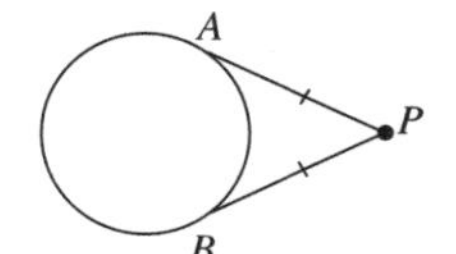

If $\overline{PA}$ and $\overline{PB}$ are tangents to a circle at A and B, then $PA = PB$.

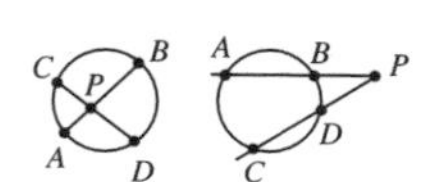

If two chords (or secants) $\overline{AB}$ and $\overline{CD}$ of a circle meet at P, then $AP \cdot BP = CP \cdot DP$.

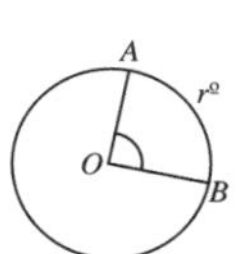

$m\angle AOB = r$
(O center)

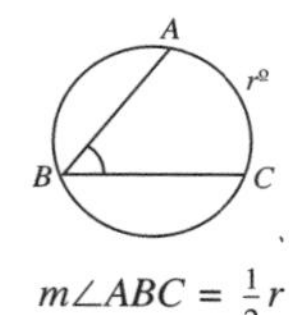

$m\angle ABC = \frac{1}{2}r$

$m\angle APB = \frac{1}{2}(r + s)$

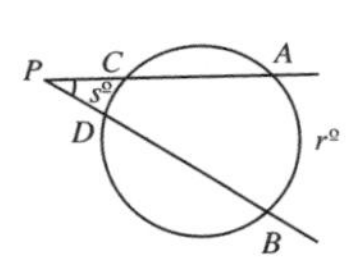

$m\angle APB = \frac{1}{2}(r - s)$

HIGH SCHOOL TRIGONOMETRY

Pythagorean Relation: $a^2 + b^2 = c^2$

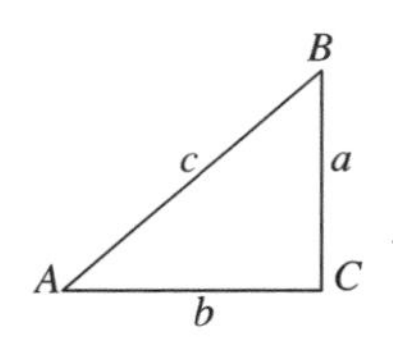

The Six Trigonometric Ratios

$\sin A$ = opposite/hypotenuse = a/c	$\csc A = c/a$
$\cos A$ = adjacent/hypotenuse = b/c	$\sec A = c/b$
$\tan A$ = opposite/adjacent = a/b	$\cot A = b/a$

Basic Identities: $A, B \leq 180$

$\sin^2 A + \cos^2 A = 1$

$1 + \tan^2 A = \sec^2 A$

$\sin (A + B) = \sin A \cos B + \cos A$

$\cos (A + B) = \cos A \cos B - \sin$

$\tan (A + B) = (\tan A + \tan B)/(1$

$\sin 2A = 2 \sin A \cos A$

$\cos 2A = \cos^2 A - \sin^2 A$

$= 1 - 2 \sin^2 A$

$= 2 \cos^2 A - 1$

$\tan 2A = (2 \tan A)/(1 - \tan^2 A)$

Miscellaneous Formulas

AREA $\triangle ABC$: $\frac{1}{2}bh$

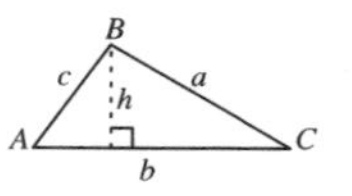

AREA $\triangle ABC$: $\sqrt{s(s-a)(s-b)(s-c)}$ where $s = \frac{1}{2}(a + b + c)$

AREA (PARALLELOGRAM): $ABCD = bh$

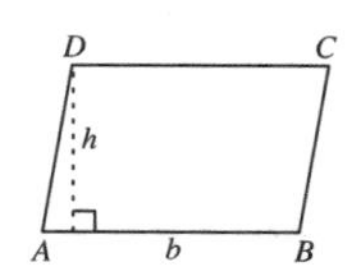

AREA (CIRCLE): $K = \pi r2$

CIRCUMFERENCE (CIRCLE): $C = 2\pi r$

AREA (SECTOR): $K = \frac{1}{2}r^2\theta$ ARC LENGTH: $s = r\theta$

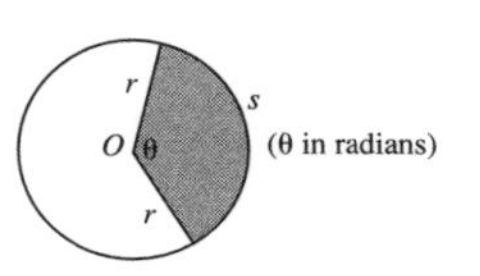

STANDARD CONSTRUCTIONS IN GEOMETRY

CONGRUENT SEGMENTS:
($AB = CD$)

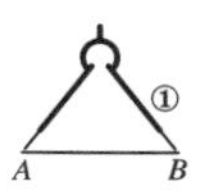

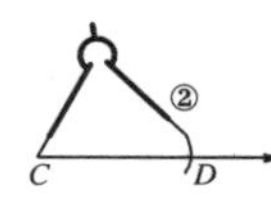

CONGRUENT ANGLES:
($BK = DL$, $PQ = RS$)

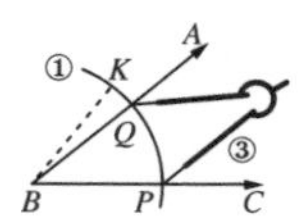

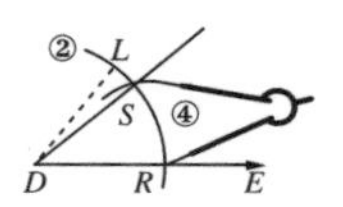

MIDPOINT AND PERPENDICULAR BISECTOR OF SEGMENT:
($AP = BP = AQ = BQ$)

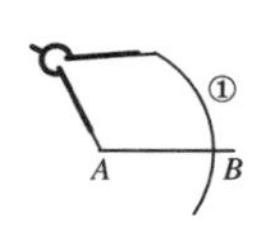

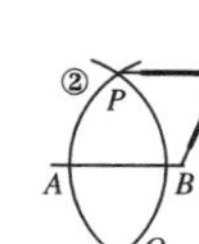

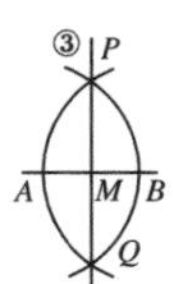

BISECTOR OF ANGLE:
($BP = BQ$, $PR = QR$)

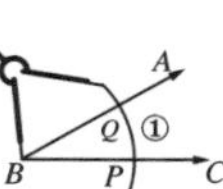

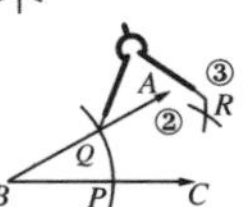

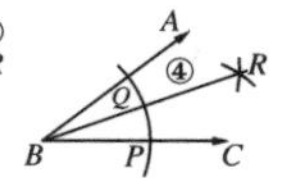

PERPENDICULAR TO LINE AT POINT ON LINE:
AQ, $PR = QR > AP$)

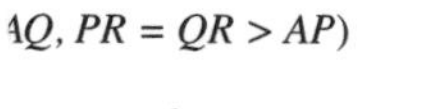

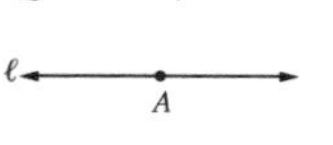

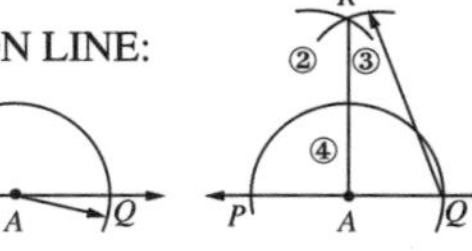

NDICULAR TO LINE FROM EXTERNAL POINT:
Q, $PR = QR$)

A

ℓ

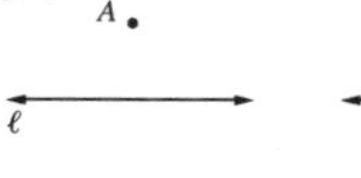

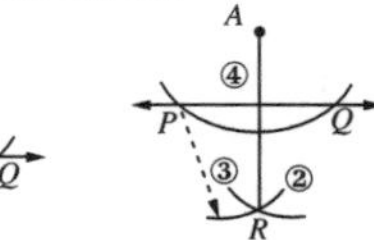

College Geometry

A DISCOVERY APPROACH

with *The Geometer's Sketchpad*

SECOND EDITION

David C. Kay

Formerly, Professor and Chairman of Mathematics

University of North Carolina at Asheville

Boston San Francisco New York
London Toronto Sydney Tokyo Singapore Madrid
Mexico City Munich Paris Cape Town Hong Kong Montreal

Sponsoring Editors: Carolyn Lee-Davis/Laurie Rosatone
Assistant Editor: RoseAnne Johnson
Marketing Manager: Michael Boezi
Managing Editor: Karen Guardino
Associate Production Supervisor: Cindy Cody
Cover Designer: Barbara T. Atkinson
Cover Photo: Stone ©2000/Steven Weinberg
Photo Research: Beth Anderson
Prepress Service Manager: Caroline Fell
Manufacturing Buyer: Evelyn Beaton
Project Coordination, Text Design, and Electronic Page Makeup: Nesbitt Graphics, Inc.

Photo Credits: page 21, Milt and Joan Mann–Cameramann Intl.; *page 58,* Courtesy of the Granger Collection; *pages 79, 128, 188, 256, 392, 424,* Courtesy of AIP Emilio Segrè Visual Archives; *page 143,* Bohdan Hrynewych/Stock Boston; *page 217,* Steve Monti/Bruce Coleman, Inc.; *page 227,* Michael George/Bruce Coleman, Inc.; *page 259,* Adam Woolfit/Corbis.

Library of Congress Cataloging-in-Publication Data
Kay, David C., 1933—
College geometry: a discovery approach/David C. Kay.–2nd ed.
Includes bibliographical references and index.
ISBN 0-321-04624-2
1. Geometry. I. Title.
QA445.K383 2000
516—dc21 00-032293

Please visit our website at http://www.awl.com

2 3 4 5 6 7 8 9 10—CRW—030201

To my wife and best friend,

Katie

INTERDEPENDENCE OF TOPICS

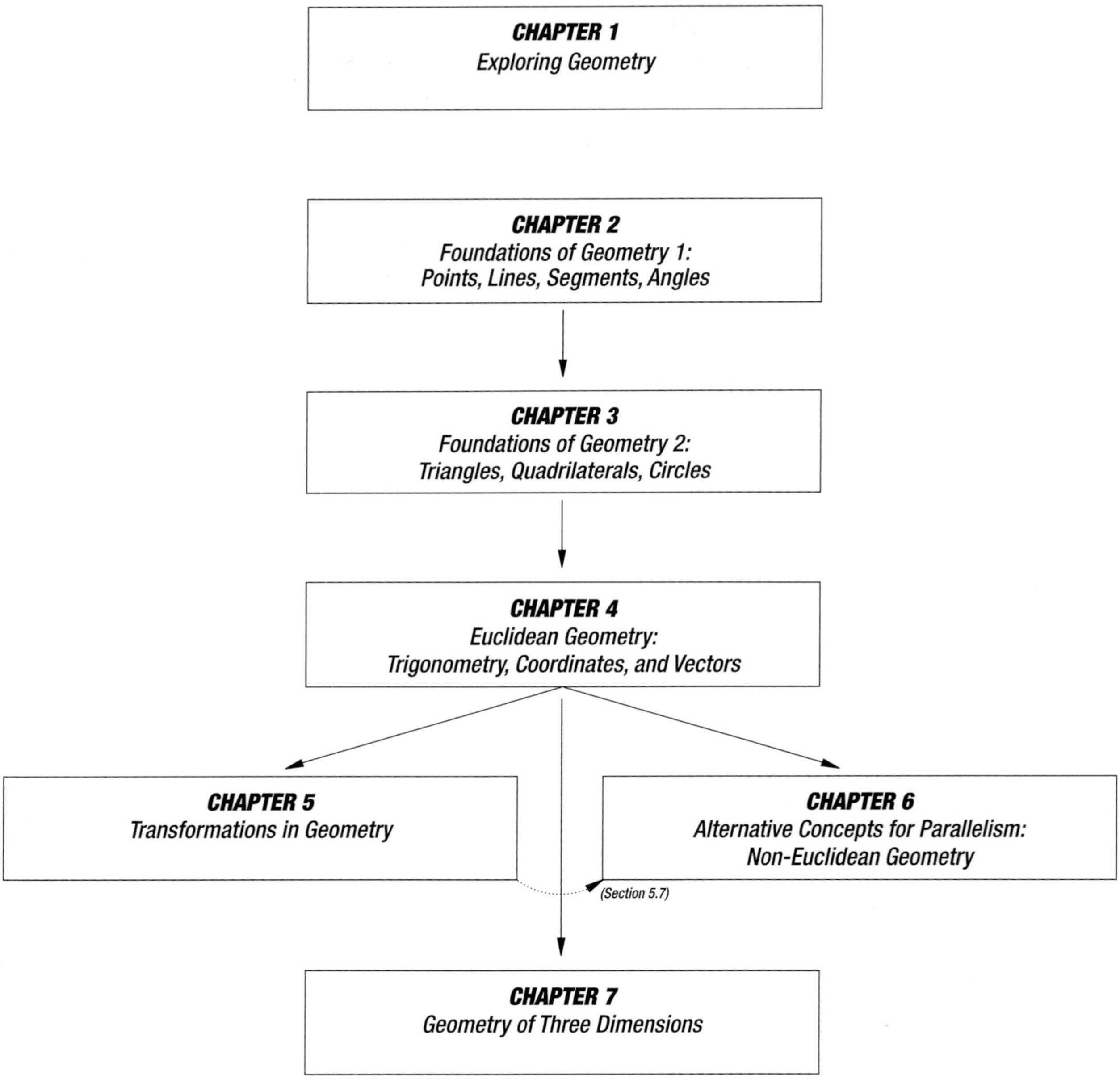

CHAPTER 1
Exploring Geometry

CHAPTER 2
Foundations of Geometry 1:
Points, Lines, Segments, Angles

CHAPTER 3
Foundations of Geometry 2:
Triangles, Quadrilaterals, Circles

CHAPTER 4
Euclidean Geometry:
Trigonometry, Coordinates, and Vectors

CHAPTER 5
Transformations in Geometry

CHAPTER 6
Alternative Concepts for Parallelism:
Non-Euclidean Geometry

(Section 5.7)

CHAPTER 7
Geometry of Three Dimensions

Contents

*Optional sections

CHAPTER 7

An Introduction to Three-Dimensional Geometry 493

Appendixes

Special Topics

An Introduction to Projective Geometry
An Introduction to Convexity Theory

Either of these topics will be available upon request to adopters of *College Geometry,* Second Edition, by contacting your local Addison Wesley sales consultant or via *math@awl.com.*

Preface

This book is an introductory, college-level geometry text written for students who are mathematics majors or education majors seeking teacher certification in secondary school mathematics. For the students who are education majors, we have developed mathematical concepts along traditional lines and have included numerous problems and examples from current secondary school textbooks.

GUIDING PHILOSOPHY

The following three underlying principles direct the design of this book:

1. The many aspects of geometry should be as visual as possible, and students should be inspired to develop a questioning, curious nature about geometry.
2. Students should be encouraged to discover many principles and phenomena of geometry for themselves.
3. The gentle art of proof-writing, as a *tool* for answering questions in geometry and discovering what is actually true, should be effectively conveyed to students.

In short, the focus of this book is not only to provide an introduction to geometry, but to help the instructor achieve a high level of student participation in and outside the classroom. The writing style is itself intended to maintain student interest, and the presentation of geometry is designed to promote student participation, interest, satisfaction, and fulfillment. The main objective in this whole endeavor is to help students master the subject of geometry at this introductory level.

CHANGES IN THE SECOND EDITION

We have improved the presentation of the material, aiming for both better clarity and for more material that is appropriate for mathematics education courses. Many changes directly address requests made by reviewers of this edition and users of the first edition. A special review section was included in the appendix, and a chapter on three-dimensional geometry was added.

- ***An improved course opener (Chapter One)***
  The opening chapter has been shortened and rendered more suitable for beginners. Because many instructors prefer to start the development of geometry in Chapter 2, Chapter 1 is designed as optional material, as in the first edition, and can either be handled as a reading assignment, as a source of special topics to be covered at a later point in the course, or skipped altogether. For those who do wish to use Chapter 1 to start the course, it is rewritten to be both appealing and easy to read and includes an exploratory discussion about some old topics (pi and the Pythagorean Theorem) and a

few topics almost certain to be new to students (the Morley Triangle, Pappus' generalization of the Pythagorean Theorem, and the Pedal Triangle). Chapter 1 also includes examples and instructions for the use of *The Geometer's Sketchpad.* The experiments considered here can easily lead to some excitement on the part of the student at this early stage. The problems for exercises in Chapter 1 are all introductory in nature, particularly the "Group A" problems designed for the average student. Some "Group B" and "C" problems, however, are designed for the more advanced student and might be assigned later in the course as a special challenge.

- ***Enhancement by* The Geometer's Sketchpad**
  The second edition has added specific experiments via *The Geometer's Sketchpad* as a tool for self-discovery. This material is marked with a special icon for easy location and to indicate that other computer software besides *Sketchpad* may be substituted with minimal alteration. Some *Sketchpad* experiments occur in examples, many in the Moments for Discovery, and the remainder in the problems. Virtually all of them feature new viewpoints and innovative techniques not currently found in other sources of *Sketchpad* projects. At the end of each experiment students are prompted to make their own discoveries and to note various geometric features. They are also encouraged to apply the theoretical results of the particular section involved to explain the phenomenon they have just witnessed. This feature meets one of the major standards for geometry and teacher education recommended by the National Council for Teachers of Mathematics.
- ***Moments for Discovery***
  Many Moments for Discovery sections have been added or improved from the first edition to further enhance self discovery. This is a natural place for *Sketchpad* experiments, so many of these units are devoted to constructions achieved by computer software. In a few cases these units occur in identical pairs—one for students who are not using a computer, and one specifically designed for the use of *Sketchpad.*
- ***New problems added***
  Many new and interesting problems have been added to the problem sections, and many which appeared in the first edition have been greatly improved. The grading of problems (into Groups A, B, or C) has also been improved. New problems have been added to Group A in most sections to encourage greater student success and interest.
- ***Testing Your Knowledge sections improved***
  Instead of true-false questions, bonafide problems at level A have been put together at the end of Chapters 2 through 7 for use by students. These may be used as "practice tests," or just as self-tests for monitoring progress in the course. The answers or solutions to these problems are located in Appendix E with the other problem solutions.
- ***Three-dimensional geometry added***
  Although available through the publisher as a special topic in the first edition, this material was redesigned as a new Chapter 7 in the second edition, recognizing another major NCTM recommendation. At several reviewer's

requests, a unit on spherical geometry was included. The development of spherical trigonometry appears, with a few real-life applications for finding distances on the earth's surface.

- ***Review section included***
  A special review section was added to the appendix which reflects a typical high school geometry course. It was designed to be short, and to include just enough topics to allow the typical student to recall enough background with which to begin the course. (It should be pointed out that the inside front cover includes the most common concepts in geometry, which will probably be adequate for most students.) In spite of its ordinary purpose, the unit was carefully written to enhance student interest, with 40 diagrams, several unusual examples, and four special problems at that level to be worked out by the student. This material should help most students quickly recall the pertinent facts they will need from their high school geometry course.

DISTINGUISHING FEATURES

The book has special units for self-discovery, frequent examples for proof-writing involving the concepts being presented, and graded problem sets for each section.

Moments for Discovery. Appearing in most sections of the text, these units are specifically designed to encourage students to make discoveries on their own. Many of them are purely experimental in nature, directing the student to conduct a drawing experiment or investigate an example. Others involve abstract reasoning, guiding the student in a step-by-step procedure to make warranted conclusions regarding certain geometric phenomena. These units can be used in a variety of ways by the instructor: as a device to arouse classroom discussion, as team projects, or just for outside reading assignments on which to report.

Topical and Graded Problem Sets. The problem sets have also been designed to foster self-discovery, and at the same time allow every student to have successful problem-solving experiences. The wide variety and large quantity of problems also serves to stimulate greater interest and participation. Occasionally, problems which correspond to major named theorems in geometry appear. These are presented with a topical heading to catch the reader's attention. Longer problems thought to be appropriate for undergraduate research, either partly or totally unsolved in the literature, or suitable for an expository treatment, are so labelled.

To make it easier for the instructor, each problem set is graded according to difficulty.

Group A: Largely of a computational nature, or involve short proofs.
Group B: Involve a higher level of sophistication, requiring formal multiple-step proof-writing and some original thinking.
Group C: Require yet a higher level of original thinking, and includes longer projects and topics for undergraduate research.

Short answers and problem solutions for the odd-numbered problems appear in Appendix E.

Visual Impact. The text is heavily illustrated, with well over 1200 separate drawings. In presenting a concept, often several stages of a line of development pursued in the text appear in the illustrations. Frequently, the steps of a proof are illustrated separately, and each major case of a proof appears with a separate figure. Sometimes a figure is even duplicated across non-facing pages to make it more convenient for the reader.

Geometric Content. Although most texts at this level concentrate on either axiomatics (excluding transformation theory and modern applications), or transformation theory (excluding axiomatics), this text contains a careful axiomatic treatment beginning with Chapter 2 and continuing through Chapter 3, then some modern geometry, including the use of coordinates and vectors in Euclidean geometry (Chapter 4), transformations and the group concept (Chapter 5), and finally, returning again to the axiomatic foundation, an introduction to Lobachevskian geometry and models for hyperbolic geometry (Chapter 6). A final chapter on three-dimensional geometry appears. (Chapter 7).

Computer Friendly. This text is compatible with a parallel use of the latest in computer technology to make the subject more lively in the minds of students. *The Geometer's Sketchpad* is an integral part of the text by intention, but the text stands on its own if such software is unavailable.

Cross-Referencing. For the further convenience of the instructor (and student), the cross-referencing of problems and text material, backward and forward, appears. If a problem is used in any significant way in some future development, it is starred, with the forward reference stated, and problems involving a previous concept or problem include a reference with the appropriate information and location in the text.

PEDAGOGICAL FEATURES

This text was written primarily for the student, although it contains much material which the average instructor should find interesting, and even exciting when it is used to engage the minds of students.

Historical Notes. The book includes numerous biographical anecdotes to reveal the human side of mathematics. Rather than present a complete chronological account of events as would be found in a more formal treatment of the history of mathematics, special or unusual features concerning the lives of geometers and their contributions were singled out.

Our Geometric World. Excerpts that relate geometric concepts to the physical world in which we live are liberally sprinkled throughout.

Chapter Overviews and Summaries. Each chapter contains an overview (where we intend to go) and a summary (where we have been). The overviews briefly state what we intend to study, and why, often with helpful reminders of what one should look out for or to avoid. The summaries at the end of each chapter discuss the major results covered in that chapter in broad outline, but detailed enough to provide pertinent information. At the very end, there is a short test included for student self-evaluation on general mastery of the concepts.

Location of Previous Results. The problem of quickly locating previous results has been eliminated by placing the statements of all theorems and corollaries, by section number, in Appendix F. To find the idea or theorem you are looking for, a quick glance is often all it takes. A complete glossary of definitions, also arranged by section, is included.

CLASS TESTED

Classes which field-tested a preliminary version of the manuscript (first edition) include the students of James Spencer, of The University of South Carolina at Spartanburg, the students of David Dezern, of The University of North Carolina at Asheville, and the students of James Tattersall, Providence College. Comments from students were carefully considered in making revisions or deletions of material found confusing or ambiguous. Because of this feedback, the book has been greatly improved.

PREREQUISITES

It is not necessary for a student to have a fresh memory of the content of high school geometry in order to succeed. (A cursory review appears in the inside front cover for the convenience of students, as well as a more complete review in Appendix B.) However, it is recommended that at least one semester of calculus be required of students entering a course using this text. This will guarantee a fairly recent encounter with trigonometry which is assumed here to be common ground for all students. Informal references are made to limits and continuity (mostly confined to Group C problems). We do not assume an ability to use abstract reasoning, but sooner or later students will be expected to make some progress in this area, and a foundations course could be considered as another prerequisite in colleges where this is a viable option.

COURSE FLEXIBILITY

The core text contains 28 sections, each designed for one or two 50-minute classroom sessions, in addition to 16 optional sections (starred) which the instructor can use for enrichment. After Chapters 1 through 4 are covered, Chapters 5 through 7, or combination of topics from these chapters, can be covered independently (one brief reference involving the half-plane model in Chapter 6 which uses a previous discussion in Chapter 5 is the only exception). It is recommended that students at least read Chapter 1 on their own. Some topics introduced in Chapter 1 are dealt with later, at a more appropriate stage of development.

Axiomatic Geometry. A course emphasizing axiomatics might include Chapters 1 through 4 and Chapters 6 and 7 (24 sections, not counting optional sections).

Modern Geometry. A course stressing primarily the techniques of geometry could be based on Chapters 2 through 5, with emphasis on Chapter 5 and concentrating on solving the problems occurring there (19 sections). This would allow the addition of several optional sections, such as Sections 3.2, 4.6, 4.8, 7.4, and 7.6. Material from the topic *An Introduction to Convexity Theory* could also be included.

Survey Course. A course intending to give students a broader view of geometry at this level can be created using Chapter 2, (including Sections 2.4 and 2.5), Chapter 3 (including Sections 3.2, 3.4, and 3.6), Chapter 4 (omitting Sections 4.1, 4.2, 4.3), Chapter 5 (omitting Sections 5.1 and 5.2), Chapter 6 (including only Section 6.4), Chapter 7 (Section 7.6), for a total of 18 sections, and content from one of the available topics (mentioned below).

SPECIAL TOPICS

Since opinions vary about the specific topics that should be included in an introductory course in geometry, we make the following topics available to anyone adopting this text:

- An Introduction to Projective Geometry
- An Introduction to Convexity Theory

Each of these is about the same length as the previous chapters, and includes the same features (units for self-discovery, historical notes, graded problem sets). Solutions to selected problems for each topic appear at the end of that topic. These may be obtained by contacting your local Addison Wesley representative, or by contacting the home office: Addison Wesley, 1 Jacob Way, Reading, MA 01867 (*math@awl.com*)

SUPPLEMENTS

Complete worked-out solutions to all discovery units and all problems in the text (except undergraduate research projects) is available in the *Instructor's Solutions and Resource Manual.* Sample tests have been prepared covering the individual sections of each chapter, except Chapter 1.

ACKNOWLEDGMENTS

I would like to express my deep appreciation to those who helped make this project possible: to George Duda, my editor and good friend who made the first edition possible, to Paul Patten at North Georgia College and State University for helping me to double check the problem solutions for accuracy and correctness, as well as making suggestions for other parts of the book and checking that the *Sketchpad* experiments work as intended, and to John Eggers, also at North Georgia College, for additional proofreading.

Reviewers. Thanks and appreciation are due to the following reviewers of the first and/or second edition: Debra Carney, University of Texas, Austin; Ray Carry, University of Texas, Austin; Jon Clauss, Augustana College; Kenneth Evans, Southwest Texas State University; Barbara Ferguson, Kennesaw State University; William Fitzgerald, Michigan State University; Karen Graham, University of New Hampshire; Paul Green, University of Maryland; J. Taylor Hollist, SUNY Oneonta; Mark Hughes, Florida State University; Jerry Johnson, Western Washington University; Mark Kon, Boston University; Kathryn Lenz, University of Minnesota; Jeanie McGehee, University of Central Arkansas; Eric Milou, Rown University; Rick Mitchell, University of Wisconsin; R. Padmanabhan, University of Manitoba;

Stephen Pennell, University of Massachusetts; Peter Sandburg, University of Minnesota.

Editorial Staff. The author deeply appreciates the expert staff at Addison Wesley Longman—true professionals, who are largely responsible for whatever success this venture may have. In particular, I would like to single out the one who originally encouraged me to make revisions for a second edition, Carolyn Lee-Davis. Her advice was quite valuable in helping me decide what approach should be taken to address the wide variety of suggestions from the reviews. I also wish to acknowledge the book team, including RoseAnne Johnson, assistant editor at Addison Wesley, Cindy Cody, associate production supervisor at Addison Wesley, Barbara Atkinson, who designed the cover, Michael Boezi, who is responsible for marketing the text, and Lois Lombardo of Nesbitt Graphics, Inc. for overseeing the production.

We welcome comments and criticism from the readers and user of this book. Feedback from varying classroom situations would be most important for making improvements in the future that could lead to an even more effective text. It is recommended that all such matters be referred to your local Addison Wesley representative at *math@awl.com.*

David C. Kay

To the Student

You may be wondering what to expect from a college level geometry course. Indeed, you might be thinking "If I had a hard time with high school geometry, how will I be able to learn something of value here?" Or you might wonder if a college course in geometry is totally different from high school geometry, with a far-ranging subject matter. (Just to put your mind at ease, a special review section is offered in Appendix B to help you get started.) These are natural questions for you to ask as you embark on a new journey. We are going to try to give you some honest answers here.

Many of you are obliged to take a college course in geometry because it is required for teacher certification. If you are in that group, this course is designed especially for you. It will prepare you for the time when you will face your own class in geometry. However, our aim is not to show you explicitly how to teach geometry, but rather to emphasize the concepts and inherent beauty of geometry so you will be inclined to teach geometry as an exciting subject yourself—one which your own students will be able to enjoy, just as you have, rather than as a subject to endure.

We also believe in the old saw "teachers should know more than they teach." Mathematics teachers should always be in a position of strength, able to field any unpredictable question which a student may ask. An unexpected question from a student can often catch a teacher off guard, and the better prepared you are as a teacher, the better you will be able to respond to those students in a helpful and effective manner.

You may be taking this course as a major elective and not as a prospective teacher. This book is also written for you, because the presentation is introductory in nature, and the variety of subjects covered, particularly the problems, will show you some very unusual aspects of geometry. We think you too will come to enjoy the subject of geometry, and may even get excited at times, when you find that you can really "do geometry."

The "Moments for Discovery" are designed for everybody. It is strongly recommended that you work through each of these sections as you come to them in your reading of the text. In each one, you will be carefully led through a logical sequence of steps, sometimes consisting of just a few simple calculations, drawing experiments, or elementary logic, based on what you already know. When you get to the end of the material, the chances are very good that you will have "discovered" a really neat concept of geometry on your own. As you grow more mature in the subject you will gradually acquire the ability to generalize and discover geometric properties on your own, or at least make good guesses about what may be true. As you can see, this course is really very creative, and we are going to encourage your best creative thinking.

In many of the "Moments for Discovery," as well as in many examples and problems for homework, an icon will appear that simulates a computer screen. That will signal the need to use a computer (and special software to be mentioned later) in order to work the problem or accomplish the goals of the discovery section. If you do not have access to such computer software, don't worry. Plenty of discovery sections and homework problems remain which do not require a computer. The first of those that do takes place in Section 1.3, where some detailed instructions are given. That section, and the next cannot be fully appreciated without a computer, so if this equipment is not accessible, just omit those two sections (and all future material in the text so indicated).

At the end of each section a set of problems has been carefully crafted. These are divided into three groups. Group A contains those problems everyone should be able to do; they are of the pocket-calculator, quick recognition, or one-step proof variety, mainly involving definitions, and generally do not require much thinking. They are just to test your basic understanding of the material in the section just preceding. Group B consists of a little higher level of abstraction, requiring the writing of short, multiple-step proofs. Then there is Group C, consisting of problems requiring a great deal more thought and involvement on your part. Occasionally, ideas for undergraduate projects will appear there. Hints appear frequently throughout all three groups of problems.

Finally, if you wish to test yourself on basic understanding, a practice test appears at the end of each chapter, starting with Chapter 2. Answers to these appear with the problem solutions in Appendix E.

In summary, we have attempted to present a good introduction to geometry within the bounds of a one-semester, undergraduate course, which not only provides a body of knowledge in mathematics that has a host of practical applications, but reveals a fascinating area of active, on-going research in which it is possible for almost anyone to participate. We hope that you will derive great benefit, personal satisfaction and accomplishment in using this textbook.

David C. Kay

CHAPTER ONE

Exploring Geometry

OVERVIEW

Geometry may be loosely defined as the systematic study of the physical world we live in. The word "geometry" ("geo-metric") means to *measure the earth*. The earliest studies in geometry were devoted to just such endeavors. A prime example is Eratothsenes' measurement of the circumference of the earth, where he used the sun's rays as they projected light at the bottom of the well in the ancient city of Syene, and properties of the circle. But geometry was used in other ways, too. During the Renaissance in the seventeenth and eighteenth centuries, artists began developing a new kind of geometry, called *projective geometry*, to help them understand how to draw images on a canvas. And engineers, craftsmen, and construction workers have always used various aspects of Euclidean geometry in their work.

As our world has advanced, our viewpoint of geometry has continually changed, and the subject now encompasses a great many widely differing branches of mathematics. In fact, due to the constant state of change of all aspects of the earth (climate, continents, landscape, soil, and inhabitants), the case could be made that the only useful kind of geometry for our modern age is one that can portray change—*geometry in motion*, if you will. In this book, we emphasize the visual and dynamic aspects of geometry, as well as the traditional. The elegant system of Euclid is prominent throughout, but the use of computers is also recognized, and from its power we can see instantly, with our own eyes, *geometry in motion*.

In this first chapter, we look at geometry in an exploratory manner that will set the stage for later work, in which you will be continually encouraged to discover geometric relations for yourself. If you feel you need a "refresher course" in geometry, first check the inside front cover and facing page. If you feel that more details are needed, a more complete review of school geometry, reflecting typical textbooks written for high school courses, can be found in Appendix B. That material includes a few sample problems, with answers included, to make it easier for you to digest the material you will need later.

1.1 Discovery in Geometry

Most elementary properties in geometry can be discovered if presented in a discovery mode. That is, instead of passively reading about it, we can often discover a geometric property by means of a figure or from a drawing experiment, if led to do so. In this manner, it is possible to participate in what can be a rather exciting venture, the goal of which is a personal observation, or "discovery," of geometric relationships.

Let us look at a few examples.

EXAMPLE 1 The medians of a triangle are concurrent, that is, they pass through a common point. (The point of concurrency is called the **centroid**.)

(a) Suppose you take two or three triangles at random (as illustrated in Figure 1.1) and carefully construct the three medians (lines joining the vertices to the midpoints of the opposite sides). Subject to experimental error, in each case you can observe that the medians pass through a common point.

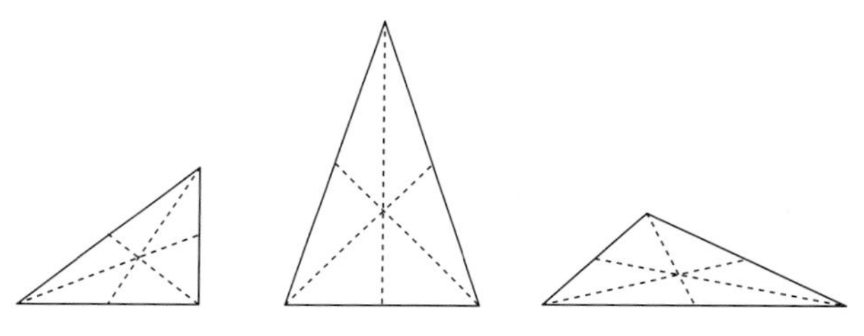

Figure 1.1

(b) Another method, also experimental, is based on the physical attributes of the center of mass. Since the medians of a triangle divide it into triangles having equal areas, a cardboard model of a triangle will perfectly balance on a knife edge along any one of its medians (Figure 1.2). If we consider two medians, their intersection is a point of equilibrium for the entire triangle—its center of mass. Because this is true of any two medians, and because there is only one center of mass, *all three* medians must pass through this center, which is the common point of intersection. ■

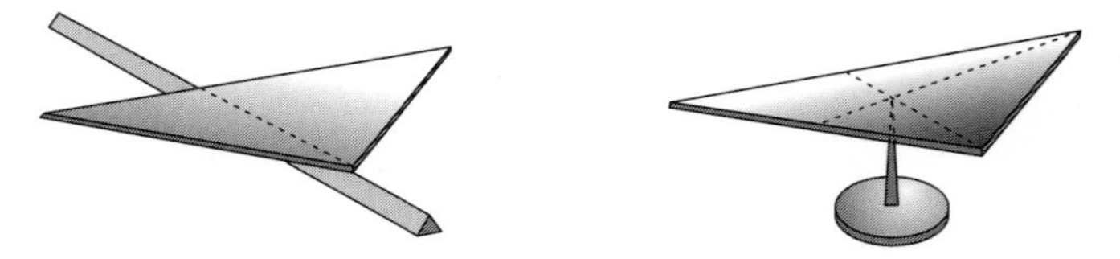

Figure 1.2

EXAMPLE 2 The square of the hypotenuse of a right triangle equals the sum of the squares of the two legs.

Start with any right triangle, and construct three other right triangles congruent to the given one. By trial and error, you could discover that in moving the triangles about, they may be made to form a square (as shown in Figure 1.3). By calculating the area of the resulting square in two different ways, the following analysis could be discovered:

I. Area of Square from Standard Formula: $c \cdot c = c^2$

II. Area of Individual Parts of Square:

(a) Small square inside $= (a - b)^2 = a^2 - 2ab + b^2$

(b) Area of four triangles $= 4 \cdot \frac{1}{2}ab = 2ab$

Total: $a^2 - 2ab + b^2 + 2ab = a^2 + b^2$

III. Therefore,

$$c^2 = a^2 + b^2. \quad \blacksquare$$

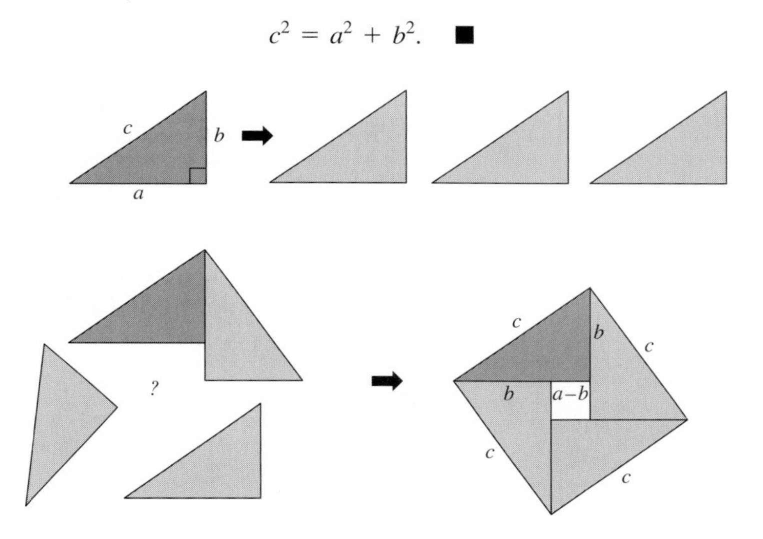

Figure 1.3

EXAMPLE 3 The sum of the measures of the angles of a triangle equals 180.

Make a paper model $\triangle ABC$ and tear off a portion of the triangle that includes vertex A. Repeat this for vertices B and C (as shown in Figure 1.4). Reassemble the torn pieces so that A, B, and C coincide and the pieces are flush. Take a ruler or straight object and place it alongside the configuration to check for the desired result. ■

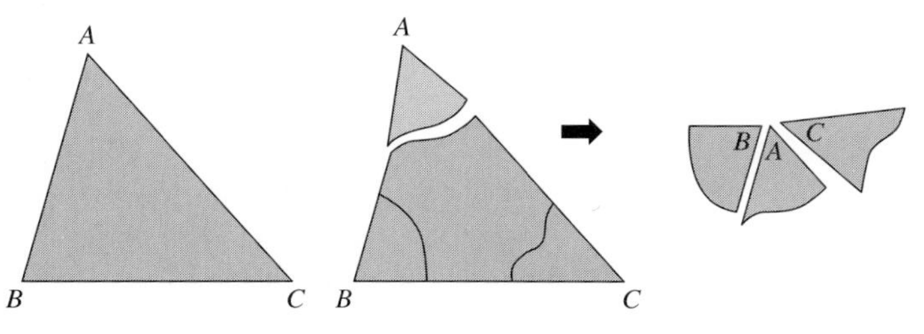

Figure 1.4

Moment for Discovery

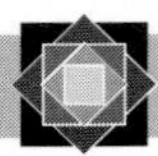

Dissection of Parallelogram to Form a Square

Do you believe that the parallelogram and its interior in Figure 1.5 can be cut into just three pieces and the pieces rearranged to form a perfect square?

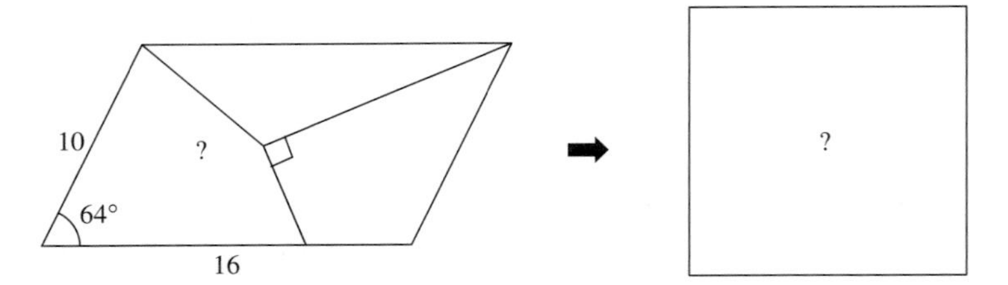

Figure 1.5

Obviously, *any* parallelogram can be dissected into *two* pieces to form a *rectangle*. But a *square*? If you follow these steps, you will be able to discover something that may surprise you. (Refer to Figure 1.6 for everything that follows.)

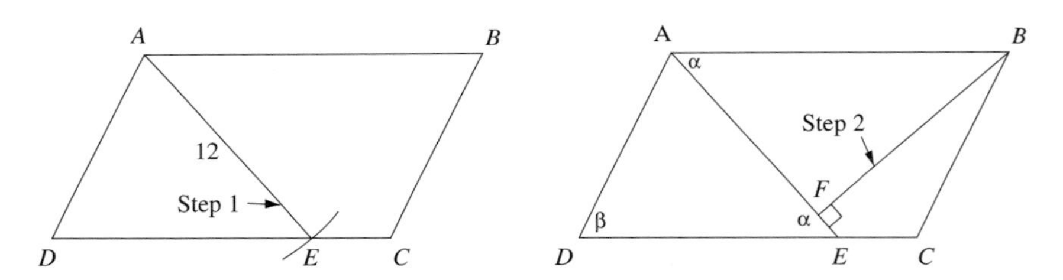

Figure 1.6

1. Carefully duplicate the figure on paper or cardboard (poster board is ideal) by using a protractor to measure the 64° angle, and a centimeter ruler to measure the sides, 10 cm and 16 cm. (If you do not have a centimeter ruler, you can use dimensions of 5 in. by 8 in.) Notice that the altitude of the parallelogram will be 9 cm (or 4.5 in.) and that the area of the parallelogram is $9 \times 16 = 144 \text{ cm}^2$.
2. The first step in the dissection procedure is to measure off from point A segment $\overline{AE}$ with E on side $\overline{DC}$ so that $AE = \sqrt{144} \text{ cm} = 12 \text{ cm}$ (or 6 in.), the length of one side of the square. (See Figure 1.6.) This is achieved by compass, or you can just use a ruler. However, accuracy is important.
3. The next step is to construct the perpendicular $\overline{BF}$ from point B to line $\overleftrightarrow{AE}$. (Construct it with compass and straight edge if you know how, or use a protractor.)
4. Now cut along the lines you have drawn, dividing the parallelogram into three pieces. See if you can arrange these pieces to form a square. Can you verify that the construction actually works? For this purpose, see the angles

marked α and β in Figure 1.6. Would this construction work for an arbitrary parallelogram? (The calculation of $\sqrt{144} = 12$ would be replaced by $\sqrt{bh}$, where b is the base of the parallelogram, and h is the altitude.)

Two very intriguing theorems of geometry involve special figures constructed from arbitrary triangles, one a circle and the other an equilateral triangle. The one involving a circle was discovered in the early 1800s; the other was discovered about a century later. They are presented here because we will show a little later (in Section 1.4) that these figures are connected with a most unusual result in geometry known as *Steiner's Theorem*. Both results are occasionally introduced in high school courses, particularly in laboratory exploration sessions using the computer.

Moment for Discovery

A Special Circle

1. Draw a large nonisosceles triangle, $\triangle ABC$, on a sheet of paper, taking care not to let any two sides appear to be congruent (see Figure 1.7). Either by compass/straight-edge or centimeter ruler, accurately locate the midpoints L, M, and N of the three sides of $\triangle ABC$.

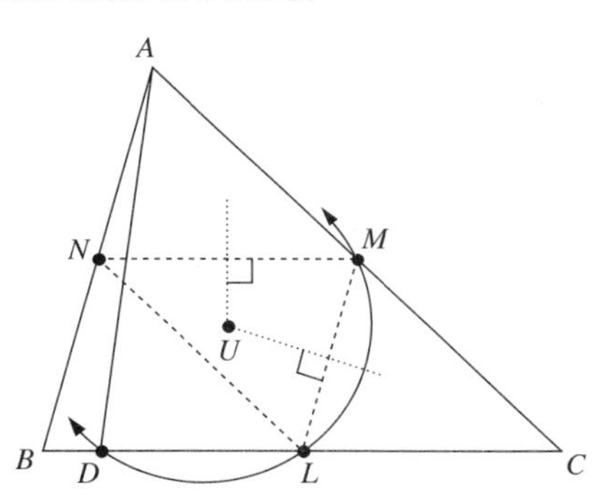

Figure 1.7

2. Construct the perpendicular bisectors of two of the sides of $\triangle LMN$, then locate the point of intersection, U.
3. Draw a circle with center U and passing through midpoints L, M, and N. This circle will meet the sides of $\triangle ABC$ at the additional points D, E, and F. Draw lines $\overline{AD}$, $\overline{BE}$, and $\overline{CF}$, and then measure, by protractor, the angles that these lines make with the respective sides of $\triangle ABC$. Did anything in particular happen?
4. Locate point H, the orthocenter of $\triangle ABC$ (point of intersection of the altitudes), and construct the midpoints X, Y, and Z of segments $\overline{AH}$, $\overline{BH}$, and $\overline{CH}$, respectively. Did anything happen?
5. Write down any general conclusions you might come to based on this experiment.

The circle you constructed in the preceding discovery unit is known as the **Nine-Point Circle** (its center is called the **Nine-Point Center**), so named by the French mathematician Jean-Victor Poncelet (1788–1867). Karl W. Feuerbach (1800–1834) in Germany (where it is called the **Feuerbach** circle) made several remarkable discoveries concerning it. These properties are so unique that we shall mention a few of them. (Some are proved in Section 5.7.)

Properties of Nine-Point Circle and Nine-Point Center

1. The Nine-Point Circle of $\triangle ABC$ with orthocenter H passes through the three midpoints L, M, and N of the sides, the feet of the altitudes D, E, and F to those sides, and the **Euler Points** X, Y, and Z, which are the midpoints of segments $\overline{AH}$, $\overline{BH}$, and $\overline{CH}$, respectively. (See Figure 1.8.)
2. The Nine-Point Center U lies on the **Euler Line** of $\triangle ABC$—the line passing through the orthocenter H, the circumcenter O, and the centroid G of the triangle. (See Figure 1.9.)
3. **Feuerbach's Theorem** The Nine-Point Circle is tangent to the **incircle** of $\triangle ABC$ and to its three **excircles**, which are the circles outside of $\triangle ABC$ that are tangent to the three sides. (See Figure 1.10.)

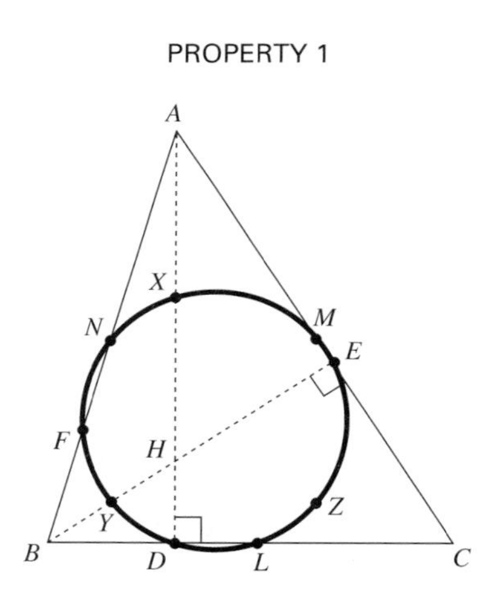

Figure 1.8

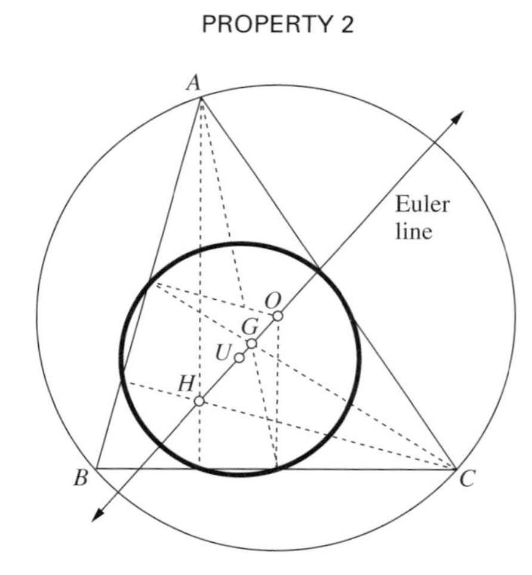

Figure 1.9

FEUERBACH'S THEOREM (PROPERTY 3)

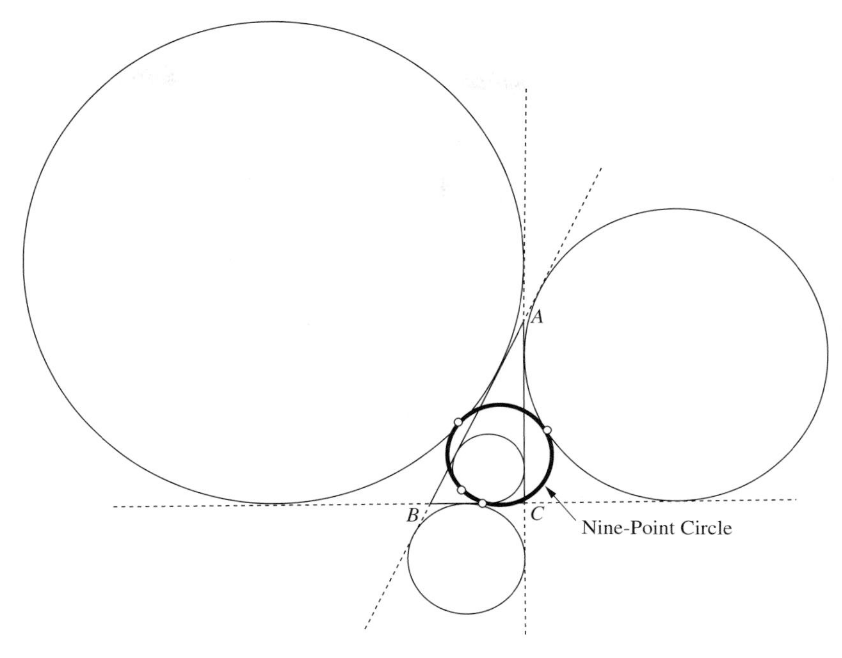

Figure 1.10

The second property about a triangle to be mentioned involves the **angle trisectors** of a triangle. These are the rays that divide each angle of the triangle into three congruent angles. In Figure 1.11 you will find these angle trisectors shown for a typical triangle $\triangle ABC$. Now, if the intersections of those pairs of trisectors lying closest to the sides are the points P, Q, and R, respectively, the result is a *perfect equilateral triangle*, $\triangle PQR$. This feature is independent of the shape of the original triangle! (We are going to look at this later using computer software, but if you want to try constructing this triangle with a protractor, we suggest that you start with a triangle $\triangle ABC$ having integer angle measures *divisible by three*, such as 30°, 45°, and 105°.)

MORLEY TRIANGLE OF $\triangle ABC$

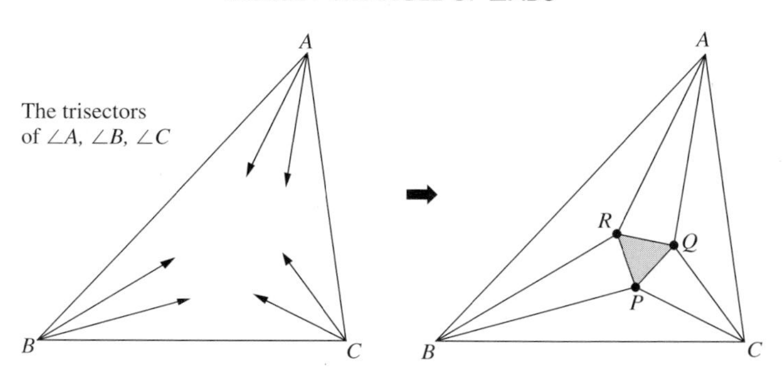

Figure 1.11

This discovery was first made by an English mathematician, Frank Morley (1860–1937), the father of novelist Christopher Morley. The equilateral triangle is for this reason known as the **Morley Triangle** of the given triangle. Its existence emerged as the result of a more general theory published by Morley in a paper about 1900. Friends of Morley became fascinated with the result and spread word of it throughout Europe. Its actual direct proof (without the more general theory originally presented by Morley) did not appear until 1914. For further historical details, see the informative article by C. Oakley and J. Baker, "The Morley Trisector Theorem," *American Mathematical Monthly*, Vol. 85, No. 9 (1987), pp. 737–745. Most proofs are either in some way indirect, use unusual methods, or employ trigonometry and special identities. We shall indicate a direct elementary proof later (Problem 18 in Section 4.1) using only high school geometry. This problem is optional, however, and is intended for more advanced work in geometry for those who wish to pursue it. For the specific case of the isosceles right triangle, a proof involving some trigonometry can be constructed with a little less work. (See Problem 14.)

PROBLEMS (1.1)

GROUP A

1. Try the experiment described in Example 3 for a quadrilateral. Construct a quadrilateral of your own, using paper or cardboard, and repeat the experiment indicated here. What seems to take place?

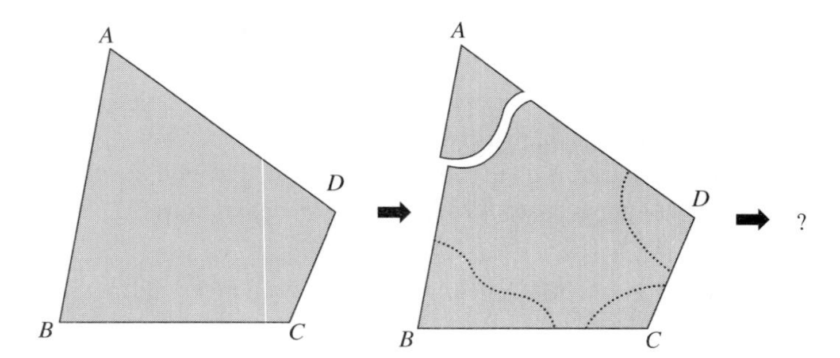

2. A square is constructed externally on a side of a regular octagon forming the figure shown at right. Can this pattern be used to tile a bathroom floor? What pattern results? Draw a diagram.

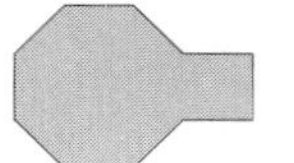

3. Draw a circle and locate a point A outside it. (See figure at top of next page.) Now draw any two lines through A, intersecting the circle at P, Q, R, and S. Draw lines $\overleftrightarrow{PR}$, $\overleftrightarrow{QS}$, $\overleftrightarrow{PS}$, and $\overleftrightarrow{QR}$, which determine the points of intersection U and V. Finally, draw line $\overleftrightarrow{UV}$, which intersects the circle at B and C. Do you observe anything special about points B and C? (Draw lines $\overleftrightarrow{AB}$ and $\overleftrightarrow{AC}$. What do you observe?) Try this experiment two more times, with point A closer to the circle, then with point A further away.

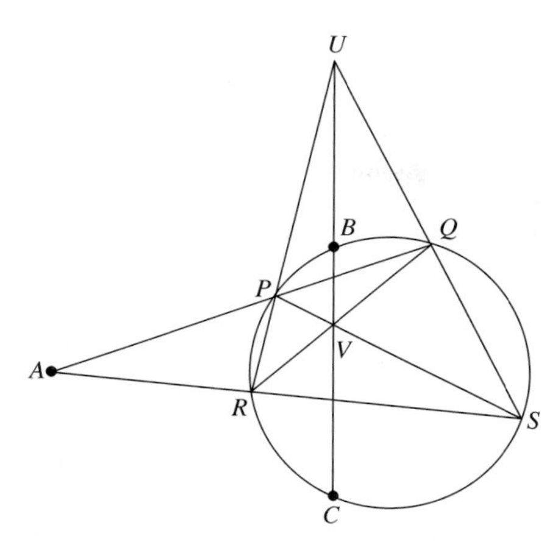

*4. Draw a circle and one of its diameters, $\overline{AB}$. Choose any point $P \neq A$ or B on that circle and draw segments $\overline{AP}$ and $\overline{BP}$. Carefully measure $\angle APB$ with a protractor. What did you find? Try the experiment several times with different locations for P to see if the phenomenon persists. What theorem in geometry does this experiment reveal?

*See Corollary A to Theorem 1, Section 4.5.

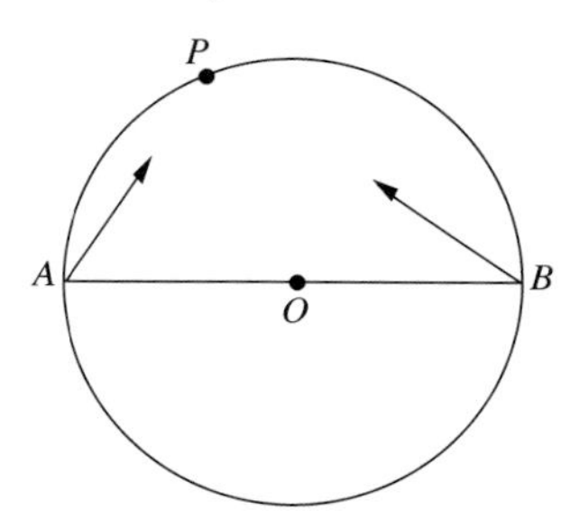

5. **Conjecture** The midpoint M of the hypotenuse $\overline{AB}$ of right triangle $\triangle ABC$ is equidistant from the vertices A, B, and C. Conduct a drawing experiment either to support or to deny this claim. What did you find? (For further analysis, draw the perpendicular from M to side $\overline{AC}$; what kind of triangle is $\triangle MAC$?)

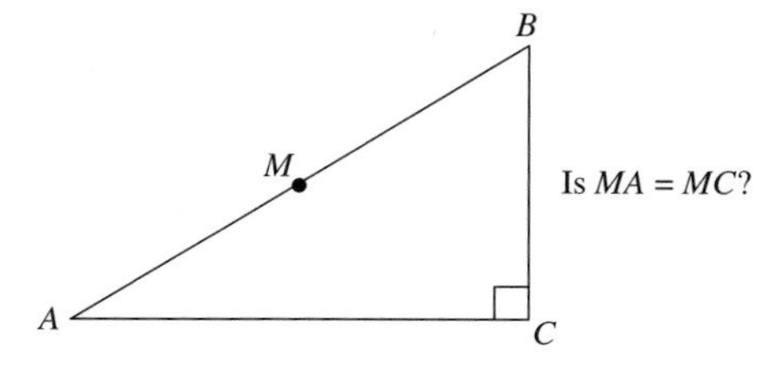

6. The first letter of the Kunif alphabet, *ali*, has the form shown at the right, with the stems having uniform thickness of one unit, and all angles of measure 60 or 120. Find, by paper experiment or drawing experiments, how to assemble six of these letters to form a perfect regular hexagon and interior. What is the length of each side of the hexagon?

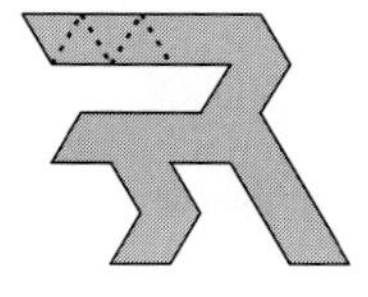

GROUP B

7. **Pick's Theorem** The points in the coordinate plane having integer coordinates are called **lattice points**. If a polygon has lattice points as vertices, it is called a **lattice polygon**. A theorem discovered in 1900 by an English mathematician, George Pick, provides a formula for the area (K) of a lattice polygon in terms of the number of lattice points on the boundary (B) and the number inside (I). It is reasonable to assume that the area of a lattice polygon is *linear* in B and I; hence, there exist constants x, y, and z such that $K = xB + yI + z$ for all polygons.

(a) Complete the table below, and find various entries for K, B, and I, to be substituted into the expression $K = xB + yI + z$. (Using the table, one equation is $2 = x \cdot 6 + y \cdot 0 + z$ or $6x + z = 2$.)

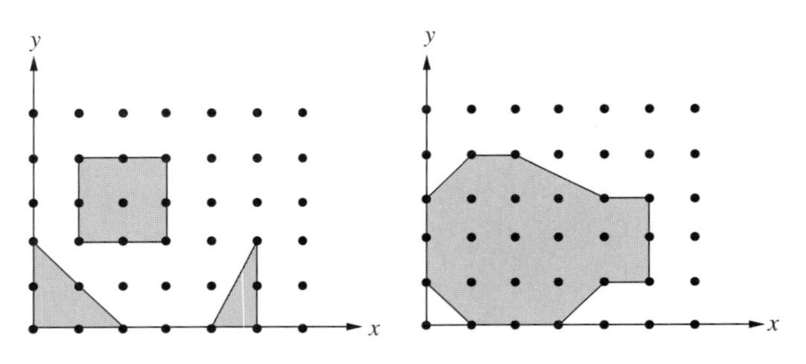

	Triangle	Square	Triangle
Area = K	2		
Boundary points = B	6		
Interior points = I	0		

(b) Solve the system of three equations you found in (a) for x, y, and z. You now have a plausible formula for the area of a lattice polygon, $K = xB + yI + z$ (after substituting the values you found for x, y, and z). Try it out on the remaining polygon. Does the formula work? (The actual proof of Pick's Theorem may be found in Coxeter, *Introduction to Geometry*, p. 34.[1])

8. Perform an experiment *in reasoning* to see if the theorem illustrated in Problem 4 is actually true. Try an argument beginning with the statement, "If P lies on the circle with diameter $\overline{AB}$ and M is the center of that circle, then $MA = MB = MP$, and two isosceles triangles are formed." What follows from this?

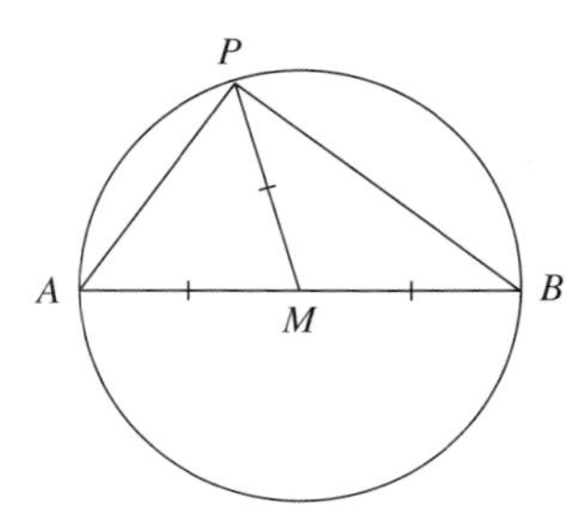

[1]Exact sources for all references to books and articles appear in the Bibliography (Appendix A).

9. Dudeney Dissection An equilateral triangle and interior is dissected according to the specifications shown. Make a paper or cardboard model, using a carefully constructed figure, and having the exact dimensions specified.

$AP = PB = AQ = QC = 5$ cm
$QR = 5\sqrt[4]{3} \approx 6.58$ cm
$RS = 5$ cm $\quad (BR > SC)$

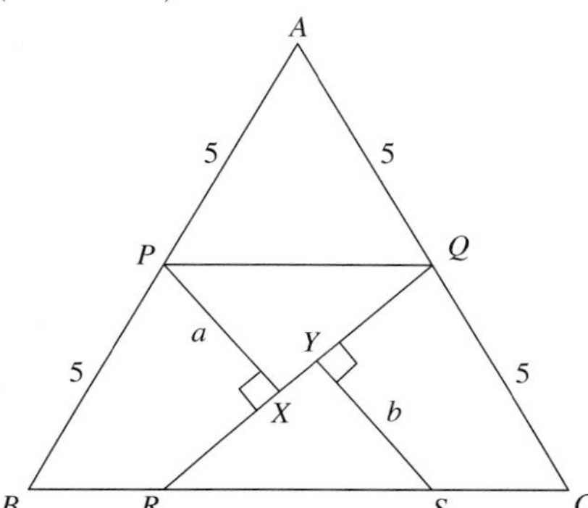

Find a way to reassemble these four pieces to form a square, and try to prove the validity of this dissection. (***Hint:*** Think of quadrilateral *BPXR* and triangle *CQR* as pivoting about points *P* and *Q*, respectively, and triangle *RYS* about *S*.)

NOTE: This construction is due to H. E. Dudeney (1857–1931) who discovered many ingenious dissections, some of which are considered here.

10. Based on numerical examples involving rectangles, the following conjecture is made: If *P* is any point on the diagonal $\overline{AC}$ of parallelogram *ABCD*, the parallel lines through *P* to the sides form two parallelograms having equal areas. Try proving it (there is a simple proof that does not involve ratios or similar triangles, only areas of parts of the given parallelogram). Do you agree with the conjecture in the case of a square?

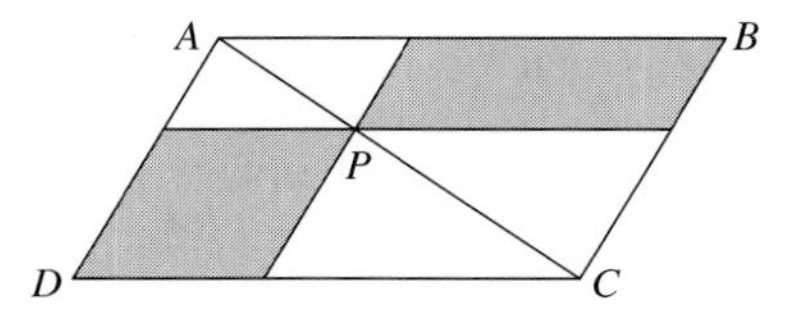

11. Find a way to decompose $\triangle ABC$ and its interior into five subtriangles, each similar to the original triangle $\triangle ABC$, if

(a) the given triangle is a right triangle

(b) the given triangle is isosceles, with angles of measure 30, 30, and 120.

NOTE: A recent theorem due to Werner Raffke ["Partitions of Triangles," *Beitrage Algebra Geometry*, No. 32 (1991), pp. 87–93] shows that the triangles described in (a) and (b) above are the *only ones* that allow such a decomposition.

12. Another Dudeney Dissection Here we consider Dudeney's six-piece dissection of a regular pentagon into a square. The first step is to dissect the pentagon to form a parallelogram. Because we know how to dissect a parallelogram into a square, the two resulting dissections of the parallelogram are superimposed to yield pieces that can be made into a square. This process is shown in the figure. Give your own explanation of how to perform the desired dissection. Verify that it works, according to basic geometric properties.

DISSECTION OF REGULAR PENTAGON TO FORM A SQUARE

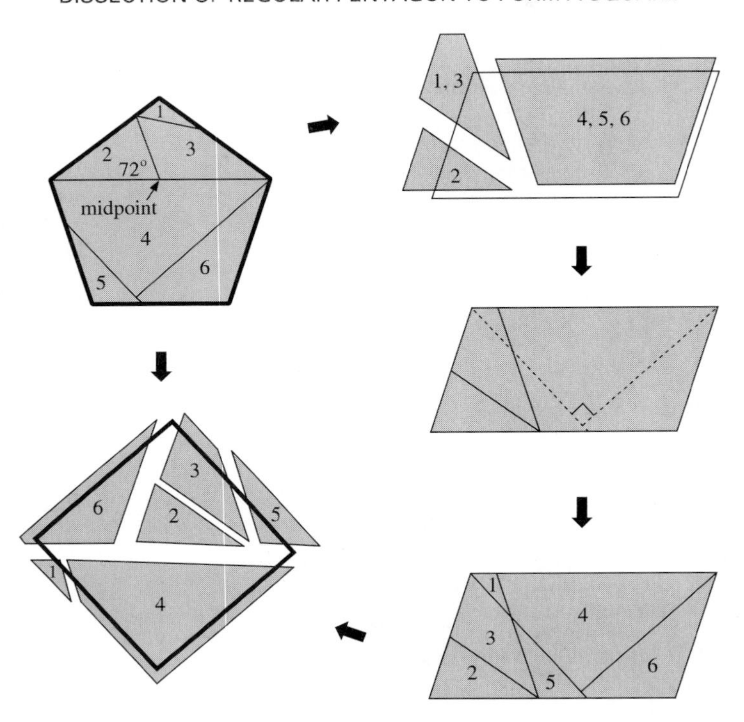

GROUP C

13. Existence of the Nine-Point Circle Observe the figure for this problem, where $\triangle ABC$ is a given triangle, L, M, N are the midpoints of the sides, D, E, and F are the feet of the three altitudes on the sides, H is the orthocenter, and X, Y, and Z are the midpoints of segments $\overline{AH}$, $\overline{BH}$, and $\overline{CH}$. Is there any reason you can think of that would make quadrilateral $MNYZ$ a parallelogram? A rectangle? In the middle figure in the sequence, the diagonals are drawn, intersecting at point U. Show that a circle centered at U passes through M, N, Y, and Z. Finally, use the third figure in the sequence to show that a circle centered at U' passes through M, N, Y, Z, X, and L. (Why does $U' = U$?) For what reason do points D, E, and F lie on this circle?

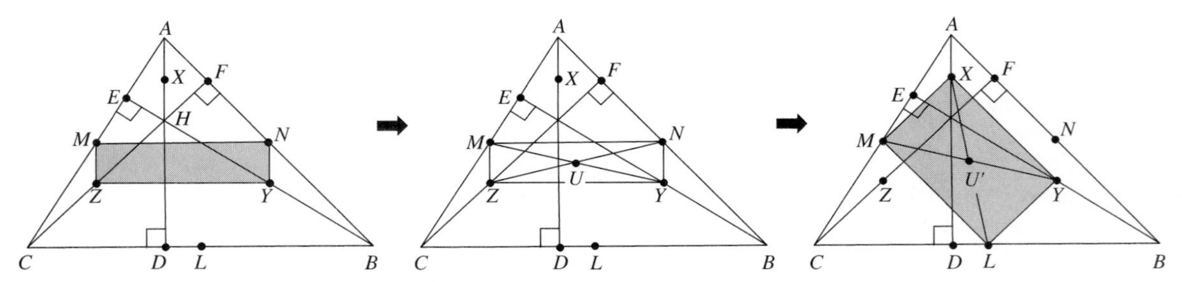

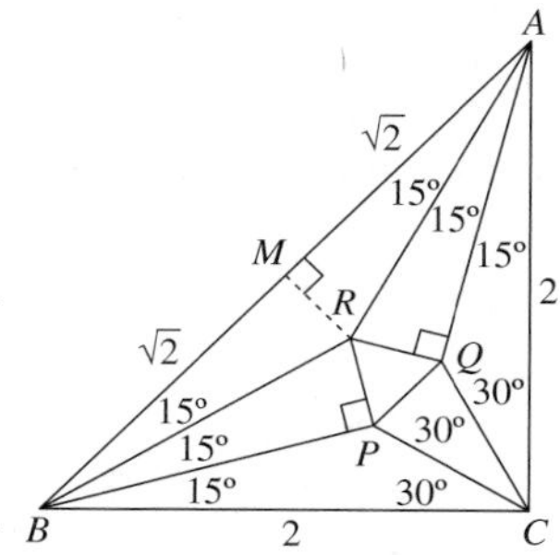

14. Special Case of Morley's Theorem The Law of Sines for $\triangle AQC$ in the figure yields $AQ/\sin 30° = AC/\sin 135°$. From this, derive the value $AQ = \sqrt{2}$. If M is the midpoint of side AB, show that $AM = \sqrt{2} = AQ$. Thus, $\triangle ARQ$ is congruent to $\triangle ARM$, and $\angle RQA$ is a right angle. (Complete these details.) Compute the resulting angle measures of $\triangle PQR$ to show it is equiangular, hence, equilateral.

15. An Ingenious Dudeney Dissection This problem will show how to dissect the regular *heptagon* (seven-sided polygon) to form a square.

(a) The figure below shows a dissection of the heptagon to form a parallelogram. Point C is located on a diagonal so that $AB = AC$, $AL = LD$, and $NE = NF$; L, M, and N are midpoints on their respective segments. It follows that the triangles labeled 1 and 2, 5, and 12 are similar isosceles triangles having angle measures 77 1/7°, 77 1/7°, and 25 5/7° respectively. Verify this, and complete your explanation of how the pieces formed will make a parallelogram.

(b) The square dissection of a parallelogram (discussed earlier) is superimposed onto the preceding dissection. Complete these details and provide your own explanation of the desired dissection of the given heptagon.

DISSECTION OF REGULAR HEPTAGON TO FORM A SQUARE

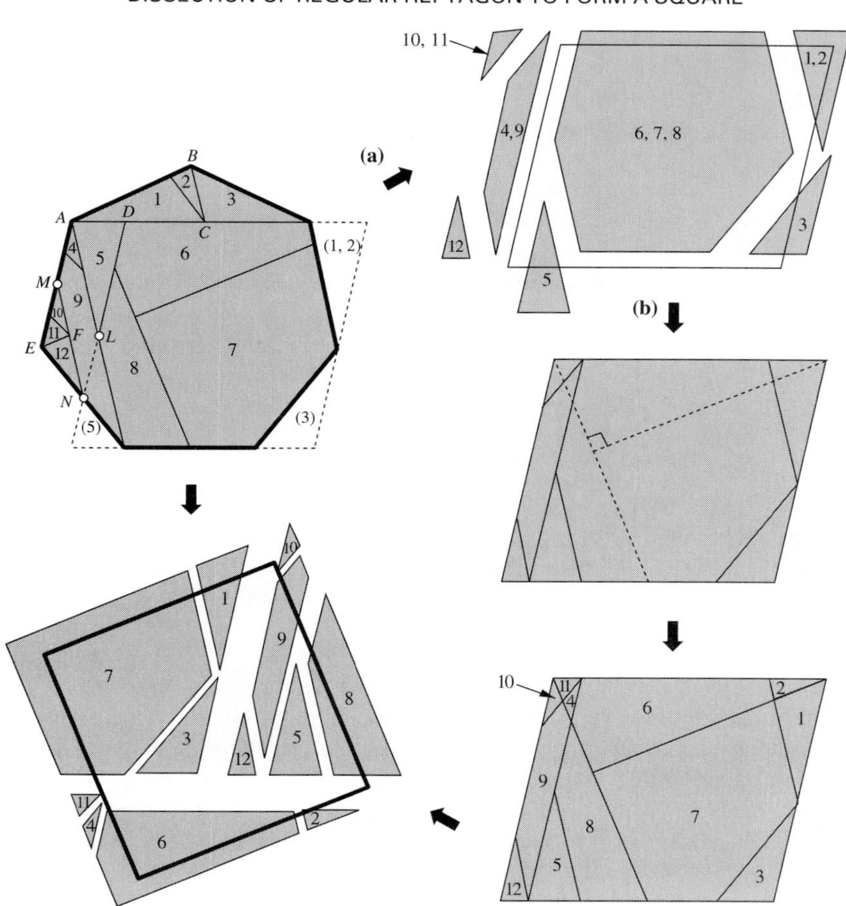

16. Undergraduate Research Project For some integer n, find an n-piece dissection of a *regular pentagon* to form a *regular heptagon*, making use of the dissections of Problems 12 and 15. (***Hint:*** Remember to use the device of *superimposing* one dissection onto another; superimposed cut lines will create additional pieces in the dissection.)

NOTE: A famous theorem of geometry asserts that any two polygons having equal areas may be dissected into the same number of pairwise congruent triangles. Thus, for any two positive integers m and n (≥ 3), a regular m-gon may be cut into triangular pieces, which can be reassembled to form a regular n-gon. This theorem is due to J. Bolyai, one of the discoverers of hyperbolic geometry, discussed later in Chapter 6.

*1.2 Variations on Two Familiar Geometric Themes

We are going to consider here a few unusual ideas involving two fundamental geometric concepts—the Pythagorean Theorem and the number π.

ORIGIN / PROOF OF THE PYTHAGOREAN THEOREM

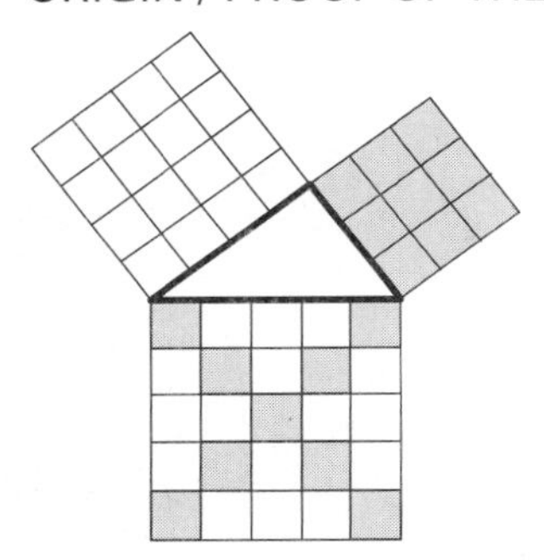

Figure 1.12

There is evidence that the early Egyptians knew of the special Pythagorean relation for the 3–4–5 right triangle, as illustrated in Figure 1.12. But they were probably unaware of the general relationship that holds for any right triangle, and were certainly unable to prove it. The Chinese were also aware of the special case, perhaps as early as 2,000 B.C.E. But the Greek school of geometers founded by Pythagoras, which flourished in the period 600–500 B.C.E., could demonstrate the general theorem, and the theorem took on its current name. Later, in 300 B.C.E., Euclid gave a more sophisticated proof of the theorem in the *Elements*.

The Pythagorean relation is one of the most pervasive formulas in mathematics. A variety of unique proofs appeared in antiquity, a few of which we will explore, and both amateurs and professionals have been deriving new proofs ever since. In his book, *The Pythagorean Proposition*, E. S. Loomis presents a collection of over 370 different proofs of the theorem!

EUCLID'S METHOD OF PROOF

The earliest versions of the Pythagorean Theorem were all stated and demonstrated in terms of *area*. Euclid himself developed the concept of area before he tackled the theorem, and both his statement and proof of it involved area, rather than our modern approach of using ratios and similar triangles. As illustrated in Figure 1.13, Euclid's demonstration consists of the clever idea that the square on the hypotenuse can be split into two rectangles whose areas are, respectively, equal to those of the other two squares.

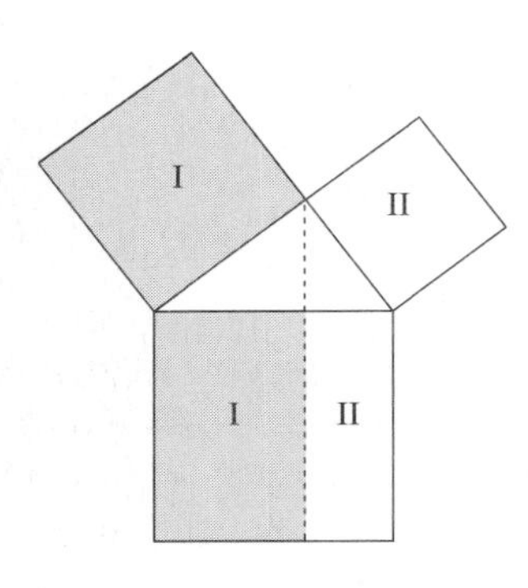

Figure 1.13

A dynamic pictorial proof (Figure 1.14) was given by H. Baravalle (found in the *Eighteenth Yearbook of the NCTM*). This sequence of diagrams is practically self-explanatory. However, it might be helpful to point out that the areas indicated by shading remain constant as one proceeds from one figure to the next—owing to the basic property that two parallelograms having equal bases and equal altitudes have equal areas. Another "proof without words" was invented by the Hindu mathematician Bhaskara (C.E. 1150), which he supposedly drew in the sand, then proclaimed to his students, "Behold!" This was the proof we introduced in Example 2, Section 1.1.

PROOF BY PICTURE: Pythagorean Theorem

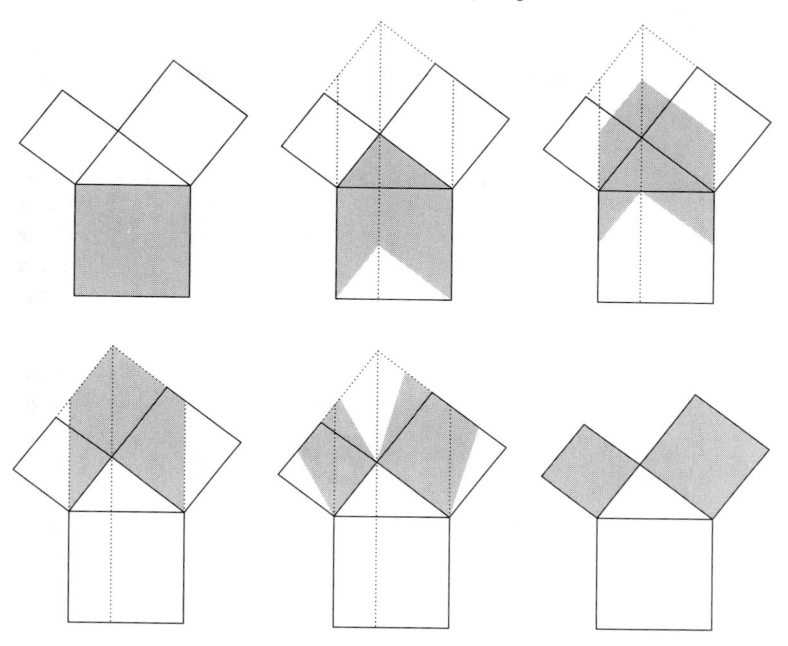

Figure 1.14

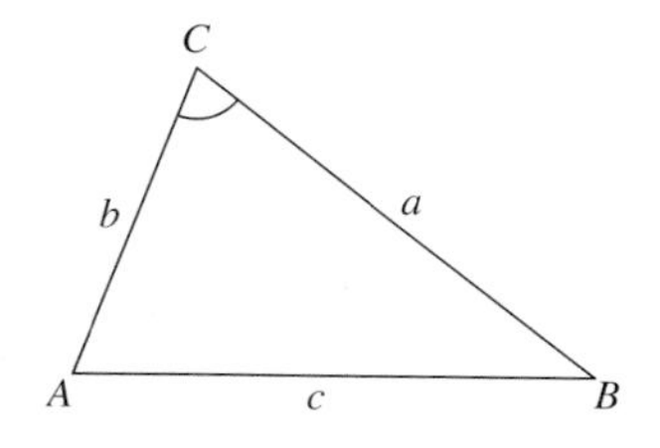

Figure 1.15

There are also a wide variety of *generalizations* of the Pythagorean Theorem, that is, theorems that have the latter as a corollary. You are no doubt already aware of one: the Law of Cosines for any triangle (Figure 1.15):

$$\textbf{(1)} \qquad c^2 = a^2 + b^2 - 2ab\cos C$$

where C represents the measure of $\angle ACB$. When the angle at C measures 90 (degrees), then, since $\cos 90° = 0$, **(1)** reduces to the Pythagorean Theorem.

A PYTHAGOREAN-LIKE THEOREM

A strictly geometric generalization of the Pythagorean Theorem that you probably do *not* know about is very intriguing. It is doubtful that it would ever have been discovered from a strictly algebraic version of the Pythagorean Theorem. In fact, the same sequence of diagrams in Figure 1.14 extends to this theorem, as indicated in Problem 7. It is due to Pappus (C.E. 300), another Greek geometer. We start with *any* triangle, say $\triangle ABC$, as shown in Figure 1.16. Then, instead of squares on the sides, we construct *parallelograms*. Parallelograms I and II on sides $\overline{BC}$ and $\overline{AC}$ can be arbitrary, and point P is the intersection of the lines containing the sides of I and II *opposite* sides $\overline{BC}$ and $\overline{AC}$, respectively. Pappus' Theorem states:

If line segments $\overline{PC}$ and $\overline{AQ}$ are both congruent and parallel, then Area I + Area II = Area III.

PAPPUS' THEOREM: Area I + Area II = Area III

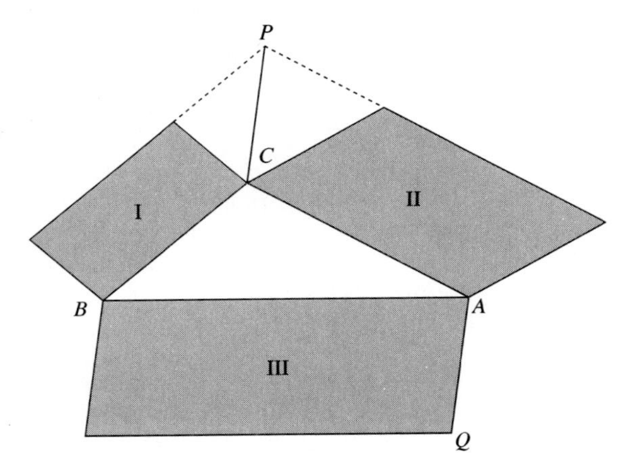

Figure 1.16

Note that in Figure 1.16, there is not a single right angle! Later, we see how this theorem implies the Pythagorean Theorem as a corollary.

EARLY HISTORY OF π

The Greek geometers early recognized a fundamental property of circles, which leads directly to the number π: The circumference (total length) of a circle divided by its diameter is a constant for all circles. To understand why this is true on an informal basis, recall that an inscribed regular polygon of a circle has a perimeter that approximates the circumference of the circle as close as we please by taking a large enough number of sides. Let C and C' be the circumferences of any two circles (Figure 1.17), and r and r' their radii. If we take an imaginary 10,000-sided, regular inscribed polygon for each circle, then, to all intents and purposes, the perimeters of the polygons are equal to the circumferences of the circles. Thus, we have (with incredibly small error), with s and s' as the lengths of the sides of the polygons,

$$C = P = 10{,}000s \qquad \text{and} \qquad C' = P' = 10{,}000s'$$

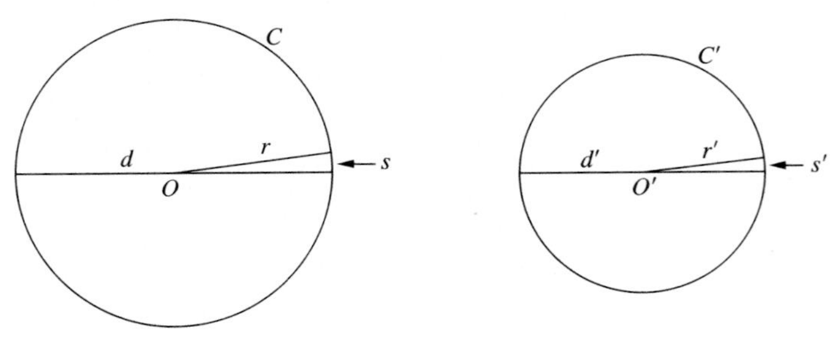

Figure 1.17

That is,

$$\frac{C}{C'} = \frac{10{,}000s}{10{,}000s'} = \frac{s}{s'}$$

But the triangles with vertices at O and O' are isosceles triangles, with vertex angles each having measures equal to 360/10,000, so they are similar triangles, with $s/s' = r/r'$. Hence,

$$\frac{C}{C'} = \frac{s}{s'} = \frac{r}{r'} = \frac{2r}{2r'} = \frac{d}{d'}$$

where d and d' are the lengths of the diameters of the two circles. Because $C/C' = d/d'$, then, by algebra, $C/d = C'/d'$ for any two circles. The Greeks then defined the constant C/d as the number π, whatever its value. Thus we come to the first fundamental formula for a circle involving its radius r and circumference C:

(2) $$C = 2\pi r$$

Bible verses indicate that the value of π is 3; Archimedes knew the value to be somewhat larger, and ultimately gave his famous estimate of $22/7$ ($= 3.14$) accurate to two decimals. Any calculus student has no doubt seen π estimated to at least eight decimal places, 3.141592654···. But a long and interesting history ensued between the early estimates and our present-day technology, along with several key mathematical developments. A survey of this bit of history concerning π reveals the following milestones (see H. Eves, *An Introduction to the History of Mathematics* for a more complete discussion).

A Time Line for π

2500 B.C.E.	Babylonians and Chinese use the value $\pi = 3$.
1650 B.C.E.	Egyptians use the value $\left(\frac{4}{3}\right)^4 = 3.1604\cdots$.
240 B.C.E.	Archimedes' inequality **(7)** below and his approximation $\pi \approx \frac{22}{7}$.
480 C.E.	Chinese mathematician Tsu Ch'ung-chih proposes the value $\frac{355}{113} = 3.1415929\ldots$ (See Problem 19 in this connection.)
1706	John Machin obtained π to 100 decimals using infinite series.
1760	Comte de Buffon devised his famous **needle problem** from which π may be determined by the laws of probability and theoretically tossing a needle.
1767	J. H. Lambert showed that π is irrational.
1853	William Rutherford of England calculated π to 400 decimals.
1882	F. Lindemann showed that π is **transcendental**. That is, π is not the solution of any polynomial equation having rational coefficients (and thus, there is no compass/straight-edge construction of π).
1948	Two mathematicians, D. F. Ferguson and J. W. Wrench, published the value of π to 808 decimals using infinite series.

A Time Line for π *(continued)*

1949 The first use of the electronic computer to calculate π was made. The ENIAC at the Army Ballistic Research Laboratories in Maryland calculated π to 2,037 decimals.

1961 Using an IBM 7090, J. W. Wrench and Daniel Shanks computed π to 100,265 decimals.

1967 Wrench and Shanks computed π to 500,000 decimals.

1989 Two Columbia professors, Gregory V. and David V. Chudnovsky, using a Cray 2 and an IBM 3090, calculated π to over a billion decimals (to be exact, 1,011,196,691 decimal places).

1999 A team headed by Yasumasa Kanada, at the Computer Science Division of the Information Technology Center, University of Tokyo, calculated π to more than 200 billion decimal places. (The 200 billionth digit of π turns out to be a 2.)

Moment for Discovery

Calculating π

The following steps will reveal a way to calculate π to progressively more accurate values, as n increases.

1. Compute the following values on your pocket calculator:

$$\sqrt{2}, \qquad \sqrt{2+\sqrt{2}}, \qquad \text{and} \qquad \sqrt{2+\sqrt{2+\sqrt{2}}}$$

2. To keep track of the number of radicals in the expressions you found in Step 1, write, symbolically,

$$\sqrt{2} = 2_1, \qquad \sqrt{2+\sqrt{2}} = 2_2, \quad \text{and} \quad \sqrt{2+\sqrt{2+\sqrt{2}}} = 2_3.$$

(In general, 2_n would be the same type expression having n radicals.) Record the values you found in Step 1 for 2_1, 2_2, and 2_3.

3. Continue calculating, and recording (or storing in memory) the values for

$$2_4, \qquad 2_5, \qquad 2_6 \qquad \text{and} \qquad 2_7$$

(Note that to calculate each successive value in this sequence, one can simply take the value just calculated, add 2, and take the square root. For example, $2_3 = 1.9615706$, so 2_4 is found by adding 2 and taking the square root:

$$1.9615706 \to 3.9615706 \to \sqrt{3.9615706} = 1.9903695 = 2_4.)$$

4. Was your value for 2_7 rather close to 2.00? (As a matter of fact, it can be shown that $\lim\limits_{n\to\infty} 2_n = 2$.)

5. Find the value of $\sqrt{2 - 2_7}$. (Note that this is *not* equal to 2_8.)

6. Calculate the product $2^8 \cdot \sqrt{2 - 2_7}$. Did anything happen?
7. (**Optional**) If you have a calculator that has an accuracy of 9 or more decimals, continue these calculations on up through 2_{10}, and find the value of $2^{11} \cdot \sqrt{2 - 2_{10}}$ to see what happens. (***Hint:*** The decimal expansion for 2_{10} will have a string of *5 nines*!)

If you are curious about the strange radical expressions for π in the Discovery Unit, in a moment we will show how they may be derived from elementary geometry. The original idea for these formulas came from Archimedes, a contemporary of Euclid. Euclid used inscribed and circumscribed polygons having many sides to deduce properties of circles, and Archimedes elaborated on this, but in a very unique way.

HISTORICAL NOTE

Archimedes (287–212 B.C.E.), sometimes called the **Father of Physics,** was the greatest scientific and mathematical intellect in the ancient world, and, according to some, even up through modern times. He was born of low estate in the ancient Greek city of Syracuse, but because of his ingenious inventions, he was soon noticed by the king. Tradition has it that King Hieron once asked Archimedes to test his gold crown as to its authenticity. As the story goes, a flash of insight struck Archimedes as he was bathing, and he immediately rushed out into the street naked, proclaiming "Eureka! I have found it!" His method was to submerge the crown in water and thereby determine its density, a method known today as **Archimedes' Principle.** Although he was brilliant in physics (having discovered the law of levers and other important concepts), Archimedes was proudest of his achievements in pure mathematics (geometry, in those days). He considered his greatest accomplishment to be the development and proof of the formula for the volume of a sphere, a problem that Euclid never solved. During a siege on Syracuse, the order of the commander Marcelus to spare Archimedes was ignored, and a Roman soldier is said to have murdered him while he was working on a problem—after Archimedes had admonished him not to destroy his drawing. Archimedes' tomb bore the inscription of a sphere inscribed in a cylinder, representing his result that the ratio of these two volumes equals $\frac{2}{3}$, established by pure logic.

ARCHIMEDES' CALCULATION

Since $C = 2\pi$ for a circle of radius 1, the value of π lies between one-half the perimeter of a regular inscribed and circumscribed polygon of n sides for all integers $n \geq 3$. If we start with an inscribed square (Figure 1.18) and begin doubling the number of sides to obtain a regular octagon, then a regular 16-sided polygon, and so on, we know that the perimeter of these polygons will tend to a limit of twice the value of π. The lengths of the sides of these polygons can be derived strictly by applying the Pythagorean Theorem. For convenience, let s and t be the sides of two successive regular polygons, one having twice as many sides as the other. Showing just a part of those polygons, Figure 1.19 indicates that $OA = OM = OB = 1$,

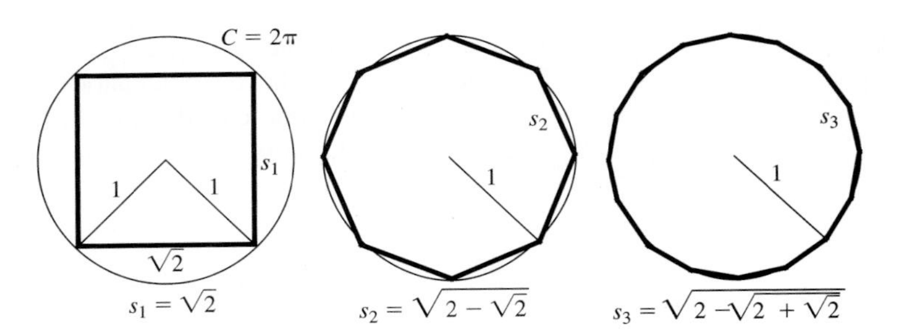

Figure 1.18

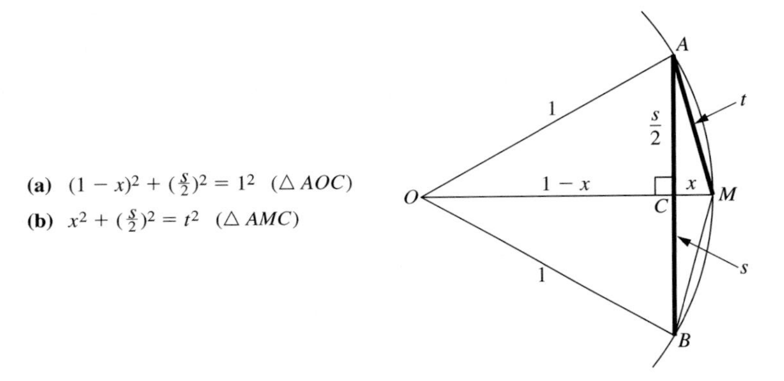

Figure 1.19

$AB = s$, and $AM = t$, where M is the midpoint of arc $\widehat{AB}$. Squaring out the binomial in **(a)**, we have

$$1 - 2x + x^2 + \frac{s^2}{4} = 1 \qquad \text{or} \qquad -2x + \left(x^2 + \frac{s^2}{4}\right) = 0$$

Now use **(b)** to make a substitution for the expression inside the parentheses in the preceding step:

$$-2x + (t^2) = 0 \qquad \text{or} \qquad x = \frac{1}{2}t^2$$

On the other hand, **(a)** yields

$$(1 - x)^2 = 1 - \frac{s^2}{4} = \frac{1}{4}(4 - s^2)$$
$$1 - x = \frac{1}{2}\sqrt{4 - s^2}$$

$$1 - \frac{1}{2}\sqrt{4 - s^2} = x = \frac{1}{2}t^2 \qquad \text{or} \qquad \frac{t^2}{2} = 1 - \frac{1}{2}\sqrt{4 - s^2}$$

Therefore, after we multiply both sides by 2, we obtain

$$t^2 = 2 - \sqrt{4 - s^2} \qquad \text{or} \qquad t = \sqrt{2 - \sqrt{4 - s^2}}$$

Finally, with $t = s_{n+1}$ and $s = s_n$, we obtain the formula for the side of the $(n + 1)$st polygon in terms of that of the nth polygon:

$$s_{n+1} = \sqrt{2 - \sqrt{4 - s_n^2}} \tag{3}$$

Now, s_1 is the side of the inscribed square in Figure 1.18, so we obtain

$$s_1 = \sqrt{2}$$

$$s_2 = \sqrt{2 - \sqrt{4 - \sqrt{2}^2}} = \sqrt{2 - \sqrt{2}}$$

$$s_3 = \sqrt{2 - \sqrt{4 - (2 - \sqrt{2})}} = \sqrt{2 - \sqrt{2 + \sqrt{2}}}$$

and so on. In general,

$$s_n = \sqrt{2 - \sqrt{2 + \sqrt{2 + \cdots \sqrt{2 + \sqrt{2}}}}} \qquad (n \text{ radicals}) \tag{4}$$

Since the perimeter of the nth polygon is $p_n = 2^{n+1} \cdot s_n$, an approximation to π is $\frac{1}{2}p_n = 2^n \cdot s_n$, or

$$\pi_n = 2^n \cdot \sqrt{2 - \sqrt{2 + \sqrt{2 + \cdots \sqrt{2 + \sqrt{2}}}}} \qquad (n \text{ radicals}) \tag{5}$$

If we use the notation suggested in the preceding discovery unit, this formula can be written in the more convenient form

$$\pi_n = 2^n\sqrt{2 - 2_{n-1}} \tag{5'}$$

OUR GEOMETRIC WORLD

The Great Pyramid at Gizeh, Egypt, has a square base with sides 755.8 ft in length and rises 481.2 ft. The ratio of the perimeter of the base to its height is very close to 2π. Some scholars have conjectured that the Egyptians must have made a peculiar use of the circle to obtain measurements, but most believe that this connection with 2π is mere coincidence.

Actually, Archimedes' analysis of π started with a regular inscribed and circumscribed *hexagon* instead of the square that we used in our calculations. This results in a slightly different expression (for its derivation, see Problem 9):

$$\textbf{(6)} \qquad \pi'_n = 3 \cdot 2^{n-1}\sqrt{(2 - \sqrt{2 + \sqrt{2 + \cdots \sqrt{2 + \sqrt{3}}}}} \ (n \text{ radicals})$$

Archimedes used a 96-sided regular inscribed and circumscribed polygon, together with an extensive analysis of square root calculations to obtain the inequality

$$\textbf{(7)} \qquad 3\frac{10}{71} < \pi < 3\frac{1}{7}\left(= \frac{22}{7}\right)$$

or, in decimals,

$$3.140845070\ldots < \pi < 3.142857142\ldots$$

This estimate of π was totally adequate for engineering and construction purposes for many centuries to come.

PROBLEMS (1.2)

GROUP A

1. **A Dissection Proof of the Pythagorean Theorem** Given any right triangle with sides equal to a, b, and c, construct a square having sides of length $a + b$. Then, dissect this square, as shown, in two different ways. Explain how this proves that $a^2 + b^2 = c^2$.

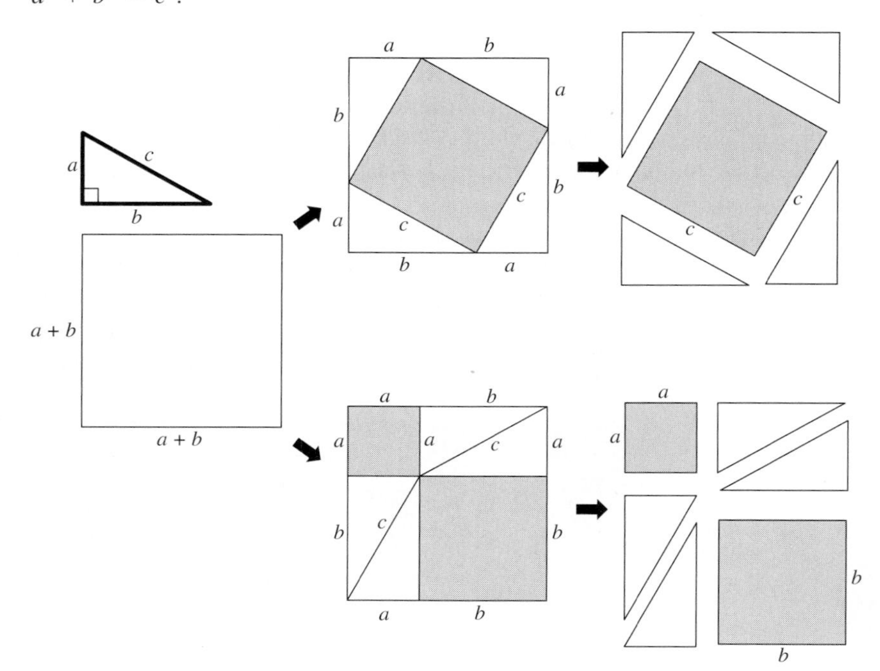

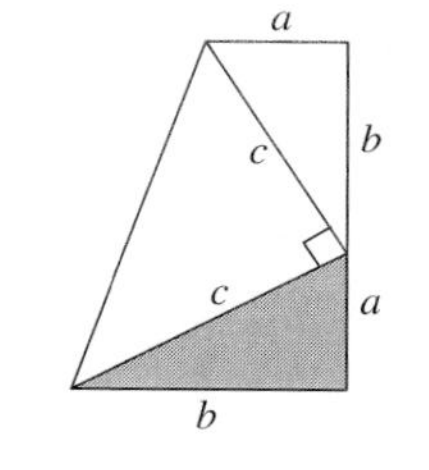

2. **President Garfield's Proof** The twentieth president of the United States, James A. Garfield, presented his own proof of the Pythagorean Theorem in 1876. He was the only president to have published a mathematical proof. It is based on the area formula for a trapezoid having bases b_1 and b_2, and height h: $A = \frac{1}{2}(b_1 + b_2)h$. In the figure, a right triangle having sides of length a, b, and c, and one congruent to it are constructed adjacent to each other, to form part of a square (why is this so?). Explain why a trapezoid is formed, and how the area formula for a trapezoid can be used to prove, algebraically, that $a^2 + b^2 = c^2$.

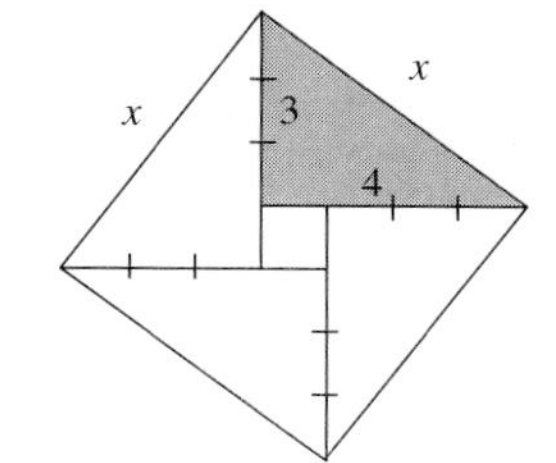

3. Without using the Pythagorean Theorem, show from the figure to the right how the value 5 must be the length of the hypotenuse of a right triangle having legs of length 3 and 4 units. (***Hint:*** Use area.)

4. It can be observed that the lengths of the altitudes of a 3–4–5 right triangle are, respectively, 4, 3, and $2.4 = 12/5$ units. (The altitude to the side of length 5 units is found by using the area formula $A = \frac{1}{2}bh$.)
   (a) Show that the reciprocals of these altitudes, $\frac{1}{4}$, $\frac{1}{3}$, and $\frac{5}{12}$, satisfy the Pythagorean relation $c'^2 = a'^2 + b'^2$.
   (b) Using the area formula for a right triangle in general, show that if h_1, h_2, and h_3 are the altitudes to the sides of length a, b, and c, respectively, then
$$\frac{1}{h_3{}^2} = \frac{1}{h_1{}^2} + \frac{1}{h_2{}^2}$$

5. In connection with Archimedes' formula **(6)** for π, evaluate the following numbers using your pocket calculator:
   (a) $\pi'_3 = 3 \cdot 2^2 \cdot \sqrt{2 - \sqrt{2 + \sqrt{3}}}$
   (b) $\pi'_4 = 3 \cdot 2^3 \cdot \sqrt{2 - \sqrt{2 + \sqrt{2 + \sqrt{3}}}}$
   (c) $\pi'_7 = 3 \cdot 2^6 \cdot \sqrt{2 - \sqrt{2 + \sqrt{2 + \sqrt{2 + \sqrt{2 + \sqrt{2 + \sqrt{3}}}}}}}$

6. Let $x_1 = 2$, $x_2 = x_1 + \sin x_1$, $x_3 = x_2 + \sin x_2, \ldots$. Using the radian mode on your calculator, compute x_2 and x_3. What did you find? What happens if $x_1 = 0.2$? Try other values for x_1 (such as $x_1 = 10, 8, 6.3 > 2\pi$, and $6.2 < 2\pi$). What discoveries did you make? (See the note by Daniel Shanks in *American Mathematical Monthly*, Vol. 99, No. 3 (1992), p. 263, where it is shown that if P is an approximation to π correct to n decimals, $P + \sin P$ is an approximation correct to $3n$ places.)

GROUP B

7. **Dynamic Proof of Pappus' Theorem** The following sequence of figures will prove Pappus' generalization of the Pythagorean Theorem. Explain how it works. (What parallelograms have equal areas in the figures?)

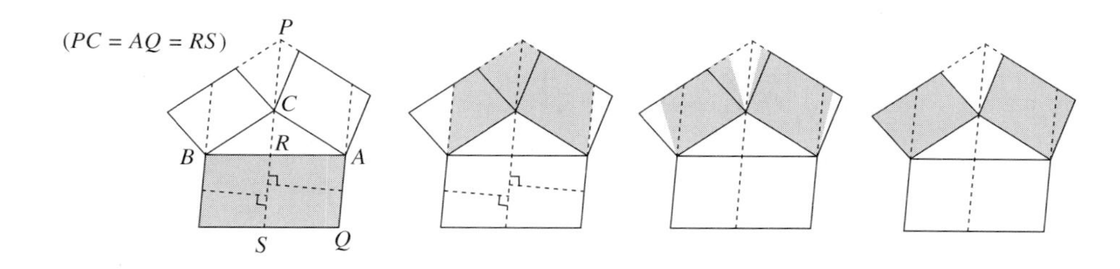

8. An area proof for the Law of Cosines when $C > 90°$ is exhibited in the figure below (found in H. Eves, *Survey of Geometry*, Vol. I, p. 268). Analyze the figure and provide a proof. (From trigonometry, the standard area for a parallelogram with two sides of length a and b and the angle between having measure θ is $A = ab \sin \theta$.)

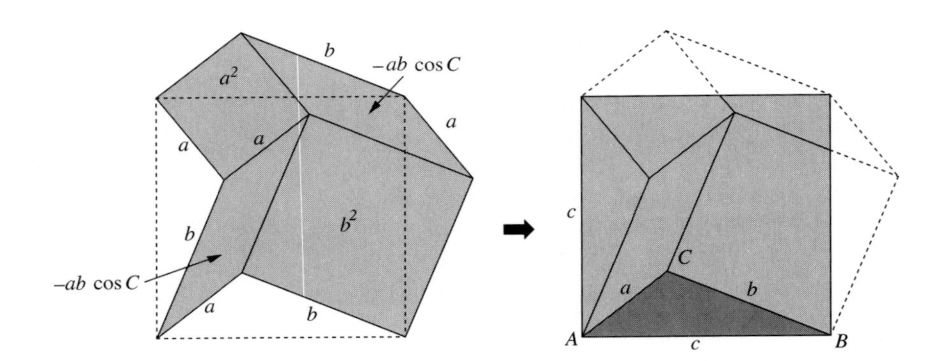

9. Use the equation **(4)** for s_{n+1} to find the length of the side of the inscribed regular dodecagon (12 sides) from the regular hexagon (for which $s_1 = 1$), then find the side of the regular 24-sided inscribed polygon from this. Generalize, to obtain Archimedes' expression for π'_n [equation **(6)**].

10. Archimedes ended his investigation of the value for π with a regular 96-sided inscribed polygon. Use equation **(6)** to obtain the approximation to π that a 96-sided polygon provides. (***Hint:*** In equation **(6)**, $n = 2$ corresponds to a 12-sided polygon. What value for n corresponds to 96 sides?)

11. Show that a parallelogram with sides of length 5 and 8 units, having an altitude of 4 units, can be dissected into three triangles to form a square.

12. Dissect an isosceles triangle with a 2-in. base, with legs each of length $\sqrt{5}$ in., into four triangles that can be reassembled to form a square.

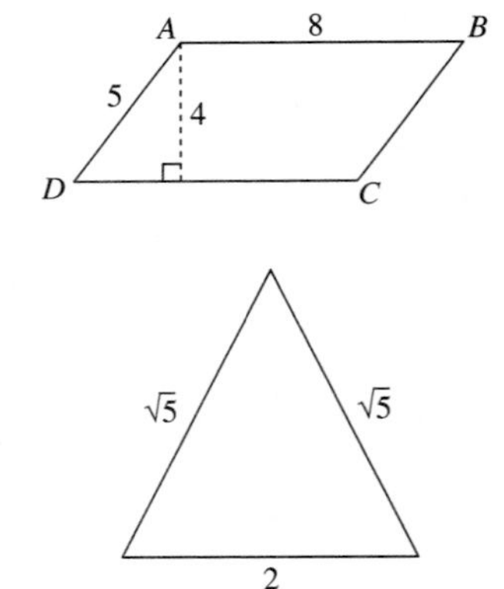

*** 13. Euclidean Construction of the Golden Ratio** The Pythagorean relation leads to a simple construction of the **Golden Ratio** τ, a number that may be defined in terms of the radical expression $\tau = \frac{1}{2}(1 + \sqrt{5}) \approx 1.618$.

(a) Let a be the length of a given segment $\overline{AB}$. Construct the midpoint M of segment $\overline{AB}$, and a square $ABCD$ on $\overline{AB}$ as side. With M as center and $\overline{MC}$ as radius, swing an arc, cutting line $\overleftrightarrow{AB}$ at point E. Show that the ratio AE/AD is the Golden Ratio.

*See Problem 15, Section 4.3 and Problems 16, 17, Section 4.4.

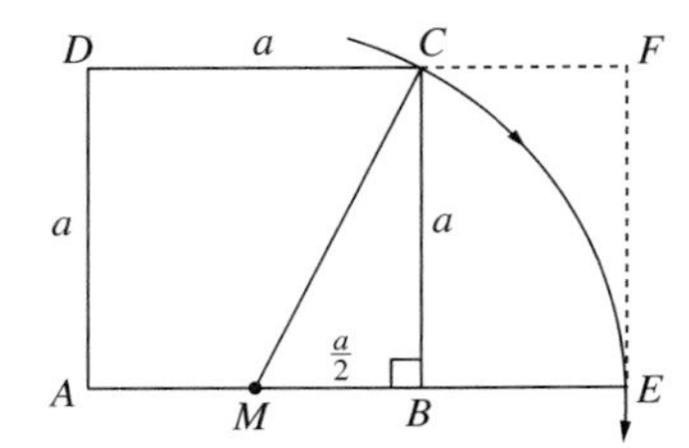

(b) Show that the Golden Ratio τ satisfies the defining relation $\tau^2 = \tau + 1$. (Show that τ is the positive root of the quadratic equation $x^2 - x - 1 = 0$.)

(c) Show that $\tau^{-2} = 2 - \tau$.

14. The Golden Rectangle If a rectangle has two sides whose lengths are in the Golden Ratio (that is, if $AB/BC = \tau$ in rectangle $ABCD$), then it is called a **Golden Rectangle**. (Rectangle $AEFD$ in the figure for Problem 13 provides an example, showing that such rectangles can be constructed with compass and straight-edge.)

A GOLDEN RECTANGLE

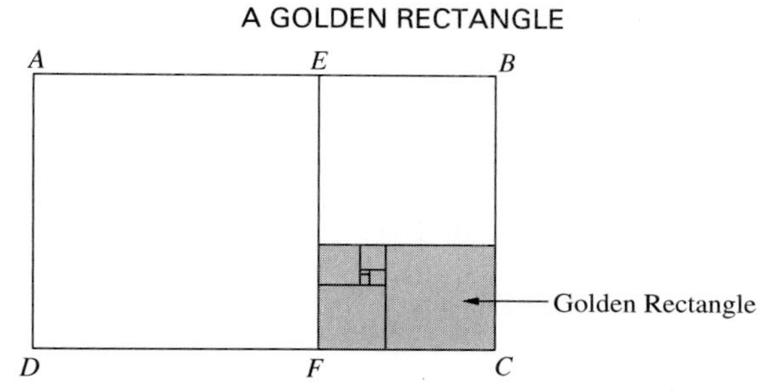

(a) Show that a Golden Rectangle $ABCD$ has the characteristic property that if a square $AEFD$ is cut off one end, there still remains a Golden Rectangle $EBCF$. Thus, the dissection process can be continued, cutting off a square from rectangle $EBCF$, which leaves another Golden Rectangle, and so forth. (***Hint:*** Use the relation from Problem 13: $\tau^2 = \tau + 1$.)

(b) Show that if $BC = 1$, the squares obtained in the dissection in **(a)** have areas 1, $\tau^{-2}, \tau^{-4}, \tau^{-6}, \ldots$.

(c) Give an area proof that $1 + \tau^{-2} + \tau^{-4} + \cdots = \tau$.

GROUP C

15. In 1579, the eminent French mathematician Francois Viète discovered the formula

$$\frac{2}{\pi} = \sqrt{\frac{1}{2}} \cdot \sqrt{\frac{1}{2} + \frac{1}{2}\sqrt{\frac{1}{2}}} \cdot \sqrt{\frac{1}{2} + \frac{1}{2}\sqrt{\frac{1}{2} + \frac{1}{2}\sqrt{\frac{1}{2}}}} \cdots$$

Find how this formula may be derived using the procedure which follows.

(a) Show, by the algebraic device of "rationalizing the denominator" that

$$2 \cdot \frac{1}{\sqrt{2}} = \sqrt{2}, \quad \frac{1}{\sqrt{2 + \sqrt{2}}} = \frac{\sqrt{2 - \sqrt{2}}}{\sqrt{2}} \quad \text{and}$$

$$\frac{1}{\sqrt{2 + \sqrt{2 + \sqrt{2}}}} = \frac{\sqrt{2 - \sqrt{2 + \sqrt{2}}}}{\sqrt{2 - \sqrt{2}}}$$

(b) Generalize the results in **(a)** and show that

$$2 \cdot \frac{2}{\sqrt{2}} \cdot \frac{2}{\sqrt{2 + \sqrt{2}}} \cdot \frac{2}{\sqrt{2 + \sqrt{2 + \sqrt{2}}}} \cdots =$$

$$\lim_{n \to \infty} 2^n \sqrt{2 - \sqrt{2 + \sqrt{2 + \cdots \sqrt{2 + \sqrt{2}}}}} = \pi$$

Thus,

$$\frac{2}{\pi} = \frac{\sqrt{2}}{2} \cdot \frac{\sqrt{2+\sqrt{2}}}{2} \cdot \frac{\sqrt{2+\sqrt{2+\sqrt{2}}}}{2} \cdots$$

(c) Derive Viète's expression.

16. Prove the formula, for $n > 1$,

$$\sin\frac{90^\circ}{2^n} = \frac{1}{2}\sqrt{2 - 2_{n-1}} \equiv \frac{1}{2}\sqrt{2 - \sqrt{2 + \sqrt{2 + \cdots \sqrt{2 + \sqrt{2}}}}}$$

(n radicals)[2]

17. Consider the two expressions for π introduced previously—as limits of the sequences $\{\pi_n\}$ and $\{\pi'_n\}$ in equations **(5)** and **(6)**. We could combine them into the single expression

$$\pi = \lim_{n\to\infty} a \cdot 2^{n-1} \cdot \sqrt{2 - \sqrt{2 + \sqrt{2 + \cdots \sqrt{2 + \sqrt{a}}}}} \text{ (n radicals)}$$

where $a = 2$ and 3, using the two equations **(5)** and **(6)**, respectively. This leads to the conjecture that the above relation holds for all positive values of a. Verify or disprove the conjecture using specific values of a and testing the formula with your calculator. If you find that the conjecture is false, try to prove that the above relation holds *only* for the values $a = 2$ or 3. (In connection with this, see Problem 18.)

18. To what does the following pseudo-Archimedean sequence converge, if a is a real number, $0 < a < 4$?

$$u_n = 2^n\sqrt{2 - \sqrt{2 + \sqrt{2 + \cdots \sqrt{2 + \sqrt{a}}}}} \qquad \text{(n radicals)}$$

What if $a \geq 4$? (***Hint:*** The key lies in a geometric interpretation of the length of an arc of the unit circle, using regular inscribed polygons to estimate the length. Start with a *chord* of length $s_1 = \sqrt{4 - a}$.)

19. Ruler–Compass "Construction" of π Verify the end result of the following details of a compass/straight-edge construction for a very close approximation to π. Let $\overline{AB}$ be a line segment having unit length, and let line $\overleftrightarrow{AC}$ be perpendicular to $\overleftrightarrow{AB}$, with $AC = AB(= 1)$. The remaining steps are as follows:

(1) $CD = \frac{AC}{8} = \frac{1}{8}$ (obtainable by successive bisections)

(2) $BE = MB = \frac{1}{2}$ (construct the midpoint M of $\overline{AB}$ and swing an arc)

(3) $\overleftrightarrow{EF} \perp \overleftrightarrow{AB}$ (standard construction—see inside front cover)

(4) $\overleftrightarrow{EG} \parallel \overleftrightarrow{DF}$ (standard construction)

(5) $AB = BH = HK$, $KL = GB$ (segment duplication)

(6) Result: $AL \approx \pi$, with an error of less than 0.0000003. Show this by calculating $AL = GB + 3$ from applications of similar triangles and the Pythagorean Theorem.

---

[2]Use of $\equiv$ designates a definition or identity in this text.

EUCLIDEAN "CONSTRUCTION" FOR π

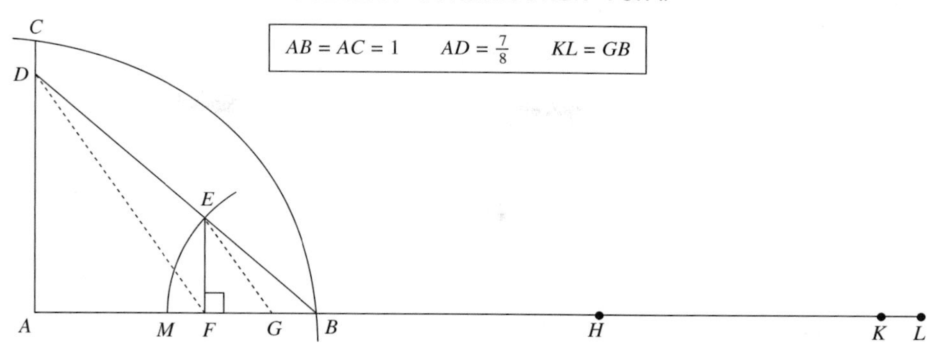

NOTE: This construction was discovered in 1849 by a German mathematician, Jakob de Gelder. It makes use of the Chinese approximation of 480 C.E. mentioned above.

1.3 Discovery via the Computer

There are a variety of software packages available that make it possible to observe virtually any geometric property on the computer. We mention two of these, which are commonly used in computer laboratories in numerous schools, colleges, and universities. They are

THE GEOMETER'S SKETCHPAD™ (Key Curriculum Press, Berkeley, California)

CABRI™ (Texas Instruments, Inc., Dallas, Texas)

For this textbook we have adopted the former, so the commands and menus referred to here are from that software (they can be readily adapted to fit other programs as well). The basic techniques pertinent to the concepts we explore may be found in Appendix C. For the most part, we expect you to already know how to carry out procedures such as constructing a segment between two points, finding the midpoint of a segment, constructing perpendicular and parallel lines, bisecting an angle, and so on. To help you get started in this venture, we will give rather explicit directions at first, which will gradually taper off as you progress, until finally, only the desired constructions are stated, without explicit reference to the software.

In order to alert you to the use of computer graphics from this point on, an icon ( ) will be inserted in the text at the appropriate places.

EXAMPLE 1 Suppose we want to examine the proposition that the sum of the measures of the angles of a triangle equals 180. First, choose Preferences in the DISPLAY menu and check Autoshow Labels for points. (This will automatically label your figure in agreement with references made here.) Next, use the SEGMENT TOOL to draw a triangle, and then use the MEASURE command to measure each angle (select three vertices in sequence with the shift key depressed, then choose Angle from the MEASURE menu). Finally, choosing Calculate from the MEASURE menu, we can display the angle-sum on the screen and observe the effect of dragging a vertex of the triangle to change its shape. This sequence of operations would appear as in Figure 1.20. (You might actually try this for warm-up if you do not have much experience with *Sketchpad*.) ■

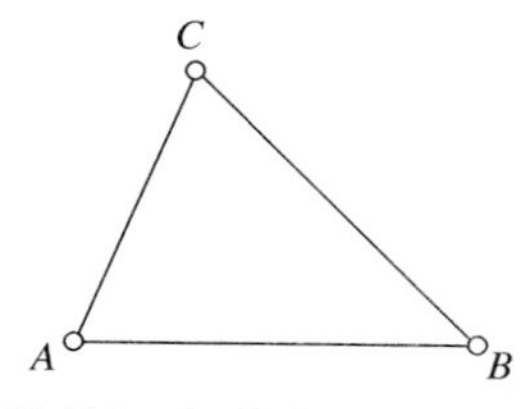

(1) Triangle *ABC* constructed and automatically labeled

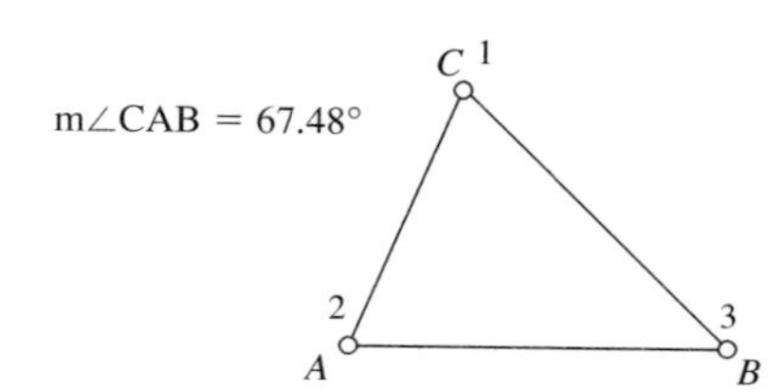

(2) Points selected, in order, to find measure of angle *A*

m∠CAB = 67.48°
m∠ABC = 45.00°
m∠BCA = 67.52°
m∠CAB + m∠ABC + m∠BCA = 180.00°

(3) All three angle measures displayed, with their sum

m∠CAB = 61.87°
m∠ABC = 30.67°
m∠BCA = 87.46°
m∠CAB + m∠ABC + m∠BCA = 180.00°

(4) Vertex *B* is dragged to observe the effect on angle measures

Figure 1.20

EXAMPLE 2 The following steps will show how to construct a parallelogram and observe some of its properties. (See Figure 1.21.)

m∠CAB = 63.65°
m∠ABB′ = 116.35°
m∠BB′C = 63.65°
m∠B′CA = 116.35°
m∠CAB + m∠B′CA = 180.00°
AC = 1.294 inches
BB′ = 1.294 inches
Area ABB′C = 2.331 inches²

Figure 1.21

(1) With the Autoshow Labels for points turned on (using Preferences from the DISPLAY menu), construct segments $\overline{AB}$, and $\overline{AC}$, dragging A to B, then A to C, using the SEGMENT tool.

(2) Select point A, then point C, with the shift key depressed. Choose Mark Vector from the TRANSFORM menu. Then, select both segment $\overline{AB}$ and point B, and choose Translate from the TRANSFORM menu.

(3) Construct segment $\overline{BB'}$ to complete the parallelogram, using the procedure of Step (1).

(4) Select the vertices in sequence A, B, B', C) and choose Polygon Interior from the CONSTRUCT menu.

(5) Display the following measurements on the screen using MEASURE, and from its menu, Angle, Distance, Calculate, and Area: $m\angle A$ ($\angle CAB$), $m\angle B$, $m\angle B'$, $m\angle C$, $m\angle A + m\angle C$, AC, BB', and Area $ABB'C$ (accessed by first selecting the interior of the parallelogram, then choosing Area).

NOTE: Since segment $\overline{AB}$ was translated to $\overline{CB'}$, we already know that $AB = CB'$ and that lines $\overleftrightarrow{AB}$ and $\overleftrightarrow{CB'}$ are parallel. What features do you observe that remain constant as point C is dragged about? (For area, keep C on a fixed line parallel to $\overleftrightarrow{AB}$.) What basic theorems concerning parallelograms are illustrated? ■

You might find it more interesting to execute the steps suggested in the next example to demonstrate a property that is less familiar (which was mentioned earlier in Section 1.2).

EXAMPLE 3 Verifying Pappus' Theorem

Follow these steps to demonstrate Pappus' Theorem on the computer (introduced in Section 1.2).

(1) With the Autoshow Label option for points turned on, use SEGMENT tool to construct $\triangle ABC$.

(2) Use the procedure of Example 2 to construct a parallelogram $ACD'D$ on side $\overline{AC}$, and another parallelogram $BCE'E$ on side $\overline{BC}$. (See Figure 1.22.)

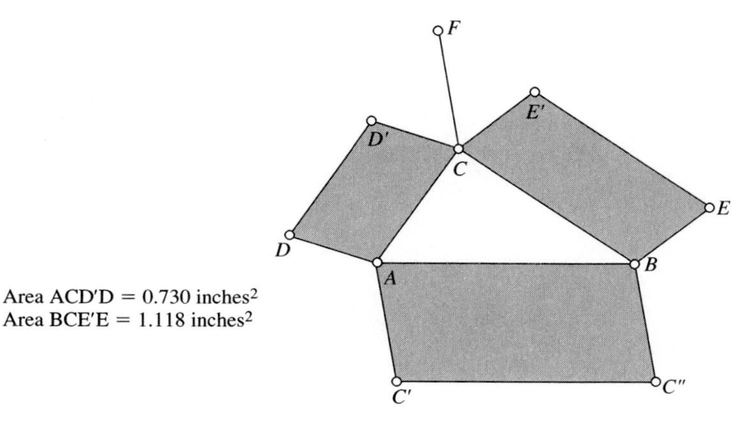

Figure 1.22

(3) Construct lines $\overleftrightarrow{DD'}$ and $\overleftrightarrow{EE'}$ using Line from the CONSTRUCT menu (you have to select the LINE tool for this operation ahead of time), then point and click on the intersection F of these two lines. Hide the lines.

(4) Construct segment $\overline{CF}$, and translate it, along with point C, from point F to point A (use Mark Vector "F->A" from the TRANSFORM menu). Point C will translate to point C'.

(5) Construct a parallelogram $AC'C''B$ on side $\overline{AB}$.

(6) As in Example 2, display the areas of the three parallelograms on the screen, and Area $ACD'D$ + Area $BCE'E$. Compare the last line with Area $ABC''C'$. Drag points C and F to observe the effect on the figure and the above calculations. ■

Now it is time for you to make a discovery of your own, involving a concept not yet introduced in this book. The answer will not be revealed or discussed until the next section. We first need to define a concept that will be used here.

If $\triangle ABC$ is any triangle and P any point, suppose we were to construct the perpendiculars from P to the sides $\overline{AB}$, $\overline{BC}$, and $\overline{AC}$ (extended, if necessary), and then locate the points of intersections D, E, and F with the sides of the triangle (as illustrated in Figure 1.23). The triangle DEF thus formed is called the **Pedal Triangle of $\triangle ABC$ with respect to point** P. (The points D, E, and F are called the **feet** of the perpendiculars from P on the sides.) As P assumes various positions in the plane, the pedal triangle will assume different shapes. Investigating this feature is the purpose of the next discovery unit.

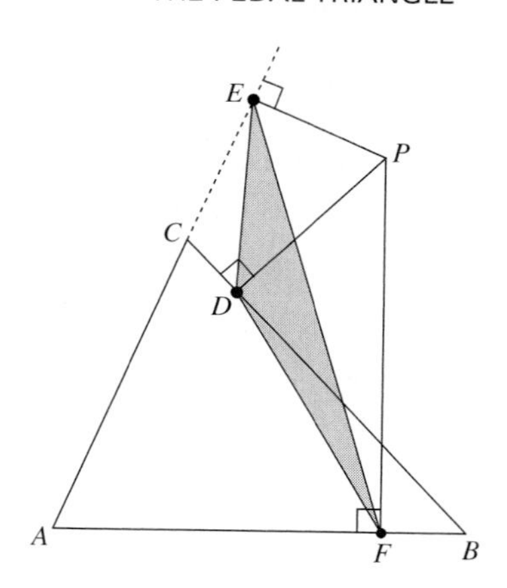

Figure 1.23

Moment for Discovery

Pedal Triangles and the Simpson Line

Follow these steps for an investigation of the pedal triangle of a given triangle.

1. With the automatic labeling device activated, construct $\triangle ABC$ and locate point D. (For a better demonstration, make the angles at A and B have measures as close to 30 and 45, as possible.) Construct line $\overleftrightarrow{AB}$, selected, and select point D, then choose Perpendicular Line from the CONSTRUCT menu. Select the point of intersection E (the foot) of the perpendicular and line $\overleftrightarrow{AB}$.
2. As in Step 1, construct the feet F and G of the perpendiculars from D to lines $\overleftrightarrow{BC}$ and $\overleftrightarrow{AC}$. Hide all lines and perpendiculars, leaving just the feet E, F, and G on the sides of $\triangle ABC$, and, of course, point D. Construct the pedal triangle $\triangle EFG$.
3. Using Polygon Interior from the CONSTRUCT menu, construct the interior of $\triangle EFG$. With the interior selected, display the area of $\triangle EFG$ on the screen using Area from the MEASURE menu.
4. Drag point D to make Area $\triangle EFG$ change value. How large can you make it? How small? Is it ever zero? Is there a position for which the pedal triangle appears to be equilateral, and is there more than one position for this to occur? Can you make the pedal triangle similar to (the same shape as) $\triangle ABC$?
5. Construct the circumcircle of $\triangle ABC$ as follows: Construct the midpoints H and I of segments $\overline{AB}$ and $\overline{BC}$, then the perpendiculars to $\overline{AB}$ at H and to $\overline{BC}$ at I, and select the point of intersection, J. Hide these objects of construction. With points J and A selected, choose Circle By Center And Point from the CONSTRUCT menu. The circumcircle of $\triangle ABC$ will appear. Now drag point D to point J. Did anything happen? Can you see a geometric reason for this?
6. Drag point D to any point on the circumcircle. Do you notice anything? Drag D about the circumcircle. What takes place? Also, change the shape of $\triangle ABC$ to see if this phenomenon depends on the shape or size of the given triangle. Write down the conclusions you think must be true from this last step.

Although working with *Sketchpad* can be gratifying (allowing you to make ruler/compass constructions effortlessly), and although its power may help you to see geometry in motion (objects can be moved about at the slightest whim), your learning should not stop there. It is easy to lose sight of the fundamental principles of geometry that make it all possible if our entire focus is on experimental drawing. As a future teacher or mathematician, you should form the habit when using the computer of always asking the question "why" or "what makes this or that feature work." And then you should make a concerted effort to find the answer. Even if you avoid doing this now, one day your students—or colleagues—may well be asking that same question of you and expecting an informative answer. At such times you will want to be in a position to not only know the answer, but (as a teacher) to be able to time your answer for the most effective teaching strategy.

It should also be recognized that you cannot truly appreciate a theorem in geometry unless you first recognize it visually for several different cases, and then understand what makes it true from basic principles (its *proof*). If we totally ignored proofs, learning geometry would be like trying to learn chemistry merely by performing experiments from the laboratory manual, with no explanation about chemical principles as to what was going on. The same thing is true about learning geometry without the theory.

This is not to say that in this book we intend to prove every theorem or relationship we encounter, but rather, our focus will be to teach you how to establish certain fundamental principles from axioms, and then to take you a short distance beyond. The proofs of some concepts actually belong in a graduate-level geometry course, and are not appropriate for an introductory textbook. However, when they are interesting, we will occasionally *mention* geometric results just to show you what can be achieved in geometry, without being obliged to prove them. A case in point is the subject of the next section in this book (Section 1.4), an optional section. It involves more advanced work with *Sketchpad*, so if you feel you are not ready for it, skip this section for now, and resolve to return to it later when you have more experience. You definitely do not want to miss it.

PROBLEMS (1.3)

All problems of this section, except Problem 9, require the use of a computer and *Sketchpad* software.

GROUP A

1. Execute the following steps in *Sketchpad* concerning the way the area and perimeter of a triangle may vary.

[1] Construct segment $\overline{AB}$ and any line ℓ parallel to $\overleftrightarrow{AB}$. (One way to do this is to construct line $\overleftrightarrow{AB}$, translate it using Translate from the TRANSFORM menu, and finally, hide the line leaving just the segment.)

[2] Select a point C on line ℓ, and with C still selected, select A and B (with shift key depressed). Choose Polygon Interior from the CONSTRUCT menu. This will enable you to measure the area and perimeter of $\triangle ABC$.

[3] Select the interior of $\triangle ABC$ and choose Area and Perimeter from the MEASURE menu. (At this point Area $\triangle ABC$ and Perimeter $\triangle ABC$ will be displayed on the screen for observation.)

By dragging point C along line ℓ, find what behavior concerning area and perimeter occurs. When is the area/perimeter maximum or minimum? If you can, explain the phenomena from basic geometric principles.

Area CAB = 1.215 inches ^{2}
Perimeter CAB = 5.244 inches

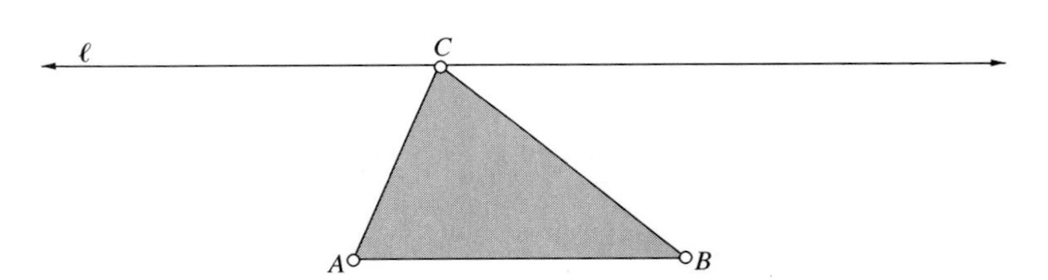

2. Repeat the steps in Problem 1 when line ℓ is *not* parallel to line $\overleftrightarrow{AB}$. (To provide more insight regarding the perimeter, construct line m perpendicular to ℓ at C, with D on m. Display $m\angle ACD$ and $m\angle BCD$ on the screen.)

3. Follow these steps in *Sketchpad* (see figure).

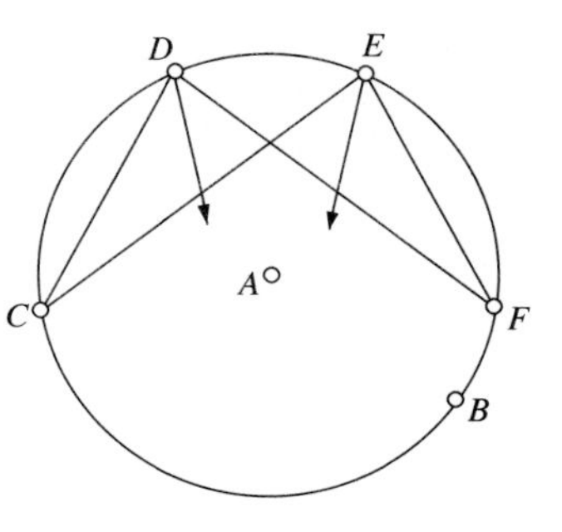

[1] Using the Circle Tool, construct a circle AB and locate points C, D, E, and F on that circle.

[2] Construct segments $\overline{CD}$, $\overline{DF}$, $\overline{CE}$, and $\overline{EF}$.

[3] Construct the bisectors of $\angle CDF$ and $\angle CEF$. (Choose Angle Bisector from the CONSTRUCT menu after selecting C, D, and F, and C, E, and F.)

[4] Construct the point G of intersection of the bisectors from Step 3.

[5] Drag both D and E to various locations on the circle to observe the effect.

What did you find? Can you back up your observation on the basis of geometric principles? (***Hint:*** Review the facts concerning inscribed angles of a circle; see Appendix B or inside front cover.)

4. Carry out the following experiment using *Sketchpad*.

[1] Construct $\triangle ABC$ and locate point D on side $\overline{AB}$ (shown in the figure below).

[2] Construct segment $\overline{AD}$, and select segments $\overline{AD}$ and $\overline{AB}$ in that order. Choose Mark Ratio from the TRANSFORM menu. (This will store the ratio AD/AB.)

[3] Select point B for the choice Mark Center on that menu, then select point C for a dilation along segment $\overline{BC}$, as follows: Choose Dilate from the TRANSFORM menu, with marked ratio checked. Point C will be mapped to C' such that $BC' = BC \cdot (AD/AB)$.

[4] Repeat Step 3 to dilate point A along segment $\overline{AC}$ to A' such that $CA' = AC \cdot (AD/AB)$. (Thus, the ratios AD/AB, BC'/BC, and CA'/CA are equal.)

[5] Construct the segments $\overline{AC'}$, $\overline{BA'}$, and $\overline{CD}$. Obtain the points of intersection E, F, and G of these segments, and construct the interior of $\triangle EFG$ displaying its area on the screen using Polygon Interior from the CONSTRUCT menu and Area from the MEASURE menu.

Drag point D on segment $\overline{AB}$. What fundamental theorem in geometry did you "discover"?

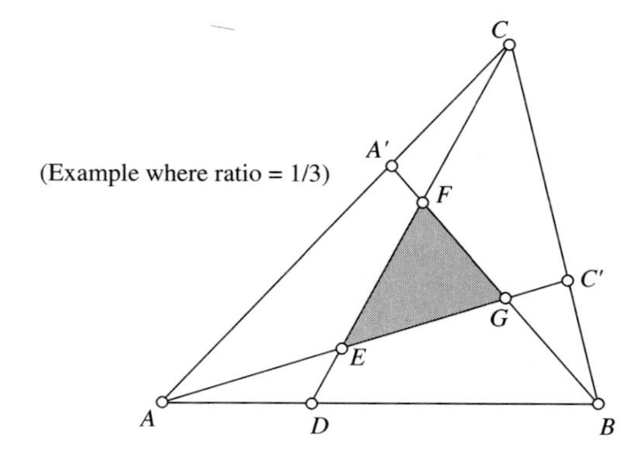

***5. Ptolemy's Theorem** The Greek astronomer Claudius Ptolemy (c. 100–170 C.E.), who invented sine tables to be used for calculations in astronomy, discovered an interesting formula relating the sides and diagonals of a cyclic quadrilateral (one whose vertices lie on a circle). Use these instructions in *Sketchpad* to reveal this theorem.

[1] Construct $\triangle ABC$.

[2] Construct the circumcircle of $\triangle ABC$ by selecting points A, B, and C and using Arc Through Three Points from the CONSTRUCT menu, then selecting points A, C, and B and repeating this command.

[3] Locate point D at random (not on the circle). Construct segments $\overline{AD}$, $\overline{BD}$, and $\overline{CD}$, then, select each of the six segments and display their lengths on the screen. (Use Length from the MEASURE menu.)

[4] Using Calculate from the MEASURE menu, display the value

$$x = AB \cdot CD + BC \cdot AD - AC \cdot BD \qquad (\text{where } AB = m\overline{AB}, \text{etc.})$$

Observe the value of x as you drag point D. Is this value ever negative? Would you conjecture that perhaps $x \geq 0$? Is x ever zero? (Be sure to move D on arc $\overset{\frown}{AC}$.) What do you think Ptolemy's Theorem might be for a cyclic quadrilateral whose sides are of length a, b, c, and d, and diagonals, m and n, as shown?

*See Problem 10, Section 1.4.

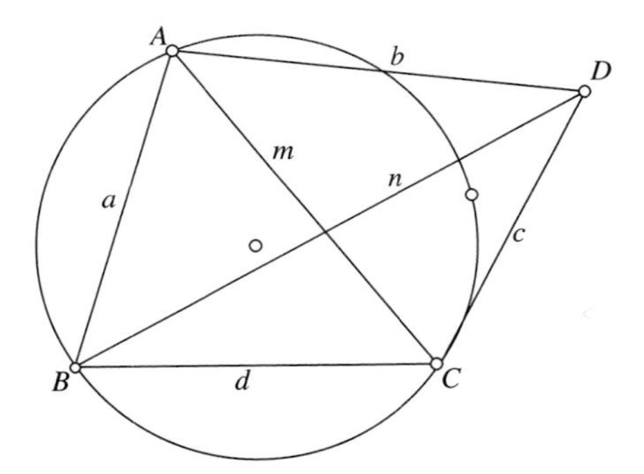

6. Using *Sketchpad*, construct an equilateral $\triangle ABC$, and locate P any interior point. Measure the distances PD, PE, and PF using Distance from the MEASURE menu. Call these distances x, y, and z, as shown in the figure. Calculate, and display, $x + y + z$.

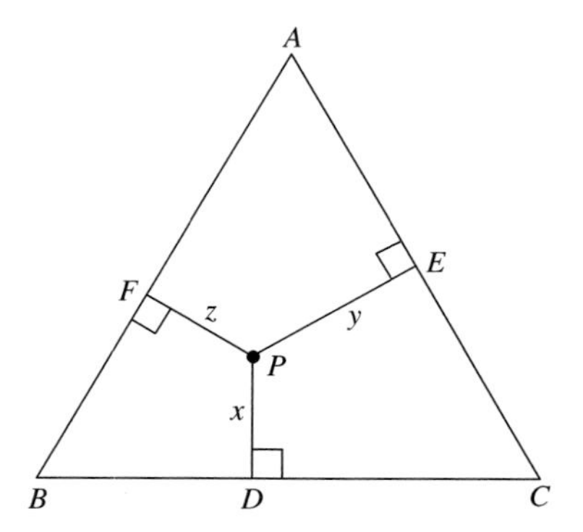

(a) Drag P inside $\triangle ABC$ and find what location(s) make the sum $x + y + z$ a maximum and minimum. What behavior do you notice. Can you prove it?

(b) Examine the sum $x + y + z$ when P lies outside $\triangle ABC$.

7. Let $\triangle ABC$ be an equilateral triangle, and construct its incircle (circle tangent to the sides of $\triangle ABC$, with center at the point of intersection of the angle bisectors).

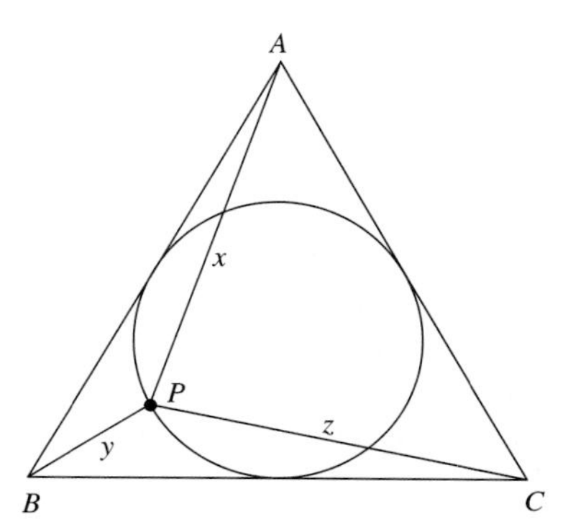

Locate P on this circle, and measure x, y, and z, as shown in the figure. Investigate the maximum/minimum properties of

(a) $x + y + z$

(b) $x^2 + y^2 + z^2$

8. Morley's Theorem The steps below will direct you in an illustration of Morley's Theorem, mentioned in Section 1.1.

[1] Construct $\triangle ABC$ using the Segment Tool.

[2] Select vertices C, A, and B in that order (with shift key depressed) and choose Angle from the MEASURE menu. Then, choosing Calculate from that menu, find the ratio $m\angle CAB/3$, which will be displayed on the screen.

[3] Select the displayed ratio ($m\angle CAB/3$ and choose Mark Angle Measurement. Then from the TRANSFORM menu, select point A and choose Mark Center. Construct ray $\overrightarrow{AB}$, selected, and choose Rotate, with Rotate By Marked Angle checked. Click OK and the first angle trisector will be displayed. With that trisector still selected, choose Rotate again, and the second angle trisector will appear. Hide ray $\overrightarrow{AB}$.

[4] Repeat the procedure of Step 3 at vertices B and C, producing the angle trisectors of the other two angles. (Select ray $\overrightarrow{BC}$, *not* ray $\overrightarrow{BA}$, to be rotated about B for the trisectors of $\angle B$, and select ray $\overrightarrow{CA}$ for the trisectors of $\angle C$.)

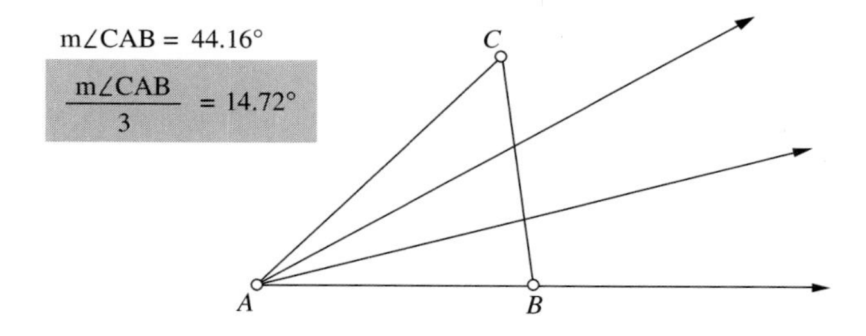

[5] Select the points of intersection of the angle trisectors closest to the sides of $\triangle ABC$ and, with the three points selected, choose Segment from the CONSTRUCT menu. This will produce $\triangle DEF$. For extra effect, display the interior of $\triangle DEF$ using Polygon Interior from the CONSTRUCT menu, with D, E, and F selected.

[6] Measure the sides of $\triangle DEF$, displaying these measurements on the screen. (Select D and E, and choose Distance from the MEASURE menu, to display the value DE, and repeat this procedure for pairs E, F and F, D.)

Drag vertex B to various positions to see the effect on these measurements and the shape of $\triangle DEF$.

GROUP B

9. An Experiment in Reasoning Let $\triangle ABC$ be given, and construct lines through the vertices parallel to the opposite sides, as shown in the figure. If P, Q, and R are the resulting points of intersection of these lines, several parallelograms are formed, one of them shown as a shaded region. (Recall that the opposite sides of a parallelogram are congruent.) Attempt to answer all the questions below, then see if your conclusion proves an important theorem in geometry.

(1) Is quadrilateral $QABC$ a parallelogram? Is $QA = CB$?

(2) Is quadrilateral $ACBR$ a parallelogram, and is $CB = AR$?

(3) Does it follow that A is the midpoint of segment $\overline{QR}$?

(4) What reasoning tells you whether B and C might be the midpoints of segments $\overline{PR}$ and $\overline{PQ}$?

(5) Would the perpendiculars $\overline{AT}$, $\overline{BU}$, and $\overline{CV}$ to segments $\overline{QR}$, $\overline{RP}$, and $\overline{QP}$ have to meet at some common point? What is that point, relative to $\triangle PQR$?

(6) Recall that if a line is perpendicular to one of two parallel lines, it is also perpendicular to the other. Does the point of concurrency of Step 5 have any bearing on an important theorem about $\triangle ABC$?

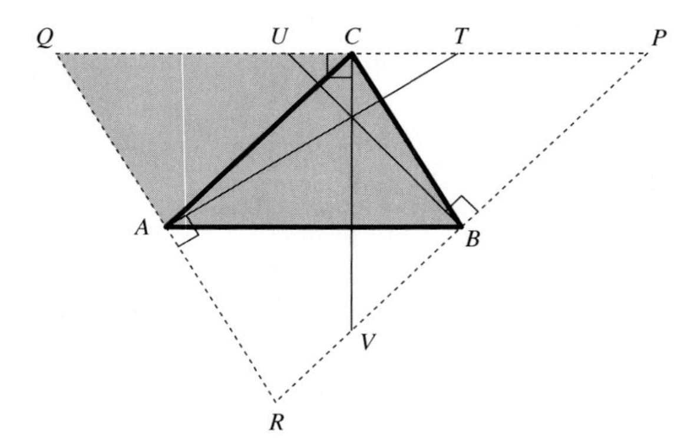

10. The Ellipse on *Sketchpad* The instructions in this problem will construct an ellipse strictly by geometry, using *Sketchpad*.

[1] Construct segment $\overline{AB}$ and locate points C and D on it (in the order A–C–D–B).

[2] Construct a circle centered at A passing through C. (Use Circle By Center and Point from the CONSTRUCT menu.)

[3] Construct segment $\overline{CB}$ using Segment from the CONSTRUCT menu. With $\overline{CB}$ still selected, select point D and choose Circle By Center and Radius.

[4] Select, or construct, the points of intersection E and F of the two circles. (If the circles do not intersect, move point D closer to point C.) Hide the circles.

[5] Select points E and F and choose Trace Points under DISPLAY.

As C is moved along segment $\overline{AB}$, points E and F will trace out an ellipse with foci at A and D. Explain why this is so, using the definition for an ellipse. Move D closer to A, then further away from A. What effect does this have on the ellipse?

NOTE: Save this construction for the next problem.

11. Using the ellipse constructed in Problem 10, construct segments $\overline{AE}$ and $\overline{DE}$, and display Area $\triangle ADE$ and Perimeter $\triangle ADE$ on the screen. Observe these two measurements as C is dragged along segment $\overline{AB}$. Where does the maximum and minimum of these two measurements occur, and why?

GROUP C

12. **The Euler Segment of a Triangle** In this problem we will demonstrate an intriguing property of the important points associated with any triangle: the circumcenter O, the centroid G, and the orthocenter H.

[1] Construct $\triangle ABC$.

[2] Construct two perpendicular bisectors of the sides of $\triangle ABC$ (locate two midpoints of sides using Point At Midpoint from the CONSTRUCT menu, then Perpendicular from that menu). Select the point of intersection F $(= O)$, which is the circumcenter of $\triangle ABC$, and hide the two lines.

[3] Using the midpoints D and E from Step 2, construct segments from the opposite vertices, and select the point of intersection, G. This is the centroid of $\triangle ABC$. Hide the two segments used to find G, and the midpoints D and E.

[4] Construct two altitudes of $\triangle ABC$ (select A, then segment $\overline{BC}$, and choose Perpendicular Line from the CONSTRUCT menu, to get the altitude from A). Select the point H of intersection. This is the orthocenter of $\triangle ABC$. Hide the altitudes just constructed.

[5] Construct segment $\overline{OH}$ $(\cong \overline{FH})$.

(a) Does it appear that the points O $(= F)$, G, and H are collinear? To provide an accurate test, calculate the distances FG, GH, and FH, and then the value $x = FG + GH$. How does x compare with FH? Drag vertex B and observe the effect.

(b) For a further relationship, examine the distances FG and FH. Do you find that $FH = 3 \cdot FG$? To find out, exhibit the value $FH - 3 \cdot FG$ on the screen. Again drag B to see if this affects that relationship.

(c) State the general theorem you illustrated here.

NOTE: Save this construction for the next problem.

13. **Euler Line and the Nine-Point Center** Construct $\triangle ABC$ and the following nine points (continuing from the construction of Problem 12): midpoints of the sides (D, E, I), feet of the altitudes (J, K, L), and the Euler points (M, N, O) (the midpoints of segments $\overline{AH}$, $\overline{BH}$, and $\overline{CH}$). Does it appear that these nine points lie on a circle? Construct the circumcenter $U(= R)$ of $\triangle DEI$ by finding midpoints P and Q of segments $\overline{DE}$ and $\overline{EI}$ and perpendiculars at these points. Hide objects of construction.

(a) Does it appear that U lies on the Euler Line of points O, G, and H?

(b) Measure the distances $HU(=HR)$, $UO(=RF)$, and $HO(=HF)$. Did you discover anything else peculiar? State a general theorem concerning the points H, U, G, and O (orthocenter, Nine-Point Center, centroid, and circumcenter) of any triangle $\triangle ABC$.

NOTE: Save the Euler segment $\overline{HO}$ and the four points H, U, G, and O for the next problem.

14. Harmonic Relation and the Euler Segment Using directed distance, the **Cross Ratio** of four collinear points A, B, C, and D is defined by the number

$$(AB, CD) \equiv \frac{AC/AD}{BC/BD} = \frac{AC \cdot BD}{AD \cdot BC}$$

If that Cross-Ratio has the value -1, then the four points A, B, C, and D are said to be **in harmonic relation**. An example is provided by taking the four points on the x-axis: $A{:}x_1 = 0$, $B{:}\ x_2 = 15$, $C{:}\ x_3 = 12$, and $D{:}\ x_4 = 20$ (as indicated in the figure). Using the formula for directed distance $P_1P_2 = x_2 - x_1$, we find

$$AC = 12 - 0 = 12 \quad BD = 20 - 15 = 5$$

$$AD = 20 - 0 = 20 \quad BC = 12 - 15 = -3$$

and

$$\frac{AC \cdot BD}{AD \cdot BC} = \frac{12 \cdot 5}{20 \cdot (-3)} = \frac{60}{-60} = -1$$

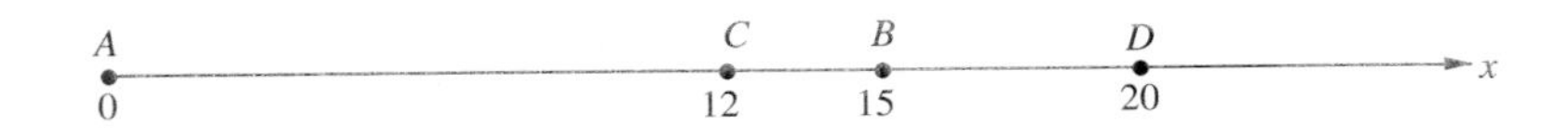

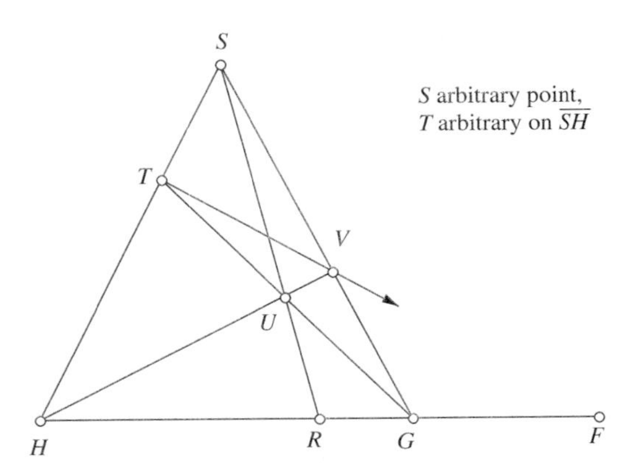

(a) Make a copy of the Euler segment from the construction of Problem 13 by using EDIT, Copy, and Paste, after selecting segment $\overline{HF}$ and points H, F, G, and R. Translate (by dragging) to a new position on the screen and delete the rest of the figure. (If segment $\overline{HF}$ is too small, select point H and use the Select, Dilation tool of *Sketchpad* to enlarge it; this will not affect any ratios of distances on that segment.) Calculate the Cross-Ratio $(HG, RF) = (HR \cdot GF)/(HF \cdot GR)$. You should obtain 1 for the answer. (The Distance feature on *Sketchpad* does not recognize directed distance, so the answer will be positive.)

(b) Follow these steps in *Sketchpad* to obtain an interesting geometric construction, involving the harmonic relation:

[1] Select point S not on line $\overleftrightarrow{HF}$ and construct segments $\overline{SH}$, $\overline{SR}$, and $\overline{SG}$.

[2] Locate T, any point on segment $\overline{SH}$, and construct segment $\overline{GT}$. Select the point U of intersection of segments $\overline{GT}$ and $\overline{SR}$.

[3] Construct line $\overleftrightarrow{HU}$ and locate its intersection, V, with segment $\overline{SG}$. Hide this line, and construct segment $\overline{HV}$.

[4] Construct line $\overleftrightarrow{TV}$.

Did anything happen? Drag points S and T about and observe the effect? State any theorem (or theorems) you think might be true [$HRGSTUV$ is sometimes referred to as the **harmonic configuration** (or **construction**) relative to H, R, G, and F].

NOTE: This so-called harmonic configuration is valid for any four collinear points that are in the harmonic relation (Cross-Ratio $= -1$). The proof of this belongs more appropriately to that area of geometry known as *Projective Geometry*.

15. Neuberg's Theorem: The Third Pedal Triangle of a Triangle Using *Sketchpad*, construct $\triangle ABC$. Choose any point D, and construct the *first* pedal triangle $\triangle EFG$ with respect to D. (Use the procedure of Steps 1 and 2 from the Discovery Unit above—*Pedal Triangles and Simpson's Line*.) From D, construct the pedal triangle $\triangle HIJ$ of $\triangle EFG$, and finally, from D, construct the pedal triangle $\triangle KLM$ of $\triangle HIJ$ (the *second* and *third* pedal triangles of $\triangle ABC$ with respect to D). If you know how, create a Script for the process of constructing the first pedal triangle, selecting A, B, C, and D, in order, and use this for the second and third pedal triangles.

(a) Drag the vertices A, B, and C in such a way that you make $\triangle ABC$ equilateral (or as close as you can make it). Did you observe anything peculiar happening to $\triangle KLM$?

(b) Make $\triangle ABC$ a distinctive right triangle, say with one rather long leg. What happened this time? In both cases, drag D and observe the effect.

NOTE: The theorem being demonstrated here is due to J. Neuberg. Its proof may be found in Coxeter and Greitzer, *Geometry Revisited*, p. 24.

*1.4 Steiner's Theorem

Our purpose here is to present one of the more beautiful theorems of geometry, and you are encouraged to actively participate in its observation. Its intricate relationships with several other geometric theorems, such as Morley's Triangle and the Nine-Point Circle (unrelated by themselves), make it a rare gem in the annals of geometric results. Discovered by Jacob Steiner in the 1800s, it is to mathematics what a famous poem or essay would be to literature.

Let us begin by discussing the components that lead up to Steiner's Theorem (which, by the way, we do not attempt to prove[3]). They will be stated as facts, some

[3] A proof of Steiner's Theorem can be found in the article by E. H. Lockwood, "Simson's Line and It's Envelope," *Mathematical Gazette,* Vol. 37 (1953), pp. 124–125.

of which you may already have discovered for yourself from previous sections. Two such facts were introduced in Section 1.1, as summarized below. (See Figure 1.24).

Nine-Point Circle

The midpoints of the sides of any triangle, the feet of the altitudes on the sides, and the midpoints between the vertices and orthocenter all lie on a circle.

Morley's Theorem

Corresponding angle trisectors of any triangle intersect at points that determine the vertices of an equilateral triangle.

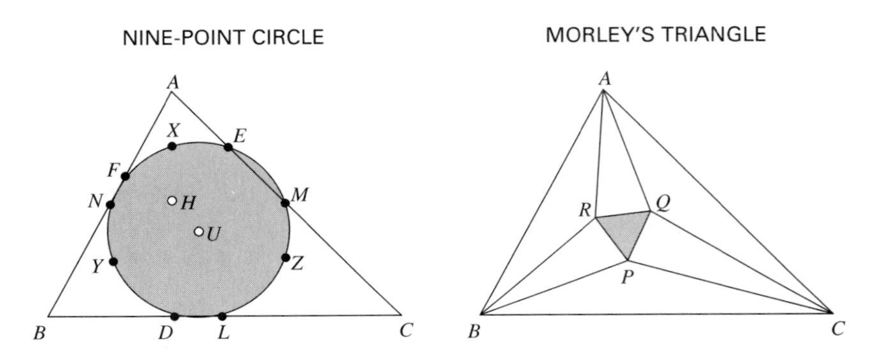

Figure 1.24

HISTORICAL NOTE

Jacob Steiner was born in 1796 in Germany. Although he did not learn to write until he was 14 years old, he made up for it later with his prolific writing in geometry. He became a serious student of mathematics at the age of 18 and soon began his studies at the universities of Heidelberg and Berlin. Steiner was one of the leading contributors to the celebrated *Crelle Journal of Mathematics*—his articles appeared in practically every issue, and everything he submitted for publication was published. A chair in geometry was established for him at Berlin in 1834, a position he held until his death in 1867.

Steiner and M. Chasles, another outstanding nineteenth century geometer, independently gave a full development of (synthetic) projective geometry, including a construction of the classical conics by projective methods. All of Steiner's work was **synthetic** (that is, in the style of Euclid, without the use of coordinates). In spite of this, his arguments in maxima and minima surpassed in power the analytic methods then in vogue. Even today, Steiner's methods and developments remain models of elegance in geometry. It was once said that Steiner hated analysis as thoroughly as Lagrange disliked geometry. But, inversely, Steiner must have liked geometry as much as Lagrange liked analysis.

A third fact we will need involves the Simson Line of a triangle, which was introduced in Section 1.3. If P is any point, recall that the pedal triangle DEF of $\triangle ABC$ with respect to P varies as P moves in the plane of the given triangle. With a little work, the following theorem can be proven (if you are interested in the proof, Problems 8–11 below are devoted to that task). If you completed the discovery unit of Section 1.3 successfully, then you have already discovered this property for yourself.

Existence of the Simson Line

The pedal triangle of a triangle with respect to a point on its circumcircle, as shown in Figure 1.25, degenerates to a line segment, and the feet of the perpendiculars from that point to the three sides of the triangle are collinear.

THE SIMSON LINE

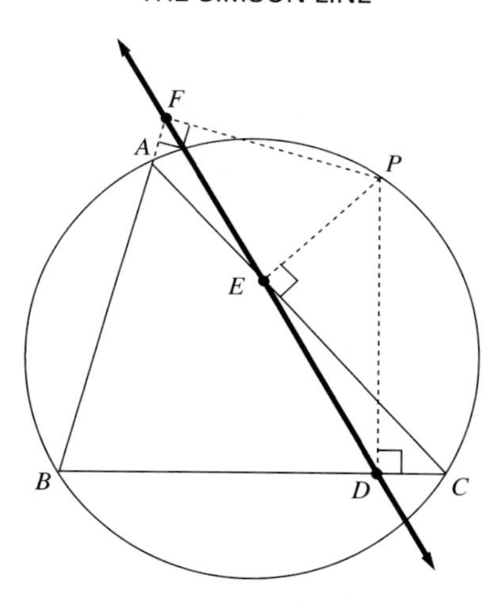

Figure 1.25

The line of collinearity in the preceding theorem is called the **Simson Line**, after Robert Simson (1687–1768), an anomaly since the line was never discovered by Simson, but rather by William Wallace in 1797. There are an infinity of Simson Lines, one for each position of P on the circumcircle. The motion of the Simson Line as P revolves about the circle is what we are most interested in for the remainder of this section. Its demonstration is made to order for the Trace/Animate feature of *Sketchpad*.

HYPOCYCLOIDS: ASTROID AND DELTOID

One last concept will be needed to fully appreciate Steiner's Theorem. You may have studied in calculus the curve we need to consider here, called the **hypocycloid**. This is the path that a point takes on one circle as it rolls inside another circle (without slipping). That point will come into contact with the base circle a number of times, and these points are called the **cusps** of the hypocycloid, as illustrated in

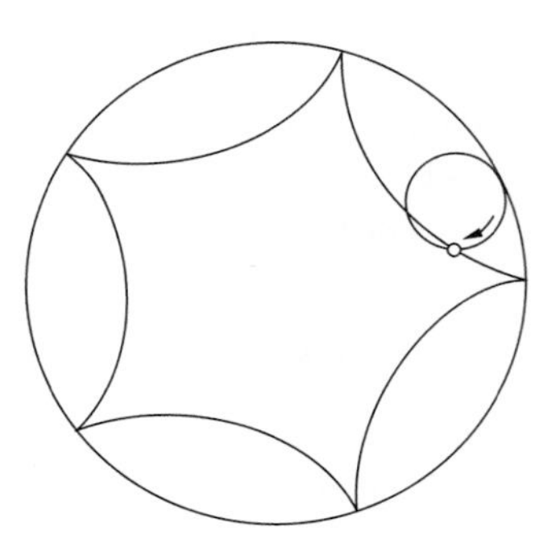

Figure 1.26

Figure 1.26. If the radius of the rolling circle has the appropriate value, the curve will be periodic and repeat itself after one cycle. Figure 1.26 shows such a hypocycloid, which has five cusps. The hypocycloid of four cusps (Figure 1.27) is called an **astroid**. It has a particularly simple coordinate equation:

$$x^{2/3} + y^{2/3} = a^{2/3} \tag{1}$$

The curve involved in Steiner's Theorem is the **deltoid**—a hypocycloid having three cusps, also shown in Figure 1.27. Its equation is very complicated. Because we do not intend to go into any proofs here, the equations for the general hypocycloid will not be considered—they can be found in almost any calculus text.

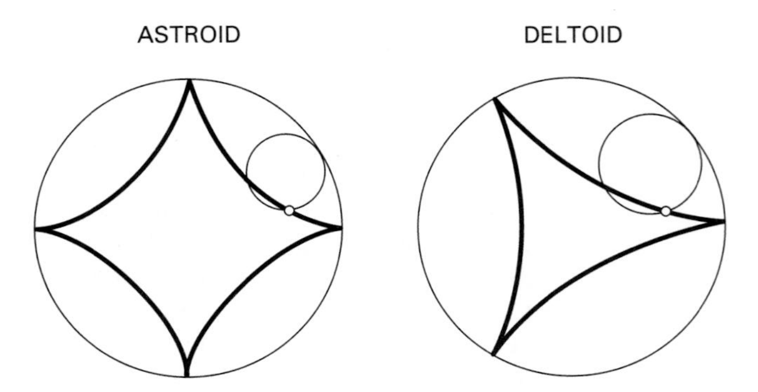

Figure 1.27

EXAMPLE 1 The mechanical definition for the hypocycloid (given above) provides the best way to graph it using *Sketchpad*. Here's how.

[1] Construct a circle AB (as radius) and locate an arbitrary point C on that circle. (This will be the base, or stationary, circle; see Figure 1.28.)

Circumference ⊙AB = 7.286 inches
Circumference ⊙DC = 2.996 inches

$$\frac{(\text{Circumference } \odot\text{AB})}{(\text{Circumference } \odot\text{DC})} = 2.432$$

Arc angle $\widehat{\text{c1BC}}$ = 44.85°

$$(-1) \cdot (\text{Arc angle } \widehat{\text{c1BC}}) \cdot \left(\frac{\text{Circumference } \odot\text{AB}}{\text{Circumference } \odot\text{DC}}\right) = -109.06°$$

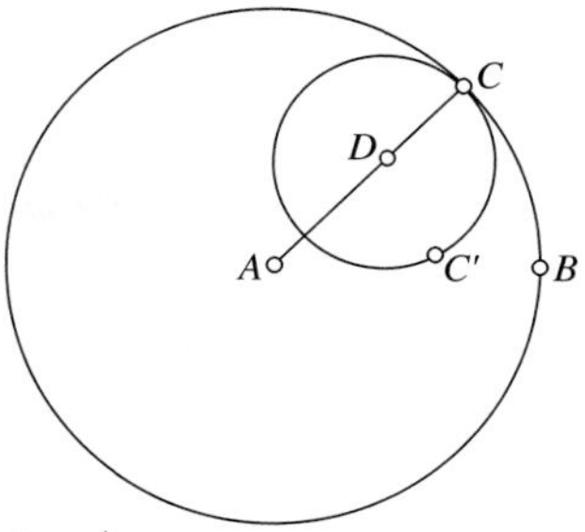

Figure 1.28

[2] Locate an arbitrary point D on segment $\overline{AC}$, and selecting points C and D, choose Circle By Center and Point under CONSTRUCT. A circle appears, centered at point D, and tangent to the base circle at point C. (This will be the rolling circle.)

[3] Select circle AB and choose Circumference under MEASURE. Repeat for circle DC. Using Calculate under MEASURE, display the ratio $x = (\text{Circumference circle } AB)/(\text{Circumference circle } DC)$.

[4] Select point B, then C, and circle AB. Then choose Arc on Circle under CONSTRUCT. With the arc still selected, choose Arc Angle under MEASURE.

NOTE: It is important to check to see if Arc Angle actually measures the amount of rotation, up to 360°. To do this, drag point C around the circle. If you are only getting arc measure only up to 180°, you will need to go back and select Arc Angle again, or make other appropriate adjustments.

[5] Calculate the quantity $[(-1) \cdot (\text{Arc angle } \widehat{c1BC} \cdot x]$ using Calculate, and selecting the displayed values appropriately. With this calculation selected (on screen) choose Mark Angle Measurement under TRANSFORM.

[6] Select point D and choose Mark Center under TRANSFORM (or double click on point D). Select C and choose Rotate, with By Marked Angle checked. Point C will be rotated clockwise about D with a rotation of $x°$ to point C'.

[7] Select point C', then C, and choose Locus under CONSTRUCT. A curve will appear, part of which should resemble a hypocycloid.

You can now adjust this curve by moving D on segment $\overline{AC}$. To get a "closed" hypocycloid, the ratio x must be very close to an integer (e.g., $x = 5.997$ or $x = 6.005$ will suffice). If necessary, enlarge the figure by moving point B, then you can adjust the value of x with greater accuracy, and shrink it down again. The value of x is the number of cusps. If you drag C counterclockwise about circle AB (slowly), the small circle actually appears to roll inside the larger circle as C' traces the curve! ■

OUR GEOMETRIC WORLD

One of the components of an automatic transmission is known as a *planetary gear train*. This is an arrangement of three rotating gears within a larger interior gear. A single tooth of one of the inside gears will travel along the path of a hypocycloid. On the command to shift gears, an electronic signal causes the gears to lock in place, and the outer gear, which is attached to the drive shaft, and rear axle begin to rotate at the same speed as the inner shaft, which is attached to the engine. The vehicle is now in "high gear."

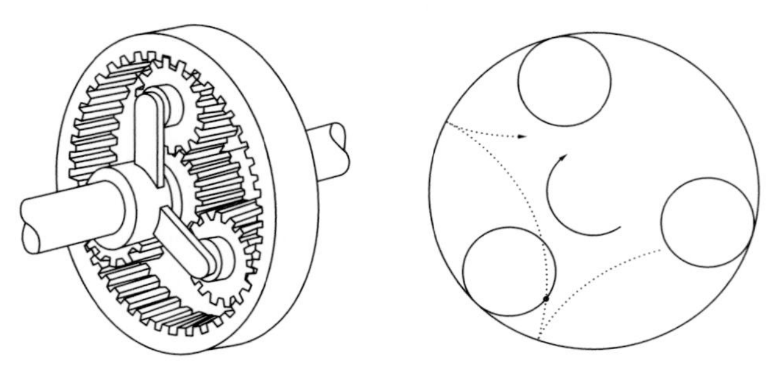

Figure 1.29

STEINER'S THEOREM

So far we have discussed the Nine-Point Circle of a triangle, the Morley Triangle, Simson's Line, and the deltoid—seemingly altogether unrelated topics. It is amazing that they all may be tied together in a beautiful theorem of geometry. We refer you to Figure 1.30 for illustration.

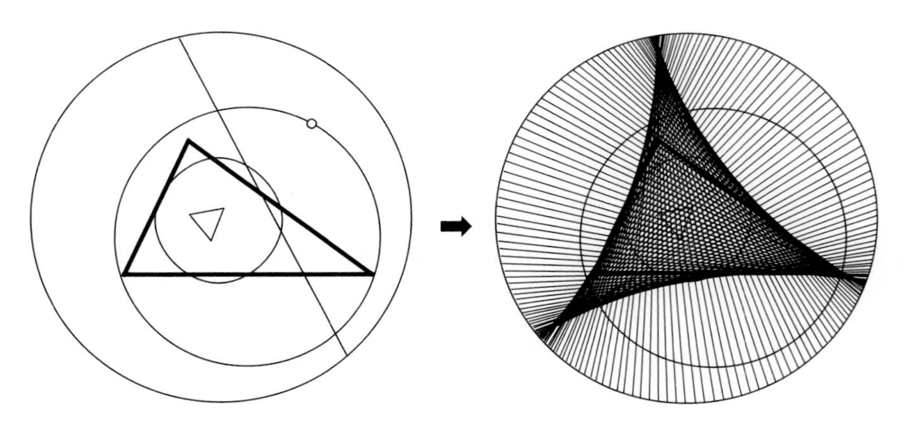

Figure 1.30

Steiner's Theorem

As a point revolves about the circumcircle of a triangle, the Simson Line with respect to that point generates the family of tangents to a deltoid, whose center is the Nine-Point Center, whose inner tangent circle is the Nine-Point Circle of the triangle, and whose cusps are located on the three lines passing through the Nine-Point Center that are perpendicular to the sides of the Morley Triangle

You can produce the diagram of Figure 1.30 yourself using *Sketchpad*. That is the objective of Problems 2 and 3 below.

NOTE: In dual recognition of this discovery of Steiner and the remarkable property that the Simson line has, the deltoid was originally named **Steiner's Curve**.

Moment for Discovery

The Ladder Problem

A classical problem in calculus is the so-called *ladder problem:* A ladder leans against the wall of a building; calculate the velocity at which the bottom slides away from the wall in the horizontal direction as the top slides down the wall at a given velocity. Here we are looking for strictly geometric features. *Sketchpad* provides an effective way to analyze this motion.

1. Construct a perfectly horizontal segment $\overline{AB}$ and a perfectly vertical segment $\overline{AC}$ about 2 in. long (no jagged appearance on the screen along either segment). Select point D on segment $\overline{AC}$. (D simulates the top of the ladder.)
2. Translate point D horizontally by the amount 1.5 in. (use Translate under TRANSFORM and check By Rectangular Vector, choosing components of 1.5 and 0, respectively). Figure 1.31 illustrates this procedure.

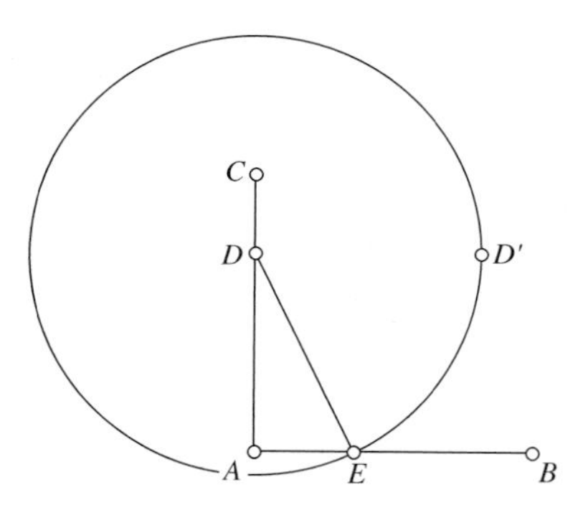

Figure 1.31

3. Select point D, then point D', and choose Circle By Center and Point under CONSTRUCT. This will produce a circle of radius 1.5 in. centered at D. Move D, if necessary to allow the circle to intersect segment $\overline{AB}$, then select the point of intersection, E. Hide D' and the circle. (Segment $\overline{DE}$ now simulates the ladder; move point D to see if the resulting action resembles a falling ladder.)
4. To cause the ladder to move automatically and at the same time trace its motion, select segment $\overline{DE}$ and choose Trace Segment under DISPLAY. Select point D and segment $\overline{AC}$, and choose Animate. You will be prompted for the parameters of animation; choose Bidirectionally (along segment k) and Quickly, then click OK. If Animate appears on the screen, point to it and double-click.

What effect did you notice? Does the ladder appear to generate the tangents to a curve? What curve? (Such a curve is called the **envelope** of the lines.)

PROBLEMS (1.4)

GROUP A

1. **Tangents to Hypocycloids** To generate a hypocycloid by its tangents, use the following instructions in *Sketchpad*. For a specific value of n that is useful, take $n = 3$. Its value can be changed to obtain other hypocycloids.

   [1] Construct a small circle AB, and any point C on that circle.

   [2] Select B, C, and circle AB. Choose Arc On Circle under CONSTRUCT, then Arc Angle under MEASURE.

   [3] Calculate the quantity $x = (1-n) \cdot$ (Arc angle $\widehat{c1BC}$] displaying it on screen.

   [4] Select x and choose Mark Angle Measurement under TRANSFORM. Select A and choose Mark Center "A." Select B and rotate about A through the angle $x°$ to yield the point B'. (Choose Rotate By Marked Angle.)

   [5] Construct line $\overleftrightarrow{B'C}$ and put a tracer on it (choose Trace under DISPLAY). Try moving C to examine its motion. Hide all points except C. Select C, then circle AB, and choose Animate under DISPLAY, checking Once (around the circle) and Normally. Does this produce the tangents to a deltoid?

   Try this for several other values of n, say $n = 4$ and 5. The value $n = 10$ yields a rather spectacular display.

2. **Displaying the Simson Lines of a Triangle** Using *Sketchpad*, construct $\triangle ABC$, its circumcircle, and a point G on that circumcircle. Then follow these steps:

   [1] Construct the lines containing two sides of the triangle and construct the perpendiculars from G to those two lines, selecting the points of intersection H and I between the perpendiculars and lines containing the sides. Hide the lines.

   [2] Construct the line $\overleftrightarrow{HI}$ (the Simson Line with respect to G).

   [3] Select $\overleftrightarrow{HI}$ and choose Trace under DISPLAY. Select G and the circumcircle on which it lies, and choose Animate, checking the items Once, and Quickly.

   NOTE: Save your construction for the next problem.

3. **Steiner's Theorem** Using the construction from the previous problem, construct these additional objects before you animate the tracing of the Simson Line: The Nine-Point Circle and the Morley Triangle. Now animate the Simson Line, and observe the relationship of the deltoid with the Nine-Point Circle and Morley Triangle, as stated in Steiner's Theorem.

4. **The Ladder Problem Revisited** Go back and reconstruct the moving ladder under *Sketchpad* as in the Discovery Unit above. Construct the midpoint M of segment $\overline{DE}$, then trace the midpoint (instead of the ladder). What did you find? Try to prove that your discovery is valid. (***Hint:*** See Problem 5, in Section 1.1.)

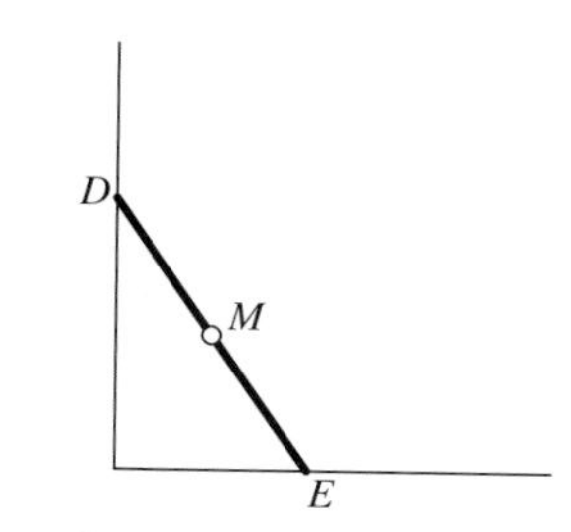

GROUP B

5. Use trigonometry to show that each side of the Morley Triangle of an equilateral triangle is precisely q times as large as that of the given triangle, where

$$q = \sin 10° \sec 20°$$

(***Hint:*** The identity $\sin^2 \frac{1}{2}\theta = \frac{1}{2}(1 - \cos\theta)$ may be useful; for convenience, assume that the side of the given equilateral triangle equals 1.)

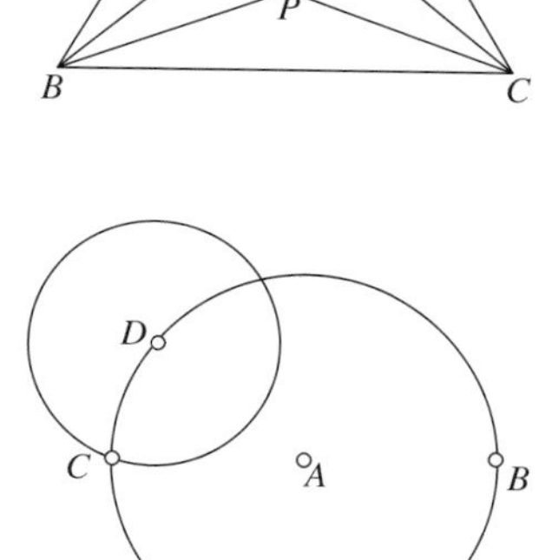

6. This experiment first appeared in *Book of Curves* by E.H. Lockwood. It will show that circles, as well as lines, can be used to generate classical curves. This is another example of the power of *Sketchpad*, which was unavailable at the time the remarkable *Book of Curves* was published.

[1] Draw a (base) circle AB, and select any point C not on that circle, but drag it to a position somewhere on circle AB. (Doing this will allow you to relocate point C to obtain other envelopes.) Locate point D on circle AB so that dragging it will not pull it off the circle.

[2] Using Circle By Center and Point under CONSTRUCT, construct a circle centered at D and passing through C.

[3] Trace the circle you just constructed, and animate with D on the base circle. Did you obtain a familiar curve? What curve is it?

NOTE: Save this construction for the next problem.

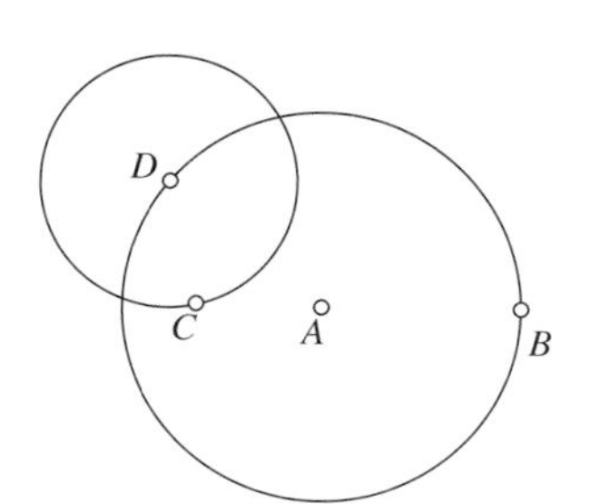

7. Try the experiment of Problem 6 again, first with point C inside, then with point C outside the base circle. What do you find? To complete this problem satisfactorily, look up the curves you found in a calculus book and accurately identify them, unless you already know what they are.

GROUP C

The following sequence of problems (8–11) provides a proof of the validity of the Simson Line for a triangle. Each problem independently contains a useful, significant geometric result.

8. If the circumradius of $\triangle ABC$ is R, then, using standard notation, these formulas can be established:

$$\sin A = \frac{a}{2R}, \qquad \sin B = \frac{b}{2R}, \qquad \sin C = \frac{c}{2R}$$

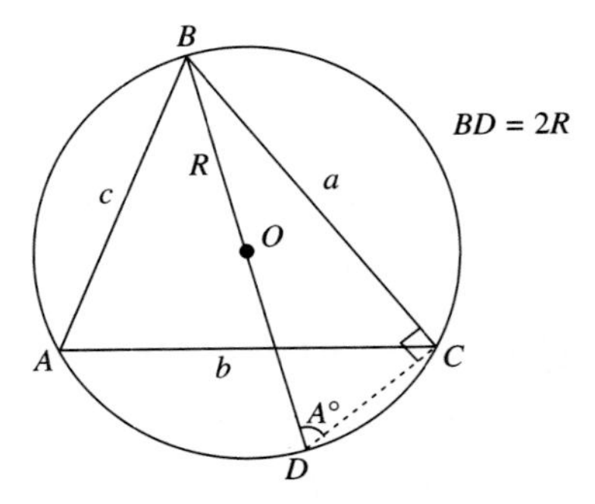

Prove this. (***Hint:*** See the figure. From elementary theorems in high school geometry, $A = m\angle BAC = \frac{1}{2}m(\text{Arc } \widehat{BC}) = m\angle BDC$; $\angle BCD$ is a right angle, so $\sin \angle BDC = BC/BD$.)

9. Formula for the Sides of a Pedal Triangle Suppose that $PA = x$, $PB = y$, and $PC = z$, and that the pedal triangle of $\triangle ABC$ with respect to P is $\triangle DEF$, where D, E, and F are the feet of the perpendiculars from P to sides $\overline{BC}$, $\overline{AC}$, and $\overline{AB}$, respectively. Then if R is the circumradius of $\triangle ABC$,

$$EF = \frac{ax}{2R}, \qquad DF = \frac{by}{2R}, \qquad DE = \frac{cz}{2R}$$

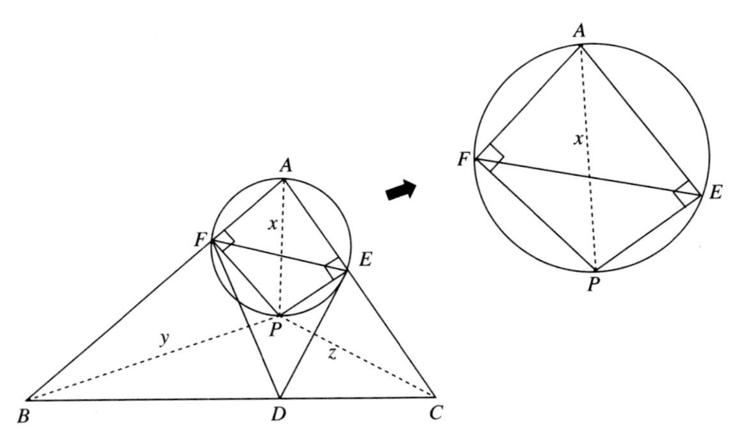

Prove this. (***Hint:*** Explain why a circle must pass through A, E, P, and F, then use the result of Problem 8 on the circumcircle of $\triangle AFE$, having circumradius $AP = x$. What does $\sin A$ equal?)

10. Ptolemy's Theorem If the sides of a cyclic quadrilateral are of length a, b, c, and d and the diagonals have lengths m and n, as shown in the figure that follows, then

$$ac + bd = mn$$

This is proven using similar triangles, as follows (you are to fill in the details).

(1) Construct line $\overleftrightarrow{BE}$ so that $\angle ABE \cong \angle CBD$. (See figure below.)

(2) $m\angle EAB = m\angle CDB$.

(3) $\triangle ABE \sim \triangle CBD \rightarrow a/m = x/c$, or $ac = mx$.

(4) $\triangle ABD \sim \triangle EBC \rightarrow m/b = d/y$, or $bd = my$.

(5) $\therefore\ ac + bd = mn$.

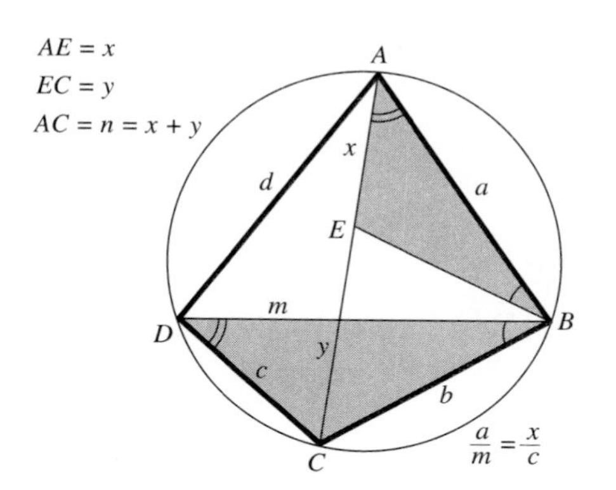

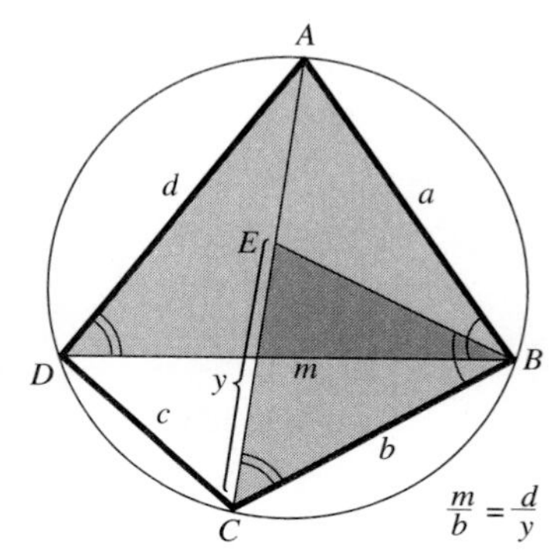

11. Existence of Simson Line For any point P on the circumcircle of $\triangle ABC$ we have, by Ptolemy's Theorem (Problem 10)

$$AB \cdot PC + BC \cdot AP = AC \cdot BP$$

Convert this to standard form, with $x = PA$, $y = PB$, $z = PC$, $a = BC$, etc. Use the formulas from Problem 9 to show that D, E, and F are collinear. (See figure below.)

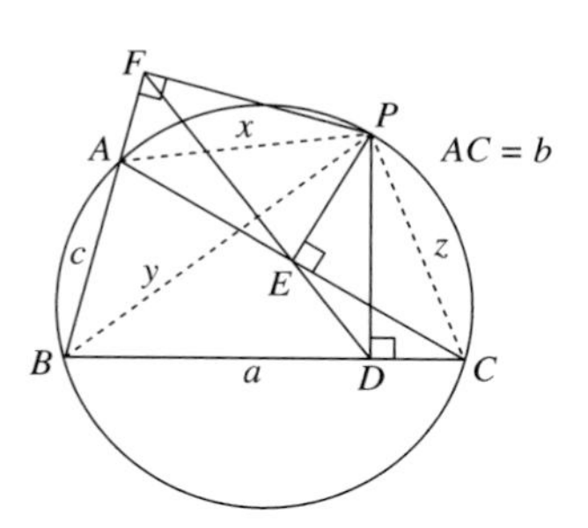

12. Undergraduate Research Project Find the polar equation of the envelope of circles in the experiment of Problem 6 using polar coordinates and the following note.

NOTE: The subject of envelopes generated by a one-parameter system of curves $F(x, y, t) = 0$ is sometimes found in the older books on differential equations. The main result is the following (under certain hypotheses): The envelope of $F(x, y, t) = 0$ is given by the solution of the following system of equations in x and y, where t is eliminated algebraically:

$$\frac{\partial F}{\partial t} = 0 \qquad \text{and} \qquad F(x, y, t) = 0$$

You should reference this result as part of the project. These equations should be converted to polar coordinates in order to work Problem 6, as mentioned. (Use the conversion equations $x = r\cos\theta$, $y = r\sin\theta$; the partial derivative is not affected since the differentiation is with respect to t.)

Example: Let $F(x, y, t) = x/t + ty - 1$ (family of lines with x- and y- intercepts t and $1/t$, respectively, which will generate a hyperbola). The system $\{\partial F/\partial t = x(-t^{-2}) + y = 0, F = x/t + ty - 1 = 0\}$ reduces to $4xy = 1$.

13. Use the note in the preceding problem to prove that in the Ladder Problem, the moving ladder is the envelope of the astroid $x^{2/3} + y^{2/3} = a^{2/3}$, where a is the length of the ladder. (To get started, let the x- and y-intercepts of the line representing the ladder be t and $\sqrt{a^2 - t^2}$, respectively; the equation of the line is then $x/t + y/\sqrt{a^2 - t^2} = 1$.)

CHAPTER TWO

Foundations of Geometry 1

POINTS, LINES, SEGMENTS, ANGLES

OVERVIEW

A good understanding of the basic principles of geometry cannot be had without knowing the source of those principles—the axioms. The axioms, and their immediate consequences, provide the components of the foundation on which everything else rests. So it is important to have an appreciation and some knowledge of the foundation of the geometric "house" we are building.

It might seem possible to start with the coordinate plane and use vector analysis and algebra to develop geometric concepts, and have some power left over. However, to do it correctly, some of the simplest ideas in geometry (e.g., congruent triangles and angle measurement) require relatively difficult algebraic developments. Thus Euclid's style remains the most appropriate, and the simplest of all for beginners. It should be pointed out, however, that although the prevailing spirit in Euclid's *Elements* was axiomatic, Euclid provided virtually *no foundation* by today's standards of rigor. For example, Euclid never defined what was meant by *line segment,* even though its use in the *Elements* was prolific. He seemed to imply that it was part of a line, because he frequently referred to "extending a line segment" in one, or both, "directions." But nowhere is to be found exactly what that meant.

Certainly by the turn of the first millennium, mathematicians had found deficiencies in Euclid's work and as time went on they realized that some revision was necessary. Most scholars, however, wanted to preserve Euclid's elegance of style, so attempts were made to bolster the development rather than to revamp it. In 1898 one of the greatest mathematicians of all time, David Hilbert, proposed an axiomatic foundation in his book *Grundlagen der Geometrie* (Foundations of Geometry) that allowed Euclid's geometry to live in a completely rigorous environment. His work dominated geometric research for many decades.

Hilbert's dominance is felt even today. Because it is really the best way to develop elementary geometry synthetically, the axioms found in many books on geometry, including most high school geometry texts, are adaptations of Hilbert's original system. One particular adaptation that has had a great deal of influence in geometry education in this country is found in Moise, *Elementary Geometry from an Advanced Standpoint*. It is the formal system we adopt in this book.

The great difficulty in studying foundations is that work must be done strictly from the axioms. Any prior geometric knowledge is disavowed. In fact, too much knowledge can be a handicap. This problem is recognized, and much help is provided by way of examples and proofs worked out for you, as well as hints for problems. (See Appendix B for review and examples of proofs found in typical high school geometry courses.) However, just because we are studying foundations of geometry does not mean we cannot have some fun along the way. You will see.

*2.1 An Introduction to Axiomatics and Proof

Let us begin by taking a look at the simplest of all logical discourses—the *syllogism*—which is the basis for all mathematical proofs. The abstract form is

1. All A is B.
2. X is A.
3. $\therefore$ X is B.

An example of this for geometry is

1. All right triangles have a pair of complementary angles.
2. Triangle ABC is a right triangle.
3. $\therefore$ Triangle ABC has a pair of complementary angles.

Thus, the syllogism provides the basis for moving from the general to the particular, a process called *deductive logic*. (*Inductive logic*, on the other hand, moves from the particular, that is, from individual cases, to the general—a method much used in scientific research.) The syllogism is really the essence of any mathematical argument, geometric or not. For each time we use a general theorem or principle already established, we are using a syllogism. When we put together a chain of such syllogisms leading from Condition A to Condition B, the result is a *proof* that A implies B.

DISCOVERING RELATIONSHIPS FROM AXIOMS

In Chapter 1 we emphasized discovery in geometry. So how would one go about making discoveries from an axiomatic system? Here is an example.

EXAMPLE 1 Consider the following three axioms.

AXIOM 1: Each line is a set of four points.

AXIOM 2: Each point is contained by precisely two lines.

AXIOM 3: Two distinct lines that intersect do so in exactly one point.

QUESTION: *Do parallel (nonintersecting) lines exist?*

SOLUTION

As with most axiomatic systems in geometry, our ideas about axioms can be more clearly understood by representative diagrams. Suppose we use an ordinary line segment for each "line" mentioned in the axioms, and dots for "points." When three or more dots in the diagram appear on a line segment, it means that those dots represent points on a line in the axiomatic system. From Axiom 1, we could then represent a line as shown in Figure 2.1 (where we assume that a line actually exists). Now Axiom 2 states that there are two lines containing each of the four points which we have represented on the "line" in Figure 2.1; our reasoning then takes us in the direction indicated by Figure 2.2.

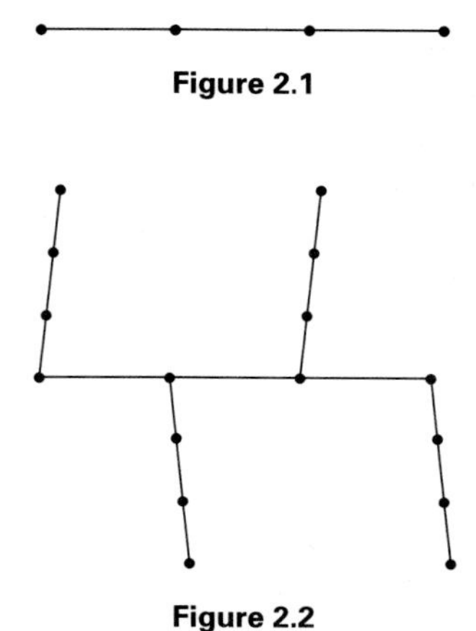

Figure 2.1

Figure 2.2

If we continue in this manner indefinitely, we can see that we are generating an example for the axiom system that has an infinite number of points and lines. We also find that there are infinitely many pairs of nonintersecting, or parallel, lines. The next several stages in the diagram were generated by computer, as shown in Figure 2.3. ■

NOTE: The resulting diagram in Figure 2.3 is a **fractal**, a geometric object that has the property that a window that includes an appropriate part of the diagram, when enlarged, is an exact copy of the original. (See Figure 2.4 for a more familiar example of a fractal.) A few problems on fractal geometry will be included later (in Section 4.6), when you have the tools available for handling the area and perimeter of complicated regions—the kind often generated by fractals.

A FRACTAL GENERATED BY AXIOMS

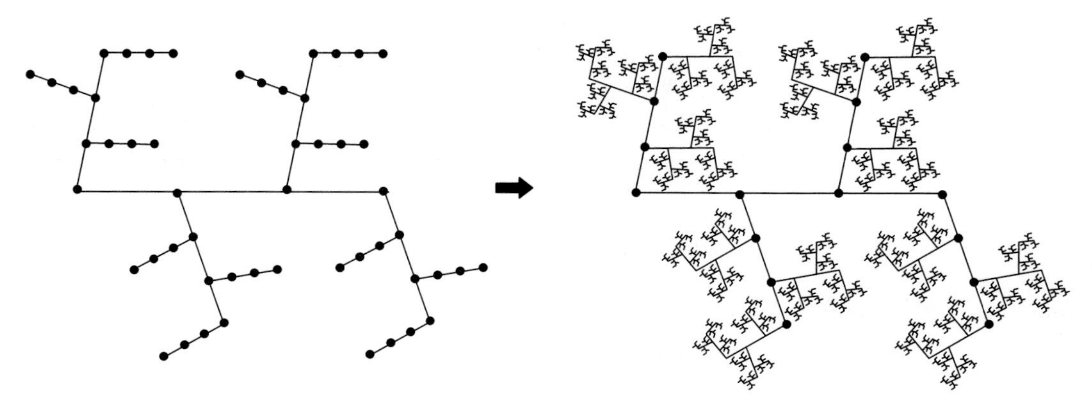

Figure 2.3

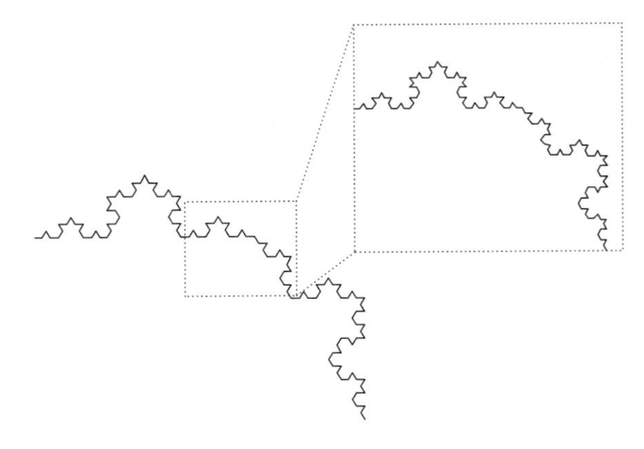

Figure 2.4

IF-THEN STATEMENTS, CONDITIONALS

The most common statement we encounter in geometry (and mathematics in general) is the conditional "if-then" proposition, which may be written symbolically as

$$p \to q \qquad \text{(read: "}p\text{ implies }q\text{," or "If }p\text{, then }q\text{")}.$$

The condition p is called the **hypothesis**, the *Given* part, and q, the **conclusion**—the objective of proof—the *Prove* part.

Given one conditional, several others may be defined. Because they are important, we focus on just two of these: the **converse** and the **contrapositive**, tabulated below for convenience. The prefix $\sim$ stands for the **negation** of a statement: if p is any condition, $\sim p$ means *not p*.

	Given Conditional	Its Converse	Its Contrapositive
Prose form	If p, then q	If q, then p	If not q, then not p
Symbolic form	$p \to q$	$q \to p$	$\sim q \to \sim p$

Each of these propositions plays its own role in proofs. The contrapositive is logically equivalent to the original conditional, but the converse is not. For example, consider the true statement

Right angles are congruent.

That is, stated as a conditional,

If two angles are right angles, then they are congruent.

The converse of this would be the patently false statement

If two angles are congruent, then they are right angles.

However, a proposition whose converse is true is the familiar

If two sides of a triangle are congruent, then the angles opposite are congruent.

The converse "If two angles of a triangle are congruent, then the sides opposite are congruent" is equally valid. Thus we have both $p \to q$ and $q \to p$. Symbolically,

$$p \to q \qquad \text{and} \qquad p \leftarrow q \qquad \text{or} \qquad p \leftrightarrow q$$

In prose form, we have

If p, then q (or, p ONLY IF q) and If q, then p (or p, IF q)

which leads to the abbreviation

$$p \text{ iff } q$$

This is read "*p if and only if q*." Here we say that p and q are **logically equivalent**, or we can also say that q **characterizes** p. In the above example, then, isosceles triangles are said to be *characterized* by having congruent base angles.

DIRECT PROOFS

The simplest form of argument is the **direct proof**, which is the embodiment of the *propositional syllogism*:

1. p implies q
2. q implies r
3. $\therefore$ p implies r

This could be regarded as a sort of *transitivity law* for logic, but we do not need to get too fancy here in our terminology. Furthermore, in proofs we just go ahead and use this logical procedure without making reference to it. We then have the basis for the following:

General Form of a Direct Proof

Suppose that $p \to q$ and $q \to r$ are valid propositions. If one assumes p as hypothesis, then q. Since q, then r. Therefore $p \to r$.

This logical procedure also allows us to insert any number of intermediate propositions, if necessary. Here is an example of a typical direct proof, dressed up so it resembles what we might encounter in a typical high school geometry course.

Outline Form of a Direct Proof

List of previously proven theorems to be used in the proof:

Theorem 1: $p \to q$
Theorem 2: $q \to r$
Theorem 3: $r \to s$

Outline Form of a Direct Proof *(continued)*

THEOREM: If p, then s.

PROOF

Given: p

Prove: s

CONCLUSIONS	JUSTIFICATIONS
(1) p	Given
(2) q	*Theorem 1*
(3) r	*Theorem 2*
(4) $\therefore s$	*Theorem 3*

Sometimes it is necessary to show the use of the definition of a particular term to write a lucid proof. For example, the proof of the proposition "All right angles are congruent" would go as follows, after recalling the two definitions:

DEFINITION 1: Two angles are **congruent** iff they have equal measures.

DEFINITION 2: A **right angle** is an angle having measure 90 (degrees).

PROOF

Given: $\angle A$and $\angle B$ are right angles

Prove: $\angle A$ is congruent to $\angle B$.

CONCLUSIONS	JUSTIFICATIONS
(1) The measures of $\angle A$ and $\angle B$ are each 90°.	***Definition*** (of right angle)
(2) The measures of $\angle A$ and $\angle B$ are equal.	Algebra (if $x = 90$ and $y = 90$, then $x = y$)
(3) $\therefore \angle A$ is congruent to $\angle B$.	***Definition*** (of congruence)

There is a growing trend to encourage the paragraph (or prose) style of proof in high school geometry courses. This is good, because that method is closer to the original style of Euclid, allows more freedom of expression, and is the accepted style of mathematical proof in other areas of mathematics (why should geometry be any different?). Hence, we will from time to time write our proofs in paragraph form, with the two-column (T-proof) being gradually phased out altogether. The outline form of proof does have the advantage of forcing us to come up with all the reasons for a particular argument, which is a good habit to form. It also provides a good visual analysis for a proof, if sometimes a bit tedious.

INDIRECT PROOF

Another kind of mathematical argument was introduced by Euclid. It is often described by the phrase *reductio ad absurdum* ("to reduce to an absurdity"). The logical pattern in this kind of argument is to assume that the conclusion you are trying to prove is false, and then show that this leads to a contradiction of the hypothesis. In other words, to show that p implies q, one assumes p and *not* q, and then proceeds to show that *not* p follows, which is a contradiction. This means that we prove the *contrapositive* of the given proposition instead of the proposition itself.

Still another line of argument, a little more involved, is the analysis of all possible logical cases in a given situation. For example, if we want to prove that $x = 3$, we could start with the three possible cases involving the numbers x and 3: Either $x < 3$, $x = 3$, or $x > 3$. Then, if we are able to show that $x < 3$ and $x > 3$ each leads to a contradiction, it follows that $x = 3$. The property of numbers we used here is known as the **Trichotomy Property** for the reals. It is useful enough to state by itself as a property that can be used in proofs from now on.

Trichotomy Property of Real Numbers

For any two real numbers a and b, either $a < b$, $a = b$, or $a > b$.

Let us look at an example from the *Elements*, in which Euclid employs the very principle we just mentioned. There is a passage from *Volume I, Book I* that immediately follows the proof of Proposition 18 (that if one side of a triangle is greater than a second, the angle opposite the first side is greater than that opposite the second). This passage argues the truth of Proposition 19, which is the *converse* of Proposition 18. Here it is, as it appears in Heath's *The Thirteen Books of Euclid's Elements*:

> PROPOSITION 19: *In any triangle the greater angle is subtended by the greater side.*
>
> Let ABC be a triangle having the angle ABC greater than the angle BCA;
> I say that the side AC is also greater than the side AB.
> For, if not, AC is either equal to AB or less.
> Now AC is not equal to AB;
> for then the angle ABC would also have been
> equal to the angle ACB; [I.5][1]
> but it is not;
> therefore AC is not equal to AB
> Neither is AC less than AB,
> for then the angle ABC would also have been less than the angle ACB; [I.18]
> but it is not;
> therefore AC is not less than AB.
> And it was proved that it is not equal either.
> Therefore AC is greater that AB.

This type of argument was used often by Greek mathematicians, and was referred to as the *method of exhaustion*, that is, the method of *exhausting all possibili-*

[1]This is Euclid's proposition about isosceles triangles: If two sides of a triangle are equal, the angles opposite are equal.

ties. In such arguments, two things are critical, and must be adhered to in order to be valid:

(1) You must consider *all* the logical possibilities.
(2) All the logical possibilities must lead to a contradiction *except* the one you are trying to prove.

The mathematical justification for the Method of Exhaustion is stated below in a more explicit and complete manner. It will be referred to frequently in proofs from now on.

HISTORICAL NOTE

About all we know concerning the man Euclid is that he was a professor of mathematics around 300 B.C.E., he taught at the University of Alexandria in Greece, and he was the author of the celebrated *Elements.* The stories told of Euclid show that he was a serious, uncompromising scholar. When asked by King Ptolemy whether there was an easier way to learn geometry (besides studying the *Elements*), Euclid is supposed to have responded, "There is no royal road to geometry!" Euclid's great contribution to mathematics was the elegant, perfectly logical arrangement of the *Elements.* This monumental work consists of 13 books containing 465 propositions, with detailed arguments for each, all based on only 10 basic axioms. The work was an immediate success, so much so that all other works of mathematics soon disappeared; only the Bible has been printed in more editions. Although the propositions found in the *Elements* were themselves not new, their logical arrangement and ingenious arguments were so effective that many of them have survived to this day as the simplest method of proof. The *Elements* have had a great and lasting impact on science, its influence extending to modern times.

Rule of Elimination

Suppose that $r, s, t, \ldots,$ and u is a complete (finite) set of logical possibilities, that some (or none) of them imply $\sim p$, and all the rest (at least one) imply q. Then $p \rightarrow q$ is valid.

A good symbol to use in proofs when a contradiction has been reached is $\rightarrow\leftarrow$. This signals the end of one part of the argument and the beginning of another, where the next case is considered.

To conclude our discussion of the indirect proof, we present an argument in the abstract that explicitly uses the Rule of Elimination to show what such a proof looks like.

Outline Form of an Indirect Proof

Previously proven propositions:
Theorem 1: $r \rightarrow \sim p$
Theorem 2: $s \rightarrow \sim p$
Complete set of logical possibilities: r, s, q

THEOREM: If p, then q.

PROOF

Given: p

Prove: q

CONCLUSIONS	JUSTIFICATIONS
(1) p	Given
(2) r, s, or q	All the logical cases
(3) Assume r.	First logical case
(4) $\sim p$ $\rightarrow\leftarrow$	*Theorem 1*
(5) Assume s.	Second logical case
(6) $\sim p$ $\rightarrow\leftarrow$	*Theorem 2*
(7) $\therefore q$	Rule of Elimination

PROBLEMS (2.1)

GROUP A

1. If p = "John wears his raincoat," q = "it rains," and r = "Mary is happy," translate the following propositions of logic into prose if-then statements:

(a) $q \rightarrow p$

(b) $p \rightarrow r$

(c) $q \rightarrow \sim r$

(d) $\sim r \rightarrow q$

2. Suppose that q = "it rains" and s = "there no are clouds in the sky." Translate the following propositions of logic into prose if-then statements:

(a) $q \rightarrow \sim s$

(b) $\sim s \rightarrow q$

(c) $\sim q \rightarrow s$

(d) Is the statement of **(b)** logically the same as that of **(a)**?

(e) Is the statement of **(c)** logically the same as that of **(b)**?

In each of the following problems (3–6) you are given a statement that is implicitly an "if-then" statement. Treat each of them as a proposition to be proven, and write down what is given (the "if" part) and what is to be proven (the "then" part). Also, state the converse of each proposition, and decide whether it is valid for each case.

3. Two angles whose measures are each 30° are congruent.

4. Every equilateral triangle is equiangular.

5. The medians of a triangle are concurrent lines.

6. **Chinese Proverb:** Success can be achieved only by hard work. (**Warning:** To obtain the correct if-then statement, the usage of the word "only" must be carefully considered.)

7. An economist claims that inflation causes high unemployment.

(a) Write this in the form of $p \to q$, identifying the components p and q.

(b) Under what circumstances would $p \to q$ be shown to be false?

8. Consider:

Definition: a is said to be b iff $p \to q$.
Theorem 1: $p \to r$
Theorem 2: $r \to q$

Fill in the missing reasons in the two-column proof that follows.

Theorem: If p, then a is b.
Proof
Given: p
Prove: a is b

CONCLUSIONS	JUSTIFICATIONS
(1) p	Given
(2) r	**(a)** __________?
(3) q	**(b)** __________?
(4) $\therefore$ a is b	**(c)** __________?

9. Consider:

Axiom 1: $s \to t$
Theorem 1: $p \to q$
Theorem 2: $q \to \sim t$

Fill in the missing reasons in the two-column proof that follows.

Theorem: If p, then s is impossible.
Proof
Given: p
Prove: $\sim s$

CONCLUSIONS	JUSTIFICATIONS
(1) p	Given
(2) q	**(a)** __________?
(3) $\sim t$	**(b)** __________?
(4) $\therefore \sim s$	**(c)** __________?

10. Rewrite the following paragraph proof in two-column form: John lost his locker key either in his homeroom or on his way home from school. Since he did not stop in his homeroom, he must have lost the key on his way home from school. (Which proof do you find easier to understand?)

GROUP B

11. Consider the following:

Definition: a is said to be b iff $r \to s$.
Axiom 1: $r \to q$
Theorem 1: If a is c then $q \to t$.
Theorem 2: $t \to s$

Fill in the missing reasons in the two-column proof that follows:

Theorem: If a is c then a is b.
Proof
Given: a is c
Prove: a is b (i.e., $r \to s$)

CONCLUSIONS	JUSTIFICATIONS
(1) a is c	Given
(2) r	Assumption for $r \to s$
(3) q	*Axiom 1*
(4) t	**(a)** ________?
(5) s	**(b)** ________?
(6) $\therefore$ a is b	**(c)** ________?

12. Suppose that the following propositions have been proven: (1) $\sim q \to r$, and (2) if r, then either $\sim p$ or q. Give a paragraph proof of the proposition $p \to q$.

13. Suppose it has been established that p implies either r or s, and

Theorem 1: $y \to \sim p$ *Theorem 3:* $s \to y$
Theorem 2: $r \to x$ *Theorem 4:* $x \to q$

Prove, in outline form, that $p \to q$.

GROUP C

14. Suppose the following have been established or assumed:

Axiom 1: $p \to \sim y$
Axiom 2: $\sim q \to r$
Theorem 1: $p \to \sim z$
Theorem 2: $x \to$ either q or z
Theorem 3: $r \to$ either x *or* y

Write a two-column proof of the proposition $p \to q$.

***15.** In the axiom system of Example 1, discover the consequences of the following additional axiom, assuming a point or line exists: *Axiom 4:* Each two lines intersect at exactly one point. Can there be infinitely many points in the resulting axiomatic system?

*See Problem 7, Section 2.2.

16. Research the usage of the term "exhaustion" in mathematics, or "method of exhaustion." Eudoxus (370 B.C.E.) was one of the first to have used the term.

*2.2 The Role of Examples and Models

Without examples, very little new mathematics would ever be discovered. Examples literally point the way for modern mathematical research, showing the possibilities of what may be true, and sometimes what is not. A model is like an example, but it is usually more elaborate and for the purpose of simulating a system of axioms. (The currently popular area known as "mathematical modeling" involves an entirely different purpose for models; in this field, models are constructed in some mathematical system to simulate real-world situations which occurs outside of mathematics.)

EXAMPLES AND FALSE PROOFS IN MATHEMATICS

As often happens, an argument we use to prove a theorem may have unsuspected flaws, and sometimes a concrete example will shed light on the error.

EXAMPLE 1 Consider the following theorem and the reasoning used to prove it.

THEOREM: The sum of the measures of the angles of a parallelogram equals 360°.

PROOF

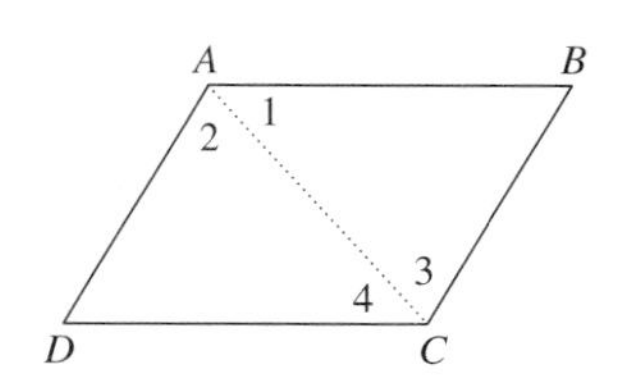

Figure 2.5

Consider Figure 2.5, where the parallelogram is divided into two triangles, each of whose angle measures sum to 180°. Hence in the original parallelogram,

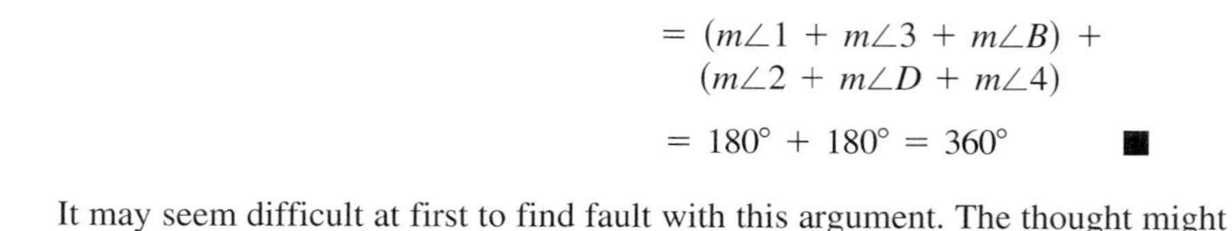

$$m\angle A + m\angle B + m\angle C + m\angle D = m\angle 1 + m\angle 2 + m\angle B + m\angle 3 + m\angle 4 + m\angle D$$
$$= (m\angle 1 + m\angle 3 + m\angle B) + (m\angle 2 + m\angle D + m\angle 4)$$
$$= 180° + 180° = 360° \quad \blacksquare$$

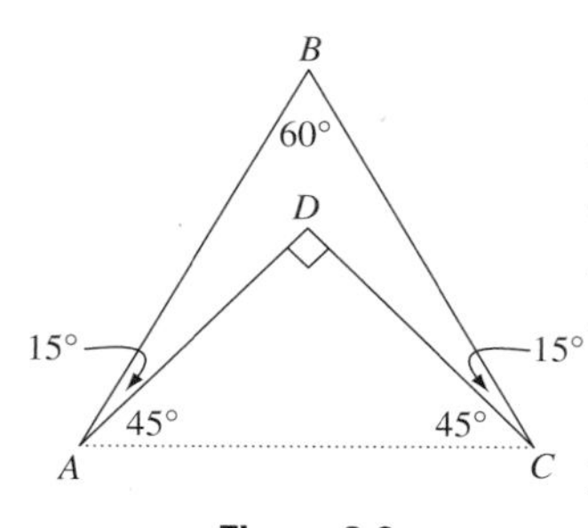

Figure 2.6

It may seem difficult at first to find fault with this argument. The thought might even occur to us to try our luck at generalizing it. Since we did not actually use the fact that the figure was a parallelogram, maybe we could prove this for any quadrilateral. But consider the example illustrated in Figure 2.6. Here we have a quadrilateral *ABCD* with angles at *A*, *B*, and *C* having measures 15°, 60°, and 15°, respectively, and with a right angle at *D*. Since a right angle has measure 90°, we obtain the sum

$$m\angle A + m\angle B + m\angle C + m\angle ADC = 15° + 60° + 15° + 90° = 180°!$$

Hence, something is wrong with the argument, since it should also apply to the quadrilateral in Figure 2.6.

NOTE: It may seem that a different measure for the angle at D should be used, namely, 270° instead of 90°, since the *interior* of the quadrilateral is on the opposite "side" of $\angle D$. But this is contrary to the essential property of an angle (*defined only by its sides*), which can have only one unique angle measure. We will have more to say about this later, when angle measure is introduced.

The next example establishes a blatantly false theorem, and serves as a warning about the danger of misusing diagrams to aid in proof-writing.

EXAMPLE 2 Consider the following result, with an apparently rigorous argument to prove it. (See Figure 2.7.)

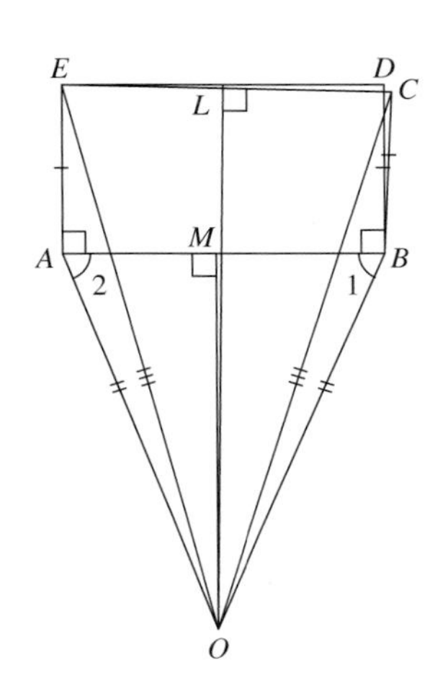

Figure 2.7

THEOREM (!): Any obtuse angle $\angle ABC$ is a right angle.

PROOF

Construct a rectangle $ABDE$, as shown, so that $AE = BD = BC$, and complete the quadrilateral $ABCE$. Then construct the perpendicular bisectors of segments $\overline{EC}$ and $\overline{AB}$ at L and M, respectively. Since lines $\overleftrightarrow{AB}$ and $\overleftrightarrow{EC}$ are not parallel, their perpendiculars at M and L will not be parallel and must meet at some point O. Construct the remaining line segments, $\overline{OA}$, $\overline{OE}$, $\overline{OC}$, and $\overline{OB}$. Now $OA = OB$ and $OE = OC$, since O is equidistant from A and B, and from E and C. Then by SSS $\triangle OBC$ is congruent to $\triangle OAE$, and

$$m\angle OBC = m\angle OAE$$

But these two angles are made up of their parts, $\angle 1$ and $\angle ABC$, and $\angle 2$ and $\angle BAE$. Hence

$$m\angle ABC + m\angle 1 = 90° + m\angle 2$$

But since $\triangle ABO$ is isosceles, $m\angle 1 = m\angle 2$. Then the above equation reduces to

$$m\angle ABC = 90°$$

Therefore, $\angle ABC$ is a right angle. ■

Can you find the fatal flaw with this reasoning? Problem 9 at the end of this section will provide you with some ideas.

AXIOMS, AXIOMATIC SYSTEMS

An axiomatic system always contains statements which are assumed without proof—the **axioms**. These axioms are chosen

(a) for their convenience and efficiency
(b) for their consistency

and, in some cases, (but not always)

(c) for their plausibility

UNDEFINED TERMS

Every axiom must, of necessity, contain some terms that have been purposely left without definitions—the **undefined terms**. To attempt a precise definition for every significant term that might appear in a particular development would result in failure. For, as in a dictionary, we might define A in terms of B and C, then B and C in terms of D, E, and F, and so on. But inevitably, we will find that at some point the term W will have been defined in terms of one defined earlier, $A, B, C, \ldots$, resulting in a circular chain.

In geometry, the most common undefined terms are "point" and "line." We may think of a point as a dot on a sheet of paper, computer screen, or chalkboard, but in reality a dot has physical (albeit microscopic) dimensions, and does not exactly fit our idealistic viewpoint. Euclid defined a point as "that which has no part" and a line as "breadthless length" (i.e., "length without width"). However, these statements mainly served as a reminder to a reader of the *Elements* about the obvious. Euclid made no attempt to prove properties of points and lines precisely from these definitions.

The modern axiomatic approach is to decide at the beginning what the undefined terms shall be, then set down the properties those objects are to have in the axioms. The choice of undefined terms and axioms can be quite arbitrary—as long as no contradictions result. Regarding this arbitrary nature of the axioms and use of undefined terms, the great philosopher Bertrand Russell once remarked, "Mathematics may be defined as the subject in which we never know what we are talking about, nor whether what we are saying is true."

MODELS FOR AXIOMATIC SYSTEMS

A **model** for an axiomatic system is a realization of the axioms in some mathematical setting in which the universe is specified in that system, all undefined terms are interpreted, and all the axioms are true. The model may "live" or exist in coordinate geometry, or in some well-known, or even contrived, system, but it is always separate from the axioms themselves. This idea actually occurred earlier, in Example 1 of Section 2.1: The fractal of Figure 2.3 served as a *model* for the three stated axioms. This particular model could be regarded as taking place in the coordinate plane by merely assigning coordinates to all the points in the diagram of Figure 2.3 in some systematic way.

Two further examples of axioms and models are now given. In these examples, each of the models illustrated by the diagrams represents a finite number of points in the Euclidean plane and certain line segments containing those points. These diagrams (models) depict only the essential features of the axioms regarding the number of points and lines and the collinearity or noncollinearity relationships. Nothing can be inferred from the pictures that cannot be derived from the axioms.

EXAMPLE 3

ABSTRACT SYSTEM

1. There exist two points.
2. There exists a line containing those two points. ■

MODEL

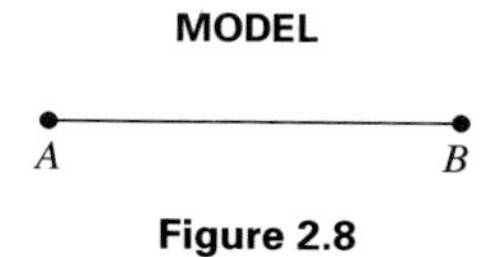

Figure 2.8

EXAMPLE 4

ABSTRACT SYSTEM

1. There exist five points.
2. Each line is a subset of those five points.
3. There exist two lines.
4. Each line contains at least two points. ■

MODELS

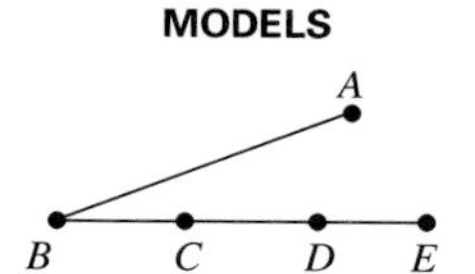

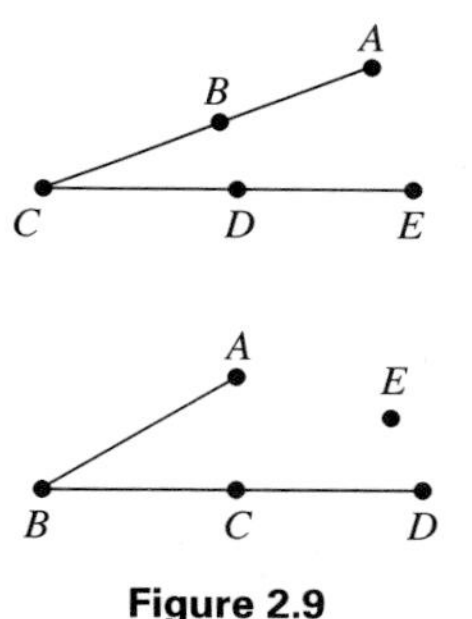

Figure 2.9

We can see that any theorem logically provable from the axioms must also be true in any model for those axioms, but not conversely. This can be seen quite clearly in Example 4, in which *three* different models are possible. A theorem true in one model but not in another (such as "every point belongs to a line" which is true in the first two models but not in the last) would amount to a theorem *not provable from the axioms*, since it is not true in all the models.

INDEPENDENCE AND CONSISTENCY IN AXIOMATIC SYSTEMS

The only things of virtue about an axiomatic system are **independence** (every axiom is essential—none is a logical consequence of the others), and **consistency** (freedom from contradictions). Models play an important role in deciding these two issues.

If a model can be constructed in a mathematical setting that is known to be free of contradictions, then we know that no contradiction can arise from the axioms themselves. Thus, a model for a set of axioms in some concrete system always establishes (at least) **relative consistency**—the axiomatic system is as consistent as the mathematical environment of the model. On the other hand, if a model can be constructed for all but one of the axioms in a certain axiomatic system, and if the one excluded axiom is false in that model, then the excluded axiom cannot be proven from the rest; thus, it is **independent**. If this can be done for each of the axioms, then the system as a whole is independent.

One of the earliest serious debates in mathematics arose over the independence of Euclid's Fifth Postulate of Parallels. (See Section 4.1 for its statement.) For centuries this postulate was thought to be unnecessary, and mathematicians, including the great C. F. Gauss, tried to use the other Euclidean assumptions to prove it. It was not until E. Beltrami constructed a model in Euclidean geometry that the issue was completely settled. More will be said about this later—in fact, in Chapter 6 we will go through the details necessary to construct and understand two of Beltrami's models (one of them known today as Poincaré's Disk Model).

A very interesting feature about Euclidean geometry is that it has *only one* model, namely, three-dimensional coordinate geometry, or the equivalent. This is the model we use so often and are familiar with from calculus. If an axiomatic system has only one model, it is called **categorical**. Thus, *Euclidean geometry (and each of the two non-Euclidean geometries) are categorical systems.*

EXAMPLE 5 Consider the following axiomatic system:

Undefined terms: member, committee

AXIOMS:

1. Every committee is a collection of at least two members.
2. Every member is on exactly one committee.

Find two distinctly different models for this set of axioms, and discuss how it might be made categorical.

SOLUTION

Let one model be:

Members:	John, Dave, Robert, Mary, Kathy, and Jane
Committees:	Committee A: John and Robert
	Committee B: Dave and Jane
	Committee C: Kathy and Mary

Another model is:

Members:	$\{a, b, c, d, e, f, g, h\}$
Committees:	Committee A: $\{a, b, c, d\}$
	Committee B: $\{e, f\}$
	Committee C: $\{g, h.\}$

Because we have found models that are different (committees having different sizes exist), the system is noncategorical. Suppose we add the axiom

3. There exist three committees and six members.

Now we have a categorical system—every model quite obviously consists of three committees, with two members on each committee. All models would be entirely similar; only the names or identities of individual committee members could differ. ■

PROBLEMS (2.2)

GROUP A

1. The rectangle shown in the figure has an area of $5 \cdot 13 = 65$ square units. The four-piece dissection of the rectangle indicated can apparently be reassembled to form a perfect square having an area of $8 \cdot 8 = 64$ square units. Where did the missing square go? Explain the fallacy in this well-known puzzle.

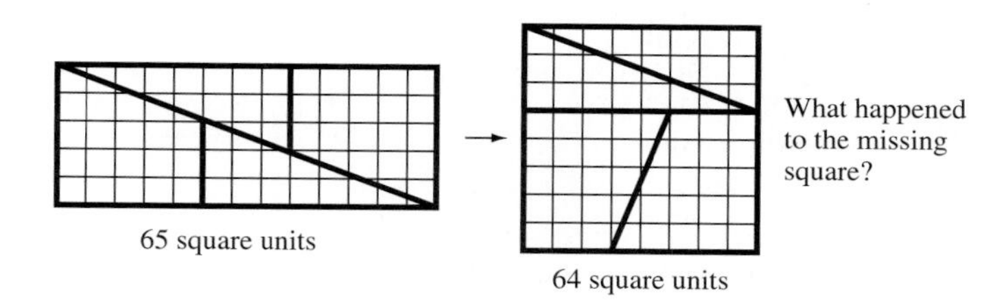

2. A not-so-well-known puzzle is called a **Curry Triangle**, but it is even more mysterious than the one in Problem 1. However, it contains the same type of fallacy. (See if you can find it.) Start with an isosceles triangle having a base of 10 units and height 12. Hence, its area is $\frac{1}{2} \cdot 10 \cdot 12 = 60$ square units. When the six pieces in the dissection of the triangle are rearranged face up as shown in **(a)**, the result is an apparent *loss* of two square units. However, in **(b)**, some of the pieces are turned face down, and the resulting loss of area is *only one* square unit!

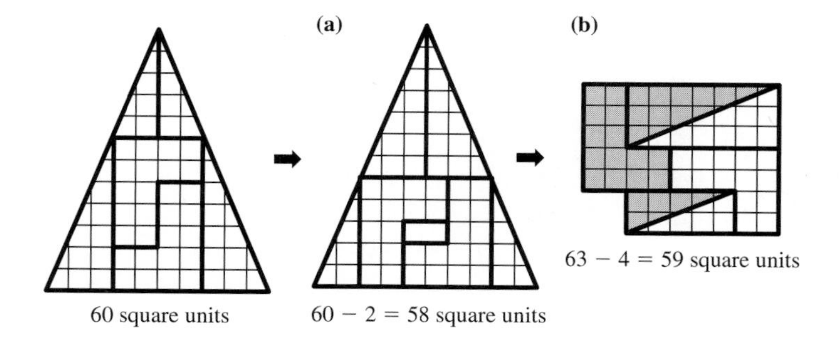

3. Trace the word *noise* through a standard dictionary until a circular chain occurs.

4. Show that the following model violates the axioms in Example 5 (concerning committees):

Members:	John, Dave, Robert, Mary, Kathy, and Jane.
Committees:	Executive Committee: Mary and Robert
	Steering Committee: John and Robert
	Nomination Committee: Jane, Kathy, and Dave.

5. Find a third model for the axioms in Example 5 (on committees) that is distinctly different from the models given there. What is different about your model, precisely?

6. Find two different models for the following set of axioms in which "point" and "line" are undefined terms:

Axiom 1: Every line is a set of at least two points.
Axiom 2: Each two lines intersect in a unique point.
Axiom 3: There are precisely three lines.

7. (a) Show that the diagram constitutes a model in the Euclidean Plane for the axioms given in Example 1, Section 2.1, which has only 10 points. We repeat the axioms here for convenience:

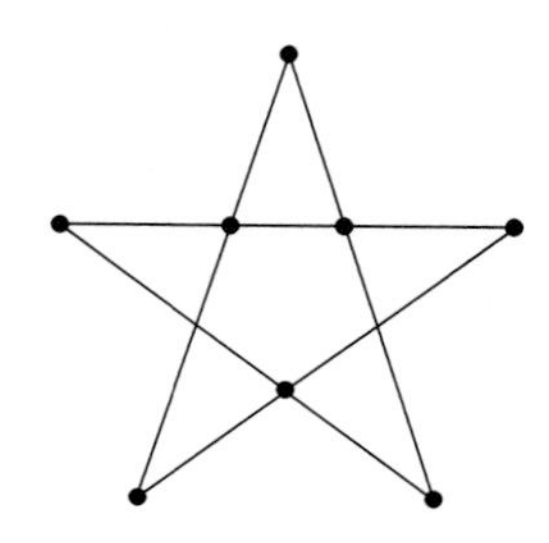

Axiom 1: Each line is a set of four points.
Axiom 2: Each point is contained by precisely two lines.
Axiom 3: Two distinct lines which intersect do so in exactly one point.

(b) Is this axiom system categorical?

GROUP B

8. (a) A **convex quadrilateral** is a quadrilateral having the property that each diagonal lies between the two sides adjacent to it. Show that the sum of the measures of the angles of any convex quadrilateral is 360.

(b) Is the quadrilateral in Figure 2.6 convex? Why not? Is a parallelogram a convex quadrilateral?

9. Locate the points A, B, C, D, E in the coordinate plane, as shown, and verify that $ABDE$ is a rectangle and $BD = BC$. Find the coordinates of the midpoint L, slope of line $\overleftrightarrow{CE}$, and the equations of the perpendiculars $\overleftrightarrow{LO}$ and $\overleftrightarrow{MO}$. Then find the coordinates of the point of intersection of these two perpendiculars. Make a careful sketch revealing your computations, and draw lines $\overleftrightarrow{OA}$, $\overleftrightarrow{OE}$, $\overleftrightarrow{OE}$, and $\overleftrightarrow{OC}$. What do you observe?

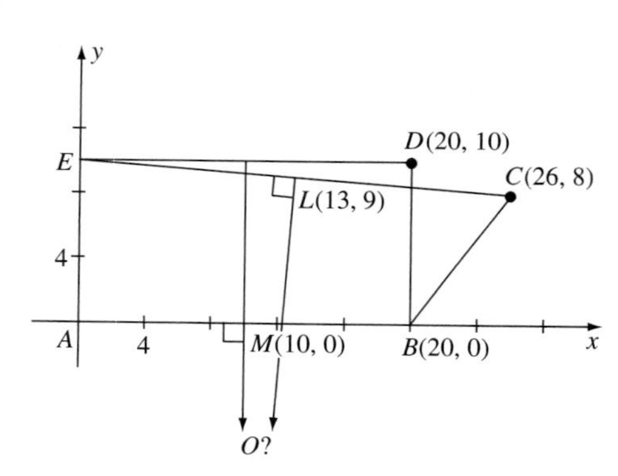

10. Find an example to explain the fallacy in the following.

THEOREM (!): Every triangle is isosceles.

PROOF

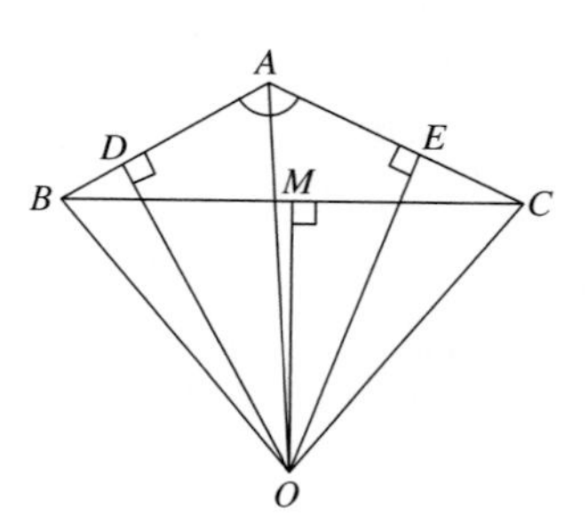

Assuming that $\triangle ABC$ is *NOT* isosceles, then $AB \neq AC$ and the angle bisector at A and perpendicular bisector of side $\overline{BC}$ are not parallel and must therefore meet at some point O. Drop perpendiculars $\overline{OD}$ and $\overline{OE}$. Then $AO = AO$ and $m\angle DAO = m\angle OAE$ so that (by Euclid) right triangles $\triangle ADO$ and $\triangle AEO$ are congruent. Therefore, $AD = AE$ and $DO = EO$.

Because O is equidistant from B and C, $OB = OC$ and the right triangles $\triangle BDO$ and $\triangle CEO$ are also congruent. Hence, in addition to $AD = AE$, we also have $BD = EC$. Then, $AD + DB = AE + EC$, or, since $AB = AD + DB$ and $AC = AE + EC$, $AB = AC$.

***11.** There are seven standing committees on the *Flying Aviators Club*. In drawing up a set of by-laws for committee membership, the members wanted to maximize both the distribution of its members serving on committees, and communication between committees. They decided to adopt the following rules:

(1) There shall be seven committees of the *Flying Aviators Club,* including the Executive, Program, and Nominating Committees. The other committees shall be designated A, B, C, and D.

(2) The Executive Committee consists of just the President, Vice-President, and Secretary/Treasurer.

(3) The President shall serve on the three named committees mentioned in Rule 1.

(4) Each member-at-large of the *Flying Aviators Club* must serve on at least three committees.

(5) Two or more members are prohibited from belonging to the same two committees.

(6) Each two committees must have a member in common.

Show that the members have unwittingly adopted a list of rules that will work only if the club has *precisely seven members.* (***Hint:*** Use "points" for members and let "lines" represent committees.)

*See Problem 6, Section 2.3.

GROUP C

12. Show that there are essentially only two models for the following axiom system:

Undefined Terms: point, adjacent to, and color.

AXIOMS:

(1) There are exactly five points.

(2) If point A is adjacent to point B, then point B is adjacent to point A.

(3) If point A is not adjacent to point B, then there exists a point C to which A and B are mutually adjacent.

(4) Each point is assigned a color, red or green.

(5) Any two adjacent points are assigned to different colors.

13. As an interesting excursion befitting Bertrand Russell's comment about mathematics, consider the following axiom system (adapted from Howard Eves, *Survey of Geometry, Volume 1,* p. 390).

Undefined terms: abbas, dabbas

AXIOMS:

(1) Every abba is a collection of at least two dabbas.

(2) There exist at least two dabbas.

(3) If d and d' are two dabbas, then there exists one and only one abba containing both d and d'.

(4) If a is an abba, then there exists a dabba d not in a.

(5) If a is an abba, and d is a dabba not in a, then there exists one and only one abba containing d and not containing any dabba that is in a.

(a) Deduce the following theorems from this postulate set:

(1) Every dabba is contained in at least two abbas.

(2) There exist at least four distinct dabbas.

(3) There exist at least six distinct abbas.

(b) Find two models for this postulate set, one having four dabbas and six abbas, and the other having nine dabbas, twelve abbas, and three dabbas on each abba.

2.3 Incidence Axioms for Geometry

The axioms we adopt for geometry in this book are based on the familiar adaptation of Hilbert's axioms found in Moise, *Elementary Geometry from an Advanced Standpoint,* and used in many high school geometry texts. To facilitate the statement of the axioms and the later development of geometry, free use of the fundamental notations of set theory will be made. Basic set theory notation (such as $x \in S$ for set membership, $S \subseteq T$ for subsets, and $S \cap T$ for intersection will be used). Remember that prior knowledge of geometry can be used only for constructing examples or models. All else must rest firmly on the axioms, and only the axioms. No outside assumptions are allowed.

We begin with those axioms which govern how points, lines, and planes interact—the so-called **incidence axioms**. The undefined terms will be

point, line, plane, space

(the latter two indicate axioms for three-dimensional space). We let $\boldsymbol{S}$ denote the **universal set**, that is, the set of *all points in space,* and we envision lines and planes as being certain subsets of $\boldsymbol{S}$, as illustrated in Figure 2.10. Sometimes a line ℓ can be a **subset** of a plane $\boldsymbol{P}$, when all the points of ℓ belong to $\boldsymbol{P}$ ($\ell \subseteq \boldsymbol{P}$). Or ℓ can meet $\boldsymbol{P}$ at just one point A ($\ell \cap \boldsymbol{P} = \{A\}$), and finally, ℓ can be **parallel** to $\boldsymbol{P}$, where there are no points at all in common ($\ell \cap \boldsymbol{P} = \varnothing$).

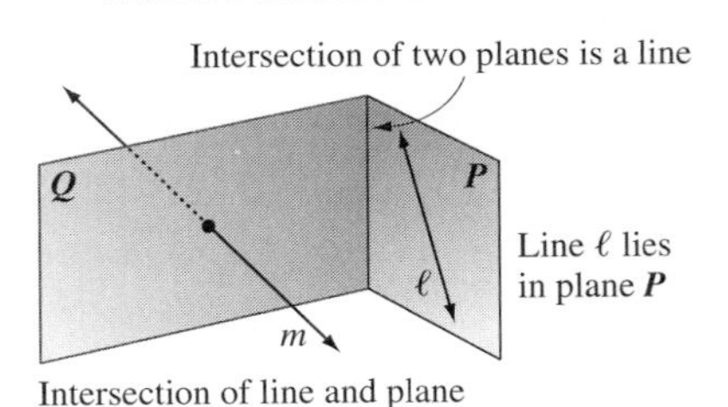

Figure 2.10

In the interest of more colorful language and smoother discourse, we shall modify the stilted language of set theory. This will actually make the axioms shorter and easier to read. For example, instead of saying that the intersection of two lines is a "nonempty set," we shall simply say that the two lines **meet**, or **pass through** a certain point. A point that belongs to a line or plane will be said to **lie on** that line or plane. Two or more lines or planes that meet at the same point are called **concurrent** lines or planes, and two or more points are said to be **collinear** if they lie on the same line. Finally, two points are said to **determine** a line provided there exists a unique line passing through those two points, and three points **determine** a plane if there exists a unique plane passing through them.

AXIOMS FOR POINTS, LINES, AND PLANES

The first axiom is basic to almost any geometry worth considering.

> AXIOM I-1: Each two distinct points determine a line.

Note that this axiom allows us to designate a line by any pair of its points, say A, and B, and we use the common notation

$$\overleftrightarrow{AB}$$

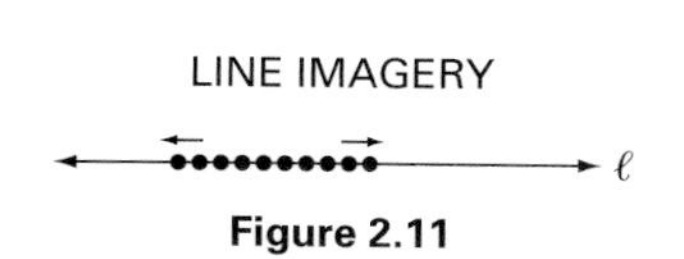

Figure 2.11

The two-headed arrow used in this notation is suggestive of the familiar imagery of a line, consisting of an infinite number of points covering the line and extending indefinitely in both directions. However, until we get to the Ruler Postulate in Section 2.4, we cannot assume that there are even an infinite number of points on a line, let alone a "continuum" (i.e., "continuous infinite stream") of points, as suggested by the diagram in Figure 2.11. At this point such notions are mere imagery.

Since $\overleftrightarrow{AB}$ is a set of points, we have (by definition)

$$A \in \overleftrightarrow{AB}, \qquad B \in \overleftrightarrow{AB}, \qquad \therefore \{A, B\} \subseteq \overleftrightarrow{AB}$$

An obvious restatement of Axiom I-1 using the notion just introduced is therefore

> THEOREM 1: If $C \in \overleftrightarrow{AB}$, $D \in \overleftrightarrow{AB}$ and $C \neq D$, then $\overleftrightarrow{CD} = \overleftrightarrow{AB}$.

Because we are interested in axiomatic geometry of three dimensions, we need to postulate how planes are to behave. The first such axiom guarantees the uniqueness of a plane passing through three noncollinear points.

> AXIOM I-2: Three noncollinear points determine a plane.

OUR GEOMETRIC WORLD

When the four legs of a chair do not rest squarely on the floor, the reason is clear from Axiom I-2: the fourth leg need not lie in the plane of the other three, and if too short or too long, creates instability. But a surveyor's transom, a camera tripod, and a child's tricycle are all highly stable objects, illustrations of Axiom I-2 at work.

Figure 2.12

A simple model at this point could consist of a set of three points, A, B, and C, three subsets for lines $\{A, B\}$, $\{B, C\}$, and $\{A, C\}$, and a single plane $\{A, B, C\}$. Note that we cannot conclude that a given plane contains three noncollinear points, nor that a line always contains two distinct points. A later axiom will take care of that.

The nature of a line to be "straight" or not "curved," or a plane to be "flat" (as illustrated in Figure 2.13) prevents a line from intersecting a plane in two different points unless the line lies completely in the plane. Another desirable property of planes is that two of them must intersect in a line. Thus, our next two incidence axioms are quite predictable.

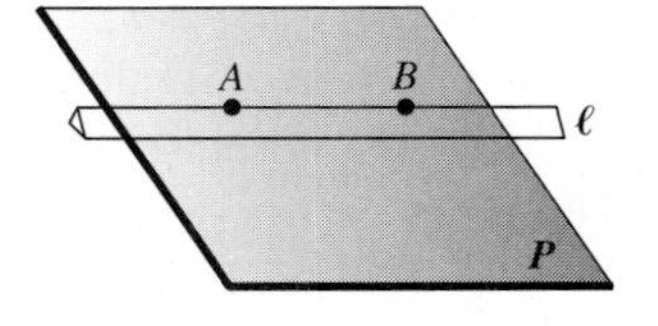

Figure 2.13

AXIOM I-3: If two points lie in a plane, then any line containing those two points lies in that plane.

AXIOM I-4: If two distinct planes meet, their intersection is a line.

The last axiom on incidence is of a technical nature, guaranteeing certain "obvious facts"—desirable properties we could not prove otherwise. (See Problem 12.)

AXIOM I-5: Space consists of at least four noncoplanar points, and contains three noncollinear points. Each plane is a set of points of which at least three are noncollinear, and each line is a set of at least two distinct points.

MODELS FOR THE AXIOMS OF INCIDENCE

The model with the smallest number of points has four points, six lines, and four planes. This will establish the consistency of the axioms of incidence we have introduced. We use three-dimensional coordinate space to aid in its construction, and we encourage you to participate in its construction in the discovery unit at the end of this section.

An interesting problem for you to work on next is to construct a model for Axioms I-2, I-3, I-4, and I-5 in which Axiom I-1 *fails to hold* (Problem 9). This would show the independence of Axiom I-1. A model can be constructed with just five points. The independence of the remaining axioms can also be considered. This explains the language used in Axiom I-3: the words *any line* refer to the possibility of more than one line passing through two points, if Axiom I-1 is not valid.

The following theorem can now be proven.

THEOREM 2: If two distinct lines ℓ and m meet, their intersection is a single point. If a line meets a plane and is not contained by that plane, their intersection is a point.

PROOF

I. *Given:* Distinct lines ℓ and m, and point A on $\ell \cap m$.
Prove: No other point, besides A, lies in $\ell \cap m$.

CONCLUSIONS	JUSTIFICATIONS
(1) Assume false. Assume point $B \neq A$ lies on both ℓ and m.	Assumption for indirect proof
(2) A and B belong to both ℓ and m.	Given and Step **(1)**
(3) $\ell = m \quad \rightarrow\leftarrow$	Axiom I-1
(4) $\therefore A$ is the only point on both ℓ and m.	Rule of Elimination

II. *Given:* Line ℓ not lying in plane $\boldsymbol{P}$, $A \in \ell \cap \boldsymbol{P}$.
Prove: No other point, besides A, lies in $\ell \cap \boldsymbol{P}$.

(The proof will be left as Problem 10.)

Moment for Discovery

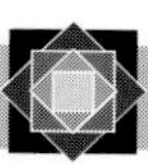

A Test of the Axioms of Incidence

According to Axiom I-5, there must be at least four points. Let these be labeled A, B, C, and D. We explore here whether the axioms hold with only these four points.

1. Arrange A, B, C, D in ordinary xyz-space (shown in Figure 2.14), where A is the origin, and B, C, and D are at a unit distance from A on each of the three coordinate axes. This forms what is known as a *tetrahadron ABCD*.

MODEL FOR AXIOMS I-1 TO I-5

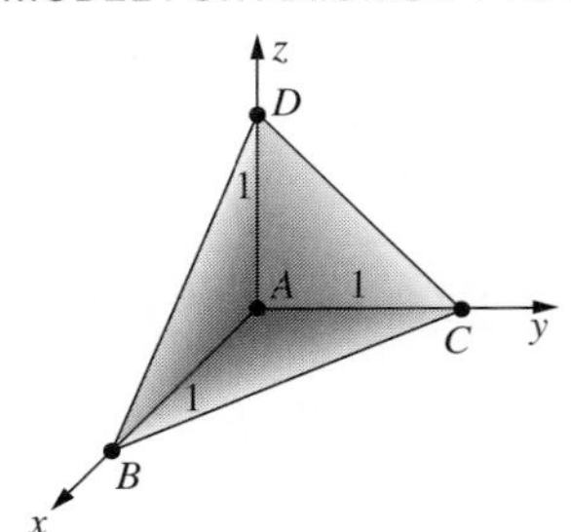

Figure 2.14

2. Using actual lines and planes in xyz-space to help you, and using set theory notation, make a list of
   (a) all the lines possible
   (b) all the planes possible

   (You should get, for example, that two of the lines are $\ell_1 = \{A, B\}$, $\ell_2 = \{A, C\}$, and one of the *planes* is $\boldsymbol{P}_1 = \{A, B, C\}$.)
3. After you determine the lines, examine whether you think Axiom I-1 holds. Can you verify this without referring to the diagram? Look at your list of lines and the points you identified for them.
4. Does Axiom I-2 hold?
5. Is Axiom I-3 obvious? Verify Axiom I-4, again using set theory, and not coordinate geometry. Does Axiom I-5 work?
6. Can you think of a way to construct a model having five points? Six points? n points, where n is an integer > 4? (Consider adding more points to line $\overleftrightarrow{BC}$ in the model.)

PROBLEMS (2.3)

GROUP A

1. Verify that the following system is a model for the incidence axioms (Axioms I-1–I-5) (show that each axiom is true):

   Points: $\boldsymbol{S} = \{1, 2, 3, 4, 5\}$
   Lines: $\{1, 2, 3\}$, $\{1, 4\}$, $\{1, 5\}$, $\{2, 4\}$, $\{2, 5\}$, $\{3, 4\}$, $\{3, 5\}$, $\{4, 5\}$
   Planes: $\{1, 2, 3, 4\}$, $\{1, 2, 3, 5\}$, $\{1, 4, 5\}$, $\{2, 4, 5\}$, $\{3, 4, 5\}$

   Draw a three-dimensional illustration of this model. (If done correctly, the picture serves as an immediate visual verification of Axioms I-1–I-5 in the model.)

2. Show that the following system is *not* a model for Axioms I-1–I-5:

Points: $\mathbf{S} = \{1, 2, 3, 4, 5\}$
Lines: $\{1, 2, 3\}, \{1, 4\}, \{1, 5\}, \{2, 4\}, \{2, 5\}, \{3, 4\}, \{3, 5\}, \{4.5\}$
Planes: $\{1, 2, 3, 4\}, \{1, 2, 3, 5\}, \{2, 3, 4, 5\}, \{1, 4, 5\}$

Which axiom is not satisfied?

3. Consider the following system of points, lines, and planes:

Points: $\mathbf{S} = \{1, 2, 3, 4, 5\}$
Lines: $\{1, 2, 3\}, \{1, 4\}, \{1, 5\}, \{2, 4\}, \{2, 5\}, \{3, 5\}, \{4, 5\}$
Planes: $\{1, 2, 3, 4\}, \{1, 2, 3, 5\}, \{1, 2, 3\}, \{1, 4, 5\}, \{2, 4, 5\}, \{3, 4, 5\}$

Which incidence axioms are satisfied here?

4. Consider the following system of points, lines, and planes:

Points: $\mathbf{S} = \{1, 2, 3, 4\}$
Lines: $\{1, 2, 4\}, \{1, 3\}, \{2, 3\}, \{3, 4\}$
Planes: $\{1, 2, 3\}, \{1, 3, 4\}, \{2, 3, 4\}$

Which incidence axioms are satisfied in this model?

GROUP B

5. Consider the infinite set of positive integers $\{1, 2, 3, \ldots\}$. Find a way to name definite subsets as lines and other definite subsets as planes in such a manner that the incidence axioms will automatically hold as in the manner of the previous tetrahedron model. (**Hint:** Let one line consist of the points 3, 4, 5, . . .)

6. The 7-Point Projective Plane Consider the following model, in which there is only one plane, the universal set ***S*** itself.

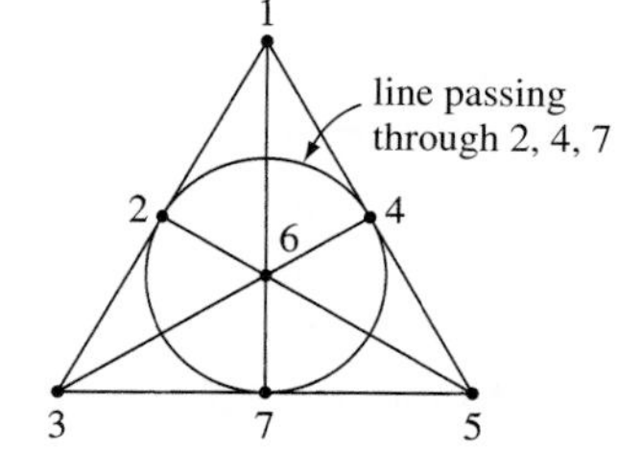

Points: $\mathbf{S} = \{1, 2, 3, 4, 5, 6, 7\}$
Lines: $\ell_1 = \{1, 2, 3\}, \ell_2 = \{1, 4, 5\}, \ell_3 = \{3, 7, 5\}, \ell_4 = \{1, 6, 7\},$
$\ell_5 = \{2, 5, 6\}, \ell_6 = \{3, 4, 6\}, \ell_7 = \{2, 4, 7\}$
Plane: $\mathbf{P} = \{1, 2, 3, 4, 5, 6, 7\}$

Which incidence axioms are satisfied? (Compare this model with the committee structure in the Flying Aviators Club of Problem 11, Section 2.2.)

7. The Four-Point Affine Plane Show that the model depicted here for four points satisfies the axioms for an **affine plane**, consisting of Axiom I-1, the pertinent part of Axiom I-5 dealing with a single plane, and the **Parallel Postulate** (a unique line exists that passes through a given point parallel to a given line).

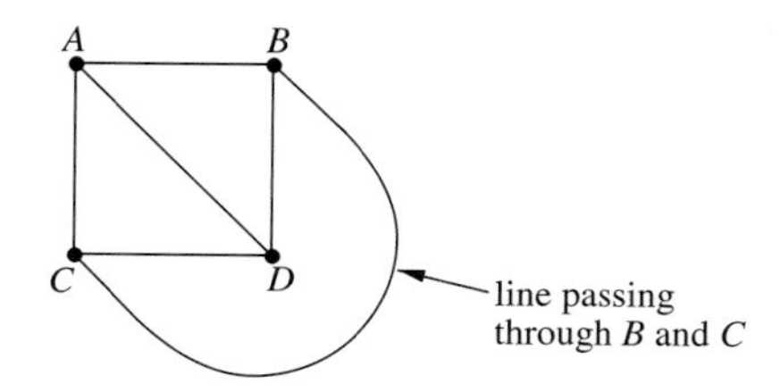

8. **The Nine-Point Affine Plane** Consider the model depicted in the figure that illustrates a set of 9 points and 12 lines:

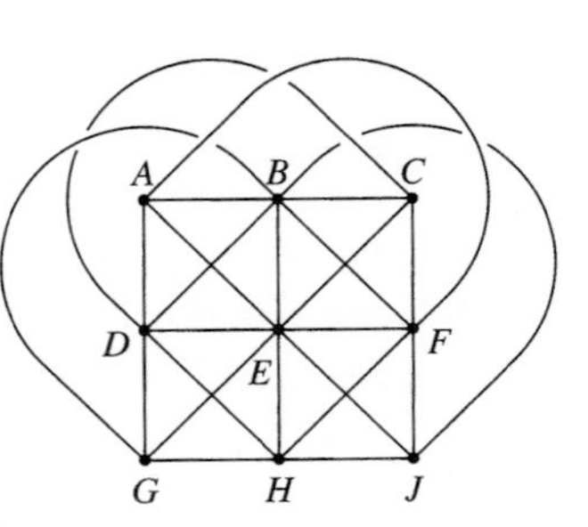

Points: $\mathbf{S} = \{A, B, C, D, E, F, G, H, J\}$
Lines: $\ell_1 = \{A, B, C\}, \ell_2 = \{D, E, F\}, \ell_3 = \{G, H, J\}, \ell_4 = \{A, D, G\},$
$\ell_5 = \{B, E, F\}, \ell_6 = \{C, F, J\}, \ell_7 = \{A, E, J\}, \ell_8 = \{B, F, G\},$
$\ell_9 = \{C, D, H\}, \ell_{10} = \{C, E, G\}, \ell_{11} = \{A, F, H\}, \ell_{12} = \{B, D, J\}$

Show that this geometry satisfies the axioms for an affine plane, defined in Problem 7.

NOTE: These figures cannot represent the models faithfully in every respect; they are intended to show only the incidence relationships defined by the given sets.

9. Construct a model having five points in which Axioms I-2, I-3, I-4, and I-5 are satisfied, but not I-1.

10. Prove the second part of Theorem 2.

GROUP C

11. (**NOTE:** A basic knowledge of algebra is needed for a feasible solution of this problem, including a basic knowledge of so-called *clock arithmetic*.) Use a finite field (clock arithmetic) to generate a finite affine plane of any prime order from the following algebraic model: A *point* is any ordered pair (x, y) of elements from the field; a *line* is, for each a, b, and c in the field, the set of ordered pairs (x, y) satisfying the equation $ax + by + c = 0$, where a and b are not both zero. **Example:** When the field consists of just the elements 0, 1, the result is the four-point affine plane of Problem 7. You are to construct the points and lines, in theory, and prove that the axioms are satisfied in general (in terms of any finite field).

12. Prove explicitly from the incidence axioms that

    (a) a plane cannot be a line

    (b) each line is contained by at least two planes, whose intersection is that line. (Do you see the necessity for all the parts of Axiom I-5?)

13. **Axioms for Affine Space** An **affine space** is any system of points, lines, and planes satisfying Axioms I-1–I-5, and, in addition, the following **Parallel Postulate** for space, with the same meaning of parallel as before (objects which do not meet):

    AXIOM I-6: Given any two noncoplanar lines in space, there exists a unique plane through the first line that is parallel to the second.

    An affine space (or affine plane) having a maximal of $n \geq 2$ points on each line is said to be of **finite order** n. Find an argument for each of the following:

    (a) Every plane in an affine space is an affine plane. (See Problem 7 for definition.)

    (b) Given a plane $\boldsymbol{P}$ and a point A not on it, there exists one and only one plane through A that is parallel to $\boldsymbol{P}$.

    (c) An affine plane of order n has n points on each line, $n^2 + n$ lines, and n^2 points altogether.

    (d) An affine space of order n has n points on each line, $n^3 + n^2 + n$ planes, and n^3 points altogether.

14. Construct a model in which Axioms I-1, I-2, I-4, and I-5 are valid, but not I-3.

15. A 27-point Affine Space The model we will consider has 27 points, 117 lines, and 39 planes. To facilitate its construction, we resort to a series of figures that are three-dimensional drawings of cubes, with the points of the space labeled accordingly. Each horizontal and vertical face, midface, and each diagonal plane indicated in the figure below contains nine points of the geometry; each of the three cubes in the figure thereby contains 13 of the 39 planes. (See Problem 8 for the geometry of each plane.) The lines are obtained by locating any of the 27 points that lie on an *ordinary Euclidean line* in each cube. (They may also be determined as lines of affine planes, 12 lines for each plane, as in Problem 8.) Lines will be duplicated as one passes from one cube to the next, but the planes will not. Examples are, for planes, $\{A, B, C, D, E, F, G, H, I\}$ and $\{F, J, Z, A, Q, X, H, O, S\}$; and for lines, $\{A, B, C\}$, $\{A, N, \Sigma\}$, and $\{B, K, T\}$.

Verify the properties of the model by performing the following valid geometric operations for any affine space.

(a) Find the unique line (the three points on that line) passing through each of the pairs of points (A, K), (V, H), and (S, I).

(b) Find the unique plane (the nine points on that plane) passing through each of the point triples (G, H, R) and (K, F, Σ).

(c) Find whether the planes through (R, S, T) and (B, F, L) intersect, and if they do, find their line of intersection.

(d) Find the unique plane through P parallel to the plane through (B, F, L).

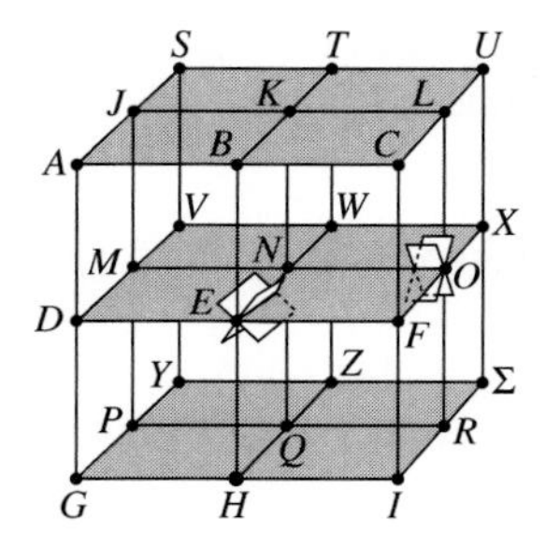

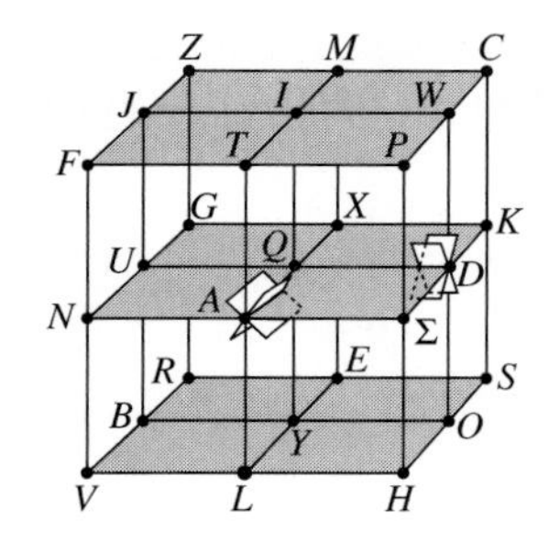

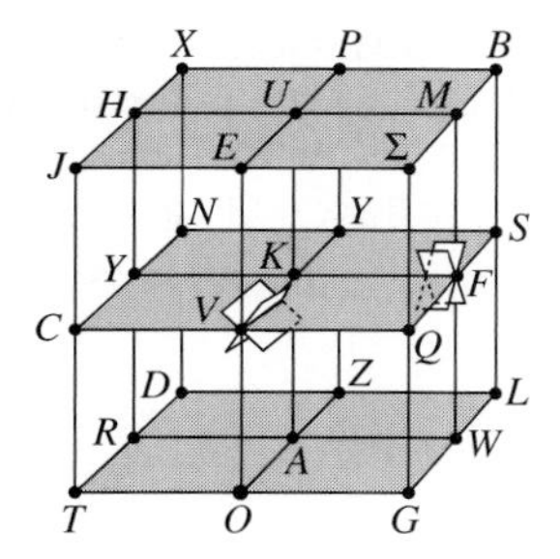

16. Project How many *lines* exist in an affine space of order 2? Find the general formula for an affine space of order n, and verify your answer for the case of $n = 2$, and that of the case $n = 3$ in the previous problem.

2.4 Distance, Ruler Postulate, Segments, Rays, and Angles

As a practical matter, we commonly assume that the distance between any two points can always be *measured* in some way (often easier said than done). Therefore, it is natural to take as the basis for distance, or *metric*, the following axioms.

Metric Axioms

AXIOM D-1: Each pair of points A and B is associated with a unique real number, called the **distance** from A to B, denoted by AB.

AXIOM D-2: For all points A and B, $AB \geq 0$, with equality only when $A = B$.

AXIOM D-3: For all points A and B, $AB = BA$.

In so many words, the metric axioms guarantee that we can measure the distance from one point to another, that the distance measured is positive if the points are distinct, and that the distance between them is not affected by the order in which the points occur (sometimes referred to as *symmetry*). The usual definition of a metric includes an additional axiom, called the **triangle inequality** (which we do not need to assume since this will ultimately be proven as a theorem): Given A, B, and C, as in Figure 2.15,

$$AB + BC \geq AC$$

If it so happens that the inequality reduces to *equality*, that is, if

$$AB + BC = AC$$

then it is only natural to assume that B is somehow "between" A and C—that the ordering of the points on the line is either A–B–C or C–B–A. In the foundations of geometry, betweenness is often taken as an undefined operation, and the desired properties postulated, *without the concept of distance*. This was the approach taken by David Hilbert (1862-1943) in his *Foundations of Geometry*, in keeping with Euclid's *Elements* that never used or mentioned real numbers. However, as a matter of convenience, we have chosen to adopt a notion for distance first, from which we can then define the property of betweenness. This will save us much hard work.

THE TWO CASES OF THE TRIANGLE INEQUALITY

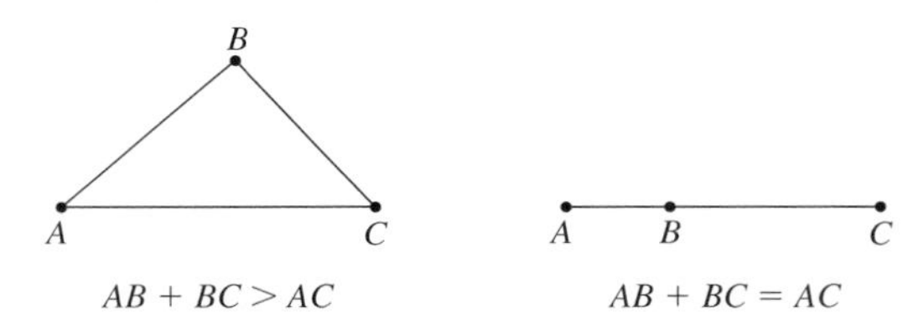

Figure 2.15

HISTORICAL NOTE

David Hilbert solved outstanding problems in both mathematics and physics, and significantly extended virtually every field of mathematics. Born in 1862 in Königsberg, East Prussia (the famous city whose arrangement of seven bridges was the source of the well-known graph theory problem), he was the son of a district judge who wanted him to become a lawyer. But Hilbert soon saw that his greatest interests, and gifts, were in mathematics. He received his doctorate from Königsberg University in May 1885, and arrived at Göttingen University to teach mathematics in 1895, almost exactly 100 years after Gauss had been a student there. Hilbert remained at Göttingen as professor of mathematics the rest of his life, and during this time directed 96 doctoral students. Hilbert's research was so far-reaching that it has been estimated that every significant mathematical discovery made in the twentieth century was involved in some way with at least one of his results. Hilbert's byword, which was engraved on his tombstone, was *We Must Know; We Shall Know.* After having solved a major problem in number theory that beautifully extended Gauss's Reciprocity Theorem, he turned to geometry and wrote his *Foundations of Geometry* in 1898. He adopted the familiar style and terminology of Euclid. That, and its logical perfection, made it a great success (like the *Elements*), quickly going through seven editions. Hilbert's favorite comment was: "Instead of points, lines and planes, one must be able to say at all times tables, chairs, and beer mugs," to emphasize the abstract nature of the axiomatic development. Other notable attempts to correct the logical shortcomings of Euclid during that period included the work of Moritz Pasch, which was actually the basis for Hilbert's work, and the development given by G. Peano (who was famous for his space-filling curve).

DEFINITION AND PROPERTIES OF BETWEENNESS

Definition of Betweenness

For any three points A, B, and C, we say that B is **between** A and C, and we write A–B–C, iff A, B, and C are distinct, collinear points, and $AB + BC = AC$.

One basic theorem concerning betweenness is the following, and we include a two-column proof of it.

THEOREM 1: If A–B–C then C–B–A, and neither A–C–B nor B–A–C.

PROOF

Given: A–B–C.

Prove: **(a)** C–B–A **(b)** A–C–B and B–A–C are impossible.

CONCLUSIONS	JUSTIFICATIONS
(a)	
(1) $AB + BC = AC$ and A, B, C are distinct and collinear points.	Definition of betweenness
(2) $BC + AB = AC$.	Algebra
(3) $CB + BA = CA$.	Symmetry (Axiom D-3)
(4) $\therefore$ C–B–A.	Definition of betweenness
(b)	
(5) Assume A–C–B. Then $AC + CB = AB$.	Assumption for indirect argument
(6) $(AC + CB) + BC = AC$	Substitution: Steps **(1)** and **(5)**
(7) $2BC = 0$, or $BC = 0$ and $B = C$ $\rightarrow\leftarrow$	Algebra ($B \neq C$)

(The case B–A–C is similar, and is left as Problem 15.)

Quite often the ordering of *four* points on a line is needed. For that purpose, we make another definition.

DEFINITION: If A, B, C, and D are four distinct collinear points, then we write A–B–C–D iff the composite of all four betweenness relations A–B–C, B–C–D, A–B–D, and A–C–D are true.

Proving the basic result regarding this concept provides a nice problem for you (Problem 17); just pattern your proof after that given above for Theorem 1.

THEOREM 2: If A–B–C, B–C–D, and A–B–D hold, then A–B–C–D is true.

EXAMPLE 1 Suppose that in a certain metric geometry satisfying Axioms D-1–D-3, points A, B, C, and D are collinear and

$$AB = 2, \quad AC = 3, \quad AD = 4, \quad BC = 5,$$
$$BD = 6, \quad \text{and} \quad CD = 7$$

What betweenness relations follow, *by definition*, among these points? Can these four points be placed in some order such that a betweenness relation for all four points holds for them, as in the above definition?

SOLUTION

Trying to draw a representative diagram of the situation is useless here because the distances given cannot be represented on an ordinary Euclidean line where the Ruler Postulate holds (stated below). We must abstractly examine each of the four possible point triples one by one to test for the possible betweenness relations:

$\{A, B, C\} \rightarrow AB = 2,\ AC = 3,$ and $BC = 5.$ Therefore $B\text{–}A\text{–}C$ holds.

$\{A, B, D\} \rightarrow AB = 2,\ AD = 4,$ and $BD = 6.$ Therefore $B\text{–}A\text{–}D$ holds.

$\{A, C, D\} \rightarrow AC = 3,\ AD = 4,$ and $CD = 7.$ Therefore $C\text{–}A\text{–}D$ holds.

$\{B, C, D\} \rightarrow BC = 5,\ BD = 6,$ and $CD = 7.$ No betweenness relation holds.

Thus, we conclude that just the relations $B\text{–}A\text{–}C$, $B\text{–}A\text{–}D$, and $C\text{–}A\text{–}D$ hold, and that no "quadruple" betweenness is valid. ■

SEGMENTS, RAYS, AND ANGLES

The concept of betweenness was all that Euclid would have needed to make segments and rays meaningful mathematically. It is not necessary to have to resort to drawing a picture and pointing. The definitions we give are set theoretic for greater clarity and convenience, illustrated in Figure 2.16. (Although a line is an undefined object and does not require definition, we include a useful formula for it.) While we are at it we define *angle,* as it is merely the union of two rays.

SEGMENT AB: $\overline{AB} = \{X : A\text{–}X\text{–}B,\ X = A, \text{ or } X = B\}$

RAY AB: $\overrightarrow{AB} = \{X : A\text{–}X\text{–}B,\ A\text{–}B\text{–}X,\ X = A, \text{ or } X = B\}$

LINE AB: $\overleftrightarrow{AB} = \{X : X\text{–}A\text{–}B,\ A\text{–}X\text{–}B,\ A\text{–}B\text{–}X,\ X = A, \text{ or } X = B\}$

ANGLE ABC: $\angle ABC = \overrightarrow{BA} \cup \overrightarrow{BC}$ (A, B, and C noncollinear)

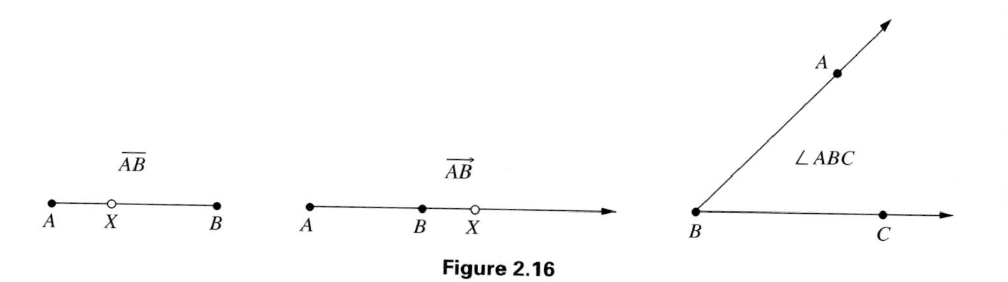

Figure 2.16

The points A and B are called the **end points** of segment $\overline{AB}$, and point A is the **end point** (also, sometimes called the **origin**) of ray $\overrightarrow{AB}$. Point B in $\angle ABC$ is called the **vertex** of the angle, and rays $\overrightarrow{BA}$ and $\overrightarrow{BC}$, its **sides**.

EXTENDED SEGMENTS, RAYS

Frequently the term *extended segment* or *ray* is used in geometry, to mean the following:

> DEFINITION: The **extension** of segment $\overline{AB}$ is either the ray $\overrightarrow{AB}$ (in the direction B), the ray $\overrightarrow{BA}$ (in the direction A), or the line $\overleftrightarrow{AB}$ (in both directions). The **extension** of ray $\overrightarrow{AB}$ is just the line $\overleftrightarrow{AB}$.

Finally, at times we need to mention a segment or ray *without its end points*. A segment without its end points is called an **open segment**, and a ray without its end point (origin) is called an **open ray**. Any point of a segment (or ray) that is not an end point is referred to as an **interior point** of that segment (or ray).

OUR GEOMETRIC WORLD

Euclid's approach to geometry could be described by the term **caliper geometry**, a geometry in which the only means of comparing lengths of segments are by a pair of mechanical calipers. In such a geometry it is possible to determine only whether one segment is larger than another, or if they are equal (congruent). Euclid's *Elements* provided no concept for the "length" of a segment (in terms of a real number apart from the segment itself, as we are assuming here). One reason for this was that the real number system had not yet been invented in Euclid's day. Another is that this is not in the true spirit of synthetic geometry, which allows no algebra or the use of numbers.

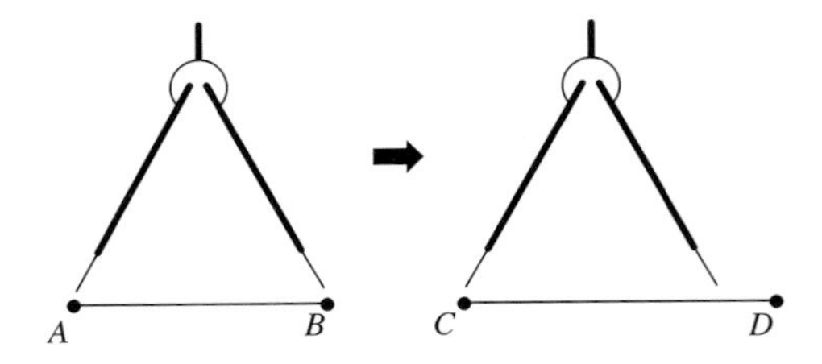

Figure 2.17

RULER POSTULATE

Consider the practice of measuring, by ruler or yardstick the distance between two points. In effect, we set up A as the origin of a coordinate system, and read off the number opposite point B, as illustrated in Figure 2.18. If A happens not to be located at the origin (i.e., the end of the ruler), then we must take *two* readings for the two points, and *subtract*, as shown in Figure 2.19. Thus, in this example, $AB = 10\frac{1}{8} - 4 = 6.125$ (in.). This leads to the important **Ruler Postulate**.

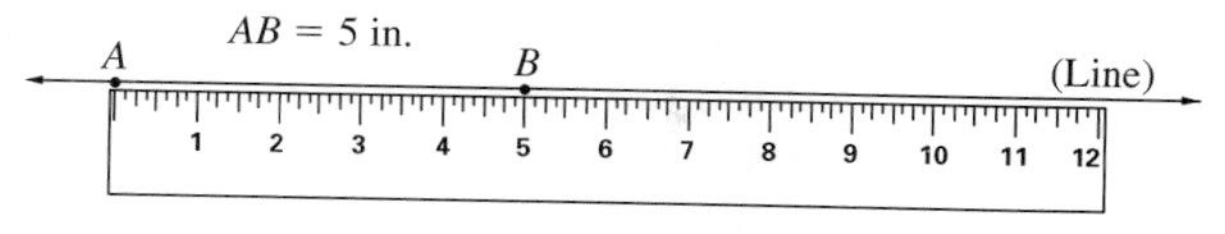

Figure 2.18

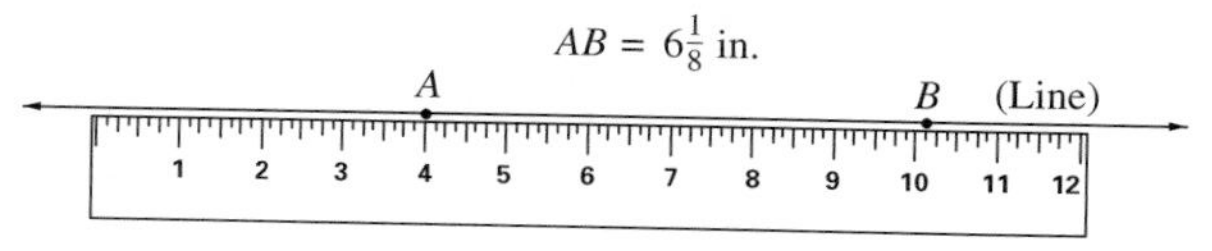

Figure 2.19

AXIOM D-4: RULER POSTULATE
The points of each line ℓ may be assigned to the entire set of real numbers x, $-\infty < x < \infty$, called **coordinates**, in such a manner that
(1) each point on ℓ is assigned to a unique coordinate
(2) no two points are assigned to the same coordinate
(3) any two points on ℓ may be assigned the coordinates zero and a positive real number, respectively
(4) if points A and B on ℓ have coordinates a and b, then $AB = |a - b|$.

As a result, notice the direct correlation between the ordering of the geometric points on a line and the ordering of their coordinates, as real numbers. Thus, in Figure 2.20 we have

$$A\text{–}B\text{–}C\text{–}D$$

for the points on line ℓ, and

$$-3 < 1 < 2 < 5$$

for their coordinates. This obvious property is stated in general for later use. We introduce the useful notation $A[a]$, to denote the (geometric) point A, with its coordinate (real number) a, as guaranteed by the Ruler Postulate.

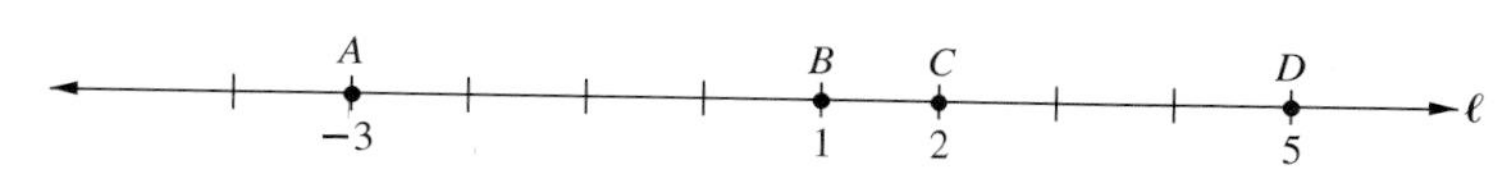

Figure 2.20

THEOREM 3: For any line ℓ and any coordinate system under the Ruler Postulate, if $A[a]$, $B[b]$, and $C[c]$ are three points on line ℓ, with their coordinates, then A–B–C iff either $a < b < c$ or $c < b < a$.

It is evident the Ruler Postulate imposes on each line in our geometry or ray a **continuum of points,** like the real number line, that has infinitely many points, is everywhere dense (each two points have a point between them), and is without gaps or "holes."

EXAMPLE 2 Prove that if A–B–C holds, then the two rays from B passing through A and C make a line. That is, prove that $\overrightarrow{BA} \cup \overrightarrow{BC} = \overleftrightarrow{AC}$.

SOLUTION

This is a problem about two sets: (1) $\overrightarrow{BA} \cup \overrightarrow{BC}$ (all points on ray $\overrightarrow{BA}$ or $\overrightarrow{BC}$) and (2) line $\overleftrightarrow{AC}$. To prove that two sets are the same, it is necessary to show that each is a subset of the other.

(a) It will be proven first that $\overrightarrow{BA} \cup \overrightarrow{BC} \subseteq \overleftrightarrow{AC}$. Observe by definition that the points of ray $\overrightarrow{BA}$ are collinear with A and B, hence, by Theorem 1, Section 2.3, $\overrightarrow{BA} \subseteq \overleftrightarrow{BA} = \overleftrightarrow{AC}$. Similarly, $\overrightarrow{BC} \subseteq \overleftrightarrow{BC} = \overleftrightarrow{AC}$. Since both sets $\overrightarrow{BA}$ and $\overrightarrow{BC}$ are subsets of $\overleftrightarrow{AC}$, their union is a subset, so we are finished with this part.

(b) Now for the reverse inclusion, $\overleftrightarrow{AC} \subseteq \overrightarrow{BA} \cup \overrightarrow{BC}$: Let X be any point on the line $\overleftrightarrow{AC}$. It helps to think of point X as moving from left to right on the line $\overleftrightarrow{AC}$ to obtain all possible positions of X as related to the given betweenness relation, A–B–C. Thus, by the Ruler Postulate, there are seven possible cases (X–A–B, $X = A$, A–X–B, . . .). But in each case it follows either $X \in \overrightarrow{BA}$ or $X \in \overrightarrow{BC}$. Hence, $X \in \overrightarrow{BA} \cup \overrightarrow{BC}$, and every point of $\overleftrightarrow{AC}$ is in $\overrightarrow{BA} \cup \overrightarrow{BC}$, as desired. ■

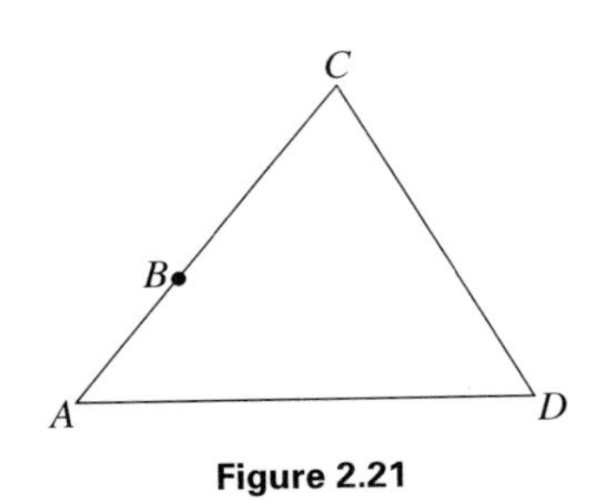

Figure 2.21

One useful result about angles often overlooked in geometry courses needs to be included, because it is used so often. In Figure 2.21 it appears that it is a question of how one "reads the angle," whether we write $\angle BAD$ or $\angle CAD$. However, because $\angle BAD = \overrightarrow{AB} \cup \overrightarrow{AD}$ and $\angle CAD = \overrightarrow{AC} \cup \overrightarrow{AD}$, this relies on the result that rays $\overrightarrow{AB}$ and $\overrightarrow{AC}$ are the same set of points whenever A–B–C. Certainly the latter is intuitively clear, but we state this result formally, with a paragraph proof for it.

THEOREM 4: If $C \in \overrightarrow{AB}$ and $A \neq C$, then $\overrightarrow{AB} = \overrightarrow{AC}$.

PROOF

As shown in Figure 2.22, let a coordinate system be determined for line $\ell = \overleftrightarrow{AB}$, with coordinates given by $A[0]$ and $B[b]$, $b > 0$, as guaranteed by the Ruler Postulate. Now by Theorem 3, the points of ray $\overrightarrow{AB}$ are precisely the points $P[x]$ on ℓ such that $x \geq 0$. Because $C[c]$ lies on $\overrightarrow{AB}$, $c \geq 0$, and because $C \neq A$, then $c > 0$. Using the betweenness definition for points on ray $\overrightarrow{AC}$, those points are, again, precisely the points $P[x]$ on ℓ such that $x \geq 0$. That is, the points of rays $\overrightarrow{AB}$ and $\overrightarrow{AC}$ are the same and $\overrightarrow{AB} = \overrightarrow{AC}$.

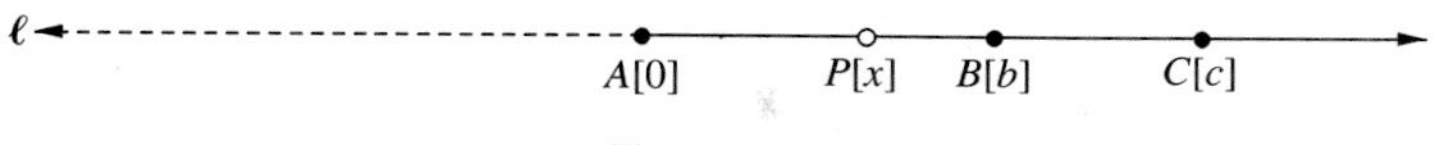

Figure 2.22

SEGMENT CONSTRUCTION THEOREM

We end this section on distance by including two rather obvious conclusions from the Ruler Postulate that we will not prove (and you need not write a proof unless you are so inclined). However, it is important to be familiar with these results because they will be used almost constantly in later work. They represent the theoretical counterpart of the Euclidean constructions of duplicating a line segment (or "moving" a segment from one place to another) and locating the midpoint of a segment. We define the **midpoint** of segment $\overline{AB}$ to be any point M on $\overline{AB}$ such that $AM = MB$. Any geometric object passing through M is said to **bisect** the segment $\overline{AB}$. Also, the **measure** of a segment $\overline{AB}$ is just the distance from A to B: $m\overline{AB} = AB$.

THEOREM 5: SEGMENT CONSTRUCTION THEOREM
If $\overline{AB}$ and $\overline{CD}$ are two segments and $AB < CD$, then there exists a unique point E on ray $\overrightarrow{CD}$ such that $AB = CE$, and C–E–D. (See Figure 2.23.)

SEGMENT CONSTRUCTION THEOREM

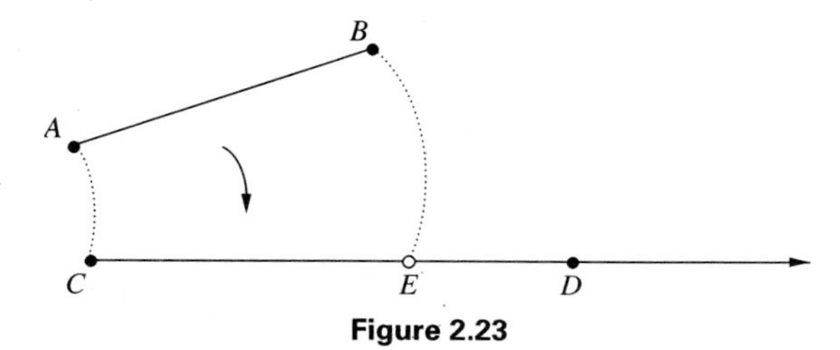

Figure 2.23

COROLLARY: The midpoint of any segment exists, and is unique.

Moment for Discovery

An Unusual Metric

An astronaut from the planet Xenon has just landed on earth, and reveals to us the science and mathematics with which he is familiar. In the area of geometry,

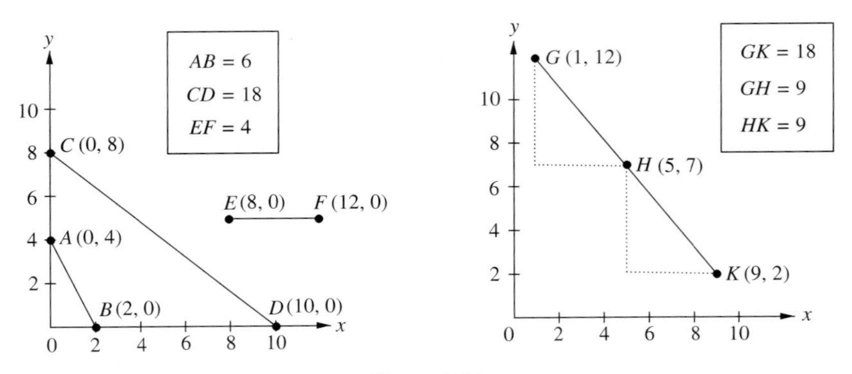

Figure 2.24

he shows us the diagram in Figure 2.24 (strangely familiar!) of a coordinate system and his calculations of distance between various points.

1. Based on your observations of these results, can you tell what the visitor's concept for distance is? If you can, write down a general formula for the *Xenon distance* from $P(x_1, y_1)$ to $Q(x_2, y_2)$
2. If you have access to *Sketchpad* software and are familiar with Scripts, set up a program for calculating Xenon distance between any two points and displaying it on the screen. Then go to the next step. If not, then skip this step.
3. Calculate the Xenon distances *TU*, *UV*, *VW*, *TV*, and *TW* for the points *T*, *U*, *V*, and *W*, as shown in Figure 2.25, and record the results. Use *Sketchpad* if possible, or use hand calculations, not overly tedious here. Note that these points all lie on a line, having equation $y = 1 + x/3$. Finally, compute the values $TU + UV$ and $TU + UV + VW$. Did anything happen?

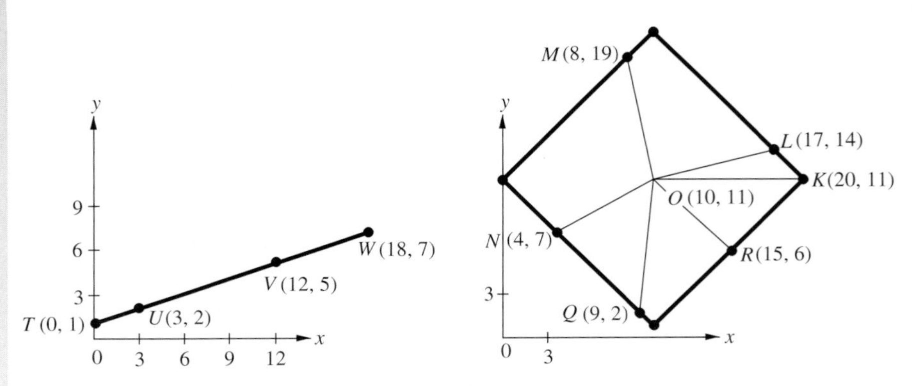

Figure 2.25

4. Again, using your formula from Step 1, calculate the distances from point $O(10, 11)$ to the points *K*, *L*, *M*, *N*, *Q*, and *R*, lying on a square, as shown. Did anything happen?

5. Based on your results in Steps 3 and 4, what are some features of Xenon Geometry? Is the triangle inequality valid?

PROBLEMS (2.4)

GROUP A

1. Suppose that in a certain metric geometry satisfying Axioms D-1–D-4, points A, B, C, and D are collinear, and

$$AB = 4 = AC, \quad BC = 8, \quad BD = 3, \quad CD = 5, \quad \text{and} \quad AD = 1.$$

What betweenness relations follow (by definition) among these points? (A diagram works here since the Ruler Postulate holds.)

2. Suppose that in a certain metric geometry satisfying Axioms D-1–D-3, points A, B, C, and D are collinear, and

$$AB = 4 = AC, \quad BC = 6, \quad BD = 3 = CD \quad \text{and} \quad AD = 2.$$

What betweenness relations follow (by definition) among these points?

3. Suppose that in a certain metric geometry satisfying Axioms D-1–D-3, points A, B, C, and D are collinear, and

$$AB = 4 = AC, \quad AD = 6, \quad BC = 8, \quad BD = 9, \quad CD = 1.$$

(a) What betweenness relations follow (by definition) among these points?

(b) Show that the Triangle Inequality is *not* satisfied.

4. **Matching** The expressions on the left represent certain geometric objects that are pictured in their totality in the figures on the right. (For example, Figure (1) shows the geometric object $\overrightarrow{SR}$.) For each expression, enter the number of the matched geometric object. The same response may be used more than once.

(a) $\overline{RS}$ _____

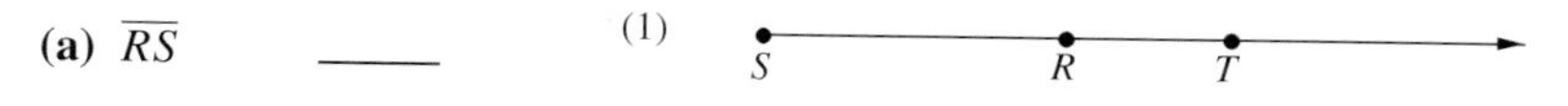

(b) $\overrightarrow{RT}$ _____

(2) R T S

(c) $\overrightarrow{ST}$ _____

(3) R S T

(d) $\overleftrightarrow{RS}]$ _____

(4) T S R

(e) $\overline{RT}$ _____

(f) $\overleftrightarrow{ST}$ _____

(5) T S R

5. Identify by the correct symbol each of the following sets appearing in the figure, without proof:

(a) $\overline{BC} \cap \overrightarrow{BD}$ (Answer: $\overline{BC}$)

(b) $\overline{BC} \cup \overline{CD}$

(c) $\overrightarrow{BC} \cup \overrightarrow{DC}$

(d) $\overrightarrow{AC} \cap \overrightarrow{BD}$

(e) $\overrightarrow{CA} \cup \overrightarrow{CD}$

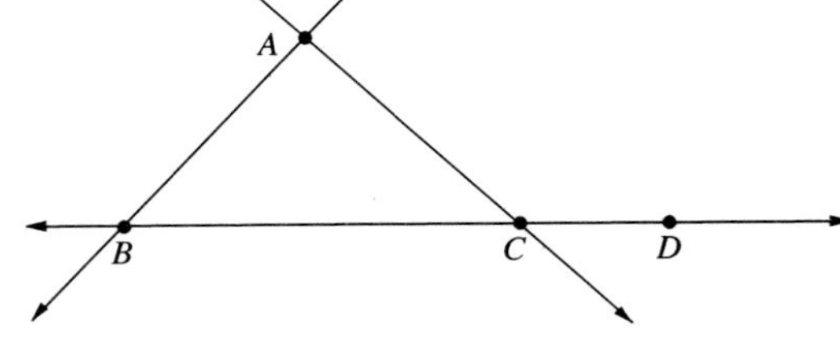

6. Identify by the correct symbol each of the following sets appearing in the figure, without proof:

(a) $\angle ABC \cap \angle ACD$

(b) $\angle ABC \cap \angle ACB$

(c) $\overrightarrow{AB} \cap \angle ABC$

7. Name all segments and rays determined by the points A, B, C, and D exhibited by illustration in the figure that do *not* contain point E.

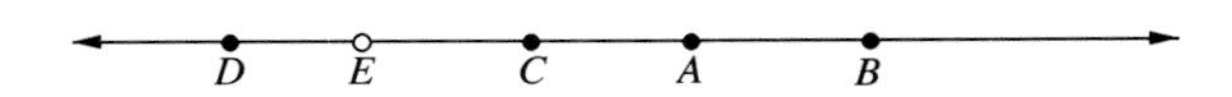

8. Let $A[3]$ and $B[-6]$ be two points on a line (with their coordinates). Find the values of x allowable if point $P[x]$ is any point on

(a) $\overline{AB}$ (Answer: $-6 \leq x \leq 3$)

(b) $\overrightarrow{AB}$

(c) $\overleftrightarrow{AB}$.

9. Let $A[5]$, $B[10]$, $C[3]$, and $D[-1]$ denote four collinear points, with their coordinates. If $P[x]$ denotes any other point on $\overleftrightarrow{AB}$, find the range of x allowable if P lies on

(a) $\overrightarrow{CB}$

(b) $\overline{AD}$

(c) $\overrightarrow{AD} \cap \overrightarrow{CA}$.

10. Let $A[-3]$, $B[-1]$, $C[6]$, and $D[5]$ denote four collinear points, with their coordinates. If $P[x]$ denotes any other point on $\overleftrightarrow{AB}$, find the range of x allowable if P lies on

(a) $\overrightarrow{AB}$

(b) $\overrightarrow{BD}$

(c) $\overline{DC}$

(d) $\overrightarrow{AB} \cup \overrightarrow{CD}$.

11. On the line indicated in the accompanying figure, if $WY = 17$, $WZ = 23$, and $XZ = 21$, find XY. (Use the betweenness relations evident from the figure.) [UCSMP,[2] p. 50]

[2]University of Chicago School Mathematics Project: Geometry.

12. The points A, B, C, D, E, F, and G are defined to lie in the coordinate xy-plane in terms of their coordinates, given in the same order, as follows: $(0,1)$, $(\pm 1, 2)$, $(\pm 2, 3)$, and $(\pm 3, 3)$. Name the betweenness relations that exist among these points, using the ordinary concept of Euclidean distance normally assumed in coordinate geometry.

13. Let $A[a]$ and $B[b]$ designate two points on a line that are 8 units apart. If the Ruler Postulate holds, what are the possible values for b in terms of a?

14. If on some line ℓ we plot all points $P[x]$ such that $x < 3$, what geometric object best describes the given set? What exactly describes it?

GROUP B

15. Complete the proof of Theorem 1. (Assume B–A–C, then obtain a contradiction.)

16. Prove that for $A \neq B$, $\overline{AB} \subseteq \overrightarrow{AB} \subseteq \overleftrightarrow{AB}$.

17. Prove that if A–B–C, B–C–D, and A–B–D then A–B–C–D (Theorem 2).

18. Prove that if A, B, and C are any three distinct, collinear points, then either A–B–C, A–C–B, or B–A–C.

19. Prove that a segment cannot be a ray.

GROUP C

20. In Problem 11, what answers are possible for XY if no betweenness relations are assumed from a figure, but it is still assumed that X, Y, Z, and W are collinear, and that the Ruler Postulate holds? (***Hint:*** Draw a diagram, placing point W[0] on the line first.)

21. The Round-Up Metric For all points on the x-axis, represented conveniently by the coordinates themselves, a concept for distance is defined as follows: $d(x, y) = \{\text{the number } |x - y| \text{ } \textit{rounded up}\}$, that is, either $|x - y|$ itself if it is an integer, or the *next higher integer* if it is not. [Examples: $d(2, 5) = 3$, $d(2, 5.5) = 4$, and $d(-2, 5.5) = 8$.]

(a) Does this concept of distance satisfy Axioms D-1, D-2, and D-3? D-4?

(b) Identify the segment $\overline{25}$. (***Hint:*** The answer is $\{2, 3, 4, 5\}$ and no other points. Explain.)

(c) Identify the segment $\overline{2a}$ where $a = 5.5$.

(d) Identify the segment $\overline{ab}$ where $a = 2.5$ and $b = 5.5$. Generalize to arbitrary numbers a and b.

(e) Prove the triangle inequality for this metric.

22. Undergraduate Research Project What is the least number of distances among n distinct, collinear points satisfying Axioms D-1–D-3 and the Triangle Inequality that will uniquely determine the ordering of those points (i.e., P_1–P_2–P_3– $\cdots$ –P_n, where the obvious extension of the betweenness concept A–B–C–D is indicated). Is there a formula $d(n)$ for this number? Start with the simplest case $n = 3$: Assume that $AB = d_1$, $BC = d_2$, and $AC = d_3$, where $d_1 = d_2 + d_3$; will this force A–C–B? Can d_1, d_2, and d_3 represent only *two* distances? If so, then this would prove that $d(3) = 2$. Now move on to $n = 4$. Note that for any ordering of n

points P_1–P_2–P_3– $\cdots$ –P_n the least number of distances seems to occur when those points are equally spaced on the line.

2.5 Angle Measure and the Protractor Postulate

The first two axioms for angle measure will assert that (1) we can measure any angle, and (2) when two angles are placed adjacent to one another and they are not too large, we can sum the measures of the smaller angles to obtain the measure of the larger angle.

AXIOM A-1: EXISTENCY OF ANGLE MEASURE
Each angle $\angle ABC$ is associated with a unique real number between 0 and 180, called its **measure** and denoted $m\angle ABC$. No angle can have measure 0 or 180.

The degree mark, as in 120°, is not normally used in the foundations of geometry because angle measure is just a real number, needing no particular embellishment. (Otherwise, we would be obliged to define *degree* and prove some theorems about it.) However, degree signs are necessary in figures to distinguish angle measure from lengths of segments, or in trigonometric expressions to distinguish degrees from radians.

NOTE: In axiomatic geometry, it is customary to assume that all angles have measure less than 180. If this seems to contradict what you studied in trigonometry, try defining (rigorously) an *angle of measure* 270. Remember that the measure of an angle cannot have *two* values—it can only depend on its sides. The concept of *rotation* is not available because our axioms do not include it. However, it is possible to allow angles to have measure 180 (when they are *straight angles*), but we are following Moise's foundation for geometry that disallows straight angles. This is largely a whim, but if straight angles were allowed we would have to precede every theorem or corollary involving the interior of an angle with the disclaimer "if the angle is not a straight angle."

INTERIOR OF AN ANGLE

To state the next axiom, we are going to need a definition for *interior point* of an angle. We will give a definition that is correct in Euclidean geometry, but it will be replaced by a better one in the next section that is valid in absolute geometry.

DEFINITION: A point D is an **interior point** of $\angle ABC$ (Figure 2.26) iff there exists a segment $\overline{EF}$ containing D as an interior point that extends from one side of the angle to the other ($E \in \overrightarrow{BA}$ and $F \in \overrightarrow{BC}$, $E \neq B, F \neq B$).

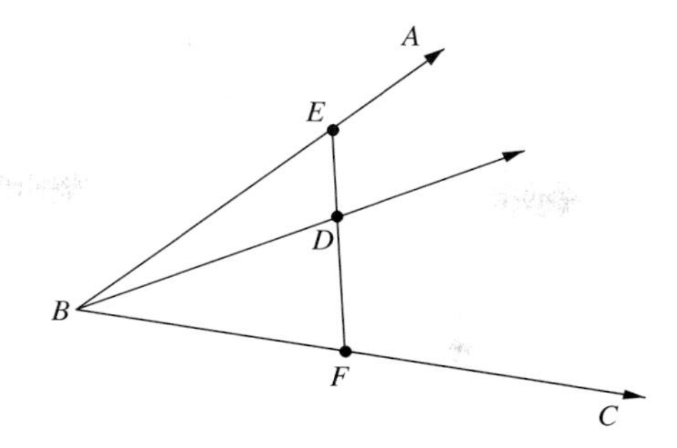

Figure 2.26

AXIOM A-2: ANGLE ADDITION POSTULATE

If D lies in the interior of $\angle ABC$, then $m\angle ABD + m\angle DBC = m\angle ABC$. Conversely, if $m\angle ABD + m\angle DBC = m\angle ABC$, then ray $\overrightarrow{BD}$ passes through an interior point of $\angle ABC$.[3]

BETWEENNESS FOR RAYS

Axiom A-2 provides the basis for a betweenness relation for rays having the same end point. Since we will be making constant use of this concept, we define it formally.

DEFINITION (BETWEENNESS FOR RAYS)

For any three rays $\overrightarrow{BA}$, $\overrightarrow{BD}$, and $\overrightarrow{BC}$ (having the same end point), we say that ray $\overrightarrow{BD}$ lies between rays $\overrightarrow{BA}$ and $\overrightarrow{BC}$, and we write $\overrightarrow{BA}$-$\overrightarrow{BD}$-$\overrightarrow{CD}$, iff the rays are distinct and $m\angle ABD + m\angle DBC = m\angle ABC$.

EXAMPLE 1 Suppose that in Figure 2.27 the betweenness relation C–E–D–B holds on line $\overleftrightarrow{BC}$. Show that $\overrightarrow{AC}$–$\overrightarrow{AE}$–$\overrightarrow{AD}$–$\overrightarrow{AB}$ holds (where the definition of betweenness for four rays is directly analogous to that of four points), and that $m\angle CAB = m\angle 1 + m\angle 2 + m\angle 3$.

---

[3]This converse is assumed for convenience only. It can, however, be established as a theorem making extensive use of the material in the next section. This will appear as a Group C problem (with hints) in the next section, for those who may wish to tackle it.

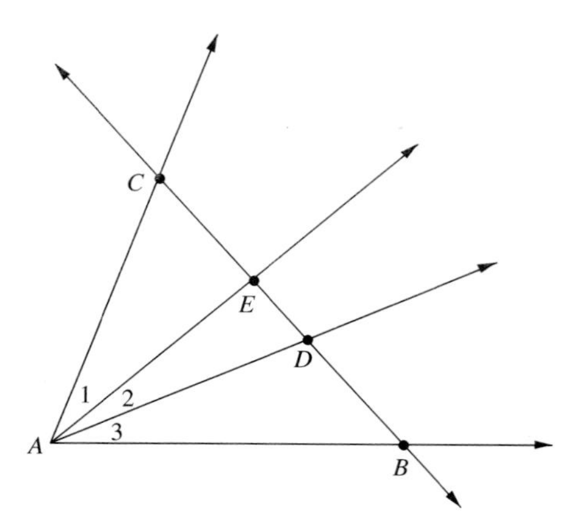

Figure 2.27

SOLUTION

Since E is interior to $\angle CAD$, $\overrightarrow{AC}$–$\overrightarrow{AE}$–$\overrightarrow{AD}$ follows from Axiom A-2. Similarly, $\overrightarrow{AC}$–$\overrightarrow{AE}$–$\overrightarrow{AB}$, $\overrightarrow{AC}$–$\overrightarrow{AD}$–$\overrightarrow{AB}$ and $\overrightarrow{AE}$–$\overrightarrow{AD}$–$\overrightarrow{AB}$. It follows by definition that $\overrightarrow{AC}$–$\overrightarrow{AE}$–$\overrightarrow{AD}$–$\overrightarrow{AB}$. Therefore, $m\angle CAB = m\angle CAD + m\angle 3 = (m\angle 1 + m\angle 2) + m\angle 3$. ■

The same kind of results follow for betweenness among rays as did for betweenness among points—the proofs are almost the same, word for word. This feature shows that there is a *duality* between concurrent rays (having the same end point) and collinear points—what is true for one, is true for the other. An important case is the Protractor Postulate, stated below. It is the dual of the Ruler Postulate.

Any angle can be measured by a protractor, much as we can measure the distance between two points with a ruler: we simply put the protractor in place along the "initial" side of the angle and read off the number opposite the "terminal" side, as shown in Figure 2.28. If we happen to have *two* rays, like $\overrightarrow{AC}$ and $\overrightarrow{AD}$ in the figure, then to find $m\angle CAD$ we must take the two readings opposite rays $\overrightarrow{AC}$ and $\overrightarrow{AD}$, and subtract. Thus, in this example, $m\angle CAD = 52.5 - 30 = 22.5$.

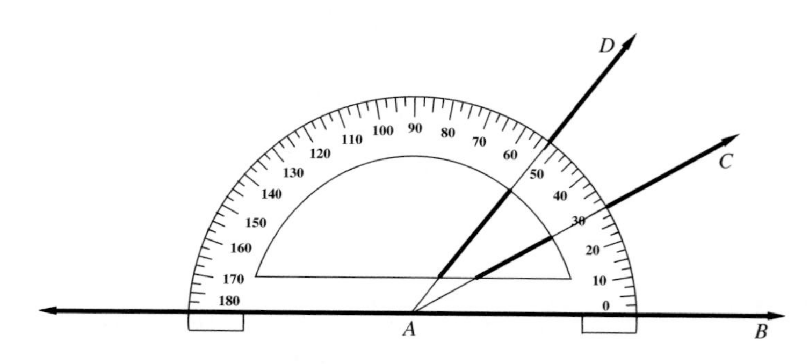

Figure 2.28

As in the preceding example, the Protractor Postulate assigns numbers between 0 and 180 to all the rays lying on *one side* of line $\overleftrightarrow{AB}$, that is, given point C not on $\overleftrightarrow{AB}$ and D–A–B, to all rays $\overrightarrow{AX}$ which meet segment $\overline{BC}$ or $\overline{CD}$ and not passing through B or C. (See Figure 2.29.) This set of rays, excluding rays $\overrightarrow{AB}$ and $\overrightarrow{AC}$, can be said to lie *on the C-side* of line $\overleftrightarrow{AB}$.

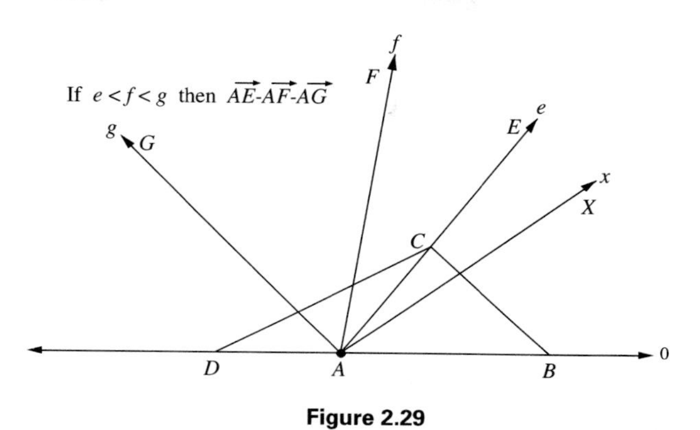

Figure 2.29

PROTRACTOR POSTULATE

> **AXIOM A-3: PROTRACTOR POSTULATE**
> The set of rays $\overrightarrow{AX}$ lying on one side of a given line $\overleftrightarrow{AB}$, including ray $\overrightarrow{AB}$, may be assigned to the entire set of real numbers x, $0 \le x < 180$, called **coordinates**, in such a manner that
> **(1)** each ray is assigned to a unique coordinate
> **(2)** no two rays are assigned to the same coordinate
> **(3)** the coordinate of $\overrightarrow{AB}$ is 0
> **(4)** if rays $\overrightarrow{AC}$ and $\overrightarrow{AD}$ have coordinates c and d, then $m\angle CAD = |c - d|$.

We can immediately prove the theorem that is the dual of Theorem 3 in Section 2.4, where the ordering of the rays correlate to the ordering of the reals, as indicated in Figure 2.29. Thus, for example, we can conclude that if rays $\overrightarrow{AB}$ and $\overrightarrow{AC}$ are between $\overrightarrow{AP}$ and $\overrightarrow{AQ}$, either the order $\overrightarrow{AP}$–$\overrightarrow{AB}$–$\overrightarrow{AC}$ or the order $\overrightarrow{AP}$–$\overrightarrow{AC}$–$\overrightarrow{AB}$ occurs. We regard all these results intuitively obvious, and we choose not to belabor the point by stating and proving all of them formally.

ANGLE CONSTRUCTION THEOREM

One result that does need a formal treatment is that which allows us to construct angle bisectors and to "copy" an angle—ideas we will use over and over. We define an **angle bisector** for $\angle ABC$ to be any ray $\overrightarrow{BD}$ lying between the sides $\overrightarrow{BA}$ and $\overrightarrow{BC}$ such that $m\angle ABD = m\angle DBC$.

THEOREM 1: ANGLE CONSTRUCTION THEOREM
For any two angles $\angle ABC$ and $\angle DEF$ (Figure 2.30) such that $m\angle ABC < m\angle DEF$, there is a unique ray $\overrightarrow{EG}$ such that $m\angle ABC = m\angle GEF$ and $\overrightarrow{EF}$–$\overrightarrow{EG}$–$\overrightarrow{ED}$.

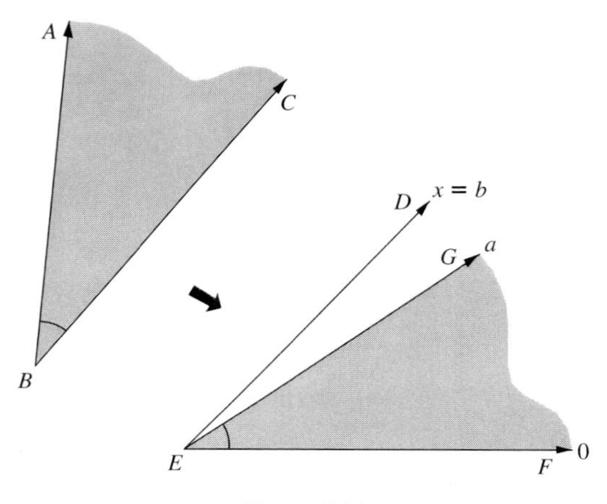

Figure 2.30

COROLLARY: The bisector of any angle exists and is unique.

Because the theorem and its corollary are important, we give an outline proof of the theorem. The corollary will be left as a problem (Problem 12).

PROOF OF THEOREM 1 (See Figure 2.30.)

Given: $m\angle ABC < m\angle DEF$

Prove: There exists a unique ray $\overrightarrow{EG}$ such that $m\angle ABC = m\angle GEF$ and $\overrightarrow{EF}$–$\overrightarrow{EG}$–$\overrightarrow{ED}$.

CONCLUSIONS	JUSTIFICATIONS
(1) Let the real numbers $a = m\angle ABC$ and $b = m\angle DEF$ be defined. Then $0 < a < b < 180$.	Given, Axiom A-1
(2) Set up a coordinate system for the rays from E on the D-side of line $\overleftrightarrow{EF}$, with 0 assigned to ray $\overrightarrow{EF}$.	Protractor Postulate
(3) If the coordinate of ray $\overrightarrow{ED}$ is x, then $b = m\angle DEF = \lvert 0 - x \rvert = x$. That is, b is the coordinate of ray $\overrightarrow{ED}$.	Protractor Postulate
(4) Let ray $\overrightarrow{EG}$ be the unique ray having coordinate a. Then $\overrightarrow{EF}$–$\overrightarrow{EG}$–$\overrightarrow{ED}$.	Protractor Postulate and the dual of Theorem 3, Section 2.4 $(0 < a < b)$
(5) $m\angle GEF = \lvert a - 0 \rvert = a = m\angle ABC$.	Protractor Postulate

We end this section by discussing the basic notion of perpendicularity of lines. Intuitively, if two lines ℓ and m are perpendicular, as shown in Figure 2.31, then those lines must form four right angles at their point of intersection. But what makes this true? It turns out that we will need an additional axiom about angle measure to justify this seemingly trivial idea.

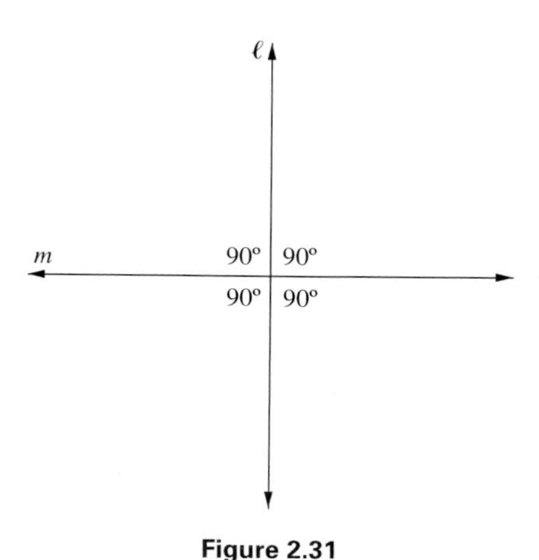

Figure 2.31

SUPPLEMENTARY, COMPLEMENTARY PAIRS

To get started, we know from Example 2, Section 2.4, that if A–B–C, then the union of the two rays $\overrightarrow{BA}$ and $\overrightarrow{BC}$ is a line, namely, line $\overleftrightarrow{AC}$. We call any two such rays **opposite** or **opposing** rays. A routine argument shows that if we are given a ray $\overrightarrow{BA}$, then its opposing ray $\overrightarrow{BC}$ exists and is unique, as indicated in Figure 2.32.

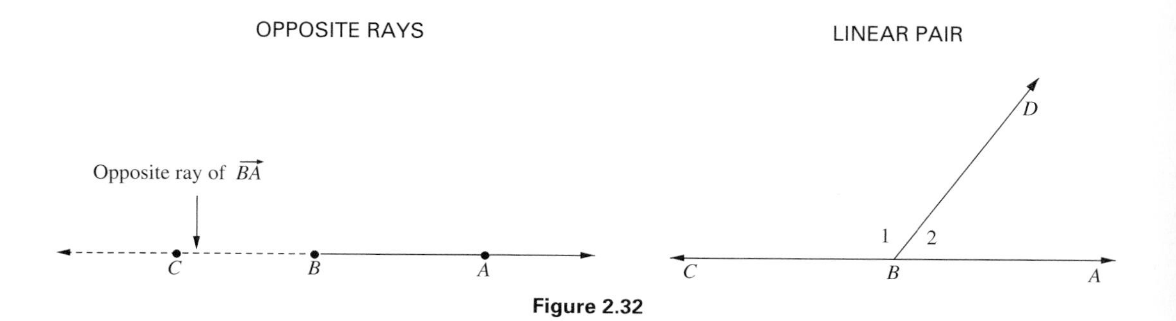

Figure 2.32

DEFINITION: Two angles are said to form a **linear pair** iff they have one side in common and the other two sides are opposite rays, as shown in Figure 2.32. We call any two angles whose angle measures sum to 180 a **supplementary pair**, or simply, **supplementary**, and two angles whose angle measures sum to 90, **complementary**.

It often happens in geometry that two angles are supplementary (or complementary) to a third angle. If for example, as illustrated in Figure 2.33, $\angle 1$ and $\angle 2$ are both supplementary to $\angle 3$, then by definition,

$$m\angle 1 + m\angle 3 = 180 \qquad \text{and} \qquad m\angle 2 + m\angle 3 = 180$$

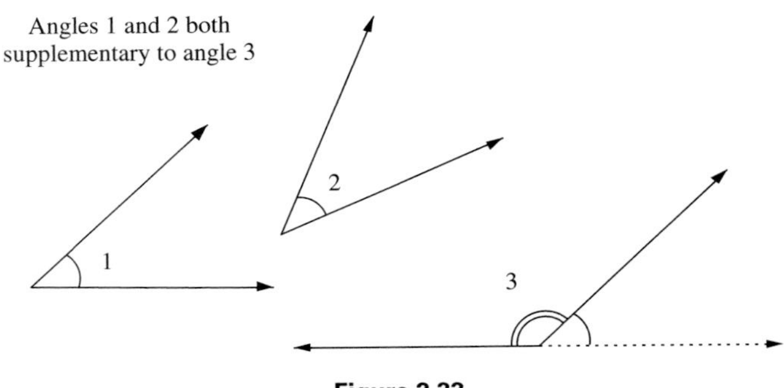

Figure 2.33

Subtraction produces the result $m\angle 1 = m\angle 2$. A similar argument can be given for two angles complementary to a third angle. Thus

> THEOREM 2: Two angles that are supplementary, or complementary, to the same angle have equal measures.

Now we are ready for the axiom mentioned above.

> AXIOM A-4: LINEAR PAIR AXIOM
> A linear pair of angles is a supplementary pair.

EXAMPLE 2 Certain rays on one side of line $\overleftrightarrow{BF}$ have their coordinates as indicated in Figure 2.34, and $m\angle GBF = 80$. Ray $\overrightarrow{BA}$ is opposite ray $\overrightarrow{BC}$, and $\overrightarrow{BG}$ is opposite $\overrightarrow{BE}$. Using the betweenness relations evident from the figure, find:

(a) $m\angle ABG$
(b) $m\angle DBG$
(c) The coordinate of ray $\overrightarrow{BE}$.

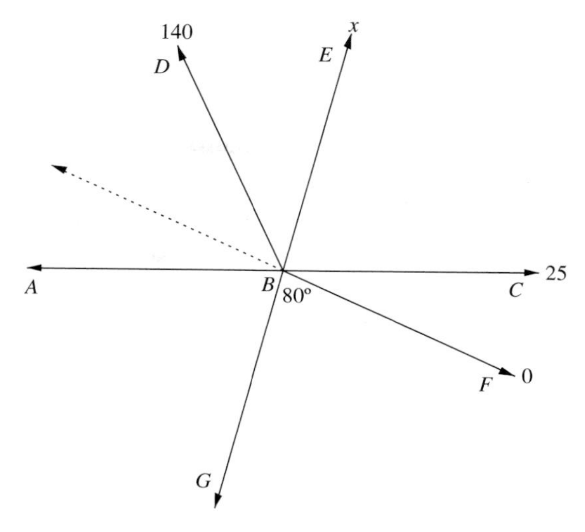

Figure 2.34

SOLUTION

(a) Because $m\angle CBF = 25$ by the Protractor Postulate, and $\overrightarrow{BF}$ lies between $\overrightarrow{BG}$ and $\overrightarrow{BC}$,

$$m\angle GBC = 80 + 25 = 105$$

Because $\angle ABG$ and $\angle GBC$ consist of a linear pair, then by the Linear Pair Axiom

$$m\angle ABG = 180 - 105 = 75$$

(b) Again, by the Linear Pair Axiom and Protractor Postulate,

$$m\angle ABD = 180 - m\angle DBC = 180 - (140 - 25) = 180 - 115 = 65$$

Hence, $m\angle DBG = 65 + 75$ (from **(a)**) $= 140$.

(c) The Linear Pair Axiom implies that $m\angle EBF = 180 - 80 = 100$. $\therefore x = 100$. ■

RIGHT ANGLES, PERPENDICULARITY

The formal definition for right angles and perpendicular lines is next.

DEFINITION: A **right angle** is any angle having measure 90. Two (distinct) lines ℓ and m are said to be **perpendicular**, and we write $\ell \perp m$, iff they contain the sides of a right angle. (For convenience, **segments** are **perpendicular** iff they lie, respectively, on perpendicular lines. Similar terminology applies to a segment and ray, two rays, and so on.)

While we are at it, let us define the further terms **acute angle**—any angle whose measure is less than 90, and **obtuse angle**—any angle whose measure is greater than 90. (In this book, an angle of measure 180 *does not exist.*)

If line ℓ is perpendicular to line m, Figure 2.35 shows the four ways in which ℓ and m can contain the sides of a right angle (in bold). Picking out any one of the cases (e.g., the second case illustrated in Figure 2.35) clearly leads to the following result via the Linear Pair Axiom: Since $\angle 1$ is supplementary to $\angle 2$ and $\angle 1$ is a right angle, $m\angle 1 + m\angle 2 = 180 = 90 + m\angle 2$, so it follows that $m\angle 2 = 90$ and $\angle 2$ is a right angle; because $\angle 3$ is supplementary to $\angle 2$ and $\angle 2$ is a right angle, $\angle 3$ is a right angle, and so on. We have justified what we wanted to. It will be stated as a **lemma** because it is a minor result used to prove other theorems—the mathematical meaning of the term. (Observe that we could not have accomplished this without the Linear Pair Axiom.)

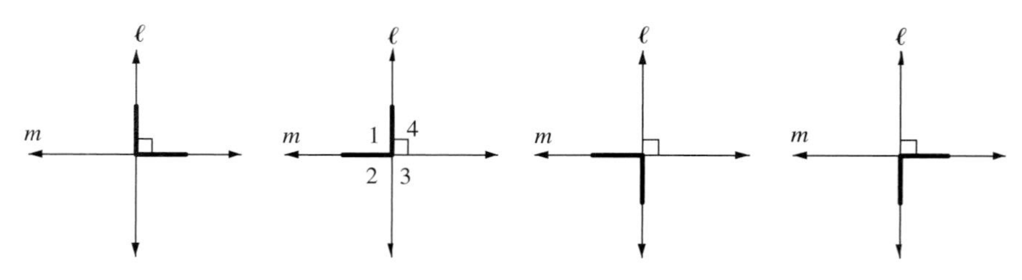

Figure 2.35

LEMMA: If two lines are perpendicular, they form four right angles at their point of intersection.

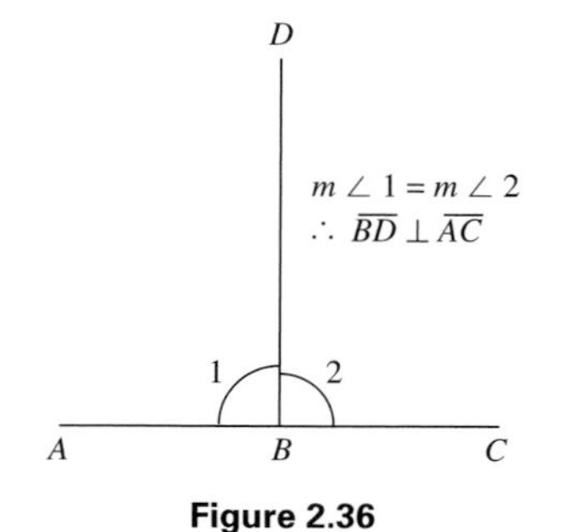

Figure 2.36

We now prove a theorem about perpendicular lines that is as useful as it is obvious, but this will save us going through the same steps repeatedly in later proofs. We define **adjacent angles** as being two distinct angles with a common side and having no interior points in common.

THEOREM 3: If line $\overleftrightarrow{BD}$ meets segment $\overline{AC}$ at an interior point B on that segment (Figure 2.36), then $\overleftrightarrow{BD} \perp \overline{AC}$ iff the adjacent angles at B have equal measures.

PROOF

(1) If $\overleftrightarrow{BD} \perp \overline{AC}$, then by the lemma above, $\angle ABD$ and $\angle DBC$ are right angles, hence have equal measures. (2) If $m\angle ABD = m\,\angle DBC$, then by the Linear Pair Axiom and a little algebra, it follows that $m\angle ABD = 90$ and, by definition, $\overleftrightarrow{BD} \perp \overline{AC}$.

The Protractor Postulate implies the following basic result; the uniqueness part will be left as Problem 19.

> THEOREM 4: Given a point A on line ℓ, there exists a unique line m perpendicular to ℓ at A.

VERTICAL ANGLES

A final useful result involves what are called **vertical angles**, or a **vertical pair** of angles, as shown in Figure 2.37. These are two angles having the sides of one opposite the sides of the other, using the previous definition of *opposite rays*.

VERTICAL PAIR

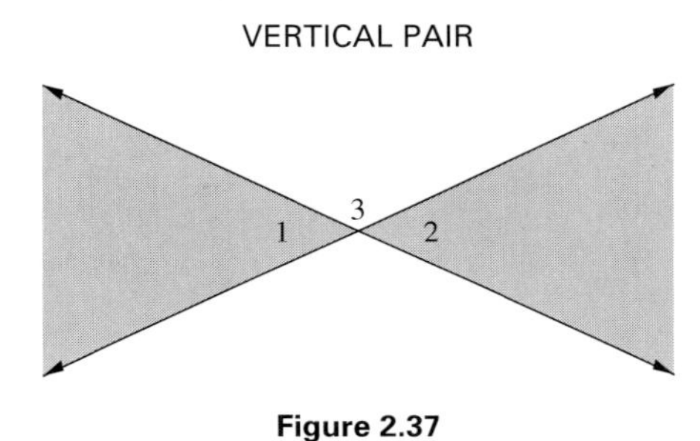

Figure 2.37

> THEOREM 5: VERTICAL PAIR THEOREM
> Vertical angles have equal measures.

The proof makes a nice problem for you (Problem 15). Just use the Linear Pair Axiom twice and Theorem 2 above.

PROBLEMS (2.5)

GROUP A

1. Coordinates of certain rays are shown in the figure (according to the Protractor Postulate). Find $m\angle JFH$.

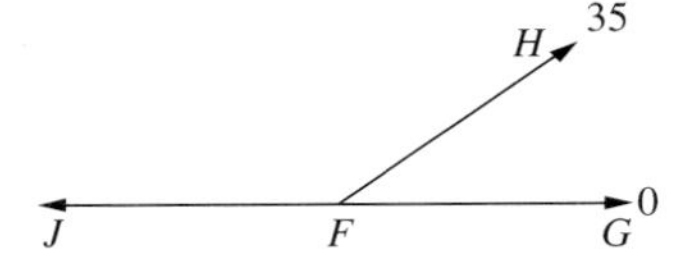

2. As in Problem 1, find

(a) $m\angle KPW$

(b) $m\angle KPT$

(c) $m\angle TPW$

(d) Why is $\overrightarrow{PT}$ the bisector of $\angle KPW$?

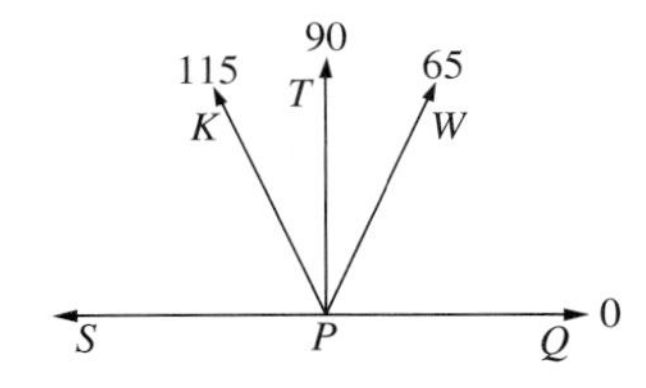

3. As in Problem 1, find

(a) $m\angle COA$

(b) $m\angle COB$

(c) $m\angle DOC$

(d) the coordinate of the bisector of $\angle DOB$,

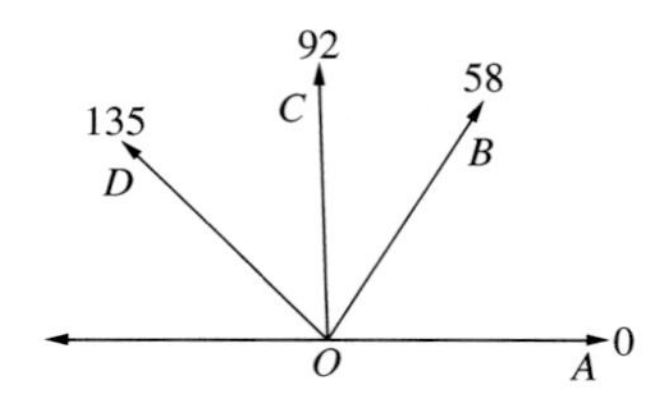

4. Coordinates for certain rays on the T-side of $\overrightarrow{RS}$ are as indicated in the figure. If angles $\angle YRS$ and $\angle MRV$ are vertical angles and $\overrightarrow{RM}$ is between $\overrightarrow{RT}$ and $\overrightarrow{RV}$, find $m\angle TRV$.

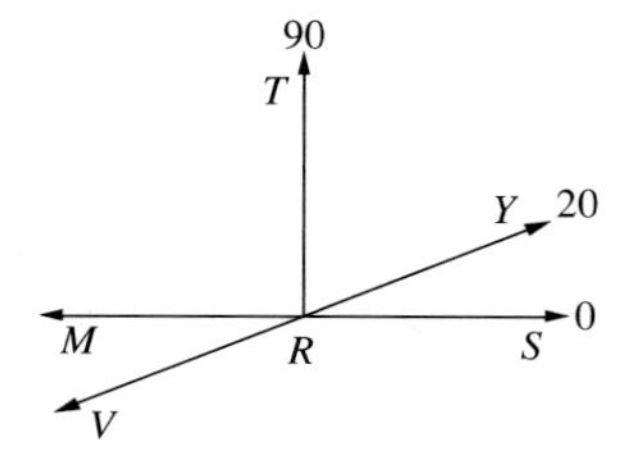

5. Are there any betweenness relations among the rays $\overrightarrow{MN}$, $\overrightarrow{MP}$, $\overrightarrow{MQ}$? Can you *prove* that $m\angle PMQ = 120$?

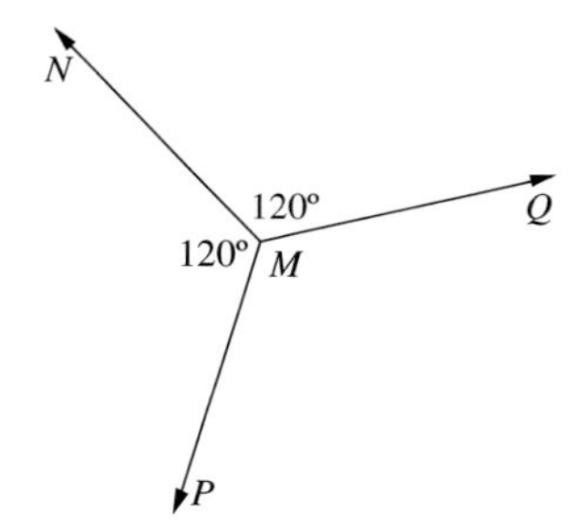

6. Rays on one side of line $\overleftrightarrow{BF}$ have their coordinates as indicated. Ray $\overrightarrow{BA}$ is opposite ray $\overrightarrow{BC}$, and $\overrightarrow{BG}$ is opposite $\overrightarrow{BE}$. Using the betweenness relations evident in the figure, find

(a) $m\angle ABG$

(b) $m\angle GBD$

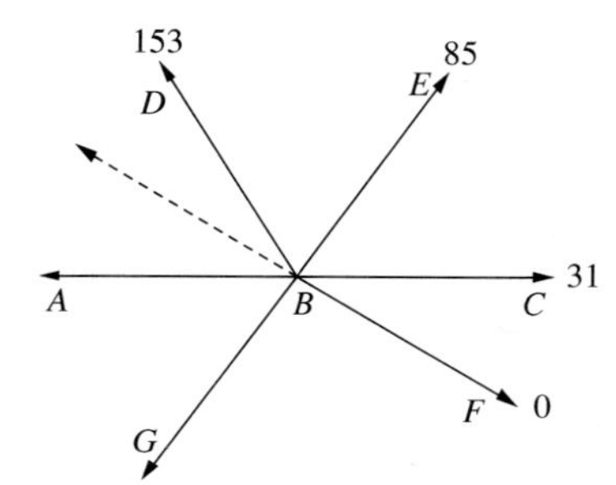

7. Ray $\overrightarrow{OC}$ is opposite $\overrightarrow{OA}$, and $\overrightarrow{OB}$ is opposite $\overrightarrow{OD}$. If $m\angle AOB = 53$ and $m\angle DOE = 1$, find $m\angle EOA$ if

(a) $\overrightarrow{OA}$–$\overrightarrow{OE}$–$\overrightarrow{OD}$

(b) $\overrightarrow{OA}$–$\overrightarrow{OD}$–$\overrightarrow{OE}$

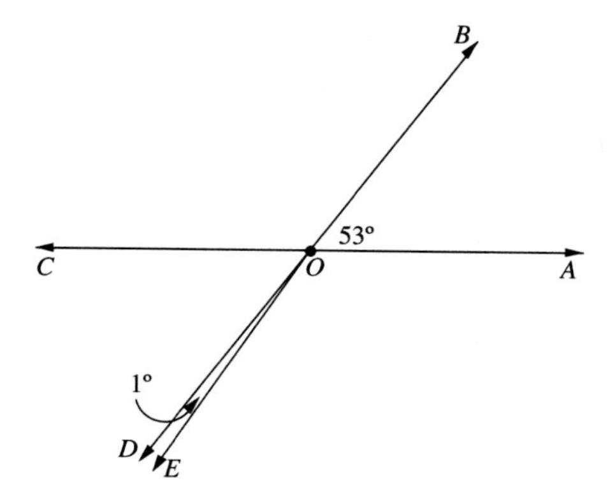

8. $\angle QMN$ and $\angle QMP$ are a linear pair of angles, with $m\angle QMN = 40$. Rays $\overrightarrow{MX}$ and $\overrightarrow{MY}$ bisect $\angle QMN$ and $\angle QMP$, respectively. Find

(a) $m\angle QMX$

(b) $m\angle QMY$

(c) Did you notice anything?

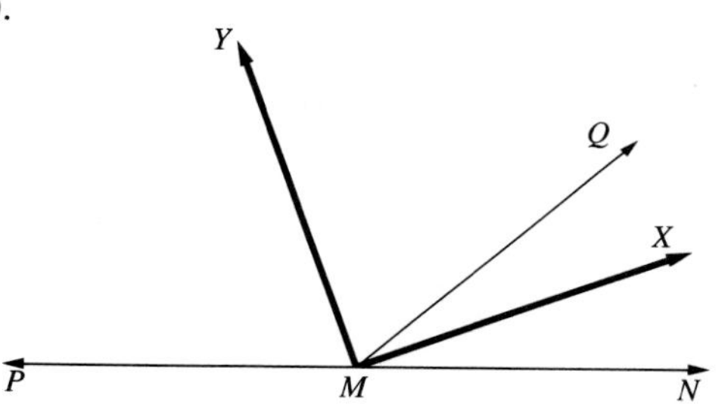

9. Using *Sketchpad*, follow these steps:

[1] Construct segment $\overline{AB}$ and locate a point C on it.

[2] Construct an arbitrary segment $\overline{CD}$

[3] Using Angle Bisector under CONSTRUCT, construct the bisectors of $\angle DCB$ and $\angle ACD$, and locate points E and F on them.

[4] Display $m\angle FCD$, $m\angle DCE$, and their sum on the screen.

Drag point D and observe the effect. Does this reveal what may be a theorem?

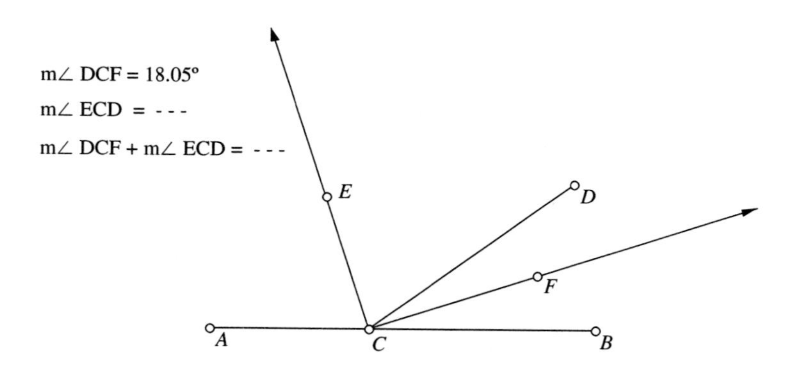

GROUP B

10. Point B lies on $\overline{CD}$, and you are given that $m\angle ABC = 40$ and $m\angle ABE = 160$. Using the figure, what values are possible for $m\angle DBE$? (You do not have to prove any betweenness relations.)

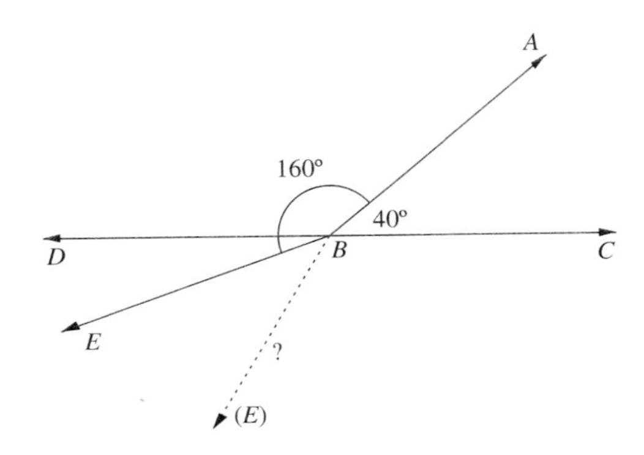

11. If in Problem 10 you are given the additional information that the opposite ray of $\overrightarrow{BE}$ bisects $\angle ABC$, what values are possible for $m\angle DBE$?

12. Prove the corollary to Theorem 1.

13. Prove that the bisectors of a linear pair of angles are perpendicular. Establish any betweenness relations used in the proof. (***Hint:*** Use the Protractor Postulate to help you establish these betweenness relations.)

14. Use the Protractor Postulate to prove that if $\overrightarrow{OA}$, $\overrightarrow{OB}$, $\overrightarrow{OC}$ are any three rays on one side of a line and having the same end point, then either $\overrightarrow{OA}$–$\overrightarrow{OB}$–$\overrightarrow{OC}$, $\overrightarrow{OA}$–$\overrightarrow{OC}$–$\overrightarrow{OB}$, or $\overrightarrow{OB}$–$\overrightarrow{OA}$–$\overrightarrow{OC}$.

15. Prove Theorem 5 using the Linear Pair Axiom.

16. Prove that if two angles have a side in common that passes through an interior point of the angle formed by the other two sides, the other two sides are perpendicular iff the given angles are complementary.

17. **Using *Sketchpad* to construct complementary angles** (see figure).

[1] Construct perfectly horizontal and perfectly vertical segments $\overline{AB}$ and $\overline{CA}$ and locate D any point inside $\angle BAC$. (If you have trouble, just rotate $\overline{AB}$ 90° about A.)

[2] Display angle measures $x = m\angle DAB$ and $y = m\angle CAD$ on the screen. (Does this sum equal 90°?)

[3] Construct any segment $\overline{EF}$ elsewhere on the screen.

[4] Construct an angle of measure x at E, as follows: (1) Double click on E making it a center for rotation. (2) Select x and choose Mark Angle Measurement under TRANSFORM. (3) Select segment $\overline{EF}$ and rotate x degrees in a counterclockwise direction. (Move point D and observe the effect on the segment you just constructed.)

[5] Construct an angle at F having measure $-y$, following the procedure in Step 4. (In order to display $-y$ on the screen for selection, choose Calculate under MEASURE and compute $(-1) \cdot y$; double-click on F to make it the center of rotation.)

[6] Select the point of intersection G of the two rotated images of segment $\overline{EF}$,

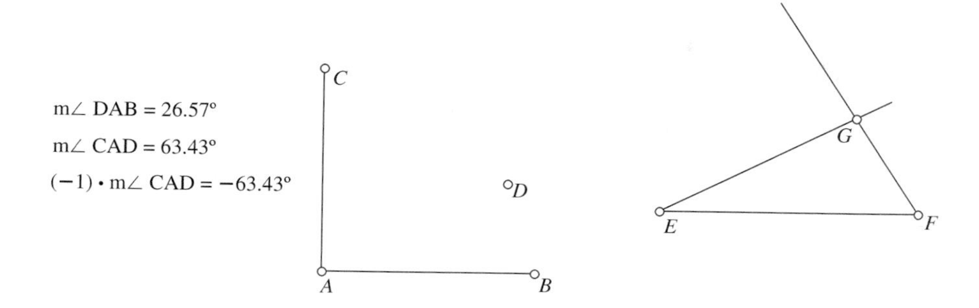

(a) Drag D and see what happens to point G.

(b) Trace point G (under DISPLAY) and again move point D. What curve seems to emerge? Does your knowledge of Euclidean geometry to this point enable you to explain this phenomenon?

GROUP C

18. Define angle trisectors and prove that any angle can be trisected in axiomatic geometry.

NOTE: A famous problem going back to ancient times is finding a compass, straight-edge construction for the trisectors of a given angle. It is known that no such construction exists in general; the proof of this fact involves abstract algebra. (See Corollary of Theorem 3, Section 4.4; also see Problem 11, Section 4.7.)

19. Prove that the perpendicular to a line at some point on that line is unique.

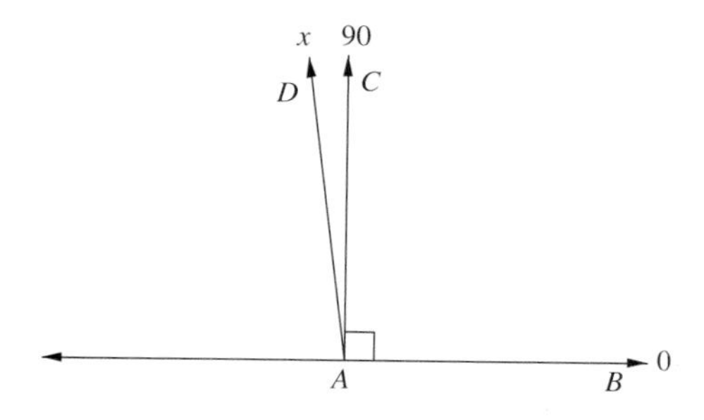

*2.6 Plane Separation, Interior of Angles, Crossbar Theorem

When you draw a line across a sheet of paper, you have effectively divided that sheet into two regions—the two *sides* of the line. This simple idea is very useful in geometry, but in order to develop a precise statement, it is helpful to analyze the experiment just described.

CONVEX SETS

Consider the points A, B, and C (Figure 2.38). It appears that when two points lie on the *same side* of line ℓ (as do A and B), then the segment joining them does not intersect ℓ and all the points of that segment lie entirely on that side. Hence, that side of ℓ is what we call a *convex set*. However, if two points lie on *opposite sides* of ℓ (as do A and C), then the segment joining them intersects ℓ at some interior point of the segment. This forms the basis for the next axiom for foundations, which establishes the basis for the idea of *half-planes* and the two *sides* of a line.

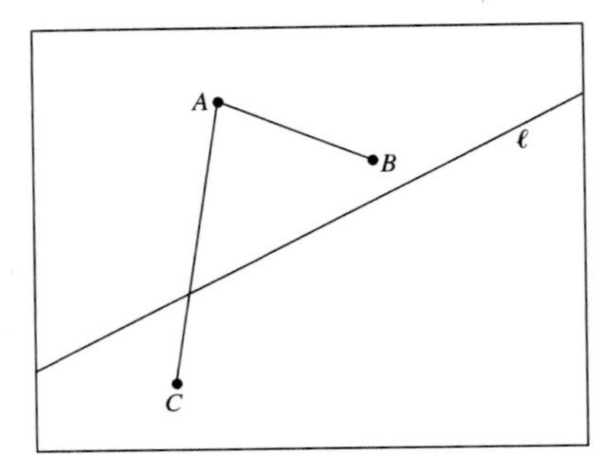

Figure 2.38

First, let us introduce formally the concept of convexity, which is used frequently in geometry.

> DEFINITION: A set K in $\mathbf{S}$ is called **convex** provided it has the property that for all points $A \in K$ and $B \in K$, the segment joining A and B lies in K ($\overline{AB} \subseteq K$). (See Figure 2.39 for illustration.)

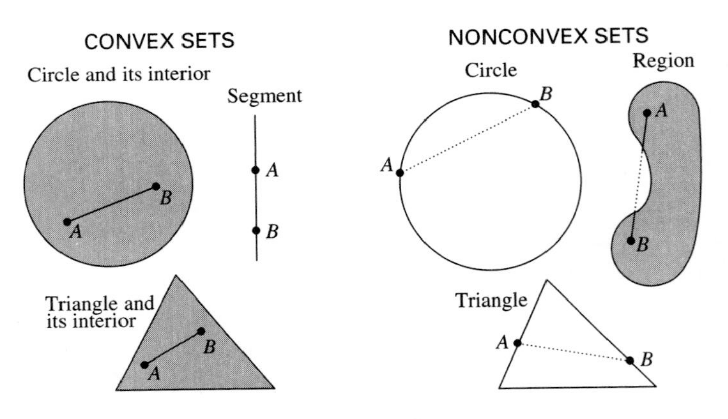

Figure 2.39

The simplest and most basic general result that can be obtained directly from the definition of convexity is the following:

If K_1 and K_2 are any two convex sets, $K_1 \cap K_2$ is also convex.

PROOF

If A and B are any two points of the set $K_1 \cap K_2$, then A and B both belong to K_1 and K_2, and by convexity of K_1 and K_2 individually, $\overline{AB} \subseteq K_1$ and $\overline{AB} \subseteq K_2$. Hence, $\overline{AB} \subseteq K_1 \cap K_2$.

PLANE SEPARATION POSTULATE

Now we are ready to introduce the axiom mentioned above.

AXIOM H-1: PLANE SEPARATION POSTULATE
Let ℓ be any line lying in any plane $\boldsymbol{P}$. The set of all points in $\boldsymbol{P}$ not on ℓ consists of the union of two subsets H_1 and H_2 of $\boldsymbol{P}$ such that
(1) H_1 and H_2 are convex sets.
(2) H_1 and H_2 have no points in common.
(3) If A lies in H_1 and B lies in H_2, the line ℓ intersects the segment $\overline{AB}$.

The two sets H_1 and H_2 are called the two **sides** of ℓ, or also, **half-planes** determined by ℓ. To identify which side of a line we are talking about, if A and B lie, respectively, in H_1 and H_2, we shall speak of the **A-side** of ℓ for H_1, and the **B-side** of ℓ for H_2. (See Figure 2.40.) A neat notational device (shown in Figure 2.40) is the following:

$H(A, \overleftrightarrow{CD}) = H_1$ (the A-side of $\overleftrightarrow{CD}$)
$H(B, \overleftrightarrow{CD}) = H_2$ (the B-side of $\overleftrightarrow{CD}$)

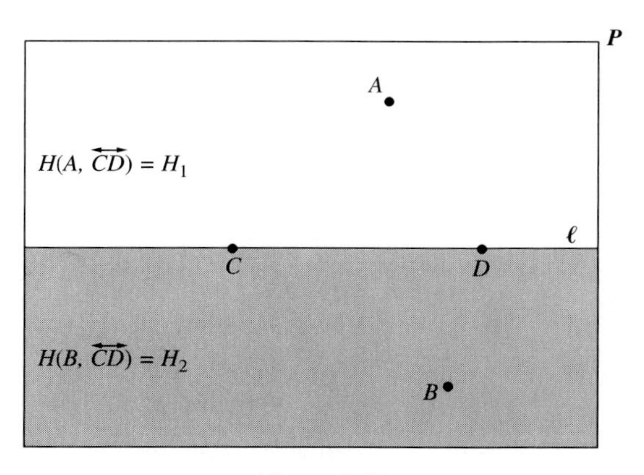

Figure 2.40

This notation will help us write down a kind of "formula" for the interior of an angle a little later. The first result we prove about half-planes is an aid to proving other results, again, a *lemma*.

PROPERTIES OF HALF-PLANES

LEMMA: If A–B–C holds and a line ℓ passes through point B but not point A (Figure 2.41), then A and C lie on opposite sides of line ℓ (that is, one point in H_1 and the other in H_2).

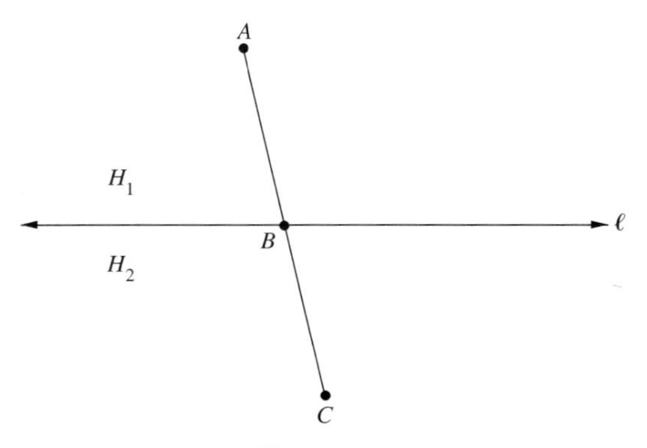

Figure 2.41

PROOF

Suppose point A lies in the half-plane H_1, as in Figure 2.41; we must show that C lies in H_2. But if C *does not* lie in H_2, it would lie either on line ℓ (impossible because $A \notin \ell$) or in H_1. Thus, A and C would then both lie in H_1. But by convexity of H_1,

$$\overline{AC} \subseteq H_1$$

and, with $B \in \overline{AC}$, then B would lie in H_1. We have reached a contradiction since, by hypothesis, B lies on line ℓ and no points of ℓ can lie in either H_1 or H_2. $\therefore C \in H_2$.

Another result, that looks obvious from a figure but requires proof, is the following.

THEOREM 1: If point A lies on line ℓ and point B lies in one of the half-planes determined by ℓ then, except for A, the entire segment $\overline{AB}$ or ray $\overrightarrow{AB}$ lies in that half-plane.

This theorem can be proven much as the lemma was, and will be left as Problem 15.

COROLLARY: Let B and F lie on opposite sides of a line ℓ and let A and G be any two distinct points on ℓ. Then segment $\overline{GB}$ and ray $\overrightarrow{AF}$ have no points in common (Figure 2.42).

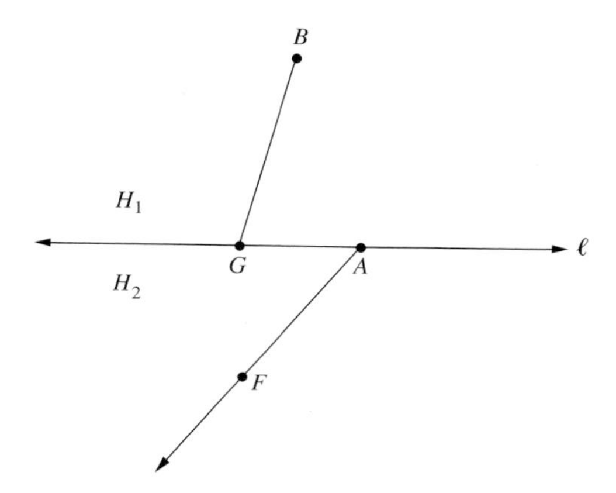

Figure 2.42

PROOF

By Theorem 1, all the points except G of $\overline{GB}$ lie in H_1, and all except A of $\overrightarrow{AF}$ lie in H_2. Because H_1 and H_2 are disjoint sets (having no points in common) and because $A \neq G$, $\overline{GB}$ is disjoint from $\overrightarrow{AF}$, as we were to prove.

In the early development of the foundations of geometry, the following proposition was taken as an axiom and the Plane Separation Postulate proved as a theorem. It is a little easier and technically less tedious to do it the other way around, which is the approach we have taken here. Although it is a theorem, we will continue to refer to it as a postulate for historic reasons. This relationship is named after Moritz Pasch (1843-1930) who was the first to recognize its significance to geometry, and the first also to treat Euclid's unstated assumptions about betweenness explicitly.

THE POSTULATE OF PASCH

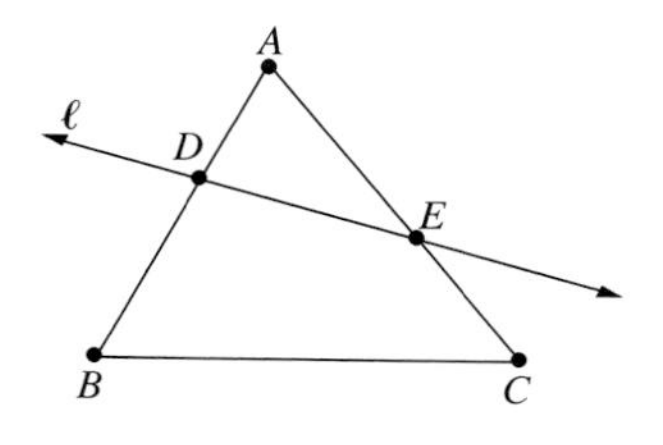

Figure 2.43

THEOREM 2: POSTULATE OF PASCH
Suppose A, B, and C are any three distinct noncollinear points in a plane, and ℓ is any line in that plane that passes through an interior point D of one of the sides, $\overline{AB}$, of the triangle determined by A, B, and C. Then line ℓ meets either $\overline{AC}$ at some interior point E, or $\overline{BC}$ at some interior point F, the cases being mutually exclusive (Figure 2.43). (Proof is left as Problem 12.)

INTERIOR OF AN ANGLE

We all know where the "interior" of an angle is, at least by picture; even a child could identify it. However, it is challenging to describe this set of points geometrically. (One such description appeared in the previous section.) It is shown in Figure 2.44 how half-planes can be used to determine the desired interior of $\angle ABC$. Informally, the interior of $\angle ABC$. consists of all points that lie strictly "between" its sides, rays $\overrightarrow{BA}$ and $\overrightarrow{BC}$. But formally, we take as our definition the following.

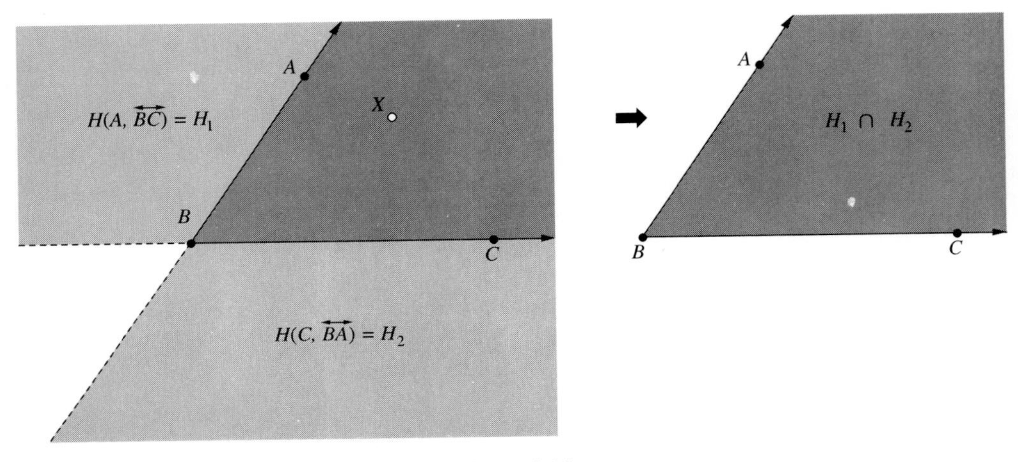

Figure 2.44

DEFINITION: The **interior** of $\angle ABC$ is the set of all points X that simultaneously lie on the A-side of $\overleftrightarrow{BC}$ and on the C-side of $\overleftrightarrow{BA}$

A "formula" for the interior of an angle is the following, simply the result of using the correct symbols for the previous definition and using notation introduced earlier.

$$\text{Interior } \angle ABC = H(A, \overleftrightarrow{BC}) \cap H(C, \overleftrightarrow{BA})$$

Because half-planes are convex sets, the very definition of the interior of an angle—as the intersection of two half-planes—makes it convex. This simple fact is not altogether useless; we often use it without realizing it. The next theorem is a more specialized result than this, but is the basic embodiment of the convexity property.

THEOREM 3: If A and C lie on the sides of $\angle B$, then, except for end points, segment $\overline{AC}$ is a subset of the interior of $\angle B$. If $D \in$ Interior $\angle B$, then, except for B, ray $\overrightarrow{BD} \subseteq$ Interior $\angle B$. (See Figure 2.45.)

PROOF

(for ray $\overrightarrow{BD}$ only): Let $X \in \overrightarrow{BD}, X \neq B$, as in Figure 2.45. Then by Theorem 1, $X \in H(A, \overleftrightarrow{BC})$ (as D lies in that half-plane). Similarly, $X \in H(C, \overleftrightarrow{BA})$. Therefore, $X \in H(A, \overleftrightarrow{BC}) \cap H(C, \overleftrightarrow{BA}) =$ Interior $\angle ABC$. Then, except for point B, ray $\overrightarrow{BD}$ is a subset of Interior $\angle B$.

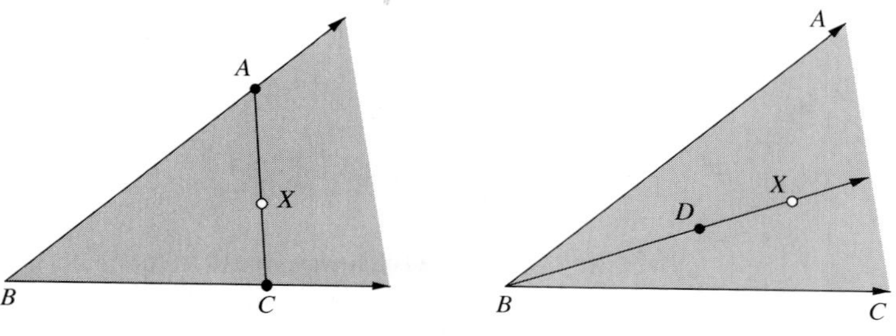

Figure 2.45

CROSSBAR THEOREM

One of the deeper results of axiomatic geometry is the so-called **Crossbar Principle** (Figure 2.46): If a segment reaches from one side of an angle to the other, then any ray lying between the sides of the angle (like the angle bisector) must intersect that segment. Let us see if we can write a foundations proof for it. The above corollary is the key, as well as the Postulate of Pasch.

THEOREM 4: CROSSBAR THEOREM
If D lies in the interior of $\angle BAC$, then ray $\overrightarrow{AD}$ meets segment $\overline{BC}$ at some interior point E. (See Figure 2.46.)

CROSSBAR THEOREM

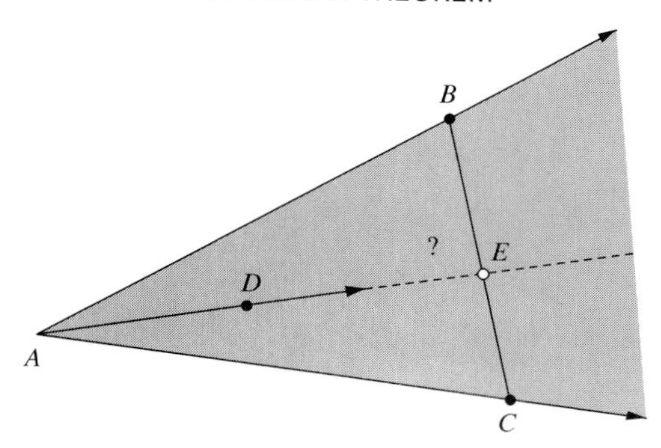

Figure 2.46

If we are to use the Postulate of Pasch, we must arrange things so that line $\overleftrightarrow{AD}$ passes through an interior point (probably point A) of the side of some triangle. Thus, we must add some points to the diagram in Figure 2.46. Suppose we locate, by the

Ruler Postulate, points F and G such that D–A–F and C–A–G; then draw segment $\overline{BG}$ and ray $\overrightarrow{AF}$, as shown in Figure 2.47. Now it looks promising: Observing the three points B, C, and G (triangle BCG), line $\overleftrightarrow{AD}$, which is the set $\overrightarrow{AD} \cup \overrightarrow{AF}$, must meet either segment $\overline{BC}$ or $\overline{BG}$, as that line does not pass through B, C, or G (because D is an

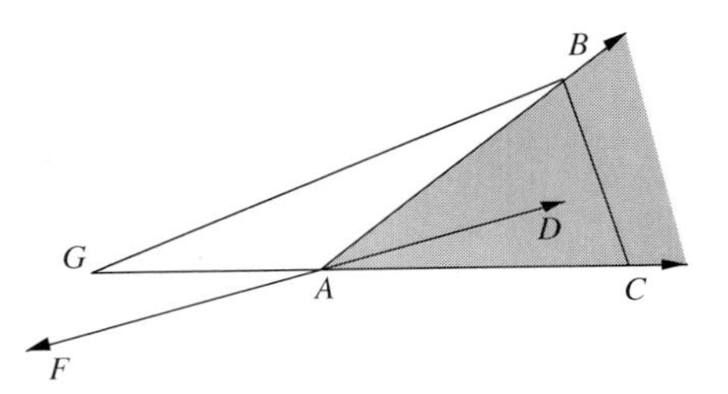

Figure 2.47

interior point of $\angle BAC$). We are close, but now it is necessary to carefully examine the figure from several points of view. To make it easier for you, we have included special diagrams of the cases we need (Figure 2.48). Is this a *proof by picture*? It is important to note here that in each diagram we are applying the result of the corollary of Theorem 1 and *not* the Postulate of Pasch.

PROOF OF THE CROSSBAR THEOREM

By the Postulate of Pasch applied to the three points B, C, and G, line $\overrightarrow{AD} \cup \overrightarrow{AF}$ either meets $\overline{BG}$ at some interior point J or $\overline{BC}$ at some interior point E. But because of Views 1 and 2 in Figure 2.48, the corollary of Theorem 1 rules out $\overrightarrow{AD} \cup \overrightarrow{AF}$ meeting $\overline{BG}$ at J. Hence $\overrightarrow{AD} \cup \overrightarrow{AF}$ meets $\overline{BC}$ at an interior point E, and E must either lie on $\overrightarrow{AD}$ or $\overrightarrow{AF}$. Because of View 3 in Figure 2.48 and the corollary, E cannot lie on ray $\overrightarrow{AF}$. Hence $\overrightarrow{AD}$ meets segment $\overline{BC}$ at an interior point E, as desired.

PROOF OF CROSSBAR THEOREM

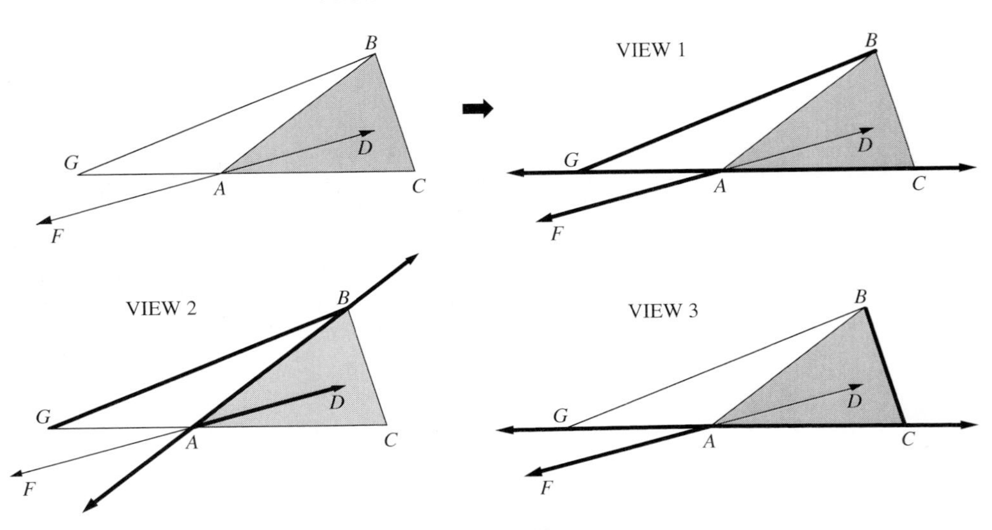

Figure 2.48

PROBLEMS (2.6)

GROUP A

1. Use the notation for half-planes introduced in this section to name the shaded regions in the figure:

(a) Region 1

(b) Region 2

(c) Region 3

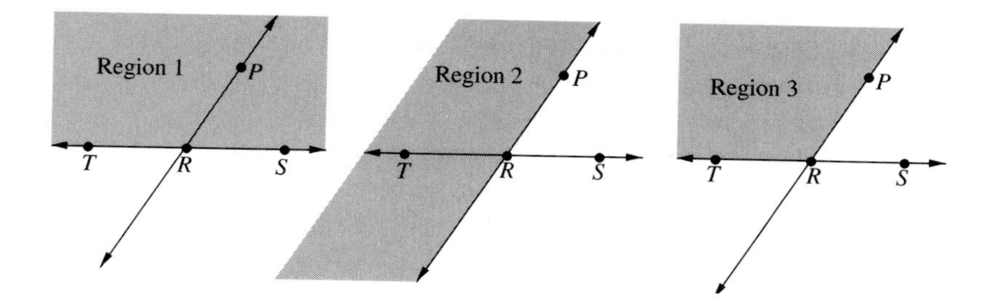

2. Use the notation for half-planes introduced in this section to name the shaded regions in the figure:

(a) Region 1

(b) Region 2

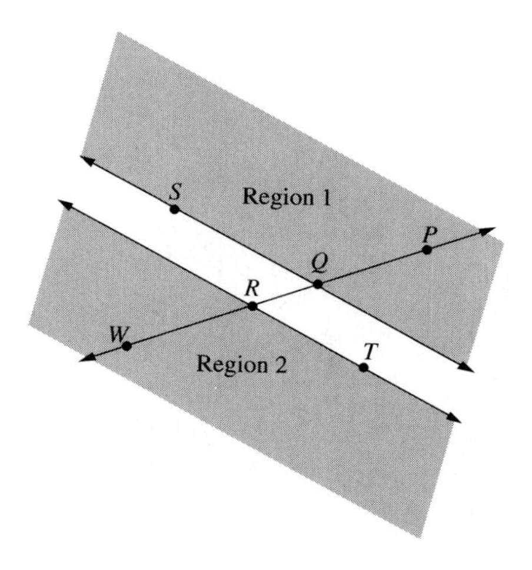

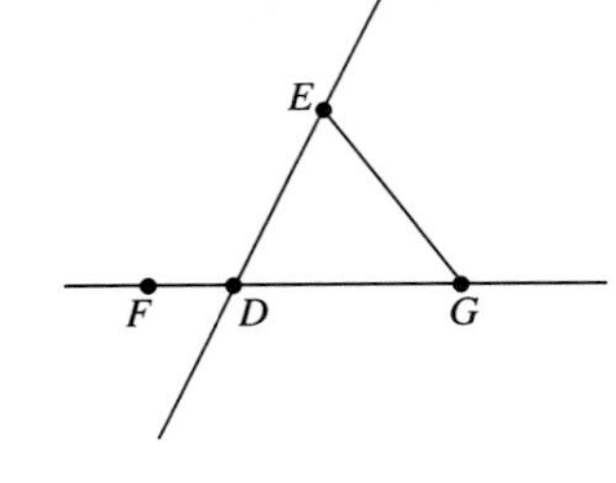

3. Name the half-plane that
   (a) contains the open segment $\overline{DE}$ but does not contain points G or F.
   (b) contains the open segment $\overline{DF}$ and opposite ray of $\overrightarrow{EG}$, but not D.
4. If $D \in H(A, \overleftrightarrow{BC})$, why is $H(D, \overleftrightarrow{BC}) = H(A, \overleftrightarrow{BC})$?
5. The rays in both half-planes H_1 and H_2 have been assigned coordinates according to the Protractor Postulate. Rays $\overrightarrow{AB}$ and $\overrightarrow{AC}$ have coordinates 70 and 110 in H_1 and ray $\overrightarrow{AD}$ has coordinate 90 in H_2 (as shown). Using betweenness relations evident from the figure, determine the following values:
   (a) $m\angle BAC$
   (b) $m\angle BAD$
   (c) $m\angle CAD$
   (d) What betweenness relations among the rays $\overrightarrow{AB}$, $\overrightarrow{AC}$, and $\overrightarrow{AD}$ hold, if any?

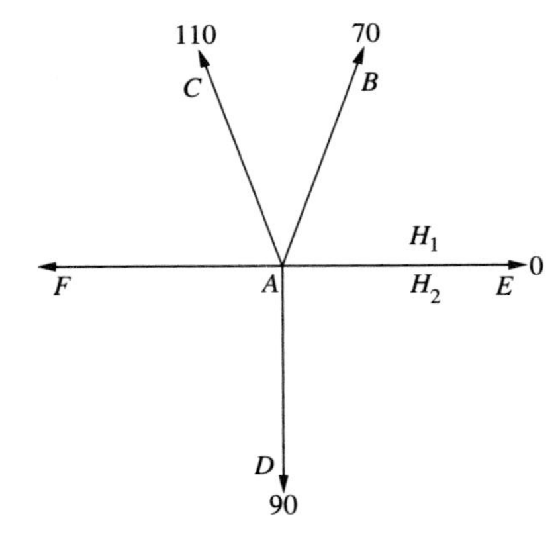

6. Rays $\overrightarrow{AE}$ and $\overrightarrow{AF}$ have coordinates 31 and 150, as indicated, relative to coordinates in H_1 and H_2, respectively, with N–A–M.
   (a) Find $m\angle EAF$.
   (b) Which of the rays $\overrightarrow{AM}$ or $\overrightarrow{AN}$ lies between $\overrightarrow{AE}$ and $\overrightarrow{AF}$? (No proofs necessary.)

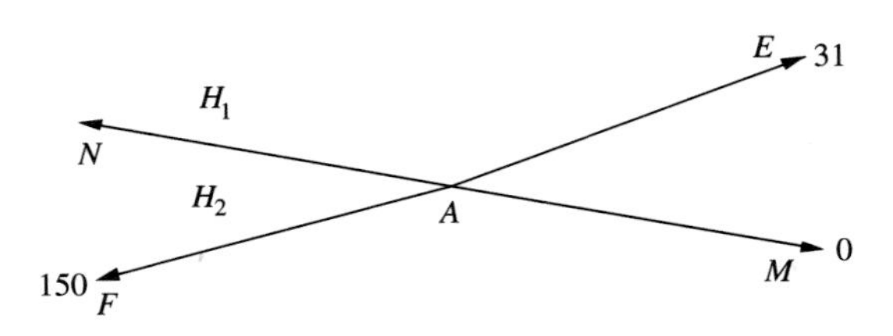

7. What sorts of figures are possible from the intersection of three half-planes? Four? Five?

GROUP B

8. Prove that if $B \notin \overleftrightarrow{AD}$ and $A–B–C$, then $C \in H(B, \overleftrightarrow{AD})$. (For an indirect proof, assume that C lies on the opposite side of $\overleftrightarrow{AD}$ as B; what must be true then?)

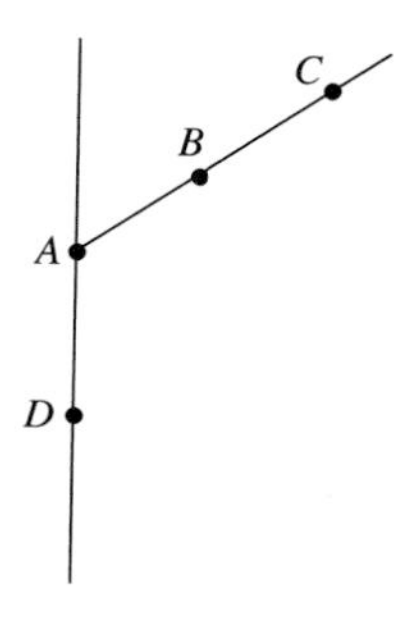

9. Using a few examples of your own and various values for x and y, verify (you need not prove) the following formula for $m\angle XAY$ when $\overrightarrow{AX}$ and $\overrightarrow{AY}$ lie on opposite sides of line ℓ, and x and y are the coordinates of $\overrightarrow{AX}$ and $\overrightarrow{AY}$, respectively:

$$m\angle XAY = x + y, \qquad \text{if } x + y < 180$$
$$= 360 - x - y, \qquad \text{if } x + y > 180$$

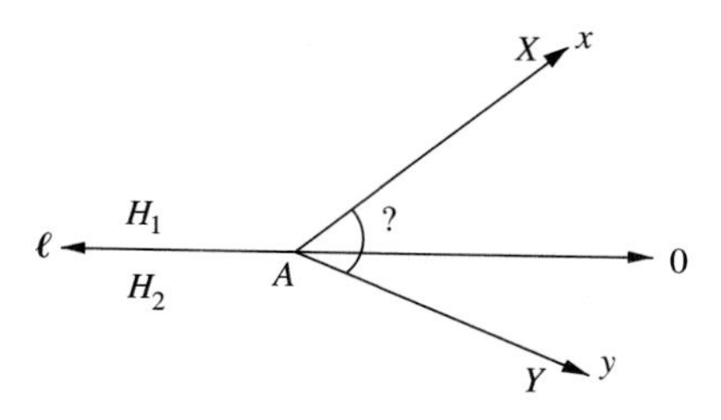

10. Prove or disprove: If two half-planes meet, their intersection is the interior of some angle.

11. Prove that segments and rays are convex sets, but an angle is not.

12. Prove the Postulate of Pasch as a theorem. (***Hint:*** Consider the two half-planes determined by line ℓ; what must be true of points A, B, and C, since they do not lie on ℓ?)

GROUP C

13. Suppose that A, B, and C are distinct, noncollinear points and that A–C–D and B–E–C. Prove there exists a unique point F such that D–E–F and A–F–B.

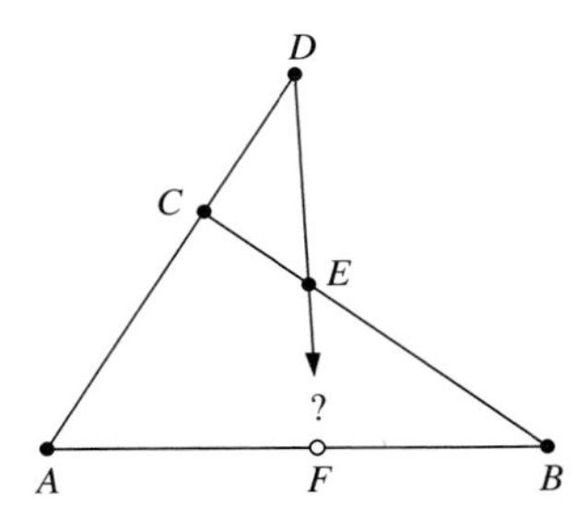

14. If $\overrightarrow{BA}$–$\overrightarrow{BD}$–$\overrightarrow{BC}$, then must ray $\overrightarrow{BD}$ pass through an interior point of $\angle ABC$? The affirmative answer to this is the essence of the converse statement of Axiom A-2: If $m\angle ABD + m\angle DBC = m\angle ABC$, then ray $\overrightarrow{BD}$ passes through an interior point of $\angle ABC$. Fill in the details of the following indicated steps in its proof: Locate points K, L, and M on rays $\overrightarrow{BA}$, $\overrightarrow{BD}$, $\overrightarrow{BC}$, respectively. *Case 1:* When K, L, and M are collinear and lie on line ℓ (not passing through B), then either K–L–M, K–M–L, or L–K–M. Use Axiom A-2, first part, to gain a contradiction if either K–M–L or L–K–M. *Case 2:* When K, L, and M are noncollinear, the cases are (a) B lies in the interior of all three angles of the triangle KLM (show you get a contradiction here by using the Linear Pair Axiom and by extending ray $\overrightarrow{BD}$ backward to form ray $\overrightarrow{BE}$, intersecting $\overline{KM}$ by the Crossbar Theorem), or (b) B lies on the opposite side of one of the lines $\overleftrightarrow{KL}$, $\overleftrightarrow{LM}$, $\overleftrightarrow{MK}$ as the respective points M, K, or L. Show this case reverts back to *Case 1,* already proven.

15. Prove Theorem 1. [Outline of proof (fill in all details): If C is any point of the segment $\overline{AB}$, then A–C–B. If $C \in \ell$ or H_2 then $B \in \ell$ $(\rightarrow\leftarrow)$ or B–A–C $(\rightarrow\leftarrow)$. (You need to *prove* that B–A–C holds in that case.) $\therefore C \in H_1$ and segment $\overline{AB} \subseteq H_1$, except for A. Similarly, if $D \in$ ray $\overrightarrow{AB}$, and A–B–D, then it follows that $D \in H_1$.]

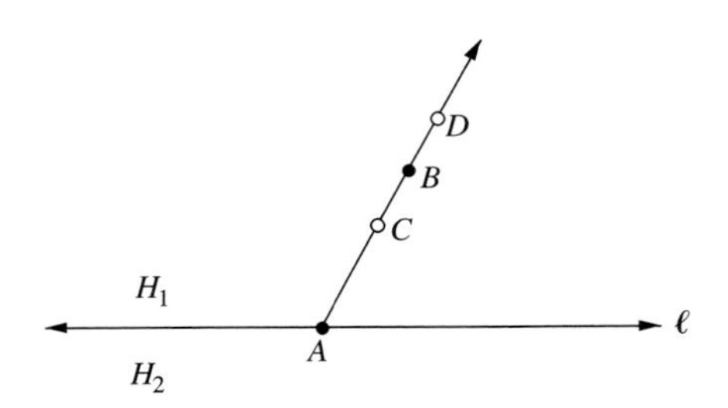

16. Prove the following form of the Angle Construction Theorem: For any two angles $\angle ABC$ and $\angle DEF$, there is a unique ray $\overrightarrow{EG}$ from E on the D-side of line $\overleftrightarrow{EF}$ such that $m\angle ABC = m\angle GEF$.

17. It is given that R–M–O, $m\angle PMO = 78$, $m\angle OMQ = 101$, and that rays $\overrightarrow{MP}$ and $\overrightarrow{MQ}$ lie on opposite sides of line $\overleftrightarrow{RO}$.

(a) Find the measure of $\angle PMQ$, establishing all the necessary betweenness relations without using the figure. (You might find the result of Problem 16 useful.)

(b) Find $m\angle PMQ$ if $m\angle PMO = 80$, and everything else is the same. Again, establish all betweenness relations used.

(c) If $m\angle PMO = 79$ and everything else is the same, prove that the pair of angles $\angle PMO$ and $\angle OMQ$ are linear.

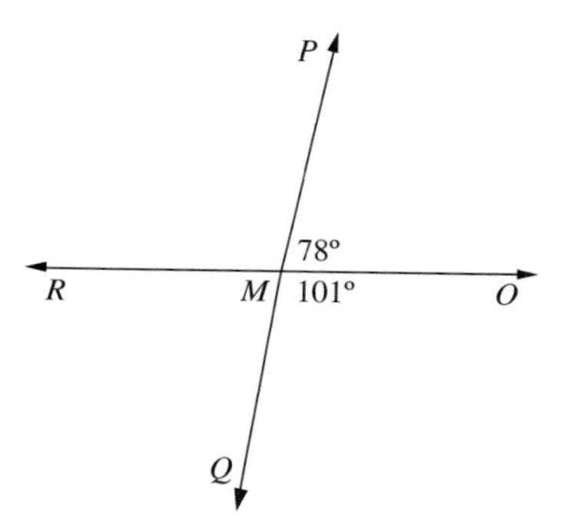

18. Prove that every half-plane is a nonempty set.

19. Two convex sets K_1 and K_2 make up the entire plane $\boldsymbol{P}$ (that is, $K_1 \cup K_2 = \boldsymbol{P}$). What must the two convex sets look like?

Chapter 2 Summary

This chapter began with a discussion of axiomatic mathematics, logic and proof, and how models play a role in determining whether a set of axioms is consistent and independent. This paved the way for beginning our study of axiomatic geometry, with *point*, *line*, and *plane* as undefined terms.

The axioms in the first group, I-1 through I-5, describe the interaction of points, lines, and planes in three-dimensional space, called *incidence* properties. There are no surprises here—points, lines, and planes behave just like we imagine them: A line will either "pierce" a plane at precisely one point, or be wholly contained by it (unless the line and plane are parallel), and given a line and point in space such that the point does not lie on the line, there is precisely one plane that contains both the line and the point. Finally, two nonparallel planes will always intersect in a straight line.

The next group of axioms, D-1 through D-3, establishes the metric concept, or distance between points. From this, betweenness ideas were explored, and *segments*, *rays*, and *angles* were defined. Axiom D-4, the Ruler Postulate, provides us with the basis for thinking of lines as a *continuum* of points, as we normally do; there is a system of coordinates for the points on each line such that if the coordinate of point A is a and that of B is b, then

$$AB = |a - b|$$

Angle measure emerges from the basic existence axiom, Axiom A-1; the Angle Addition Postulate, Axiom A-2, provides the connection between the interior of an angle and betweenness of rays, which would not otherwise exist. The Protractor Postulate, Axiom A-3, establishes for all the rays having a common end point (or origin) and lying on one side of a line, a system of coordinates between 0 and 180 such that if ray $\overrightarrow{OA}$ has coordinate a and ray $\overrightarrow{OB}$ has coordinate b, then

$$m\angle AOB = |a - b|$$

The final axiom on angle measure is the Linear Pair Axiom, Axiom A-4, which enables us to deal with perpendicular lines, right angles, and vertical angles in the customary way.

The Plane Separation Postulate, Axiom H-1, makes explicit the assumptions usually made regarding the behavior of the two regions into which a line divides a plane. Everything we take for granted about such concepts as the "side of a line" is given a formal basis here in terms of *half-planes*. The concept of half-planes makes it possible to define, and deal with, the interior of an angle without using pictures or diagrams—thereby adhering to one of the maxims of axiomatic geometry. The Crossbar Theorem was proven from basic properties of half-planes and the Postulate of Pasch. A completely rigorous proof could have been written that does not depend on diagrams in any way, but those we used did help to organize the many logical possibilities that occur in this theorem.

This completes the presentation and development of 14 of the 16 axioms needed for the development of Euclidean geometry. The SAS axiom concerning congruent triangles and the Parallel Postulate are the only ones left.

Testing Your Knowledge

SPECIAL NOTE FOR PRACTICE TESTS: In the practice tests at the end of each chapter we will assume that you have access to a list of the definitions, axioms, and theorems/corollaries as found in Appendix F. Otherwise, these tests are intended as "closed-book" tests. A few other "ground rules" should be mentioned. In any given chapter, the results proven in later chapters are forbidden in solutions or proofs for the current chapter. We remind you also that ordinary knowledge of coordinate geometry, trigonometry, calculus, or any other source of information can be used only for constructing or dealing with *models* or *examples*, and cannot be used for proofs in axiomatic geometry.

1. Which of the axioms on incidence would prevent the kind of behavior shown here for lines?

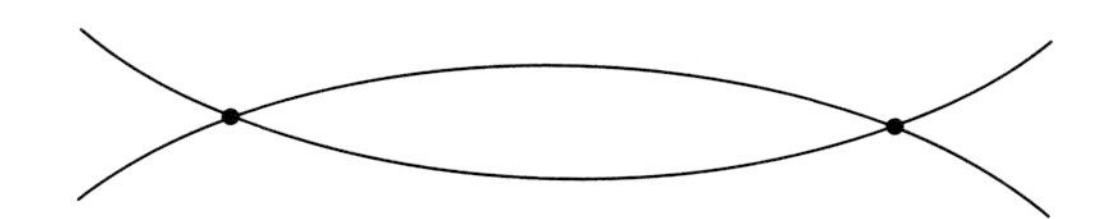

2. Construct a proof strictly from the axioms on incidence that if a line intersects a plane and does not lie in that plane, then the line and the plane have exactly one point in common.

3. Given that K, P, Q, and T are points on a line such that $P\text{–}K\text{–}Q$ and $P\text{–}K\text{–}T$, Q is the midpoint of $\overline{KT}$, $PK = 9$, and $PT = 15$, find QT.

4. The coordinates of A, B, and C on line ℓ are, respectively, 3, 5, and 8. If $A\text{–}B\text{–}C\text{–}D$ and $BD = 6$, **(a)** find the coordinate of D, and **(b)** show that C is the midpoint of $\overline{BD}$.

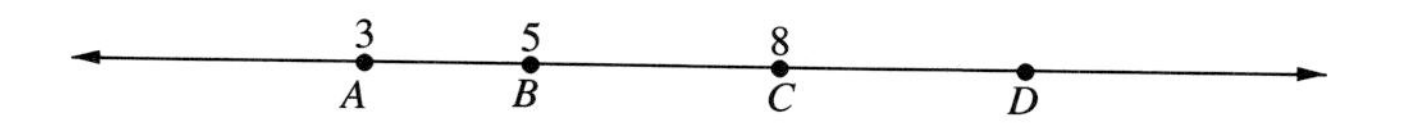

5. Consider the following relationships among three angles: $\angle 1$ is supplementary to $\angle 2$ and complementary to $\angle 3$, and $\angle 2$ and $\angle 3$ are a linear pair. Find $m\angle 1$.

6. Under the Protractor Postulate, the coordinates of rays $\overrightarrow{CM}$ and $\overrightarrow{CN}$, are, respectively, 48 and 115 with respect to some half-plane containing $\overrightarrow{CM}$ and $\overrightarrow{CN}$. What must the coordinate of the bisector of $\angle MCN$ be?

7. Suppose that points R, S, and T are collinear, and that $RS = 3$, $ST = 4$, $RT = 7$. Furthermore, $\overrightarrow{TP} \perp \overrightarrow{PS}$ and $m\angle TPR = 135$. Find $m\angle SPR$.

8. Among the rays on one side of line ℓ, rays $\overrightarrow{AG}$ and $\overrightarrow{AH}$ have coordinates 45 and 120. If $\overrightarrow{AF}$ and $\overrightarrow{AG}$ are opposite rays, find $m\angle FAH$.

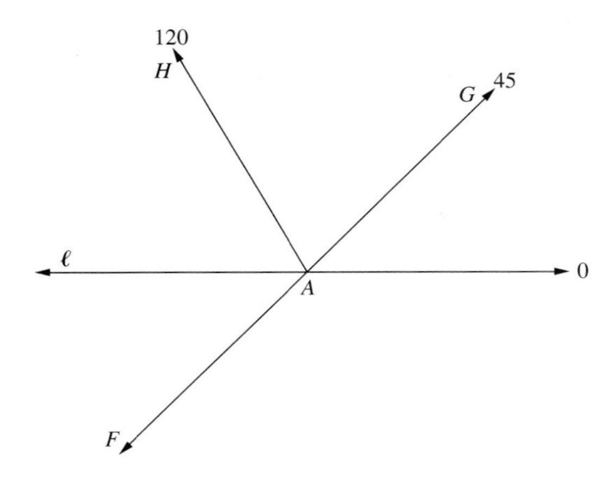

9. Point T is an interior point of acute $\angle EFG$, point R lies on ray $\overrightarrow{FG}$ such that $\overline{TR} \perp \overrightarrow{FG}$, and T–R–S holds. Point S will then lie in half-planes that are determined by lines $\overleftrightarrow{EF}$ and $\overleftrightarrow{FG}$. Draw a diagram of this situation, and name the two half-planes.

10. Using the Crossbar Theorem and definition of angle bisector, write a proof that the bisectors of any two angles of a triangle intersect at a point that is an interior point of both angles.

CHAPTER THREE

Foundations of Geometry 2

TRIANGLES, QUADRILATERALS, CIRCLES

OVERVIEW

In his organization of geometry, Euclid developed the basic properties of triangles before developing the basic properties of quadrilaterals and rectangles. Thus, the postulate of parallels and its consequences (such as *the angle sum of a triangle equals* 180) was postponed until late in the development. We consider here that same body of material, which provides a foundation not only for Euclidean geometry but for non-Euclidean geometry as well. This development is known as **absolute geometry,**[1] and it consists of that part of Euclidean geometry that includes all the axioms we have introduced in the previous chapter and the one we shall introduce in this chapter, but excludes all references to parallel lines, or results therefrom.

It is impossible to understand the basis for non-Euclidean geometry and its historical impact unless we follow Euclid's method of development. It is surprising what can be derived *without* parallelism, and it is most interesting what happens when we finally do assume the Parallel Postulate (which we do in Chapter 4). As you learn this material and write some proofs of your own, we remind you once again that you cannot use anything you might be familiar with from your high school

[1]This term was used by J. Bolyai to describe a general development of three-dimensional space, or what *any basic geometric study must include,* without a commitment regarding parallelism. Any logical consequence thereof would be an *absolute truth* about physical space, thus the term.

geometry course, or from trigonometry or calculus—only what can be derived from the axioms or previously proven theorems. The only exception is when we construct examples or models from Euclidean geometry. Naturally, in those cases we will be using our knowledge of Euclidean geometry, just as we did in the previous chapter when we constructed various models.

3.1 Triangles, Congruence Relations, SAS Hypothesis

TRIANGLES

The terminology commonly associated with triangles will be introduced formally.

> DEFINITION: A **triangle** is the union of three segments (called its **sides**), whose end points (called its **vertices**) are taken, in pairs, from a set of three noncollinear points. Thus, if the vertices of a triangle are A, B, and C, then its sides are $\overline{AB}$, $\overline{BC}$, and $\overline{AC}$, and the triangle is then the set defined by $\overline{AB} \cup \overline{BC} \cup \overline{AC}$, denoted by $\triangle ABC$. The **angles** of $\triangle ABC$ are $\angle A \equiv \angle BAC$, $\angle B \equiv \angle ABC$, and $\angle C \equiv \angle ACB$.

(See Figure 3.1 for a pictorial definition of these and other terms often associated with triangles; their formal statements will be omitted.)

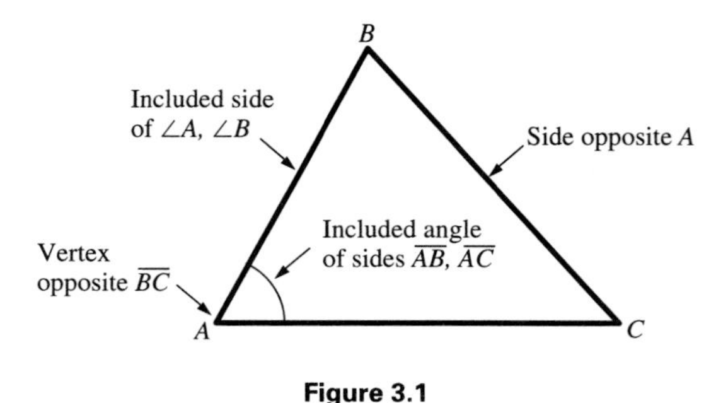

Figure 3.1

OUR GEOMETRIC WORLD

The triangle is a convenient and sturdy form used in engineering and construction. The illustration in the following figure shows a type of bridge construction known as the *Burr Truss*. It combines several triangles and one arch for extra strength to support the load, allowing longer spans to be achieved. Due to certain beams having equal lengths, and the presence of right angles, the congruence theorems we study in Section 3.3 show that several pairs of triangles in this drawing are congruent.

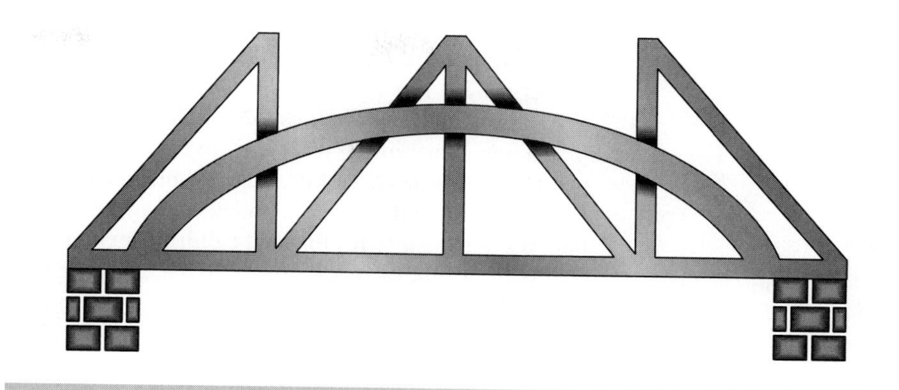

EUCLID'S CONCEPT FOR CONGRUENCE

Euclid had no word reserved exclusively for what we today call **congruence.** He used the term "equal" to describe not only congruence of triangles, but also equality of areas. Two polygons were said to be "equal" if they had the same area. Euclid attributed to congruent (equal) triangles the property that one triangle could be placed *precisely on top of another.* The act of placing one figure on top of another has been called **superposition** (a term not used by Euclid).

If we were to use superposition as a criterion for congruence, then somehow the properties of motions or isometries (distance-preserving maps) would have to be dealt with in the axioms. This can, in fact, be done, making isometries the basis for congruence, and this is not a bad way to study geometry. We shall, however, take the more traditional approach here. (In Chapter 5, we will discuss Euclid's superposition proof in the context of transformation theory.)

CONGRUENCE AS AN EQUIVALENCE RELATION

Modern usage of the term "equal" in mathematics is restricted to mean "identically the same as," as in equal sets, equal numbers, or equal algebraic expressions. For example, if we say that two points are equal, as in $A = B$, then we mean they coincide, or are the *same* point. We write $\overline{AB} = \overline{CD}$ only if the *set* of points $\overline{AB}$ is the exact same set of points denoted by $\overline{CD}$ (which can presumably happen only if either $A = C$ and $B = D$, or $A = D$ and $B = C$). In arithmetic, $6/(-3)^2 = \frac{2}{3}$ because these two numbers are identically the same, or have the *same value*, although their representations differ. Equality among people is similar. We cannot say that "Bill equals Sam" in the mathematical sense unless the person named Bill is the same identical person as the one named Sam. Now, "Bill = Sam" could mean, however, that Bill and Sam have the same height, or that they have the same weight, or that they belong to the same personality type. But this usage of *equals* is entirely different, and gives rise to a general type of equality in mathematics called an **equivalence relation,** a concept of equality that applies to many different situations. Its defining properties are just the three familiar properties of equality: *Reflexive Law, Law of Symmetry*, and the *Transitive Law* (to be defined later).

In geometry, if we want "=" to be used as it is in all other branches of mathematics (and we shall), then when a segment can be made to fit exactly on top of another (as in Euclid) but the segments are *different sets of points,* we need a new symbol. We shall use $\cong$ for this and call it **congruence.** Under this kind of equality, it is necessary for the two segments to have the *same length,* but they do not have to be identically the same set of points. Similarly, two *angles* are said to be congruent if and only if their measures are equal.

Thus, we require

Congruence for Segments and Angles

$\overline{AB} \cong \overline{XY}$ iff $AB = XY$

$\angle ABC \cong \angle XYZ$ iff $m\angle ABC = m\angle XYZ$

CONGRUENCE FOR TRIANGLES

The concept for congruence in triangles is more complicated. What do *you* think a good criterion would be? Certainly the concept of area comes to mind. But obviously two triangles in Euclidean geometry can have the same area without having the same size or shape. Perhaps both area and perimeter might help, but here, again, it is possible to have two triangles with noncongruent sides but with their areas and perimeters the same. (See Problem 14.)

The accepted definition is actually simpler as it does not require a prior development for area and perimeter. It involves a **correspondence** between the vertices of two triangles—congruent or not. If we are given any two triangles, $\triangle ABC$ and $\triangle XYZ$ we write

$$ABC \leftrightarrow XYZ$$

to mean that, in the order written, A corresponds to X, B to Y, and C to Z, as illustrated in Figure 3.2. That is,

$$A \leftrightarrow X, \qquad B \leftrightarrow Y, \qquad C \leftrightarrow Z$$

This induces a correspondence between the sides and angles of the two triangles (e.g., $\overline{BC} \leftrightarrow \overline{YZ}$ and $\angle BAC \leftrightarrow \angle YXZ$).

CORRESPONDENCE BETWEEN TWO TRIANGLES

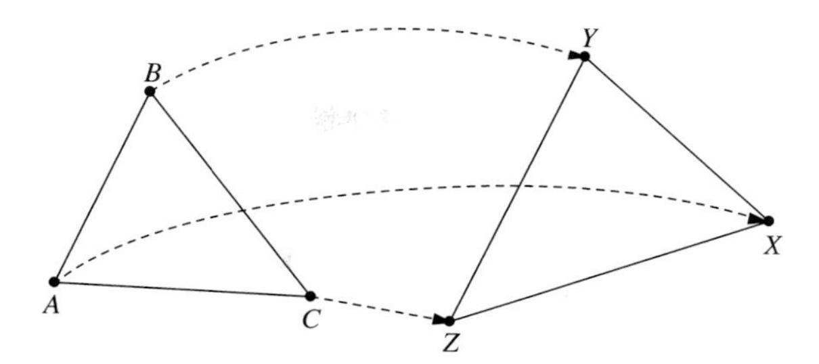

Figure 3.2

Note that there are potentially six ways for one triangle to correspond to another, simply because there are six ways to make the vertices correspond:

$$ABC \leftrightarrow XYZ \qquad ABC \leftrightarrow XZY \qquad ABC \leftrightarrow YXZ$$

$$ABC \leftrightarrow YZX \qquad ABC \leftrightarrow ZXY \qquad ABC \leftrightarrow ZYX$$

DEFINITION: CONGRUENCE FOR TRIANGLES

If, under some correspondence between the vertices of two triangles, corresponding sides and corresponding angles are congruent, the triangles are said to be **congruent**. Thus we write $\triangle ABC \cong \triangle XYZ$ whenever $\overline{AB} \cong \overline{XY}$, $\overline{BC} \cong \overline{YZ}$, $\overline{AC} \cong \overline{XZ}$, $\angle A \cong \angle X$, $\angle B \cong \angle Y$, and $\angle C \cong \angle Z$, as illustrated in Figure 3.3.

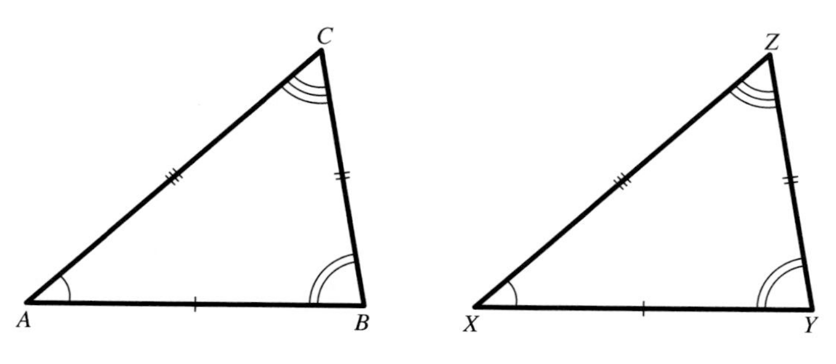

Figure 3.3

It is important to take note of the fact that the correct *order* of vertices in naming the triangles in a congruence is essential; changing the order changes the congruence. This brings up the question of whether a triangle can be *congruent to itself* under various orderings of its vertices. (See Problem 15 and the discovery unit at the end of this section.) The symbol CPCF will be used quite often in proofs as a time saver. It is an abbreviation for *corresponding parts of congruent figures* (*are congruent*), which comes from the definition of congruence.

The three basic properties of congruence for segments, angles, and triangles can be stated and established in a straightforward manner, which will be left as an exercise. We state these properties in terms of triangles only.

Properties of Congruence

1. Reflexive Law: $\triangle ABC \cong \triangle ABC$
2. Symmetry Law: If $\triangle ABC \cong \triangle XYZ$, then $\triangle XYZ \cong \triangle ABC$
3. Transitive Law: If $\triangle ABC \cong \triangle XYZ$ and $\triangle XYZ \cong \triangle UVW$, then $\triangle ABC \cong \triangle UVW$.

SAS HYPOTHESIS

Earlier we raised the issue of how much information would be needed to guarantee the congruence of triangles. Can we get by with fewer than six sets of congruent pairs, as required in the definition? Obviously, we could not merely require that the three sets of *angles* in two triangles be congruent, because similar noncongruent triangles have this feature. Euclid's *Elements* provides some clues. Euclid presents an argument that two triangles are congruent if the following hypothesis is satisfied:

The SAS Hypothesis

Under the correspondence $ABC \leftrightarrow XYZ$, let two sides and the included angle of $\triangle ABC$ be congruent, respectively, to the corresponding two sides and the included angle of $\triangle XYZ$. That is, for example, $\overline{AB} \cong \overline{XY}$, $\overline{BC} \cong \overline{YZ}$, and $\angle ABC \cong \angle XYZ$.

Euclid's familiar result actually cannot be established with the current set of axioms (those presented in the last chapter). In the next section, we develop an interesting model that shows this. Thus, if we want this property for congruence, we must assume it as an axiom.

Moment for Discovery

The Reflexive Law of Congruence

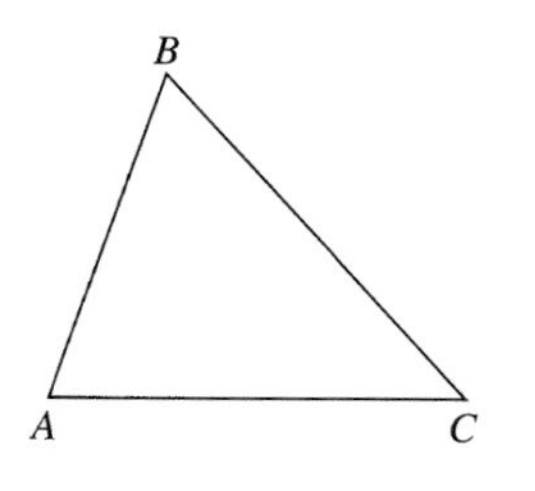

Figure 3.4

Suppose $\triangle ABC$ is a given triangle (Figure 3.4). Because the triangles $\triangle ABC$ and $\triangle BAC$ are identical sets, $\triangle ABC = \triangle BAC$. Therefore, the Reflexive Law of Congruence demands that $\triangle ABC$ be congruent to $\triangle BAC$. Can this congruence be written as $\triangle ABC \cong \triangle BAC$? Let us explore the implications of this congruence in detail.

1. What correspondence between A, B, C and A, B, C is indicated *by definition* if $\triangle ABC \cong \triangle BAC$?
2. Name the corresponding angles (i.e., $\angle A \leftrightarrow \angle B$, etc.).

3. Name the corresponding sides.
4. Using CPCF, write down the pairs of *congruent* sides and angles under this correspondence.
5. Do you still think that $\triangle ABC \cong \triangle BAC$ in general? If not, what kind of triangle would lead to this situation?

PROBLEMS (3.1)

GROUP A

1. Suppose $RS = VW$ and $RT = UW$.

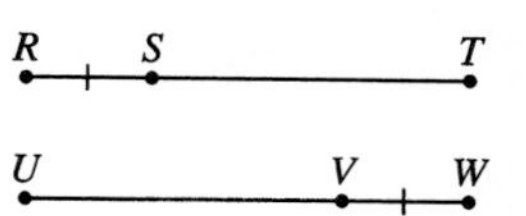

(a) Name all congruent pairs of segments as guaranteed *by definition.*

(b) Prove that $\overline{ST} \cong \overline{UV}$ using betweenness relations as evident in the figure.

2. (a) Name the congruent pairs of angles as indicated, and as guaranteed *by definition.*

(b) Prove that $\angle SVW \cong \angle UVR$ using betweenness relations from the figure.

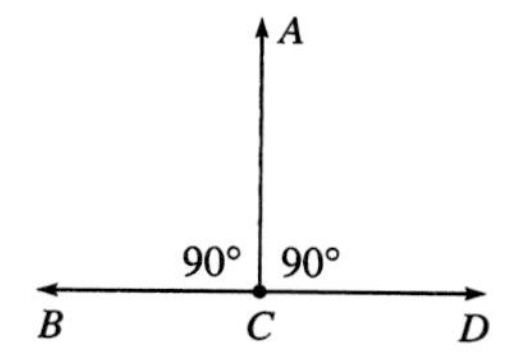

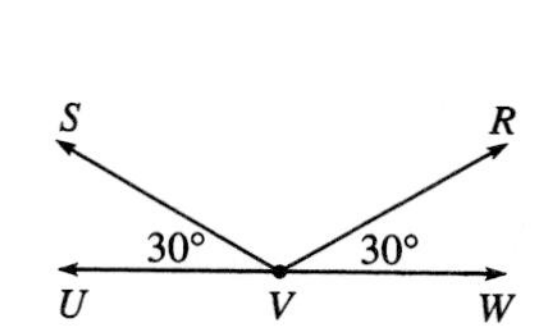

3. Name the congruent pairs of segments and angles justified by our axioms thus far using betweeness from the figure. (Segments have their measures as indicated in the figure; perpendicularity is indicated by small squares in the customary manner.)

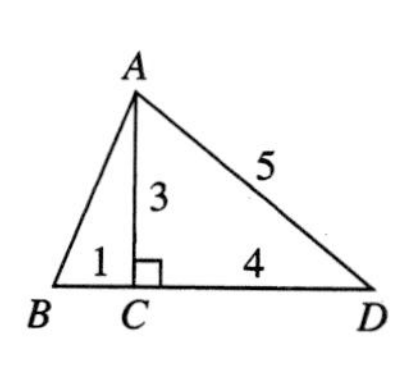

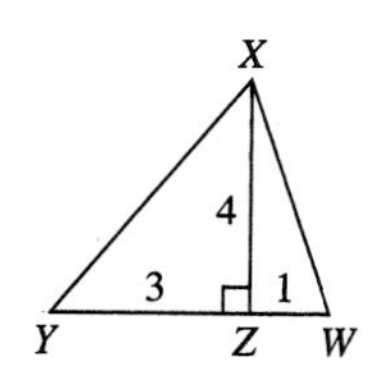

4. Name the congruent pairs of segments if $\triangle RST \cong \triangle LMN$.

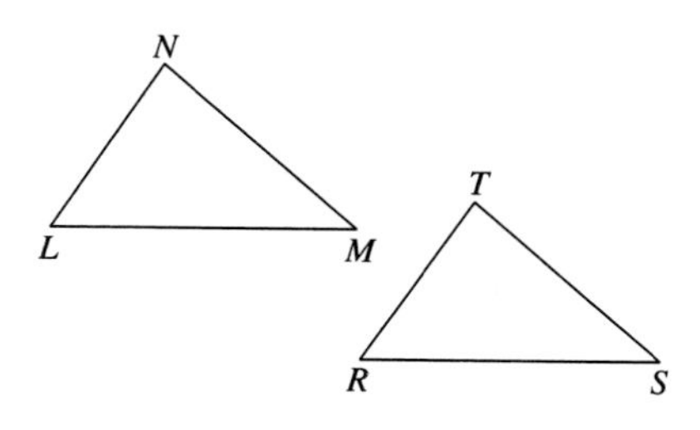

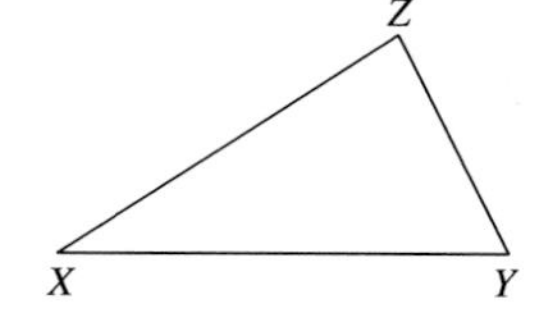

5. Name the congruent pairs of (distinct) angles if $\triangle XYZ \cong \triangle XZY$.

6. Under the correspondence $LMN \leftrightarrow VUW$, the triangles $\triangle LMN$ and $\triangle UVW$ are congruent, and $\triangle LMN$ has $LN = LM$. Name the congruent segments.

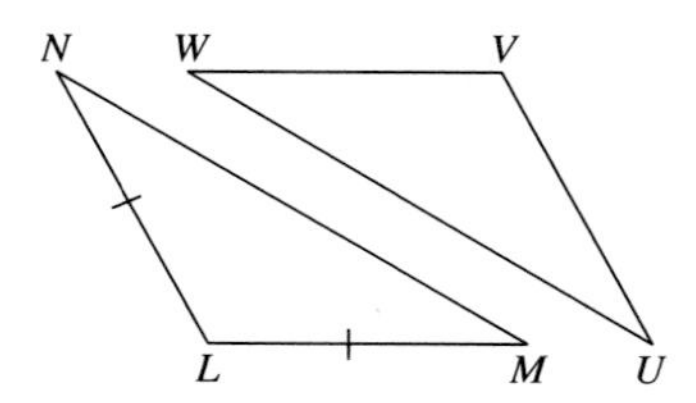

7. If $\triangle ABC \cong \triangle XYZ$, under what circumstances is $\triangle BCA \cong \triangle XZY$?

8. Which pairs of triangles in the figure (with measures as indicated) would satisfy the SAS Hypothesis? (Do not assume prior knowledge of congruent triangles.)

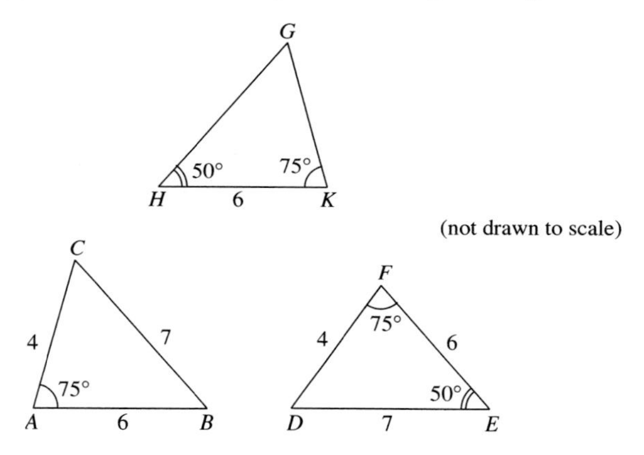

9. Which pairs of triangles in the figure would satisfy the SAS Hypothesis? (Do not assume prior knowledge of congruent triangles.)

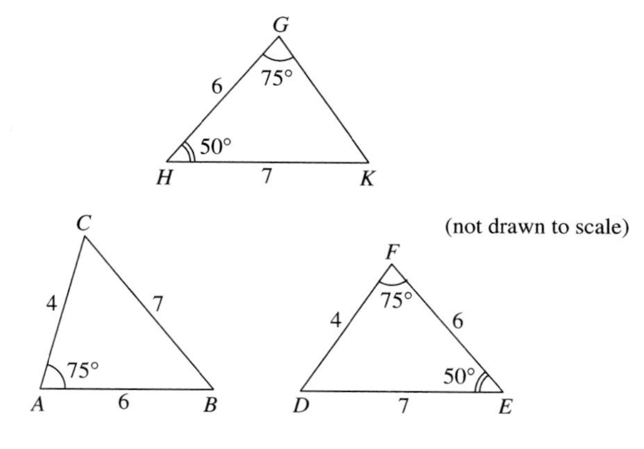

10. What reason(s) would you give for the (valid) general statements

(a) $AB = BA$

(b) $\overline{AB} \cong \overline{BA}$

11. What reason(s) would you give for the (valid) general statements

(a) $\triangle ABC = \triangle BCA$

(b) $\triangle ABC \cong \triangle ABC$

(c) Under what circumstances is $\triangle ABC \cong \triangle BCA$?

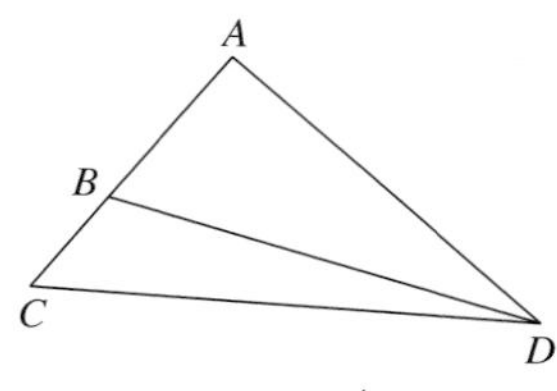

12. In the figure, what reason(s) would you give for the statement $\angle DAB \cong \angle DAC$?

GROUP B

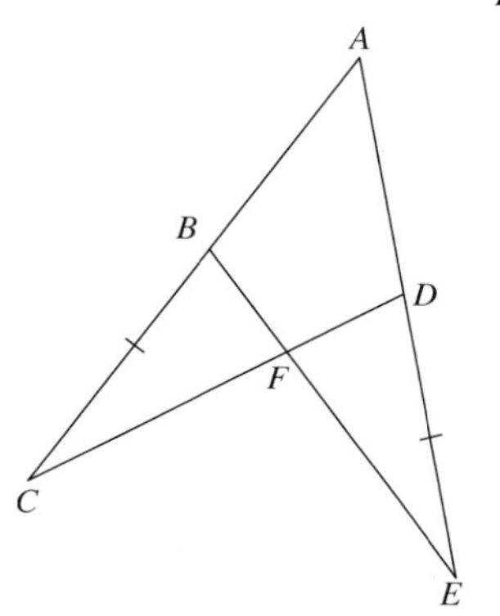

13. It is given that A–B–C and A–D–E, that $\triangle ACD$ and $\triangle ABE$ have equal perimeters (sum of the measures of the three sides) and $BC = DE$. Prove that $CD = BE$. Note that this problem has absolutely nothing to do with congruence of angles or triangles.

14. The triangles in the figure have the dimensions as indicated. Show, by *Sketchpad* or by direct calculation, that both the areas and perimeters of the two triangles are the same. (***Hint:*** Recall Heron's formula: $A = \sqrt{s(s-a)(s-b)(s-c)}$; here it is permissible to use prior knowledge since this is an example from Euclidean geometry.)

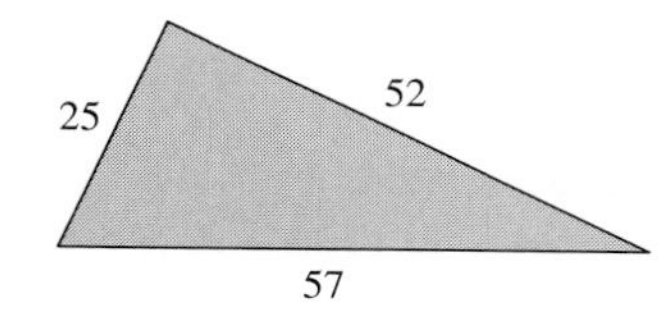

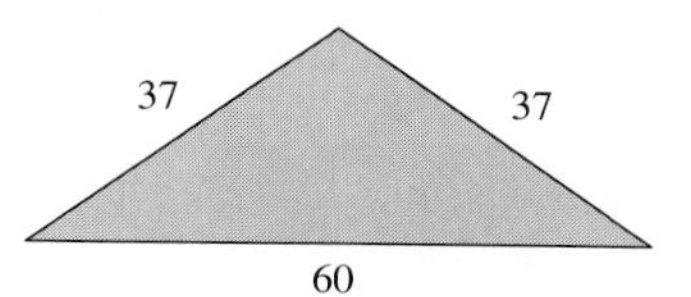

15. How many correspondences exist between the vertices of a triangle and itself? For a *scalene triangle* (no two sides congruent) how many of these are congruences?

16. Prove the Transitive Law for triangles using the Transitive Law for ordinary equality in the real number system.

GROUP C

17. Prove that if $\overline{AB} = \overline{CD}$ and $A \neq B$, then either $A = C$ and $B = D$, or $A = D$ and $B = C$.

18. Conjecture Two triangles are congruent if any five of the six parts of one are congruent, respectively, to five parts of the other, in some order. (That is, three pairs of angles and two pairs of sides, or two pairs of angles and three pairs of sides, are required to be congruent.) Use Euclidean geometry/trigonometry to prove or disprove.

*3.2 Taxicab Geometry: Geometry without SAS Congruence

We are now going to look at a rather unusual geometry. While it is something you can have some fun with, the purpose is entirely serious. This geometry serves as a model for all the axioms thus far assumed, but violates the SAS Postulate that we shall introduce in Section 3.3. What does this imply logically about the relationship of SAS to our other axioms?

To construct a true model for the axioms thus far assumed, which were three dimensional in character, the points of our model will be the ordinary points of three-dimensional coordinate geometry: $P(x, y, z)$, where x, y, and z are real numbers. The lines and planes will be defined exactly as in coordinate/vector geometry. That is, a *plane* is the set of points whose coordinates satisfy a linear equation $ax + by + cz = d$ for a, b, and c not all zero, and a *line* is the intersection of two planes. In this setting, we know that the axioms of incidence (Axioms I-1–I-5) will all be satisfied.

Moving on to the concept of distance and angle measure, instead of taking the distance between two points $P(x_1, y_1, z_1)$ and $Q(x_2, y_2, z_2)$ to be the usual metric given by the distance formula

$$PQ = \sqrt{(x_1 - x_2)^2 + (y_1 - y_2)^2 + (z_1 - z_2)^2}$$

take the following metric (which will satisfy Axioms D–1, D–2, and D–3)

$$\textbf{(1)} \qquad PQ^* = |x_1 - x_2| + |y_1 - y_2| + |z_1 - z_2|$$

HISTORICAL NOTE

The Taxicab Metric was originally discovered by Hermann Minkowski (1864–1909) as a special case of a metric defined in terms of an arbitrary convex set centered at the origin, whose boundary is a "circle" of the geometry. Minkowski, like Hilbert, grew up in Königsberg. In 1879, when he won a prize for a competition in mathematics, he met Hilbert for the first time, and they became close friends. At Hilbert's urging, Minkowski accepted a teaching position at Göttingen in 1902, and they were colleagues there until Minkowski's untimely death of appendicitis at the age of 45. Besides Minkowski's monumental work in the theory of numbers (using his metric), he is also known for his mathematical foundation for the Theory of Relativity, which Einstein—a student of Minkowski—used extensively in his work. The idea of the four-dimensional space/time framework commonly used for relativity is due to Minkowski. The physicist Max Born, also a student of Minkowski, once said: "Minkowski laid out the whole arsenal of relativity mathematics . . . as it has been used ever since by every theoretical physicist."

TAXICAB METRIC

When restricted to the plane, this metric is commonly known as the **Taxicab** (or **Manhattan**) **Metric.** We shall soon learn why.

The definition for betweenness will be the same as that for Euclidean betweenness: If A, B, and C are distinct collinear points, and $AB^* + BC^* = AC^*$, then A–B–C^*. It will be seen later that even though the two metrics for the plane are very different, taxicab betweenness and Euclidean betweenness coincide. If we adopt

Euclidean angle measure for this geometry, all the axioms on angle measure (Axioms **A**-1, **A**-2, **A**-3, and **A**-4) hold, and this geometry becomes a model for the previous axioms, assuming we can verify the Ruler Postulate.

At this point, let us confine our attention to a single plane. For that purpose, we can assume that $z = 0$ and that in **(1)**, $z_1 = z_2 = 0$. Hence, the distance formula reduces to

(2) $$PQ^* = |x_1 - x_2| + |y_1 - y_2|$$

In Figure 3.5, where we have illustrated the components of this formula, $|x_1 - x_2|$ is the length of one leg of the right triangle $\triangle PRQ$ shown, and $|y_1 - y_2|$ is that of the other. That is, the "distance" from P to Q is the Euclidean distance *around the corner* along the two legs of the triangle, as if to get from P to Q, you had to go through point R!

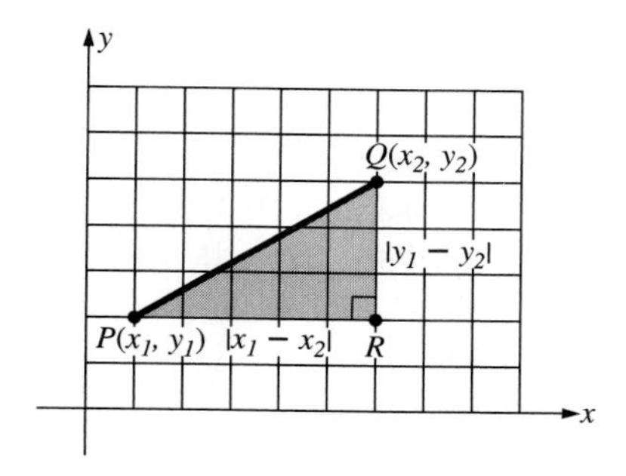

Figure 3.5

To make this more graphic, Figure 3.6 shows a portion of a fictitious city map. The route we must take to get from point A to B is $AB^* = AC + CB$. But a variety

THE TAXICAB METRIC:
$AB^* = AC + CB$

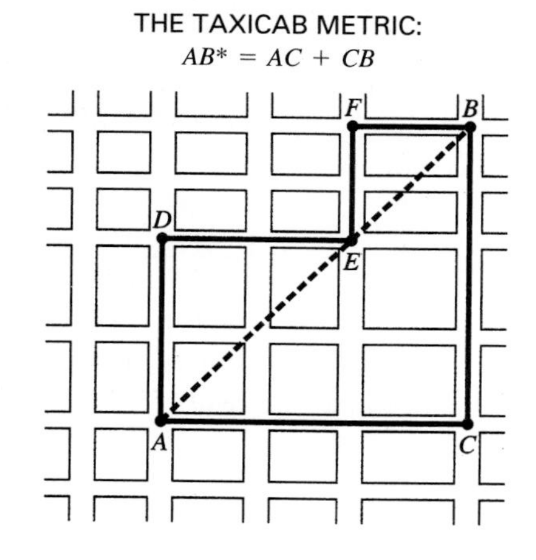

Figure 3.6

of other routes would also take us from A to B with the same total distance traveled, such as

$$AD + DE + EF + FB$$

Point E is the only point of the route $ADEFB$ that lies *between* A and B, but many other points, such as C, D, and F, are **metrically between** A and B, in the sense that equality in the triangle inequality holds for them. For example, since $AD^* = AD$ and $DB^* = DE + EF + FB$, we have

$$AB^* = AD^* + DB^*$$

If we are given a line ℓ in this geometry, such as $y = mx + b$, then for any two points along this line, say $P(x_1, y_1)$ and $Q(x_2, y_2)$, we have, by substitution into the taxicab distance formula **(2),**

$$\textbf{(3)} \qquad \begin{aligned} PQ^* &= |x_1 - x_2| + |(mx_1 + b) - (mx_2 + b)| \\ &= |x_1 - x_2| + |m|\,|x_1 - x_2| = k|x_1 - x_2|, \end{aligned}$$

where $k = 1 + |m|$. Since k is a constant, the Ruler Postulate will be cleared up if we take as the correspondence between real numbers and points on ℓ:

$$\textbf{(4)} \qquad P(x, y) \leftrightarrow kx \qquad (k = 1 + |m|)$$

That is, the taxicab **line coordinate** of $P(x, y)$ on ℓ is kx, or $P[kx]$ as in the earlier notation. The value k depends on the line chosen, but it is a *fixed constant for each line.* Thus, if $A[a]$ and $B[b]$ are any two points on ℓ with line coordinates $a = kx_1$ and $b = kx_2$, where $A = (x_1, y_1)$ and $B = (x_2, y_2)$, as two points in the coordinate plane, then by **(3)**,

$$\textbf{(5)} \qquad AB^* = k|x_1 - x_2| = |kx_1 - kx_2| = |a - b|$$

Thus, the Ruler Postulate is valid for the Taxicab Metric. Now we want to establish:

THEOREM 1: On any given line ℓ, betweenness under the Euclidean and Taxicab Metrics coincide.

The proof requires only a simple observation: Points on lines and their single coordinates in Euclidean geometry correspond to their coordinate pairs according to the formula

$$\textbf{(4')} \qquad P(x, y) \leftrightarrow k'x$$

where $k' = \sqrt{1 + m^2}$. This follows from the ordinary Euclidean distance formula: $\sqrt{(x_1 - x_2)^2 + (y_1 - y_2)^2}$; substitute $mx_1 + b$ for y_1 and $mx_2 + b$ for y_2, respectively, for two points $P(x_1, y_1)$ and $Q(x_2, y_2)$ on the line $y = mx + b$, and use a little algebra to arrive at

$$\begin{aligned} \sqrt{x_1 - x_2)^2 + m^2(x_1 - x_2)^2} &= \sqrt{(x_1 - x_2)^2(1 + m^2)} = \\ k'\sqrt{(x_1 - x_2)^2} &= k'|x_1 - x_2| \end{aligned}$$

[as in the derivation of **(3)** for the Taxicab Metric]. Then, $PQ = k'|x_1 - x_2|$, so the Ruler Postulate on ℓ becomes, for the Euclidean metric,

$$P[x] \leftrightarrow k'x$$

for any real x. Now compare this with **(4)** for the Taxicab Metric. Points therefore have the same order on line ℓ (even though k and k' are not the same real number, they are constant for each particular line). This finishes the proof.

We can therefore use the Euclidean definitions for segments, rays, angles, and half-planes in the new geometry under the Taxicab Metric. Because angle measure has been defined as in Euclidean geometry, the rest of the axioms for geometry up to this point will be satisfied, including the Plane Separation Postulate. The geometry we obtain will be called **taxicab geometry** from now on.

Since collinearity means the same thing in the two geometries, Theorem 1 shows that *geometric* betweenness for the Euclidean and for the Taxicab Metrics are the same. But *metric* betweenness is an entirely different matter. As mentioned before, we could have $AB^* = AD^* + DB^*$ under the Taxicab Metric and $AB < AD + DB$ under the Euclidean metric (for the same three points), as shown in Figure 3.6. Consequently, there are some unusual features of the Taxicab Metric, as we might imagine. For starters, circles are diamond shaped, ellipses are crystal-like hexagons or octagons, and "perpendicular bisectors" of line segments can be irregular curves (see Figure 3.7 for some illustrations). Establishing these unusual facts will be left as entertaining problems. Problems 13–17 address some of these features.

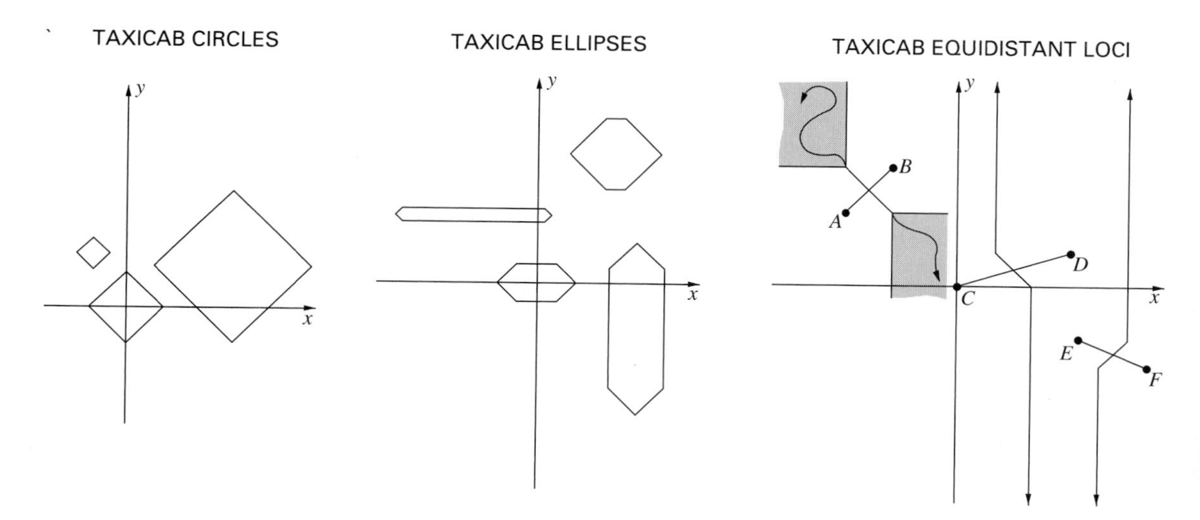

Figure 3.7

A less spectacular example is in connection with the SAS hypothesis. Because this was the reason for introducing taxicab geometry in the first place, we state it as a theorem. The proof makes an interesting problem.

THEOREM 2: In taxicab geometry, the SAS Hypothesis for two triangles does not always imply that the triangles are congruent.

The statement to be proven is of the form *not p*, so proving it consists of just coming up with one example showing that p is false (a **counterexample**). (See Problem 1.)

EXAMPLE 1 Show that the coordinate equation of the taxicab circle centered at (0, 0) having radius r is $|x| + |y| = r$ (Figure 3.8).

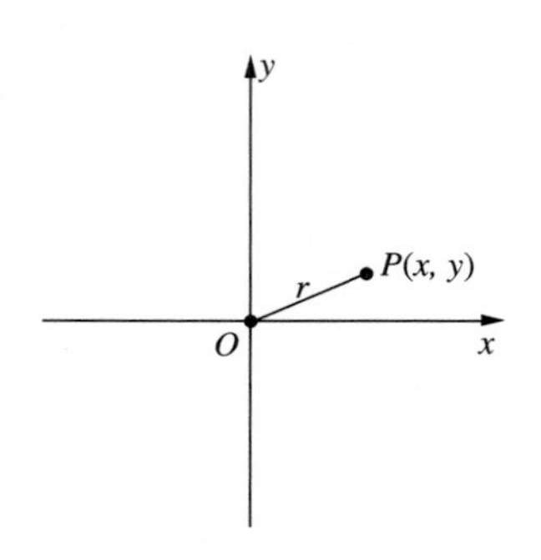

Figure 3.8

SOLUTION

Using the distance formula **(2)**, we have, with $O = (0, 0)$ and $P = (x, y)$,

$$r = OP^* = |x - 0| + |y - 0|$$
$$\therefore |x| + |y| = r. \quad \blacksquare$$

EXAMPLE 2 The triangle whose vertices are $A(2, 2), B(-1, 1)$, and $C(1, -1)$ in the xy-plane is equilateral under the Taxicab Metric. Verify this. (See Figure 3.9.)

SOLUTION

$$AB^* = |2 - (-1)| + |2 - 1| = 3 + 1 = 4,$$
$$BC^* = |-1 - 1| + |1 - (-1)| = 2 + 2 = 4,$$
$$AC^* = |2 - 1| + |2 - (-1)| = 1 + 3 + 4$$

Thus $AB^* = BC^* = AC^*$ under the Taxicab Metric, and the triangle is a taxicab equilateral triangle having sides of measure 4 units. ■

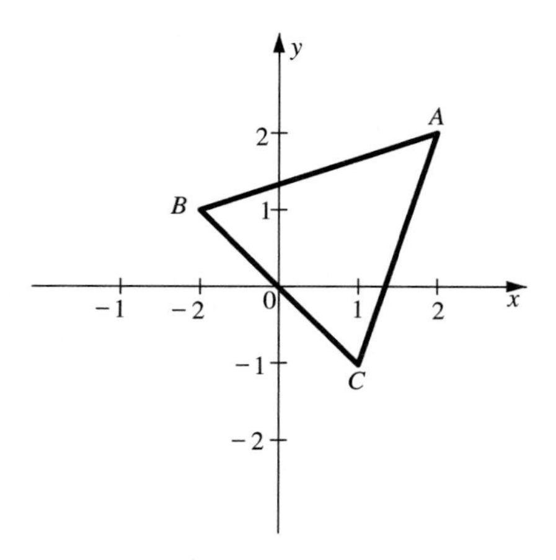

Figure 3.9

EXAMPLE 3 In taxicab geometry consider the hexagon border of the crystal-shaped figure shown in Figure 3.10. ($\overline{AB} \perp \overline{AG}$ and $\overline{CD} \perp \overline{DE}$.) Define the two points F_1 and F_2 (foci) to be the centers of the squares formed by the two ends of the

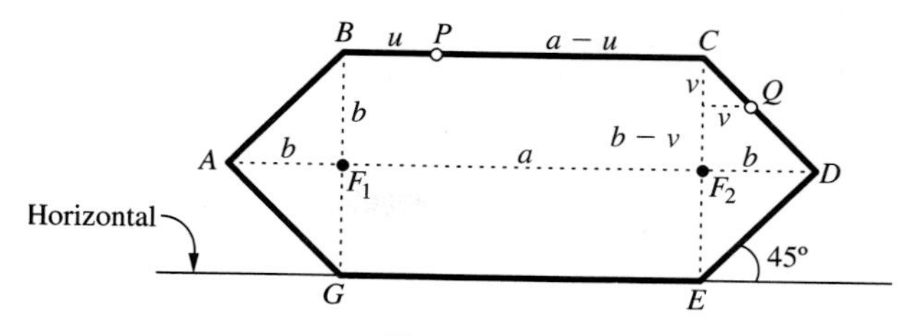

Figure 3.10

hexagon. Show that under the two-foci definition of an ellipse, the hexagon is an ellipse in taxicab geometry. That is, show that for all points P on the hexagon, $PF_1^* + PF_2^* = \text{(constant)}$.

SOLUTION

First, if P lies on either the top or bottom side, we have

$$PF_1^* + PF_2^* = (u + b) + (a - u + b) = a + 2b = AD^*$$

Next, if Q is a point on segment $\overline{CD}$, then

$$QF_1^* + QF_2^* = [v + (b - v) + a] + [v + (b - v)] = [b + a] + [b] = a + 2b = AD^*$$

By symmetry of the argument and figure, we can deduce the same result for a point P that lies on any other side of the hexagon. Therefore, in all cases,

$$PF_1^* + PF_2^* = AD^* \quad \blacksquare$$

EXAMPLE 4 The analogue of the Euclidean parabola $y = \frac{1}{4}x^2$ in taxicab geometry, having "focus" $F(0, 1)$ and "directrix" $y = -1$, is illustrated in Figure 3.11. Verify this by finding conditions for the coordinates x and y of a variable point

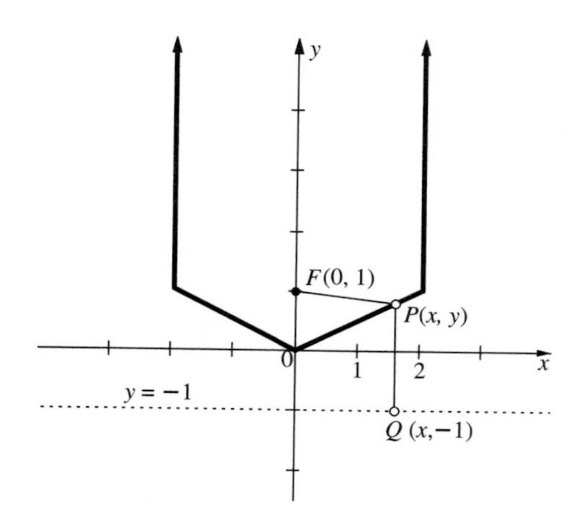

Figure 3.11

$P(x, y)$ on this parabola using the defining property $PF^* = PQ^*$, and the distance formula **(2)**.

SOLUTION

By **(2)**, $PF^* = |x - 0| + |y - 1|$ and $PQ^* = |y - (-1)|$, so we must have

$$|x| + |y - 1| = |y + 1|$$

There are three possibilities for y: $y < -1$, $-1 \leq y < 1$, and $y \geq 1$. We take each case separately.

(1) $y < -1$: Because $y + 1 < 0$, and hence $y - 1 < 0$, the above equation becomes $|x| - (y - 1) = -(y + 1)$ or $|x| - y + 1 = -y - 1$. This gives us $|x| = -2 \rightarrow\leftarrow$ (absolute value cannot be negative).

(2) $-1 \leq y < 1$: In this case, $y + 1 \geq 0$ and $y - 1 < 0$, so the previous equation becomes $|x| - (y - 1) = y + 1$ or $|x| - y + 1 = y + 1$ or $|x| = 2y$. $\therefore y \geq 0$. Hence, P lies on either of the two lines $y = \pm\frac{1}{2}x$, $0 \leq y < 1$.

(3) $y \geq 1$: Here, $y - 1 \geq 0$ and therefore $y + 1 > 0$, so the above equation becomes $|x| + y - 1 = y + 1$ or $|x| = 2$. Hence, P lies on either of the two lines $x = \pm 2$ for $y \geq 1$. ■

Using Sketchpad to Measure Taxicab Distance

Program to Create Script

[1] Open New Sketch and New Script, placed side by side. In Sketch, locate any two points A and B. Access recording mode by selecting (REC) in Script.

[2] Select A and translate it 1 unit in the horizontal direction. (Under Translate, use Polar Vector having coordinates of 0.00° and 1.000 unit.)

[3] Construct a line passing through A and A', the point you found in Step 2.

[4] With the line still selected, select point B and choose Perpendicular Line under CONSTRUCT.

[5] Select the point C of intersection of the perpendicular lines. Hide lines and points except A, B, and C.

[6] Using Distance and Calculate under MEASURE, display the quantities AC, BC, and $AC + BC$ on the screen, then hide the measurements Distance AC, Distance BC, and point C. Turn off Record by selecting STOP in Script. Save the script under a name like TAXIDIST.

You are now ready to use the script you just created. To activate, select any two points P and Q and choose FAST in TAXIDIST script. You can drag the points in any manner you wish and the displayed taxicab distance PQ^* will change accordingly.

Moment for Discovery

Square Circles

Investigate a phenomenon that certain squares have in taxicab geometry using the following procedure. Observe the line segment $\overline{AB}$ in Figure 3.12, which makes an angle of 45° with the x-axis.

1. Verify that $OP^* = OQ + QP = OA$ for $P \in \overline{AB}$. Thus OP^* is constant for $P \in \overline{AB}$.
2. Similarly, verify that if P is any point on $\overline{BC}$, then $OP^* = OB = OA$.
3. What about the behavior of OP^* when P varies on segments $\overline{CD}$ and $\overline{AD}$?

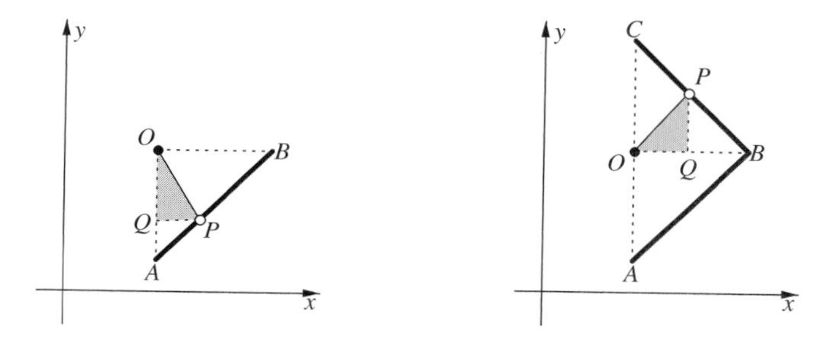

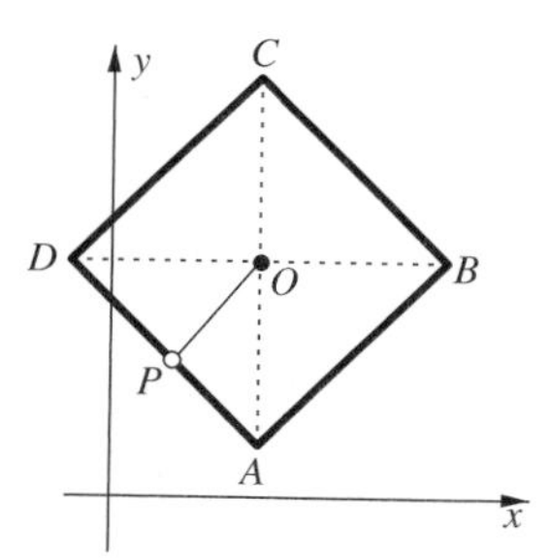

Figure 3.12

4. Do you have any conclusion to state about "circles" in taxicab geometry?
5. (Optional) Using *Sketchpad*, construct the square of Figure 3.12 and locate points on each side, say P, Q, R, and S, then display the four distances OP^*, OQ^*, OR^*, and OS^* on the screen, using the TAXIDIST script. Drag the four points and observe the effect on their "distances" from O.

PROBLEMS (3.2)

Many of the problems in this section are investigative in nature and are suitable for *Sketchpad* and the TAXIDIST script introduced above. Any problem of a general nature, however, such as 9, 12, and 14, needs a formal proof or explanation.

GROUP A

1. Find the lengths of the sides of right triangles $\triangle ABC$ and $\triangle XYZ$ using the Taxicab Metric (without coordinates). Do these two triangles satisfy the SAS hypothesis? (Are $\overline{AB} \cong \overline{XY}$, $\angle B \cong \angle Y$, $\overline{BC} \cong \overline{YZ}$ true?) Are they congruent?

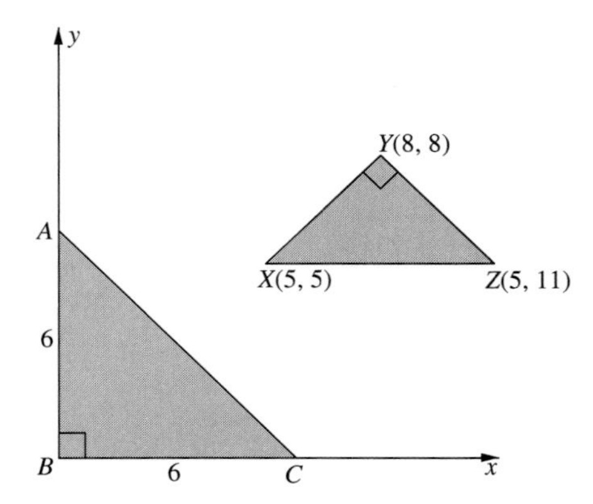

2. Is the triangle $\triangle ABC$ of Example 2 equilateral under the Euclidean metric? Isosceles?

3. A triangle in the coordinate plane has vertices $A(0, 0)$, $B(2, 0)$, and $C(1, \sqrt{3})$. This triangle is equilateral under the Euclidean metric. For the Taxicab Metric, find

(a) the lengths of the sides of $\triangle ABC$.

(b) the angle measures.

(c) Based on these calculations, is $\triangle ABC$ equilateral in taxicab geometry? Equiangular? Isosceles?

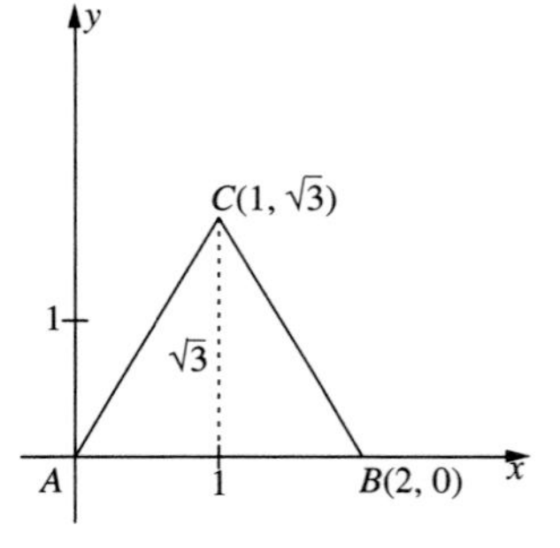

4. Three points R, S, and T are placed in a unit grid (where each square represents unit length).

(a) Show that $\triangle RST$ is equilateral under the taxicab metric.

(b) Is $\triangle RST$ equiangular under the taxicab metric?

(c) Is $\triangle RST$ equilateral under the Euclidean metric? Isosceles?

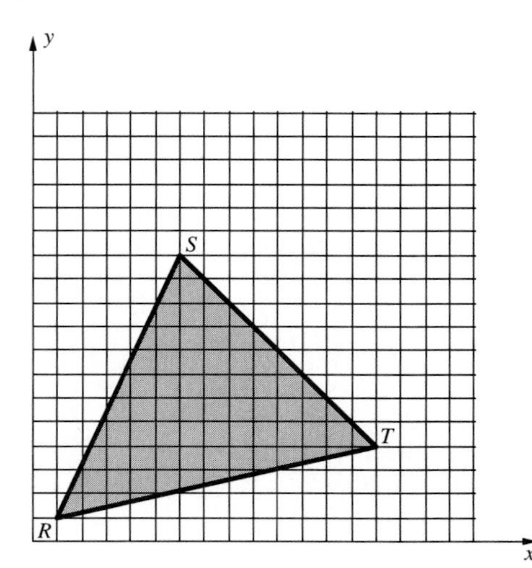

5. Consider the taxicab triangle $\triangle ABC$, with its vertices as indicated in the coordinate plane.

(a) Show that $\triangle ABC$ is a right triangle. (Use slopes to show that $\overline{AC} \perp \overline{BC}$.)

(b) Find the taxicab lengths of the three sides $a = BC^*$, $b = AC^*$, and $c = AB^*$ using the distance formula **(2)**.

(c) Does the Pythagorean relation hold for this right triangle in taxicab geometry?

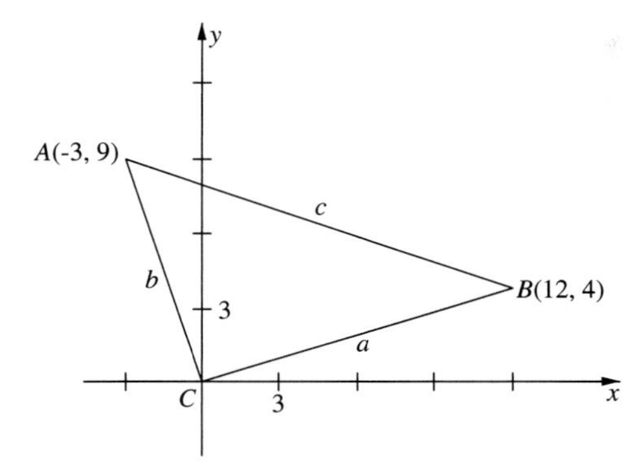

6. A 3–4–5 right triangle in the Euclidean plane can be placed in such a position in the xy-plane so its vertices have coordinates $A(0, 0)$, $B(4, 3)$ and $C(4, 0)$.

(a) Find the taxicab hypotenuse $AB^* = c$, and taxicab legs $BC^* = a$ and $AC^* = b$.

(b) Compute c^2 and $a^2 + b^2$.

(c) Does the Pythagorean relation hold for this right triangle in taxicab geometry?

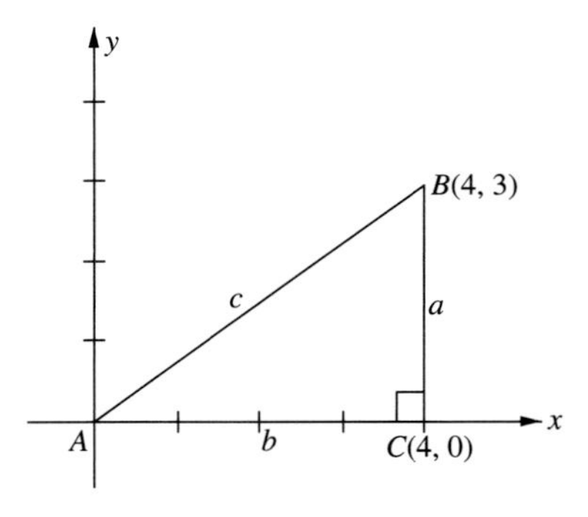

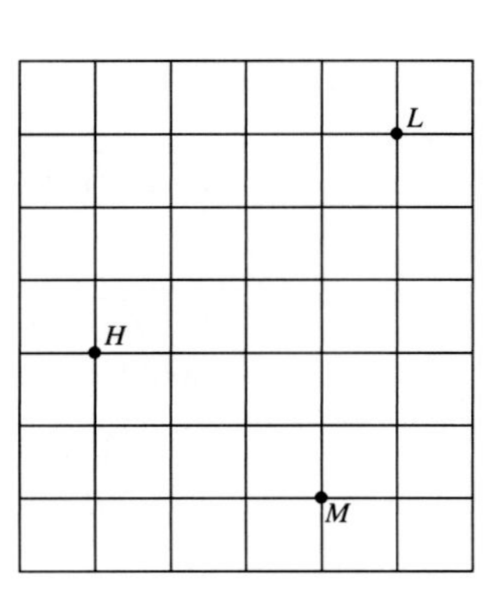

7. Members of a family on vacation in New York City were staying at a hotel (H). They decide to walk from their hotel to visit a museum (M), and then from there, walk to a library (L), and finally return to their hotel (H). If the grid in the figure simulates the city streets, find the distances (in city blocks) they must walk: HM^*, ML^*, and LH^*.

8. Find the Euclidean distances HM, ML, and LH of Problem 7 (in terms of city blocks as unit), and compare with the results HM^*, ML^*, and LH^* obtained in Problem 7.

9. Suppose that a rectangle in taxicab geometry is defined as a *four-sided polygon having four right angles*. Are the opposite sides of a rectangle congruent in taxicab geometry?

GROUP B

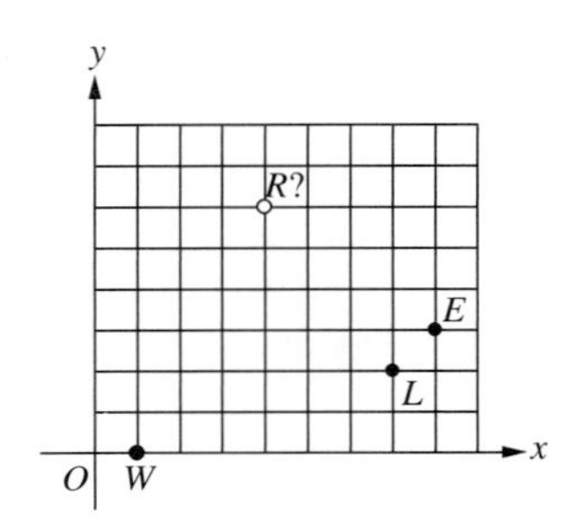

10. A man has a newspaper stand at $W = (1, 0)$, eats regularly at a cafeteria located at $E = (8, 3)$, and does his laundry at a laundromat at $L = (7, 2)$.

    (a) If he wants to find a room R using the Taxicab Metric so as to be at the same *walking* distance from each of these points, where should R be located? (Give the coordinates.)

    (b) How many blocks does he have to walk from his room to each of the three points, assuming he finds such a room?

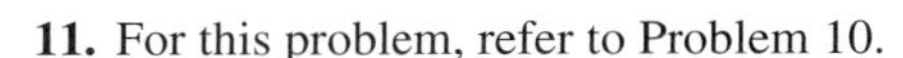

11. For this problem, refer to Problem 10.

    (a) If all conceivable shortcuts are possible (i.e., using the Euclidean metric), where should the man's room be? (Find the perpendicular bisectors of $\overline{WL}$ and $\overline{LE}$.)

    (b) Answer part (b) of Problem 10 if his room is located *using the Euclidean metric*. [Note that the distance is $\approx 8\frac{1}{2}$ blocks compared to 5 blocks in Problem 5(b)!]

12. Do three noncollinear points always determine a unique taxicab circle? Consider all cases.

13. Verify to your own satisfaction that *all points* in the shaded region are equidistant from A and B in taxicab geometry.

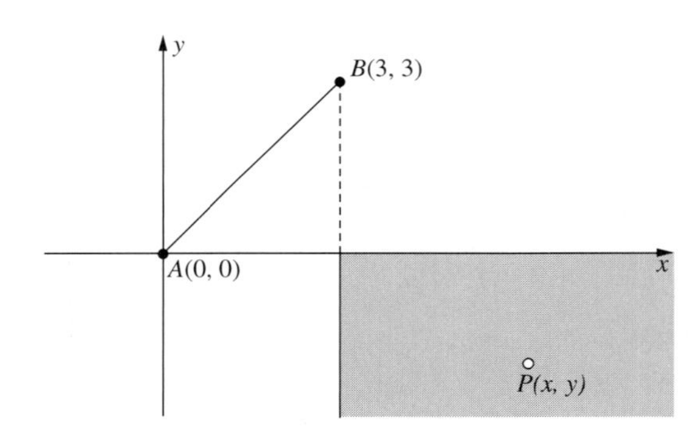

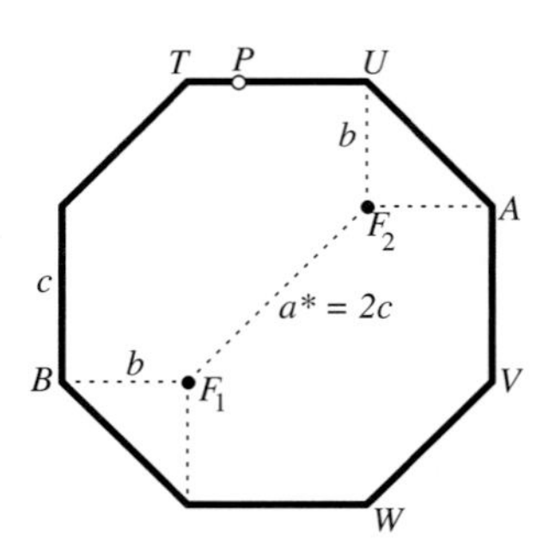

14. Show that a regular octagon in Euclidean geometry with two horizontal sides is an ellipse in taxicab geometry, having foci located inside the polygon and on a line making a 45° angle with the horizontal. (***Hint:*** Show that $PF_1{}^* + PF_2{}^* = AB^*$ for any point P on the octagon.)

15. Consider two points $A(0, 0)$ and $B(30, 20)$, and two other points $C(5, 20)$ and $D(25, 0)$, and join points A and B. Draw the two lines $x = 5$ for $y \geq 20$, and $x = 25$ for $y \leq 0$, and join the points C and D. This will give you a broken line. Make the necessary calculations to satisfy yourself that for any point P on this broken line, $PA^* = PB^*$. [***Hint:*** The coordinate approach would look like this: Any point P on $x = 5$ has coordinates $P(5, y)$ for $y \geq 20$, so $PA^* = |5 - 0| + |y - 0| = 5 + y$, while $PB^* = |5 - 30| + |y - 20| = y + 5$. If P lies on $\overline{CD}$, its coordinates are $(x, 25 - x)$ for $5 \leq x \leq 25$, and $PA^* = |x - 0| + |(25 - x) - 0| = x + 25 - x = 25$, etc.]

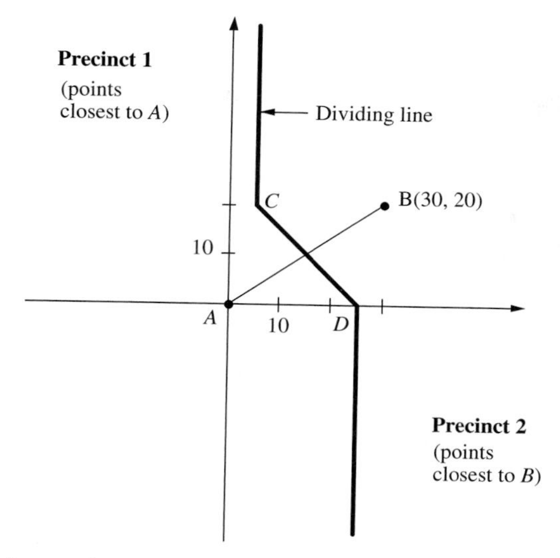

16. In a town having perfect square blocks and equally spaced streets running north and south, east and west, two police stations are to be located at $A(0, 0)$ and $B(30, 20)$. The town officials want to divide the town into two precincts—Precinct 1 served by Station A and Precinct 2 served by Station B. How should the boundary be drawn? (***Hint:*** First work Problem 15.)

GROUP C

17. Use *Sketchpad* to verify that a hexagon in the coordinate plane with two opposite sides lying on vertical lines and with the top and bottom consisting of isosceles right triangles (as shown) is a taxicab ellipse with foci F_1 and F_2 as indicated in the figure.

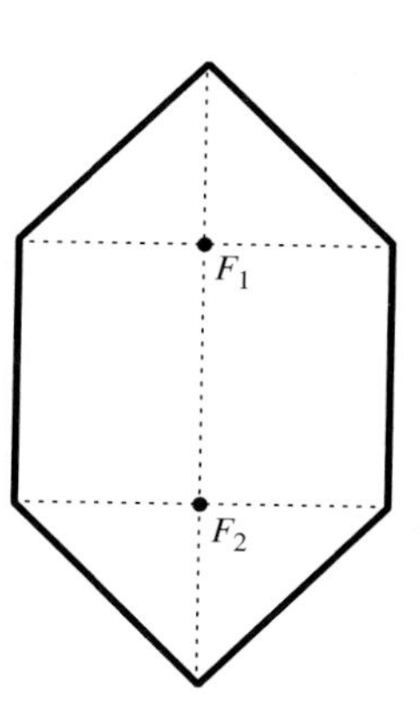

18. Rework Problem 16 if there are to be *three* precincts: At $A(0, 0)$, $B(30, 20)$, and $C(10, 40)$.

19. Investigate the focus/directrix property of a parabola having an oblique axis. (See Example 4.) What do such curves look like? Explore how the computer might help.

***20.** Using the two-focus property of hyperbolas in Euclidean geometry, investigate taxicab hyperbolas. What curves would be the "asymptotes" of these hyperbolas?

*Readable references to Problems 14, 17, 19 and 20 include E. Krause, "Taxicab Geometry," *Mathematics Teacher*, Vol. 66 (1973), pp. 695–706, and R. Laatsch, "Pyramidal Sections in Taxicab Geometry," *Mathematics Magazine*, Vol. 55, No. 4 (1982), pp. 205–212.

3.3 SAS, ASA, SSS Congruence and Perpendicular Bisectors

The postulate anticipated in the preceding sections represents the last axiom we need for absolute geometry. This postulate leads to all the familiar properties of Euclidean geometry that do not involve the concept of parallel lines. Thus, all the results we obtain are also true for non-Euclidean geometry as well. We can be sure that this new axiom is independent from the rest due to the validity of taxicab geometry as a model for the axioms we have already assumed.

> **AXIOM C-1: SAS POSTULATE**
> If the SAS Hypothesis holds for two triangles under some correspondence between their vertices, then the triangles are congruent.

A major application of the SAS Postulate is to establish the other two fundamental congruence criteria for triangles, namely the ASA and SSS theorems. (Secondary school texts frequently take these results as axioms also.) This, and the basic properties of isosceles triangles, will be our goal for this section.

THE ASA CONGRUENCE CRITERION

Euclid's proof of the ASA Theorem relied heavily on the undefined operation of superposition—placing or "applying" one triangle onto another. However, a logical proof based on the development thus far is not difficult to construct.

First, suppose that in Figure 3.13 $\angle A \cong \angle X$, $\overline{AB} \cong \overline{XY}$, and $\angle B \cong \angle Y$. If we knew that $\overline{AC} \cong \overline{XZ}$, then $\triangle ABC \cong \triangle XYZ$ by the SAS Postulate. But if *not,* then either $AC > XZ$ or $AC < XZ$. Let us see where this will lead. If it is assumed that $AC > XZ$, then we can locate D on $\overline{AC}$ so that A–D–C and $\overline{AD} \cong \overline{XZ}$ by the Segment Construction Theorem.

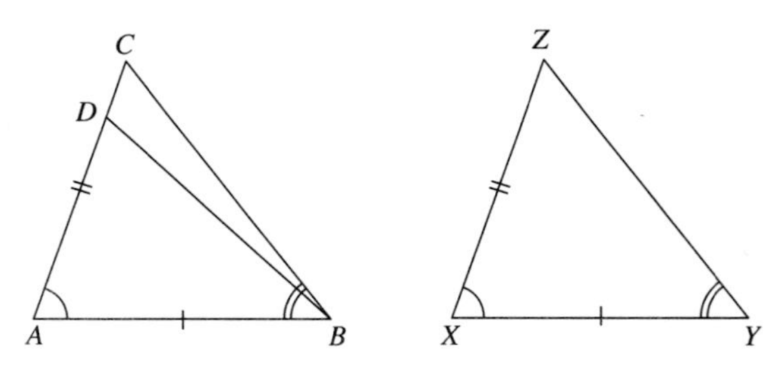

Figure 3.13

To encourage your participation, we now ask a few pointed questions for you to answer:

1. What do you observe about $\triangle ABD$ and $\triangle XYZ$? Why?
2. Does this make $\angle ABD \cong \angle Y$?
3. We were given that $\angle Y \cong \angle ABC$. What about $\angle ABD$ and $\angle ABC$? Do you reach a contradiction? What properties are used? What if $AC < XZ$?

The contradiction inherent in the above line of reasoning then proves:

> THEOREM 1: ASA THEOREM
> If, under some correspondence, two angles and the included side of one triangle are congruent to the corresponding angles and included side of another, the triangles are congruent under that correspondence.

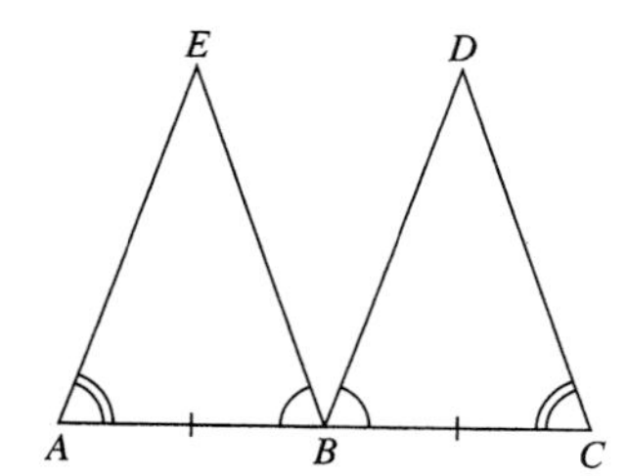

Figure 3.14

EXAMPLE 1 (See Figure 3.14.)
Given: $\angle EBA \cong \angle CBD$, $\overline{AB} \cong \overline{BC}$, $\angle A \cong \angle C$.
Prove: $\overline{EB} \cong \overline{DB}$.

CONCLUSIONS	JUSTIFICATIONS
(1) $\triangle ABE \cong \triangle CBD$.	ASA Theorem
(2) $\therefore \overline{EB} \cong \overline{DB}$.	CPCF

[UCSMG, p. 317] ■

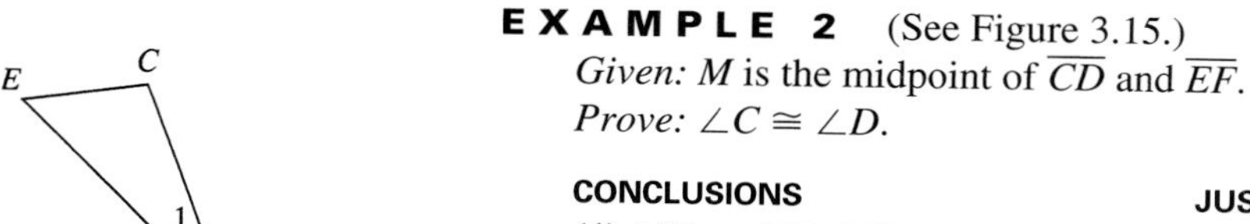

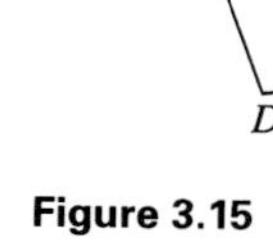

Figure 3.15

EXAMPLE 2 (See Figure 3.15.)
Given: M is the midpoint of $\overline{CD}$ and $\overline{EF}$.
Prove: $\angle C \cong \angle D$.

CONCLUSIONS	JUSTIFICATIONS
(1) $MC = MD$, $ME = MF$.	Definition of midpoint
(2) Angles 1 and 2 form a vertical pair, hence $\angle 1 \cong \angle 2$.	Definition and Vertical Pair Theorem
(3) $\triangle CME \cong \triangle DMF$.	SAS Postulate
(4) $\therefore \angle C \cong \angle D$.	CPCF

[UCSMG, p. 318] ■

EXAMPLE 3 Prove that a line that bisects an angle also bisects any segment perpendicular to it that joins two points on the sides of that angle (Figure 3.16).

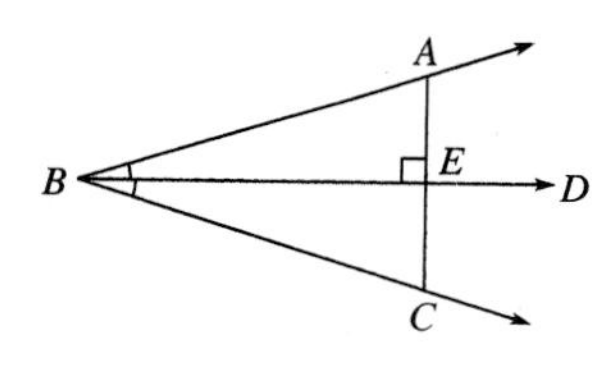

Figure 3.16

PROOF

Given: $\angle ABC$, line $\overleftrightarrow{BD}$ bisects $\angle ABC$, $\overleftrightarrow{AC} \perp \overleftrightarrow{BD}$.
Prove: Line $\overleftrightarrow{BD}$ bisects segment $\overline{AC}$.

CONCLUSIONS	JUSTIFICATIONS
(1) Ray $\overrightarrow{BD}$ intersects segment $\overline{AC}$ at some interior point E.	Crossbar Theorem
(2) $\angle ABE \cong \angle EBC$.	Definition (angle bisector)
(3) $\overline{BE} \cong \overline{BE}$.	Reflexive Property
(4) $\angle BEA \cong \angle BEC$.	Theorem 3, Section 2.5
(5) $\triangle ABE \cong \triangle CBE$.	ASA
(6) $\overline{AE} \cong \overline{EC} \rightarrow AE = EC$.	CPCF
(7) $\therefore E$ is the midpoint of segment $\overline{AC}$ and the line $\overleftrightarrow{BD}$ bisects $\overline{AC}$.	Definition of midpoint ■

ISOSCELES TRIANGLE THEOREM

A basic result that makes use of SAS and ASA involves an **isosceles triangle**—a triangle having two sides congruent. (The congruent sides are called the **legs** of the triangle, and the third side is called the **base,** with the angles along the base called the **base angles.** The vertex opposite the base is called the **vertex,** and the angle at that vertex, the **vertex angle.**) The result may be stated as follows: If two sides of a triangle are congruent, the angles opposite those sides are congruent.

Euclid went to a great deal of trouble to prove this using a special construction method that later became a test of geometric ability (the *pons asinorum*—"bridge of asses"). Any student who could not "cross this bridge" was considered unable to go further in the study of geometry. (See Problem 22.)

However, there is a very efficient, streamlined proof, first given by Pappus, used by Hilbert, and reportedly rediscovered by a computer in the 1950s when programmers were conducting early experiments in artificial intelligence. The reader has only to look back to the discovery unit of Section 3.1 and examine the correspondence $ABC \leftrightarrow BAC$ in connection with that. Hence, there is a one-line proof!

LEMMA A: In $\triangle ABC$, if $\overline{AC} \cong \overline{BC}$, then $\angle A \cong \angle B$. (See Figure 3.17.)

PROOF

By the SAS Postulate, $\triangle CAB \cong \triangle CBA$. Therefore, corresponding angles $\angle CAB$ and $\angle CBA$ are congruent. That is, $\angle A \cong \angle B$.

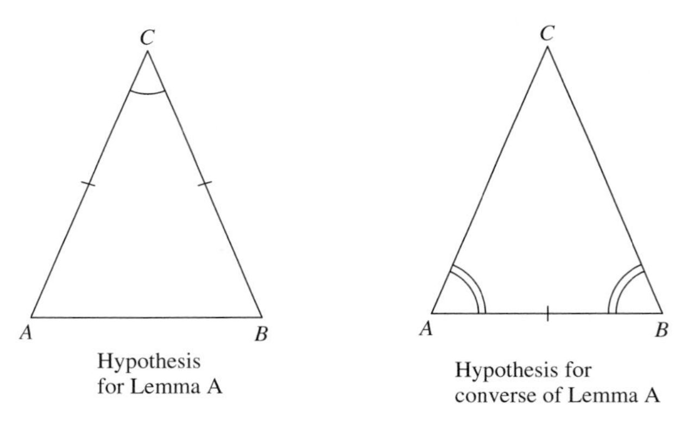

Figure 3.17

THEOREM 2: ISOSCELES TRIANGLE THEOREM

A triangle is isosceles iff the base angles are congruent.

PROOF

In view of Lemma A, we have only to prove its converse: If $\angle A \cong \angle B$, then $\overline{AC} \cong \overline{BC}$. (See Figure 3.17.) By the ASA Theorem, $\triangle CAB \cong \triangle CBA$. Therefore, $\overline{AC} \cong \overline{BC}$.

EXAMPLE 4 Solve the following problem in proof writing.

Given: In Figure 3.18, $GJ = KM = 2$, $JK = HJ = HK = 10$, with betweenness relations as evident from the figure.

Prove: $HG = HM$.

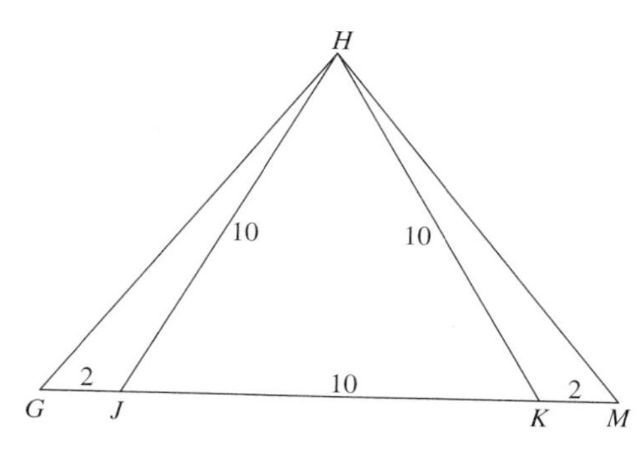

Figure 3.18

CONCLUSIONS	JUSTIFICATIONS
(1) $\angle HJK \cong \angle HKJ$; $m\angle HJK = m\angle HKJ$.	Isosceles Triangle Theorem, definition of $\cong$
(2) $m\angle GJH = 180 - m\angle HJK$, and $m\angle MKH = 180 - m\angle HKJ$.	Linear Pair Axiom
(3) $m\angle GJH = m\angle MKH$ or $\angle GJK \cong \angle MKH$.	Algebra (Steps 1 and 2)
(4) $\triangle HGJ \cong \triangle HMK$.	SAS
(5) $\therefore \overline{HG} \cong \overline{HM}$, or $HG = HM$.	CPCF, definition of $\cong$ ■

OUR GEOMETRIC WORLD

The center beam in the photograph is used to support the beams at the top. Depending on whether certain other supporting beams are of equal length, it may or may not follow that the center support is erect ($\perp$ to the base). A theoretical basis for this question is provided by Theorem 3.

SYMMETRY

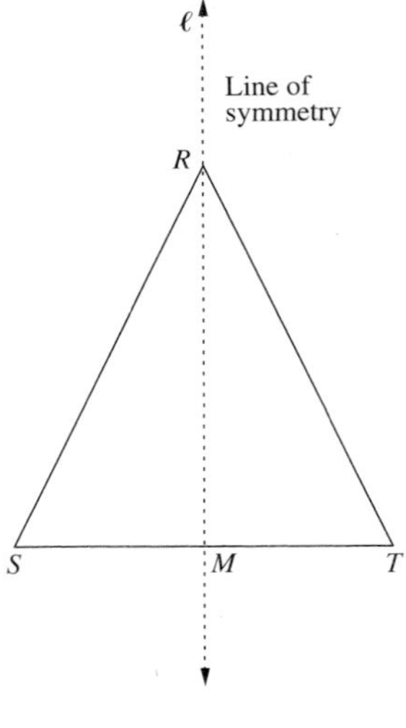

Figure 3.19

As in nature, symmetry is a feature that occurs repeatedly in geometry. Our first encounter with it involves the isosceles triangle. The line passing through the vertex of an isosceles triangle and the midpoint of the base essentially divides the triangle into two smaller triangles that are reflections of each other in that line (they are congruent). That line, ℓ in Figure 3.19, is perpendicular to the base, and bisects the vertex angle. In turn, if a line bisects the base and is perpendicular to it, then it passes through the vertex and coincides with that line of symmetry, ℓ. There are a number of good problems implied in these statements, which provide excellent practice in proof writing. They appear as the next three results. They will be left for you to prove (Problems 14, 15, and 16).

LEMMA B: If M is the midpoint of segment $\overline{AB}$ and line $\overleftrightarrow{PM}$ is perpendicular to $\overline{AB}$, then $PA = PB$. (See Figure 3.20.)

LEMMA C: If $PA = PB$ and M is the midpoint of segment $\overline{AB}$, then line $\overline{PM}$ is perpendicular to segment $\overline{AB}$. (See Figure 3.21.)

LEMMA D: If $PA = PB$ and M is the midpoint of segment $\overline{AB}$, then ray $\overrightarrow{PM}$ bisects $\angle APB$. (See Figure 3.21.)

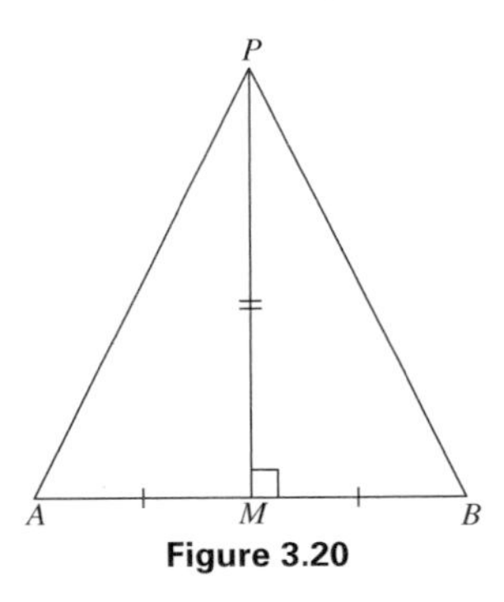

Figure 3.20

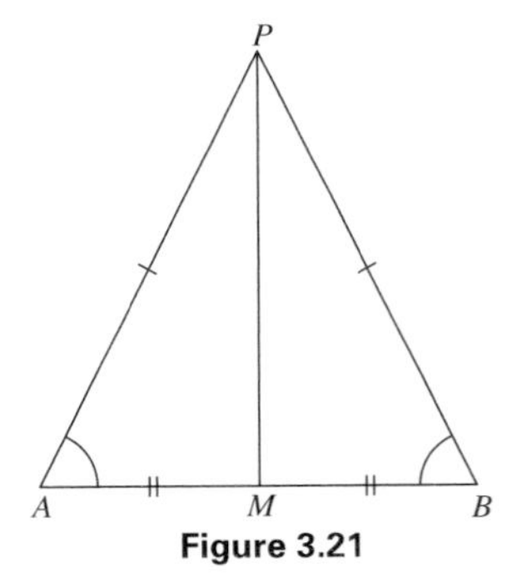

Figure 3.21

PERPENDICULAR BISECTORS, LOCUS

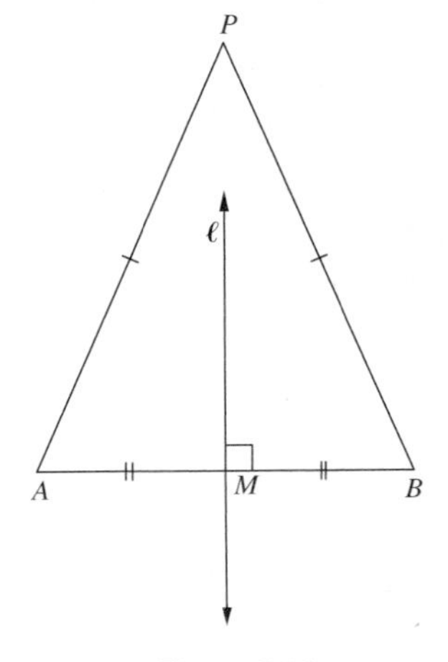

Figure 3.22

One of the first encounters with the notion of *locus* in geometry is in the next result we will prove. We define the **perpendicular bisector** of a segment $\overline{AB}$ to be the line that both bisects $\overline{AB}$ and is perpendicular to it.

> THEOREM 3: PERPENDICULAR BISECTOR THEOREM
> The set of all points equidistant from two distinct points A and B is the perpendicular bisector of the segment $\overline{AB}$. (See Figure 3.22.)

The traditional language sounds more dynamic: *The locus of a point that remains at equal distances from two fixed points is the perpendicular bisector of the segment joining those two fixed points.* Both theorems say essentially the same thing, but the latter suggests that a point somehow *moves* in a plane in a certain manner. Although the axioms do not address the motion of a point, it can be a way of *interpreting* a geometric fact, however, and can be helpful at times.

All locus problems state that a certain point describes, or *lies on*, a certain curve or line, that is, a *set of points*. Accordingly, two things must always be proved for such *locus problems*.

(1) That every point of the curve, or locus, must satisfy the given conditions.
(2) That every point that satisfies the given conditions must lie on the curve.

If we think of the curve or line as the set $\boldsymbol{C}$, and the points that satisfy the given properties as set $\boldsymbol{P}$, then the above two steps will prove

(1) $$\boldsymbol{C} \subseteq \boldsymbol{P}$$

and

(2) $$\boldsymbol{P} \subseteq \boldsymbol{C}$$

This is the standard way you prove that two sets $\boldsymbol{C}$ and $\boldsymbol{P}$ are equal in set theory.

PROOF OF THEOREM

Applying steps (1) and (2) to the above "locus" theorem, first assume that point P is any point of ℓ, the perpendicular bisector of segment $\overline{AB}$. By definition, ℓ passes through M, the midpoint of $\overline{AB}$. Then by Lemma B, $PA = PB$ and P is equidistant from A and B. To reverse, suppose that P is equidistant from A and B. By Lemma C, $\overleftrightarrow{PM} \perp \overline{AB}$. But line ℓ is also perpendicular to $\overline{AB}$ at M, and there can be only one perpendicular to $\overline{AB}$ at M. Therefore, $\overleftrightarrow{PM} = \ell$ and $P \in \ell$.

We now come to the familiar SSS theorem on congruence, whose proof follows from the Perpendicular Bisector Theorem. Instead of just presenting the proof for you to read, it might be more interesting to let you "discover" the proof on your own. Thus, a discovery unit on kites and darts was designed for just that purpose, appearing at the end of this section.

THEOREM 4: SSS THEOREM
If, under some correspondence between their vertices, two triangles have the three sides of one congruent to the corresponding three sides of the other, then the triangles are congruent under that correspondence.

EXAMPLE 5 Name all congruent pairs of distinct segments and angles in Figure 3.23, which follow logically from previous concepts, or from the congruence theorems thus far developed.

SOLUTION

(a) $\overline{PT} \cong \overline{TQ}$
$\angle P \cong \angle Q$
$\overline{PW} \cong \overline{WQ}$
$\angle PWT \cong \angle QWT$
$\angle PTW \cong \angle WTQ$

(b) $\angle U \cong \angle V$
$\overline{RW} \cong \overline{WS}$
$\angle WSR \cong \angle WRS$
$\angle SPU \cong \angle RPV$
(cannot conclude $\triangle SUR \cong \triangle RVS$ nor $\triangle UWR \cong \triangle VWS$) ■

(a)

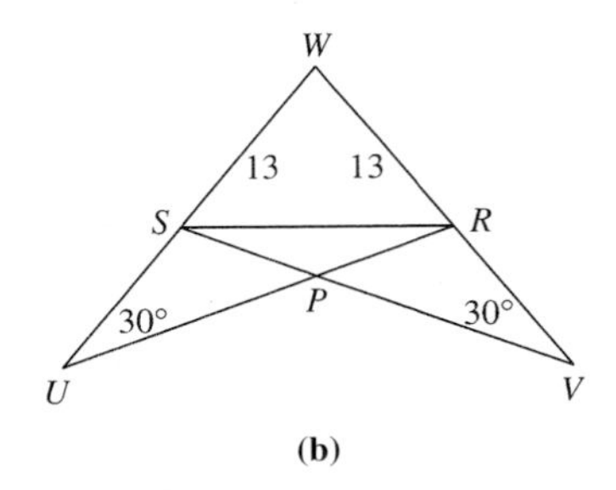

(b)

Figure 3.23

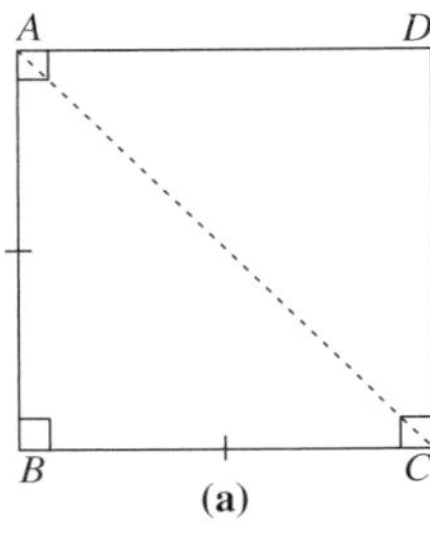

(a)

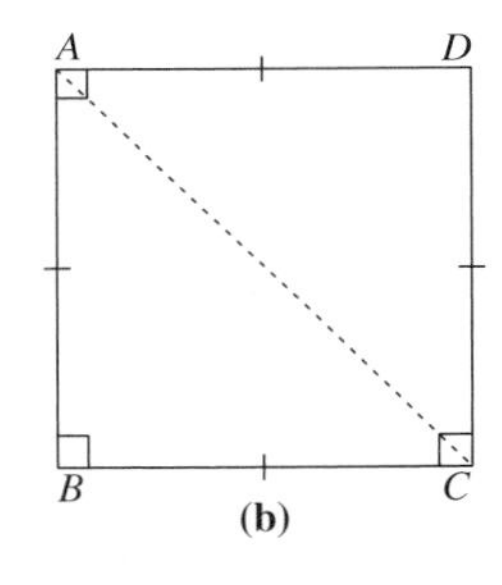

(b)

Figure 3.24

EXAMPLE 6 In Figure 3.24(a), we have a four-sided figure with the congruent sides and right angles as marked.

(a) Prove that $\overline{AD} \cong \overline{CD}$.

(b) If, in addition, $\overline{BC} \cong \overline{CD}$ prove $\angle D$ is a right angle. (Presumably, $ABCD$ is then a square.) We shall see later that it is impossible, with what we know, to prove that $\angle D$ is a right angle with just the assumption $\overline{AB} \cong \overline{BC}$.

SOLUTION

(a)

CONCLUSIONS	JUSTIFICATIONS
(1) Construct $\overline{AC}$.	Axiom I-1
(2) $\triangle ABC$ is isosceles, so $\angle BAC \cong \angle BCA$.	Definition, Isosceles Triangle Theorem
(3) $m\angle CAD = 90 - m\angle BAC$ $= 90 - m\angle BCA$ $= m\angle ACD$.	Betweenness considerations
(4) $\therefore \overline{AD} \cong \overline{DC}$.	Isosceles Triangle Theorem

(b)

CONCLUSIONS	JUSTIFICATIONS
(1) $\overline{AD} \cong \overline{DC}$.	Part (a) above
(2) $\overline{AB} \cong \overline{AD}$.	Transitive Law
(3) $\overline{AC} \cong \overline{AC}$.	Reflexive Law
(4) $\triangle ABC \cong \triangle ADC$.	SSS
(5) $m\angle D = m\angle B = 90$ $\therefore \angle D$ is a right angle. ■	CPCF, definition

The final result takes care of an axiomatic detail that is intuitively obvious, but requires proof, nevertheless. It is important since it is used so often.

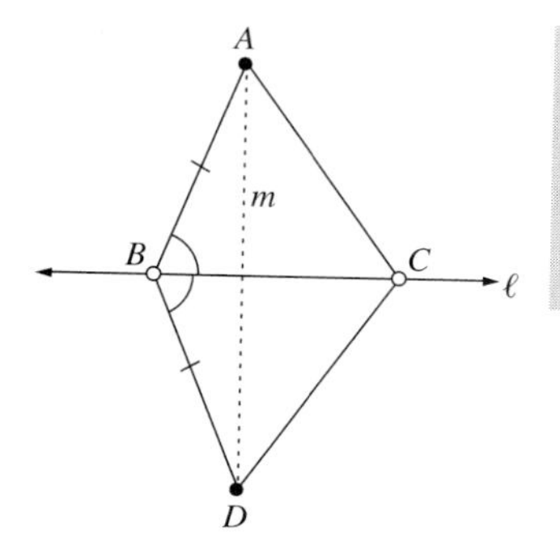

Figure 3.25

THEOREM 5: EXISTENCE OF PERPENDICULAR FROM AN EXTERNAL POINT
Let the line ℓ and point A not on ℓ be given. Then there exists a unique line m perpendicular to ℓ and passing through A.

PROOF

Given A not on line ℓ locate arbitrary points B and C on ℓ, as shown in Figure 3.25, and let $\angle DBC \cong \angle ABC$, $\overline{BD} \cong \overline{BA}$. It follows that B and C are both equidistant from A and D. (Why?) Hence, $\overline{AD} \perp \overline{BC}$ by the Perpendicular Bisector Theorem. (The uniqueness will be left as Problem 24.)

Moment for Discovery

SSS Theorem Via Kites and Darts

Two geometric figures, the *kite* and *dart*, though elementary, are quite useful. The figures we have in mind are shown in Figure 3.26, where it is assumed that $AB = AD$ and $BC = CD$. The dart is distinguished from the kite by virtue of the eight angles at A and C formed with the diagonals $\overline{AC}$ and $\overline{BD}$ being either all acute angles (for the kite), or two of them, such as the angles at C, being obtuse or right angles (for the dart). Because $AC = AC$, the two triangles on either side of line $\overline{AC}$ are congruent by SSS. However, this fact will follow from results preceding the SSS Theorem, thereby providing a way to *prove* the SSS criterion from kites and darts. Write out a proof, with the following steps to guide you.

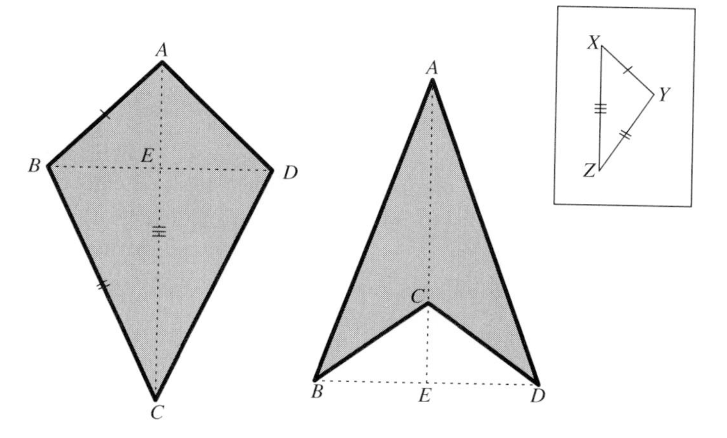

Figure 3.26

1. Given $\triangle ABC$ and $\triangle XYZ$ with congruent sides (under $ABC \leftrightarrow XYZ$), construct $\triangle ADC$ congruent to $\triangle XYZ$ on the opposite side of line $\overleftrightarrow{AC}$ as point B, forming a kite or dart. (Show that a kite or dart is actually formed.)
2. We now have A equidistant from B and D, and C equidistant from B and D. What can be said now? (See the theorem about equidistant loci.)
3. What theorem tells us that ray $\overrightarrow{AC}$ will bisect $\angle BAD$? Be sure to consider both the kite and dart in your argument, and be careful not to use SSS, which you are trying to prove. (Note that the case C–A–E not pictured is actually covered by this same argument, by interchanging A and C.)
4. Now finish the argument that $\triangle ABC \cong \triangle XYZ$. Try writing a polished, completely rigorous proof of SSS along these lines.

PROBLEMS (3.3)

GROUP A

1. Name all congruent pairs of distinct segments in each figure that follow logically either from previous concepts, or from the congruence theorems thus far developed.

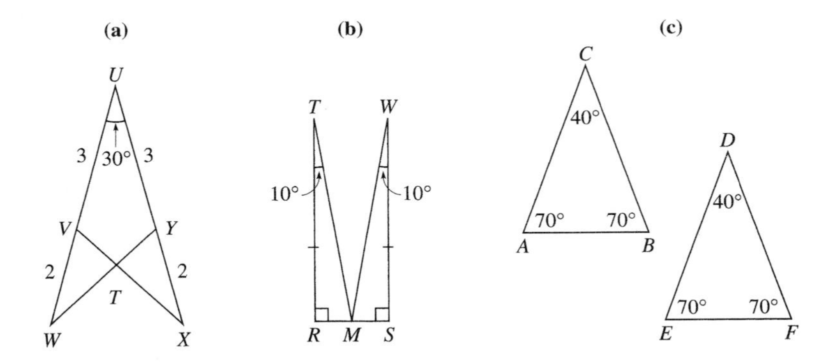

2. Name all congruent pairs of distinct angles in the figure of Problem 1 that follow logically either from previous concepts, or from the congruence theorems thus far developed.
3. Using the figure, find the missing lengths x and y, and state which congruence criterion you used.

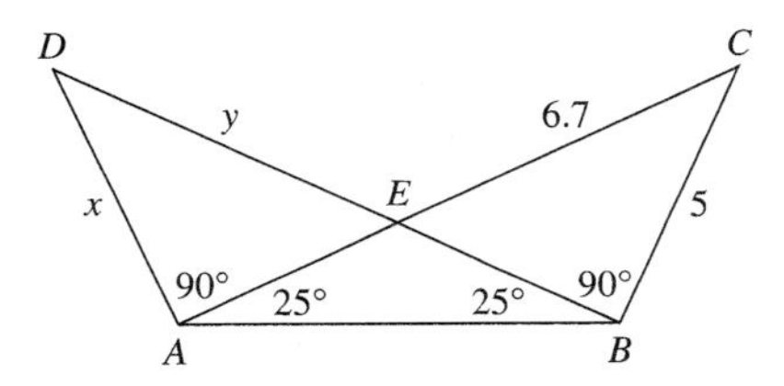

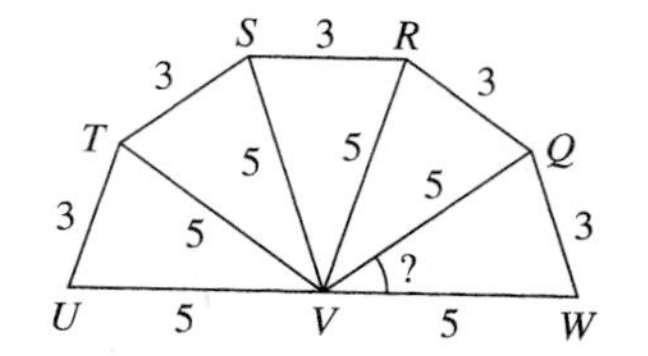

4. Find the measure of angle $\angle QVW$ and state what congruence criterion you used. (Assume that U–V–W, $\overrightarrow{VW}$–$\overrightarrow{VQ}$–$\overrightarrow{VR}$–$\overrightarrow{VS}$, and $\overrightarrow{VR}$–$\overrightarrow{VS}$–$\overrightarrow{VT}$.)

5. Find the missing angle measures x, y, z using the congruence criteria of this section. Explain your solution.

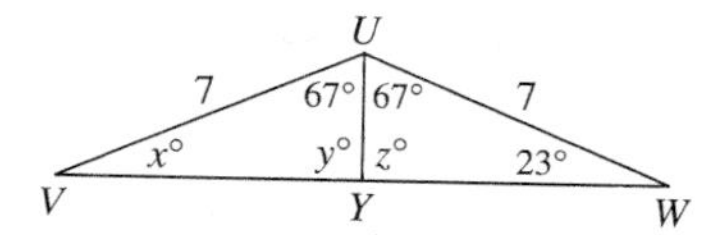

6. **(a)** List the eight triangles shown.

(b) From the marked information, there are three pairs of triangles that can be proved congruent. Name them, *with vertices in corresponding order.* [UCSMP, p. 321]

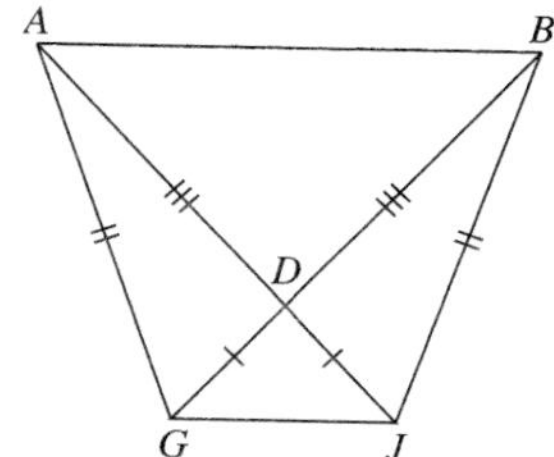

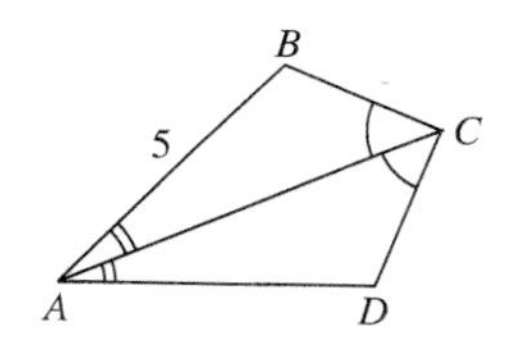

7. The following proof applies to the figure for this problem, with the information as indicated. Supply the missing justifications.

CONCLUSIONS	JUSTIFICATIONS
(1) $\overline{AC} \cong \overline{AC}$	**(a)** ____?
(2) $\triangle ABC \cong \triangle ADC$	**(b)** ____?
(3) $\overline{AD} \cong \overline{AB}$	**(c)** ____?
(4) $AD = 5$	**(d)** ____?

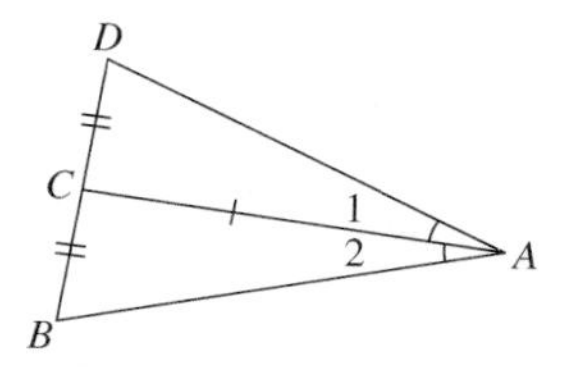

8. Explain why the SAS Postulate cannot be used to prove $\triangle ADC \cong \triangle ABC$. (Assume $\overline{BC} \cong \overline{CD}$ and $\angle 1 \cong \angle 2$.)

9. A support brace has the shape illustrated in the figure, with the approximate dimensions as indicated. What principle in geometry requires that line $\overleftrightarrow{ST}$ be perpendicular to line $\overleftrightarrow{QR}$?

S
10
10
9.3
T
9.3
Q
R

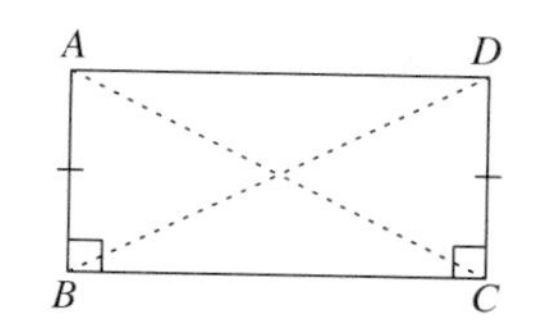

***10.** Suppose $\overline{AB} \cong \overline{CD}$, $\overline{AB} \perp \overline{BC}$, and $\overline{CD} \perp \overline{BC}$.

(a) Why must $\overline{AC} \cong \overline{BD}$? (What congruence criterion guarantees this?)

(b) Prove $\angle A \cong \angle D$.

*See Theorem 2, Section 3.7.

11. Prove that the standard compass, straight-edge construction for the bisector of an angle is valid, that it actually produces the bisector of a given angle.

12. Prove that the standard compass, straight-edge construction for the perpendicular to a line either from a point on the line or from an external point actually produces a line that is perpendicular to the given line.

GROUP B

13. **A problem on Locus** Here is a locus problem you can either attempt to prove theoretically, or try it out on *Sketchpad*. There is actually an item in the CONSTRUCT menu called Locus, and it can be used to solve many locus problems. Consider this: Start with any isosceles triangle, $\triangle ABC$, and points X and Y on its legs $\overline{AB}$ and $\overline{AC}$ such that $AX = AY$. Find the locus of P, the point of intersection of $\overline{XC}$ and $\overline{YB}$ as X varies on $\overline{AB}$. An outline of the steps needed in *Sketchpad* are as follows:

[1] (Turn Autoshow Label off.) Construct any segment $\overline{BC}$, locate its midpoint, and construct the perpendicular at that midpoint. Select a point A on the perpendicular, and construct segments $\overline{AB}$ and $\overline{AC}$. Hide the perpendicular and the midpoint. The vertex of the isosceles triangle created can be dragged, as well as points B and C, to create any shape for the isosceles triangle.

[2] Select any point X on $\overline{AB}$, and draw segment $\overline{CX}$.

[3] Construct the Circle By Center and Point having center A and passing through X. Select the point of intersection of this circle with $\overline{AC}$. This will be point Y in the locus problem mentioned.

[4] Obtain the point P of intersection of segments $\overline{XC}$ and $\overline{YB}$. To activate the Locus command properly, select point X, then point P (in that order), and choose Locus under CONSTRUCT.

Did anything predictable happen? Can the result be proven geometrically? What do you *think* the locus is? Is this feature invariant when you drag points A and B? For further investigation, try this with a nonisosceles triangle.

14. Prove Lemma B (Figure 3.20). (***Hint:*** Why is $\triangle PMA \cong \triangle PMB$?)

15. Prove Lemma C (Figure 3.21).

16. Prove Lemma D (Figure 3.21).

17. Use *Sketchpad* to construct a triangle $\triangle ABC$ having these measurements: $AB = 1.5$, $AC = 2$, and $m\angle ACB = 30$:

[1] Locate A and translate it 1.5 units horizontally to $A' = B$. Translate it 2 units horizontally to A''.

[2] Construct Circle With Center A and Radius $\overline{AA''}$. (Since $AB = 1.5$ and $AA'' = 2$, then any point C on the circle will have $AC = 2$, as desired, but $\angle ACB$ probably needs to be adjusted.)

[3] Locate point C on circle AA'' and construct segments $\overline{AC}$ and $\overline{BC}$, then display $m\angle ACB$ on the screen. By dragging C, you can make $m\angle ACB = 30$ (or nearly so). Accuracy to within 0.05 units will be acceptable.

(a) Did you find more than one triangle having the required measurements? If so, construct a second point D on the circle with $m\angle ADB = 30$. (There are *four* positions for point P on the circle for which $m\angle APB = 30$.)

(b) Try to find something significant about the two noncongruent triangles you should have obtained. (If you have trouble finding anything, try locating C and D on opposite sides of line $\overleftrightarrow{AB}$.)

(c) Invent a *Sketchpad* construction that will produce triangles having the *exact* measurements given.

18. If $WR = WU$ and $RS = ST = TU$, with R–S–T–U, prove that $\angle RWS \cong \angle TWU$.

***19.** A common misconception related to Problem 18 is assuming that joining the vertex of an isosceles triangle with the trisection points on the base will trisect the vertex angle, thereby providing a compass/straight-edge construction of the trisectors of any angle. Create a convincing argument to dispel this notion. A simple example would be good. (Suggestion: If the construction were valid, then a right angle could be trisected this way. What are the angle measures involved, if such is the case? You can use your knowledge of Euclidean geometry here.)

*See Problem 13, Section 3.5.

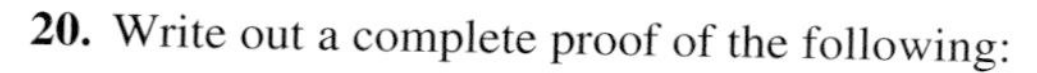

20. Write out a complete proof of the following:

> THEOREM: The perpendicular bisectors of the sides of a triangle, if two of them meet, are concurrent at a point that is the center of a circle **(circumcircle)** passing through the vertices. (The point of concurrency is called the **circumcenter** of the triangle.)

NOTE: The conditional is needed here because in absolute geometry it is possible for two such perpendicular bisectors to be nonintersecting lines.

21. (a) Prove that an isosceles triangle exists with a given segment as base.

(b) Prove that equilateral triangles exist. (***Hint:*** Do not start with a given base.)

NOTE: Later we develop the tools necessary for proving that an equilateral triangle exists with a given base—Euclid's first proposition in Volume I of the *Elements*.

22. In the accompanying figure, $\triangle ABC$ is given with $\overline{AB} \cong \overline{AC}$. Can you construct a proof based on the diagram shown that $\angle ABC \cong \angle ACB$? (This was Euclid's *pons asinorum.*)

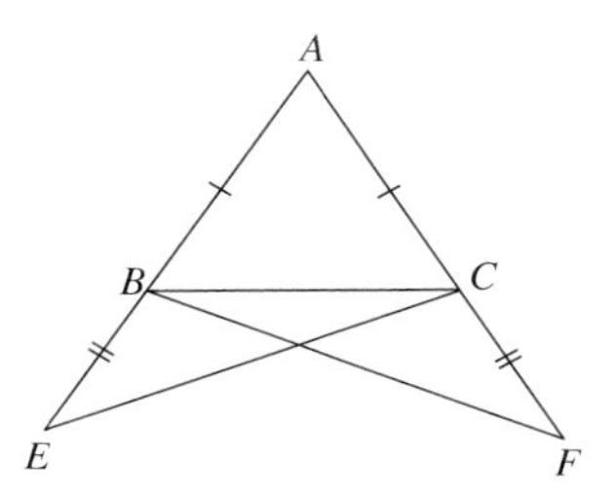

GROUP C

23. Establish all betweenness relations needed to write a rigorous proof for the following geometry problem. (Recall the definition of vertical angles and how it involves betweenness relations.)

Given: $DF = FE$, $BF = FC$, A–D–B, and C–F–D.

Prove: $AD = AE$.

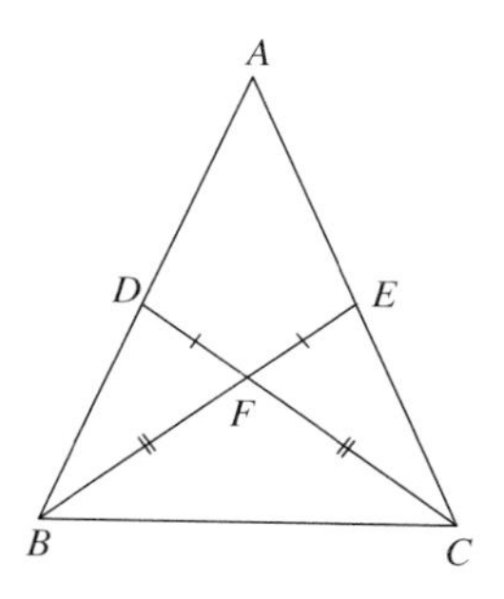

(***Hint:*** To prove B–F–E, draw ray $\overrightarrow{CF}$ and show that ray $\overrightarrow{CF}$ passes through an interior point of $\angle ACB$, then use the Crossbar Theorem. A similar trick works for the proof that A–E–C.)

24. Prove that there can be only one perpendicular from a given point to a given line. (***Hint:*** Can a triangle $\triangle ABC$ have two right angles, say at B and C? Locate D on the opposite side of line $\overleftrightarrow{BC}$ as A so that $BA = BD$ and $m\angle ABC = m\angle CBD$. What can be proven from this figure to get a contradiction?)

25. To prove that $AB + BC > AC$ (the Triangle Inequality for three noncollinear points), assume that (1) $AB + BC = AC$ or (2) $AB + BC < AC$, and obtain a contradiction. Case (1) is shown in the figure. Obtain a contradiction for this case first, then attempt the second case. [For Case (2) you will need the fact that the base angles of an isosceles triangle are acute, which can be proven with what we already know; see Problem 24.]

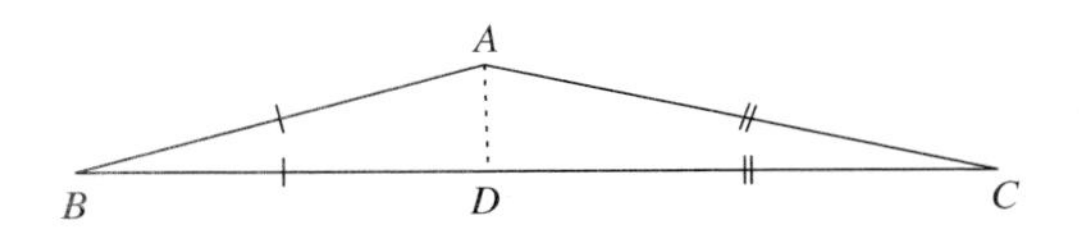

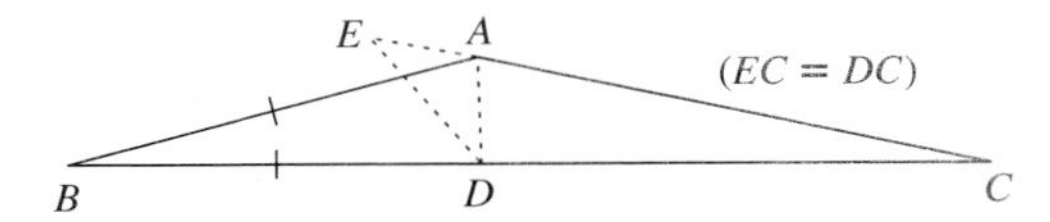

3.4 Exterior Angle Inequality

EXTERIOR ANGLE OF A TRIANGLE

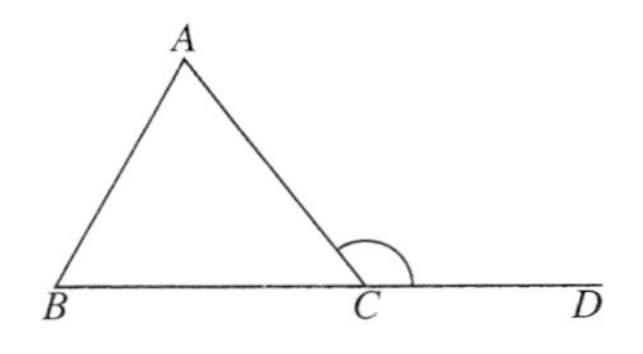

Figure 3.27

A result that plays a major role in the development of absolute geometry is the Exterior Angle Theorem. Our proof is virtually the same as that found in the *Elements*, with betweenness arguments added to make it complete.

DEFINITION: Let $\triangle ABC$ be given, and suppose D is a point on $\overleftrightarrow{BC}$ such that the betweenness relation B–C–D holds (Figure 3.27). Then $\angle ACD$ is called an **exterior angle** of the given triangle. The angles at A and B of $\triangle ABC$ are called **opposite interior angles** of $\angle ACD$.

OUR GEOMETRIC WORLD

When sailors use *Polaris* (the North Star) for sighting, it is routinely assumed that their line of sight is parallel to the **polar axis** (the line through the north and south poles, which points directly to *Polaris*). This assumption implies that the angle of elevation gives their precise latitude θ (since $\theta = 90 - \phi_1 = 90 - \phi_2$). Although the error is small enough to justify the procedure, these lines are, in fact, not parallel. The inequality we study in this section shows that $\phi_2 > \phi_1$. (See inset to Figure 3.28.)

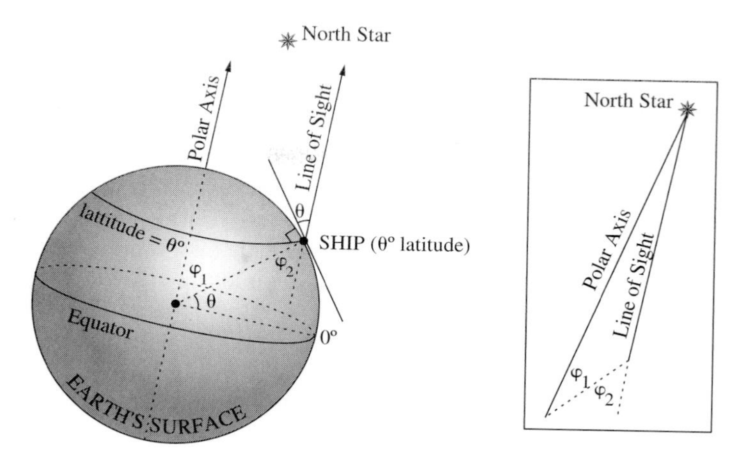

Figure 3.28

You may remember that in Euclidean geometry the measure of an exterior angle of any triangle in Euclidean geometry equals the sum of the measures of the other two opposite interior angles. In absolute geometry, this relationship is not valid, in general. In 1882 the French mathematician Henri Poincaré discovered a model for geometry that satisfies the axioms we have thus far introduced, yet violates many of the geometric concepts we take for granted and use frequently. Let us explore this model briefly. More details will be presented in Chapter 6.

EXTERIOR ANGLES IN THE POINCARÉ MODEL

Let $\boldsymbol{C}$ be a circle with center O (Figure 3.29), and consider all points inside. These points will become the "points" of our geometry (a point on or outside the circle is

POINCARÉ MODEL FOR ABSOLUTE GEOMETRY

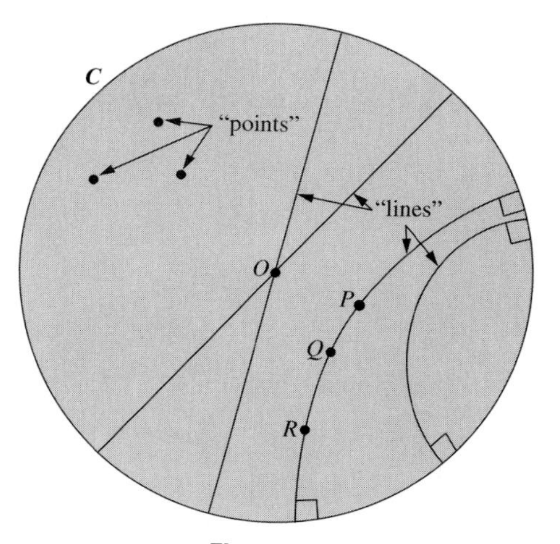

Figure 3.29

not a "point"). To construct a "line," draw either an ordinary straight line through point O, cut off by $\boldsymbol{C}$, or construct a *circle that makes right angles with* $\boldsymbol{C}$, and take the arc of that circle that lies inside $\boldsymbol{C}$. We need not concern ourselves just yet with precisely *how* to draw circles perpendicular **(orthogonal)** to $\boldsymbol{C}$.

As for betweenness and the definitions for segments, rays, and angles, we merely use the betweenness that already exists in the plane: If on an arc of a circle point Q is between points P and R (as shown in Figure 3.29), then we define P–Q–R in the new geometry. Thus, an example of a "segment" would be the set of points shown in Figure 3.30 (at left), "segment $\overline{AB}$."

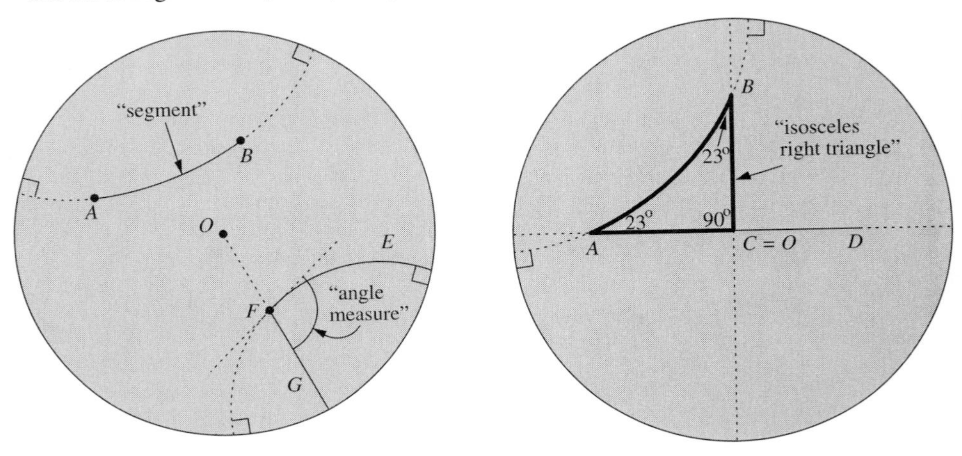

Figure 3.30

For angle measure, we take that used in calculus: the angle between two curves (defined as the angle measure of the angle formed by the tangents to the two curves at the point of intersection). We ask you to be patient since we will not show you, just yet, what the concept of distance will be—that is much more complicated and will be taken up in Chapter 6.

Now it is most intriguing that this geometry inside $\boldsymbol{C}$ *satisfies all the axioms for absolute geometry* in a plane. (You might think about some of these axioms now, such as *Two "points" determine a unique "line"*; can you see how this principle would work here?) Furthermore, if we consider one of the "triangles" of this geometry, say $\triangle ABC$ as shown in Figure 3.30, we find what could be a disturbing fact. Carefully measuring the angles A and B with a protractor, we have

$$m\angle A = m\angle B \approx 23 \qquad \text{and} \qquad m\angle C = 90$$

Thus,

$$m\angle A + m\angle B + m\angle C = 23 + 23 + 90 = 136$$

Also, from this we find that

$$m\angle A + m\angle B = 46 \neq m\angle BCD\ (= 90)$$

This geometry has the characteristic property that the **angle sum** of a triangle (that is, the sum of the measures of its angles) is always *less than* 180.

EXTERNAL ANGLES IN SPHERICAL GEOMETRY

It is interesting to also consider a geometry where the angle sum of a triangle is always *greater than* 180. Spherical geometry has this property. Here, we take the unit sphere ***S*** (a sphere of radius one in three-dimensional space) and consider as "points" all the points on the surface of the sphere (the *shell* only), and as "lines" all great circles of that sphere (Figure 3.31). It is often helpful to use familiar geographical terms on the earth's surface for ***S***. For example, the "equator" is that great circle on ***S*** that lies in a horizontal plane. Thus the meridian lines (great circles passing through the north and south poles) are "lines," as is the equator. For the concept of distance, use ordinary (Euclidean) arc length (thus a great circle on ***S*** has length 2π), and for angle measure, use the measure of the angle between the tangents to the sides at the vertex of the angle.

The sphere is definitely not a model for absolute geometry because it has a bounded metric. However, if we restrict ourselves to an open hemisphere, the axioms of absolute geometry work, including the principle *two points determine a line.* A "triangle" on the sphere, that is, a *spherical triangle*, is simply one made up of arcs of great circles, pairwise connected at the endpoints. Even the SAS Postulate holds for spherical geometry, as it does for the Poincaré model.

SPHERICAL GEOMETRY

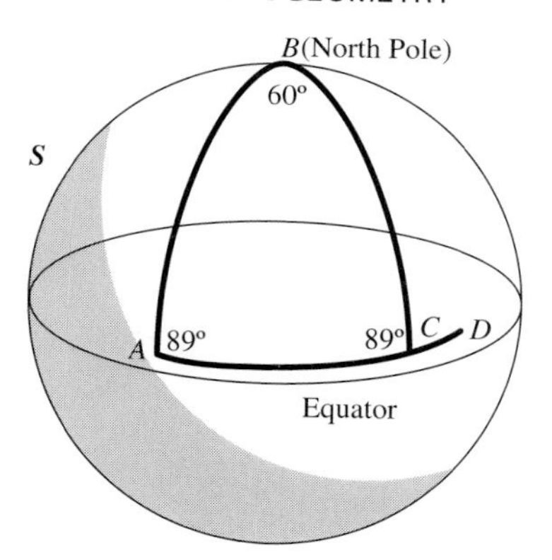

Figure 3.31

But there is one major exception where a Euclidean principle is violated. Observe Figure 3.31, in which an isosceles triangle $\triangle ABC$ has been constructed in the Northern Hemisphere. Here, the angle-sum of $\triangle ABC$ has the value

$$m\angle A \neq m\angle B + m\angle C = 2 \cdot 89 + 60 = 178 + 60 = 238$$

Moreover,

$$m\angle A + m\angle B = 89 + 60 = 149 \neq m\angle BCD.$$

The following theorem is the best we can do without introducing properties of parallelism into our system of axioms. Although it is a very weak form of its Euclidean counterpart, it is extremely useful in establishing all kinds of important inequalities in absolute geometry, as we shall see.

EXTERIOR ANGLE OF A TRIANGLE IN ABSOLUTE GEOMETRY

THEOREM 1: EXTERIOR ANGLE INEQUALITY

An exterior angle of a triangle has angle measure greater than that of either opposite interior angle.

PROOF

Given: Exterior $\angle ACD$ of $\triangle ABC$, as in Figure 3.32.

Prove: **(a)** $m\angle ACD > m\angle A$.
(b) $m\angle ACD > m\angle B$.

(a)

CONCLUSIONS	JUSTIFICATIONS
(1) Locate M, the midpoint of $\overline{AC}$.	Existence of midpoint
(2) Extend $\overline{BM}$ its own length to point E such that M is the midpoint of $\overline{BE}$.	Segment Construction Theorem
(3) A–M–C and $AM = MC$; B–M–E and $BM = ME$.	Definition of midpoint
(4) $\angle AMB$ and $\angle CME$ form a vertical pair.	Definition
(5) $\angle AMB \cong \angle CME$	Vertical Pair Theorem
(6) $\triangle AMB \cong \triangle CME$	SAS
(7) $m\angle ACE \equiv m\angle MCE = m\angle MAB \equiv m\angle A$	CPCF
(8) A, M, and E are on the same side of $\overleftrightarrow{BC} \equiv \overleftrightarrow{CD}$, hence $E \in H(A, \overleftrightarrow{CD})$.	Theorem 1, Section 2.6
(9) Similarly, $E \in H(D, \overleftrightarrow{CA})$ and hence $E \in$ Interior $\angle ACD$.	Definition of the interior of an angle

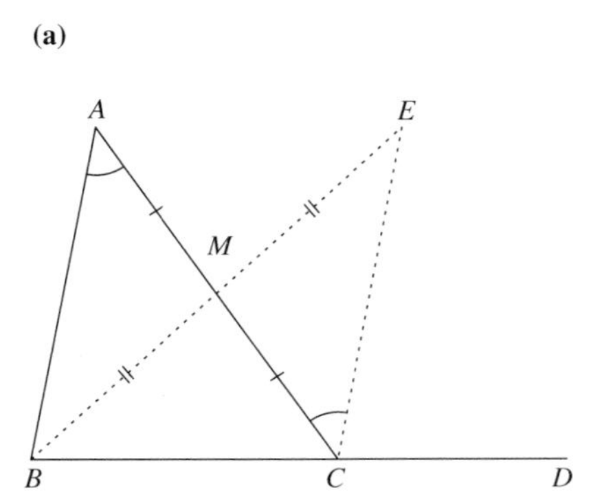

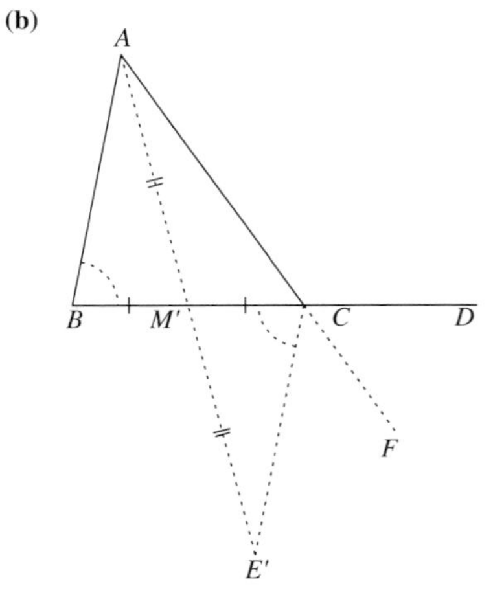

Figure 3.32

(10) $m\angle ACD = m\angle ACE + m\angle ECD$ $> m\angle ACE$	Angle Addition Postulate, algebra
(11) $\therefore m\angle ACD > m\angle A$	Steps 7 and 10

(b)
For the second part, merely locate F so that $A\text{–}C\text{–}F$; then $m\angle BCF = m\angle ACD$ by the Vertical Pair Theorem, and from the first part, $m\angle BCF > m\angle B$. By substitution, $m\angle ACD > m\angle B$.

APPLICATIONS

The Exterior Angle Inequality has far-reaching effects in geometry. For example, it puts a limit on the number of obtuse or right angles that a triangle can have. A short list of the simpler consequences follows. (Figure 3.33 indicates how the proof of the first can be derived; the others are corollaries.)

- The sum of the measures of any two angles of a triangle is less than 180.
- A triangle can have at most one right or obtuse angle.
- The base angles of an isosceles triangle are acute.

METHOD FOR PROVING $m\angle A + m\angle B < 180$

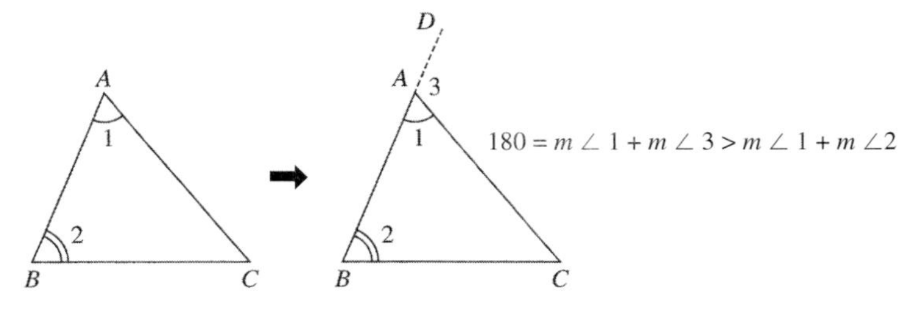

Figure 3.33

EXAMPLE 1 Consider the triangle shown in Figure 3.34, with certain angle measures indicated.

(a) Use the Exterior Angle Inequality to show that $\angle C$ has measure less than 131.

(b) In this example, is the angle sum of $\triangle ABC$ equal to, or less than 180? (Do not use Euclidean geometry.)

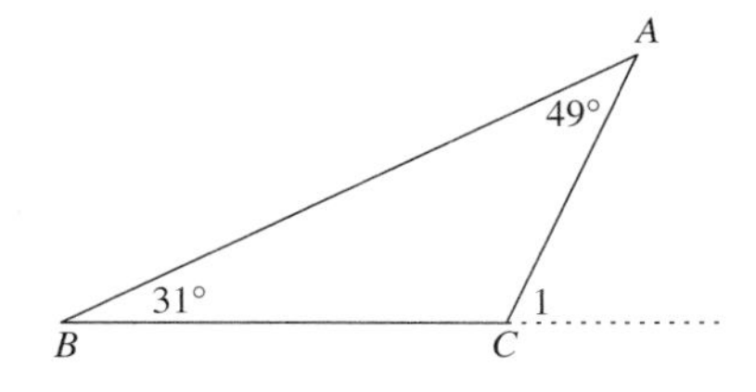

Figure 3.34

SOLUTION

(a) If we extend side $\overline{BC}$ to form an exterior angle, we obtain

$$m\angle 1 > 31 \qquad \text{and} \qquad m\angle 1 > 49$$

It gives us a stronger inequality if we choose $m\angle 1 > 49$. So this means that the supplement of $\angle 1$ has measure *less than* $180 - 49 = 131$. Hence, $m\angle C < 131$.

(b) No, because if we use only what we have derived in geometry so far, all we can say is

$$m\angle A + m\angle B + m\angle C < 31 + 49 + 131 = 211. \quad \blacksquare$$

EXAMPLE 2 If $\angle ECD$ is an exterior angle of $\triangle EAC$ (Figure 3.35) and A–B–C–D holds, use the Exterior Angle Inequality to find upper and lower bounds for

(a) $m\angle EBC = x$
(b) $m\angle BEC = y$.

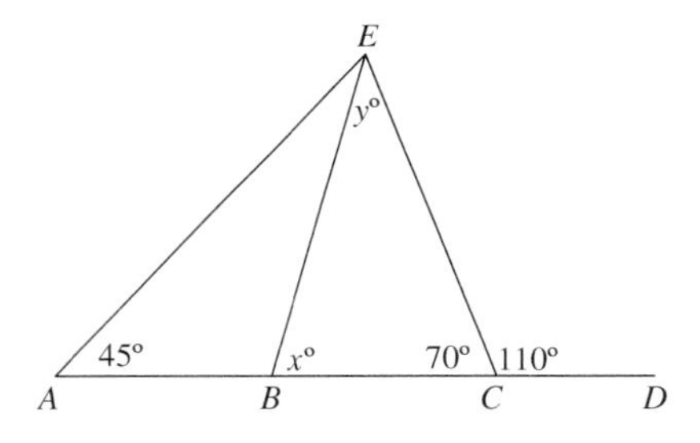

Figure 3.35

SOLUTION

(a) By the Exterior Angle Inequality applied to $\triangle ABE$, then to $\triangle EBC$, we obtain

$$x > 45 \qquad \text{and} \qquad 110 > x$$

That is,

$$45 < x < 110$$

(b) In $\triangle EBC$ we have $y < 110$. On the other hand, y is always positive. Therefore,

$$0 < y < 110. \quad \blacksquare$$

A major geometric result will now be considered. It is the famous Saccheri–Legendre Theorem: *Every triangle in absolute geometry has angle sum* $\leq$ *180.* This can be proven with what is known up to this point—no additional assumptions are needed. The same construction used to prove the Exterior Angle

Inequality (Figure 3.32) is used over and over again to prove the Saccheri–Legendre Theorem. A special discovery unit at the end of this section has been designed to allow you to discover for yourself a key step in that proof. This ingenious discovery was first made by Saccheri, an Italian geometer we mention later in more detail, and improved on by Legendre.

Let $\triangle ABC$ be any given triangle. As in the proof of the Exterior Angle Inequality, locate the midpoint M of $\overline{AC}$ then extend $\overline{BM}$ its own length to point E such that B–M–E (see Figure 3.36). Repeat this construction in $\triangle BEC$: Locate the midpoint N of $\overline{CE}$ and extend $\overline{BN}$ its own length to point F such that B–N–F. Continue this process ad infinitum. Here is what happens. First, the angle sums of all the new triangles we constructed in the process ($\triangle BCE, \triangle BCF, \triangle BCG, \ldots$) remain constant, as you will see in the discovery unit. Now let us suppose that, contrary to what we are trying to prove, the angle-sum of $\triangle ABC$ is *greater than* 180. Hence there is some positive constant t such that

$$m\angle A + m\angle ABC + m\angle BCA = 180 + t$$

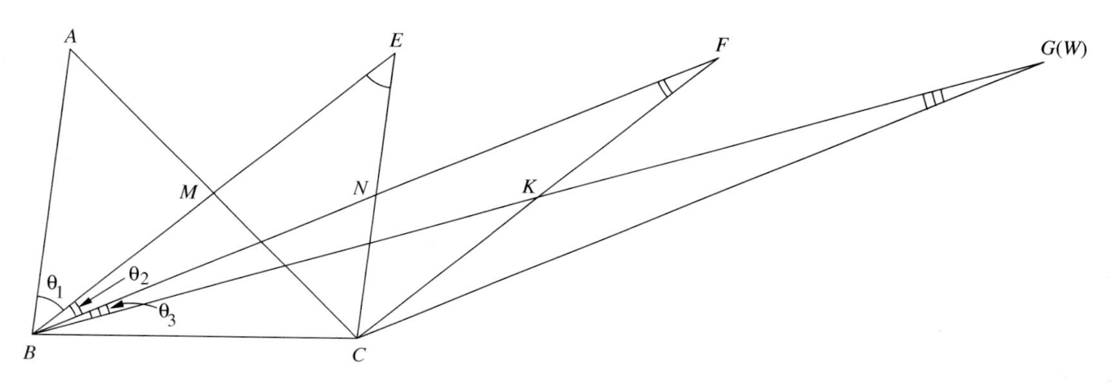

Figure 3.36

Then all the other triangles we constructed will also have angle sum $= 180 + t$. Next, we establish that the angles at $E, F, G, \ldots$ are decreasing in size and that they eventually have measure $< t$. To obtain an actual proof of this, observe the angles marked $\theta_1, \theta_2, \theta_3, \ldots$ in Figure 3.36. From the proof of the Exterior Angle Inequality, it can be recalled that $\triangle AMB \cong \triangle CME$ so that $m\angle E = \theta_1, \triangle ENB \cong \triangle CNF$ so that $m\angle F = \theta_2$, $\triangle FKB \cong \triangle CKG$ so that $m\angle G = \theta_3$, and so on. But also observe that by betweenness considerations,

$$\theta_1 < m\angle ABC, \qquad \theta_1 + \theta_2 < m\angle ABC, \qquad \theta_1 + \theta_2 + \theta_3 < m\angle ABC.$$

In fact, for any n,

$$\theta_1 + \theta_2 + \theta_3 + \theta_4 + \cdots + \theta_n < m\angle ABC.$$

This means ultimately that for at least one n taken large enough, as will be shown,

$$\theta_n < t$$

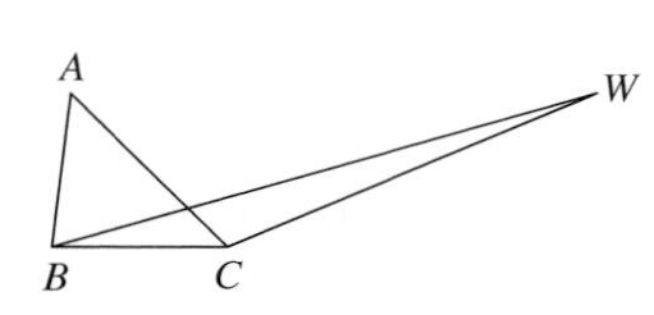

If this were **not** true, then $\theta_n \geq t$ would hold for every value of n and it would follow

$$\theta_1 \geq t, \qquad \theta_1 + \theta_2 \geq t + t = 2t, \qquad \theta_1 + \theta_2 + \theta_3 \geq 3t, \qquad \cdots$$

That is, in general, $m\angle ABC > \theta_1 + \theta_2 + \theta_3 + \theta_4 + \cdots + \theta_n \geq nt$. But since the number nt grows without bound as n gets larger and larger, $m\angle ABC > 180$, which is impossible. This means that $\theta_n < t$ for some n, and one of the angles at $E, F, G, \ldots$ has measure less than t. Suppose, for convenience, that the triangle for which this occurs is labeled BCW and that $m\angle W < t$. But the angle sum of $\triangle BCW = 180 + t$, so it follows that

$$180 + t = m\angle WBC + m\angle BCW + m\angle W < m\angle WBC + m\angle BCW + t$$

or, canceling t from both sides,

$$180 < m\angle WBC + m\angle BCW,$$

a contradiction, since the sum of the measures of two angles of a triangle is *always less than 180*. Hence, the following theorem has been proven:

> THEOREM 2: SACCHERI–LEGENDRE THEOREM
> The angle sum of any triangle cannot exceed 180.

Note that we have not yet proven that the angle sum of a triangle *equals* 180; that will come later. We are showing you a bit of geometric history, as well as how much can be done, and how close we seem to be getting to Euclidean geometry without actually introducing the powerful Parallel Postulate. If we could only somehow at the same time prove that the angle-sum of a triangle ≥ 180, then equality would follow from Theorem 2 and we could do away with the Parallel Postulate altogether.

Moment for Discovery

Central Idea in Proof of Saccheri–Legendre Theorem

Suppose the angles of $\triangle ABC$ are as marked in Figure 3.37: $m\angle A = 56$, $m\angle ABC = 50$, and $m\angle C = 70$. Points M and E are constructed so that M is the midpoint of both $\overline{AC}$ and $\overline{BE}$. Thus, $\triangle AMB \cong \triangle CME$, and $m\angle MCE = m\angle A$, $m\angle E = m\angle ABM$.

1. Calculate the angle sum for $\triangle ABC$: $m\angle A + m\angle B + m\angle C = ?$ (Note that this answer is not 180, which is quite possible in non-Euclidean geometry.)
2. Suppose that $m\angle EBC = 26$. Find the measures of the angles at C and E. (Did you get $m\angle BCE = 126$?)
3. Record the following angle measures thus far determined:

$$m\angle EBC, \qquad m\angle BCE, \qquad m\angle BEC$$

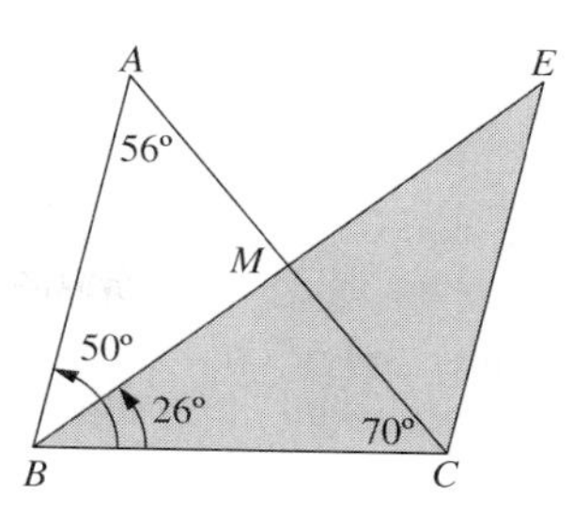

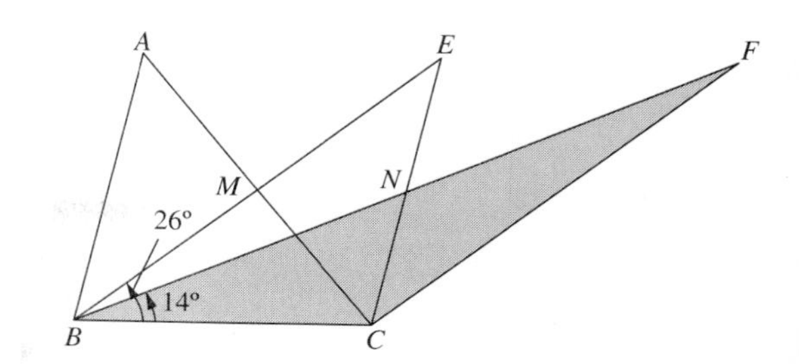

Figure 3.37

4. Record the angle sums of:
   (a) $\triangle ABC$
   (b) $\triangle BCE$
   Did anything happen?
5. In the second diagram of Figure 3.37, $\triangle BCF$ is constructed from $\triangle BCE$ in the same way that $\triangle BCE$ was constructed from $\triangle ABC$ (N is the midpoint of $\overline{CE}$ and $\overline{BF}$). Suppose $m\angle FBC = 14$, for example. Calculate

$$m\angle FBC, \qquad m\angle BCF, \qquad m\angle BFC$$

6. Find the angle sum of $\triangle BCE$. Did you notice anything?
7. Try to write a proof, in general, that under this type of geometric construction the angle sum of $\triangle BCE$ equals that of $\triangle ABC$, as are those of $\triangle BCE$ and $\triangle BCF$. This would prove that the phenomenon persists in all the constructed triangles, as needed in the proof of the Saccheri–Legendre Theorem.

Moment for Discovery

Using *Sketchpad* for Calculations in the Poincaré Model

It is interesting that we can calculate the angle sum for many triangles at once in the Poincaré Model.

1. Construct the circle **C**, with center O, with Autoshow Label off.
2. Draw two radii to any two points C and D on circle **C**. Display the angle measure between these two radii on the screen (call this quantity x).
3. Draw another radius to some point E on circle **C**, then construct a line perpendicular to this radius at E; select any point F on this line. Hide radius $\overline{OE}$.
4. Construct the Circle By Center and Point having center F and passing through E. This circle will be orthogonal to **C** and can be dragged or adjusted to any position in the figure. Cause this circle to intersect the previous radii $\overline{OC}$ and $\overline{OD}$ at, say, points G and H.
5. Construct the segments $\overline{FG}$ and $\overline{FH}$, and lines perpendicular to them at G and H. Select points G' and H' on these lines interior to the "triangle" in the

model ($\triangle OGH$) and display the angle measures $m\angle G'GO$ and $m\angle H'HO$ on the screen. Call these values y and z.

6. Display the sum $x + y + z$ on the screen. This is the angle sum of $\triangle OGH$ in the Poincaré Model. See what you can discover about this angle sum as you drag various points in the figure and change the triangle. What is the largest you can make this sum, and when does this happen? What is the smallest you can make it and when does this happen?

PROBLEMS (3.4)

GROUP A

1. Three triangles are shown in the figure with certain angle measures indicated. Which of them cannot exist in absolute geometry?

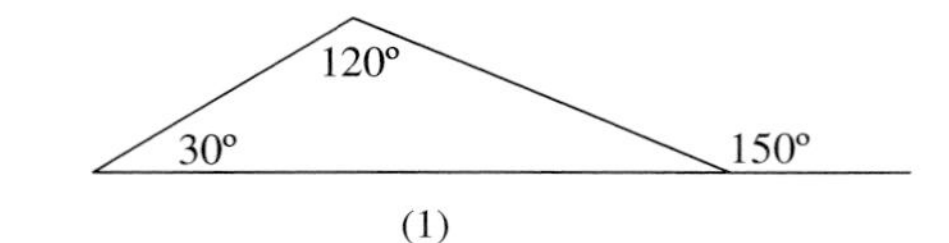

(1)

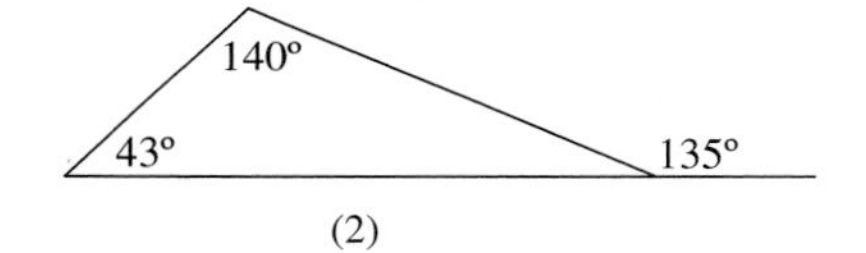

(2)

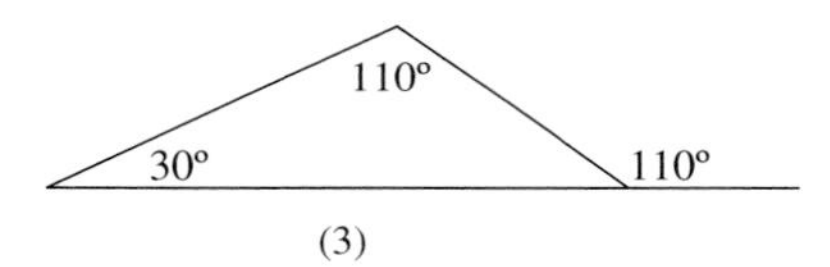

(3)

2. As in Problem 1, which of the triangles in the figure cannot exist in absolute geometry?

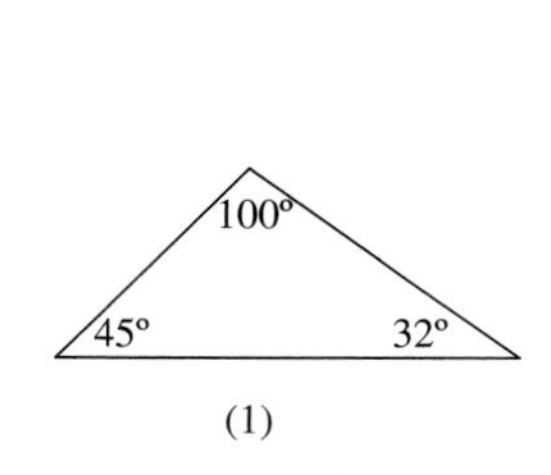

(1)

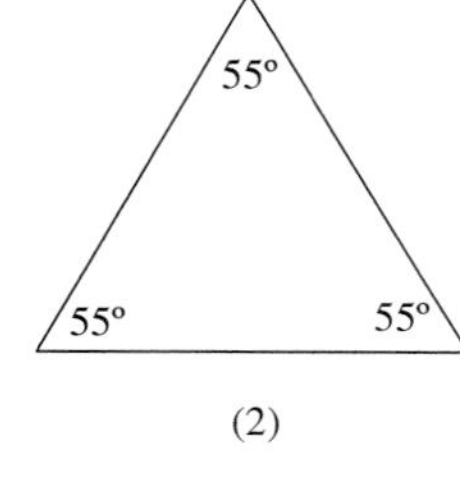

(2)

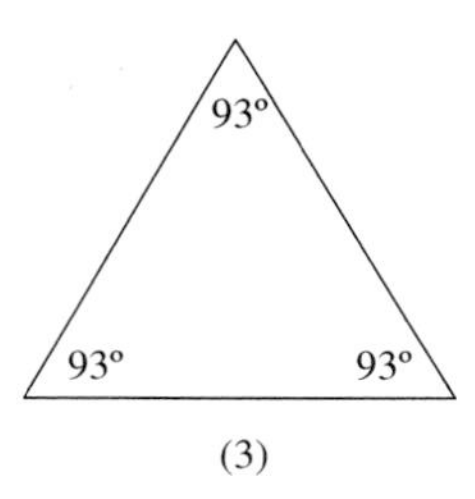

(3)

3. As in Problem 1, which of the indicated isosceles triangles cannot exist in absolute geometry?

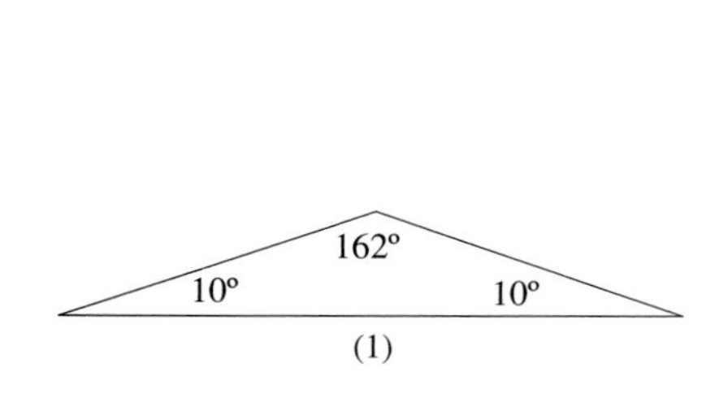

(1)

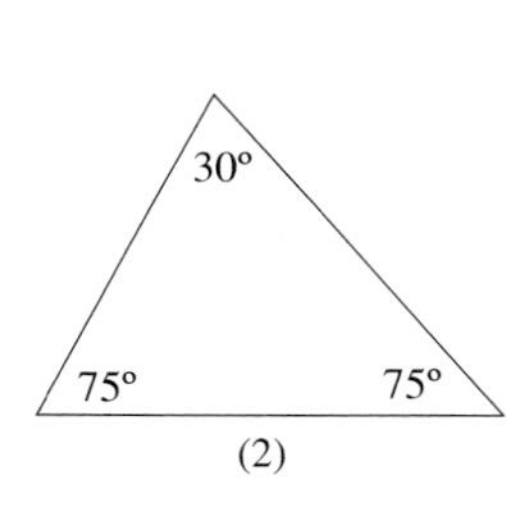

(2)

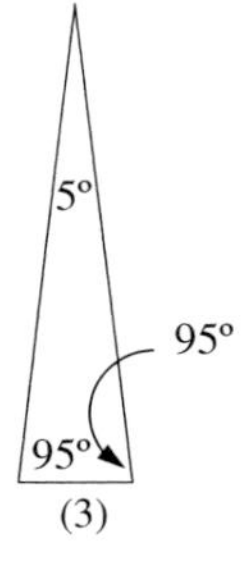

(3)

4. Use the Exterior Angle Inequality to find an upper bound for the measure of $\angle Q$ in $\triangle PQR$.

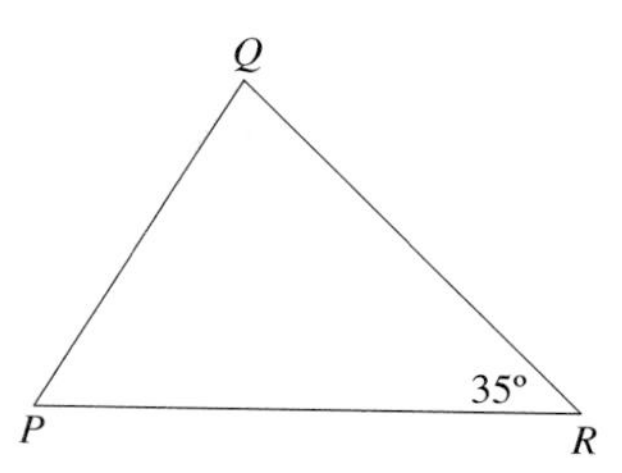

5. Use the Exterior Angle Inequality to find an upper bound for the measure of $\angle K$ in $\triangle KLW$.

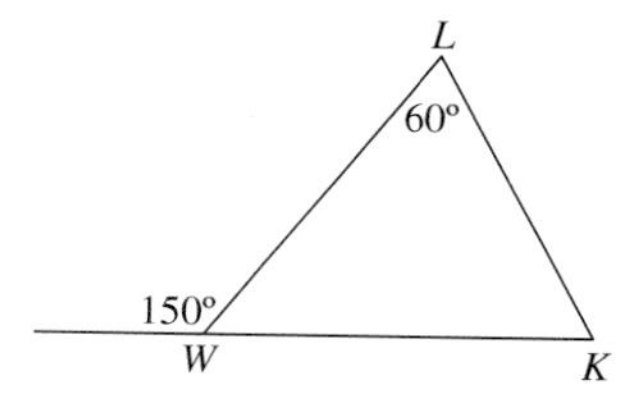

6. Use the Exterior Angle Inequality to find an upper and lower bound for $m\angle PSR = x$.

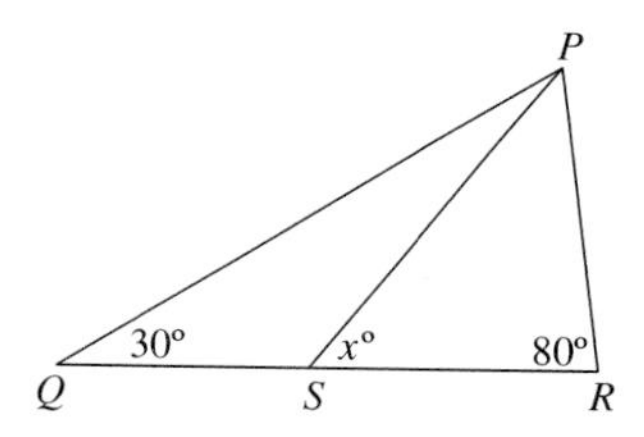

7. Use the Exterior Angle Inequality to find an upper and lower bound for $\angle MLK = y$.

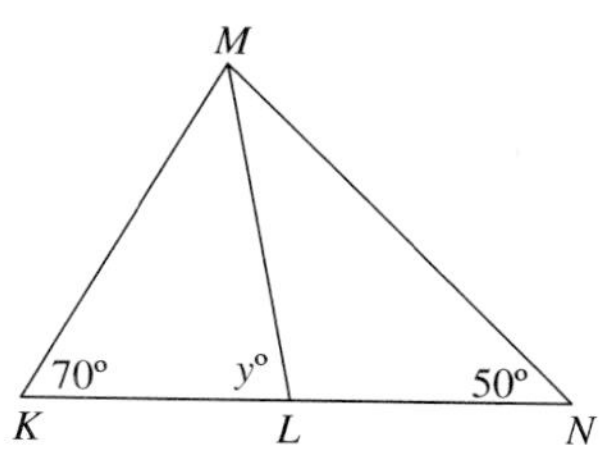

8. Given that $\triangle UVW$ is isosceles, with base angles at U and V each of measure 56. What is the (least) upper bound for the measure of the vertex angle at W?

9. Use the Exterior Angle Inequality to prove that any right triangle $\triangle ABC$ has only one right angle, and the other two angles must both be acute angles.

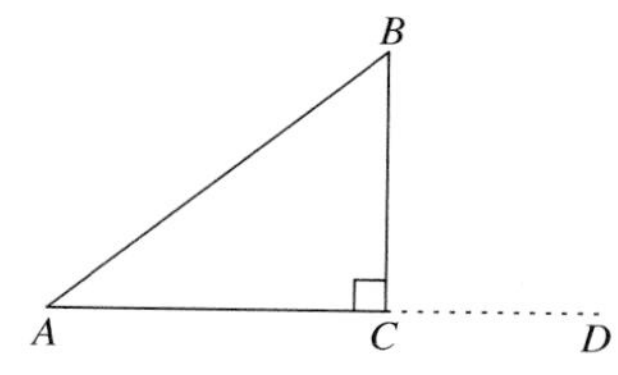

10. Prove that the base angles of any isosceles triangle are acute. (See figure.)

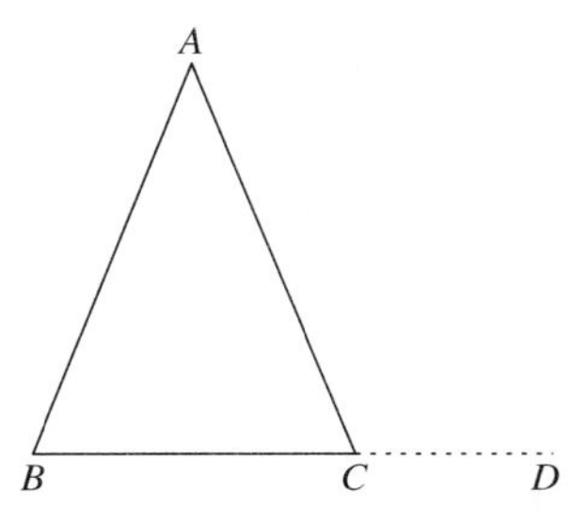

11. The measure of an exterior angle of a triangle is 160. Show that the other five exterior angles must each have measure ≥ 20.

12. In $\triangle MNT$, $MN = MT$ and M–T–K, with certain angle measures as indicated. Show that $m\angle K < 30$.

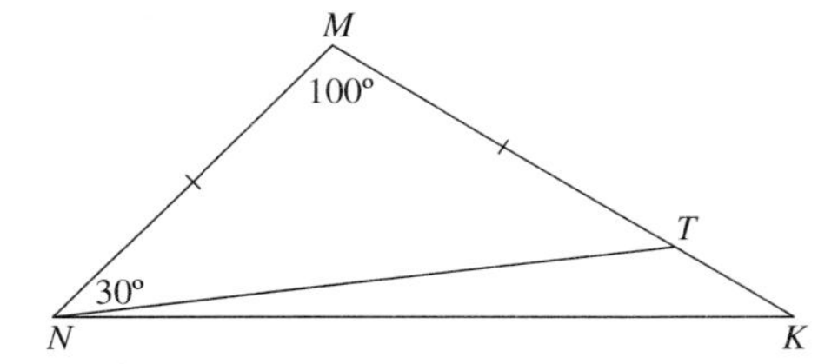

13. Give an appropriate inequality for x, y, and z if in right triangle $\triangle MLK$, K–D–E–L.

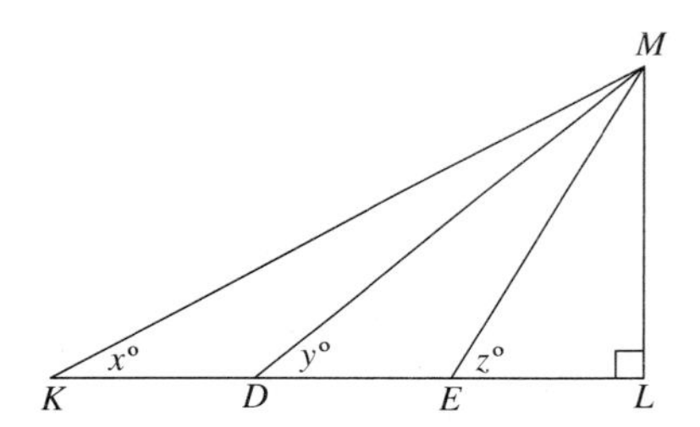

14. Point D varies on ray $\overrightarrow{AK}$, and $\angle BCK$ is an exterior angle of $\angle ABC$, with $m\angle A = 25$ and $\overline{BC} \perp \overline{AK}$. What can be said about the range of possible values for $x = m\angle BDC$ if

(a) A–D–C (as shown)

(b) A–C–D?

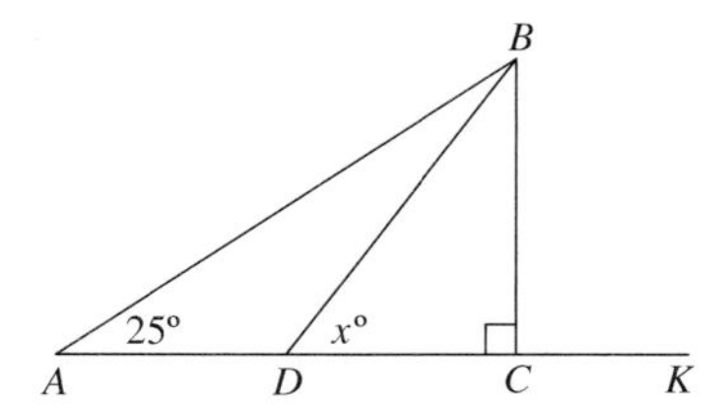

GROUP B

15. Prove that the sum of the measures of any two angles of a triangle is less than 180 without using the Saccheri–Legendre Theorem. (See Figure 3.33.)

16. Point B varies on ray $\overrightarrow{AD}$, A–C–D and $m\angle ECD < 90$. What can be said about the range of values possible for $y = m\angle BEC$ if

(a) A–B–C (as shown)?

(b) A–C–B?

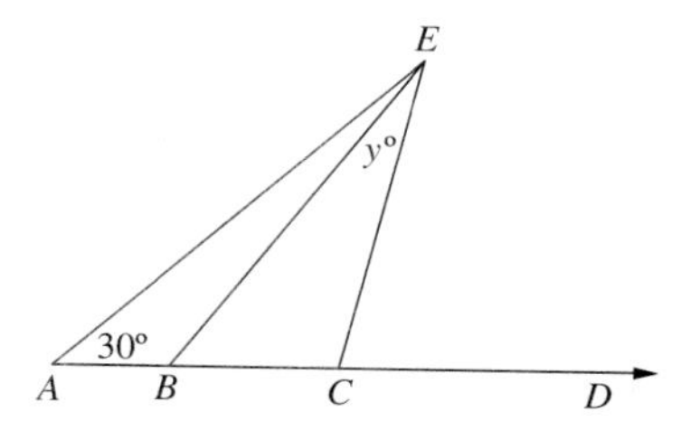

***17.** If $\triangle WKT$ has acute angles at K and T, and if $\overline{WR} \perp \overline{KT}$, show that R lies between K and T. (If R does *not* lie between K and T, then either K–T–R or R–K–T; show in either case that a contradiction to the Exterior Angle Theorem occurs.)

*An important result of absolute geometry often taken for granted.

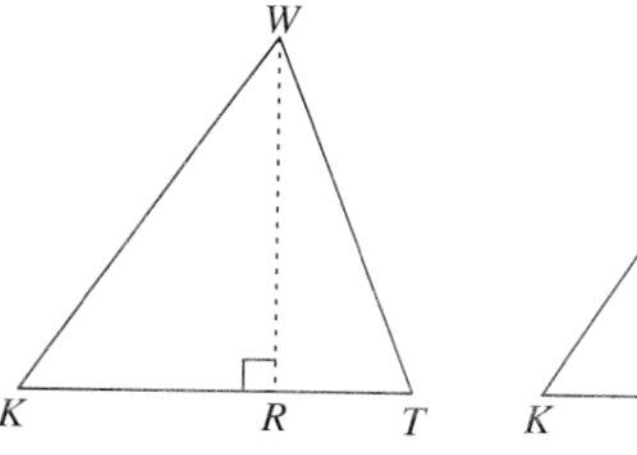

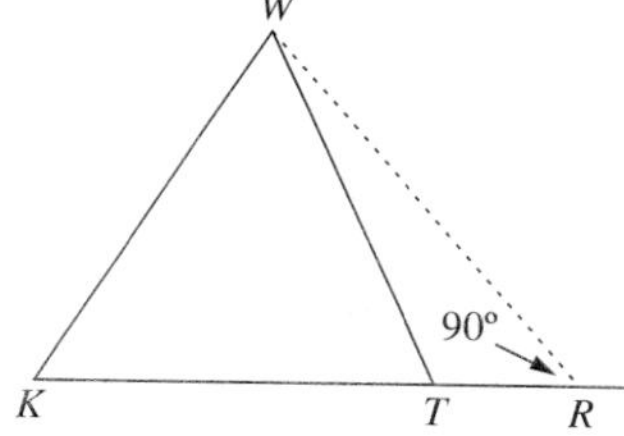

18. In $\triangle SKL$ we know that $m\angle 1 + m\angle 2 < 180$ by the corollary of the Exterior Angle Inequality mentioned in this section. Show that if the rays at K and L that "double" $\angle 1$ and $\angle 2$ meet at W (i.e., rays $\overrightarrow{KS}$ and $\overrightarrow{LS}$ bisect $\angle WKL$ and $\angle WLK$), then $m\angle 1 + m\angle 2 < 90$.

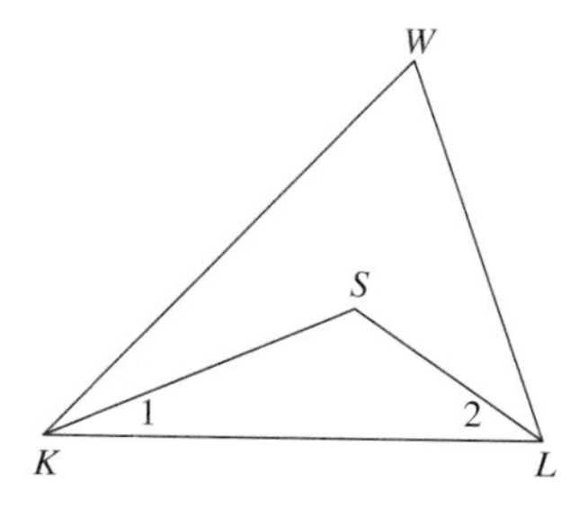

19. Rays $\overrightarrow{AN}$ and $\overrightarrow{MT}$ bisect $\angle PAM$ and $\angle AMQ$. Show that even though $\angle 1$ is not an exterior angle of $\triangle AMN$, $m\angle 1 > m\angle 2$.

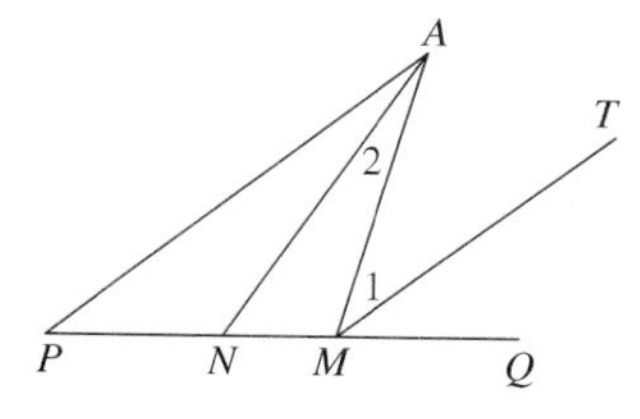

GROUP C

20. If C lies between B and C' and $AC = CC'$, prove that $y \leq x/2$.

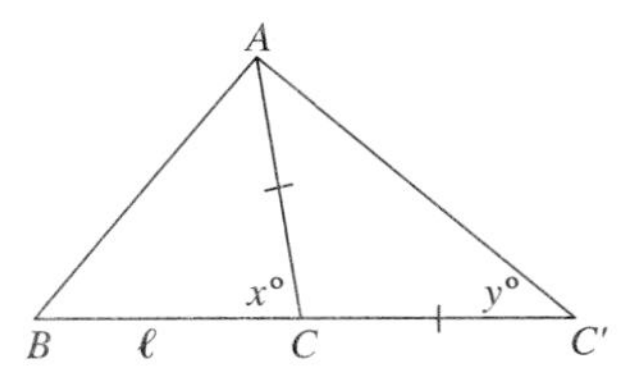

***21.** An important principle in geometry involves the question of whether, in the figure for Problem 20, $m\angle AC'B$ can be made smaller than a preassigned constant by moving C' out far enough on ray $\overrightarrow{BC}$. This seems intuitively clear. To prove it, follow this construction (discovered by Legendre): Let $\epsilon > 0$ be the given constant. We must show that for some point C' on line ℓ, $m\angle AC'B < \epsilon$. Locate points $C_1, C_2, C_3, \ldots, C_n$ on ray $\overrightarrow{BC}$ such that B–C_1–C_2 and $C_1C_2 = AC_1$, B–C_2–C_3 and $C_2C_3 = AC_2$, and so on. Show that $x_2 \leq x_1/2$, $x_3 \leq x_2/2 \leq x_1/4, \ldots$, and, in general, $x_{n+1} \leq x_1/2^n$. (See previous problem.) Now finish the argument to show that for some n, $m\angle AC_nB < \epsilon$.

*Used in the proof of Theorem 1, Section 6.3.

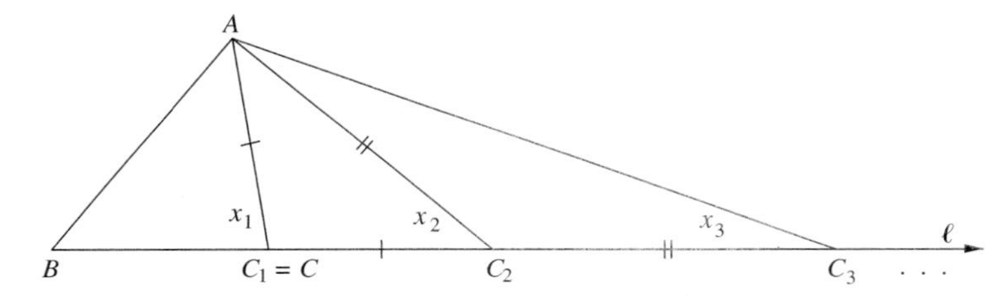

3.5 The Inequality Theorems

The Exterior Angle Inequality leads to numerous other inequality theorems for triangles. These, in turn, are very useful in proving certain "equality" theorems, identities, and congruences. The first group of inequalities involves relationships *within a single triangle.*

THEOREM 1: SCALENE INEQUALITY

If one side of a triangle has greater length than another side, then the angle opposite the longer side has the greater angle measure, and, conversely, the side opposite an angle having the greater measure is the longer side.

PROOF

(1) In $\triangle ABC$ it is given that $AC > AB$ (Figure 3.38). Locate D on $\overline{AC}$ so that $AD = AB$, and join points B and D. By betweenness properties, $m\angle ABC > m\angle 1$, and by the Isosceles Triangle Theorem, $m\angle 1 = m\angle 2$. Thus, by the Exterior Angle Inequality for $\triangle BDC$,

$$m\angle ABC > m\angle 1 = m\angle 2 > m\angle C$$

or, $m\angle ABC > m\angle C$, as was to have been proven.

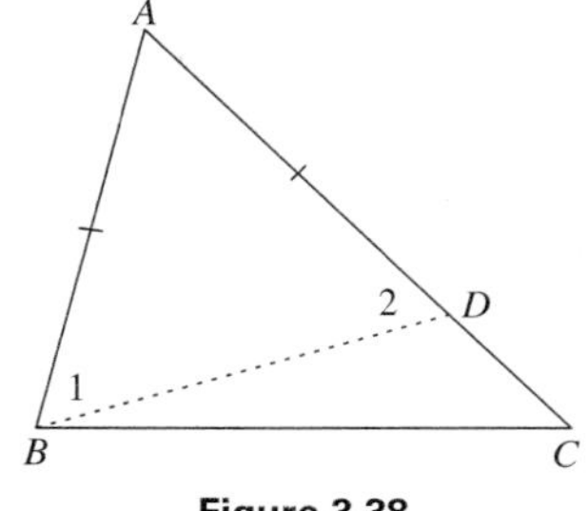

Figure 3.38

(2) For the converse, Euclid's old argument is perfectly valid, as quoted in Section 2.1: We are given $m\angle B > m\angle C$ and want to show that $AC > AB$. By the result just obtained, we cannot have $AC \leq AB$, or else $m\angle B \leq m\angle C$, a contradiction. Hence, $AC > AB$ is the only remaining possibility.

Two immediate corollaries will be stated informally; the proofs are merely a matter of recalling the definitions of the terms involved.

- If a triangle has an obtuse or right angle, then the side opposite that angle has the greatest measure.
- The hypotenuse of a right triangle has measure greater than that of either leg.

The next result, promised earlier, is the Triangle Inequality. It embodies practically all the significant results of absolute geometry proven thus far. (We might note, however, that a proof was already proposed in Problem 25, Section 3.3, using only congruence theorems for triangles.) The argument presented now is one of the more standard arguments.

The statement to be proven is that if A, B, and C are any three distinct points, then $AB + BC \geq AC$. A trivial case is taken care of first, that of three collinear points. If A, B, and C lie on some line, then we know by the Ruler Postlate that either B–A–C, A–B–C, or A–C–B. If B–A–C, then $AB + AC = BC$. That is, $BC > AC$ and therefore, $AB + BC > AC$. If A–B–C, then $AB + BC = AC$ (the only case where equality holds). Finally, if A–C–B then $AC + CB = AB$ and hence $AB > AC$ or $AB + BC > AC$. This proves that for any three collinear points A, B, and C,

$$AB + BC \geq AC$$

with equality only when A–B–C.

THEOREM 2: TRIANGLE INEQUALITY

If A, B, and C are any three distinct points, then $AB + BC \geq AC$, with equality only when the points are collinear, and A–B–C.

PROOF

Because we have already taken care of the case in which the points are collinear, we need only consider the noncollinear case, which means that we must now prove $AB + BC > AC$ (the strict inequality). In Figure 3.39 is

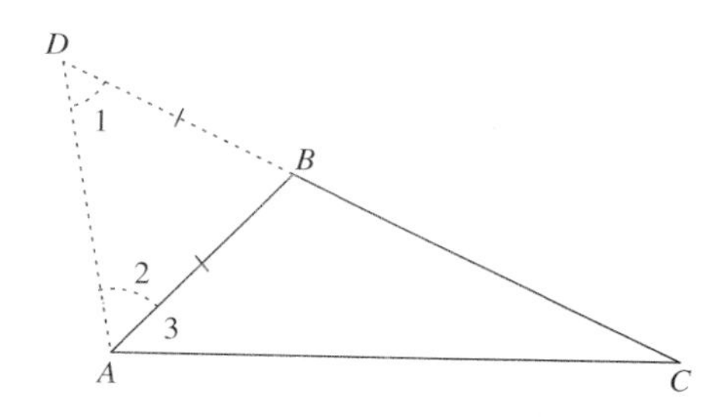

Figure 3.39

shown $\triangle ABC$ and the construction of point D on ray $\overrightarrow{CB}$ such that $BD = BA$. Now observe the angles formed in the figure, denoted by $\angle 1$, $\angle 2$, and $\angle 3$. Because $\triangle ABD$ is isosceles, $\angle 1 \cong \angle 2$. But by betweenness properties, $m\angle DAC = m\angle 2 + m\angle 3$. Hence, $m\angle DAC > m\angle 2 = m\angle 1 = m\angle ADC$. Then in $\triangle ADC$, by the Scalene Inequality,

$$CD > AC \rightarrow AB + BC = BD + BC = CD > AC$$

as desired.

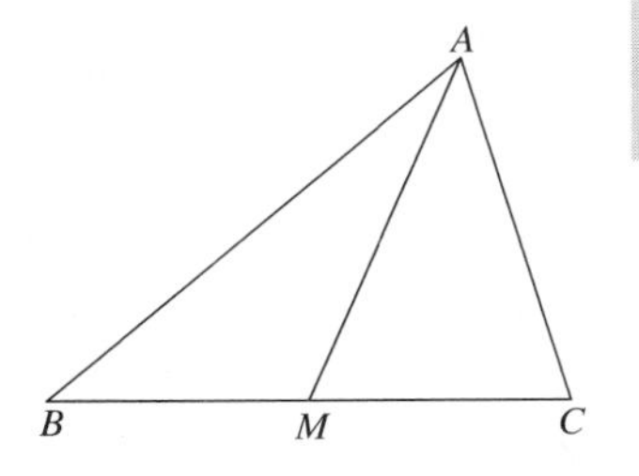

Figure 3.40

COROLLARY: MEDIAN INEQUALITY

Suppose that $\overline{AM}$ is the median to side $\overline{BC}$ of $\triangle ABC$ (Figure 3.40). Then

$$AM < \tfrac{1}{2}(AB + AC)$$

(The proof will be left as Problem 18.)

EXAMPLE 1 Find which of the angle measures x, y, z, r, and s indicated in Figure 3.41 is the least. Prove your answer.

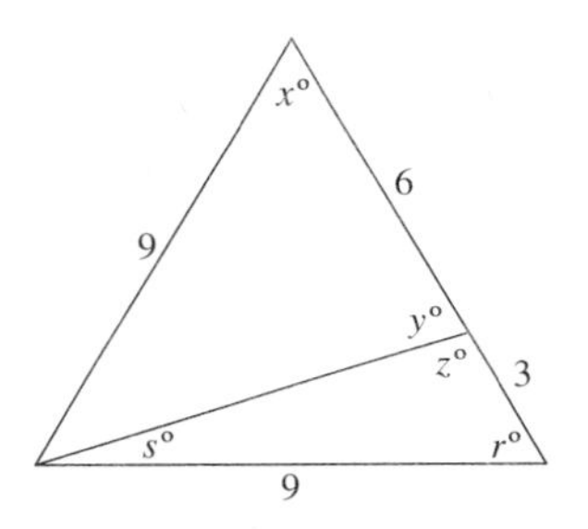

Figure 3.41

SOLUTION

It seems clear that s should be the smallest. We must then show that $s <$ each of the other values x, y, z, and r. Because s is opposite a shorter side of its triangle, by the Scalene Inequality,

$$s < z$$

Now because the large triangle is equilateral, therefore equiangular, $x = r =$ measure of the angle of which s is a part, so

$$s < x \qquad \text{and} \qquad s < r$$

Finally, by the Exterior Angle Inequality,

$$s < y. \quad \blacksquare$$

The next result is sometimes referred to as the "Hinge" Theorem, or "Alligator" Theorem (the wider the angle, the larger the opening). We shall give it a more formal name, however. Note that this time the inequality involves *two different* triangles, unlike the Scalene Inequality, which involved only one triangle.

THEOREM 3: SAS INEQUALITY THEOREM
If in $\triangle ABC$ and $\triangle XYZ$ we have $AB = XY$, $AC = XZ$, but $m\angle A > m\angle X$, then $BC > YZ$, and conversely if $BC > YZ$, then $m\angle A > m\angle X$. (See Figure 3.42.)

PROOF

CONCLUSIONS	JUSTIFICATIONS
(1) Construct ray $\overrightarrow{AD}$ such that $\overrightarrow{AB}$–$\overrightarrow{AD}$–$\overrightarrow{AC}$ and $\angle BAD \cong \angle X$, with $AD = XZ = AC$.	Angle and Segment Construction Theorems
(2) $\triangle ABD \cong \triangle XYZ$.	SAS
(3) $\overline{BD} \cong \overline{YZ}$.	CPCF
(4) Construct the bisector of $\angle DAC$; this bisector will cut segment $\overline{BC}$ at an interior point E.	Angle Construction Theorem, Crossbar Theorem
(5) $\angle DAE \cong \angle EAC$, $\overline{AE} \cong \overline{AE}$.	Definition of bisector, Reflexive Property
(6) $\triangle DAE \cong \triangle CAE$.	SAS
(7) $DE = EC$.	CPCF
(8) $BC = BE + EC = BE + DE > BD$, $\therefore BC > YZ$.	B–E–C, substitution from Step 7, Triangle Inequality and Step 3
(9) Converse argument similar to previous converse argument for Theorem 1.	

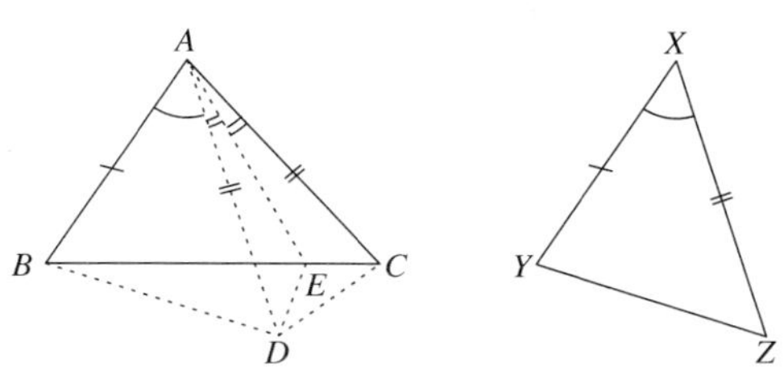

Figure 3.42

EXAMPLE 2 In Figure 3.43, a circle with diameter $\overline{QR}$ and center O is shown. If P varies on the circle on either side of $\overleftrightarrow{QR}$, and if $\theta = m\angle POQ$, define the function

$$f(\theta) = PQ, \qquad 0 < \theta < 180$$

Explain why $f(\theta)$ is an increasing function. [That is, if $\theta_1 < \theta_2$, then $f(\theta_1) < f(\theta_2)$.] Observe points P and P' in the figure. We need to prove that PQ, which is $f(\theta_1)$, is less than $P'Q$, which is $f(\theta_2)$.

Figure 3.43

SOLUTION
In $\triangle OPQ$ and $\triangle OP'Q$, $\theta_1 = m\angle POQ < m\angle P'OQ = \theta_2$ and $f(\theta_1) = PQ$, $f(\theta_2) = P'Q$. Because $OP = OP'$ and $OQ = OQ$, by the SAS Inequality Theorem, $PQ < PQ'$, or $f(\theta_1) < f(\theta_2)$. ■

PROBLEMS (3.5)

GROUP A

1. Order the quantities x, y, and z from least to greatest. What geometric principle did you use?

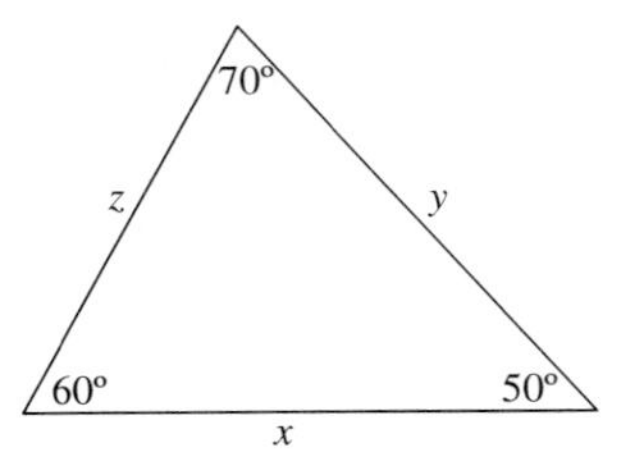

2. Order the quantities x, y, and z from least to greatest. What geometric principle did you use?

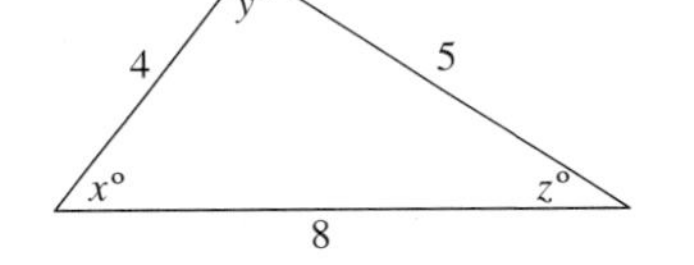

3. If a triangle has a right angle, why must the side opposite have the greatest length? (Your explanation constitutes a proof that the hypotenuse of a right triangle is of greater length than either leg.)

4. Name the longest and shortest segments in the figure and state your reasons.

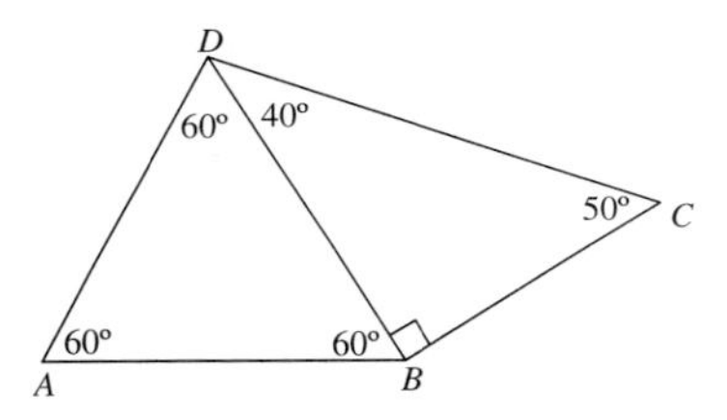

5. Given $m\angle ACB < 90$, name the smallest and largest angle, and state your reasons.

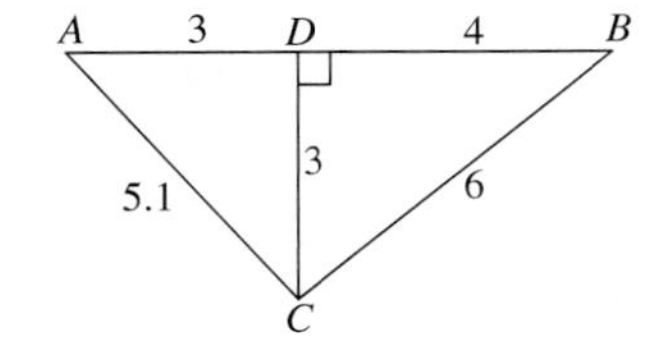

6. Which angle in the figure is smallest, and why?

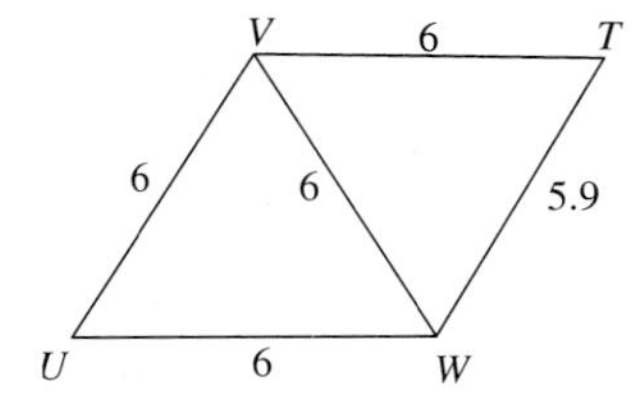

7. In the figure for this problem, which of the angles at A, B, and C is the largest, and why?

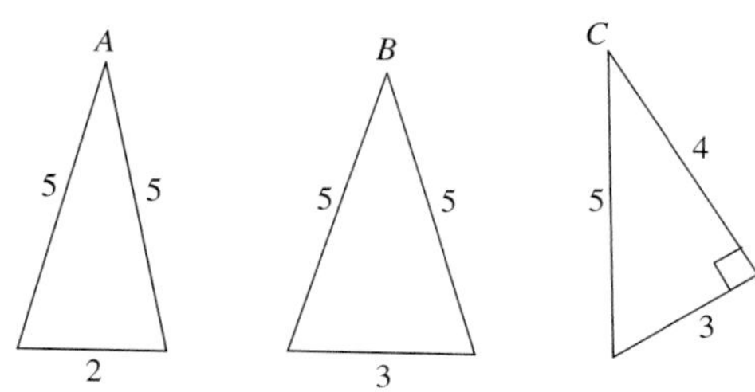

8. Use the Triangle Inequality to prove that for any three points A, B, and C,

$$AB - BC \leq AC \leq AB + BC$$

9. Certain inequality problems lend themselves to the *Sketchpad* software, as the following project will show.

   [1] Construct segment $\overline{AB}$, then rays $\overrightarrow{AB}$ and $\overrightarrow{BA}$.

   [2] By rotating ray $\overrightarrow{AB}$ about A and $\overrightarrow{BA}$ about B, construct angles at A and B having measures 43 and 36, respectively (for rotation about B, use $-36°$).

   [3] Select the point of intersection C of the two rotated rays, then hide all four rays. Finally, construct segments $\overline{AC}$ and $\overline{BC}$. You should now have a triangle $\triangle ABC$ with $m\angle A = 43$ and $m\angle B = 36$.

   [4] Construct a point D on segment $\overline{AB}$, and join points C and D.

   [5] Display the angle measures $x = m\angle ADC$ and $y = m\angle DCB$.

   Now drag point D, beginning at A and ending at B. As you do so, watch the values of x and y change.

   (a) What are the smallest and largest values that x seems to attain. Write down the resulting inequality.

   (b) Repeat for the value of y.

   (c) Use the inequalities established in this section to prove the inequalities you obtained in (a) and (b).

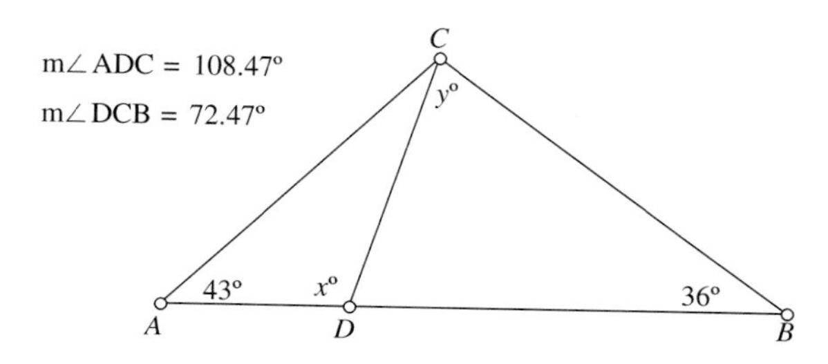

GROUP B

10. Prove or disprove: In $\triangle RST$ with angle measures as indicated, $RS \geq RT$.

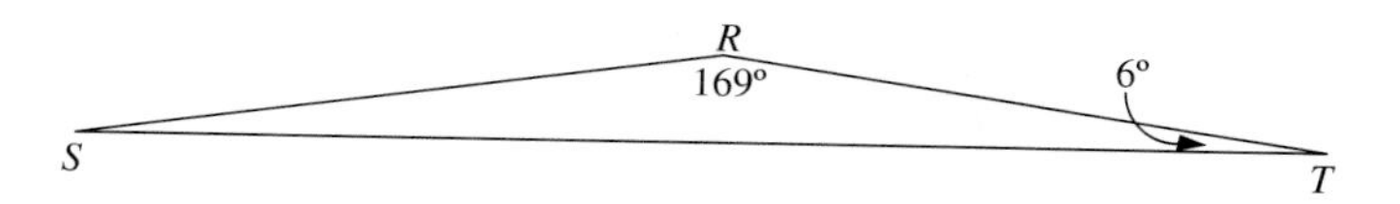

11. Suppose that $\triangle ABC$ has sides $AB = 31$, $AC = 35$ and M is the midpoint of $\overline{BC}$. The object is to obtain the inequality $2 < x < 33$, where $x = AM$. Construct the auxiliary lines shown, with M the midpoint of $\overline{AE}$.

(a) Find $y = CE$. Then show that $2x = AE < 66$.

(b) Show that $AM < \frac{1}{2}(66) = 33$.

(c) Show that $2 < x < 33$. (Use the Triangle Inequality in $\triangle AEC$.)

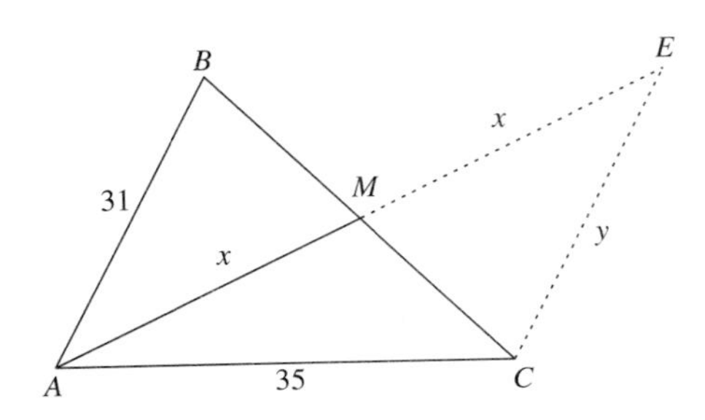

12. If $QR = QT$ and certain angle measures are as indicated, which segment in the figure is the shortest, and prove your answer. (Do not assume $m\angle QST = 60$.)

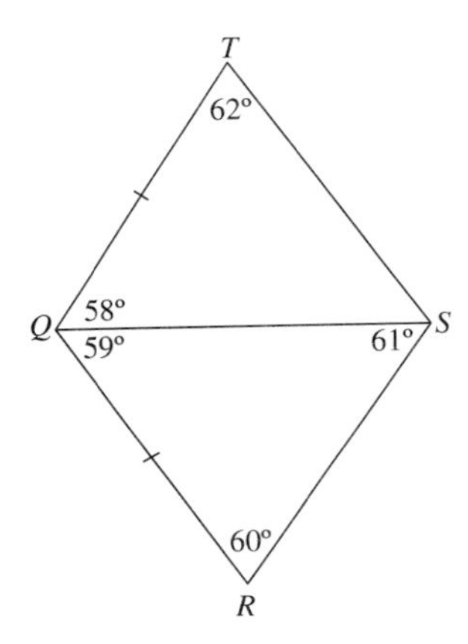

13. As mentioned in an earlier problem (Problem 19, Section 3.3), a common incorrect proposal often advanced by amateur mathematicians to solve the ancient problem of constructing the trisectors of an angle with the Euclidean tools is to lay out equal distances AD and AE on the sides of the angle to be trisected, then trisect the segment joining D and E, as shown. Prove that this *never* works, for any angle, that, in fact,

$$m\angle LAM > m\angle DAL = m\angle MAE$$

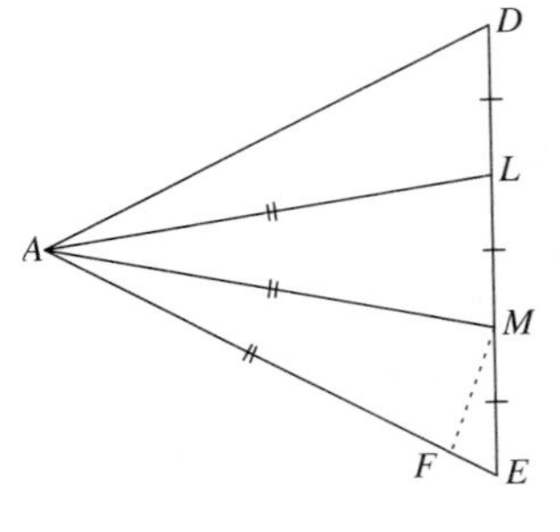

14. A mechanical device for trisecting an angle called the *tomahawk* appeared in a book in 1835 (its inventor is unknown). It may be constructed and tested efficiently using *Sketchpad*.

[1] Construct point R and translate it horizontally 0.75 units (inches) three times, to obtain points S, T, and U, equally spaced. Construct segment $\overline{RU}$.

[2] Construct the perpendicular bisector of segment $\overline{RT}$ at S, locate a point V on it, then hide the line and construct segment $\overline{SV}$.

[3] With T as center, construct a semicircle with diameter $\overline{SU}$ as base. (Use Construct Arc on Circle technique.)

[4] (Optional) If desired, fill in with shading using Polygon Interior and Arc Segment Interior under CONSTRUCT—inessential to the function of the Tomahawk as a trisector.

(a) To test this device, construct by rotation, an angle $\angle ABC$ having measure 60—an angle that cannot be trisected with the Euclidean tools. Using the Translate Tool and by dragging one point of the ray $\overrightarrow{BA}$, position the angle so that B lies on $\overline{SV}$, ray $\overrightarrow{BA}$ passes through R, and ray $\overrightarrow{BC}$ is tangent to the semicircle. Construct segments $\overline{BS}$ and $\overline{BT}$. These will be the angle trisectors. Measure the angles $\angle RBS$, $\angle SBT$, and $\angle TBC$ to check.

(b) Does this instrument seem to work? Why? Try to prove its validity. (We are in Euclidean geometry now.)

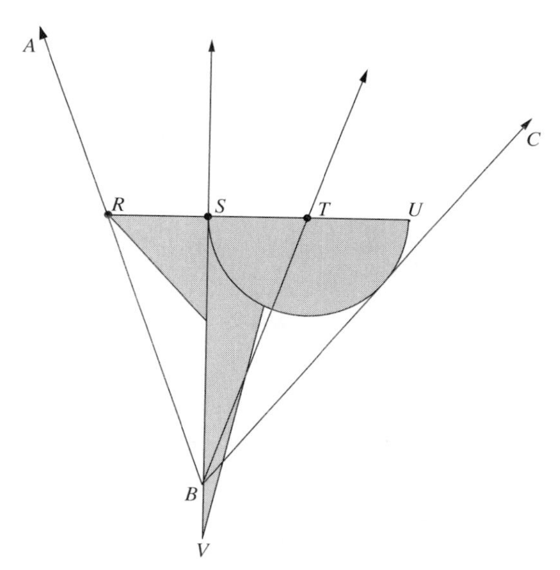

15. Prove the SSS Congruence Theorem using inequality theorems.

16. Prove in absolute geometry that the total angle sum of two adjacent triangles with their bases lying on a common line exceeds the angle sum of the large triangle by 180. That is, if B–D–C and k_1 and k_2 are the angle sums of $\triangle BAD$ and $\triangle BDC$, then $k_1 + k_2 = m\angle BAC + m\angle B + m\angle C + 180$.

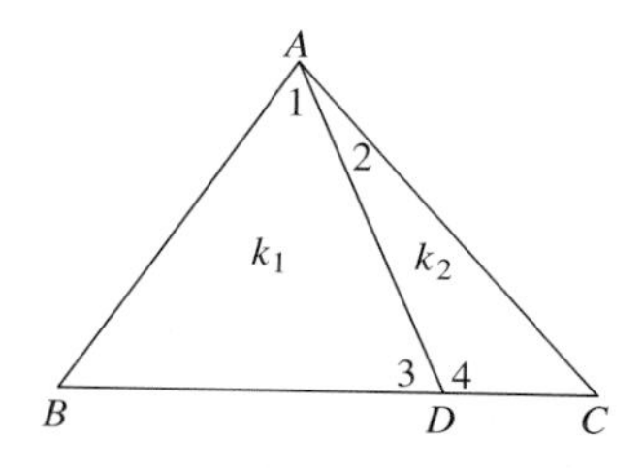

GROUP C

17. Using the result of Problem 16, prove in absolute geometry that if the value for the angle sum of triangles does not vary (a reasonable but incorrect assumption), then the angle sum for all triangles equals 180.

18. Prove the Median Inequality: $AM = \frac{1}{2}(AB + AC)$, where M is the midpoint of side $\overline{BC}$ of $\triangle ABC$. (See Problem 11 for ideas.)

3.6 Additional Congruence Criteria

The inequality theorems established in the previous section allow us to obtain two further congruence criteria for arbitrary triangles, which, in turn, lead to special criteria for right triangles.

AAS AND SSA CONGRUENCE CRITERIA

The first to be considered involves triangles having two pairs of congruent angles and a pair of congruent sides. Consider the situation illustrated in Figure 3.44 in which $\triangle ABC$ and $\triangle XYZ$ have $\angle A \cong \angle X$, $\angle B \cong \angle Y$, and $\overline{AC} \cong \overline{XZ}$. It would be straightforward to conclude from this that the third pair of angles (at C and Z) are also congruent *if* we knew that the angle sums of both triangles equal 180. Then ASA could be applied to complete the proof. But this is not the case in absolute geometry. The proper way to handle this problem is the following.

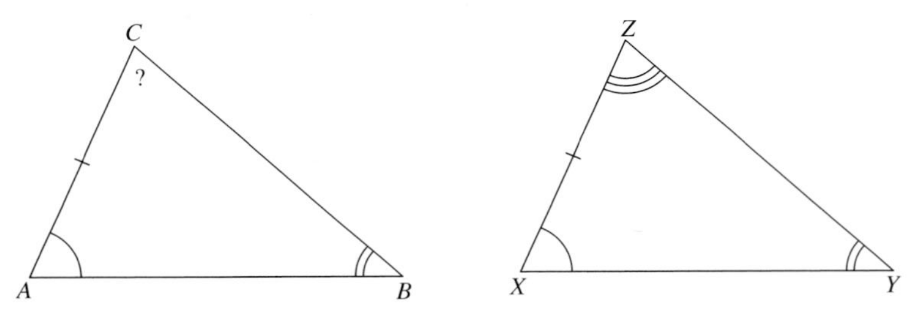

Figure 3.44

THEOREM 1: AAS CONGRUENCE CRITERION

If under some correspondence between their vertices, two angles and a side opposite in one triangle are congruent to the corresponding two angles and side of a second triangle, then the triangles are congruent.

PROOF

Suppose, as in Figures 3.44 and 3.45, that $\angle A \cong \angle X$, $\angle B \cong \angle Y$, and $\overline{AC} \cong \overline{XZ}$. To prove $\triangle ABC \cong \triangle XYZ$, we show first that $m\angle C = m\angle Z$. To that end, suppose that $m\angle C \neq m\angle Z$, and, for sake of argument, that

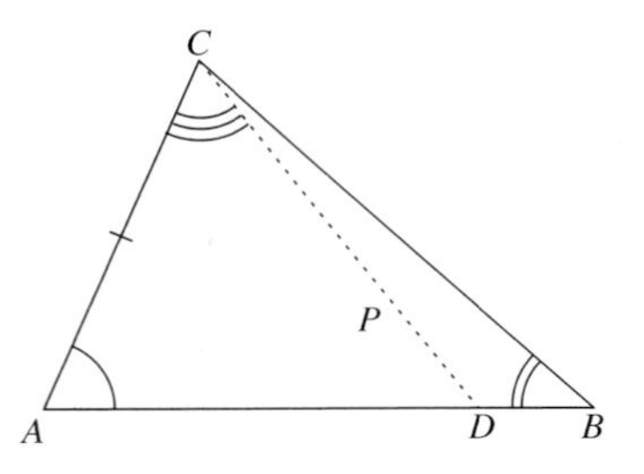

Figure 3.45

$m\angle C > m\angle Z$. By the Angle Construction Theorem, there exists a unique ray $\overrightarrow{CP}$ lying between $\overrightarrow{CA}$ and $\overrightarrow{CB}$ such that $m\angle ACP = m\angle Z$ (Figure 3.45). By the Crossbar Theorem, ray $\overrightarrow{CP}$ meets segment $\overline{AB}$ at an interior point D such that A–D–B. By ASA, $\triangle ADC \cong \triangle XYZ$. Hence $m\angle ADC = m\angle XYZ = m\angle B$ (by hypothesis). But $\angle ADC$ is an exterior angle of $\triangle CDB$ and $m\angle ADC > m\angle B$. $\rightarrow\leftarrow$ A similar contradiction follows if we assume $m\angle C < m\angle Z$. Therefore, $m\angle C = m\angle Z$ and $\triangle ABC \cong \triangle XYZ$ by ASA.

You might recall from trigonometry the so-called "ambiguous" case in solving for triangles specified by various sides and angles. This particular case involves solving for the remaining parts of a triangle when the measures of two sides and an angle opposite one of them are given. The problem is ambiguous because there can either be two solutions, one solution, or no solution at all—depending on circumstances. Because of this, we do not obtain the congruence criterion implied by the symbol SSA (see Figure 3.46 for a specific counterexample; see also Problem 17, Section 3.3, in this connection). In fact, the example in the figure shows the *only* way that two triangles can satisfy the SSA Hypothesis without being congruent. This is because we can reassemble the triangles, by congruence, into the first diagram in Figure 3.46, where $BC = BD$, and thus show that $\angle ACB$ is supplementary to $\angle BCD$, which is congruent to $\angle BDA$. This result can actually be put to good use, so it will be stated formally.

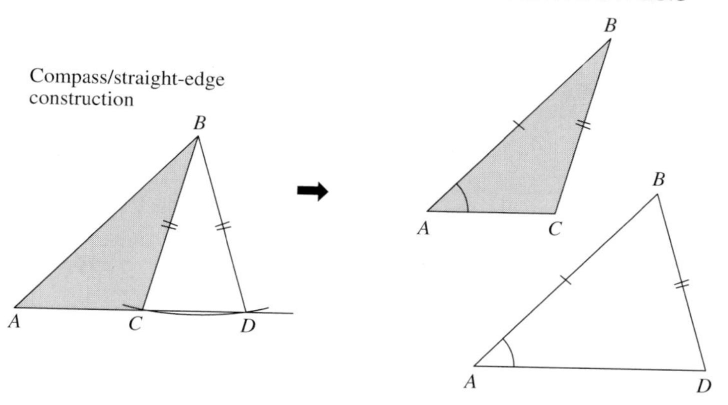

Figure 3.46

THEOREM 2: SSA THEOREM

If, under some correspondence between their vertices, two triangles have two pairs of corresponding sides and a pair of corresponding angles congruent, and if the triangles are *not* congruent under this correspondence, then the remaining pair of angles not included by the congruent sides are supplementary angles. (See Figure 3.47.)

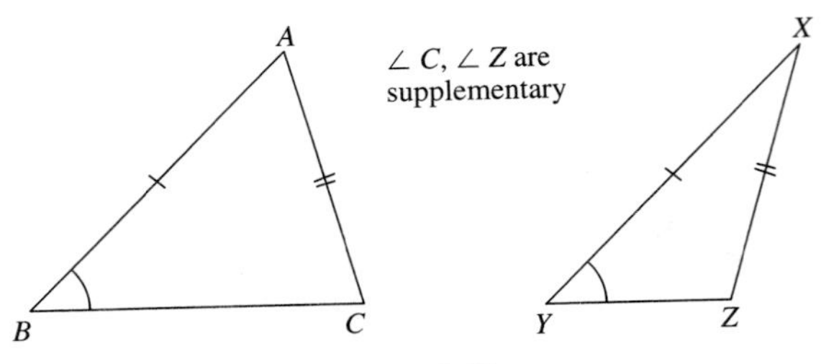

Figure 3.47

An obvious corollary is the following.

COROLLARY A: If under some correspondence of their vertices, two acute angled triangles have two sides and an angle opposite one of them congruent, respectively, to the corresponding two sides and angle of the other, the triangles are congruent.

Another corollary applies to right triangles, and establishes the useful HL (Hypotenuse/Leg) congruence criterion.

THE RIGHT TRIANGLE CONGRUENCE CRITERIA

COROLLARY B: HL THEOREM
If two right triangles have the hypotenuse and leg of one congruent, respectively, to the hypotenuse and leg of the other, the right triangles are congruent. (See Figure 3.48.)

THE HL HYPOTHESIS

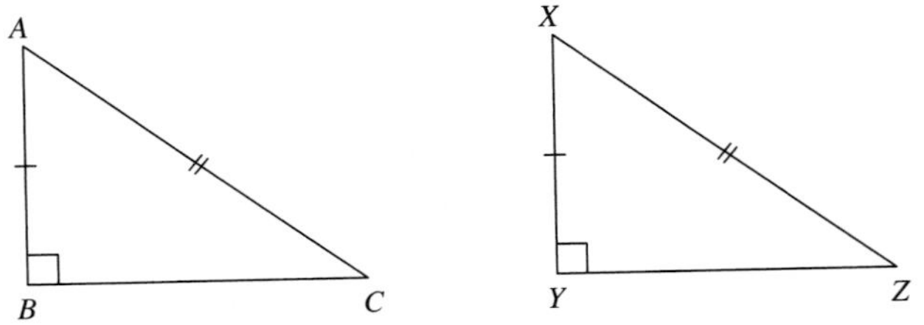

Figure 3.48

PROOF

Suppose $\overline{AB} \cong \overline{XY}$, $\overline{AC} \cong \overline{XZ}$ and $\angle B \cong \angle Y$ are right angles. By Theorem 2, if the triangles are not congruent, then $\angle C$ and $\angle Z$ are supplementary. But this is impossible since both angles $\angle C$ and $\angle Z$ must be acute angles, by properties of right triangles. Therefore, $\triangle ABC \cong \triangle XYZ$.

The other two congruence criteria for right triangles are corollaries of the AAS Theorem.

COROLLARY C: HA THEOREM

If two right triangles have the hypotenuse and acute angle of one congruent, respectively, to the hypotenuse and acute angle of the other, the triangles are congruent. (See Figure 3.49.)

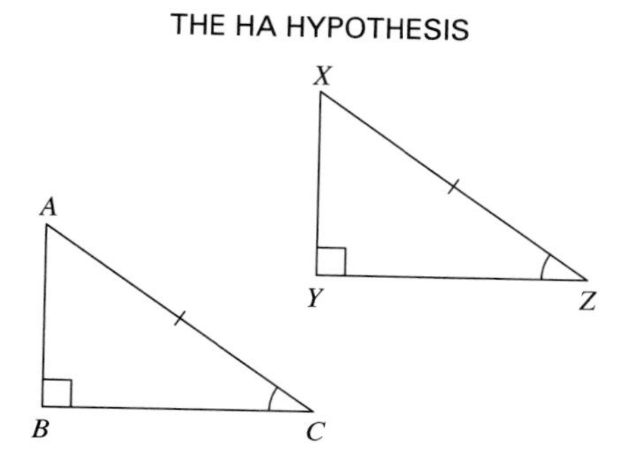

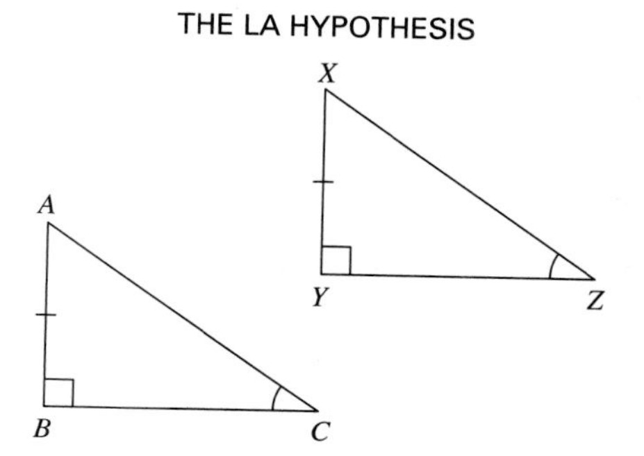

Figure 3.49

COROLLARY D: LA THEOREM

If under some correspondence between their vertices, two right triangles have a leg and acute angle of one congruent, respectively, to the corresponding leg and acute angle of the other, the triangles are congruent. (See Figure 3. 49.)

The final result of this section is included primarily because it appears in at least one high school geometry text.

COROLLARY E: SsA CONGRUENCE CRITERION

Suppose that in $\triangle ABC$ and $\triangle XYZ$, $\overline{AB} \cong \overline{XY}$, $\overline{BC} \cong \overline{YZ}$, $\angle A \cong \angle X$, and $BC \geq BA$. Then $\triangle ABC \cong \triangle XYZ$. (See Figure 3.50.)

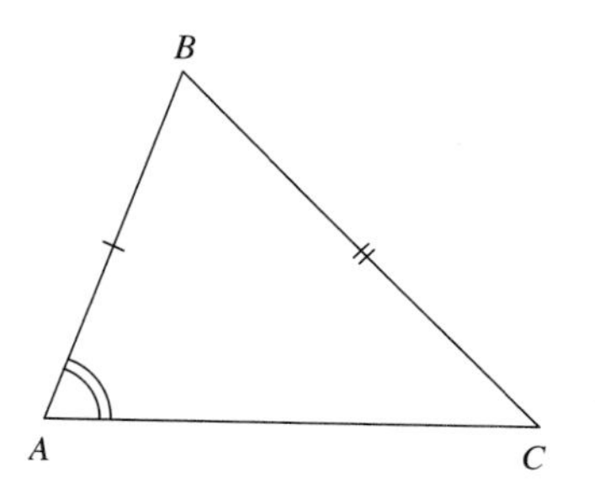

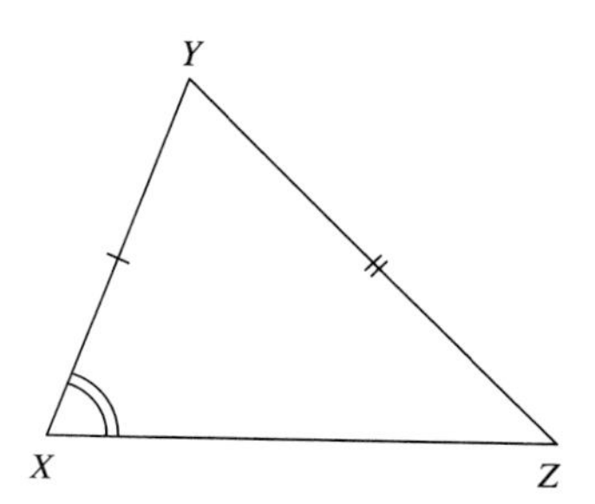

Figure 3.50

The proof follows immediately from the SSA Theorem; it will be left as Problem 4.

EXAMPLE 1

Given: $\overline{PQ} \perp \overline{PR}$, $\overline{QS} \perp \overline{SR}$, and $\overline{PQ} \cong \overline{QS}$, as shown in Figure 3.51.
Prove: $\overline{PR} \cong \overline{RS}$.

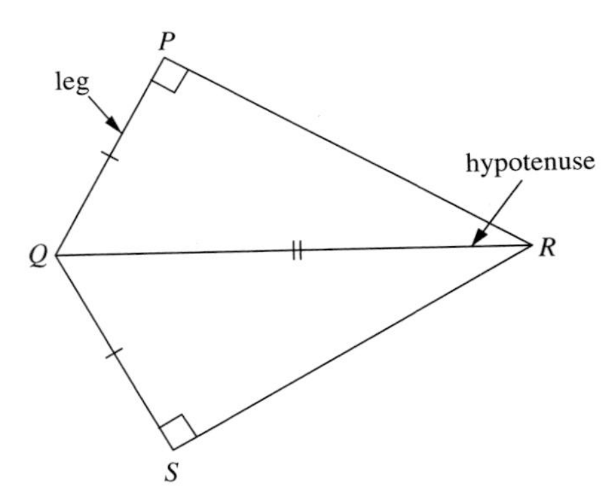

Figure 3.51

SOLUTION

Since $\overline{QR} \cong \overline{QR}$, by the HL Theorem $\triangle PQR \cong \triangle SQR$. $\therefore$ $\overline{PR} \cong \overline{SR}$ by CPCF. ■

Some definitions depend on proven results in order to make good sense. A case in point is the following, involving the *distance* from a point to a line. The "distance" between two geometric objects (i.e., sets of points) would presumably be the distance between the closest points in the two sets, as illustrated in Figure 3.52. We are interested in the distance from a point to a line. The result that should be proven first is that the distance from P to any point Q in line ℓ is least when $\overline{PQ} \perp \ell$. (See Figure 3.53.) This argument will be left for the reader.

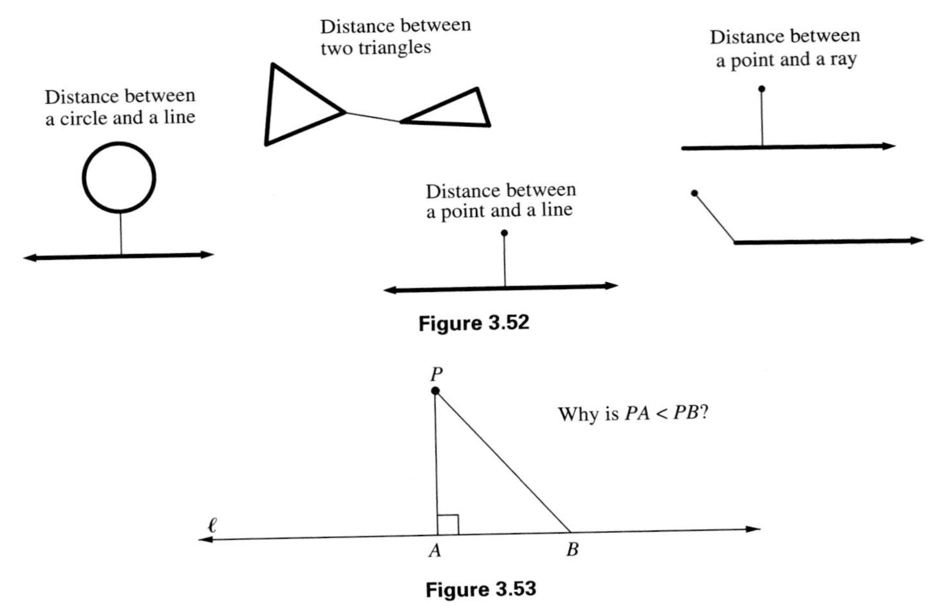

Figure 3.52

Figure 3.53

> DEFINITION: The **distance** from any point P to a line ℓ not passing through P is the distance from P to the foot of the perpendicular Q from P to line ℓ. A point is **equidistant** from two lines iff the distances from the point to the two lines are equal.

EXAMPLE 2 Prove that if, in Figure 3.54, $PA = PB$ and M is the midpoint of $\overline{AB}$, then M is equidistant from rays $\overrightarrow{PQ}$ and $\overrightarrow{PR}$.

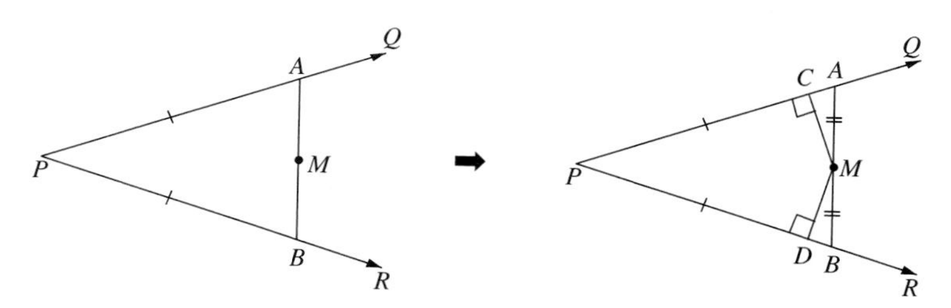

Figure 3.54

SOLUTION

Drop perpendiculars $\overline{MC}$ and $\overline{MD}$ to rays $\overrightarrow{PQ}$ and $\overrightarrow{PR}$. Because $MA = MB$ and $\angle CAM \cong \angle DBM$, by the Isosceles Triangle Theorem, $\triangle MAC \cong \triangle MBD$ by the HA Theorem. Hence $MC = MD$. ■

PROBLEMS (3.6)

GROUP A

1. Draw two separate noncongruent triangles of your own that satisfy the SSA Hypothesis. Can you draw these triangles in such a way that neither of them has an obtuse angle? Can both triangles be obtuse triangles?

2. From the information given in the figure, determine what triangle congruence theorem tells you that the triangles are congruent, if any. Write any valid congruence using the proper notation.

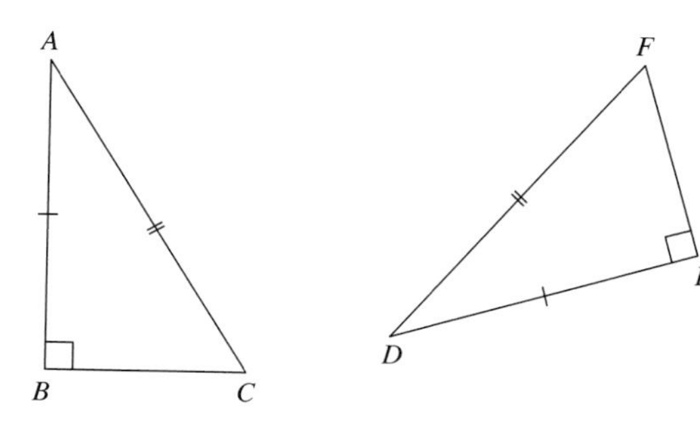

3. From the information given in the figure, determine what triangle congruence theorem tells you that the triangles are congruent, if any. Write any valid congruence using the proper notation.

(a)

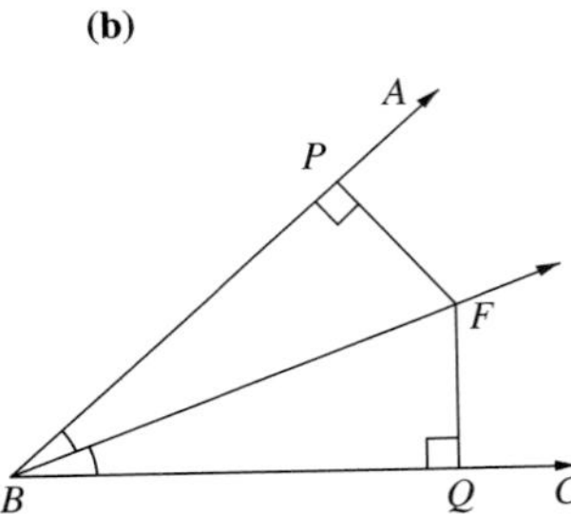

(b)

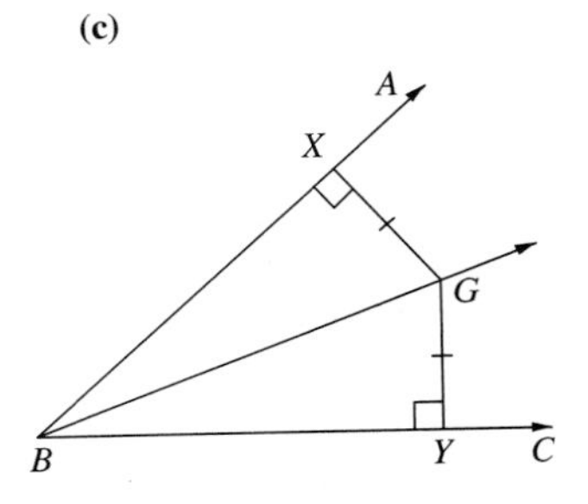

(c)

4. Provide the missing parts in the following proof of the SSA Congruence Theorem.
Given: $\triangle ABC$ and $\triangle XYZ$ with $AB = XY$, $BC = YZ$, $\angle A \cong \angle X$, and $BC \geq BA$.
Prove: $\triangle ABC \cong \triangle XYZ$.

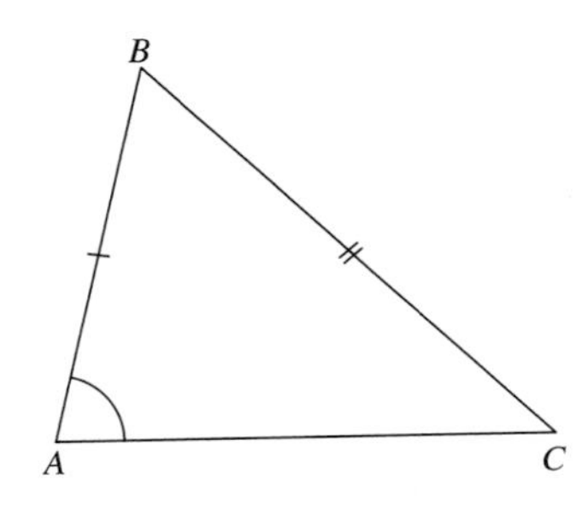

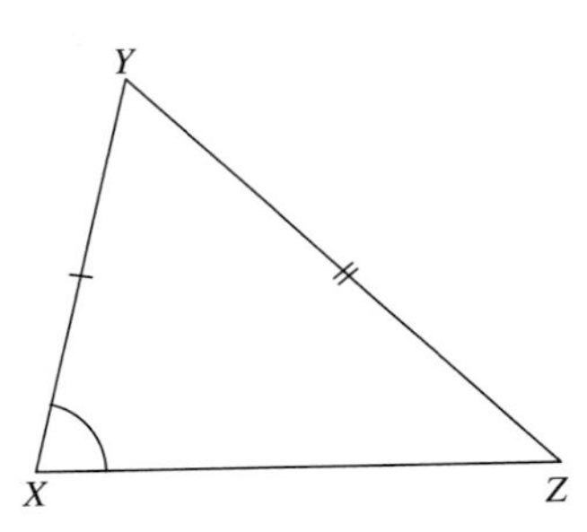

CONCLUSIONS	JUSTIFICATIONS
(1) Suppose the triangles are *not* congruent.	Given
(2) Then $\angle C$ and $\angle Z$ are supplementary angles.	**(a)** ________?
(3) One of the angles, say $\angle Z$, is either a right angle, or an obtuse angle.	Definition of supplementary angles
(4) $m\angle Z > m\angle X$	A triangle has at most one right angle or obtuse angle
(b) ________?	Scalene Inequality
(5) $AB > BC$. $\rightarrow\leftarrow$	**(c)** ________?
(6) $\therefore\ \triangle ABC \cong \triangle XYZ$.	The only remaining possibility

5. In elementary geometry texts in which the SSA theorems are *not* included, a different proof of the HL Congruency Theorem is required. Find such a proof. (See figure for hint.)

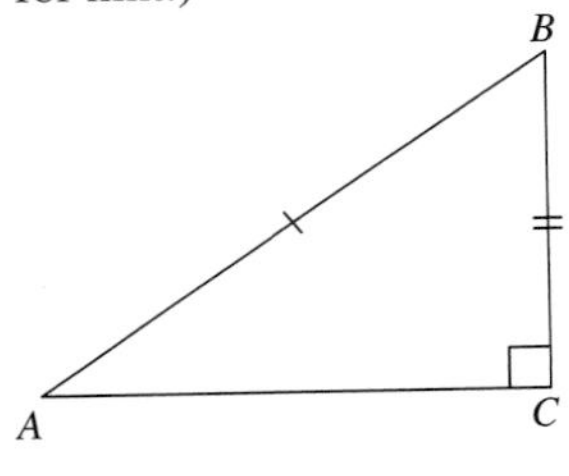

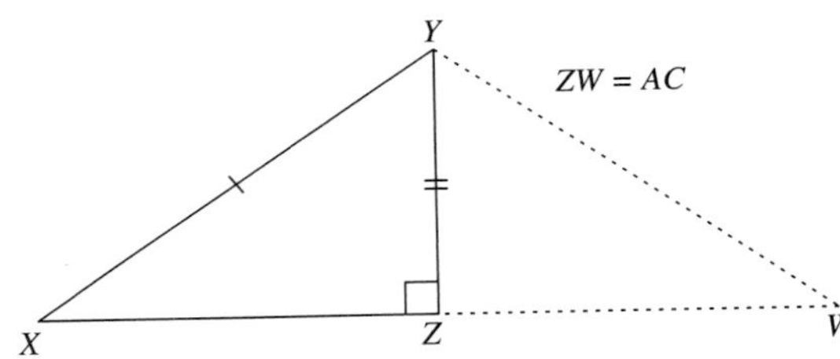

6. **(a)** List all eight triangles occurring in the figure below.

(b) From the marked information, there are three pairs of triangles that can be proved congruent. Name them, with vertices in corresponding order.

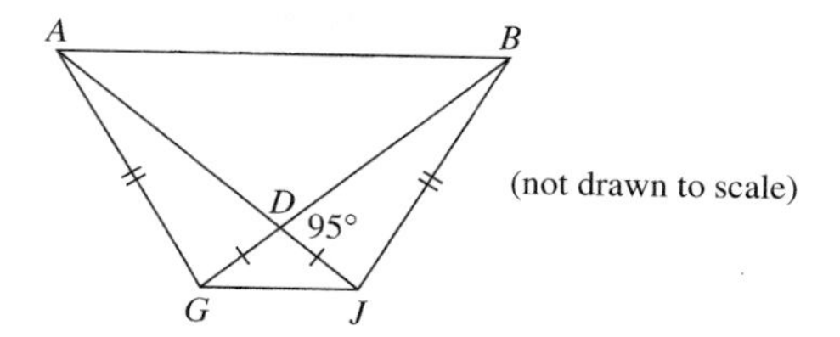

7. Follow these steps to draw (construct) a certain triangle.

a. Draw a ray $\overrightarrow{XY}$.

b. Draw $\angle WXY$ with measure (approximately) 50.

c. Draw $\overline{XZ}$ on ray $\overrightarrow{XW}$ such that $ZX = 11$ cm.

d. Draw circle Z with radius 9 cm. Let P be a point of intersection (circle $Z \cap \overleftrightarrow{XY}$).

e. Consider $\triangle XPZ$. Will everyone else who does this correctly have a $\triangle XPZ$ congruent to yours? Discuss.

[UCSMP, p. 331]

8. Explain why the SsA Theorem cannot be used to prove that $\triangle ADC \cong \triangle ABC$ in the accompanying figure.

[UCSMP, p. 321]

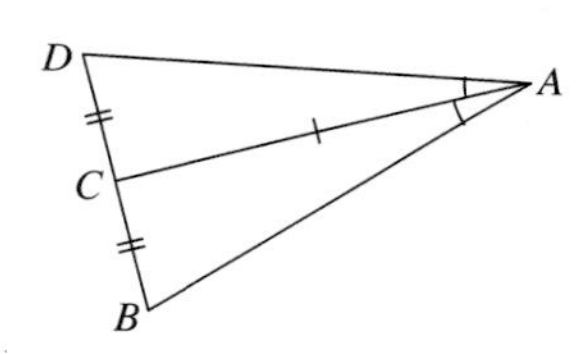

GROUP B

9. The following steps in *Sketchpad* should lead you to discover two very different geometric properties at once.

[1] Construct any two congruent segments $\overline{AB}$ and $\overline{AB'}$ such that $\angle BAB'$ is acute. (Rotate $\overline{AB}$ and end point B about A through an acute angle of your choice.)

[2] Construct ray $\overrightarrow{BC}$ making an acute angle with ray $\overrightarrow{BA}$, and passing through the interior of $\angle BAB'$.

[3] Using MEASURE, display $m\angle CBA$, and while it is selected, choose Mark Angle Measurement under TRANSFORM.

[4] Construct ray $\overrightarrow{B'A}$ and double-click on B' to make it a center for rotation (or select B' and use Mark Center under TRANSFORM). Rotate ray $\overrightarrow{B'A}$ about B' to ray $\overrightarrow{B'Q}$ choosing Rotate By Marked Angle under TRANSFORM. (This will make $\angle QB'A \cong \angle CBA$.) ***Note:*** It is important that both rays pass through the interior of $\angle BAB'$. The particular orientation of your figure may require you to go back to Step 3 and select $(-1) \cdot m\angle CBA$ instead. (Use Calculate to multiply by -1.) Then repeat Steps 3 and 4.

[5] Hide ray $\overrightarrow{B'A}$, then select, or construct, the point of intersection D of rays $\overrightarrow{BC}$ and $\overrightarrow{B'Q}$.

(a) Drag C and observe the behavior of point D. Use trace under DISPLAY to exhibit the path of point D. What is it? Try to prove your conjecture.

(b) Drag C to such a position that ray $\overrightarrow{BC}$ lies *exterior* to $\angle BAB'$. What happens to the locus of D in this case? Is your knowledge of Euclidean geometry sufficient for you to explain this behavior?

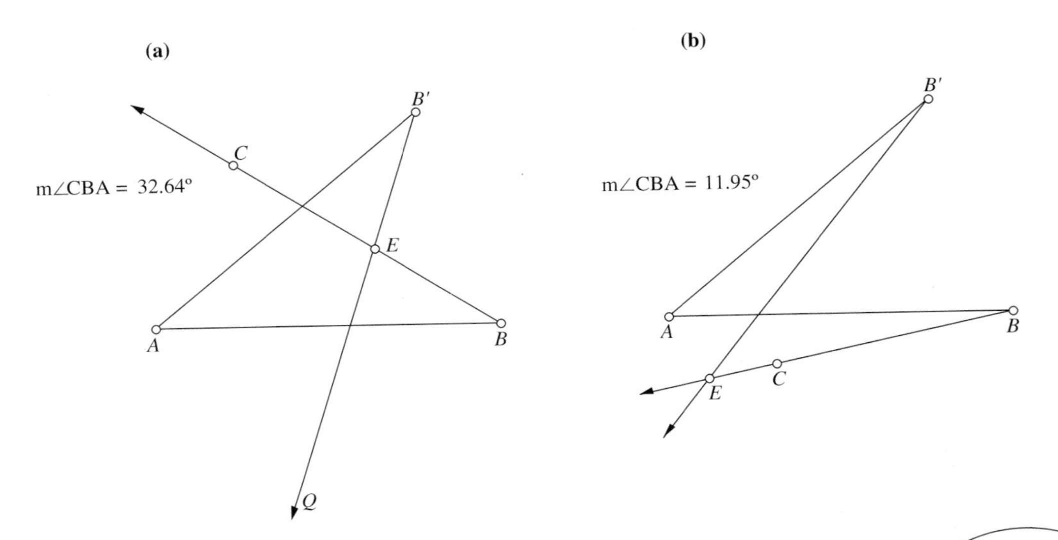

10. Assume the usual definition for circles. Consider the concentric circles in the figure, and show that if $\overline{OA}$ and $\overline{OB}$ are radii of the smaller of the two circles, and $\angle OAC \cong \angle OBD$, then $\overline{AC} \cong \overline{BD}$.

11. Prove that if I is any point on the bisector $\overrightarrow{BD}$ of $\angle ABC$, then I is equidistant from the sides of the angle, and conversely.

12.** Prove that the angle bisectors of any triangle are concurrent at a point I, called the **incenter**, that is equidistant from the three sides of the triangle. (Hint:*** Use the result of Problem 11; the argument is virtually the same as that for the concurrence of the perpendicular bisectors of the sides of a triangle, Problem 20, Section 3.3.)

*This result is needed for Problem 14, Section 3.8.

GROUP C

13. Steiner–Lehmus Theorem If two angle bisectors of a triangle are congruent, the triangle is isosceles. Prove, using the inequality theorems. (***Hint:*** In $\triangle ABC$, suppose bisectors $\overline{BD}$ and $\overline{CE}$ are congruent, but that $AC > AB$ and hence, $m\angle ABC > m\angle ACB$. Construct F on $\overline{AD}$ so that $\angle FBD \cong \angle ACE$, and gain a contradiction.)

$FC > BF \rightarrow CG > BD$

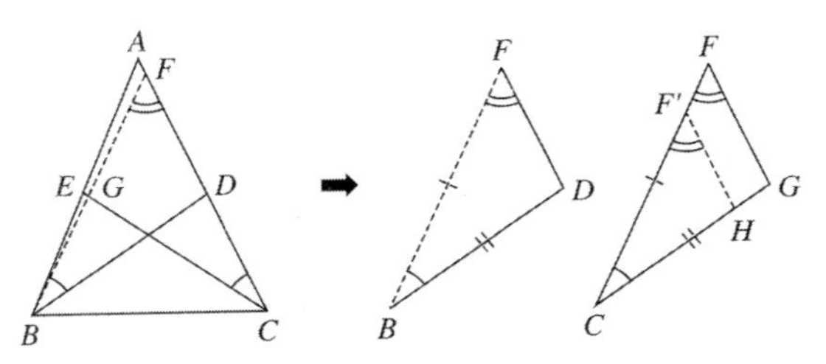

3.7 Quadrilaterals

This section will develop the subject of quadrilaterals formally. The symbol $\diamond$ or $\square$ is often used to denote quadrilaterals; we shall adopt the former for the general case and reserve the latter for squares and rectangles; $\varDelta$ is used for parallelograms.

> DEFINITION: If A, B, C, and D are any four points lying in a plane such that no three of them are collinear, and if the points are so situated that no pair of open segments determined by each pair of points taken in the order A, B, C, D ($\overline{AB}$, $\overline{BC}$, etc.) have points in common (see Figure 3.55), then the set
>
> $$\diamond ABCD \equiv \overline{AB} \cup \overline{BC} \cup \overline{CD} \cup \overline{DA}$$
>
> is a **quadrilateral,** with **vertices** A, B, C, D, **sides** $\overline{AB}, \overline{BC}, \overline{CD}, \overline{DA}$, **diagonals** $\overline{AC}, \overline{BD}$, and **angles** $\angle DAB, \angle ABC, \angle BCD, \angle CDA$.

QUADRILATERALS

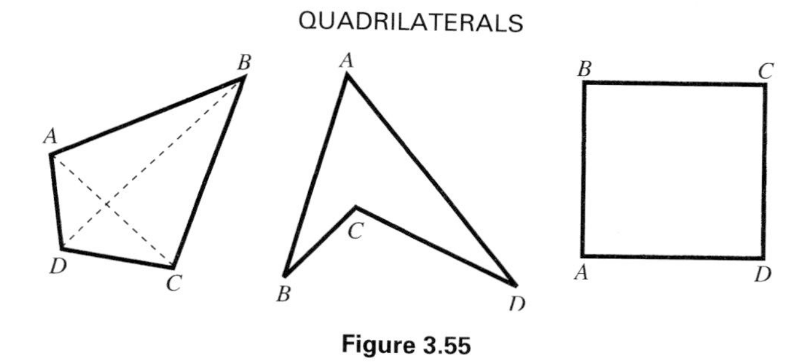

Figure 3.55

A few more terms associated with quadrilaterals are used frequently. Sides like $\overline{BC}$ and $\overline{CD}$ with a common end point are called **adjacent** (or sometimes **consecutive**), and angles containing a common side, like $\angle BCD$ and $\angle CDA$ (where $\overline{CD} \subseteq \overrightarrow{CD} \cap \overrightarrow{DC}$), are also called **adjacent** (or **consecutive**). Two sides (or angles) that are not adjacent are called **opposite** (such as $\overline{AB}$ and $\overline{CD}$).

CONVEX QUADRILATERALS

In Figure 3.55, one of the quadrilaterals has the peculiar property that its diagonals do not intersect, and two vertices (A and C) lie on one side of the diagonal joining the other two vertices (B and D). In most work with quadrilaterals in elementary geometry, we want the diagonals to intersect at a point that lies between opposite vertices. Such quadrilaterals are called **convex.** We leave fundamental details concerning this, as well as the formal definition, as problems. A few useful properties of convex quadrilaterals may be listed.

- The diagonals of a convex quadrilateral intersect at an interior point on each diagonal.
- If $\diamond ABCD$ is a convex quadrilateral, then D lies in the interior of $\angle ABC$ (and similarly for the other vertices).
- If A, B, C, and D are consecutive vertices of a convex quadrilateral, then $m\angle BAD = m\angle BAC + m\angle CAD$.

CONGRUENCE CRITERIA FOR CONVEX QUADRILATERALS

A few basic congruence properties will be used repeatedly throughout the rest of this text.

DEFINITION: Two quadrilaterals $\diamond ABCD$ and $\diamond XYZW$ are **congruent** under the correspondence $ABCD \leftrightarrow XYZW$ iff all pairs of corresponding sides and angles under the correspondence are congruent (i.e., CPCF). Such congruence will be denoted by $\diamond ABCD \cong \diamond XYZW$. That is,

$\diamond ABCD \cong \diamond XYZW$ means $\overline{AB} \cong \overline{XY}$, $\overline{BC} \cong \overline{YZ}$, $\overline{CD} \cong \overline{ZW}$, $\overline{DA} \cong \overline{WX}$, $\angle A \cong \angle X$, $\angle B \cong \angle Y$, $\angle C \cong \angle Z$, $\angle D \cong \angle W$

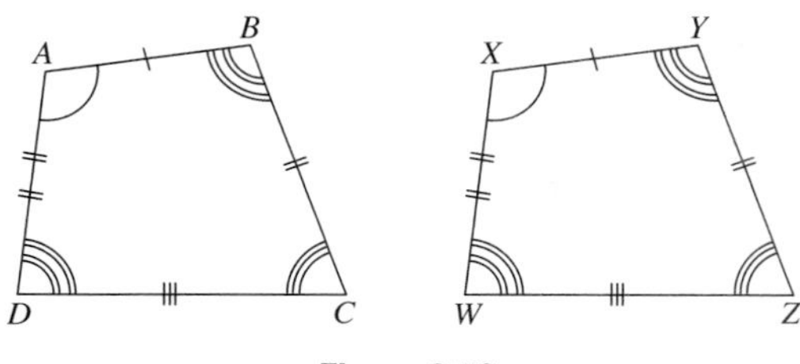

Figure 3.56

THEOREM 1: SASAS CONGRUENCE

Suppose that two convex quadrilaterals $\diamond ABCD$ and $\diamond XYZW$ satisfy the **SASAS Hypothesis** under the correspondence $ABCD \leftrightarrow XYZW$. That is, as shown in Figure 3.57, three consecutive sides and the two angles included by those sides of $\diamond ABCD$ are congruent, respectively, to the corresponding three consecutive sides and two included angles of $\diamond XYZW$. Then $\diamond ABCD \cong XYZW$.

PROOF

We must prove that the remaining corresponding sides and angles of the two quadrilaterals are congruent. That is, it is to be proven that (1) $\overline{AD} \cong \overline{XW}$, (2) $\angle D \cong \angle W$, and (3) $\angle BAD \cong \angle YXW$. As shown in Figure 3.57, draw the diagonals $\overline{AC}$ and $\overline{XZ}$. Then by SAS, $\triangle ABC \cong \triangle XYZ$. Hence $m\angle 1 = m\angle 3$, $m\angle 5 = m\angle 7$, and $AC = XZ$. Because rays $\overrightarrow{CA}$ and $\overrightarrow{ZX}$ lie between the sides of their respective quadrilaterals (by convexity), we can subtract certain angle measures in the figure to obtain $m\angle 2 = m\angle 4$, so by SAS, $\triangle ACD \cong \angle XZW$. Therefore, (1) $\overline{AD} \cong \overline{XW}$, (2) $\angle D \cong \angle W$, and $m\angle 6 = m\angle 8$. Hence, $m\angle BAD = m\angle 5 + m\angle 6 = m\angle 7 + m\angle 8 = m\angle YXW$ and (3) $\angle BAD \cong \angle YXW$.

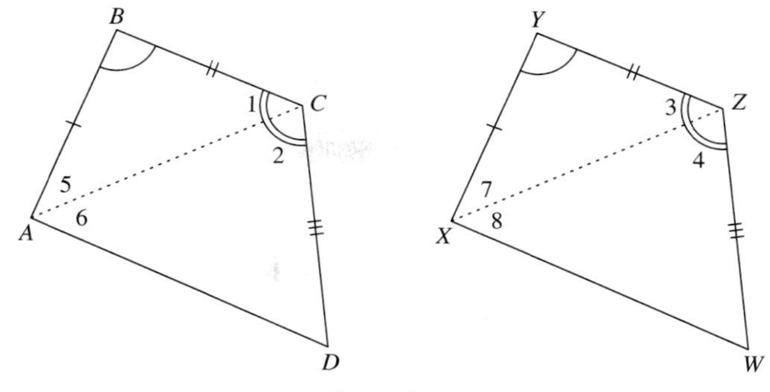

Figure 3.57

Other congruence theorems for convex quadrilaterals, symbolized in the usual manner, may also be derived (they make good problems for you and good exercises in proof writing):

- ASASA Theorem
- SASAA Theorem
- SASSS Theorem

QUESTION: *Do you think ASAAA is a valid congruence criterion for convex quadrilaterals?*

OUR GEOMETRIC WORLD

Unlike triangles, quadrilaterals are unstable geometric designs. This is reflected in the fact that a quadrilateral can be made rigid only by fixing one of its angles, in addition to its four sides (the SASSS congruence criterion is valid, but SSSS is not). If a single brace is added (dotted line in Figure 3.58), then the whole assembly becomes (theoretically) stable.

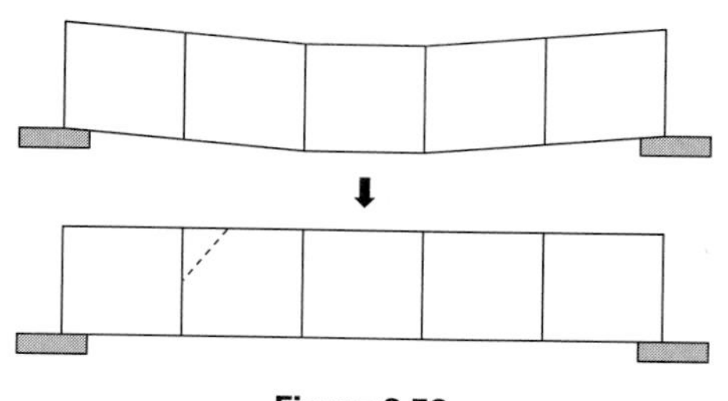

Figure 3.58

SACCHERI, LAMBERT QUADRILATERALS

Our main goal now is to develop the basic properties of the so-called **Saccheri Quadrilateral.** This figure is named after the first geometer to pursue seriously the consequences of a non-Euclidean hypothesis for parallels, Girolamo Saccheri (pronounced "Sack-er'-ee") (1667–1733), a Jesuit priest. The Saccheri Quadrilateral is about the nearest thing we can get to a rectangle in absolute geometry. In defining a rectangle, we have to be careful not to presume the Euclidean Postulate for Parallels and its consequences; our definition must be meaningful in absolute geometry.

DEFINITION: A **rectangle** is a convex quadrilateral having four right angles.

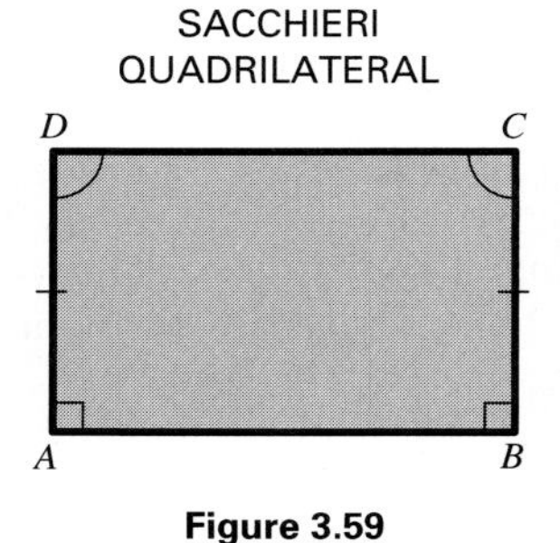

Figure 3.59

DEFINITION: Let $\overline{AB}$ be any line segment, and erect two perpendiculars at the endpoints A and B. (See Figure 3.59.) Mark off points C and D on these perpendiculars so that C and D lie on the same side of line $\overleftrightarrow{AB}$, and $BC = AD$. The resulting quadrilateral is a **Saccheri Quadrilateral.** Side $\overline{AB}$ is called the **base,** $\overline{BC}$ and $\overline{AD}$ the **legs,** and side $\overline{CD}$ the **summit.** The angles at C and D are called the **summit angles.**

To see what a Saccheri Quadrilateral might look like in non-Euclidean geometry, consider the models previously introduced: Poincaré's Model and the unit sphere, as shown in Figure 3.60. For the Poincaré Model, we construct a segment $\overline{AB}$ having O (center of $\boldsymbol{C}$) as midpoint, then construct circles at A and B perpendicular to $\overline{AB}$ and to $\boldsymbol{C}$. Finally, construct another circle orthogonal to $\boldsymbol{C}$ centered directly above O cutting the other two circles at C and D. The quadrilateral $ABCD$ can be seen to have the required features to make it a Saccheri Quadrilateral. On the sphere, take $\overline{AB}$ on the Equator and draw two great circles intersecting at the north and south poles, marking

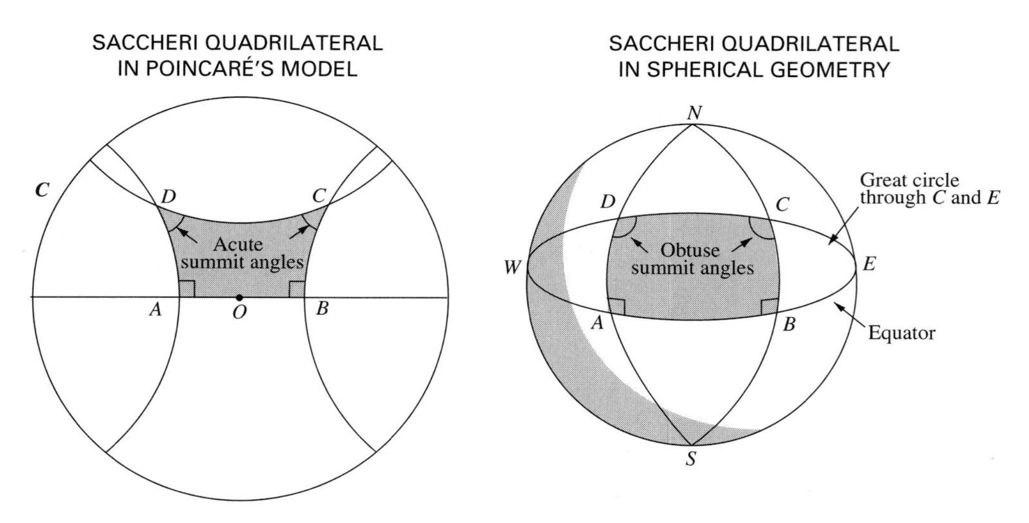

Figure 3.60

equal arcs from A and B along those circles, to obtain C and D. Observe that a Saccheri Quadrilateral in the Poincaré Model has *acute* summit angles (the angles at C and D), whereas the one on the unit sphere has *obtuse* summit angles.

By reviewing the previously stated properties of convex quadrilaterals, and the fact that because of the uniqueness of perpendiculars from an external point in absolute geometry, lines $\overline{BC}$ and $\overline{AD}$ cannot meet, we conclude that

LEMMA: A Saccheri Quadrilateral is convex.

The basic result concerning a Saccheri Quadrilateral in absolute geometry is next.

THEOREM 2: The summit angles of a Saccheri Quadrilateral are congruent.

PROOF

(1) In Figure 3.61, $\overline{DA} \cong \overline{CB}$, $\angle A \cong \angle B$, $\overline{AB} \cong \overline{BA}$, $\angle B \cong \angle A$ and $\overline{CB} \cong \overline{DA}$ (definition)

(2) $\diamond DABC \cong \diamond CBAD$, under the correspondence $DABC \leftrightarrow CBAD$. (SASAS Theorem)

(3) $\therefore \angle C \cong \angle D$. (CPCF)

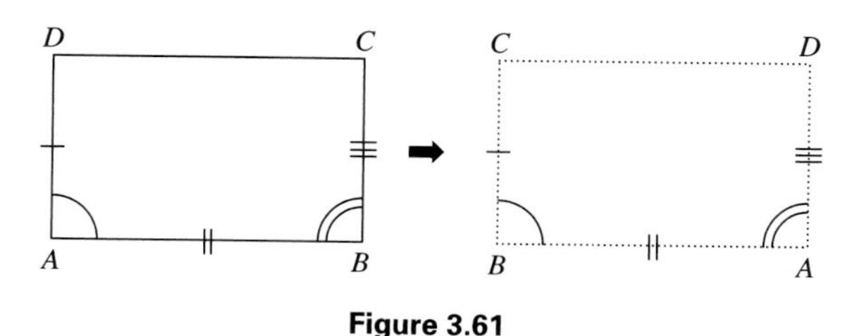

Figure 3.61

Numerous properties of Saccheri Quadrilaterals may be proven as corollaries to Theorem 2, which we simply list for possible additional exercises. They are all intuitively obvious if we appeal to geometric symmetry.

- The diagonals of a Saccheri Quadrilateral are congruent.
- The line joining the midpoints of the base and summit of a Saccheri Quadrilateral is the perpendicular bisector of both the base and summit.
- If each of the summit angles of a Saccheri Quadrilateral is a right angle, the quadrilateral is a rectangle, and the summit is congruent to the base.

LAMBERT QUADRIALTERAL

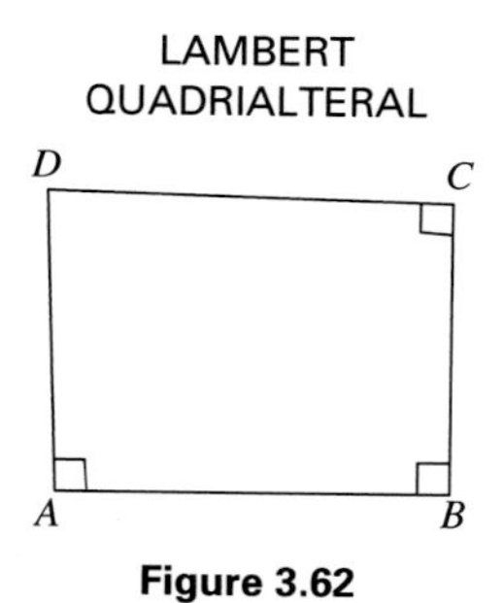

Figure 3.62

NOTE: The existence of a common perpendicular bisector of the base and summit of a Saccheri Quadrilateral proves the existence of a quadrilateral in absolute geometry having *three* right angles, called a **Lambert Quadrilateral,** after J.H. Lambert (1728–1777), another early pioneer in the development of non-Euclidean geometry. (See Figure 3.62.)

ATTEMPTS TO REPAIR EUCLID'S ELEMENTS

Saccheri dreamed of providing a proof of Euclid's Parallel Postulate. His method was to investigate systematically the three possible cases for a given Saccheri Quadrilateral.

Hypothesis of the Obtuse Angle:	Summit angles of a Saccheri Quadrilateral are obtuse angles.
Hypothesis of the Right Angle:	Summit angles of a Saccheri Quadrilateral are right angles.
Hypothesis of the Acute Angle:	Summit angles of a Saccheri Quadrilateral are acute angles.

Saccheri titled his work *Euclides ab omne naevo vindicatus* ("Euclid freed of every flaw"), which was published in 1733, the year of his death. However, had he succeeded in his effort, namely, to eliminate the Hypothesis of the Acute Angle, he would have destroyed rather than repaired Euclidean geometry on logical grounds. For this would have meant that Euclidean geometry is self-contradictory due to the validity of models for non-Euclidean geometry that can be constructed with the Euclidean axioms. (In Chapter 6, we shall study two of those models in detail.)

HISTORICAL NOTE

Saccheri and Lambert (pictured) were the forerunners of non-Euclidean geometry. They were the first to start with a hypothesis contrary to Euclidean geometry and derive results from that hypothesis. Having shown that the Hypothesis of the Obtuse Angle is impossible, Saccheri next assumed the Hypothesis of the Acute Angle, hoping to reach another contradiction. (This would have shown that the Hypothesis of the Right Angle was the only valid one.)

His work in 1733 established important facts about absolute geometry, some of which appear in this section, but the sought-for contradiction never showed up. In desperation, thinking there had to be a contradiction, Saccheri appealed to the "intuitive nature of the straight line." Thirty years later, in 1766, Lambert undertook a similar investigation, revealing unusual flashes of insight. Among them was a plausibility argument showing that if the angle sums of triangles are not 180, then there exists an absolute unit of measure. This was to be proven 60 years later by both Bolyai and Lobachevski, the recognized founders of non-Euclidean geometry.

As we have said before, one of the reasons we are postponing the study of parallel lines for so long is chiefly historical—so we can appreciate some of the great ideas of geometry that were invented in order to prove that, among other things, the angle sum of a triangle equals 180.

ELIMINATING THE HYPOTHESIS OF THE OBTUSE ANGLE

One of Saccheri's ingenious methods was to take a given triangle and construct a quadrilateral in such a way that two of its angles are right angles and the other two angles together have the same measure as the angle sum of the given triangle. You might guess that this quadrilateral turns out to be a Saccheri quadrilateral.

The construction is as follows (see Figure 3.63). Locate the midpoints M and N of sides $\overline{AB}$ and $\overline{AC}$ of any triangle ABC, and draw line $\ell = \overleftrightarrow{MN}$. Then drop perpendiculars $\overline{BB'}$ and $\overline{CC'}$ from B and C to line ℓ. You will be given some direction in the problem set that follows for proving that (1) $\diamond BCC'B'$ is a Saccheri Quadrilateral, with $BB' = CC'$ and congruent summit angles at B and C (the quadrilateral is, so to speak, turned *upside down*), and (2) the angle sum of $\triangle ABC$ has twice the value of the measure x of each summit angle. The resulting quadrilateral is called the **Saccheri Quadrilateral associated with** $\triangle ABC$. (See Problems 16 and 17.)

Does this idea of Saccheri prove anything important? Since we know that the angle sum of a triangle ≤ 180, then

$$2x \leq 180 \qquad \rightarrow \qquad x \leq 90$$

TRIANGLE AND ITS ASSOCIATED SACCHERI QUADRIALTERAL

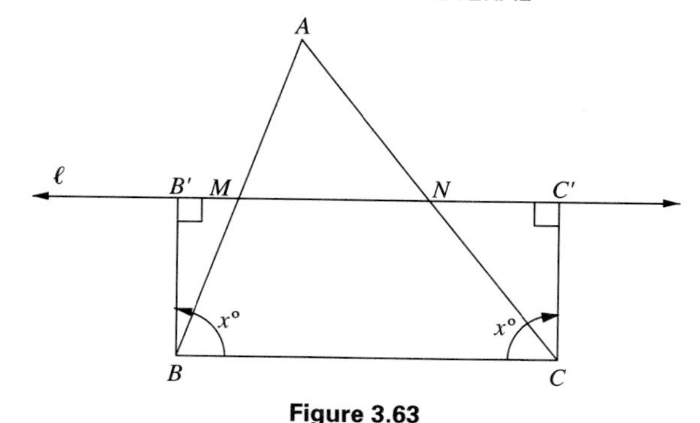

Figure 3.63

That is, the summit angles of this Saccheri quadrilateral are either right angles, or acute, and we have eliminated the Hypothesis of the Obtuse Angle. (This brings up the question whether we could have started with an arbitrary Saccheri Quadrilateral and constructed an associated triangle like the one above; can you think of a way?)

THEOREM 3: The Hypothesis of the Obtuse Angle is not valid in absolute geometry.

This left only the Hypothesis of the Acute Angle for Saccheri to show to be false. He did not succeed in this venture, and today we know that this is impossible to accomplish with only the axioms of absolute geometry.

One further fact of interest can be deduced from the construction in Figure 3.63: The length of the base $\overline{B'C'}$ of the Saccheri Quadrilateral associated with $\triangle ABC$ equals twice the length of the line segment $\overline{MN}$ (Problem 17). Put this together with the fact that $m\angle B'BC \leq m\angle BB'C' = 90$, and we can prove two more results:

- The summit of a Saccheri Quadrilateral has length greater than or equal to that of the base. (See Problem 21.)
- The line segment joining the midpoints of two sides of a triangle has length less than or equal to one-half that of the third side.

After properties of parallel lines have been introduced in Chapter 4, we prove the more familiar result of Euclidean geometry that the line segment mentioned in the second result above has measure *equal* to one-half that of the third side.

PROBLEMS (3.7)

GROUP A

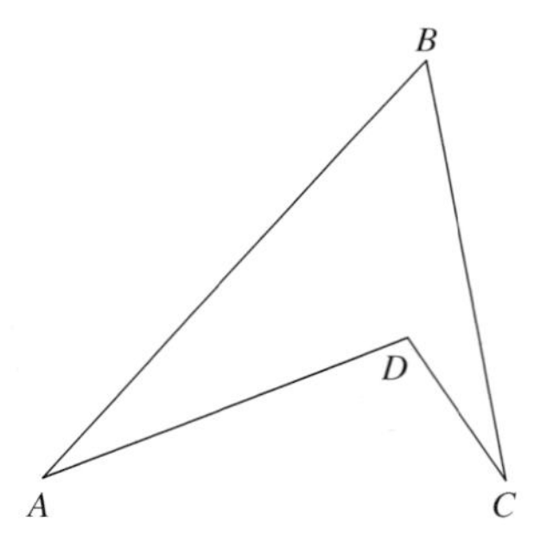

1. In terms of the criterion for a convex quadrilateral given in this section, why is $\diamond ABCD$ shown here not convex?

2. In $\diamond ABCD$ of Problem 1, point D lies in the interior of $\angle ABC$. Answer each of the following:

(a) Does A lie in the interior of $\angle BCD$?

(b) Does B lie in the interior of $\angle ADC$?

(c) Does C lie in the interior of $\angle BAD$?

3. In $\diamond ABCD$ of Problem 1, are the following statements true or false?

(a) $m\angle ABC = m\angle ABD + m\angle DBC$.

(b) $m\angle BAD = m\angle BAC + m\angle CAD$.

4. A convex quadrilateral $\diamond ABCD$ has right angles at A, B, and C, as shown. By drawing a diagonal, show that in absolute geometry, $m\angle D \leq 90$.

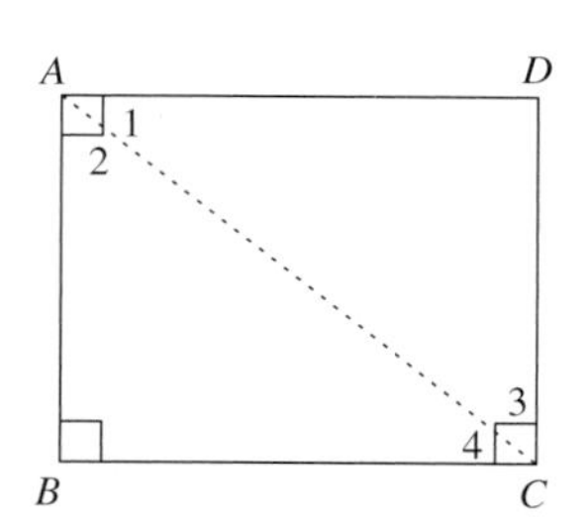

5. In convex quadrilateral $\diamond PQRS$, $\angle PQR$ is a right angle, $m\angle R = 50$ and $m\angle PSR = 150$. Show that $m\angle P \le 70$.

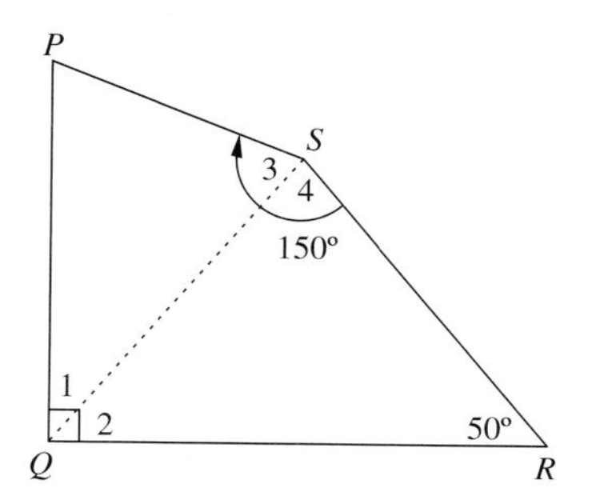

6. State, without proof, the ASASA Congruence Theorem in terms of convex quadrilaterals $\diamond ABCD$ and $\diamond XYZW$.

7. State, without proof, the SASAA Congruence Theorem in terms of convex quadrilaterals $\diamond ABCD$ and $\diamond XYZW$.

8. In the Lambert Quadrilateral $\diamond ABCD$ shown, prove that $AD = CD$ and ray $\overrightarrow{BD}$ bisects $\angle ABC$.

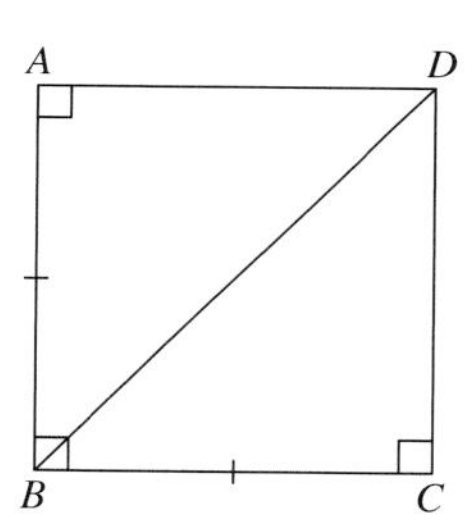

9. Quadrilateral $\diamond AEFD$ is convex, with midpoints B and C of sides $\overline{AE}$ and $\overline{FD}$ located, and segment $\overline{BC}$ drawn.

(a) If $\diamond ABCD$ is a rectangle, show that $\diamond BEFC$ and $\diamond AEFD$ are rectangles.

(b) If $\diamond AEFD$ is a rectangle, show that $\diamond ABCD$ is a rectangle. (For this part you may "borrow" the result, proven later in Problem 18, that the opposite sides of a rectangle are congruent.)

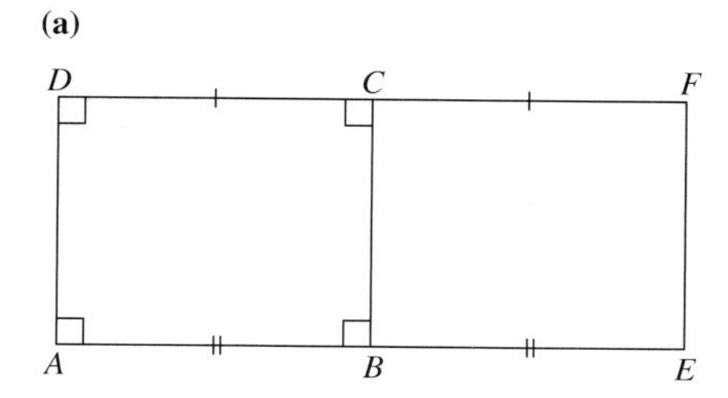

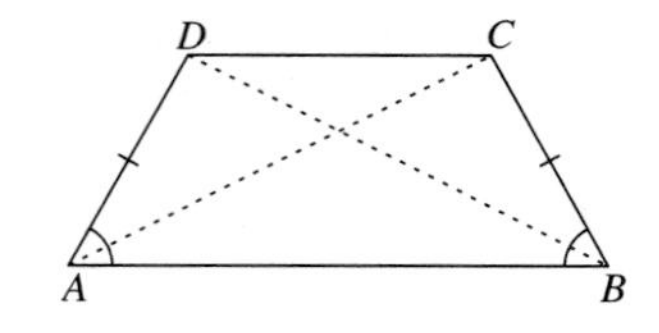

10. In quadrilateral $\diamond ABCD$, $\overline{AD} \cong \overline{BC}$ and $\angle A \cong \angle B$.

(a) Prove $\overline{AC} \cong \overline{BD}$.

(b) Use this result to prove that the diagonals of a Saccheri Quadrilateral are congruent.

11. Prove that if the summit and base of a Saccheri Quadrilateral have equal lengths, the quadrilateral is a rectangle.

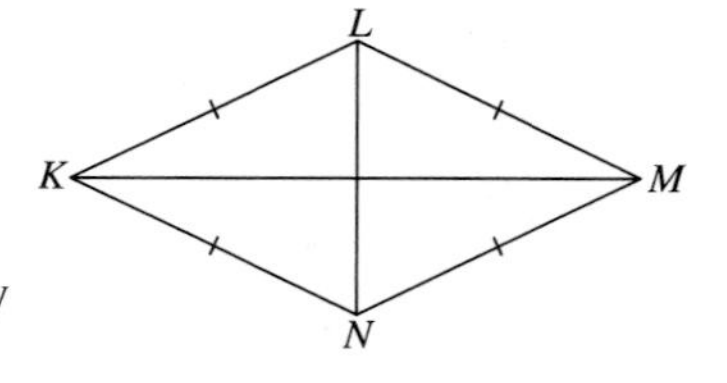

12. An equilateral convex quadrilateral is given ($\diamond KLMN$). Show:

(a) The diagonals are perpendicular.

(b) Opposite angles are congruent.

13. Show that if two Lambert Quadrilaterals $\diamond ABCD$ and $\diamond GEFH$, with $BC = GH$, $AB = GE$ and right angles at A, B, C, E, G, and H, are placed side by side, as shown, the resulting quadrilateral $\diamond AEFD$ is a Saccheri Quadrilateral.

D C(H) F

A B(G) E

GROUP B

14. Prove the ASASA Congruence Theorem. (You must prove that the remaining three corresponding parts of the two quadrilaterals are congruent.)

15. Prove the SASSS Congruence Theorem. (See comment in Problem 14.)

16. In Figure 3.63 (duplicated here for this problem) the perpendicular $\overline{AQ}$ to line ℓ is drawn. Prove by congruent right triangles that $BB' = AQ$ and $AQ = CC'$. What conclusion can be drawn?

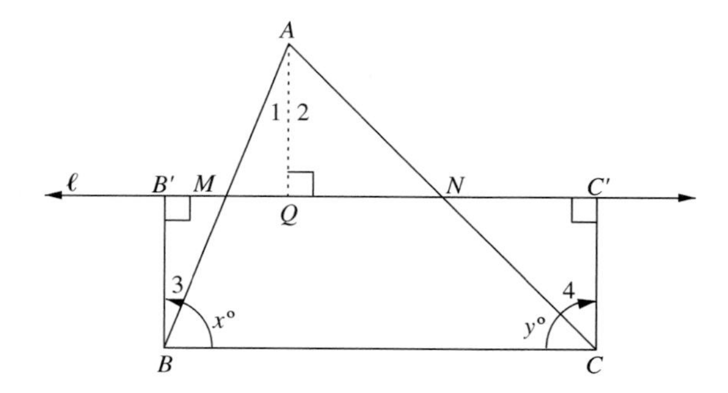

17. In the figure for Problem 16, prove

(a) $m\angle A + m\angle ABC + m\angle BCA = m\angle B'BC + m\angle C'CB = 2x$

(b) $x \leq 90$

(c) $B'C' = 2MN$.

18. Prove that the opposite sides of a rectangle are congruent. (***Hint:*** Assume one side, say $\overline{AB}$, has length greater than that of the opposite side $\overline{CD}$. What construction can you make on segment $\overline{AB}$? Use the Exterior Angle Inequality.)

19. Use Euclidean geometry to draw a figure to show that, unlike triangles, specifying the measures of the sides does not uniquely determine a convex quadrilateral. (Thus the saying, *quadrilaterals are not rigid* becomes meaningful.)

20. Use Euclidean geometry to draw a figure to show that, unlike triangles, specifying the measures of the angles and one side does not determine a unique convex quadrilateral. (Thus, there is no ASAAA Congruence Criterion.)

GROUP C

21. Prove that the base of a Saccheri Quadrilateral in absolute geometry has length less than or equal to that of the summit. (***Hint:*** Show that $m\angle 1 + m\angle 2 = 90 \geq m\angle 2 + m\angle 3$, which implies that $m\angle 1 \geq m\angle 3$; apply the Hinge Theorem in Section 3.5 to $\triangle CAD$ and $\triangle ACB$.)

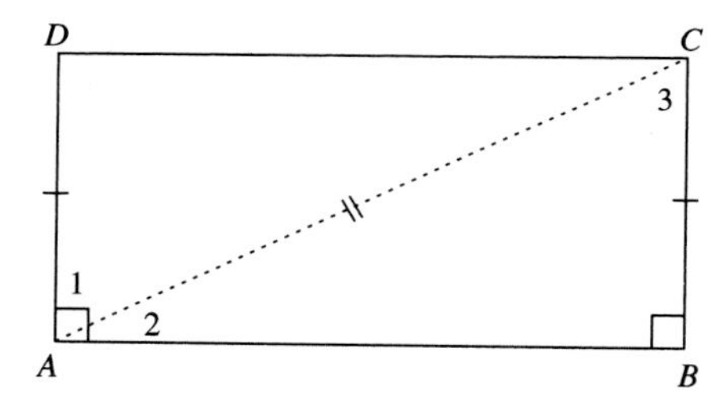

22. Prove that the segment joining any two points of opposite sides of a convex quadrilateral forms two adjacent convex quadrilaterals. (Prove that the new quadrilaterals each have the property that each diagonal lies between opposite vertices.)

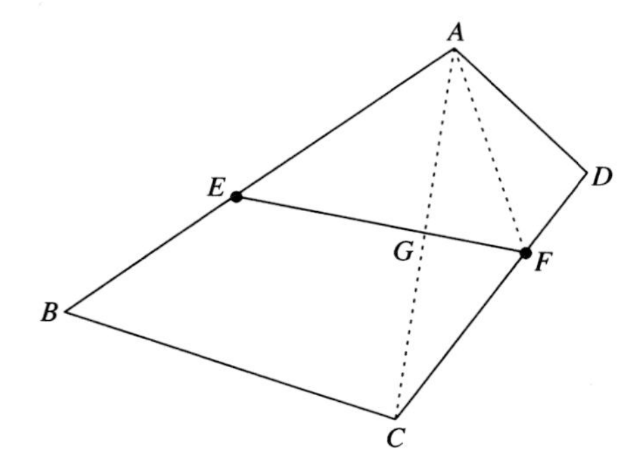

23. Determine whether ASAASA denotes a valid congruence criterion for convex quadrilaterals. (If you decide to look for a counterexample, Euclidean geometry may be used.)

24. As in Problem 23, determine whether SSASA denotes a valid congruence criterion for convex quadrilaterals.

25. Undergraduate Research Project Write a paper on all the possible congruence criteria for convex quadrilaterals

(a) in absolute geometry

(b) in Euclidean geometry (involving, for example, Problem 22, Section 4.2).

3.8 Circles

Many familiar properties of circles carry over to absolute geometry. We need a definition to introduce the subject formally.

DEFINITION: A **circle** is the set of points in a plane that lies at a positive, fixed distance r from some fixed point O. The number r is called the **radius** (as well as any line segment joining point O to any point on the circle), and the fixed point O is called the **center** of the circle. A point P is said to be **interior** to the circle, or an **interior point,** whenever $OP < r$; if $OP > r$, then P is said to be an **exterior point.**

The numerous other terms commonly associated with circles, such as **diameter, chord,** and **tangent,** *will be defined by a pictorial glossary.* From this you should be able to write formal definitions of these terms if it ever becomes necessary. (See Figure 3.64.)

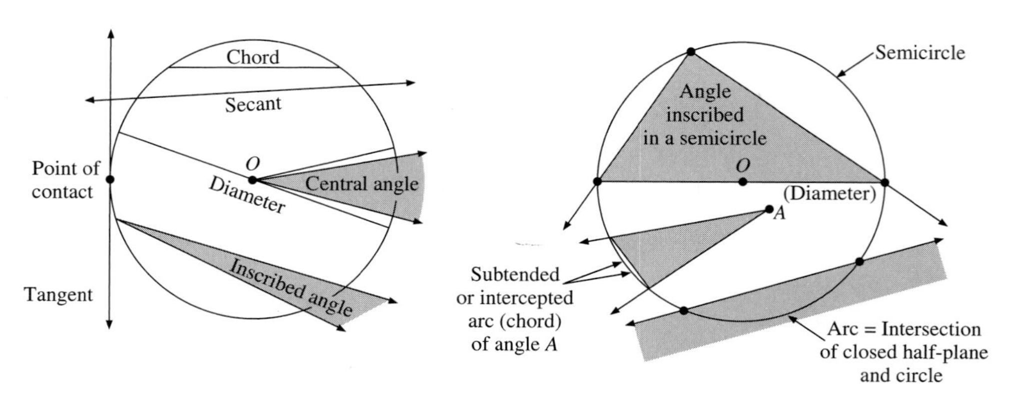

Figure 3.64

THE ELEMENTARY PROPERTIES

All the basic theorems for circles may be proven by the use of the congruence theorems given so far, so most of these results will be left as problems. A few examples are stated as follows.

- The center of a circle is the midpoint of any diameter.
- The perpendicular bisector of any chord of a circle passes through the center.
- A line passing through the center of a circle and perpendicular to a chord bisects the chord.
- Two congruent central angles subtend congruent chords, and conversely.
- Two chords equidistant from the center of a circle have equal lengths, and conversely.

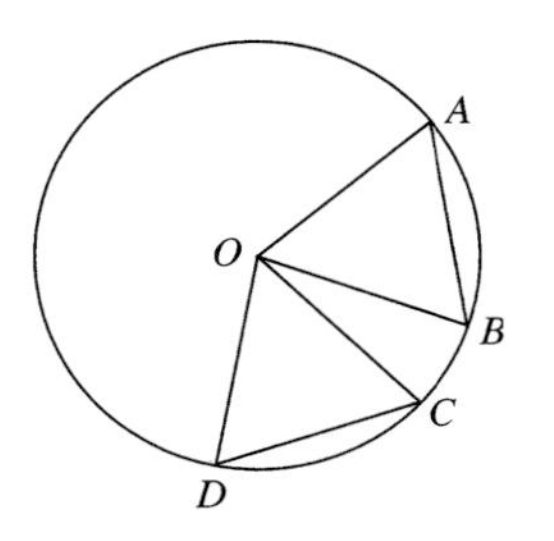

Figure 3.65

EXAMPLE 1 Prove the property of circles that two congruent central angles subtend congruent chords, and conversely.

SOLUTION

Let the circle be as shown in Figure 3.65. By hypothesis, we have $\angle AOB \cong \angle COD$ and must prove that $\overline{AB} \cong \overline{CD}$. But, by the definition of a circle, $\overline{OA} \cong \overline{OB} \cong \overline{OC} \cong \overline{OD}$, so $\triangle AOB \cong \triangle COD$ by SAS. Hence $\overline{AB} \cong \overline{CD}$ by CPCF. For the converse, we are given that $\overline{AB} \cong \overline{CD}$ and must prove that $\angle AOB \cong \angle COD$. This time SSS applies to prove $\triangle AOB \cong \angle COD$, and therefore $\angle AOB \cong \angle COD$. ■

EXAMPLE 2 Prove that if two chords of a circle are equidistant from the center, the chords are congruent. Conversely, prove that congruent chords are equidistant from the center. (See Figure 3.66.)

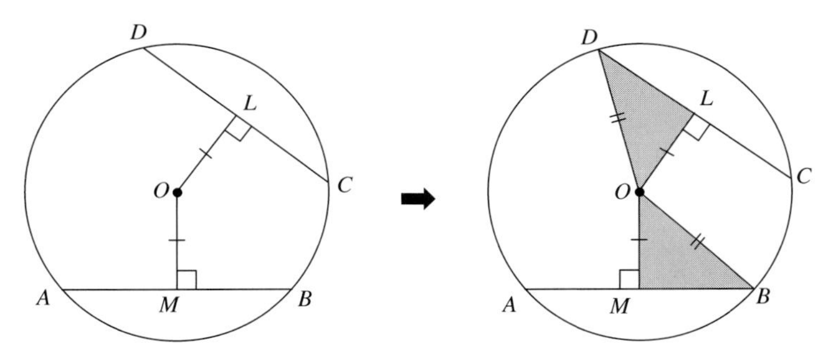

Figure 3.66

SOLUTION

(1) It is given that $OM = OL$, where $\overline{OM} \perp \overline{AB}$ and $\overline{OL} \perp \overline{CD}$. It must be shown that $\overline{AB} \cong \overline{CD}$. Draw radii $\overline{OB}$ and $\overline{OD}$, as in Figure 3.66. Then $OB = OD$ and $OM = OL$ (given). Since $\triangle BOM$ and $\triangle DOL$ are right triangles, by the HL Theorem $\triangle BOM \cong \triangle DOL$. $\therefore BM = DL$. Since M and L are midpoints of chords $\overline{AB}$ and $\overline{CD}$, it follows that $AB = CD$.

(2) For the converse, $AB = CD$ is given, and it must be shown that $OM = OL$. But this argument is essentially the reverse of the one just given, hence it follows that $OM = OL$ and the chords are equidistant from O. ■

CIRCULAR ARC MEASURE

A natural idea in geometry, and a useful one, is that of circular arc measure and its property of additivity (the measure of the union of two nonoverlapping arcs equals the sum of the measures of the given arcs). This can be easily introduced into absolute geometry in the conventional way. This development will be used later when we want to establish the properties of inscribed angles of circles in Euclidean geometry (Chapter 4).

DEFINITION: As shown in Figure 3.67, consider the three types of arcs of a circle with center O, along with their measures. A **minor arc** is the intersection of the circle with a central angle and its interior, a **semicircle** is the intersection of the circle with a closed half-plane whose edge passes through O, and a **major arc** of a circle is the intersection of the circle and a central angle and its exterior (that is, the complement of a minor arc, plus end points). If the end points of an arc are A and B, and C is any other point of the arc (which must be used in order to uniquely identify the arc), then we define the **measure** $m\widehat{ACB}$ of the arc as follows.

MINOR ARC	SEMICIRCLE	MAJOR ARC
$m\widehat{ACB} = m\angle AOB$	$m\widehat{ACB} = 180$	$m\widehat{ACB} = 360 - m\angle AOB$

THE THREE TYPES OF ARCS AND THEIR MEASURES

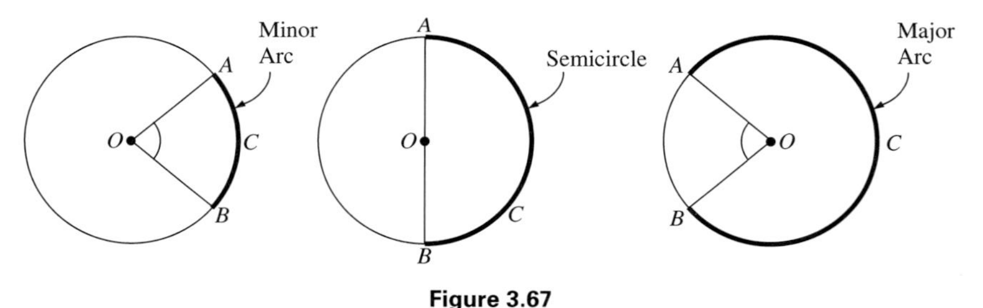

Figure 3.67

Having three separate definitions for the measure of a circular arc makes proving the additive property problematic. It helps to carefully organize the numerous possible cases. First, observe that if we are given two arcs on a circle, one of them has to be a minor arc (there is not enough room on a circle to hold *two* nonoverlapping major arcs). So, throughout the proof that follows, we will assume that arc $\widehat{APB}$ is a minor arc, as in Figure 3.68. The cases to prove are then determined merely by the kind of arc that the combined arc $\widehat{ABC} = \widehat{APB} \cup \widehat{BQC}$ is. Further, we shall arrange things so that arc $\widehat{APB}$ always has its end point A on a diameter of the circle and the rest of the arc lies in a half-plane H_1 determined by that diameter.

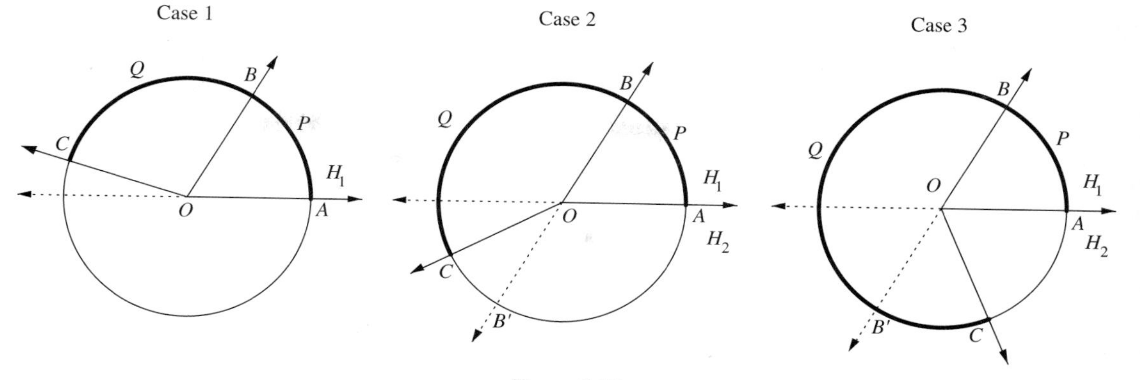

Figure 3.68

> **THEOREM 1: ADDITIVITY OF ARC MEASURE**
> Suppose arcs $\boldsymbol{A}_1 = \widehat{APB}$ and $\boldsymbol{A}_2 = \widehat{BQC}$ are any two arcs of circle O having just one point B in common and such that their union $\boldsymbol{A}_1 \cup \boldsymbol{A}_2 = \widehat{ABC}$ is also an arc. Then $m(\boldsymbol{A}_1 \cup \boldsymbol{A}_2) = m\boldsymbol{A}_1 + m\boldsymbol{A}_2$.

PROOF (OPTIONAL)

CASE 1. When $\widehat{ABC}$ is a minor arc or a semicircle.

First, if $\widehat{ABC}$ is a minor arc, then, except for end points, it lies in the interior of $\angle AOC$. Hence B is an interior point of $\angle AOC$. Therefore $\overrightarrow{OA}$–$\overrightarrow{OB}$–$\overrightarrow{OC}$ and we have, by definition,

$$m\widehat{ABC} = m\angle AOC = m\angle AOB + m\angle BOC = m\widehat{APB} + m\widehat{BQC}$$

as desired. If arc $\widehat{ABC}$ is a semicircle, by the Linear Pair Axiom, $m\angle AOB + m\angle BOC = 180$ and we have

$$m\widehat{ABC} = 180 = m\angle AOB + m\angle BOC = m\widehat{APB} + m\widehat{BQC}$$

CASE 2. When $\widehat{ABC}$ is a major arc and $\widehat{BQC}$ is either a minor arc or a semicircle. Let $\overrightarrow{OB'}$ be the ray opposite ray $\overrightarrow{OB}$, so that both $\overrightarrow{OB'}$ and $\overrightarrow{OC}$ lie in H_2, the opposite half-plane of H_1. Then if $\widehat{BQC}$ is a minor arc, $\overrightarrow{OC}$–$\overrightarrow{OB'}$–$\overrightarrow{OA}$. Hence,

$$\begin{aligned} m\widehat{ABC} &= 360 - m\angle AOC \\ &= 180 - (m\angle AOB' + m\angle B'OC) + 180 \\ &= (180 - m\angle AOB') + (180 - m\angle B'OC) \\ &= m\angle AOB + m\angle BOC \text{ (by the Linear Pair Axiom)} \\ &= m\widehat{APC} + m\widehat{BQC}. \end{aligned}$$

If $\widehat{BQC}$ is a semicircle, a similar argument is valid.

CASE 3. When both $\widehat{ABC}$ and $\widehat{BQC}$ are major arcs.

In this case, $\overrightarrow{OC}$ and $\overrightarrow{OA}$ lie on the same side of line $\overleftrightarrow{BB'}$ and $\overrightarrow{OB}$–$\overrightarrow{OA}$–$\overrightarrow{OC}$. Then

$$\begin{aligned} m\widehat{ABC} &= 360 - m\angle AOC \\ &= 360 - (m\angle BOC - m\angle BOA) \\ &= (360 - m\angle BOC) + m\angle AOB \\ &= m\widehat{BQC} + m\widehat{APC}. \end{aligned}$$

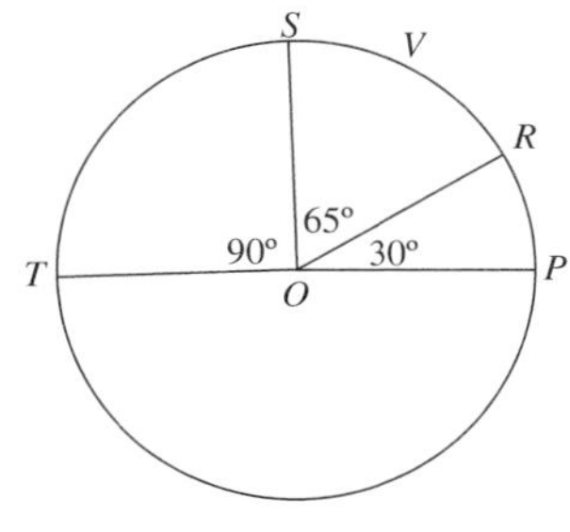

Figure 3.69

EXAMPLE 3 Arc $\widehat{SPT}$ shown in Figure 3.69 is a major arc and is the union of arc $\widehat{SVR}$ (a minor arc) and arc $\widehat{RPT}$ (a major arc). Using the angle measures shown in the figure, determine the measures of each of the three arcs, and verify the property of additivity in this case.

SOLUTION

By definition, we have

$$m\widehat{SVR} = 65$$

Because $\widehat{RPT}$ is a major arc and $m\angle TOR = 90 + 65 = 155$,

$$m\widehat{RPT} = 360 - 155 = 205$$

Finally,

$$m\widehat{SPT} = 360 - 90 = 270$$

Hence,

$$m\widehat{SVR} + m\widehat{RPT} = 65 + 205 = 270 = m\widehat{SPT}. \quad \blacksquare$$

Because we are going to prove theorems about them, we define both a secant and tangent of a circle formally.

DEFINITION: A line that meets a circle in two distinct points is a **secant** of that circle. A line that meets a circle at only one point is called a **tangent** to the circle, and the point in common between them is the **point of contact**, or **point of tangency**.

There is a discovery unit designed to reveal the following basic theorem and its proof at the end of this section.

THEOREM 2: TANGENT THEOREM

A line is tangent to a circle iff it is perpendicular to the radius at the point of contact.

One of the useful corollaries of the tangent theorem is the following.

COROLLARY: If two tangents $\overline{PA}$ and $\overline{PB}$ to a circle O from a common external point P have A and B as the points of contact with the circle, then $\overrightarrow{PA} \cong \overline{PB}$ and $\overrightarrow{PO}$ bisects $\angle APB$ (Figure 3.70).

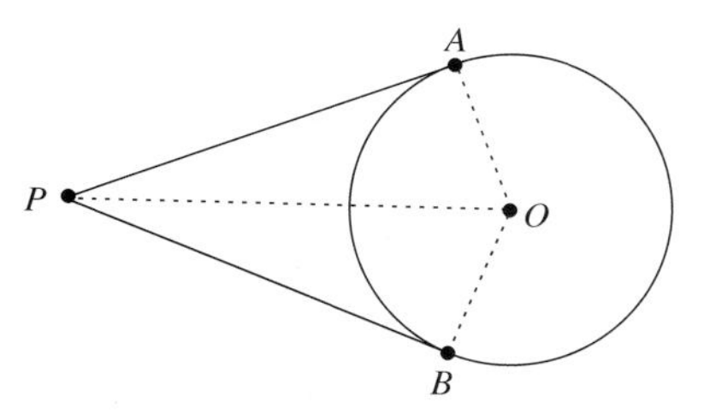

Figure 3.70

(For the proof, just consider the radii $\overline{OA}$, $\overline{OB}$, and the line of center $\overleftrightarrow{PO}$.)

THE SECANT THEOREM

We shall concentrate our remaining efforts on a somewhat deeper result for absolute geometry. If a line passes through an interior point of a circle, it should be obvious that it must intersect that circle and be one of its secants. We may either take this as an additional axiom, or prove it is a theorem. We naturally prefer the latter approach.

Unfortunately, a proof involves recalling a bit of analysis from calculus. (Because some users of this book may not have taken calculus, the proof is optional; if you are in that group, nothing will be lost if you skip to the discovery units and problems at the end of the section.) The preliminary result is a hybrid of both geometry and analysis, and, consequently, so is our proof of the Secant Theorem. A strictly geometric proof is much longer and requires even more preliminary work. One such proof can be found in Moise, *Elementary Geometry from an Advanced Standpoint*.

LEMMA: Given ray $\overrightarrow{AB}$ and any point O, as shown in Figure 3.71, define a function $d(x)$ for any real $x \geq 0$ as the distance from O to P on $\overrightarrow{AB}$, where x is the coordinate of P. That is,

$$d(x) = OP \qquad \text{iff} \qquad x = AP$$

Then $d(x)$ is continuous.

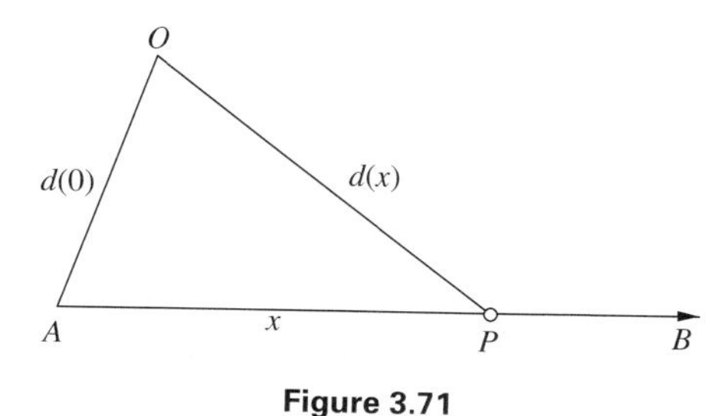

Figure 3.71

Since continuity is intuitively clear here [small changes in x produce small changes in $d(x)$], we leave that part as a problem (see Problem 19). The key to our proof of the Secant Theorem is in recalling the Intermediate Value Theorem of calculus as it applies to the function $d(x)$: *If r lies between $d(a)$ and $d(b)$, then there exists a value x_0 between a and b such that $d(x_0) = r$.*

THEOREM 3: SECANT THEOREM

If a line ℓ passes through an interior point A of a circle, it is a secant of the circle and intersects that circle in precisely two points. (See Figure 3.72.)

PROOF

Since A is an interior point of the circle, $OA < r =$ radius of the circle. Set up the function $d(x)$ for the distance from O, the center, to any point P on ray $\overrightarrow{AB}$ on line ℓ. Again, $d(x) = OP$ iff $x = AP$. Note that because $AP = 0$ for $P = A$, then $d(0) = OA < r$. On the other hand, we can prove that $d(2r) > r$ using the Triangle Inequality: If $AN = 2r$, then $AO + ON > AN$, or $AO + ON > 2r$, and therefore, $ON > 2r - AO$. But $AO < r$, so $ON = d(2r) > 2r - r = r$. Hence, $d(0) < r$ and $d(2r) > r$. By the Intermediate Value Theorem, there exists x_0 between 0 and $2r$ such that $d(x_0) = r$. Let C be that point on ray $\overrightarrow{AB}$ such that $AC = x_0$. Then $OC = d(x_0) = r$, and C lies on the circle. Thus C is a point of intersection of circle O and line ℓ, and an elementary geometric construction will then yield a second point D of intersection. Therefore, ℓ is a secant of circle O, as desired.

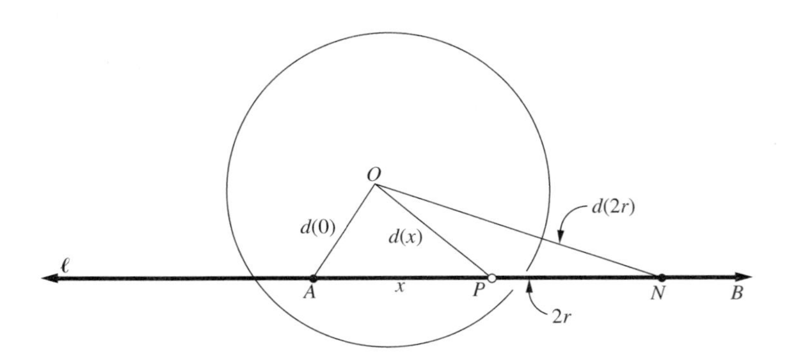

Figure 3.72

NOTE: The Secant Theorem proves, in effect, that a line segment joining a point inside a circle with a point outside must intersect the circle. This is a special case of a more general theorem that applies to a much larger class of closed curves than the circle, including triangles, polygons, and ellipses, for example. They are called **simple closed curves**. Such a curve can be described as *any continuous path beginning and ending at the same point that never crosses itself.* The theorem states that (1) every simple closed curve C in the plane has a well-defined interior and exterior, and (2) if A lies in the interior of C and B lies in the exterior, an arc of another curve joining A and B must intersect C. This is a remarkable result when one realizes that there is no requirement that C enclose a convex region. So it could be quite complex, like the one shown in Figure 3.73. This diagram consists of a map of the hedge that forms the famous maze in the gardens at Hampton Court Palace, England. In this example, it is even difficult to determine which points of the plane lie interior to C and which lie in its exterior. In fact, the diagram actually consists of *two separate closed curves*! Can you find them? The tools of geometry seem not to be of much help in proving a theorem of this type. It took the clever work of C. Jordan (1828–1922) to find a proof. His theorem is a landmark for topologists, and it is appropriately called the **Jordan Closed Curve Theorem**.

Figure 3.73

Moment for Discovery

Property of the Tangent to a Circle

The following steps in *Sketchpad* will lead you to "discover" an important theorem in geometry.

1. Construct an arbitrary circle AB (Figure 3.74).
2. Construct a line ℓ intersecting the circle at points C and D.
3. Construct segment $\overline{AC}$.
4. Display $m\angle ACD$ on the screen.

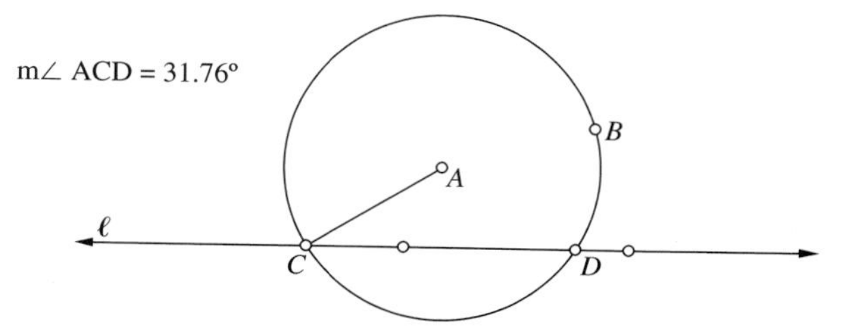

Figure 3.74

Now drag line ℓ parallel to itself (select ℓ). Does $m\angle ACD$ change? What do you discover as points C and D come very close to each other? Make a conjecture, then try to prove it. (***Hints:*** In Figure 3.75, suppose $\overline{AC} \perp \ell$, and let E be any other point on ℓ. Why is $AE > AC$ = radius of circle A? Then is every point of ℓ except C an exterior point? This would prove that ℓ is tangent to circle A. Now reverse this: Prove that if ℓ is tangent to circle A, then $\ell \perp$ radius $\overline{AC}$. Assume $m\angle ACE \neq 90$ and show that ℓ meets the circle at a *second* point D, contradicting that ℓ was a tangent.)

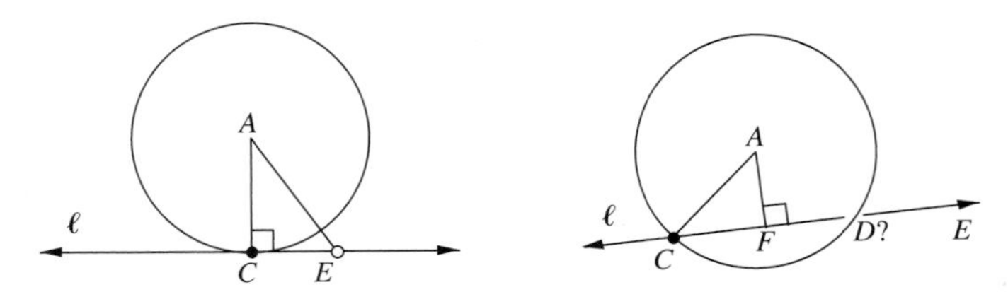

Figure 3.75

Moment for Discovery

The Secant Theorem

If a line passes through an interior point of a circle, the question is whether that line can, perhaps, "roam around" inside the circle and never "get out." (See Figure 3.76.) Even if we could prove that there exist points on the line that are arbitrarily far from the center, we still have not proven that the line actually *intersects the circle.* Obviously, the line cannot just "slip" through the circle, can it? (Actually, it *can* if we are working in the rational plane. See Problem 17 in this connection.) To be specific, consider a circle of radius 5, with A on line ℓ inside the circle.

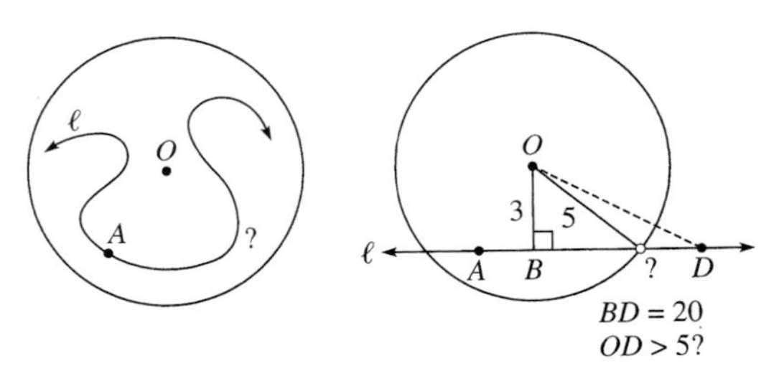

Figure 3.76

1. Locate the foot of the perpendicular, B, on the given line, from A. In the example, $OB = 3$ units. How far from B should the points of intersection of the line and circle be? If this were the Euclidean plane, how might the Pythagorean Theorem be applied?

2. Since we cannot use the Pythagorean Theorem in absolute geometry, locate point D on ℓ so that $BD = 20$. Can we be certain that $OD > 5$? Why?
3. Generalize Step 2 to any circle of radius r and any line ℓ passing through any interior point A.
4. This gives us two points on ℓ, one in the interior of the circle (point A) and one exterior to it (point D). The Jordan Curve Theorem mentioned above then implies that $\overline{AD}$ meets circle O.

PROBLEMS (3.8)

GROUP A

1. Prove that the perpendicular bisector of any chord of a circle passes through its center.
2. Find the measures x and y of the two arcs in the figure, then find $m\widehat{ABC}$ and check the additive property in this case.

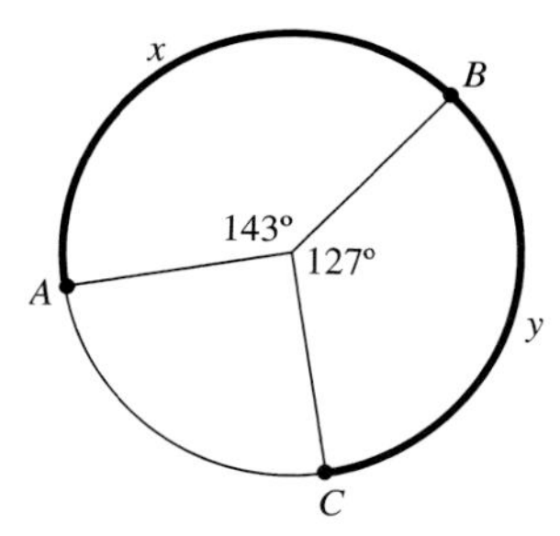

3. Prove two chords of a circle are congruent iff they subtend arcs of equal measure. (You must establish $m\widehat{AB} = m\widehat{CD}$ iff $AB = CD$. Recall that $m\widehat{AB} = m\angle AOB$.)

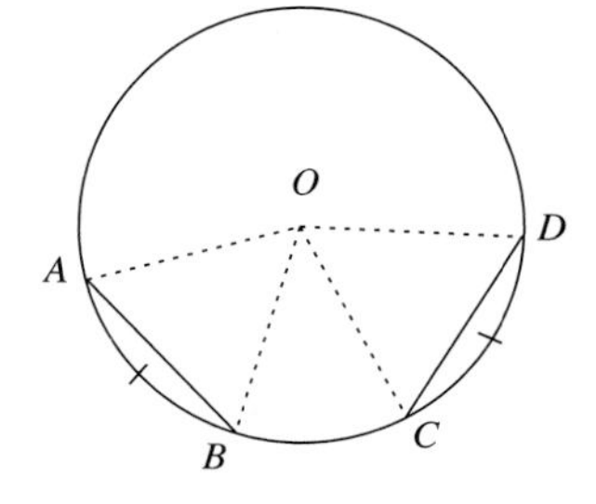

4. Line ℓ passes through the center of a circle. Prove that if ℓ is perpendicular to a chord of that circle, then ℓ bisects the chord. (You must prove that if $\ell \perp \overline{AB}$ in the figure, then M is the midpoint of chord $\overline{AB}$.)

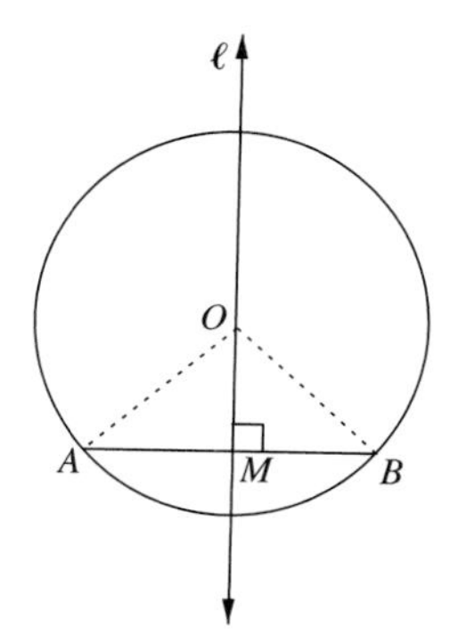

5. Prove or disprove: A chord half as far from the center of a circle as another chord has twice the length.

6. Secants from P are drawn, intersecting a circle to form segments having the measures as indicated. Find x, and prove your answer is correct.

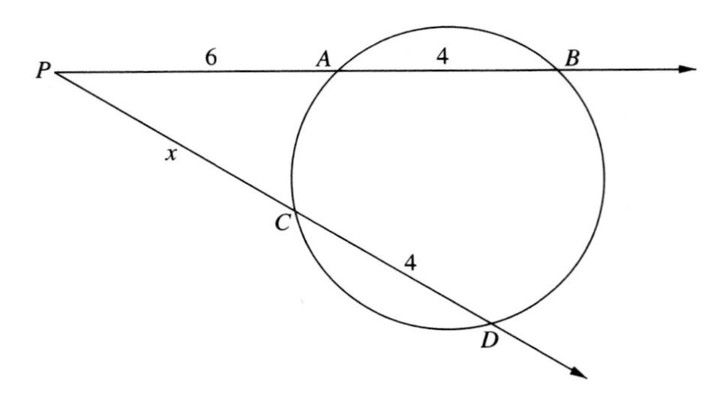

7. This *Sketchpad* project will show the animated motion of a circle tangent to the sides of an angle.

[1] Construct two rays $\overrightarrow{AB}$ and $\overrightarrow{AC}$. Locate point D on ray $\overrightarrow{AB}$.

[2] Construct the bisectors of $\angle ADC$ and $\angle ACD$.

[3] Select, or construct, the intersection E of the two bisectors. Hide the bisectors.

[4] With E selected, select ray $\overrightarrow{AB}$ and choose Perpendicular Line under CONSTRUCT.

[5] Select the point of intersection F of the perpendicular and ray $\overrightarrow{AB}$, then Construct a circle with center E passing through F. (This circle should be tangent to both rays.) Hide the perpendicular.

[6] Select the circle and its center E to trace. (Use Trace under DISPLAY.) What happens when you drag point D?

[7] Select point D and ray $\overrightarrow{AB}$ to animate the circle and its center. (Choose Animate under DISPLAY, and choose the options Bidirectionally and Quickly.)

What happened? What conjecture could be made? Try to prove it.

8. Circles K and T intersect, forming a common chord $\overline{MN}$. Prove that the line of centers $\overleftrightarrow{KT}$ is the perpendicular bisector of $\overline{MN}$.

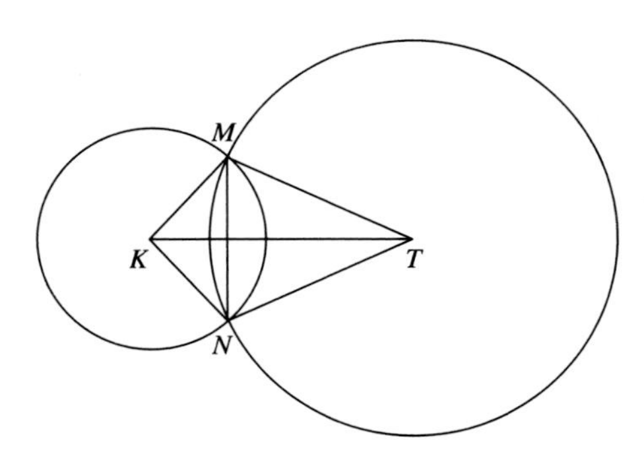

9. In Problem 8 prove that $\angle KMT \cong KNT$.

10. Answer the question posed in Figure 3.73. Try to find the two curves mentioned by imagining that you are gradually pulling on the curve, changing only its shape, to straighten it out. The diagram is topologically equivalent to that shown here. Where are points A and B in this diagram?

GROUP B

11. A circle passes through the vertices of $\diamond ABCD$, and $AB = CD$. Prove that $m\angle A = m\angle D$. (***Hint:*** Use Theorem 1 and the results of Example 1 and Problem 3.)

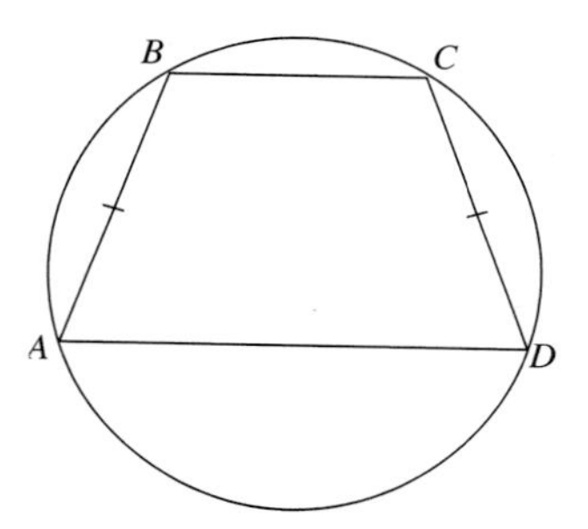

12. Prove the Corollary of the Tangent Theorem. $\overline{PA} \cong \overline{PB}$ and $\overrightarrow{PO}$ bisects $\angle APB$. (***Hint:*** Draw radii $\overline{OA}$ and $\overline{OB}$.)

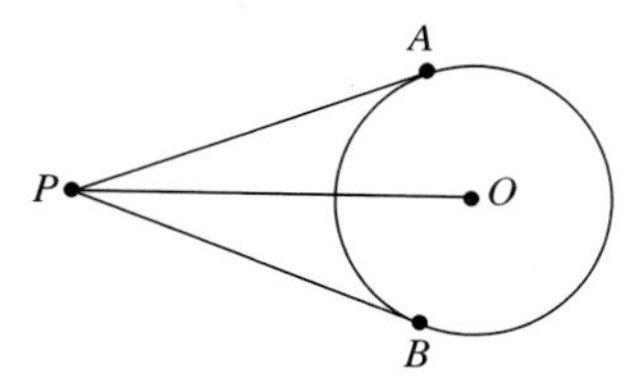

13. Tangents are drawn to circle O from points R and S, which lie on a line passing through the center O. If M, N, P, and Q are the points of contact, prove $m\overset{\frown}{MP} = m\overset{\frown}{NQ}$. (***Hint:*** Use SASAS and the result of Problem 3.)

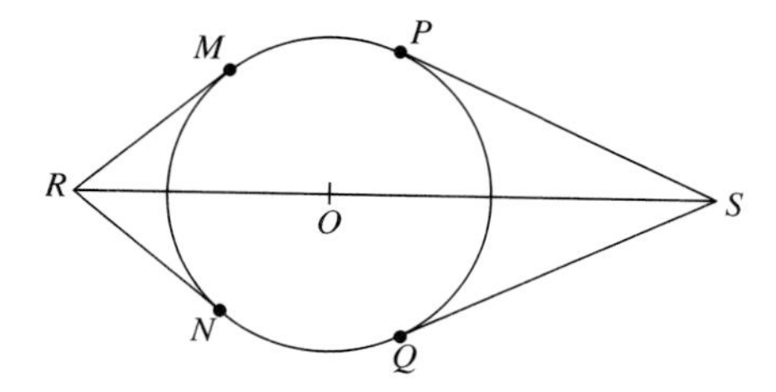

***14. The Incircle of a Triangle** A circle is called the **incircle** of a triangle iff it is tangent to all three sides (and accordingly, lies inside the triangle). The radius and center of the incircle are called the **inradius**, denoted r, and **incenter**, denoted I.

(a) Prove that the incircle exists for $\triangle ABC$. (See Problem 12, Section 3.6.)

(b) Show that, in standard notation, and with **semiperimeter** $s = \frac{1}{2}(a + b + c)$, the measures x, y, and z in the figure are given by

$$x = s - a, \qquad y = s - b, \qquad z = s - c$$

*See Problem 9, Section 4.8.

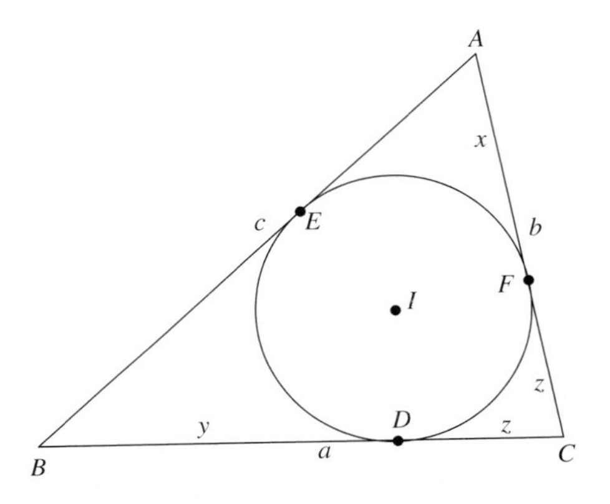

15. Two circles are tangent externally at C, with the common external and internal tangents $\overleftrightarrow{AB}$ and $\overleftrightarrow{CD}$ drawn. Prove that line $\overleftrightarrow{CD}$ bisects segment $\overline{AB}$. (You may assume that $\overleftrightarrow{CD}$ intersects $\overline{AB}$.)

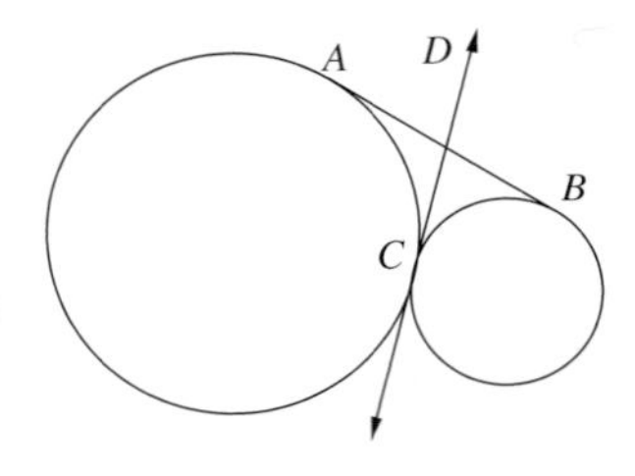

16. Observe the Euclidean four-step construction of the tangent to a circle from a given outside point (P). What well-known theorem in Euclidean geometry do you need to justify it? (It was referred to in Problem 4, Section 1.1; see also Section 4.5.)

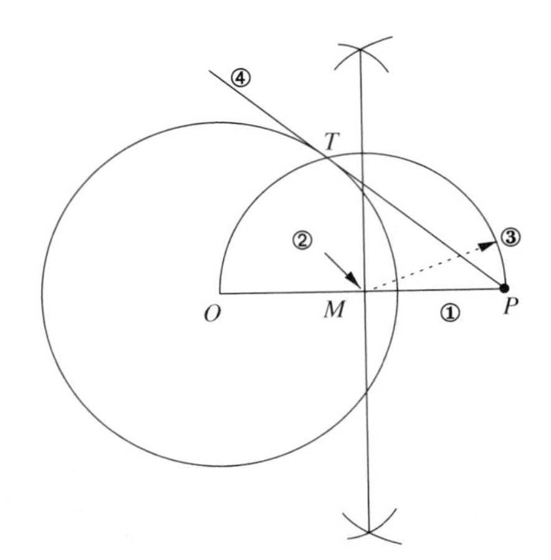

17. If it were possible for the **rational plane** to be a model for absolute geometry, then the Secant Theorem would not be valid. To be specific, consider the coordinate plane where all points $P(x, y)$ have only rational coordinates x and y ($\sqrt{2}$, $\sqrt{3}$, π, etc. are disallowed values for x and y). Consider the circle whose equation is $x^2 + y^2 = 25$ and the line whose equation is $y = 1$ (all variables x and y are rational). Show that this line and this circle have no points in common. Does the line pass through an interior (rational) point of the circle?

GROUP C

18. Prove that the set of interior points of any circle is convex.

19. Prove the lemma that was used for the Secant Theorem, using the triangle inequality. (***Hint:*** You will find the result of Problem 8, Section 3.5 helpful. Applied to $\triangle OPP_0$, it reads: $OP_0 - PP_0 \leq OP \leq OP_0 + PP_0$, which can be transformed into the inequality $-PP_0 \leq OP - OP_0 \leq PP_0$, or $|OP - OP_0| \leq PP_0$. Start your proof like this: Let x_0 be any positive real and $\epsilon > 0$ be given. Choose $\delta = \epsilon$. If $|x - x_0| < \delta$, then $\cdots$)

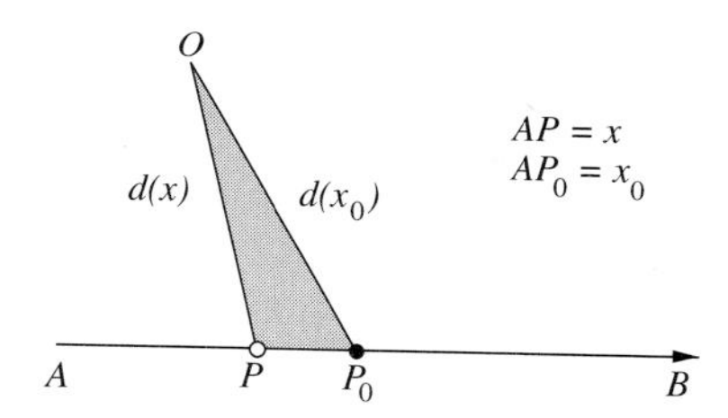

20. Use circles and circular arc measure to prove:

THEOREM: The sum of the measures of all the angles about a point is 360 provided those angles and their interiors contain the entire plane and their interiors are mutually disjoint.

21. Prove the **Two Circle Theorem:** If circles O and O' having radii $r \geq r'$, respectively, have their centers at a distance d apart, where $r - r' < d < r + r'$, then the circles will meet at two distinct points. (***Hint:*** First define for any circle O' and diameter $\overline{AB}$ the function $d(x) = OP$ where O is any fixed point on line $\overleftrightarrow{AB}$ and P varies on circle O' such that $x = m\widehat{AP}$. Prove that $d(x)$ is continuous using an argument similar to that of Problem 19, then use the given inequalities and the Intermediate Value Theorem to prove there exists point P on circle O' such that $d(x) = OP = r$, hence P lies on both circles.)

22. **Converse of Triangle Inequality Theorem** Use the Two Circle Theorem (Problem 21) to prove in absolute geometry that if three numbers a, b, and c satisfy the (strict) Triangle Inequality (that is, $a + b > c$, $a + c > b$, and $b + c > a$), then there exists a triangle $\triangle ABC$ having a, b, and c as the lengths of its sides. Thus,

COROLLARY: Given a segment $\overline{AB}$, there exists an equilateral triangle $\triangle ABC$ having $\overline{AB}$ as base.

Chapter 3 Summary

The heart of absolute geometry is the development in this chapter resulting from the notion of congruent triangles. We saw at first, after looking at taxicab geometry, that for congruence to play a role it would be necessary to add one additional axiom—the SAS Postulate, Axiom C–1 (thus bringing the total number of axioms to 15 so far). The early consequences of the SAS Postulate were then established: The ASA Congruence Theorem and the standard results concerning isosceles triangles and related theorems on equidistant loci. The remaining congruence theorem SSS was then proven, using kites and darts.

The primary inequality theorem in absolute geometry is the Exterior Angle Inequality in triangles. Euclid's proof was revised to acceptable standards of rigor, which then led to many other important inequality theorems, notably the Triangle Inequality ($AB + BC > AC$ if A, B, and C are noncollinear), the Scalene Inequality, involving parts of the *same* triangle (in $\triangle ABC, AB > AC$ iff $m\angle C > m\angle A$), and the so-called Hinge Theorem, or the SAS Inequality, involving parts from *different* triangles (in $\triangle ABC$ and $\triangle XYZ$, if $AB = XY$ and $AC = XZ$, then $m\angle A > m\angle X$ iff $BC > YZ$). Further congruence criteria then result, including the very useful HL, HA, and LA theorems for right triangles.

Finally, quadrilaterals and circles were introduced and basic results proven. The congruence criterion SASAS for convex quadrilaterals was proven—analogous to SAS for triangles. This led to an important concept, the Saccheri Quadrilateral—a convex quadrilateral having two consecutive right angles at the base and two congruent legs. Its line of symmetry divides it into two congruent quadrilaterals, each known as a Lambert Quadrilateral, and having *three* right angles, with the fourth being a right angle only in Euclidean geometry. The Saccheri–Legendre Theorem showing that the angle sum of any triangle does not exceed 180 disallows Saccheri's Hypothesis of the Obtuse Angle in absolute geometry. The basic properties of circles are (1) arc measure is additive, (2) a line is tangent to a circle if and only if it is perpendicular to the radius at the point of contact, and (3) a line is a secant of a circle if and only if it passes through an interior point of the circle.

Testing Your Knowledge

You are expected to take this test using only the list of axioms, definitions, and theorems in Appendix F as references.

1. What is the purpose of taxicab geometry in the study of axiomatic geometry?
2. (a) Find x in the figure below.
   (b) What reason justifies your answer in (a)?

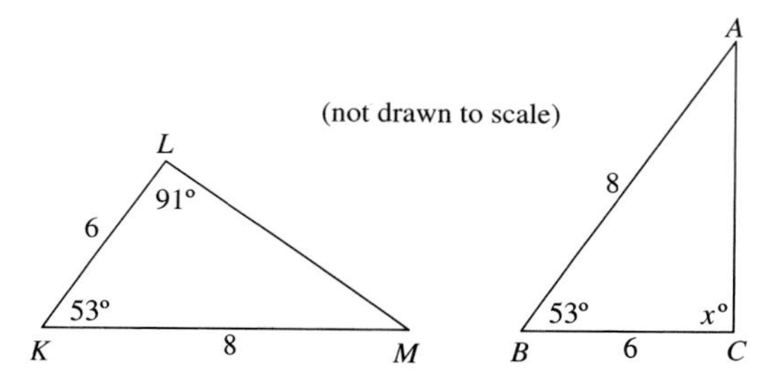

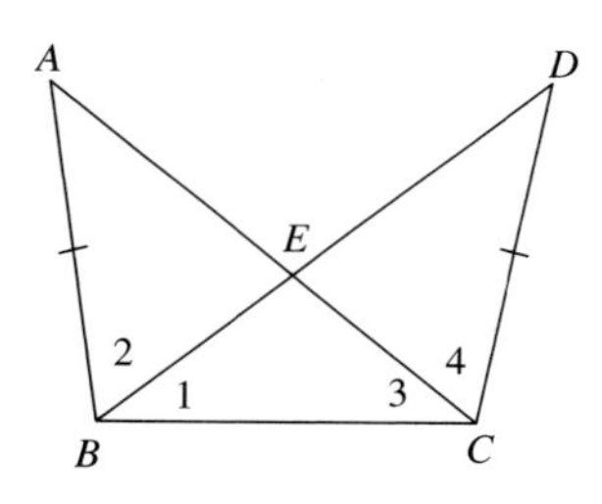

3. *Given:* $m\angle ABC = 100$, $m\angle BCD = 100$, $\overline{AB} \cong \overline{DE}$. with A–E–C, B–E–D.
   Prove: $\overline{AE} \cong \overline{DE}$.

   Fill in the missing blanks in the outline proof that follows.

CONCLUSIONS	JUSTIFICATIONS
(1) $\overline{BC} \cong \overline{CB}$	Reflexive Law of Congruence
(2) $\triangle ABC \cong \triangle DCB$	(a) ________?
(3) $m\angle 1 = m\angle 3$	(b) ________?
(4) $m\angle ABC = m\angle 1 + m\angle 2$ $m\angle BCD = m\angle 3 + m\angle 4$	Angle Addition Postulate
(5) $m\angle 1 + m\angle 2 = m\angle 3 + m\angle 4$	Given, substitution from Step 4
(6) $m\angle 2 = m\angle 4$	Algebra (Steps 3 and 5)
(7) $m\angle AEB = m\angle DEC$	(c) ________?
(8) $\triangle AEB \cong \triangle DEC$	(d) ________?
(9) (e) ________?	CPCF

4. In isosceles triangle UVW (at right) with $UV = UW$ and K on side $\overline{UW}$, explain why $m\angle 1 > m\angle 2$.

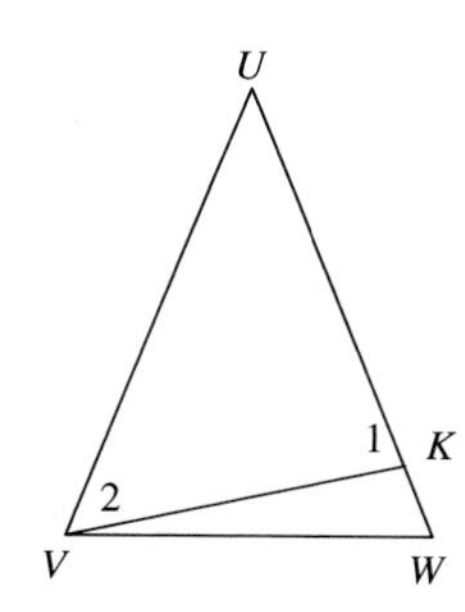

5. Which of the three triangles shown in the figure below cannot exist in absolute geometry?

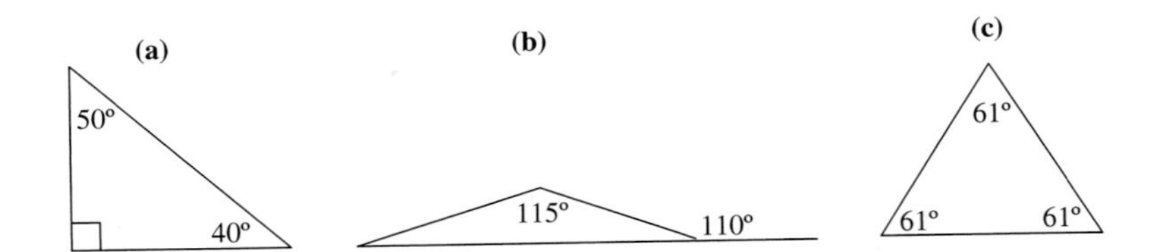

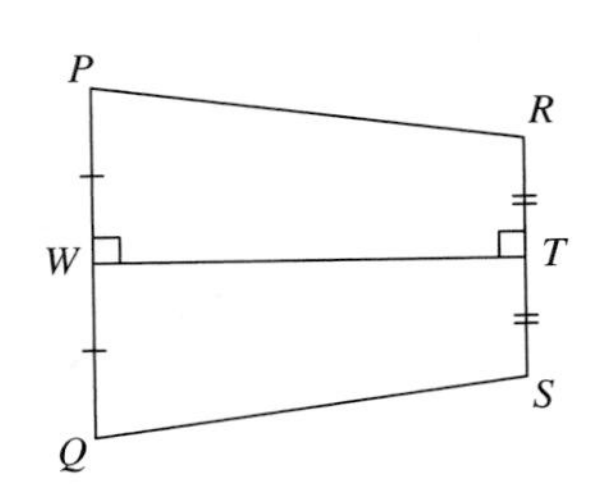

6. *Given:* Convex quadrilateral $\diamond RSQP$ with

$$PW = WQ,\ RT = TS,\ \overline{PQ} \perp \overline{TW}$$
$$\text{and } \overline{RS} \perp \overline{TW}.$$

Prove: $PR = QS$.

7. By observation, is it possible for $\triangle DEF$ and $\triangle RSU$ (with certain measurements indicated in the figure) to satisfy
   (a) the AAS hypothesis?
   (b) the SSA hypothesis?
   (c) If your answer to (b) was "yes," what sides and angles would be congruent? If your answer was "no," state the reason.

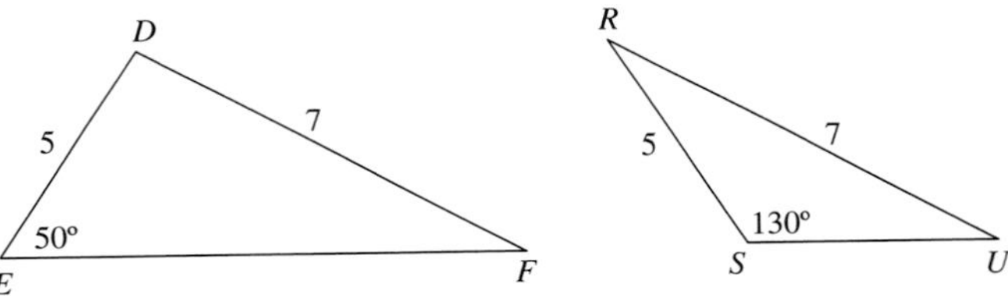

8. **(a)** Find a in the figure below.
   (b) What theorem justifies your answer in (a)?

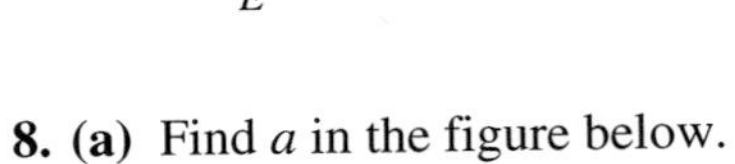

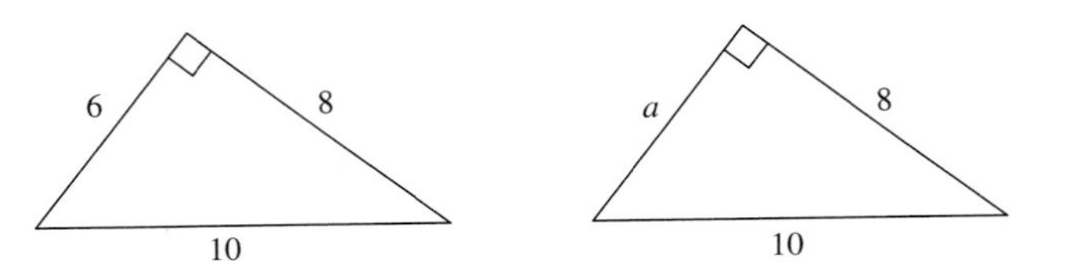

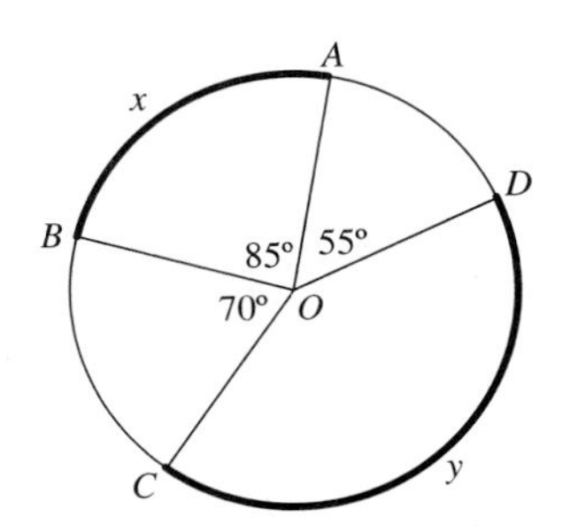

9. Find the measures x and y of the two arcs shown in the figure.

10. Prove that the center of a circle lies on the bisector of the angle formed by two intersecting tangents of that circle.

CHAPTER FOUR

Euclidean Geometry

TRIGONOMETRY, COORDINATES, AND VECTORS

OVERVIEW

It is the purpose of this chapter to adopt the Parallel Postulate for Euclidean geometry and to develop the basic concepts of classical geometry—rectangles, regular polygons, and the circle theorems, including coordinates and vectors. All this is accomplished by adding to our list of 15 axioms one further axiom, the parallel postulate for Euclidean geometry. Properties of parallelism have a dramatic effect on absolute geometry, as we shall see.

4.1 Euclidean Parallelism, Existence of Rectangles

Before we get started, we need to establish what we mean by parallel lines. Euclid's Definition 23 in Book I of the *Elements* is quite descriptive:

> Parallel lines are lines which, being in the same plane and being produced indefinitely in both directions, do not meet one another in either direction.

Our statement is similar, but will omit the unnecessary references to "extending a line" and "direction." Such features have already been taken care of by the axioms.

> DEFINITION: Two distinct lines ℓ and m are said to be **parallel** (and we write $\ell \parallel m$) iff they lie in the same plane and do not meet (Figure 4.1).

For convenience of terminology, we often speak of two segments (or rays) being parallel; this is taken to mean that such objects lie on lines that are parallel.

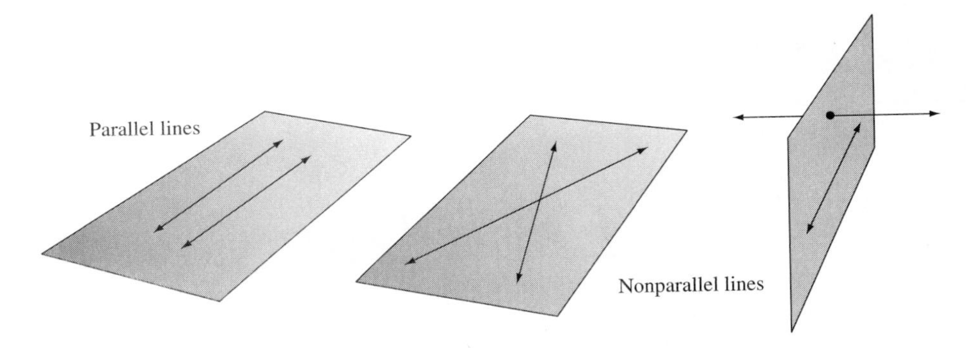

Figure 4.1

TRANSVERSALS

Another preliminary detail is introducing the usual terminology associated with the angles created when one line, a **transversal,** intersects two other lines (parallel or not). We let Figure 4.2 speak for itself—formal definitions using betweenness considerations can be formulated from the figure when necessary.

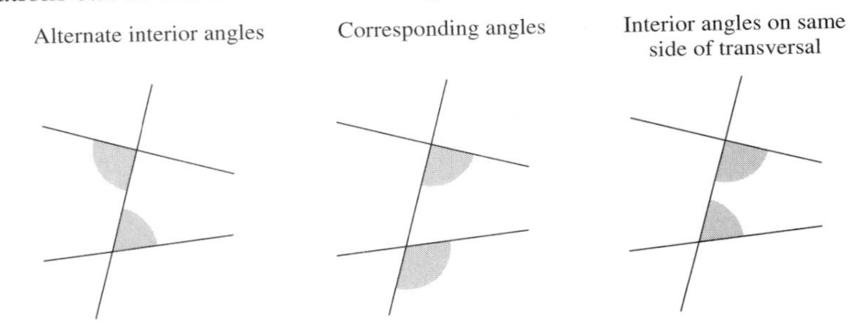

Figure 4.2

The following result was essentially the last theorem proved by Euclid before he began using his parallel postulate. Although it involves parallel lines, it in no way uses the parallel postulate and is valid in absolute geometry.

THEOREM 1: PARALLELISM IN ABSOLUTE GEOMETRY

If two lines in the same plane are cut by a transversal so that a pair of alternate interior angles are congruent, the lines are parallel.

PROOF

Given: Lines ℓ and m, transversal t meeting line ℓ at A and m at B, and $\angle 1 \cong \angle 2$ (Figure 4.3).

Prove: $\ell \parallel m$.

Suppose that ℓ is not parallel to m and that, therefore, ℓ and m meet at some point R. Then $\angle 1$ (or $\angle 2$) is an exterior angle of $\triangle ABR$. Hence $m\angle 1 > m\angle 2$ (or $m\angle 2 > m\angle 1$) by the Exterior Angle Inequality. But this contradicts the given ($\angle 1 \cong \angle 2$). $\therefore\ \ell \parallel m$.

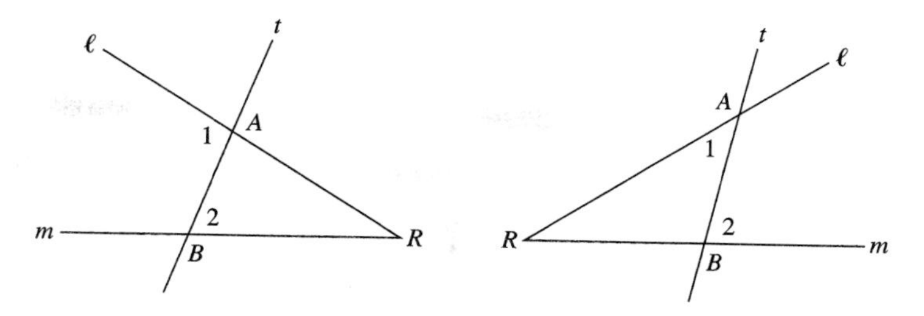

Figure 4.3

EUCLID'S FIFTH POSTULATE OF PARALLELS

It is the *converse* of this proposition that is the Euclidean Parallel Postulate, a rather curious phenomenon. We now know that this converse cannot be proven, which is in itself of great significance. This fact is the legacy of a long, unprecedented struggle in mathematics. It represents the defeat of efforts of many first-rate mathematicians, including Gauss and Legendre, who attempted to prove the postulate, but failed. Euclid is obviously not among those who failed because he stated the property as a postulate and made no attempt to prove it. Whether this was a matter of luck, mere convenience, or deep insight on his part can only be speculated.

We paraphrase Euclid's parallel postulate for later reference.

> **Euclid's Fifth Postulate of Parallels**
>
> If two lines in the same place are cut by a transversal so that the sum of the measures of a pair of interior angles on the same side of the transversal is less than 180, the lines will meet on that side of the transversal.

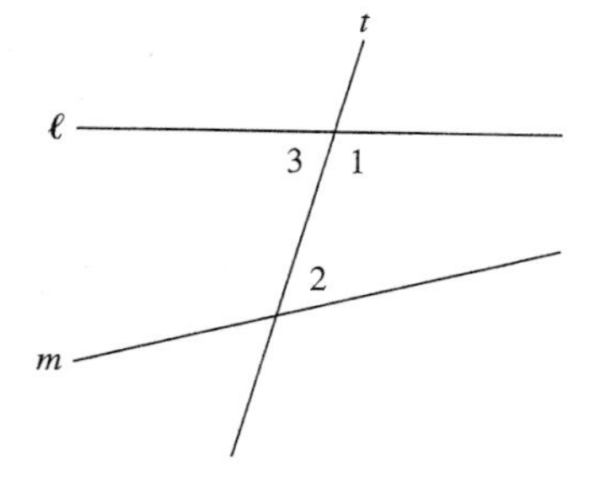

Figure 4.4

In effect, this says that in Figure 4.4:

$$\text{If } m\angle 1 + m\angle 2 \neq 180, \text{ then } \ell \text{ is not parallel to } m.$$

The contrapositive, logically equivalent to this statement, is

$$\text{If } \ell \parallel m, \text{ then } m\angle 1 + m\angle 2 = 180 \text{ (or } m\angle 2 = m\angle 3).$$

In this form, Euclid's postulate is more easily spotted as an equivalent form of the parallel postulate we adopt, and as the *converse* of Theorem 1 (which stated that if $m\angle 2 = m\angle 3$, then $\ell \parallel m$).

THREE POSSIBLE NOTIONS OF PARALLELISM

Consider in a single fixed plane a line ℓ and a point P not on it (Figure 4.5). There are three logical cases:

(1) There exists no line through P parallel to ℓ.
(2) There exists exactly one line through P parallel to ℓ.
(3) There exists more than one line through P parallel to ℓ.

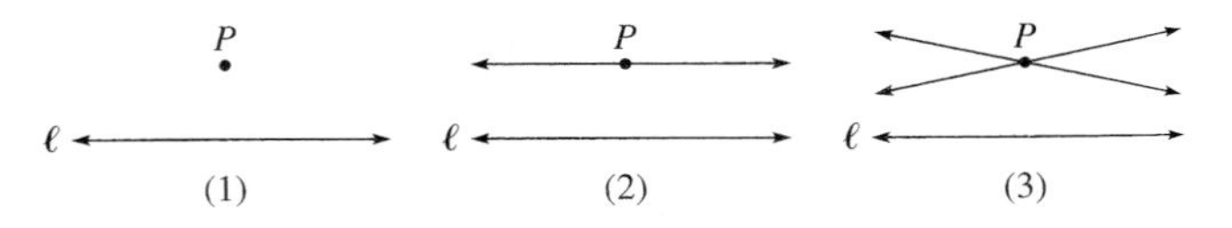

Figure 4.5

Which of these do you think is valid? The obvious answer, no doubt, is **(2).** That is what most people who are familiar only with Euclidean geometry would say. But, in fact, both projective geometry and spherical geometry satisfy case **(1),** and hyperbolic geometry satisfies case **(3),** which we study later, in Chapter 6. Since we want to study Euclidean geometry in this and the following chapter, we shall adopt hypothesis **(2)** at this time—the **Euclidean Hypothesis**—formally stated as our last axiom for Euclidean geometry.

AXIOM P-1: EUCLIDEAN PARALLEL POSTULATE
If ℓ is any line and P any point not on ℓ, there exists in the plane of ℓ and P one and only one line m that passes through P and is parallel to ℓ.

This form of the parallel postulate is due to John Playfair (1748–1819), an English mathematician who made important contributions to the foundations of geometry in several editions of a book on geometry, first published in 1795.

NOTE: The remaining major results of this text up through Chapter 6 will be restricted to a *single plane,* so at this point we no longer make references to three-dimensional space in statements of definitions or theorems.

The first consequence of this axiom will almost immediately imply a host of familiar ideas about triangles, rectangles, and parallelograms. So that you can participate in establishing this key result, we revert back, temporarily, to the two-column proof. We are going to ask you to fill in the missing parts, and also to determine precisely where the Parallel Postulate is used.

THEOREM 2: If two parallel lines are cut by a transversal, then each pair of alternate interior angles are congruent. (See Figure 4.6.)

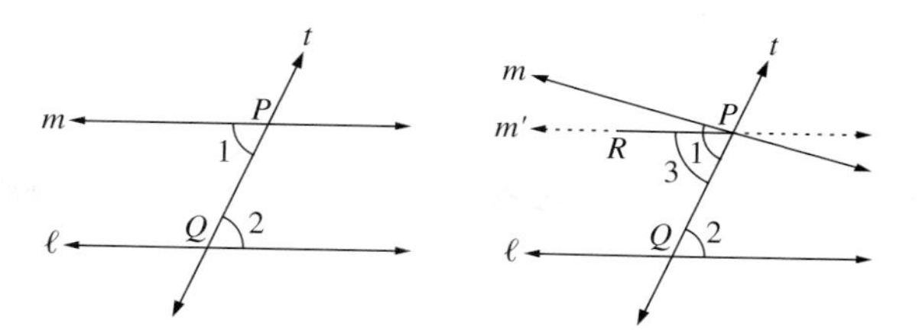

Figure 4.6

PROOF

CONCLUSIONS	**JUSTIFICATIONS**
(1) Parallel lines ℓ and m intersect a transversal t at P and Q, with alternate interior angles $\angle 1$ and $\angle 2$.	Given
(2) Suppose $m\angle 1 \neq m\angle 2$, and, for sake of argument, $m\angle 1 > m\angle 2$. (The proof is virtually the same if $m\angle 1 < m\angle 2$.)	Assumption for indirect proof
(3) Construct ray $\overrightarrow{PR}$ on the opposite side of t as $\angle 2$ so that $m\angle QPR = m\angle 2$.	Angle Construction Theorem
(4) Then line $\overleftrightarrow{RP}$ $(= m')$ is parallel to ℓ.	________?
(5) Thus, $m' \parallel \ell$ and $m \parallel \ell$. $\rightarrow\leftarrow$	________?
(6) $\therefore m\angle 1 = m\angle 2$ or $\angle 1 \cong \angle 2$.	________?

THE TRANSVERSAL THEOREMS FOR PARALLELISM

Theorem 2 has three important corollaries that are used constantly in Euclidean geometry. You are no doubt familiar with them, but we present them in a way that makes them both easier to remember and to use. Their proofs make good problems—just remember, they are direct consequences of Theorems 1 and 2. Each of them is illustrated in Figure 4.7.

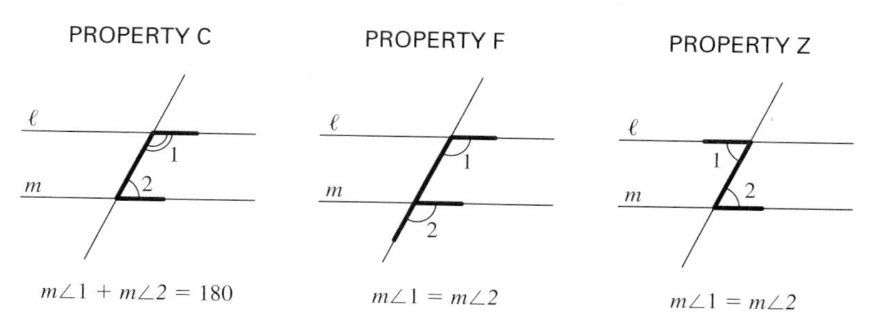

Figure 4.7

COROLLARY A: THE C PROPERTY If two lines in the same plane are cut by a transversal, then the two lines are parallel iff a pair of interior angles on the same side of the transversal are supplementary.

COROLLARY B: THE F PROPERTY If two lines in the same plane are cut by a transversal, then the two lines are parallel iff a pair of corresponding angles are congruent.

COROLLARY C: THE Z PROPERTY If two lines in the same plane are cut by a transversal, then the two lines are parallel iff a pair of alternate interior angles are congruent.

(Note that Corollary C is merely the composite of Theorems 1 and 2.) We include one additional corollary, which is itself a corollary to any one of those preceding.

COROLLARY D: If a line is perpendicular to one of two parallel lines, it is perpendicular to the other also.

EXAMPLE 1 Write a proof for Corollary B, the F-Property.

SOLUTION

Using the middle diagram in Figure 4.7, first suppose $\angle 1 \cong \angle 2$. But $\angle 2$ and the angle formed by the opposite rays are congruent vertical angles, so $\angle 1$ is congruent to the other angle in the vertical pair, thereby showing that $\angle 1$ is congruent to an alternate interior angle. Therefore, by Theorem 1, $\ell \parallel m$. Conversely, suppose that $\ell \parallel m$. By Theorem 2, $\angle 1$ is congruent to an alternate interior angle, which, in turn is congruent to $\angle 2$. Therefore, $\angle 1 \cong \angle 2$. ■

In working geometry problems involving parallel lines, it is often helpful to look for configurations resembling the letters *C*, *F*, or *Z* in the diagrams involved. Here is an example.

EXAMPLE 2 Prove in Euclidean geometry that if $\angle ACD$ is an exterior angle of $\triangle ABC$ (Figure 4.8), then $m\angle ACD = m\angle A + m\angle B$.

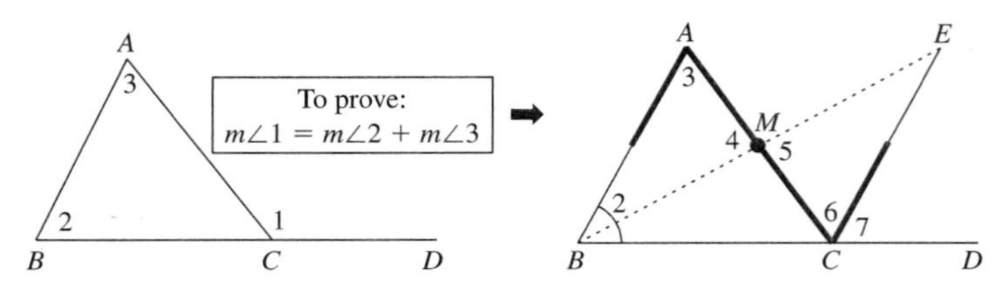

Figure 4.8

SOLUTION

Not surprisingly, the same construction used to establish the Exterior Angle Inequality can be used here. Recall that we constructed segment $\overline{BE}$ so that the midpoint M of $\overline{AC}$ was also the midpoint of $\overline{BE}$. (We gave a rigorous proof that $\overrightarrow{CA}$–$\overrightarrow{CE}$–$\overrightarrow{CD}$ was valid, so we need not repeat that here.) Now by the Vertical Pair Theorem, $\angle 4 \cong \angle 5$, $\overline{AM} \cong \overline{CM}$, and $\overline{MB} \cong \overline{ME}$. Therefore (as before), $\triangle AMB \cong \triangle CME$, and $\angle 3 \cong \angle 6$ by CPCF. By the Z-Property, $\overleftrightarrow{CE} \parallel \overleftrightarrow{AB}$, and by the F-Property, $\angle 2 \cong \angle 7$. (Do you see a letter F in the figure?) Therefore,

$$m\angle ACD = m\angle 6 + m\angle 7 = m\angle 2 + m\angle 3$$

as desired. ■

The result just established is so important it will be stated formally.

> THEOREM 3: EUCLIDEAN EXTERIOR ANGLE THEOREM
> The measure of an exterior angle of any triangle equals the sum of the measures of the two opposite interior angles.

OUR GEOMETRIC WORLD

The great mathematician Carl F. Gauss, who in the 1800s was an official cartographer for the German Government, once led an incredible expedition. He set out to test the Euclidean Hypothesis by measuring the angles of a triangle formed by the lines of sight between three distant mountain peaks. Beyond experimental error, no discrepancy from 180° was found. Of course, as Gauss himself realized, this experiment does not *prove* that our world is Euclidean, and no experiment of this type could ever prove it because of the possibility of experimental error.

ANGLE SUM THEOREM

A direct result of the preceding theorem is a key characteristic of Euclidean geometry.

COROLLARY: The sum of the measures of the angles of any triangle is 180.

Can you prove this? (If we pick any angle of a triangle, it is supplementary to its adjacent exterior angle, whose measure equals the sum of the measures of the remaining two angles.)

A few other basic results are also readily obtained.

- The acute angles of any right triangle are complementary angles.
- The sum of the measures of the angles of any convex quadrilateral is 360.
- Rectangles exist in Euclidean geometry, and any Saccheri Quadrilateral or Lambert Quadrilateral is a rectangle.
- Squares exist in Euclidean geometry. (See Section 4.2.)

EXAMPLE 3 In Figure 4.9, the line segment joining the midpoints L and M of two sides of $\triangle ABC$ is extended to point P, such that M is also the midpoint of $\overline{LP}$. Prove that line $\overleftrightarrow{PC}$ is parallel to $\overleftrightarrow{AB}$.

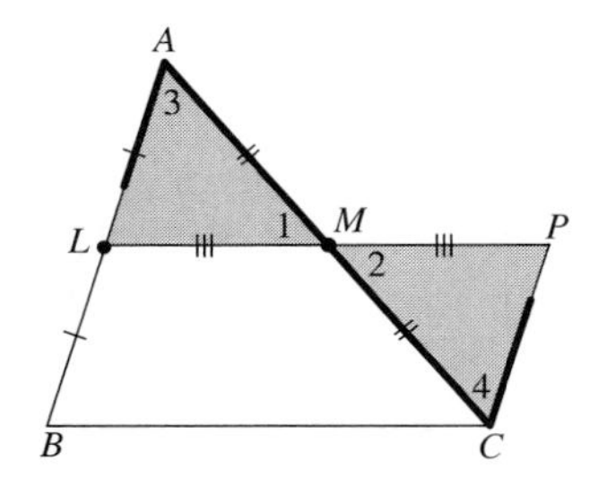

Figure 4.9

SOLUTION

First it is proved that $\triangle ALM \cong \triangle CPM$. Then it will follow by Property Z that $\overleftrightarrow{PC} \parallel \overleftrightarrow{AB}$.

(1) $AM = MC$ and A–M–C by definition of midpoint, and, similarly, $LM = MP$ and L–M–P. Since $\angle 1$ and $\angle 2$ are vertical angles, $\angle 1 \cong \angle 2$. $\therefore \triangle ALM \cong \triangle CPM$, by SAS.

(2) $\angle 3 \cong \angle 4$ (CPCF) $\therefore$ by Property Z, $\overleftrightarrow{PC} \parallel \overleftrightarrow{LA}$. ■

MIDPOINT CONNECTOR THEOREM

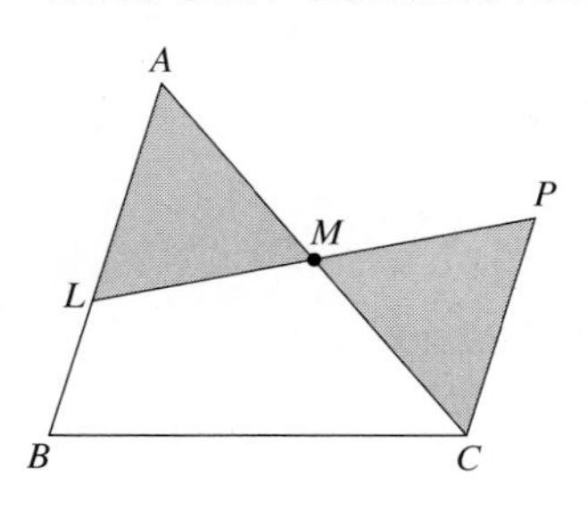

Figure 4.10

The previous example is somewhat puzzling. Where did we use the fact that L was the midpoint of segment $\overline{AB}$? As a matter of fact, this had nothing to do with the proof. As shown in Figure 4.10, the only essential feature was that M be the midpoint of $\overline{AC}$, and P be located so that M is the midpoint of $\overline{LP}$, where L is *any* point on $\overline{AB}$. The shaded triangles are still congruent, and $\overleftrightarrow{PC} \parallel \overleftrightarrow{AB}$. This is a good illustration of what often happens in mathematics: a feature that one thinks, at first, is essential in a proof turns out not to be so, and thus a more general theorem can been proven. But in this case, there is another reason to require that L be the midpoint of $\overline{AB}$; from this a very significant result can be established.

> THEOREM 4: THE MIDPOINT CONNECTOR THEOREM
> The segment joining the midpoints of two sides of a triangle is parallel to the third side and has length one-half that of the third side. (See Figure 4.11.)

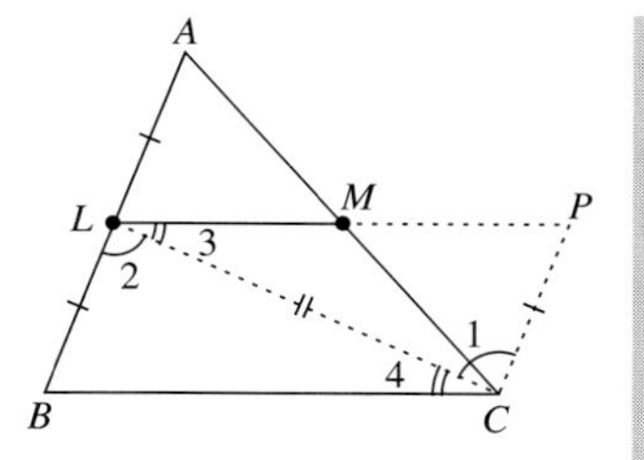

Figure 4.11

PROOF

Extend line segment $\overline{LM}$ to point P such that M is the midpoint of $\overline{LP}$ (Segment Construction Theorem). Then, as in Example 4, $\triangle ALM \cong \triangle CPM$ and $\overleftrightarrow{PC} \parallel \overleftrightarrow{AB}$. If we also draw segment $\overline{LC}$, then we have $PC = AL = LB, LC = CL$, and $\angle 1 \cong \angle 2$ (the Z-property). By SAS, $\triangle LPC \cong \triangle CBL$. Then $LM = \frac{1}{2}LP = \frac{1}{2}BC$ by CPCF, and $\angle 3 \cong \angle 4$ making $\overleftrightarrow{LP} \parallel \overleftrightarrow{BC}$ by Property Z.

It should be pointed out that Theorem 4 could have been obtained rather quickly by using the Saccheri Quadrilateral associated with a triangle. That will be pursued in a problem in this section (Problem 19) for those who have studied that concept and are familiar with it. However, that method uses a few properties we did not prove about Saccheri Quadrilaterals (which did appear in some of the problems in Section 3.7, however).

A final result is clear once we realize that the line $\overleftrightarrow{LM}$ in Theorem 4 joining the midpoints L and M and the parallel to $\overleftrightarrow{BC}$ passing through L are one and the same line (only *one* line through L can be parallel to $\overleftrightarrow{BC}$).

COROLLARY: If a line bisects one side of a triangle and is parallel to the second, it also bisects the third side.

PROBLEMS (4.1)

GROUP A

1. In the slanted block letter A, $m\angle 1 = m\angle 2$. Prove: $m\angle 3 = m\angle 4$.

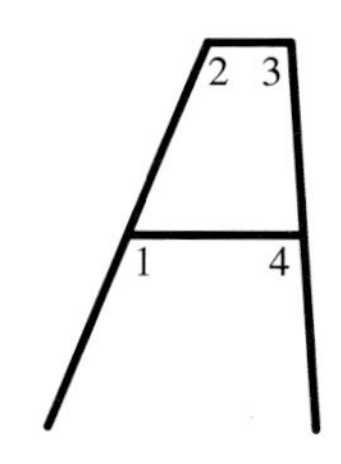

2. If $\overline{AB} \perp \overline{BD}$, $\overline{ED} \perp \overline{BD}$ and $\overline{AC} \perp \overline{CE}$, with the measure of $\angle ACB$ as indicated, find the measures of $\angle 1$, $\angle 2$, and $\angle 3$.

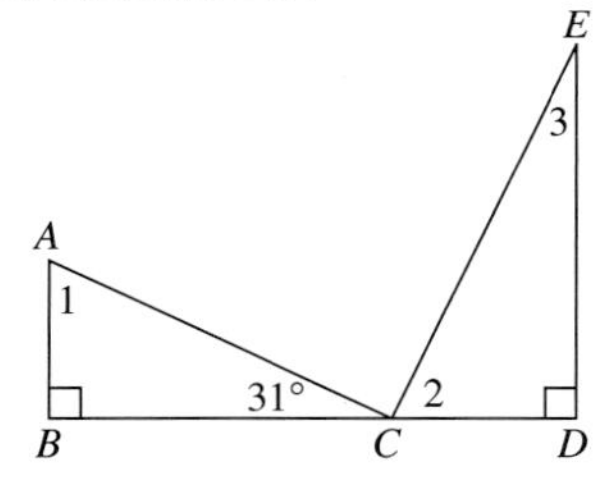

3. Given right triangle $\triangle ABC$ with angle measures as indicated in the figure, find x, y, and z.

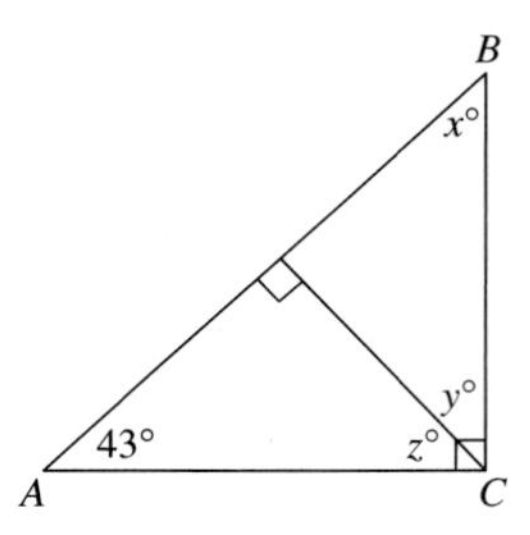

4. In the figure, $\square ABCD$ and $\square RSTU$ are rectangles. If $m\angle CBT = 10$, find the measures of $\angle 1$ and $\angle 2$.

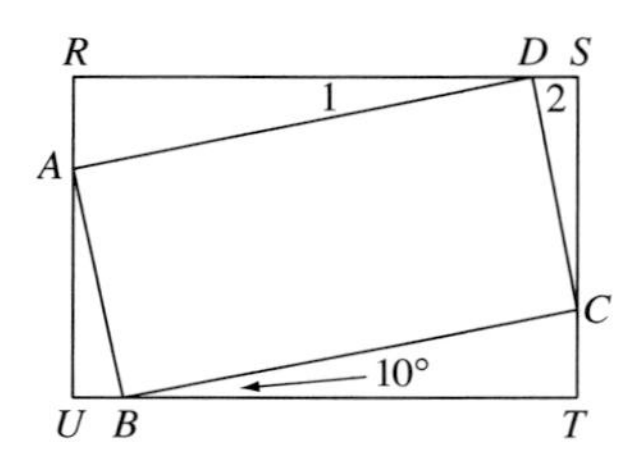

5. It is given that $\triangle ABC$ is equilateral and that $\overline{DE} \perp \overline{BC}$. Prove that $\triangle ADF$ is isosceles.

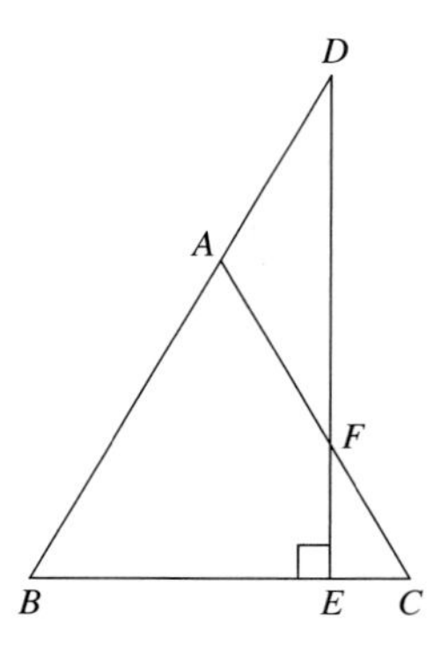

6. Use Corollary A to give another proof that the perpendicular to a line ℓ from a given external point A is unique.

7. Investigate the phenomenon that results from the following procedure using *Sketchpad*.

[1] Construct segment $\overline{AB}$, its midpoint C, and the perpendicular bisector of $\overline{AB}$. Locate point D on that perpendicular. (What kind of triangle is $\triangle ABD$?)

[2] Construct ray $\overrightarrow{AD}$ and segment $\overline{BD}$.

[3] Locate point E on ray $\overrightarrow{AD}$ so that A–D–E, then select points E, D, and B, and construct the angle bisector $\overrightarrow{DF}$ of $\angle EDB$ using Angle Bisector under CONSTRUCT.

Now drag points B and D and observe the effect on the figure. What do you notice? Have you discovered a theorem? Try to write a theoretical proof for it.

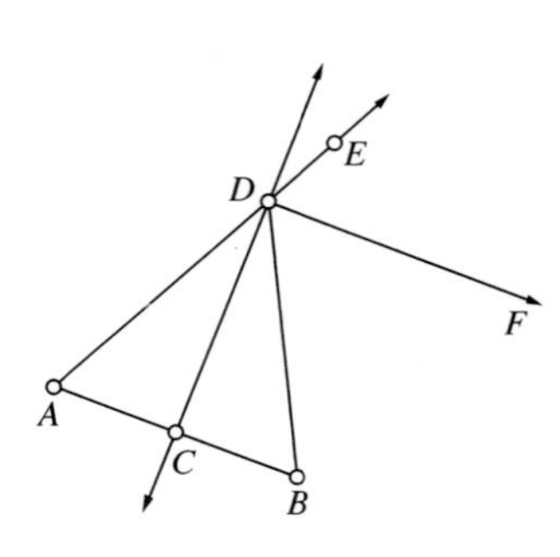

GROUP B

8. Prove your choice of Corollary A or D of Theorem 2. (Corollaries B and C have already been established.)

9. Prove from the theory of this section that parallel lines are everywhere equidistant.

10. It is given that L and M are midpoints of $\overline{AB}$ and $\overline{AC}$, and D and E are midpoints of $\overline{AL}$ and $\overline{AM}$.

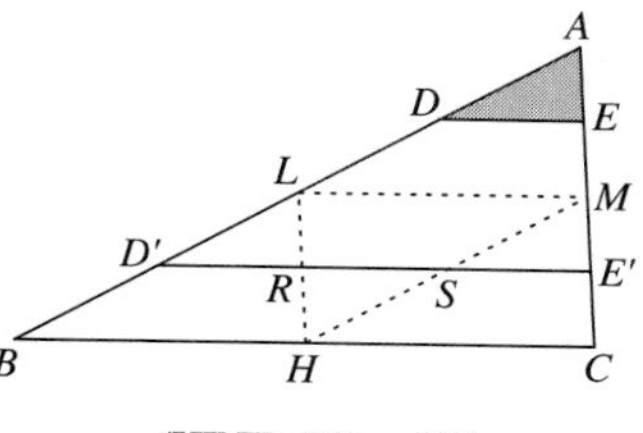

(HINT: $BH = HC$)

(a) Prove by the Midpoint Connector Theorem that the lengths of the sides of $\triangle ADE$ are one-fourth those of $\triangle ABC$.

(b) Show that if D' and E' are the midpoints of $\overline{LB}$ and $\overline{MC}$, then the lengths of the sides of $\triangle AD'E'$ are three-fourths those of $\triangle ABC$.

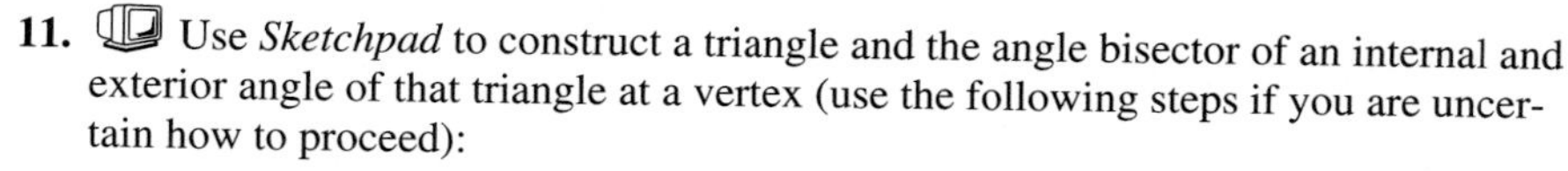

11. Use *Sketchpad* to construct a triangle and the angle bisector of an internal and exterior angle of that triangle at a vertex (use the following steps if you are uncertain how to proceed):

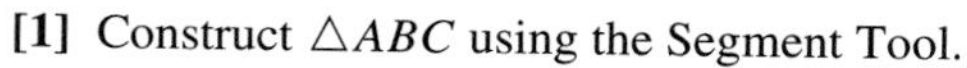

[1] Construct $\triangle ABC$ using the Segment Tool.

[2] Construct ray $\overrightarrow{AC}$ and locate D on that ray so that A–C–D.

[3] Using Angle Bisector under CONSTRUCT, select points A, C, B to construct the bisector $\overline{CE}$ of $\angle ACB$. Repeat this for the bisector $\overline{CF}$ of $\angle DCB$.

(a) Drag point A, keeping it on (or parallel) to a fixed line through B. What happens to $\angle FCE$? Does it change position and measure?

(b) What seems to be true of $\triangle ABC$ when ray $\overrightarrow{CF}$ is parallel to $\overline{AB}$? (See in this connection Problem 7.)

(c) Prove theoretically each phenomenon you noticed in (a) and (b).

12. Using *Sketchpad*, investigate the following phenomenon in connection with the remarks made concerning Example 3.

[1] Construct $\triangle ABC$ using the Segment Tool.

[2] Locate point D on side $\overline{AB}$, and E on side $\overline{AC}$.

[3] Construct the midpoint F of $\overline{AE}$.

[4] Construct $\overline{DF'}$ so that F is its midpoint. (Mark the vector D to F, then translate F by Marked Vector, producing point F'. Then join D and F'.)

[5] Join points E and F', and display the distances DF and BC. (For extra effect, shade in the triangles $\triangle ADF$ and $\triangle FEF'$ using Polygon Interior under CONSTRUCT.)

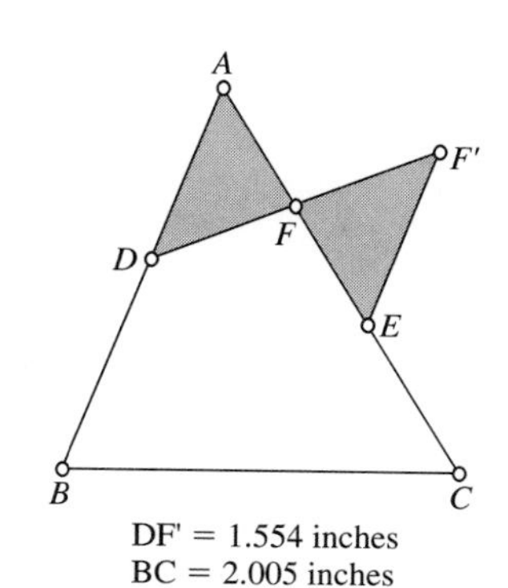

DF' = 1.554 inches
BC = 2.005 inches

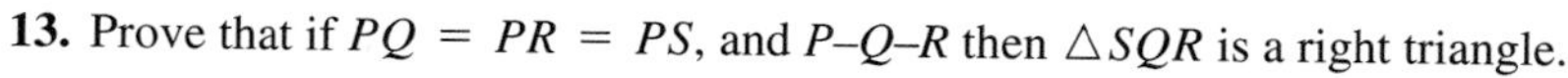

Drag points D and E and write down your observations (there should be several). (Be sure to move E on top of point C in your experiment.) Did you rediscover the Midpoint Connector Theorem?

13. Prove that if $PQ = PR = PS$, and P–Q–R then $\triangle SQR$ is a right triangle.

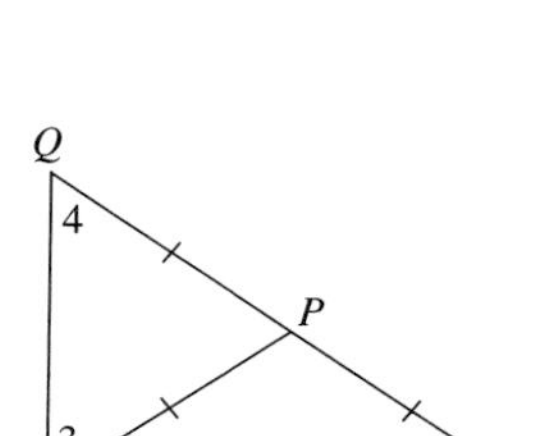

14. The interior of a chapel is built in the shape of an equilateral triangle, with the roof extending upward to an additional height equal to the lengths of the sides of the equilateral triangle. Find the angle measure x of the peak of the roof. Generalize.

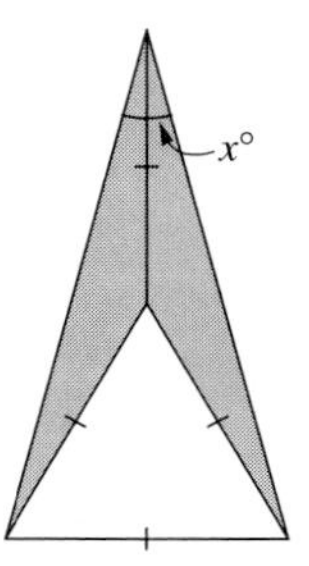

15. Transitivity of Parallelism in Euclidean Geometry Prove that for three distinct lines ℓ, m, and n, if $\ell \parallel m$ and $m \parallel n$, then $\ell \parallel n$.

16. If $\triangle ABC \cong \triangle DCE$ and B–C–E, prove that $\overleftrightarrow{AD} \parallel \overleftrightarrow{BE}$.

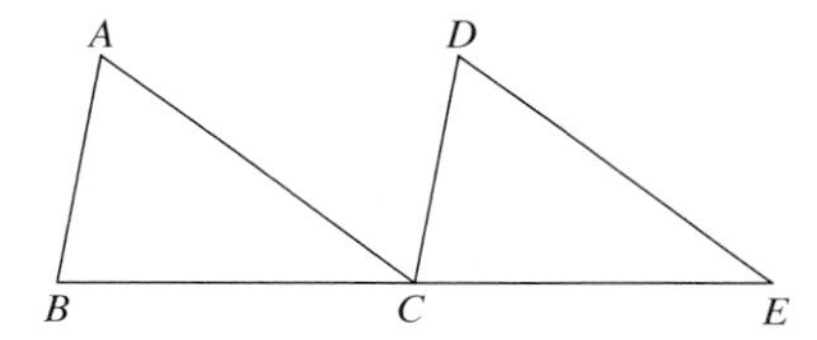

GROUP C

17. Point P is located on leg $\overline{UV}$ of isosceles triangle $\triangle UVW$ so that $m\angle PWV = 70$. If the base angles of $\triangle UVW$ at V and W are each of measure 80, show that $UP = VW$. (***Hint:*** Construct an equilateral triangle.)

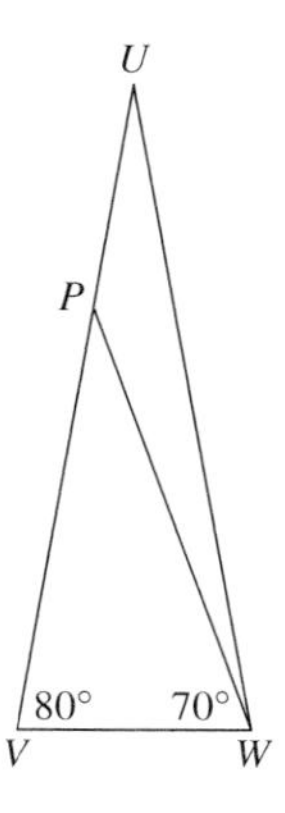

18. Formal Proof of Morley's Theorem The following four steps will guide you in proving Morley's Theorem referred to in Sections 1.1 and 1.4. Construct points X, Q', R' as shown in the figure, where $m\angle XPQ' = m\angle XPR' = 30$ and $\triangle PQ'R'$ is equilateral. (Note that P is the incenter of $\triangle BCX$ and $\overrightarrow{XP}$ bisects $\angle BXC$.) Let Y

and X be the intersections of the altitudes of $\triangle PQ'R'$ from Q' and R' with rays $\overrightarrow{CP}$ and $\overrightarrow{BP}$. From the angle measures to be obtained, $m\angle YR'Q' + m\angle ZQ'R' > 180$, so rays $\overrightarrow{YR'}$ and $\overrightarrow{ZQ'}$ will meet at some point A'. Now complete the details in the following steps:

(1) $m\angle PQ'Z = 60 - C/3$ and $m\angle PR'Y = 60 - B/3$. (**Hint:** $m\angle PQ'Z = m\angle Q'PZ = 180 - m\angle Q'PB = 180 - (m\angle XPB + 30)$; $m\angle XPB = 180 - B/3 - \frac{1}{2}m\angle BXC = 90 + C/3$.)

(2) $m\angle A' = A/3$.

(3) $m\angle A'ZB = 60 + 2C/3$ and $m\angle A'YC = 60 + 2B/3$. (**Hint:** $m\angle A'ZB = 2 \cdot m\angle R'ZQ' = 2 \cdot (90 - m\angle PQ'Z.)$

(4) $m\angle A'BC = B$ and $m\angle A'CB = C$, hence $A' = A$, $Q' = Q$, $R' = R$, and $\triangle PQR \equiv \triangle PQ'R'$ is equilateral. (To show $m\angle A'BC = B$, construct ray $\overrightarrow{BK}$ such that $\overrightarrow{BC}$–$\overrightarrow{BX}$–$\overrightarrow{BK}$ and $m\angle KBC = B$; show that ray $\overrightarrow{BK}$ meets $\overrightarrow{ZA'}$ at some point W, and that $m\angle R'WZ = A/3 = m\angle R'A'Z$. $\therefore W = A'$. Similarly, $m\angle A'CB = C$.)

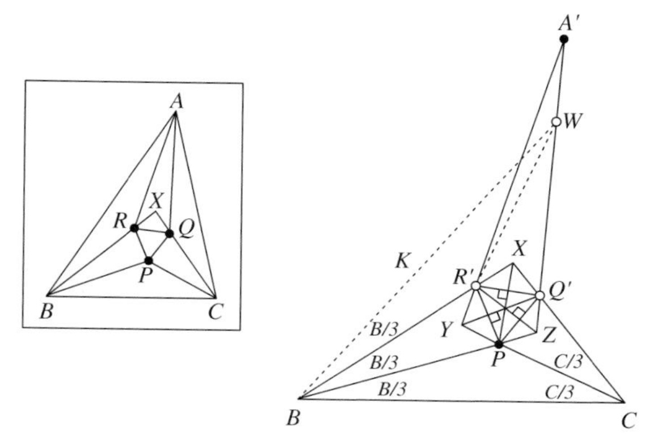

19. Prove the Midpoint Connector Theorem using the associated Saccheri Quadrilateral of a triangle. In particular, the following ideas must be established (using the notation of Figure 3.63): (1) $B'C' = 2MC$, (2) every Saccheri Quadrilateral is a rectangle, and (3) the opposite sides of a rectangle are congruent and parallel. See Problems 16, 17, and 18 in Section 3.7.

20. Prove that Euclid's Fifth Postulate is logically equivalent to Axiom P–1. This proof has two parts:

(a) Prove that Euclid's Postulate implies Axiom P–1. (Assume there are *two* lines through point P parallel to line ℓ in Figure 4.5, drop a perpendicular from P, and gain a contradiction using Euclid's Fifth Postulate.)

(b) Prove that Axiom P–1 implies Euclid's Postulate. (Any of the results of this section can be used because they are logical consequences of Axiom P–1.)

21. If two angles have their respective sides parallel, the angles are either congruent or supplementary. Prove.

22. If two angles have their respective sides perpendicular, the angles are either congruent or supplementary. Prove.

23. Work the following problem in absolute geometry: If lines ℓ and m are cut by a transversal so that a pair of alternate interior angles are congruent, then ℓ and m have a common perpendicular. Do not use any of the results of this section that depend on the Euclidean Parallel Postulate. (***Hint:*** In the accompanying figure, the angles at A and B are congruent, as indicated. Drop perpendiculars from the midpoint M to ℓ and m at C and D.

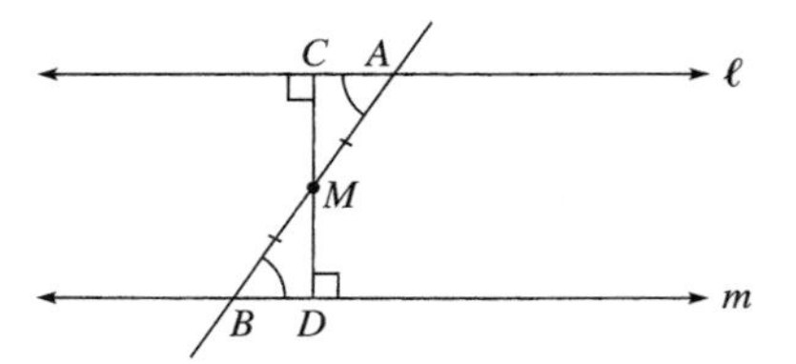

4.2 Parallelograms and Trapezoids: Parallel Projection

The first sequence of theorems in this section is an elementary application of the basic results of parallelism that appeared in the last section. After we prove the first of these, you should have no difficulty with the rest. Our intention here is to cover only the highlights and major ideas in an area containing literally hundreds of intricate relationships and theorems.

DEFINITION: A convex quadrilateral $\diamond ABCD$ is called a **parallelogram** if the opposite sides $\overline{AB}$, $\overline{CD}$ and $\overline{BC}$, $\overline{AD}$ are parallel (Figure 4.12). A **rhombus** is a parallelogram having two adjacent sides congruent. A **square** is a rhombus having two adjacent sides perpendicular.

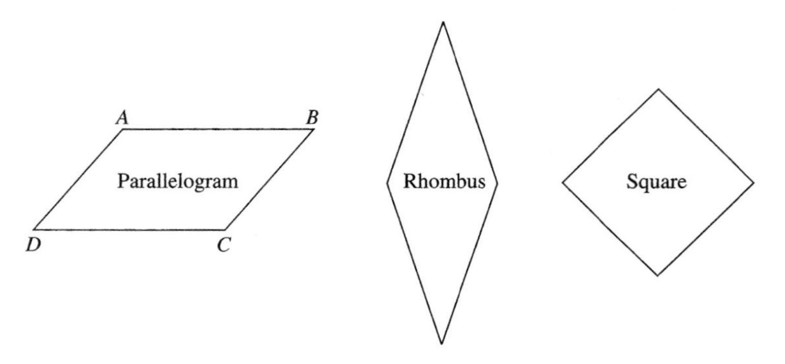

Figure 4.12

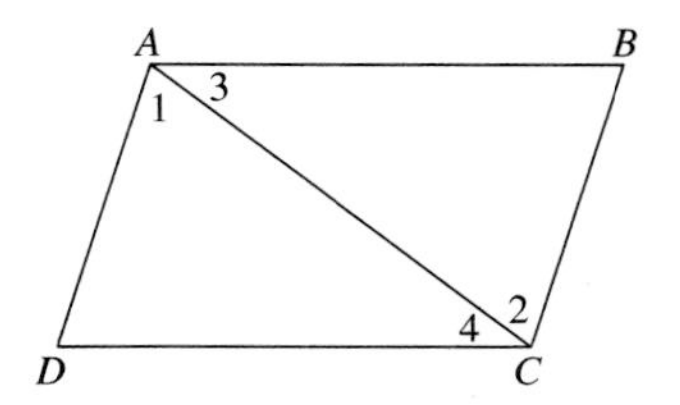

Figure 4.13

THEOREM 1: A diagonal of a parallelogram divides it into two congruent triangles.

PROOF (See Figure 4.13)

We want to show that $\triangle ABC \cong \triangle CDA$. But, by the Z-property applied twice, $\angle 1 \cong \angle 2$ and $\angle 3 \cong \angle 4$, with $\overline{AC} \cong \overline{CA}$. By ASA, $\triangle ABC \cong \triangle CDA$.

COROLLARY: The opposite sides of a parallelogram are congruent. Opposite angles are also congruent, while adjacent angles are supplementary (due to Property C.)

The following list of further properties of parallelograms includes corollaries of the preceding, or to each other. Any of them make good exercises (to be taken in the order given for easier proofs).

- If a convex quadrilateral has opposite sides congruent, it is a parallelogram.
- If a convex quadrilateral has a pair of sides that are both congruent and parallel, it is a parallelogram. (Problem 17.)
- The diagonals of a parallelogram bisect each other. Conversely, if the diagonals of a convex quadrilateral bisect each other, the quadrilateral is a parallelogram.

Further results are:

- A parallelogram is a rhombus iff its diagonals are perpendicular.
- A parallelogram is a rectangle iff its diagonals are congruent.
- A parallelogram is a square iff its diagonals are both perpendicular and congruent. (Problem 10.)

OUR GEOMETRIC WORLD

Carpenters and construction crews routinely use one of the geometric principles in this section to check on their measurements. When a rectangular component is called for, such as a door frame or cabinet top, the lengths of diagonals $\overline{AC}$ and $\overline{BD}$ are measured. If they are off by a half inch or more, workers know the angles are not true.

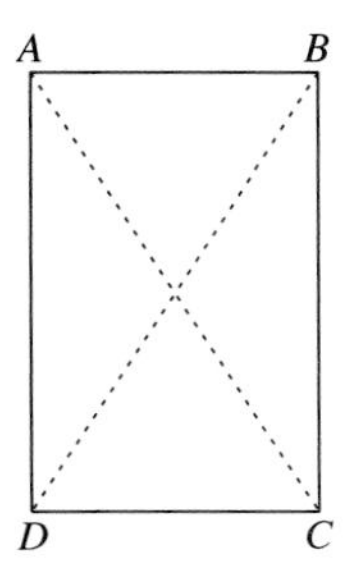

Figure 4.14

TRAPEZOIDS AND MEDIANS

DEFINITION: A **trapezoid** is a (convex) quadrilateral with a pair of opposite sides parallel, called the **bases,** and the other two sides, the **legs.** (See Figure 4.15.) The segment joining the midpoints of the legs of a trapezoid is called the **median.** A trapezoid is said to be **isosceles** iff its legs are congruent *and if it is not an oblique parallelogram* (one having no right angles).

PARTS OF A TRAPEZOID

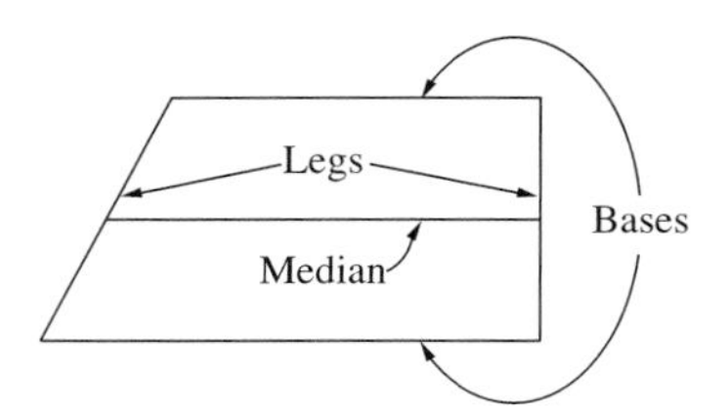

Figure 4.15

NOTE: Our definition of a trapezoid allows parallelograms to be trapezoids. This means that any result proven for a trapezoid is automatically true for a parallelogram, and we do not have to give separate proofs.

THEOREM 2: MIDPOINT-CONNECTOR THEOREM FOR TRAPEZOIDS
If a line segment bisects one leg of a trapezoid and is parallel to the base, then it is the median and its length is one-half the sum of the lengths of the two bases. Conversely, the median of a trapezoid is parallel to each of the two bases, whose length equals one-half the sum of the lengths of the bases.

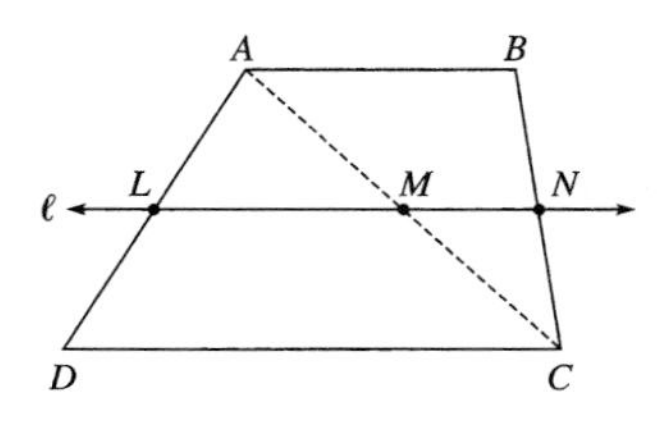

Figure 4.16

PROOF

(Figure 4.16) Let line ℓ bisect leg $\overline{AD}$ at L and be parallel to base $\overline{CD}$; hence $\ell \parallel \overleftrightarrow{AB}$ by definition of trapezoid and Transitivity of Parallelism (Problem 15, Section 4.1). Construct diagonal $\overline{AC}$.

(1) ℓ bisects $\overline{AC}$ at some point M (corollary to Midpoint Connecter Theorem).
(2) $\therefore$ ℓ bisects $\overline{BC}$ at some point N (same reason).
(3) $LM = \frac{1}{2}DC$ and $MN = \frac{1}{2}AB$ (Midpoint Connector Theorem).
(4) L–M–N ($\overrightarrow{AB}$–$\overrightarrow{AC}$–$\overrightarrow{AD}$, $\overrightarrow{AB}$–$\overrightarrow{AN}$–$\overrightarrow{AC}$, and Crossbar Theorem).
(5) $\therefore LN = LM + MN = \frac{1}{2}(AB + DC)$ [Steps (3) and (4)].

(The converse will be left as Problem 18.)

OUR GEOMETRIC WORLD

The base of the Statue of Liberty has the shape of a **truncated square pyramid** (a square pyramid with the top cut off). The outer edge of each lateral face is an isosceles trapezoid.

PARALLEL PROJECTIONS

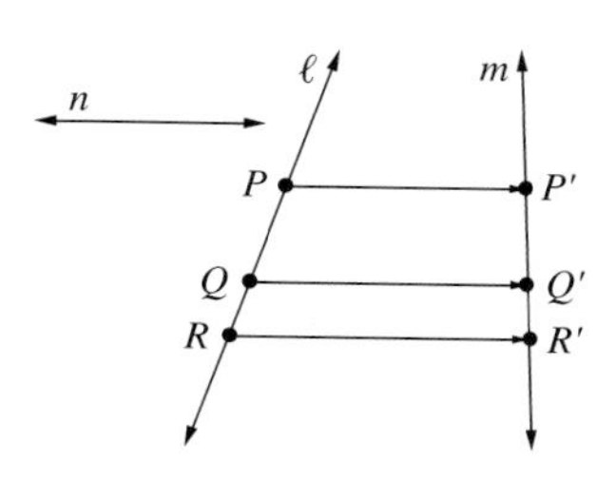

Figure 4.17

The basis for ideas about ratio and proportion needed for the theory of similar polygons is an extension of Theorem 2. A certain mapping from one line to another will play a prominent role in later developments.

In Figure 4.17, we are given two lines, ℓ and m, and point P on ℓ. We set up a correspondence $P \leftrightarrow P'$ between the points of ℓ and m by requiring $\overleftrightarrow{PP'} \parallel n$ for all P on ℓ, where n is some fixed line. It is clear that this mapping is **one-to-one** (each distinct P and Q on ℓ leads to distinct **images** P' and Q' on m). This correspondence has a number of **invariant properties** (properties that are unchanged by the mapping). One of these is betweenness on ℓ (obvious because parallel lines cannot cross[1]), and the other, which we want to focus on here, is *ratios of segments* on ℓ. We claim that, in general,

$$\frac{PQ}{QR} = \frac{P'Q'}{Q'R'}$$

[1]A rigorous proof makes use of the Postulate of Pasch.

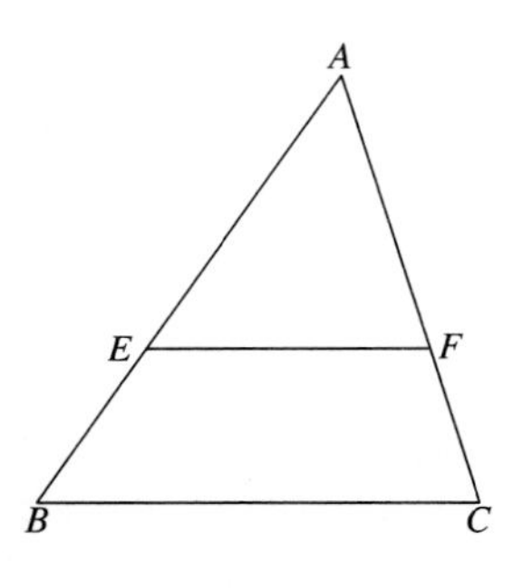

Figure 4.18

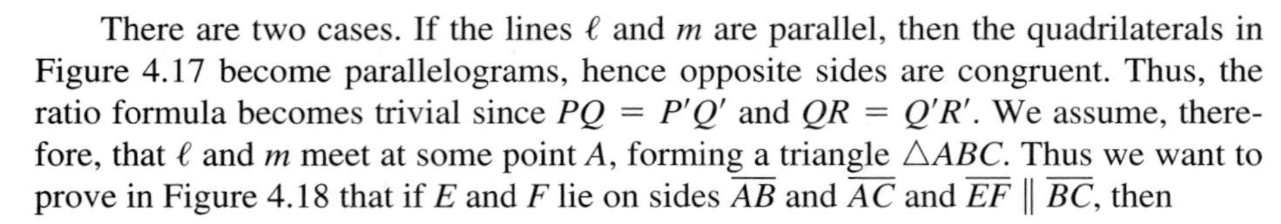

There are two cases. If the lines ℓ and m are parallel, then the quadrilaterals in Figure 4.17 become parallelograms, hence opposite sides are congruent. Thus, the ratio formula becomes trivial since $PQ = P'Q'$ and $QR = Q'R'$. We assume, therefore, that ℓ and m meet at some point A, forming a triangle $\triangle ABC$. Thus we want to prove in Figure 4.18 that if E and F lie on sides $\overline{AB}$ and $\overline{AC}$ and $\overline{EF} \parallel \overline{BC}$, then

$$\frac{AE}{AB} = \frac{AF}{AC}$$

Our method of proof is to assume that $\overline{EF}$ *is not parallel to* $\overline{BC}$, then show that $AE/AB \neq AF/AC$. This is the contrapositive of what we are trying to prove, and hence is logically equivalent to it. Thus, once again, establishing an *inequality* leads to the proof of an *equality*. The proof given here is in the style of Archimedes and does not make use of limits, which is the usual method of proof for this result.

LEMMA: In $\triangle ABC$, with E and F points on the sides, if $\overline{EF}$ is *not* parallel to $\overline{BC}$, then $AE/AB \neq AF/AC$.

PROOF

Construct the parallels $\overline{EH}$ and $\overline{GF}$ to $\overline{BC}$ through E and F, respectively. This forms with $\overline{EF}$ either a Z or backward Z depending on whether A–G–E or A–E–G (the case actually depicted in Figure 4.19). Let us assume for the sake of argument that A–E–G holds; since betweenness is preserved under parallel projection, we also have A–H–F. The rest of the argument appears in the following steps.

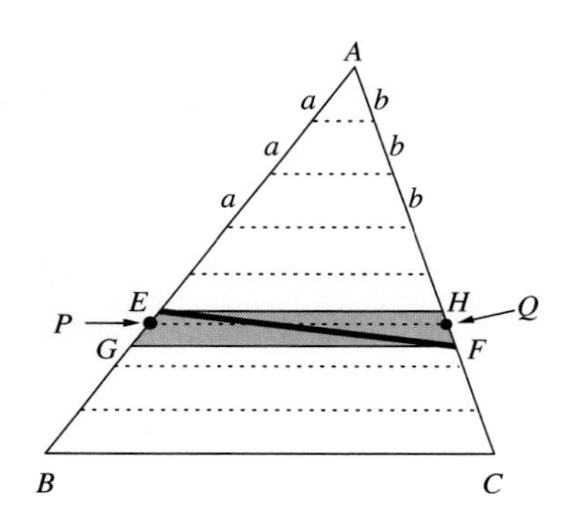

Figure 4.19

(1) Locate the midpoints of segments $\overline{AB}$ and $\overline{AC}$, then bisect the segments that these midpoints determine, and continue the bisection process indefinitely. It is clear that at some stage we will have one of the midpoints, say P, falling on segment $\overline{EG}$, and, correspondingly, midpoint Q falls on $\overline{HF}$.[2] The illustration in the figure shows that it takes *three* bisections to achieve the desired point P in this case.)

(2) The lines joining corresponding midpoints are parallel to $\overline{BC}$ by the Midpoint Connector Theorem for trapezoids and triangles.

(3) Suppose there are k parallel lines between A and $\overline{PQ}$, counting $\overline{PQ}$, and n between A and $\overline{BC}$, counting $\overline{BC}$. Since betweenness is preserved, and if a and b are the lengths of the congruent segments on $\overline{AB}$ and $\overline{AC}$, we obtain

$$AP = ka \qquad AQ = kb$$

$$AB = na \qquad AC = nb$$

[2]This depends on the so-called **Archimedean Principle** of the real numbers, which states that there exists an integer n such that $n \cdot EG > AB$, or since $2^n > n$, $2^n \cdot EG > AB$, and $AB/2^n < EG$. At this stage in the bisection process, segment $\overline{EG}$ is too large for two consecutive midpoints to straddle it.

(4) By algebra,

$$\frac{AP}{AB} = \frac{ka}{na} = \frac{k}{n} \quad \text{and} \quad \frac{AQ}{AC} = \frac{kb}{nb} = \frac{k}{n}$$

$\therefore AP/AB = AQ/AC$.

(5) Also, by betweenness, $AE < AP$ and $AQ < AF$, so

$$\frac{AE}{AB} < \frac{AP}{AB} = \frac{AQ}{AC} < \frac{AF}{AC} \quad \text{or} \quad \frac{AE}{AB} < \frac{AF}{AC}$$

(6) The case A–G–E leads to the inequality $AE/AB > AF/AC$ by the same argument.

(7) $\therefore AE/AB \neq AF/AC$.

COROLLARY: If $AE/AB = AF/AC$, then $\overline{EF} \parallel \overline{BC}$.

THEOREM 3: THE SIDE-SPLITTING THEOREM

If a line parallel to the base $\overline{BC}$ of $\triangle ABC$ cuts the other two sides $\overline{AB}$ and $\overline{AC}$ at E and F, respectively, then

$$\frac{AE}{AB} = \frac{AF}{AC}$$

and by algebra,

$$\frac{AE}{EB} = \frac{AF}{FC}$$

(See Figure 4.20.)

PROOF

In case it is not true that $AE/AB = AF/AC$, then construct a point F' on $\overline{AC}$ so that $AE/AB = AF'/AC$ by the Segment Construction

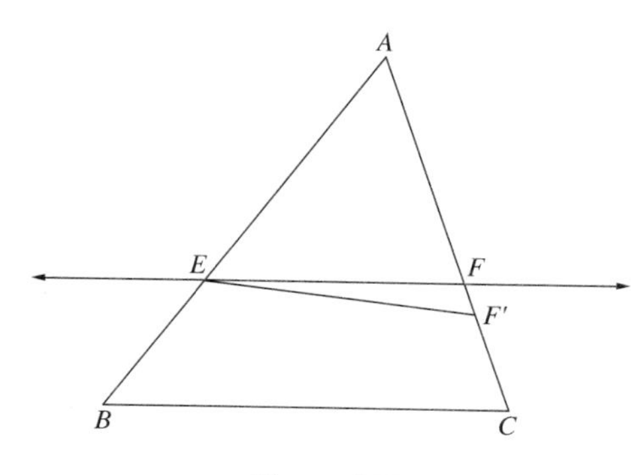

Figure 4.20

Theorem. The above corollary then tells us that $\overline{EF'} \parallel \overline{BC}$. But $\overline{EF} \parallel \overline{BC}$. Therefore $\overline{EF'}$ must coincide with $\overline{EF}$ and $F' = F$, with

$$\frac{AE}{AB} = \frac{AF}{AC}$$

To prove the other ratio, since $AB = AE + EB$ and $AC = AF + FC$, then

$$\frac{AE}{AE + EB} = \frac{AF}{AF + FC} \qquad \text{or} \qquad \frac{AE + EB}{AE} = \frac{AF + FC}{AF}$$

which reduces, by algebra, to $EB/AE = FC/AF$ and this implies the desired result, $AE/EB = AF/FC$.

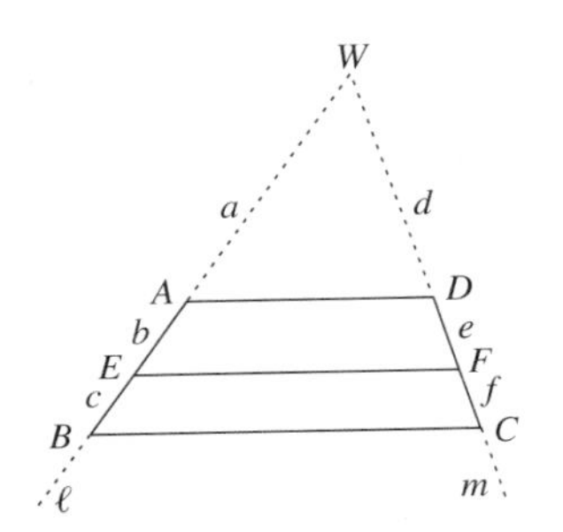

Figure 4.21

NOTE: The above logical sequence could just as well have been applied to a trapezoid $ABCD$ (Figure 4.21). The result would have been: If $\overline{AD} \parallel \overline{EF} \parallel \overline{BC}$ then $AE/EB = DF/FC$. This is also an algebraic result involving the quantities a, b, c, d, e, and f shown in the figure, where W is the point of intersection of lines $\overleftrightarrow{AB}$ and $\overleftrightarrow{CD}$ and the Side-Splitting Theorem is used (Problem 24). This finally completes the proof of the result that *ratios are preserved under parallel projection:* $PQ/QR = P'Q'/Q'R'$ in Figure 4.17.

EXAMPLE 1 In Figure 4.22, $\overline{DE} \parallel \overline{BC}$ and certain measurements are as indicated in each case.

(a) Find x

(b) Find y and z if you are given that $AD = DE$. Prove your answer.

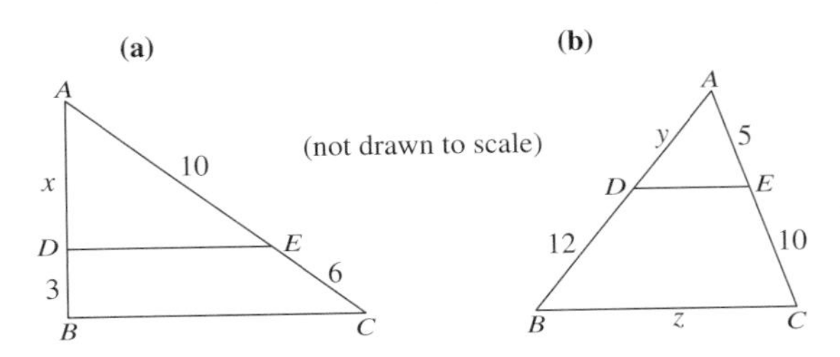

Figure 4.22

SOLUTION

(a) By the Side-Splitting Theorem,

$$\frac{x}{3} = \frac{10}{6} \qquad \text{or} \qquad 6x = 3 \cdot 10$$

Thus, $x = 5$.

(b) Here $y/12 = 5/10 = \frac{1}{2}$ so $y = 12 \cdot \frac{1}{2} = 6$. We need to show that $\triangle ABC$ is isosceles to conclude that $z = BC = BA$. But, $\overline{DE} \parallel \overline{BC}$ so by Property F, $\angle BCA \cong \angle DEA \cong \angle DAE \cong \angle BAC$. Therefore, $AB = BC$, as desired, and $z = BA = y + 12$. Since $y = 6$ we obtain $z = y + 12 = 18$. ■

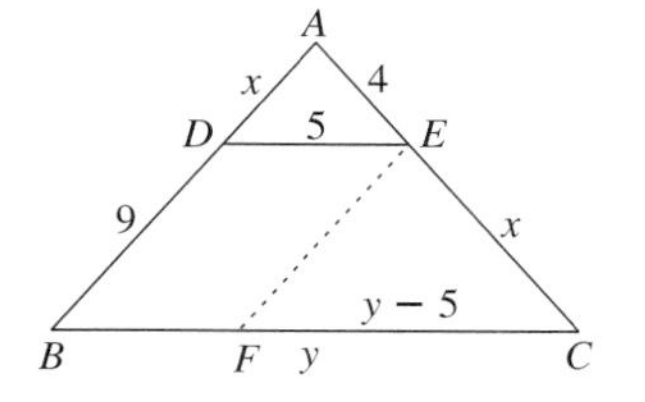

Figure 4.23

EXAMPLE 2 In Figure 4.23, $\overline{DE} \parallel \overline{BC}$, with certain measurements as indicated. Find x and y.

SOLUTION

For the value of x we have:

$$\frac{x}{9} = \frac{4}{x} \qquad \text{or} \qquad x^2 = 36 \rightarrow x = 6$$

(since $x > 0$). To find y, we proceed as follows. Draw a line $\overleftrightarrow{EF}$ through point E parallel to line $\overleftrightarrow{AB}$. This forms a parallelogram with sides of length 5 and 9, as shown. Thus, by the Side-Splitting Theorem applied to $\triangle CAB$ and segment $\overline{EF}$, $CE/EA = CF/FB$, or

$$\frac{x}{4} = \frac{y-5}{5} \qquad \text{or} \qquad 5x = 4(y-5)$$

Since $x = 6$, we find $5 \cdot 6 = 4y - 20$ or $4y - 20 = 30$. $\therefore y = 50/4 = 12.5$. ■

Moment for Discovery

A Quadrilateral Within a Quadrilateral

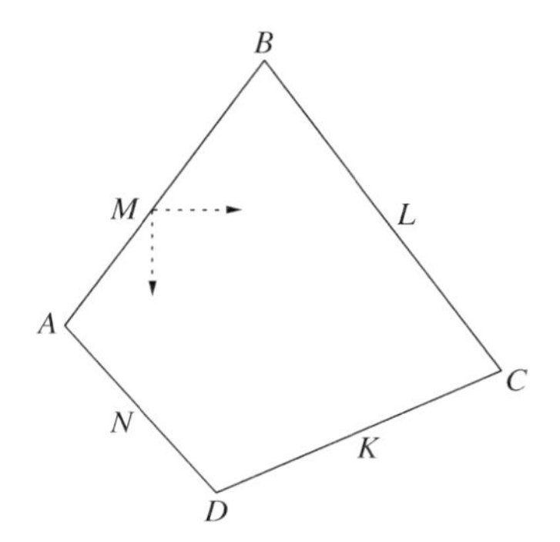

Figure 4.24

(**NOTE:** This discovery unit can also be performed without using *Sketchpad*.)

1. Draw any quadrilateral $\diamond ABCD$.
2. Locate, either by actual construction or by careful measuring, the midpoints M, L, K, and N of the sides of the quadrilateral.

3. Join the points M, L, K, and N, in order, forming a second quadrilateral. Do you observe anything in particular? (If you are using *Sketchpad*, drag A or B to obtain quadrilaterals of different shapes, including one that is not convex. Also, see if you can cause the smaller quadrilateral to be a rectangle or square.)
4. Construct a kite $\diamond ABCD$ (with $AB = BC$ and $AD = CD$) and locate the midpoints M, L, K, and N as before. Join these points, and observe the resulting quadrilateral. Do you have a conjecture? Try a proof.
5. Construct an isosceles trapezoid $\diamond ABCD$ with $AB = CD$, and construct the quadrilateral joining the midpoints of the sides as before. Did anything new occur?
6. Under what circumstances would $\diamond KLMN$ be a square?
7. State and prove any theorems you may have discovered above. (NOTE: A "theorem" is only a *conjecture* unless it can be proven theoretically.)

PROBLEMS (4.2)

GROUP A

1. In $\triangle ABC$, $\overleftrightarrow{DE} \parallel \overleftrightarrow{BC}$. Find x and y.

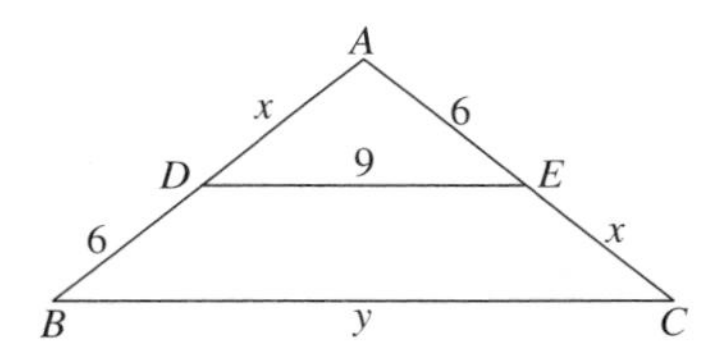

2. If $\angle A \cong \angle B$ and $\overleftrightarrow{DE} \parallel \overleftrightarrow{BC}$, find x, y, and z.

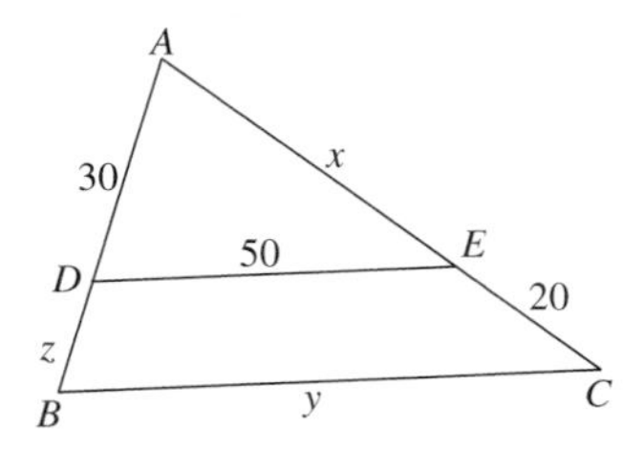

3. In the figure, $\overleftrightarrow{DE} \parallel \overleftrightarrow{BC}$. Find x, $y = BC$, and z if $z = \frac{2}{3}x$.

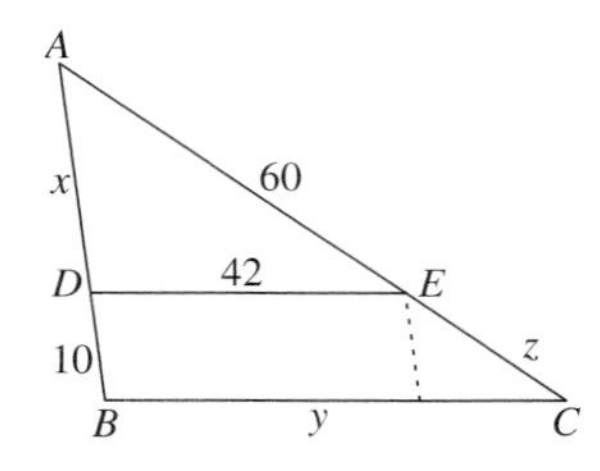

4. Four congruent triangular gardens are bordered by a walkway and, together they form a larger triangle, as shown. If the sides of the larger triangle are of length 8, 10, and 12 yards, and a person makes a tour of the gardens by walking along the path indicated, how many total yards did that person walk?

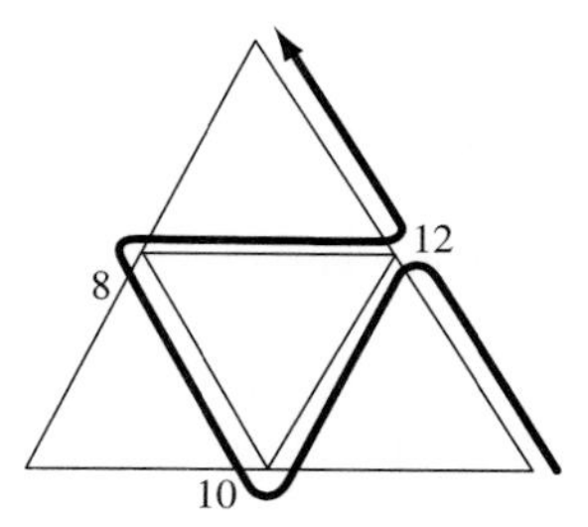

5. In trapezoid $ABCD$, E and F are the midpoints of $\overline{AD}$ and $\overline{BC}$, $AB = 20$, $DC = 28$, and $\overleftrightarrow{AK} \parallel \overleftrightarrow{BC}$.

(a) Find x and y.

(b) Find the length of the median $\overline{EF}$.

(c) Is your answer for EF the average of AB and DC? Try to prove this relation for any trapezoid using this method. (Compare with the above proof given for Theorem 2.)

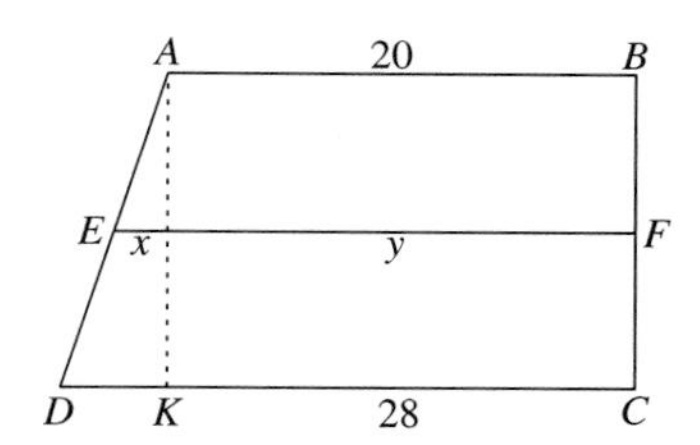

6. Equally spaced parallel lines cut two transversals at corresponding points A, B, C, D, and E on one transversal, and A', B', C', D', and E' on the other, forming several trapezoids, with $AE = 24$, $A'E' = 32$, $AA' = 12$, and $EE' = 34$. Find x, y, z, w, and u. (For y, z, and w, see Problem 5 for ideas. You do not have to prove your answer.)

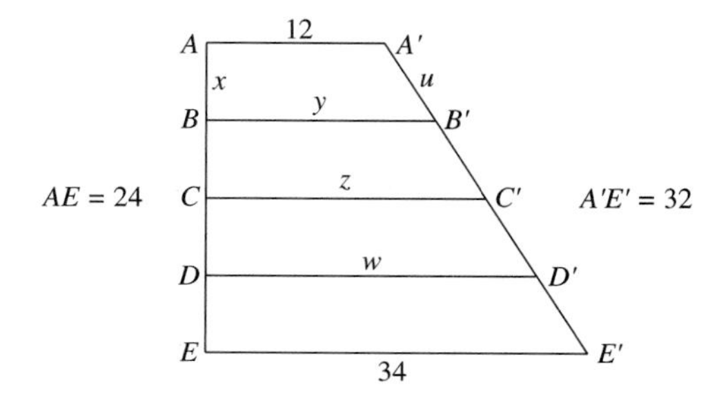

7. You have a given line segment $\overline{AB}$ of length > 6 in. drawn on an ordinary $8\frac{1}{2}$ by 11-in. sheet of lined paper, and you want to use the lines of the paper to divide $\overline{AB}$ into five congruent segments. How should you proceed? Prove the validity of your method in terms of the properties given in this section.

8. If $\diamond WXYZ$ is an isosceles trapezoid with base angles X and Y. (Problem 23 shows that $\angle X \cong \angle Y$.) If $m\angle X = -2q + 71$ and $m\angle Y = -5q + 32$ for some quantity q, find $m\angle X$.
[UCSMP, p. 239]

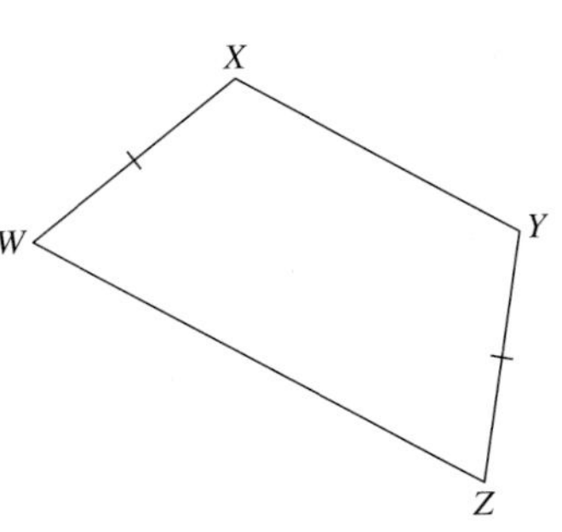

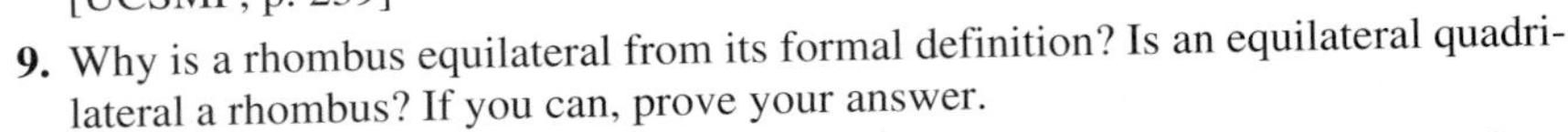

9. Why is a rhombus equilateral from its formal definition? Is an equilateral quadrilateral a rhombus? If you can, prove your answer.

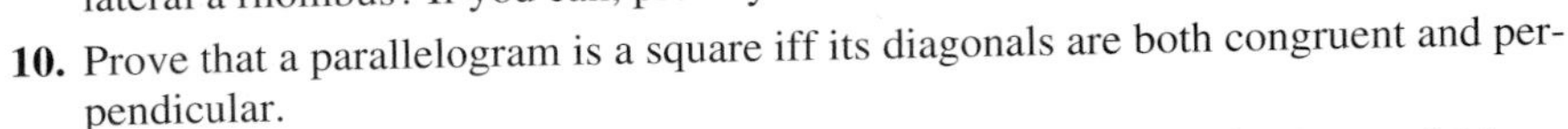

10. Prove that a parallelogram is a square iff its diagonals are both congruent and perpendicular.

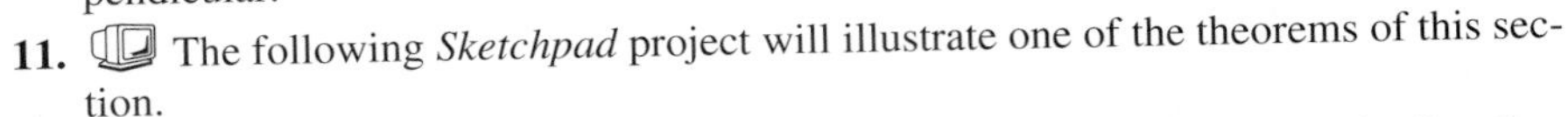

11. The following *Sketchpad* project will illustrate one of the theorems of this section.

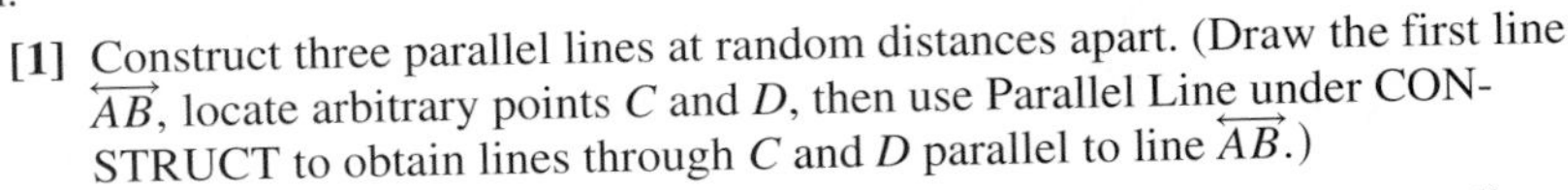

[1] Construct three parallel lines at random distances apart. (Draw the first line $\overleftrightarrow{AB}$, locate arbitrary points C and D, then use Parallel Line under CONSTRUCT to obtain lines through C and D parallel to line $\overleftrightarrow{AB}$.)

[2] Construct a point E on one line and point F on another, then construct line $\overleftrightarrow{EF}$. Select the point G of intersection of $\overleftrightarrow{EF}$ with the third line.

[3] Display the ratio EF/FG on the screen.

As you drag E to change the slope of line $\overleftrightarrow{EF}$, observe what happens to the ratio EF/FG. What theorem explains this behavior?

GROUP B

12. In $\triangle ABC$, $\overleftrightarrow{DE} \parallel \overleftrightarrow{BC}$ and certain measurements are as indicated. Find x and y.

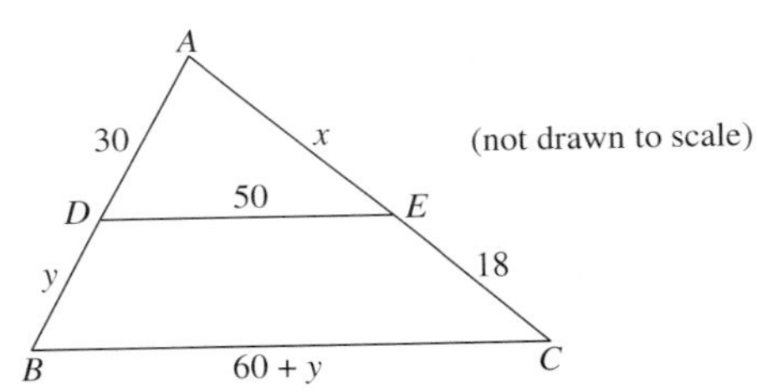

13. Two congruent isosceles triangles having angles of measure 30, 75, and 75, respectively, are placed vertex to vertex at point B, with the two congruent sides $\overline{DE}$ and $\overline{AC}$ opposite B lying on parallel lines as shown. Line segments $\overline{CD}$ and $\overline{AE}$ are then drawn. Prove:

(a) $\triangle BCD$ is an equilateral triangle.

(b) $\diamond ACDE$ is a square.

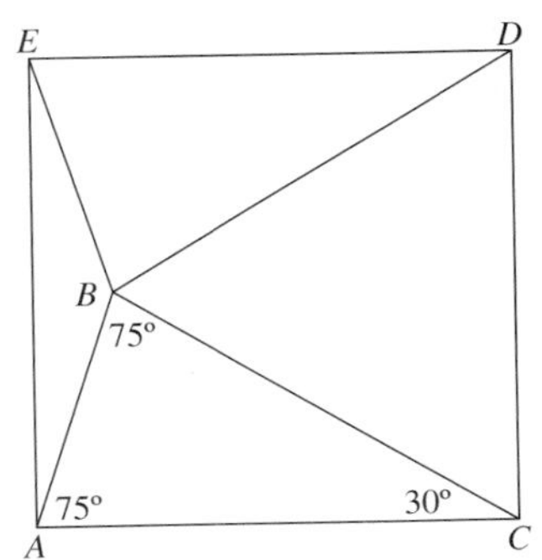

14. Investigate the phenomenon resulting from the following steps in *Sketchpad*:

[1] Construct $\triangle ABC$.

[2] Locate D any point on side $\overline{AC}$.

[3] Construct a line through D parallel to $\overline{AB}$, then select its intersection E with $\overline{BC}$. (What geometric principle predicts that this line must intersect $\overline{BC}$?)

[4] Display the ratios CD/CA and DE/AB side by side.

As you drag D on $\overline{AC}$, what happens to these ratios? Try proving your observation is correct based on geometric principles so far (note that we have not yet come to similar triangles).

15. In trapezoid $\diamond ABCD$ we are given $AB = AD$. Show that $\overrightarrow{BD}$ bisects $\angle ABC$.

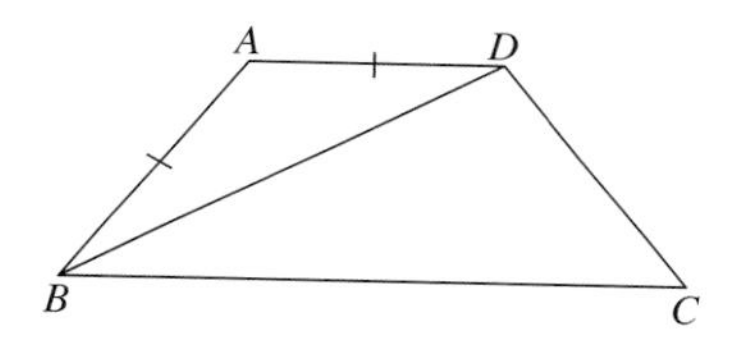

16. Perpendiculars $\overline{AE}$ and $\overline{CF}$ are dropped from the vertices to the diagonal $\overline{BD}$ of parallelogram $\square ABCD$. Prove: $AE = CF$.

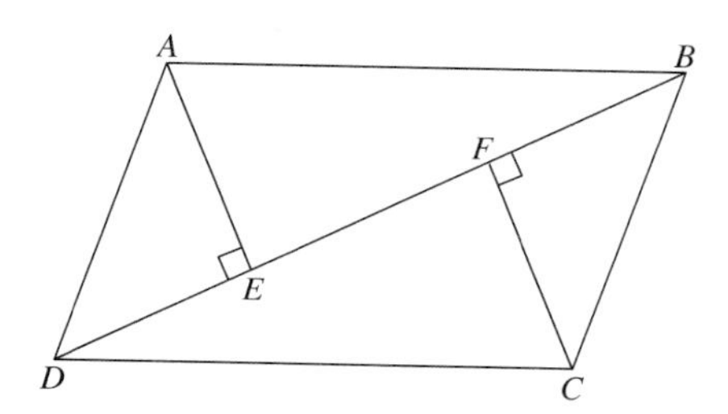

17. If two opposite sides of a convex quadrilateral are congruent and parallel, prove that the quadrilateral is a parallelogram.

18. Prove the converse part of the Midpoint Connector Theorem for Trapezoids: The median of a trapezoid is parallel to either base. (***Hint:*** Suppose the median is *not* parallel to the base; construct a line passing through the midpoint of one leg parallel to the base, and intersecting the other leg. What must happen?)

19. Euclidean Construction of Parallel Lines Given line ℓ and point A not on it, explain, with proofs, how the sequence of steps illustrated in the figure will produce the unique parallel m to line ℓ through A.

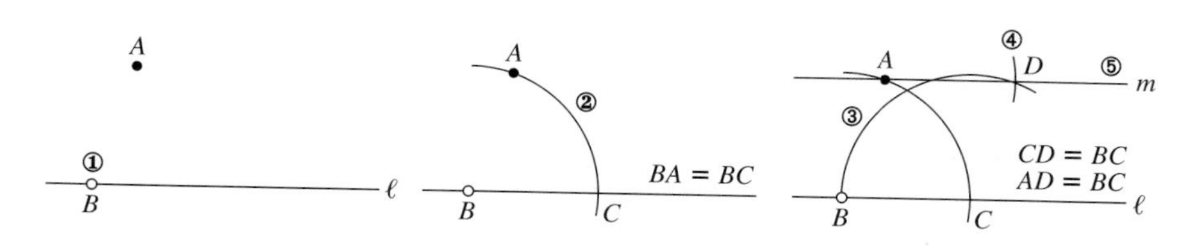

20. Design your own compass/straight-edge construction of

(a) a parallelogram with two adjacent sides given.

(b) an isosceles trapezoid.

(c) How would these constructions proceed under *Sketchpad*?

21. The figure shows an example for Step 1 in the proof of the lemma preceding the Side-Splitting Theorem, where successive bisections are performed. Here, $AB = 45, AE = 17\sqrt{2}$, and $EG = 1$. How many times must segments be bisected for one of the midpoints P to fall on $\overline{EG}$? (***Hint:*** This is strictly a calculator problem, not requiring much geometric reasoning.)

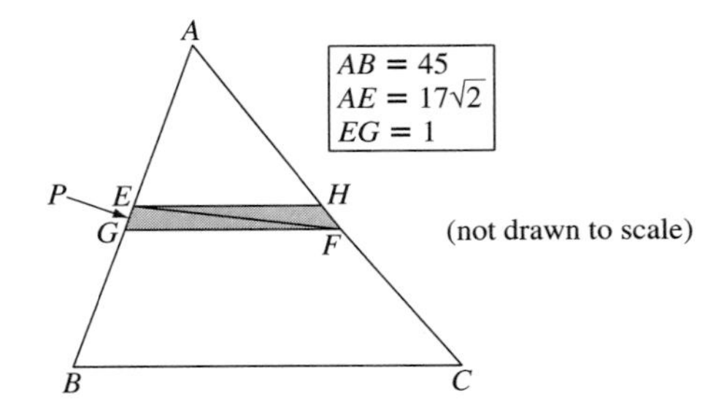

22. State and prove the SSAAA Congruence Criterion for convex quadrilaterals in Euclidean Geometry.

GROUP C

23. Prove each of the following:

(a) The base angles of an isosceles trapezoid are congruent. (Note that this property is not valid if the trapezoid is an oblique parallelogram.)

(b) If the base angles of a trapezoid are congruent, it is isosceles.

(c) The diagonals of an isosceles trapezoid are congruent.

24. Provide the necessary algebra to prove the relation $b/c = e/f$ using the other values shown in Figure 4.21 and Theorem 3.

4.3 Similar Triangles, Pythagorean Theorem, Trigonometry

The development of Euclidean geometry will now continue with the concept of similarity and similar triangles, of obvious importance.

DEFINITION: Two polygons P_1 and P_2 are said to be **similar,** denoted $P_1 \sim P_2$, iff under some correspondence between their vertices, corresponding angles are congruent, and the ratio of the lengths of corresponding sides is constant $(= k)$. The number k is called the **constant of proportionality,** or **scale factor,** for the similarity.

SIMILAR POLYGONS

The diagrams in Figure 4.25 illustrate two pairs of similar polygons, and one pair that is not. The basic idea of similarity, as opposed to congruence, is that of **shape.** Polygons having both the same size and shape are, of course, congruent. But they are merely similar if they have the same shape, but not the same size.

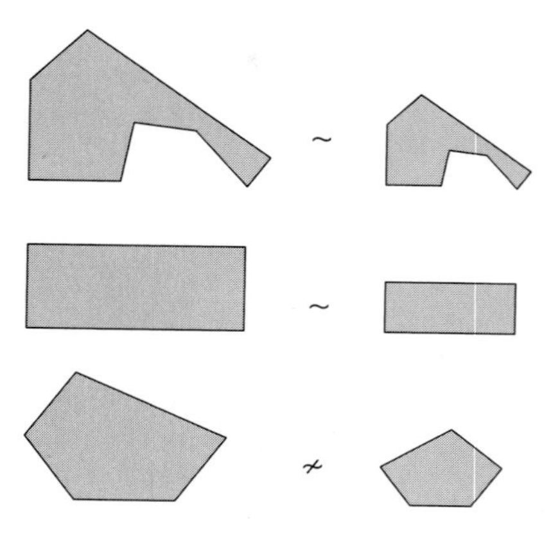

Figure 4.25

An important special case is similar triangles. The same care that was necessary for correspondence in congruent triangles will also be needed for similar triangles. Thus, if k is the constant of proportionality, then, as illustrated in Figure 4.26,

$$\triangle ABC \sim \triangle XYZ \quad \text{iff} \quad \begin{array}{ll} \angle A \cong \angle X, & AB = k \cdot XY \\ \angle B \cong \angle Y, & BC = k \cdot YZ \\ \angle C \cong \angle Z, & AC = k \cdot XZ \end{array}$$

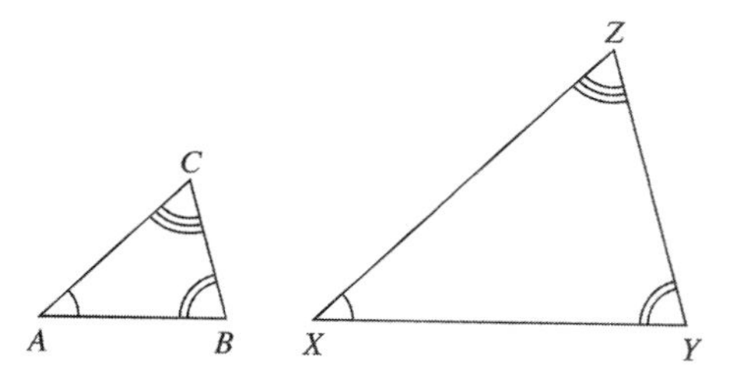

Figure 4.26

(Note that $k = 1$ corresponds to congruent triangles.)

While similar triangles may seem very familiar to you, their very existence is characteristic of Euclidean geometry. In non-Euclidean geometry similar triangles do not exist (unless they are congruent). So in an axiomatic treatment, the following lemma is crucial and plays a key role in our development.

LEMMA: Consider $\triangle ABC$ with D and E on sides $\overline{AB}$ and $\overline{AC}$ (Figure 4.27). If $\overline{DE} \parallel \overline{BC}$ then

$$\frac{AD}{AB} = \frac{AE}{AC} = \frac{DE}{BC}$$

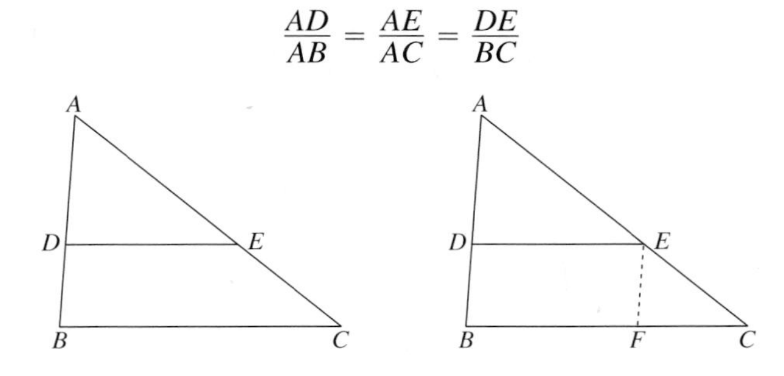

Figure 4.27

PROOF

The first equality, $AD/AB = AE/AC$, is due to the Side-Splitting Theorem. To prove the second, just draw line $\overleftrightarrow{EF} \parallel \overleftrightarrow{AB}$. Then, by the Side-Splitting Theorem,

$$\frac{CE}{EA} = \frac{CF}{FB}$$

We need to change the form of this equation, using betweenness and algebra. First, add 1 to both sides, then combine fractions:

$$1 + \frac{CE}{EA} = 1 + \frac{CF}{FB} \quad \text{or} \quad \frac{EA + CE}{AE} = \frac{FB + CF}{FB}$$

$$\frac{AC}{AE} = \frac{BC}{BF}$$

But $\diamond DEFB$ is a parallelogram, so $DE = BF$. Therefore, by substitution into the last expression, $AC/AE = BC/DE$, or $AE/AC = DE/BC$, as desired.

Notice how the last result immediately proves the existence of similar triangles: By the lemma the corresponding sides of $\triangle ABC$ and $\triangle ADE$ have measures in the same ratio (alternatively, if $k = AD/AB$, then $AD = kAB, AE = kAC$, and $DE = kBC$), and by the F-Property, the corresponding angles are congruent. Thus,

$$\triangle ABC \sim \triangle ADE$$

With just a little more thought, it can be seen that the lemma implies all three of the following useful similarity criteria (by constructing a figure like Figure 4.27 from the given triangles). The actual proofs will be left as problems.

THEOREM 1: AA SIMILARITY CRITERION
If, under some correspondence, two triangles have two pairs of corresponding angles congruent, the triangles are similar under that correspondence.

THEOREM 2: SAS SIMILARITY CRITERION
If in $\triangle ABC$ and $\triangle XYZ$ we have $AB/XY = AC/XZ$ and $\angle A \cong \angle X$, then $\triangle ABC \sim \triangle XYZ$.

THEOREM 3: SSS SIMILARITY CRITERION
If in $\triangle ABC$ and $\triangle XYZ$ we have $AB/XY = BC/YZ = AC/XZ$, then $\angle A \cong \angle X$, $\angle B \cong \angle Y$, $\angle C \cong \angle Z$, and $\triangle ABC \sim \triangle XYZ$.

EXAMPLE 1 Using Figure 4.28, determine as many missing angles or sides (their measures) as possible without using trigonometry. [UCSMP, p. 607]

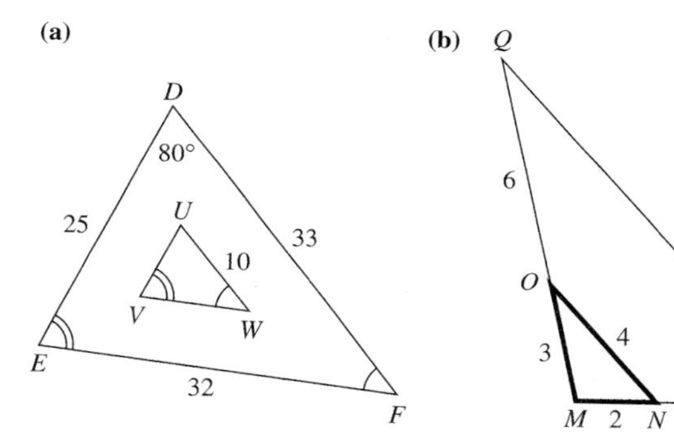

Figure 4.28

SOLUTION

(a) The AA Similarity Criterion applies to $\triangle DEF$ and $\triangle UVW$, with scale factor $k = \frac{10}{33}$.

$$\therefore\ UV = k \cdot DE = \left(\frac{10}{33}\right) \cdot 25 = \frac{250}{33}$$

$$VW = \left(\frac{10}{33}\right) \cdot 32 = \frac{320}{33}$$

$$m\angle U = m\angle D = 80$$

(b) For $\triangle MPQ$ and $\triangle MNO$, the SAS Similarity Criterion applies, with $MP/MN = k = \frac{6}{2} = 3$. Therefore, $PQ = k \cdot ON = 3 \cdot 4 = 12$.

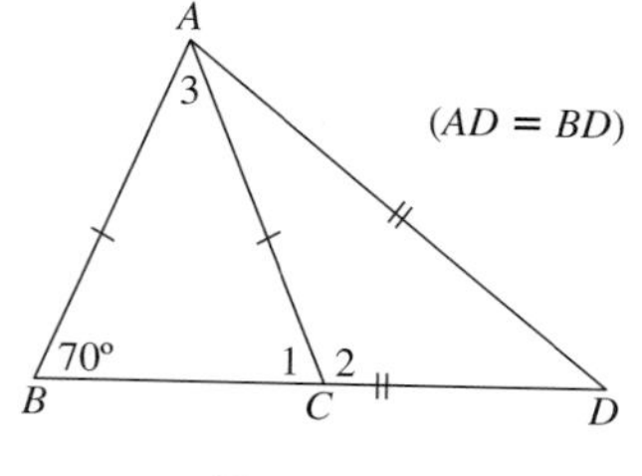

Figure 4.29

EXAMPLE 2 In Figure 4.29, $\triangle ABC$ and $\triangle ADB$ are both isosceles triangles ($AB = AC$ and $AD = BD$).

(a) Prove the two triangles are similar and establish the relation $AB^2 = BC \cdot BD$.

(b) If $m\angle B = 70$, find $m\angle D$.

SOLUTION

(a) Since $\angle B$ is in common between the two triangles, they have one pair of congruent angles; since they are isosceles, $\angle 1 \cong \angle B \cong \angle BAD$ $\therefore \triangle ABC \sim \triangle DBA$ by the AA Similarity Criterion. Hence

$$\frac{AB}{DB} = \frac{BC}{BA} \quad \text{or} \quad AB^2 = DB \cdot BC$$

(b) $m\angle D = m\angle 3 = 180 - (m\angle B + m\angle 1) = 180 - 140 = 40.$ ■

THE PYTHAGOREAN THEOREM DERIVED FROM SIMILAR TRIANGLES

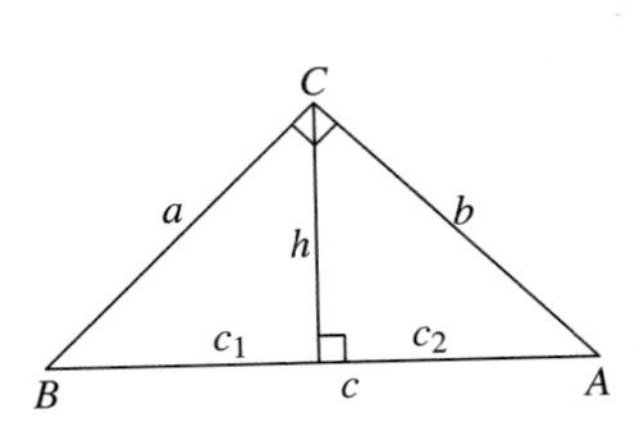

Figure 4.30

Although we have more than ample evidence of the validity of the Pythagorean Theorem from numerous points of view (as discussed in Section 1.2), it is interesting to see how its proof is handled using similar triangles, a method conducive to axiomatic geometry (without additional concepts being necessary, such as area). The actual details will be left for you in an interesting discovery unit at the end of this section, where you will receive guidance for splitting a given right triangle into two smaller triangles, each similar to the given triangle. The whole analysis will end up as an exercise in algebra where you can rather quickly come up with most (if not all) of the following relationships (Figure 4.30):

(1) $a^2 + b^2 = c^2$ (Pythagorean Theorem)

(2) $h^2 = c_1 c_2$ (First relation for altitude to hypotenuse)

(3) $h = \dfrac{ab}{c}$ (Second relation for altitude to hypotenuse)

Three lesser known formulas can also be derived:

(4) $$\frac{a^2}{b^2} = \frac{c_1}{c_2}$$

(5) $$a^2 - b^2 = {c_1}^2 - {c_2}^2$$

Relation **(2)** proves the well-known geometric principle

- The measure of the altitude of a right triangle to the hypotenuse equals the geometric mean of the measures of the two segments it cuts off on the hypotenuse. (See Appendix B.5 for the definition of *geometric mean.*)

EXAMPLE 3 In Figure 4.31 is shown three adjacent right triangles, with the measurements of their sides as indicated.

(a) Show that the three triangles are each similar to a 3–4–5 right triangle.
(b) Show that several numbers in the figure are geometric means of other numbers appearing.

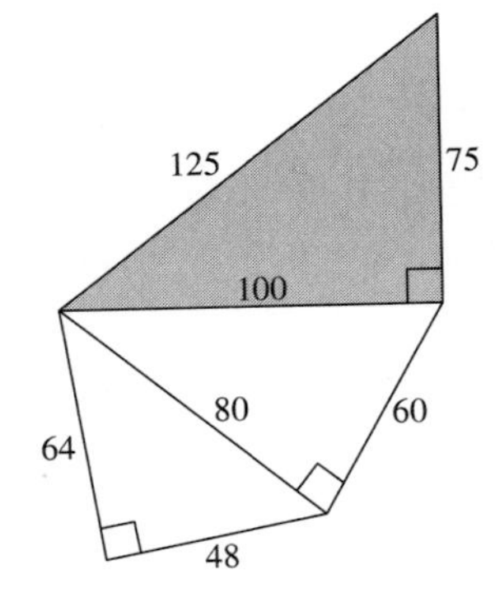

Figure 4.31

SOLUTION

(a) First triangle: 75:100:125 = 3:4:5 ($k = 25$)
Second triangle: 60:80:100 = 3:4:5 ($k = 20$)
Third triangle: 48:64:80 = 3:4:5 ($k = 16$)

(b) 100 is the geometric mean of 80 and 125: $100^2 = 80 \cdot 125 = 10{,}000$
60 is the geometric mean of 48 and 75: $60^2 = 48 \cdot 75 = 3600$
80 is the geometric mean of 64 and 100: $80^2 = 64 \cdot 100 = 6400$. ■

A problem analogous to the next item in Our Geometric World series appears in a high school geometry text, *Geometry: An Integrated Approach*, by Larson, Boswell, and Stiff. At first glance, it appears to have little to do with similar triangles. But a second look shows there is similarity in action, and it is possible to actually obtain three triangles similar to each other like those shown in Figure 4.31 (except that they are not right triangles). One triangle is indicated in the figure; see if you can determine the other two, along with their dimensions and geometric mean relationships.

OUR GEOMETRIC WORLD

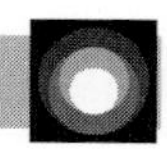

Two gears are said to have **gear ratio** $a/b = k$ if the first gear has p teeth, the second gear has q teeth, and $p/q = k$. The gear ratio determines how fast the gears turn with respect to each other. For example, a gear ratio of $3/2 = 1.5$ means that the smaller gear will make 1.5 revolutions for every one revolution made by the larger gear. Three gears in tandem (as shown in Figure 4.32) are such that the relative turning speed of Gear A to Gear B equals that of Gear B to Gear C (which is itself undetermined). If Gear A has 18 teeth and Gear C has 8 teeth, the number of teeth in Gear B *is the geometric mean of the numbers for gears A and C*. This is because, using gear ratio as defined above,

$$\frac{18}{x} = \frac{x}{8}$$

or, by algebra, $x^2 = 18 \cdot 8 = 144$, which yields the answer $x = 12$ teeth for Gear B. The relative turning speeds can now be found: That of Gear C to Gear A is $18/8 = 2.25$, whereas those of Gear C to Gear B and of Gear B to Gear A (stated to be equal) are each equal to $18/12 = 1.5$. Note that $2.25 = (1.5)^2$. An interesting fact for gear trains in linear sequence (such as this one) is that if there are $n + 1$ gears in tandem, and all engaged pairs have equal relative turning speeds $k > 1$, then the smaller gear has a turning speed of k^n with respect to the larger gear.

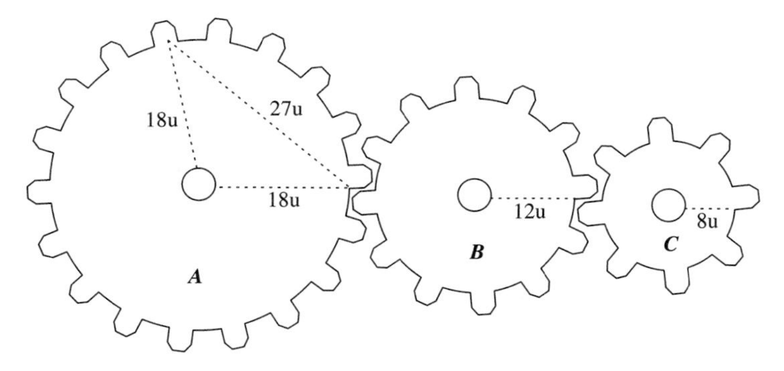

Figure 4.32

THE TRIGONOMETRY OF RIGHT TRIANGLES

If each of the three right triangles in Figure 4.33 has an acute angle congruent to $\angle A$, they are similar, and the ratio of a particular pair of side lengths, say the *opposite over the hypotenuse,* is a constant for the three right triangles pictured, and for all other right triangles having acute angle congruent to $\angle A$. The same is true for other pairs of sides. Hence, we define the following **trigonometric ratios** (which depend only on the measure of the acute angle A):

$$\sin A = \frac{a}{c} = \frac{a'}{c'} = \frac{a''}{c''} \qquad \cos A = \frac{b}{c} \qquad \tan A = \frac{a}{b}$$

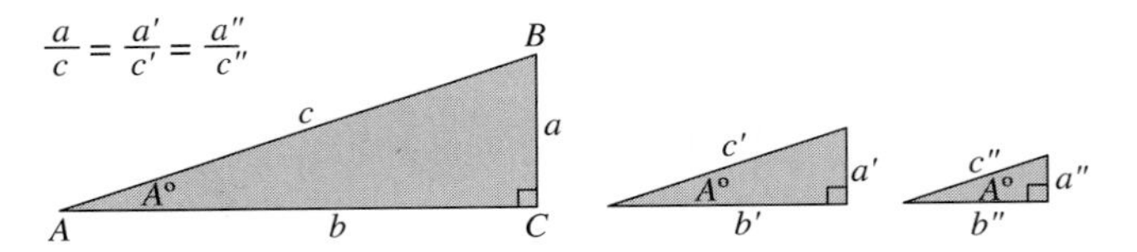

Figure 4.33

called **sine, cosine,** and **tangent,** respectively. Thus, the **cosine** of an acute angle of measure A in any right triangle is the ratio of the lengths of the *adjacent leg to the hypothenuse,* and its **tangent** is the ratio of the lengths of the *opposite leg to the adjacent leg,* etc. The other three ratios are the reciprocals of those already defined, and are denoted csc A, sec A, and cot A (which are abbreviations of their full names, **cosecant, secant,** and **cotangent**).

NOTE: We have introduced the six trigonometric functions from the viewpoint of similar triangles. The form for these definitions should be reminiscent of your introduction to trigonometry. The only reason for repeating this here is to show the fundamental importance of similar triangles to trigonometry now that you have seen a more sophisticated treatment of geometry.

The following summarizes the definitions and results involving right triangles introduced so far.

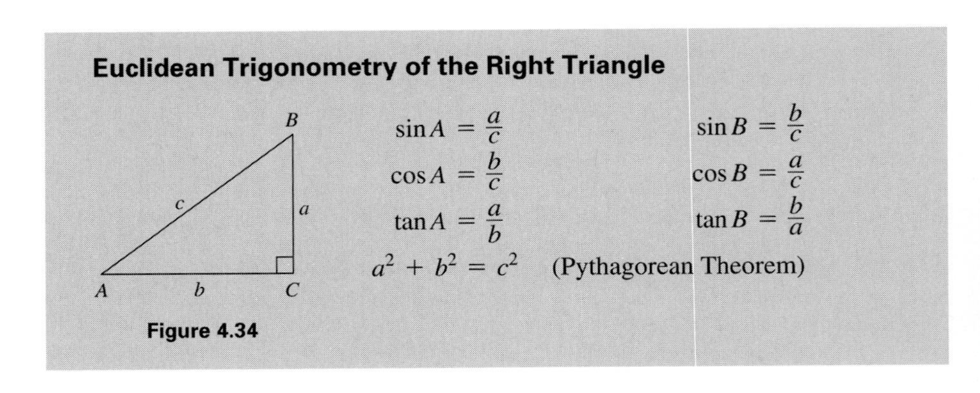

Euclidean Trigonometry of the Right Triangle

$$\sin A = \frac{a}{c} \qquad \sin B = \frac{b}{c}$$

$$\cos A = \frac{b}{c} \qquad \cos B = \frac{a}{c}$$

$$\tan A = \frac{a}{b} \qquad \tan B = \frac{b}{a}$$

$$a^2 + b^2 = c^2 \quad \text{(Pythagorean Theorem)}$$

Figure 4.34

To illustrate the geometric nature of these definitions, you should be able to use geometry to find the exact values of sin 30, cos 45, and tan 60. (***Hint:*** For $A = 30$, consider an equilateral triangle, etc.)

The preceding definitions are obviously valid only for $0 < A < 90$. To make the usual applications to geometry, it is necessary to extend them to the range $0 \leq A \leq 180$. To that end, we make the following definitions for $0 < A < 180$.

(6)
$$\sin(180° - A) = \sin A$$

(7)
$$\cos(180° - A) = -\cos A$$

(The definitions for the remaining functions then follow, although we will not need them for the present.)

The next set of definitions are motivated by continuity considerations:

(8)
$$\sin 0° = 0 \qquad \sin 90° = 1 \qquad \sin 180° = 0$$

(9)
$$\cos 0° = 1 \qquad \cos 90° = 0 \qquad \cos 180° = -1$$

The familiar **Pythagorean Identity** can be derived as follows, for $0 < A < 90$:

$$a^2 + b^2 = c^2 \qquad \rightarrow \qquad a^2/c^2 + b^2/c^2 = c^2/c^2$$

or

$$(a/c)^2 + (b/c)^2 = 1$$

That is,

(10)
$$\sin^2 A + \cos^2 A = 1$$

which can then be extended to the range $0 \leq A \leq 180$ using the definitions **(6)–(9).**

Finally, we come to the geometric proofs of the Law of Sines and Law of Cosines. (Remember that we are showing how ideas, which may already be familiar to you, can be derived from the axioms.) Each of these so-called laws requires three separate cases in geometry; we consider only one of them and let you try writing the very similar proofs for the other two cases. In all the proofs, an altitude of length h is drawn, as shown. It should be pointed out that the proof for a single angle A automatically makes it valid for B and C by arbitrariness of choice.

Law of Sines for Any Triangle

$$\frac{a}{\sin A} = \frac{b}{\sin B} = \frac{c}{\sin C}$$

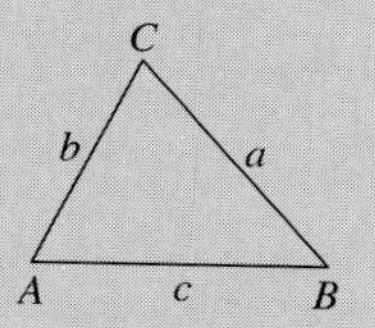

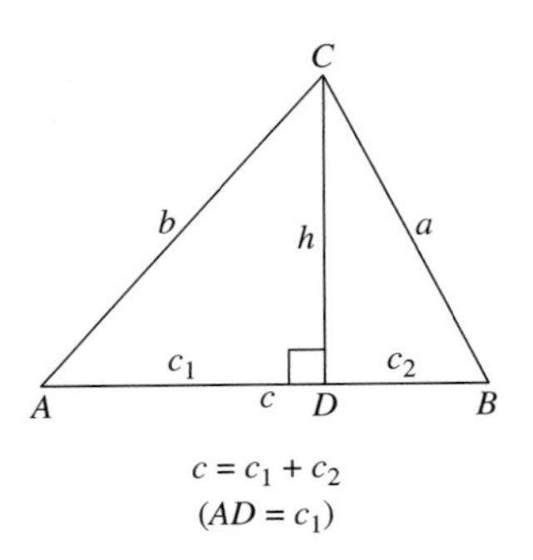

Figure 4.35

PROOF

CASE 1. $A < 90$ and $B < 90$. (See Figure 4.35.)

?

CASE 2. $A < 90$ and $B \geq 90$. (See Figure 4.36.) *Proof:*

$\sin A = h/b$

$\sin(180° - B) = h/a$ or $\sin B = h/a$

$\therefore \sin A/\sin B = (h/b) \cdot (a/h) = a/b \rightarrow a/\sin A = b/\sin B$

CASE 3. $A \geq 90$ and $B < 90$. (See Figure 4.37.)

?

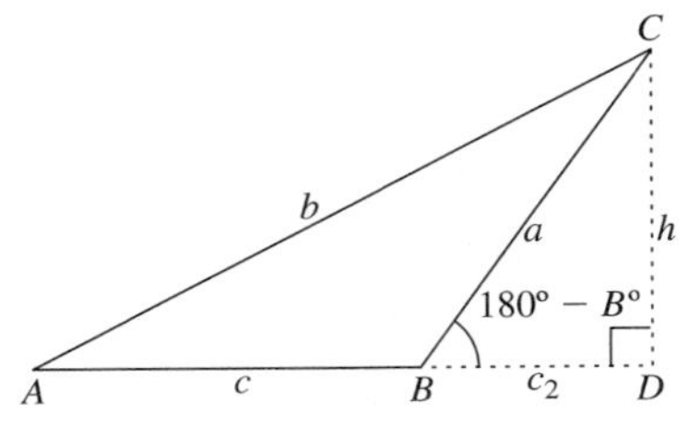

Figure 4.36

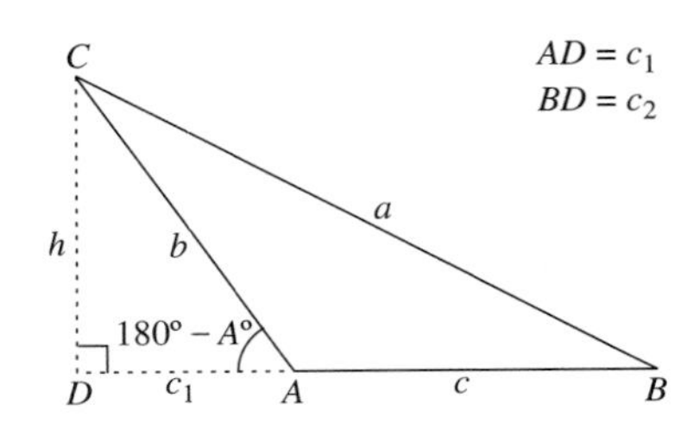

Figure 4.37

Law of Cosines for Any Triangle

$$a^2 = b^2 + c^2 - 2bc\cos A$$
$$b^2 = a^2 + c^2 - 2ac\cos B$$
$$c^2 = a^2 + b^2 - 2ab\cos C$$

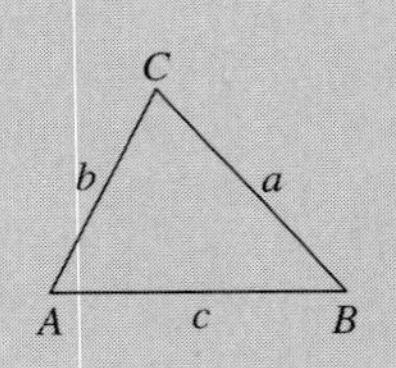

PROOF

CASE 1. $A < 90$ and $B < 90$. (See Figure 4.35.)

?

CASE 2. $A < 90$ and $B \geq 90$. (See Figure 4.36.)

?

CASE 3. $A \geq 90$ and $B < 90$. (See Figure 4.37.) *Proof:*

$\cos(180° - A) = c_1/b$

$-\cos A = c_1/b$ (definition)

$-2bc\cos A = 2cc_1$ (algebra)

$a^2 = h^2 + c_2^2$ (Pythagorean Relation for $\triangle BCD$)

$= h^2 + (c + c_1)^2$ ($c_2 = c + c_1$, from the figure)

$= h^2 + c^2 + 2cc_1 + c_1{}^2$ (algebra)

$= (h^2 + c_1{}^2) + c^2 - 2bc\cos A$ (substitution: $2cc_1 = -2bc\cos A$)

$= b^2 + c^2 - 2bc\cos A$ (Pythagorean Relation for $\triangle ACD$)

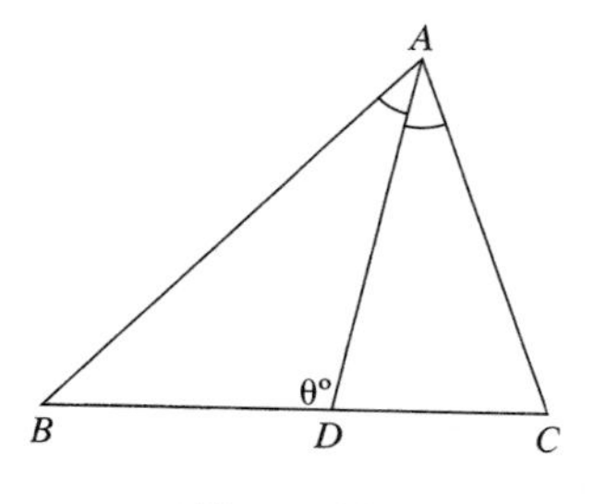

Figure 4.38

EXAMPLE 4 Use the Law of Sines to prove that if $\overrightarrow{AD}$ is the angle bisector of $\angle A$ in $\triangle ABC$ (Figure 4.38), then

$$\frac{AB}{AC} = \frac{BD}{DC}$$

SOLUTION
From the Law of Sines in $\triangle ADB$, we have

$$\frac{AB}{\sin\theta} = \frac{BD}{\sin A/2}$$

From the Law of Sines in $\triangle ADC$,

$$\frac{AC}{\sin(180° - \theta)} = \frac{AC}{\sin\theta} = \frac{CD}{\sin A/2}$$

That is,

$$\frac{AB}{\sin\theta} \cdot \frac{\sin\theta}{AC} = \frac{BD}{\sin A/2} \cdot \frac{\sin A/2}{CD}$$

or

$$\frac{AB}{AC} = \frac{BD}{DC}$$

The result in the next example is a useful formula, and can be used from this point on.

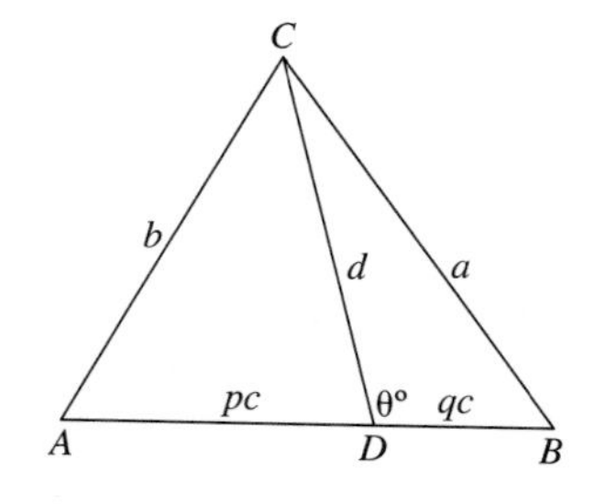

Figure 4.39

EXAMPLE 5 **The Cevian Formula** (See Figure 4.39.) Use the Law of Cosines to prove that for any triangle $\triangle ABC$, if d is the length of the line segment $\overline{CD}$ (called a **cevian**) and $p = AD/AB$, $q = DB/AB$, then d is given in terms of the side lengths a, b, and c by the formula

(11) $$d^2 = pa^2 + qb^2 - pqc^2$$

SOLUTION
Note that by definition of p and q, $AD = p \cdot AB = pc$ and $DB = q \cdot AB = qc$, as shown in Figure 4.39. Hence, the Law of Cosines for $\triangle CDB$ and $\triangle ADC$ yields

$$a^2 = d^2 + (qc)^2 - 2d(qc)\cos\theta$$
$$b^2 = d^2 + (pc)^2 - 2d(pc)\cos(180° - \theta)$$

Now multiply the first equation by p, the second equation by q, and use the fact that $\cos(180 - \theta) = -\cos\theta$:

$$pa^2 = pd^2 + pq^2c^2 - 2dpqc\cos\theta$$

$$qb^2 = qd^2 + qp^2c^2 + 2dpqc\cos\theta$$

Sum these two equations and the terms involving $\cos\theta$ will cancel. With a little factoring we have

$$pa^2 + qb^2 = (p + q)d^2 + pq(q + p)c^2$$

But $p + q = AD/AB + DB/AB = 1$, so this becomes

$$pa^2 + qb^2 = d^2 + pqc^2 \qquad \text{or} \qquad d^2 = pa^2 + qb^2 - pqc^2. \quad \blacksquare$$

The Cevian Formula is quite useful in working with triangles in geometry. For example, suppose we wanted to find the length of a median of a triangle and we know the lengths of the three sides. We just let $p = \frac{1}{2}$ and $q = \frac{1}{2}$ and place these two values into **(11)**. The result is the **formula for the length of a median** of a triangle in terms its sides:

$$\textbf{(12)} \qquad d = \sqrt{\frac{1}{2}a^2 + \frac{1}{2}b^2 - \frac{1}{4}c^2}$$

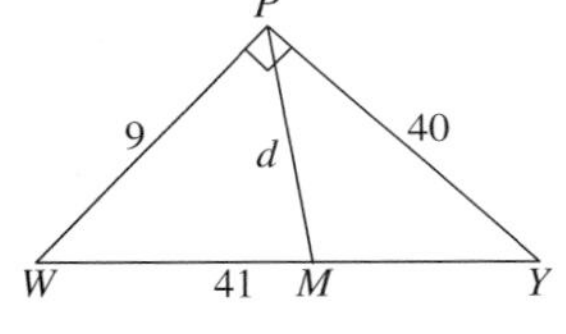

Figure 4.40

EXAMPLE 6 Use **(12)** to verify that in the case of the right triangle $\triangle PWY$ shown in Figure 4.40, the midpoint of the hypotenuse of a right triangle is equidistant from the three vertices.

SOLUTION

By **(12)** we have

$$\begin{aligned} d &= \sqrt{\frac{1}{2}(40)^2 + \frac{1}{2}(9)^2 - \frac{1}{4}(41)^2} \\ &= \sqrt{\frac{1}{2}(1600 + 81) - \frac{1}{4}(1681)} \\ &= \sqrt{\frac{1}{2}(1681) - \frac{1}{4}(1681)} \\ &= \sqrt{\frac{1}{4}(1681)} = 41/2 \end{aligned}$$

Thus, $MP = MW = MY = \frac{1}{2}(41)$. $\blacksquare$

Moment for Discovery

Discovering the Proof of the SSS Similarity Criterion

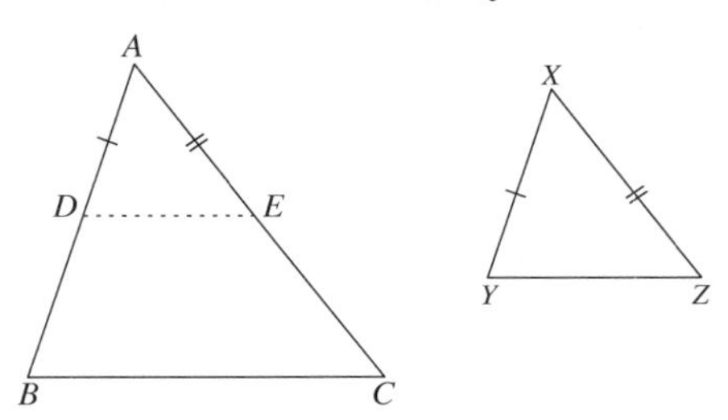

Figure 4.41

Observe Figure 4.41, where we are given: $\triangle ABC$ and $\triangle XYZ$ with the three sides in proportion, $AB/XY = BC/YZ = AC/XZ\ (= k)$, with $k > 1$. We are to prove $\triangle ABC \sim \triangle XYZ$. To that end, provide answers to the following questions:

1. Is $AB > XY$? Then can you construct segment $\overline{AD} \cong \overline{XY}$ on $\overline{AB}$? How about $\overline{AE} \cong \overline{XZ}$ on $\overline{AC}$?
2. Is $\triangle ADE \cong \triangle XYZ$? (Remember, we cannot assume $\angle A \cong \angle X$; this is what we want to prove.)
3. What reason might be given for $\overline{DE} \parallel \overline{BC}$? Is $AD/AB = AE/AC$?
4. Does it follow that $DE = YZ$? (Recall the lemma in this section.)
5. Now finish the proof from this point.

Moment for Discovery

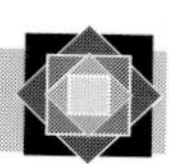

The Pythagorean Theorem from Similar Right Triangles

Let $\triangle ABC$ be any right triangle (Figure 4.42), with the measures of the sides shown, and the altitude to the hypotenuse, of length h. Since the angles at A and B are acute, we must have A–D–B by the result of Problem 17, Section 3.4. Observe the triangles formed by drawing this altitude, $\overline{CD}$, which we have separated and rearranged into similar positions.

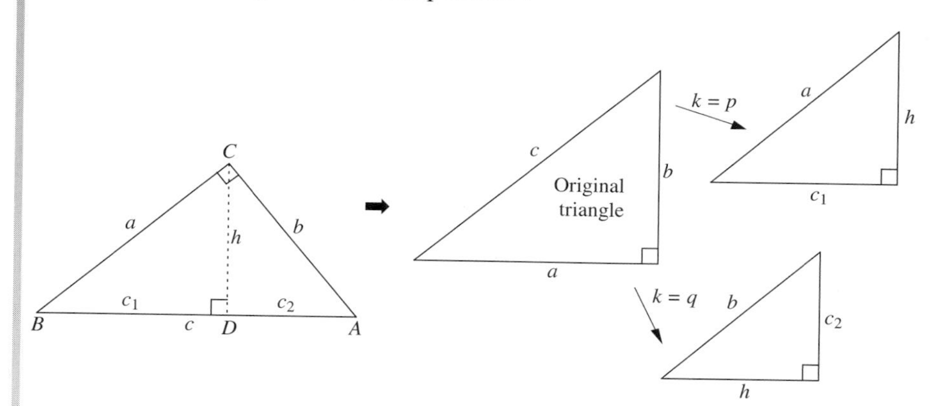

Figure 4.42

1. Why are the three right triangles to the right in the figure similar to each other? What similarity criterion is used?
2. Then, being similar, there are constants of proportionality k (which we denote by p and q, respectively) such that

$$\begin{aligned} a &= pc_1 = qh \\ b &= ph = qc_2 \\ c &= pa = qb \end{aligned}$$

(verify this).

3. The rest is algebra. See how many relations among **(1)–(5)** you can find. In each case, the goal is to *eliminate*, by algebra, the constants p and q.

PROBLEMS (4.3)

Since trigonometry has been derived axiomatically, that method of proof is valid for any of the following problems, as well as in the future. However, for most problems in this book, trigonometry is not the most appropriate method of solution.

GROUP A

1. *Given:* X is the midpoint of $\overline{WY}$. V is the midpoint of $\overline{WZ}$.
   Prove: $\triangle WXV \sim \triangle WYZ$.
   [UCSMP, p. 608]

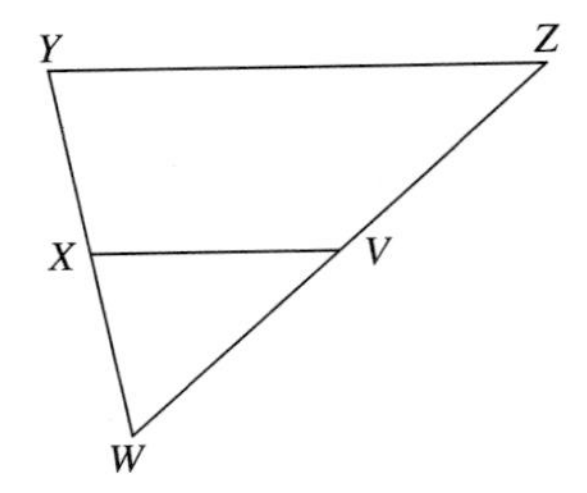

2. Consider $\triangle ABC$ and $\triangle XYZ$, with the sides and approximate angle measures as indicated in the figure. Determine whether the triangles are similar, and if so write the similarity in the proper manner and determine the measures of $\angle A$, $\angle B$, and $\angle C$. If not, state why not.

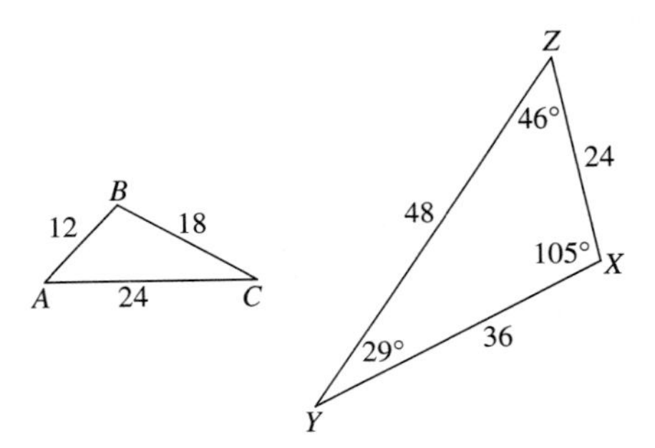

3. As in Problem 2, determine whether $\triangle PQR$ and $\triangle UVW$ are similar in **(a)** and **(b)**. If similar, write the similarity using the proper notation and determine the remaining sides and angles; if not, state why not.

(a)

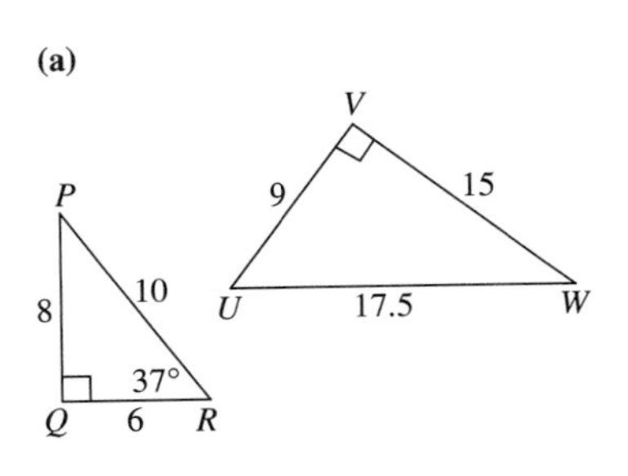

(b)

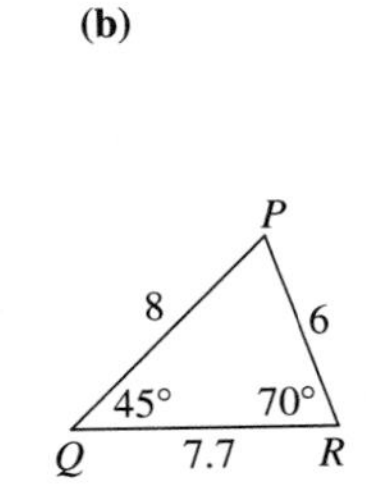

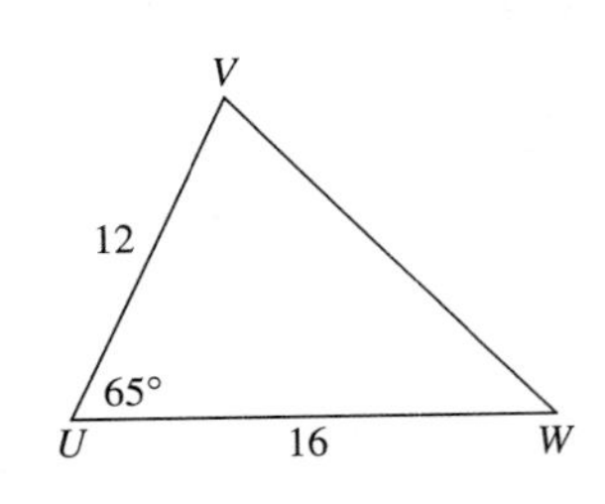

4. If $\overleftrightarrow{DE} \parallel \overleftrightarrow{BC}$ and $AC = BC$ in $\triangle ABC$, find x, y, and z.

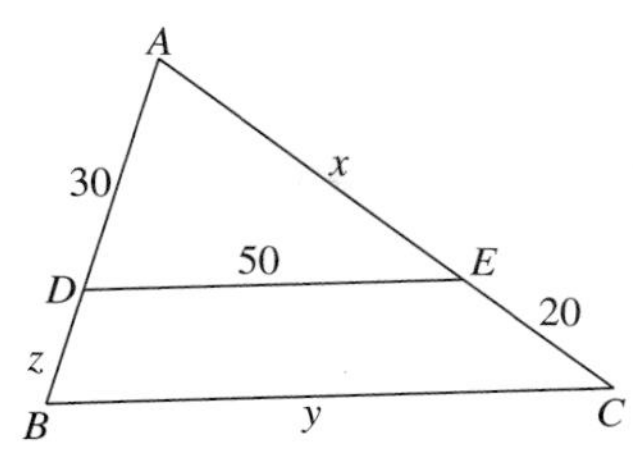

5. If $\triangle ABC$ is isosceles $(AB = AC)$, $m\angle BAC = 28$, $\overline{AM} \perp \overline{BC}$, and $\overline{MN} \perp \overline{AC}$, find $m\angle CMN$.

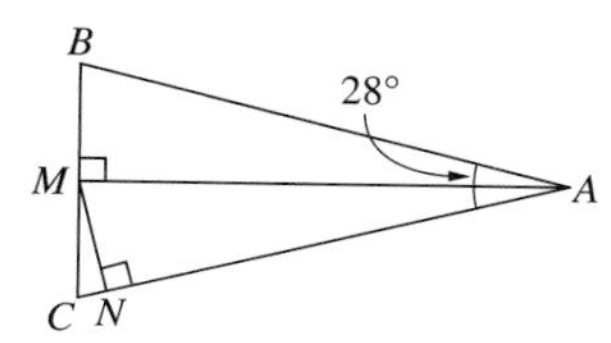

6. If $\triangle MNP \sim \triangle SRT$ with proportionality factor k, show that if $\overline{MA}$ and $\overline{SB}$ are altitudes of the given triangle, then $MA = kSB$.

In Problems 7 and 8, the two triangles in the figures are similar. For each problem, perform the following tasks:

(a) Determine the proper correspondence for the vertices.

(b) Find the scale factor k.

(c) Determine the measures of as many unknown angles and segments as possible, without trigonometry.

[UCSMP, p. 612]

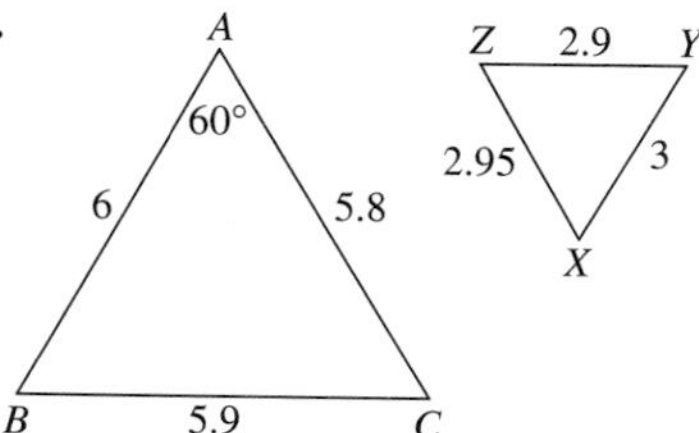

8.

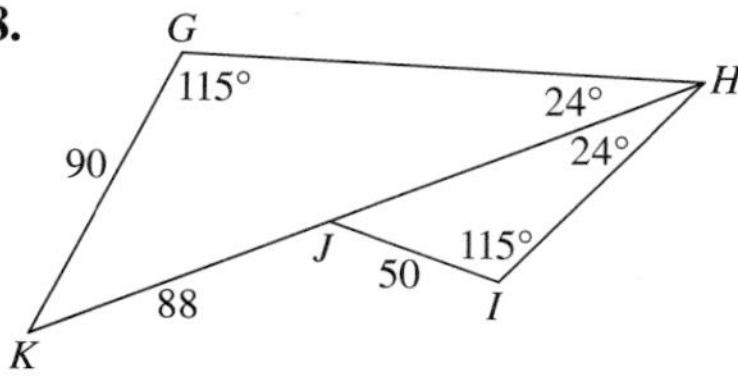

9. The figure shows a triangle that is very nearly isosceles. Show, in fact, that it is a right triangle. (Where is the right angle?) Verify all conclusions made. Note the use of the *converse* of the Pythagorean Theorem, considered in Problem 24.

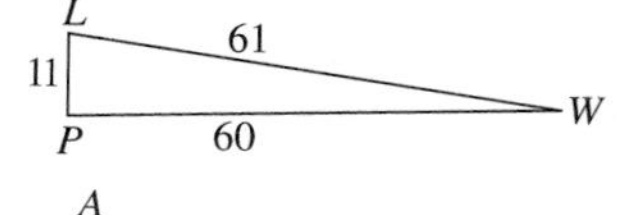

10. If $AB = AC$, $AD = BD$, and $m\angle D = 30$, find:

(a) $m\angle B$.

(b) Show that x is the geometric mean of BC and BD, then solve for x. (Refer to Example 2 if you need help.)

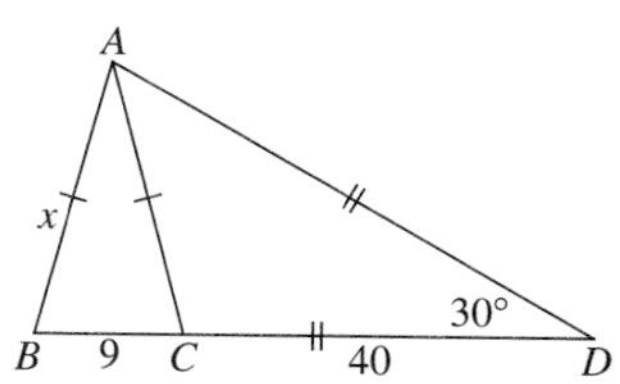

11. Algebraic Lemma for Ratio and Proportion Prove that if $a/b = c/d$ then

$$\frac{a+c}{b+d} = \frac{a}{b} \quad \text{and} \quad \frac{a-c}{b-d} = \frac{a}{b}$$

12. The diagonals of a rhombus have lengths 10 and 24 units, respectively. Find the lengths of the sides.

GROUP B

13. Suppose that in $\triangle ABC$ points D, E, and F are located such that D–E–F and $\triangle ADE$ is equilateral. If other measurements are as indicated in the figure, find x.

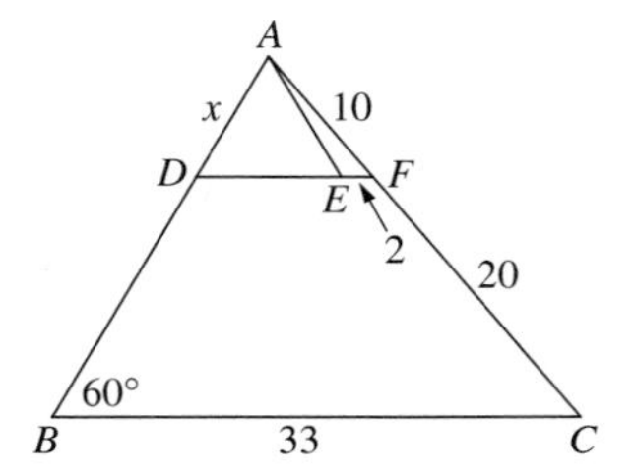

14. In a sequence of three squares, the second and third squares are symmetrically inscribed within their predecessors, as indicated. Find the ratio of the side lengths of the largest square to the smallest.

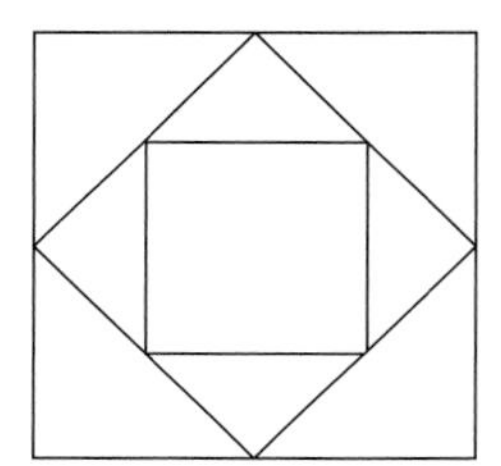

***15. Golden Isosceles Triangle** Isosceles triangle $\triangle ABC$ lies within another isosceles triangle $\triangle ABD$ ($AB = AC$ and $AD = BD$), as in Example 2. In addition, it is given that $\triangle ACD$ is isosceles with $AC = CD$.

(a) Show that this can happen (and will happen) precisely when $\triangle ABC$ is a **Golden Isosceles Triangle**, that is, when $AB = \tau BC$, where $\tau = \frac{1}{2}(1 + \sqrt{5})$— the Golden Ratio. (See Problems 13 and 14 of Section 1.2.) (***Hint:*** Use similar triangles to show that $AB^2 = BC \cdot BD = BC^2 + BC \cdot AB$, which can be converted into the relation $x^2 = x + 1$ where $x = AB/BC$.)

(b) Show that $\theta = 72$ and $\phi = 36$ in the figure.

(c) Using trigonometry in $\triangle ABC$, show that $\sec 72° = 2\tau$ and $\cos 36° = \tau/2$.

*Used in the construction of a regular pentagon. See Problem 16, Section 4.4.

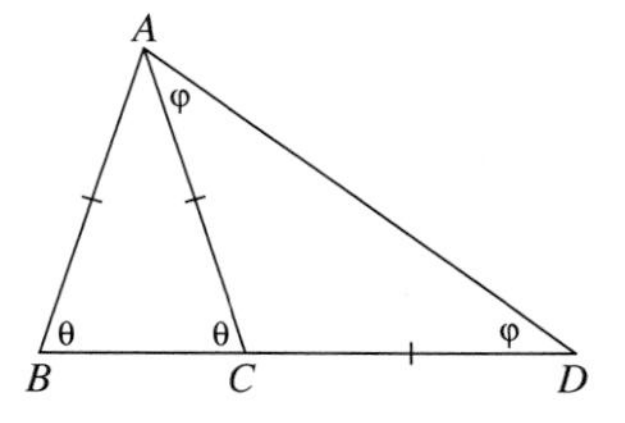

16. **(a)** In the square shown, three line segments join midpoints of two sides of the square to vertices opposite, forming a triangle (shaded). If the sides of the square are of measure $2\sqrt{5}$, find the lengths of the sides of the shaded triangle.

(b) Several line segments have been drawn that join the vertices of a square with the midpoints of the sides opposite, as shown. The shaded triangle is similar to a very familiar right triangle. What is it? (Prove your answer.)

(c) The figure for **(b)** contains a number of triangles similar to the shaded triangle. How many can you find? (***Hint:*** There are actually 24 of them.)

(a)

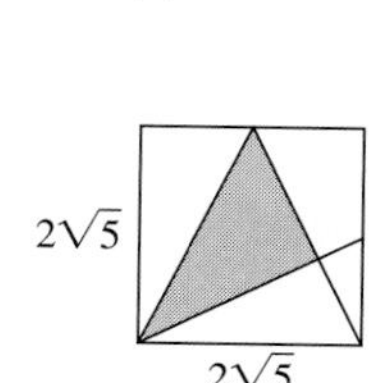

(b)

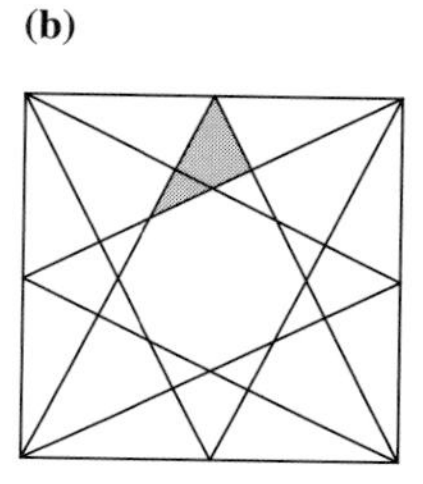

The next four problems, 17–20, are intended as a unit. They make an excellent mini-project for the reader and end in a significant result in geometry.

17. Recall from a previous result that if $AE/AB = AF/AC$, then $\overleftrightarrow{EF} \parallel \overleftrightarrow{BC}$ and $\triangle AFE \sim \triangle ABC$. However, if segment $\overline{EF}$ is "turned around" to the position shown in the figure, where $AF/AC = AE/AB$, then $\triangle AEF \sim \triangle ABC$. Show each of the following in general, regarding points F and E on sides $\overline{AB}$ and $\overline{AC}$ in $\triangle ABC$:

(a) If $\angle AEF \cong \angle ABC$ then $AF{\cdot}AB = AE{\cdot}AC$ (segment $\overline{EF}$ is called **antiparallel** to segment $\overline{AB}$ in this case).

(b) If $\triangle AFE \sim \triangle ACB$, then $\triangle ABE \sim \triangle ACF$.

(c) Conversely, if $\triangle ABE \sim \triangle ACF$, then $\triangle AFE \sim \triangle ACB$.

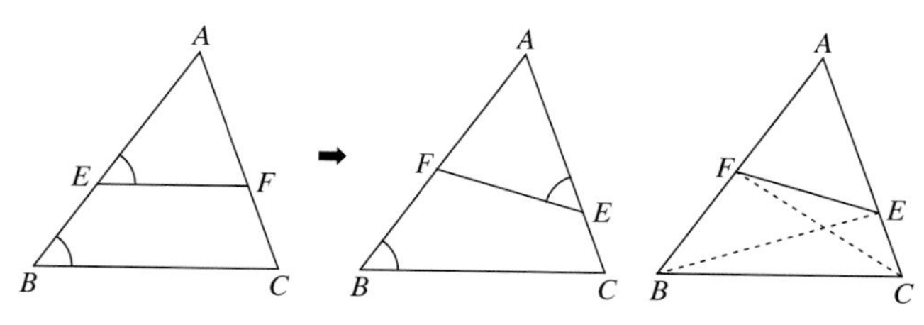

18. If $\overline{BE}$ and $\overline{CF}$ are altitudes of $\triangle ABC$ to sides $\overline{AC}$ and $\overline{AB}$, respectively, then prove:

(a) $\triangle ABE \sim \triangle AFC$.

(b) $\triangle AFE \sim \triangle ACB$. (See Problem 17.)

19. (Optional) Using *Sketchpad*, perform the following operations.

[1] Construct $\triangle ABC$.

[2] Construct segment $\overline{AD} \perp \overline{BC}$ and $\overline{BE} \perp \overline{AC}$.

[3] Locate F on side $\overline{AB}$ and draw segments $\overline{FC}$, $\overline{FD}$, and $\overline{FE}$.

[4] Display these measurements on screen:

$$m\angle C \equiv m\angle ACB,\ m\angle AFC,\ m\angle AFE, \text{ and } m\angle BFD.$$

[5] (Optional) Shade the interiors of $\triangle AFE$ and $\triangle BFD$.

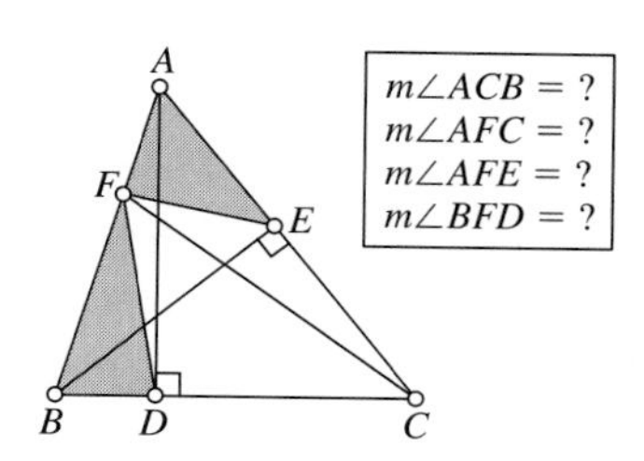

(a) Drag F on $\overline{AB}$ and watch the effect on the above measurements. Does $m\angle AFE$ ever become equal to $m\angle BFD$? When does that happen? Write down all the discoveries you make.

(b) Try proving these discoveries. (The results of Problems 17 and 18 should help. See the next problem for more on this.)

***20. Characterization of the Orthic Triangle** Based on the results of Problems 17 and 18, prove the following theorem about the orthic triangle of a given triangle $\triangle ABC$.

THEOREM: If D, E, and F are the feet of the altitudes of an acute-angled triangle $\triangle ABC$ on the sides $\overline{BC}$, $\overline{AC}$, and $\overline{AB}$, respectively, then $m\angle AFE = m\angle BFD = m\angle C$, and similarly for the remaining angles, as indicated in the figure. Conversely, if $\overline{CF} \perp \overline{AB}$ and $m\angle AFE = m\angle BFD = m\angle C$, then $\overline{BE}$ and $\overline{AD}$ are altitudes of $\triangle ABC$.

COROLLARY: The altitudes of a triangle are the angle bisectors of its orthic triangle.

*This result is used in the proof of Theorem 2, Section 5.7.

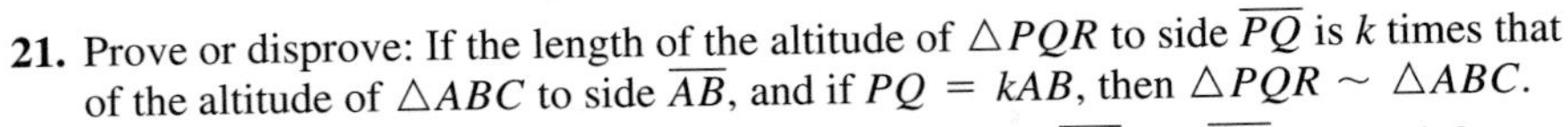

21. Prove or disprove: If the length of the altitude of $\triangle PQR$ to side $\overline{PQ}$ is k times that of the altitude of $\triangle ABC$ to side $\overline{AB}$, and if $PQ = kAB$, then $\triangle PQR \sim \triangle ABC$.

22. In trapezoid $\diamond ABCD$, the median $\overline{MN}$ cuts diagonals $\overline{AC}$ and $\overline{BD}$ at P and Q, respectively, with $BC = 6$ and $AD = 10$. Find PQ.

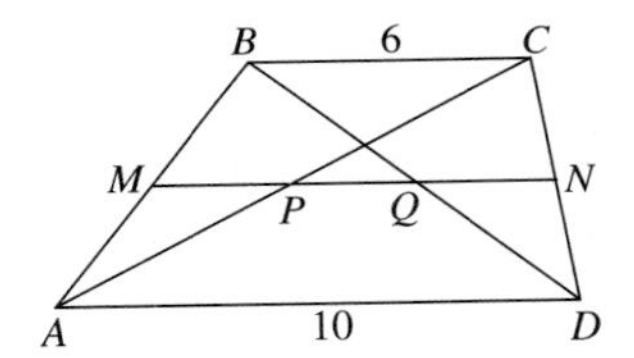

23. Prove the SAS Criterion for Similar Triangles.

24. State and prove the converse of the Pythagorean Theorem. (Use trigonometry.)

GROUP C

The problems in this group constitute a unit, intended to be worked in sequence. It culminates in an interesting research problem for an advanced undergraduate.

25. Directed Distance The formulas of geometry are often greatly simplified by the use of directed distances: If ℓ is any line coordinatized as in the Ruler Postulate, for any two points $A[a]$, $B[b]$ with their coordinates, define the **directed distance**

$$AB = b - a$$

Prove the following fundamental properties of directed distance (for arbitrary A, B, C on ℓ):

(a) $AB = -BA$ (or $AB + BA = 0$)

(b) $AB + BC = AC$ (or $AB + BC + CA = 0$)

26. Stewart's Theorem Given three collinear points A, B, and C, P any other point, and Q the foot of the perpendicular from P to line $\overleftrightarrow{AB}$ (using directed distance as introduced in Problem 20), show that

(a) $QA^2 \cdot BC + QB^2 \cdot CA + QC^2 \cdot AB + BC \cdot CA \cdot AB = 0$

(b) $PA^2 \cdot BC + PB^2 \cdot CA + PC^2 \cdot AB + BC \cdot CA \cdot AB = 0$

[***Hint:*** For **(a),** replace QB by $QA + AB$ and QC by $QA + AC$ and expand; for **(b),** let $h = PQ$ and replace PA^2 by $QA^2 + h^2$, etc.]

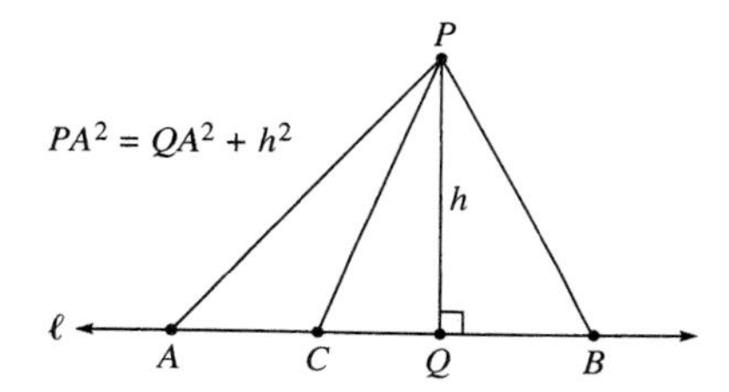

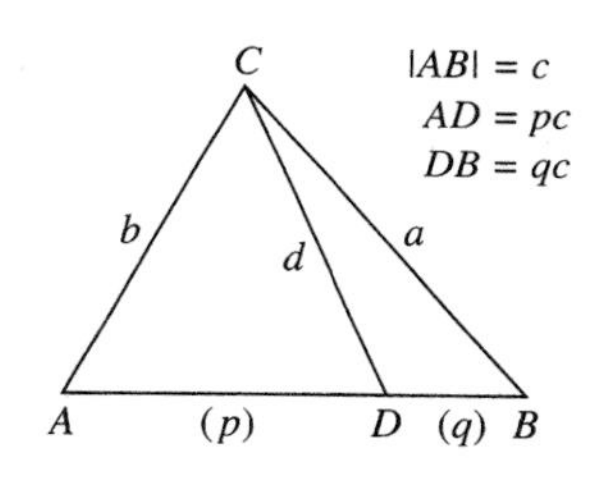

27. Comprehensive Cevian Formula Assuming directed distance on line $\overleftrightarrow{AB}$, define $p = AD/AB$ and $q = DB/AB$ (where p can be an arbitrary real number), and $CD = d$. Thus, $p + q = 1$ in all cases. Use Stewart's (Theorem Problem 21) to prove *without trigonometry* that for all points D on line $\overleftrightarrow{AB}$, $d^2 = pa^2 + qb^2 - pqc^2$.

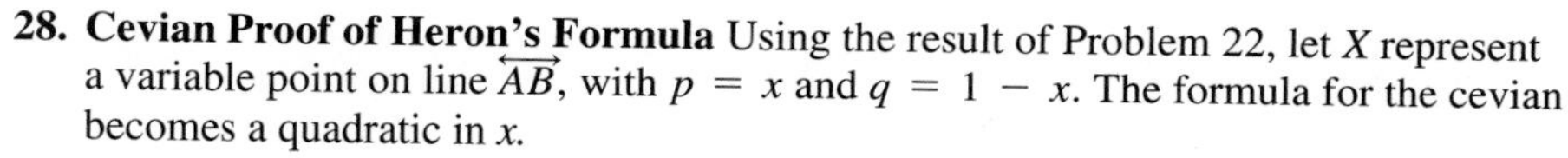

28. Cevian Proof of Heron's Formula Using the result of Problem 22, let X represent a variable point on line $\overleftrightarrow{AB}$, with $p = x$ and $q = 1 - x$. The formula for the cevian becomes a quadratic in x.

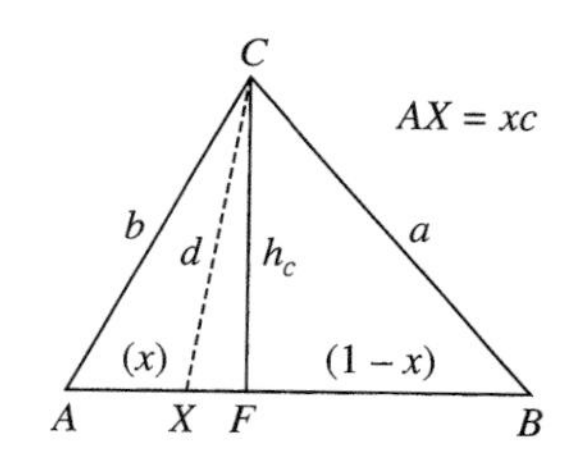

(a) Fix the value $d = h_c$ = length of the altitude $\overline{CF}$ to $\overleftrightarrow{AB}$ (F can fall exterior to side $\overline{AB}$), and show that $AF/AB = x$ must satisfy the equation

$$c^2x^2 + (a^2 - b^2 - c^2)x + (b^2 - h_c{}^2) = 0$$

What must be true of the possible values for x?

(b) That being the case, the discriminant Δ of the quadratic equation in x must equal zero. Write this out and solve the resulting equation $\Delta = 0$ for d (or h_c) in terms of a, b, and c.

(c) Put the formula you found in **(b)** in the form

$$h_c = \frac{2}{c}\sqrt{s(s - a)(s - b)(s - c)}$$

(d) Prove Heron's Formula.

29. Undergraduate Research Project Find a formula for the *sides* of a triangle in terms of

(a) its *medians* (Can these lengths be given arbitrarily to determine a triangle, and is the triangle uniquely determined?).

(b) its *altitudes* (see Problem 28).

(c) Is there a characteristic relationship for the medians of a right triangle? For the altitudes of a right triangle?

(d) Establish a Heron-like formula for the area of a triangle *in terms of its three altitudes.* [***Hint:*** Let $1/s_h = \frac{1}{2}(1/h_a + 1/h_b + 1/h_c)$.]

4.4 Regular Polygons and Tiling

An old problem in geometry, going back to Gauss, is how to construct a 17-sided regular polygon with the Euclidean tools, the compass and straight-edge. We consider the problem of both the existence of regular polygons, and their construction, and then we discuss briefly the interesting concept of tiling.

REGULAR POLYGONS IN ABSOLUTE GEOMETRY

We first tackle the problem of existence for regular polygons with an arbitrary number of sides (≥ 3). Since our construction does not use the Parallel Postulate or any of its consequences, it is valid in absolute geometry.

DEFINITION: An n-sided **polygon,** or **n-gon,** for any integer $n \geq 3$, is the closed union of line segments

$$\overline{P_1P_2} \cup \overline{P_2P_3} \cup \overline{P_3P_4} \cup \ldots \cup \overline{P_{n-1}P_n} \cup \overline{P_nP_1}$$

with **vertices** $P_1, P_2, P_3, \ldots, P_n$, and **sides** $\overline{P_iP_{i+1}}, \overline{P_nP_1}$ for $1 \leq i \leq n - 1$, where no three **consecutive vertices** $P_i, P_{i+1}, P_{i+2} (1 \leq i \leq n$, with $P_{n+1} = P_1$ and $P_{n+2} = P_2)$ are collinear and no two sides meet except at vertices. (Two sides are **adjacent** if they share a common endpoint.) An **angle** of the polygon is any of the n angles formed by three consecutive vertices, $\angle P_iP_{i+1}P_{i+2}$. If the polygon is completely contained by the sides and interior of each of its angles, it is said to be **convex.** A **regular polygon** is a convex polygon having congruent sides and congruent angles. The **radius** of a regular polygon is the radius of the circle O passing through its vertices, called the **circumscribed circle,** and point O is called the **center** of the polygon.

It is clear that since we are free to "construct" angles with any given measure (that is, such angles *exist*), then a regular n-gon may be constructed for each integer $n \geq 3$, and with any given length as radius. For example, to prove that a regular decagon (10-sided polygon) exists, with radius $r > 0$, let O be any point and P_1 any other point such that $OP_1 = r$ (Figure 4.43). Construct ray $\overrightarrow{OP_2}$ such that $m\angle P_1OP_2 = 360/10 = 36$ and $OP_2 = OP_1 = r$. Continue this process of con-

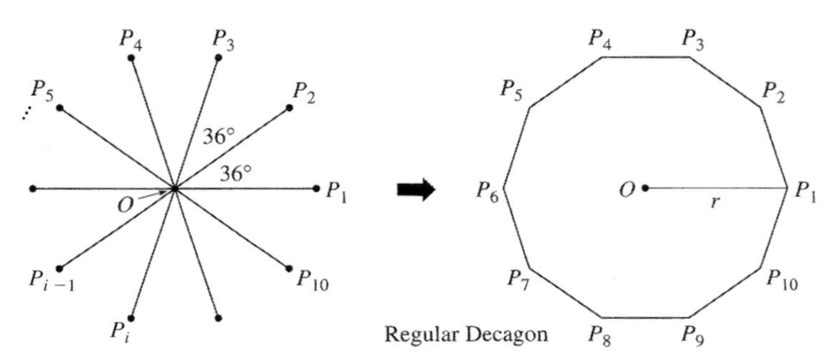

Figure 4.43

structing rays $\overline{OP_i}$ such that $m\angle P_{i-1}OP_i = 36$ and $OP_i = r$ for $3 \le i \le 10$. Finally, join the points $P_1, P_2, P_3, \ldots, P_{10}$ by segments $\overline{P_1P_2}, \overline{P_2P_3}, \overline{P_3P_4}, \ldots, \overline{P_9P_{10}}$, and $\overline{P_{10}P_1}$. The result is a regular decagon inscribed in a circle of radius r. Actually proving that the sides and angles of this polygon are congruent is a bit of a chore, but it can be done. The proof involves some work with betweenness properties of rays and additivity of circular arcs, along with the result of Problem 20 , Section 3.8.

The generalization of the preceding procedure for a decagon will give us the following basic result:

THEOREM 1: Any regular polygon having n sides, $n \ge 3$, exists, and may be inscribed in any given circle (has its vertices lying on that circle).

REGULAR POLYGONS IN EUCLIDEAN GEOMETRY

The above ideas are valid in absolute geometry. In Euclidean geometry, we can say a lot more, including exact formulas for the interior and exterior angles of regular polygons—formulas not valid in spherical or hyperbolic geometry.

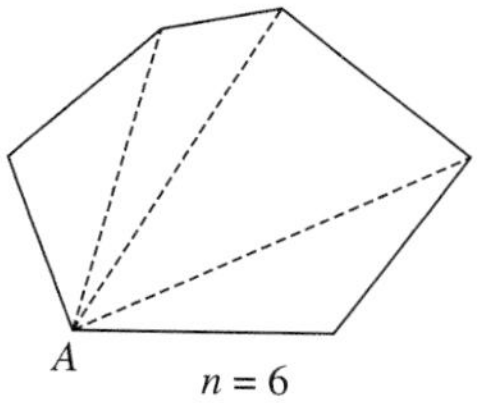

A

$n = 6$

Angle sum = 4·180 = 720

Figure 4.44

THEOREM 2: The angle sum of a convex n-gon is $180(n - 2)$. (See Figure 4.44.)

PROOF

Since the polygon is convex, the diagonals from any vertex $P_1 = A$ to the remaining $n - 2$ vertices not adjacent to A will occur in order, and betweenness considerations allow us to sum the angles of the $n - 2$ triangles thus formed, which is $180(n - 2)$, to give us the total sum of the measures of the angles of the polygon.

COROLLARY A: Each interior angle of a regular polygon (Figure 4.44) has measure

$$\phi = \frac{180(n - 2)}{n}$$

COROLLARY B: The sum of the measures of the exterior angles of any convex n-gon, taken in the same direction, is 360. (See Figure 4.45.)

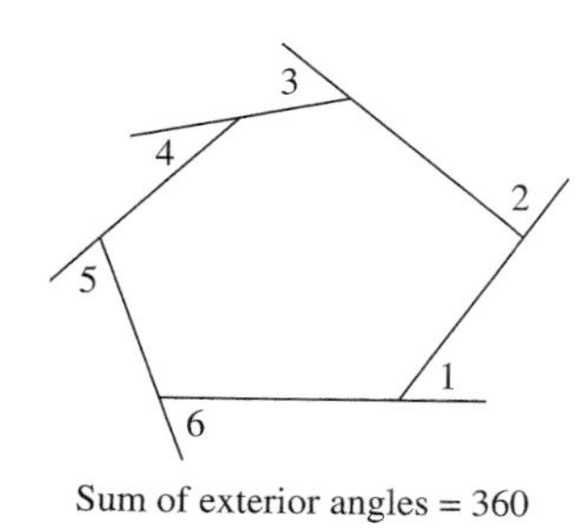

Sum of exterior angles = 360

Figure 4.45

The last corollary is intuitively clear. As you proceed around the polygon, the measure of each exterior angle gives the amount of rotation necessary in passing from one side to the next; since the total rotation necessary to return to your initial starting point is 360, then the sum of the exterior angles must be 360. There is a *Sketchpad* project (Problem 8) that presents a very convincing and clever argument appearing in some of the training manuals for that software; it is based on lines through a common point parallel to the sides of the polygon, again using the result of Problem 20, Section 3.8. This could lead to a rigorous proof, but some of the details get rather tedious. The simplest *axiomatic* proof seems to be as a corollary to our Theorem 2, where the Linear Pair Axiom is used to write $m\angle 1 + m\angle 1' = 180$,

$m\angle 2 + m\angle 2' = 180, \ldots$, summing, and using angle sum $(m\angle 1' + m\angle 2' + \ldots) = (n - 2)180$ and some algebra. You should have no difficulty finishing the details.

CLASSICAL CONSTRUCTION OF REGULAR POLYGONS

Although we have discussed how to prove the *existence* of a regular polygon having $n \geq 3$ sides, the problem of actually constructing such a polygon with only a compass and straight-edge is quite another matter. Only those regular polygons that belong to a select group are so constructible. The following theorem is quite remarkable; it was discovered by Gauss when he was only 19.

THEOREM 3: THEOREM OF GAUSS ON REGULAR POLYGONS
A regular n-gon may be constructed with the Euclidean tools iff n is either a power of two, or the product of a power of two and distinct **Fermat primes,** that is, distinct primes of the form

$$F_m = 2^{2^m} + 1$$

NOTE: Fermat Primes were named after the great number theorist Pierre de Fermat (1601–1665) because he conjectured around 1640 that F_m is prime regardless of the value of m. These numbers become incredibly large very quickly with increasing m. For example, $F_5 = 4{,}294{,}967{,}297 = 641 \cdot 6{,}700{,}417$—a product of two primes (thus, Fermat's conjecture is false). It is known that F_m is a composite for $5 \leq m \leq 16$, and it is believed that for $m > 16$, F_m is never prime. The largest values for which F_m is known to be prime are $F_3 = 257$ and $F_4 = 65{,}537$, which therefore satisfy Gauss's theorem. According to H.S.M. Coxeter, *Introduction to Geometry,* p. 27, Richelot and Schwendennwein constructed the 257-gon about 1898. J. Hermes spent 10 years on the regular 65,537-gon and deposited the manuscript in a large box at the University of Göttingen for others to inspect. (In view of this discussion, you would be advised not to start working on constructing a 4,294,967,297-sided regular polygon! Can you see why?)

HISTORICAL NOTE

Carl Frederick Gauss, known as the *prince of mathematicians,* is regarded by some scholars to be the greatest mathematician of all time, making outstanding contributions to all fields of mathematics. His work fills many volumes, including his desk drawer collections, where he kept unpublished work. The latter was a point of contention with several prominent mathematicians of his day, notably Jacobi and Legendre. The story is told that Jacobi was once describing to Gauss the details of a recent discovery of his, whereupon Gauss reached in his desk drawer and pulled out his own work which contained the essentials of Jacobi's results. Jacobi responded angrily, in essence, "It is a pity you did not publish this paper, for it is so much better than many other papers which you have published!" Gauss was a perfectionist, however, and he would not publish material that did not meet his own high standards of rigor and thorough checking. Gauss was born in Germany in 1777 into a poor and unlettered family. His father was, from time to time, a bricklayer, gardener, and construction worker. Having noticed Gauss's genius, the Duke of Brunswick helped pay for Gauss to attend Caroline College, then later, the University of Göttingen, from 1795 to 1798. In spite of his other accomplishments of greater importance, Gauss seemed proudest of the result concerning the 17-sided regular polygon (the first advance since the time of Euclid), which he had discovered while he was a student at Göttingen. He asked that this figure be carved on his tombstone, a request that was never carried out. In 1806, he was appointed director of the newly built observatory at the University of Göttingen, a position he held until his death nearly a half-century later (in 1885). His interests had, by that time, turned to astronomy, and he devoted three years to observing and calculating the orbit of Ceres—an asteroid—with amazing accuracy. His least squares method for minimizing experimental error allowed him to accurately calculate the orbits of a series of planetoids that lie between Mars and Jupiter as soon as they were discovered by astronomers, and he made significant new contributions to the theory of astronomy. Many scholars feel that this activity amounted to genius wasted, and that such efforts might have been more effectively devoted to mathematics and theoretical physics.

Since 3, 5, and 17 are Fermat primes (corresponding to $m = 0, 1$, and 2), $n = 34 = 2 \cdot 17$ and $n = 15 = 3 \cdot 5$ are of the required form, and we can construct a 34-sided and 15-sided regular polygon with the Euclidean tools. But $n = 45 = 3^2 \cdot 5$ is not (it is not the product of *distinct* Fermat primes), so we cannot construct a 45-sided regular polygon. For $n \leq 20$, the values of n which yield constructible n-gons are

$$n = 3, 4, 5, 6, 8, 10, 12, 15, 16, 17, 20$$

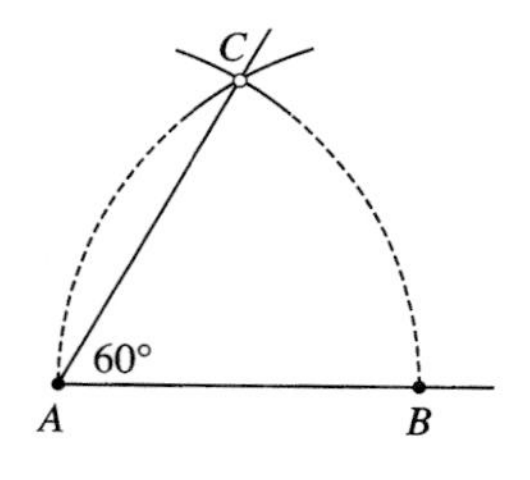

Figure 4.46

We can get through this list fairly quickly with what we already know. For $n = 3$ (corresponding to the equilateral triangle), we merely swing two circular arcs centered at A and B, each having radius $\overline{AB}$ as shown in Figure 4.46. These two arcs will meet at the third vertex C of the desired triangle (Euclid's original construction, which also yields the construction for an angle of measure 60). From this we can construct a regular hexagon $n = 6$) and, after bisecting an angle of measure 60, we can construct a regular dodecagon $n = 12$). The regular n-gons for $n = 4, 8, 16$ are obtained by constructing a right angle, then bisecting a right angle and the half-angle obtained from that. The construction of the regular pentagon involves the Golden Ratio (introduced in Problem 13 of Section 1.2), and was also given by Euclid (see Problems 16 and 17 following this section). The essence of this construction is the fact that $\cos 36 = \frac{1}{2}\tau$ (Problem 15, Section 4.3). From that we can obtain the regular pentagon, regular decagon, and the regular 20-sided polygon. This leaves the values 15 and 17. For $n = 17$, we refer the reader to Eves, *Survey of Geometry, Volume I,* pp. 218–224, or Coxeter's *Introduction to Geometry,* p. 27. The value $n = 15$ can be based on the numerical fact

$$\frac{1}{15} = \left(\frac{2}{5} - \frac{1}{3}\right)$$

Using Figure 4.47, construct, with compass/straight-edge, an equilateral triangle with side $\overline{AC}$, and a regular pentagon with sides $\overline{AB}$ and $\overline{BD}$. Then

$$m\widehat{CD} = m\widehat{ABD} - m\widehat{ABC} = 2\left(\frac{360}{5}\right) - \frac{360}{3} = 360\left(\frac{2}{5} - \frac{1}{3}\right) = 360 \cdot \frac{1}{15} = 24$$

the required angle measure for the central angle of a 15-sided regular polygon.

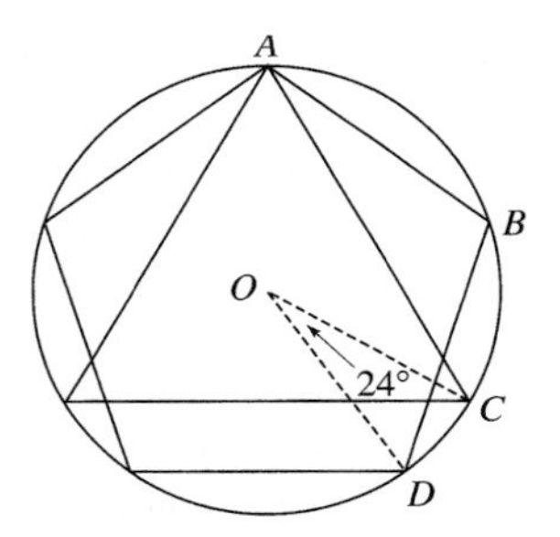

Figure 4.47

Note that $n = 7$ (corresponding to the **regular heptagon**) is the least number of sides for which there is no Euclidean construction. This was recognized in antiquity by Archimedes, who showed it could be constructed by the aid of a special transcendental curve.

An interesting corollary to Theorem 3 is the following, which was suggested by a college instructor at the University of Wisconsin at Superior, David Beran.

COROLLARY: An angle of measure 60 cannot be trisected using a compass/straight-edge.

PROOF

Suppose $m\angle A = 20$ and that it has a finite-step construction by compass/straight-edge. Then an 18-sided regular polygon could be constructed. But $18 = 2 \cdot 3^2$, a prime factorization with the Fermat prime 3 *repeated*, not allowed by Theorem 3.

TILING AND PLANE TESSELATIONS

The subject of **tiling theory** is very old, which will be apparent from some of our illustrations. However, recently this area has had a surge of new interest, aroused in part by the publication of the definitive book by B. Grünbaum and G.C. Sheppard, *Tiling and Patterns,* which has brought a semblance of order to an unwieldy collection of seemingly unrelated ideas (viz, wallpaper patterns, and ancient classical designs), and this work is a monumental achievement in terms of both number of results and the high quality of the art work. Here, we will just introduce you to this topic.

OUR GEOMETRIC WORLD

An unusual tiling pattern decorates a museum in Grenada, Spain. Built in the 1300s, the artist is unknown. The structure is used today as a mosque.

Johann Kepler (1571–1630) was the first to apply geometry systematically to the subject of plane tilings when he observed in 1619 that the only regular polygons that can be used to tile the plane are the equilateral triangle, square, and regular hexagon, as shown in Figure 4.48—the so-called **regular tilings**. He also discovered the unusual tiling shown in Figure 4.49, which uses the regular pentagon and decagon, and includes a "fused" double decagon and five-pointed star.

(Tiles continue indefinitely in all directions)

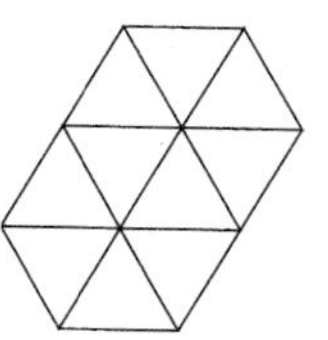

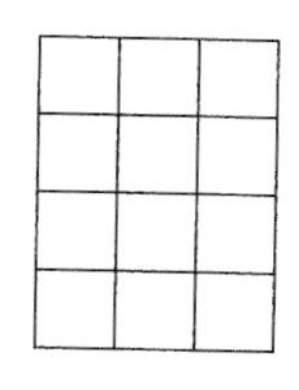

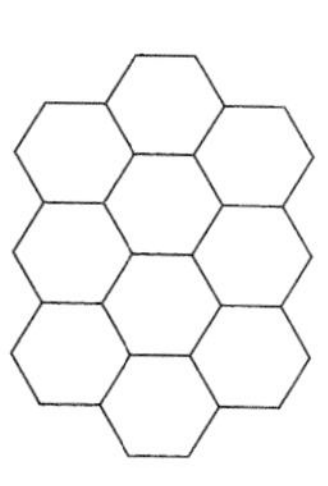

Figure 4.48

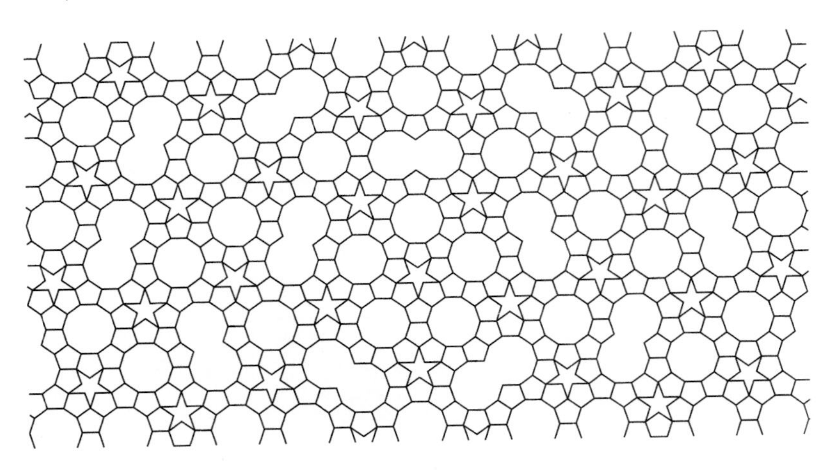

Figure 4.49

In general, a **tiling** or **tesselation** of the plane is a collection of regions $T_1, T_2, \ldots, T_n, \ldots$, called **tiles** (usually polygons and their interiors), such that

(1) no two of the tiles have any interior points in common

(2) the collection of tiles completely covers the plane.

Here, we shall require that all tilings by polygons be **edge to edge,** that is, each edge of a polygon in the tiling must be an edge of one of its neighbors.

To make it more interesting, a tiling is normally required to have certain regularity properties, such as restricting all tiles to be regular polygons, requiring the tiles to be congruent to a single tile, etc. The question of the *existence* of such tilings (or proving they do not exist) becomes a very interesting, and sometimes very challenging, problem.

DEFINITION: If all the tiles in a plane tiling are congruent to a single region, the tiling is said to be **elementary of order one** (or, simply, **elementary**) and the single region is called the **fundamental region** or **prototile** of the tiling. If the tiles are each congruent to one of n tiles $T_1, T_2, \ldots, T_n$ (also called **fundamental regions**), the tiling is called **elementary of order *n*.** If the adjective **semiregular** is added to any of the preceding terms, it means that all the fundamental regions are regular polygons.

A major problem in the theory of tiling is to determine whether a given two-dimensional figure, such as a polygonal region, can serve as a fundamental region for an elementary tiling (a tiling of order one). For example, we could ask what regular polygons can serve as the fundamental region for such a tiling. Suppose that an n-sided regular polygon, $n \geq 3$, is such a fundamental region. If there are k of these polygons at each vertex, then with θ the measure of each interior angle of the regular polygon, we have

$$k\theta = 360, \qquad \text{where } \theta = \frac{n-2}{n}180 \quad \text{(Corollary A of Theorem 2)}$$

Then, by substitution,

$$k \cdot \frac{n-2}{n} 180 = 360 \qquad \text{or} \qquad k = \frac{2n}{n-2}$$

Since we are seeking integer solutions for k, observe that the right side of the last equation above must be an integer. But integer values for

$$\frac{2n}{n-2}$$

occur only for $n = 3$, 4, and 6—that is, when the regular polygon has either 3, 4, or 6 sides, precisely the cases found by Kepler (the regular tilings).

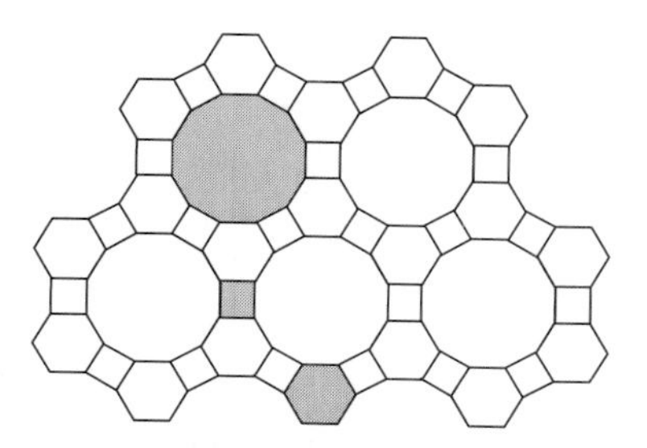

Figure 4.50

The next logical place to look for tilings involving regular polygons is in higher order semiregular tilings (having more than one fundamental region). Many more types of regular polygons can occur in such tilings. For example, the regular dodecagon (12-sided), together with a regular hexagon and square, are the fundamental regions for the tiling illustrated in Figure 4.50. A few other examples are shown in Figures 4.51 and 4.52. A systematic way to find all possible tesselations of this type is pursued in Problem 12. (See also Problem 24.) In this type of problem, one must consider only semiregular tilings whose vertices are all of the *same type,* that is, the set of polygons surrounding each vertex is the same.

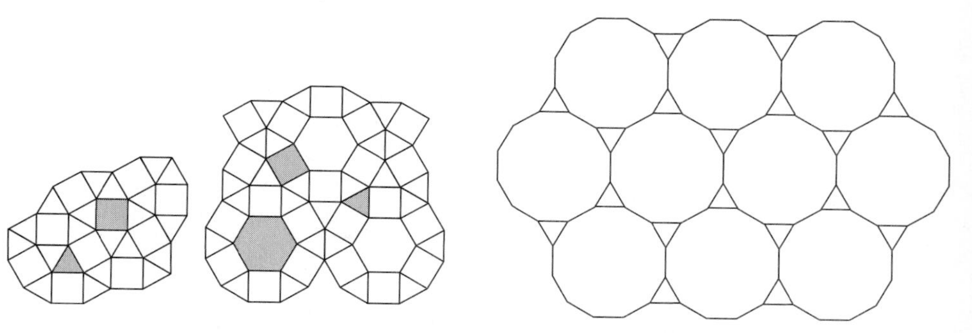

Figure 4.51

Figure 4.52

We still have not found a tiling that utilizes the pentagon, heptagon, nonagon, and decagon. To involve these, we must allow a different type of fundamental region other than regular polygons. The pentagon can be used in a tiling if we also allow a rhombus to be a fundamental region, thus providing us with a tiling of order two. (See Figure 4.53.) It is more difficult to obtain a tiling that involves a regular decagon: this one requires not only a rhombus and regular pentagon to serve as fundamental regions, but also the regular **pentagonal star** (the famous five-pointed star that the Pythagoreans wore as a badge of identity). The result is a tiling of order four as shown in Figure 4.54. Note that it has *five types* of vertices.

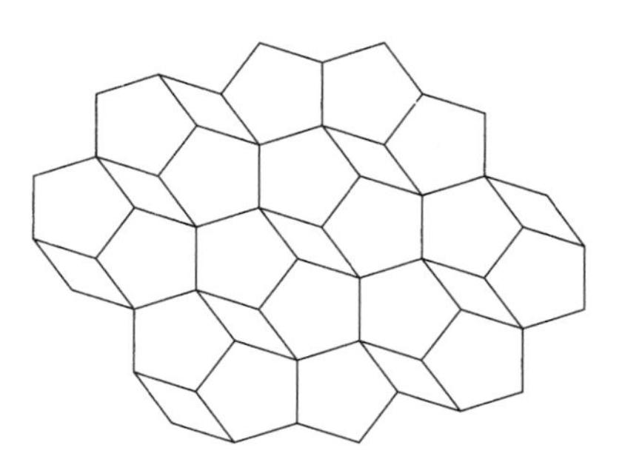

Figure 4.53

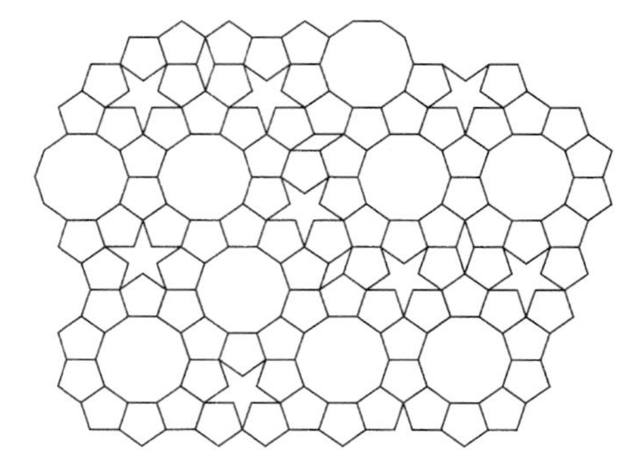

Figure 4.54

Moment for Discovery

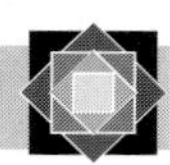

Tiling With Regular Octagons

1. Construct a regular octagon (stop sign) on a piece of cardboard or extra heavy paper (e.g., poster board). To make this construction, draw a circle of radius 1 in., use a protractor and ruler and, starting at the center of the polygon and, measuring angles of 45°, intersect the sides of these angles with the circle, then join these points of intersection to complete the regular polygon. (The entire polygon should not measure over 2 in.)
2. Cut out the polygon you have drawn in Step 1 to make a template to be used to draw congruent octagons.
3. Using the template, draw two regular octagons on a sheet of paper edge to edge. Will another octagon fit in the space next to the end point of the common edge? (Obviously not.)
4. Draw as many more edge-to-edge pairs of octagons as you can so that no two interiors of those octagons overlap, filling up most of the page.
5. What spaces are not covered by the octagons? Do you see a semiregular tiling? What is its order, and what are the fundamental regions?

Moment for Discovery

Tiling With Regular Octagons (using *Sketchpad*)

1. Use *Sketchpad* to construct a regular octagon, as follows. Construct a circle AB and double-click on A, select B, and rotate by 45° once, producing point

B'. Construct segment $\overline{BB'}$ and select B' and $\overline{BB'}$, then repeatedly rotate by 45°. [Use the keyboard shortcuts ALT-T (to choose TRANSFORM), the Down Arrow once (to choose Rotate), then Enter twice; repeat this procedure until the polygon closes.] To downsize the polygon to an overall dimension of about 2 in., drag point B toward point A.

2. Select the vertices of the octagon in sequence, then choose Polygon Interior from the CONSTRUCT menu. Hide all points and the circle of Step 1. (Shortcut: Use Select All Points under EDIT after choosing the POINT TOOL, then choose Hide Points under DISPLAY.)
3. Select the polygon (interior) and choose EDIT, Copy, Paste. Translate the copy and place it beside the original octagon edge to edge. Continue this process several times in order to fill up as much of the screen as possible. (More accuracy is achieved by using Translation under TRANSFORM.)

NOTE: The above procedure for constructing a regular octagon with *Sketchpad* can be adapted to any regular polygon. You can then experiment with other tilings, such as those discussed in this section.

PROBLEMS (4.4)

GROUP A

For Problems 1 and 2, determine which of the polygonal regions (a), (b), and (c) in the figures shown can serve as a fundamental region for an elementary tiling of the plane.

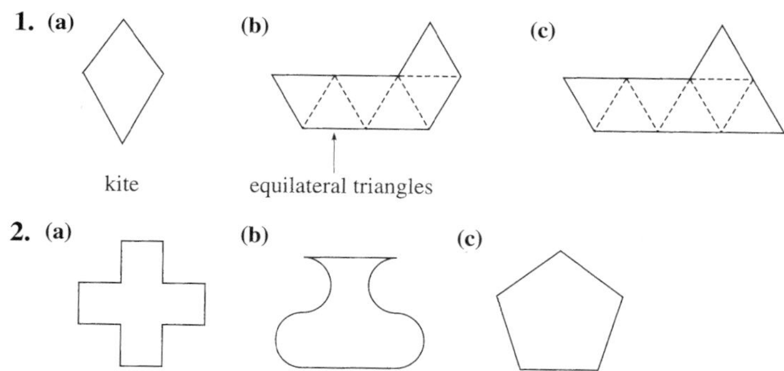

3. Verify with a sketch that an isosceles trapezoid is the fundamental region of an elementary tiling of the plane.
4. Show that any convex pentagon and interior which has a pair of opposite sides parallel and congruent, as shown here, may be used to tile the plane.

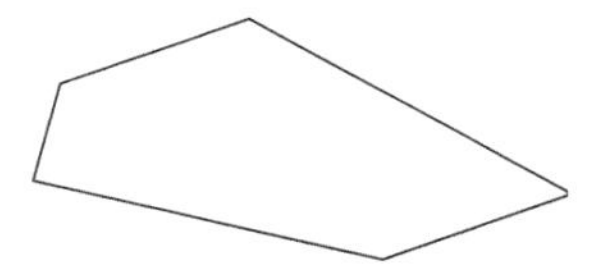

5. Show that a dart can be used to tile the plane. (A sketch of several tiles edge to edge will suffice; no proofs are necessary.)

6. If the two nontriangular polygons are regular, find the measures of the angles of $\triangle PQR$.

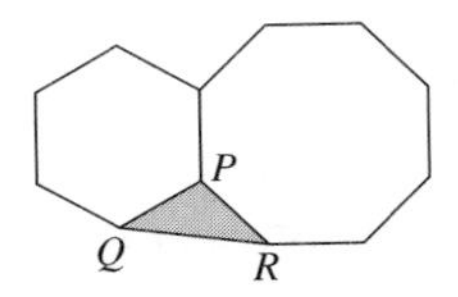

7. If the two nontriangular polygons are regular, find the measures of the angles of $\triangle WST$.

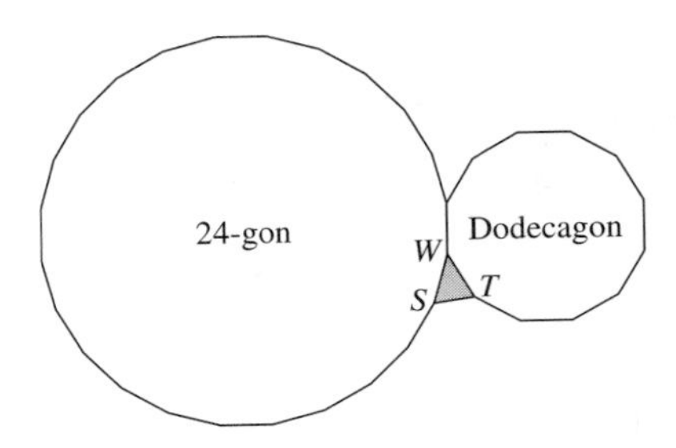

8. Construct an arbitrary convex polygon $ABCDE \cdots$ having at least five sides using *Sketchpad*, then construct consecutive rays through those sides all in the same direction about the polygon (as in Figure 4.45). Using the DILATE TOOL (opposite the ARROW TOOL), and double-clicking on A, drag vertex B toward A in order to shrink the polygon to a very small size. Does the result of Corollary B stand out?

9. Verify the validity of the tiling shown in Figure 4.52 by analyzing angle measures at each vertex.

GROUP B

10. Show that each of the regions illustrated in the following figure is a fundamental region for an elementary regular tiling of the plane.

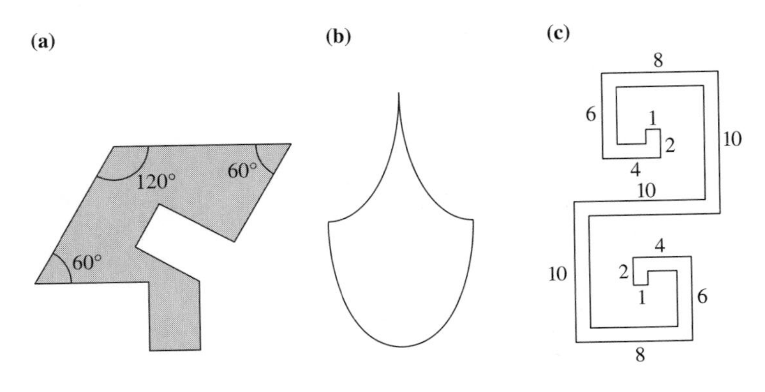

11. Using the method of proof of the Corollary of Theorem 3, prove that

(a) an angle of measure 72 can be trisected with compass, straight-edge

(b) an angle of measure 72 cannot be divided into nine equal parts using compass, straight-edge. (That is, show that an angle of measure 8 cannot be constructed.)

12. Show that there are only 10 semiregular tilings of order ≤ 2 having only one type of vertex (3 of order 1, and 7 of order 2). (The three tilings of order 1 are, of course, Kepler's regular tilings shown in Figure 4.48.) If $n_1, n_2, \ldots, n_k$ are the numbers of sides of regular polygons permitted at a vertex, it can be shown from Corollary A that $\frac{1}{n_1} + \frac{1}{n_2} + \cdots + \frac{1}{n_k} = \frac{k-2}{2}$.

13. **A Tiling Experiment on *Sketchpad*[3]** Follow these steps for a drawing experiment involving an arbitrary polygon that is a tile for a tessellation.

[1] Construct segment $\overline{AB}$ in a corner of the screen, then construct point C above this segment.

[2] Construct a parallelogram $\square ABCC'$ by using Marked Vector A to B and translating point C to C', then joining the vertices.

[3] Construct several (two for openers) connected segments from point A to point C, and call this broken line segment an *irregular edge*.

[4] Select all the segments and points of irregular edge AC and translate them by the marked vector (AB).

[5] Make an irregular edge from A to B.

[6] Mark the vector from point A to point C and translate all the parts of irregular edge AB by the marked vector.

[7] Construct the polygon interior of the irregular figure. This is the desired tile.

[8] Translate the polygon interior by the marked vector AC already marked.

[9] Repeat this process until you have a column of tiles all the way up your sketch. By translating the column, you can then complete a part of a tessellation having the irregular polygon and interior as the fundamental tile.

Explore this construction with other types of tiles that can be created, perhaps even using arcs of circles. Also, you may animate your sketch in various ways to make it more dynamic.

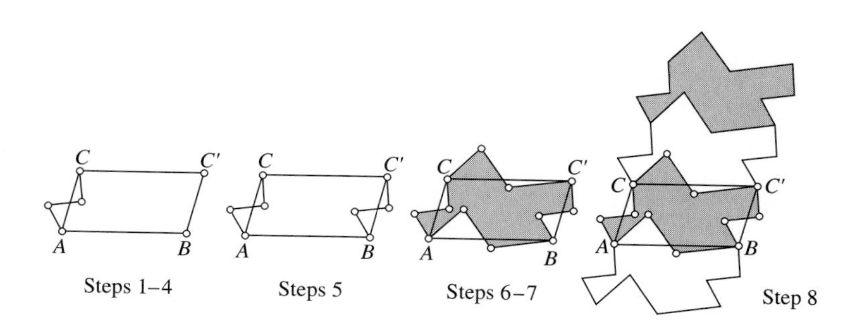

[3]Taken from *Teaching Geometry with the Geometer's Sketchpad,* Key Curriculum Press, pp. 21, 22, 1998; used by permission.

14. The Cairo Tessellation Follow these steps to create the **Cairo tile**, and use it to make a tessellation.

(1) Construct segment $\overline{AB}$ of length 2 units (inches). Determine its midpoint, M.

(2) At M, construct rays $\overrightarrow{MX}$ and $\overrightarrow{MY}$ so that $m\angle AMX = m\angle BMY = 45$.

(3) Locate C on ray $\overrightarrow{MY}$ such that C lies on the circle having center B and radius $AB = 2$.

(4) Locate E on ray $\overrightarrow{MX}$ such that E lies on the circle having center A and radius $AB = 2$.

(5) Locate D, the intersection of the circle having center C and radius AB, and the circle having center E and radius AB.

The resulting pentagon $ABCDE$ is the Cairo tile. See if you can find the tessellation it generates.

NOTE: This tile was used to decorate the streets of Cairo, Egypt. To verify the correctness of this tessellation we need to closely consider the features of the Cairo tile. (See Problem 15.)

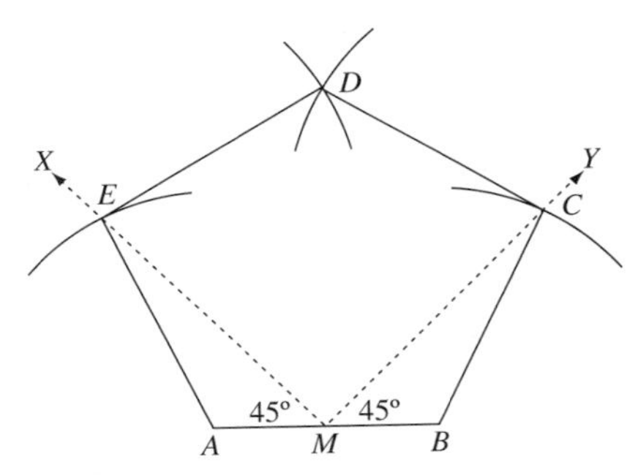

15. The Geometry of the Cairo Tile Using the same notation as Problem 14, show from the Laws of Sines and Cosines that $MD = \sqrt{7}$, $MC = (\sqrt{7} + 1)/\sqrt{2}$, and $EC = \sqrt{7} + 1$. It then follows that $m\angle ABC$ is approximately 114° and $m\angle EDC \approx 132°$. Find $m\angle BCD$ and $m\angle DEA$ using these estimations.

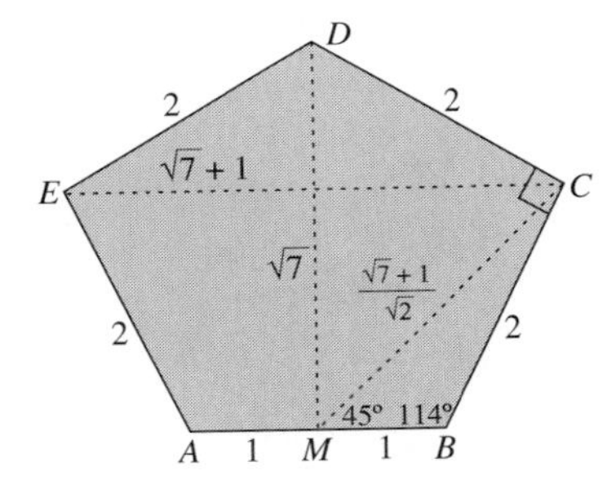

16. Euclidean Construction of the Regular Pentagon Using trigonometry in right triangle $\triangle ABF$ in unit circle O with diameter $\overline{AB}$ (of length 2), show that $AF = \tau$ and $\cos\theta = \tau/2$, so that $\theta = 36$, $m\angle FOB = 72$, and the construction yields a regular pentagon. The steps of the construction are as follows:

(1) Starting with diameter $\overline{AB}$, construct $\overline{CO} \perp \overline{AB}$.

(2) Construct the midpoint D of $\overline{AO}$.

(3) Construct point E on $\overline{AB}$ so that $DE = DC$.

(4) Construct point F on the circle so that $AF = AE$.

(5) Draw segment $\overline{BF}$; this is a side of the regular pentagon.
(***Hint:*** Show that $AF = AE = \tau$ using the Pythagorean Theorem, then use $\cos\theta = AF/AB$; see Problem 15, Section 4.3.)

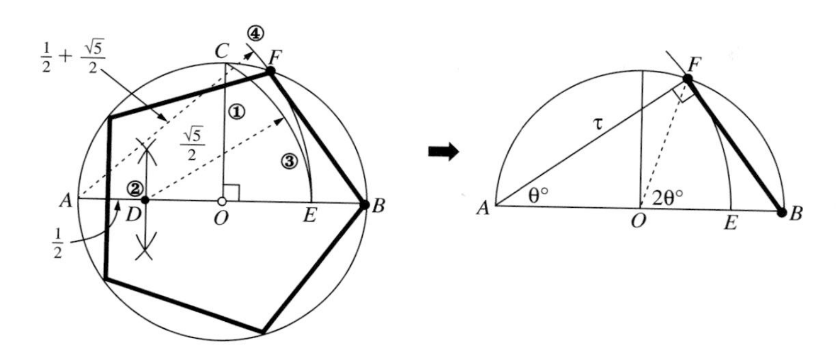

17. Euclidean Construction of the Regular Decagon This simple construction of the regular decagon appeared in an old issue of the *Mathematics Teacher* (December 1960). Starting with a unit circle O and diameter $\overline{AB}$ (of length 2), follow these steps:

(1) Construct $\overline{AP} \perp \overline{AB}$.

(2) Construct $AC = AO = 1$ on $\overrightarrow{AP}$.

(3) Draw $\overline{CB}$ and construct $CD = CA$ on $\overline{CB}$.

(4) Construct the midpoint E of segment $\overline{DB}$.

(5) Draw segment $\overline{BF} \cong \overline{BE}$; this is a side of the regular decagon. Prove that this construction is valid by showing that $m\angle BOM = 18$, where $\overline{OM} \perp \overline{BF}$. [***Hint:*** Show that $\csc\theta = 2\tau$ so that $\theta = 18$; see Problem 15, Section 4.3 where it was shown that $\sec(90° - 18°) = 2\tau$.)

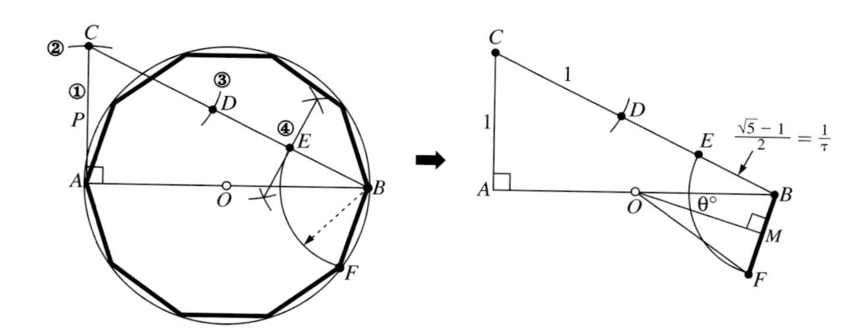

18. The midpoints of the sides of a regular pentagon are joined, in order, forming a second regular pentagon. Show that the lengths of the sides of the new pentagon are precisely $\frac{1}{2}\tau$ times those of the original polygon.

19. The Star of Pythagoras The Pythagoreans were known by the regular five-pointed pentagonal star that they wore as a badge for identification in their elite society. As part of their initiation rites, members were required to know how to demonstrate the Pythagorean Theorem. Show that this star, derived from the diagonals of a regular pentagon, has congruent sides, each having length τ times the side of the regular pentagon that defines it, and that the points form an angle of measure 36 each.

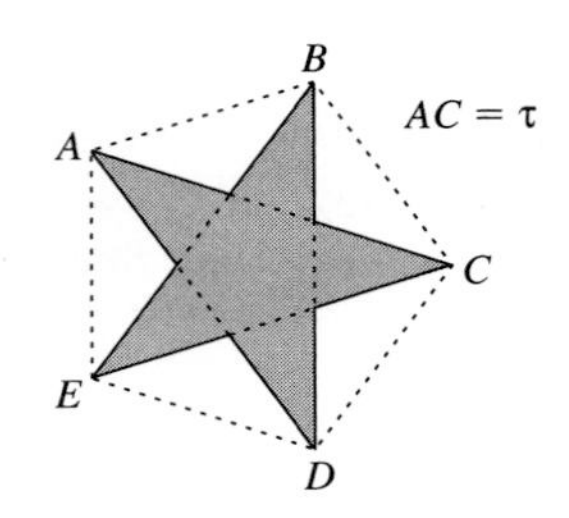

20. The regular hexagon can be used in two distinctly different semiregular tilings of the plane of order 2. Find the other polygons, and explain how this may be accomplished.

GROUP C

21. Use Gauss's Theorem to determine which of the regular n-gons for $21 \leq n \leq 100$ may be constructed with compass and straight-edge.

22. The 24 regions illustrated in the following figure were proposed by T.H. O'Bierne as possible fundamental regions for elementary (order one) plane tesselations. Each region proposed consists of the union of 7 equilateral triangles and interior. One of the 24 regions is not such a fundamental region. Can you find it?

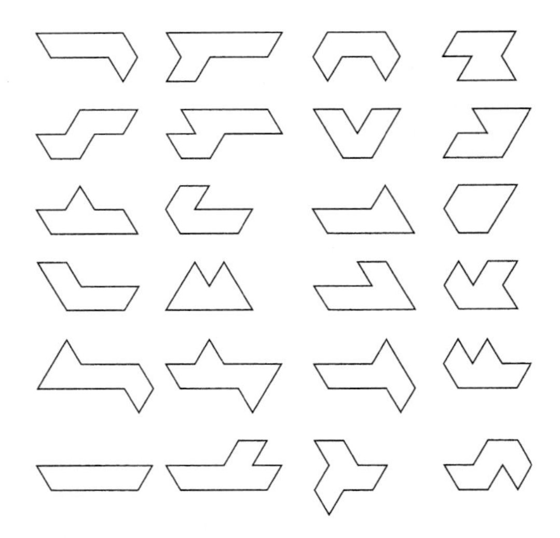

23. Niven's Theorem A 1978 theorem due to I. Niven states that if the areas of all tiles of an arbitrary plane tiling are to be greater than some positive constant α, and if the tiles consist of convex polygons having seven or more sides, then there exist tiles whose perimeters β are *arbitrarily large* (see Klamken and Liu, "Note on a Result of Niven on Impossible Tesselations," *American Mathematical Monthly,* Vol. 87, No. 8 (1980), pp. 651–653). Show that if the convexity requirement is dropped, the polygons in the accompanying figure are the fundamental regions of a tiling of order two, for which the perimeters of the tiles are bounded. (This would be a counterexample to the theorem if convexity requirements were deleted.)

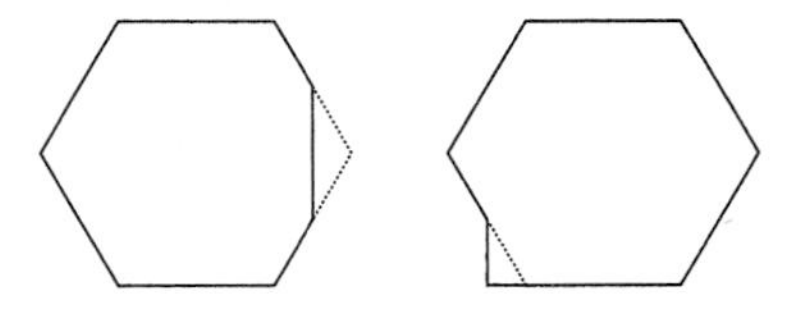

24. Undergraduate Research Project Make a list of the regular polygons and interiors that can be used in a tiling of the plane of order three, where other regular polygons and semiregular nth-order **star polygons** (stars whose n points are distributed uniformly on a circle—including the rhombus of order two used in Figures 4.53 and 4.54) are allowed. (***Hint:*** If $n_1, n_2, \ldots, n_k$ are the numbers of sides of regular polygons permitted at a vertex, and if θ is the common measure of the angles of a semiregular star polygon, assuming only one of these is permitted at each point, then show that these numbers must satisfy

$$\frac{2}{n_1} + \frac{2}{n_2} + \cdots + \frac{2}{n_k} = \frac{\theta}{180} + k - 2$$

4.5 The Circle Theorems

Much of the power of synthetic geometry derives from the study of circles. We present here the basic development of circles in Euclidean geometry. A fact about circles you can discover from a discovery unit at the end of this section is one of the most amazing, and useful, theorems in Euclidean geometry. It defies intuition, and, indeed, it is totally false in non-Euclidean geometry. It is also difficult to prove by coordinates or vectors. The synthetic geometry of Euclid remains the most appropriate method of proof. We begin with a special case.

LEMMA: If $\angle ABC$ is an inscribed angle of circle O and the center of the circle lies on one of its sides, then $m\angle ABC = \frac{1}{2}m\widehat{AC}$. (See Figure 4.55.)

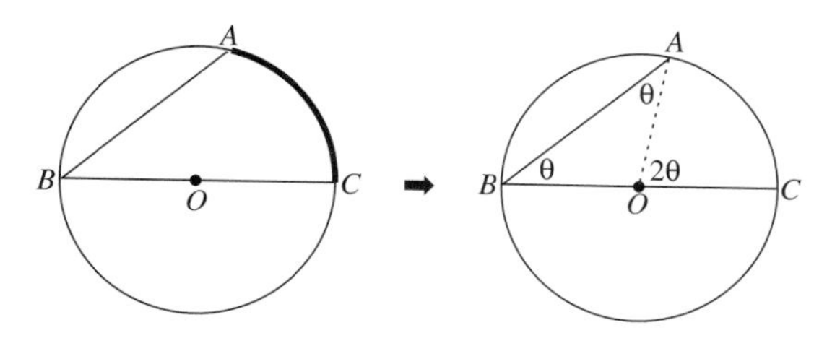

Figure 4.55

PROOF

Suppose O lies on side $\overline{BC}$, as shown in Figure 4.55. Draw radius $\overline{OA}$, and observe that $OA = OB$. Hence $\triangle AOB$ is isosceles, with $m\angle ABO = m\angle BAO \equiv \theta$. By the Euclidean Exterior Angle Theorem,

$$m\angle AOC = \theta + \theta = 2\theta$$

or

$$m\angle ABC = \theta = \tfrac{1}{2}m\angle AOC = \tfrac{1}{2}m\widehat{AC}.$$

THEOREM 1: INSCRIBED ANGLE THEOREM

The measure of an inscribed angle of a circle equals one-half that of its intercepted arc.

PROOF

Let $\angle ABC$ be an inscribed angle of circle O. There are three cases: (1) When O lies on one of the sides of the angle (as in the lemma), (2) when $O \in$ Interior $\angle ABC$, or (3) when $O \in$ Exterior $\angle ABC$. Since Case 1 is covered by the lemma, consider Cases 2 and 3. In each, construct the diameter $\overline{BD}$, as shown in Figure 4.56.

Case 2: $O \in$ Interior $\angle ABC$.

By the Angle Addition Postulate, the lemma, and the additivity of circular arcs,

$$m\angle ABC = m\angle ABD + m\angle DBC = \frac{1}{2}m\widehat{AD} + \frac{1}{2}m\widehat{DC} = \frac{1}{2}m\widehat{AC}$$

Case 3: $O \in$ Exterior $\angle ABC$.

Since rays $\overrightarrow{BA}$ and $\overrightarrow{BC}$ must lie on one side of line $\overleftrightarrow{BD}$ (otherwise, points A and C would be on opposite sides and $O \in$ Interior $\angle ABC$ reverting back to Case 1), we either have $\overrightarrow{BA}$–$\overrightarrow{BC}$–$\overrightarrow{BD}$ or $\overrightarrow{BC}$–$\overrightarrow{BA}$–$\overrightarrow{BD}$, cases which are logically equivalent. Thus, consider just the case $\overrightarrow{BA}$–$\overrightarrow{BC}$–$\overrightarrow{BD}$, as shown in Figure 4.56.

$$m\angle ABC = m\angle ABD - m\angle CBD = \frac{1}{2}\widehat{ACD} - \frac{1}{2}\widehat{CD} = \frac{1}{2}\widehat{AC}.$$

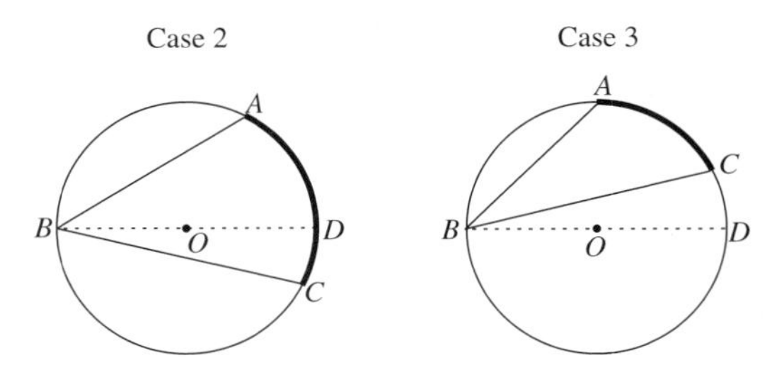

Figure 4.56

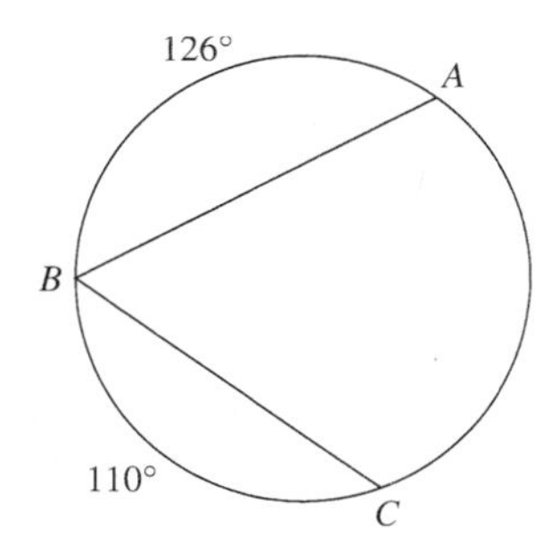

Figure 4.57

EXAMPLE 1 In Figure 4.57, find $m\angle ABC$.

SOLUTION

Observe: $m\widehat{AC} = 360 - 126 - 110 = 124$. Hence, by Theorem 1, $m\angle ABC = \frac{1}{2}(124) = 62$. ■

EXAMPLE 2 Find $m\angle ABC$ in Figure 4.58.

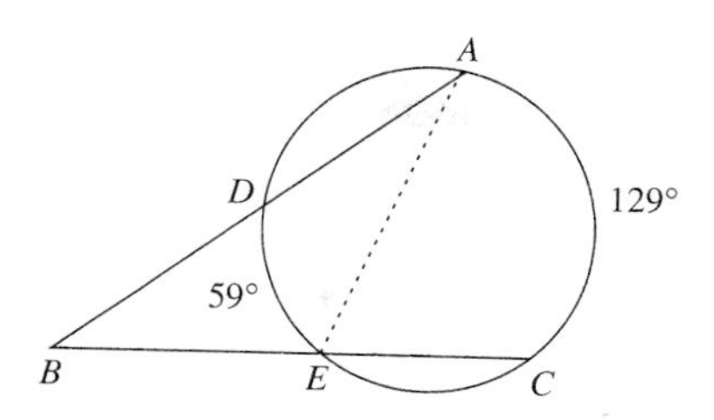

Figure 4.58

SOLUTION

Construct chord $\overline{AE}$. Then $\angle AEC$ is an exterior angle of $\triangle ABE$, so $m\angle AEC = m\angle B + m\angle BAE$. But angles $\angle AEC$ and $\angle DAE$ are both inscribed angles of the circle, so by the Exterior Angle Theorem,

$$
\begin{aligned}
m\angle B &= m\angle AEC - m\angle BAE \\
&= \frac{1}{2}(m\widehat{AC} - m\widehat{DE}) \\
&= \frac{1}{2}(129 - 59) = 35 \quad \blacksquare
\end{aligned}
$$

A special case of Theorem 1 is important enough to state separately.

COROLLARY A: An angle inscribed in a semicircle is a right angle.

The converse of this result has already been mentioned several times; it is a simple property of right triangles: The midpoint of the hypotenuse of a right triangle is the center of a circle passing through its vertices. Hence, a right angle is always an inscribed angle of a semicircle with diameter joining any two points on its sides.

OUR GEOMETRIC WORLD

A carpenter's square is hung on a wall between two nails, as shown. As the square slides along the two nails, the corner will trace the arc of a perfect circle. The reason is the converse of the corollary of Theorem 1.

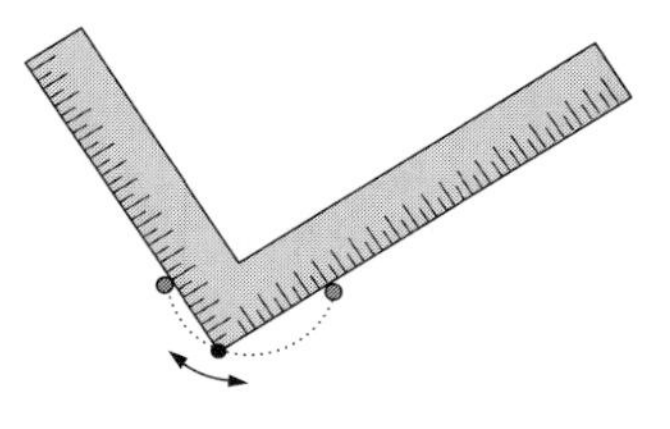

Figure 4.59

The preceding illustration of the carpenter's square leads to the question of what happens to the vertex of an angle other than a right angle that "slides" between two fixed points, as shown in Figure 4.60. You might want to try the *Sketchpad* experiment at the end of this section before you go any further. The result is a corollary to the Inscribed Angle Theorem.

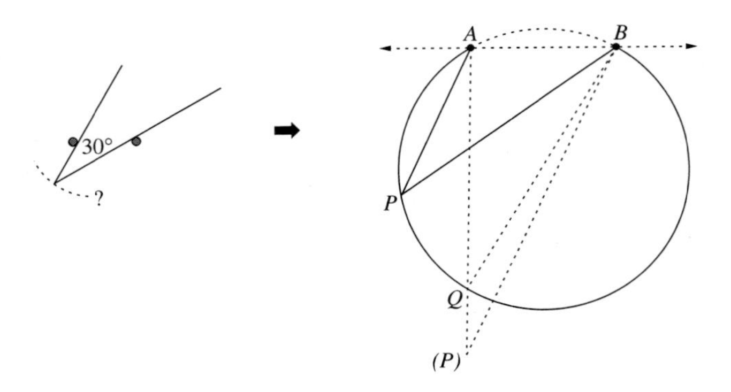

Figure 4.60

COROLLARY B: The locus of a point P that lies on one side of a line $\overline{AB}$ such that $m\angle APB$ remains constant is the arc of a circle with end points A and B.

A proof is indicated in Figure 4.60. The first part, showing that if P lies on the arc of a circle then $m\angle APB$ = constant, follows from Theorem 1; the second part, showing that if $m\angle APB = m\angle AQB$ then P lies on the arc, uses the Exterior Angle Inequality (indicated in Figure 4.60). This interesting result provides the explanation needed in a few previous *Sketchpad* experiments.

The Inscribed Angle Theorem also has several other important corollaries, which we will leave for you to prove as exercises (based on the idea advanced in Example 2. (See Figure 4.61.)

- An angle whose vertex lies inside a circle and is formed by intersecting chords of the circle (intercepting arcs of measure x and y) has measure $\theta = \frac{1}{2}(x + y)$.

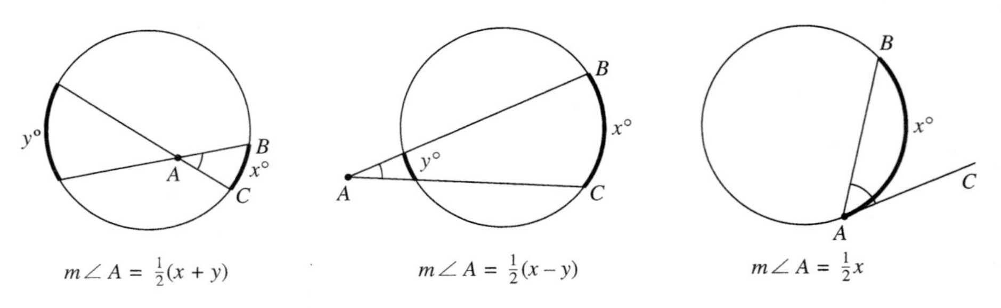

Figure 4.61

- An angle whose vertex is exterior to a circle and is formed by intersecting secants of the circle (intercepting arcs of measure x and y) has measure $\theta = \frac{1}{2}|x - y|$.
- An angle formed by a chord and tangent of a circle, with its vertex at the point of tangency and intercepting an arc of measure x on that circle, has measure $\theta = \frac{1}{2}x$. (See Problem 19.)

$AP \cdot PB = CP \cdot PD$

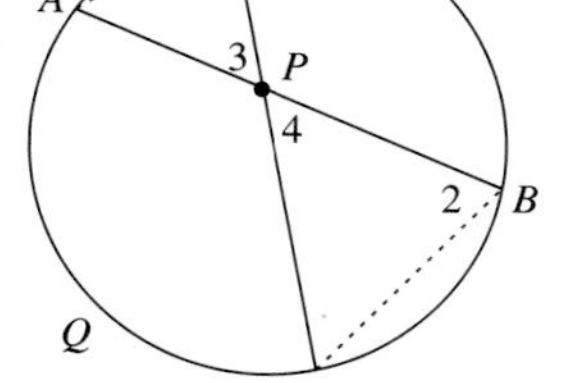

Figure 4.62

Two other important applications of the Inscribed Angle Theorem are (1) the **Two-Chord Theorem** and (2) the **Secant–Tangent Theorem**, sometimes referred to as the **power theorems** in geometry (see Problems 22 and 23 where the term *power* is defined and some interesting results appear). These theorems will be explored informally as experiments at the end of the section, and their actual proofs are relegated to Group B problems.

THEOREM 2: TWO-CHORD THEOREM

When two chords of a circle intersect, the product of the lengths of the segments formed on one chord equals that on the other chord. That is, in Figure 4.62,

$$AP \cdot PB = CP \cdot PD$$

$PC^2 = PA \cdot PB$

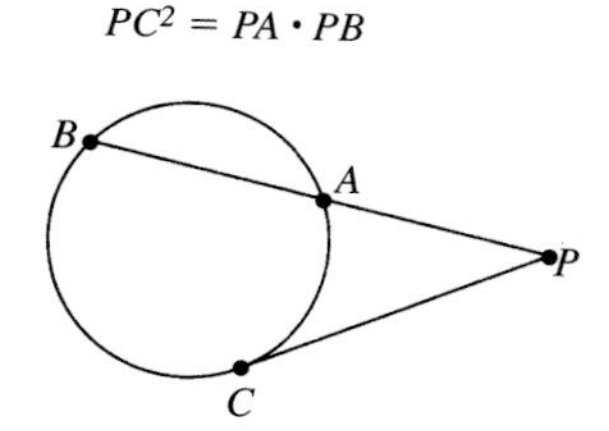

Figure 4.63

THEOREM 3: SECANT–TANGENT THEOREM

If a secant $\overleftrightarrow{PA}$ and tangent $\overleftrightarrow{PC}$ meet a circle at the respective points A, B, and C (point of contact), then (Figure 4.63)

$$PC^2 = PA \cdot PB$$

COROLLARY: TWO-SECANT THEOREM

If two secants $\overleftrightarrow{PA}$ and $\overleftrightarrow{PC}$ of a circle meet the circle at A, B, C, and D, respectively (Figure 4.64), then

$$PA \cdot PB = PC \cdot PD$$

PROOF

Draw a tangent from P and apply the Secant–Tangent Theorem to both secants: $PA \cdot PB = PE^2 = PC \cdot PD$.

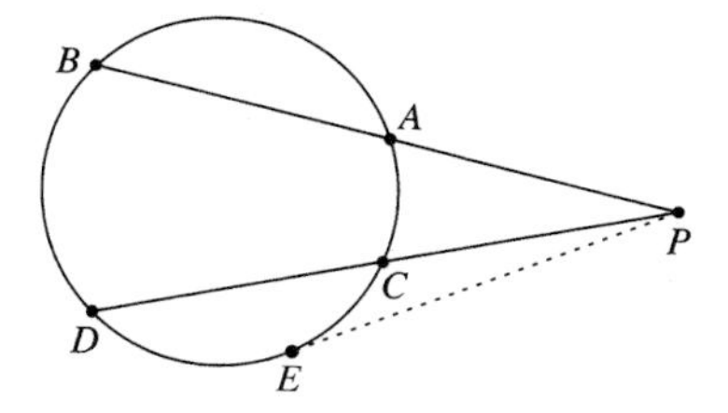

Figure 4.64

EXAMPLE 3 A kite $\Diamond ABCD$ has perpendicular struts meeting at E such that $AE = 2$ and $EC = 8$ (Figure 4.65). What must the length of strut $\overline{BD}$ equal for the corners of the kite to exactly fit on a circle?

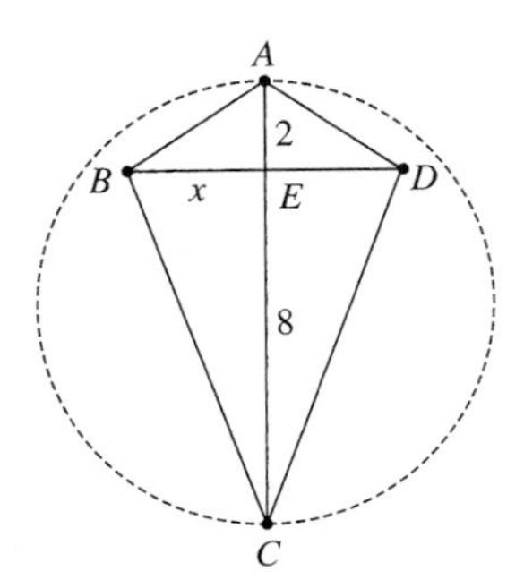

Figure 4.65

SOLUTION

According to Theorem 2, if the four corners lie on a circle, we must have $BE \cdot ED = AE \cdot EC$, or $x^2 = 16$. Therefore, $x = 4$ and $BD = 2x = 8$. ■

Moment for Discovery

Generalized Carpenter's Square Problem

If an angle has measure 30 and its sides are constrained to pass through two fixed points, the vertex still has some freedom of movement. To see what its locus might be, follow these steps in *Sketchpad.*

1. Locate two points A and B. To one side, construct a small circle CD with point E on it and construct segment $\overline{CE}$.
2. Rotate segment $\overline{CE}$ and point E about center C through an angle of 30° to segment $\overline{CE'}$.
3. Select A and segment $\overline{CE}$, and construct a line parallel to $\overline{CE}$ passing through A; Repeat for point B and $\overline{CE'}$, obtaining a line parallel to $\overline{CE'}$ passing through B. Select the point F of intersection of the two lines. Hide the lines, then construct segments $\overline{AF}$ and $\overline{BF}$.
4. Select F and choose Trace under DISPLAY.
5. Select E and circle CD, then choose Animate under DISPLAY with options Once Around Circle and Quickly checked.

What result did you get? Note that this construction will yield the entire circle, while the locus result in Corollary B yields only an arc of a circle. Can you explain why? Drag point B to see what effect this has on the phenomenon. Try the experiment using a different angle measure, such as 75°.

Moment for Discovery

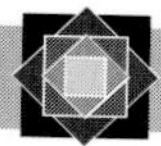

Inscribed Angles of a Circle

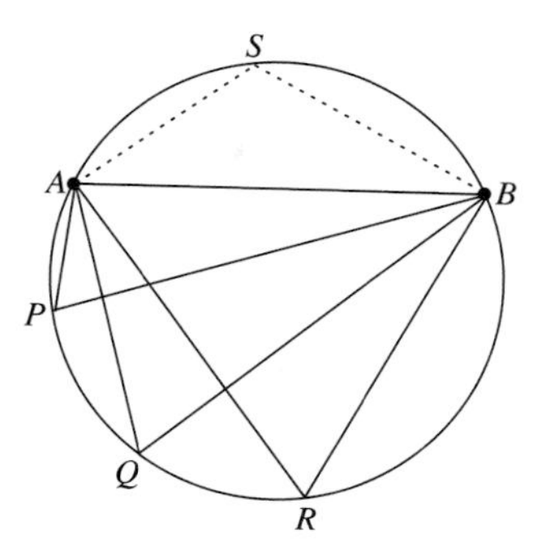

Figure 4.66

Draw a fairly large circle and chord $\overline{AB}$, as shown in Figure 4.66. Draw several inscribed angles $\angle APB$, $\angle AQB$, $\angle ARB$ on one side of the chord, and one, $\angle ASB$, on the other side.

1. Carefully measure the angles at P, Q, and R with a protractor. Did you notice anything?
2. Measure the angle at S. Does it bear any relation to the other measurements you obtained? Find a theoretical reason for this, if you can.

Moment for Discovery

Inscribed Angles of a Circle (using *Sketchpad*)

1. Construct circle AB and an arbitrary chord $\overline{CD}$.
2. Locate E and F on the circle on opposite sides of $\overline{CD}$.
3. Draw segments $\overline{CE}$, $\overline{DE}$, $\overline{CF}$, and $\overline{DF}$.
4. Display $m\angle CED$, $m\angle CFD$, and $m\angle CED + m\angle CFD$.

Drag E and F to see what you can observe. What general theorem seems to emerge? Can you prove it?

Moment for Discovery

The Two-Chord Theorem

Point P is an interior point of a circle having radius 20 units, located 16 units from the center O (as shown in Figure 4.67). Chord $\overline{AB}$ passes through P, and the distance from O to that chord is precisely $\sqrt{175} \approx 13.2$ units.

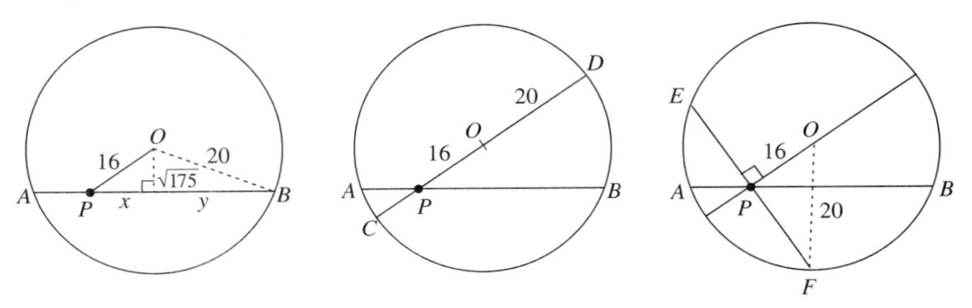

Figure 4.67

1. Determine x and y, using the Pythagorean Theorem.
2. Find AP and PB. (Note that $AB = 2y$.)
3. A second chord $\overline{CD}$ passing through P also passes through the center O of the circle. Find CP and PD in this case.
4. A third chord $\overline{EF}$ passing through P is perpendicular to chord $\overline{CD}$. Find EP and PF. (Note that $EP = PF$.)
5. Evaluate the three products

$$AP \cdot PB, \qquad CP \cdot PD, \qquad \text{and} \qquad EP \cdot PF,$$

Did anything unusual happen? What conjecture might be made regarding any chord $\overline{XY}$ of circle O that passes through point P?

Moment for Discovery

Secant–Tangent Theorem

Follow these steps in a *Sketchpad* experiment that illustrates an interesting geometric phenomenon.

1. Draw circle AB and radius $\overline{AB}$.
2. Construct a line perpendicular to $\overline{AB}$ at B and choose point C on it. Construct segment $\overline{BC}$ and hide the perpendicular.
3. Locate any point D on the circle and construct line $\overleftrightarrow{CD}$ intersecting the circle at a second point E.
4. Display the values BC^2, DC, EC, and $DC \cdot EC$ on the screen.

Do you see a relationship? What is it? Drag D on the circle. Does anything happen? Finally, drag C and observe what changes take place. Is there a general unchanging relationship (called an **invariant**)?

PROBLEMS (4.5)

GROUP A

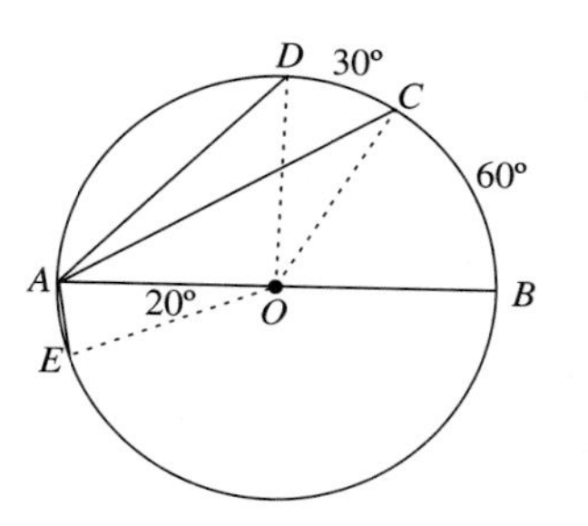

1. Several arcs of circle O and an angle have their measures as indicated in the figure.
   (a) Find the measures of $\angle CAB$, $\angle COB$, $\angle DAC$, and $\angle DOC$.
   (b) Find the measure of $\angle BAE$ in two ways: by finding the measure of minor arc $\widehat{BE}$ and using the Inscribed Angle Theorem, and by using $OA = OE$ and properties of isosceles triangles.

2. If $\overline{TA}$ is tangent to circle O, $\overline{AB}$ is a diameter, and C is a point on the circle such that $m\widehat{CB} = 100$, determine the correct value for x, evaluate $m\angle CAB$, and deduce $m\angle TAC$ from the Angle Addition Postulate. Does your result agree with the statement about such angles in this section?

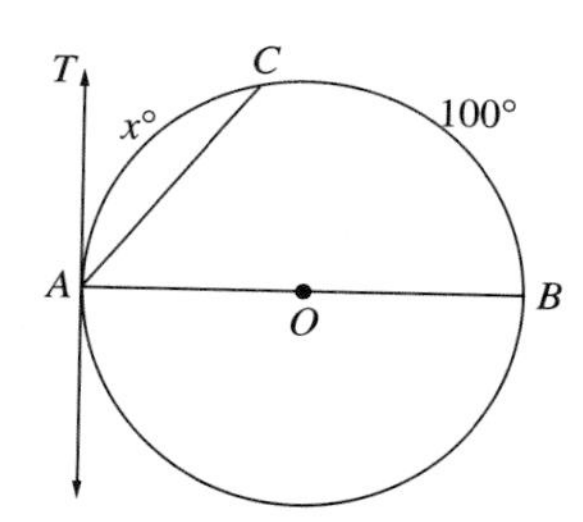

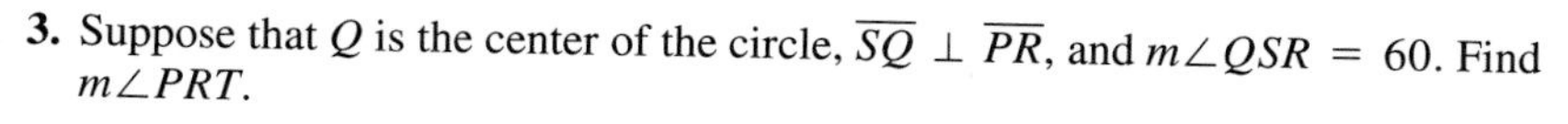

3. Suppose that Q is the center of the circle, $\overline{SQ} \perp \overline{PR}$, and $m\angle QSR = 60$. Find $m\angle PRT$.

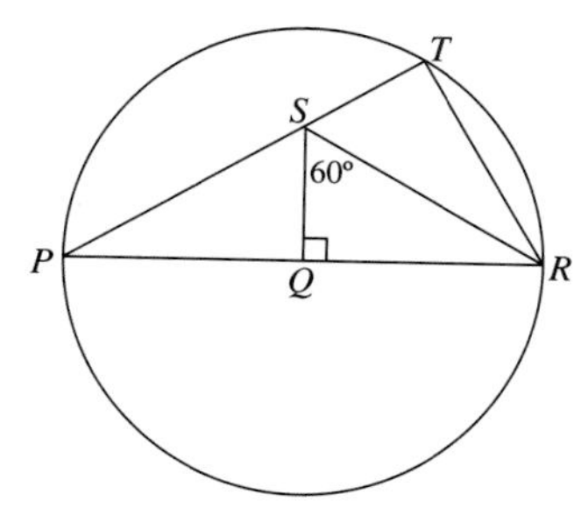

4. Find the length of chord $\overline{TR}$ in Problem 3 if the radius of the circle equals 4 units.

5. A circle has radius 6 units. A tangent is drawn from an external point P that lies at a distance of 10 units from the center. If A is the point of contact of the tangent with the circle, find PA.

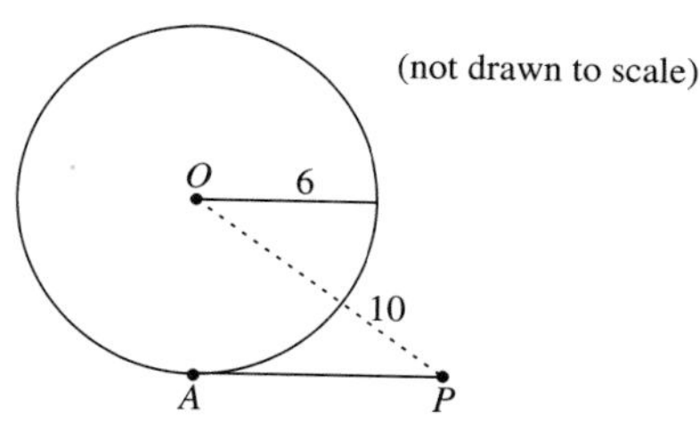

6. *Given:* Isosceles $\triangle ABC$ inscribed in a circle, with base $\overline{BC}$.
Prove: If P is any point on minor arc $\overset{\frown}{BC}$, then ray $\overrightarrow{PA}$ bisects $\angle BPC$.

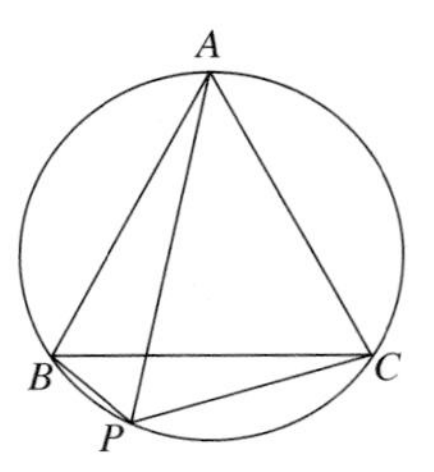

7. Polygon FANCYPROBLEM is a regular dodecagon inscribed in a circle.

(a) Find the measure of each of the arcs $\overset{\frown}{FA}$, $\overset{\frown}{AN}$, $\overset{\frown}{NC}$, . . .

(b) What argument would prove these measures are equal?

(c) Find $m\angle RPY$ in two different ways, using the concepts from this section, and those from the previous one involving properties of regular polygons. Compare your answers.

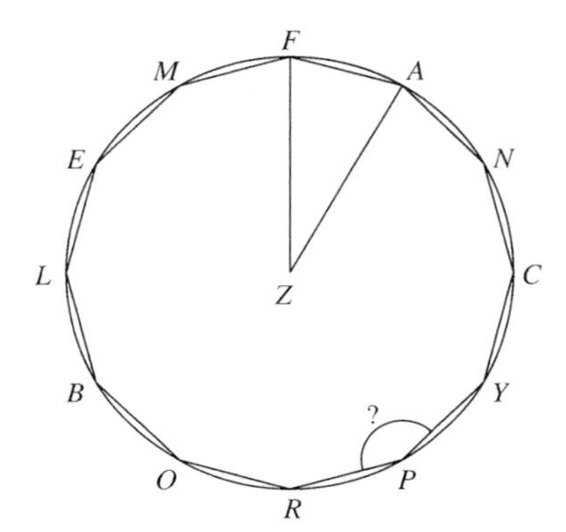

8. Two chords $\overline{KV}$ and $\overline{QR}$ of circle O are perpendicular at point P, with $PQ = 6$ and $PR = 8$. If the radius of circle O is $\sqrt{65}$, find KP and PV.

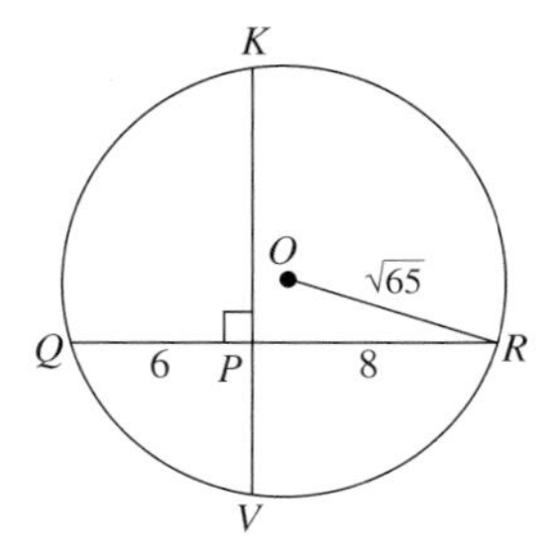

9. Two chords $\overline{RT}$ and $\overline{KW}$ of circle O are perpendicular at point S, with $RS = 6$, $ST = 32$, $KS = 12$, and $SW = 16$. Find the radius of the circle.

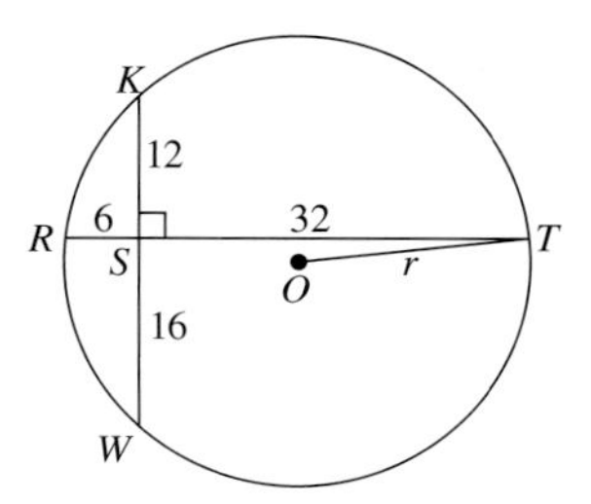

10. Two secants of a circle are perpendicular at P and meet the circle at the points A, B, C, and D, with $CD = 1$, $PA = 2$, and $PC = 3$. Find the radius r. (***Hint:*** First show that $AB = 4$. Observe the resulting dimensions of the half-dotted rectangle in the figure.)

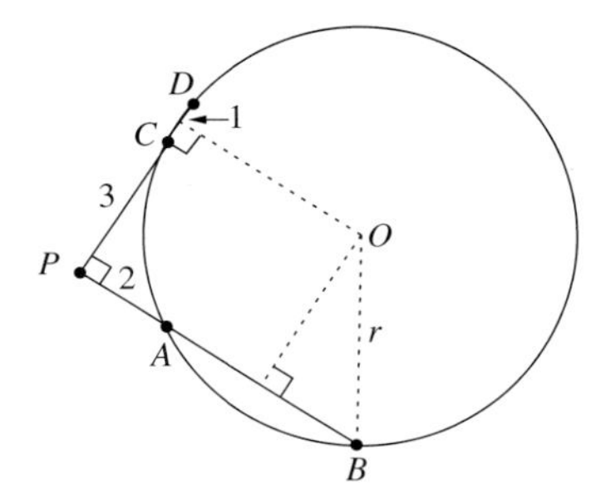

11. The Coast Guard has determined that the waters near a certain point O on shore are treacherous due to rocks hidden below the surface. The warning has been issued to all ships sailing in these waters to stay outside of a region marked by two warning lights at A an B (on shore) at a distance of 2 miles on either side of O and all other points within 2 miles of O. What simple method of sighting could be used by the captain of a ship to guarantee that he obeys this order? Prove your answer. (Assume a relatively straight shore line.)

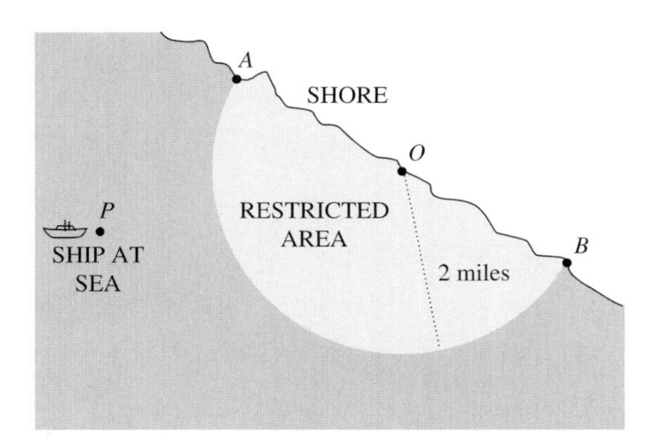

12. **(a)** Suppose $\triangle ABC$ is an isosceles triangle inscribed in the circle, with $AB = AC$. If $\overline{CD}$ is a diameter, find the measures of $\angle 1$, $\angle 2$, $\angle 3$, and $\angle 4$ in terms of $\theta \equiv m\angle ABC$.

(b) What is the value of θ if $\triangle OBC$ is equilateral?

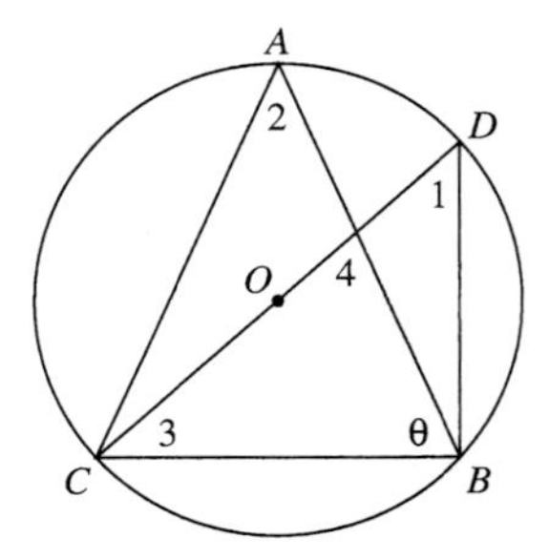

13. A circle is tangent to the sides of an angle of measure 28. Find the measures of the two arcs on the circle subtended by that angle.

14. If $\overline{MT} \perp \overline{RW}$, $VN = VM = VW = 3$, $RN = 1$, and $m\widehat{MW} = 75$ find

(a) VT

(b) NS

(c) $m\widehat{RS}$

(d) $m\widehat{ST}$.

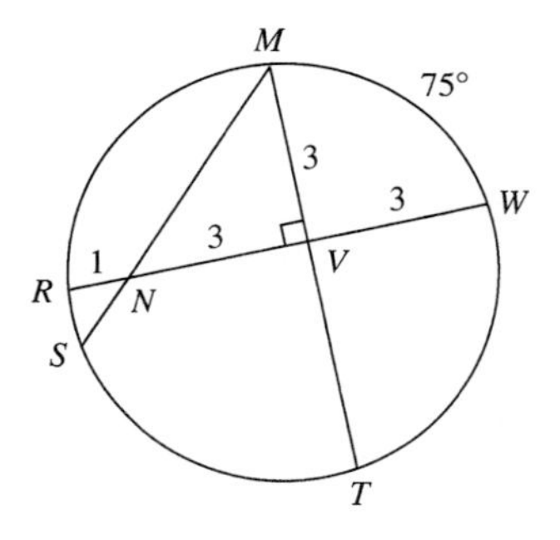

15. A circle P of radius 10 units passes through the center Q of another circle, of radius $r > 10$, cutting off an arc of measure 90 on the larger circle. Find r.

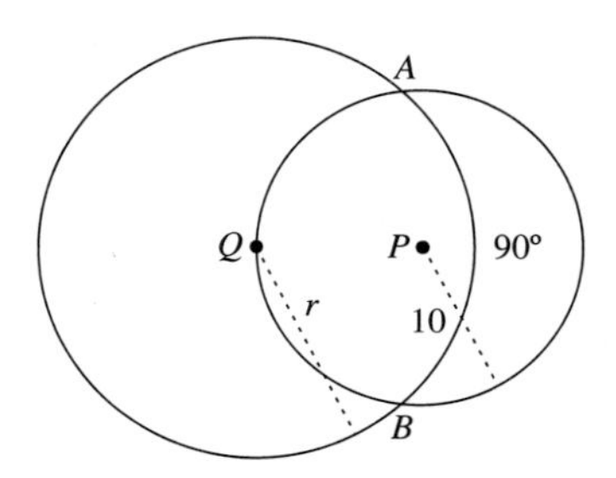

GROUP B

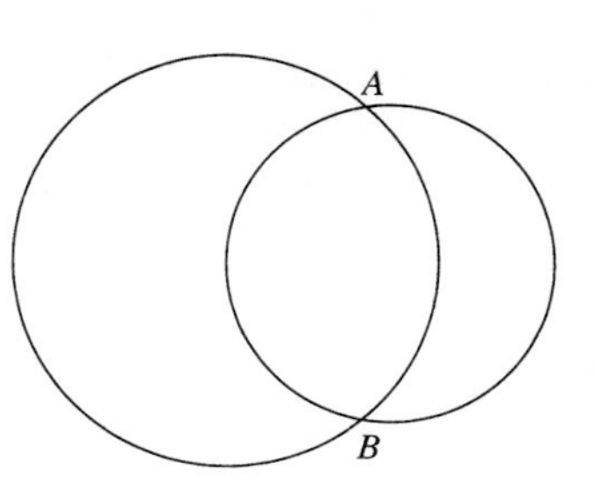

16. In the figure for this problem, the center of one circle lies on the other.

(a) What is the relationship between the measures of the two minor arcs having end points at A and B?

(b) Find $m\widehat{AB}$ if the circles have equal radii.

17. Two circles intersect at A and B. If $m\widehat{AB} = 40$ on the larger circle and $m\widehat{AB} = 170$ on the smaller, and if the radius of the larger circle is 12, find the radius of the smaller circle. Use trigonometry and geometric facts about circles.

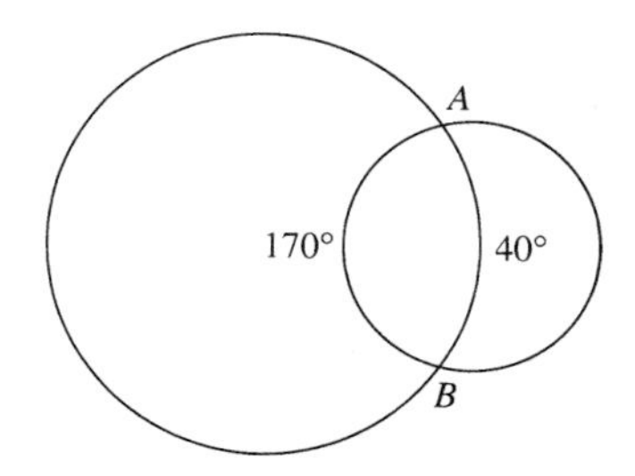

18. Prove the Two-Chord Theorem by first showing that $\triangle PAC$ and $\triangle PBD$ in the figure have two sets of congruent angles, hence are similar triangles with proportional side lengths. Then complete the proof that $PA \cdot PB = PC \cdot PD$.

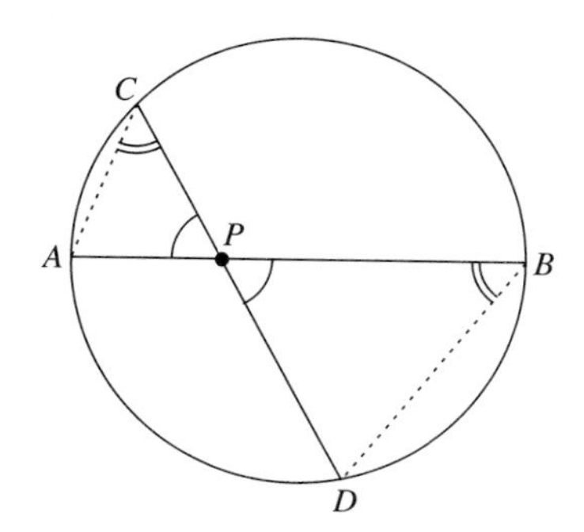

19. Prove that an angle formed by a chord and tangent of a circle has measure one-half that of the intercepted arc (in the figure, $m\angle CAT = \frac{1}{2}x$). (***Hint:*** Draw diameter $\overline{AB}$, then use the Inscribed Angle Theorem for $\angle BAC$. Part of the proof involves explaining why $m\widehat{BC} = 180 - x$.)

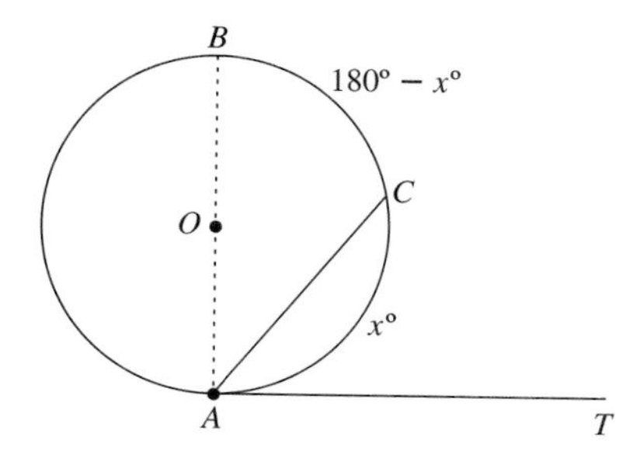

20. In the figure, secant $\overline{PA}$ and tangent $\overline{PC}$ are drawn, forming triangles $\triangle PBC$ and $\triangle PCA$. (Then, $\angle B \cong \angle ACP$. Why?) Use similar triangles to prove the Secant–Tangent Theorem: $PC^2 = PA \cdot PB$.

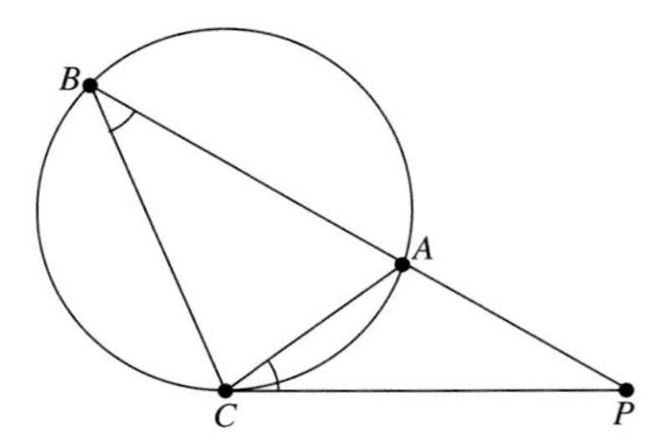

21. In right $\triangle ABC$, $\overleftrightarrow{MR}$ is the perpendicular bisector of the hypotenuse $\overline{AB}$, and $MR = MA$. Prove: Ray $\overrightarrow{CR}$ bisects $\angle ACB$. (***Hint:*** Draw a circle through A, B, and C.)

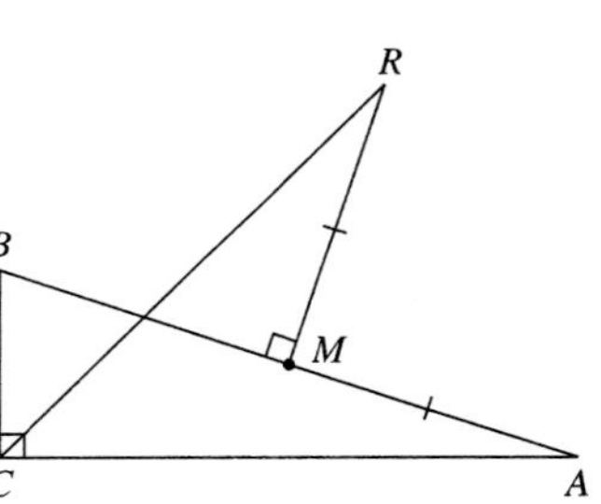

The next two problems are devoted to the interesting relationship between the *power* of a point P with respect to a circle (defined in Problem 22), and the constant products $AP \cdot PB$ relative to variable chords and secants through P.

22. The Power of a Point The **power** of point P with respect to a circle with center O and radius r is the real number

$$\text{Power } (P) = PO^2 - r^2$$

Note that the number is positive if P lies outside the circle, is zero if P lies on the circle, and is negative if P lies interior to the circle.

(a) Prove that if P lies outside circle O and $\overleftrightarrow{PT}$ is tangent at T, then

$$\text{Power } (P) = PT^2$$

(b) Identify the set of all points P for which

$$\text{Power } (P) = k \text{ (a constant)}$$

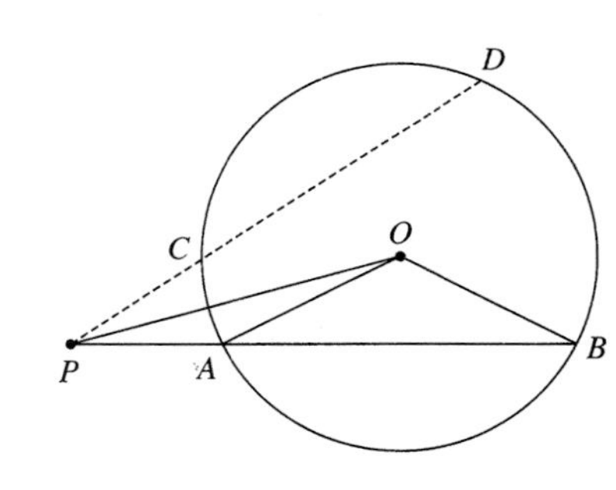

23. **(a)** In the following figure, use the Cevian Formula and directed distance (as introduced in Problems 25–27, Section 4.3) to prove that, regardless of the location of P inside or outside the circle,

$$\text{Power } (P) = PA \cdot PB$$

(b) Prove Theorems 2, 3, and the corollary to Theorem 3 as immediate corollaries of this result.

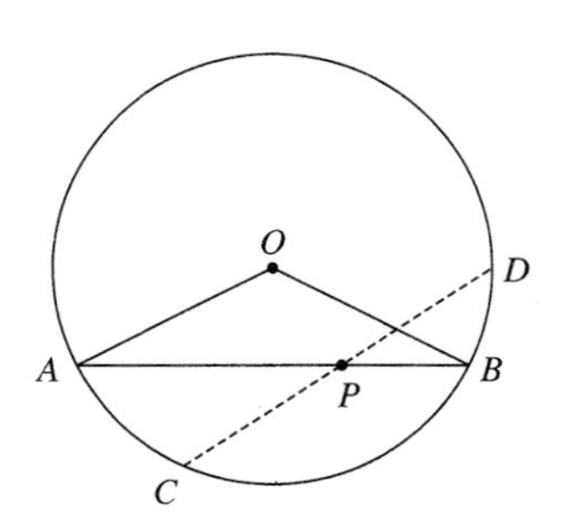

24. Using only the result of Problem 23, prove that $m\angle C = m\angle D = \frac{1}{2}m\widehat{AB}$, where C is any point on the circle for which $\angle ACB$ subtends the minor arc $\widehat{AEB}$, and $\overline{DE}$ is the diameter of the circle perpendicular to chord $\overline{AB}$. (The point of this and the preceding problem is to show that the circle theorems may all be derived without using additivity of arc measure.)

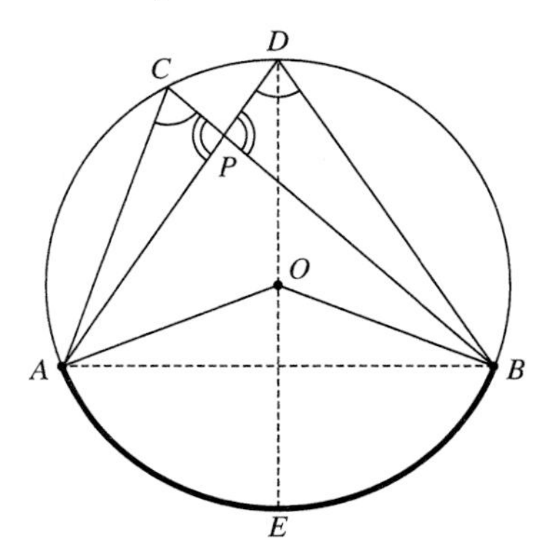

GROUP C

25. Suppose $\overline{BC}$ and $\overline{CD}$ are congruent chords of circle O, and $\triangle ABC$ is isosceles with base $\overline{BC}$. Find the angles of $\triangle OAE$ in terms of $\theta \equiv m\angle OCD$ and $\phi \equiv m\angle ADC$.

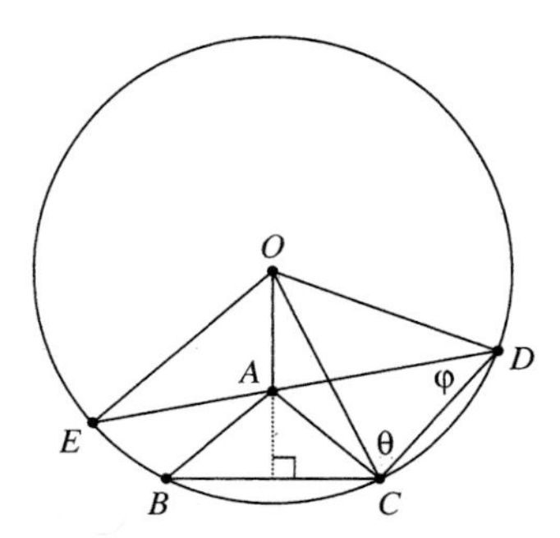

26. How to Find the Radius of a Circle Given a circle O, construct an equilateral triangle $\triangle ABC$ with one of its sides as a chord of the circle (side $\overline{BC}$ in the figure). Construct chord $\overline{CD} \cong \overline{BC}$, then draw line $\overline{DA}$ cutting the circle again at point E. Then AE will equal the radius of circle O. Provide a proof of this fact, based on the result of Problem 25. (***Hint:*** With $m\angle OCD = \theta$, obtain the angles of $\triangle OAE$ in terms of θ, as in Problem 25.)

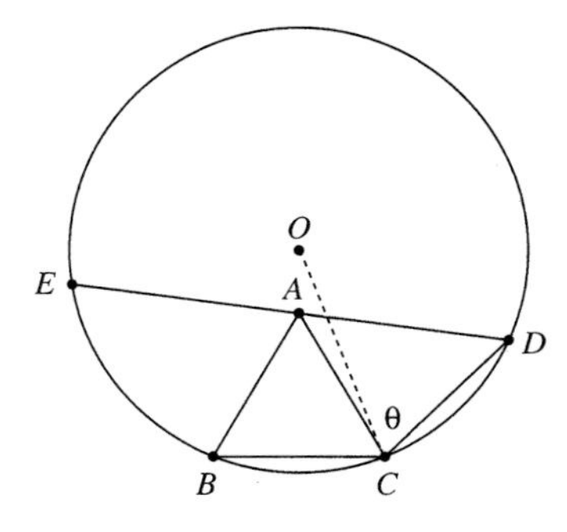

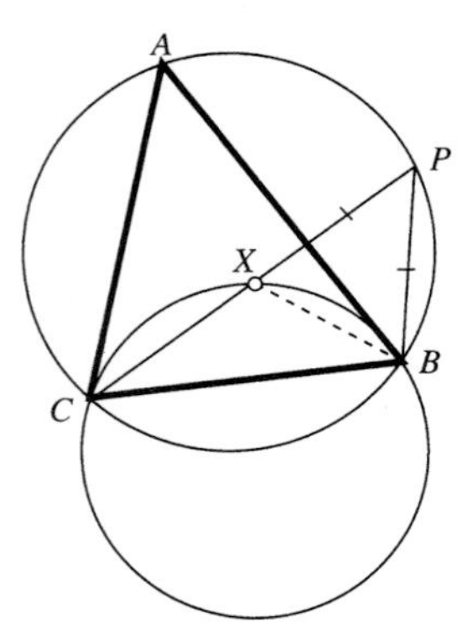

27. A point P varies on the circumcircle of $\triangle ABC$, and for each position of P a point X is located on segment $\overline{PC}$ such that $PX = PB$.

(a) Find the locus of point X.

(b) Do you notice anything peculiar about the configuration when $m\angle A = 60$?

(c) Study this phenomenon using *Sketchpad.*

28. Problem of Appolonius Let three circles having equal radii be given. Find a Euclidean construction for a fourth circle that is tangent to the other three.

Note: The general problem of constructing a circle tangent to three given circles is the famous **Problem of Appolonius.** Appolonius was a Greek geometer who lived around 250 B.C.

29. Using *Sketchpad,* explore the following locus problem: Find the locus of the center C of a circle $\boldsymbol{C}$ that is tangent (externally or internally) to two other circles $\boldsymbol{C}_1$ and $\boldsymbol{C}_2$, which are themselves tangent externally. Try to prove your conjecture using properties of circles and tangents.

*4.6 Euclid's Concept of Area and Volume

An enduring topic for geometry and applied mathematics, yet one that is as troublesome as it is obvious, is that of area and volume. We begin with a question. Which do you think has the greater area: a 1-in. tall isosceles triangle with a 2-in. base, or a rectangle 1/64,000 of an inch wide and a mile long? Could you rely on your intuition to answer this question conclusively? Does the area even *exist* for this very long and human-hair-width rectangle?

Our intuitive concept of area and volume no doubt involves the idea of simple counting—counting the number of unit squares, or fractional parts thereof, (or unit cubes) which are contained by the region concerned. This works fine for an integer-sided or even rational-sided rectangle, which involves only a finite number of unit squares or fractional parts (like the first two examples shown in Figure 4.68), and the formula $K = bh$ is pretty obvious. But it quickly becomes problematical in the case of a rectangle with irrational sides, or a triangle, even though these figures in geometry are quite elementary (by contrast, finding the area of a circle would certainly seem insurmountable). The ancient Greeks struggled with the problem of area for the so-called **incommensurable case** of the rectangle (when the ratio of length to width is irrational)—and ultimately conquered that problem, as well as that of the circle.

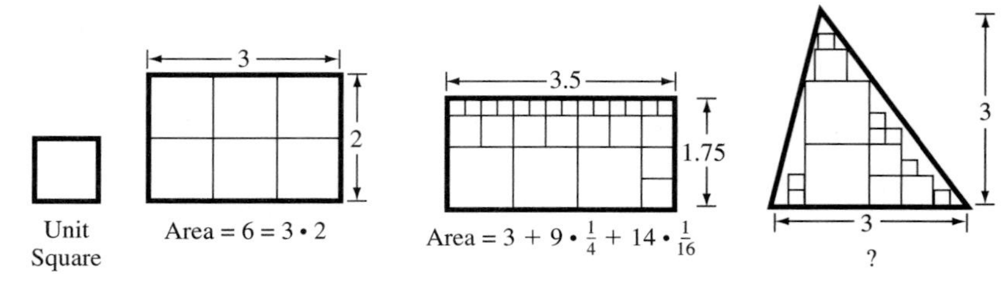

Figure 4.68

The modern approach to this idea of counting squares or cubes involves **measure theory.** This topic is devoted essentially to the study of area and volume as applied to arbitrary sets in the plane or in three-dimensional space. For a region in the plane, one essentially covers the given region by a grid of squares, "counts" the number of squares that just cover the region, and then multiplies by the area of each square in the grid (as shown in Figure 4.69). By making the grid finer and finer, a converging sequence whose limit is taken as the area of the region is obtained. A similar idea is used for the volume of a solid in three-dimensional space.

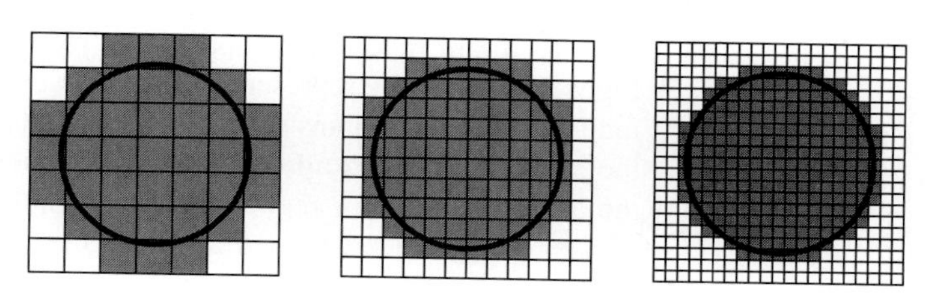

Figure 4.69

Although perhaps appealing to our basic instincts, the idea of "counting" squares remains elusive, to say the least. Since it requires using a complicated process involving limits, actual calculations are unwieldy, and useful area and volume formulas are difficult to develop. Thus we come face to face with the recurring problem in mathematics of trying to formulate a natural and workable definition, and then using the definition to derive the basic theory.

To further illustrate the dilemma facing us, we propose a short drama. It is predicated on the assumption that the only part you remember about area in geometry are the basic formulas for rectangles and triangles. As a promotion, suppose a large firm has decided to sponsor a contest on mathematics. The rules involve a $500 entry fee, coming to the site of the contest to participate, and working on the problem without recourse to textbooks or calculators. The prize is a million dollars. Thus, the motivation for you to derive from scratch the formulas you have forgotten is guaranteed. Now suppose the problem is to find which of the following geometric figures (Figure 4.70)—a trapezoid, circle, regular hexagon, and a square with the four corners cut out along the curve indicated in the coordinate system, as shown in the figure—has the largest area. Tie breakers consist of the degree of accuracy of the answers (exact answers are obviously desirable), and the inclusion of a mathematical development for the formulas used. No doubt the contestants will try very hard to capture the essence of the concept of area in order to derive the formulas they need.

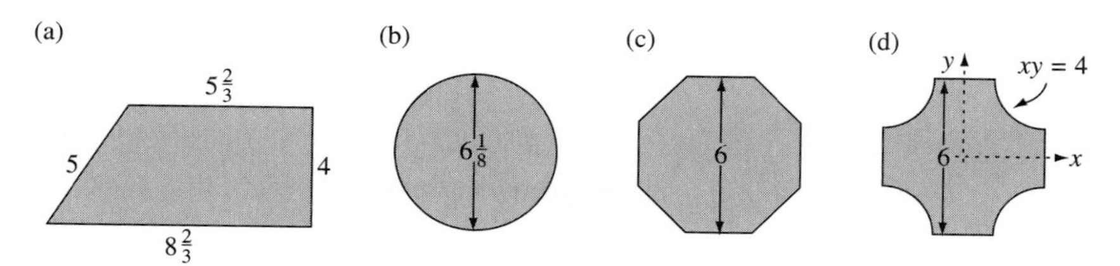

Figure 4.70

This example points up the nature of the area problem confronting ancient scholars. They, too, were trying to find formulas for the area and volume of certain geometric objects, along with proofs or plausibility arguments for those formulas. They had no books in which to look up the answer, nobody who could give them a hint. They had only their intellect and ingenuity. It is sometimes surprising to learn how extensive and ingenious those early developments of area and volume were, particularly since the tools of calculus and the theory of limits had not yet been invented. A prime example is the marvelous development given by Archimedes in his monumental derivation of the volume for a sphere.

We might note that the particular regions chosen for our imaginary contest in Figure 4.70 actually represent major plateaus of achievement in the history of mathematics. The Babylonians (2000–1600 B.C.) could have found the correct formula for the area of the trapezoid [region (a)], but they did not have the general formula for the area of a circle or regular polygon. Euclid (300 B.C.) conquered the circle and regular polygon [regions (b) and (c)], but it requires some form of calculus (A.D. 1790) to find the area of region (d), which, in terms of the usual coordinate system, requires calculating the area under the curve $y = 4/x$.

It is strangely illuminating, in modern times, to examine Euclid's ancient approach to the problem of area and volume. In the *Elements,* Euclid never bothered to define area and volume. Rather, he *identified* each object in the plane with its area and each object in space with its volume; area and volume were not regarded as real numbers, separate from the objects themselves. When he stated that two objects in the plane were "equal," he meant that they had *equal areas or volumes.* For example, his Proposition 35, Book I, states: "Parallelograms which are on the same base and in the same parallels are equal to one another."

Indeed, what Euclid did is logically equivalent to our modern approach, a development that appears in many high school geometry textbooks. One assumes the *existence* of the area and volume of all regions under consideration, and then states certain desirable laws that area and volume should obey (the *axioms*).

We present the axioms usually assumed, and show how to derive some of the basic formulas from them. In this axiom system, we take as an undefined term **region.** We may think of a region as just a set of points (like a circle or triangle and their interiors), whose boundary is not too pathological. (There do exist bounded regions in the plane whose areas do not exist, called **nonmeasurable sets** in mathematics.)

	AREA	VOLUME
1. Existence Postulate	To each region M in the plane, there corresponds a real number Area $M \geq 0$, called its area.	To each region T in space, there corresponds a real number Vol $T \geq 0$, called its volume.
2. Dominance Postulate	For regions M_1 and M_2 in a plane, if $M_1 \subseteq M_2$, then Area $M_1 \leq$ Area M_2.	For regions T_1 and T_2 in space, if $T_1 \subseteq T_2$, then Vol $T_1 \leq$ Vol T_2.

3. Postulate of Additivity	For any two planar regions M_1 and M_2 such that Area $(M_1 \cap M_2) = 0$ then Area $(M_1 \cup M_2) =$ Area $M_1 +$ Area M_2.	For any two regions T_1 and T_2 in space such that Vol $(T_1 \cap T_2) = 0$ then Vol $(T_1 \cup T_2) =$ Vol $T_1 +$ Vol T_2.
4. Congruence Postulate	Congruent regions in a plane have equal areas.	Congruent regions in space have equal volumes.
5. Unit of Measure	The area of the unit square is one.	The volume of the unit cube is one.
6. Cavalieri's Principle[4]	If all the lines parallel to some fixed line that meet the plane regions M_1 and M_2 do so in line segments having equal lengths, whose end points lie in the boundaries of the two regions, then Area $M_1 =$ Area M_2. (See Figure 4.71.)	If all the planes parallel to some fixed plane that meet the solid regions T_1 and T_2 do so in plane sections having equal areas, whose boundaries lie in the boundaries of the two regions, then Vol $T_1 =$ Vol T_2.

CAVALIERI'S PRINCIPLE FOR AREA

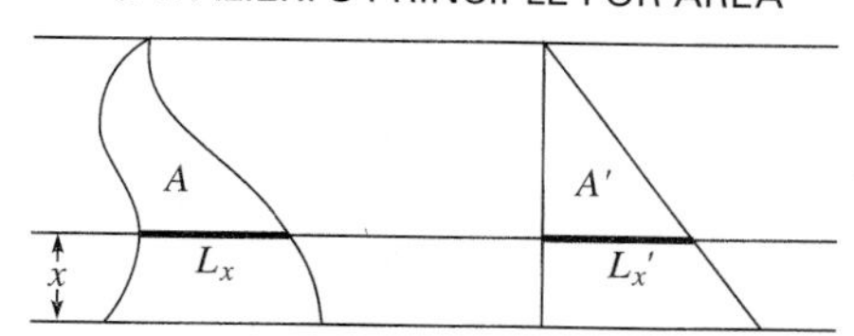

If $L_x = L_x'$ then Area A = Area A'

CAVALIERI'S PRINCIPLE FOR VOLUME

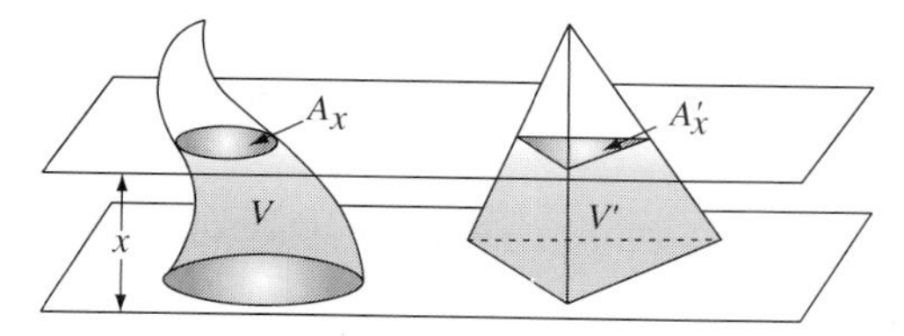

If Area A_x = Area A_x' then Volume V = Volume V'

Figure 4.71

[4]This ingenious axiom is due to one of the early pioneers of calculus, B. Cavalieri (1598–1647). Euclid and Archimedes obviously had to do without this labor-saving concept.

NOTE: Cavalieri's principle may actually be proven from the axioms that precede it using the theory of integration.

OUR GEOMETRIC WORLD

Designer scratch pads with a twist provide a perfect illustration of Cavalieri's Principle. A cubical pile of square pages is twisted to form the more interesting three-dimensional figure below. If we started with this solid and asked for its volume, the problem might seem unmanageable. But, since, page for page, the solids have equal cross-sectional area, the volume of the twisted solid equals that of the cube, whose volume is quite elementary.

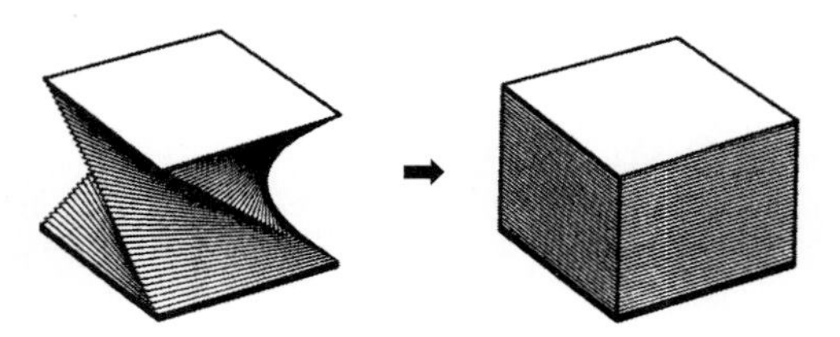

Figure 4.72

It is common in elementary treatments to assume as axioms the basic formulas for the area of a rectangle and the volume of a "box." But these may be derived by logic, using the previous axioms. To illustrate, we show this for a rectangle. Let two rectangles R and R' be given, with bases of length b and b', and altitudes h and h'.

We show first that if $h = h'$,

$$\frac{\text{Area } R'}{\text{Area } R} = \frac{b'}{b}$$

Let m/n be a rational approximation of b'/b in positive integers m and n, such that

(1) $$\frac{m}{n} < \frac{b'}{b} \leq \frac{m+1}{n}$$

If we multiply throughout by the positive quantity nb, then

(2) $$mb < nb' \leq (m+1)b$$

Now consider three rectangles R_1, R_2, R_3 (Figure 4.73) each having altitude h and bases of length mb, nb', and $(m+1)b$, respectively. By the Dominance Postulate (since the rectangle plus interior having the smaller length would fit inside the one having the greater length by congruence, etc.), the inequality **(2)** implies that

(3) $$\text{Area } R_1 < \text{Area } R_2 \leq \text{Area } R_3$$

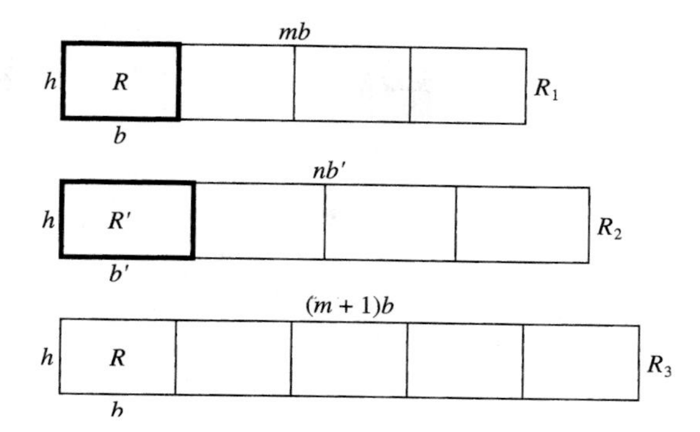

Figure 4.73

By the Postulate of Additivity, since exactly m rectangles congruent to R make up R_1, n congruent to R' make up R_2, and $m + 1$ congruent to R make up R_3, we get

$$m \text{ Area } R < n \text{ Area } R' \leq (m + 1) \text{ Area } R$$

or

(4)
$$\frac{m}{n} < \frac{\text{Area } R'}{\text{Area } R} \leq \frac{m + 1}{n}$$

Therefore, by comparing the inequalities **(1)** and **(4)** (as $n \to \infty$)

(5)
$$\frac{\text{Area } R'}{\text{Area } R} = \frac{b'}{b}$$

Similarly, if the original rectangles R and R' have $b = b'$, then

(6)
$$\frac{\text{Area } R'}{\text{Area } R} = \frac{h'}{h}$$

From **(5)** and **(6)** it now follows that if R' is a rectangle with height h and base 1, and U is the unit square, then

$$\frac{\text{Area } R}{\text{Area } R'} = \frac{b}{1} = b \qquad \text{and} \qquad \frac{\text{Area } R'}{\text{Area } U} = \frac{h}{1} = h$$

Simply multiply these two equations, and use Postulate 5 to obtain the following:

THEOREM 1: If R is a rectangle with base of length b units and height of length h units, then

(7)
$$\text{Area } R = bh$$

The formula Vol $P = Bh$ for a rectangular box P having base area B and height h may be derived similarly. (See Problem 18.) The formula for the general parallelogram then follows from Cavalieri's Principle. Triangles, trapezoids and regular polygons come next.

We do not attempt a rigorous axiomatic treatment for the area of a circle. That may be found in sources such as Moise, *Elementary Geometry from an Advanced Viewpoint*. Instead, we present two *plausibility* arguments any prospective teacher of geometry should know about. The first shows how to come up with the following relationship between the *area* of a circle and its *circumference*:

(8) $$2A = rC$$

where A is the area of a circle of radius r, and C is its circumference. Using the standard formula $C = 2\pi r$ already introduced (in Section 1.2), the relation **(8)** may then be readily converted into the desired formula

(9) $$A = \pi r^2.$$

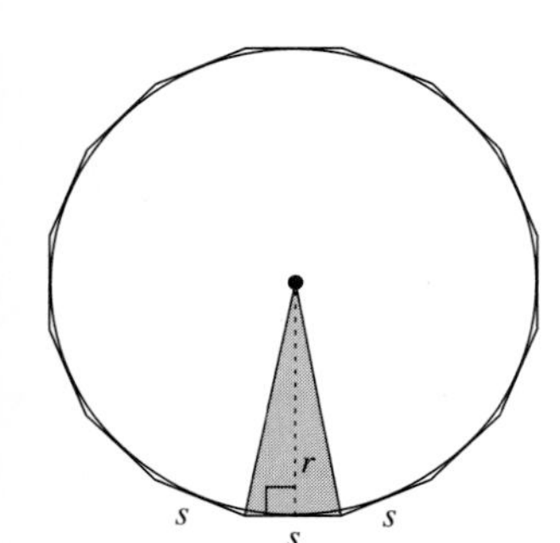

Figure 4.74

To obtain **(8)**, consider a regular polygon with many sides circumscribing the circle, as shown in Figure 4.74 (for a 16-sided regular polygon.). The interior of the polygon subdivides into 16 congruent isosceles triangles with altitude r and base s. Each of these has area $\frac{1}{2}rs$. The total area of the polygon is therefore

$$K = 16 \cdot \frac{1}{2}rs = \frac{1}{2}r(16s) = \frac{1}{2}rP$$

where P is the perimeter of the polygon. Thus

(10) $$K = \frac{1}{2}rP \qquad \text{or} \qquad 2K = rP$$

But for a polygon having 1000 or more sides, K would be very nearly the area of the circle, A, and P would be nearly equal to C, the circumference. Hence,

$$2A = rC.$$

The second plausibility argument is quite interesting and presents a very unique type of construction often found in high school geometry texts. It will be left as a discovery unit at the end of this section.

EXAMPLE 1 Show how Cavalieri's Principle can be used to prove the area formula $K = bh$ for the area of a parallelogram having base b and altitude h.

SOLUTION

Let the given parallelogram be $ABCD$, with base $AB = b$ and altitude h, as shown in Figure 4.75. On line $\overleftrightarrow{AB}$ construct a rectangle $\square EFGH$ with $EF = b$ and $EH = h$. If

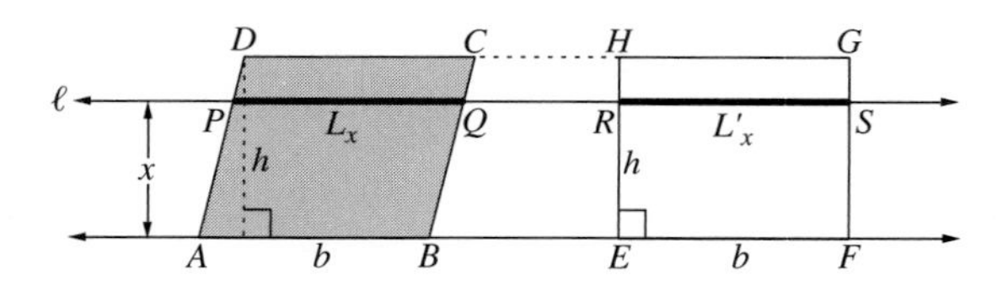

Figure 4.75

a line parallel to line $\overleftrightarrow{AB}$ lies at a distance x from $\overleftrightarrow{AB}$, that parallel will cut the sides of the parallelogram and rectangle at the points P, Q, R, and S, creating segments $\overline{PQ}$ and $\overline{RS}$ having lengths L_x and L_x'. In order to apply Cavalieri's Principle, we must show that $L_x = L_x'$. But $\overline{PQ} \parallel \overline{AB}$, so $\diamond PQBA$ is a parallelogram. Hence $PQ = AB$. Similarly, $RS = EF$, and thus we have

$$L_x = PQ = AB = b = EF = RS = L_x'$$

Since this is true for $0 < x < h$, by Cavalieri's Principle,

$$K = \text{Area } \square ABCD = \text{Area } \square EFGH = bh$$

as desired. ■

EXAMPLE 2 Establish the area formula

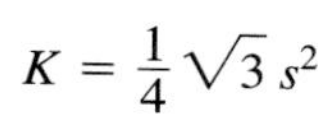

$$K = \frac{1}{4}\sqrt{3}\, s^2$$

for an equilateral triangle having side s. (See Figure 4.76.)

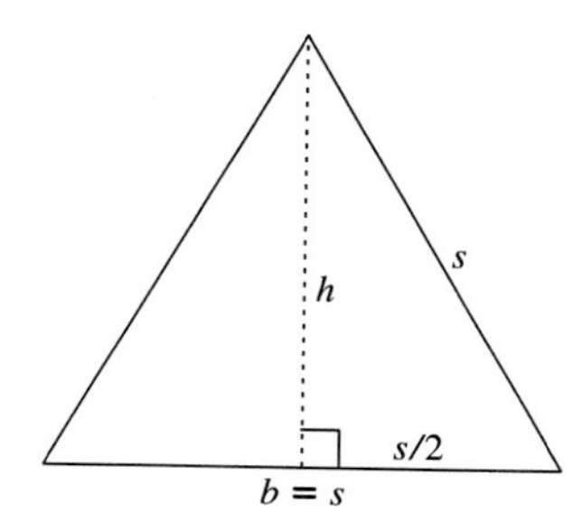

Figure 4.76

SOLUTION
Using the area formula for a triangle $K = \frac{1}{2}bh$ (which will be established in Problem 6), we have $K = \frac{1}{2}sh$. By the Pythagorean Theorem,

$$h^2 + (s/2)^2 = s^2 \qquad \text{or} \qquad h^2 = s^2 - \frac{1}{4}s^2 = \frac{3}{4}s^2$$

$$h = \sqrt{\frac{3}{4}s^2} = (\sqrt{3}/2)s$$

Substituting this into the preceding equation for K, we have

$$K = \frac{1}{2}s(\sqrt{3}/2)s = \frac{1}{4}\sqrt{3}s^2 \quad ■$$

EXAMPLE 3 Show that the area of a sidewalk that borders a square (shown in Figure 4.77) is simply its width (w) times the total length of its *midline* (L):

$$K = wL$$

(The **midline** is defined as the locus of midpoints of all line segments joining two adjacent borders of the shaded region.)

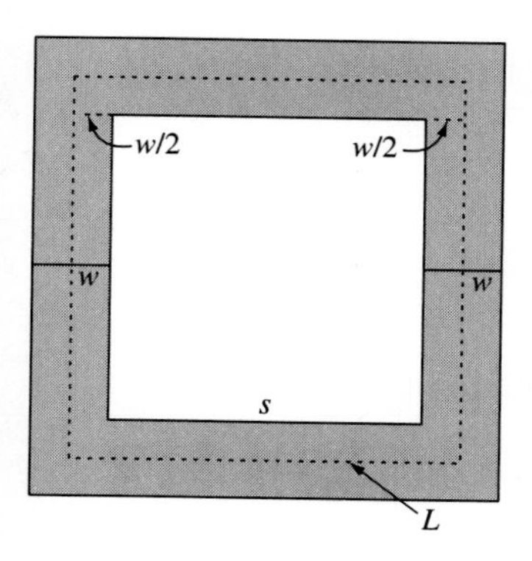

Figure 4.77

SOLUTION
Let s = side of the inner square; then the side of the outer square will be $s + 2w$. The area of the sidewalk is the difference of the areas of these two squares, so

$$\begin{aligned} K &= (s + 2w)^2 - s^2 \\ &= s^2 + 4sw + 4w^2 - s^2 \\ &= 4sw + 4w^2 \end{aligned}$$

On the other hand, the length of one side of the square midline is

$$w/2 + s + w/2 = s + w$$

so its total length is

$$L = 4(s + w) = 4s + 4w$$

Hence going back to the equation for K,

$$K = 4sw + 4w^2 = w(4s + 4w) = wL. \quad \blacksquare$$

Moment for Discovery

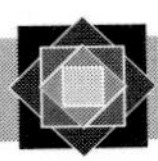

The Area of a Circle

The following diagrams depict a unique way to deduce the area of a circle of radius r.

1. A circle is cut into a sequence of pie-shaped pieces, then reassembled into rectangular shapes of dimensions r and L_1, r and L_2, r and L_3, as shown in Figure 4.78.

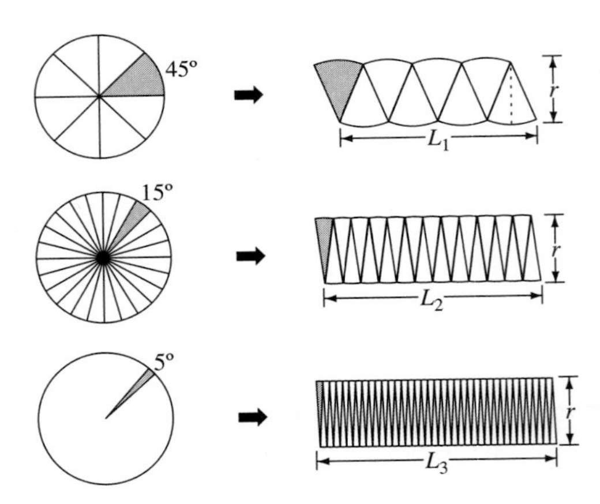

Figure 4.78

2. Find what number L the sequence $L_1, L_2, L_3, \ldots$ seems to be approaching. (Give a formula for L in terms of r.)
3. To what value, then, do the areas of these figures seem to be converging? (Give a formula.)
4. What have you deduced about the area of a circle?

Moment for Discovery

Cavalieri's Principle

This *Sketchpad* experiment will provide a dynamic demonstration of Cavalieri's Principle. Follow these steps (illustrated in Figure 4.79):

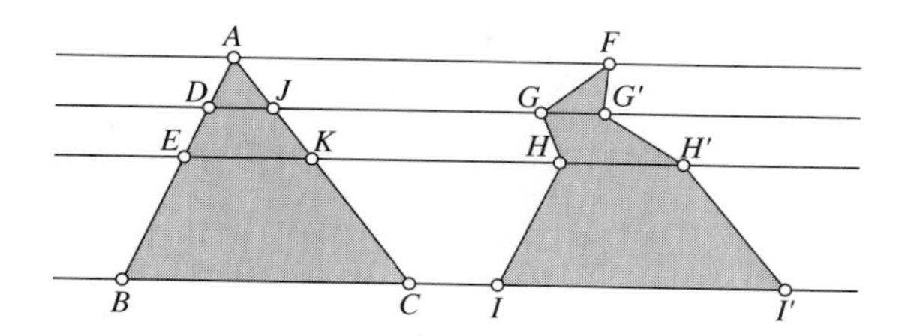

Figure 4.79

1. Construct a triangle $\triangle ABC$ and locate points D and E on side $\overline{AB}$.
2. Construct line $\overleftrightarrow{BC}$, then through points A, D, and E construct lines parallel to $\overleftrightarrow{BC}$.
3. Locate arbitrary points F, G, H, and I on the four lines of Step 2.
4. Select the points of intersection J and K of the parallels $\overleftrightarrow{GD}$ and $\overleftrightarrow{HE}$ with segment $\overline{AC}$. Hide the lines.
5. Mark Vector D to J, then Translate point G to G' By Marked Vector, so that $DJ = GG'$.
6. Repeat Step 5 for points H and I, yielding H' and I' such that $EK = HH'$ and $BC = II'$.
7. Select Interior ABC and Interior $FGHII'H'G'$ and calculate their areas.

Did anything happen? Drag points F, G, H, I before and after changing the shape of $\triangle ABC$. Also, drag D and E along segment $\overline{AB}$. Did you notice anything? Discuss, with proofs, if possible.

PROBLEMS (4.6)

The problems in this section that are prefixed by the letter C require calculus for their solution.

GROUP A

1. Find the areas of the first three regions proposed for the contest in Figure 4.70.

C2. One quadrant of region (d) of Figure 4.70 is shown in the figure for this problem (the shaded region). Part of this region is a rectangle, and the rest consists of the area under the curve $y = 4/x$ between A and B. Find the coordinates of points A and B and use calculus to find the area under the curve, then determine the area of region (d). Finally, find which of the regions of Figure 4.70 has the greatest area.

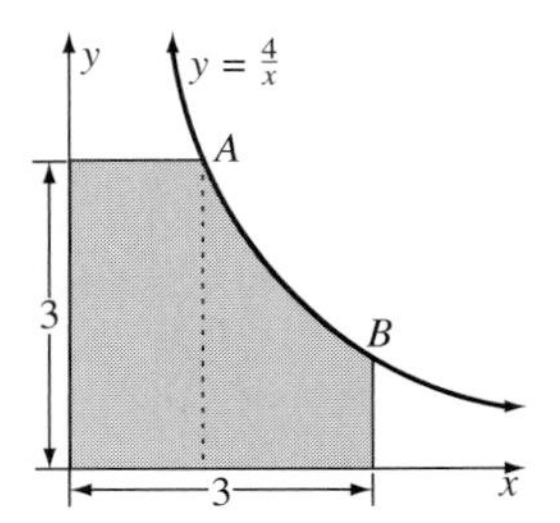

3. Use Cavalieri's Principle to find the area of the shaded region in the figure.

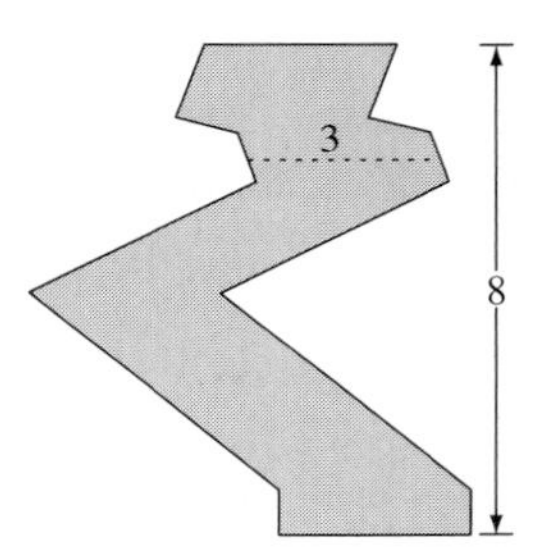

4. Use Cavalier's Principle to find the total area covered by the letters in the word *Cavalieri* if the width of the downstroke of each letter is 2 mm and the height of the letters is 8 mm. (Assume that the gap in "C" has height equal to that of the hook in the letter "L.")

5. These steps in *Sketchpad* lead to an interesting phenomenon concerning area.

[1] Construct any triangle ABC and line $\overleftrightarrow{AB}$. Locate D any point on line $\overleftrightarrow{AB}$, with an initial position between A and B.

[2] Calculate the ratio AD/AB, displayed on the screen.

[3] Select the interiors of $\triangle ACD$ and $\triangle ACB$ and display the ratio

$$(\text{Area } ACD) \,/\, (\text{Area } ACB)$$

Do you notice anything? Drag D on line $\overleftrightarrow{AB}$ to see if the phenomenon persists. Is there a conjecture you can make based on this experiment? State it, and try proving it using the results of this section.

6. (a) Use the axioms for area to prove the formula

$$\text{Area } T = \frac{1}{2}bh$$

for a triangle T having base of length b and height h. (See figure below.)

(b) Prove the formula

$$K = \frac{1}{2}(b_1 + b_2)h$$

for the area of a trapezoid having bases of length b_1 and b_2, and height h.

(a)

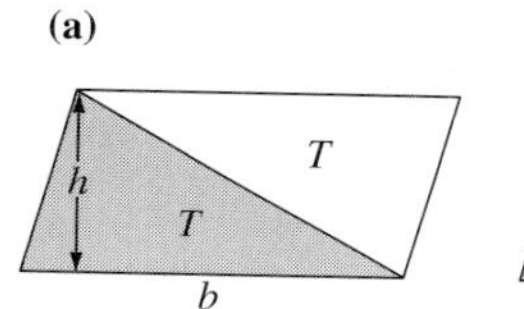

(b)

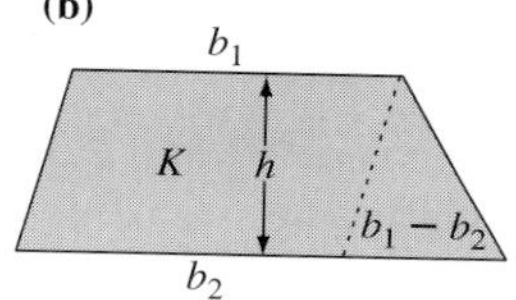

7. A window design consists of a rectangle and semicircle at the top, and is to be placed into a rectangular opening having a perimeter of 20 ft. To increase the amount of light the window will allow, a carpenter has decided that the maximal area for such a window is obtained when the rectangle is a square with 4 ft. on each side, as shown. Do you think the carpenter is correct? Base your answer on the area of the carpenter's window as compared to one having a base of

(a) 6 ft.

(b) 5 ft.

(***Hint:*** If x is the height of the rectangle in the window having a base of 6 ft., the perimeter of the opening is given by $12 + 2(x + 3) = 20$ or, solving for x, $x = 1$ ft.)

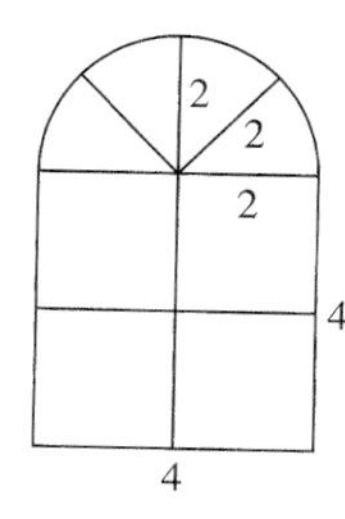

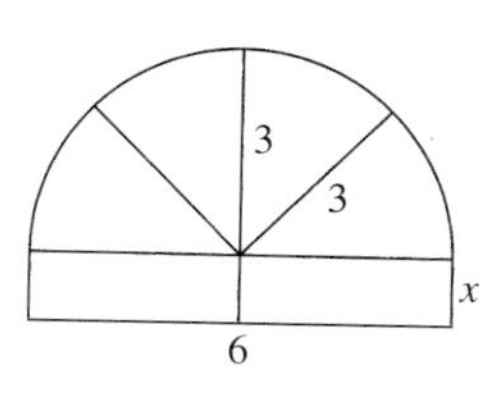

8. Use the classical area formulas to determine the ratio of the side of a square to the radius of a circle if the square and circle have the same area.

GROUP B

C9. Show by calculus that the window of Problem 7 in fact has maximal area when the rectangle has a base of approximately 4.52 ft. Find the maximal area.

10. A farmer has 30 yd. of fencing and wants to enclose an area beside his barn. What are the dimensions of the region of maximal area, and what is that maximal area, if the region is

(a) a rectangle?

(b) an isosceles triangle with its base along the side of the barn?

(c) a semicircle with diameter along the side of the barn?

(Refer to the figure below. Calculus is not needed for this problem.)

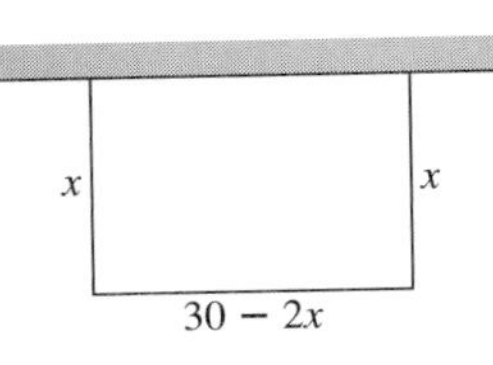

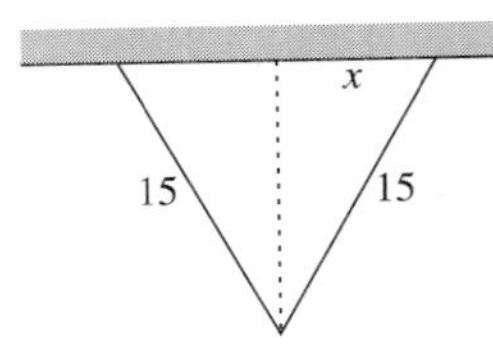

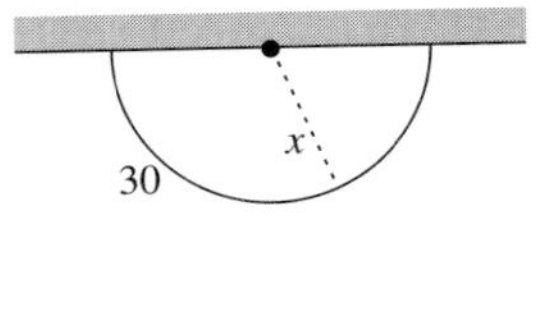

11. Show that the area of the washer shown in the figure having width w and a midline of length L is given by $A = wL$. (***Hint:*** L = circumference of a circle of radius $r + w/2$.)

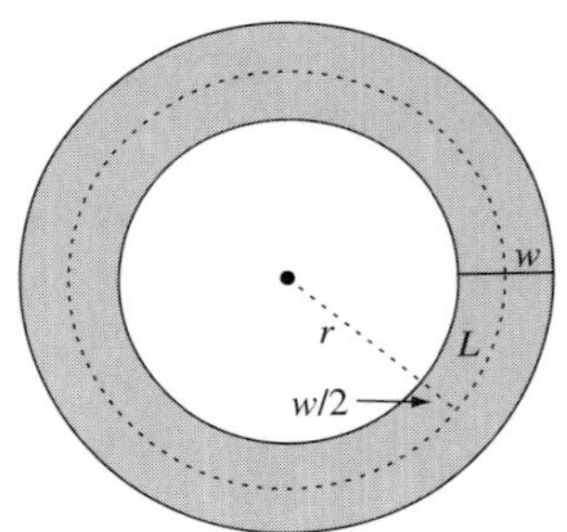

12. **Ratios of Areas of Similar Figures** Use *Sketchpad* to investigate an important phenomenon concerning area by performing these steps:

[1] Construct an arbitrary pentagon $ABCDE$. (See figure below.)

[2] Select segment $\overline{AB}$ and choose Length under MEASURE to find $m\overline{AB}$.

[3] Select points A, B, C, D, and E and choose Polygon Interior to calculate the area of the pentagon.

[4] Calculate the quantities

$$(\text{Area } ABCDE) / (m\overline{AB}) = x \quad \text{and} \quad (\text{Area } ABCDE) / (m\overline{AB})^2 \equiv y$$

displaying these two values on the screen.

[5] Double-click on A and using the DILATE TOOL, shrink the polygon to a smaller size (select all five sides of the pentagon). As you do so, watch what happens to x and y.

[6] Return to ARROW TOOL and radically change the shape of the pentagon (for example, making it nonconvex). Then repeat Step 5 to see if a relationship prevails.

State the general relationship you discovered and apply it to state a conjecture about the ratios of the areas of any two similar polygons $\boldsymbol{P}$ and $\boldsymbol{Q}$ as related to the measures of two corresponding sides.

$$\frac{(\text{Area p1})}{(\text{m}\overline{\text{AB}})} = 1.444 \text{ inches}$$

$$\frac{(\text{Area p1})}{(\text{m}\overline{\text{AB}})^2} = 1.202$$

D C E A B

13. (a) A replica of a Model T Ford has a scale of 1:12 (1 in. on the model equals 1 ft. on the car). The front of the windshield on the model has an area 6.3 in.^{2}. What is the corresponding area of the windshield of the car in square inches?

(b) A bedroom measures 10 by 12 ft. How many square yards of carpet will be needed to cover the floor?

(c) Two fishermen each catch a Salmon on a lake, and the two fish have similar shapes in all respects. One fish is twice as long as the other. How do the weights compare? (***Hint:*** Weight is proportional to volume.)

14. Area of Regular Polygon If a is the **apothem** of a regular n-sided polygon (distance from center to any side), prove that its area is given by $K = \frac{1}{2}ap$, where p is the perimeter. (See the figure below.)

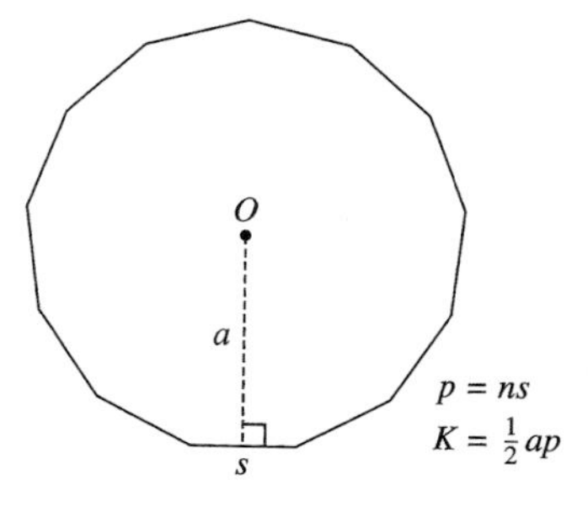

15. Find an explicit formula for the area of a regular hexagon and regular octagon in terms of the length s of a side.

16. Use Cavalieri's Principle to find the areas of each of the regions indicated in the figure.

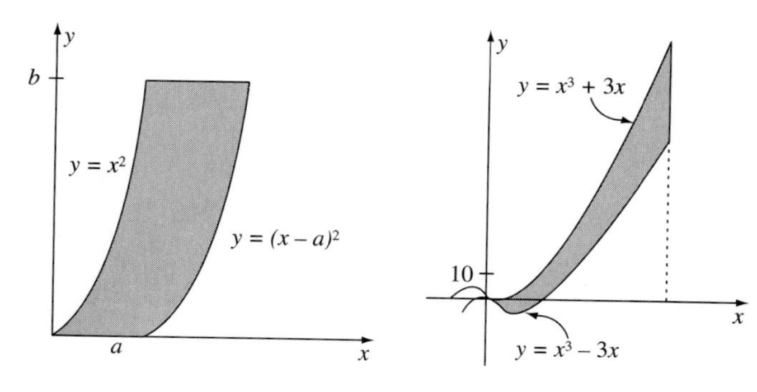

GROUP C

17. Find the area of the shaded region in the figure without calculus. (***Hint:*** Cavalieri's Principle applies—show that at height y the generating segment of the region has length y. You will need to solve the quadratic equation $x^2 - 2xy + 2y^2 - 2y = 0$ for x.)

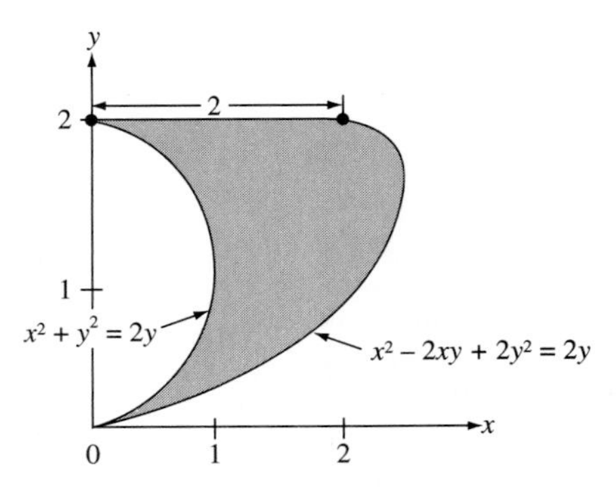

18. Using the same procedure as that used for rectangles, prove the volume formula $V = \ell wh$ (or $V = Bh$) for a parallelepiped having dimensions ℓ, w, h, and B = area of base $= \ell w$. (***Hint:*** First prove that Vol P'/Vol $P = \ell'/\ell$ for boxes P' and P have the same width and height, and lengths ℓ' and ℓ, respectively.)

19. Alternate Euclidean Proof of Pyramid Formula See the figure below and fill in the details of the following analysis.

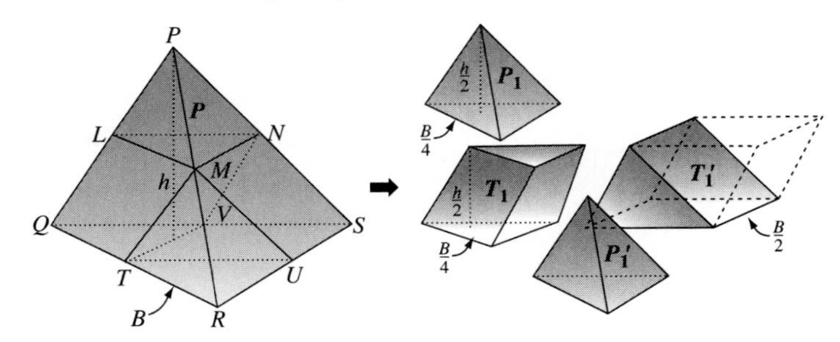

(1) Dissect the original pyramid $\boldsymbol{P}$ into two triangular prisms $\boldsymbol{T}_1$ and $\boldsymbol{T}_1'$ and two congruent pyramids $\boldsymbol{P}_1$ and $\boldsymbol{P}_1'$ similar to $\boldsymbol{P}$ and half its size. Why does Vol $\boldsymbol{T}_1 = \frac{1}{8}Bh$ and Vol $2\boldsymbol{T}_1' = \frac{1}{4}Bh$ (where $2\boldsymbol{T}_1'$ = parallelepiped shown, having base $\frac{1}{2}B$)? Thus,

$$\text{Vol}\,\boldsymbol{P} = \frac{1}{4}Bh + 2\,\text{Vol}\,\boldsymbol{P}_1$$

(2) Dissect the pyramid $\boldsymbol{P}_1$ in like manner and obtain

$$\begin{aligned}\text{Vol}\,\boldsymbol{P} &= \frac{1}{4}Bh + 2\left(\frac{1}{4}B_1h_1 + 2\,\text{Vol}\,\boldsymbol{P}_2\right)\\ &= Bh\left(\frac{1}{4} + \frac{1}{16}\right) + 4\,\text{Vol}\,\boldsymbol{P}_2\end{aligned}$$

where B_1 and h_1 are the base area and altitude, respectively, of $\boldsymbol{P}_1$.

(3) Continue the process and obtain the result

$$\text{Vol } \boldsymbol{P} = Bh\left(\frac{1}{4} + \frac{1}{16} + \frac{1}{64} + \cdots\right) \equiv Bh \sum_{n=1}^{\infty} \frac{1}{4^n}$$

(4) Evaluate the geometric series in **(3)**.

NOTE: Euclid used this same decomposition of the pyramid to prove that two triangular pyramids having congruent bases and equal altitudes were "equal," which then enabled him to finish his proof of the relation $V = \frac{1}{3}Bh$ which appears in Section 7.4. Cavalieri's Principle was not available to Euclid, so he had to find a different method of proof.

The following problems are on the topic of fractals, discussed briefly in Section 2.1. They are optional, and are not graded according to difficulty.

20. Sierpinski Triangle For us the working definition of the term **fractal** (whose Latin derivative *fractus* means "to break") is: a geometric figure or set of points in the plane having the property that a window, if properly placed, contains a replica similar to the original figure. A good example to begin with is Sierpinski's Triangle, which can be created by the following initial stages using *Sketchpad*. The process continues ad infinitum and a fractal results. Determine which of the windows shown in the figures below contains a reduced version of the entire fractal.

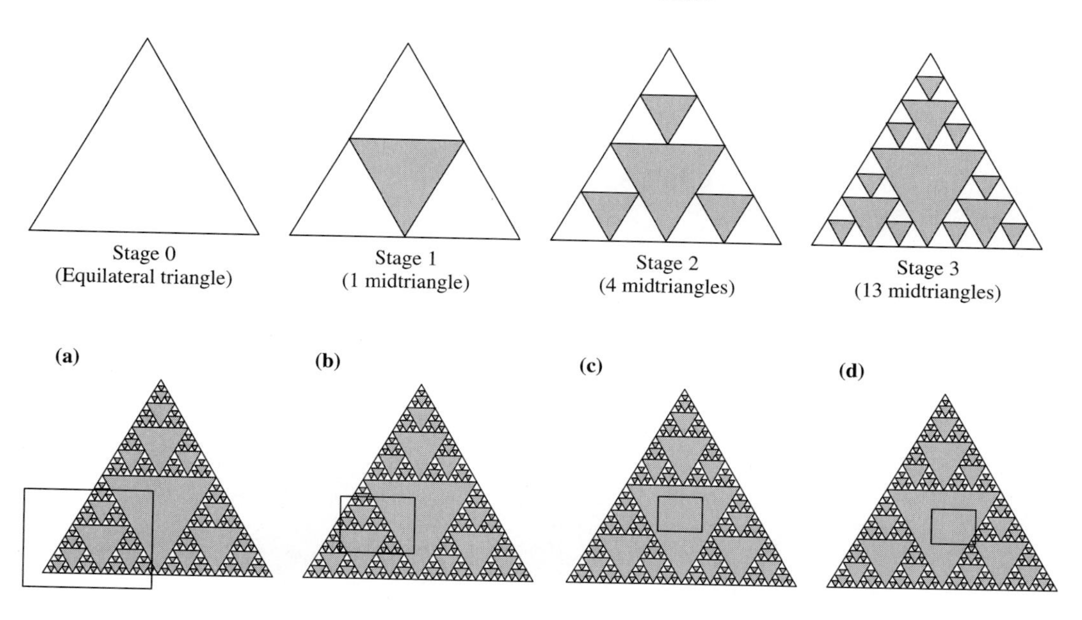

21. If the side of the equilateral triangle in Stage 0 of Sierpinski's Triangle (Problem 20) is unity, calculate the following:

(a) The total perimeters of the shaded triangles (i.e., *boundary*) in Stages 1–3.

(b) The total areas of the shaded triangles in Stages 1–3.

(c) Based on your answers in **(a)** and **(b)** find the **perimeter** (= total boundary length) and **area** of Sierpinski's triangle. In terms of area, do the shaded regions "fill up" the equilateral triangle of Stage 0?

22. A Path Fractal The following stages define a one-dimensional fractal, a particular type called a **hat fractal**. Sketch the very next stage (Stage 3) and determine the total path length of each of the three stages, if the length of Stage 0 equals unity. Is the length increasing without bound? Does the fractal possess a finite length?

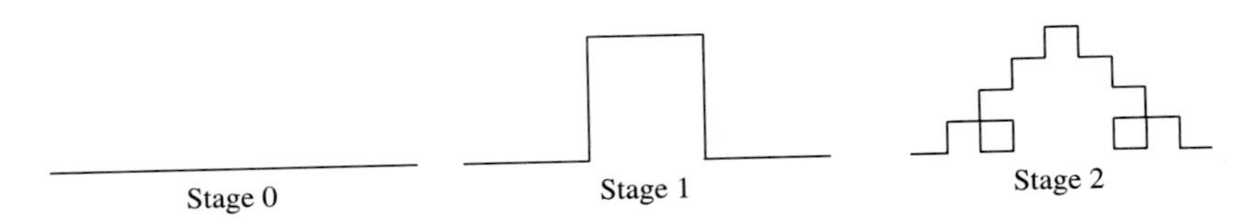

In Problems 23 and 24, Stages 0, 1, 2, and 3 of a fractal are shown. Stage 0 has an area of 1. Find the area of the shaded regions of Stages 1, 2, and 3.[5]

23.

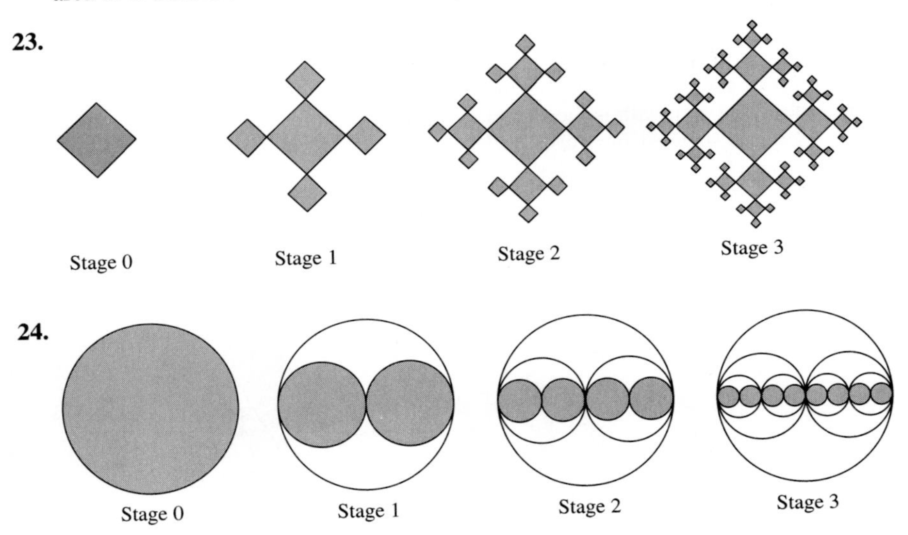

25. Stages 0, 1, 2, and 3 of a fractal are shown below. Stage 0 consists of one regular hexagon. How many regular hexagons are in Stage 1? Stage 2? Stage 3?

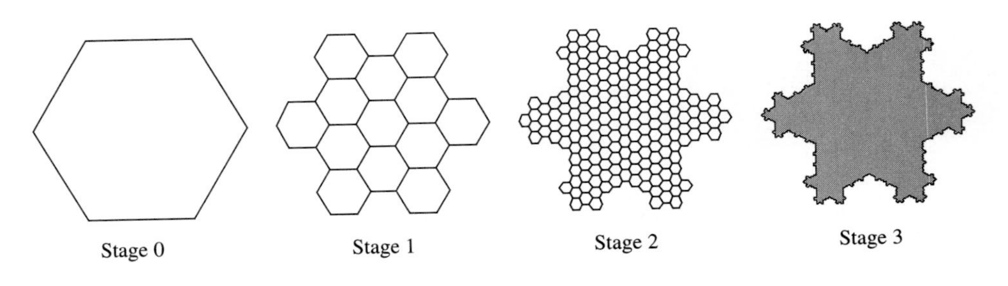

---

[5]Problems 24–26 are from the high school text *Geometry: An Integrated Approach* by R. E. Larson, L. Boswell, and L. Stiff, D.C. Heath and Company, 1995. They are used by permission.

26. A Pentagonal Fractal Observe the fractal defined by the stages shown in the figure. Find its area if the pentagon has unit side. [It can be shown from trigonometry that the side of the square is $a = 1/(1 + 2\tan 36°) \approx .408$.]

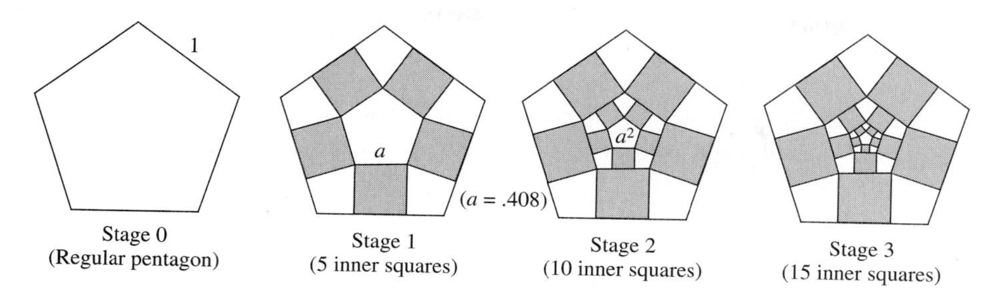

27. A Stair Step Fractal The first stages of a path fractal are shown in the figure. Show that the perimeter of Stage n of the fractal equals 2 (constant).

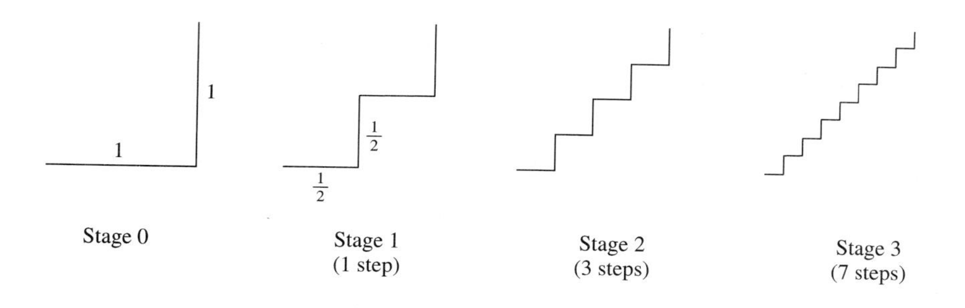

*4.7 Coordinate Geometry and Vectors

In 1637, the great mathematician/philosopher René Descartes made a discovery that would revolutionize geometry. He found that geometric configurations could be described entirely by coordinated real number pairs, and two-variable equations. From this concept ultimately emerged the entire field of real analysis, vectors, linear algebra, and matrix theory. Without it, calculus as we know it could not have developed. No doubt Descartes was one of the "giants" Newton was referring to in his famous quote.

We shall run quickly through the major steps in creating what are called **Cartesian coordinates**—a system that is also known as **coordinate geometry**—and **vectors,** or **vector geometry.** The previous development is distinguished from the present one by the term **synthetic,** versus the **analytic method** we are about to introduce. It is expected that you will be familiar with most of these ideas. *As you read through this material, you should focus on the deeper question of validity and the existence of a "linear" coordinate system for Euclidean geometry.* Non-Euclidean geometry does not have such a coordinate system.

THE COORDINATE PLANE: THE EQUATION OF A LINE

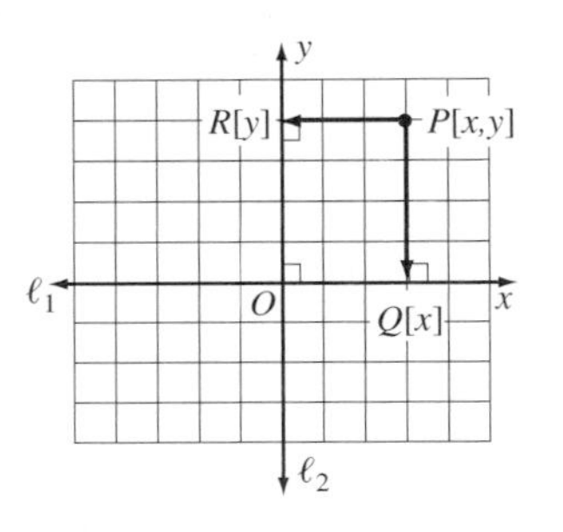

Figure 4.80

Let ℓ_1 and ℓ_2 be two perpendicular lines intersecting at point O, and on each line take a coordinate system as guaranteed by the Ruler Postulate so that O has the zero coordinate on each line. (See Figure 4.80.) These lines are, respectively, the **coordinate axes** for the system. For any point P in the plane, drop perpendiculars to each line ℓ_1 and ℓ_2; if the feet of the perpendiculars meet ℓ_1 and ℓ_2 at the points $Q[x]$ and $R[y]$, respectively, then assign to P the **coordinate pair** (x, y). Conversely, every ordered pair of real numbers (x, y) is a coordinate pair of a unique point P. We designate this relation by

$$P(x, y)$$

Thus, ℓ_1 is called the ***x*-axis** and ℓ_2 the ***y*-axis.**

One major result at this point is that the lines of the plane are sets of number pairs whose coordinates satisfy first-degree equations. To see this, let ℓ be an **nonvertical line** that makes an angle of measure $\theta \neq 90$ with the x-axis (Figure 4.81), and suppose ℓ intersects the y-axis at $B[b] \equiv B(0, b)$. If $P(x, y)$ is any point on ℓ, then, in right triangle $\triangle PQB$ (since $\theta \neq 90$),

$$\tan\theta = m = \frac{y - b}{x}$$

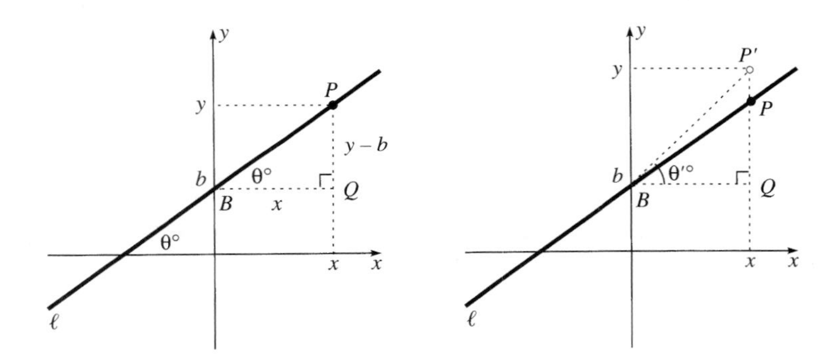

Figure 4.81

Conversely, if the coordinates of P' satisfy this relation, then

$$\tan\theta' = \frac{y - b}{x} = \tan\theta$$

hence $m\angle P'BQ = \theta = m\angle PBQ$ so that $P' = P$ and P' lies on ℓ. Solving the preceding equation for y, we deduce that $P(x, y)$ lies on ℓ iff $y = mx + b$, where $m = \tan\theta$ (the **slope** of ℓ); θ is the **inclination** of ℓ, and b the ***y*-intercept** of ℓ. The equation of a **vertical line,** one that is parallel to the y-axis (with inclination $\theta = 90$), is clearly $x = a$, where a is some real constant. Thus, all lines have an equation of the form $ax + by + c = 0$, and hence the term **linear equation.**

SLOPES AND GEOMETRIC RELATIONSHIPS BETWEEN LINES

To establish by geometry that *two nonvertical lines are parallel iff they have equal slopes,* note that if $m_1 = m_2$ and hence $\tan\theta_1 = \tan\theta_2$, or $\theta_1 = \theta_2$, then by Property F, $\ell_1 \| \ell_2$. A good problem for you is to prove, by geometry, that *two nonvertical, nonhorizontal lines are perpendicular iff the product of their slopes is −1.* [See Figure 4.82 for a hint; if $P = (1, m_1)$, $Q = (1, 0)$, and $R = (1, m_2)$, where $m_2 < 0$, why is $PQ \cdot QR = 1$ and $m_1 m_2 = -1$?]

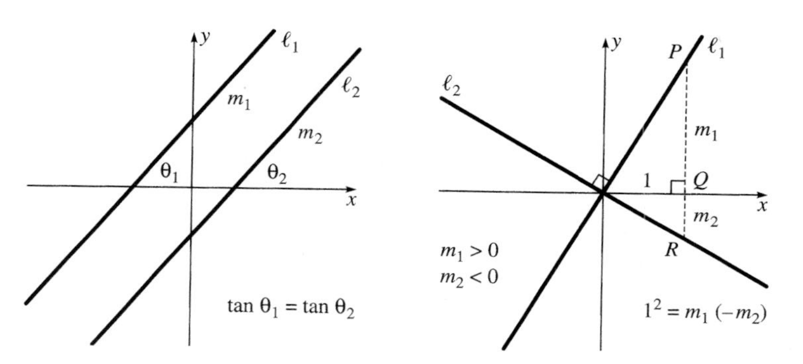

Figure 4.82

FORMULAS FOR SLOPE AND DISTANCE

Finally, we derive the analytic formulas for slope and distance. Let $P(x_1, y_1)$ and $Q(x_2, y_2)$ be two points in the plane (Figure 4.83). Construct the lines parallel to the coordinate axes through P and Q, as shown, which then determines a third point $R(x_2, y_1)$. Note that since $\overleftrightarrow{PR}$ and $\overleftrightarrow{QR}$ are, respectively, horizontal and vertical lines, $PR = |x_2 - x_1|$ and $QR = |y_2 - y_1|$. Since $\triangle PQR$ is a right triangle

$$PQ^2 = PR^2 + RQ^2$$

or

(1) $$PQ = \sqrt{(x_1 - x_2)^2 + (y_1 - y_2)^2}$$

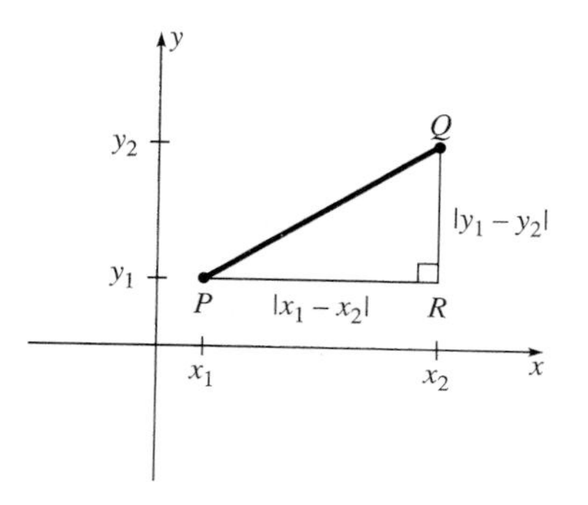

Figure 4.83

We leave the proof of the following standard slope formula as a problem: If $x_1 \neq x_2$, the slope of the line $\overleftrightarrow{PQ}$ (with the same coordinates as before) is given by

(2) $$m = \frac{y_1 - y_2}{x_1 - x_2}$$

EXAMPLE 1 Use coordinate geometry to prove that the diagonals of a rhombus are perpendicular.

SOLUTION

It is immaterial where we locate the coordinate system, so we can suppose that the vertices of the rhombus have the coordinates as shown in Figure 4.84. Since the sides of the parallelogram have equal lengths, it follows from **(1)** that $a^2 = OP^2 = PQ^2 = (a + b - a)^2 + (c - 0)^2$ or $a^2 = b^2 + c^2$. Also, by **(2)**,

$$m_1 = \{\text{Slope } \overleftrightarrow{OQ}\} = \frac{c - 0}{a + b - 0} = \frac{c}{a + b}$$

$$m_2 = \{\text{Slope } \overleftrightarrow{PR}\} = \frac{0 - c}{a - b} = -\frac{c}{a - b}$$

We must show that $m_1 m_2 = -1$.

$$m_1 m_2 = -\frac{c^2}{(a + b)(a - b)} = -\frac{c^2}{a^2 - b^2} = \frac{-c^2}{c^2} = -1$$

Therefore, $\overleftrightarrow{OQ} \perp \overleftrightarrow{PR}$. ■

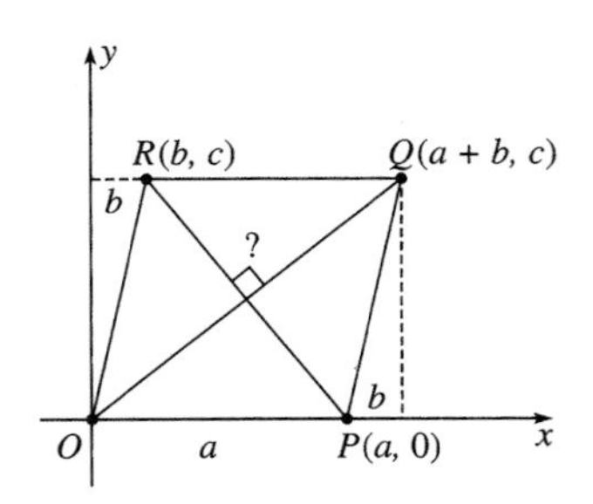

Figure 4.84

VECTORS

Coordinate geometry provides the basis for introducing another tool having tremendous power in the study of geometry, with far-reaching applications in mathematics and science. We define a **vector** $\boldsymbol{v}$ to be a *column of two real numbers,* called its **components,** as in

$$\boldsymbol{v} = \begin{bmatrix} x \\ y \end{bmatrix}$$

also written as $[x, y]$. We concentrate first on the algebraic properties of vectors.

VECTOR, SUM, LINEAR COMBINATION OF VECTORS

We can define a **zero** for vectors, we can **add** two vectors and **multiply** a real number times a vector, and form arbitrary **linear combinations** of vectors, according to the following definitions (let $\boldsymbol{u} = [x, y]$ and $\boldsymbol{v} = [z, w]$):

ZERO VECTOR $$O = \begin{bmatrix} 0 \\ 0 \end{bmatrix}$$

SUM OF TWO VECTORS $$\boldsymbol{u} + \boldsymbol{v} = \begin{bmatrix} x \\ y \end{bmatrix} + \begin{bmatrix} z \\ w \end{bmatrix} = \begin{bmatrix} x + z \\ y + w \end{bmatrix}$$

SCALAR MULTIPLICATION $$a\boldsymbol{u} = a\begin{bmatrix} x \\ y \end{bmatrix} = \begin{bmatrix} ax \\ ay \end{bmatrix}$$

LINEAR COMBINATION $$a\boldsymbol{u} + b\boldsymbol{v} = a\begin{bmatrix} x \\ y \end{bmatrix} + b\begin{bmatrix} z \\ w \end{bmatrix} = \begin{bmatrix} ax + bz \\ ay + bw \end{bmatrix}$$

INNER PRODUCT OF TWO VECTORS, LENGTH OR NORM OF A VECTOR

We can also multiply two vectors, although the kind of product we consider produces a *real number* (a **scalar**) instead of another vector: If $\boldsymbol{u} = [x, y]$ and $\boldsymbol{v} = [z, w]$, then we define the **inner** or **scalar product,** sometimes also called the **dot product,** as follows.

INNER PRODUCT $$\boldsymbol{u} \cdot \boldsymbol{v} = \begin{bmatrix} x \\ y \end{bmatrix} \cdot \begin{bmatrix} z \\ w \end{bmatrix} = xz + yw$$

For convenience, we shall use the notation $\boldsymbol{uv}$ for $\boldsymbol{u} \cdot \boldsymbol{v}$ and $\boldsymbol{u}^2$ for $\mathbf{u} \cdot \mathbf{v}$ when $\mathbf{u} = \mathbf{v}$.

The resulting **commutative** and **distributive** laws are obvious: For arbitrary vectors $\boldsymbol{u}, \boldsymbol{v}, \boldsymbol{w}$:

(3) $$\boldsymbol{uv} = \boldsymbol{vu}$$

and

(4) $$\boldsymbol{u}(\boldsymbol{v} + \boldsymbol{w}) = \boldsymbol{uv} + \boldsymbol{uw}$$

Also, we have the obvious **associative** laws for scalar and dot products, $(a\boldsymbol{u})\boldsymbol{v} = \boldsymbol{u}(a\boldsymbol{v}) = a(\boldsymbol{uv}) \equiv a\boldsymbol{uv}$. It then follows that

(5) $$(a\boldsymbol{u} + b\boldsymbol{v})^2 = (a\boldsymbol{u} + b\boldsymbol{v}) \cdot (a\boldsymbol{u} + b\boldsymbol{v}) = a^2\boldsymbol{u}^2 + 2ab\boldsymbol{uv} + b^2\boldsymbol{v}^2$$

This rule has the same appearance as the ordinary algebraic rule for squaring a binomial, but *do not be misled.* There is a lot more to this rule than appears, and it has a great deal of geometric power built into it, as will soon be evident.

Finally, we introduce the **length** or **norm** of a vector:

LENGTH OF A VECTOR $$\|\boldsymbol{v}\| = \sqrt{\boldsymbol{v}^2}$$

This too, is misleading; it looks like the familiar rule for absolute values of real numbers $|a| = \sqrt{a^2}$. But a closer examination shows that, in reality, there is also hidden information in this equation, for it can be expanded by writing

$$\boldsymbol{v}^2 = \boldsymbol{vv} = \begin{bmatrix} x \\ y \end{bmatrix} \cdot \begin{bmatrix} x \\ y \end{bmatrix} = x^2 + y^2, \qquad \text{or} \qquad \|\boldsymbol{v}\| = \sqrt{x^2 + y^2}$$

Thus, from the Distance Formula **(1)**, we find that $\|\boldsymbol{v}\|$ is just the distance from $O(0, 0)$ to $P(x, y)$. Note the following basic properties, obvious from the definition.

$$\|\boldsymbol{v}\| \geq 0, \text{ with equality only when } \boldsymbol{v} = 0$$

$$\|a\boldsymbol{v}\| = |a| \, \|\boldsymbol{v}\| \text{ for any scalar } a$$

The first inequality implies an important property of the dot product:

$$\boldsymbol{v}^2 > 0 \text{ unless } \boldsymbol{v} = 0.$$

VECTORS AS DIRECTED SEGMENTS; TRIANGLE ADDITION LAW

To actually use vectors in geometry, we must show how they are to be connected to objects in geometry.

> DEFINITION: A **directed line segment** from A to B, denoted $\boldsymbol{AB}$, is the ordinary segment $\overline{AB}$ with a **direction,** defined as the ordered pair (A, B), where point A is called the **initial point** and B the **terminal point** of the directed segment.

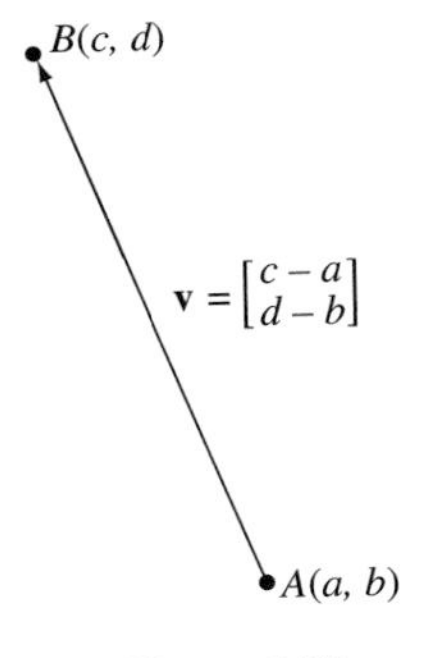

Figure 4.85

(Thus, each segment $\overline{AB}$ as previously defined has two directed segments associated with it, namely $\boldsymbol{AB}$ and $\boldsymbol{BA}$.)

The **vector representation** of the directed line segment $\boldsymbol{AB}$, where $A(a, b)$ and $B(c, d)$ are any two points, is now defined as the vector $\boldsymbol{v}$ whose components are precisely the numbers $c - a$ and $d - b$. (See Figure 4.85.) That is,

> **Vector Representation of Directed Line Segment**
>
> **(6)** $\quad \boldsymbol{v} = \boldsymbol{v}(\boldsymbol{AB}) = \begin{bmatrix} c - a \\ d - b \end{bmatrix} \quad$ where $A = (a, b)$, $B = (c, d)$

In reality, we are defining a mapping or function *from the set of all directed line segments to the set of all vectors.* We could ask if the mapping is one to one, that is, for two distinct directed segments $\boldsymbol{AB}$ and $\boldsymbol{CD}$, is it possible to have

$$\boldsymbol{v}(\boldsymbol{AB}) = \boldsymbol{v}(\boldsymbol{CD})?$$

It follows by definition and a little algebra that this is true iff C and D have the coordinates as indicated in Figure 4.86, which means that the quadrilateral $\diamond ABDC$ is a parallelogram. We omit the proof; although simple, it is in no way a trivial matter—this is the basis for virtually all the applications of vectors to engineering mathematics and physics.

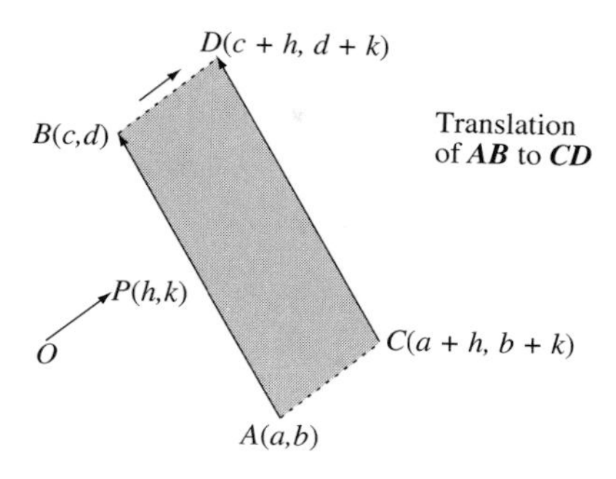

Figure 4.86

Although it is inaccurate to do so, it is customary to *identify* the vector with its geometric representation, as in

$$\boldsymbol{v} = \boldsymbol{AB}$$

No great harm is done as long as we understand that equality here is taken as an *equivalence up to translation.* Under this identification, note that if

$$\boldsymbol{AB} = \boldsymbol{u} \qquad \text{and} \qquad \boldsymbol{BC} = \boldsymbol{v}$$

then

(7)
$$\boldsymbol{AB} + \boldsymbol{BC} = \boldsymbol{u} + \boldsymbol{v} = \boldsymbol{AC}$$

(**PROOF:** Suppose that $A = (a, b), B = (c, d), C = (e, f)$; if $\boldsymbol{AB} = \boldsymbol{u}$ and $\boldsymbol{BC} = \boldsymbol{v}$, then $\boldsymbol{u} = [c - a, d - b]$ and $\boldsymbol{v} = [e - c, f - d]$, and therefore $\boldsymbol{u} + \boldsymbol{v} = [c - a + e - c, d - b + f - d] = [e - a, f - b] = \boldsymbol{AC}$.)

The relation **(7)** is called the **Triangle Law for Addition,** and may also be used to generate an equivalent **Parallelogram Law for Addition,** as indicated in Figure 4.87, which also shows the geometric interpretation for scalar multiplication.

VECTOR ADDITION

SCALAR MULTIPLICATION

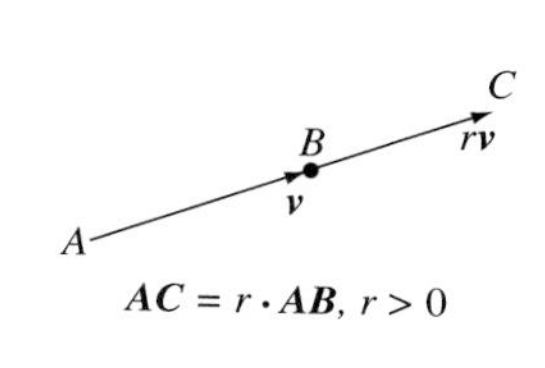

Figure 4.87

We have noted that each vector may be associated with an *infinite number* of directed line segments, parallel to one another and having equal lengths. If we consider only directed segments of the form $\boldsymbol{OP}$, where O is the origin, then there is a *one-to-one correspondence* between vectors and such directed segments. These are called **position vectors,** a misnomer because they are not vectors at all but directed line segments. The association between position vectors and the algebraic vectors they represent is given by

$$\boldsymbol{v} = \begin{bmatrix} x \\ y \end{bmatrix} \leftrightarrow \boldsymbol{OP}, \qquad \text{where} \qquad P = (x, y)$$

LAW OF COSINES

An important question in both geometry and analysis is: What meaning does $\boldsymbol{uv} = 0$ have geometrically? First we note from the Law of Cosines that for any triangle $\triangle ABD$ (Figure 4.87), where we let $\boldsymbol{u} = \boldsymbol{AB}$, $\boldsymbol{v} = \boldsymbol{AD}$, and $\boldsymbol{DB} = \boldsymbol{DA} + \boldsymbol{AB} = -\boldsymbol{v} + \boldsymbol{u}$, with $\theta = m\angle BAD$,

$$BD^2 = AB^2 + AD^2 - 2AB \cdot AD \cdot \cos\theta$$

or

$$\|\boldsymbol{u} - \boldsymbol{v}\|^2 = \|\boldsymbol{u}\|^2 + \|\boldsymbol{v}\|^2 - 2\|\boldsymbol{u}\|\,\|\boldsymbol{v}\|\cos\theta$$

and, from **(5),**

$$\boldsymbol{u}^2 - 2\boldsymbol{uv} + \boldsymbol{v}^2 = \boldsymbol{u}^2 + \boldsymbol{v}^2 - 2\|\boldsymbol{u}\|\|\boldsymbol{v}\|\cos\theta$$

This last expression simplifies to

(8) LAW OF COSINES FOR VECTORS $\quad \boldsymbol{uv} = \|\boldsymbol{u}\|\,\|\boldsymbol{v}\|\cos\theta$

where θ is the measure of the angle between $\boldsymbol{u}$ and $\boldsymbol{v}$ (as represented geometrically). Thus, after setting $\theta = 90$, we deduce the following result.

THEOREM 1: Two nonzero vectors $\boldsymbol{u}$ and $\boldsymbol{v}$ are perpendicular (as directed line segments) iff $\boldsymbol{uv} = 0$. (In this case, the vectors $\boldsymbol{u}$ and $\boldsymbol{v}$ are called **orthogonal.**)

VECTOR EQUATIONS FOR LINES AND CIRCLES

Suppose we regard the vector $\boldsymbol{a} = [a_1, a_2]$ as a **direction** in the plane (actually, it represents the position vector $\boldsymbol{OA}$, where the coordinates of A are the vector components of $\boldsymbol{a}$). Then in Figure 4.88, from the addition law for vectors a line parallel to $\boldsymbol{a}$ and passing through $B(x_0, y_0)$ would have the vector form for some real t, $\boldsymbol{OX} = \boldsymbol{OB} + \boldsymbol{BX} = \boldsymbol{OB} + t\boldsymbol{a}$, or with $\boldsymbol{x}_0 = \boldsymbol{OB}$,

(9) VECTOR EQUATION FOR A LINE $\quad \boldsymbol{x} = t\boldsymbol{a} + \boldsymbol{x}_0$

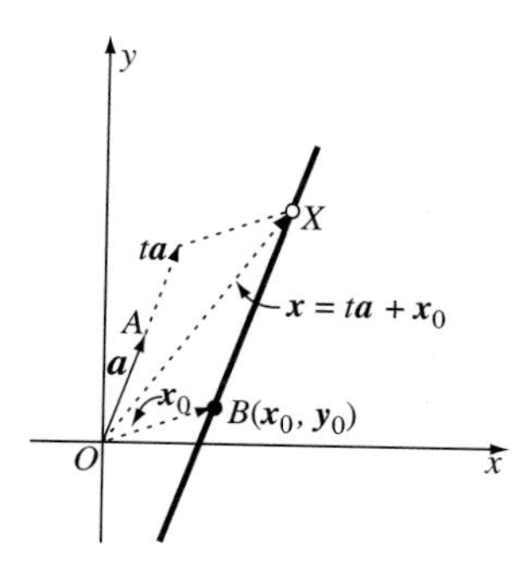

Figure 4.88

In terms of vector components, this is

$$\begin{bmatrix} x \\ y \end{bmatrix} = t\begin{bmatrix} a_1 \\ a_2 \end{bmatrix} + \begin{bmatrix} x_0 \\ y_0 \end{bmatrix} = \begin{bmatrix} ta_1 + x_0 \\ ta_2 + y_0 \end{bmatrix}$$

where $\boldsymbol{x}_0 = [x_0, y_0]$ as before. Thus, by reading off the components, we have found the **parametric (coordinate) equations** for ℓ:

$$\begin{cases} x = ta_1 + x_0 \\ y = ta_2 + y_0 \end{cases} \qquad (t \text{ real})$$

For the vector equation of a circle (Figure 4.89), observe that if $X(x, y)$ is any point on the circle having radius r and centered at $C(h, k)$, then with $\boldsymbol{x}$ as the vector $\boldsymbol{OX}$, and $\boldsymbol{c} = \boldsymbol{OC}$, we have $\boldsymbol{OX} - \boldsymbol{OC} = \boldsymbol{CX}$, *or* $\boldsymbol{x} - \boldsymbol{c} = \boldsymbol{r}$. That is, if $\boldsymbol{x} = [x, y]$ and $\boldsymbol{c} = [h, k]$,

(10) **VECTOR EQUATION FOR A CIRCLE** $\|\boldsymbol{x} - \boldsymbol{c}\| = r$

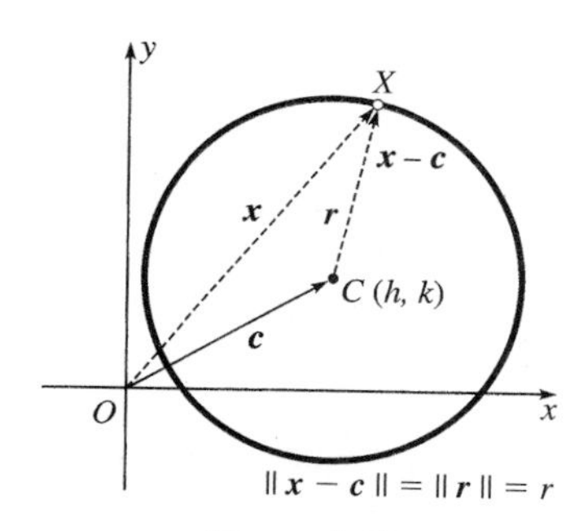

Figure 4.89

This may be converted into coordinate form in one step using the definition of $\|\boldsymbol{v}\|$ and $\boldsymbol{x} - \boldsymbol{c}$:

$$\sqrt{(x - h)^2 + (y - k)^2} = r$$

or

$$(x - h)^2 + (y - k)^2 = r^2$$

ELEMENTARY APPLICATIONS

As an example of what we can do with vectors, we are going to give a vector proof of the Triangle Inequality. This does not sound like much, but you should recall that this was a theorem in Chapter 3 that took practically all the major theorems we had at our disposal to use at the time (in particular, it depended on the Exterior Angle Inequality, which itself involved many key geometric principles, including the SAS Postulate). We think you will be struck with the elegance and efficiency of the vector proof. It definitely demonstrates the power of linear algebra and its usefulness to geometry.

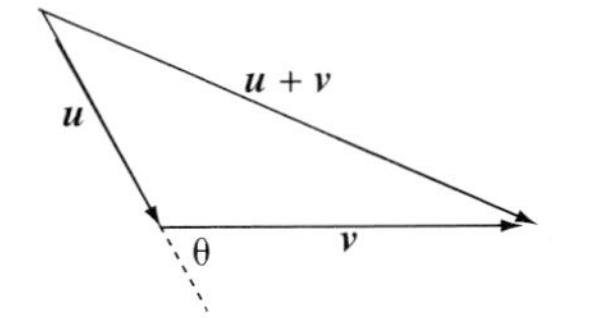

Figure 4.90

THEOREM 2: TRIANGLE INEQUALITY, VECTOR FORM

For any two vectors $\boldsymbol{u}$ and $\boldsymbol{v}$, $\|\boldsymbol{u} + \boldsymbol{v}\| \leq \|\boldsymbol{u}\| + \|\boldsymbol{v}\|$, with equality only when $\boldsymbol{u}$ and $\boldsymbol{v}$ represent two collinear directed line segments (Figure 4.90).

PROOF

Since all quantities are nonnegative, let's square the desired inequality.

$$\textbf{(11)} \qquad \|\boldsymbol{u} + \boldsymbol{v}\|^2 \leq (\|\boldsymbol{u}\| + \|\boldsymbol{v}\|)^2 = \|\boldsymbol{u}\|^2 + 2\|\boldsymbol{u}\|\,\|\boldsymbol{v}\| + \|\boldsymbol{v}\|^2$$

We also have

$$\|\boldsymbol{u} + \boldsymbol{v}\|^2 = (\boldsymbol{u} + \boldsymbol{v})^2 = (\boldsymbol{u} + \boldsymbol{v}) \cdot (\boldsymbol{u} + \boldsymbol{v})$$
$$= \boldsymbol{u}^2 + \boldsymbol{u} \cdot \boldsymbol{v} + \boldsymbol{v} \cdot \boldsymbol{u} + \boldsymbol{v}^2 \leq \|\boldsymbol{u}\|^2 + 2\boldsymbol{u}\boldsymbol{v} + \|\boldsymbol{v}\|^2$$

By substitution of this expression into **(11)** and making the obvious cancellation, we find that the desired Triangle Inequality is equivalent to

$$\textbf{(12)} \qquad \mathbf{uv} \leq \|\boldsymbol{u}\|\,\|\boldsymbol{v}\|$$

But using the vector Law of Cosines **(8)**, we have, since $\cos\theta \leq 1$,

$$\textbf{(13)} \qquad \mathbf{uv} = \|\boldsymbol{u}\|\,\|\boldsymbol{v}\| \cos\theta \leq \|\boldsymbol{u}\|\,\|\boldsymbol{v}\|$$

Thus we have proven the desired inequality. As for equality, this holds only if it holds in both **(12)** and **(13)**, which can happen only when $\cos\theta = 1$ or $\theta = 0$, or when $\boldsymbol{u}$ and $\boldsymbol{v}$ are collinear.

NOTE: The inequality **(12)** in the previous proof is the famous **Cauchy–Schwarz Inequality** of vector analysis.

EXAMPLE 2 Use vectors to prove that the medians of a triangle are concurrent, and that the point of concurrency is located on each median at a point that is two-thirds the distance from the vertex to the midpoint of the opposite side.

SOLUTION

It suffices to prove that any two medians, say $\overline{AL}$ and $\overline{BM}$, meet at G such that $AG = \frac{2}{3}AL$ and $BG = \frac{2}{3}BM$ (Figure 4.91). We will prove this by *defining* G to be the point on $\overline{AL}$ such that $AG = \frac{2}{3}AL$, and G' that point on $\overline{BM}$ such that $BG' = \frac{2}{3}BM$, then showing that $G = G'$. Using vectors, we have

$$\boldsymbol{AL} = \boldsymbol{AB} + \boldsymbol{BL} = \boldsymbol{AB} + \tfrac{1}{2}\boldsymbol{BC} \quad \rightarrow \quad \boldsymbol{AG} = \tfrac{2}{3}\boldsymbol{AL} = \tfrac{2}{3}\boldsymbol{AB} + \tfrac{1}{3}\boldsymbol{BC}$$

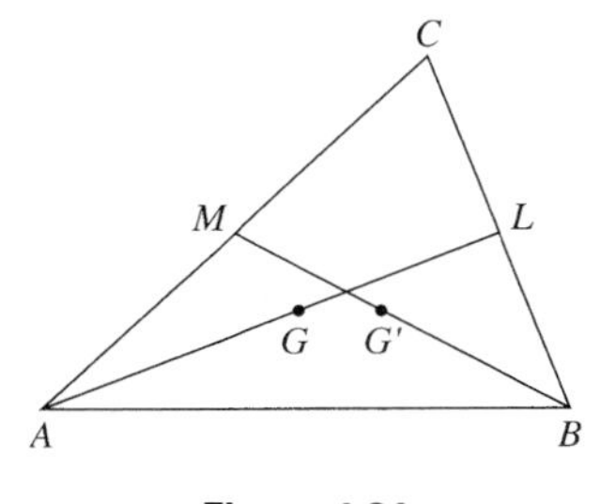

Figure 4.91

Similarly,

$$\boldsymbol{BM} = \boldsymbol{BA} + \boldsymbol{AM} = \boldsymbol{BA} + \tfrac{1}{2}\boldsymbol{AC} \quad \rightarrow \quad \boldsymbol{BG'} = \tfrac{2}{3}\boldsymbol{BM} = \tfrac{2}{3}\boldsymbol{BA} + \tfrac{1}{3}\boldsymbol{AC}$$

Then, by substitution,

$$\begin{aligned} \boldsymbol{AG'} &= \boldsymbol{AB} + \boldsymbol{BG'} = \boldsymbol{AB} + \tfrac{2}{3}\boldsymbol{BA} + \tfrac{1}{3}\boldsymbol{AC} \\ &= \boldsymbol{AB} - \tfrac{2}{3}\boldsymbol{AB} + \tfrac{1}{3}(\boldsymbol{AB} + \boldsymbol{BC}) \\ &= \tfrac{2}{3}\boldsymbol{AB} + \tfrac{1}{3}\boldsymbol{BC} = \boldsymbol{AG} \end{aligned}$$

That is,

$$\boldsymbol{AG'} = \boldsymbol{AG} \qquad \text{or} \qquad G' = G.$$

as desired. Thus, the "two-thirds point," G on any two medians is the same, and we are finished.

EXAMPLE 3 Give a vector proof that the diagonals of a parallelogram are perpendicular iff it is a rhombus.

SOLUTION

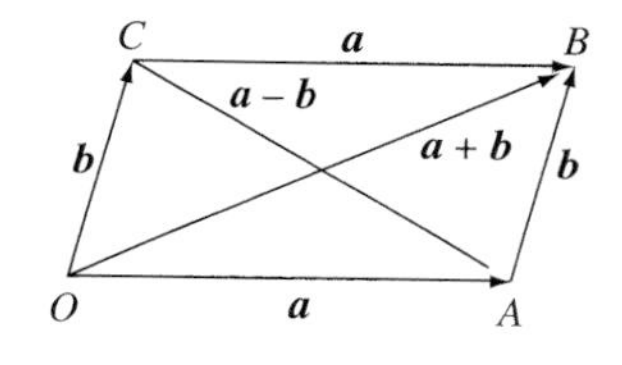

Figure 4.92

The sides and diagonals may be labelled as shown in Figure 4.92. Now $\boldsymbol{OB} \perp \boldsymbol{AC}$ iff $\boldsymbol{OB} \cdot \boldsymbol{AC} = 0$, or

$$(\boldsymbol{a} + \boldsymbol{b}) \cdot (\boldsymbol{a} - \boldsymbol{b}) = 0$$

or

$$\boldsymbol{a}^2 - \boldsymbol{b}^2 = 0$$

That is, iff

$$\boldsymbol{a}^2 = \boldsymbol{b}^2 \qquad \text{or} \qquad \|\boldsymbol{a}\|^2 = \|\boldsymbol{b}\|^2 \leftrightarrow \|\boldsymbol{a}\| = \|\boldsymbol{b}\| \leftrightarrow OA = AB,$$

precisely when $\diamond OABC$ is a rhombus. ■

EXAMPLE 4 Use vectors to prove the midpoint formula for the midpoint $M(x_3, y_3)$ of the segment joining $P(x_1, y_1)$ and $Q(x_2, y_2)$:

(14)
$$x_3 = \frac{x_1 + x_2}{2}, \qquad y_3 = \frac{y_1 + y_2}{2}$$

SOLUTION

With O = origin,

$$\begin{aligned} \begin{bmatrix} x_3 \\ y_3 \end{bmatrix} &= \boldsymbol{OM} = \boldsymbol{OP} + \boldsymbol{PM} = \boldsymbol{OP} + \tfrac{1}{2}\boldsymbol{PQ} = \begin{bmatrix} x_1 \\ y_1 \end{bmatrix} + \frac{1}{2}\begin{bmatrix} x_2 - x_1 \\ y_2 - y_1 \end{bmatrix} \\ &= \begin{bmatrix} \frac{1}{2}(x_1 + x_2) \\ \frac{1}{2}(y_1 + y_2) \end{bmatrix} \end{aligned}$$

■

PROBLEMS (4.7)

GROUP A

All problems in this section are to be worked in the coordinate plane. In most situations, it is best to choose a coordinate system so that a given triangle $\triangle ABC$ has vertices $A(a, 0)$, $B(b, 0)$, and $C(0, c)$, where $b > 0$, $b > a$, and $c > 0$ ($a < 0$ unless $m\angle A \geq 90$). An arbitrary parallelogram can be assumed to have vertices: $A(a, 0)$, $B(b, 0)$, $C(b - a, c)$, and $D(0, c)$.

1. If $A(-2, 5)$, $B(1, 7)$, $C(4, -4)$ are the vertices of a triangle,

(a) determine the numerical components of $\boldsymbol{u} = \boldsymbol{AB}$, $\boldsymbol{v} = \boldsymbol{AC}$, and $\boldsymbol{w} = \boldsymbol{BC}$. Verify that $\boldsymbol{u} + \boldsymbol{w} = \boldsymbol{v}$.

(b) Show that $\triangle ABC$ is a right triangle.

(c) Find cos B using the Vector Law of Cosines.

2. Give a coordinate proof that the diagonals of a parallelogram bisect each other.

3. Give a coordinate proof that a parallelogram is a rectangle iff its diagonals are congruent.

4. Give a coordinate proof that the segment joining the midpoints of two sides of a triangle is parallel to the third side and has length equal to one-half that of the third side.

5. Give a coordinate proof that the midpoints of the sides of any quadrilateral are the vertices of a parallelogram. Note that three vertices, A, B, and D can be taken as above, but C must be taken as a general point, $C(d, e)$.

6. Give a vector proof that the midpoints of the sides of a rhombus are the vertices of a rectangle.

7. State and prove the vector form of the Pythagorean Theorem. Is the converse clear from your proof? (***Hint:*** Use $\boldsymbol{u} + \boldsymbol{v}$ for the hypotenuse and $\boldsymbol{u}$ and $\boldsymbol{v}$ for the two legs. In the proof, use $(\boldsymbol{u} + \boldsymbol{v})^2 = (\boldsymbol{u} + \boldsymbol{v}) \cdot (\boldsymbol{u} + \boldsymbol{v})$.

GROUP B

8. **How to graph functions using *Sketchpad*** The standard functions (algebraic, trigonometric, and logarithmic) can be graphed using the tools of *Sketchpad*. In addition, an extensive amount of interplay can be created between geometric problems and graphs. For a standard example, follow these steps:

[1] Choose Create Axes under GRAPH, and locate a point C on the x-axis.

[2] With C selected, choose Coordinates under MEASURE. The coordinates of point C will appear as $C(x_C, 0)$ where x_C is some real number.

[3] Choose Calculate under MEASURE, and with $C(x_C, 0)$ selected, choose Values in the Sketchpad Calculator. Choose Point C, then x. The number x_C will appear in the Calculator display. Check OK to display x_C on the computer screen.

[4] Calculate ln x_C on the Sketchpad Calculator, and place this value on the screen. (You will find the function "ln" in the list of standard functions in the *Sketchpad* calculator. This function could be replaced by any other in the list, or your own calculation if it leads to a specific value.)

[5] Select x_C, then ln x_C with the shift key depressed. Now choose Plot As (x, y) in the GRAPH menu. The point $D(x_C, \ln x_C)$ will be plotted in the coordinate plane. To obtain a curve, either trace the point and animate, with C on the x-axis, or choose C, then D, and Locus under CONSTRUCT.

NOTE: To obtain the graphs for sine and cosine, you need to convert x_C from degrees to radians by calculating, say, $\sin(180 \cdot x_C/\pi)$ instead of just $\sin x_C$. Also, you can adjust the units of measure on the axes by dragging point B toward or away from the origin, A. Most graphs need minor adjustments such as these to obtain a good display.

9. Vector Problem: "Blowin' In the Wind" A small aircraft with a cruising speed of 145 mph proceeds to a destination 425 miles due north, taking $3\frac{1}{8}$ hours. On the return trip, due south, with the same prevailing wind conditions, the trip took only $2\frac{5}{6}$ hours. In what direction was the wind blowing, and with what velocity? (***Hint:*** Let $\mathbf{w} = [x, y]$ be the wind vector and $\mathbf{v} = [r, s]$ the vector of the aircraft *heading,* not *bearing*. Then by hypothesis, $r^2 + s^2 = 145^2$. Continue.)

10. Using the coordinates for the vertices of $\triangle ABC$ suggested at the head of this problem section, show the following.

(a) If H is the point having coordinates $(0, -ab/c)$, then $\overline{AH} \perp \overline{BC}$, $\overline{BH} \perp \overline{AC}$, and $\overline{CH} \perp \overline{AB}$. What theorem in geometry does this prove?

(b) If D, E, and F are the feet of the altitudes from A, B, and C on $\overline{BC}$, $\overline{AC}$, and $\overline{AB}$, respectively (assuming that $\triangle ABC$ is acute-angled), show that $\overleftrightarrow{FD}$ and $\overleftrightarrow{FE}$ have respective equations

$$y = \frac{\pm(bc - ac)x}{(ab + c^2)}$$

What does this prove about the sides of $\angle EFD$ in relation to altitude $\overline{CF}$?

(c) What is the ultimate consequence regarding H and the orthic triangle $\triangle DEF$ for acute-angled triangles $\triangle ABC$?

11. Using a Hyperbola to Trisect An Angle The following construction was discovered by Pappus of Alexandria about 300 C.E. Justify and explain each of the following steps in the construction (as shown in the figure):

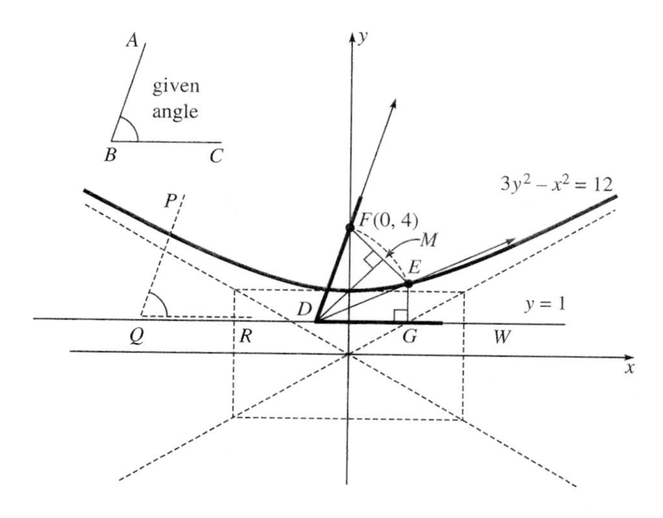

(1) Let $\angle ABC$ be a given angle. In any xy-coordinate system, consider the graph of the hyperbola $3y^2 - x^2 = 12$, and the line $y = 1$, or $\overleftrightarrow{QW}$.

(2) Construct a line through $F(0, 4)$ (the focus of the hyperbola), cutting the line $\overleftrightarrow{QW}$ at some point D such that $\angle FDW \cong \angle ABC$. (This may be accomplished by first constructing $\angle PQR \cong \angle ABC$ on $\overleftrightarrow{QW}$, then constructing $\overleftrightarrow{FD} \parallel \overleftrightarrow{PQ}$.)

(3) With D as center and DF as radius, draw a circle, cutting the hyperbola at $E(x, y)$.

(4) Ray $\overrightarrow{DE}$ is a trisector of $\angle FDW \cong \angle ABC$.

(***Hint:*** Drop perpendicular $\overline{EG}$ to line $\overleftrightarrow{QW}$; use coordinate geometry to show that $FE = 2EG$, then consider $DM \perp FE$.)

12. Give a coordinate proof of the Secant Theorem proven in Section 3.8. [***Hint:*** With no loss of generality, we may assume that the circle has equation $x^2 + y^2 = r^2$ and the interior point A given in the theorem has coordinates $(a, 0)$, where $0 < a < r$. Let ℓ have equation $x = a$ (if ℓ is vertical) or $y = mx + b$ (otherwise), passing through A [thus, $0 = ma + b$ or $b = -ma$ and $y = m(x - a)$]. Now solve the system of two equations (for the circle and the line) and prove that there are always two points of intersection.]

13. The Parallelogram Law For Vectors A famous law of vectors, known as the **Parallelogram Law,** is useful in topology and real analysis. It is the identity for any two vectors $\boldsymbol{u}$ and $\boldsymbol{v}$,

$$\|\boldsymbol{u} + \boldsymbol{v}\|^2 + \|\boldsymbol{u} - \boldsymbol{v}\|^2 = 2\|\boldsymbol{u}\|^2 + 2\|\boldsymbol{v}\|^2$$

Interpret this geometrically, and prove in two ways:

(a) by vectors and inner products

(b) by geometry.

[***Hint:*** For the vector proof, simply expand $\|\boldsymbol{u} \pm \boldsymbol{v}\|^2 \equiv (\boldsymbol{u} \pm \boldsymbol{v})^2$ using properties of the inner product, then sum. For the geometric proof, use the formula for the median of a triangle. See Equation **(12)** following Example 6 in Section 4.3.]

14. The Euler Line Let the vertices of $\angle ABC$ have the coordinates suggested at the beginning of this problem section.

(a) Show that if $d = ab/c$, the orthocenter H, circumcenter O, centroid G, and Nine-Point Center U have the coordinates $H(0, -d)$ (see Problem 10), $O(\frac{1}{2}a + \frac{1}{2}b, \frac{1}{2}c + \frac{1}{2}d)$, $G(\frac{1}{3}a + \frac{1}{3}b, \frac{1}{3}c)$ (see Problem 15), and $U(\frac{1}{4}a + \frac{1}{4}b, \frac{1}{4}c - \frac{1}{4}d)$. (***Hint:*** U is the midpoint of $\overline{XL}$, where X and L are the midpoints of $\overline{AH}$ and $\overline{BC}$, respectively.)

(b) Show that the **Euler Line** $\overleftrightarrow{HG}$ has equation $y = -d + (3d + c)x/(a + b)$, and verify that H, O, G, and U all lie on the Euler line of $\triangle ABC$.

15. Prove by vector methods that the centroid $G(x, y)$ of a triangle whose vertices are $A(a_1, a_2)$, $B(b_1, b_2)$ and $C(c_1, c_2)$ has coordinates

$$x = \frac{1}{3}(a_1 + b_1 + c_1), \qquad y = \frac{1}{3}a_2 + b_2 + c_2)$$

(***Hint:*** Use the fact that if L is the midpoint of $\overline{BC}$, G occupies the point of $\overline{AL}$ two-thirds the distance from A to L, as shown in Example 2.)

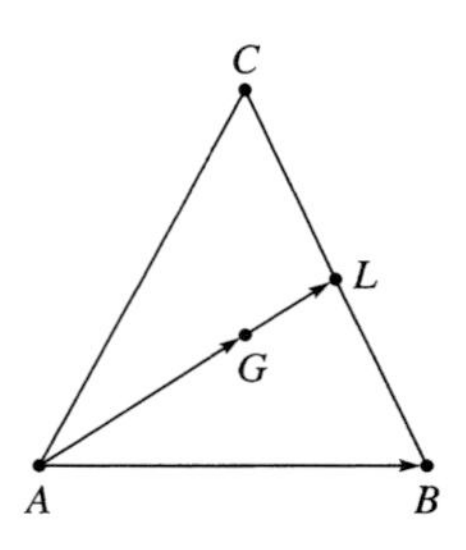

GROUP C

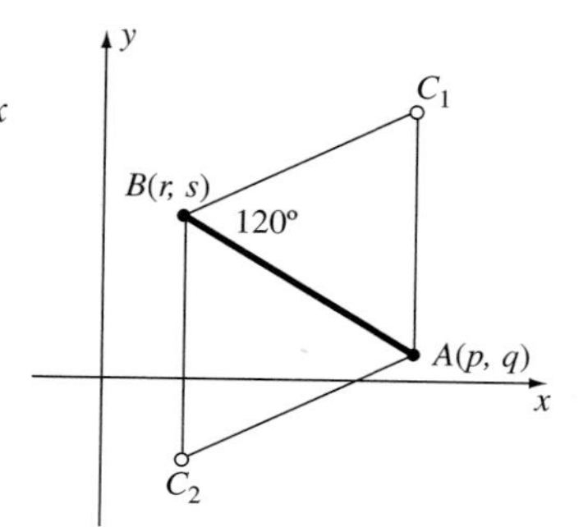

16. Show that if the end points of a segment are $A(p, q)$ and $B(r, s)$, the coordinates x and y of the vertices C_1 and C_2 of the two equilateral triangles having $\overline{AB}$ as base are given by

$$2x = p + r \pm \sqrt{3}\,(q - s)$$
$$2y = q + s \pm \sqrt{3}\,(r - p)$$

where the + or − signs are taken together in the two equations (***Hint:*** Use the parametric equations for the perpendicular bisector of $\overline{AB}$.)

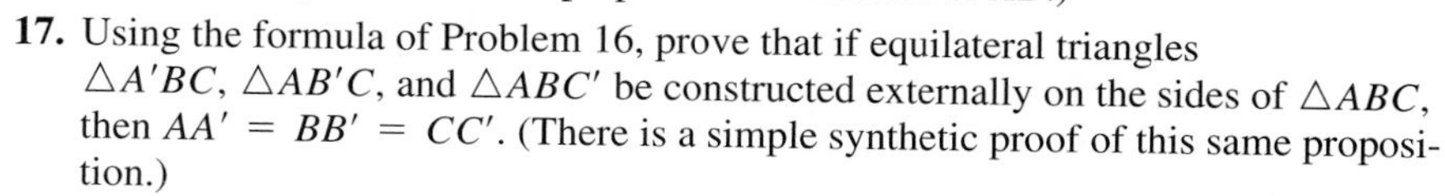

17. Using the formula of Problem 16, prove that if equilateral triangles $\triangle A'BC$, $\triangle AB'C$, and $\triangle ABC'$ be constructed externally on the sides of $\triangle ABC$, then $AA' = BB' = CC'$. (There is a simple synthetic proof of this same proposition.)

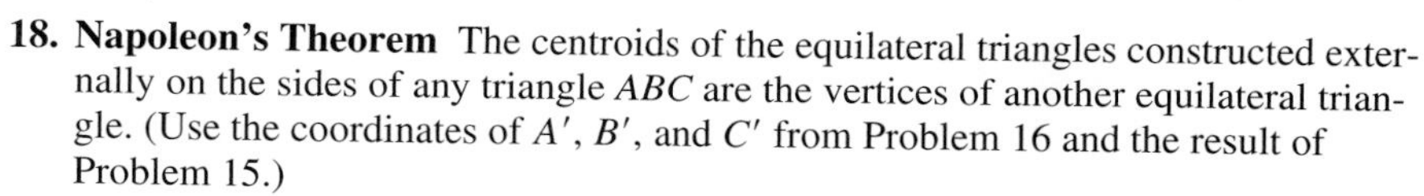

18. Napoleon's Theorem The centroids of the equilateral triangles constructed externally on the sides of any triangle ABC are the vertices of another equilateral triangle. (Use the coordinates of A', B', and C' from Problem 16 and the result of Problem 15.)

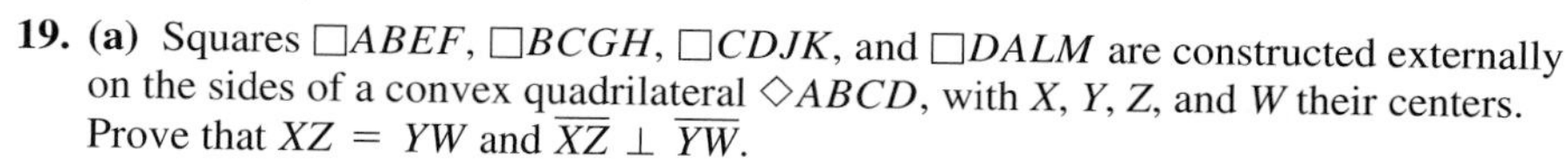

19. **(a)** Squares $\square ABEF$, $\square BCGH$, $\square CDJK$, and $\square DALM$ are constructed externally on the sides of a convex quadrilateral $\diamond ABCD$, with X, Y, Z, and W their centers. Prove that $XZ = YW$ and $\overline{XZ} \perp \overline{YW}$.

(b) Prove that if $\diamond ABCD$ is a parallelogram, $\diamond XYZW$ is a square.

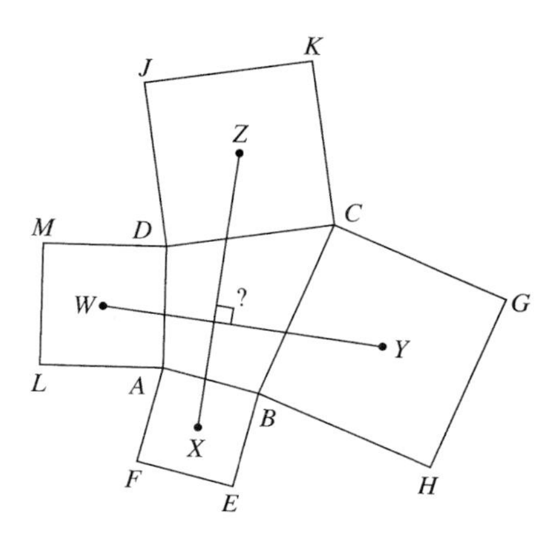

*4.8 Some Modern Geometry of the Triangle

An area known as *modern geometry,* which has developed over approximately the past 300 years, includes material that extends Euclidean geometry far beyond the self-imposed bounds of the ancient Greek geometers. This literature is extensive, involving hundreds of articles on scores of different topics, such as the Nine-Point Circle, Euler Line, Gergonne Point, and Simson Line, to name a few.

Since many of these results involve collinear points or concurrent lines obtained in a variety of ways, we discuss here two major tools for such results, the theorems of Ceva and Menelaus. Dual to each other, these are two of the most intriguing theorems outside the mainstream of the body of material surrounding Euclid's *Elements.* Curiously, their discoveries were separated by more than a thousand years. The theorem bearing the name Ceva (from which the term "cevian" originates) was discovered by an Italian mathematician Giovanni Ceva (1647–1736), who noted that it complemented the theorem proved approximately 1600 years earlier by the Greek Astronomer Menelaus of Alexandria (around A.D. 100).

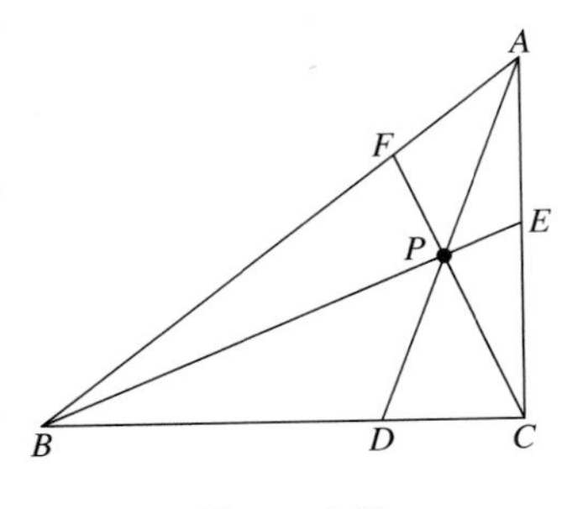

Figure 4.93

Ceva's Theorem involves a simple criterion for three cevians of an arbitrary triangle to be concurrent, as shown in Figure 4.93. Certainly if $P = G$ (the centroid)—when the cevians are the medians of $\triangle ABC$—we have

$$\frac{BD}{DC} = 1, \qquad \frac{CE}{EA} = 1, \qquad \text{and} \qquad \frac{AF}{FB} = 1$$

But suppose we are not dealing with the medians. For example, we might have

$$\frac{BD}{DC} = \frac{1}{3} \qquad \text{and} \qquad \frac{CE}{EA} = \frac{2}{3}$$

What must the value of AF/FB then be in order that the cevians all pass through a common point P? One could experiment with *Sketchpad* to determine the answer. In fact, such an experiment is designed for you appearing at the end of this section. You may want to try it and come up with your own discovery before you continue with our discussion.

It seems inevitable that in some way, the three ratios BD/DC, CE/EA, and AF/FB are going to be involved. Thus, for convenience, we define the **linearity number** of the points D, E, and F with respect to $\triangle ABC$ as the *product* of these three ratios, denoted by the symbol $\begin{bmatrix} ABC \\ DEF \end{bmatrix}$. Points D, E, and F always lie on the extended sides of $\triangle ABC$ opposite A, B, and C, respectively, and are distinct from A, B, and C. Thus, we take as the definition for this linearity number,

$$\textbf{(1)} \qquad \begin{bmatrix} ABC \\ DEF \end{bmatrix} = \frac{AF}{FB} \cdot \frac{BD}{DC} \cdot \frac{CE}{EA}$$

LINEARITY NUMBER PROFILE

A B C

D E F

WORKING WITH THE LINEARITY NUMBER

Note the mnemonic device in the diagram to the right of the definition of the linearity number to aid in the construction of the desired ratios. Thus, one can change the labels and still obtain the correct ratios, with a little practice. For example, if $\overline{PR}$, $\overline{QT}$, and $\overline{WS}$ are cevians in $\triangle PQW$, the linearity number would be

$$\begin{bmatrix} PQW \\ RTS \end{bmatrix} = \frac{PS}{SQ} \cdot \frac{QR}{RW} \cdot \frac{WT}{TP}$$

EXAMPLE 1 A triangle has sides $MN = 9$, $NO = 5$, and $OM = 8$, with cevians $\overline{MU}$, $\overline{NV}$, and $\overline{OW}$ drawn to the respective sides to form segments of various lengths, as indicated by Figure 4.94. Find the linearity number $\begin{bmatrix} MNO \\ UVW \end{bmatrix}$. (Note that the cevians are *not* concurrent.)

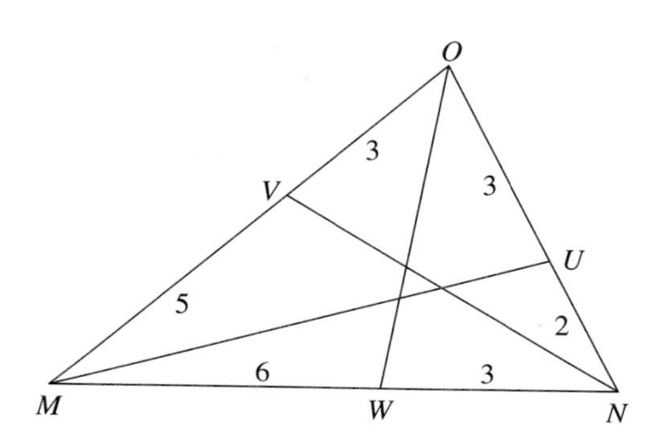

Figure 4.94

SOLUTION

By definition, we have

$$\begin{bmatrix} MNO \\ UVW \end{bmatrix} = \frac{MW}{WN} \cdot \frac{NU}{UO} \cdot \frac{OV}{VM} = \frac{6}{3} \cdot \frac{2}{3} \cdot \frac{3}{5} = \frac{4}{5} \quad \blacksquare$$

EXAMPLE 2 Show that the linearity number satisfies the properties

(a) $\begin{bmatrix} ACB \\ DFE \end{bmatrix} = \begin{bmatrix} ABC \\ DEF \end{bmatrix}^{-1}$

(b) $\begin{bmatrix} ABC \\ DEF \end{bmatrix} = \begin{bmatrix} ABC \\ XEF \end{bmatrix}$ iff $F = X$.

SOLUTION

(a) By definition,

$$\begin{bmatrix} ACB \\ DFE \end{bmatrix} = \frac{AE}{EC} \cdot \frac{CD}{DB} \cdot \frac{BF}{FA}$$

which is the reciprocal (inverse) of $\begin{bmatrix} ABC \\ DEF \end{bmatrix}$. (Note from this result that the *odd* permutations of the columns in the symbol for the linearity number will transform it into the reciprocal of the original number, but the *even* permutations leave it unchanged.)

(b) By hypothesis, we have

$$\frac{AF}{FB} \cdot \frac{BD}{DC} \cdot \frac{CE}{EA} = \frac{AF}{FB} \cdot \frac{BX}{XC} \cdot \frac{CE}{EA}$$

All ratios cancel except

$$\frac{BD}{DC} = \frac{BX}{XC}$$

involving two points D and X on line $\overleftrightarrow{BC}$, Adding 1 to both sides of the above equation we obtain

$$1 + \frac{BD}{DC} = 1 + \frac{BX}{XC} \qquad \text{or} \qquad \frac{DC + BD}{DC} = \frac{XC + BX}{XC}$$

and, assuming B–D–C and B–X–C, this gives us

$$\frac{BC}{DC} = \frac{BC}{XC} \qquad \text{or} \qquad DC = XC$$

Since such a point X is unique on $\overline{BC}$, we have $D = X$. (The converse is trivial since $D = X$ obviously implies that the above linearity numbers are the same.) ■

Notice that we had to hedge a bit in the preceding argument. We used a fact that was not given, that both D and X belong to segment $\overline{BC}$. Such arguments can be greatly simplified if we agree to work with *directed distance* on each specific line involved. In fact, the theorem of Menelaus requires the use of directed distance for its correct statement and proof. Although the concept of directed distance has been introduced in previous problems, it is now time to introduce this concept formally.

DEFINITION OF DIRECTED DISTANCE: For any line ℓ, the Ruler Postulate guarantees a coordinate system for the points of ℓ. Let $P[x_1]$ and $Q[x_2]$ be any two points of ℓ, with their coordinates. The **directed distance** from $P[x_1]$ to $Q[x_2]$ is the number

(2)
$$PQ = x_2 - x_1$$

(Note the reversal of the coordinates x_1 and x_2 in this distance formula.) Among several interesting and useful phenomena, we have the following basic rules, for arbitrary points P, Q, R, and X on ℓ:

(a) $PQ = -QP$ (or $PQ + QP = 0$)
(b) $PQ = PR + RQ$ (or $PQ + QR + RP = 0$)
(c) $PX = PQ \rightarrow X = Q$ $(X \in \ell)$
(d) $PQ/QR > 0$ iff P–Q–R holds.

It is important that you take note of the condition **(d)** above:

$$\frac{PQ}{QR} > 0 \qquad \text{iff} \qquad P\text{–}Q\text{–}R \text{ holds}$$

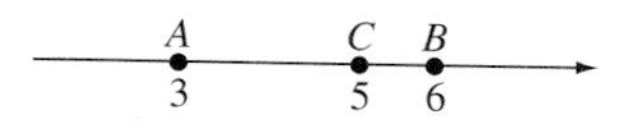

Figure 4.95

Suppose we had three points with coordinates $A[3]$, $B[6]$, and $C[5]$ (Figure 4.95). By the Ruler Postulate we would have $A–C–B$ instead of $A–B–C$, and in this case $AB = 6 - 3 = 3$, $BC = 5 - 6 = -1$. Hence

$$\frac{AB}{BC} = \frac{3}{-1} = -3$$

and the ratio is negative, as it should be. Also, note that property **(c)** now takes care of the error in Example 2 we mentioned earlier: it is not necessary to worry about order of points on a line when dealing with directed distance. The proof of property **(c)** may be had by first writing $P[a]$, $X[x]$, and $Q[b]$ for the three points involved, and then we have

$$PX = PQ \quad \rightarrow \quad x - a = b - a \qquad \therefore x = b \text{ or } X = Q.$$

THE THEOREMS OF CEVA AND MENELAUS

We now turn our attention to the actual statements and proofs of the two theorems mentioned earlier.

THEOREM 1: CEVA'S THEOREM

The cevians $\overline{AD}$, $\overline{BE}$, $\overline{CF}$ of $\triangle ABC$ are concurrent iff $\begin{bmatrix} ABC \\ DEF \end{bmatrix} = 1$.

PROOF

(1) First observe that if P is a point of concurrency of the cevians (Figure 4.96) and all three points D, E, and F are on the sides of the triangle, then the linearity number is positive.

(2) If $B–C–D$ (other cases similar), then looking at the three possible cases $A–P–D$, $P–A–D$, and $A–D–P$, we find that *exactly one* other cevian falls outside the triangle: For example, if $A–P–D$, then ray $\overrightarrow{CP}$ is interior to $\angle ACD$, which puts $F \in \overrightarrow{CP}$ on the same side of $\overleftrightarrow{AC}$ as D, hence on the opposite side of A as B, or $B–A–F$, and further, ray $\overrightarrow{BP}$ falls interior to $\angle ABC$, so $A–E–C$ follows.

(3) Thus, by Property **(d),** *if there is a point of concurrency* the linearity number is *positive.* (The next part of the proof is to show it equals 1.)

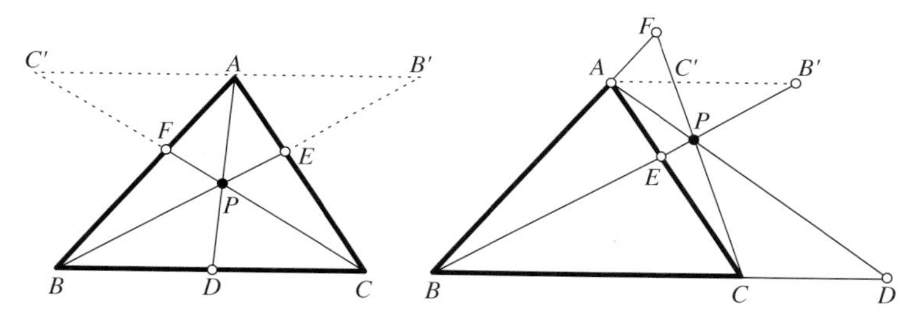

Figure 4.96

(4) Construct line $\overleftrightarrow{B'C'} \parallel \overleftrightarrow{BC}$ passing through A. By central projection through P (and using similar triangles), we obtain, in magnitude only,

$$\frac{BD}{B'A} = \frac{DP}{PA}, \qquad \frac{DP}{PA} = \frac{DC}{AC'} \quad \text{or} \quad \frac{BD}{DC} = \frac{B'A}{AC'}$$

(5) By similar triangles,

$$\frac{CE}{EA} = \frac{BC}{B'A} \quad \text{and} \quad \frac{AF}{FB} = \frac{AC'}{BC}$$

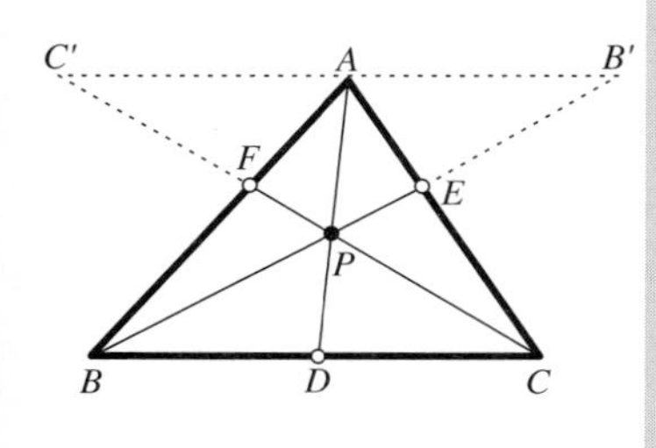

(6) Hence

$$\left|\frac{AF}{FB} \cdot \frac{BD}{DC} \cdot \frac{CE}{EA}\right| = \left|\frac{AC'}{BC} \cdot \frac{B'A}{AC'} \cdot \frac{BC}{B'A}\right| = 1$$

and since the linearity number is positive, it must equal 1.

(7) Conversely, suppose the linearity number is 1. If the first two cevian lines meet at P, let X be the point where the third cervian through P meets the remaining side of $\triangle ABC$, say $\overleftrightarrow{AP} \cap \overleftrightarrow{BC} = X$. Then by the first case, and by hypothesis,

$$\begin{bmatrix} ABC \\ XEF \end{bmatrix} = 1 = \begin{bmatrix} ABC \\ DEF \end{bmatrix}$$

By the properties of linearity numbers, $X = D$ and $\overleftrightarrow{AD}$ coincides with $\overleftrightarrow{AX}$. Therefore, the given cevians are concurrent.

THEOREM 2: MENELAUS' THEOREM

If points D, E, and F lie on the sides of $\triangle ABC$ opposite A, B, and C, respectively, then D, E, and F are collinear iff

$$\begin{bmatrix} ABC \\ DEF \end{bmatrix} = -1$$

PROOF (See Figure 4.97.)

(1) First assume that D, E, and F are collinear, lying on line ℓ. Pasch's postulate implies that if one of the points D, E, and F lies on a side of $\triangle ABC$, then a second point lies on another side and the remaining point lies exterior to the third side, or else all three points lie exterior

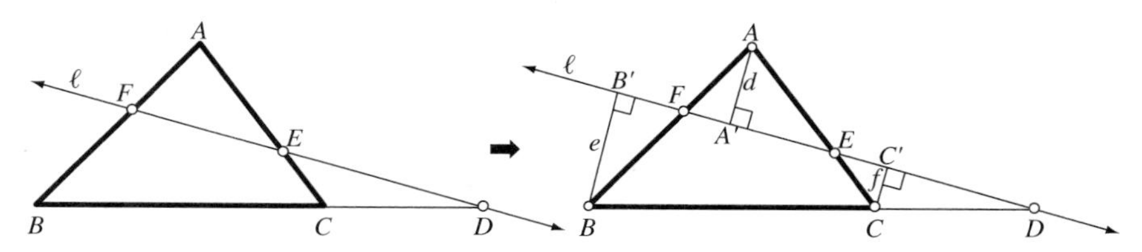

Figure 4.97

to the sides of the triangle. Since the ratio AF/FB is positive iff $A\text{–}F\text{–}B$, and similarly for the other ratios, this proves that

$$\begin{bmatrix} ABC \\ DEF \end{bmatrix} < 0$$

(2) For the algebraic part of the proof, let A', B', and C' be the feet of the perpendiulars from A, B, and C on ℓ. Then from similar triangles (in magnitude only):

$$\frac{AF}{FB} = \frac{d}{e}, \qquad \frac{BD}{DC} = \frac{e}{f}, \qquad \frac{CE}{EA} = \frac{f}{d}$$

Hence, since the linearity number is negative, we have

$$\begin{bmatrix} ABC \\ DEF \end{bmatrix} = \frac{AF}{FB} \cdot \frac{BD}{DC} \cdot \frac{CE}{EA} = -\left(\frac{d}{e} \cdot \frac{e}{f} \cdot \frac{f}{d}\right) = -1$$

(3) Conversely, suppose the linearity number is -1. We prove first that $\overleftrightarrow{DE}$ meets $\overleftrightarrow{AB}$.

(4) Suppose that $\overleftrightarrow{DE} \parallel \overleftrightarrow{AB}$ (Figure 4.98). Since $B\text{–}D\text{–}C$ iff $A\text{–}E\text{–}C$, BD/DC is positive iff CE/EA is positive, and hence, these two ratios have like signs. Therefore, by the Side Splitting Theorem,

$$\frac{BD}{DC} \cdot \frac{CE}{EA} = 1$$

The hypothesis then implies that

$$\frac{AF}{FB} = -1$$

or $AF = -FB = BF$. By Property **(c)** of directed distance, $A = B \quad \rightarrow\leftarrow$

(5) Therefore, $\overleftrightarrow{DE}$ meets $\overleftrightarrow{AB}$ at some point X, and by hypothesis and the first part of the proof,

$$\begin{bmatrix} ABC \\ DEX \end{bmatrix} = -1 = \begin{bmatrix} ABC \\ DEF \end{bmatrix}$$

By permutations and properties of the linearity number, $X = F$, so that D, E, and F are collinear.

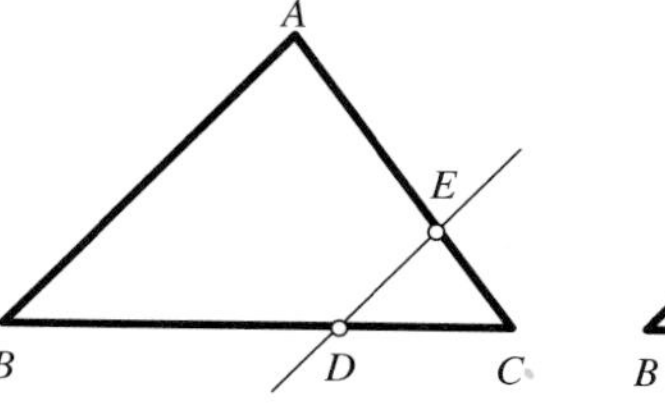

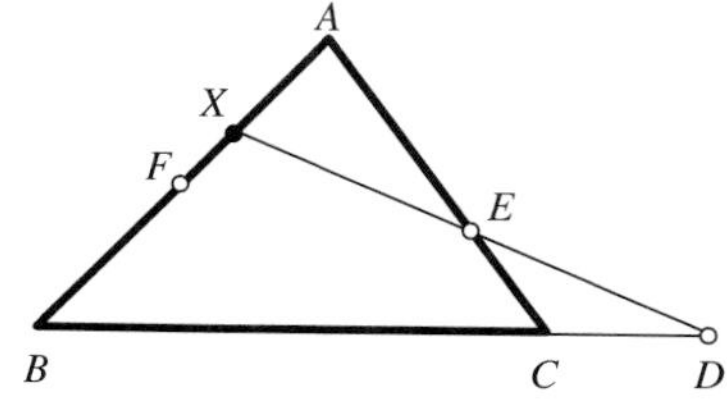

Figure 4.98

The theorems of Menelaus and Ceva provide an excellent example of the duality between collinearity and concurrency in elementary geometry and they can be used to prove a variety of incidence theorems.

EXAMPLE 3 Prove that the medians of a triangle are concurrent using the linearity number.

SOLUTION

Let L, M, and N be the midpoints of sides $\overline{BC}$, $\overline{AC}$, and $\overline{AB}$, respectively, and consider the linearity number of L, M, and N with respect to $\triangle ABC$:

$$\begin{bmatrix} ABC \\ LMN \end{bmatrix} = \frac{AN}{NB} \cdot \frac{BL}{LC} \cdot \frac{CM}{MA} = 1 \cdot 1 \cdot 1 = 1$$

Therefore, by Ceva's Theorem, the medians $\overline{AL}$, $\overline{BM}$, and $\overline{CN}$ are concurrent. ■

EXAMPLE 4 If two of the ratios in the linearity number for the cevians $\overline{AD}$, $\overline{BE}$, and $\overline{CF}$ of Figure 4.93 are $BD/DC = 1/3$ and $CE/EA = 2/3$, what value must the ratio AF/FB have for the cevians to be concurrent?

SOLUTION

Since we must have

$$\frac{AF}{FB} \cdot \frac{BD}{DC} \cdot \frac{CE}{EA} = 1$$

then

$$\frac{AF}{FB} \cdot \frac{1}{3} \cdot \frac{2}{3} = 1$$

$$\therefore AF/FB = 9/2 = 4.5 \quad ■$$

EXAMPLE 5 The Fermat Point of a Triangle

Let $\triangle ABC$ be a given triangle with no angle having measure greater than 120. Construct equilateral triangles externally on each of the sides. (See Figure 4.99.) If the outer vertices of these three equilateral triangles are joined to the vertices of $\triangle ABC$ opposite them, the resulting cevian lines $\overline{AA'}$, $\overline{BB'}$, and $\overline{CC'}$ will be concurrent. The point of concurrency is called **Fermat's Point**.[6] Prove that this point of concurrency exists using Ceva's Theorem.

SOLUTION

Observe that, by SAS, $\triangle ACA' \cong \triangle B'CB$. The congruent angles resulting from this are designated by the same number in Figure 4.99. By the same procedure, two other

---

[6] See the discussion in Section 5.7, where a further property of the Fermat Point is derived.

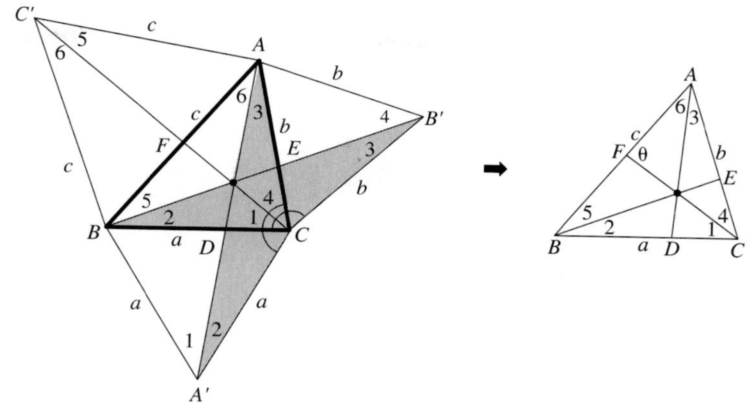

Figure 4.99

pairs of congruent triangles can be observed and the resulting congruent angles labeled by the same number in Figure 4.99. (This also proves the interesting side result that $AA' = BB' = CC'$.) Repeated use of the Law of Sines is now made in the subtriangles of $\triangle ABC$ (see diagram to the right in the figure). First, in $\triangle ACF$ we have

$$\frac{AF}{AC} = \frac{\sin\angle 4}{\sin\theta} \quad \text{or} \quad \frac{AF}{b} + \frac{\sin\angle 4}{\sin\theta} \quad \rightarrow \quad AF = \frac{b\sin\angle 4}{\sin\theta}$$

and in $\triangle BFC$

$$\frac{FB}{BC} = \frac{\sin\angle 1}{\sin(180 - \theta)} \quad \text{or} \quad \frac{FB}{a} = \frac{\sin\angle 1}{\sin\theta} \quad \rightarrow \quad FB = \frac{a\sin\angle 1}{\sin\theta}$$

The preceding two equations give us the first ratio AF/FB needed in the linearity number:

$$\frac{AF}{FB} = \frac{b\sin\angle 4}{a\sin\angle 1}$$

The same procedure is used in triangles $\triangle ABD$, $\triangle ACD$, and $\triangle BCE$, $\triangle BAE$ with the result

$$\frac{BD}{DC} = \frac{c\sin\angle 6}{b\sin\angle 3} \quad \text{and} \quad \frac{CE}{EA} = \frac{a\sin\angle 2}{c\sin\angle 5}$$

Therefore, the linearity number is given by

$$\begin{bmatrix} ABC \\ DEF \end{bmatrix} = \frac{b\sin\angle 4}{a\sin\angle 1} \cdot \frac{c\sin\angle 6}{b\sin\angle 3} \cdot \frac{a\sin\angle 2}{c\sin\angle 5} = \frac{\sin\angle 4\sin\angle 6\sin\angle 2}{\sin\angle 1\sin\angle 3\sin\angle 5}$$

We still have to prove that the last expression = 1. But observe the Law of Sines applied to $\triangle BAB'$, $\triangle CBC'$, and $\triangle BCB'$:

$$\frac{\sin\angle 4}{\sin\angle 5} = \frac{c}{b}, \qquad \frac{\sin\angle 6}{\sin\angle 1} = \frac{a}{c}, \qquad \frac{\sin\angle 2}{\sin\angle 3} = \frac{b}{a}$$

When you multiply these three equations, the right side cancels, leaving 1, and the left side is what we had in the above linearity number. This completes the proof. ■

Moment for Discovery

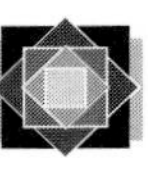

Values of The Linearity Number

Follow these steps in *Sketchpad*.

1. Construct $\triangle ABC$ and line $\overleftrightarrow{AB}$.
2. Locate points D, E, and F on segments $\overline{BC}$ and $\overline{CA}$ and on *line* $\overleftrightarrow{AB}$, respectively. (Drag F to a position on segment $\overline{AB}$ initially.)
3. Construct segments $\overline{AD}$, $\overline{BE}$, and $\overline{CF}$.
4. Construct line $\overleftrightarrow{DE}$ and select the point of intersection G on line $\overleftrightarrow{AB}$. Arrange points D and E so that G falls on the B side of A, that is, on ray $\overrightarrow{AB}$.
5. Calculate and display the ratios

$$\frac{BD}{DC}, \qquad \frac{CE}{EA}, \qquad \text{and} \qquad \frac{AF}{AB - AF}\left(\text{which equals } \frac{AF}{FB}\right)$$

Finally, display the following value on the screen:

$$\frac{AF}{FB} \cdot \frac{BD}{DC} \cdot \frac{CE}{EA} = \begin{bmatrix} ABC \\ DEF \end{bmatrix}$$

Drag F until the cevians $\overline{AD}$, $\overline{BE}$, and $\overline{CF}$ are concurrent. What do you observe about the value of the bracket expression? Drag F toward point G. What happens to the linearity number when D, E, and F are collinear? Drag C, D, and F and repeat the experiment. What did you discover?

PROBLEMS (4.8)

GROUP A

1. In right triangle ABC, with certain segment measures indicated, line $\overleftrightarrow{DF}$ meets side $\overline{AC}$ at E. Use Menelaus' Theorem to determine the ratio CE/EA.

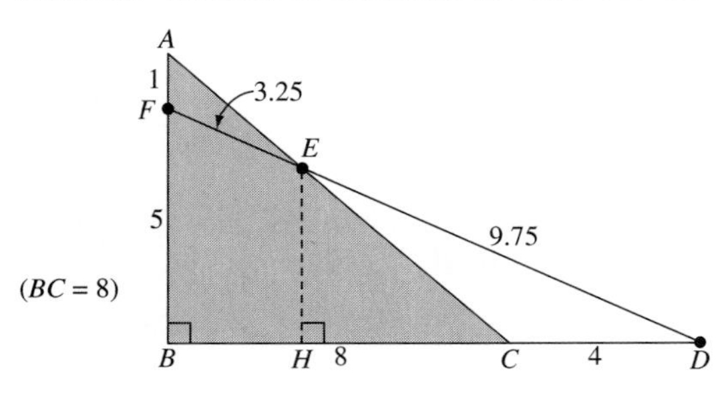

2. Using the result of Problem 1 and similar right triangles, show that $DE = 9.75$ and $EF = 3.25$, and show that, relative to $\triangle BDF$

$$\begin{bmatrix} BDF \\ EAC \end{bmatrix} = -1$$

3. In $\triangle KMW$, $MW = 6$ in., side $\overline{KM}$ is divided into three equal parts and $\overline{KW}$ into four equal parts, with points E and F located as shown. Decide if line $\overline{EF}$ meets line $\overline{MW}$, and if so, where on that line (in terms of distances from M or W).

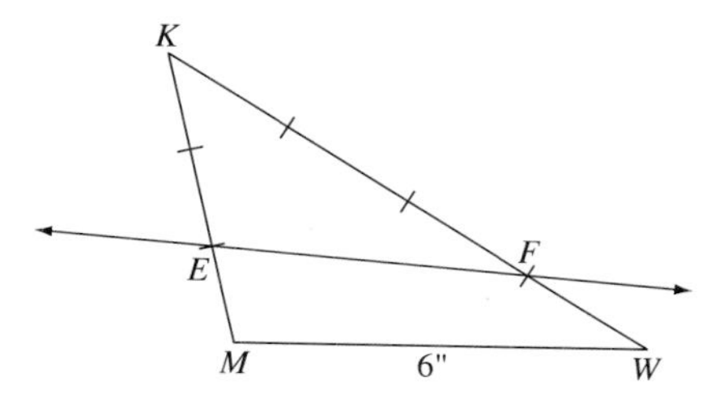

4. Use directed distance to prove from the definition of linearity number that if

$$\begin{bmatrix} ABC \\ DEF \end{bmatrix} = \begin{bmatrix} ABC \\ DEX \end{bmatrix}$$

where X is some point on line $\overleftrightarrow{AB}$, then $F = X$.

5. Using the property $c/b = a_1/a_2$ (proved in Example 4, Section 4.3) for angle bisectors (in the figure, $\overrightarrow{AD}$ is the bisector of $\angle CAB$ and $BD = a_1$, $DC = a_2$), use Ceva's Theorem to prove that the angle bisectors of a triangle are concurrent.

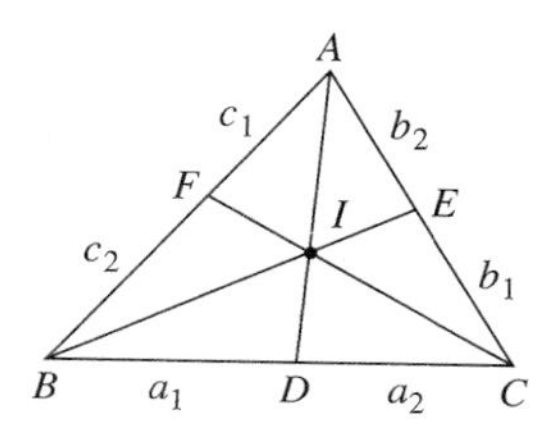

6. Use Menelaus' Theorem in $\triangle ABE$ to prove that medians $\overline{BE}$ and $\overline{CF}$ meet at G, the two-thirds point on $\overline{BE}$ from B to E. (***Hint:*** Use the linearity number $\begin{bmatrix} ABE \\ GCF \end{bmatrix}$.)

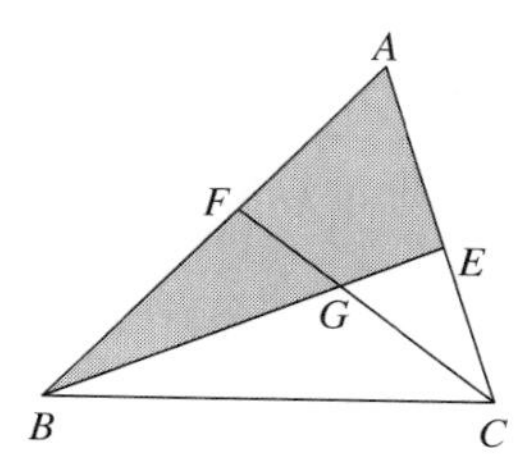

7. If the sides of a quadrilateral are cut by a transversal at the points X, Y, Z, W (as shown in the accompanying figure), show that

$$\frac{AX}{XB} \cdot \frac{BY}{YC} \cdot \frac{CZ}{ZD} \cdot \frac{DW}{WA} = 1$$

(***Hint:*** Draw $\overline{BD}$ and use Menelaus' Theorem on $\triangle ABD$ and $\triangle BCD$.)

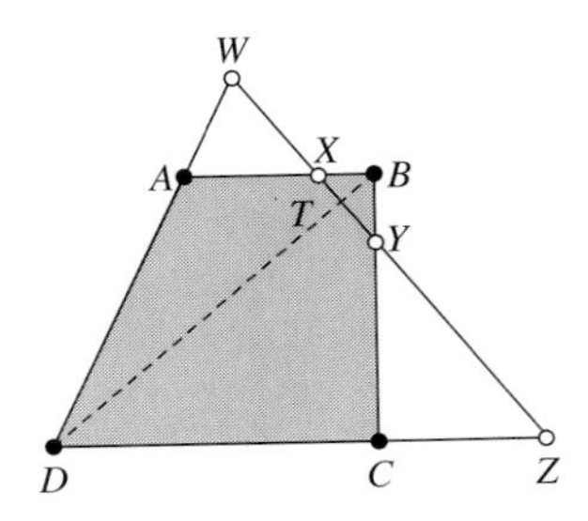

8. Use Ceva's Theorem to prove that the altitudes of an acute angled triangle are concurrent. (***Hint:*** First establish that $\triangle BEC \sim \triangle ADC$, etc., what does the ratio DC/CE then equal? Continue in like manner around the sides of $\triangle ABC$.)

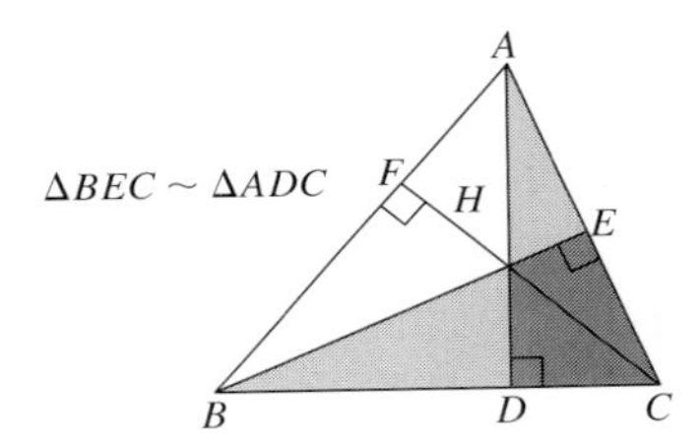

GROUP B

9. Gergonne Point The cevians joining the vertices of a triangle with the points of contact of the incircle on the opposite sides are concurrent. Prove. [This point of concurrency is called the **Gergonne Point** of the triangle, after J.D. Gergonne (1771–1859).] (***Hint:*** Use the result of Problem 14, Section 3.8.)

10. Nagel Point Let D, E, and F be the points of contact of the sides opposite A, B, and C with the three **excircles** of triangle $\triangle ABC$ (the circles tangent to the three sides that lie outside the triangle). The cevians $\overline{AD}$, $\overline{BE}$, and $\overline{CF}$ are concurrent in a point called the **Nagel Point** of the triangle. Prove the concurrency of these cevians. (***Hint:*** Show that $BD = s - c$, $DC = s - b$, $CE = s - a$, etc.)

11. A point P is selected at random on the median $\overline{AD}$ of triangle $\triangle ABC$. If E and F are the points of intersection of $\overleftrightarrow{BP}$ and $\overleftrightarrow{CP}$ with sides $\overline{AC}$ and $\overline{AB}$, respectively, show that $\overleftrightarrow{FE} \parallel \overleftrightarrow{BC}$.

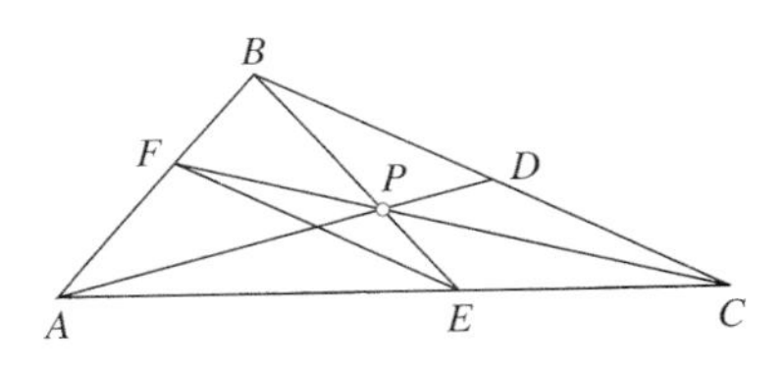

GROUP C

12. Prove the existence of the Simson Line, defined in Section 1.4, using Menelaus' Therem. (See figure for hints.)

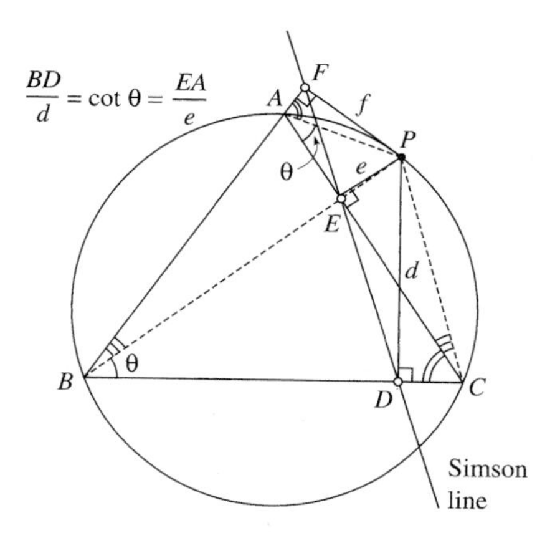

13. The Theorem of Desargues Use the linearity number to prove the following:

THEOREM: If under some correspondence of their vertices two triangles have their corresponding vertices lying on concurrent lines, then corresponding sides meet in three collinear points, if they are not parallel. (See figure for this problem.)

(***Hint:*** Work with the following linearity numbers:

$$\begin{bmatrix} PAB \\ NB'A' \end{bmatrix}, \begin{bmatrix} PBC \\ LC'B' \end{bmatrix}, \begin{bmatrix} PCA \\ MA'C' \end{bmatrix}$$

and show by algebra that the product of these three linearity numbers equals the linearity $\begin{bmatrix} ABC \\ LMN \end{bmatrix}$.)

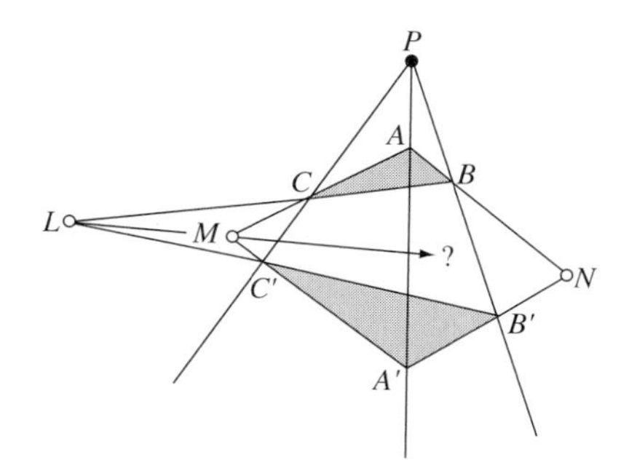

Chapter 4 Summary

This chapter began with a discussion about various parallel postulates, and the one adopted, the Euclidean parallel postulate, essentially requires a pair of alternate interior angles to be congruent along any transversal cutting two parallel lines. From this logically follows the familiar properties of rectangles, parallelograms, and trapezoids. The median of a trapezoid (the line joining midpoints of the two nonparallel sides, or the legs) is always parallel to the two bases, and, conversely, a line passing through the midpoint of a leg and parallel to the bases will bisect the other leg. This leads to the important properties of parallel projection and the useful similarity criteria for triangles.

Similar polygons, having corresponding angles congruent and corresponding sides proportional, are pervasive in Euclidean geometry. This concept leads to everything else of importance: the Pythagorean Theorem, the trigonometry of the triangle, the circle theorems, the concept of area, and coordinates and vectors, to name a few.

Regular polygons and their basic properties were introduced next, along with the usual terms and formulas connected with that topic. Gauss's Theorem concerning the constructibility of certain regular polygons was presented and discussed, after which some work with regular and semiregular tilings was presented.

The key concepts for circles were (1) the measure of an angle inscribed in a circle is one-half that of its intercepted arc, and (2) the chord–secant relationships (products of lengths of corresponding segments on them are equal).

Then an axiomatic system for area appeared, and it was shown how to prove some of the basic area formulas. Coordinates and vectors were introduced to show how these concepts may be constructed in an axiomatic system. How to use these tools to solve problems in geometry was then shown.

Finally, the last section, an optional section, showed how to derive an important linearity number that is related to the concurrence of cevians of a triangle (lines joining each vertex with a point on the opposite side) and the collinearity of points lying on the sides (extended) of a triangle.

Testing Your Knowledge

You are expected to take this test using only the list of axioms, definitions, and theorems in Appendix F as references.

1. If $\overleftrightarrow{BD} \parallel \overleftrightarrow{EF}$ and the angles at A, C, and B have measures as indicated, find

(a) $m\angle 1$ **(b)** $m\angle 2$ **(c)** $m\angle 3$

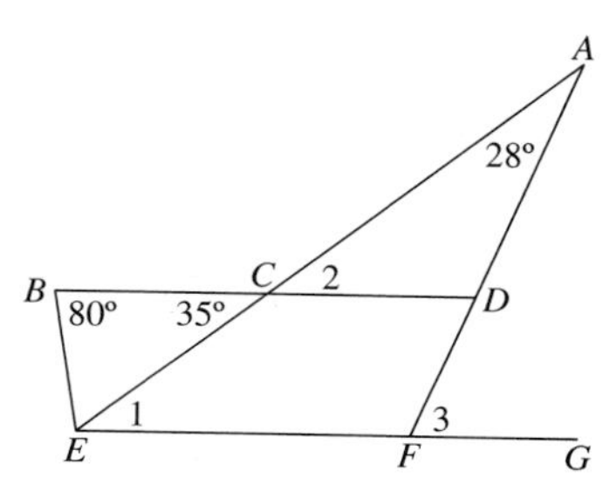

2. The figure shows two rectangles with their diagonals on the same line and with certain segment measures indicated. Find the dimensions x and y of the large rectangle.

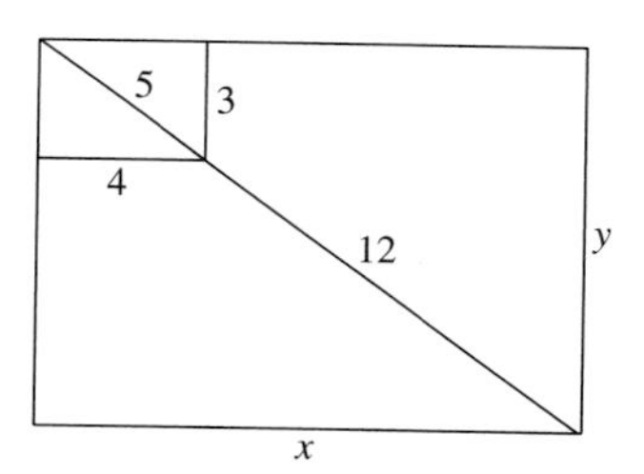

3. Sketch a figure and give a paragraph demonstration for the following geometry problem:

Given: P–R–W, Q–R–T, and $\overleftrightarrow{PQ} \parallel \overleftrightarrow{TW}$

Prove: $\dfrac{PR}{RW} = \dfrac{QR}{RT}$

4. In the 3–4–5 right triangle $\triangle ABC$, $\overline{BC} \perp \overline{CA}$, $\overline{CK} \perp \overline{AB}$, and $\overline{KL} \perp \overline{BC}$, creating a right triangle $\triangle KLA$ that is similar to $\angle ABC$. How much smaller is $\triangle KLA$ than $\triangle ABC$? Express your answer as a fraction $(= x)$reduced to lowest terms.

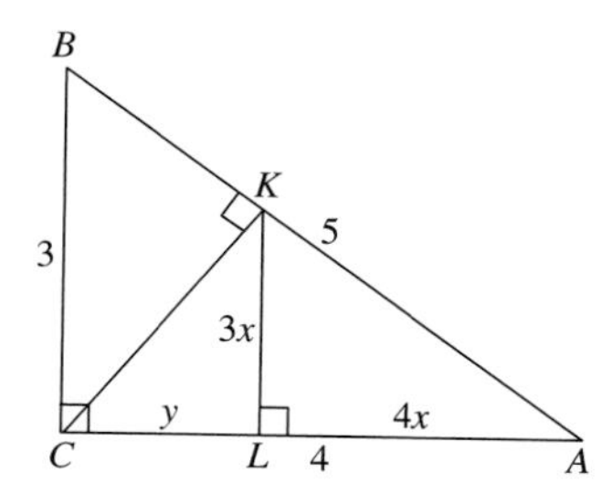

5. True or False? If p is a prime number, then a regular polygon having $n = p$ as the number of sides may be constructed with a compass, straight-edge. Discuss.

6. Which of the pairs of tiles shown in **(a)**, **(b)**, and **(c)** can be used in a tessellation of the plane of order 2? (The sides of all polygons have equal lengths.)

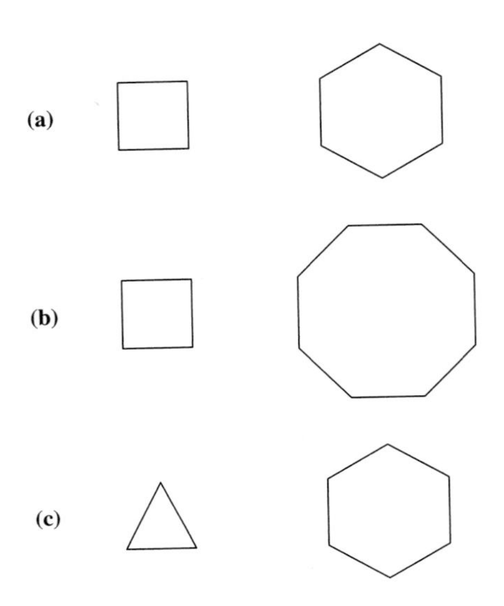

7. Consider a nine-sided regular polygon having radius 3 units.

(a) Find the measure of each interior angle.

(b) Find a trigonometric expression which gives the length of a side of the polygon.

8. The shadow of a silo on the ground (shown in the figure with dimensions marked) is in the shape of a rectangular base and semicircular top (like the silo), with the diameter of the semicircle parallel to the base. Find its area to one decimal accuracy.

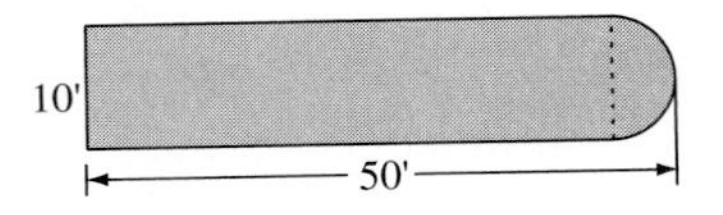

9. Describe how you would investigate, with coordinates, whether, and under what circumstances, the midpoint of a side of a triangle is equidistant from the vertices of the triangle.

10. Give a paragraph proof that if the diagonals $\overline{AC}$ and $\overline{BD}$ of a parallelogram $\square ABCD$ intersect at M, then M is the midpoint of both diagonals.

CHAPTER FIVE

Transformations in Geometry

OVERVIEW

In geometry, as in all branches of mathematics, transformations are of key importance and provide a powerful tool for both discovering and proving new theorems. We first introduce synthetically the two major geometric transformations, the translation and rotation, then branch out to other types such as dilations and similitudes. After making a study of the basic properties these mappings have, we show how they may be used in a variety of ways to prove some very unusual results (such as Napoleon's Theorem), and also to provide some elegant proofs of a few results we have seen before (for example, involving the Nine-Point Circle).

*5.1 Euclid's Superposition Proof and Plane Transformations

The origin of the concept for transformation theory came from Euclid himself. This is curious, because in his own writings Euclid never referred to it directly, and, in fact, abandoned it entirely after the first several propositions. In the proof of Proposition 4, however, he in effect assumed that one can "move" certain figures around in space without changing their size or shape. He was taking for granted, without saying so, that there exists a certain "rigid motion" that maps one triangle to another.

It is instructive to study the passage in Euclid's *Elements* where the idea of motions is first used to prove a theorem. The following is Euclid's argument for Proposition I.4 (our Axiom C-1, the SAS Postulate), as quoted from Heath's *The Thirteen Books of Euclid's Elements, Book I,* pp. 247–248. We are given $\triangle ABC$ and $\triangle DEF$ with $AB = DE, AC = DF$, and $m\angle A = m\angle D$, and we want to prove that $BC = EF, m\angle B = m\angle E$, and $m\angle C = m\angle F$. In the following passage, we have italicized Euclid's reference to motions.

I say that the base BC is also equal to the base EF, the triangle ABC will be equal to the triangle DEF, and the remaining angles will be equal to the remaining angles, respectively, namely those which the equal sides subtend, that is, the angles ABC to the angle DEF, and the angle ACB to the angle DFE.

For, if the triangle *ABC be applied to the triangle DEF and if the point A be placed on the point D and the straight line AB on DE,* then the point B will also coincide with E, because AB is equal to DE.

Again, AB coinciding with DE, the straight line AC will also coincide with DF, because the angle BAC is equal to the angle EDF; hence the point C will also coincide with the point F, because AC is again equal to DF.

But B also coincided with E; hence the base BC will coincide with the base EF . . . and will be equal to it.

Thus the whole triangle ABC will coincide with the whole triangle DEF . . .

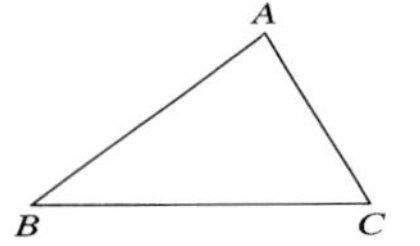

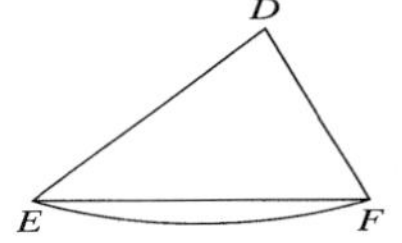

This bit of mathematical poetry contains the germ of the idea for Euclidean motions. We are going to first develop this concept from the axioms (synthetically), then coordinates will be brought in, which will show how these transformations fit into a much larger and more general class of mappings. We begin with the definition of a *linear transformation.*

DEFINITION: A **transformation** in absolute geometry is a function (or mapping) f that associates with each point P in some plane with some other point in that plane, denoted by $f(P) = P'$, such that

(1) f is **one-to-one**: If $P \neq Q$ then $f(P) \neq f(Q)$, as indicated in Figure 5.1;

(2) f is **onto**: Every point R in the plane has a **preimage** under f, that is, there exists a point S such that $f(S) = R$.

If a transformation preserves collinearity, that is, if it maps any three collinear points to three other collinear points, it is said to be **linear**.

Thus, a linear transformation maps lines into lines. A further property of a transformation [from properties (1) and (2)] is that it always has an **inverse**: If P is a given point, let its unique preimage Q be denoted $f^{-1}(P)$, where f^{-1} now becomes a new

TRANSFORMATION

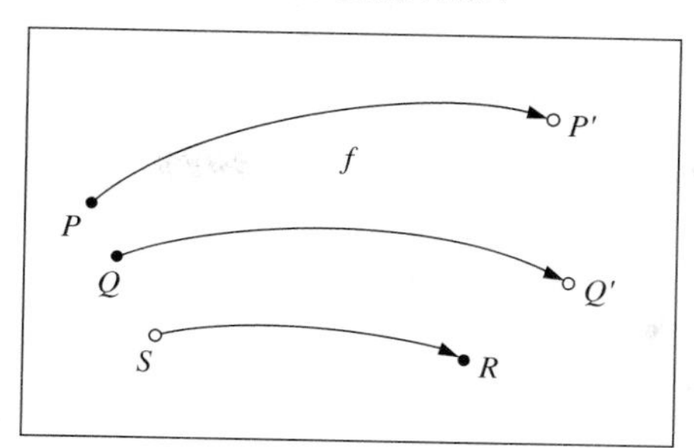

LINEAR TRANSFORMATION

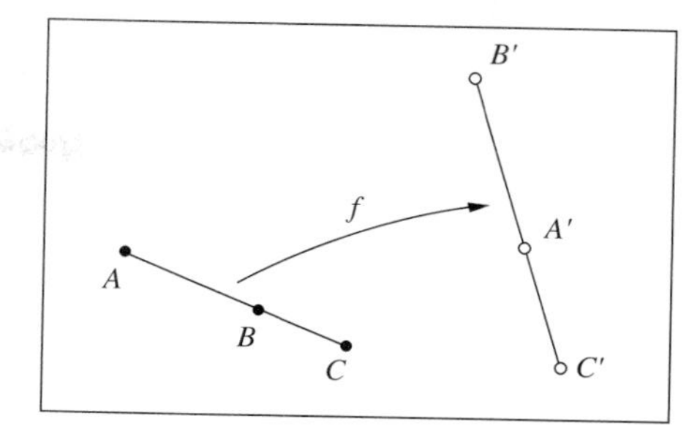

Figure 5.1

mapping (transformation?). For each point P, if f takes P to P', then f^{-1} takes P' back to P. It seems clear enough (but needs a formal proof) that the inverse of a linear transformation is also a linear transformation. (See theorem below.)

NOTATION

Coordinates will frequently be used to convey examples of transformations in coordinate geometry. Instead of $f(P) = P'$, one can write

$$f(x, y) = (x', y')$$

where $P = (x, y)$ and $P' = (x', y')$. This is equivalent to having *two separate equations*—one for x' and one for y'—to specify the coordinates of P', as in

(1) $$f: \begin{cases} x' = f_1(x, y) \\ y' = f_2(x, y) \end{cases}$$

Under this scheme, which is essentially algebraic in nature, the inverse of the transformation is found by solving the system of equations for x and y in terms of x' and y', resulting in equations of the form

(2) $$f^{-1}: \begin{cases} x = f_3(x', y') \\ y = f_4(x', y') \end{cases}$$

EXAMPLE 1 A transformation is given in coordinate form by

$$f: \begin{cases} x' = 2x + y + 1 \\ y' = x + y - 2 \end{cases}$$

(a) Find the images of the three points $A(3, 1)$, $B(4, 6)$, and $C(5, 11)$ under f; that is, find $f(A)$, $f(B)$, and $f(C)$ using the above coordinate equations for x' and y'.

(b) Find the unique preimage of the point $D(2, 2)$; that is, find a point $P(x, y)$ such that $f(P) = D$.

(c) Find the inverse transformation, f^{-1}.

SOLUTION

(a) By substituting the given coordinates into the equations for f, we find, as shown in Figure 5.2:

$$f(3, 1): \begin{cases} x' = 2(3) + (1) + 1 \\ y' = (3) + (1) - 2 \end{cases} \text{ or } \begin{matrix} x' = 6 + 2 = 8 \\ y' = 4 - 2 = 2 \end{matrix} \quad \therefore f(3, 1) = (8, 2)$$

$$f(4, 6): \begin{cases} x' = 2(4) + (6) + 1 \\ y' = (4) + (6) - 2 \end{cases} \text{ or } \begin{matrix} x' = 8 + 7 = 15 \\ y' = 10 - 2 = 8 \end{matrix} \quad \therefore f(4, 6) = (15, 8)$$

$$f(5, 11): \begin{cases} x' = 2(5) + (11) + 1 \\ y' = (5) + (11) - 2 \end{cases} \text{ or } \begin{matrix} x' = 10 + 12 = 22 \\ y' = 16 - 2 = 14 \end{matrix} \quad \therefore f(5, 11) = (22, 14)$$

The points A, B, C and their images A', B', and C' have been plotted in Figure 5.2. For convenience, we have depicted two separate planes, the xy- and $x'y'$-planes, even

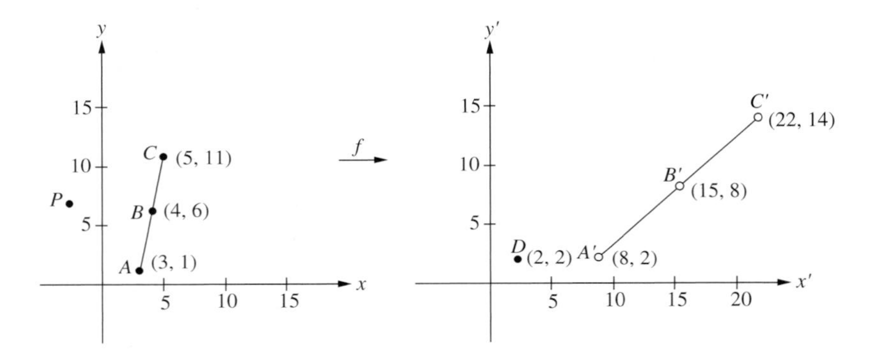

Figure 5.2

though the transformation is taking place in a single plane.

(b) Since $x' = 2$ and $y' = 2$, we must solve the following system for x and y:

$$\begin{cases} 2 = 2x + y + 1 \\ 2 = x + y - 2 \end{cases}$$

But this is equivalent to the system of equations

$$\begin{cases} 2x + y = 1 \\ x + y = 4 \end{cases}$$

Using the usual methods, we find (subtracting the two equations)

$$\begin{matrix} x + 0 = -3 & \text{or} & x = -3 \\ -3 + y = 4 & \text{or} & y = 7 \end{matrix}$$

Thus, the preimage of $D(2, 2)$ is $P(-3, 7)$, shown in Figure 5.2.

(c) Here we solve the system for x and y in terms of x' and y':

$$\begin{cases} x' = 2x + y + 1 \\ y' = x + y - 2 \end{cases} \quad \text{or} \quad \begin{cases} 2x + y = x' - 1 \\ x + y = y' + 2 \end{cases}$$

Again we have

$$x + 0 = (x' - 1) - (y' + 2)$$
$$x = x' - y' - 1 - 2 = x' - y' - 3$$

and, by substituting this for x in the second equation,

$$(x' - y' - 3) + y = y' + 2$$
$$y = y' + 2 - x' + y' + 3 = -x' + 2y' + 5$$

Hence, the inverse of f is given by

$$f^{-1}: \begin{cases} x = x' - y' - 3 \\ y = -x' + 2y' + 5 \end{cases} \quad \blacksquare$$

NOTE: It can be shown that the transformation f in **(1)** is linear iff the functions f_1 and f_2 are linear in x and y, that is, iff they each have the form $Ax + By + C$, as they were in Example 1.

EXAMPLE 2 Show synthetically that a linear transformation f maps three noncollinear points A, B, and C into three noncollinear points A', B', and C'. (See Figure 5.3.)

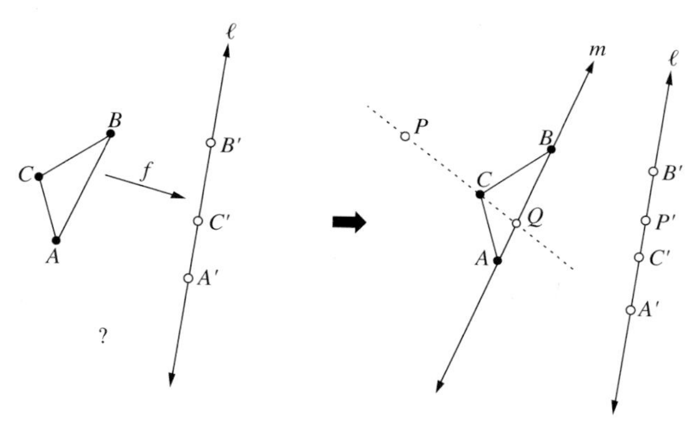

Figure 5.3

SOLUTION

Suppose, on the contrary, that A', B', and C' lie on some line ℓ. We are going to show that every point in the plane maps to line ℓ. To that end, let P be any point besides A, B, and C. If P lies on either line $\overleftrightarrow{AC}$, $\overleftrightarrow{BC}$, or $m = \overleftrightarrow{AB}$, then since f preserves collinearity, $P' = f(P) \in \overleftrightarrow{A'C'} = \overleftrightarrow{B'C'} = \overleftrightarrow{A'B'} = \ell$. Otherwise, consider line $\overleftrightarrow{PC}$. If $\overleftrightarrow{PC}$ is not parallel to m, then $\overleftrightarrow{PC}$ meets m at some point Q, and again, since $Q' \in \ell$ and $C' \in \ell$, then $P' \in \ell$. Finally, if $\overleftrightarrow{PC} \parallel m$, certainly line $\overleftrightarrow{PD}$ is not parallel to m if D is some other point on line $\overleftrightarrow{AC}$ besides A and C, and $\overleftrightarrow{PD}$ intersects m at Q. Then D' and Q' lie on ℓ and again it follows that $P' \in \ell$. Thus, we have shown that f maps the entire plane into line ℓ, which is a contradiction that f is onto. Therefore, A', B', and C' are not collinear. $\blacksquare$

The argument given in Example 2 proves an important theorem.

THEOREM: The inverse of a linear transformation is also a linear transformation.

PROOF

It is already clear that if f is one-to-one and onto, then f^{-1} will also be one to one and onto. This leaves the linearity property. Suppose f^{-1} were *not* linear. Then there would exist collinear points A', B', and C' such that $f^{-1}(A') = A, f^{-1}(B') = B$, and $f^{-1}(C') = C$ are not collinear. Then f would map the noncollinear points A, B, and C into the collinear points A', B', C', which contradicts the result of the argument of Example 2. Therefore, f^{-1} is a linear transformation.

EXAMPLE 3 Show that under the mapping

$$f: \begin{cases} x' = -x + y \\ y' = -2x + y \end{cases}$$

(a) the inverse is given by

$$f^{-1}: \begin{cases} x = x' - y' \\ y = 2x' - y' \end{cases}$$

(b) Use this to show that the images of the two parallel lines $y = 3x$ and $y = 3x + 2$ are parallel lines.

SOLUTION

(a) To reverse the defining equations (as in Example 1) write as a 2×2 linear system in x, y and solve for x and y:

$$\begin{cases} -x + y = x' \\ -2x + y = y' \end{cases}$$

Subtract the two equations to find that $x = x' - y'$, then substitute this into the first equation and solve for y:

$$-(x' - y') + y = x' \quad \rightarrow \quad y = 2x' - y'$$

(b) Substitute the equations found in **(a)** into $y = 3x$ and $y = 3x + 2$:

$$2x' - y' = 3(x' - y') \quad \rightarrow \quad y' = \frac{1}{2}x'$$

$$2x' - y' = 3(x' - y') + 2 \quad \rightarrow \quad y' = \frac{1}{2}x' + 1$$

Thus, the two image lines are also parallel (having slope $\frac{1}{2}$). ■

For later reference, we define two more fundamental terms in transformation theory.

DEFINITION: A transformation f of the plane is said to have A as a **fixed point** iff $f(A) = A$.

DEFINITION: A transformation of the plane is called the **identity mapping** iff every point of the plane is a fixed point. This transformation is denoted e.

EXAMPLE 4 Find those values of the parameter a for which the linear transformation

$$f: \begin{cases} x' = 3ax + y \\ y' = x - ay \end{cases}$$

has a *nontrivial* fixed point (a point different from the origin (0, 0)), and characterize those fixed points (describe geometrically) for each value of a.

SOLUTION

We must have $x' = x$ and $y' = y$ for any fixed point. That is,

$$\begin{cases} x = 3ax + y \\ y = x - ay \end{cases} \quad \text{or} \quad \begin{cases} (3a - 1)x + y = 0 \\ x - (a + 1)y = 0 \end{cases}$$

Since both equations must be satisfied simultaneously for certain $(x, y) \neq (0, 0)$, we must have, by substitution of $x = (a + 1)y$ from the last equation into the preceding equation,

$$\textbf{(3)} \qquad (3a - 1)(a + 1)y + y = 0$$

If $y = 0$, then these equations force $x = 0$, so we conclude that $y \neq 0$ and, dividing both sides of **(1)** by y,

$$(3a - 1)(a + 1) + 1 = 0, \qquad \text{or} \qquad 3a^2 + 2a = 0$$

with roots $a = 0, -\frac{2}{3}$. For $a = 0$, we get

$$(3 \cdot 0 - 1)x + y = 0, \qquad \text{or} \qquad y = x,$$

which shows that for $a = 0$, every point of the line $y = x$ is a fixed point. If $a = -\frac{2}{3}$, then

$$[3 \cdot (-\tfrac{2}{3}) - 1]x + y = 0, \qquad \text{or} \qquad y = 3x$$

again a line of fixed points. ■

Moment for Discovery

Are Linear Transformations Midpoint Preserving?

Consider the transformation given by

$$f: \begin{cases} x' = 3x + 2y - 1 \\ y' = -x + 4y + 2 \end{cases}$$

and the line segments $\overline{AB}$ and $\overline{CD}$ with endpoints $A(1,1)$, $B(3, 5)$, $C(2, 3)$, and $D(-4, -3)$.

1. Using the midpoint formula **(14)** in Section 4.7, find the coordinates of the midpoints M and N of $\overline{AB}$ and $\overline{CD}$, respectively.
2. Carefully calculate the images of the points A, B, and M under f.
3. Independently, calculate the midpoint of $\overline{A'B'}$ using the midpoint formula. Did anything happen?
4. Repeat Steps 2 and 3 for segment $\overline{CD}$ and the images C', D', and N'.

Do you observe anything significant? If so, make a conjecture. (See Problems 14 and 15 in the problem section that follows.

PROBLEMS (5.1)

GROUP A

1. Let the point (x, y) be associated with (x', y') iff

f: $x' = 2x$ and $y' = 3x$

Find

(a) the images of $(1, -1)$, $(\frac{1}{2}, \frac{1}{3})$, and $(-3, 4)$

(b) the preimages of $(2, -3)$ and $(4, 6)$.

(c) Is the pairing a transformation?

2. For the mapping given by

$$f: \begin{cases} x' = x + y + 2 \\ y' = x - y - 3 \end{cases}$$

Find the

(a) fixed points

(b) inverse map.

3. (a) Show that the image of the parabola $y = x^2$ under the mapping $(x, y) \to (x', y')$ defined by

$$f\colon x' = y \qquad \text{and} \qquad y' = x^2 + y$$

is part of a straight line. Identify it.

(b) Is this mapping a transformation? Show why, or why not.

4. For the mapping $g(x, y) = (x^2, y)$,

(a) give an example to show g is not one-to-one

(b) tell why g is not onto.

5. Show that the image of the curve $x^3 + y^2 = 1$ (as shown in the figure) is a broken line under the mapping defined by

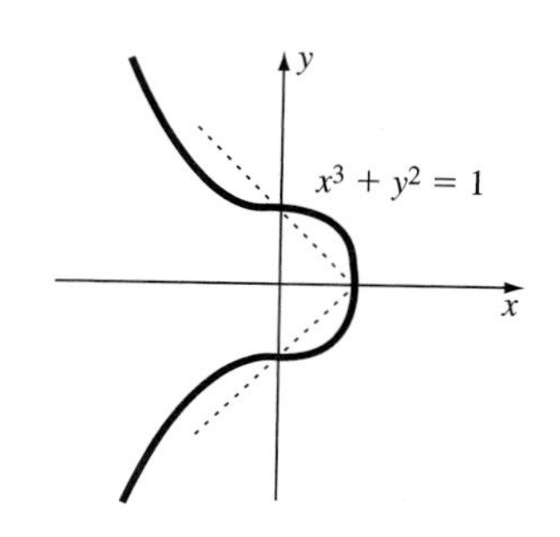

$$f\colon x' = x^3 \qquad \text{and} \qquad y' = \delta y^2$$

where $\delta = +1$ if $y \geq 0$ and $\delta = -1$ if $y < 0$. Is this mapping a transformation? Show why or why not.

6. Find the image of $\triangle ABC$ and its interior (and sketch) under the transformation defined by

$$f\colon x' = 2x \qquad \text{and} \qquad y' = 2y$$

if $A = (0, 0)$, $B = (0, 1)$, and $C = (1, 1)$.

GROUP B

7. Find the line into which all points of the plane $\boldsymbol{P}$ map, under the mapping

$$f\colon x' = 2x - y \qquad \text{and} \qquad y' = -6x + 3y$$

Is the mapping a transformation?

8. Show, by substitution, that the perpendicular lines $y = 2x + 1$ and $y = (-1/2)x + 3$ map to perpendicular lines under the transformation f whose inverse is given by

$$f^{-1}\colon x = 2x' - 3y' \qquad \text{and} \qquad y = 3x' + 2y'$$

(After making the substitution, simplify, solve the equations for y' and find the two slopes.)

9. Repeat Problem 8 for the transformation whose inverse is

$$f^{-1}\colon x = 3x' + 4y' \qquad \text{and} \qquad y = -4x' + 3y'$$

10. If a transformation preserves all distances (that is, $PQ = P'Q'$ for all points P and Q), then it necessarily preserves angle measure as well.

(a) Prove this synthetically.

(b) Give a counterexample to show that the converse of this statement is not true. (For ideas, see Problem 6.)

11. **How to Work with Linear Transformations Using *Sketchpad*** Consider the linear transformation

$$f: \begin{cases} x' = x - y + 1 \\ y' = 2x - y - 1 \end{cases}$$

[1] Choose Create Axis from the GRAPH menu.

[2] Construct a circle CD of radius 1 (using CIRCLE TOOL), and locate E, any point on it.

[3] Under MEASURE, choose Coordinates to display $E(x_E, y_E)$ on the screen.

[4] Select E (x_E, y_E) and choose Calculate under MEASURE, then calculate and display the value $x' = x_E - y_E + 1$, where x_E and y_E are found under Values, Point E.

[5] Calculate and display the value $y' = 2x_E - y_E - 1$, as in Step 4 for x'.

[6] Select the values for x' and y', in order (with Shift key depressed), then choose Plot As (x, y) under GRAPH to locate $E' = F$ on the screen.

[7] Select point E, then F'. and choose Locus under CONSTRUCT. The image of the unit circle under f will appear.

Adjust the circle by shrinking, enlarging, and dragging to observe the effect on its image under f. What observations do you make about this image? Does it change shape, or just size and position? When the circle intersects its image, is the point of intersection a fixed point?

[8] Locate any point $G(x, y)$ and, using x_G and y_G calculate the values

$$x' = x_G - y_G + 1$$
$$y' = 2x_G - y_G - 1$$

and then use Plot Point $G' = H(x', y')$ as in Step 6.

Drag G and H and see what happens. Maneuver G and H until they coincide. Estimate the coordinates of $G = H$. Is this a fixed point of f? Is it the only one? Solve for the exact coordinates of the fixed point(s), and compare this with your experimental results.

GROUP C

12. Characterize geometrically the set of all fixed points of each of the following transformations.

(a) $f: \begin{cases} x' = x^3 \\ y' = y^3 + y \end{cases}$

(b) $g: \begin{cases} x' = ax + y - a \\ y' = ax + a \end{cases}$

13. Show without coordinates that if a linear transformation f preserves perpendicularity, then its inverse f^{-1} also preserves perpendicularity.

14. Show without coordinates that any linear transformation maps parallel lines to parallel lines. (***Hint:*** If ℓ' meets m', then where did the point $P' = \ell' \cap m'$ come from?)

15. Using Problem 14, show that a linear transformation maps a parallelogram and its diagonals to another parallelogram and its diagonals. Hence, prove that *midpoints of line segments are preserved* under any linear transformation.

16. Can a transformation of the form

$$f: \begin{cases} x' = ax + by \\ y' = cx + dy \end{cases}$$

ever have *exactly* one other fixed point besides (0, 0) by choosing the appropriate values for a, b, c, and d? Investigate.

17. The result of Problem 15 leads to a synthetic version of the issue raised in Problem 16. If a linear transformation has *two* fixed points A and B, then the midpoint M_1 of $\overline{AB}$ is a fixed point. Thus the midpoints M_2 of $\overline{AM_1}$ and M_3 of $\overline{M_2B}$ are fixed points, and so on. Show, by continuity of linear transformations (a property you may assume without proof), that every point of segment $\overline{AB}$ is a fixed point.

18. Use Problem 17 to show that if a linear transformation fixes three noncollinear points A, B, and C, then it fixes every point of the plane, and hence is the identity.

5.2 Reflections: Building Blocks for Isometries

We are concerned here with two basic types of reflections: reflections in lines and reflections in points.

DEFINITION OF REFLECTIONS IN EUCLIDEAN PLANE

DEFINITION: If a transformation f has the property that some fixed line ℓ is the perpendicular bisector of each segment $\overline{PP'}$ for any point P in the plane, where $P' = f(P)$, then f is a **reflection** in the line ℓ (called the **line of reflection**). We denote a reflection in line ℓ by s_ℓ. On the other hand, if point C is always the midpoint of segment $\overline{PP'}$ for all P, then f is a **reflection** in point C (called the **center of reflection**), to be denoted by s_C. (See Figure 5.4.)

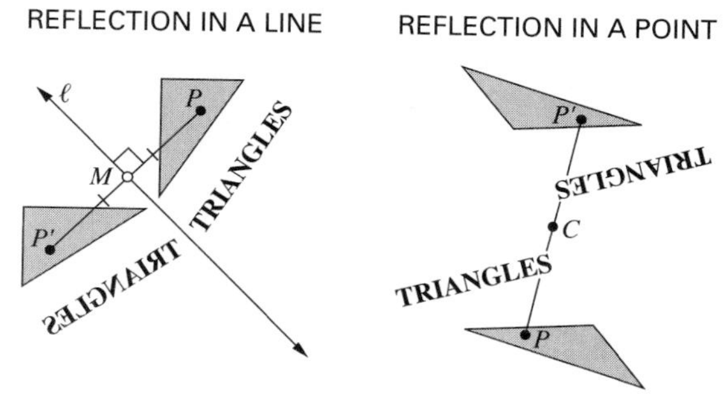

Figure 5.4

To distinguish between the two kinds of transformations in the above definition, we refer to them, respectively, as *line reflections* and *central reflections*. A majority of the development in this book involves line reflections.

To construct the image of a given point P under a reflection in line ℓ, first define $P' = P$ if P lies on ℓ; otherwise, construct the perpendicular $\overline{PM}$ from P to line ℓ, then extend the segment $\overline{PM}$ to point P', making M the midpoint of $\overline{PP'}$. Similarly, if C is a given point and P any other, define $P' = P$ if $P = C$; otherwise, extend segment $\overline{PC}$ to P', making C the midpoint of $\overline{PP'}$. This proves the *existence* of the mappings s_ℓ and s_C. From these constructions it is apparent that s_ℓ and s_C are one-to-one and onto. (As a simple application, can you see why $m \parallel \ell$ implies $s_\ell(m) \parallel \ell$?)

REFLECTIONS AS ISOMETRIES AND THE ABCD PROPERTY

The fundamental properties of reflections are all collected into one result. It will establish the invariance of angle measure, betweenness, collinearity, and distance. These same invariant properties are shared by a more general kind of mapping.

DEFINITION: Any mapping of the plane that preserves distances (and is therefore one-to-one and a transformation) is called an **isometry** (also **motion, rigid motion,** or **Euclidean motion**). Thus, f is an isometry iff for each P and Q, with $P' = f(P)$ and $Q' = f(Q)$,

$$PQ = P'Q'$$

LEMMA A: An isometry preserves collinearity, betweenness, and angle measure.

PROOF

Let A', B', and C' be the images of points A, B, and C, respectively, under a given distance-preserving transformation, and suppose that A–B–C. Then $AB + BC = AC$. Since the mapping is distance preserving, $A'B' + B'C' = A'C'$. Therefore, A'–B'–C' (definition of betweenness and the Triangle Inequality). Hence, betweenness and collinearity of any three points are preserved. Finally, consider $\angle ABC$ and $\triangle A'B'C'$. By SSS, $\triangle ABC \cong \triangle A'B'C'$. Therefore, $\angle ABC \cong \angle A'B'C'$, and angle measure is preserved.

THEOREM 1: ABCD PROPERTY
Reflections are angle-measure preserving (A), betweenness preserving (B), collinearity preserving (C), and distance preserving (D).

PROOF

In view of the lemma, it suffices to prove that a reflection s_ℓ or s_C is distance preserving. Suppose that A and B are any two points not both on line ℓ (non-trivial case). Consider the case when A and B are on the same side of ℓ (Figure 5.5). Since ℓ is the perpendicular bisector of $\overline{AA'}$ and $\overline{BB'}$ at midpoints M and N, respectively, then $\diamond AMNB$ and $\diamond A'MN'B$ are convex, and by SASAS, $\diamond AMNB \cong \diamond A'MNB'$. Hence, $AB = A'B'$. (Note that the quadrilateral $\diamond ABB'A'$ is an isosceles trapezoid for this case.) The case when A or B lies on ℓ simply involves congruent right triangles, so will be omitted. Finally, when A and B lie on opposite sides of ℓ (Figure 5.6), then A and B' lie on the same side, so that, with $s_\ell(A) = A'$ and $s_\ell(B') = B$, by the first case just proven we have $AB' = A'B$. Hence $\diamond AB'BA'$ is an isosceles trapezoid with congruent diagonals $\overline{AB}$ and $\overline{A'B'}$ (see Problem 23, Section 4.2.) Therefore, $AB = A'B'$ in all cases. (The proof for point reflections will be left as an exercise.)

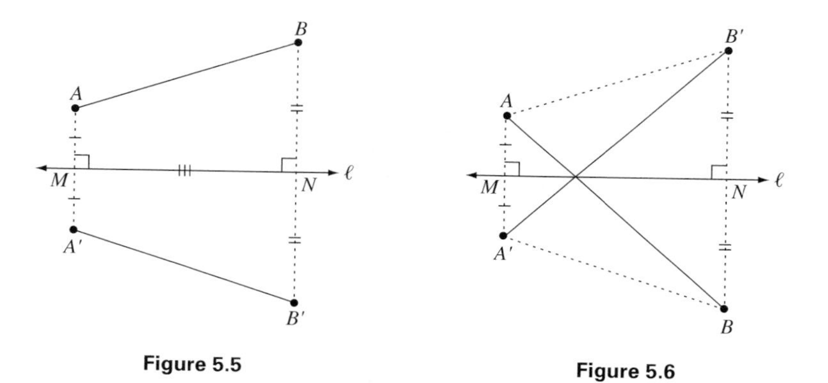

Figure 5.5

Figure 5.6

APPLICATIONS OF REFLECTIONS, ORIENTATION

A common everyday experience illustrates the geometric principle of a line reflection. A mirror placed along line ℓ will reflect images in the manner of the transformation s_ℓ. A point reflection can be illustrated by the occurrence of light rays passing through a small opening.

It may seem surprising, however, that the two types of reflections have altogether different orientation-preserving properties. When we observe printed letters in a mirror, they are reversed, and a right hand will appear as a left hand in the mirror (Figure 5.7). But for a camera, the image is merely inverted, not reversed. To deal with this concept mathematically, we make a couple of definitions.

Right hand is reflected as left hand

Figure 5.7

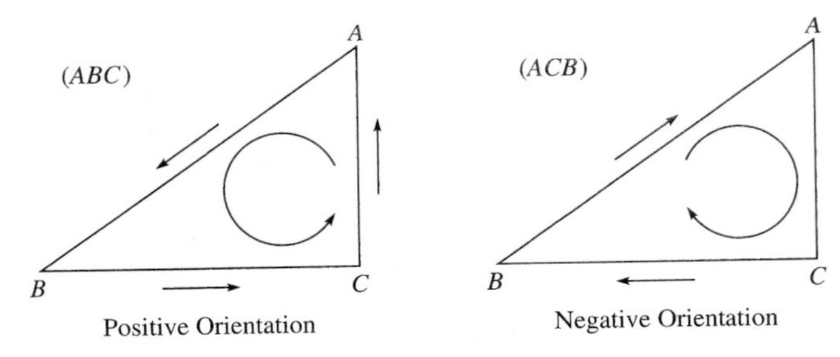

Figure 5.8

DEFINITION: Given a triangle $\triangle ABC$ in the plane, the counterclockwise direction (the path ABC in Figure 5.8) is called the **positive orientation** of its vertices, while the clockwise direction (the path ACB) is the **negative orientation.**

We encourage you to experiment with orientations relative to line and point reflections in the discovery unit at the end of this section before proceeding further. The idea of orientation can easily be extended to polygons and simple, closed curves. The formal definitions will be omitted since we will not need these more general concepts here.

DEFINITION: A linear transformation of the plane is called **direct** iff it preserves the orientation of any triangle, and **opposite** iff it reverses the orientation of each triangle.

OUR GEOMETRIC WORLD

When we look across a lake, the principle of a line reflection is at work. The reflected image of the scenery in the water is the image of the objects on land under a reflection s_ℓ, where ℓ is a line running along the surface of the lake. The image we see is always upside down (inverted). In a similar manner, the pinhole effect of a camera illustrates the mapping s_C, where C is the lens opening. The image on the film is also inverted.

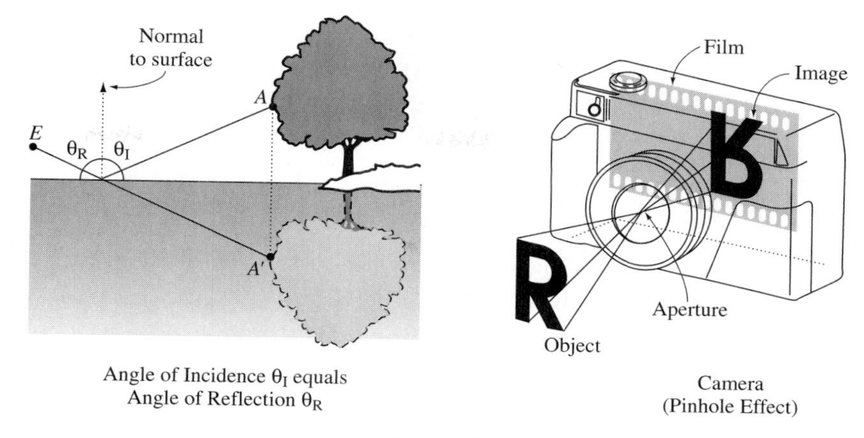

Angle of Incidence θ_I equals Angle of Reflection θ_R

Camera (Pinhole Effect)

Figure 5.9

PRODUCTS OF TWO OR MORE REFLECTIONS

An obvious theorem about direct and opposite mappings will be stated after we introduce the necessary idea of the **composition** or **product** of two transformations. If f and g are any two transformations, their *product,* denoted by gf, is just the ordinary functional composition of $g \circ f$ of f and g. That is, for each point P,

$$gf(P) = g \circ f(P) = g[f(P)].$$

This means that we must first apply the mapping f, and then follow this with the mapping g, going from any point P to the point $P' = f(P)$, then on to $P'' = g(P')$; the product gf is the mapping that takes P directly to P''. This can be easily illustrated in an example using coordinates.

EXAMPLE 1 Consider the two linear transformations f and g given in terms of coordinates, as follows.

$$f: \begin{cases} x' = 2x - 3y \\ y' = 3x + 4y \end{cases} \qquad g: \begin{cases} x' = x - y \\ y' = 2x + y \end{cases}$$

Find the coordinate form of the product transformation gf, then repeat for the product fg, and compare the two results.

SOLUTION

For convenience, change the form of g to

$$g: \begin{cases} x'' = x' - y' \\ y'' = 2x' + y' \end{cases}$$

To find gf, merely substitute the equations for f (above) into those just given for g.

$$x'' = (2x - 3y) - (3x + 4y) = 2x - 3y - 3x - 4y = -x - 7y$$
$$y'' = 2(2x - 3y) + (3x + 4y) = 4x - 6y + 3x + 4y = 7x - 2y$$

Hence,

$$gf: \begin{cases} x' = -x - 7y \\ y' = 7x - 2y \end{cases}$$

where we have changed the double primes back to primes. Similarly, if we reverse the mappings we have

$$x'' = 2(x - y) - 3(2x + y) = 2x - 2y - 6x - 3y = -4x - 5y$$
$$y'' = 3(x - y) + 4(2x + y) = 3x - 3y + 8x + 4y = 11x + y$$

or

$$fg: \begin{cases} x' = -4x - 5y \\ y' = 11x + y \end{cases}$$

Thus, $gf \neq fg$, and we would say that f and g **do not commute.** ■

By applying the definitions given previously, we can obtain the next two results.

THEOREM 2: The product of an even number of opposite linear transformations is direct, and the product of an odd number is an opposite transformation.

LEMMA B: The product of any number of isometries is an isometry.

THEOREM 3: TRANSLATION MAPPING

The product of two line reflections s_ℓ and s_m, where ℓ and m are parallel lines, is distance and slope preserving, and maps a given line n to one that is parallel to it.

PROOF

The lemma proves the first part; for the second part, we have:

(1) If line n is parallel to ℓ and m, then $n' \equiv s_\ell(n)$ is parallel to ℓ, hence to m, and $n'' = s_m(n')$ is parallel to m, hence to both m and n.

(2) Otherwise, n meets both ℓ and m (Figure 5.10). Let $n' = s_\ell(n)$ and $n'' = s_m(n') \equiv s_m s_\ell(n)$, as before. By the *ABCD* Property, $\angle ABC \cong \angle A'B'C' \equiv \angle A'BC$ and therefore $\angle 1 \cong \angle 2$.

(3) Similarly, $\angle 4 \cong \angle 5$ [because $s_m(n') = n''$].

(4) By the F- and Z-Properties of Parallelism, respectively, $\angle 1 \cong \angle 3$ and $\angle 2 \cong \angle 4$.

(5) Therefore, $\angle 3 \cong \angle 1 \cong \angle 2 \cong \angle 4 \cong \angle 5$, and $\angle 3 \cong \angle 5$.

(6) By the Z-Property of Parallelism, $n \parallel n''$.

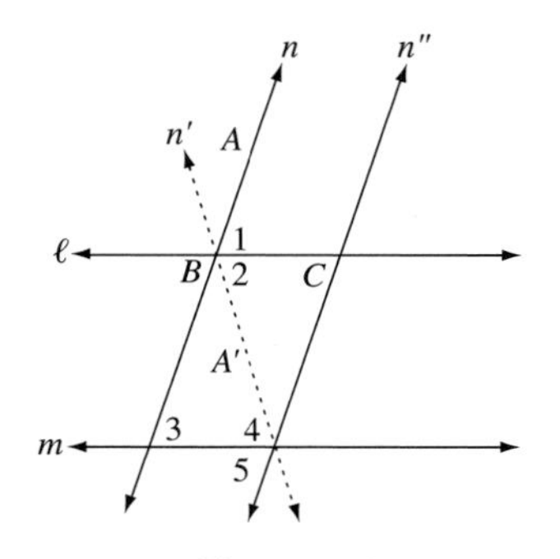

Figure 5.10

QUESTION: Is Theorem 3 true if lines ℓ and m are not parallel? (You should get a negative answer because of Step (4) in the above proof.)

AN APPLICATION IN PHYSICS

A law of optics which governs the effect of light rays reflecting from a surface, such as a plane mirror (Figure 5.11), is the following: The light ray OA travels a broken straight line path $OABCE$ lying in a plane that is perpendicular to the reflecting surface, such that the **angle of incidence** (the measure of the angle that the incoming ray $\overrightarrow{AB}$ makes with the normal to the reflecting surface) equals the **angle of reflection** (the measure of the angle that the outgoing ray $\overrightarrow{BC}$ makes with the normal).

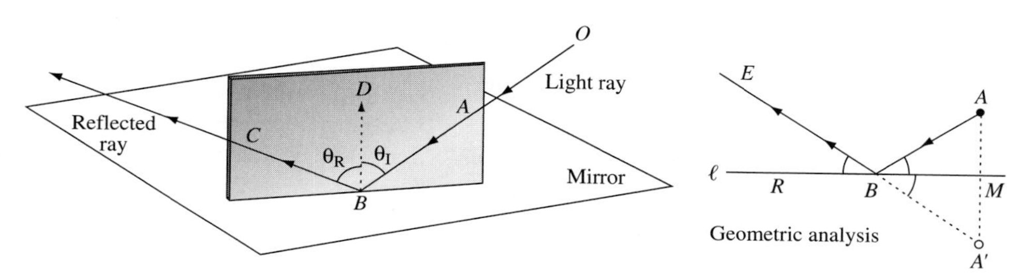

Figure 5.11

Thus, in Figure 5.11, $m\angle ABD = m\angle DBC$ ($\theta_I = \theta_R$). For example, if the light ray comes in at an angle of 55° with the normal (or 35° with the surface of the mirror), it leaves the mirror at an angle of 55° with the normal (or 35° with the mirror).

A geometric analysis of this shows that if the eye of an observer is at E, which sees an object at A through the mirror along the line ℓ, the position of the object appears to the observer to be at the point A', the image of A *under the geometric line reflection* s_ℓ, *where* $\ell = \overleftrightarrow{RM}$.

PROOF

The previously mentioned law of optics implies that $\angle ABM \cong \angle EBR \cong \angle MBA'$ (Vertical Angle Theorem). Assuming that the depth of the image from B as seen in the mirror equals the distance from B to the object at A, that is $BA' = BA$, then $\triangle ABM \cong \triangle A'BM$ by SAS, and $\overleftrightarrow{MB}$ is the perpendicular bisector of segment $\overline{AA'}$. Hence, $A' = s_\ell(A)$.

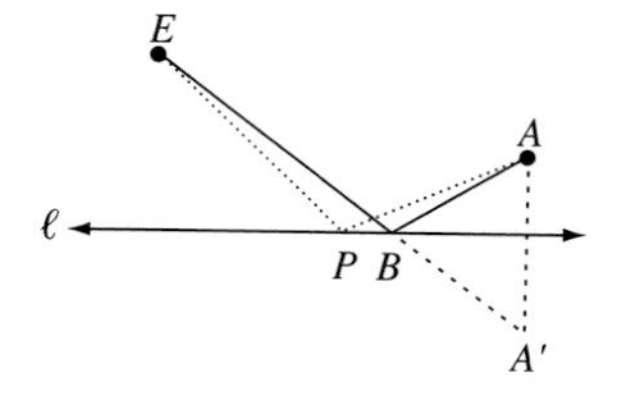

Figure 5.12

A mathematical reason for this law of optics has been proposed: The ray of light follows the path of *shortest distance* from A to E, by way of a point on the mirror (that is, from A to B to E). In other words, the ray of light takes a path so as to minimize the total distance of travel $AP + PE$ from A to E, where P is any point on line ℓ (Figure 5.12). Can you prove that $EB + BA < EP + PA$? Write $EB + BA = EB + BA' = EA' < EP + PA' = EP + PA$.)

The law of optics previously mentioned also governs the motion of a projectile when it strikes a resilient surface, as in billiards, and bounces off it or **caroms,** as the phenomenon is sometimes called. Again, the angle of incidence equals the angle of reflection, as illustrated in Figure 5.13. Further ideas on this are taken up in the problem section that follows.

THE GAME OF BILLIARDS

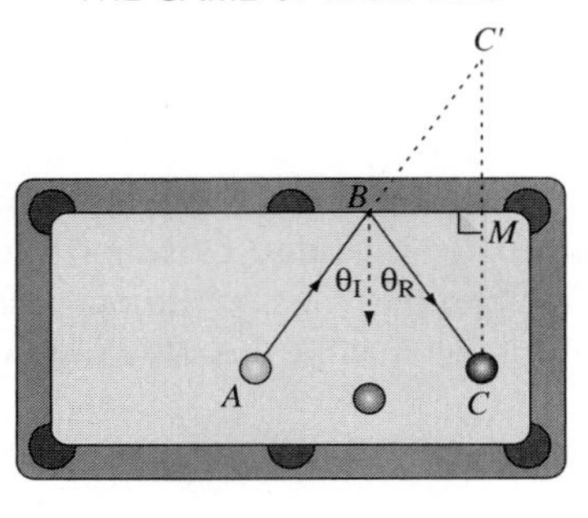

Figure 5.13

y
A(1, 3)
B(5, 1)
1
P(t, 0)
4
5
x
B'(5, –1)

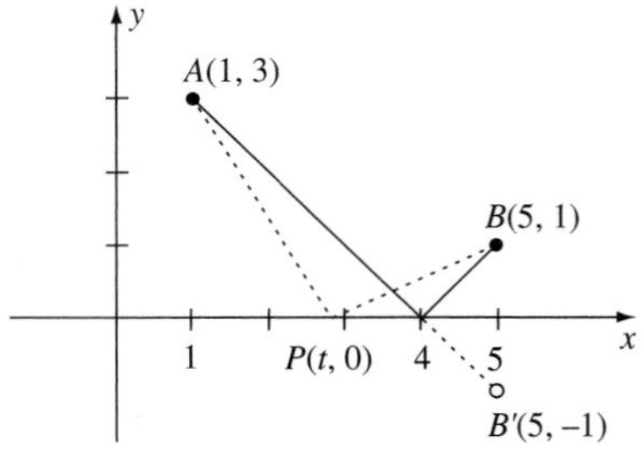

Figure 5.14

C EXAMPLE 2[1] Show that the answer obtained by calculus for minimizing the total distance $AP + PB$, where $A = (1, 3)$, $B = (5, 1)$, and $P = (t, 0)$ for some real t, $1 \le t \le 5$, is the same as that obtained by geometry, taking P as the intersection of the line $\overleftrightarrow{AB'}$ with the x-axis, where $B' = (5, -1)$, the reflection of B in the x-axis. (See Figure 5.14.)

SOLUTION

(a) By the distance formula,

$$AP = \sqrt{(t-1)^2 + (0-3)^2} = \sqrt{t^2 - 2t + 10}$$
$$BP = \sqrt{(t-5)^2 + (0-1)^2} = \sqrt{t^2 - 10t + 26}$$

We want to minimize the function

$$F(t) = AP + BP = \sqrt{t^2 - 2t + 10} + \sqrt{t^2 - 10t + 26}$$

so we take its derivative. After simplifying, we find

$$F'(t) = \frac{t-1}{\sqrt{t^2 - 2t + 10}} + \frac{t-5}{\sqrt{t^2 - 10t + 26}}$$

Setting $F'(t) = 0$ and squaring, we obtain, again with a little algebraic manipulation,

$$\frac{t^2 - 2t + 1}{t^2 - 2t + 10} = \frac{t^2 - 10t + 25}{t^2 - 10t + 26}$$
$$1 - \frac{9}{t^2 - 2t + 10} = 1 - \frac{1}{t^2 - 10t + 26}$$

or, after canceling the ones and cross-multiplying,

$$t^2 - 11t + 28 = 0 \quad \text{or} \quad (t-4)(t-7) = 0$$

---

[1]Makes use of calculus.

The only root lying on [1, 5] is $t = 4$, which produces the answer $P(4, 0)$.
(b) The equation of $\overleftrightarrow{AB'}$ is

$$y - 3 = -(x - 1) = -x + 1 \quad \text{or} \quad y = 4 - x$$

This line cuts the x-axis where $y = 0$, or $x = 4$, in agreement with **(a)**. (*Geometry is better!*) ■

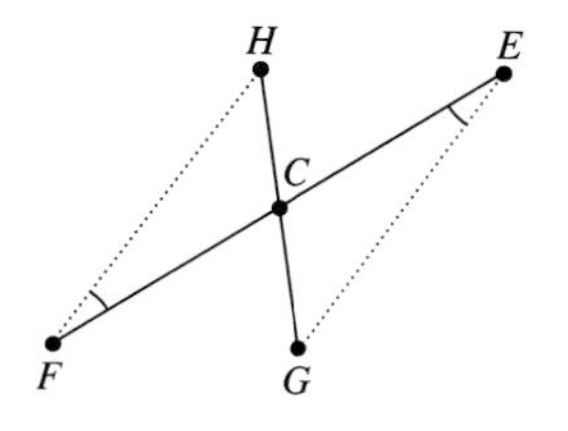

Figure 5.15

EXAMPLE 3 In Figure 5.15, $s_C(E) = F$ and $s_C(G) = H$. Prove that $\angle GEC \cong \angle HFC$ without using any part of Theorem 1 (*ABCD* Property).

SOLUTION
By definition of s_C, C is the midpoint of segments $\overline{EF}$ and $\overline{GH}$, and $\angle GCE \cong \angle HCF$ by the Vertical Pair Theorem. By SAS, $\triangle GCE \cong \triangle HCF$ and by CPCF, $\angle GEC \cong \angle HFC$. ■

Moment for Discovery

Do Reflections Preserve Orientation?

Suppose $\diamond ABCD$ is a quadrilateral that has been positively oriented. Consider the image quadrilateral $\diamond A'B'C'D'$ under a line reflection with axis ℓ, then under a point reflection with center L.

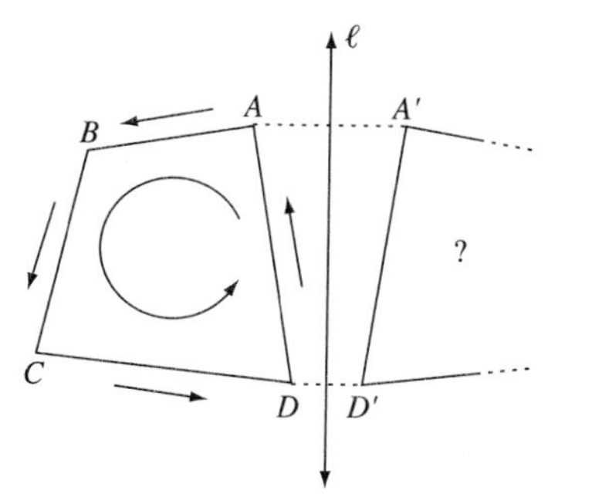

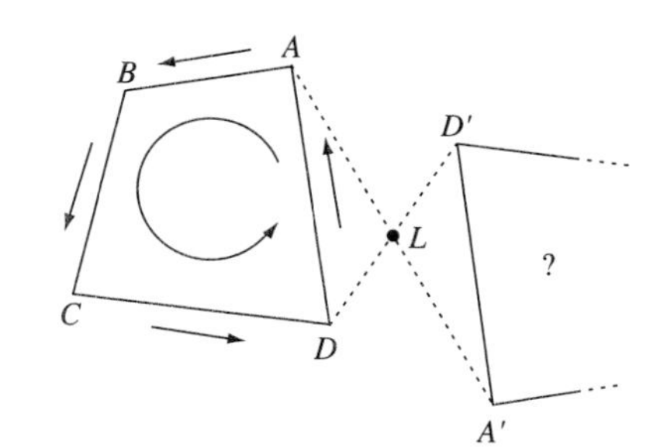

Figure 5.16

1. On your own paper, finish the construction of the image quadrilateral $\diamond A'B'C'D'$ as carefully as you can in each case.
2. Look at the order of the image vertices, and decide whether the orientation of the image is clockwise or counterclockwise in each case.
3. Did you detect any difference in the orientation-preserving properties of the two types of reflections?

PROBLEMS (5.2)

GROUP A

1. The block letter A of the alphabet has a vertical line of symmetry.
   (a) Which of the block letters K, M, L, and X have vertical lines of symmetry?
   (b) Which letters of the alphabet have some line of symmetry?
2. (a) Decipher the message in the figure below.
   (b) Which letter is written incorrectly? [UCSMP, p. 161]

Help! I'm trapped
inside this page!

3. In the accompanying figure we are given that $s_\ell(C) = D$. Prove that $CQ = DQ$ without using any part of Theorem 1. [UCSMP, p. 184]

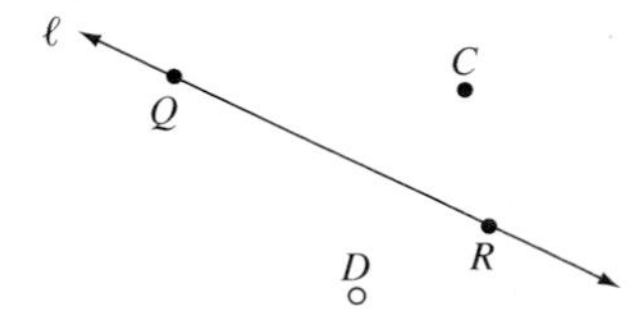

4. Use the information in the figure and prove that $VC = CX$, making use of reflections. [UCSMP, p. 184]

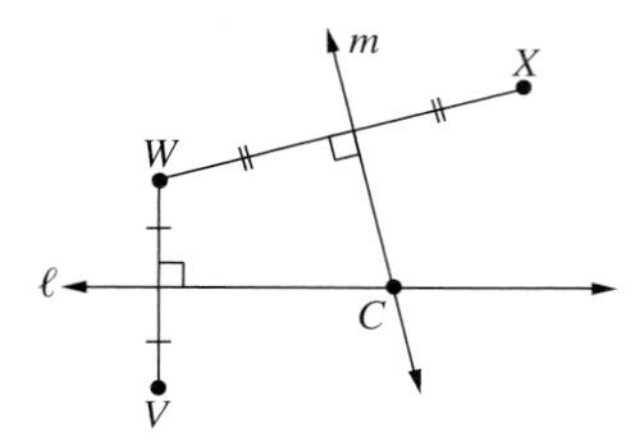

5. Find the equation of the line of reflection ℓ if $s_\ell(-4, 5) = (6, 7)$.
6. Show that the mapping given below is
   (a) one to one
   (b) is not an isometry.

$$f:\begin{cases} x' = 3x \\ y' = y + 2 \end{cases}$$

7. Find the coordinate equation for each of the composition mappings gf and fg if

$$f:\begin{cases} x' = y \\ y' = 2x \end{cases} \qquad g:\begin{cases} x'' = 5x' + 3y' \\ y'' = 6x' + 5y' \end{cases}$$

(***Hint:*** For fg you will need to switch primes and double primes before making substitutions, as in Example 1.)

8. The following transformation f represents the reflection in the line $y = x$, while g is the reflection in the parallel line $y = x + 2$. (Verify this by trying to calculate the images of a few choice points in each case.) Find, by substitution, the coordinate forms for gf and fg.

$$f:\begin{cases} x' = y \\ y' = x \end{cases} \qquad g:\begin{cases} x'' = y' - 2 \\ y'' = x' + 2 \end{cases}$$

NOTE: Any coordinate mapping of the form $x' = x + a$, $y' = y + b$ is a *translation*, which maps the origin to the point (a, b) and displaces every other point by an equal amount.

9. Using the definitions for s_ℓ and s_C, prove that s_ℓ^2 and s_C^2 are each the identity mapping. That is, show that for any point P,

$$s_\ell^2(P) \equiv s_\ell[s_\ell(P)] = P \qquad \text{and} \qquad s_C^2(P) = P.$$

10. Derive a compass/straight-edge construction for the image P' of a point P that is reflected in

(a) a line

(b) a point.

GROUP B

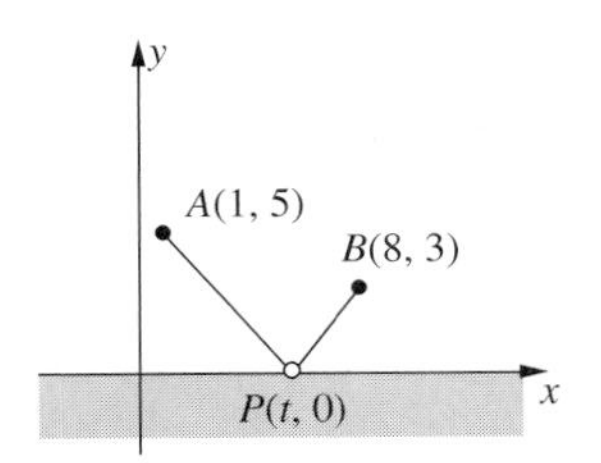

C11. Taking the x-axis as a straight river bank, at what point $P(t, 0)$ along the x-axis should pipe be laid to join two towns $A(1, 5)$ and $B(8, 3)$ to a pumping station at P so as to minimize the amount of pipe used? (Units are in miles.) Work this in two ways:

(a) by calculus

(b) by geometry, using reflections.

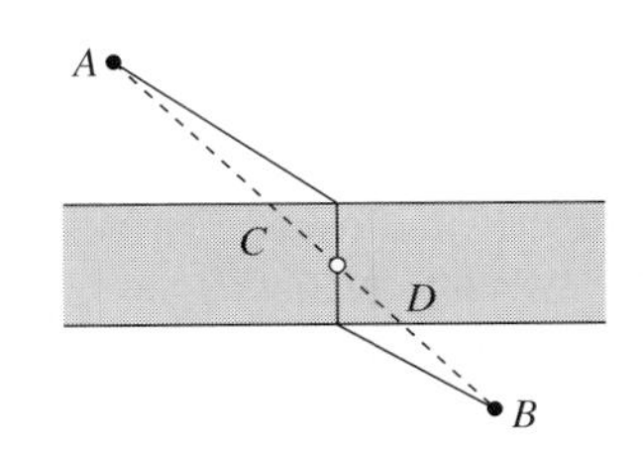

12. The Bridge Problem A road and bridge combination is being planned joining two towns, A and B, separated by a river. Assume that the river banks are straight parallel lines, and that the bridge will be built perpendicular to the river banks. Where should the bridge be built so that the total distance across the bridge and along the straight roadway between the towns and bridge be minimized?

NOTE: The seemingly logical solution depicted in the figure, where the midpoint of the bridge is the midpoint between points C and D lying on the straight line joining A and B, does *not* give the minimum distance. (The solution does not directly involve reflections.)

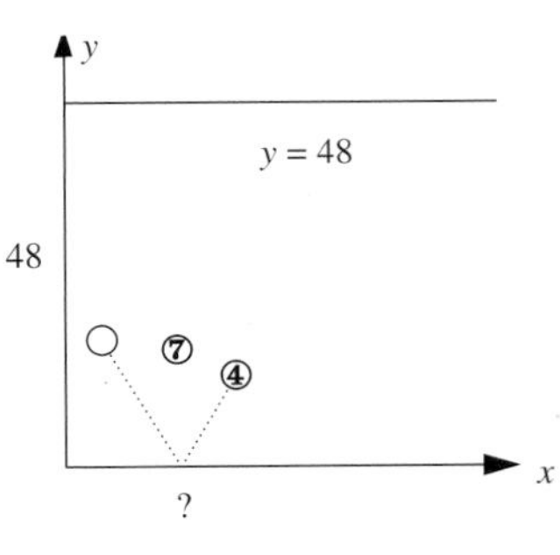

13. With the coordinate system shown in the figure, the 7-ball is blocking the cue ball at $A(5, 15)$ from the 4-ball at $B(25, 10)$. In the game of snooker, the balls must be pocketed in numerical order. At what point

(a) along the x-axis (cue ball can also strike opposite side, $y = 48$)

(b) along the y-axis (cue ball can also strike the x-axis)

(c) on the opposite side of the table $y = 48$

should the cue ball be aimed to squarely strike the 4-ball after one carom off that particular side, or sides, of the table? (Units are in inches.)

14. Darlene is playing miniature golf. Her golf ball ended up at point A, and a barrier blocks a direct shot into the hole at B. By making a geometric construction, show how to locate precisely the desired point P (as illustrated) so the golf ball will bounce to point Q, and then into the hole at B.

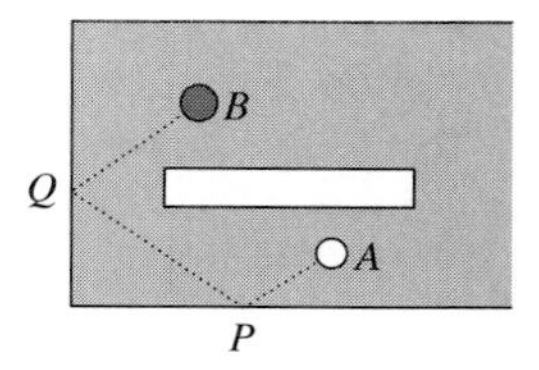

***15.** Where should a cue ball be located on an empty, triangular pool table with no pockets, and in what direction should it be hit, so that the ball will return to its original position after three caroms, one off each side?

*See Problem 20, Section 4.3.

16. Work Problem 15 assuming the fictitious pool table is a rectangle, allowing four caroms instead of three.

GROUP C

17. The given transformation f has a line ℓ of fixed points and is a reflection in that line. Find the coordinate equation of ℓ, and then verify that ℓ is actually the line of reflection. (That is, verify that ℓ is the perpendicular bisector of $\overline{PP'}$ for any point P. (***Hint:*** Find the fixed points.)

$$f: \begin{cases} x' = \frac{3}{5}x + \frac{4}{5}y + 2 \\ y' = \frac{4}{5}x - \frac{3}{5}y - 4 \end{cases}$$

18. Prove that any isometry with a line of fixed points (and no others) is either the identity or a reflection in that line. What is true if the isometry has just one fixed point?

19. Kaleidoscope Effect, 90 Degrees Using two mirrors along perpendicular lines ℓ and m, the eye at E will see four copies of triangle T, symmetrically distributed about O, as shown. Explain this phenomenon.

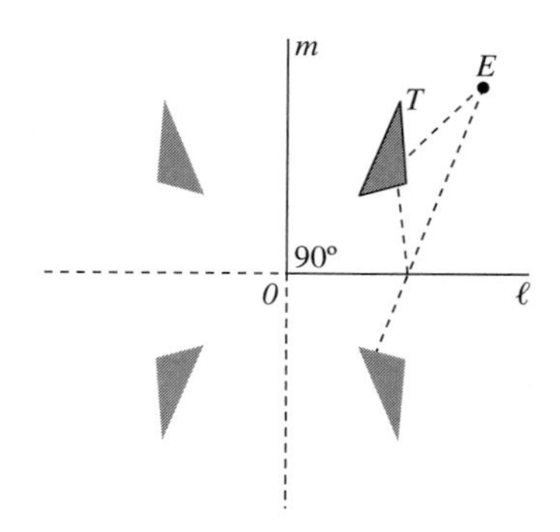

20. Kaleidoscope Effect, 60 Degrees Using two mirrors along lines ℓ and m, forming an angle of measure 60, the eye at E will see six copies of triangle T, symmetrically distributed about O, as shown. Explain this phenomenon.

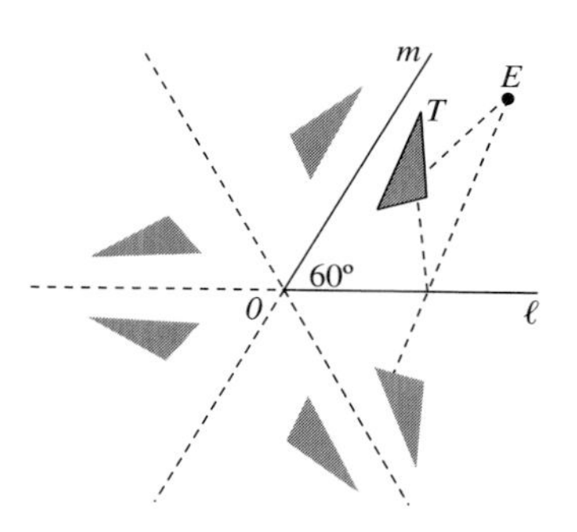

5.3 Translations, Rotations, and Other Isometries

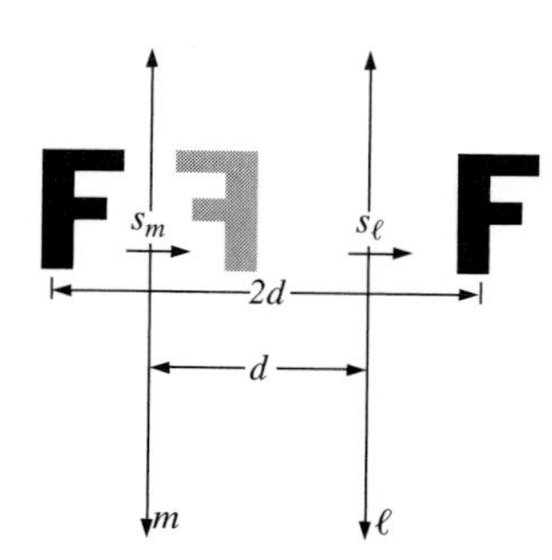

Figure 5.17

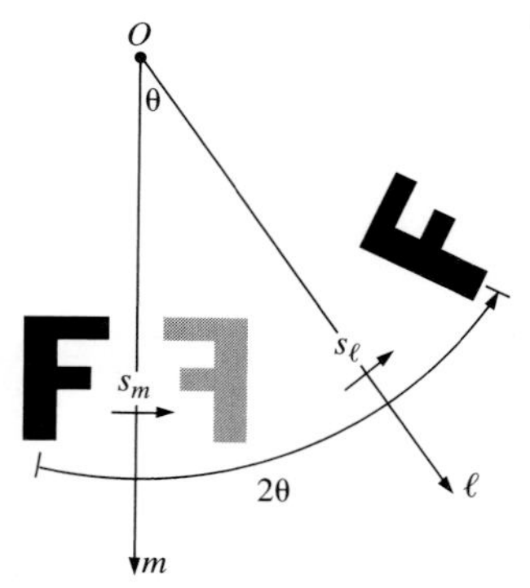

Figure 5.18

We now take a brief look at the two basic Euclidean motions, the translations and rotations. In this section, a *reflection will always be with respect to some line.* Point reflections will not be used.

Figure 5.17 is an example of the product of two reflections over parallel lines. The image of the original configuration (the block letter "F") looks as though it could be obtained by directly sliding the figure to the new position a distance $2d$. In the example of Figure 5.18, the same configuration ("F") has been reflected through two *intersecting lines.* Here it appears as though the letter "F" has been *rotated about point O*, the point of intersection of the two lines.

DEFINITION: A **translation** in the plane is the product of two reflections $s_\ell s_m$, where ℓ and m are parallel lines. A **rotation** is the product of two reflections $s_\ell s_m$, where ℓ and m are lines that meet, and the point of intersection is called the **center** of the rotation.

Since each individual reflection defining a translation or rotation has the *ABCD* property (angle measure, betweenness, collinearity, and distance are invariant), clearly the translation or rotation itself has the *ABCD* property. But a translation or rotation will have the additional property of *preserving orientation* as well, according to Theorem 2 of Section 5.2.

To work effectively with translations, it is important to know how to construct the correct parallel lines ℓ and m so that $s_\ell s_m$ will translate one figure to another. The two discovery units at the end of this section were designed to allow you to discover the secret for yourself.

EXAMPLE 1 Show that the usual definition of a translation (as the displacement of all points at equal distances along a direction parallel to a fixed vector) is in agreement with the synthetic definition using line reflections, and that the coordinate form of a translation is

$$f: \begin{cases} x' = x + a \\ y' = y + b \end{cases}$$

where a is the "change in x" and b is the "change in y."

SOLUTION

In Figure 5.19, we want to show that in terms of vectors, the displacement from P to P' equals the given displacement from O to $C(a, b)$ for all points $P(x, y)$. It suffices to show that $\diamond OCP'P$ is a parallelogram. Recall Theorem 3, Section 5.2, where it was

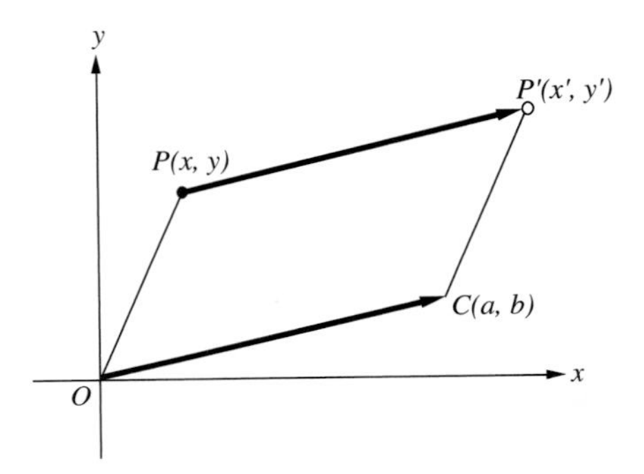

Figure 5.19

shown that a translation maps any line to one that is parallel to itself. In Figure 5.19, O is mapped to $C(a, b)$ and $P(x, y)$ is mapped to $P'(x', y')$, so line $\overleftrightarrow{CP'}$ must be parallel to line $\overleftrightarrow{OP}$. Also, since a translation is distance preserving,

$$CP' = OP$$

That is, $\diamond OCP'P$ has two sides that are both parallel and congruent. Therefore, $\diamond OCP'P$ is a parallelogram. Hence, $\overleftrightarrow{OC} \| \overleftrightarrow{PP'}$ and $OC = PP'$. Thus, it follows that directed line segments $\boldsymbol{PP'}$ and $\boldsymbol{OC}$ are represented by the same vector:

$$v(\boldsymbol{PP'}) = v(\boldsymbol{OC})$$

which, by **(4)** of Section 4.7, gives us

$$[x' - x, y' - y] = [a - 0, b - 0]$$

Hence,

$$x' - x = a, \qquad y' - y = b \qquad \text{or} \qquad x' = x + a, \qquad y' = y + b. \qquad \blacksquare$$

NOTE: A useful observation about translations, made clear by the previous example, is that a translation is completely determined by specifying one point A and its corresponding image $A' = B$. Thus an appropriate notation for such a translation would be t_{AB}. A rotation r is likewise completely determined by specifying its center C and angle of rotation θ. Hence, we can denote this by $r_\theta[C]$.

EXAMPLE 2 Find the rotation $s_m s_\ell = r_{64}[C]$ that maps the letter "G" to its new position in Figure 5.20 (left diagram), if line ℓ has already been located, as shown. That is, locate line m. Show how the two line reflections actually work with respect to one of the points of G being rotated.

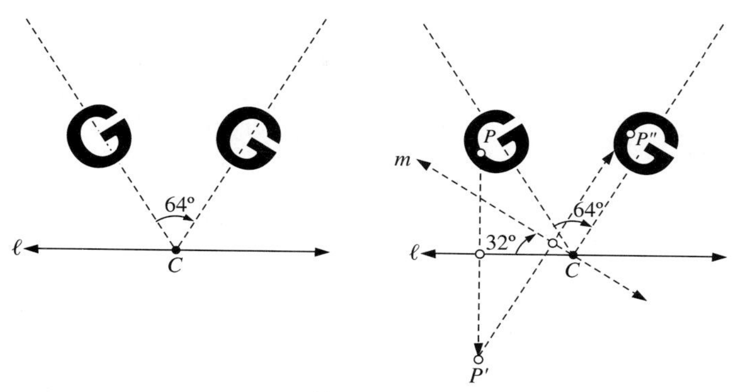

Figure 5.20

SOLUTION

Line m must make an angle of measure $\frac{1}{2} \cdot 64 = 32$ with line ℓ, so we draw line m through point C accordingly. To see how this works, let point P be the point opposite the "hook" in the letter "G," as shown, and reflect P to P', then P' to P''. ■

ISOMETRIES AS PRODUCTS OF REFLECTIONS, TRANSITIVITY PROPERTY

An interesting sequence of results ultimately shows that *any isometry is the product of at most three reflections.* The translation and rotation are special cases, requiring two reflections each to characterize them. The first step is to show that an isometry exists that maps a given triangle into one that is congruent to it.

> THEOREM 1: Given any two congruent triangles $\triangle ABC$ and $\triangle PQR$, there exists a unique isometry that maps the first triangle onto the second.

The proof of this theorem has two major parts:

(1) prove that the isometry *exists.*

(2) prove that it is *unique*.

We tackle the uniqueness property first.

LEMMA: The only isometry that has three noncollinear fixed points is the identity mapping e (which fixes all points in the plane).

PROOF

(1) First we show that if T is an isometry and $A' = T(A) = A$, $B' = T(B) = B$ for $A \neq B$, then every point P on segment $\overline{AB}$ obeys the property $P' = T(P) = P$, and is a fixed point. (See Figure 5.21.)

Suppose that A–P–B. Then A'–P'–B' or A–P'–B, since an isometry preserves betweenness (Lemma A of Section 5.2). But $AP' = AP$, and by the Segment Construction Theorem, there is only one such point P, hence $P' = P$.

(2) Now assume T also fixes some point C not on line $\overleftrightarrow{AB}$. Thus ℓ and m pass through a given point P in the plane so that both lines cut two of the sides of $\triangle ABC$ at certain points D, E and F, G, respectively. Since, by **(1)**, $D' = D$, $E' = E$, $F' = F$, and $G' = G$, T fixes the two lines, hence T fixes P, their point of intersection. Since P was arbitrary, T fixes every point of the plane and is the identity.

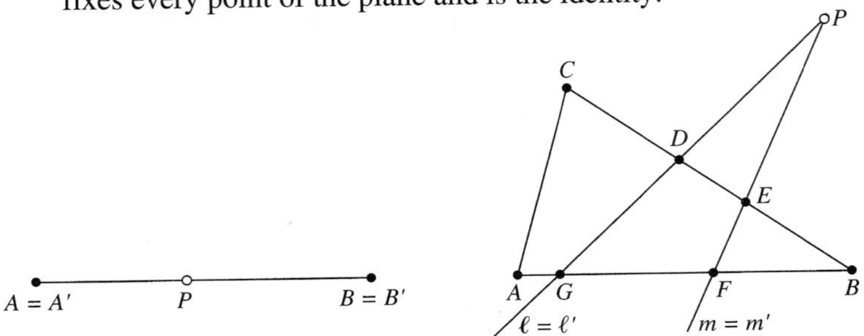

Figure 5.21

COROLLARY: If f and g are two isometries and they both map $\triangle ABC$ to $\triangle PQR$, then $f = g$.

PROOF

Consider the product $h = g^{-1}f$. By the preceding Lemma, $g^{-1}f$ is the identity mapping since it takes $\triangle ABC$ to itself (A goes to P, then P goes back to A, etc.). Hence if X is any point,

$$X = g^{-1}[f(X)]$$

$$g(X) = g\{g^{-1}[f(X)]\} = f(X)$$

That is, $g(X) = f(X)$ for all X, hence $f = g$.

It remains to be shown how to construct the unique isometry that maps a given triangle $\triangle ABC$ onto another given triangle $\triangle PQR$ congruent to it ($\triangle ABC \cong \triangle PQR$). If $A \neq P$, construct the perpendicular bisector ℓ of segment $\overline{AP}$ and, via ℓ, reflect A to $A_1 = P$, B to B_1, and C to C_1. (If $A = P$, then just apply this argument to any pair of distinct, corresponding vertices; if none exists, then the identity map does the job trivially.) If $B_1 \neq Q$, construct line m, the perpendicular bisector of $\overline{B_1Q}$ (which passes through $A_1 = P$ since $A_1B_1 = AB = PQ$), and reflect A_1 to $A_2 = P$, B_1 to $B_2 = Q$ and C_1 to C_2. If $C_2 = R$, we are finished; otherwise, use one more reflection in line $\overleftrightarrow{PQ}$ to map C_2 to R. (Why is $\overleftrightarrow{PQ}$ the perpendicular bisector of segment $\overline{RC_2}$?)

Figure 5.22 shows the successive reflections we have constructed in the proof; in the example illustrated in Figure 5.22, three reflections are actually required. Thus, we have completed the proof of Theorem 1. Indeed, using uniqueness, we have also proven

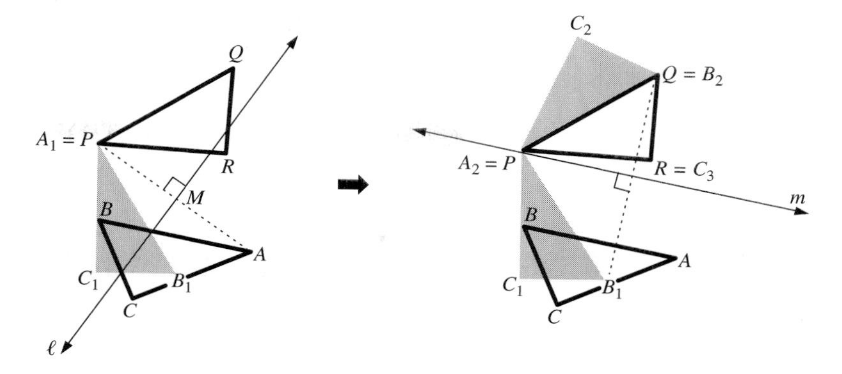

Figure 5.22

> THEOREM 2: FUNDAMENTAL THEOREM ON ISOMETRIES
> Every isometry in the plane is the product of at most three reflections, exactly two if the isometry is direct and not the identity.

COROLLARY: A nontrivial direct isometry is either a translation or rotation.

Moment for Discovery

Translations By Reflections

In Figure 5.23, a triangle $\triangle ABC$ is shown, and we want to find a translation $s_m s_\ell$ which will map $\triangle ABC$ to a new *given* position, $\triangle DEF$. It is pretty clear that since the direction of the desired translation is the directed line segment $\overline{AD}$, the two lines ℓ and m need to be *perpendicular* to line $\overleftrightarrow{AD}$. The next job is to find out where to locate these two parallel lines. Try the following experiment.

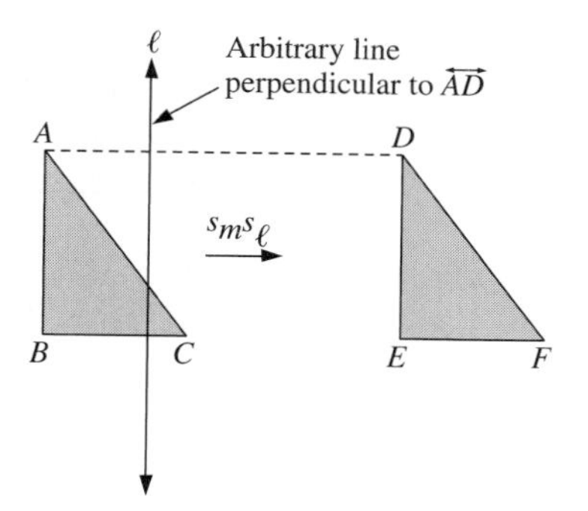

Figure 5.23

1. Draw any line ℓ perpendicular to $\overleftrightarrow{AD}$, such as that shown here. (In your figure, use a different position for ℓ, but closer to $\triangle ABC$ than $\triangle DEF$.)

2. Using your best estimate, or by construction, draw line m parallel to line ℓ at the distance $d = \frac{1}{2}AD$ from ℓ.
3. Carefully locate, according to your best estimates, the points A', B', and C', which are the reflected images of A, B, and C, respectively, in line ℓ.
4. Carefully locate, according to your best estimates, the points A'', B'', and C'', the reflected images of A', B', and C', respectively, in line m.
5. What seems to be the result?

Moment for Discovery

Translations By Reflections (using *Sketchpad*)

1. Construct a right triangle ABC, locate point D outside it, then translate $\triangle ABC$ to $\triangle DB'C'$ using Mark Vector A to D. (See Figure 5.24.)

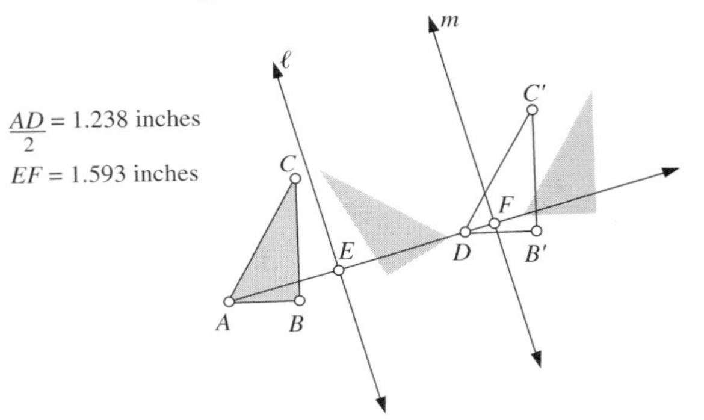

Figure 5.24

2. Construct ray $\overrightarrow{AD}$ and locate point E on it. Construct a perpendicular line ℓ to $\overrightarrow{AD}$ at E.
3. Locate a second point F on ray $\overrightarrow{AD}$ and construct a second perpendicular m at F.
4. Construct the interior of $\triangle ABC$ using the CONSTRUCT command and Polygon Interior.
5. Using the TRANSFORM command, reflect Interior ($\triangle ABC$) in line ℓ [double click on ℓ, select Interior ($\triangle ABC$), and choose Reflect under TRANSFORM.]
6. While still selected, reflect the image of Interior ($\triangle ABC$) in line m.
7. Measure the distance AD and display $AD/2$. Measure the distance EF. Hide Distance AD.

Now drag line m to various positions. Does anything happen? Can you make the shaded region match up with $\triangle DB'C'$? Are the distances EF and $AD/2$ ever equal? Also drag line ℓ as well as line m, and see if you can still match the shaded region with $\triangle DB'C'$.

8. Investigate a similar phenomenon with respect to rotations. Find out whether the rotation $s_m s_\ell$ depends on the choice of lines ℓ and m passing through the center of rotation if the angle between ℓ and m remain the same.

Moment for Discovery

Constructing a Rotation

Follow these steps in *Sketchpad* to find a rotation $s_m s_\ell$ that will map a given figure onto another congruent to it.

1. Construct any triangle $\triangle ABC$. Locate any point, choose it as a center for rotation by double-clicking on it, then rotate $\triangle ABC$ to $\triangle A'B'C'$. Make note of the angle of rotation you used. Finally, translate $\triangle A'B'C'$ to some other position using the Copy/Paste technique under FILE. The new triangle will have the same labels.
2. Hide all objects of construction, including the intermediate triangle. You should now have $\triangle ABC \cong \triangle A'B'C'$ on the screen, with the same orientation, the result of a rotation followed by a translation. The Fundamental Theorem (Theorem 2) states that just two line reflections, that is, either a translation or rotation, will map $\triangle ABC$ to $\triangle A'B'C'$.
3. Construct line $\overleftrightarrow{AA'}$, locate a point on it, and construct a line $\ell \perp \overleftrightarrow{AA'}$ at that point.
4. Select the perpendicular line you just constructed as a line of reflection by double-clicking on it, then reflect $\triangle ABC$ through that line. The new triangle is labeled $\triangle A'B'C'$; we refer to it as $\triangle EFG$.
5. Repeat the procedure of Step 4 with vertex $B' = F$ of the last triangle you constructed and B'. That is, construct line $\overleftrightarrow{FB'}$, locate a point on it, construct $m \perp \overleftrightarrow{FB'}$, and reflect $\triangle EFG$ in line m to a new triangle we call $\triangle PQR$. Is $\triangle PQR$ a rotation or translation of $\triangle ABC$?
6. Select the vertices of the last triangle you constructed ($\triangle PQR$) and choose Polygon Interior under CONSTRUCT to shade in this triangle. Hide the vertices and sides of the triangle, leaving just the interior shading. Hide $\triangle EFG$ (sides and vertices).

By dragging lines ℓ and m, see if you can position the shaded region to exactly fit inside $\triangle A'B'C'$. What has to happen? Select the point of intersection S of lines ℓ and m, and measure $\triangle ASA'$. Is this angle of rotation the same as that used originally? What theorem did you discover about the product of a rotation and translation?

PROBLEMS (5.3)

GROUP A

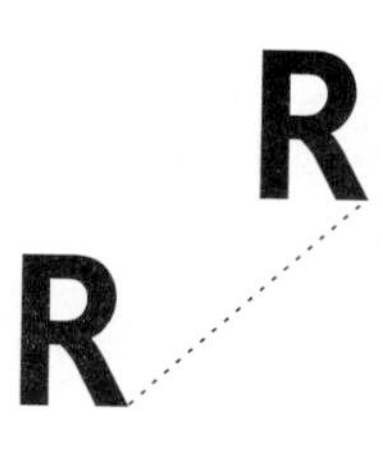

1. On your paper, sketch the diagram in the accompanying figure showing a translation of the letter "R." Draw a pair of lines ℓ and m such that "R" maps to its translated image under $s_m s_\ell$.

2. On your paper, sketch the following figure showing a rotation of the letter "R." Draw a pair of lines ℓ and m such that "R" maps onto its rotated image. Estimate, by protractor, the angle of rotation. (***Hint:*** Begin by choosing corresponding points and drawing the perpendicular bisector of the segment joining them, following some of the steps of the proof of Theorem 1.)

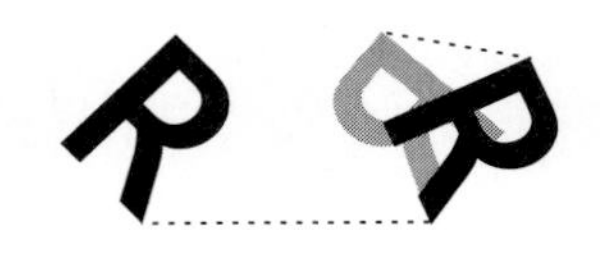

3. When the letter "S" is reflected over line ℓ, and then some line m, its final image is that shown in the accompanying figure. But someone erased line m. Trace the figure and put line m back in. [UCSMP, p. 264]

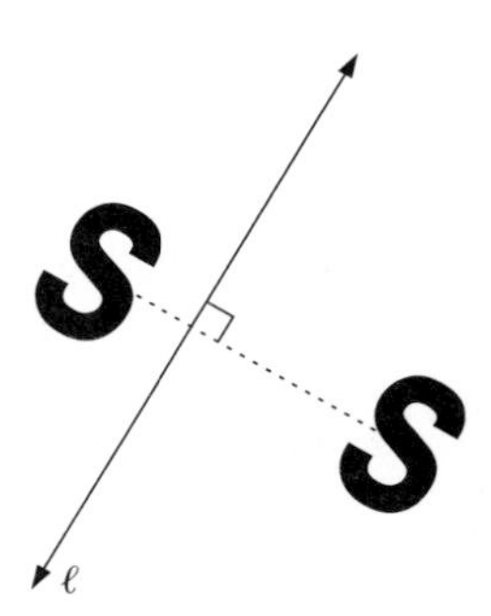

4. Carla has just had her hair cut. With her back to a mirror on the wall, she holds a hand mirror in front of her face. If Carla's eyes are 1 ft. from the hand mirror, her head is 8 in. thick, and the back of her head is 3 ft. from the wall mirror, about how far from her eyes will the image of the back of Carla's head appear in her mirror? [UCSMP, p. 264]

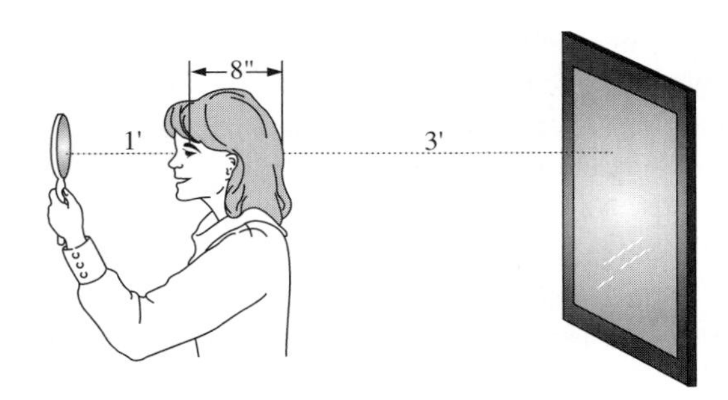

5. The transformation with the rule $T(x, y) = (x + 2, y + 6)$ is a translation.
   (a) Graph (7, 3) and $T(7, 3)$.
   (b) Find the slope of the line through (7, 3) and $T(7, 3)$, its image. (The slope helps to indicate the direction of the translation.)
   (c) Describe in words the effect of T on a figure. [UCSMP, p. 264]
6. Find the coordinate equations for the translation that maps (2, 5) to $(6, -4)$.
7. If one translation is followed by another, the result is a translation. Verify this by finding $t_2 t_1$ if

$$t_1: \begin{cases} x' = x + 2 \\ y' = y + 6 \end{cases} \qquad t_2: \begin{cases} x'' = x' + 2 \\ y'' = y' + 2 \end{cases}$$

8. Reflect the point (4, 2) over the y-axis. Reflect its image over the line $y = x$ (interchange coordinates).

(a) What is the final image?

(b) What rotation has taken place? Through what angle?

9. Derive the coordinate equations for a half-turn about $C(a, b)$. (***Hint:*** Use the midpoint formula.)

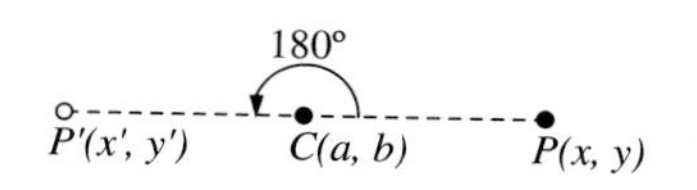

10. Give a compass, straight-edge construction of the rotational image of segment $\overline{AB}$ by the given angle in the figure.

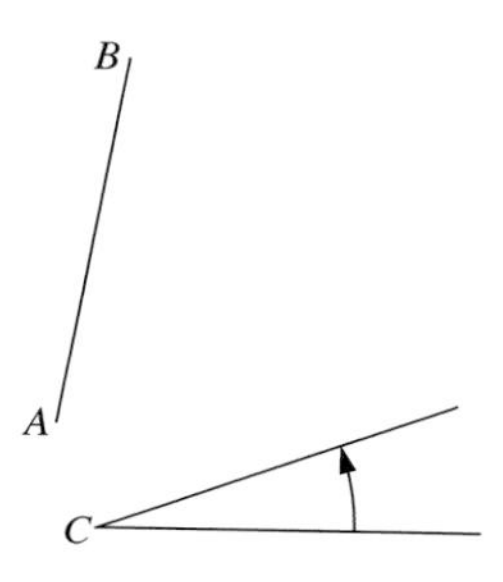

11. Show the compass, straight-edge construction of the parallelogram needed to obtain the translation of a point P by a given position vector $\boldsymbol{OA}$ when

(a) P does not lie on line $\overleftrightarrow{OA}$

(b) P lies on line $\overleftrightarrow{OA}$.

12. Given two points P and C in the plane, can a compass, straight-edge construction be given that will produce the image of P when it is rotated about C through an angle of 45° in the clockwise direction?

13. Use the coordinate equations of a translation to show that a translation is an isometry.

GROUP B

14. Consider the rotation $f = s_m s_\ell$, where ℓ and m intersect at point C, and $\angle ACB$ is either of the acute (or right) angles formed by ℓ and m, with $\theta = m\angle ACB$.

(a) Prove synthetically that if f maps point P to point P', then $m\angle PCP' = 2\theta$. Be sure to examine the cases when P is on either side of ℓ.

(b) What happens when $\theta = 90$? Prove your answer.

15. Point B is the image of A under a rotation in the figure to the right.

(a) Identify a point that could be the center of the rotation.

(b) Identify another point that could be the center.

(c) Identity a third point that could be the center.

(d) Generalize parts **(a)**, **(b)**, and **(c)**. [UCSMP, p. 272]

16. Point B is the image of A under a clockwise rotation through an angle of measure 40. Show that this information uniquely identifies the center of rotation C, and give its exact location in terms of A, B and line segment $\overline{AB}$.

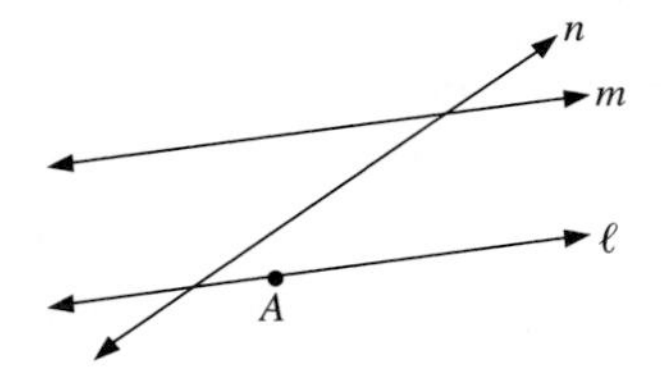

17. Suppose that $t = s_m s_\ell$ is a translation and that $r = s_\ell s_n$ is a rotation. Identify the composite (product) mapping tr. Start with a point A (as shown) and trace its movement under the reflections involved to reach your conclusion. Determine and describe the exact location of the final image of A, then describe the composition mapping tr in general. (***Hint:*** Simplify the composition $tr = s_m s_\ell s_\ell s_n$ using the result of Problem 9, Section 5.2.)

18. In the figure, $\triangle PQR \cong \triangle UVW$. Sketch this figure on your paper, then draw the lines of reflection needed to map $\triangle PQR$ onto $\triangle UVW$ in each case.

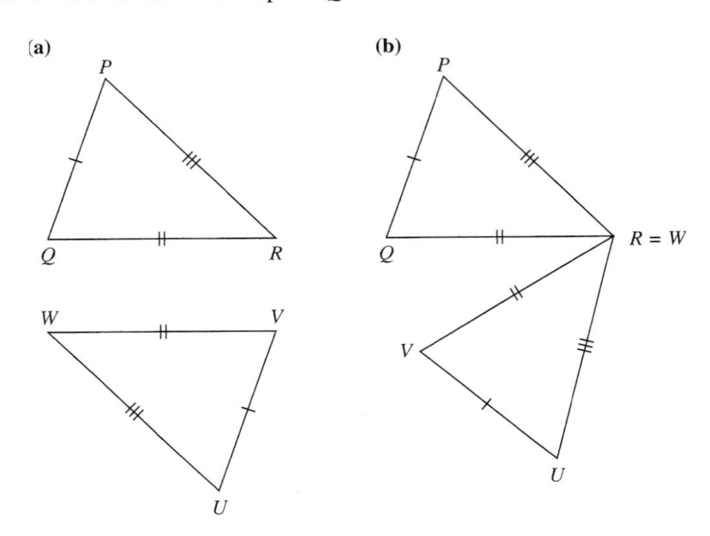

GROUP C

19. Prove the Corollary of Theorem 2: A nontrivial direct isometry is either a translation or rotation.

20. Conjecture: The only isometry with exactly one fixed point is a rotation. If you can, settle this conjecture (prove, if true, or give a counterexample, if false).

21. What can be said, in general, about the mappings

(a) $f = s_m s_\ell s_m$, if $\ell \perp m$?

(b) rt or tr, where r and t are, respectively, a rotation and a translation?

5.4 Other Linear Transformations

We will give a brief account of other important mappings used in geometry, looking at them from the synthetic point of view.

GLIDE REFLECTIONS

The simple act of walking is itself a demonstration of the mapping to be defined next.

DEFINITION: A **glide reflection** is the product of a reflection s_ℓ and a translation t_{AB} in a direction parallel to the line of reflection (that is, $\overleftrightarrow{AB} \parallel \ell$).

Thus, any glide reflection $g = t_{AB}s_\ell$ is the product of *three* line reflections, since a translation is itself the product of two reflections in parallel lines, which are, in this case, each perpendicular to ℓ. The analogy of walking is made clear by observing that each step is the image of the previous step under a glide reflection (Figure 5.25). Since it is immaterial whether the translation or the reflection is performed first, we easily find that

$$s_\ell t_{AB} = t_{AB}s_\ell = g$$

GLIDE REFLECTION

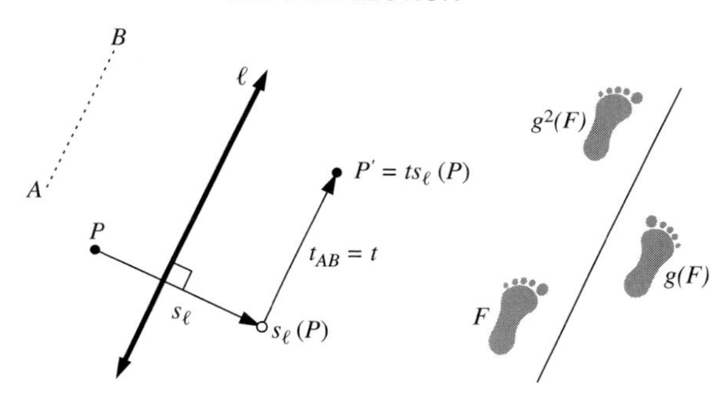

Figure 5.25

DILATIONS AND SIMILITUDES

We next consider linear transformations that preserve angle measure, but not necessarily distance.

DEFINITION: A **similitude**, or **similarity transformation**, is a linear transformation s such that for some positive constant k,

$$P'Q' = k\,PQ$$

for all pairs of points P, Q and their images $P' = s(P)$ and $Q' = s(Q)$. The number k is called the **dilation factor** for the similitude.

The important special case $k = 1$ for the dilation factor of a similitude reduces it to an isometry ($P'Q' = PQ$ for all P, Q). Thus, every isometry is a similitude, but not conversely.

THEOREM 1: A similitude is angle preserving. That is, if A, B, and C are any three points in the plane and A', B', and C' their images under a similitude, then

$$\angle A'B'C' \cong \angle ABC.$$

In Problem 15 you are asked to prove this theorem. If you have trouble getting started, think back to the SSS Similarity Criterion for triangles; what would that mean in this situation?

Similitudes affords a means of defining *similar figures* in geometry ("a figure" meaning an arbitrary set of points). We say that two sets S_1 and S_2 are **similar** iff there exists a similitude that maps S_1 onto S_2. For example, two polygons and their interiors are similar iff one can be mapped to the other by means of a similitude. The distances between all corresponding point pairs are shrunk or expanded (the correct term is *dilated*) by the same amount, k.

Since a similitude need not preserve distances, we obtain only a reduced version of the so-called ABCD Property that isometries enjoy.

COROLLARY: ABC PROPERTY
A similitude preserves angle measure, betweenness, and collinearity.

PROOF

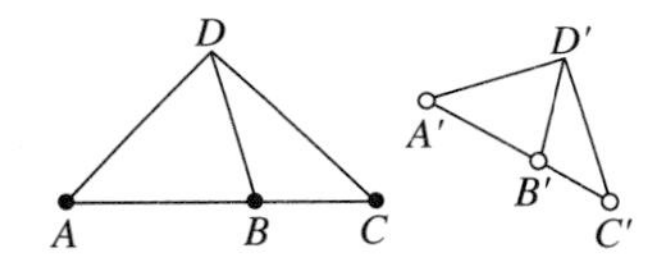

Figure 5.26

In view of Theorem 1, we need only prove a similitude has properties B and C (betweenness and collinearity preserving). We can take care of both of these at once by observing in Figure 5.26 that, if A–B–C holds and D is any point not on line $\overleftrightarrow{AC}$, then by Theorem 1

$$m\angle D'A'B' = m\angle DAB = m\angle DAC = m\angle D'A'C'$$

so C' must lie on ray $\overrightarrow{A'B'}$ (hence A', B', and C' are collinear). Also, since A–B–C then

$$m\angle ADB + m\angle BDC = m\angle ADC \;\rightarrow\; m\angle A'D'B' + m\angle B'D'C = m\angle A'D'C'$$

and A'–B'–C' must hold, finishing the proof.

EXAMPLE 1 Show that a similitude f preserves equidistant loci. Specifically, show that if line ℓ is the equidistant locus of points A and B, then f maps ℓ to the equidistant locus ℓ' of the images $A' = f(A)$ and $B' = f(B)$.

SOLUTION
Since ℓ is the perpendicular bisector of segment $\overline{AB}$ at M (Figure 5.27) and f is angle preserving, ℓ' will be perpendicular to segment $\overline{A'B'}$ at some point N. Since a linear

transformation preserves midpoints of segments, M maps to N, the midpoint of $\overline{A'B'}$. Hence ℓ' is the perpendicular bisector of $\overline{A'B'}$, or equidistant locus. ■

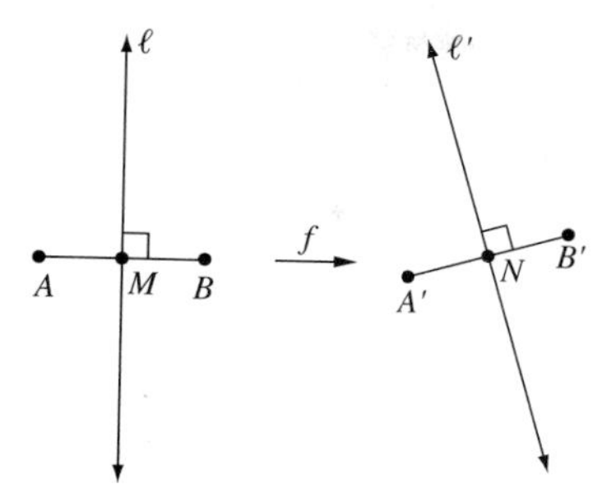

Figure 5.27

A special kind of similitude that is useful in geometry is characterized by its slope preserving property.

DEFINITION: A **dilation**[2] with **center** O and **dilation factor** $k > 0$ is that transformation d_k that leaves O fixed and maps any other point P to that point P' on ray $\overrightarrow{OP}$ such that $OP' = k \cdot OP$, as in Figure 5.28. (If $k < 0$, we agree to locate P' on the ray $\overrightarrow{OQ}$ opposite $\overrightarrow{OP}$, with $OP' = |k|OP$.) Two configurations are said to be **homothetic** iff one is the image of the other under some dilation map.

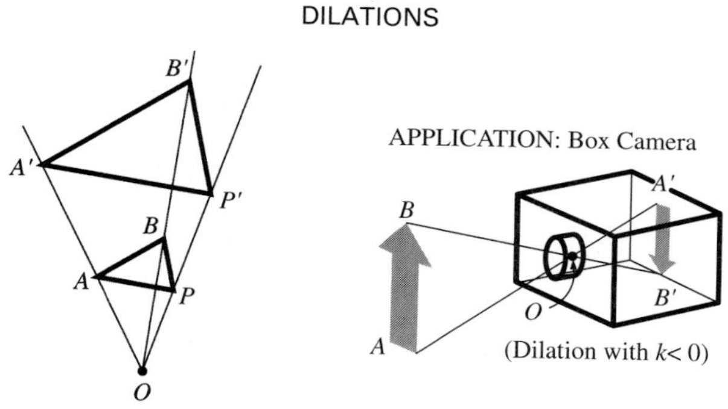

DILATIONS

APPLICATION: Box Camera

(Dilation with $k < 0$)

Figure 5.28

[2]Here the term "dilation" will designate either shrinking ($k < 1$) *or* stretching ($k > 1$). Another term for this that frequently appears is "dilitation," which we will not use here.

OUR GEOMETRIC WORLD

The operation of a movie projector, slide projector, or TV picture tube are illustrations of dilation mappings. The image in the film, slide, or cathode tube is enlarged by the action of light rays or electrons from some point projected toward the screen, after passing through the image in the film, slide, or electronic signal. An overhead projector and a model train or car randomly placed are illustrations of the more general similitude mapping.

EXAMPLE 2 Find the equations of a dilation d with center $C(-5, 9)$ and $k = 3$. Then use these equations to find the image of $\triangle CAB$ if $A = (0, 10)$ and $B = (0, 5)$.

SOLUTION

By using directed segments and the definition of a dilation, for any point $P(x, y)$ and its image $P'(x', y')$ we have, as shown in Figure 5.29

$$\mathbf{CP'} = 3\mathbf{CP}$$

or, using the vector representation of this,

$$[x' + 5, y' - 9] = 3[x + 5, y - 9] = [3(x + 5), 3(y - 9)]$$

That is,

$$x' + 5 = 3x + 15 \quad \rightarrow \quad x' = 3x + 10$$
$$y' - 9 = 3y - 27 \quad \rightarrow \quad y' = 3y - 18$$

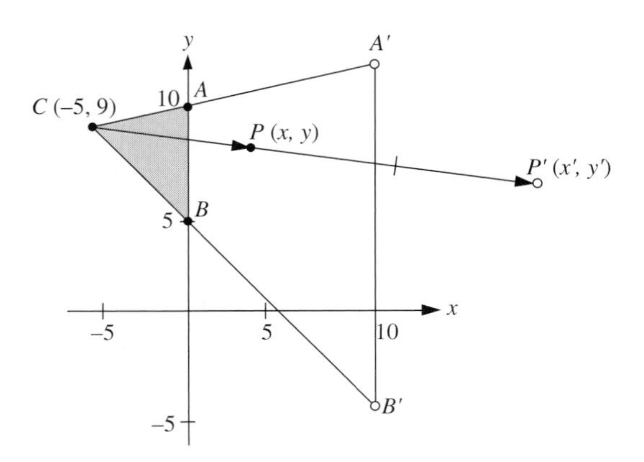

Figure 5.29

(As a check, C should be a fixed point of the mapping, so if we substitute -5 for x and 9 for y we ought to get -5 for x' and 9 for y', as we do.) The coordinates of A' and B' are found as follows:

$$d(0, 10): \begin{cases} x' = 3(0) + 10 = 10 \\ y' = 3(10) - 18 = 12 \end{cases} \qquad \therefore A' = (10, 12)$$

$$d(0, 5): \begin{cases} x' = 3(0) + 10 = 10 \\ y' = 3(5) - 18 = -3 \end{cases} \qquad \therefore B' = (10, -3) \quad \blacksquare$$

In the coordinate plane, a dilation is a slope-preserving similitude, just as a translation is a slope-preserving isometry, and this property is characteristic of these mappings if we exclude central reflections. The dilations also belong to the class of building block transformations, along with the reflections, because a similitude can be shown to be *the product of a dilation and three, or fewer, reflections.*

EXAMPLE 3 In Figure 5.30 a triangle ($\triangle ABC$) is shown that has been reflected in a vertical line to produce $\triangle LEM$, which, in turn, is rotated clockwise about E through an angle of 20°, to yield $\triangle GEH$. Finally, $\triangle GEH$ is acted on by a dilation with factor $k = \frac{1}{2}$, to give us $\triangle DEF$. Thus, $\triangle ABC \sim \triangle DEF$ and the unique similitude that takes $\triangle ABC$ to $\triangle DEF$ is the product of the first reflection, the rotation (two more reflections), and a dilation. No fewer mappings will suffice. ■

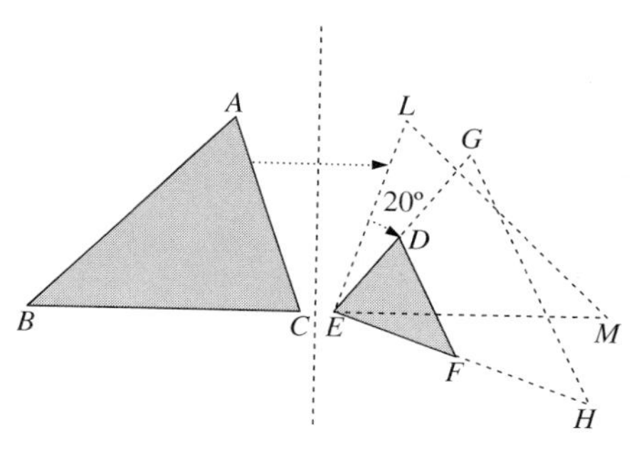

Figure 5.30

THEOREM 2: A dilation is a similitude, and, accordingly, has the *ABC* Property.

PROOF

From the definition and SAS Criterion for similar triangles, it follows that for any two points P and Q, $\triangle POQ \sim \triangle P'OQ'$ and $P'Q' = |k|PQ$.

THEOREM 3: Given any two similar triangles $\triangle ABC$ and $\triangle PQR$, there exists a unique similitude that maps the first triangle onto the second.

PROOF

See Problem 21.

COROLLARY: Any similitude is the product of a dilation and at most three reflections.

PROOF

See Problem 17.

A few examples will demonstrate some applications of similitudes in geometry.

EXAMPLE 4 Problem: Given a triangle, the object is to find a construction that will inscribe a square inside it such that one side of the square lies on the base of the triangle and the other two vertices of the square lie on the other two sides of the triangle.

SOLUTION

We can start with *any* square having its base on $\overrightarrow{BC}$ and opposite vertex on $\overrightarrow{BA}$, then "pull it down" to the desired position by means of a dilation, as shown in Figure 5.31. **Justification:** The construction defines a dilation (how?) and dilations preserve angle measure and ratios of distances, hence a square maps to a square. ■

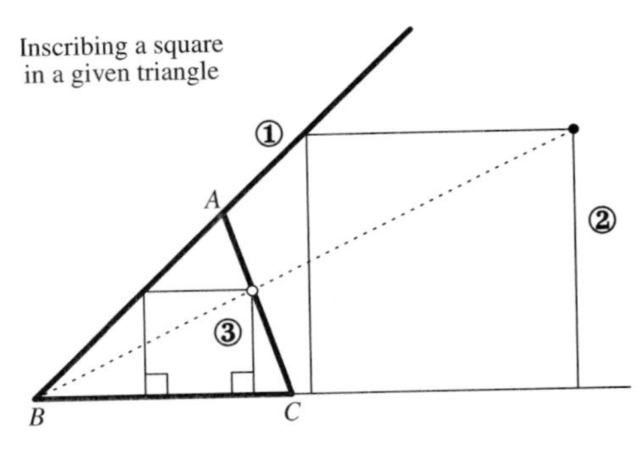

Figure 5.31

A more challenging problem is illustrated next.

EXAMPLE 5 Solve the following, using similitudes: Given a triangle $\triangle ABC$ and three circles **C**, **D**, and **E**, find points P, Q, and R on the three circles, respectively, such that $\triangle PQR \sim \triangle ABC$.

SOLUTION

Locate P on circle **C** (P can sometimes be chosen arbitrarily for the purpose), as illustrated in Figure 5.32. Rotate circle **D** to **D′** about P as center through an angle congruent to $\angle A$. Use the dilation d_k, where $k = AB/AC$, to map **D′** to **D″**. If **D″** meets **E** at Q, there will be a solution with the original choice of point P; otherwise, one must change the location of P. (If no such location for P exists, then there is no solution.) Let line $\overleftrightarrow{PQ}$ meet **D′** at R', the image of a unique point R on **D** under the preceding rotation. Then $\triangle PQR$ is the desired triangle. **Justification:** We must

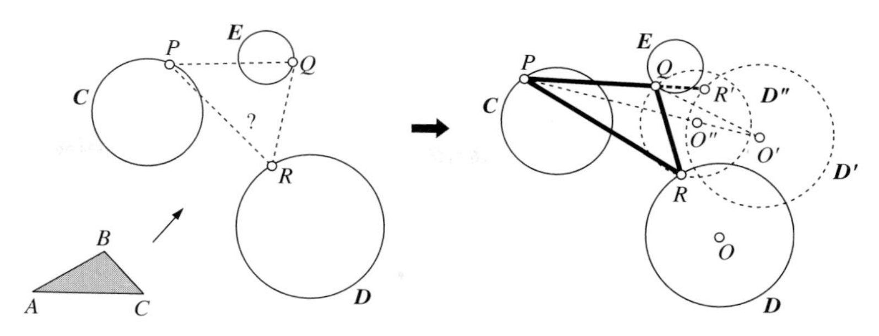

Figure 5.32

show that $\triangle PQR \sim \triangle ABC$. We have $PQ/PR = PQ/PR' = k = AB/AC$, and $m\angle QPR = m\angle R'PR = m\angle A$ by our choice of angle of rotation. Hence, by the SAS Similarity Criterion, $\triangle PQR \sim \triangle ABC$. ■

NOTE: The famous *Problem of Appolonius* also begins with three given circles, and the object is to find (construct) a circle mutually tangent to those three circles. This is a difficult problem, requiring 10 cases for its solution. (For further information, see Eves, *A Survey of Geometry, Volume I*, p. 192.)

EXAMPLE 6 Use similitudes to show that the medians of a triangle are concurrent.

SOLUTION

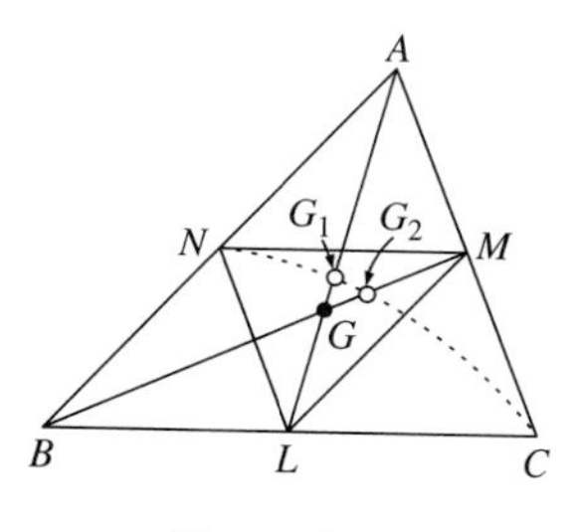

Figure 5.33

Let $\triangle ABC$ be any triangle, with medians $\overline{AL}$, $\overline{BM}$, and $\overline{CN}$ (Figure 5.33). Since $\triangle LMN \sim \triangle ABC$ by the SSS Similarity Criterion, there exists a unique similitude f that maps $\triangle ABC$ to $\triangle LMN$ (by Theorem 2). Now consider what f must do to each median. Since f preserves midpoints (in fact, any linear transformation does so), it maps L to the midpoint of the image of $\overline{BC}$, or the midpoint of $\overline{B'C'} = MN$. But since $\overline{AL}$ cuts $\overline{MN}$ at the midpoint of $\overline{MN}$ (note the two diagonals of $\square AMLN$), that midpoint must lie on $\overline{AL}$, and f maps $\overleftrightarrow{AL}$ to itself. Then any two medians map to themselves, and their point of intersection, G, is a fixed point of f. If G does not lie on the third median, then that third median cuts the other two in distinct points G_1 and G_2, which are also fixed points of f. Then f has three noncollinear fixed points, forcing it to be the identity. But this is impossible. Hence G lies on all three medians. ■

LINEAR TRANSFORMATIONS

A more general transformation preserves only betweenness and collinearity (the *BC* property). It is, accordingly, called a **linear transformation,** meaning, *line preserving,* which was defined in Section 5.1. A convenient way to construct such a transformation without using coordinates is to use three-dimensional geometry—simply project orthogonally from plane $\boldsymbol{P}$ to plane $\boldsymbol{Q}$ (Figure 5.34), then orthogonally from $\boldsymbol{Q}$ back to $\boldsymbol{P}$. The resulting composite mapping from $\boldsymbol{P}$ to $\boldsymbol{P}$ is a one-to-one mapping that clearly preserves collinearity and betweenness, but it does not map circles into circles, which, as indicated in Problem 22 below, is a property that characterizes similitudes among all linear transformations.

Circle maps to ellipse in plane P

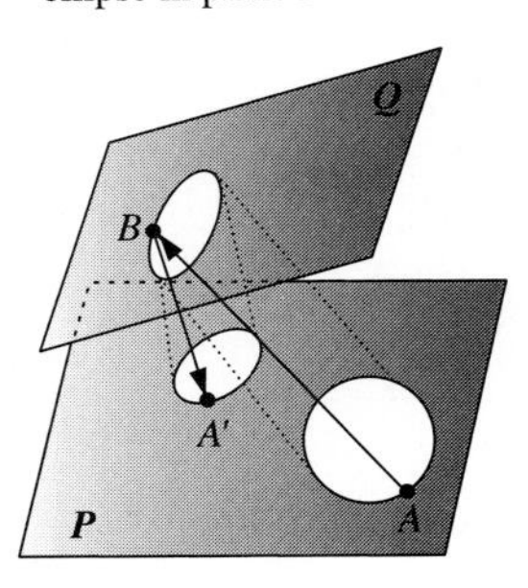

Figure 5.34

We shall state three interesting facts about linear transformations without proof. Although they can be proved synthetically, coordinate proofs are much simpler, and we will defer such proofs until the next section.

- A linear transformation preserves parallelism, betweenness, and ratios of lengths of parallel or collinear segments.
- Given any two triangles $\triangle ABC$ and $\triangle A'B'C'$, there exists a unique linear transformation that takes $\triangle ABC$ onto $\triangle A'B'C'$. (See Problem 16, Section 5.5)
- A linear transformation is a similitude iff it maps circles onto circles. (See Problem 17, Section 5.5)

Moment for Discovery

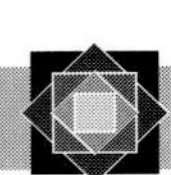

Dilations and Similitudes

1. Consider the mapping given in coordinate form $f(x, y) = (3x, 3y)$. By plotting points, discover what geometric properties this mapping seems to have. Figure 5.35 shows an example. (Is f distance preserving? Is it slope preserving? Is it a similitude or dilation?)

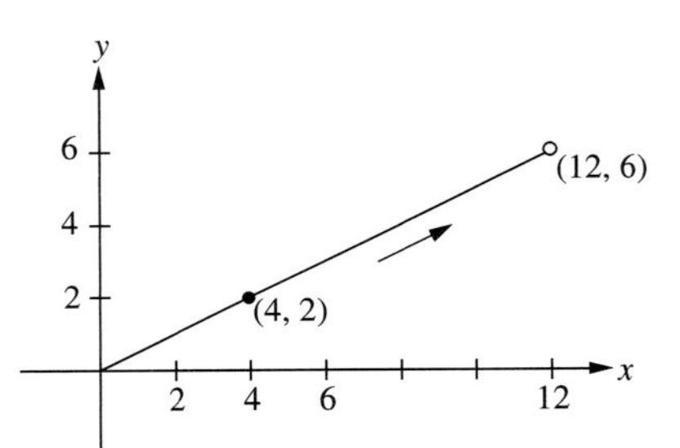

Figure 5.35

2. If we follow the mapping f defined in Step 1 by an isometry (like a rotation or reflection in some line), what would be the overall effect? What type of mapping would you have now?
3. What do you think the relationship is between similitudes and dilations in terms of transformation products?

PROBLEMS (5.4)

GROUP A

1. Find the dilation factor for each of the following dilations by finding the images of the points $A(0, 0)$ and $B(1, 2)$.

(a) $\begin{cases} x' = 3x \\ y' = 3y \end{cases}$ **(b)** $\begin{cases} x' = -2x \\ y' = -2y \end{cases}$

2. Find the dilation factor for each of the following dilations by finding the images of the points $A(0, 0)$ and $B(1, 2)$ in each case.

(a) $\begin{cases} x' = 3x - 2 \\ y' = 3y - 4 \end{cases}$ (b) $\begin{cases} x' = -x + 2 \\ y' = -y + 4 \end{cases}$

3. Find the image of $\triangle ABC$ under the transformation

$$\begin{cases} x' = 4x + 3 \\ y' = -4y \end{cases}$$

if $A = (0, 0)$, $B = (3, 0)$, and $C = (2, 2)$. Show that $\triangle ABC \sim \triangle A'B'C'$.

4. Show that if $\ell \perp m$ and $\ell \perp n$, then $s_n s_m s_\ell$ is either s_ℓ or a glide reflection.

5. Find the equation for the reflecting line in the glide reflection

$$\begin{cases} x' = -y + 6 \\ y' = -x - 6 \end{cases}$$

(***Hint:*** Find the inverse map and substitute into $y = mx + b$; find m and b such that this line is fixed by this mapping.)

6. If the y-axis is the reflecting line in a glide reflection and each "step" is 5 units ahead (parallel to the y-axis), derive the equations for this transformation. ($x' = ?, y' = ?$)

7. Derive the equations for the dilation with center (2, 3) and $k = 5$. ($x' = ?, y' = ?$)

8. If $\triangle ABC$ and $\triangle ADE$ in the figure are related by a dilation, which of the points A, B, C, D, or E is the center for the dilation, and why?

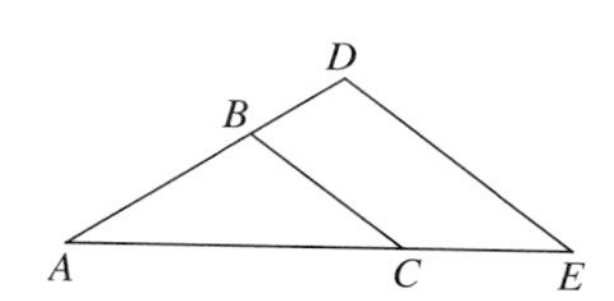

9. If $\triangle UVW \sim \triangle RST$ and the base $\overline{RS}$ of $\triangle RST$ lies on the base $\overline{UV}$ of $\triangle UVW$, how would you locate the center for a dilation that maps $\triangle UVW$ to $\triangle RST$?

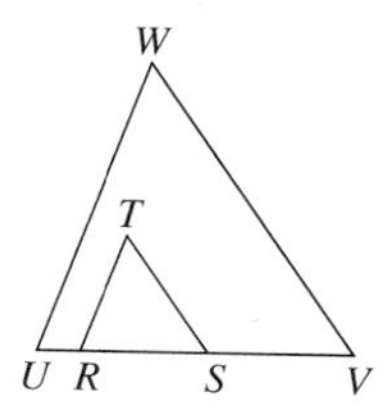

10. A point C is located inside a square that does not lie on either diagonal. Make a sketch to show the result of applying a dilation with center C and $k = 1/3$ to the square, showing both the original square and the dilated square. (If you have access to *Sketchpad*, work this problem using that software.)

11. A dilation having center A maps P to P', as shown. Find a construction that gives the image of Q' under this dilation. (***Hint:*** Draw ray $\overleftrightarrow{AQ}$ and segment $\overline{PQ}$.)

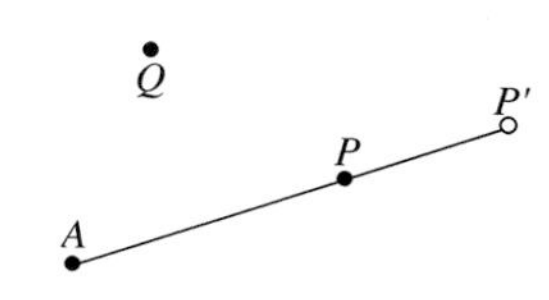

12. Find the dilation factor k and center C for the dilation d if $d(2, 2) = (2, 12)$ and you are told that C lies on the line $y = -3$.

GROUP B

13. If you know that a certain mapping is a direct similitude, that it has exactly one fixed point, and a line through that fixed point is also fixed by the mapping (with individual points on that line allowed to move), what else can definitely be said about this mapping?

14. Prove that a similitude with dilation factor k maps a circle centered at O and radius r to a circle centered at O' and radius kr.

15. Prove that a similitude is angle preserving.

16. Prove that a similitude having two fixed points is an isometry.

17. If a similitude with dilation factor k is followed by the dilation $d_{1/k}$ centered at some point C, what is the behavior of the composite mapping (product)? Thus, prove that a similitude is the product of a dilation and an isometry, and therefore it is the product of a dilation and three or fewer line reflections. (Use the fact to be established later that if three transformations f, g, and h satisfy the equation $f = gh$ then $g^{-1}f = h$.)

18. Use a dilation to inscribe a Golden Rectangle in a given acute-angled triangle $\triangle ABC$ (with one side of the rectangle lying on the base $\overline{AB}$). [Recall that a Golden Rectangle is one in which two sides are in the ratio 1 to τ, where $\tau = \frac{1}{2}(1 + \sqrt{5})$.]

19. Prove the converse of Theorem 1, that an angle-preserving mapping is a similitude. (***Hint:*** You may find the figure for this problem helpful; define $k = A'B'/AB$.)

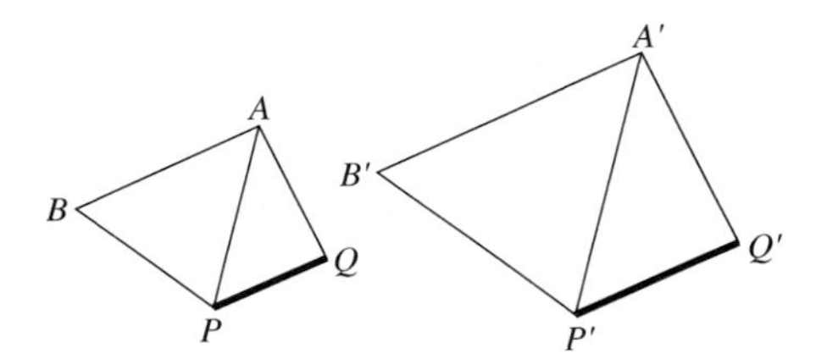

20. Given three parallel lines ℓ, m, and n, find a construction that will

(a) locate the vertices of a 45° right triangle $\triangle PQR$ on those three lines, as indicated in the figure

(b) locate the vertices of an equilateral triangle $\triangle PQR$ on those three lines.

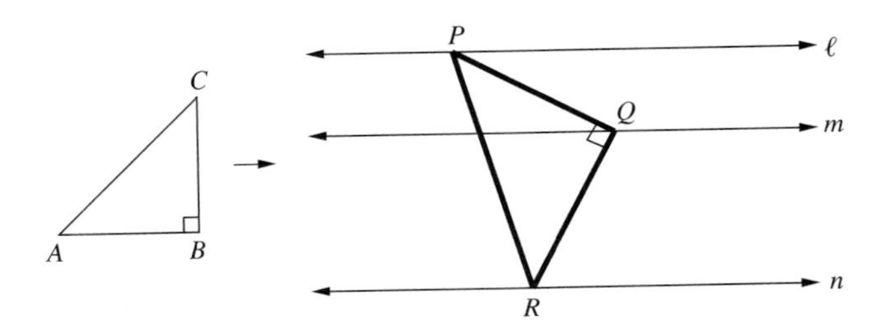

GROUP C

21. Prove that there exists a unique similitude that maps a given triangle to one that is similar to it. (***Hint:*** See Problem 17 for ideas. See also Theorem 1, Section 5.3.)

22. Find a synthetic proof that a linear transformation is a similitude iff it maps circles into circles. (You may use the fact that a linear transformation preserves ratios of distances between collinear points.)

23. Project for Undergraduate Research Given a similitude f with dilation factor $k < 1$ and some point A, form the sequence $A_1, A_2, A_3, \ldots$ by defining $A_1 = f(A), A_2 = f(A_1), A_3 = f(A_2)$, and so on. Show that $\{A_n\}$ is a **Cauchy sequence,** that is, given $\epsilon > 0$, there exists n_0 such that for all $m, n > n_0$.

$$A_mA_n < \epsilon$$

As such, the limit

$$\lim_{n \to \infty} A_m = B(b, c)$$

exists. Find a formula for b and c in terms of the parameters of the mapping f, if possible. Illustrate in the coordinate plane when f is defined explicitly by

$$x' = \frac{2}{3}x + 9, \qquad y' = \frac{2}{3}y + 27$$

and $A = (0, 0)$. Find the coordinates, by computer if necessary, of the limiting point $B(b, c)$ for this case. Explore whether a general formula for this limit exists in terms of the coordinate formula for f (as given in the next section).

5.5 Coordinate Characterizations

It can be shown that any transformation that maps lines onto lines in the coordinate and vector plane must preserve parallelograms, thus vector addition, and ultimately scalar multiplication of vectors as well (this latter result is a basic theorem of linear algebra whose proof can be found in some texts on the subject). Accordingly, if $f(x, y) = (x', y')$ is any linear transformation, then for certain constants a, b, c, d, h, and k we must have

Coordinate Form of a General Linear Transformation

(1)
$$f: \begin{cases} x' = ax + by + h \\ y' = cx + dy + k \end{cases} \qquad (ad \neq bc)$$

NOTE: The condition $ad \neq bc$ characterizes the one-to-one property of f. If f were, for example, $x' = 2x + 2y$ and $y' = 3x + 3y$, then the *entire plane* would be mapped into a *single line,* namely, $3x' = 2y'$, and every point on the line $y = -x$ would map to the single point $(0, 0)$.

We observe that the mapping f in **(1)** is actually a composition of two simpler linear transformations:

$$f^*: \begin{cases} x' = ax + by \\ y' = cx + dy \end{cases} \qquad t: \begin{cases} x'' = x' + h \\ y'' = y' + k \end{cases}$$

The latter mapping t is the translation that takes the origin to (h, k), and we call it the **translational component** of f. Since the geometric properties of f^* and f will be virtually the same (but for a translation), we can concentrate our efforts on f^*. Accordingly, we look at the **matrix** defining f^*

$$A = \begin{bmatrix} a & b \\ c & d \end{bmatrix}$$

and its **determinant**

$$\det A \equiv |A| \equiv \begin{vmatrix} a & b \\ c & d \end{vmatrix} = ad - bc$$

(which we require to be nonzero). The geometric behavior of f is thereby completely determined by the four numbers in this 2×2 matrix.

EXAMPLE 1 By merely substituting algebraic expressions and simplifying, find what distortions the linear transformation

$$f: \begin{cases} x' = 4x + 2y \\ y' = -x + 2y \end{cases} \qquad \text{matrix} = \begin{bmatrix} 4 & 2 \\ -1 & 2 \end{bmatrix}$$

imposes on

(a) a square having vertices $A(1, 1), B(2, 1), C(2, 2), D(1, 2)$
(b) the unit circle $x^2 + y^2 = 1$, parameterized as $x = \cos t, y = \sin t, t$ real (Figure 5.36).

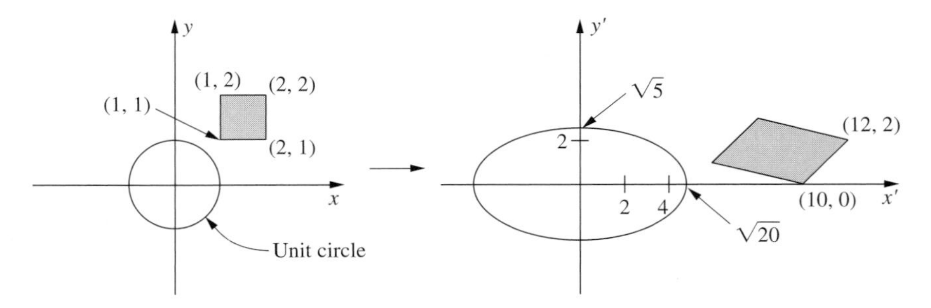

Figure 5.36

SOLUTION

(a) We know $\square ABCD$ maps to at least a parallelogram. To find its vertices, merely substitute the coordinates of A B, C, and D into the above equations.

$$\begin{array}{lll} A(1, 1): & x' = 4 \cdot 1 + 2 \cdot 1 = 6, & y' = -1 + 2 \cdot 1 = 1 \\ B(2, 1): & x' = 4 \cdot 2 + 2 \cdot 1 = 10, & y' = -2 + 2 \cdot 1 = 0 \\ C(2, 2): & x' = 4 \cdot 2 + 2 \cdot 2 = 12, & y' = -2 + 2 \cdot 2 = 2 \\ D(1, 2): & x' = 4 \cdot 1 + 2 \cdot 2 = 8, & y' = -1 + 2 \cdot 2 = 3 \end{array}$$

The results are shown in Figure 5.36. The image of $\square ABCD$ is then an oblique-angled parallelogram.

(b) Substitute $x = \cos t$ and $y = \sin t$ into the transformation equations, then eliminate t by squaring and summing.

$$\begin{aligned} x' &= 4\cos t + 2\sin t \\ x'^2 &= 16\cos^2 t + 16\sin t\cos t + 4\sin^2 t \\ y' &= -\cos t + 2\sin t \\ y'^2 &= \cos^2 t - 4\sin t\cos t + 4\sin^2 t \\ 4y'^2 &= 4\cos^2 t - 16\sin t\cos t + 16\sin^2 t \end{aligned}$$

Sum the resulting squared equations to obtain

$$x'^2 + 4y'^2 = 20\cos^2 t + 20\sin^2 t = 20$$

or

$$\frac{(x')^2}{20} + \frac{(y')^2}{5} = 1$$

which is an ellipse. (See graph in Figure 5.36.) ■

It is interesting, and instructional, to see what geometric patterns are effected by the numbers in the matrix defining a transformation. Here are a few examples (Figure 5.37) where a uniform test pattern is used throughout (unit square plus cross hairs).

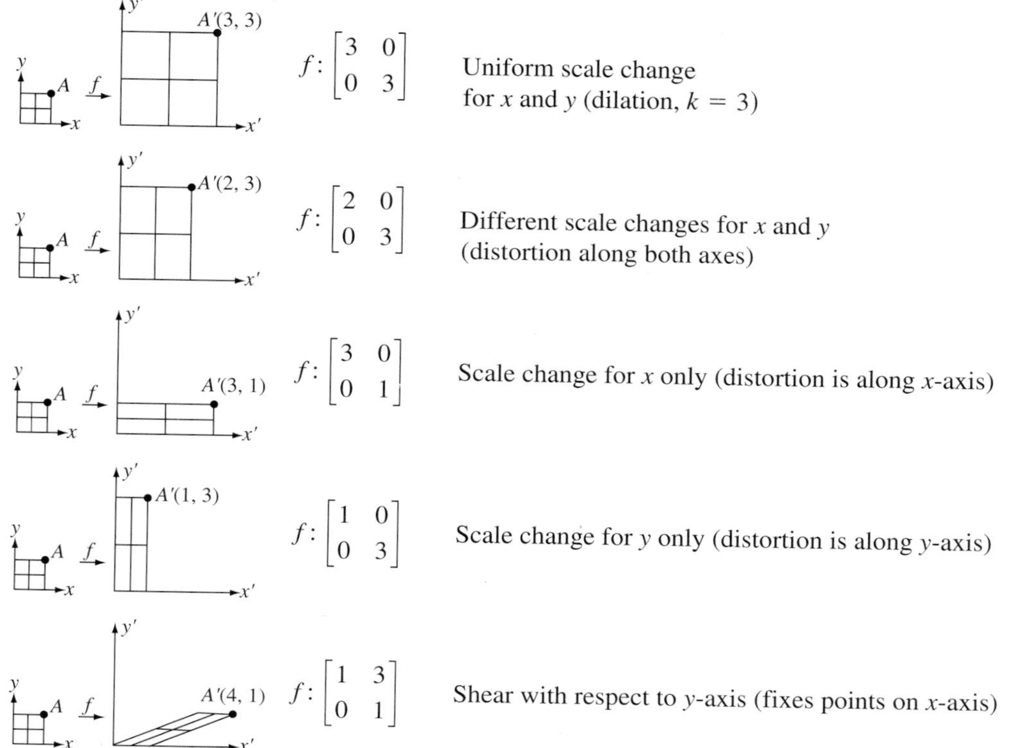

(Figure continued on next page)

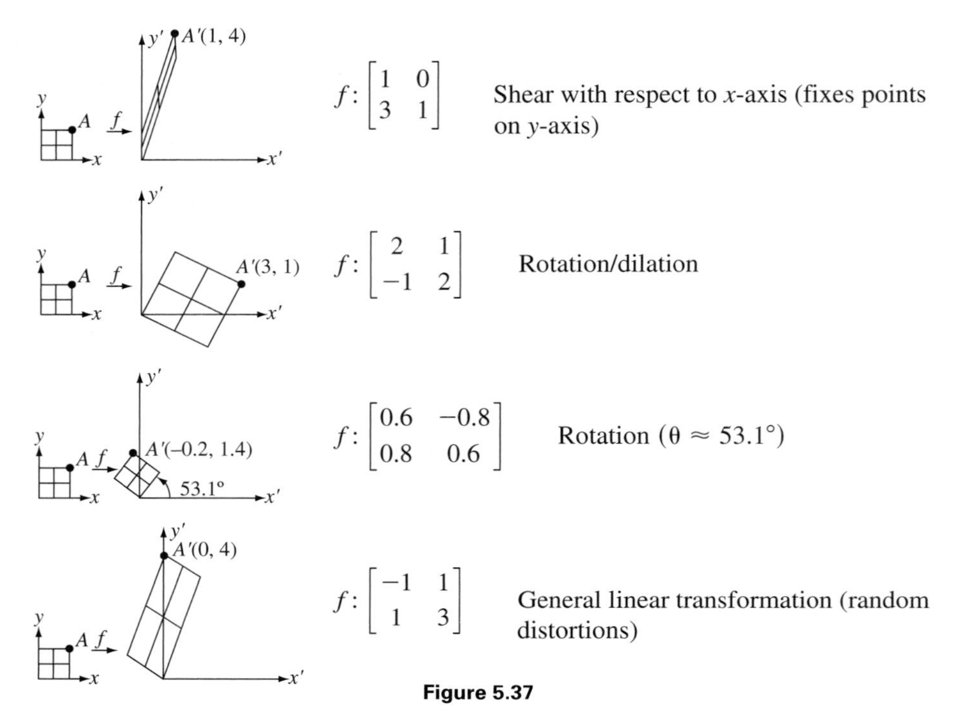

Figure 5.37

In working with transformations, it is helpful to have one more tool at our disposal besides coordinates and vectors. It is the theory of matrix products. You may already be familiar with matrix multiplication. In case you are not, the product of two 2×2 matrices is given by the rule

$$\begin{bmatrix} a & b \\ c & d \end{bmatrix} \cdot \begin{bmatrix} x & z \\ y & w \end{bmatrix} = \begin{bmatrix} ax + by & az + bw \\ cx + dy & cz + dw \end{bmatrix}$$

(The rule is to regard each *row* of A and each *column* of B as vectors, then calculate their scalar products for each entry of the product.)

The more general product of a 2×2 matrix and a $2 \times n$ (for $n = 4$) matrix is

$$\begin{bmatrix} a & b \\ c & d \end{bmatrix} \cdot \begin{bmatrix} x_1 & x_2 & x_3 & x_4 \\ y_1 & y_2 & y_3 & y_4 \end{bmatrix} = \begin{bmatrix} ax_1 + by_1 & ax_2 + bx_2 & ax_3 + bx_3 & ax_4 + by_4 \\ cx_1 + dy_1 & cx_2 + dy_2 & cx_3 + dy_3 & cx_4 + dy_4 \end{bmatrix}$$

It should be clear from the general form that the matrix product represents a labor-saving device in dealing with linear transformations. To find the images of two or more points under a linear transformation given in coordinate form, one can use the matrix product as a schematic to organize the various computations.

EXAMPLE 2 Calculate the images of the four points $P(1, 2)$, $Q(1, -2)$, $R(-1, -2)$, and $S(-1, 2)$ in Figure 5.38 under the transformation

$$f: \begin{cases} x' = 2x + 3y \\ y' = x + 2y \end{cases} \qquad \text{matrix: } \begin{bmatrix} 2 & 3 \\ 1 & 2 \end{bmatrix}$$

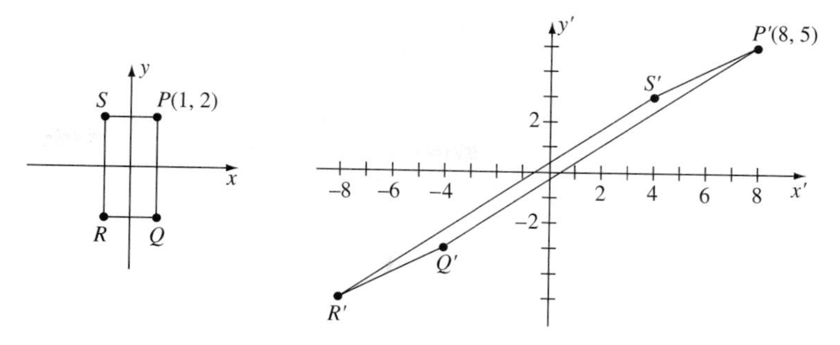

Figure 5.38

SOLUTION

$$\begin{matrix} & & P & Q & R & S & & P' & Q' & R' & S' \end{matrix}$$
$$\begin{bmatrix} 2 & 3 \\ 1 & 2 \end{bmatrix}\begin{bmatrix} 1 & 1 & -1 & -1 \\ 2 & -2 & -2 & 2 \end{bmatrix} = \begin{bmatrix} 8 & -4 & -8 & 4 \\ 5 & -3 & -5 & 3 \end{bmatrix}$$

$\therefore P' = (8, 5), Q' = (-4, -3), R' = (-8, -5)$, and $S' = (4, 3)$.) ■

In general, the image of $P(x, y)$ under the transformation

$$f: \begin{cases} x' = ax + by + h \\ y' = cx + dy + k \end{cases} \qquad \text{matrix} : A = \begin{bmatrix} c & b \\ c & d \end{bmatrix}$$

may be written in the matrix/vector form

(2) $\qquad f: \boldsymbol{x}' = A\boldsymbol{x} + \boldsymbol{b} \quad |A| \neq 0$

where $\boldsymbol{b}$ is the vector $[h, k]$, $\boldsymbol{x} = [x, y]$, and $\boldsymbol{x}' = [x', y']$. A more explicit formulation of the same thing is

(2′) $\qquad \begin{bmatrix} x' \\ y' \end{bmatrix} = \begin{bmatrix} a & b \\ c & d \end{bmatrix}\begin{bmatrix} x \\ y \end{bmatrix} + \begin{bmatrix} h \\ k \end{bmatrix}, \quad \begin{vmatrix} a & b \\ c & d \end{vmatrix} \neq 0$

We are now ready to make a systematic study of the geometric properties of linear transformations as characterized by the form of matrices defining them. The first step is to obtain a general formula to determine the way a linear transformation, in terms of its matrix, alters the slope of a line under **(2)**. ■

EXAMPLE 3 Consider the linear transformation given by

$$\begin{cases} x' = 3x + 7y \\ y' = 6x - 5y \end{cases} \qquad \text{with matrix} \begin{bmatrix} 3 & 7 \\ 6 & -5 \end{bmatrix}$$

and a line whose slope is m,

$$y = mx + k$$

Find the slope m' of the transformed line under the above transformation.

SOLUTION

We will find the coordinates of two (convenient) points on the given line, transform them using the equations for x' and y', then use the slope formula to find the slope of the transformed points, which will be the slope of the transformed line. Let the two points be those for which $x = 0$ and $x = 1$; by substitution into $y = mx + k$ we find

$$A(0, k) \qquad \text{and} \qquad B(1, m + k)$$

Then, to find A' and B', write

$$x' = 3(0) + 7(k) = 7k, \qquad y' = 6(0) - 5(k) = -5k \quad \rightarrow \quad A' = (7k, -5k)$$

$$x' = 3(1) + 7(m + k) = 3 + 7m + 7k, \qquad y' = 6(1) - 5(m + k)$$
$$= 6 - 5m - 5k \quad \rightarrow \quad B' = (3 + 7m + 7k, 6 - 5m - 5k)$$

Now we use the slope formula to find the slope of $\overleftrightarrow{A'B'}$:

$$m' = \frac{y_2 - y_1}{x_2 - x_1} = \frac{6 - 5m - 5k - (-5k)}{3 + 7m + 7k - (7k)} = \frac{6 - 5m}{3 + 7m} \quad \blacksquare$$

It is not difficult to generalize the result of the preceding example, using a, b, c, and d in place of 3, 7, 6, and −5, and show

Effect of Linear Transformations on Slope

(3)
$$m' = \frac{c + dm}{a + bm}$$

An obvious immediate deduction from this formula is a key property of linear transformations in the plane, mentioned earlier.

THEOREM 1: A linear transformation maps parallel lines to parallel lines.

PROOF

If two parallel lines have slopes m_1 and m_2, then $m_1 = m_2$ and the above formula obviously yields $m_1' = m_2'$. If the lines are both vertical, then either $b \neq 0$ and the image lines are nonvertical, with slope $m_1' = d/b = m_2'$, or $b = 0$ and both image lines are vertical.

We already know that linear transformations do not always preserve perpendicularity (Example 2). If we impose this property on a transformation, a very surprising thing takes place, and we get more than we might expect. First, we will need to prove a special algebraic lemma.

LEMMA: Suppose the following conditions hold for real numbers a, b, c, and d:
(a) $ad \neq bc$
(b) $a^2 + c^2 = b^2 + d^2$
(c) $ab = -cd$.
Then either $b = -c$ and $d = a$, or $b = c$ and $d = -a$.

PROOF

(1) First, suppose that $a = 0$. Then by **(a)**, $bc \neq 0$ and both $b \neq 0$ and $c \neq 0$. Apply **(b)** and **(c)**: $0 + c^2 = b^2 + d^2, 0 = -cd$. Since $c \neq 0$, $d = 0$, and we obtain $b^2 = c^2$ or $b = \pm c$. Then either $b = -c$ and $d = a = 0$, or $b = c$ and $d = -a = 0$, as desired.

(2) Next, suppose $c = 0$. Then in this case, $a \neq 0$ and **(c)** implies $b = 0 \to a^2 = d^2$ or $d = \pm a$, and again the desired conclusion results.

(3) The last case is $a \neq 0$ and $c \neq 0$. Let $x = -b/c, y = d/a$. That is, $b = -cx$ and $d = ay$. Substitute into **(b)** and **(c)**.

$$a^2 + c^2 = c^2x^2 + a^2y^2,$$
$$-acx = -acy$$
$$\therefore x = y$$

$$a^2 + c^2 = (c^2 + a^2)x^2$$
$$x^2 = 1$$
$$\therefore x = y = \pm 1$$

If $x = 1 = y$, then $b = -c$ and $d = a$, and if $x = -1, b = c$ and $d = -a$, as desired.

NOTE: The matrix form of the lemma is: If $ad \neq bc, a^2 + c^2 = b^2 + d^2$, and $ab = -cd$, then either

$$\begin{bmatrix} a & b \\ c & d \end{bmatrix} = \begin{bmatrix} a & -c \\ c & a \end{bmatrix} \text{ or } \begin{bmatrix} a & b \\ c & d \end{bmatrix} = \begin{bmatrix} a & c \\ c & -a \end{bmatrix}$$

To summarize the two cases, we can simply write

$$\begin{bmatrix} a & b \\ c & d \end{bmatrix} = \begin{bmatrix} a & -\delta c \\ c & \delta a \end{bmatrix}, \qquad \text{where } \delta = \pm 1$$

If a linear transformation $\mathbf{x}' = A\mathbf{x}$ preserves perpendicularity, then for all slopes such that $m_1m_2 = -1$, we must have $m'_1m'_2 = -1$. We shall call such a transformation an **orthomap.** Making use of **(3)**, we have, for all $m_1m_2 = -1 = m'_1m'_2$,

$$-1 = \left(\frac{c + m_1d}{a + m_1b}\right) \cdot \left(\frac{c + m_2d}{a + m_2b}\right) = \frac{c^2 + (m_1 + m_2)\,cd - d^2}{a^2 + (m_1 + m_2)\,ab - b^2}$$

You can show from this (substituting, in turn, $m_1 = 1, m_2 = -1$ and $m_1 = 2, m_2 = -\frac{1}{2}$)

$$-1 = \frac{c^2 - d^2}{a^2 - b^2} \qquad \text{and} \qquad -1 = \frac{c^2 + \frac{3}{2}cd - d^2}{a^2 + \frac{3}{2}ab - b^2}$$

which ultimately implies that $a^2 + c^2 = b^2 + d^2$, and $ab = -cd$. Since the conditions of the lemma are therefore satisfied, it follows that

$$A = \begin{bmatrix} a & -\delta c \\ c & \delta a \end{bmatrix}$$

Thus, we have (with b replacing c):

Coordinate Form of an Orthomap

(4)
$$f:\begin{cases} x' = ax - \delta by \\ y' = bx + \delta ay \end{cases} \qquad \text{matrix}: \begin{bmatrix} a & -\delta b \\ b & \delta a \end{bmatrix}$$
$$(\delta = \pm 1)$$

EXAMPLE 4 Show that, by its form, the following mapping is an orthomap, and determine the sign of δ:

$$f:\begin{cases} x' = 2x - 3y \\ y' = -3x - 2y \end{cases} \qquad \text{matrix}: \begin{bmatrix} 2 & -3 \\ -3 & -2 \end{bmatrix}$$

SOLUTION
By direct comparison,

$$\begin{bmatrix} 2 & -3 \\ -3 & -2 \end{bmatrix} \leftrightarrow \begin{bmatrix} a & -\delta b \\ b & \delta a \end{bmatrix}$$

and we find that $a = 2$, $b = -3$, and $\delta a = -2$, or $\delta = -1$. ■

Since a similitude is angle preserving, it is certainly an orthomap, and we can conclude that its matrix form is given by **(4).** (However, is it certain that for every possible value of a, b, and δ **(4)** gives us a similitude? The answer to this is left for you in the discovery unit at the end of this section.) Thus we conclude, after replacing the translational component:

Matrix/Coordinate Form of a Similitude

$$(\text{Dilation factor } k = \sqrt{a^2 + b^2})$$

(5)
$$f:\begin{cases} x' = ax - \delta by + x_0 \\ y' = bx + \delta ay + y_0 \end{cases} \qquad \text{matrix}: \begin{bmatrix} a & -\delta b \\ b & \delta a \end{bmatrix}$$
$$(\delta = \pm 1)$$

An important special case of this occurs when $b = 0$ and $\delta = 1$.

Matrix/Coordinate Form for Dilation Centered at (*h, k*)

(Dilation factor a)

(6) $$f: \begin{cases} x' = ax + x_0 \\ y' = ay + y_0 \end{cases} \qquad \text{matrix: } \begin{bmatrix} a & 0 \\ 0 & a \end{bmatrix}$$

where $x_0 = h - ah$, $y_0 = k - ak$

Now let us consider the distance-preserving property. If f is an isometry, then as we have seen, f preserves angle measure and hence is a similitude, having the form in **(5)**. Substitute the coordinates $O(0, 0)$ and $P(1, 0)$ into **(5)**; since distance is preserved, $O'U' = OU = 1$, and we obtain the additional requirement

$$a^2 + b^2 = 1$$

Thus

Matrix/Coordinate Form of an Isometry

(7) $$f: \begin{cases} x' = ax - \delta by + h \\ y' = bx + \delta ay + k \end{cases} \qquad \text{matrix: } \begin{bmatrix} a & -\delta b \\ b & \delta a \end{bmatrix}$$

where $a^2 + b^2 = 1$ and $\delta = \pm 1$

NOTE: It is interesting that the sign of δ in the above formulas characterizes the direct/opposite nature of the mappings: If $\delta = 1$, the mapping is direct, and if $\delta = -1$, opposite. (See Problem 12.)

The remaining special forms of transformations we introduced synthetically can now be found. In particular, if $a = \cos\theta$, $b = \sin\theta$, $\delta = 1$, $h = 0$, and $k = 0$, we obtain the *rotation about the origin through an angle of measure* θ.

Matrix/Coordinate Form of Rotation about Origin

(8) $$f: \begin{cases} x' = x\cos\theta - y\sin\theta \\ y' = x\sin\theta + y\cos\theta \end{cases} \qquad \text{matrix: } \begin{bmatrix} \cos\theta & -\sin\theta \\ \sin\theta & \cos\theta \end{bmatrix}$$

(θ = measure of angle of rotation)

The form in **(8)** can be generalized to give the rotation about an arbitrary point $C(h, k)$, which will be important later. First, translate the center $C(h, k)$ to the origin O under

(9) $$f: \begin{cases} x' = x - h \\ y' = y - k \end{cases}$$

[That is, the new coordinates of $C(h, k)$ are (0, 0).] Now perform the rotation about (0, 0) in the prime coordinate system.

$$\textbf{(10)} \qquad g: \begin{cases} x'' = x'\cos\theta - y'\sin\theta \\ y'' = x'\sin\theta + y'\sin\theta \end{cases}$$

Now translate back to the original center $C(h, k)$ under

$$\textbf{(11)} \qquad h: \begin{cases} x''' = x'' + h \\ y''' = y'' + k \end{cases}$$

Using **(9)**, **(10)**, and **(11)**, the overall effect is the composition hgf, after changing the triple primes to ordinary single primes and by substitution,

Coordinate Form of Rotation about C(h, k)

$$\textbf{(12)} \qquad f: \begin{cases} (x' - h) = (x - h)\cos\theta - (y - k)\sin\theta \\ (y' - k) = (x - h)\sin\theta + (y - k)\cos\theta \end{cases}$$

(θ = measure of angle of rotation)

We come full circle back to the line reflection, whose matrix representation, curiously enough, is the most complicated of all. For sake of completeness, we shall include it, leaving the details of the following analysis for you to work out. First, let the equation of the line of reflection be ℓ: $ax + by = c$, with $P(x, y)$ and $P'(x', y')$ a point and its image under s_ℓ. Now ℓ is the perpendicular bisector of segment $\overline{PP'}$, so

$$\text{Slope } \overleftrightarrow{PP'} = (y' - y)/(x' - x) = -1/\{\text{Slope } \ell\} = -1/(-a/b) = b/a$$

This gives us the equation, after rearranging,

$$bx' - ay' = bx - ay$$

Also, the midpoint M of segment $\overline{PP'}$, with coordinates $u = \frac{1}{2}(x' + x)$, $v = \frac{1}{2}(y' + y)$, must lie on line ℓ, so its coordinates must satisfy the equation $au + bv = c$. After simplifying, this condition becomes

$$ax' + by' = 2c - ax - by$$

This gives us a system of equations to solve for x' and y':

$$\begin{cases} bx' - ay' = bx - ay \\ ax' + by' = 2c - ax - by \end{cases}$$

By Cramer's Rule,

$$D_1 = \begin{vmatrix} bx - ay & -a \\ 2c - ax - by & b \end{vmatrix} = (b^2 - a^2)x - 2aby + 2ac, \; D = \begin{vmatrix} b & -a \\ a & b \end{vmatrix} = b^2 + a^2$$

Then

$$x' = D_1/D \quad \rightarrow \quad kx' = (b^2 - a^2)x - 2aby + 2ac \qquad \text{where } k = a^2 + b^2$$

A similar determinant D_2 for y' leads to the equation for ky'. The final result is:

Matrix/Coordinate Form of Reflection in a Line

(Line = ℓ: $ax + by = c$, with $k = a^2 + b^2$)

s_ℓ (coordinate form):

$$\textbf{(13)} \quad \begin{cases} kx' = (b^2 - a^2)x - 2aby + 2ac \\ ky' = -2abx - (b^2 - a^2)y + 2bc \end{cases}$$

s_ℓ (matrix form):

$$\frac{1}{k}\begin{bmatrix} b^2 - a^2 & -2ab \\ -2ab & a^2 - b^2 \end{bmatrix}$$

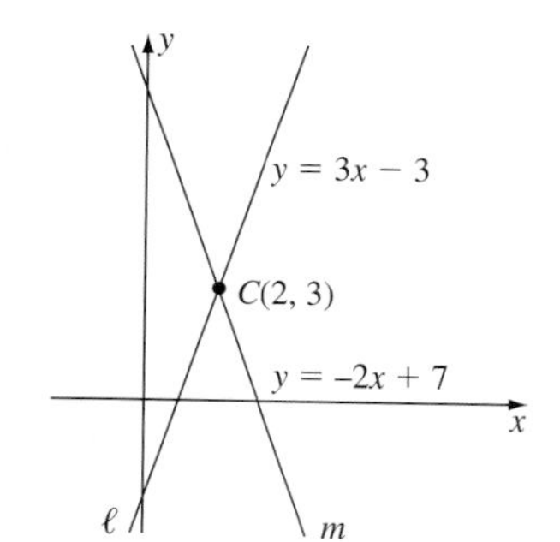

Figure 5.39

EXAMPLE 5 The lines ℓ: $y = 3x - 3$ and m: $y = -2x + 7$ meet at $C(2, 3)$, as shown in Figure 5.39.

(a) Calculate the matrices for the reflections s_ℓ and s_m and the equations in **(13)**.
(b) Verify by the appropriate substitutions that the product $s_m s_\ell$ is a rotation about C, and find the angle of rotation.
(c) Verify that this angle is twice the angle between ℓ and m.

SOLUTION

(a) We have:

$$\ell: 3x - y = 3 \rightarrow a = 3, b = -1, c = 3, \text{ and } k = 10$$

Hence, by **(13)** we have

$$10x' = (1 - 9)x - 2(3)(-1)y + 2(3)(3) = -8x + 6y + 18 \quad \text{or}$$
$$x' = -0.8x + 0.6y + 1.8 \quad \text{(exact decimals)}$$

$$10y' = 6x + 8y + 2(-1)(3) = 6x + 8y - 6 \quad \text{or} \quad y' = 0.6x + 0.8y - 0.6$$

Therefore,

$$s_\ell: \begin{cases} x' = -0.8x + 0.6y + 1.8 \\ y' = 0.6x + 0.8y - 0.6 \end{cases}$$

For s_m we have

$$m: 2x + y = 7 \rightarrow a = 2, b = 1, c = 7, \text{ and } k = 5.$$

Hence

$$5x' = (1 - 4)x - 2(2)(1)y + 2(2)(7) = -3x - 4y + 28 \text{ or}$$
$$x' = -0.6x - 0.8y + 5.6$$

$$5y' = -4x + 3y + 2(1)(7) = -4x + 3y + 14 \text{ or}$$
$$y' = -0.8x + 0.6y + 2.8$$

$$s_m: \begin{cases} x'' = -0.6x' - 0.8y' + 5.6 \\ y'' = -0.8x' + 0.6y' + 2.8 \end{cases}$$

(b) We find the equations for $s_m s_\ell$ by substitution of those of s_ℓ into those of s_m:

$$\begin{aligned} x'' &= -0.6(-0.8x + 0.6y + 1.8) - 0.8(0.6x + 0.8y - 0.6) + 5.6 \\ &= 0.48x - 0.36y - 1.08 - 0.48x - 0.64y + 0.48 + 5.6 = -y + 5 \end{aligned}$$

$$\begin{aligned} y'' &= -0.8(-0.8x + 0.6y + 1.8) + 0.6(0.6x + 0.8y - 0.6) + 2.8 \\ &= .64x - .48y - 1.44 + 0.36x + .48y - 0.36 + 2.8 = x + 1 \end{aligned}$$

That is,

$$s_m s_\ell : \begin{cases} x'' = -y + 5 \\ y'' = x + 1 \end{cases}$$

Comparing with **(12),** we find that these equations can be written as

$$\begin{aligned} x'' - 2 &= -(y - 3) = (\cos 90°)(x - 2) - (\sin 90°)(y - 3) \\ y'' - 3 &= x - 2 = (\sin 90°)(x - 2) + (\cos 90°)(y - 3) \end{aligned}$$

This is the form of a rotation about $C(2, 3)$ through the angle $\theta = 90°$.

(c) The angle between ℓ (slope 3) and m (slope -2) is found by the formula

$$\tan \phi = \frac{m_2 - m_1}{1 + m_1 m_2} = \frac{-2 - 3}{1 - 6} = 1$$

Thus, $\phi = 45°$ and $\theta = 90°$. ■

Moment for Discovery

Experimenting with an Orthomap

Every similitude is an orthomap, but is every orthomap a similitude? Recall the following formula from trigonometry, which gives the angle measure between two lines having slopes m_1 and m_2.

$$\tan \theta = \frac{m_2 - m_1}{1 + m_1 m_2}$$

The proof for this formula is actually indicated in Figure 5.40; the key is the trigonometric identity for $\tan(\theta_2 - \theta_1)$ and the fact that $m_2 = \tan \theta_2$ and $m_1 = \tan \theta_1$.

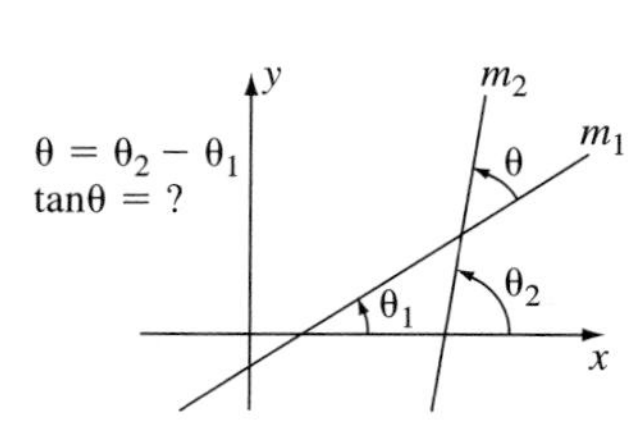

Figure 5.40

As an experiment, consider the orthomap whose matrix is

$$\begin{bmatrix} 4 & -3 \\ 3 & 4 \end{bmatrix}$$

and consider lines ℓ_1 and ℓ_2 having slopes $m_1 = 1$ and $m_2 = 3$, respectively.

1. Find $\tan\theta$ for lines ℓ_1, ℓ_2. (You should get $\frac{1}{2}$ for this.)
2. Use **(3)** to find the slopes m_1' and m_2' of the images of ℓ_1 and ℓ_2 under this transformation. (Did you get $m_1' = 7$?)
3. Calculate $\tan\theta'$ for ℓ_1' and ℓ_2'.
4. Did anything happen?
5. Repeat the experiment for $m_1 = 0$ and $m_2 = 4$.
6. Have you discovered a theorem? Try proving it in general, using algebra techniques for simplifying complex fractions.

Moment for Discovery

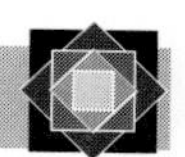

Experimenting with an Orthomap (using *Sketchpad*)

Consider the orthomap

$$\begin{cases} x' = 2x - y \\ y' = x + 2y \end{cases}$$

This experiment requires the construction given in Problem 11, Section 5.1, for locating the image point of a given point under a linear transformation. To efficiently carry out the parts of this experiment, we first create a Script for plotting the images of given points.

Script for $P'(x', y')$

1. Locate any point $C = P(x, y)$ on the screen and choose Create Axis from the GRAPH menu.
2. Start a script choosing SCRIPT, then REC (record).
3. Find the coordinates (x_C, y_C) of C using MEASURE, then calculate the values $x' = 2x_C - y_C$ and $y' = x_C + 2y_C$. Select these two values, in order, then choose Plot As (x, y) under GRAPH. Hide all coordinate measurements and calculations. Stop recording.

Experimenting with the orthomap

4. Hide points C and D, and construct a right angle $\angle PQR$ that can be dragged. (Draw a segment and rotate it 90°.)
5. Locate points G and H on the sides of $\angle PQR$, activate the Script on each of these points to locate G', Q', and H'. If necessary, adjust the figure so that these images come into view.
6. Select G and G', then Locus under CONSTRUCT, and repeat for H and H'. The image of $\angle PQR$ will appear. Try moving $\angle PQR$ to see what happens to its image.

7. Construct an arbitrary angle $\angle STU$ and, using the procedure of Steps 5 and 6, construct its image.

Does it appear that this transformation always maps right angles to right angles? Do you believe this mapping is an orthomap? Next, drag $\angle STU$ about, changing its size. (To be more certain of the result, measure both $\angle STU$ and its image.) Does this mapping appear to be angle-measure preserving? Is it really a similitude? Experiment some more by drawing a few triangles and circles, and their images. Write down all observations.

PROBLEMS (5.5)

GROUP A

1. Identify each of the following linear transformations by their coordinate forms.

$$\text{(a)}\begin{cases} x' = 2x + y \\ y' = -x + 2y \end{cases} \qquad \text{(b)}\begin{cases} x' = 3x \\ y' = 3y \end{cases} \qquad \text{(c)}\begin{cases} x' = x - 3 \\ y' = y - 5 \end{cases}$$

2. Identify each of the following linear transformations by their coordinate forms.

$$\text{(a)}\begin{cases} x' = -x + 1 \\ y' = -y + 2 \end{cases} \qquad \text{(b)}\begin{cases} x' = 2x/\sqrt{5} + y/\sqrt{5} \\ y' = x/\sqrt{5} + 2y/\sqrt{5} \end{cases} \qquad \text{(c)}\begin{cases} x' = y \\ y' = x \end{cases}$$

3. The following matrices define certain linear transformations:

$$\text{(a)}\begin{bmatrix} 3 & -2 \\ 2 & -3 \end{bmatrix} \qquad \text{(b)}\begin{bmatrix} 1 & 2 \\ -1 & 1 \end{bmatrix}$$

Find the images of the points $P(3, 0)$, $Q(5, -1)$, $R(-2, 3)$, and $S(0, 2)$ by finding the matrix products

$$\text{(a)}\begin{bmatrix} 3 & -2 \\ 2 & -3 \end{bmatrix}\begin{bmatrix} 3 & 5 & -2 & 0 \\ 0 & -1 & 3 & 2 \end{bmatrix} \qquad \text{(b)}\begin{bmatrix} 1 & 2 \\ -1 & 1 \end{bmatrix}\begin{bmatrix} 3 & 5 & -2 & 0 \\ 0 & -1 & 3 & 2 \end{bmatrix}$$

4. If T_1 and T_2 are the linear transformations defined by

$$T_1\text{:}\begin{cases} x' = 3x - 2y \\ y' = 2x - 3y \end{cases} \qquad T_2\text{:}\begin{cases} x'' = 2x' + 2y' \\ y'' = -x' - 2y' \end{cases}$$

suppose we want to find the image of $\triangle ABC$, where $A = (-1, 2)$, $B = (3, -2)$, and $C = (2, 0)$ under the product T_2T_1.

(a) Find A'', B'', C'' by direct substitution into the given coordinate equations defining T_1 and T_2, and then do this by use of matrix products.

(b) Compare your answers. Which way is more efficient?

(c) What 2×2 matrix characterizes T_2T_1?

5. Verify that

$$\begin{bmatrix} 3 & 2 \\ 2 & -3 \end{bmatrix}$$

is the matrix of a similitude by graphing $\triangle ABC$ and its altitude $\overline{BE}$, and the image triangle and altitude, if $A = (0, 0)$, $B = (2, 2)$, $C = (3, 0)$, and $E = (2, 0)$.

6. (a) Using the form **(13)** for a reflection in a line, find the equations that represent a reflection in the line ℓ: $3x + 2y = 13$. (***Hint:*** $k = 3^2 + 2^2 = 13$.)

(b) Test your answer by finding the image A' of the point $A(6, 4)$ and that of B, the foot of the perpendicular $\overline{AB}$ to ℓ (find the coordinates of B and verify that $B = B'$), then use the midpoint formula to verify that $B =$ the midpoint of segment $\overline{AA'}$.

7. (a) Using **(13)** write down the equations for the reflections through each of the parallel lines $x - 2y = 1$ and $x - 2y = 3$.

(b) Show, by coordinates, that the product of the two reflections in **(a)** is a translation.

8. (a) Repeat Problem **7(a)** for the lines $x - 2y = 0$ and $3x - y = 0$.

(b) Show, by coordinates, that the product of the two reflections in **(a)** is a rotation about the origin, and verify that the angle of rotation is twice that between the given lines as in Example 5.

9. Reflection in a Point *C*(*h*, *k*) Verify that the transformation

$$f: \begin{cases} x' = -x + 2h \\ y' = -y + 2k \end{cases} \qquad \text{matrix: } \begin{bmatrix} -1 & 0 \\ 0 & -1 \end{bmatrix}$$

is the reflection s_C in the point $C(h, k)$.

10. Three of the following mappings, given in matrix form, map squares into squares, but have decidedly different ways of doing it. Without plotting points, find which of the three mappings do so, and tell what the different ways are.

(a) $\begin{bmatrix} 5 & 2 \\ -2 & -5 \end{bmatrix}$, **(b)** $\begin{bmatrix} -4 & 3 \\ 3 & 4 \end{bmatrix}$, **(c)** $\begin{bmatrix} \frac{1}{3} & -\frac{1}{4} \\ \frac{1}{4} & \frac{1}{3} \end{bmatrix}$, **(d)** $\begin{bmatrix} -\frac{4}{5} & \frac{3}{5} \\ \frac{3}{5} & \frac{4}{5} \end{bmatrix}$

11. Find the factor k for the following similitudes:

(a) $\begin{cases} x' = 12x + 5y \\ y' = -5x + 12y \end{cases}$ **(b)** $\begin{cases} x' = -2/3x + \sqrt{5}/3y \\ y' = \sqrt{5}/3x + 2/3y \end{cases}$

Is either one an isometry? Is either one direct?

GROUP B

12. (a) Verify, by finding the images of $A(0, 0)$, $B(1, 0)$, and $C(0, 1)$ given in counterclockwise order, that under the linear transformation f whose matrix is

$$A = \begin{bmatrix} a & b \\ c & d \end{bmatrix}$$

the image points A', B', and C' will also be oriented counterclockwise iff $|A| = ad - bc > 0$.

(b) Show that, therefore, an isometry **(7)** is direct iff $\delta = 1$.

(c) Show that the determinant of the 2×2 matrix defining a line reflection **(13)** (deleting the translational component) equals -1, and thus verify that this transformation is opposite, as it should be.

13. (a) Prove that if a linear transformation f preserves the slope of any line and maps (0, 0) to (0, 0), its matrix form must be

$$A = \begin{bmatrix} a & 0 \\ 0 & a \end{bmatrix}$$

for some real a. [***Hint:*** Use the formula **(3)** for m' and the fact that m is an arbitrary real.]

(b) Show by the synthetic definition given in Section 5.3 that this is a dilation d_a centered at the origin.

14. Illustrate the result of the fundamental theorem that states that precisely one isometry exists mapping one triangle onto another congruent to it, by solving for the unique parameters a, b, δ, h, and k for the isometry that maps $A(0, 0)$, $B(3, 4)$, and $C(4, -3)$ to $A'(-1, 1)$, $B'(4, 1)$, and $C'(-1, 6)$, respectively. (***Hint:*** Plug these coordinates into the equations of the transformation, then solve for the parameters.)

15. A transformation $\boldsymbol{x}' = \boldsymbol{Ax} + \boldsymbol{b}$ is area preserving iff $|A| = \pm 1$. Verify this directly when $\boldsymbol{b} = [0, 0]$ and

$$A = \begin{bmatrix} 3 & 5 \\ 1 & 2 \end{bmatrix}$$

by finding the image of the triangle having vertices $P(0, 0)$, $Q(2, -1)$, and $R(2, 4)$. Graph and calculate the areas of both triangles directly.

GROUP C

16. Fundamental Theorem for Linear Transformations Using coordinates, prove that there exists a unique linear transformation that maps a given triad of noncollinear points onto another given triad. [***Hint:*** To get organized, prove first that there exists a unique transformation f that maps $A(0, 0)$, $B(1, 0)$, and $C(0, 1)$ to $P(p, q)$, $Q(r, s)$, and $R(u, v)$ and another, g, that maps A, B, C to $P'(p', q')$, $Q'(r', s')$, and $R'(u', v')$, respectively. Then the desired mapping is gf^{-1}.]

17. A linear transformation is a similitude iff it maps some circle onto another. Prove, using coordinates.

18. Project for Undergraduate Research A linear transformation can be represented as a product of rotations and a **simple affine map** represented in matrix form by

$$A = \begin{bmatrix} a & 0 \\ 0 & b \end{bmatrix}$$

For example,

$$\begin{bmatrix} 2 & -2 \\ 2 & 1 \end{bmatrix} = \begin{bmatrix} 2/\sqrt{5} & -1/\sqrt{5} \\ 1/\sqrt{5} & 2/\sqrt{5} \end{bmatrix} \begin{bmatrix} 3 & 0 \\ 0 & 2 \end{bmatrix} \begin{bmatrix} 2/\sqrt{5} & -1/\sqrt{5} \\ 1/\sqrt{5} & 2/\sqrt{5} \end{bmatrix}$$

Prove that any nonsingular matrix (or transformation) may be so decomposed, and that the given transformation completely determines the parameters a and b of A. (Is there a way to compute a and b directly from the given transformation?)

Conjecture: Any linear transformation f may be represented as the product $f = r_\theta A r_\phi$ where A is the transformation $x' = ax$, $y' = by$. (Recall that r_θ designates a rotation through an angle of measure θ; this part of the problem requires the consideration of translational components.)

This problem requires a knowledge of linear algebra.

*5.6 Transformation Groups

The study of geometric invariants over groups of transformations provides important insight for a deeper understanding of geometry. The structure of geometric transformation groups is rich and elegant, as we shall see. We begin with the definition.

DEFINITION: A set of transformations is called a **group,** or **transformation group,** iff

(1) the product fg of any two members f and g belongs to the set (**product closure**), and

(2) the inverse f^{-1} of each member f also belongs to the set (**inverse closure**).

The number of distinct elements in a transformation group is its **order** if it is finite; if there are an infinite number of elements, the group is said to have **infinite order.**

NOTE: The above definition requires that for any f belonging to the group, $f^{-1}f$, which is the identity map, must also belong to the group. Thus, by its definition, every transformation group must contain the identity e.

In this section we make use of the classical notation r_θ to represent a rotation about a point through an angle of measure θ (degrees), $0 < \theta < 180$. It will be necessary to extend this range to arbitrary real numbers using circular arc length (as measured in degrees), as is usually done in trigonometry. Thus, r_{180} is a half-turn about the origin in the counterclockwise direction, r_{240} is 60 degrees beyond a half-turn, and so on. The rotation r_{-360} represents a complete revolution in the clockwise direction, returning any point to its starting point (the identity transformation). Furthermore, the result of one rotation followed by another is the same as a single rotation given by the relation $r_\phi r_\theta = r_{\theta+\phi}$. Note that as a result of these conventions, we would have, for example, $r_{120}r_{300} = r_{420} = r_{60}$ (a rotation of 300 degrees, followed by a rotation of 120 degrees, resulting in a point being rotated 60 degrees from its starting point).

EXAMPLE 1 Show that the rotations $r_0 = e$, r_{90}, and r_{180} about some common point do *not* form a transformation group, and find what must be added to make it one.

SOLUTION

The set $\boldsymbol{S} = \{r_0, r_{90}, r_{180}\}$ fails to be a group on both counts: The set is not closed under products (because $r_{90}r_{180} = r_{270} \notin \boldsymbol{S}$), and $r_{90}{}^{-1} = r_{-90} \notin \boldsymbol{S}$ (it does not have inverse closure). The remedy is simply to add the rotation r_{270} to the set (which is the same as $r_{-90} = r_{90}{}^{-1}$). The result is a *group of order 4:*

$$\boldsymbol{G} = \{e_0, r_{90}, r_{180}, r_{270}\}$$

To check that this set has product and inverse closure, we will construct a multiplication table, which will allow us to systematically observe all possible products. Each entry in the table is read xy, where x is directly opposite xy, in the column to the left, and y lies in the top row, directly above xy.

	e	r_{90}	r_{180}	r_{270}
e	e	r_{90}	r_{180}	r_{270}
r_{90}	r_{90}	r_{180}	r_{270}	e
r_{180}	r_{180}	r_{270}	e	r_{90}
r_{270}	r_{270}	e	r_{90}	r_{180}

■

CYCLIC GROUP OF ORDER *n*

Because of the arrangement of the entries of the table in Example 1, where each succeeding row of products is merely a cyclic shift by one position of the row preceding it, the group in Example 1 is known as **cyclic group of order 4.** By analogy, there is a **cyclic group of order *n*** for each positive integer n, denoted $\boldsymbol{C}_n$, and an **infinite cyclic group, $\boldsymbol{C}_\infty$.** Another way to describe cyclic behavior is to observe that all the elements in the group may be generated by taking the *integer powers of a single element* of the group, called a **group generator.** For $\boldsymbol{G} = \boldsymbol{C}_4$ in Example 1, one of its generators is r_{90}, because we have

$$(r_{90})^1 = r_{90}, \quad (r_{90})^2 = r_{180}, \quad (r_{90})^3 = r_{270}, \quad \text{and} \quad (r_{90})^4 = r_{360} = e.$$

The element r_{180} is *not* a generator, since its powers do not yield all the elements of the group:

$$(r_{180})^1 = r_{180}, \quad (r_{180})^2 = r_{360} = e, \quad (r_{180})^3 = r_{180}, \cdots .$$

EXAMPLE 2 A rectangle $\square ABCD$ that is not a square has two lines of symmetry ℓ and m, as shown in Figure 5.41, and a *point* of symmetry, E, such that the isometries s_ℓ, s_m, and $r_E \equiv r_{180}[E]$ map the rectangle back into itself. These are the only isometries that do, and therefore we call them the **symmetries of the rectangle.** Counting the identity e, there are, again, four transformations that form a group of order 4. Another way to visualize this group is to think of the rectangle as an ordinary sheet of paper, which can be turned over sideways (as if turning a page in a book) and placed on top of its original position, or turned upside down from top to bottom, or both operations. (See Figure 5.42.) Thus, if these operations are denoted, s, u, and

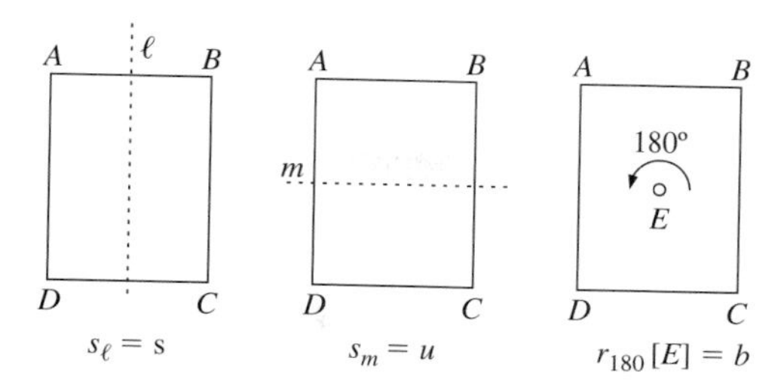

Figure 5.41

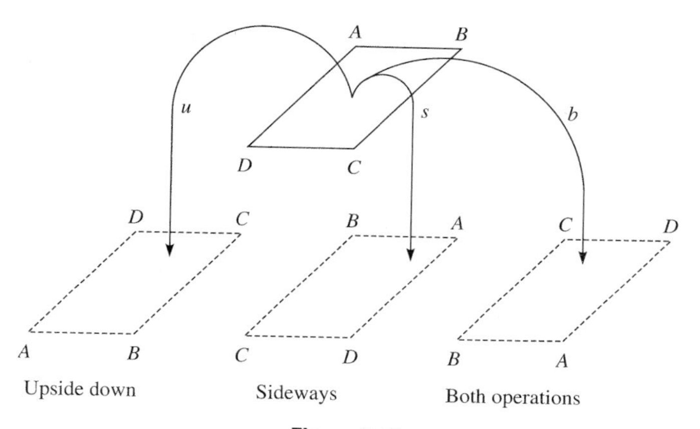

Figure 5.42

b, respectively, we obtain the following table of products, which exhibits product closure.

	e	s	u	b
e	e	s	u	b
s	s	e	b	u
u	u	b	e	s
b	b	u	s	e

It remains to check inverse closure. However, this, too may be observed from the table: Since the identity appears in each row of products, each element has an inverse (in fact, *is its own inverse,* in this particular case). ■

The two groups of order 4 presented in the two preceding examples are distinctly different. The first one was cyclic, generated by a single element, but the second has the property

$$x^2 = e$$

for each element x. Thus, the integer powers of an element will get us no further than itself and e, so it is noncyclic.

The first notable study of such groups and their relation to geometry was carried out by Felix Klein (1849–1925). The significance of the group displayed in Example 2 to geometry was discovered by him, and is sometimes called the **Klein 4-group** or simply the **Fours Group.**

HISTORICAL NOTE

When he was only 23 years of age, Felix Klein became professor of mathematics at the University of Erlangen. His inaugural lecture there made mathematical history. Known as the **Erlangen Program,** it was a bold proposal to use the group concept to classify and unify the many diverse and seemingly unrelated geometries that had developed in the 1800s. By the time he was 36, when Hilbert met him, he was already a legendary figure, and a year later he went to Göttingen where he became the leading research mathematician (and chairman of the department). He helped to restore the earlier fame enjoyed by the university in the days of Gauss and raised it to new heights. Later, both Hilbert and Minkowski joined his department, and the three of them, although having totally different personalities, worked together in rare harmony. The stimulus toward research and learning that Klein provided can be seen from the fact that he directed a total of 48 doctoral students during his lifetime. His "favorite pupil" in those days was the Englishwoman Grace Chisholm Young, the first female candidate to be granted a doctoral degree from Germany (in 1895) in any subject through the regular examination process.

THE DIHEDRAL GROUPS, SUBGROUPS

In the preceding example, we studied the symmetry groups of a rectangle with unequal sides. Here we look at the symmetry groups of a square. The square obviously has more self-mapping isometries than the rectangle. In Figure 5.43 are shown the

SYMMETRIES OF THE SQUARE

Reflections ℓ_1 ℓ_2 ℓ_3 ℓ_4

Rotations 90° 180° 270° 0° A A' $A' = A''$ (Identify)

Figure 5.43

lines of symmetry and rotations that lead to all the nonidentity isometries that map the square onto itself. The resulting group of transformations is called the **dihedral group** of the square, denoted D_4, having eight elements. To test your understanding, see if you can determine the result of a reflection in ℓ_3 followed by a 270° rotation. (You should obtain the reflection in line ℓ_2.)

> DEFINITION: When one transformation group $\boldsymbol{G}$ contains a subset that is itself another transformation group $\boldsymbol{H}$, then $\boldsymbol{H}$ is called a **subgroup** of $\boldsymbol{G}$, a relationship that is denoted by
>
> $$\boldsymbol{H} < \boldsymbol{G}$$

In the example, the set of rotations (which is the cyclic group $\boldsymbol{C}_4$) is a subset of the larger group $\boldsymbol{D}_4$, hence one writes

$$\boldsymbol{C}_4 < \boldsymbol{D}_4$$

By analogy, there is a dihedral group D_n of order $2n$ for each positive integer n, and the cyclic group $\boldsymbol{C}_n$ of order n will always be a subgroup of $\boldsymbol{D}_n$. This is the result of considering the symmetries of a regular n-gon. Because of the geometric properties of a regular polygon, the rotation $r_\theta = r$ (where $\theta = 360/n$ and the center of rotation is the center of the polygon) always maps the n-gon to itself. Thus, the symmetries of the regular n-gon always include as a subgroup the cyclic group

$$\boldsymbol{C}_n = \{r, r^2, r^3, \ldots, r^{n-1}, r^n = e\}$$

To allow you to see for yourself what one of these dihedral groups and cyclic subgroups looks like, the discovery unit at the end of this section is recommended.

GROUP DIAGRAMS

One way to study group structure is to determine the subgroups and then indicate in a diagram the relationship that these subgroups have to one another. For example, since $\boldsymbol{C}_2 < \boldsymbol{C}_4 < \boldsymbol{D}_4$, and since there are two other cyclic subgroups $\boldsymbol{C}_2'$ and $\boldsymbol{C}_2''$ of $\boldsymbol{C}_4$ (what are they?), we have the diagram shown in Figure 5.44. To indicate subgroup relationships, an arrow is directed from any group represented in the diagram to each of its subgroups.

GROUP DIAGRAM

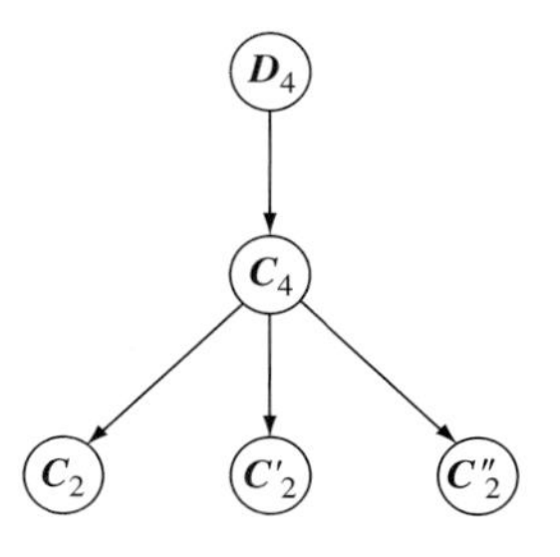

Figure 5.44

THE GENERAL LINEAR GROUP AND MATRICES

The **general linear group**—the group of all linear transformations from the plane to itself—is a tremendously large group, usually denoted by $\boldsymbol{GL}(2)$. The number in parentheses represents the dimension of the underlying space, which is, in this case, a plane. As we have seen, each element of this group may be represented in the matrix form

$$f: \boldsymbol{x}' = A\boldsymbol{x} + \boldsymbol{c}$$

where A is a nonsingular 2×2 matrix. The product of this mapping with another,

$$g: \boldsymbol{x}'' = B\boldsymbol{x}' + \boldsymbol{d},$$

may be written, at least symbolically,

$$gf: \boldsymbol{x}'' = B(A\boldsymbol{x} + \boldsymbol{c}) + \boldsymbol{d}$$

or, by the rules for matrix multiplication,

$$gf: \boldsymbol{x}' = BA\boldsymbol{x} + B\boldsymbol{c} + \boldsymbol{d},$$

where BA denotes the matrix product as defined in Section 5.5. This formula for gf can be verified by using the coordinate form introduced earlier, or by observing that $B\boldsymbol{x}'$ *means* matrix multiplication of B by any column matrix $\boldsymbol{x}'$, and the fact that $\boldsymbol{x}' = A\boldsymbol{x} + \boldsymbol{c}$ is a column matrix. The latter assumes the **associative** and **distributive laws** for matrix multiplication,

$$(AB)C = A(BC) \qquad \text{and} \qquad A(B + C) = AB + AC$$

for all matrices A, B, and C.

Another basic concept for matrices is the **identity matrix,** denoted by I and defined by

$$I = \begin{bmatrix} 1 & 0 \\ 0 & 1 \end{bmatrix}$$

It is obvious that the identity matrix represents the identity transformation $\boldsymbol{x}' = I\boldsymbol{x} = \boldsymbol{x}$. It behaves just like a unit or group identity for matrix multiplication: $AI = IA = A$ for all 2×2 matrices A.

Finally, it is useful to know the formula for the inverse of a nonsingular 2×2 matrix.

$$\textbf{(1)} \qquad A^{-1} = \begin{bmatrix} a & b \\ c & d \end{bmatrix}^{-1} = \frac{1}{k}\begin{bmatrix} d & -b \\ -c & a \end{bmatrix} = \begin{bmatrix} d/k & -b/k \\ -c/k & a/k \end{bmatrix}, \qquad k = |A|$$

Matrix theory can be used effectively to verify group structure within $\boldsymbol{GL}(2)$, as the following two examples show. The first is a numerical example to show how one can use **(1)** in connection with transformations.

EXAMPLE 3 Using matrices, find the coordinate form of the inverse of the linear transformation

$$f: \begin{cases} x' = 3x + 4y + 2 \\ y' = 4x + 5y - 1 \end{cases}$$

SOLUTION

First we transform to matrices. [The formula **(1)** for the inverse is used in the third equation.]

$$f: \begin{bmatrix} x' \\ y' \end{bmatrix} = \begin{bmatrix} 3 & 4 \\ 4 & 5 \end{bmatrix} \begin{bmatrix} x \\ y \end{bmatrix} + \begin{bmatrix} 2 \\ -1 \end{bmatrix}$$

$$\begin{bmatrix} 3 & 4 \\ 4 & 5 \end{bmatrix}^{-1} \begin{bmatrix} x' \\ y' \end{bmatrix} = \begin{bmatrix} 3 & 4 \\ 4 & 5 \end{bmatrix}^{-1} \left(\begin{bmatrix} 3 & 4 \\ 4 & 5 \end{bmatrix} \begin{bmatrix} x \\ y \end{bmatrix} + \begin{bmatrix} 2 \\ -1 \end{bmatrix} \right)$$

$$\begin{bmatrix} 3 & 4 \\ 4 & 5 \end{bmatrix}^{-1} = \frac{1}{15 - 16} \begin{bmatrix} 5 & -4 \\ -4 & 3 \end{bmatrix} = \begin{bmatrix} -5 & 4 \\ 4 & -3 \end{bmatrix}$$

$$\begin{bmatrix} -5 & 4 \\ 4 & -3 \end{bmatrix} \begin{bmatrix} x' \\ y' \end{bmatrix} = I \begin{bmatrix} x \\ y \end{bmatrix} + \begin{bmatrix} -5 & 4 \\ 4 & -3 \end{bmatrix} \begin{bmatrix} 2 \\ -1 \end{bmatrix} = \begin{bmatrix} x \\ y \end{bmatrix} + \begin{bmatrix} -14 \\ 11 \end{bmatrix}$$

or

$$\begin{bmatrix} x \\ y \end{bmatrix} = \begin{bmatrix} -5 & 4 \\ 4 & -3 \end{bmatrix} \begin{bmatrix} x' \\ y' \end{bmatrix} - \begin{bmatrix} -14 \\ 11 \end{bmatrix}$$

Therefore,

$$f^{-1}: \begin{cases} x = -5x' + 4y' + 14 \\ y = 4x' - 3y' - 11 \end{cases} \quad \blacksquare$$

EXAMPLE 4 Show that all transformations having the matrix form

$$\begin{bmatrix} 1 & a \\ 0 & 1 \end{bmatrix}$$

is a subgroup $\boldsymbol{H}$ of $\boldsymbol{GL(2)}$. [Note that this set of matrices represents all shears with respect to the y-axis introduced earlier in Section 5.1. It is assumed here that (0, 0) is a fixed point.]

SOLUTION

We must prove product and inverse closure. Take any two representative elements of $\boldsymbol{H}$ (these have to be arbitrary and different),

$$\begin{bmatrix} 1 & a \\ 0 & 1 \end{bmatrix} \quad \text{and} \quad \begin{bmatrix} 1 & b \\ 0 & 1 \end{bmatrix}$$

Their product is

$$\begin{bmatrix} 1 & a \\ 0 & 1 \end{bmatrix} \begin{bmatrix} 1 & b \\ 0 & 1 \end{bmatrix} = \begin{bmatrix} 1 \cdot 1 + a \cdot 0 & 1 \cdot b + a \cdot 1 \\ 0 & 1 \end{bmatrix} = \begin{bmatrix} 1 & a + b \\ 0 & 1 \end{bmatrix}$$

which has the desired form, hence belongs to $\boldsymbol{H}$. To calculate the inverse of any member of $\boldsymbol{H}$, write

$$\begin{bmatrix} 1 & a \\ 0 & 1 \end{bmatrix}^{-1} = \frac{1}{1 - 0} \begin{bmatrix} 1 & -a \\ -0 & 1 \end{bmatrix} = \begin{bmatrix} 1 & -a \\ 0 & 1 \end{bmatrix}$$

which also has the desired form, and belongs to $\boldsymbol{H}$. $\blacksquare$

The important subgroups of $GL(2)$ include mappings we have studied previously:

Similitudes	Dihedral groups
Dilations	Translations
Isometries	Rotations about a single point
Direct isometries	

THE ROTATION-TRANSLATION GROUP

The tools developed in this section will be particularly useful in one of our applications of group theory to geometry. It may seem surprising to learn that the set of all rotations in the plane (about different points) is *not* a subgroup, while the set of all translations is. More surprising is that, *taken together,* the family of all rotations and translations *is* a subgroup. To understand these ideas, let us go back to the basic elements of rotations and translations, the line reflections.

Recall that a translation in the plane is the product of two line reflections s_ℓ and s_m, where ℓ and m are parallel lines, each perpendicular to the direction of the translation. These perpendiculars may be chosen arbitrarily, as long as their distance apart is $\frac{1}{2}$ that of the amount of translation. The rotation $r_\theta[A]$ of θ degrees about center A is the product of two reflections s_ℓ and s_m, where lines ℓ and m intersect at A. These lines may be chosen arbitrarily, as long as they pass through A and form an angle of measure $\frac{1}{2}\theta$.

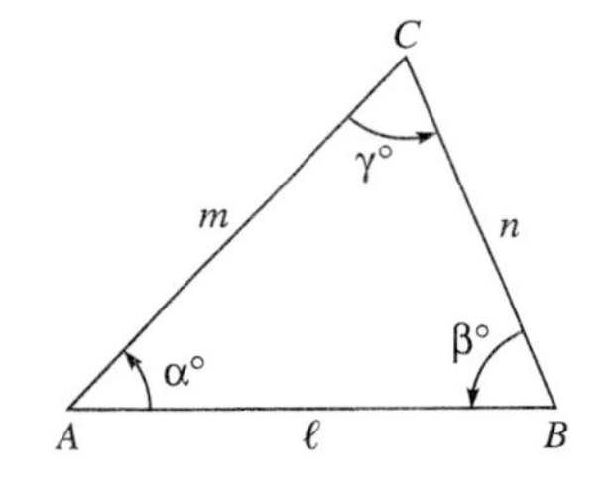

Figure 5.45

Now consider the product of two rotations having different centers, $r_{2\alpha}[A]$ preceded by $r_{2\beta}[B]$. In Figure 5.45 is shown the case when $\alpha + \beta < 180$, which determine the base angles of A and B of $\triangle ABC$. Suppose we *choose* $\ell = \overleftrightarrow{AB}$, $m = \overleftrightarrow{AC}$, and $n = \overleftrightarrow{BC}$ in representing the two rotations:

$$r_{2\alpha}[A] = s_m s_\ell, \qquad r_{2\beta}[B] = s_\ell s_n.$$

Since the product of a line reflection with itself is the identity, we have

$$r_{2\alpha}[A] \cdot r_{2\beta}[B] = (s_m s_\ell) \cdot (s_\ell s_n) = s_m {s_\ell}^2 s_n = s_m e s_n = s_m s_n$$

Thus, the product of any two rotations is either a translation (if $n \parallel m$), or another rotation (if n meets m). In the latter case, the center is $C = n \cap m$ and the angle of rotation is 2θ, where $\theta = -m\angle ACB = \alpha + \beta - 180$ (the sign denoting the clockwise direction). This proves

> **THEOREM 1:** The product of two rotations $r_{2\alpha}[A]$ and $r_{2\beta}[B]$ is either a translation or rotation. It is a translation iff $\alpha + \beta = 180$, and a rotation otherwise. If $\alpha + \beta < 180$, then
>
> $$r_{2\alpha}[A] r_{2\beta}[B] = r_{-2\gamma}[C]$$
>
> where C is the third vertex of the triangle having base $\overline{AB}$, whose vertices A, B, and C are oriented positively (counterclockwise) and whose angles are α, β and γ, respectively, each oriented in the positive direction.

COROLLARY A: If $\triangle ABC$ is any triangle in the plane, with vertices oriented in the positive (counterclockwise) direction, and α, β, and γ the three angles at A, B, and C, respectively, oriented positively, then the product

$$r_{2\alpha}[A]r_{2\beta}[B]r_{2\gamma}[C]$$

is the identity transformation.

PROOF

Use the fact that $r_{2\gamma}[C]$ is the inverse of $r_{-2\gamma}[C]$.

The converse of this corollary is also true, and it is a very interesting, but manageable problem. (See Problem 12 for hints.) We state it for later reference.

COROLLARY B: Suppose that $r_{2\alpha}[A]r_{2\beta}[B]r_{2\gamma}[C] = e$ for any triangle $\triangle ABC$ whose vertices are oriented positively, and such that α, β, and γ are each positive and less than 180, corresponding to counterclockwise rotations in each case. Then α, β, γ are the respective measures of the angles at A, B, and C.

We include one last result that is interesting and also makes a nice problem.

COROLLARY C: If $2\alpha + 2\beta + 2\gamma = 360$ and $\alpha, \beta, \gamma > 0$, then for any three noncollinear points A, B, and C the product

$$r_{2\alpha}[A]r_{2\beta}[B]r_{2\gamma}[C]$$

is either a translation or the identity.

Moment for Discovery

Symmetry Group for an Equilateral Triangle

The various members of the dihedral group $\boldsymbol{D}_3$ are illustrated in Figure 5.46.

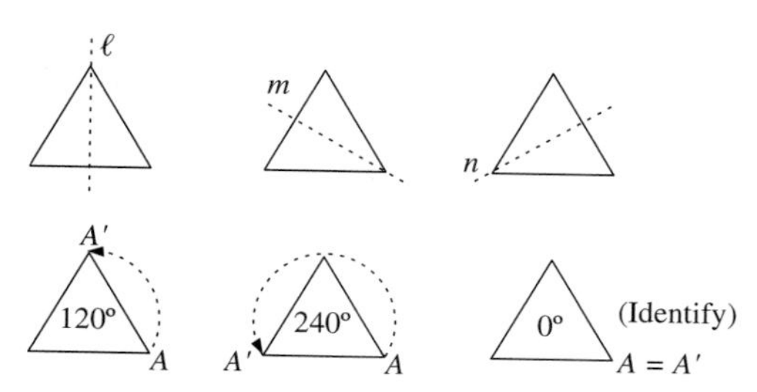

Figure 5.46

1. How many elements are there in this group?
2. What is $s_\ell s_\ell = (s_\ell)^2$? $(s_m)^2$? $(s_n)^2$?
3. What is $s_\ell s_m$? $s_\ell s_n$? $s_m s_n$? Does $s_\ell s_m = s_m s_\ell$?
4. What elements give $s_\ell{}^{-1}$, $s_m{}^{-1}$, and $s_n{}^{-1}$?

5. How do the rotations behave? What is $(r_{120})^2$?
6. Write out the complete multiplication table for $\boldsymbol{D}_3$.
7. Can you spot $\boldsymbol{C}_3$ as a subgroup of $\boldsymbol{D}_3$?
8. Find all the subgroups you can, and write their individual multiplication tables.
9. Draw a group diagram for $\boldsymbol{D}_3$ and its subgroups.

PROBLEMS (5.6)

GROUP A

1. If f and g are linear transformation as given by their matrix forms

$$f: \begin{bmatrix} 4 & 1 \\ 3 & 1 \end{bmatrix} \qquad g: \begin{bmatrix} 2 & 2 \\ 6 & -4 \end{bmatrix}$$

find the matrix representation for fg, gf, and f^{-1} in two ways:

(a) By matrix products and the formula for inverse, **(1)**.

(b) By coordinates, writing f and g in coordinate form and using primes and double primes for making substitutions, as in Section 5.5. Compare your results in **(a)** and **(b)**.

2. Show that the set of matrices of all *direct* similitudes [$\delta = 1$ in **(5)**, Section 5.5] is closed under multiplication and the taking of inverses, proving that the direct similitudes is a subgroup of the group of all similitudes. That is, show that

$$\textbf{(a)} \begin{bmatrix} a & -b \\ b & a \end{bmatrix}\begin{bmatrix} c & -d \\ d & c \end{bmatrix} \qquad \textbf{(b)} \begin{bmatrix} a & -b \\ b & a \end{bmatrix}^{-1}$$

are also matrices of direct similitudes with $\delta = 1$.

3. Show that the set of all dilations, having matrices of the form

$$\begin{bmatrix} a & 0 \\ 0 & a \end{bmatrix}, \qquad a \neq 0$$

is a subgroup of the similitudes, having matrices of the form

$$\begin{bmatrix} a & -\delta b \\ b & \delta a \end{bmatrix}, \qquad a^2 + b^2 \neq 0, \qquad \delta = \pm 1$$

4. The four matrices

$$E = \begin{bmatrix} 1 & 0 \\ 0 & 1 \end{bmatrix}, \qquad S = \begin{bmatrix} -1 & 0 \\ 0 & 1 \end{bmatrix}, \qquad U = \begin{bmatrix} 1 & 0 \\ 0 & -1 \end{bmatrix}, \qquad B = \begin{bmatrix} -1 & 0 \\ 0 & -1 \end{bmatrix}$$

form a 4-element group, which represent, respectively, the identity, the reflections in the coordinate axes, and reflection in the origin. Verify that this is a group, and show that it has identically the same table of operations as the Fours Group discussed earlier. (When this happens, the two groups are said to be **isomorphic.**)

5. In the complex numbers (where $i^2 = -1$), consider the 4-element set $\boldsymbol{S} = \{1, -1, i, -i\}$. Show that this is a group under ordinary multiplication of complex numbers. Is it the Fours Group, as in Problem 4, or the cyclic group $\boldsymbol{C}_4$?

6. In the example shown in the figure verify that both products

$$r_{260}[B]r_{100}[A] \qquad \text{and} \qquad r_{100}[A]r_{260}[B]$$

are translations.

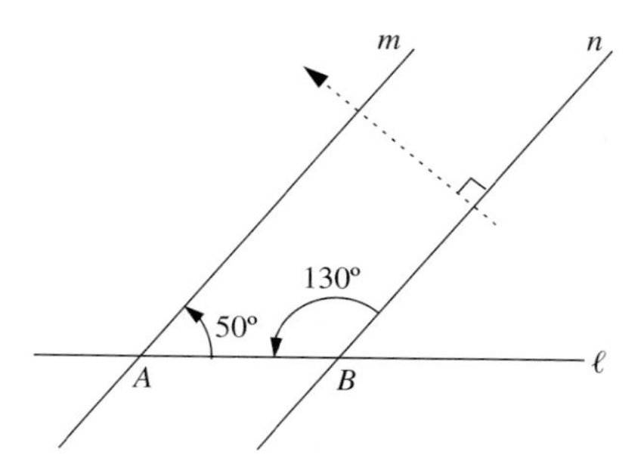

7. The dihedral group $\boldsymbol{D}_2$, containing $\boldsymbol{C}_2$, may be generated as the **symmetries of a line segment.** Let $\overline{AB}$ be any line segment, with M the midpoint, and ℓ the perpendicular bisector through M. With $m = \overleftrightarrow{AB}$, analyze the reflections s_ℓ and s_m to find the four elements of $\boldsymbol{D}_2$.

8. Find the inverse of each of the following matrices, which represent isometries, the last one representing the general case.

$$\begin{bmatrix} \sqrt{3}/2 & -1/2 \\ 1/2 & \sqrt{3}/2 \end{bmatrix}, \qquad \begin{bmatrix} 8/17 & 15/17 \\ -15/17 & -8/17 \end{bmatrix}, \qquad \begin{bmatrix} a & -\delta b \\ b & \delta a \end{bmatrix} \qquad (a^2 + b^2 = 1)$$

9. Give two proofs that the product of any two isometries is an isometry. (See Problem 8.)

GROUP B

10. If you have not already done so, analyze the subgroup structure of the dihedral group $\boldsymbol{D}_3$ explored in the discovery unit above. Make a group diagram.

11. Verify Theorem 1, making careful sketches or using compass, straight-edge. Start with the following example and track the mapping of $\triangle UVW$ under the following rotations $r_{180}[B]$ and $r_{120}[A]$, comparing this with the result of applying $r_{-60}[C]$ on the same triangle $\triangle UVW$.

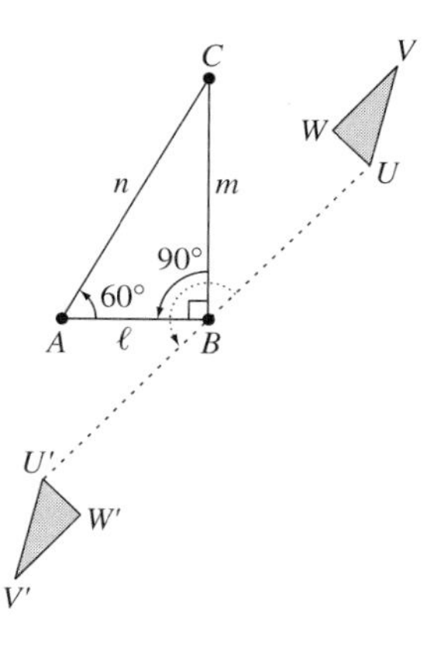

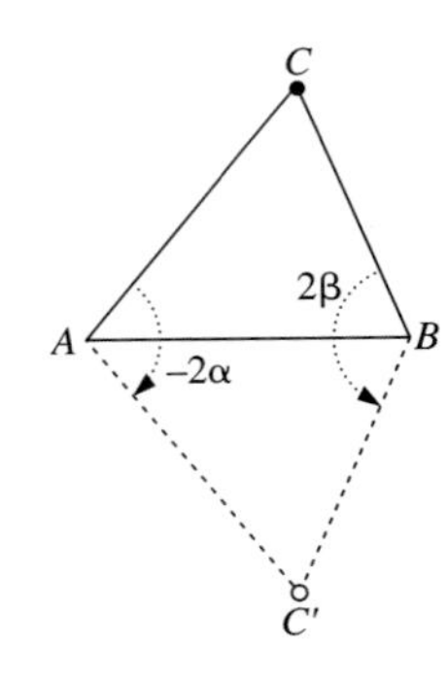

12. **(a)** Using the fact that $r_{2\gamma}[C]$ maps its center C to itself, write the identity of Corollary B in the form

$$(r_{2\alpha}[A])^{-1} = r_{2\beta}[B]r_{2\gamma}[C]$$

and analyze what the left side must do to point C (locate the point precisely).

(b) What must the right side of the above equation do to point C?

(c) Since the two sides are the same by hypothesis, then the results in **(a)** and **(b)** lead to what geometric configuration? Prove that $m\angle CAB = \alpha$, etc.

(d) Complete the proof of Corollary B.

GROUP C

13. Prove Corollary C of Theorem 1. (***Warning:*** Do not assume that α, β, and γ are the measures of the angles of $\triangle ABC$.

14. A **Generalized Lorentz Transformation** in the plane is defined as follows (a and b are arbitrary real numbers not both zero and $\delta = \pm 1$):

Coordinate Form Matrix Form

$$f: \begin{cases} x' = ax + \delta by + h \\ y' = bx + \delta ay + k \end{cases} \qquad \begin{bmatrix} a & \delta b \\ b & \delta a \end{bmatrix}$$

(a) Show that the set of matrices of the form

$$\begin{bmatrix} a & \delta b \\ b & \delta a \end{bmatrix}, \qquad a^2 \neq b^2, \qquad \delta = \pm 1$$

is closed under products and the taking of inverses, thus proving that the set of Generalized Lorentz Transformations is another subgroup of ***GL*[2].** (Note the close resemblance of these matrices with those of similitudes.)

(b) Show that instead of the invariance of perpendicularity of lines, pairs of lines having *reciprocal slopes* is an invariant. (That is, $m_1m_2 = 1 \rightarrow m_1'm_2' = 1$; see the figure.)

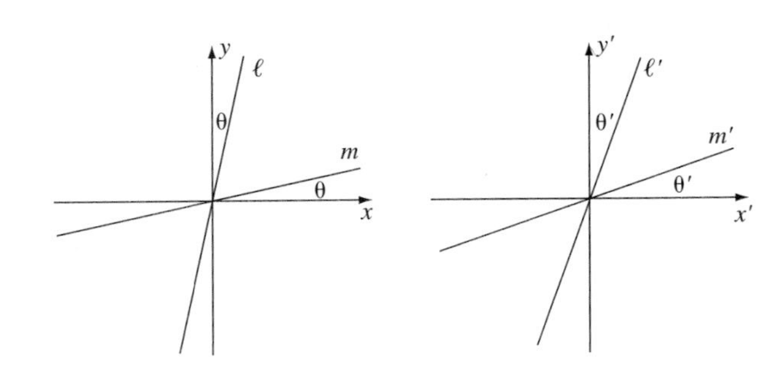

15. **Lorentz Group (Special Relativity)** If the subgroup in Problem 14 is specialized to the case $a^2 - b^2 = 1$, the result is the famous **Lorentz Group.** If $\delta = 1$, this yields the **Direct Lorentz Group,** prominent in Special Relativity. Show that the Direct Lorentz Group is, indeed, a subgroup of the General Lorentz Group.

16. Project. Complete the following diagram by adjoining the appropriate directed lines that exhibit the interrelationships of the various subgroups of $GL[2]$ we have considered previously. (Avoid redundancies in the diagram—if a chain of subgroups is indicated, then a line from the first to the last is unnecessary.)

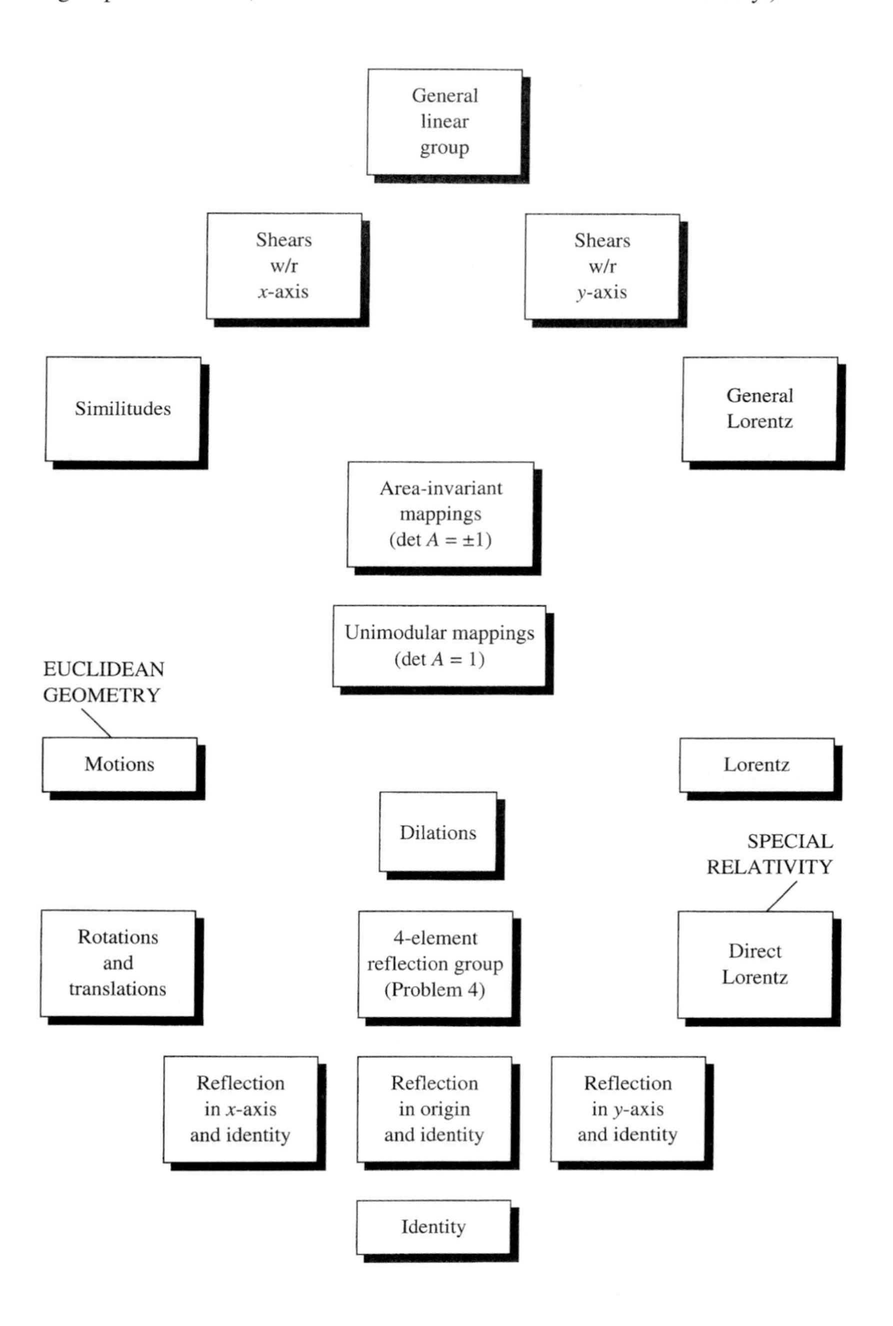

*5.7 Using Transformation Theory in Proofs

Many theorems in geometry can be made simpler and easier to prove if the theory of transformations is brought to bear. We have chosen a few examples to show you for illustration.

NINE-POINT CIRCLE

A dilation can be used not only to derive the existence of the Nine-Point Circle, but to prove other related properties as well. The center U of this famous circle lies on the **Euler Line** of the triangle—the line of collinearity of the orthocenter, centroid, and circumcenter of the triangle (Problem 3 will provide some guidance to enable you to give an elegant transformation proof for the existence of this line). To set things up properly, we will need two results about the relationship of the orthocenter of a triangle and its circumcircle.

LEMMA A: If H is the orthocenter of $\triangle ABC$, $\overline{AD}$ the altitude to side $\overline{BC}$, O the circumcenter, and L' and D' the points of intersection of $\overrightarrow{AO}$ and $\overrightarrow{AH}$ with the circumcircle, then L is the midpoint of $\overline{HL'}$ and D is the midpoint of $\overline{HD'}$ (Figure 5.47).

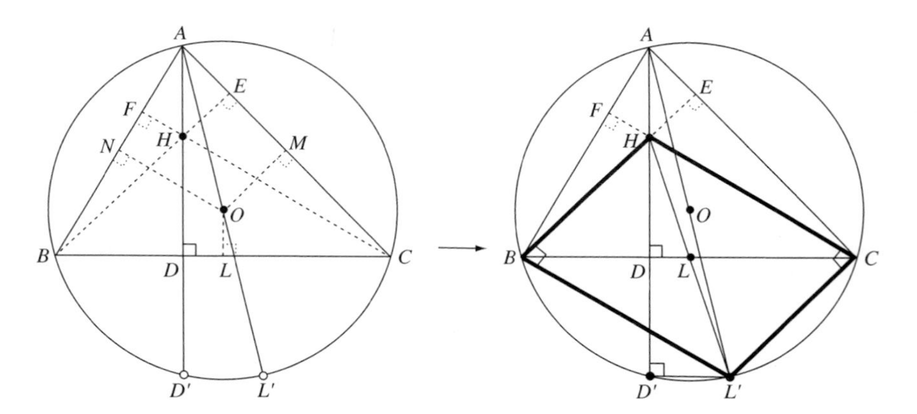

Figure 5.47

PROOF

Since $\angle ABL'$ and $\angle ACL'$ are inscribed angles of semicircles (which, together, make up the entire circumcircle of $\triangle ABC$), these angles are right angles, and hence $\overleftrightarrow{BL'} \parallel \overleftrightarrow{HC}$ (since both are perpendicular to line $\overleftrightarrow{AB}$), and similarly, $\overleftrightarrow{L'C} \parallel \overleftrightarrow{BH}$. Then $\diamond HCL'B$ is a parallelogram, with diagonals $\overline{HL'}$ and $\overline{BC}$. Hence L is the midpoint of both diagonals, and, in particular, of $\overline{HL'}$. Since $\angle AD'L'$ is also inscribed in a semicircle, $\overline{AD'} \perp \overline{D'L'}$ and $\overleftrightarrow{DL} \parallel \overleftrightarrow{D'L'}$, By the Side-Splitting Theorem, D is the midpoint of $\overline{HD'}$. This finishes the proof.

Now in Figure 5.48, consider the dilation $d_{1/2}$ with center H and scaling factor $\frac{1}{2}$. Since $d_{1/2}$ maps circles onto circles and is midpoint preserving, the circumcircle of $\triangle ABC$ maps to a circle ω whose center U is the image of O, the circumcenter of $\triangle ABC$. Hence, by Lemma A, the points D', E', F' on the circumcircle map to D, E, and F on ω, and L', M', N' on the circumcircle map to L, M, N on ω. Finally, by definition of X, Y, Z as the **Euler points** of the triangle (midpoints of $\overline{AH}$, $\overline{BH}$, and $\overline{CH}$), the vertices of $\triangle ABC$ map to these latter three points. Hence, the circle ω contains the nine points L, M, N, D, E, F, and X, Y, Z, and is the Nine-Point Circle of $\triangle ABC$. Moreover its center U lies on line $\overleftrightarrow{HO}$, the Euler Line of the triangle, and $HU = \frac{1}{2}HO$.

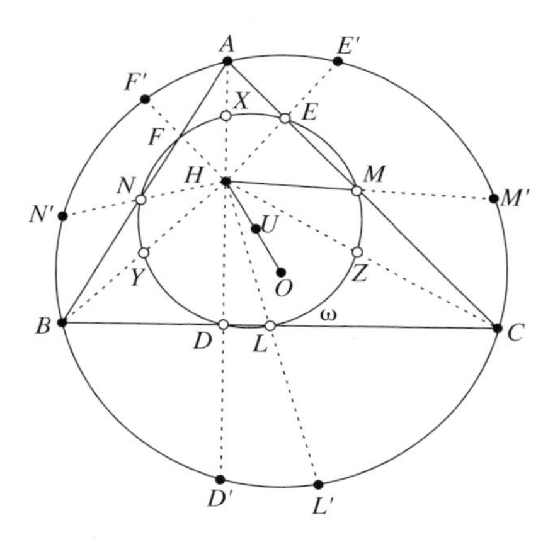

Figure 5.48

THEOREM 1: Nine-Point Circle Theorem
The midpoints of the sides of $\triangle ABC$, the feet of the altitudes on the sides, and the Euler Points all lie on a circle, the Nine-Point Circle of the triangle. The radius of the Nine-Point Circle is one-half the circumradius of $\triangle ABC$, and the center U lies on the Euler Line. Moreover, if H, U, G, and O are, respectively, the othocenter, Nine-Point Center, centroid, and circumcenter, the cross ratio $(HG, UO) = 1$. [If directed distances are used, then $(HG, UO) = -1$.]

(The proof of the last part will be left as Problem 4; use the definition of the Euler Line and the fact that $GO = \frac{1}{2}GH$. See definition of Cross Ratio in Problem 14, Section 1.3.)

FAGNANO'S THEOREM

The second example involves a singularly interesting application of reflections. The property was originally discovered by J. F. Fagnano in 1775; the reflection proof given here is due to L. Fejr. To understand the proof, you will need to recall a previous result that appeared in Problem 20, Section 4.3. It states that the orthic triangle $\triangle DEF$ of an acute-angled triangle $\triangle ABC$ forms angles with the sides of $\triangle ABC$ that

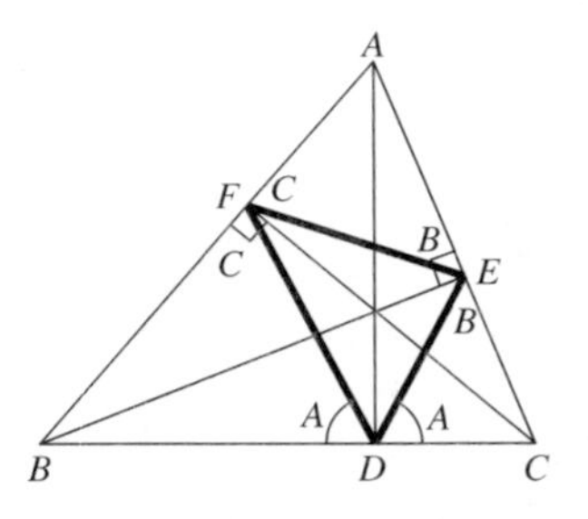

Figure 5.49

are congruent to the angles opposite; that is, in Figure 5.49, $m\angle FDB = m\angle EDC = m\angle A$, and similarly for the rest.

THEOREM 2: The orthic triangle of an acute-angled triangle has the minimum perimeter among all triangles inscribed within the given triangle.

PROOF

The proof has three major parts.

(1) Start with any inscribed triangle $\triangle PQR$ of $\triangle ABC$, as shown in Figure 5.50. Reflect $\triangle APR$ in line $\ell = \overleftrightarrow{AB}$ and $\triangle APQ$ in line $m = \overleftrightarrow{AC}$ to obtain $\triangle ASR$ and $\triangle ATQ$. Since reflections are isometries, $\triangle ASR \cong \triangle APR$ and $\triangle ATQ \cong \triangle APQ$. Using CPCF and the Triangle Inequality, we have

$$\begin{aligned} p &= \text{Perimeter}\,(\triangle PQR) = PQ + QR + RP \\ &= (TQ + QR) + RS \geq TR + RS \geq TS \end{aligned}$$

with equality only when S, R, Q, and T are collinear. Holding P fixed momentarily, the way to make p smaller is to allow R and Q to coincide with the points of intersection R' and Q' of segment $\overline{ST}$ with $\overline{AB}$ and $\overline{AC}$. Having done this, we now proceed to move P on $\overline{BC}$ to see how to further minimize p, keeping Q' and R' collinear with S and T.

(2) Observe that $AS = AP = AT$, so $\triangle AST$ is isosceles, with vertex angle $\angle SAT$ of measure

$$\begin{aligned} &m\angle 1 + m\angle 2 + m\angle 3 + m\angle 4 \\ &\qquad = 2m\angle 2 + 2m\angle 3 = 2m\angle BAC \equiv 2A \end{aligned}$$

Hence, $\triangle AST$ has a fixed vertex angle, and only the legs and base can change size. Since $AS = AP$, then if AP is made smaller, AS decreases, as well as the base ST. But $p = ST$, so p is made the smallest possible by making $\overline{AP}$ the altitude to side $\overline{BC}$ (when $P = D$, the foot of that altitude).

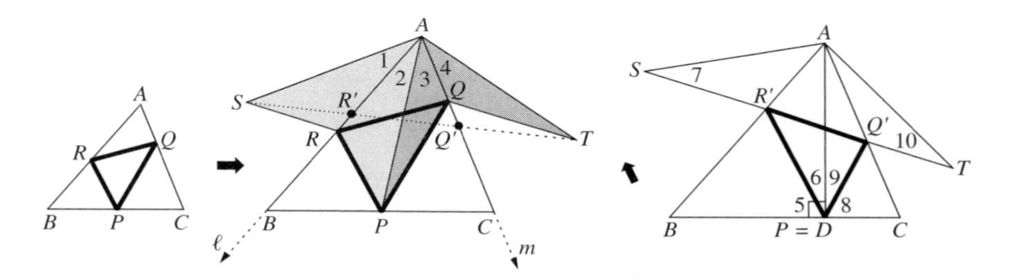

Figure 5.50

(Proof continued on next page.)

(3) It remains to show that $\triangle DQ'R'$ is the orthic triangle. But $m\angle 6 = m\angle 7$, and

$$m\angle 7 = \tfrac{1}{2}(2\, m\angle 7) = \tfrac{1}{2}(m\angle 7 + m\angle 10)$$
$$= \tfrac{1}{2}(180 - m\angle SAT) = 90 - A$$

Since $m\angle 7 = m\angle 6 = 90 - m\angle 5$, this proves that $m\angle 5 = A$. In a similar manner, $m\angle 8 = A$, and it follows from Figure 5.49 and the previous discussion that $\triangle DQ'R'$ is the orthic triangle.

OUR GEOMETRIC WORLD

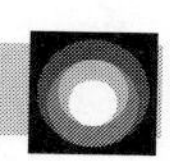

An important application of line reflections, and n-fold product of such reflections, is the technology of fiber optics. Instead of wires conducting electricity, fibers seemingly conduct light along a curved path. Actually, light rays entering one end are merely reflected back and forth along the sides, until they emerge at the other end. (See Problem 20 for the case $n = 3$.)

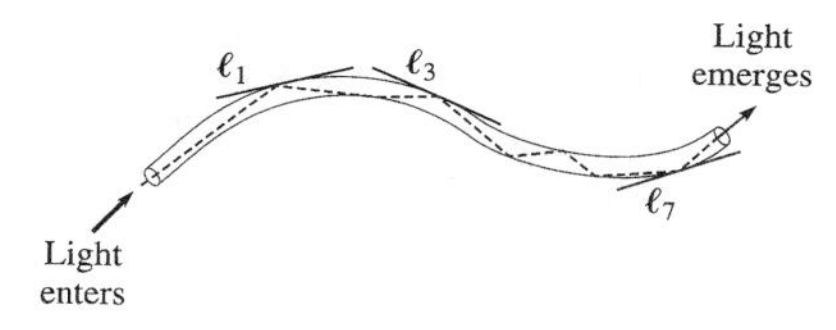

Figure 5.51

NAPOLEON-LIKE THEOREMS

NAPOLEON'S THEOREM

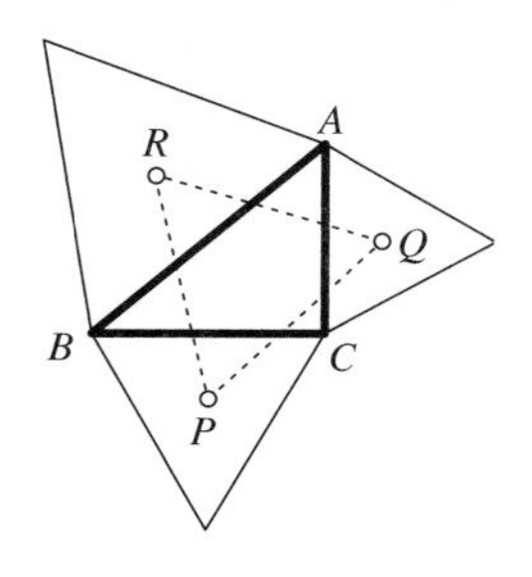

Figure 5.52

A well-known theorem attributed to an even more famous French general[4] states that if you construct equilateral triangles externally on the sides of any triangle, the centroids of those equilateral triangles also form an equilateral triangle (Figure 5.52). The proof we have in mind is an apt illustration of the power of the method of transformation theory because a coordinate proof is quite messy, and a synthetic proof is all but unmanageable. If you simply apply the theorem on the product of three rotations about the vertices of a triangle (see previous section), the proof falls right out. We are going to let you prove this classic theorem for your own enjoyment. (See Problem 10 which provides some guidance.)

Several other theorems having the same flavor as Napoleon's Theorem, some of which apply more generally to convex polygons, can be established using transformation theory. One such result follows. Before you actually read and study through the result, there is an interesting related discovery you can make for yourself. See the discovery unit at the end of this section titled *An Amazing Geometric Effect.*

---

[4]See J. Wetzel, "Converses of Napoleon's Theorem," *American Mathematical Monthly,* Vol. 99, No. 4 (1992), pp. 339–351, for an interesting discussion. According to the article, there is some doubt whether Napoleon actually discovered this theorem.

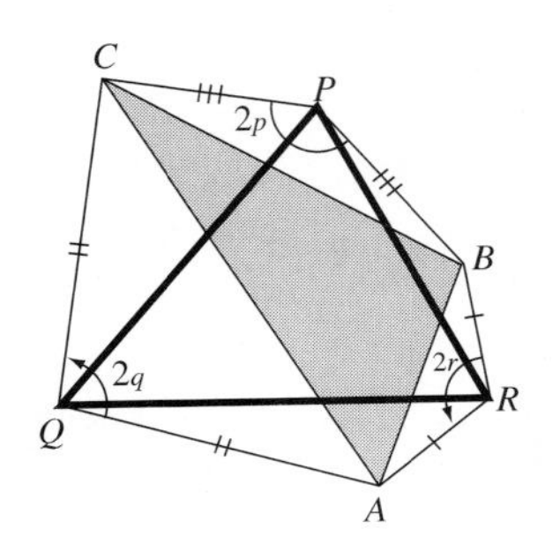

Figure 5.53

THEOREM 3: Let $\triangle PQR$ be any given acute-angled triangle in the plane, oriented positively. There exist points A, B, and C exterior to $\triangle PQR$ such that $\triangle RAB$, $\triangle QCA$, and $\triangle PBC$ are each isosceles triangles, whose vertex angles are, respectively, twice the measure of the angles of $\triangle PQR$.

PROOF

(See Figure 5.53.) Let p, q, and r denote the measures of $\angle P$, $\angle Q$, and $\angle R$ of $\triangle PQR$. By Corollary A of Section 5.6,

$$r_{2p}[P]r_{2q}[Q]r_{2r}[R] = e \text{ (identity)}$$

For less cumbersome notation, let $r_1 = r_{2r}[R]$, $r_2 = r_{2q}[Q]$, and $r_3 = r_{2p}[P]$. We can take any convenient point $B \neq P, Q, R$, and define

$$A = r_1(B), C = r_2(A)$$

Since $r_3r_2r_1(B) = B$, then $B = r_3(C)$. Thus, points A, B, and C satisfy the requirements stated in the theorem. (Do you see why?)

NOTE: The solution for points A, B, and C is clearly not unique, since B can be chosen almost at random. It would be interesting to find natural conditions for Theorem 3 that would result in a unique solution.

EXAMPLE 1 Use transformation theory to locate the vertices of the unique triangle that has a given set of three noncollinear points L, M, and N as the midpoints of its sides (Figure 5.54).

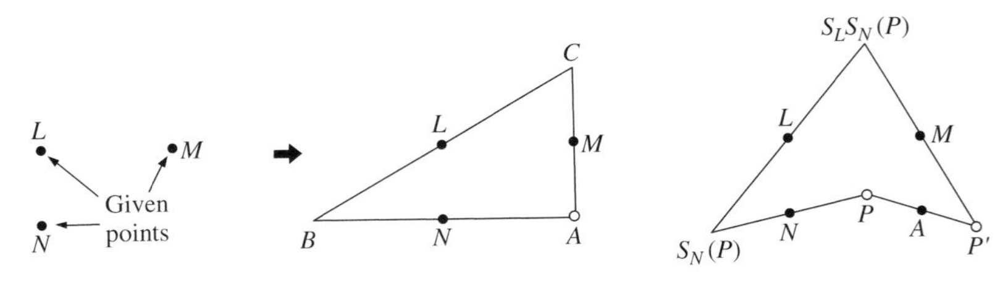

Figure 5.54

SOLUTION

We use the fact that the product of an odd number of central reflections having different centers is another central reflection (there is a short coordinate proof of this; see Problem 12). Consider the product of point reflections

$$s_Ms_Ls_N = s_A$$

(the center A is *defined* by the product). Take $B = s_N(A)$ and $C = s_L(B)$, and this will determine the desired $\triangle ABC$. (To find A constructively, simply choose some point P in the plane, and find its image P' under the above product; the midpoint of $\overline{PP'}$ will then be the desired center A, and this is one of the vertices required in the problem. Proof of the validity of this construction, which can now be converted to an elementary Euclidean construction, will be left as a problem.) ■

THEOREM 4: YAGLOM'S THEOREM

Let $\diamond ABCD$ be a parallelogram, and suppose that squares are constructed externally on the four sides of the parallelogram. Then the centers of these squares also form a square (Figure 5.55).

PROOF

Consider the rotation $r = r_{90}[P]$. The square centered at P (having $\overline{AB}$ as side) will rotate under r onto its original position, with $r(B) = A = B'$. The square centered at Q on side $\overline{BC}$ revolves 90° onto the square on side $\overline{DA}$, hence center Q maps to center S (because r preserves distance and angle measure). Thus, $r(Q) = S = Q'$. But this tells us that segment $\overline{PQ}$ rotates 90° onto segment $\overline{PS}$, and therefore $PQ = PS$ and $m\angle QPS = 90$. Since this is true at each of the other vertices Q, R, and S, $\diamond PQRS$ is a square.

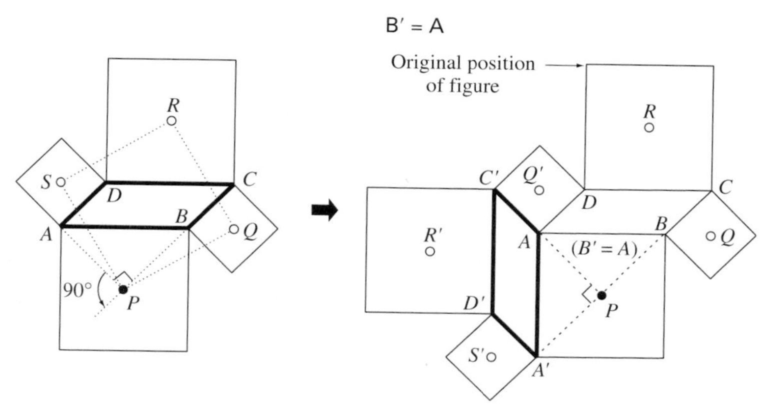

Figure 5.55

The previous proof becomes more dramatic by tiling the plane with copies of the original configuration, as shown in Figure 5.56, with copies of the points P, Q, R, and S generating a lattice. Under the same rotation we considered in the previous proof, $r_{90}[P] = r$, the entire tiling quite clearly maps onto itself (it is invariant under r), and, accordingly, so is the corresponding lattice.

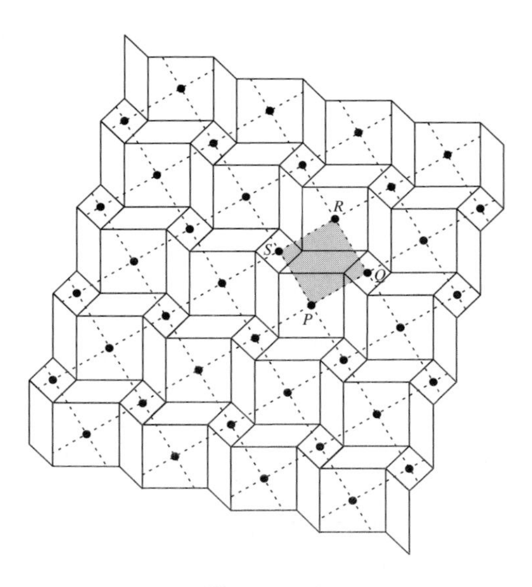

Figure 5.56

FERMAT'S POINT

A problem that often appears in calculus texts (but for which calculus is one of the more inappropriate methods of solution) goes back to Fermat. The solution presented here was given by J. E. Hofmann in 1929.

The problem is to locate a point inside a triangle so that the sum of the distances from that point to the vertices is a minimum. The solution obviously has all sorts of real-world applications. It turns out that there is a neat *geometric* solution not often realized by students of calculus. Moreover, once we find out where the point is located, it may be constructed by the ancient tools of Euclidean construction, the compass and straight-edge. The typical engineering student who struggles to work this problem by methods of calculus will no doubt completely miss the beautiful geometric aspects of this problem.

The pleasant thing about our solution is that its development is "top down," requiring no intermediate lemmas or artificial starting points. We just start with any point P inside a triangle $\triangle ABC$ (which we assume has angle measures < 120), as shown in Figure 5.57. The only clever idea required is the following: Perform the ro-

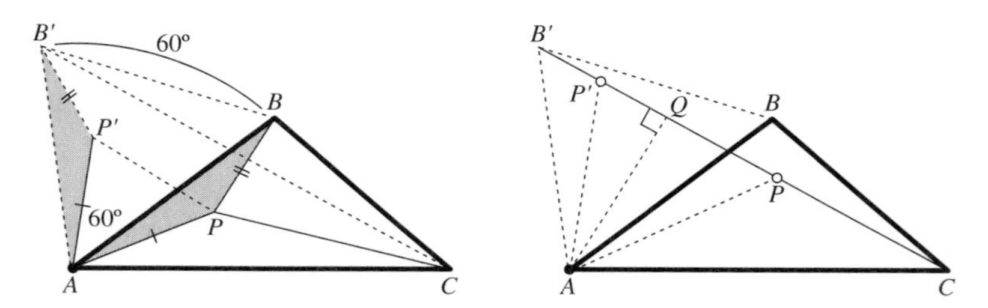

Figure 5.57

tation about vertex A through an angle of measure 60, mapping $\triangle APB$ to $\triangle AP'B'$. Since $AP = AP'$ and $m\angle P'AP = 60$, $\triangle AP'P$ is equilateral. Note that $PB = P'B'$. (Can you see why $\triangle ABB'$ is equilateral also?) Hence the sum whose minimum we seek,

$$x = PA + PB + PC$$

equals the sum

$$B'P' + P'P + PC$$

and by the Triangle Inequality

$$x \geq B'C = c.$$

Thus, we have found an absolute minimum for all sums x where P is any point inside the triangle. ($B'C$ is a fixed quantity as P varies.) If we can now prove that this sum actually takes on the value c for some choice of P inside the triangle, we shall be finished.

Drop the perpendicular $\overline{AQ}$ to $\overline{B'C}$, and locate point P on $\overline{QC}$ so that $m\angle QAP = 30$, as shown in Figure 5.57. Now perform the same rotation as before, rotating through an angle of 60 about A; again the sum x will equal the above quantity, but this time both P and P' lie on segment $B'C$, with B'–P'–P–C. Hence

$$x = B'P' + P'P + PC = B'C = c$$

NOTE: The point P of minimum sum $PA + PB + PC$ can be shown to be unique and is called the **Fermat Point** of $\triangle ABC$ (see Example 5, Section 4.8).

There is one thing about the above argument that might cause us concern. There is nothing special about our using vertex A as a center of rotation, which ultimately led to an equilateral triangle $\triangle ABB'$ on side $\overline{AB}$ and diagonal $\overline{B'C}$ of $\triangle AB'BC$. If we consider the other vertices of $\triangle ABC$, the same argument leads to equilateral triangles on the other two sides (constructed externally), and the configuration for Napoleon's Theorem reappears. (See Figure 5.58.) The argument seems to tell us that P could lie on any of the three diagonals $\overline{AD}$, $\overline{BE}$, and $\overline{CF}$.

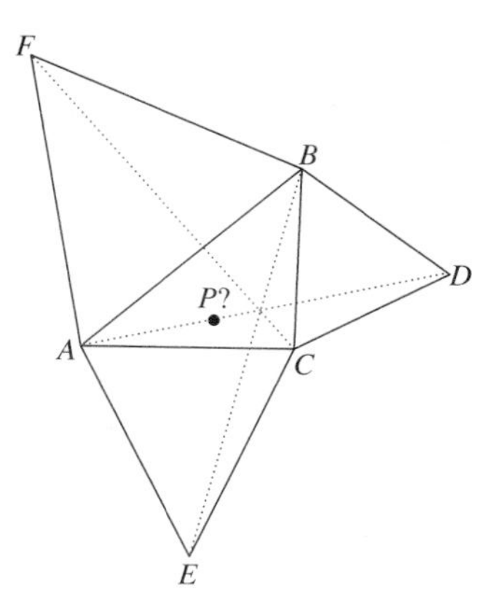

Figure 5.58

The explanation lies in Example 5, Section 4.8, where it was shown that the three diagonals $\overline{AD}$, $\overline{BE}$, and $\overline{DF}$ are concurrent, and $AD = BE = DF$. Thus the argument we presented is actually symmetric with respect to these three diagonals, and P is precisely the point of concurrency, with the minimum distance equal to the length of any one of the diagonals.

USING TRANSFORMATION THEORY TO PROVE THE SAS POSTULATE

One last application provides the foundation for textbooks that use the transformation approach in geometry. If we want to *prove* the SAS Postulate as a theorem, we must go back to the material preceding Chapter 3, and introduce transformations in a manner that is meaningful there. An extra bonus for us is that this same work will save us the trouble of having to prove the SAS Property for non-Euclidean models for which line reflections exist, which occurs in Chapter 6 (Section 5).

The danger in working with these ideas is the tendency to use properties we have been taking for granted in Euclidean geometry, such as the SSS theorem. One cannot assume in this context that an isometry (distance-preserving map) is automatically angle-measure preserving. We shall adopt a term, used only infrequently up to now as a synonym for "isometry."

> DEFINITION: A **motion** is a transformation in the plane that preserves both distance and angle measure. If a motion has a line of fixed points, that line is called its **axis.** A motion is **nontrivial** if it is not the identity.

LEMMA B: A nontrivial motion with axis ℓ has only the points of ℓ as fixed points.

PROOF

Suppose otherwise—that some point A not on ℓ is left unchanged, along with any two points B and C on ℓ (Figure 5.59). Then, since betweeness is preserved by any distance-preserving map (as argued previously), every point on segments $\overline{AB}$ and $\overline{AC}$ is fixed. It now follows that every point P in the plane is fixed, a contradiction.

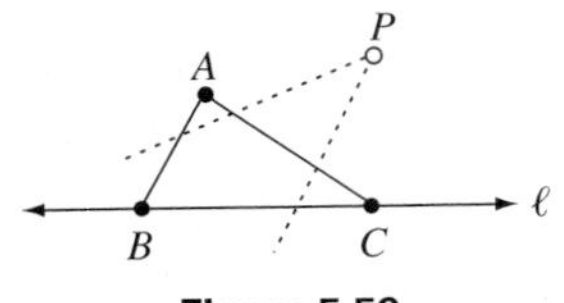

Figure 5.59

LEMMA C: A nontrivial motion with axis ℓ maps every point not on ℓ to a point on the opposite side of ℓ.

PROOF

Suppose A is any point not on ℓ, and let A' be the image of A under the given motion (Figure 5.60). Since any two points B and C on ℓ are fixed, then by definition of motions, $m\angle A'BC = m\angle ABC$. If A' and A were on the same side of ℓ, then rays $\overrightarrow{BA}$ and $\overrightarrow{BA'}$ would coincide by the Angle Construction Theorem, and by the Segment Construction Theorem, since $BA = BA'$ we have $A = A'$ (since either B–A–A' or B–A'–A must hold if $A \neq A'$). But this contradicts Lemma B. Hence, A and A' are on opposite sides of ℓ.

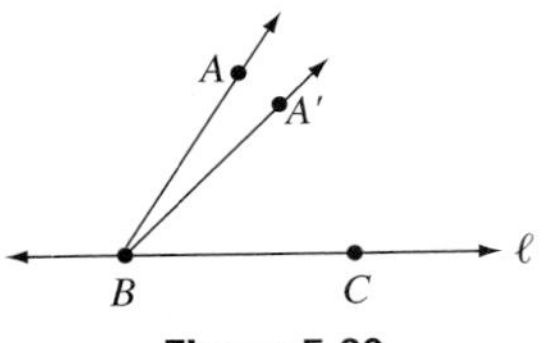

Figure 5.60

The axiom we must assume in place of the SAS Postulate is the following.

AXIOM M: There exists a nontrivial motion having any given line as an axis.

LEMMA D: Given any two angles $\angle BAC$ and $\angle YXZ$ having equal measures, there exists a motion that maps point A to point X, ray $\overrightarrow{AB}$ to ray $\overrightarrow{XY}$, and ray $\overrightarrow{AC}$ to ray $\overrightarrow{XZ}$.

(Proof left as a problem—see Problem 1.)

THEOREM 5: SAS CONGRUENCE CRITERION
Suppose that triangles $\triangle ABC$ and $\triangle XYZ$ satisfy the SAS Hypothesis under the correspondence $ABC \leftrightarrow XYZ$. Then $\triangle ABC \cong \triangle XYZ$.

PROOF

Suppose that $\overline{AB} \cong \overline{XY}$, $\overline{AC} \cong \overline{XZ}$, and $\angle A \cong \angle X$. By Lemma D, there exists a motion M mapping point A to point X, ray $\overrightarrow{AB}$ to ray $\overrightarrow{XY}$, and ray $\overrightarrow{AC}$ to ray $\overrightarrow{XZ}$. This motion must then map B to Y and C to Z by the Segment Construction Theorem. Hence, $\triangle XYZ$ is the image of $\triangle ABC$ under M, and since M preserves both distance and angle measure, the corresponding sides and angles of the two triangles are congruent. $\therefore \triangle ABC \cong \triangle XYZ$.

Moment for Discovery

An Amazing Geometric Effect

1. Start with any triangle $\triangle ABC$. On the sides of this triangle, construct external isosceles triangles $\triangle BPC$, $\triangle AQC$, and $\triangle ABR$ whose vertex angles total 360. (For this example, we have chosen isosceles triangles with vertex angles 90, 120, and 150, respectively, as shown in Figure 5.61; Use *Sketchpad* if you have access to it.)
2. Join the vertices P, Q, and R, as shown, forming a fifth triangle, $\triangle PQR$.

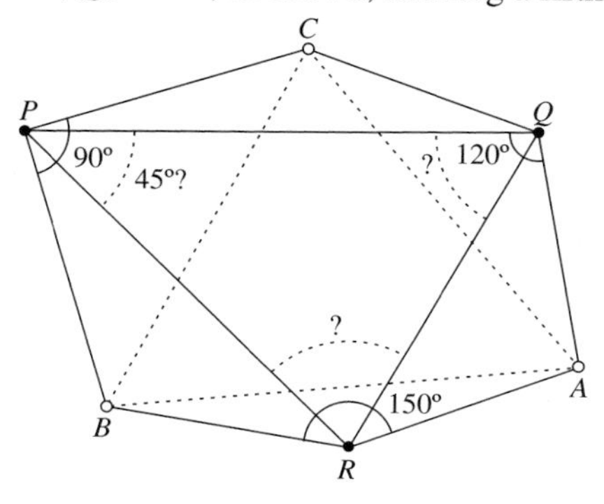

Figure 5.61

3. Using a protractor or *Sketchpad,* measure carefully the angles of $\triangle PQR$. Do you notice anything? It might be imagined that $\triangle PQR$ depends, in some way, on the original "mother" triangle $\triangle ABC$—if you change the shape of $\triangle ABC$ then the shape of $\triangle PQR$ will be affected.
4. Draw a different triangle $\triangle ABC$, making its angles radically different (say, by taking one angle to be an obtuse or right angle).
5. Construct isosceles triangles externally on the sides of your triangle *similar* to those of your original diagram (in our example, isosceles triangles having vertex angles of measure 90, 120, and 150 would be constructed).
6. Now measure the angles of the new triangle $\triangle PQR$. Did you discover anything?
7. Write down what seems like a reasonable conjecture. (You get a prize if you discover more than one property here.) Do you see any connection with Napoleon's Theorem?

Moment for Discovery

An Amazing Geometric Effect (using *Sketchpad*)

1. Construct right triangle $\triangle LMN$, with an acute angle of measure 15°. (Rotate point P to $P' = N$ about point M through 15° using Rotate under TRANSFORM, then drop perpendicular from P' to line $\overleftrightarrow{P'M}$ to obtain point L.) Construct an arbitrary triangle $\triangle ABC$.
2. On side $\overline{BC}$, construct isosceles triangle $\triangle BCP$ whose vertex angle at P has measure 30°. (Rotate ray $\overrightarrow{BC}$ about B through an angle of 75° to ray $\overline{BC'}$, then intersect this ray with the perpendicular bisector of segment $\overline{BC}$.)
3. On side $\overline{AC}$, construct isosceles triangle $\triangle ACQ$ whose vertex angle at Q has measure 150°. (Rotate ray $\overrightarrow{AC}$ about C through an angle of 15°, etc.)
4. On side $\overline{AB}$, construct isosceles triangle $\triangle ABR$ whose vertex angle at R has measure 180° (just construct the midpoint R of side $\overline{AB}$).
5. Join points P, Q, and R by line segments and construct the interior of $\triangle PQR$ using CONSTRUCT menu.

Does $\triangle PQR$ appear to be a right triangle? Is it similar to the triangle ($\triangle LMN$) you originally constructed? Drag the vertices of $\triangle ABC$ to see the effect on this phenomenon. Try this experiment with a different $\triangle LMN$. What all have you discovered? If the results you found were provable as a general theorem, would it imply Napoleon's Theorem as a special case?

Moment for Discovery

Finding Fermat's Point

Here you will investigate some features about the sum of the distances from a point to the vertices of a given triangle. Follow these steps.

1. Construct an acute-angled triangle $\triangle ABC$. On sides $\overline{AB}$ and $\overline{BC}$, construct equilateral triangles $\triangle ABD$ and $\triangle BCE$, externally.
2. Construct segments $\overline{AE}$ and $\overline{CD}$, and select the point of intersection F. Hide the segments just constructed, leaving point F.
3. Locate any point G inside $\triangle ABC$, and draw segments $\overline{GA}$, $\overline{GB}$, and $\overline{GC}$.
4. Display the sum $m\overline{GA} + m\overline{GB} + m\overline{GC} = x$ on the screen, as well as $m\angle AGB = y$. Now drag G and watch the effect on x and y. What values for y will make x as small as possible? Does F have any bearing on the problem? What theorems did you discover?

PROBLEMS (5.7)

GROUP B

1. (***Note:*** *This is the only problem in this section that will not assume the Euclidean axioms.*) Prove Lemma D in the context of a geometry satisfying the axioms of Chapter 2 and Axiom M. [***Hint:*** Let ℓ be the perpendicular bisector of line segment $\overline{AX}$ and consider the nontrivial motion M_1 with axis ℓ. Show that A maps to X, and if ray $\overrightarrow{AB}$ is not mapped to ray $\overrightarrow{XY}$, let m bisect $\angle YXB'$ and apply the motion M_2 with axis m. The composite motion M_2M_1 then maps ray $\overrightarrow{AB}$ to ray $\overrightarrow{XY}$ by the Angle Construction Theorem, etc. (a third motion M_3 may be needed).
2. How can one determine where to place the vertices of an equilateral triangle so as to inscribe it in the square shown, with one vertex of the triangle on a vertex of the square and the other two vertices lying on the sides of the square opposite that vertex?

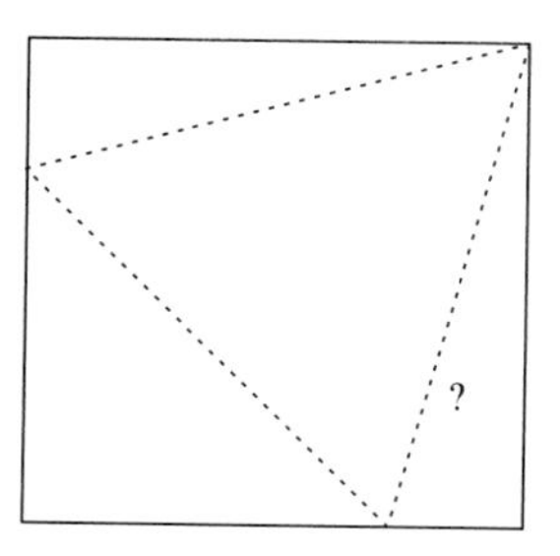

3. **Euler Line of a Triangle** A dilation that maps one triangle to another must map the altitudes and orthocenter of the first triangle to those of the second, since perpendicularity and incidence are preserved by dilations. Recall that if G is the centroid of $\triangle ABC$, then $AG = \frac{2}{3}AL = 2GL$, $BG = 2GM$, and $CG = 2GN$. Finally,

note that O is the *orthocenter of the triangle joining the midpoints of the sides of* $\triangle ABC$, that is, of $\triangle LMN$. (Prove this.) Now define the dilation $d_{-\frac{1}{2}}[G]$ with center G and negative dilation factor (hence points A, B, and C are reflected through G one-half their distance to G and must therefore map to L, M, and N). Use this to prove that O, G, and H are collinear and that $HG = 2GO$.

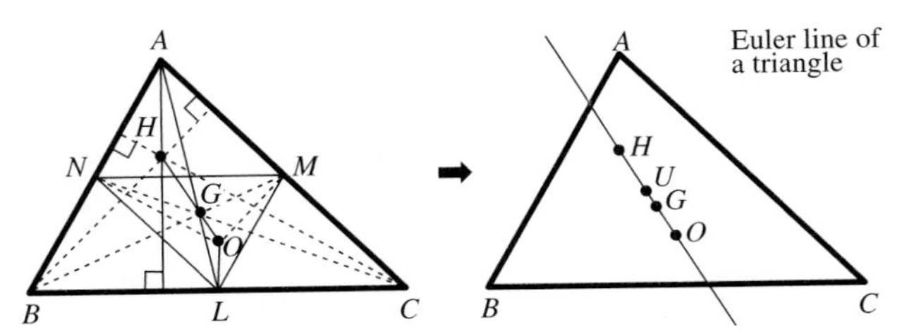

4. Referring to Figure 5.48 above, use the facts $GO = \frac{1}{2}GH$ and $HU = \frac{1}{2}HO$ to show that the segments in the above figure are correctly labeled, then prove that $(HG, UO) = 1$. (If you are familiar with directed distance, show that this cross ratio is -1.)

5. A map of Alaska and its photocopy are placed on a table at random, face up. Show that as point P traces the features of the original map and its correspondent P' traces corresponding features in the copy, the midpoint M of segment $\overline{PP'}$ traces a map that is *similar to the original map.* (Use a coordinate representation of an isometry for this endeavor.) Under what circumstances will the new map be the same size as the original?

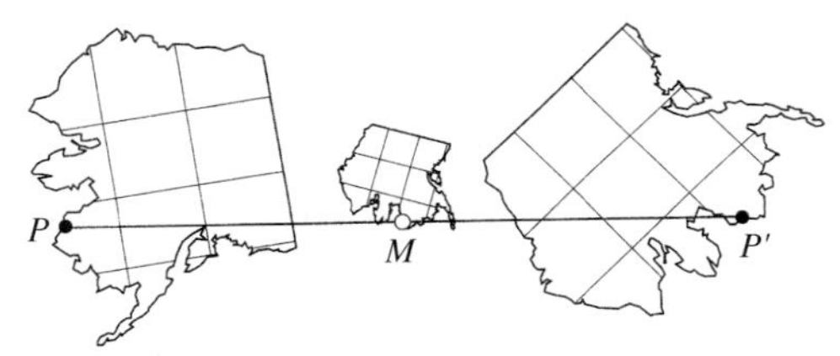

6. Scheiner's Pantograph An instrument invented in 1630 by Christolph Scheiner called the **pantograph** may be used to increase or decrease the scale of any drawing. The mechanism is fastened at a point of pivot O with the remaining rods allowed to move in the manner illustrated. The main ingredient is parallelogram $\diamond ABCA'$. Discuss the operation of this device in connection with ratios and a dilation mapping from center O.

SCHEINER'S PANTOGRAPH

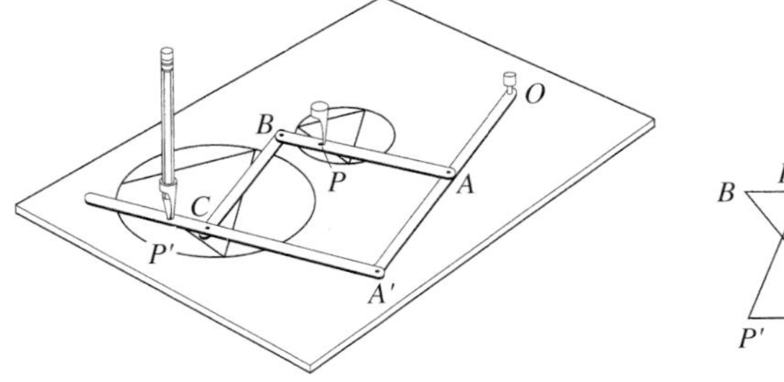

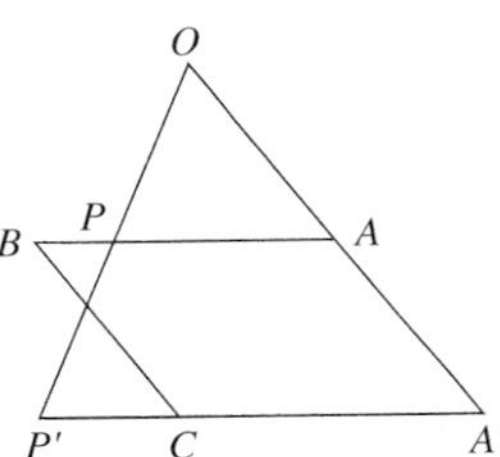

7. **Buried Treasure Problem** Without a knowledge of geometric transformations, the nature of these directions to a buried treasure would lead the average person to doubt their validity:

> *The island is uninhabited and has only two palm trees to serve as landmarks. Find a suitable place on the island where you can see both trees, and drive a stake. Proceed to the taller tree, counting paces. When you come to the tree, turn 90° to the left, and walk an equal number of paces. At that point drive a second stake. Return to your starting point and proceed to the other tree, counting your paces. When you come to the second tree, turn 90° to the right, and walk an equal number of paces. Drive a third stake. Now the treasure lies midway between the second and third stakes.*

The flaw in this "map" is that the starting point is not specified, which could land you almost anywhere on the island. However, like several problems we have seen in this section, the ending point of the directions might actually be invariant, independent of the starting point!

Simulate the directions on *Sketchpad*, or by hand, and see what happens. (A hand construction requires at least two starting points.) Then try a theoretical solution using the general rotation mapping **(12)**, Section 5.5 applied to each of the points, which can be assumed to be $A(0, 0)$ and $B(1, 0)$, where the two rotations are, respectively, $-90°$ and $90°$.

8. Analogous to the Buried Treasure Problem (preceding problem), start with three noncollinear points A, B, and C and let P be a given point. Construct the points Q, R, and S such that A is the midpoint of $\overline{PQ}$, B is the midpoint of $\overline{QR}$, and C is the midpoint of $\overline{RS}$. Let P' be the midpoint of line segment $\overline{PS}$. As P varies, what is the locus of P'? (Example 1 will provide you with some insight.) What is the result if you only use two points A and B and P' is the midpoint of $\overline{PR}$? If P describes a circle, what are the loci of R and P'?

9. Let five points J, K, L, M, and N be given. Find a way to construct by compass/straight-edge the unique pentagon $ABCDE$ such that the given points are, respectively, the midpoints of its consecutive sides. (See Example 1 for ideas.) This problem lends itself to experiment (and discovery) using *Sketchpad*.

10. **Napoleon's Theorem** In Figure 5.52, note that the angles $\angle BPC$, $\angle AQC$, and $\angle ARB$ are each of measure 120. Thus, applying the product

$$r_{120}[P]r_{120}[Q]r_{120}[R]$$

to point B, show that B returns to itself. Hence, the preceding mapping, which must either be a translation or the identity by Corollary C of Section 5.6, is the identity. Using the other material in Section 5.6, finish the proof of Napoleon's Theorem.

11. Can the plane be tiled by copies of the diagram for Napoleon's theorem in the manner of the tiling of Theorem 4 in Figures 5.55 and 5.56? If so, find it. (Make a drawing, by hand or via the computer.)

12. For the purpose of finishing the solution of Example 1, consider any three point reflections in coordinate form, s_N, s_L, and s_M, where $N = (x_1, y_1)$, $L = (x_2, y_2)$, and $M = (x_3, y_3)$.

$$s_N: \begin{cases} x' = -x + 2x_1 \\ y' = -y + 2y_1 \end{cases} \qquad s_L: \begin{cases} x' = -x + 2x_2 \\ y' = -y + 2y_2 \end{cases} \qquad s_M: \begin{cases} x' = -x + 2x_3 \\ y' = -y + 2y_3 \end{cases}$$

GROUP C

Show that the product $s_M s_L s_N$ is a point reflection (of the form $x' = -x + 2a$, $y' = -y + 2b$).

13. Generalization of Napoleon's Theorem The following result is what you might have discovered in the discovery unit (*Amazing Geometric Effect*). Prove it has the following corollary to Theorem 1, Section 5.6.

COROLLARY: Let $\triangle XYZ$ be a given acute-angled triangle, having angles of measure p, q, and r, respectively. Given any other triangle $\triangle ABC$, suppose tht isosceles triangles $\triangle PBC$, $\triangle QCA$, and $\triangle RAB$ are constructed externally on the sides of the given triangle as bases, such that the vertex angles at P, Q, and R have measures $2p$, $2q$, and $2r$, respectively. Then $\triangle PQR$ has angles of measure p, q, and r, respectively, and $\triangle PQR \sim \triangle XYZ$, regardless of the choice of $\triangle ABC$.

14. Given two circles ***C*** and ***D*** and line ℓ, as shown in the following figure, construct a line $m \parallel \ell$ that cuts off equal chords in the two circles. (***Hint:*** Think translations.)

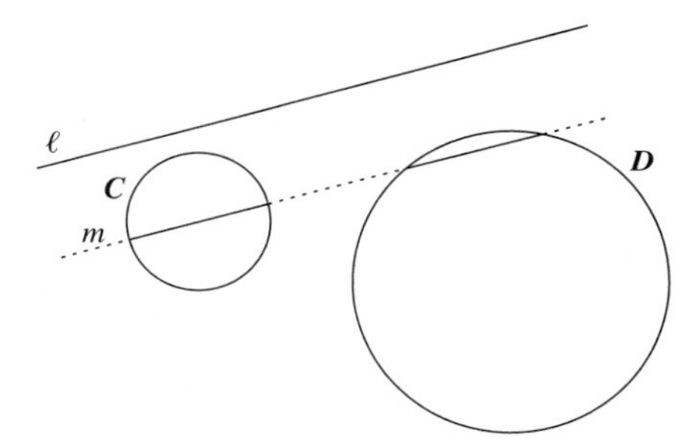

15. Use a similitude to effect the following construction (accompanying figure): Construct a circle that is tangent to the sides of an angle and passing through some given point P in its interior.

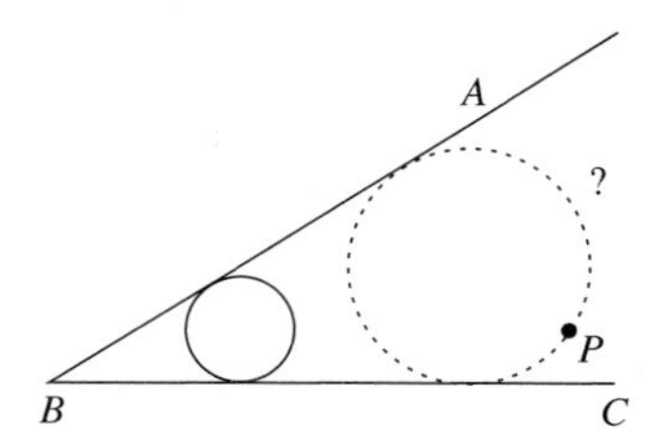

16. Given two circles ***C*** and ***D*** intersecting at A (as in figure at the right), find a line through A that cuts the circles in chords of equal lengths. (***Hint:*** Think central reflections.)

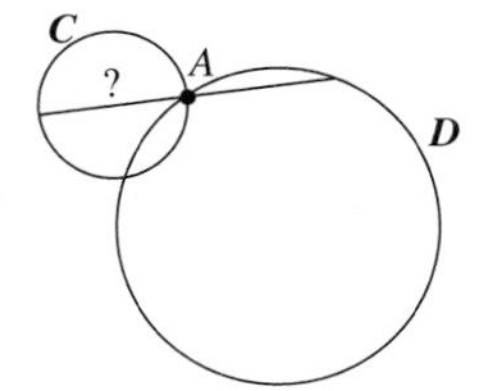

17. Given a point P in the interior of $\angle ABC$, construct a line through P that cuts the sides of the angle in two points Q and R equidistant from P. (***Hint:*** Think central reflections.)

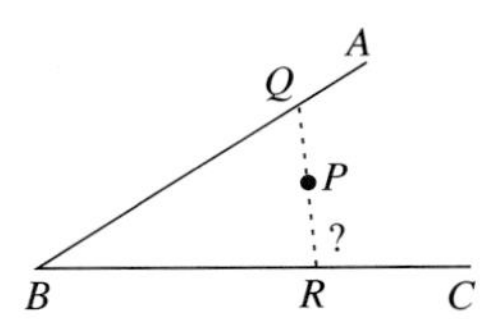

18. In a given convex quadrilateral, inscribe a parallelogram having a given point inside the quadrilateral as center. (See Problem 17.)

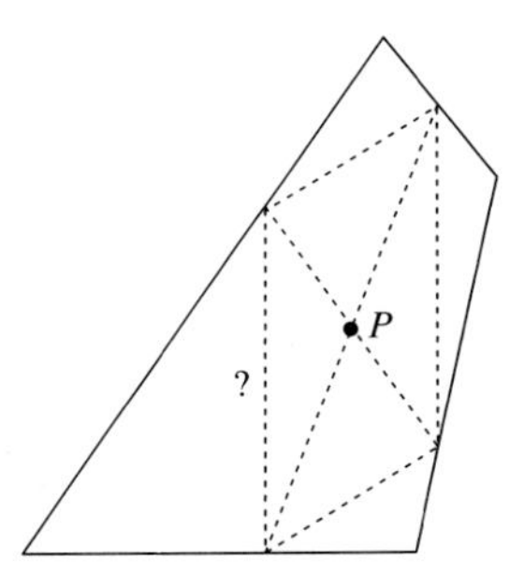

19. Consider three concentric circles. If it exists, construct an equilateral triangle having its three vertices A, B, and C on the given circles, respectively. Determine when the problem does not have a solution.

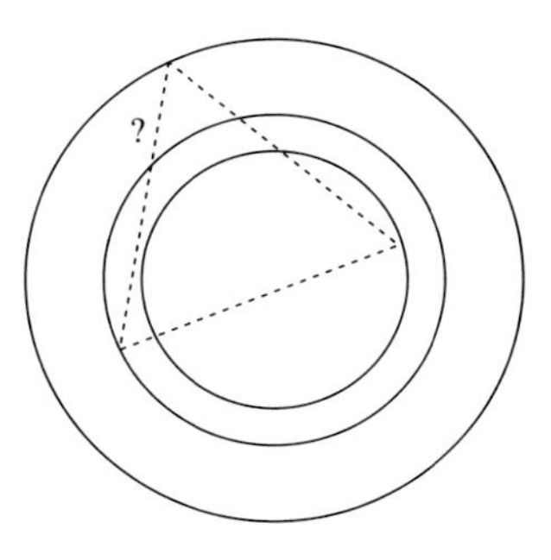

20. A Mirror Problem Three mirrors perpendicular to the plane of this page (accompanying figure) lie along three sides of a convex polygon (double lines). A laser beam is to be shot from point P and aimed in such a way toward the mirrors so as to strike the objective at point Q and stay within a narrow corridor, as shown. How can the direction of the beam be determined? (***Hint:*** First solve the problem when there are only two mirrors.)

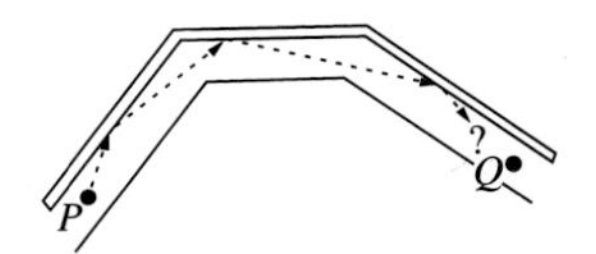

21. Project for Undergraduate Research Explore various extensions of the section on *Napolean-like Theorems* to find either further examples, conditions yielding uniqueness of certain constructions given, or extensions of these results to quadrilaterals and polygons. (For the latter, the development of the last part of Section 5.6 would have to be revamped.) For example: Does a new theory of products of four rotations exist that yields results similar to Theorem 3 and its corollaries?

Chapter 5 Summary

The concept of a transformation in geometry was developed, starting with the most basic type—the reflection. The reflection in a line—whose simple model is the common mirror—was used to define all the plane isometries. In particular, a rotation is the product of two reflections in lines that intersect, and a translation is the product of two reflections in lines that are parallel. The Fundamental Theorem states that any isometry in the plane is the product of at most three reflections. Equally important are the similitudes and dilations. A similitude is any linear transformation that preserves angle measure. A dilation also preserves angle measure, but, in addition, preserves the slopes of all lines and has a central point of projection (i.e., the image of each point lies on a line through the given point and some fixed central point). Every similitude is the product of an isometry and a dilation, making the relationship between the two clear. Glide reflections were also discussed briefly.

The chapter ended with a discussion of groups of transformations and its various subgroups (e.g., the isometries are a subgroup of the similitudes), and several examples showed how transformations can be used to solve problems in geometry.

Testing Your Knowledge

You are expected to take this test using only the list of axioms, definitions, and theorems in Appendix F as references.

1. Which pairs of figures are related by a dilation map, and which by a similitude that is not a dilation map?

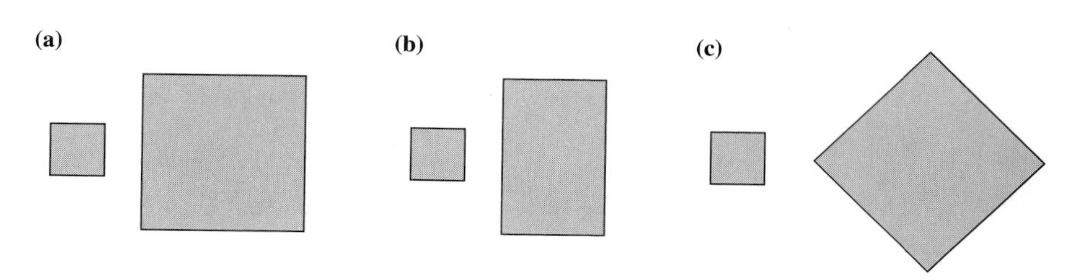

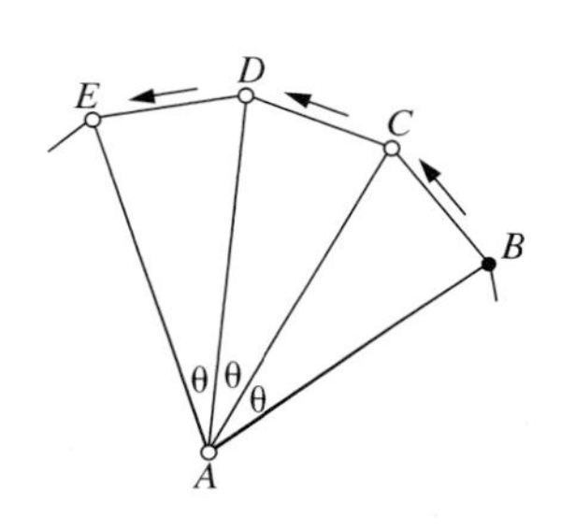

2. Point B in the plane is rotated repeatedly about point A through the same angle θ [that is, $r_\theta(B) = C$, $r_\theta(C) = D$, $r_\theta(D) = E$, and so on], where $\theta = 360/n$ for some integer n. Show that the result is a regular n-sided polygon. (For this purpose, it will suffice to show that in the figure, $BC = CD$ and $m\angle BCD = m\angle CDE$.)

3. What geometric principle justifies the conclusion that if $A' = f(A)$, $B' = f(B)$, $C' = f(C)$ where f is an isometry, then

$$\triangle A'B'C' \cong \triangle ABC?$$

4. The various pairs of right triangles shown in the figure are congruent. Find the least number of line reflections that can be used in a product of transformations that will map the first of each pair onto the second. (No proofs necessary.)

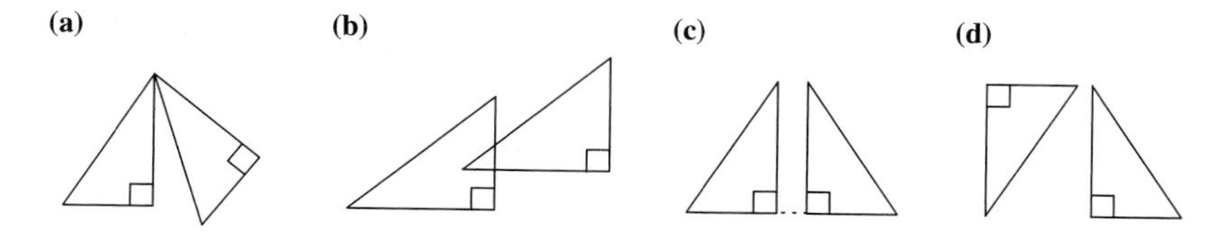

5. Find the unique value for a for which the linear transformation given below does *not* have a fixed point.

$$f: \begin{cases} x' = 2x - ay \\ y' = 3x - 2ay - 1 \end{cases}$$

6. A rotation has only one fixed point, its center, while a translation has none. Hence, a rotation followed by a translation cannot have the center of rotation as a fixed point. Show, however, in a diagram of your own choice, how a rotation followed by a translation *can* have a fixed point.

7. Find the inverses of each of the following two linear transformations:

(a) $\begin{cases} x' = x + 4 \\ y' = y + 3 \end{cases}$

(b) $\begin{cases} x' = 3x - y + 2 \\ y' = 5x - 2y + 3 \end{cases}$

8. Show that the following mapping has an infinite number of fixed *lines* (not points), that is, lines that map to themselves. Find an equation for one of them, and tell how to find the rest.

$$\begin{cases} x' = x + 2 \\ y' = y + 6 \end{cases}$$

9. Identify, by their forms, each of the following mappings as to either (1) an isometry, (2) a similitude that is not an isometry, and (3) a linear transformation that is not a similitude.

(a) $\begin{cases} x' = 5x \\ y' = 3y \end{cases}$ **(b)** $\begin{cases} x' = 5x + 1 \\ y' = 3y + 2 \end{cases}$ **(c)** $\begin{cases} x' = y + 1 \\ y' = x + 2 \end{cases}$

(d) $\begin{cases} x' = 2x - 3y \\ y' = 3x + 2y \end{cases}$ **(e)** $\begin{cases} x' = 3x + 2y \\ y' = 2x + 3y \end{cases}$ **(f)** $\begin{cases} x' = \frac{5}{13}x - \frac{12}{13}y \\ y' = \frac{12}{13}x + \frac{5}{13}y \end{cases}$

10. Prove the theorem that a linear transformation maps parallel lines to parallel lines.

CHAPTER SIX

Alternative Concepts for Parallelism

NON-EUCLIDEAN GEOMETRY

OVERVIEW

We develop here the concept of non-Euclidean geometry within the axiomatic framework of absolute geometry begun in Chapter 2. Included is a historical discussion of its origin, which accompanied a great geometric revolution in the 1800s. The mathematical side of this revolution is explored, with some samples of false proofs that have appeared in the past. We then concentrate on hyperbolic geometry and its models, derive a surprisingly simple formula for area, and show how to use the special formulas that govern the sides and angles of a right triangle in hyperbolic geometry.

6.1 Historical Background of Non-Euclidean Geometry

The first definitive study of parallelism and its effect on geometry was Euclid's, and because of the way he handled this topic, he has been called *the world's first non-Euclidean geometer.* The development of rectangles, parallelograms, area, the Pythagorean Theorem, and volume, which came after a full treatment of triangles, was based on one suspiciously complex postulate—the famous **Fifth Postulate of Parallels** (Figure 6.1):

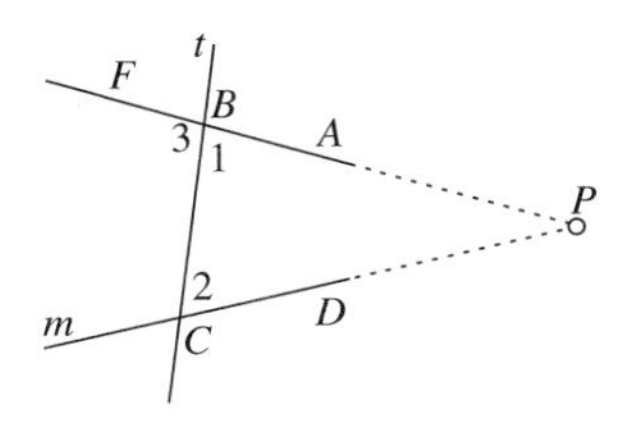

Figure 6.1

> If a straight line falling on two straight lines makes the interior angles on the same side less than two right angles, the two straight lines, if produced indefinitely, meet on that side on which are the angles less than the two right angles.

That is, in Figure 6.1, if $m\angle 1 + m\angle 2 < 180$, then lines ℓ and m meet on the A-side of line t. Euclid's earliest critic was Proclus (410–485), who wrote detailed commentaries on the works of early Greek geometers. Proclus refused to accept Euclid's postulate because he felt it was too complicated to be a postulate, and thought it could be proven from the other axioms. He constructed his own argument, which he thought would settle the issue once and for all (we will take a look at this proof later). His proof was among the first of scores of pseudoproofs of the postulate that were to emerge over the next 1400 years.

The very organization of Euclid's material invites speculation. The use of the parallel postulate is postponed until the last possible moment. Fully 28 propositions appear before that, substantially the same results we obtained in Chapter 3 for absolute geometry. The next to last of these results (Proposition 27) reads:

> If a straight line falling on two straight lines make the alternative angles equal to one another, the straight lines will be parallel to one another.

This will be recognized as Theorem 1, which we proved in Section 4.1. Euclid's poetic proof, as quoted from Heath's *Thirteen Books of Euclid* (with its own peculiar illustration, Figure 6.2), reads

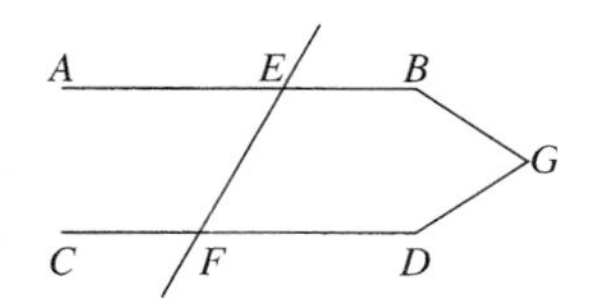

Figure 6.2

> I say that AB is parallel to CD.
> For, if not, AB, CD when produced will meet either in the direction of B, D or towards A, C.
> Let them be produced and meet, in the direction of B, D, at G.
> Then, in the triangle GEF, the exterior angle AEF is equal to the interior and opposite angle EFG: which is impossible.
> Therefore AB, CD when produced will not meet in the direction of B, D.
> Similarly, it can be proved that neither will they meet towards A, C.
> But straight lines which do not meet in either direction are parallel; [Def. 23] therefore AB is parallel to CD.

Proposition 28 merely uses a different pair of angles along the transversal, so it is a simple corollary to Proposition 27. It is highly fortuitous that the very *next* proposition in the *Elements* is the converse of Propositions 27 and 28, and in proving it, Euclid uses his Parallel Postulate for the first time.

The impressive volume of material and methods associated with proving 28 theorems, the last of which is directly related to parallel lines, has a fatal psychological effect. We begin to wonder why the Parallel Postulate is needed at all. Since there is so much to work with, the conviction emerges that we should be able to prove the postulate as a theorem, in the style of Proclus. Since geometry abounds with propositions for which a simple logical twist proves their converse, why not try a similar tactic for the converse of Proposition 27, without using the Parallel Postulate?

This was to become the question of the ages, a question whose answer led to a fascinating journey stretching over more than 2000 years, marked with worthwhile developments as well as controversy, and it ultimately revolutionized our thinking about mathematics and axiomatic systems. The problem engaged some of the greatest mathematical minds in history, including Carl Friedrich Gauss and A. Legendre. Even the great analyst Joseph Lagrange worked on the problem, but soon gave it up. (The story is told that on one occasion Lagrange was presenting a paper on parallels to the French Academy, but broke off in the middle of his presentation with the exclamation, "I must meditate further on this!" With that, he put the paper in his pocket, walked out of the meeting, and never spoke of it publicly afterward.)

During this very long period of time, very few suspected the real source of difficulty. In reality, the postulate is an *undecidable issue* in the context of absolute geometry. That is, it can neither be proven, nor can a counterexample be constructed. However, it should be realized that mathematicians in those days were not well acquainted with models and axiomatic systems. Their postulates were based on the most concrete of observations, and hence they trusted axiomatics more than we do today. With our sophisticated approach, we are much more aware of the logical pitfalls and the use of models in abstract situations. Euclid, Archimedes, and hosts of followers merely postulated what seemed to be self-evident in the world about them—what they thought they could see with their own eyes.

This was very much in line with the teachings of the eighteenth-century philosopher Immanuel Kant, who held that there can be only one valid perception of the universe. As a result, the corresponding belief that there can be only one consistent geometry—the geometry of the world in which we live—came to be firmly rooted in scholarly thinking. Leading experts believed that any other kind of hypothesis would lead to dire consequences, paradoxes, and mathematical foolishness.

The stage was set for a monumental problem to overcome. It defeated G. Saccheri, whom we have already mentioned, who dutifully concluded that his acute angle hypothesis could not be consistent with "the nature of lines." While J.H. Lambert showed great insight in his study of variable angle sums of triangles, he too did not venture far from the Euclidean hypothesis. Although Gauss was the first to recognize the true nature of the problem and to develop a consistent non-Euclidean geometry, even he was to write in a letter to F.W. Bessel in 1829:

> It may take very long before I make public my investigations on this issue; in fact this may not happen in my lifetime for I fear the scream of dullards if I make my views explicit.

So, although Gauss had worked extensively on the problem of parallels, he did not publish his work. How then were those who followed to know that he had already constructed rather extensive features of what we now call **hyperbolic geometry,** possibly as early as 1800?

This antischolarly environment was responsible for one of the great tragedies and misunderstandings in modern mathematics. It involved a young and talented scholar, János Bolyai (1802–1860) of Hungary, who was affected by the incident the rest of his life and was discouraged from further work in mathematics.

When he was only 23, Bolyai discovered a new development of geometry that eventually led to settling the problem of parallels. During the years 1823–1832, he and the Russian mathematician Nicholai Lobachevski (1793–1856) independently, and without knowledge of each other's work, developed an elaborate system of non-Euclidean trigonometry that, from its complete analogy to the well-known formulas of spherical trigonometry, strongly suggested that assuming a denial of the Fifth

Postulate will not lead to a contradiction. Instead, another very bizarre but seemingly consistent geometry (hyperbolic geometry) is obtained.

After Bolyai made this discovery, it astounded him, and he wrote of his excitement to his father, Wolfgang (Farkas) Bolyai:

> I have discovered such magnificent things that I am myself astonished at them. . . . Out of nothing I have created a strange new world.

His work was printed in 1832 as an appendix to a geometry text published by the senior Bolyai, which was a short 26 pages. (Lobachevski's work, which was incredibly similar, was first published in 1829, in Russian.) In that 26 pages, Bolyai develops the bulk of hyperbolic geometry—including the construction of a surface in hyperbolic three-dimensional space on which Euclidean geometry takes place (much like spherical geometry takes place on the surface of the sphere in Euclidean three-dimensional space), which Bolyai called a *parasphere.* (This same surface was constructed by Lobachevski, which he called a *horosphere.*) One starts with an ordinary spherical surface in absolute geometry, deduces formulas on it that are true without the Parallel Postulate, then lets the radius of the sphere become infinite. In Euclidean geometry this produces the ordinary plane, but in hyperbolic geometry a different surface is created. If in the underlying space we assume that the Parallel Postulate is false, then the ultimate result is a two-dimensional *Euclidean geometry* that lives in a three-dimensional non-Euclidean environment!

The senior Wolfgang Bolyai, who was a friend of Gauss, proudly sent this new work of his son's to him to see what he thought of it. Gauss wrote back, in so many words: "I cannot praise this work, for to do so would be to praise my own work. . . . I myself, long ago came to these same conclusions." This pontifical response was bred of a lifetime of conservative habits by a perfectionist, and the fear of being ridiculed. But his response was viewed by young Bolyai as discouraging, to say the least. Even to imagine that this work of Gauss's may have been locked away in a desk drawer since before he was born, would have been quite bizarre indeed. Bolyai feared Gauss was actually seeking to take credit for his accomplishment, or play down its importance. The incident also turned him against his father for a time, whom he suspected of being in collusion with Gauss.

HISTORICAL NOTE

Although Bolyai and Lobachevski (pictured) were heroes of the geometric revolution of the 1800s, their splendid accomplishments were met with total indifference by the mathematical community. János Bolyai was born in 1802, and, taught by his father, had already mastered calculus by the time he was 13. Bolyai was a Hungarian, with a flamboyant spirit and quick temper. When his father's request to have Gauss take him as a private student went unanswered, he began his education for the military. He became a skillful fencer, and once accepted the challenge of thirteen officers for a duel on condition that he be allowed to play his violin after each duel; he defeated them all. After his rebuff by Gauss over hyperbolic geometry, he published nothing else, although he left behind some 1,000 pages of manuscript. In his last years, he learned of Lobachevski's work. A posthumous victory was had when, in 1905, the Hungarian Academy of Science established the *Bolyai Prize* in his honor, consisting of 10,000 gold crowns awarded to the mathematician who had most greatly contributed to progress in mathematics. (The first to be awarded this prize was, appropriately, H. Poincaré; Hilbert was second, in 1910, and Einstein was third in 1915.)

In contrast to Bolyai's free spirit, Lobachevski was a reserved scholar. He was appointed professor of the University of Kasan (Russia) in 1816, and rector in 1827. His work in geometry, like Bolyai, courageously challenged the Kantian doctrine of space. It included a complete theory of parallels and the construction of an "imaginary" geometry, which he later called *pangeometry.* Although his development was published three years before Bolyai's, he received no recognition for it. The work was largely misunderstood and even stated to be incorrect by one reviewer. If the lack of recognition for his work were not enough, the administration of the University of Kasan decided to revamp its organization in 1846, and Lobachevski was dismissed after more than 30 years of faithful service. During his final 10 years he suffered blindness and had to dictate his remaining works. But, unlike Bolyai, he never gave up, anticipating the day when his work would finally win approval.

It would take approximately 40 years for mathematicians to realize the significance of the work of Bolyai and Lobachevski. In the 1850s, the concepts of differential geometry, begun by Gauss himself, were developed to the point that surfaces of constant negative curvature (called **pseudospheres**) were more clearly understood, and a variety of models for non-Euclidean geometry were subsequently discovered. After Bernhard Riemann's famous lecture in 1854 on a very general kind of geometry having arbitrary dimension and variable curvature—a development that was to play an important role in the development of the theory of General Relativity—in 1868 Eugenio Beltrami laid out the basic idea for his proof of the relative consistency of non-Euclidean geometry. At first he used the geometry of the pseudosphere as a model, then he developed both the **upper half-plane** and **circular disk** models. We will study these later in detail. In 1871, Klein gave Beltrami's disk model a new interpretation in the projective plane, which made its study more elegant. (This construction is known as the **Beltrami–Klein model.**) Later, in 1882, H. Poincaré reintroduced Beltrami's disk model in connection with transformation groups of complex numbers, traditionally known as **Poincaré's Model.**

6.2 An Improbable Logical Case

When we introduced the Euclidean Parallel Postulate in Chapter 4, we observed that there were three logical cases associated with a line ℓ, a point P not on that line, and the lines through P parallel to ℓ.

(1) There are *no* lines parallel to ℓ. (Postulate for Spherical Geometry)

(2) There is *exactly one* line parallel to ℓ. (Euclidean Parallel Postulate)

(3) There are *two or more* lines parallel to ℓ. (Lobachevskian Parallel Postulate)

With a slight modification of our postulates to accommodate a bounded metric, these three hypotheses lead to the three classical geometries, which were given the more esoteric names (1) **elliptic,** (2) **parabolic,** and (3) **hyperbolic geometry** by Felix Klein (1849–1925), in order to place them on an equal footing. Obviously, parabolic geometry is a fancy name for Euclidean geometry; other names commonly used for elliptic and hyperbolic geometry are, respectively, **spherical** (also **Riemannian**) **geometry** and **Lobachevskian geometry.**

In absolute geometry, the *existence* of parallel lines is a proven fact (Theorem 1 in Section 4.1). Thus, the set of 15 of axioms we have adopted for absolute geometry eliminates elliptic geometry. In Appendix D we show how to properly modify the axioms to allow the existence of elliptic geometry, but right now, let us continue to work in absolute geometry. Also, we will not assume Axiom P-1, the Euclidean Parallel Postulate. That is, we take up where we left off at the end of Chapter 3, omitting all the material in Chapters 4 and 5.

PLAUSIBILITY OF THE HYPERBOLIC POSTULATE

The alternative (logical) possibility for parallelism in absolute geometry is: Given a point P and a line ℓ not passing through it, there exist *at least two distinct* lines through P, in the same plane as P and ℓ, which are both parallel to line ℓ. Thus we have lines m_1 and m_2 through P parallel to ℓ (which lie in the same plane and do not intersect), as illustrated in Figure 6.3.

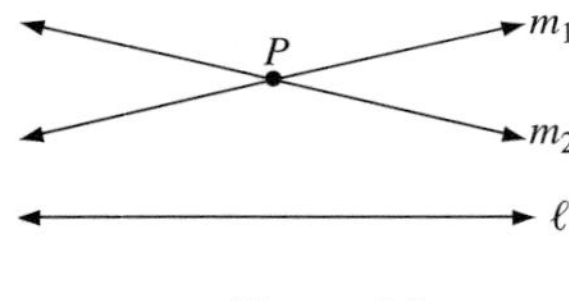

Figure 6.3

No doubt this possibility seems so bizarre that disproving it on purely logical grounds should be a simple matter. But to attempt to do so would be repeating history, and it would lead us directly to that "bottomless night, which extinguished all light and joy," as described by one man who devoted his entire life to the problem without success, Wolfgang Bolyai.

It is natural to attempt to refute this hypothesis from a "practical" standpoint by looking at the apparent behavior of lines in ordinary geometry (a more careful approach might oblige us to ask exactly what the term "ordinary geometry" means). Nevertheless, if you draw two lines through P and they are both supposed to be parallel to a third line, this would seem to force some lines to be curved rather than straight, as illustrated in Figure 6.4. We could reason that the larger the angle θ between m_1 and m_2 is, the more curved the lines have to be. However, we know that drawings and figures do not prove anything from a logical standpoint. Whether lines are "curved" or not depends on the axioms.

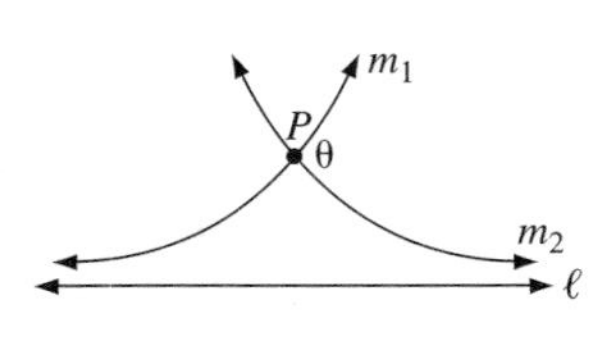

Figure 6.4

Furthermore, the argument is inconclusive because of what could really be taking place *in the world in which we live:* Suppose the angle θ between such parallels were always incredibly small, say less than one-millionth of a degree. No drawing instrument would then be capable of producing precise, distinct lines, and measuring devices would not be accurate enough to detect the small amount of curvature that might be taking place. Even modern physics and the state of the art in technology (the use of lasers, high-tech measuring devices, etc.) can only provide evidence of one hypothesis over another. Recall, as mentioned earlier, the expedition Gauss undertook to measure the angles between three distant mountain peaks, in which no discrepancy from the Euclidean result was detected. (Had such a discrepancy been found, then it certainly would have *disproven* the validity of the Euclidean Parallel Postulate in the physical world.)

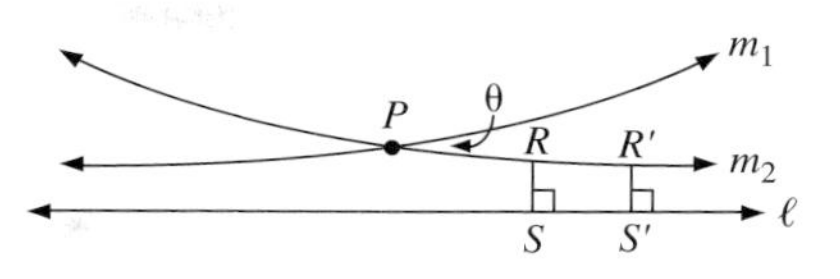

Figure 6.5

A second line of reasoning, which is based more on logic than on physical observation, is to consider the *distances* between m_2 and ℓ. Apparently, we must have $RS > R'S'$ in Figure 6.5. If not detectable by actual measuring devices, this inequality must surely be provable in fact. This would then contradict our idea of parallel lines. Parallel lines are supposedly everywhere equidistant—like the rails of a railroad track. Indeed, this point of view is the basis for several early proofs of the impossibility of hyperbolic geometry, and some early scholars used it to circumvent the problem of dealing with Euclid's Fifth Postulate. But in reality, what they were doing was assuming this property as a substitute for Euclid's Fifth Postulate (which is logically equivalent to it!) and proving the postulate from it. So nothing was actually gained. That is, Euclid's other postulates *by themselves* had not been shown to imply Euclid's Fifth Postulate.

EQUIVALENT FORMS OF EUCLID'S PARALLEL POSTULATE

Many "natural" geometric properties equivalent to Euclid's Fifth Postulate have been proposed, but each of these, like the equidistant property of parallels, has turned out to be nothing more than a substitute for the Fifth Postulate. A table of a few of these follows. Regarding the first property listed, Gauss once wrote:

> If one can prove that there exists a right triangle whose area is greater than any given number, then I am able to establish the entire system of (Euclidean) geometry with complete rigor. I am in possession of several theorems of this sort, but none of them satisfy me.

Equivalent Forms of Euclid's Fifth Postulate

- The area of a right triangle can be made arbitrarily large.
- The angle sum of all triangles is constant.
- The angle sum of a single triangle equals 180.
- Rectangles exist.
- A circle can be passed through any three noncollinear points.
- Given an interior point of an angle, a line (transversal) can be drawn through that point intersecting both sides of the angle.
- Two parallel lines are everywhere equidistant.
- The perpendicular distance from one of two parallel lines to the other is always bounded.

Each of these properties can be readily recognized as results that have either already been proven in Euclidean geometry (Chapter 4), or can be proven without difficulty assuming the Euclidean Parallel Postulate.

EXAMPLE 1 Show that if a rectangle exists in absolute geometry, then a triangle having angle sum 180 exists.

SOLUTION

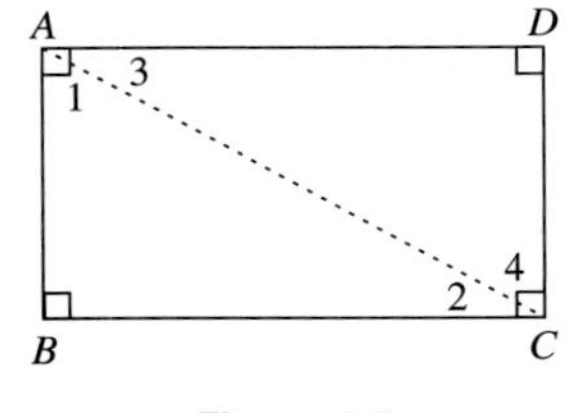

Figure 6.6

In Figure 6.6, we are given a rectangle $\square ABCD$ (all four angles are right angles). We have only to draw a diagonal $\overline{AC}$ and consider the angles of $\triangle ABC$. Since a rectangle is clearly a convex quadrilateral (reason?), diagonal $\overline{AC}$ passes through the interior of $\angle BAD$ and $\angle BCD$. Hence $\overrightarrow{AB}$–$\overrightarrow{AC}$–$\overrightarrow{AD}$ and $m\angle A = m\angle 1 + m\angle 3$. Similarly, $m\angle C = m\angle 2 + m\angle 4$. Therefore,

$$\textbf{(1)}\qquad (m\angle 1 + m\angle 2 + m\angle B) + (m\angle 3 + m\angle 4 + m\angle D) = m\angle A + m\angle B + m\angle C + m\angle D = 360$$

(by definition of a rectangle). But by the Saccheri–Legendre Theorem (Section 3.4),

$$m\angle 1 + m\angle 2 + m\angle B \le 180 \qquad \text{and} \qquad m\angle 3 + m\angle 4 + m\angle D \le 180$$

If $m\angle 1 + m\angle 2 + m\angle B < 180$, then the above equation **(1)** cannot hold. Therefore,

$$m\angle 1 + m\angle 2 + m\angle B = 180 \qquad \blacksquare$$

Another property in the preceding table will be proposed for "discovery" at the end of this section.

TWO FAMOUS "PROOFS" OF THE EUCLIDEAN PARALLEL POSTULATE

One of the earliest proofs of the Parallel Postulate was given in ancient times by Proclus (as mentioned earlier). His argument assumes

(1) that the perpendicular distance from a point on one of two intersecting lines ℓ and m to a point on the other increases without bound as the point varies on that line (that is, in Figure 6.7, $x \to \infty$ as $AP \to \infty$)

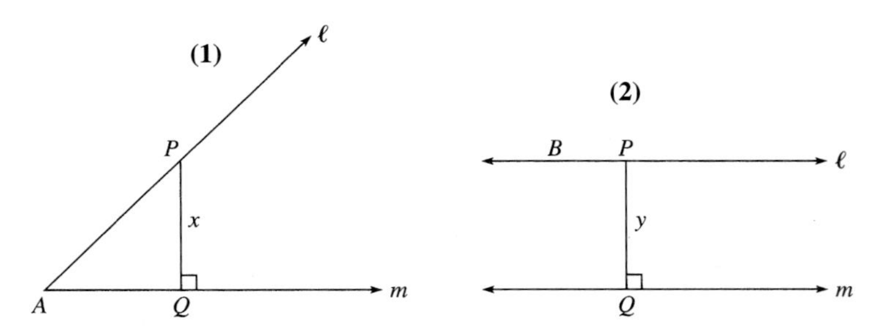

Figure 6.7

(2) that the distance from a point on one of two parallel lines to the other remains bounded ($y < b$ as $BP \to \infty$).

PROCLUS

The argument proceeds as follows, relying heavily on properties obtained from the diagram: Let lines ℓ and m be perpendicular to line $\overleftrightarrow{AB}$, as in Figure 6.8. Then by Euclid I.27, the lines are parallel. It is to be proven that ℓ is the only parallel to m through A. Suppose $n = \overleftrightarrow{AW}$ is another line through A parallel to m. Since ray $\overrightarrow{AW}$ does not meet line m, $\overrightarrow{AW}$ cuts the perpendicular $\overline{PQ}$ at some point R, and we let $x = PR$, $y = RQ$. Now let $AP \to \infty$. Then $x \to \infty$ by assumption **(1),** but $\overleftrightarrow{PQ}$ is bounded, by assumption **(2).** Hence, $x > PQ$ for sufficiently large AP, so for some previous position of P, $x = PQ$ and $y = 0$. Thus, n intersects line m at that point, a contradiction, so line ℓ is the unique parallel to m through point A.

PROCLUS' ARGUMENT

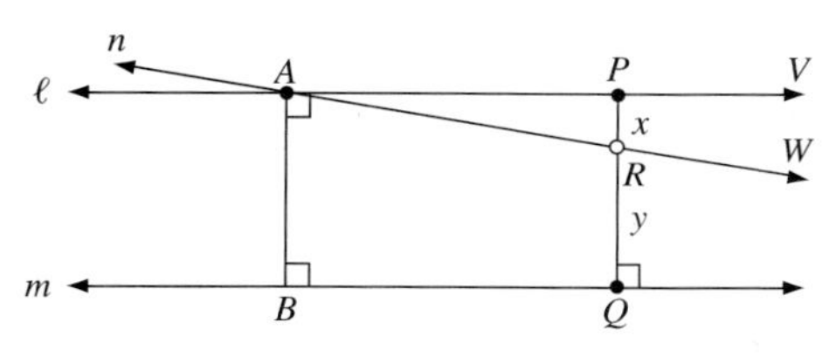

Figure 6.8

This is certainly an interesting argument. Indeed, in Problem 19 you will be asked to prove that Proclus' assumption **(1)** is valid in absolute geometry.

Question: What does Proclus' argument actually prove? Can his argument (as presented here) be made rigorous?

LEGENDRE

A somewhat more sophisticated approach was taken by Legendre (1752–1833). His proof is as follows (as found in R. Bonola's *Non-Euclidean Geometry,* pp. 58–59): Let $\triangle ABC$ have, if possible, angle sum less than 180 (Figure 6.9). Construct $\triangle A'BC$ congruent to $\triangle ABC$, with $A' \in$ Interior $\angle CAB$ and lying on the opposite side of line $\overleftrightarrow{BC}$ as A. (One way of accomplishing this construction is to repeat the construction used for the Exterior Angle Inequality, with M the midpoint of $\overline{BC}$ and $\overline{AA'}$.) Through point A', *draw a transversal meeting the sides of* $\angle CAB$ *at* B_1 *and* C_1, *respectively.* Thus, B_1C_1 forms a larger triangle $\triangle AB_1C_1$, which contains four subtriangles inside it. Define the **defect** of $\triangle ABC$ to be the number

LEGENDRE'S ARGUMENT

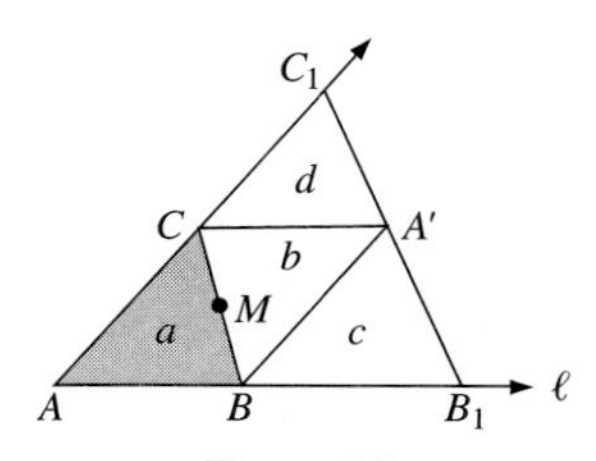

Figure 6.9

$$a = 180 - m\angle A - m\angle B - m\angle C$$

which we are assuming to be positive (where $\angle B = \angle ABC$ and $\angle C = \angle ACB$). Similarly, define the defects of the other three triangles inside $\triangle AB_1C_1$, namely $\triangle BA'C$, $\triangle BA'B_1$, and $\triangle CA'C_1$, as the numbers b, c, and d, respectively. Now, a little later, we prove the principle that *defect is additive.* That is, the defect a_1 of $\triangle AB_1C_1$ equals the *sum* of the defects of each of the four subtriangles:

$$a_1 = a + b + c + d$$

But since the angles of $\triangle A'BC$ are congruent to those of $\triangle ABC$,

$$a = b$$

By the Saccheri–Legendre theorem (Section 3.4),

$$c \geq 0 \qquad \text{and} \qquad d \geq 0$$

Thus,

$$a_1 \geq a + b = 2a$$

We have thereby constructed a new triangle, $\triangle AB_1C_1$, with defect a_1 at least twice that of $\triangle ABC$. Using an identical construction on $\triangle AB_1C_1$, we can construct $\triangle AB_2C_2$ with defect

$$a_2 \geq 2a_1 \geq 4a.$$

Continuing in this fashion, there exists a triangle with defect

$$a_n \geq 2^n a$$

Since $a > 0$, then $2^n a \to \infty$ and a triangle having defect > 180 exists, which is impossible, by its definition.

Question: What theorem of absolute geometry does Legendre's argument actually prove? Can it be made rigorous? (See italicized step in the above proof.)

A USEFUL LEMMA

The following set of problems will explore further interesting arguments that history provides for us, masquerading as proofs of Euclid's Fifth Postulate.

Another argument (this time correct) will establish a key result we will use in our development of hyperbolic geometry in the following section.

LEMMA A: Let $\triangle ABC$ be given, with M the midpoint of side $\overline{BC}$ (Figure 6.10). If the angle sum of $\triangle ABC$ is less than 180, then so is the angle sum of both $\triangle ABM$ and $\triangle AMC$.

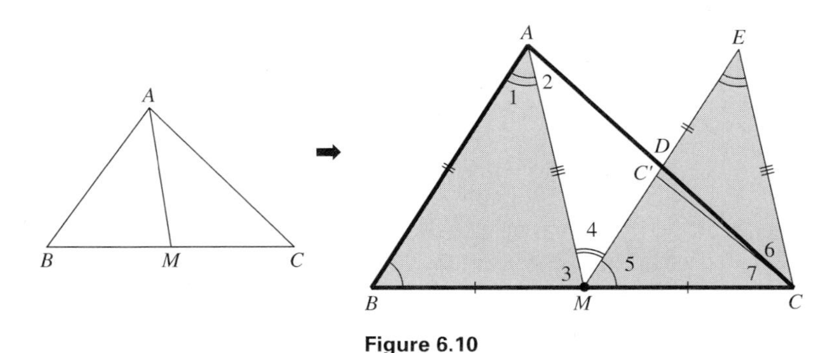

Figure 6.10

PROOF

Construct D on side $\overline{AC}$ so that $\angle DMC \cong \angle B$ and $\overline{ME} \cong \overline{AB}$. Then $\triangle EMC \cong \triangle ABM$ (the shaded triangles in Figure 6.10), and it can be shown that $ME = AB > MD$ (Problem 15), which is certainly intuitively clear. Therefore M–D–E. At this point *assume that the angle sum of* $\triangle ABM = 180$; we will obtain a contradiction. An analysis of the angles in Figure 6.10 shows that

$$m\angle B + m\angle 1 + m\angle 3 = 180 = m\angle 3 + m\angle 4 + m\angle 5$$

$$m\angle B + m\angle 1 = m\angle 4 + m\angle 5$$

$$\therefore m\angle 1 = m\angle 4 \quad \text{(because } m\angle B = m\angle 5 \text{ by CPCF)}$$

Hence, since $m\angle 1 = m\angle E$ by CPCF, $m\angle 4 = m\angle E$. Also, the vertical angles at D are congruent, and $AM = EC$ by CPCF. Hence, by AAS, $\triangle AMD \cong \triangle CED$ and $m\angle 2 = m\angle 6$. Now we have

$$\begin{aligned}\text{Angle Sum } \triangle ABC &= m\angle B + m\angle 1 + m\angle 2 + m\angle 7 \\ &= m\angle 5 + m\angle 4 + m\angle 6 + m\angle 7 \\ &= m\angle 5 + m\angle 4 + m\angle ECM \\ &= m\angle 5 + m\angle 4 + m\angle 3 = 180 \rightarrow\leftarrow\end{aligned}$$

The contradiction proves that $\triangle ABM$ does not have angle sum 180. Therefore, by the Saccheri–Legendre Theorem,

$$\text{Angle Sum } \triangle ABM < 180$$

as desired. The same proof (with different labels) works for $\triangle AMC$, and will be omitted.

We are now going to extend the lemma to cover any point D on side $\overline{BC}$, not just the midpoint. If B–D–C then D must either coincide with the midpoint M of $\overline{BC}$, or else D lies on segment $\overline{BM}$ or $\overline{MC}$. Assume for the time being that $D \in \overline{BM}$. Now let M_1 be the midpoint of $\overline{BM}$, M_2 the midpoint of $\overline{BM_1}$, M_3 the midpoint of $\overline{BM_2}$, and so on, as shown in Figure 6.11. For some n we must have $M_n \in \overline{BD}$. (Note the similarity of this argument to the one we used to prove the lemma of Section 4.2.) For convenience, let $N = M_n$ (Figure 6.12); by the preceding lemma, both $\triangle ABM$ and

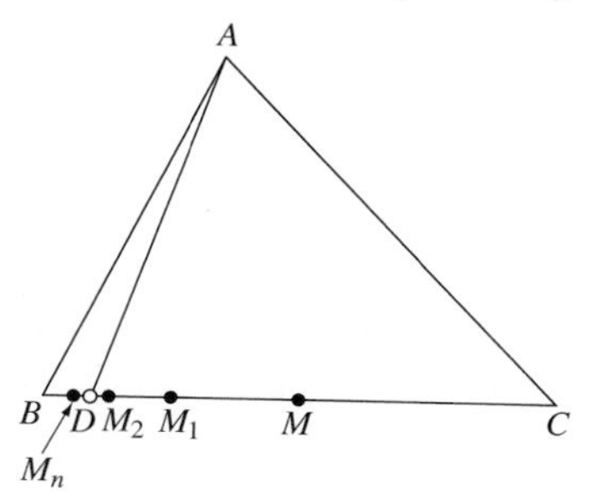

Figure 6.11

$\triangle AMC$ have angle sum less than 180: since M_1 is the midpoint of $\overline{BM}$ then $\triangle ABM_1$ has angle sum < 180, and similarly, $\triangle ABM_2$, $\triangle ABM_3, \ldots, \triangle ABN$ all have angle sums < 180. The next lemma will then clinch the argument.

LEMMA B: If B–N–D, D–M–C, and the angle sum of $\triangle ABN$ is less than 180, then that of $\triangle ABD$ is also less than 180. Likewise, if the angle sum of $\triangle AMC$ is less than 180, then that of $\triangle ADC$ is less than 180. (See Figure 6.12.)

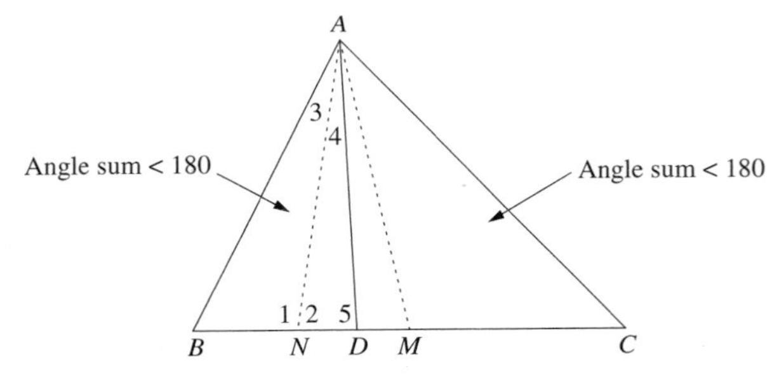

Figure 6.12

PROOF

$$
\begin{aligned}
\text{Angle Sum } \triangle ABD &= m\angle B + m\angle 3 + m\angle 4 + m\angle 5 \\
&= m\angle B + m\angle 3 + m\angle 4 + m\angle 5 + (m\angle 1 + m\angle 2 - 180) \\
&= (m\angle B + m\angle 3 + m\angle 1) + (m\angle 2 + m\angle 4 + m\angle 5) - 180 \\
&= \text{Angle Sum } \triangle ABN + (\text{Angle Sum } \triangle AND - 180) \\
&\leq \text{Angle Sum } \triangle ABN \qquad \text{(Saccheri–Legendre Theorem)}
\end{aligned}
$$

By hypothesis, this last quantity is less than 180, hence Angle Sum $\triangle ABD < 180$, as desired. A little thought should convince you that the second case is logically equivalent to the one just proven, hence a separate proof is unnecessary.

THEOREM 1[1]: If $\triangle ABC$ has angle sum less than 180 and D is any point on side $\overline{BC}$, then both $\triangle ABD$ and $\triangle ADC$ have angle sum less than 180.

PROOF

In view of Lemma A, there is nothing to prove if $D = M$, the midpoint of $\overline{BC}$. If B–D–M, obtain the sequence of midpoints in the above construction (Figure 6.11). By Lemma A, $\triangle ABN$ and $\triangle AMC$ have angle sum < 180 (Figure 6.12), so by Lemma B, $\triangle ABD$ and $\triangle ADC$ each has angle sum < 180. A similar argument is used if M–D–C.

[1]It is the *converse* of this result that is the familiar consequence of additivity of defect in absolute geometry, whose proof is much easier. (See Problem 9, Section 6.3.)

Moment for Discovery

Are Parallel Lines Everywhere Equidistant?

Suppose we assume that two parallel lines ℓ and m are everywhere equidistant in absolute geometry. This means that if A, C, and E are any three points on line ℓ and B, D, and F are the respective feet of the perpendiculars on m, then

$$AB = CD = EF$$

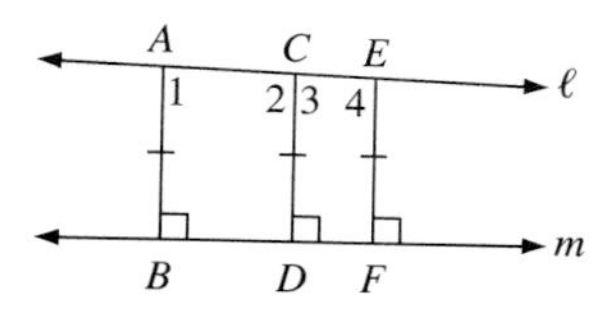

Figure 6.13

1. What kind of quadrilateral is $\diamond ABDC$? $\diamond CDFE$?
2. What can you deduce conclusively about the angles 1, 2, 3, and 4 in the figure? (Be careful not to assume $\overline{AB} \perp \ell$ from the Z property of parallels, which is derived from the Euclidean Parallel Postulate.)
3. In particular, what about $\angle 2$ and $\angle 3$? Can you prove that they are right angles?
4. What have you discovered? Does the assumption "parallel lines are everywhere equidistant" lead to another property listed in the previous table?

PROBLEMS (6.2)

GROUP A

1. If you have not already done so, prove that if the angle sum of all triangles has a constant value k, then $k = 180$. (See Problems 16, 17, Section 3.5.)
2. Satisfy yourself that the first three statements in the previous table (*Equivalent Forms of Euclid's Fifth Postulate*) are actually consequences of our Axiom P-1 (and logically equivalent to Euclid's Fifth Postulate). Find the exact theorem in Chapter 4 that implies each property, providing simple arguments where needed.
3. Points A and B are 100 miles apart. At A and B, lines $\overleftrightarrow{AC}$ and $\overleftrightarrow{BD}$ are constructed, making angles of measure 89.5 and 89.9 (degrees) with line $\overleftrightarrow{AB}$. How can you be sure that $\overleftrightarrow{AC}$ and $\overleftrightarrow{BD}$ will meet if these constructions are taking place in a Euclidean plane?

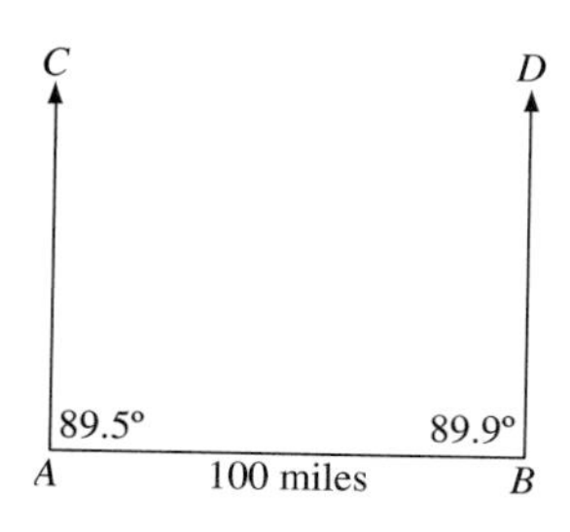

4. Suppose a line ℓ is constructed perpendicular to line $\overleftrightarrow{AB}$ at point A 245 miles away from B, line m is constructed perpendicular to line $\overleftrightarrow{BC}$ at C 658 miles away from B, and that $m\angle ABC = 179.9$ (degrees). If laser beams are shot into outer space from A and C along lines ℓ and m tangent to the earth's surface, will the beams cross at some point in outer space? (See next problem.) If the answer is affirmative, use trigonometry to find how far away the point is from A. (Assume all lines lie in a Euclidean plane.)

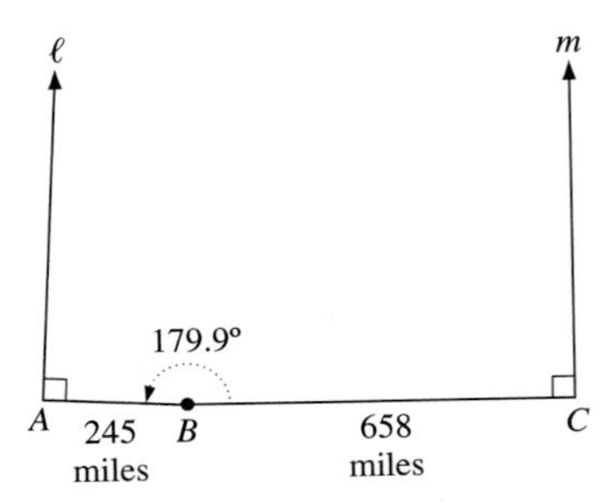

5. In the Euclidean plane, suppose that perpendiculars $\overleftrightarrow{AP}$ and $\overleftrightarrow{CQ}$ to lines $\overleftrightarrow{AB}$ and $\overleftrightarrow{BC}$, respectively, are parallel. Show that A, B, and C are collinear.

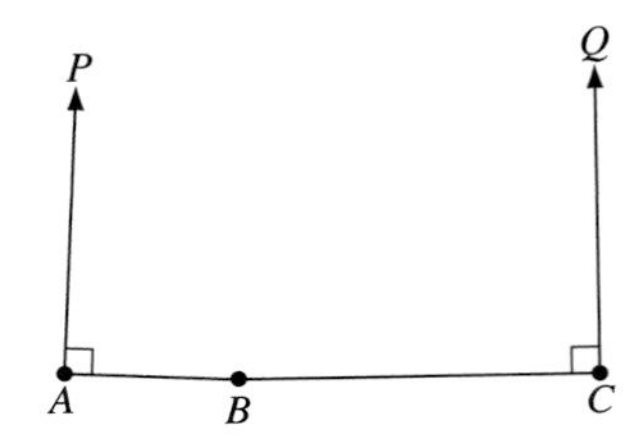

6. Prove from the Euclidean Parallel Postulate: A circle can be passed through any three noncollinear points. (***Hint:*** See previous problem.)

7. Proclus adopted the following Parallel Postulate: If a line intersects one of two parallel lines, it intersects the other also. Establish from this Euclid's Fifth Postulate of Parallels.

GROUP B

8. Prove in Euclidean geometry that if point P lies in the interior of $\angle ABC$, there can be found a line ℓ passing through P that intersects both sides of the angle. (***Hint:*** Draw a line through P parallel to $\overleftrightarrow{AB}$. Why must that line meet $\overrightarrow{BC}$ at some point D? Choose E on $\overrightarrow{BC}$ so that $B–D–E$, and consider line $\overleftrightarrow{EP}$.)

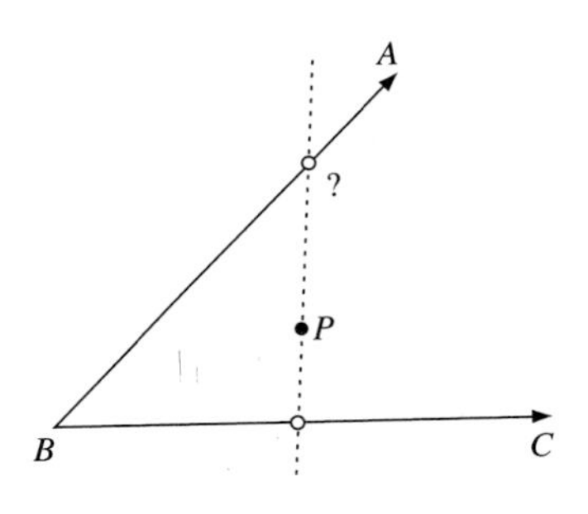

9. Prove in absolute geometry that if the summit angles of a Saccheri Quadrilateral are right angles, the summit and base are congruent. (***Hint:*** Draw a diagonal.)

10. Prove in absolute geometry that the opposite sides of a rectangle are congruent. (This is Problem 18, Section 3.7.)

11. In the figure, ℓ and m are parallel lines, $\overline{AB} \perp m$, and $\overline{CD} \perp m$. Using Euclid's Fifth Postulate, how would you prove that $AB = CD$? (***Hint:*** See Problem 10.)

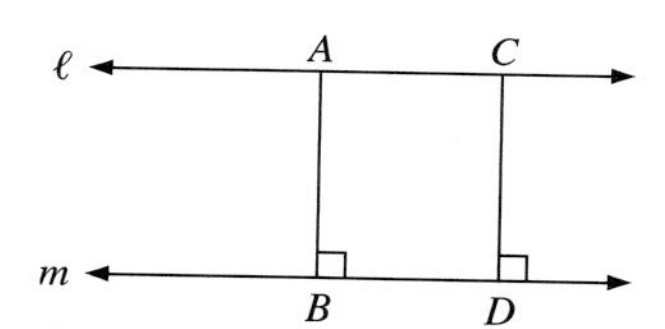

12. If you have not already done so, prove in absolute geometry that if two lines are cut by a transversal such that a pair of alternate interior angles are congruent, the two lines possess a common perpendicular. (***Hint:*** Drop perpendiculars to the lines from the midpoint of the segment joining the points of intersection of the transversal and the lines. This is Problem 23, Section 4.1.)

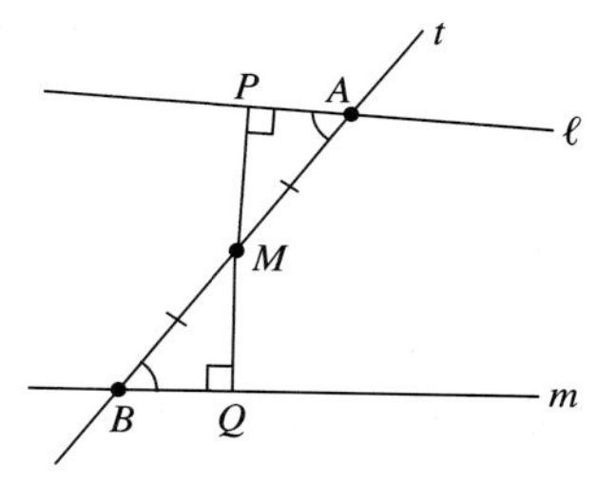

13. Prove the obvious, that $AB > MD$ in Figure 6.10. (Observe point C' constructed on segment $\overline{MD}$ in the figure, where, assuming that $AB \leq MD$, $MC' = AB$. First establish $\triangle ABM \cong \triangle C'MC$, then use CPCF to obtain a contradiction of the Exterior Angle Inequality in $\triangle AMC$.)

GROUP C

14. Prove in absolute geometry that if a quadrilateral has three right angles and a pair of opposite sides congruent, it is a rectangle.

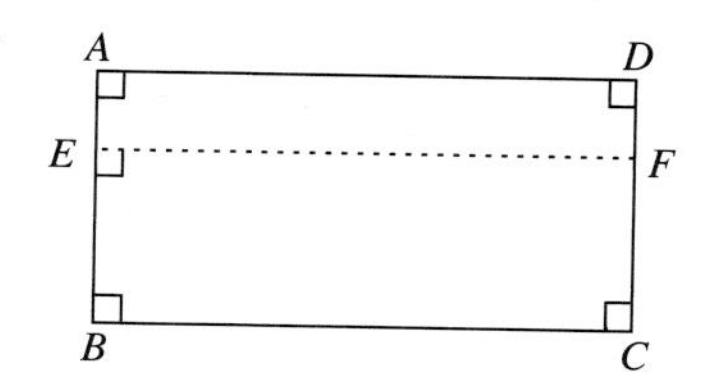

15. Prove in absolute geometry that if $\diamond ABCD$ has four right angles (is a rectangle) and $\overline{EF} \perp \overline{AB}$, then both $\diamond AEFD$ and $\diamond EBCF$ are rectangles. (***Hint:*** Use Problem 10; locate G on $\overline{CD}$ so that $CG = BE$ and draw $\overline{EG}$.)

16. Prove that if a single rectangle exists in absolute geometry, then one that is arbitrarily large (in terms of its sides) exists. (***Hint:*** To get started, double one side of the given rectangle and prove the figure obtained by filling in the remaining sides is a rectangle.)

17. In absolute geometry, prove that if a single rectangle exists, then all Lambert Quadrilaterals are rectangles and hence all Saccheri Quadrilaterals are rectangles. (See Problems 15 and 16.)

18. **Legendre's Second Theorem** From the result of the last problem, using the associated Saccheri Quadrilateral of any triangle, prove Legendre's Second Theorem (the *All-or-None Theorem*): If a single triangle exists in absolute geometry that has angle sum 180, then every triangle has angle sum 180.

19. Prove in absolute geometry that assumption (1) of Proclus' argument is justified, that indeed the distance x from P to side $\overrightarrow{AC}$ of a given angle $\angle BAC$ increases without bound as $AP \to \infty$. (See accompanying figure.)

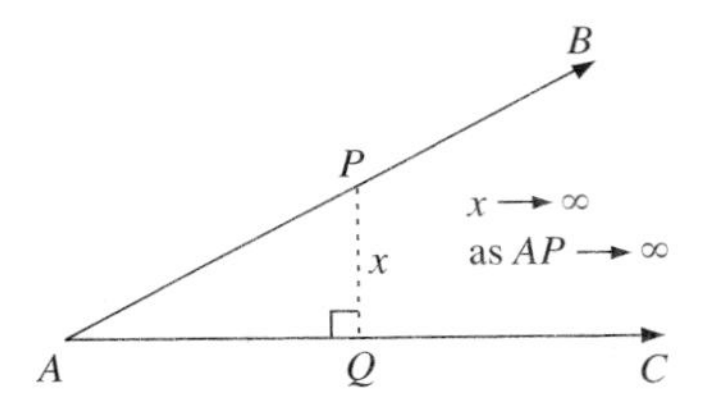

20. Wallis' Theorem[2] Wallis introduced the following axiom for Euclidean geometry:

> Given any triangle and any line segment, a triangle similar to the given triangle may be constructed with the given segment as base.

Provide the details of a proof that this axiom implies Axiom P-1 in absolute geometry. (***Hint:*** Let A be a point and ℓ a line not on A, and construct $\overline{AB} \perp \ell$ and $m \perp \overline{AB}$ at A. Then $m \parallel \ell$. To show that m is the only parallel to ℓ, consider line $n \neq \ell$ through A. Choose C on n on the B-side of m and drop the perpendicular from C to $\overline{AB}$. Then by Wallis's Axiom, there exists $\triangle ABE \sim \triangle ADC$. Finish the argument.)

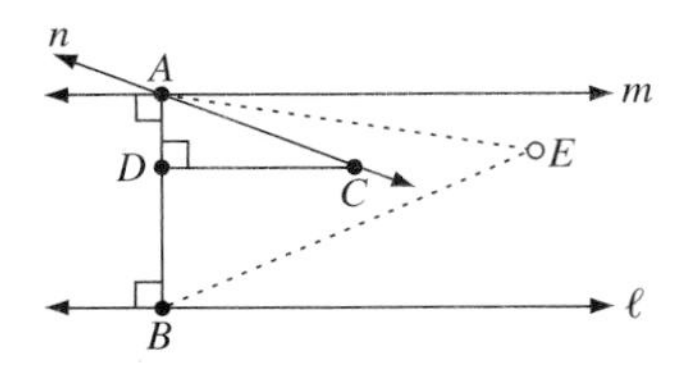

6.3 Hyperbolic Geometry: Angle Sum Theorem

In this section we will adopt the following axiom, within the context of absolute geometry.

> AXIOM P-2: HYPERBOLIC PARALLEL POSTULATE
> If ℓ is any line and P any point not on ℓ, there exists more than one line passing through P parallel to ℓ.

ANGLE SUM THEOREM FOR HYPERBOLIC GEOMETRY

A characteristic feature of hyperbolic geometry is that the sum of the measures of the angles of any triangle is always less than 180. This may be proven directly from Axiom P-2 and the theorem established in Section 6.2.

---

[2]John Wallis (1616–1703) was famous for products often mentioned in calculus, called **Wallis products,** such as

$$\frac{\pi}{2} = \frac{2}{1}\cdot\frac{2}{3}\cdot\frac{4}{3}\cdot\frac{4}{5}\cdot\frac{6}{5}\cdot\frac{6}{7}\cdots$$

THEOREM 1: The sum of the measures of the angles of any right triangle is less than 180.

PROOF

Let $\triangle ABC$ be a right triangle with right angle at B (Figure 6.14). Through A, draw the line $\overleftrightarrow{AP} \equiv m$ perpendicular to $\overleftrightarrow{AB}$, with P on the C-side of line $\overleftrightarrow{AB}$. By Theorem 1, Section 4.1, $m \parallel \overleftrightarrow{BC}$. By Axiom P-2, there is another line n through A parallel to $\overleftrightarrow{BC}$, and we may assume that one of its rays from A, ray $\overrightarrow{AQ}$, lies interior to $\angle BAP$. (For, if ray $\overrightarrow{AQ} \subseteq$ Interior $\angle BAP'$ where P–A–P', then we could apply the argument below to show that the triangle $\triangle ABC'$ congruent to $\triangle ABC$ on the other side of line $\overleftrightarrow{AB}$ has angle sum < 180.) Hence, the first major step of the proof has been established:

(1) There exists a ray $\overrightarrow{AQ}$ parallel to line $\overleftrightarrow{BC}$ such that $\overrightarrow{AB}$–$\overrightarrow{AQ}$–$\overrightarrow{AP}$. Set $t = m\angle PAQ$.

(2) Next, we want to locate a point W so far out on ray $\overrightarrow{BC}$ that $m\angle AWB$ is arbitrarily small, smaller than t. We know this can be done; recall the argument of Legendre that was introduced in Problem 21, Section 3.4. Thus, there exists $W \in \overrightarrow{BC}$ such that B–C–W and

$$m\angle AWB \equiv m\angle W < t$$

(3) Now if the order of the rays through A were $\overrightarrow{AB}$–$\overrightarrow{AQ}$–$\overrightarrow{AW}$, by the Crossbar Theorem ray $\overrightarrow{AQ}$ would meet segment $\overline{BW}$ $\rightarrow\leftarrow$. Therefore: $\overrightarrow{AB}$–$\overrightarrow{AW}$–$\overrightarrow{AQ}$. Since also $\overrightarrow{AB}$–$\overrightarrow{AQ}$–$\overrightarrow{AP}$, then we have $\overrightarrow{AB}$–$\overrightarrow{AW}$–$\overrightarrow{AQ}$–$\overrightarrow{AP}$.

(4) We can now estimate the angle sum of $\triangle ABW$:

$$\begin{aligned}\text{Angle Sum } \triangle ABW &= 90 + m\angle W + m\angle BAW \\ &< 90 + t + m\angle BAQ \\ &= 90 + m\angle BAP \\ &= 180\end{aligned}$$

(5) $\therefore$ By Theorem 1, Section 6.2, Angle Sum $\triangle ABC < 180$.

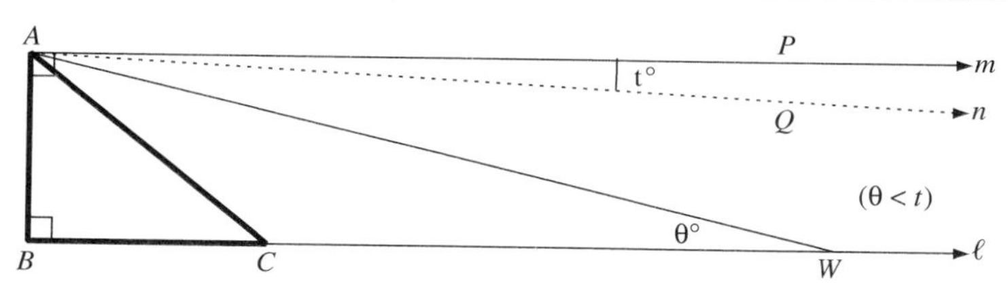

Figure 6.14

COROLLARY: The sum of the measures of the angles of any triangle is less than 180.

To prove this, just drop the perpendicular to the largest side of a triangle from the vertex opposite, forming two right subtriangles, each of whose angle sums is less than 180 by Theorem 1. The given triangle therefore has angle sum less than 180, by algebra and the Linear Pair Axiom.

AREA IN HYPERBOLIC GEOMETRY: DEFECT OF TRIANGLES AND POLYGONS

One way to use the property just proven (that every triangle has angle sum less than 180) is to define the so-called *defect* of a triangle as the *amount by which the angle sum of a triangle misses the value of 180.* That is, we define the **defect** of $\triangle ABC$ to be the value

$$\delta(\triangle ABC) = 180 - m\angle A - m\angle B - m\angle C$$

Defect may also be defined for polygons in general. (This definition includes the one for triangles as a special case, for $n = 3$.)

DEFINITION: The **defect** of the convex polygon $P_1P_2P_3 \ldots P_n$ is the number $\delta(P_1P_2P_3 \ldots P_n) = 180(n-2) - m\angle P_1 - m\angle P_2 - m\angle P_3 - \cdots - m\angle P_n$

One of the surprising properties of defect is its additivity. If a convex polygon and its interior is subdivided in any manner into convex subpolygons, the sum of the defects of the subpolygons equals that of the original polygon. Thus, all the properties of area outlined in Section 4.6 (except the postulate concerning the area of a unit square which does not exist in hyperbolic geometry) are valid, as long as the regions are restricted to be the *family of all convex polygons.* Hence, defect defines a perfectly legitimate area function for the hyperbolic plane. We may therefore define, for some constant $k > 0$, the **area** of polygon $P_1P_2P_3 \ldots P_n$ as *k times its defect.*

Concept for Area in Hyperbolic Geometry

The **area** of a convex polygon $P_1P_2P_3 \ldots P_n$ is defined by the number

$$K = k[(n - 2)180 - m\angle P_1 - m\angle P_2 - m\angle P_3 - \cdots - m\angle P_n]$$

where k is some predetermined constant for the entire plane (not depending on each given polygon). The value for k is frequently taken as $\pi/180$, which converts degree measure to radian measure.

We shall merely prove a special case of the additivity property and leave the rest for problems. This case is the one actually used in later developments.

LEMMA: If $\overleftrightarrow{AD}$ is a cevian of $\triangle ABC$ and δ_1 and δ_2 denote the defects of the subtriangles $\triangle ABD$ and $\triangle ADC$, then

$$\delta(\triangle ABC) = \delta_1 + \delta_2$$

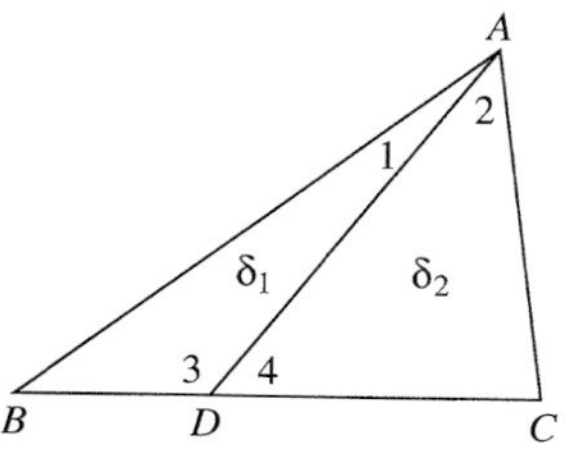

Figure 6.15

PROOF

We have (Figure 6.15)

$$\delta_1 = 180 - m\angle B - m\angle 1 - m\angle 3$$

$$\delta_2 = 180 - m\angle C - m\angle 2 - m\angle 4$$

so that, summing,

$$\begin{aligned} \delta_1 + \delta_2 &= 360 - m\angle B - m\angle 1 - m\angle 3 - m\angle C - m\angle 2 - m\angle 4 \\ &= 360 - (m\angle 1 + m\angle 2) - m\angle B - m\angle C - (m\angle 3 + m\angle 4) \\ &= 180 - m\angle A - \angle B - \angle C \\ &= \delta(\triangle ABC) \end{aligned}$$

EXAMPLE 1 Figure 6.16 shows a convex pentagon $ABCDE$ enclosing three triangles and a convex quadrilateral, with angle measures as indicated. Calculate the defect of each interior polygon, sum, and compare with the defect of the outer pentagon.

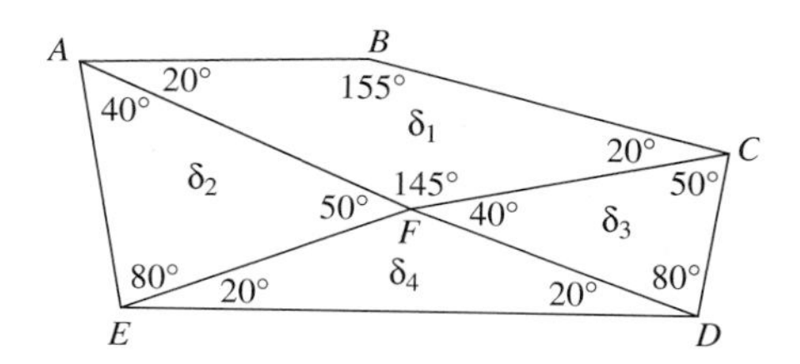

Figure 6.16

SOLUTION

First, $m\angle EFD = 360 - (145 + 40 + 50) = 125$. Now we have

$$\delta_1 = 360 - \text{Angle Sum } \Diamond ABCF = 360 - 340 = 20$$

$$\delta_2 = 180 - \text{Angle Sum } \triangle AEF = 180 - 170 = 10$$

$$\delta_3 = 180 - \text{Angle Sum } \triangle CDF = 180 - 170 = 10$$

$$\delta_4 = 180 - \text{Angle Sum } \triangle DEF = 180 - 165 = 15$$

Thus $\delta_1 + \delta_2 + \delta_3 + \delta_4 = 55$, which should then be the defect δ of the pentagon. The angle sum of $ABCDE = 60 + 155 + 70 + 100 + 100 = 485$, and from the formula for the defect for a pentagon (with $n = 5$),

$$\delta = 3 \cdot 180 - 485 = 540 - 485 = 55$$

in agreement. ■

EXAMPLE 2 In Figure 6.17 is shown a line m parallel to ℓ, with $m\angle ACD = 88$, and right $\triangle ABC$ with acute angles of measures $\frac{1}{2}$ and $89\frac{1}{4}$.

(a) Calculate the area of $\triangle ABC$ using defect.
(b) Assume that point P moves on line ℓ so that $AP \to \infty$, $\theta \to 0$, and $\phi \to 0$. Show that the area of $\triangle APC$ is *bounded*, and find a bound. (Recall in this connection the comment of Gauss, quoted in Section 6.2.)

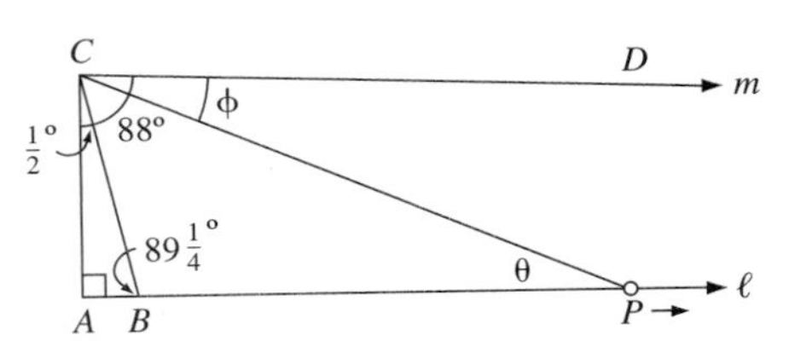

Figure 6.17

SOLUTION

(a) Area $\triangle ABC = k(180 - 90 - 89\frac{1}{4} - \frac{1}{2}) = \frac{1}{4}k.$
(b) Area $\triangle APC = k(180 - 90 - \theta - (88 - \phi)) = k(2 + \phi - \theta)$
$\leq \lim k(2 + \phi - \theta) \leq 2k$

Hence Area $\triangle APC \leq 2k$ (the function Area $\triangle APC$ is increasing as $AP \to \infty$, since $A\text{–}P\text{–}P'$ implies that Area $\triangle APC <$ Area $\triangle AP'C$). ■

NONEXISTENCE OF SIMILAR TRIANGLES IN HYPERBOLIC GEOMETRY

A major consequence of the positive value of defect for all triangles is that it abolishes all possibility of similar triangles or of a concept for similarity transformations in hyperbolic geometry. A dramatic conclusion can be your own discovery if you work out the discovery unit that follows. That conclusion proves the following incredible result in hyperbolic geometry.

THEOREM 2: AAA CONGRUENCE CRITERION FOR HYPERBOLIC GEOMETRY
If two triangles have the three angles of one congruent, respectively, to the three angles of the other, the triangles are congruent.

COROLLARY: There do not exist any similar, noncongruent triangles in hyperbolic geometry.

Moment for Discovery

An Unusual Criterion for Congruence

Suppose that $\triangle ABC$ and $\triangle PQR$ are two triangles having corresponding angles congruent: $\angle A \cong \angle P$, $\angle B \cong \angle Q$, and $\angle C \cong \angle R$. If $AB = PQ$, then the triangles are congruent by ASA. If the triangles are *not* congruent, then one of the triangles has at least two sides that are each of greater length than the corresponding sides in the other, say $AB > PQ$ and $AC > PR$.

1. Construct points D and E on $\overline{AB}$ and $\overline{AC}$ such that $AD = PQ$ and $AE = PR$ (Figure 6.18). Draw $\overline{DE}$ and $\overline{DC}$. Is $\triangle ADE \cong \triangle PQR$?

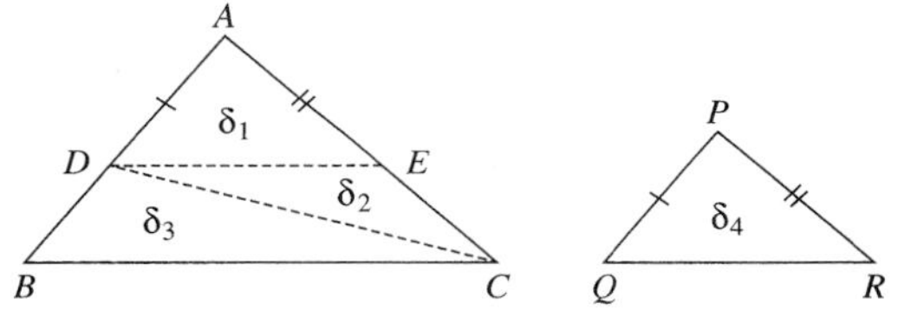

Figure 6.18

2. Label the defects of the various triangles as shown in the figure, with δ the defect of $\triangle ABC$. What must be true of δ_1 and δ_4? Of δ_4 and δ?
3. What is the sum of the defects δ_1 and δ_2? (See lemma in this section.)
4. What equals the sum of the defects δ_1, δ_2, and δ_3?
5. How is the defect of $\triangle ABC$ and δ_1 related? Do you see any contradiction here?
6. What does this prove?

PROBLEMS (6.3)

GROUP A

1. A certain equilateral triangle in hyperbolic geometry has angles of measure 55 each. Find its defect.
2. An equilateral triangle has defect 24.
   (a) What must each angle of the triangle measure?
   (b) In Euclidean geometry, two equilateral triangles having equal areas must be congruent. Is this proposition true or false in hyperbolic geometry? Prove if true or find a counterexample if false.
3. Show why congruent triangles in hyperbolic geometry have equal areas. Is the converse proposition true?

4. Find the defect δ of $\triangle PKW$ without using additivity of defect.

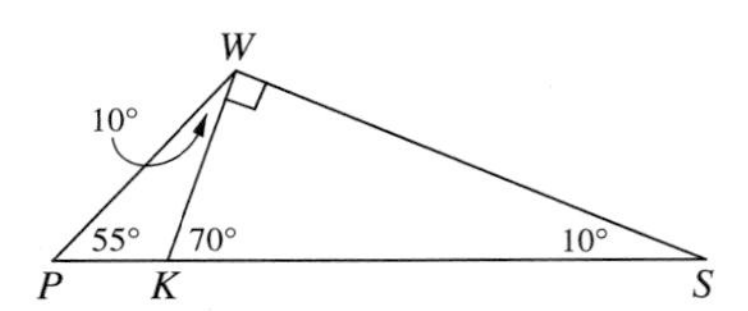

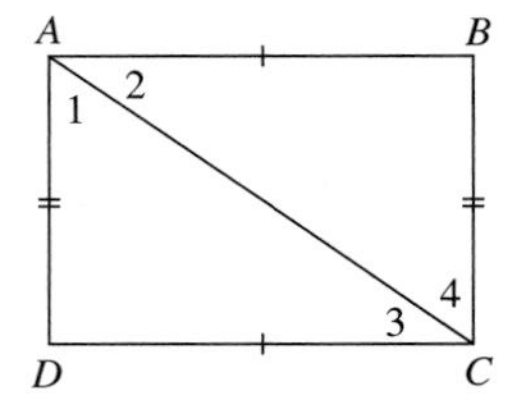

5. Work Problem 4 using the Linear Pair Axiom.

6. The defect of $\triangle ABC$ is 7.5 and $\triangle ACD \cong \triangle CAB$.

(a) Find the defect of the convex quadrilateral $\diamond ABCD$.

(b) If $\angle B$ is a right angle, find $m\angle BAD$.

7. Suppose the angles in the figure have the measures as indicated. Calculate the defects δ_1, δ_2, δ_3, and δ_4 of the interior polygons enclosed by $\diamond ABCD$, then test the validity of the additive property of defect by calculating $\delta(\diamond ABCD)$, as in Example 1.

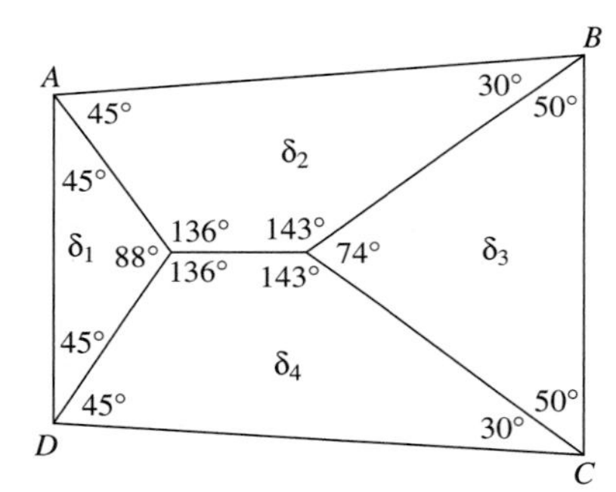

8. Isosceles triangle $\triangle DEF$, with $\overline{FD} \cong \overline{FE}$, is formed by the angle trisectors of the angles of $\triangle ABC$. If $m\angle A = m\angle B = 66$, $m\angle C = 27$, and the remaining angle measures are as indicated in the figure, find the defect of $\triangle DEF$ in two different ways. (***Hint:*** Observe that there are two pairs of congruent triangles in the figure.)

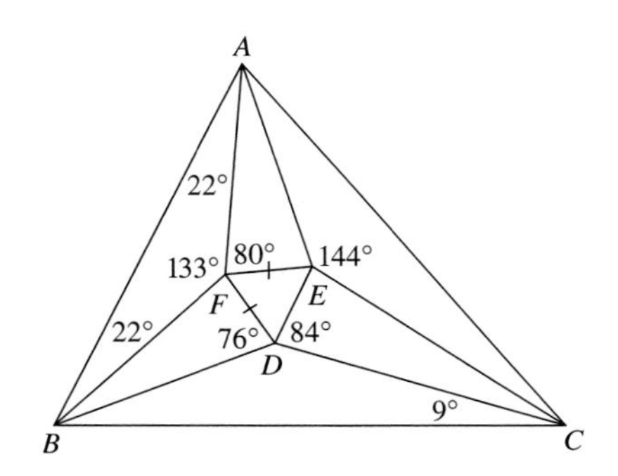

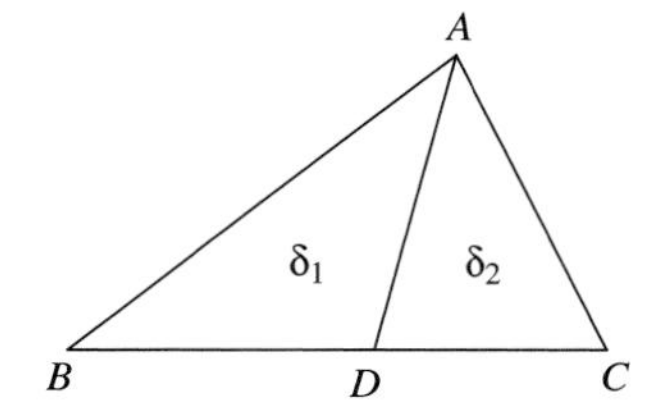

9. Additivity of defect is valid in absolute geometry because the proof of the lemma is valid. In that context, prove that if δ is the defect of $\triangle ABC$ and B–D–C holds, then $\delta = 0$ implies that $\triangle ABD$ and $\triangle ADC$ both have zero defects δ_1 and δ_2. (This is equivalent to saying that if either $\triangle ABC$ or $\triangle ADC$ has positive defect, then $\triangle ABC$ has positive defect, which is the contrapositive of the above statement. Would this prove the converse of Theorem 1, Section 6.2?)

10. The acute angle of a certain Lambert Quadrilateral has measure 83. Find its defect. What is the maximum defect of any Lambert Quadrilateral, and what has to happen if such a quadrilateral has nearly the maximum defect (area)?

11. The summit angles of a certain Saccheri Quadrilateral each has measure 83. Find the defect of the quadrilateral. (Compare with Problem 10; why should your answer to this problem be precisely twice that of Problem 10?)

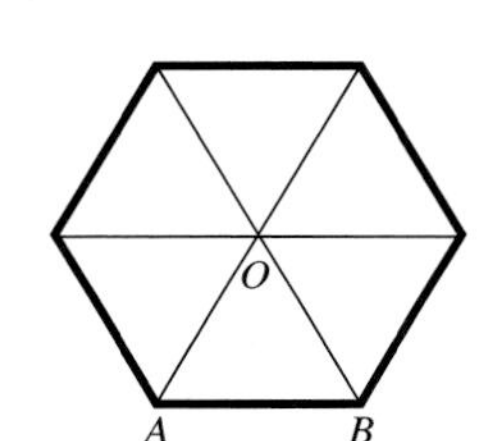

12. The defect of a certain regular hexagon in hyperbolic geometry is 12.

(a) Find the measure of each angle of the hexagon.

(b) If O is the center of the hexagon, find the defect of each subtriangle making up the hexagon, such as $\triangle ABO$ shown in the figure.

(c) Are each of these subtriangles equilateral triangles, as they would be if the geometry were Euclidean?

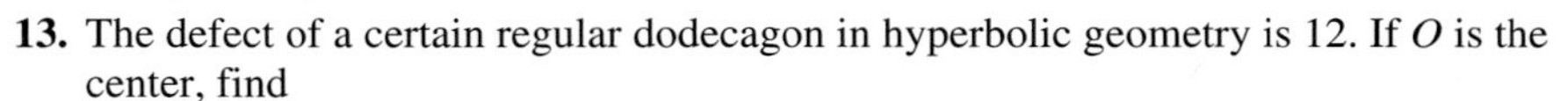

13. The defect of a certain regular dodecagon in hyperbolic geometry is 12. If O is the center, find

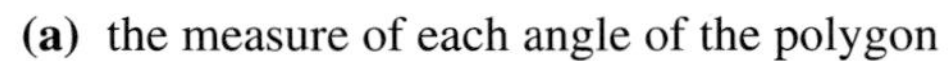

(a) the measure of each angle of the polygon

(b) the defect of the subtriangle $\triangle AOB$, where O is the center of the polygon.

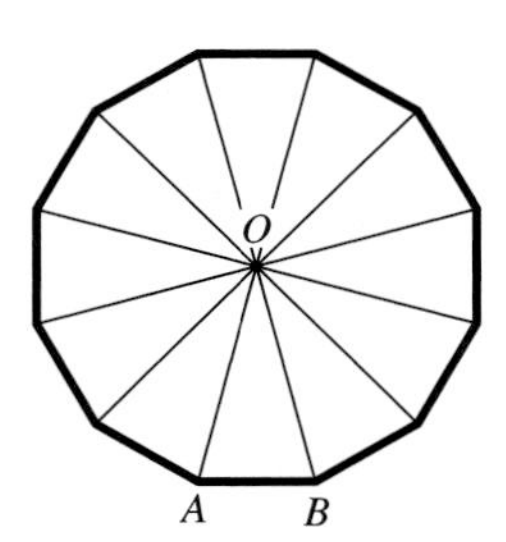

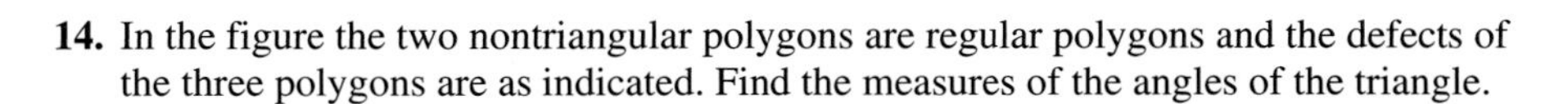

14. In the figure the two nontriangular polygons are regular polygons and the defects of the three polygons are as indicated. Find the measures of the angles of the triangle.

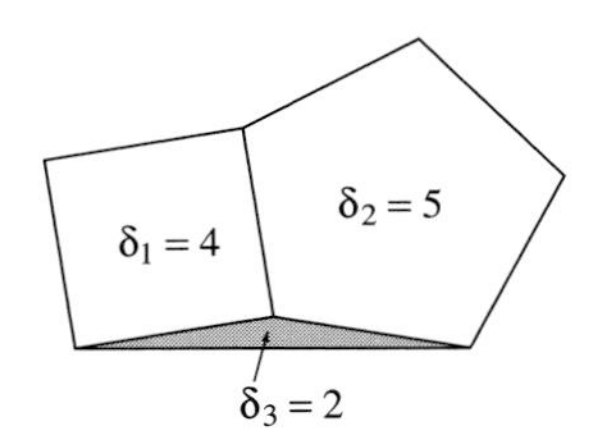

GROUP B

15. Two regular pentagons have equal defects. Show they must be congruent.

16. Points P, Q, and R are located on three lines meeting at point A and forming angles of equal measure, 120, such that $AP = AQ = AR$. (Hence, $\triangle PQR$ is equilateral.)

(a) What happens to $\delta(\triangle PQR)$, the defect of $\triangle PQR$, as AP, AQ, and AR become large without bound? (Recall the result of Problem 21, Section 3.4.)

(b) Based on the result in (a), what seems to be the area of the entire hyperbolic plane (in terms of k)?

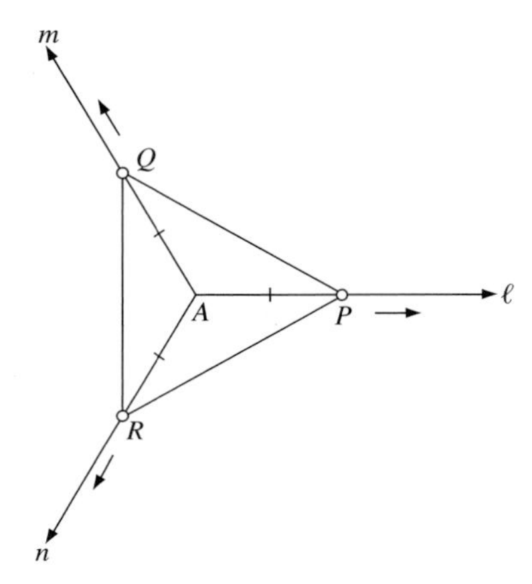

17. Lines ℓ and m are perpendicular at point O, and points at intervals of one unit apart are constructed on each of the four outgoing rays. If we draw in the regular quadrilaterals as shown, find the answer to each of these questions (with proofs):

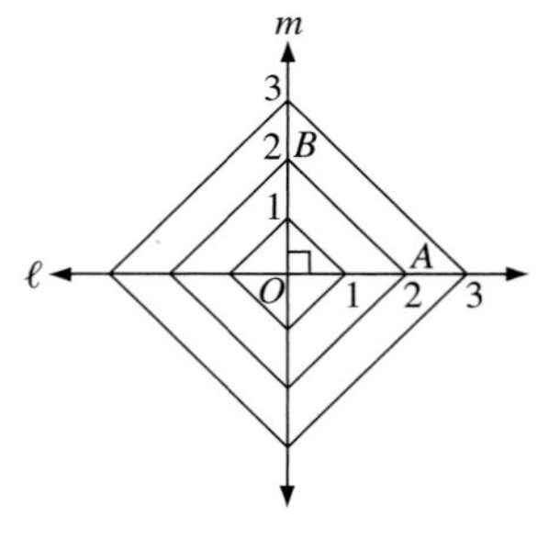

(a) Are the angles of any of the resulting quadrilaterals right angles, as they would be in Euclidean geometry?

(b) Assuming that the angles of the sequence of quadrilaterals become arbitrarily small (which is true), what seems to be the area of the hyperbolic plane, since it appears that these quadrilaterals eventually take in every point of the plane. (***Hint:*** Write k times the defect of a quadrilateral whose angle measures are nearly equal to zero.)

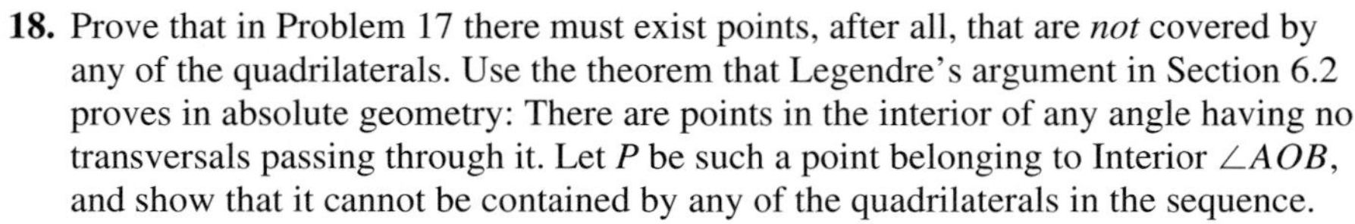

18. Prove that in Problem 17 there must exist points, after all, that are *not* covered by any of the quadrilaterals. Use the theorem that Legendre's argument in Section 6.2 proves in absolute geometry: There are points in the interior of any angle having no transversals passing through it. Let P be such a point belonging to Interior $\angle AOB$, and show that it cannot be contained by any of the quadrilaterals in the sequence.

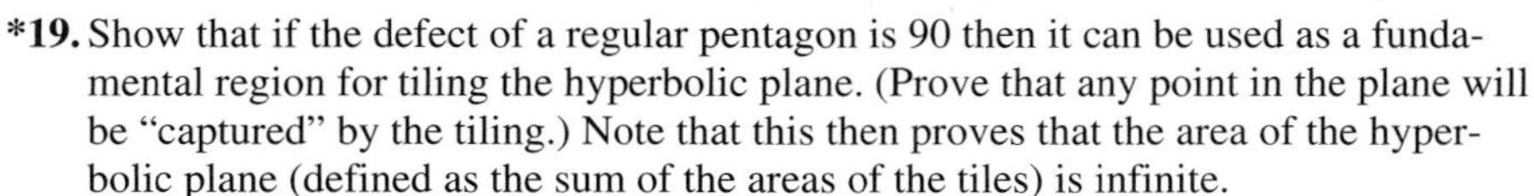

***19.** Show that if the defect of a regular pentagon is 90 then it can be used as a fundamental region for tiling the hyperbolic plane. (Prove that any point in the plane will be "captured" by the tiling.) Note that this then proves that the area of the hyperbolic plane (defined as the sum of the areas of the tiles) is infinite.

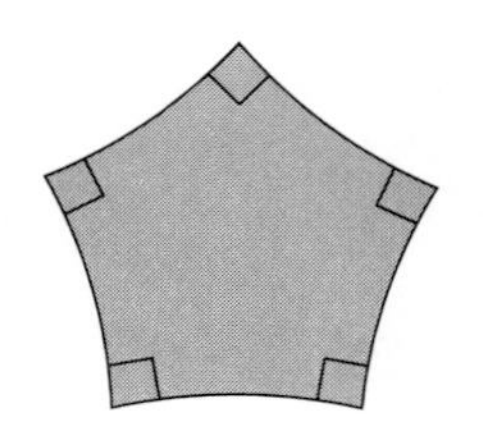

*See Problem 17, Section 6.5.

20. Suppose that $\triangle ABC$ is equilateral, and that $\delta(\triangle ABC) = 180 - 3\theta$, where $m\angle A = \theta$. If L, M, and N are the midpoints of the sides of $\triangle ABC$, and $m\angle LMN = \phi$, show that $\phi > \theta$ and $\delta(\triangle AMN) = \phi - \theta$.

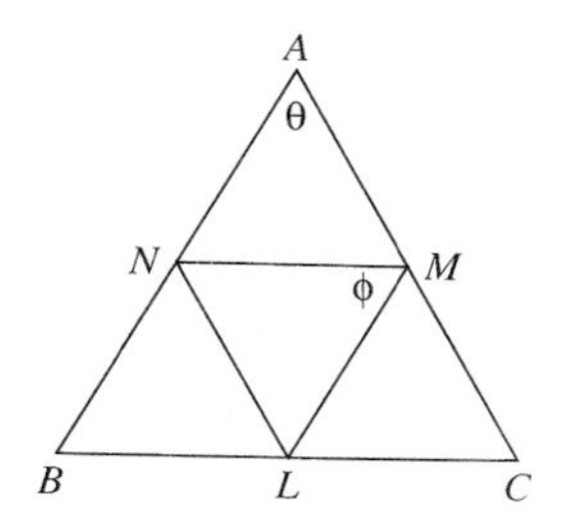

6.4 Two Models for Hyperbolic Geometry

If you were skeptical about the validity of the bizarre geometry we studied in the previous section, you are in good company. That was the reaction that many otherwise astute mathematicians had regarding the original work of Bolyai and Lobachevski, some of whom branded the work as "incorrect." Two famous models will be introduced in this section that perfectly depict axiomatic hyperbolic geometry. It may seem ironic that these models take place in the Euclidean coordinate plane, and depend on Euclidean precepts to describe them and to make deductions about them. But this is what makes the axioms for hyperbolic geometry consistent, or as consistent as Euclidean geometry itself.

OUR GEOMETRIC WORLD

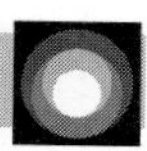

Where relatively small distances are concerned, Euclidean geometry is certainly the simplest model for the physical universe. But it is not the proper model for physicists, astronomers, and astronauts. Even the basic question of what constitutes a straight line in outer space is problematic. Consider the following demonstrated phenomenon: Light rays are actually bent by strong gravitational forces, such as the sun. Thus the visual image of a distant star may be distorted, and the apparent shortest path to it along a ray of light can actually be a curved line! (This was the first of Einstein's theoretical consequences of relativity to be verified by experiment.)

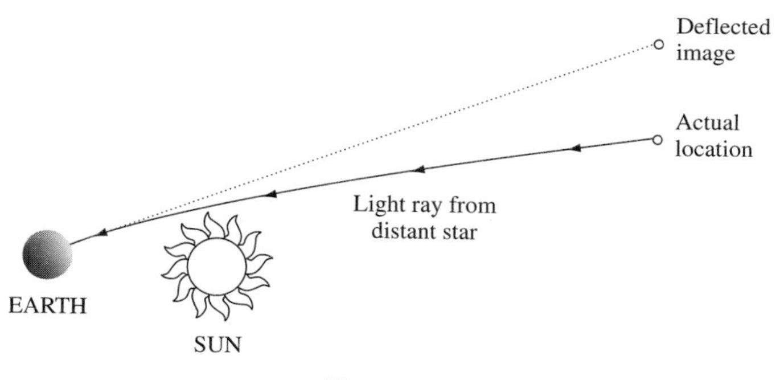

Figure 6.19

POINCARÉ'S DISK MODEL

We mentioned earlier—in Chapter 3—the disk model of Poincaré. To remind you of its details we repeat its description, and include the definition for distance not given earlier.

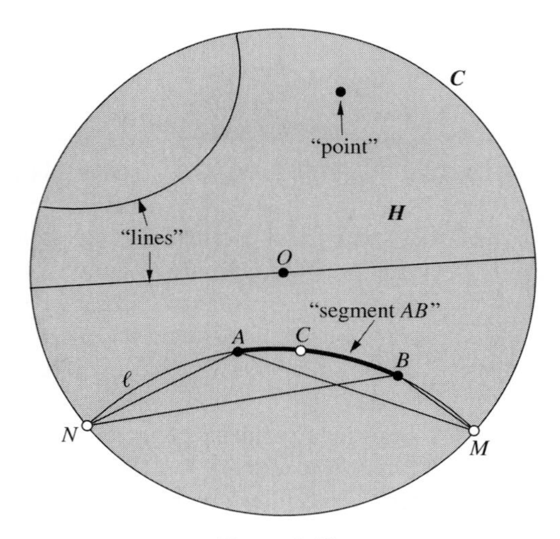

Figure 6.20

Begin with a circle $\boldsymbol{C}$ in the Euclidean plane, and its interior $\boldsymbol{H}$, as shown in Figure 6.20. The components of the geometry of Poincaré are as follows.

POINT Any interior point of circle $\boldsymbol{C}$ (the ordinary points of $\boldsymbol{H}$)

LINE Any diameter of $\boldsymbol{C}$ or any arc of a circle orthogonal to $\boldsymbol{C}$ in $\boldsymbol{H}$

DISTANCE If A and B are any two "points" and ℓ represents a "line" passing through A and B (which will be shown to be unique), let its end points on $\boldsymbol{C}$ be M and N, as shown in Figure 6.20. Define the "distance" from A to B as the real number

$$AB^* = \left| \ln\left(\frac{AM \cdot BN}{AN \cdot BM} \right) \right|$$

ANGLE MEASURE If $\triangle ABC$ is any "angle" in $\boldsymbol{H}$ consisting of two "rays" $\overrightarrow{BA}$ and $\overrightarrow{BC}$ (either part of a diameter of $\boldsymbol{C}$ or part of a circle orthogonal to $\boldsymbol{C}$, then consider the Euclidean rays $\overrightarrow{BA}'$ and $\overrightarrow{BC}'$ that are tangent to $\overrightarrow{BA}$ and $\overrightarrow{BC}$ and in the same direction (Figure 6.21). Define

$$m\angle ABC^* = m\angle A'BC'$$

(the Euclidean measure of the Euclidean angle $\triangle A'BC'$).

NOTE: To make the complicated formula for distance more convenient to work with, we give the fraction following the logarithm the name **Cross Ratio**, which has

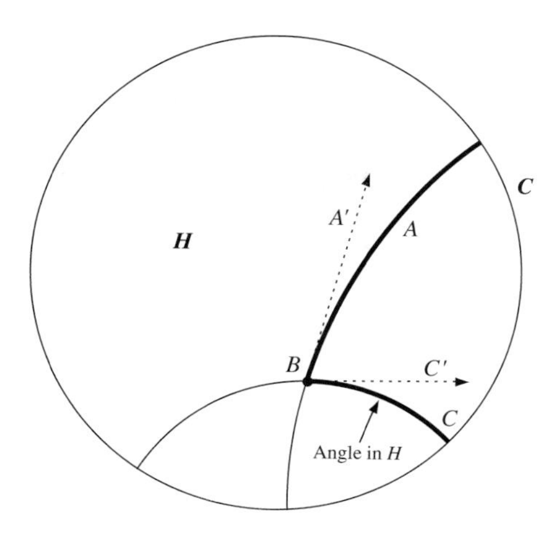

Figure 6.21

properties of interest all its own. This concept was actually introduced earlier in the text, in the problem sections. Let

$$(AB, MN) = \left(\frac{AM}{BM}\right) \Big/ \left(\frac{AN}{BN}\right) = \frac{AM \cdot BN}{AN \cdot BM}$$

The first expression on the right shows why it is called a "cross ratio" (or *ratio of ratios*); the second expression is what we use in working with it since it is simpler and easier to remember. Thus, we can reformulate the definition for distance as

$$AB^* = |\ln(AB, MN)|$$

Since one of the properties of cross-ratio is, by elementary algebra,

(1) $$(AC, MN) \cdot (CB, MN) = (AB, MN)$$

then the logarithm transforms this into the sum

(2) $$\ln(AC, MN) + \ln(CB, MN) = \ln(AB, MN)$$

and we obtain $AC^* + CB^* = AB^*$ for any point C on the Euclidean arc $\widehat{AB}$ of the "line" passing through A and B. (See solution to Example 3 that follows for algebraic details.) Thus $A–C–B$ holds in the "geometry" of $\boldsymbol{H}$, and the "segments" in $\boldsymbol{H}$ are basically their Euclidean counterparts, which makes it easy to identify them in figures and to work with them in general. This also means that rays, angles, and triangles are readily identified. (See Figures 6.20 and 6.21.)

HISTORICAL NOTE

When Hilbert met Henri Poincaré for the first time in Paris in 1885, Poincaré had already published more than 100 research papers in mathematics and physics. Poincaré was admitted to the Academy of Paris at the early age of 32. He was the outstanding mathematician of that period. Much of his work was directly connected to geometry, but he also solved some significant problems in mathematical physics associated with the so-called *three body problem.* He founded the field of algebraic topology, and he applied the theory of groups in powerful ways to provide new insight into the more advanced developments of geometry. His work with complex variables and fractional linear transformations (which are precisely the isometries of $\boldsymbol{H}$) led to a reformulation of the models for hyperbolic geometry then known, which were later named after him (but were actually first discovered by E. Beltrami in 1868).

This "geometry" satisfies all the axioms for absolute geometry. For example, to verify the axiom *two points determine a line*, consider a coordinate system where $\boldsymbol{C}$ is taken as the unit circle $x^2 + y^2 = 1$. To find circles orthogonal to $\boldsymbol{C}$, we must examine the general equation of a circle $x^2 + y^2 + ax + by = c$ and the equivalent center/radius form $(x - h)^2 + (y - k)^2 = r^2$, where, by squaring out we obtain

$$x^2 - 2hx + h^2 + y^2 - 2ky + k^2 = r^2$$
$$\rightarrow x^2 + y^2 - 2hk - 2ky = r^2 - h^2 - k^2$$

By matching coefficients,

$$\textbf{(3)} \qquad a = -2h, \qquad b = -2k, \qquad \text{and} \qquad c = r^2 - h^2 - k^2$$

With these formulas, we can transfer back and forth between the two forms. Now consider what must be true about a circle $\boldsymbol{D}$ that is orthogonal to $\boldsymbol{C}$, as in Figure 6.22. Orthogonality of $\boldsymbol{C}$ and $\boldsymbol{D}$ means that their tangents at the point of intersection P are perpendicular, and since the radii drawn to point P are perpendicular to the tangents, the tangent of one circle must pass through the center of the other. Thus $\triangle OPD$ is a right triangle. By the Pythagorean Theorem,

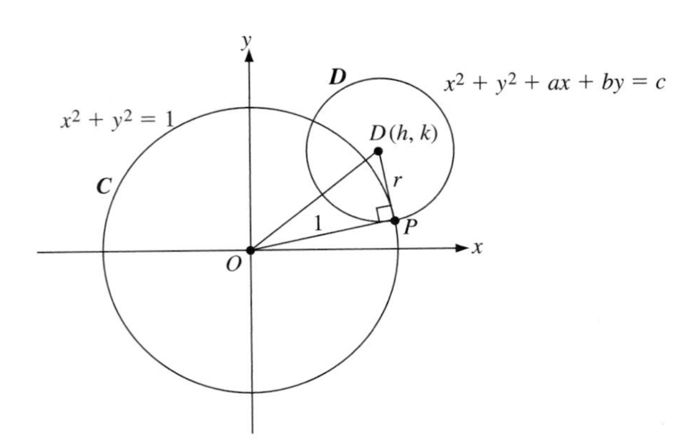

Figure 6.22

$$\textbf{(4)} \qquad OD^2 = OP^2 + PD^2 = 1 + r^2$$

where $D(h, k)$ is the center of $\boldsymbol{D}$ and r its radius. By the distance formula, $OD^2 = h^2 + k^2$ so we obtain

$$\textbf{(5)} \qquad h^2 + k^2 = 1 + r^2$$

From **(3)** we have $c = r^2 - h^2 - k^2$ or $h^2 + k^2 = r^2 - c$. Substitute this into **(5)**:

$$r^2 - c = r^2 + 1 \quad \rightarrow \quad c = -1$$

This determines the general equation of any circle $\boldsymbol{D}$ orthogonal to $\boldsymbol{C}$, namely

$$\textbf{(6)} \qquad x^2 + y^2 + ax + by = -1$$

As long as $\boldsymbol{C}$ is the unit circle, to actually find the "line" passing through two given points A and B we have only to substitute their coordinates into **(6)** and solve the resulting system for a and b. (There is no solution if the coordinates of A and B are proportional, but this means that A and B are collinear with O and the desired "line" is then the diameter of $\boldsymbol{C}$ passing through A and B.)

EXAMPLE 1 Find the equation of the "line" $\boldsymbol{D}$ in $\boldsymbol{H}$ passing through the two points $A(0.1, 0.3)$ and $B(-0.1, 0.7)$. Find its center and radius, and sketch the graph.

SOLUTION

Applying **(6)** and the above procedure, we have, by substitution,

$$(0.1)^2 + (0.3)^2 + 0.1a + 0.3b = -1$$
$$\rightarrow 0.01 + 0.09 + 0.1a + 0.3b = -1$$

Simplifying and clearing decimals, we obtain

$$0.1a + 0.3b = -1.1 \quad \rightarrow \quad a + 3b = -11$$

Similarly, for the coordinates of B:

$$0.01 + 0.49 - 0.1a + 0.7b = -1$$
$$\rightarrow 0.50 - 0.1a + 0.7b = -1 \quad \rightarrow \quad -a + 7b = -15$$

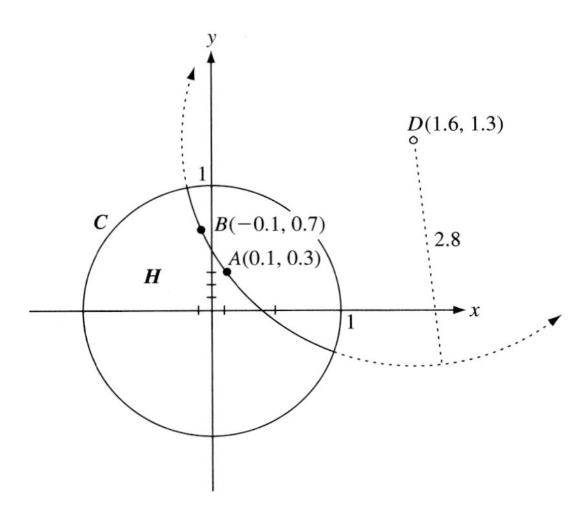

Figure 6.23

The result is the system of equations in a and b:

$$\begin{cases} a + 3b = -11 \\ -a + 7b = -15 \end{cases}$$

By adding we find that $10b = -26$ or $b = -2.6$, and substituting this into the first equation, $a + 3(-2.6) = -11$ or $a = -3.2$. Thus, the equation for $\boldsymbol{D}$ is

$$x^2 + y^2 - 3.2x - 2.6y = -1$$

By completing the square we obtain the equivalent equation

$$(x - 1.6)^2 + (y - 1.3)^2 = 3.25$$

Thus, the center of $\boldsymbol{D}$ is the point $D(1.6, 1.3)$ and its radius, $r \approx 1.8$. The graph is shown in Figure 6.23. ■

We will not pursue this model any further, except in the problems. Instead, we will work almost exclusively from this point on with another of Poincaré's models. These two models were both originally discovered by Eugenio Beltrami (1835-1900), who did extensive work in differential geometry. Using methods in this area he discovered the complicated formula for distance given above. Though different in the details, the two models are equivalent in the essentials, and that equivalence is established by means of circular inversion, a Cross Ratio-preserving transformation that is introduced in the next section. (See Problem 26, Section 6.5.) The half-plane model has the advantage of making it easier to work with a coordinate system, and many standard geometric constructions, such as obtaining the line through two points, are simpler than in the disk model.

BELTRAMI–POINCARÉ HALF-PLANE MODEL

Begin with the coordinate plane (Figure 6.24), and consider all points $P(x, y)$ for which $y > 0$. This is, of course, the **upper half-plane**. All such points will be considered the "points" in the model, and we denote this "universe" by $\boldsymbol{H}$, as in the Poincaré model. The x-axis is not part of the geometry; to be a bit melodramatic, we might think of it as the *edge of the universe*. The components of this model are then:

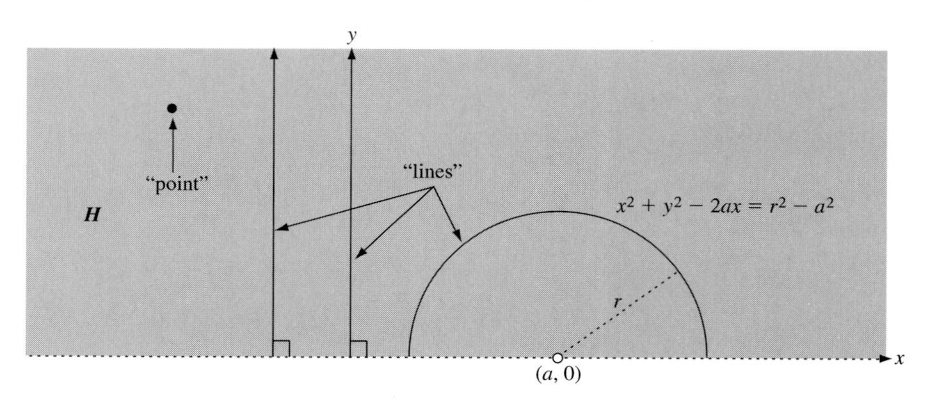

Figure 6.24

POINT Any point $P(x, y)$ such that $y > 0$

LINE Semicircles of the form

$$(x - a)^2 + y^2 = r^2, \qquad y > 0$$

with center on the x-axis, or vertical rays of the form

$$x = a \text{ (constant)}, \qquad y > 0$$

DISTANCE For any two "points" A and B and ℓ the "line" containing A and B, let the **end points** of ℓ be (if it is a semicircle with equation as above) $M(a + r, 0)$ and $N(a - r, 0)$ or if ℓ is a vertical ray, let its **end point** be $M(a, 0)$, as shown in Figure 6.25. Then, in the two cases for ℓ, define the "distance" from A to B as

(7) $$AB^* = \left| \ln \frac{AM \cdot BN}{AN \cdot BM} \right| \equiv |\ln(AB, MN)|$$

(8) $$AB^* = \left| \ln \frac{AM}{BM} \right|$$

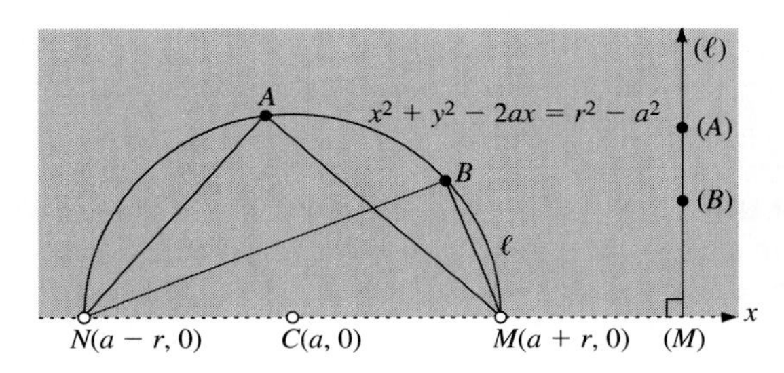

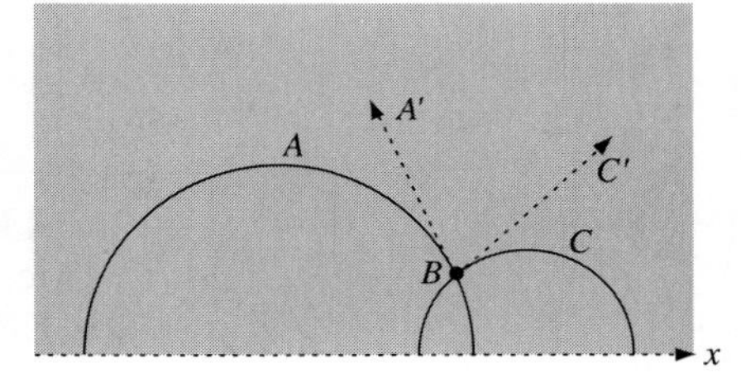

Figure 6.25

NOTE: Observe the use of the Cross Ratio symbol (as in the case of the Poincaré disk model). In looking at the two formulas for distance, we can see by using elementary limit techniques that as N recedes along the x-axis to minus infinity (i.e., $BN \to \infty$), the semicircle becomes a vertical ray and the cross-ratio $AM \cdot BN/(BM \cdot AN)$ converges to the single ratio AM/BM. Formula **(8)** could then be viewed as a special case of formula **(7)**.

ANGLE MEASURE Let $m\angle ABC^* = m\angle A'BC'$ where $\overrightarrow{BA'}$ and $\overrightarrow{BC'}$ are the Euclidean rays tangent to the "sides" of $\angle ABC$, as in the disk model.

Instead of continuing to use quotation marks for objects in the model, we are going to start using the more descriptive terms **h-line**, **h-angle**, **h-distance**, and so on, to designate objects in the model. To further clarify the distinction between objects of Euclidean geometry and those in ***H***, we also use an asterisk following any symbol representing a geometric object or concept in the model (as already done in the case of distance and angle measure).

EXAMPLE 2 Show geometrically that given any two h-points A and B, there is always a unique h-line passing though them.

SOLUTION

If A and B lie on a vertical line $x = a$, then no semicircle centered on the x-axis can pass through A and B, and the h-line $x = a, y > 0$, is the only one that passes through A and B. On the other hand, as in Figure 6. 26, suppose $\overleftrightarrow{AB}$ is not a vertical line. Then the perpendicular bisector of Euclidean segment $\overline{AB}$ will be nonhorizontal, hence will meet the x-axis at a unique point L. Take L as center and AL as radius, and draw the semicircle through A and B. This will be the unique h-line passing through A and B in this case. ■

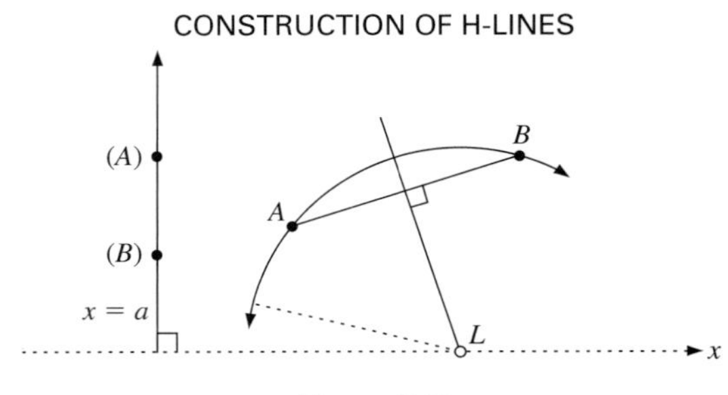

Figure 6.26

EXAMPLE 3 As shown in Figure 6.27, C is any point on arc $\overset{\frown}{AB}$ of the semicircle, or h-line ℓ. Verify the h-betweenness relation $A\text{–}C\text{–}B^*$. That is, show that $AC^* + CB^* = AB^*$. It then follows that the circular arc $\overset{\frown}{AB}$ is an h-segment.

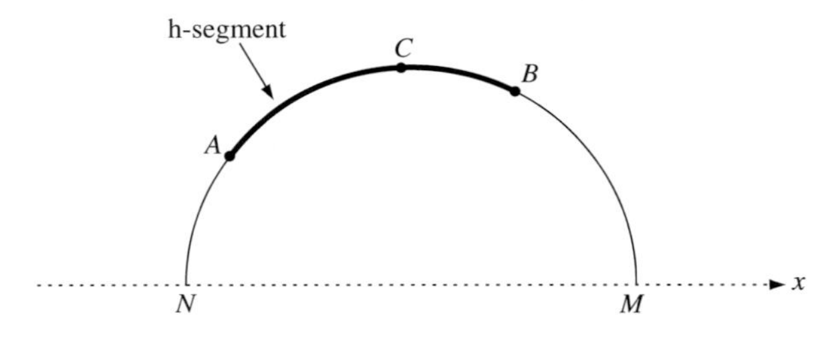

Figure 6.27

SOLUTION

By definition, and because the logarithms involved are positive,

$$AC^* = \ln(AC, MN) \qquad \text{and} \qquad \text{CB}^* = \ln(CB, MN)$$

Thus, using the property $\ln xy = 1\text{n}x + \ln y$ for the logarithm, we have

$$\begin{aligned} AC^* + CB^* &= \ln\left(\frac{AM \cdot CN}{AN \cdot CM}\right) + \ln\left(\frac{CM \cdot BN}{CN \cdot BM}\right) = \ln\left(\frac{AM \cdot CN}{AN \cdot CM} \cdot \frac{CM \cdot BN}{CN \cdot BM}\right) \\ &= \ln\left(\frac{AM \cdot BN}{AN \cdot BM}\right) = \ln(AB, MN) = AB^* \end{aligned}$$

■

NOTE: The coordinate equation of a semicircle h-line depends on only two parameters. If the center-radius form is $(x - h)^2 + (y - k)^2 = r^2$, then since the center $C(h, k)$ lies on the x-axis, $k = 0$ and the equation reduces to $(x - h)^2 + y^2 = r^2$, or equivalently, $x^2 + y^2 - 2hx = r^2 - h^2$. That is, the equation has the general form,

(9) $$x^2 + y^2 + ax = b$$

This makes it particularly simple to find the equation of an h-line passing through two points having their numerical coordinates given, by the method of substituting into **(9)** and solving for a and b. The center can then be readily found, if needed.

EXAMPLE 4 (See Figure 6.28.) Find the equation of the h-line passing through the points $A(1, 3)$ and

(a) $B(1, 6)$
(b) $C(5, 7)$

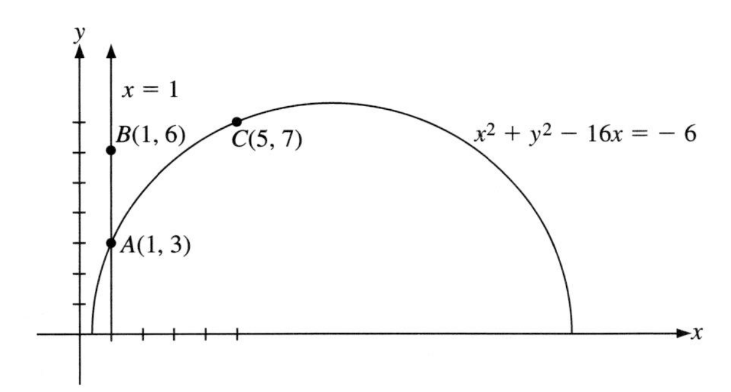

Figure 6.28

SOLUTION

(a) The h-line $\overleftrightarrow{AB}^*$ is a vertical ray, hence its equation is $x = 1, y > 0$.

(b) We use **(9)**: the desired equation is $x^2 + y^2 + ax = b$ for some a and b. By substitution of the coordinates of points A and C into this equation, we obtain

$$1^2 + 3^2 + a(1) = b \quad \rightarrow \quad 10 + a = b \quad \text{or} \quad a - b = -10$$

$$5^2 + 7^2 + a(5) = b \quad \rightarrow \quad 25 + 49 + 5a = b \quad \text{or} \quad 5a - b = -74$$

Thus we solve the following system for a and b:

$$\begin{cases} a - b = -10 \\ 5a - b = -74 \end{cases}$$

Subtracting:

$$-4a = -10 + 74 \qquad \text{or} \qquad -4a = 64 \quad \rightarrow \quad a = -16$$

Substitute this value into the first equation:

$$-16 - b = -10 \quad \rightarrow \quad b = -6$$

The equation of the h-line $\overleftrightarrow{AC}*$ is then

$$x^2 + y^2 - 16x = -6 \qquad \blacksquare$$

EXAMPLE 5 Hyperbolic Distance in Upper Half-Plane Model To three-place accuracy, find the h-distance $KW*$ if $K = (1, 4)$ and $W = (8, 3)$ (See Figure 6.29.)

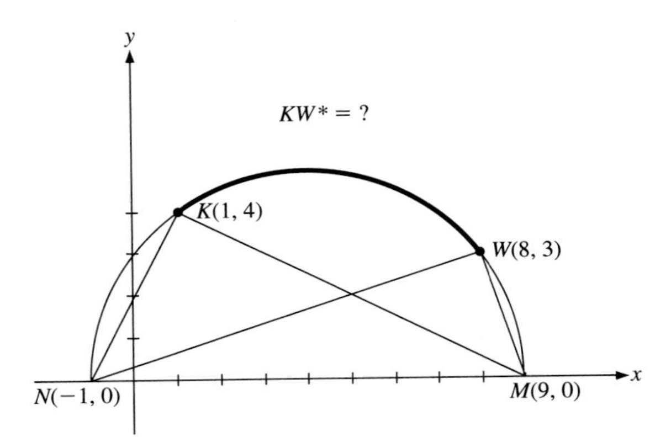

Figure 6.29

SOLUTION

We must first find the end points of $\overleftrightarrow{KW}* = \ell$. The equation of ℓ is found to be $x^2 + y^2 - 8x = 9$, the details of which are just like the preceding example. To find the end points M and N we set $y = 0$:

$$x^2 - 8x - 9 = 0 \;\rightarrow\; (x - 9)(x + 1) = 0 \;\rightarrow\; M(9, 0), N(-1, 0)$$

We use the distance formula for the required distances KM, WN, ... and the property of logarithms $\ln \sqrt{x} = \frac{1}{2}\ln x$.

$$KW* = \ln(KW, MN) = \ln\left(\frac{KM \cdot WN}{KN \cdot WM}\right) =$$

$$\frac{1}{2}\ln\frac{[(1-9)^2 + 4^2] \cdot [(8+1)^2 + 3^2]}{[(1+1)^2 + 4^2] \cdot [(8-9)^2 + 3^2]}$$

$$= \frac{1}{2}\ln\frac{[64 + 16] \cdot [81 + 9]}{[4 + 16] \cdot [1 + 9]} = \frac{1}{2}\ln\frac{80 \cdot 90}{20 \cdot 10} = \frac{1}{2}\ln 36$$

Thus, $KW* = \ln\sqrt{36} = \ln 6 \approx 1.792$. $\blacksquare$

EXAMPLE 6 Hyperbolic Angle Measure in Upper Half-Plane Model Using the formula $\tan\theta = (m_2 - m_1)/(1 + m_1 m_2)$ for the measure θ of the angle from a line having slope m_1 to a line having slope m_2, find the measure of the

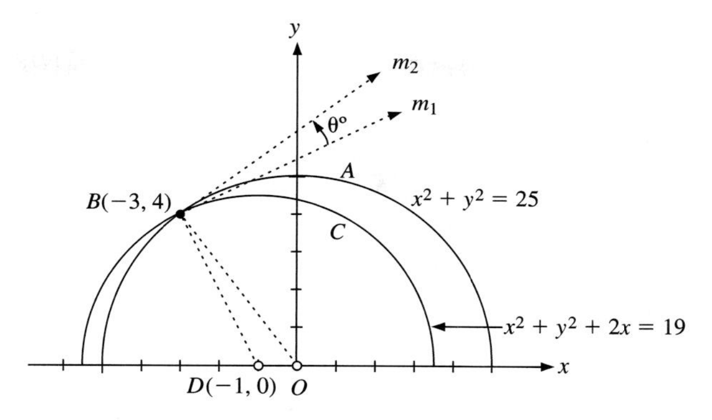

Figure 6.30

acute h-angle determined by the two h-lines $x^2 + y^2 = 25$ and $x^2 + y^2 + 2x = 19$, accurate to within 0.1 (degree). (See Figure 6.30.)

SOLUTION

We find the vertex B of the desired h-angle $\angle ABC^*$ by solving the system

$$\begin{cases} x^2 + y^2 = 25 \\ x^2 + y^2 + 2x = 19 \end{cases}$$

Subtracting yields $0 - 2x = 25 - 19$ or $x = -3$. Substitute this into the first equation and find $y = \pm 4$; since $y > 0$, we obtain $y = 4$, and the vertex is $B(-3, 4)$. The desired slopes m_1 and m_2 of the tangents to these circles at B must be found. The slope m_2 of the radius of $\overleftrightarrow{AB}^*$ $(x^2 + y^2 = 25)$ from O to B will determine m_2: $m_2' = 4/(-3) = -4/3$. Hence $m_2 = 3/4$ (the negative reciprocal of m_2'). Next, the center of the other semicircle is found by rewriting its equation:

$$(x^2 + 2x + 1) + y^2 = 19 + 1 \quad \rightarrow \quad (x + 1)^2 + y^2 = 20$$

or $D(-1, 0)$. The slope of $\overleftrightarrow{DB}$ is $m_1' = (4 - 0)/(-3 + 1) = 4/(-2) = -2$; then $m_1 = \frac{1}{2}$. Hence

$$\tan\theta^\circ = \frac{m_2 - m_1}{1 + m_1 m_2} = \frac{3/4 - 1/2}{1 + (3/4)(1/2)} = \frac{1/4}{1 + 3/8} = \frac{2}{11}$$

Therefore,

$$m\angle ABC^* = \theta \approx 10.3 \qquad \blacksquare$$

We now proceed to go through the axioms for absolute *plane* geometry and examine their validity for the half-plane model. The only incidence axioms that apply to a plane are:

(a) *Two points determine a line.*

(b) *Each plane contains three noncollinear points and each line contains at least two points.*

The first of these, **(a),** was verified in detail in Example 2. The second is clear by construction. Moving on, for the metric axioms we have:

(c) *For any two h-points A and B, $AB^* > 0$, with equality only when $A = B$.*
(d) $AB^* = BA^*$
(e) *Ruler Postulate*

Again, by the definition of distance as the absolute value of a function, distance is positive or zero, and when $A = B$ the cross-ratio reduces to unity, which yields $AB^* = \ln 1 = 0$. By the fact that $(AB, MN) = (BA, MN)^{-1}$ we obtain $\ln(AB, MN) = -\ln(BA, MN)$, and since we are taking the absolute value, we obtain the same value for both cases, and $AB^* = BA^*$. The Ruler Postulate is the most difficult to verify. To that end, let $\ell = \overleftrightarrow{AB}^*$ be any h-line, and P be any point on ℓ. Then assign the real number x to P by the formula

$$x = \ln(AP, MN) \qquad [\text{similarly, } x = \ln(PM/AM) \text{ if } \ell \text{ is a vertical ray}]$$

Notice that we have omitted the absolute values; this will allow x to be negative, and thus the range of x is the complete set of real numbers (see Figure 6.31). First, note that if $P \neq Q$ as in Figure 6.32, then the coordinates x and y of P and Q are distinct: for if $x = y$ then $(AP, MN) = (AQ, MN)$, which reduces to $PN/QN = PM/QM$, which is impossible, because one of these ratios is greater than 1 and the other is less than 1, as Figure 6.32 shows. Thus, each coordinate has been assigned to only one point, so this defines a valid coordinate system. It remains to prove

$$PQ^* = |x - y|$$

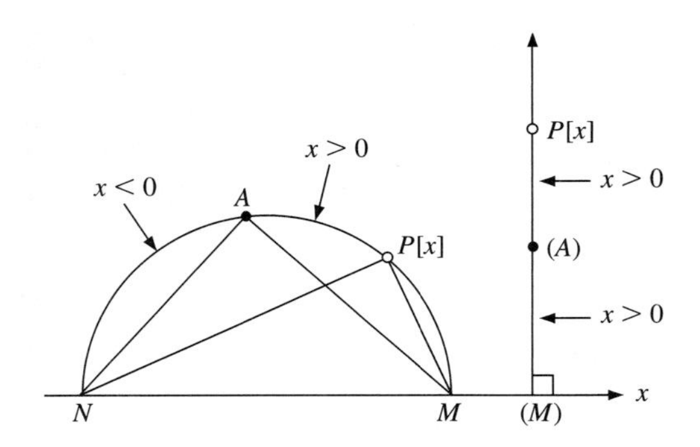

Figure 6.31

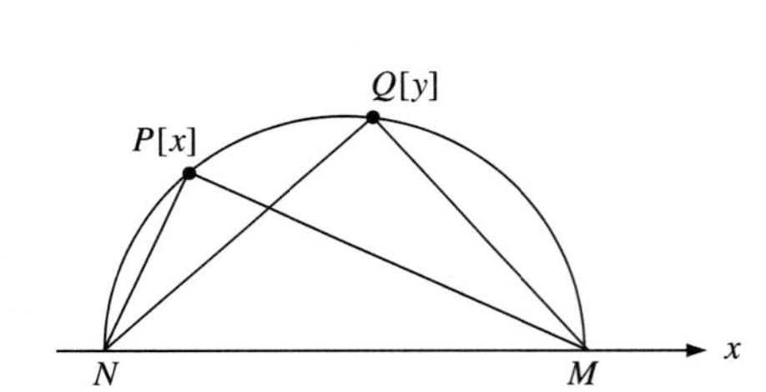

Figure 6.32

But $x = \ln(AP, MN)$ and $y = \ln(AQ, MN)$ and

$$|x - y| = |\ln(AP, MN) - \ln(AQ, MN)| = \left|\ln\frac{(AP, MN)}{(AQ, MN)}\right|$$

If you work a little with the ratios defining the quotient of the above Cross Ratios, you will find that this expression reduces to

$$|\ln(PQ, MN)^{-1}| = |-\ln(PQ, MN)| = PQ^*$$

as desired.

To proceed, since the model is **conformal** (that is, h-angle measure coincides with Euclidean angle measure), all the axioms on angle measure are valid, including the Linear Pair Axiom. (Some case checking may be necessary.) The last two axioms are the Plane Separation and SAS Postulates. We postpone the verification of the SAS Postulate to the next section since it requires the concept of circular inversion. For the Plane Separation Postulate, let ℓ be any h-line, and let the two sides of ℓ, H_1 and H_2, be determined as in the diagrams in Figure 6.33. Since the convexity of H_1 and H_2 is reasonably clear when ℓ is a vertical line, let us concentrate on the case when ℓ is a semicircle. It looks like it might be difficult to prove that the nonconvex region H_1 in Euclidean geometry is, nevertheless, h-convex. However, one can reason in the following manner. In the case shown in Figure 6.34, let A and B be any two points of H_1 and consider arc $\overset{\frown}{AB}$ of the semicircle centered on the x-axis containing A and B. It must be proven that $\overline{AB}^*$ (circular *arc* $\overset{\frown}{AB}$, not *segment* $\overline{AB}$) lies in H_1. We know that two semicircles (or line and semicircle) cannot be tangent to each other (they cannot meet at an angle of measure zero), and must cross over each other or else have no points in common. Therefore, if arc $\overset{\frown}{AB}$ is not entirely contained by H_1, there exists a first encounter (intersection) of arc $\overset{\frown}{AB}$ with line ℓ (point C) as we proceed from A to B, at which point $\overset{\frown}{AB}$ crosses over into H_2, then a *second* encounter with ℓ (point D) must occur to have $B \in H_1$. This gives us two h-points in common with h-lines $\overleftrightarrow{AB}$ and ℓ, which is impossible. Hence, $\overline{AB}^* \subseteq H_1$.

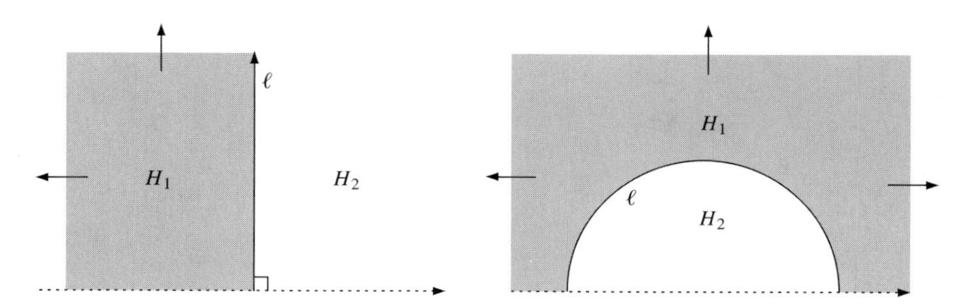

Figure 6.33

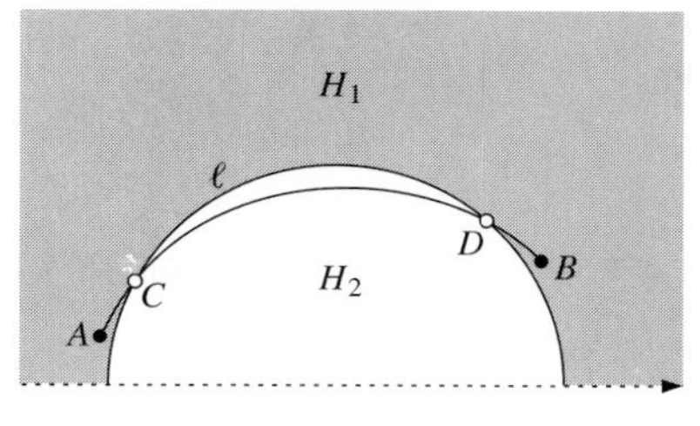

Figure 6.34

As stated above, the SAS Postulate will be postponed until the next section. But it might help to see an example.

EXAMPLE 7 Two Triangles Satisfying the SAS Hypothesis
As shown in Figure 6.35, consider the two h-triangles $\triangle ABC^*$ and $\triangle XYZ^*$, having vertices $A(0, 7)$, $B(0, 5)$, $C(3, 4)$, $X(12, 35)$, $Y(0, 37)$, and $Z(0, 74)$. It can be shown that $AB^* = XY^*$, $BC^* = YZ^*$, and $m\angle ABC^* = m\angle XYZ^* = 90$. Thus $\triangle ABC^*$ and $\triangle XYZ^*$ satisfy the SAS Hypothesis, and will be congruent in the upper half-plane model. All this can be verified by direct computation, or using *Sketchpad* scripts for h-distance and h-angle measure as introduced in Problems 14 and 15 below. ■

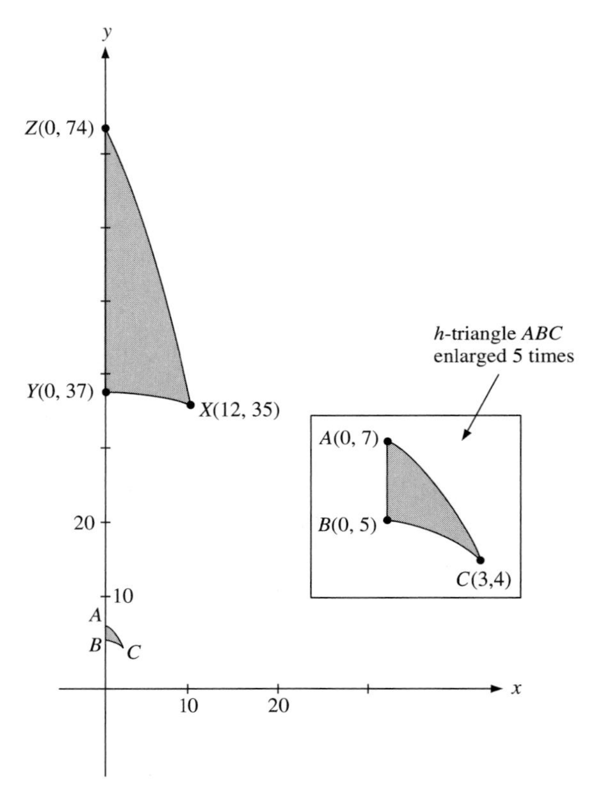

Figure 6.35

Moment for Discovery

Angle Sums for Triangles in Upper Half-Plane Model

This experiment will allow you to discover for yourself something about the angle sum of a triangle that is not easily observed by simply looking at the model. In Figure 6.36 an h-triangle $\triangle ABC$ is shown, having sides $x = 3$, $x^2 + y^2 = 25$, and $x^2 + y^2 + 2x = 19$. (Recall Example 6, where $m\angle ABC^*$ was already found.)

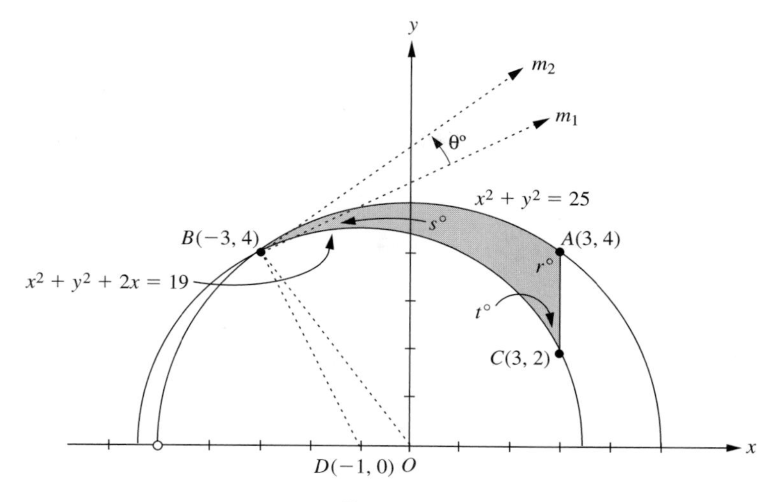

Figure 6.36

1. Show that the points of intersection of these three curves, which are the vertices of the h-triangle, are $A(3, 4)$, $B(-3, 4)$ and $C(3, 2)$.
2. Make an accurate sketch of this figure on your paper, using compass and straight-edge.
3. Measure the h-angles of $\triangle ABC^*$ with a protractor to estimate the values for r, s, and t in Figure 6.36. To save you some work, use the value for s (≈ 10.3) already determined in Example 6. [For greater accuracy, follow the method of Example 6, where the slopes of the tangents were calculated and the formula $\tan\theta = (m_2 - m_1)/(1 + m_2 m_1)$ was used.]
4. Calculate the sum $r + s + t$. What did you find?

Moment for Discovery

Parallel Postulate in the Upper Half-Plane Model

1. On a sheet of paper, draw a horizontal boundary line and a semicircle ℓ on it. Let ℓ have end point M and N, as shown in Figure 6.37.

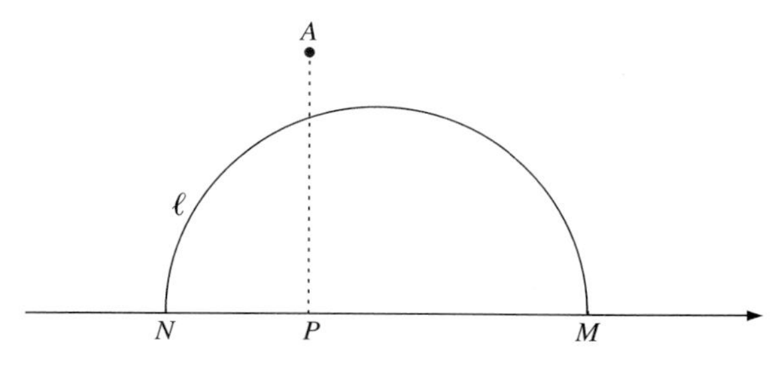

Figure 6.37

2. Locate point A somewhere above h-line ℓ in your drawing.
3. In Figure 6.37, the vertical line through A intersects ℓ. In your figure, find and draw an h-line m through A that does not intersect ℓ. (Thus, $m \parallel \ell$.)
4. If you can, draw a few more h-lines through A parallel to ℓ. What does this show about parallelism in the upper half-plane model?
5. Construct the vertical h-line n at N. Is $n \parallel \ell$? Are there other h-lines besides n having end point N that are parallel to ℓ? Draw the one passing through A, and label it n'.
6. Draw or construct the h-line m' passing through A with end point M. (*Optional*: Create a compass, straight-edge construction of the h-line passing through A with end point M.)
7. Classify all the h-lines in the model passing through A relative to the lines m' and n' of previous steps. Do you detect two major types of parallels to ℓ passing through A?

Moment for Discovery

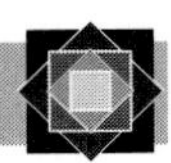

Using *Sketchpad* in the Upper Half-Plane Model

1. Construct a horizontal boundary line $\overleftrightarrow{AB}$ near the bottom of your sketch using LINE TOOL and dragging to make it perfectly horizontal. This line will be the boundary of $\boldsymbol{H}$. Hide the control points A and B. Locate three points C, D, and E.
2. Construct the Euclidean segment joining C and D, and the midpoint of that segment.
3. Construct the perpendicular (bisector) of $\overline{CD}$ through F, and select its intersection G with $\overleftrightarrow{AB}$. (See Figure 6.38.)

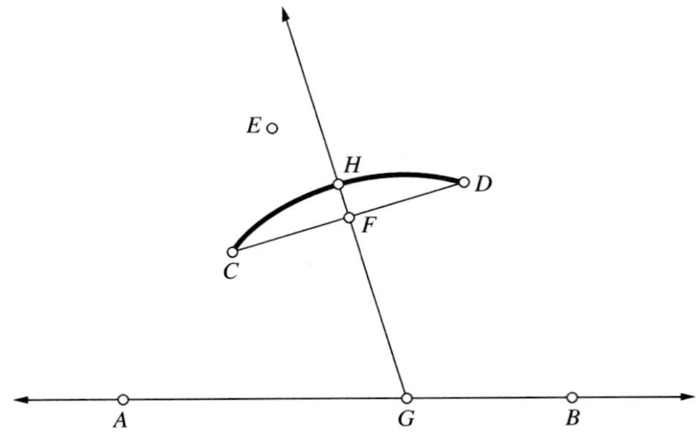

Figure 6.38

4. With G as center, construct the circle passing through C and D. Construct ray $\overrightarrow{GF}$, and select its intersection H with the circle.
5. Select C, H, and D, and choose Arc Through Three Points under CONSTRUCT to construct circular arc $\overparen{CHD}$. This will be h-segment $\overline{CD}$*.

6. Choose Thick Line under Line Style in the DISPLAY menu while arc $\widehat{CHD}$ is still selected. Hide all objects of construction except points C, D and the h-segment $\overline{CD}*$. (Also, leave point E alone.)
7. Repeat Steps 3–6 for points C, E and D, E (or make a script for Steps 3–6 and use it for this purpose) to construct h-segments $\overline{CE}*$ and $\overline{ED}*$. You now have constructed an h-triangle $\triangle CDE*$. (The construction fails if the pair of points selected lie on a vertical line.)

Drag points C, D, and E about to experiment with your h-triangle. How small or large can you make the angles? In your experiment, be sure to drag C close to the boundary line to see what happens? What if all three vertices are close to the boundary line? Verify the corollary of Theorem 1, Section 6.3.

NOTE: To make the above construction available for future use in the model, create a script for your construction of the h-segment joining a given pair of points in Steps 2–6. Simply repeat those steps with the Record (REC) option of CREATE SCRIPT activated. To play this script, you must select two (given) points and the horizontal boundary line in that order.

PROBLEMS (6.4)

In each of the problems below, ***H*** stands for the upper half-plane model of hyperbolic geometry and not the Poincaré disk model. Problems 1, 16, and 25 are the only problems dealing with Poincaré's disk model.

GROUP A

1. Make a sketch of the Poincaré disk model, and sketch a "line" ℓ and a "point" A not on ℓ. Examine the "lines" passing through A. Can you find at least two of them parallel to ℓ in the model?
2. Using coordinate geometry, find the equations for the h-line in the upper half-plane model that pass through $A(0, 2)$ and B, where
   (a) $B = (0, 5)$
   (b) $B = (1, \sqrt{3})$
   (c) $B = (4, 4)$
3. In the upper half-plane model ***H*** show that the h-line $x = 5$ $(y > 0)$ is h-parallel to the h-line $x^2 + y^2 - 6x + 5 = 0$ $(y > 0)$. Sketch the graph of these two lines.
4. Sketch a graph for the four points $A(-3, 1)$, $B(-3, 10)$, $C(10, 24)$, and $D(18, 20)$ in ***H***. Note that AB and CD (Euclidean distance) appear to be approximately the same. Use the method of Example 5 to find, and compare:
   (a) $AB*$
   (b) $CD*$
5. The h-segments $\overline{AB}*$ and $\overline{CD}*$, where $A = (2, 1)$, $B = (2, 3)$, $C = (2, 14)$, and $D = (2, 16)$, have the same Euclidean lengths (namely 2 units). Find the distances $AB*$ and $CD*$, and compare. Which h-segment is longest?

6. If $A(-9, 12)$, $B(9, 12)$, $C(-9, 40)$, and $D(9, 40)$ are four points in $\boldsymbol{H}$, find, and compare:

(a) AB and CD

(b) AB^* and CD^*

7. The points $A(-3, 4)$ and $B(3, 4)$ are transformed to $A'(-3k, 4k)$ and $B'(3k, 4k)$ by a dilation with center at the origin. If the factor k takes on the respective values $\frac{1}{2}, \frac{1}{4}, \ldots, \frac{1}{2^n}, \ldots$ then segment $\overline{AB}$ transforms to smaller and smaller segments $\overline{A'B'}$ under the Euclidean metric, converging to zero length. Calculate $A'B'^*$ under the h-metric, and determine its limit.

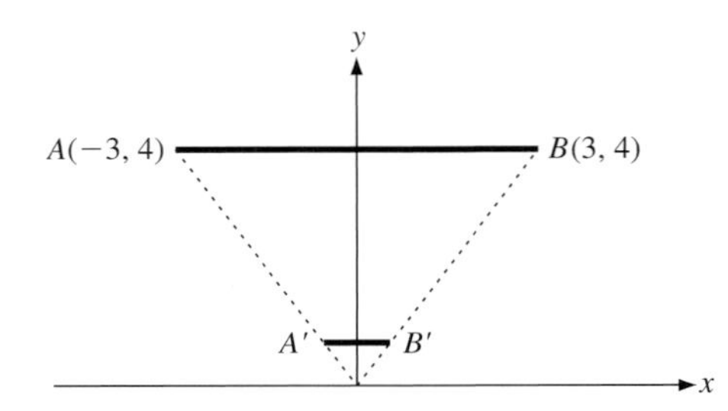

8. Show that in $\boldsymbol{H}$, the point $M(0, b)$ is the midpoint of the h-segment joining $A(0, a)$ and $B(0, c)$ iff b is the geometric mean of a and c (i.e., $b^2 = ac$).

9. In the example illustrated in the figure for this problem, explain why

(a) P is in the h-interior of $\angle ABC^*$

(b) No h-line passing through P is a transversal of $\angle ABC^*$ (intersects both sides of the angle at points different from the vertex).

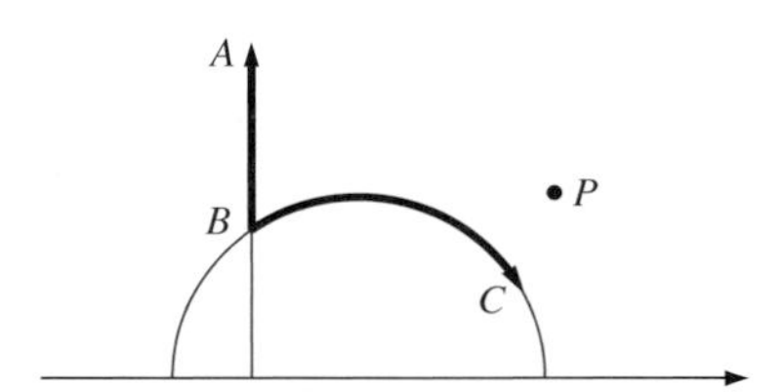

GROUP B

10. Calculate the h-distance from $C(1, 10)$ to

(a) $A(7, 8)$

(b) $B(1, 20)$

(c) Show that $\angle ABC^*$ is an isosceles right triangle in $\boldsymbol{H}$.

(d) Verify that the point $D(-24, 0)$ is the center of the semicircle $\overline{AB}^*$, and use this to calculate each of the h-angles of $\triangle ABC^*$ using the slope formula for $\tan \theta°$ as in a previous example. What do you observe about the angle sum?

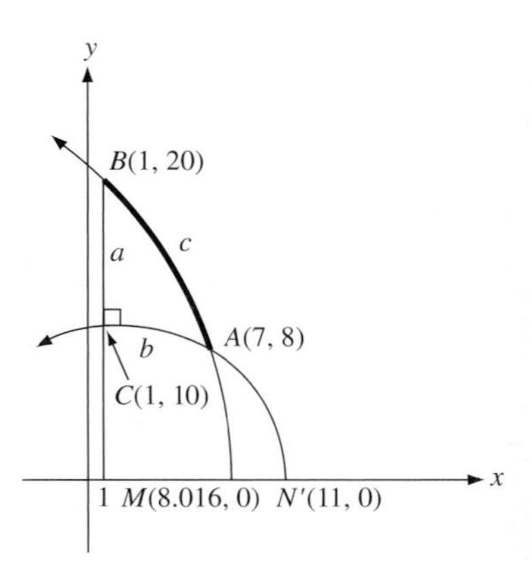

11. (a) Investigate the Pythagorean Theorem for right triangle $\triangle ABC^*$ of Problem 10, where $a = BC^* = AC^* = b = \ln 2$ and $BC^* = c \approx 1.016$, by calculating AB^* and using the results of AC^* and BC^* from Problem 10. Is the relation $c^2 = a^2 + b^2$ valid?

(b) The correct relation for a right triangle in hyperbolic geometry using standard notation (c being the length of the hypotenuse) is

$$\cosh c = \cosh a \cosh b$$

where $\cosh x$ is defined as $(e^x + e^{-x})/2$. Test this new formula for $AB^* = c$, $BC^* = a$, and $AC^* = b$ using the calculations made in (a).

12. The lines ℓ: $x^2 + y^2 = 9$ and m: $x^2 + y^2 = 21$ in $\boldsymbol{H}$ have the positive y-axis as a common h-perpendicular (line $\overleftrightarrow{AB}$ in the figure). Thus, $\ell \parallel m$. In Euclidean geometry, the semicircles ℓ and m are concentric, and are everywhere equidistant.

(a) Show that this is not the case for the h-metric. (Specifically, verify that $CD^* > AB^*$, where $\overline{CD}^*$ is the semicircle $x^2 + y^2 - 10x = -9$.)

(b) Verify that the circle $x^2 + y^2 = 9$ is orthogonal to $x^2 + y^2 = 10x = -9$. What kind of figure is $\diamond ABCD$?

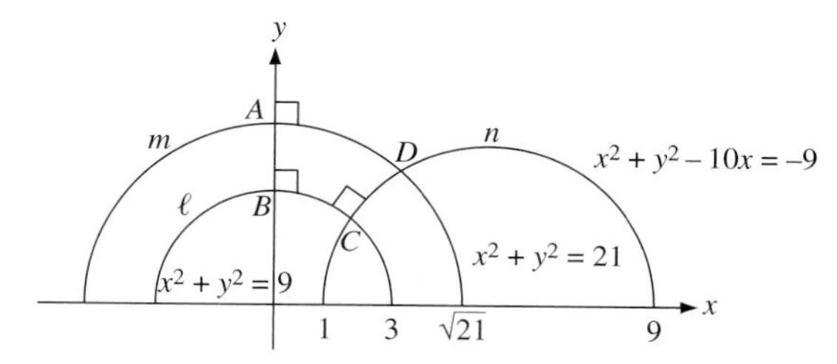

13. (Refer to Problem 12.) By symmetry, we can infer that since n': $x^2 + y^2 + 10x = 9$ is the Euclidean reflection of n: $x^2 + y^2 - 10x = 9$ in the y-axis, n' is h-perpendicular to ℓ. Suppose n' cuts ℓ and m at the points F and E. Carefully draw this figure, and observe its features. What kind of quadrilateral is $\diamond CDEF$? What can be said about its angle measures?

14. **Experimenting with h-Angles using *Sketchpad* Script** To conduct this experiment efficiently in *Sketchpad*, a script for h-angle measure needs to be created. Here are the instructions. (See figure on next page.)

[1] Construct a horizontal boundary line ℓ near the bottom of your sketch, and hide its control points.

[2] Construct three points C, D, and E.

[3] Open a new Script under FILE and select Record (REC).

[4] Construct segment $\overline{CD}$ and its midpoint F. Then construct segment $\overline{DE}$ and its midpoint G. (It is important that you select the points in this script in the order C, D, and E if you want to measure the h-angle $\angle CDE^*$.)

[5] Construct the perpendiculars at F and G to $\overline{CD}$ and $\overline{DE}$, respectively, and locate their intersections H and I with ℓ (boundary line).

[6] Construct segment $\overline{HD}$ and the perpendicular to it at D. With the perpendicular still selected, select C with shift key depressed and construct the perpendicular to the preceding line from C. Select $C'(= J)$, the intersection of the last two perpendicular lines constructed.

[7] Repeat Step 7 for segment $\overline{ID}$ and point E to obtain $E'(= K)$, the foot of the perpendicular from E to the perpendicular to $\overline{ID}$ at D.

[8] Select points J, D, and K to measure $\angle JDK$ ($= m\angle C'DE'$ = h-measure of $\angle CDE$). Hide all objects except the original three points C, D, and E and the

boundary line ℓ. Select Stop Recording. Save this script under an appropriate name such as "H-Angle" for Windows, or Hyperbolic Angle Measure for MacIntosh.

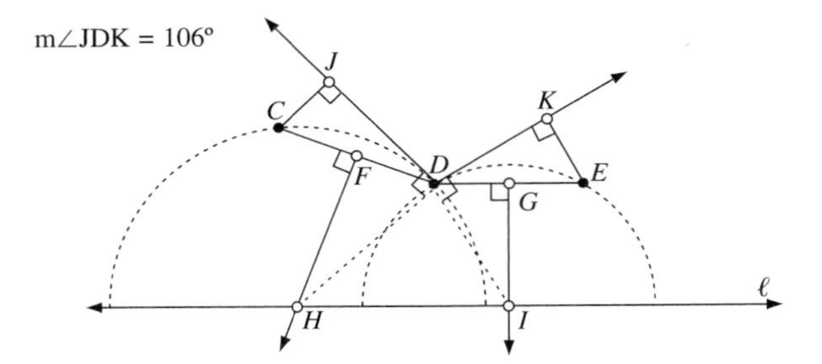

You now have a script that will yield the h-measure of any h-angle $\angle PQR^*$ by selecting P, Q, R, and ℓ, and selecting FAST in the Script menu. Using this script, and the previous one that constructs h-segments, construct an h-triangle and calculate its angle sum x. Drag the vertices to observe how the size of the triangle affects x. Write down any conclusions you find. (Be sure to drag the points to create a very small triangle, and a fairly large triangle, to see what happens in those cases.)

15. **Experimenting with h-Distance using *Sketchpad* Script** A script for h-distance can be created as follows (see figure on top of next page):

[1] Construct a horizontal boundary line ℓ near the bottom of your sketch, and hide the control points.

[2] Construct two points C and D.

[3] Open a new script under FILE and select Record.

[4] Construct segment $\overline{CD}$, its midpoint, and perpendicular bisector and select the intersection F of this perpendicular with line ℓ.

[5] Construct circle FC with F as center and passing through points C and D. Select the points of intersection of this circle, G and H, with ℓ.

[6] Calculate the distances:

$$\begin{array}{ll} CG, & DH \\ CH, & DG \end{array}$$

[7] Calculate the quantity:

$$\text{abs}\left(\ln\left(\frac{CG \cdot DH}{CH \cdot DG}\right)\right)$$

[8] Hide all but the last measurement, and all objects except C, D, and line ℓ. Stop recording.

You now have a script that will calculate the h-distance between any two points by selecting those points and line ℓ. Save under an appropriate name such as "H-Metric" or "Hyperbolic Distance." Now construct a horizontal line segment $\overline{PQ}$ and a copy of this below it, $\overline{RS}$. These two segments are congruent under the Euclidean metric. Using your script for h-distance, find PQ^* and RS^*. Do you notice anything? Drag segment $\overline{RS}$ to various points in the plane and observe the changing value of RS^*. What effect do you observe? Write down all observations you make.

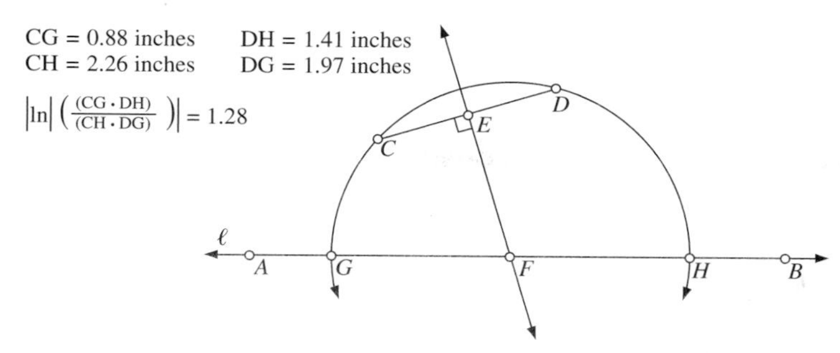

The following sequence of problems 16-22 develops, via the half-plane model, the theory of asymptotic triangles.

16. Define an **asymptotic triangle** in ***H*** as the union of an h-segment $\overline{AB}^*$ and two h-rays $\overrightarrow{AC}^*$ and $\overrightarrow{BD}^*$ that meet at the point Ω on the x-axis, as shown in the figure. Such a triangle will be denoted $\triangle AB\Omega$. Note that this "triangle" has only two vertices in ***H*** (the missing vertex is the point Ω on the x-axis). It has three sides, but only one of them is a segment. The only measurable objects, therefore, are the **base** (segment $\overline{AB}^*$) and the **base angles** ($\angle ABD^*$ and $\angle BAC^*$).

(a) Extend such a definition to the Poincaré disk model, and make a sketch of several asymptotic triangles in that model.

(b) Prove that in the upper half-plane model side $\overrightarrow{AC}^*$ has these two characteristic properties: (1) $\overrightarrow{AC}^*$ does not meet side $\overrightarrow{BD}^*$, and (2) every ray $\overrightarrow{AP}$ in the interior of $\angle BAC^*$ meets side $\overrightarrow{BD}.^*$

NOTE: In axiomatic hyperbolic geometry, the ray $\overrightarrow{AC}$ is defined as the *limiting position* of ray $\overrightarrow{AP}$, for $P \in \overrightarrow{BD}$ as BP becomes infinite.

AN ASYMPTOTIC TRIANGLE IN *H*

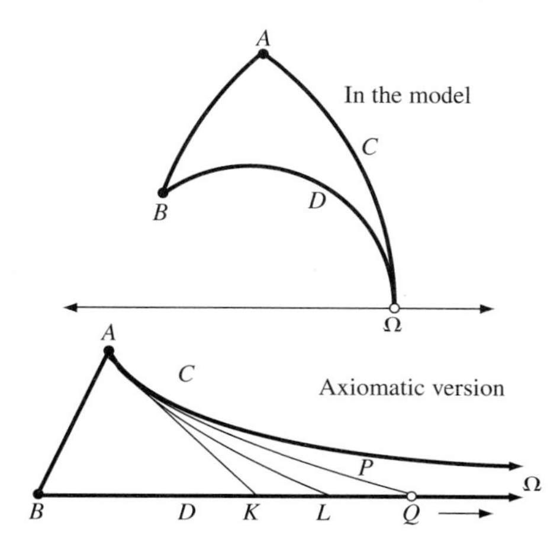

17. A **right asymptotic triangle** is an asymptotic triangle for which one of its base angles is a right angle. An **isosceles asymptotic triangle** is one for which the base angles are congruent. Show how to construct, either in theory or with a compass, straight-edge,

(a) a right asymptotic triangle

(b) an isosceles asymptotic triangle.

18. Can an asymptotic triangle in $\boldsymbol{H}$ have two right angles? Test this using *Sketchpad* and the h-angle script you created in Problem 14, or draw a right asymptotic triangle $\triangle AB\Omega$ with $\overleftrightarrow{AB}$ = y-axis and $\angle BA\Omega$ a right angle, then determine how (or whether) you could draw a semicircle $B\Omega$ orthogonal to the y-axis.

GROUP C

19. To give a conclusive answer to the question raised in Problem 18, it is more appropriate to use the synthetic method to prove the proposition that an asymptotic triangle can have at most one right angle. Suppose $\triangle AB\Omega$ is an asymptotic triangle having both $\angle A$ and $\angle B$ as right angles (see figure). Since $\overrightarrow{AC} \parallel \overrightarrow{BD}$ and there must be a second parallel ℓ to $\overrightarrow{BD}$ passing through A, assume that P is a point on that second parallel lying on the C-side of $\overleftrightarrow{AB}$.

(a) If P lies interior to $\angle BAC$, what must happen to ray $\overrightarrow{AP}$ that leads to an immediate contradiction? [See Problem 16(b).]

(b) Then $P \in$ Interior $\angle B'AC$, where B'–A–B, which makes $m\angle B'AP = m\angle QAB < 90$, where Q–A–P. In this case, construct ray $\overrightarrow{AR}$ on the C-side of $\overleftrightarrow{AB}$ so that $\angle BAR \cong \angle BAQ$. Why must ray $\overrightarrow{AR}$ meet $\overrightarrow{BD}$ at some point S? Now construct T such that T–B–S and $TB = BS$, and prove a contradiction.

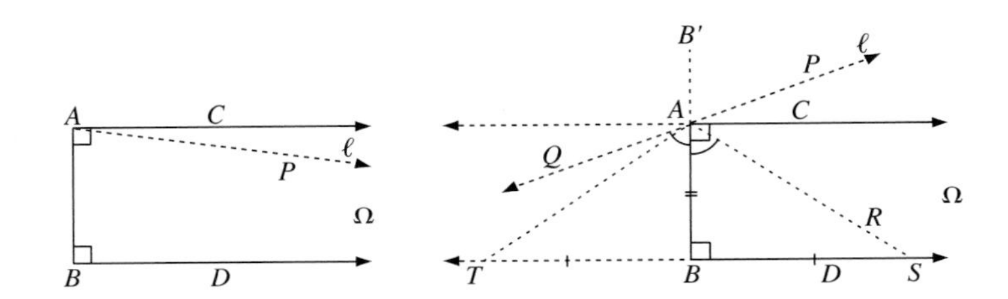

20. Prove in an asymptotic triangle $\triangle AB\Omega$, $m\angle A + m\angle B < 180$. [***Hint:*** If the proposition is false, there are two cases: (1) $m\angle A + m\angle B = 180$, and (2) $m\angle A + m\angle B > 180$. In the first case, construct the midpoint M of $\overline{AB}$, drop the perpendicular $\overline{ME}$ to line $\overleftrightarrow{BD}$, and let F be the intersection of $\overleftrightarrow{ME}$ and $\overleftrightarrow{AC}$, where F–M–E and F–A–C hold. Prove that $\overline{FE} \perp \overline{FC}$? What does this contradict? In the second case, construct $\angle BAE$ supplementary to $\angle ABD$, with $\overrightarrow{AE}$ interior to $\angle CAB$, and prove a contradiction.]

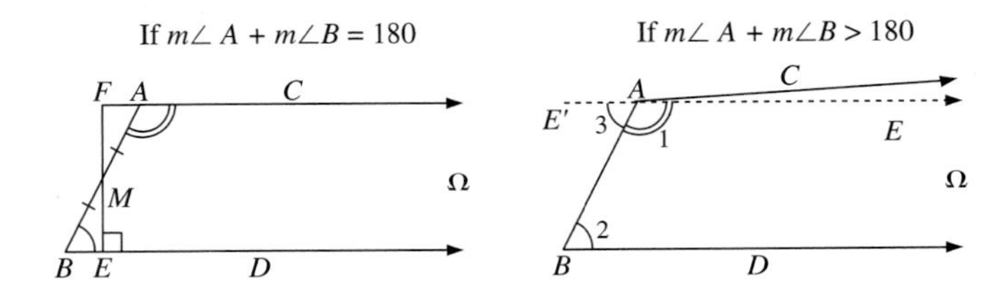

21. Prove the BA Congruence Criterion for asymptotic triangles: *If $\triangle AB\Omega$ and $\triangle XY\Gamma$ are two asymptotic triangles with $AB = XY$ and $m\angle A = m\angle X$, then $m\angle B = m\angle Y$.*

22. Prove the AA Congruence Criterion for asymptotic triangles: *If* $\triangle AB\Omega$ *and* $\triangle XY\Gamma$ *are two asymptotic triangles with* $m\angle A = m\angle X$ *and* $m\angle B = m\angle Y$, *then* $AB = XY$.

23. Criterion for h-Perpendicularity in the Coordinate Plane

(a) Using **(4)**, page 448, for the criterion for orthogonal circles that led to the criterion for a circle centered at $D(h, k)$ to be orthogonal to the unit circle **C**, show that the circle $x^2 + y^2 + ax = b$ centered at $O(-a/2, 0)$ with radius $r^2 = b + a^2/4$ is orthogonal to the circle $x^2 + y^2 + cx = d$ centered at $D(-c/2, 0)$ with radius $s^2 = d + c^2/4$ iff

$$b + d = -ac/2 \qquad \text{or} \qquad d = -b - ac/2$$

(***Hint:*** $OD^2 = OP^2 + PD^2 = r^2 + s^2$ and $OD = |a/2 - c/2|$.)

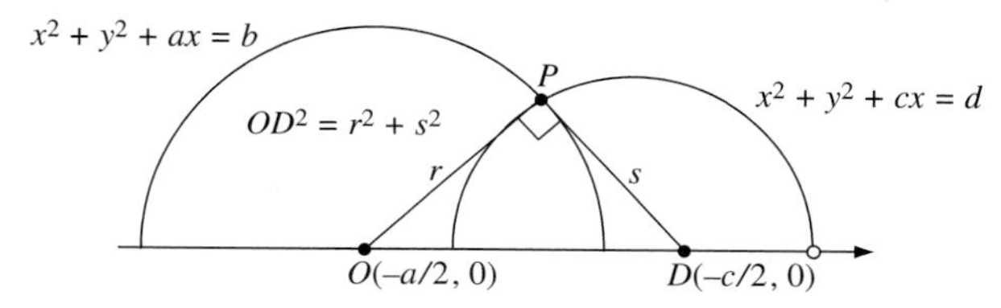

Then, by substituting $-b - ac/2$ for d in the second equation, it follows that

$$\textbf{(10)}\quad \ell: x^2 + y^2 + ax = b \quad \perp \quad m: x^2 + y^2 + cx = -b - ac/2$$

for essentially arbitrary a, b, and c. A useful special case is when ℓ has an equation of the form $x^2 + y^2 = r^2$ [with center at (0, 0)], or when $a = 0$:

$$\textbf{(10')}\qquad \ell: x^2 + y^2 = b \quad \perp \quad m: x^2 + y^2 + cx = -b$$

(b) There are infinitely many h-perpendiculars to line ℓ of **(10)**—one for each value of c. How many of these pass through a given point A?

(c) Find the equation of the h-line passing through $A(3, 2)$ perpendicular to $x^2 + y^2 = 8$.

The following sequence of problems 24–28 develops the concept of parallelism via the upper half-plane model.

24. **Parallelism in *H* using *Sketchpad*** Construct a horizontal boundary line ℓ near the bottom of your sketch, hide the control points A and B, and construct two other points C and D above ℓ. By the usual manner (using the perpendicular bisector of $\overline{CD}$), construct a semicircle through C and D. Hide all objects except h-line $\overleftrightarrow{CD}^*$ and ℓ. (Select points of intersection H and G of circle CD and ℓ, then the circle, and then choose Arc On Circle under CONSTRUCTION.) Locate any point I and J not on $\overleftrightarrow{CD}^*$ and repeat the previous construction to obtain the h-line $\overleftrightarrow{IJ}^*$. Observe the h-lines through J parallel to $\overleftrightarrow{CD}^*$. Drag I about until you observe a few. Experiment until you find two distinctly different types of parallels to line $\overleftrightarrow{CD}^*$ passing through J.

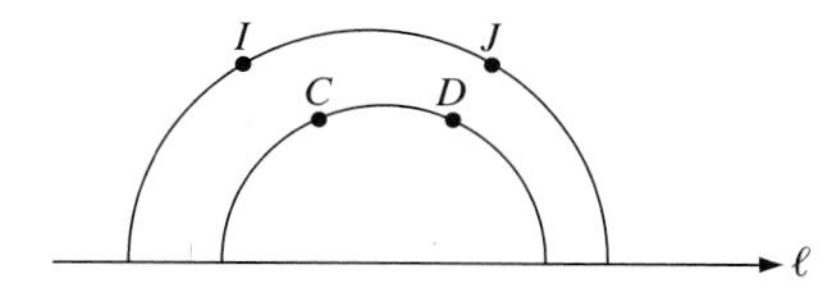

25. In axiomatic hyperbolic geometry, the two types of lines parallel to ℓ passing through A are (1) the **asymptotic parallels**, which are determined by the rays on either side of the perpendicular $\overline{AB}$ to ℓ ($B \in \ell$) that form asymptotic triangles $\triangle AB\Omega$ and $\triangle AB\Omega'$, and (2) all the rest, called **divergent parallels**. (The names **limit** and **hyperbolic** parallels are also used for these two types of parallels, respectively.) Make a sketch for these two types for

(a) the Poincaré disk model

(b) the upper half-plane model.

THE TWO TYPES OF PARALLELS TO ℓ THROUGH A

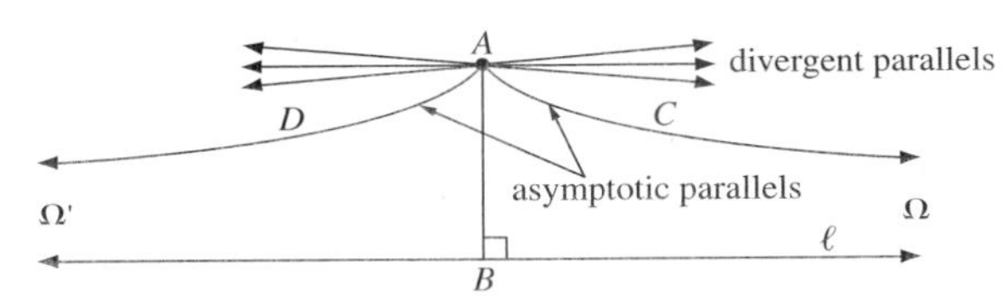

26. **(a)** Prove (or verify using *Sketchpad*) that $m\angle BAC^* = m\angle BAD^* < 90$ in the figure for Problem 25.

(b) Suppose $\overline{XY} \perp \overline{YP}$ and ray $\overrightarrow{XZ}$ is asymptotically parallel to $\overrightarrow{YP}$. Using properties of asymptotic triangles, prove that

(1) If $XY^* = AB^*$ then $m\angle YXZ^* = m\angle BAC^*$.

(2) If $m\angle YXZ^* = m\angle BAC^*$ then $XY^* = AB^*$.

NOTE: The measure of $\angle BAC^*$ in **(a)** is called the **angle of parallelism** with respect to the distance AB^*. (If you change AB^*, you change $m\angle BAC^*$.) Both Bolyai and Lobachevski independently discovered the formula

$$\tan \frac{1}{2}A = e^{-a} \tag{11}$$

relating the angle of parallelism A that corresponds to the distance $AB^* = a$. (You might want to try this formula out on a few examples using *Sketchpad*.)

27. In connection with Problem 26, can the angle of parallelism be made

(a) arbitrarily small?

(b) arbitrarily close to 90?

Investigate these questions in $\boldsymbol{H}$ using *Sketchpad*, and, if you can, back up your observations using the formula (**11**) for the angle of parallelism given in Problem 26.

28. **Metric Characterization of Parallelism** Given point A and line ℓ, let m be a line through A parallel to ℓ. Consider the following theorem that can be proven axiomatically, with the notation and construction appearing in the figure for this problem, where $\overline{AB} \perp \ell$ and $\overline{PQ} \perp \ell$.

THEOREM: As P varies on line m such that $AP^* \to \infty$, then $PQ^* \to 0$ iff m is asymptotically parallel to ℓ, and $PQ^* \to \infty$ iff m is divergently parallel to ℓ.

Verify this theorem for the following example in the upper half-plane model: Let ℓ be the y-axis, $Q = (0, u)$, $\overline{PQ}^*$ the semicircle $x^2 + y^2 = u^2$, and

(a) m: $x^2 + y^2 = x$ for m asymptotically parallel to ℓ [in this case show that $P(x, y)$ must have coordinates $x = u^2$, $y^2 = -u^4 + u^2$, that $u \to 0$, and $PQ^* \to 0$ as $P \to O$]

(b) m: $(x - 2)^2 + y^2 = 1$ or $x^2 + y^2 - 4x = -3$ for m divergently parallel to ℓ [in this case show that $P(x, y)$ must have $x = \frac{1}{4}(u^2 + 3)$, $y = \frac{1}{4}\sqrt{-u^4 + 10u^2 - 9}$, and that $u \to 1$ as $P \to C$]. The algebra for this case gets rather messy. You might prefer an axiomatic proof depicted in the figure, where line n is constructed through B asymptotically parallel to m, then Proclus' result of Problem 19, Section 6.2 is used.

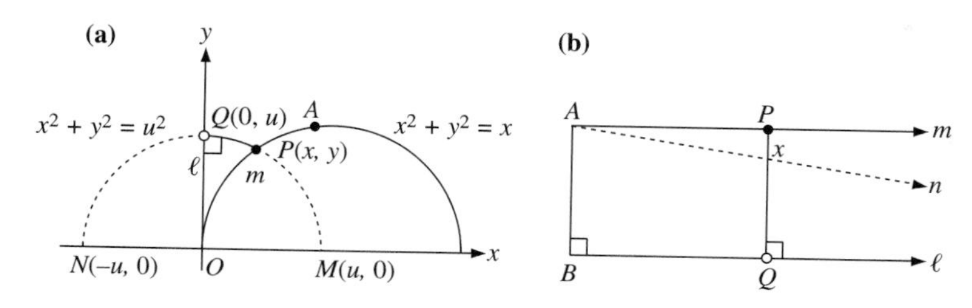

29. Hilbert's Theorem We know in absolute geometry that if two lines have a common perpendicular they are parallel (in fact, in hyperbolic geometry they must be divergently parallel). The converse of this was proven by David Hilbert (1862–1943):

THEOREM: If line ℓ is divergently parallel to line m, then there can be found a third line n that is perpendicular to both ℓ and m.

Verify this theorem in the upper half-plane model if ℓ has equation $x^2 + y^2 = r^2$ and m, $x^2 + y^2 + ax = b$, $a \neq 0$. [Find the parameters c and d in the equation of the common perpendicular n: $x^2 + y^2 + cx = d$ in terms of a, b, and r using **(10)**, Problem 23.]

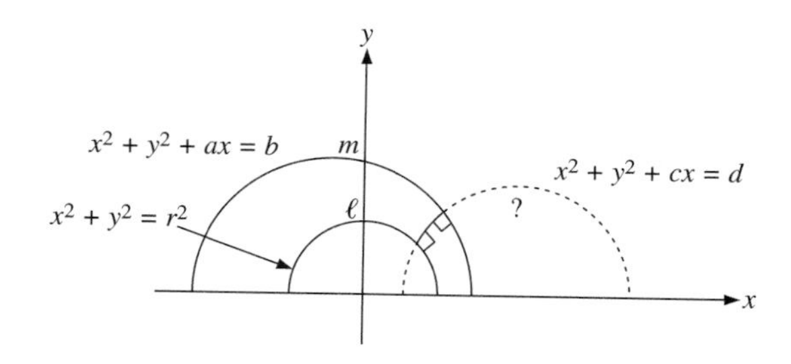

*6.5 Circular Inversion: Proof of SAS Postulate for Half-Plane Model

To prove the SAS Postulate in the Beltrami–Poincaré model, we need show only that a motion with an arbitrary axis ℓ exists. (We shall call such a motion an **h-reflection** in line ℓ.) To achieve this, we use the concept of **circular inversion,** defined below. The rest follows from our discussion in Section 5.7, for, as we shall see,

(a) a circular inversion preserves Cross Ratio, hence h-distance;

(b) a circular inversion is conformal relative to curvilinear angle measure, hence preserves h-angle measure;

(c) Euclid's superposition argument, as discussed in Section 5.1 and presented formally in Section 5.7, proves SAS in any geometry that is rich enough to allow reflections in arbitrary lines.

CIRCULAR INVERSION

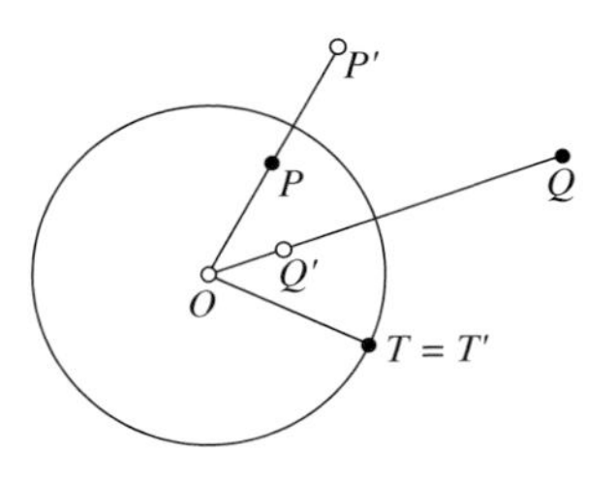

Figure 6.39

To invert a number in arithmetic usually means to "take its reciprocal." A closely related idea in geometry is that of "inverting a point." Suppose O is the origin and P is any other point (Figure 6.39). Then point P' may be located on ray $\overrightarrow{OP}$ such that the distance from P' to O is the *reciprocal* of the distance from P to O. That is,

$$OP' = \frac{1}{OP}$$

Now suppose we attempt to find the coordinates of P' assuming we know those of $P = P(x, y)$. Since $P'(x', y')$ lies on the ray $\overrightarrow{OP}$, the dilation $x' = kx, y' = ky$ must map point P to point P' for some $k > 0$. To determine k, use $OP' \cdot OP = 1$; then

$$\sqrt{x'^2 + y'^2} \cdot \sqrt{x^2 + y^2} = \sqrt{(kx)^2 + (ky)^2}\,\sqrt{x^2 + y^2}$$
$$= 1 \quad \rightarrow \quad \sqrt{k^2(x^2 + y^2)^2} = 1$$

or, since $k > 0$,

$$k(x^2 + y^2) = 1 \quad \rightarrow \quad k = 1/(x^2 + y^2)$$

Hence, substituting this value for k,

$$x' = kx = \frac{x}{x^2 + y^2} \qquad y' = ky = \frac{y}{x^2 + y^2}$$

This can be generalized by starting with the condition $OP' \cdot OP = r^2$ for any $r > 0$, in place of $OP' \cdot OP = 1$. The resulting equations would then have an extra factor of r^2.

> DEFINITION: The mapping of each point $P(x, y) \neq O(0, 0)$ in the plane to the unique *image point* $P'(x', y')$, where
>
> **(1)** $$x' = \frac{r^2 x}{x^2 + y^2} \quad \text{and} \quad y' = \frac{r^2 y}{x^2 + y^2}$$
>
> is called a **circular inversion.** Point P' is called the **inverse** of P, the circle $x^2 + y^2 = r^2$ is called the **circle of inversion,** and O and r are, respectively, the **center** and **radius** of inversion. Any curve or line $\boldsymbol{C}$ mapped by inversion to $\boldsymbol{C'}$ is called its **inverse image,** or simply, **inverse.** (See Figure 6.39.)

Note that if P lies on the circle of inversion, then $x^2 + y^2 = r^2$ and the equations in (**1**) yield $x' = r^2x/r^2$, $y' = r^2y/r^2$, or $x' = x$ and $y' = y$. Thus $P = P'$. Also, since $OP \cdot OP' = r^2$ and P' lies on ray $\overrightarrow{OP}$, the same inversion applied to P' produces P back again (an inversion *is its own inverse*). Hence, we obtain

(1′) $$x = \frac{r^2 x'}{x'^2 + y'^2} \quad \text{and} \quad y = \frac{r^2 y'}{x'^2 + y'^2}$$

One important property of inversions is the effect it has on lines and circles. From its definition, we know that a line through O will map to itself. Any other line has equation ℓ: $ax + by + c = 0$, with $c \neq 0$. By substitution of (**1′**) into this equation, we have

$$\frac{ar^2x'}{x'^2 + y'^2} + \frac{br^2y'}{x'^2 + y'^2} + c = 0 \qquad \text{or} \qquad ar^2x' + br^2y' + c(x'^2 + y'^2) = 0$$

which has the form (dividing by c)

$$x'^2 + y'^2 + dx' + ey' = 0 \qquad (d = ar^2/c,\ e = br^2/c)$$

This is the equation of a circle passing through O. If ℓ intersects the circle of inversion at A and B, then, since A and B map to themselves, ℓ maps to *the circle passing through A, B, and O*, and we could even give a compass, straight-edge construction of it. On the other hand, the inverse of a circle passing through O is a straight line. More generally, let

$$\boldsymbol{C}\colon x'^2 + y'^2 + ax' + by' = c$$

be an arbitrary circle in the plane. By substitution of (**1**),

$$\left(\frac{r^2x}{x^2 + y^2}\right)^2 + \left(\frac{r^2y}{x^2 + y^2}\right)^2 + \frac{ar^2x}{x^2 + y^2} + \frac{br^2y}{x^2 + y^2} = c$$

$$\frac{r^4(x^2 + y^2)}{(x^2 + y^2)^2} + \frac{ar^2x}{x^2 + y^2} + \frac{br^2y}{x^2 + y^2} = c$$

or

$$\frac{r^4}{x^2 + y^2} + \frac{ar^2x + br^2y}{x^2 + y^2} = c$$

which finally simplifies to

$$\boldsymbol{C}'\colon c(x^2 + y^2) - ar^2x - br^2y = r^4$$

which is another circle, provided $c \neq 0$. If the original circle passes through the origin, then $c = 0$ and the above reduces to a line, $-ax - by = r^2$ (which we already knew). Note that the slope of this line is given by $-a/b$. On the other hand, the center of the given circle is $C(-a/2, -b/2)$, so the slope of line $\overleftrightarrow{OC}$ is $(-b/2)/(-a/2) = b/a$, the negative reciprocal of $-a/b$. Thus, the line into which the circle maps is *perpendicular* to the line passing through the center of that circle and O. (See Figure 6.40.) We state this as a lemma, which will be used a little later.

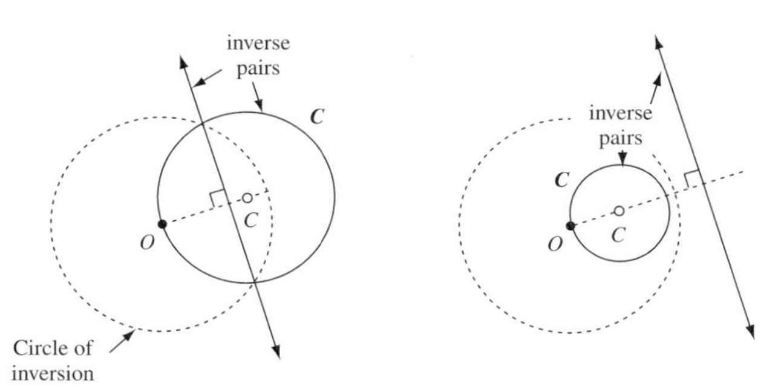

Figure 6.40

LEMMA A: A circular inversion with center O maps a line not passing through O to a circle through O, and the given line is perpendicular to the line that passes through O and the center of the circle.

The two crucial properties of inversion we need to demonstrate are (1) inversions preserve Cross Ratio, and (2) inversions preserve **curvilinear angle measure**—the angle between *tangents* to curves and their images, at the point of intersection.

LEMMA B: If A' and B' are the inverses of A and B with respect to the circle $x^2 + y^2 = r^2$, then $\triangle A'B'O \sim \triangle BAO$, and $A'B' = AB \cdot A'O/BO$. (See Figure 6.41.)

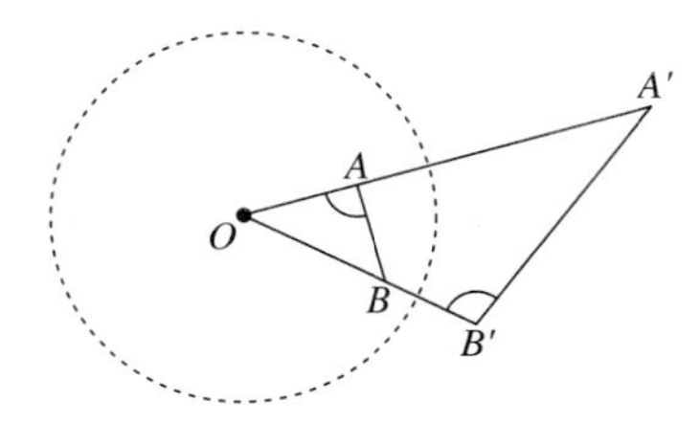

Figure 6.41

PROOF

Since $OA' = r^2/OA$ and $OB' = r^2/OB$, we have $OA'/OB' = OB/OA$. But $\angle A'OB' \cong \angle BOA$, so by the SAS Similarity Criterion, $\triangle A'B'O \sim \triangle BAO$. Therefore, corresponding side lengths are in proportion, so $A'B'/AB = A'O/BO$ or $A'B' = AB \cdot A'O/BO$.

Apply this result to the Cross Ratio of the inverse of any four points, and we have

$$\frac{A'C'}{A'D'} = \frac{AC \cdot A'O/CO}{AD \cdot A'O/DO} = \frac{AC/CO}{AD/DO} = \frac{AC \cdot DO}{AD \cdot CO}$$

and

$$\frac{B'D'}{B'C'} = \frac{BD \cdot B'O/DO}{BC \cdot B'O/CO} = \frac{BD \cdot CO}{BC \cdot DO}$$

Therefore,

$$(A'B', C'D') = \frac{A'C' \cdot B'D'}{A'D' \cdot B'C'} = \frac{AC \cdot DO}{AD \cdot CO} \cdot \frac{BD \cdot CO}{BC \cdot DO}$$

$$= \frac{AC \cdot BD}{AD \cdot BC} = (AB, CD)$$

This proves

THEOREM 1: Circular inversion preserves the Cross Ratio of any four points.

At this point, let us summarize what we know about circular inversion. For convenience, we shall include the result of the next theorem as well.

Properties of Circular Inversion (C = circle of inversion, O = center)

1. Points inside C map to points outside of C, points outside map to points inside, and each point on C is self-inverse (maps to itself).
2. Lines or circles map to lines or circles.
3. A line through O is invariant, but the individual points of that line are changed.
4. A circle through O maps to a line not through O, and the image line is perpendicular to the line that passes through O and the center of the given circle (Lemma A).
5. Cross Ratio is invariant under circular inversion (Theorem 1).
6. A circular inversion is **conformal,** that is, it preserves curvilinear angle measure (Theorem 2).

We now state and prove Theorem 2.

THEOREM 2: A circular inversion preserves curvilinear angle measure.

PROOF

We must show that if C_1 and C_2 are any two curves meeting at point P, and t_1 and t_2 are their respective tangents at P, the angle θ between those tangents is equal to that of the corresponding tangents to the curves C'_1 and C'_2 (Figure 6.42). Assuming $P \neq O$ (origin), the inversion maps t_1 and t_2 to circles t'_1 and t'_2 through O, and the angle θ' between those two circles at P' equals the angle ϕ at O. Now by Lemma A, the radii of circles t'_1 and t'_2 are perpendicular to the inverse images of those circles, lines t_1 and t_2. That is, lines t_1 and t_2 are parallel to the tangents of t'_1 and t'_2 at O. Hence $\theta = \phi = \theta'$, as desired.

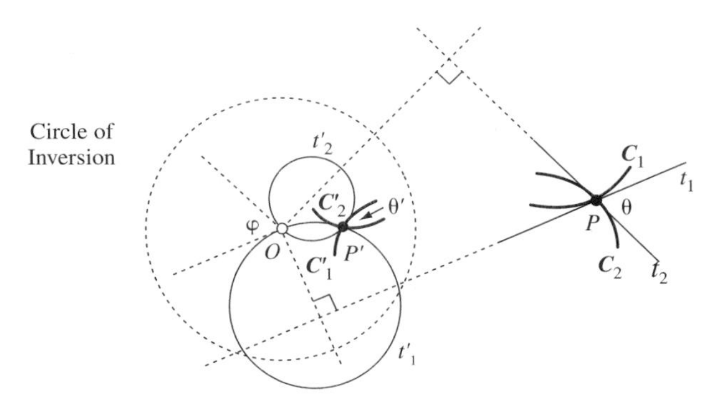

Figure 6.42

REFLECTIONS IN h-LINES

To construct an **h-reflection** in the h-line $\ell: x = a$ for some a, merely use a Euclidean reflection in that line, which clearly maps points above the x-axis to points above the x-axis (the points of $\boldsymbol{H}$), and preserves all other geometric properties of the upper half-plane model, including h-distance and h-angle measure. To construct an **h-reflection** in the h-line $\ell: x^2 + y^2 + ax = b$, first we simply make a translation of coordinates so that the new origin is the center of ℓ, and perform the inversion as before, with all the same invariants as in the above list, in particular, (5) and (6). Figure 6.43 shows a few examples of the result of such a "reflection."

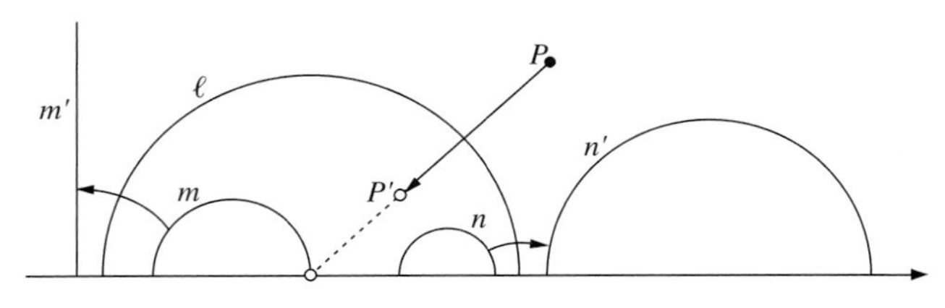

Figure 6.43

THEOREM 3: An h-reflection exists with respect to any h-line, which preserves h-distance and h-angle measure, and maps h-lines to h-lines, h-segments to h-segments, and h-angles to h-angles.

COROLLARY: The SAS Postulate holds for the upper half-plane model.

EXAMPLE 1 Points $A(2, 4)$ and $B(2, 6)$ are reflected in the h-line $\ell: x^2 + y^2 = 80$ to the points A' and B', as shown in Figure 6.44.

(a) Find the coordinates of A' and B'.
(b) Verify that $AB^* = A'B'^*$ by direct calculation

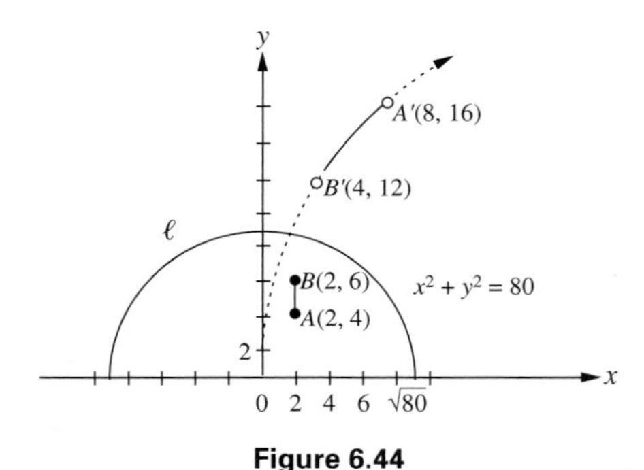

Figure 6.44

SOLUTION

(a) Since $r^2 = 80$, the inversion equations are, from **(1)**

$$x' = \frac{80x}{x^2 + y^2} \qquad y' = \frac{80y}{x^2 + y^2}$$

Substitute the coordinates of A and B into these two equations:

$$A: \quad x' = \frac{80 \cdot 2}{2^2 + 4^2} = \frac{160}{4 + 16} = \frac{160}{20} = 8 \qquad y' = \frac{80 \cdot 4}{2^2 + 4^2} = \frac{320}{20} = 16$$

$$B: \quad x' = \frac{80 \cdot 2}{2^2 + 6^2} = \frac{160}{4 + 36} = \frac{160}{40} = 4 \qquad y' = \frac{80 \cdot 6}{2^2 + 6^2} = \frac{480}{40} = 12$$

$$\therefore A' = (8, 16) \text{ and } B' = (4, 12).$$

(b) $$AB^* = \left|\ln\left(\frac{BM'}{AM'}\right)\right| = \ln\frac{6}{4} = \ln\frac{3}{2} \qquad [M' = (2, 0)]$$

To find $A'B'^*$ we must first find the equation for the h-line $\overleftrightarrow{A'B'}^*$. Let its equation be $x^2 + y^2 + ax = b$, and substitute the coordinates of A' and B' into it. The final result is $x^2 + y^2 - 40x = 0$. (You can just test the coordinates in this equation to verify it.) We must now find the end points M and N of $\overleftrightarrow{A'B'}^*$. Set $y = 0$ and solve for x: $x^2 - 40x = 0$ or $x(x - 40) = 0$. Hence $M = (0, 0)$ and $N = (40, 0)$.

$$A'B'^* = \ln(A'B', MN) = \ln\frac{A'M \cdot B'N}{A'N \cdot B'M}$$

$$= \frac{1}{2}\ln\frac{(8^2 + 16^2)[(4 - 40)^2 + 12^2]}{[(8 - 40)^2 + 16^2](4^2 + 12^2)}$$

$$= \frac{1}{2}\ln\frac{(64 + 256)(1296 + 144)}{(1024 + 256)(16 + 144)} = \frac{1}{2}\ln\frac{320 \cdot 1440}{1280 \cdot 160} = \ln\sqrt{\frac{9}{4}} = \ln\frac{3}{2}$$

$\therefore A'B'^* = AB^*$. ■

OUR GEOMETRIC WORLD

In Euclidean geometry we may change the scale of a drawing without affecting the drawing itself. A triangle three times the size of a given right triangle is still a right triangle. In hyperbolic geometry, however, a change of scale also involves a change in angle measure for all figures involved. A triangle three times the size of a given right triangle is no longer a right triangle. All three angles will decrease, as indicated in Figure 6.45. All measurements shown in the figure are the actual results of formulas in hyperbolic geometry.

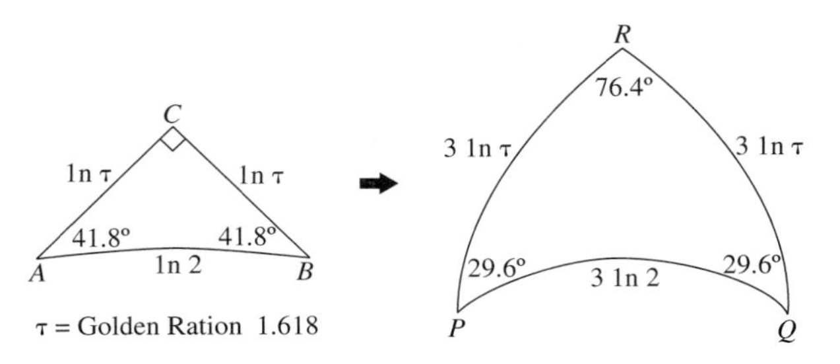

Figure 6.45

The various angle measures appearing in the preceding example (Figure 6.45) can be calculated using **hyperbolic trigonometry.** (See Problem 13.) This system of formulas was derived by both Bolyai and Lobachevski, and can also be proven, or verified, in the upper half-plane model. The formulas given below make use of the so-called **hyperbolic functions** often found in calculus texts, and defined below (x denotes a real number).

(2) $$\sinh x = \tfrac{1}{2}(e^x - e^{-x})$$

(3) $$\cosh x = \tfrac{1}{2}(e^x + e^{-x})$$

(4) $$\tanh x = \frac{\sinh x}{\cosh x} = \frac{e^x - e^{-x}}{e^x + e^{-x}} = \frac{e^{2x} - 1}{e^{2x} + 1}$$

The other three functions seem familiar from trigonometry, but they are really definitions:

(5) $\operatorname{csch} x = 1/\sinh x$, $\quad \operatorname{sech} x = 1/\cosh x$, $\quad$ and $\quad \coth x = 1/\tanh x$

Note that

$$\sinh 0 = \tfrac{1}{2}(1 - 1) = 0, \quad \cosh 0 = \tfrac{1}{2}(1 + 1) = 1, \quad \text{and } \tanh 0 = 0$$

Identities involving these functions bear a striking resemblance to trigonometric identities even though the hyperbolic functions themselves bear absolutely no resemblance to their trigonometric counterparts. A few of these are

(6) $$\cosh^2 x - \sinh^2 x = 1$$

(7) $$1 - \tanh^2 x = \operatorname{sech}^2 x$$

(8) $$\sinh(x + y) = \sinh x \cosh y + \cosh x \sinh y$$

(9) $$\sinh 2x = 2 \sinh x \cosh x$$

An important property of the hyperbolic cosine function to be used later follows from the fact that a number and its reciprocal are always > 2 unless the number itself is 1. Thus,

(10) $$\cosh x > 1 \qquad (x \neq 0)$$

Hyperbolic Trigonometry of the Right Triangle[3]

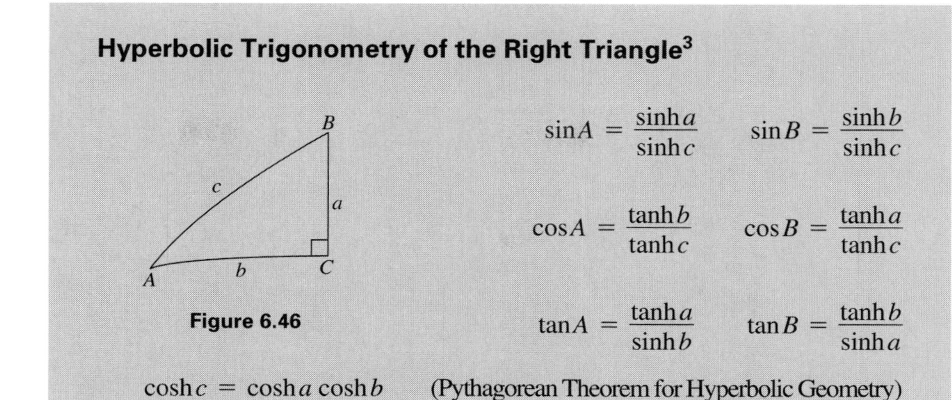

Figure 6.46

$$\sin A = \frac{\sinh a}{\sinh c} \qquad \sin B = \frac{\sinh b}{\sinh c}$$

$$\cos A = \frac{\tanh b}{\tanh c} \qquad \cos B = \frac{\tanh a}{\tanh c}$$

$$\tan A = \frac{\tanh a}{\sinh b} \qquad \tan B = \frac{\tanh b}{\sinh a}$$

$$\cosh c = \cosh a \cosh b$$ (Pythagorean Theorem for Hyperbolic Geometry)

To solve problems in hyperbolic trigonometry, the following inverse hyperbolic functions will be needed, which are frequently derived in calculus.

(11) $$\sinh^{-1} x = \ln(x + \sqrt{x^2 + 1})$$

(12) $$\cosh^{-1} x = \ln(x + \sqrt{x^2 - 1})$$

(13) $$\tanh^{-1} x = \tfrac{1}{2} \ln\left(\frac{1 + x}{1 - x}\right)$$

EXAMPLE 2 Use hyperbolic trigonometry to solve the following problems involving a right triangle in ***H***. (See Figure 6.47.)

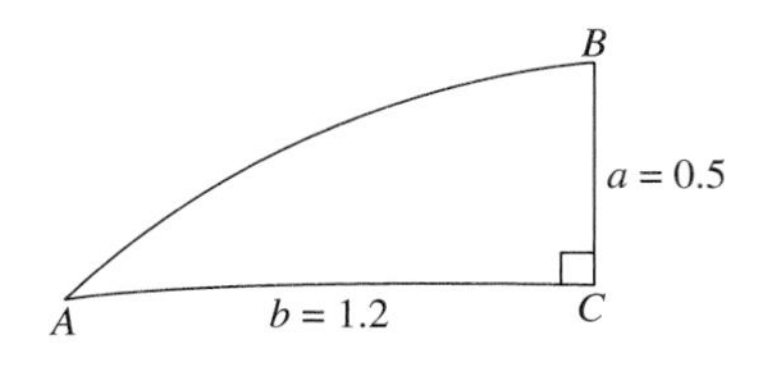

Figure 6.47

(a) Find the length of the hypotenuse of a right triangle whose legs are of length 0.5 and 1.2. Give answers that are accurate to three decimals.
(b) Find the measures of the two acute angles, accurate to 1/10 (degree).
(c) Calculate the angle sum of the triangle.

---

[3]These formulas have their direct counterpart to those of spherical trigonometry, which are remarkably similar. The latter will be derived later, in Section 7.6.

SOLUTION

(a) Set $a = 0.5$ and $b = 1.2$. Then

$$\cosh c = \cosh 0.5 \cosh 1.2 \approx (1.12763)(1.81066) \approx 2.04175$$

$$c = \cosh^{-1} 2.04175 = \ln(2.04175 + \sqrt{2.04175^2 - 1} \approx 1.34074$$

Therefore, $c = 1.341$ to three decimals.
(We might note that in Euclidean geometry the result would have been $c = \sqrt{0.5^2 + 1.2^2} = 1.3$ exactly.)

(b) $\sin A = \dfrac{\sinh a}{\sinh c} = \dfrac{\sinh 0.5}{\sinh 1.34074} \approx 0.29273$ Therefore, $A \approx 17.0°$

$\sin B = \dfrac{\sinh b}{\sinh c} = \dfrac{\sinh 1.2}{\sinh 1.34074} \approx 0.84796$ Therefore, $B \approx 58.0°$

(c) Angle sum $= 17.0 + 58.0 + 90 = 165.0$. ■

EXAMPLE 3 Derive the following formula relating the acute angles of a right triangle to the length of the hypotenuse in hyperbolic geometry, and use it to show that the angle sum of a right triangle is less than 180.

(14) $$\cosh c = \cot A \cot B$$

SOLUTION
Begin by taking the reciprocals of the formulas for $\tan A$ and $\tan B$:

$$\cot A = \frac{\sinh b}{\tanh a} = \frac{\sinh b}{\sinh a / \cosh a} = \frac{\sinh b \cosh a}{\sinh a} \qquad \cot B = \frac{\sinh a \cosh b}{\sinh b}$$

(interchanging a and b in the second formula). Multiplication yields

$$\cot A \cot B = \frac{\sinh b \cosh a}{\sinh a} \cdot \frac{\sinh a \cosh b}{\sinh b} = \cosh a \cosh b = \cosh c$$

which proves **(14)**

Since $\cosh c > 1$, then $\cot A \cot B > 1$. We know that $A + B \leq 90$, so if $A + B = 90$ then

$$\cot A \cot B = \cot A \cot(90 - A) = \cot A \tan A = 1 \rightarrow\leftarrow$$

Therefore, $A + B < 90$ and the angle sum of the triangle is < 180. ■

EXAMPLE 4 In hyperbolic geometry there is only one triangle, up to congruence, having angles of measure 20, 60, and 90. Verify this fact by actually finding the sides of such a triangle (to three-place accuracy) using hyperbolic trigonometry and **(14)**.

SOLUTION
We begin with **(14):**

$$\cosh c = \cot 60° \cot 20° \approx 1.58626$$

$$c = \cosh^{-1} 1.58626 = \ln(1.58626 + \sqrt{1.58626^2 - 1}) \approx 1.03589$$

Therefore, $c = 1.036$. Using the formula for $\cos A$ we have (with $A = 60$)

$$\frac{1}{2} = \cos A = \frac{\tanh b}{\tanh 1.03589} \approx \frac{\tanh b}{0.77626}$$

or

$$\tanh b = \frac{1}{2}(0.77626) = 0.38813$$

$$b = \frac{1}{2}\ln(1 + 0.38813)/(1 - 0.38813) \approx 0.40960$$

Use the hyperbolic Pythagorean Theorem to find a. Thus $a = 0.928, b = 0.410, c = 1.036$. ■

OTHER MODELS FOR HYPERBOLIC GEOMETRY

Having found one model, it is not surprising, given the number of mathematicians who worked on this problem, that many others were constructed. We already introduced the half-plane model, widely used for research in hyperbolic geometry.

KLEIN'S MODEL

Another model, due to F. Klein, resembles Poincaré's model, but is not conformal. It is easier to describe, but both distance and angle measure must be defined using Cross Ratios. Conceptually, it consists of all the points inside the unit circle ***C***, with "lines" consisting of all chords of ***C***, and "segments," "rays," and "angles" all being the intersection of their Euclidean counterparts with the interior of ***C*** (Figure 6.48). We shall not define distance and angle measure for this model, but simply provide you with a reference where a complete, readable, development may be found: M. J. Greenberg, *Euclidean and Noneuclidean Geometries,* Chapter 7. The reason this model cannot be conformal is obvious: Any "triangle" in the model is an ordinary Euclidean triangle, which would have angle sum 180.

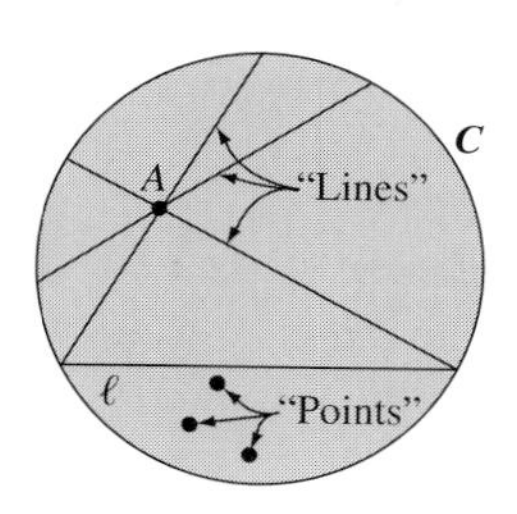

Figure 6.48

A most interesting fact is that Klein's model can be mapped onto the Poincaré Circular Disk Model ***H***. The latter model ***H*** can, in turn, be mapped into the Beltrami–Poincaré Half-Plane Model by simply taking a point T on ***C*** as a center of inversion, and inverting in some suitable circle centered at T. (See Problem 26.) As it turns out, all models of hyperbolic geometry are isomorphic, hence are logically equivalent. This bears witness to the fact that the three classical geometries (parabolic, hyperbolic, and elliptic) are **categorical**. (See Section 2.2 for a discussion of this term.)

We mention one other source of models. If you take a right circular cylinder or cone and roll it on a plane, as indicated in Figure 6.49, it will "pick up" the geometric properties of Euclidean geometry, as if we were inking all the Euclidean configurations onto the cylinder and cone. (The particular feature illustrated is the Euclidean property concerning an exterior angle of a triangle.) In fact, this rolling process suggests a way to define formally a one-to-one mapping from a portion of the Euclidean

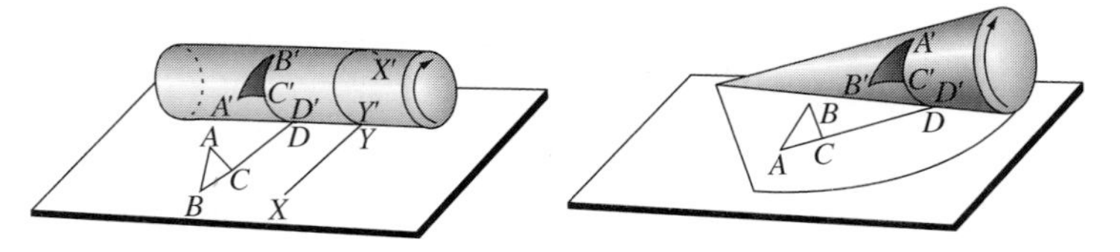

Figure 6.49

plane to the cylinder or cone. A straight line, for example, maps to either a circle or helix on the cylinder, and to a kind of spiraling helix on the cone.

What is interesting, however, is that the geometry of the cylinder and cone can be defined independently of this mapping by taking as metric the length of the shortest arc on these surfaces between two given points, and, as angle measure, the Euclidean angle measure between arcs of curves in three-dimensional space. The resulting geometry is called the **intrinsic geometry** of the cylinder or cone. In theory, this procedure can be carried out for any reasonably behaved surface S, yielding a two-dimensional **intrinsic geometry of** S. It turns out that the intrinsic geometries of both the cone and cylinder are locally isometric to the Euclidean plane, that is, small regions of the cylinder and cone are isometric to the Euclidean plane.

Analogously, the hyperbolic plane can be locally realized on a surface in Euclidean space, as discovered by Beltrami in 1872: He found a surface S having *constant negative curvature* (defined in terms of the curvatures of certain curves on S). Moreover, its intrinsic geometry was found to be locally isometric to the hyperbolic plane. This remarkable surface, called a **tractroid** or **pseudosphere**, is obtained by revolving the curve in the xy-plane

$$y = \ln\left(\frac{1 + \sqrt{1 - x^2}}{x}\right) - \sqrt{1 - x^2}, \qquad 0 < x \leq 1,$$

SURFACE MODEL FOR HYPERBOLIC GEOMETRY

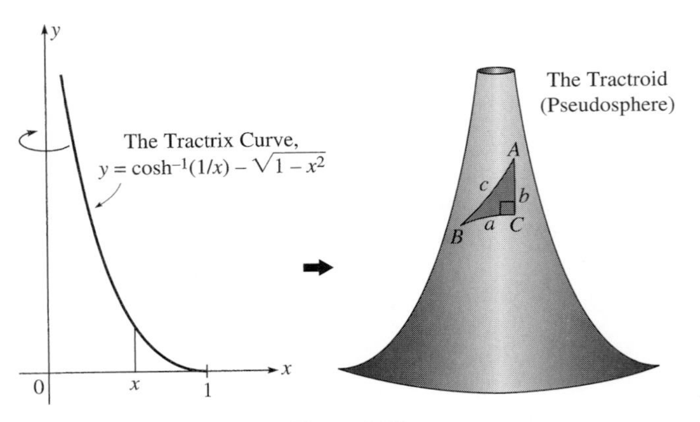

Figure 6.50

(called the **tractrix**), about the y-axis, as shown in Figure 6.50. Thus, with the proper units of measure, right triangle $\triangle ABC$, with standard notation representing the intrinsic lengths of segments, satisfies the relation

$$\cosh c = \cosh a \cosh b,$$

which is one of the formulas for a right triangle in hyperbolic geometry, as previously introduced. It should be realized, however, that the pseudosphere is only a *local* realization of hyperbolic geometry; that is, only small portions of the hyperbolic plane can be mapped isometrically onto this surface.

PROBLEMS (6.5)

GROUP A

1. Points $K(3, 3)$ and $L(3, 6)$ are reflected in the h-line ℓ: $x^2 + y^2 = 30$ to K' and L'.

(a) Find the coordinates of the image points K' and L'.

(b) Verify that $KL^* = K'L'^*$.

2. The points $A(17, 17)$, $B(25, 15)$, and $C(32, 8)$ are h-collinear and are transformed by the inversion

$$x' = \frac{9x}{x^2 + y^2}, \qquad y' = \frac{9y}{x^2 + y^2}$$

to A', B', and C'. Verify that A, B, and C, are h-collinear by showing they lie on the semicircle $(x - 17)^2 + y^2 = 289$, $y > 0$, then find the coordinates of A', B', and C' and show that these three points are also h-collinear.

3. The inversion in the circle $x^2 + y^2 = 1$ (or its inverse) will map the line $y = 1 - x$ to a circle passing through the center of inversion (origin $= O$) and the points A and B of the intersection of the circle of inversion and that line. By substituting the equation $y = 1 - x$ into the equation of the circle, find the coordinates of A and B, then by finding the equation for the image circle, using **(1)**, verify that it does pass through A, B, and O.

4. Use Theorem 2 to prove that any circle orthogonal to the circle of inversion must map back to itself under inversion.

5. Carefully sketch the semicircles ℓ: $x^2 + y^2 - 4x = 0$ and m: $x^2 + y^2 - 12x = -24$, $y > 0$, which by algebra are equivalent to $(x - 2)^2 + y^2 = 4$ and $(x - 6)^2 + y^2 = 12$, having centers (2, 0) and (6, 0) and radii 2 and $\sqrt{12} \approx 3.5$.

(a) Verify that these two h-lines are perpendicular by sketching their graphs.

(b) If these two h-lines are reflected through the h-line $x^2 + y^2 = 6$, find the equations of transformation and, by substitution and simplifying, find the equations for the image lines, and show that they are also perpendicular.

6. Show that translations of the form $x' = x + a$, $y' = y$ (which map $\boldsymbol{H}$ onto $\boldsymbol{H}$) are isometries for the hyperbolic metric, but general translations of the form $x' = x + a$, $y' = y + b$ (which, for $b > 0$, map $\boldsymbol{H}$ into $\boldsymbol{H}$) are not.

7. Consider the reflection

$$x' = \frac{3x}{x^2 + y^2}, \qquad y' = \frac{3y}{x^2 + y^2}$$

and the points $C(\frac{1}{2}, \frac{1}{2})$, $D(1, 2)$, and $E(2, 1)$. Find the coordinates of the image points C', D', and E', then perform the following experiment in *Sketchpad,* using your H-Segment and H-Angle scripts:

[1] Create a coordinate system by choosing Create Axes under GRAPH.

[2] Plot the six points C, D, E, $F = C'$, $G = D'$, and $H = E'$ in this coordinate system. [Choose Plot Points under GRAPH and enter the coordinates, in order.

[3] Use the H-Segment script to display the h-triangles $\triangle CDE$ and $\triangle FGH$.

[4] Measure $\angle CDE$ and $\angle FGH$ using the H-Angle script. Measure other pairs of angles as well, such as $\angle DCE$ and $\angle GFH$.

Did anything happen? Find a theoretical explanation.

8. **Just for Fun** Although it involves a little work, this *Sketchpad* experiment will dramatically reveal the effect of h-reflections on h-triangles. First, we want to use *Sketchpad* to construct the inverse of a point. (To save time, create a script by recording Steps 2–4 below.)

[1] Construct a horizontal boundary line $\overleftrightarrow{AB}$ near the bottom of your sketch, and construct a circle *CD* centered at *C* on this line. (This will be the circle of inversion.) Hide points *A*, *B*, and *D*.

[2] Locate an h-point *E* inside circle *CD* and construct ray $\overrightarrow{CE}$.

[3] With $\overrightarrow{CE}$ and *E* selected, construct a perpendicular line at *E*. Select the points of intersection *F* and *G* of this perpendicular with the circle.

[4] Select points *F*, *G*, and *C* in that order, and choose Arc Through Three Points under CONSTRUCT. Then select the point *H* of intersection of this arc and ray *CE*. *This will be the inverse point of E.* (Why?) Hide all objects of construction except points *E* and $E' = H$. (Stop recording.)

[5] Locate h-points *I* and *J* inside the circle and repeat the construction of steps 2–4 to obtain the inverse points $I' = K$ and $J' = L$, or use your script.

[6] Using your H-Segment script, construct h-triangles $\triangle EIJ$ (inside the circle of inversion) and its inverse image, $\triangle HKL$ (outside the circle).

Now you have a triangle and its reflected image. As you drag points *E*, *I*, and *J*, you can see the effect that an inversion has on the image triangle. Experiment until you find something interesting, then write down your observations.

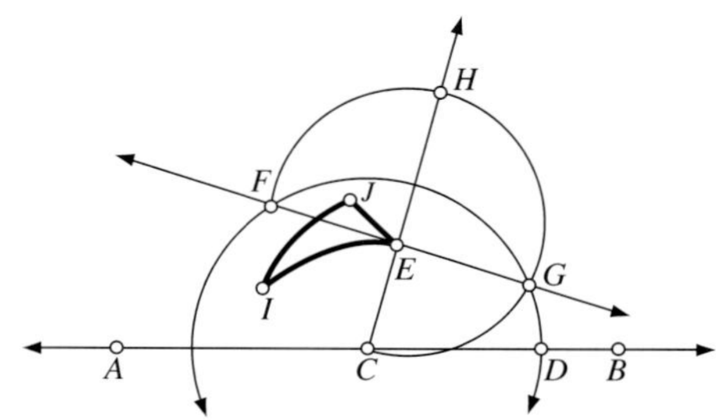

GROUP B

The following problems 10–15 involve hyperbolic trigonometry.

9. Given an h-isosceles right triangle $\triangle ABC$ with $a = b = 3$, find c accurate to three places and the measure of each of the acute angles accurate to one decimal.

10. As in Problem 9, $\triangle ABC$ is an isosceles right triangle, but with acute angles of measure just under 45 (say 44.9). What size must the sides be, compared with that of the triangle in Problem 9? Answer by finding c using equation **(14).** [*Hint:* Use formula **(14)**.]

11. The legs of a certain hyperbolic right triangle are of length $a = 3$, $b = 4$. Find c accurate to three places.

12. A hyperbolic right triangle has $a = 0.03$ and $b = 0.04$.

(a) Find c accurate to six places. (Is c approximately 0.05, as it would be in Euclidean geometry?)

(b) Find A and B, and show that $A + B \approx 90$.

13. Show that $\ln \tau$, $\ln \tau$, and $\ln 2$ are the sides of a hyperbolic isosceles right triangle, where τ is the Golden Ratio $\left[= \frac{1}{2}(1 + \sqrt{5})\right]$. (By using properties of τ, such as $\tau^2 = \tau + 1$, $\tau^{-\tau} = \tau - 1$, etc., this problem can be done without a calculator and tedious decimal approximations.) Also, find the measures of the acute angles (which are congruent).

14. Find the length of each side of an equilateral triangle having angles each of measure

(a) 45

(b) 30

(c) Show that the relationship between the side (of length a) and angle (of measure A) in any equilateral triangle is given by $\sec A = \operatorname{sech} a + 1$.

15. Proof of Bolyai–Lobachevski Formula for Angle of Parallelism (Part **(a)** of this problem involves working with elementary limits, such as $\lim_{x\to\infty} \frac{x + 1}{x - 1} = 1$.)

(a) Write down the formula for $\cos\theta$ in $\triangle ABC$ using hyperbolic trigonometry, then establish the formula for the angle of parallelism A with respect to a:

$$\textbf{(15)} \qquad \cos A = \tanh a$$

(In the figure, let $x = BC \to \infty$; then $\theta \to A$ and, since $y > x$, $y \to \infty$ and $\tanh y \to 1$.

(b) Using established identities and the result of **(a),** prove the following sequence of formulas, culminating in **(11)** of Section 6.4 (Problem 26, Section 6.4).

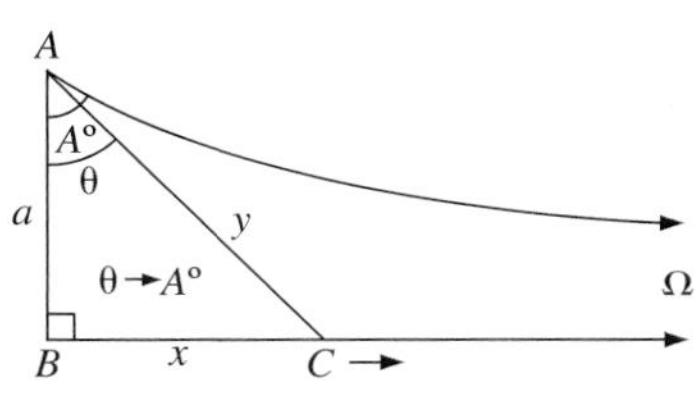

$$\sin A = \operatorname{sech} a$$

$$\tan A = \operatorname{csch} a$$

$$\tan\tfrac{1}{2}A = \cosh a - \sinh a$$

$$\tan\tfrac{1}{2}A = e^{-a}$$

Identities used are $\sin^2 A = 1 - \cos^2 A$, $1 - \tanh^2 a = \operatorname{sech}^2 a$, and $\tan\frac{1}{2}A = (1 - \cos A)/\sin A$.]

16. Let an **h-square** be defined as any quadrilateral in hyperbolic geometry having congruent sides and congruent angles. There are an infinite number of noncongruent squares (as in Euclidean geometry), but as the size of the square changes, so do its angles, becoming smaller as the sides grow larger (unlike Euclidean geometry).

(a) Using the formula for area $K = k(360 - 4A)$ where A = measure of each angle of an h-square, find an upper bound for A using the fact that the defect of any h-square is positive.

(b) The measure A of the angles of an h-square determine its sides uniquely. For what integer values of A can an h-square of angle measure A be used to tile the plane? (***Hint:*** There must be an integer number of squares placed about each vertex of the tiling, thus $nA = 360$ for some integer n. See the figure for an example when $A = 72$ in the upper plane model.)

FIVE CONGRUENT H-SQUARES ABOUT A POINT
IN UPPER HALF-PLANE MODEL

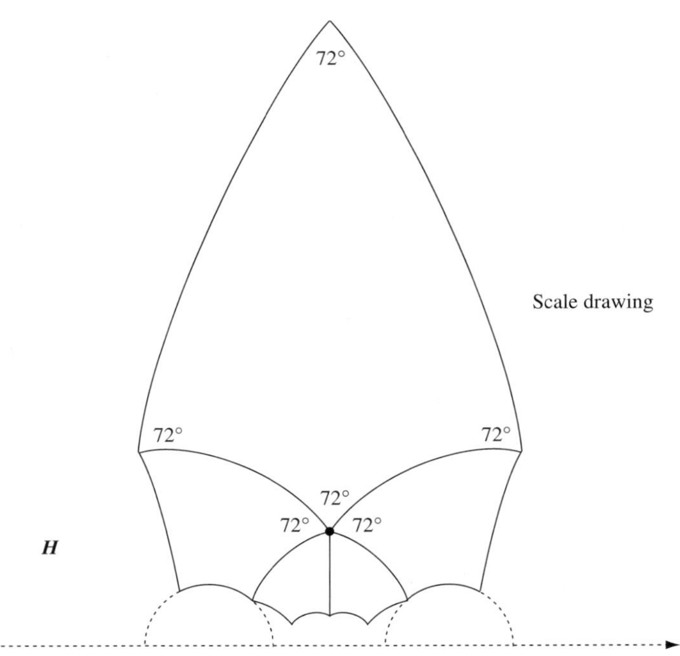

17. Show that the smallest regular pentagon that can be used to tile the plane in hyperbolic geometry must have angles of measure 90. The figure at the top of the next page shows a tiling of the upper half-plane model by such tiles. Note the abundance of hyperparallel lines. In fact, any two nonintersecting lines in the figure are hyperparallel (divergently parallel).

TILING THE UPPER HALF-PLANE MODEL WITH REGULAR 90° ANGLED PENTAGONS

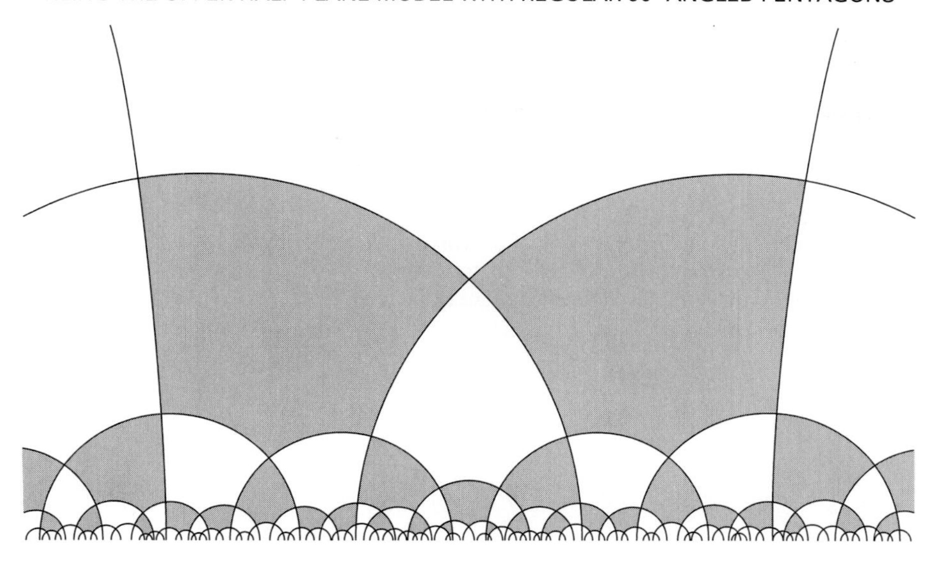

18. Show that in hyperbolic geometry, any regular polygon can be used to tile the plane provided the angles are properly chosen.

19. Using circular inversion, explain why the Euclidean construction depicted in the figure will yield the midpoint M of a given h-segment $\overline{AB}$* (lying on semicircle ℓ having center O), where $\overleftrightarrow{CM}$ is constructed tangent to circle ℓ at M.

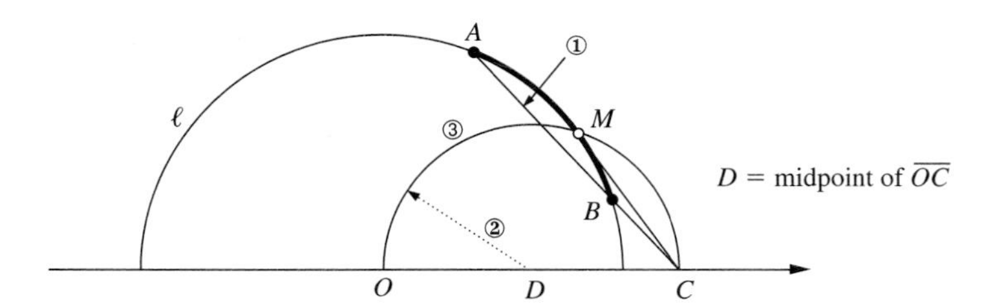

GROUP C

20. Use hyperbolic trigonometry to validate the following construction due to Beltrami for the angle of parallelism (a variation of one given by J. Bolyai). Let $AB = a$ be given, and construct the perpendicular ℓ to $\overline{AB}$ at B. Then, in axiomatic hyperbolic geometry, follow these steps:

(1) Construct a Lambert Quadrilateral $\diamond ABCD$ with $\overline{AB}$ as base and $C \in \ell$, and with right angles at A, B, and D. (The value for $x = AD$ is undetermined—it depends on where C is located on line ℓ and where the foot D of the perpendicular from C falls on line $\overleftrightarrow{AD}$.)

(2) With A as center and $BC = c$ as radius, construct a circular arc, cutting segment $\overline{CD}$ at E. ($AD < c < AC$ guarantees that the arc intersects $\overline{CD}$.)

(3) Then $\overrightarrow{AE}$ is asymptotically parallel to $\overrightarrow{BC}$, and the desired angle of parallelism is $m\angle BAE$.

(You must show that $\tan\frac{1}{2}\theta = e^{-a}$ where $\theta = m\angle BAE$, or one of the forms equivalent to this that were established in Problem 15. Use hyperbolic trigonometry for the right triangles in the figure, and use algebra to eliminate all quantities except a and θ.)

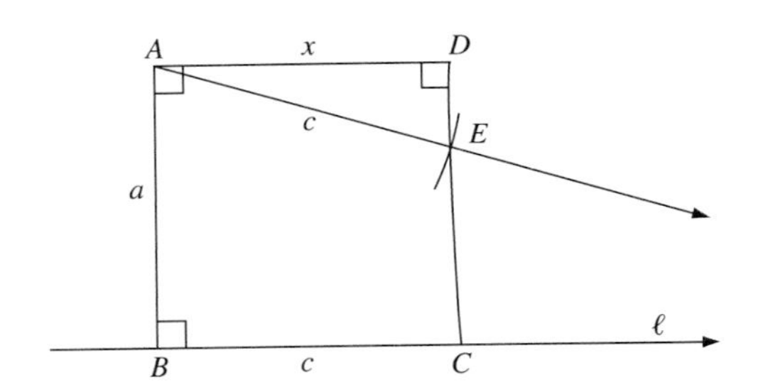

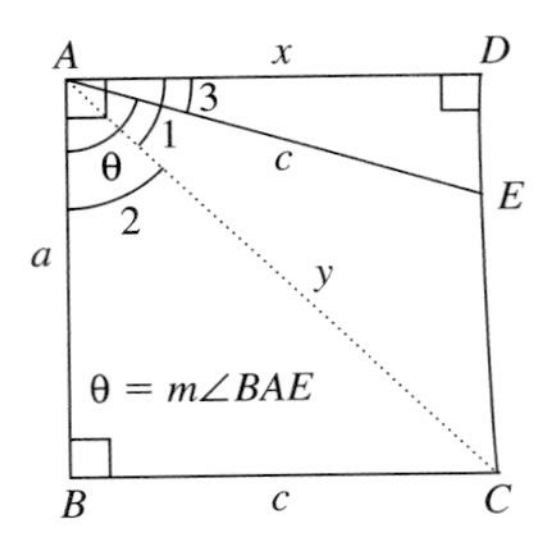

21. In axiomatic hyperbolic geometry, the locus of a point P such that its distance to a line is constant is called an **equidistant curve.** In Euclidean geometry, this locus would simply be a line parallel to the given line. Explain why this locus cannot be a line in hyperbolic geometry.

22. Ray ℓ is a Euclidean ray in $\boldsymbol{H}$, shown in the figure. Because ℓ is not an h-line, it is some curve in hyperbolic geometry. Identify it and prove your answer. (***Hint:*** See Problem 21.)

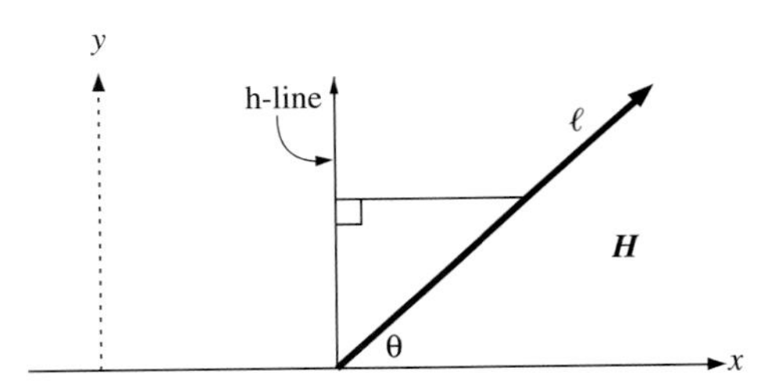

23. Circular arc ℓ is a semicircle with center C on the x-axis, so is an h-line, but circular arc $\boldsymbol{A}$ does not have its center on the x-axis, so it is not an h-line. Construct this figure using *Sketchpad*, and locate a variable point F on ℓ to make a discovery about $\boldsymbol{A}$.

[1] Construct segment $\overline{CF}$ and the perpendicular to $\overline{CF}$ at F.

[2] Select the point G of intersection of that perpendicular with the horizontal boundary.

[3] Construct circle GF with center at G passing through F, intersecting arc $\boldsymbol{A}$ at point H.

[4] Display the hyperbolic distance FH^*. (Use the H-Metric script created in Problem 15, Section 6.4.)

Drag F and observe the effect on FH^*. (For an extra project, prove this phenomenon by an inversion map.)

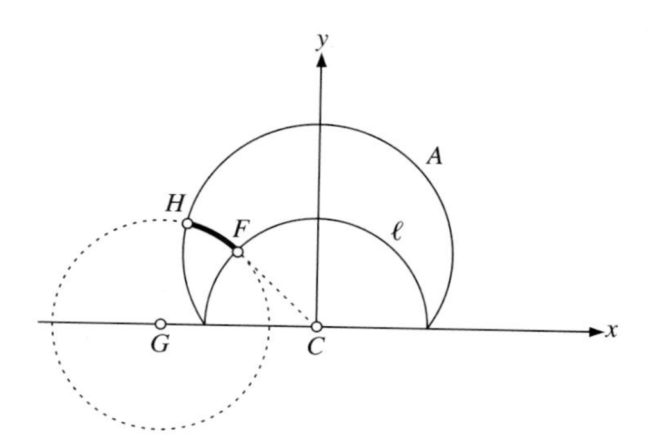

24. Law of Cosines for Hyperbolic Geometry Establish the following formula for a triangle in standard notation in hyperbolic geometry.

(16) $$\cosh a = \cosh b \cosh c - \sinh b \sinh c \cos A$$

[***Hint:*** First write $\cosh a = \cosh c_2 \cosh h = \cosh (c - c_1) \cosh h$ and use the hyperbolic addition formula $\cosh (x - y) = \cosh x \cosh y - \sinh x \sinh y$, then use $\cos A = (\tanh c_1)/(\tanh c)$.]

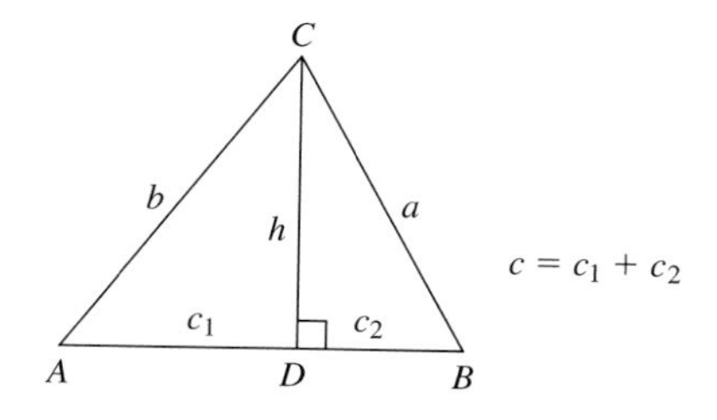

25. Cevian Formula in Hyperbolic Geometry Derive the following formula for the length d of cevian $\overline{CD}$ in the triangle:

(17) $$\cosh d = p \cosh a + q \cosh b$$

where $p = (\sinh c_1)/(\sinh c)$ and $q = (\sinh c_2)/(\text{sinch}\, c)$. (***Hint:*** Use the Law of Cosines for hyperbolic geometry given in Problem 24.)

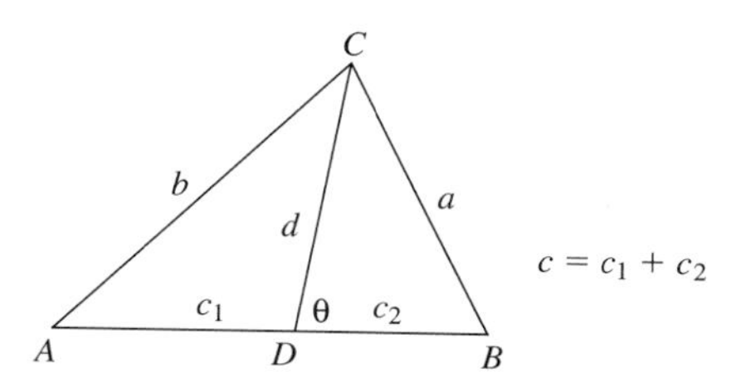

26. Show that the inversion indicated in the accompanying figure maps the Poincaré Circular Disk Model onto the Beltrami–Poincaré Half-Plane Model, with, in particular, h-lines ℓ and m mapping to ℓ' and m', respectively.

MAPPING THE POINCARÉ MODEL TO HALF-PLANE MODEL

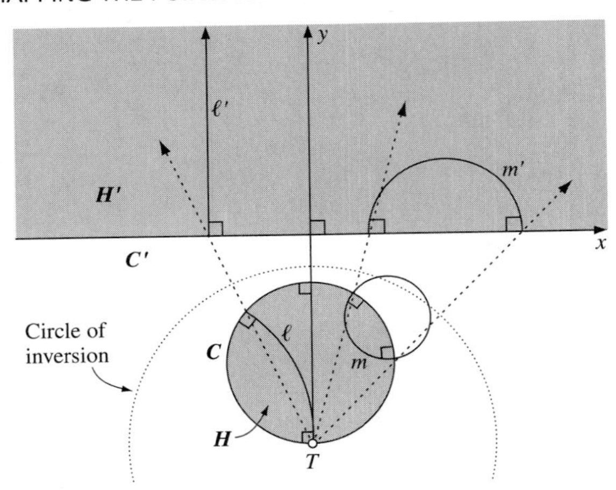

27. If the x-axis defining the upper half-plane model were actually at the "edge of the universe" and we were dealing with ordinary planetary distances, say no greater than 1 million miles, explain how we might conceive our world to be Euclidean even though we were using the definitions of h-lines, h-distance, and h-angle measure relative to the x-axis. Assume all the axioms of absolute geometry that are presumed to be derivable from observing ordinary space; you may use hyperbolic trigonometry.

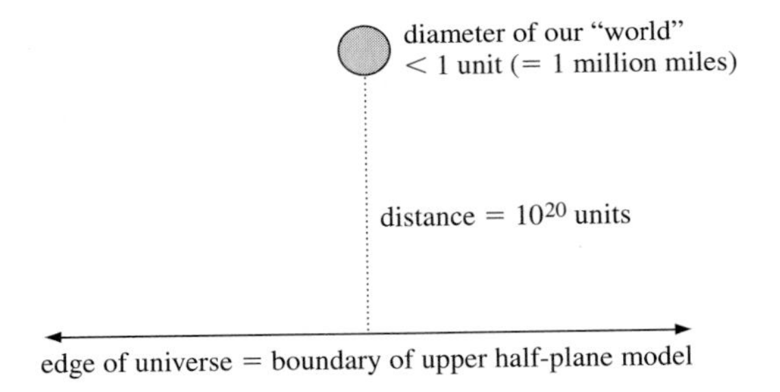

Chapter 6 Summary

The origin of the concept of non-Euclidean geometry was actually Euclid himself, who has been called the first non-Euclidean geometer. His development of parallelism naturally led to the question of whether the Fifth Postulate is provable in absolute geometry. The Half-Plane Model and Circular Disk Model discovered by E. Beltrami and H. Poincaré show conclusively that such a proof it not possible, although many very good mathematicians throughout history have gone to great effort to find a proof. János Bolyai's development and that of N. Lobachevski were the first extensive treatments of the hypothesis denying the Parallel Postulate—all without apparent contradiction. Bolyai called this development absolute science—that which is *absolutely true* whichever assumption about parallels is made. In his development, Bolyai examined the geometry of a sphere in the ordinary three-dimensional space of absolute geometry, then allowed the radius of the sphere to approach infinity. The resulting surface has as its intrinsic geometry ordinary Euclidean geometry. The formulas gleaned from this limiting case, however, were the trigonometric formulas for hyperbolic geometry.

The chief property of triangles, when there is more than one parallel to a line from a given point (the Hyperbolic Parallel Postulate), is that regarding the angle sum of a triangle: The sum of the measures of the angles of any triangle is less than 180. This theorem resulted in the unusual AAA Congruence Criterion for triangles and the fact that no noncongruent similar triangles exist in hyperbolic geometry.

The last two sections of the chapter were devoted to a study of models for hyperbolic geometry and verifying the axioms for absolute geometry. Circular inversion was introduced to show that line reflections exist, and to prove the SAS Postulate in the upper Half-Plane Model. Finally, the formulas for the trigonometry of the right triangle were given and pursued in the problems.

Testing Your Knowledge

You are expected to take this test using only the list of axioms, definitions, and theorems in Appendix F as references. The formulas needed for hyperbolic trigonometry will be provided here.

1. Arrange the following isosceles triangles in the correct order according to area (=k times the defect) from smallest to largest.

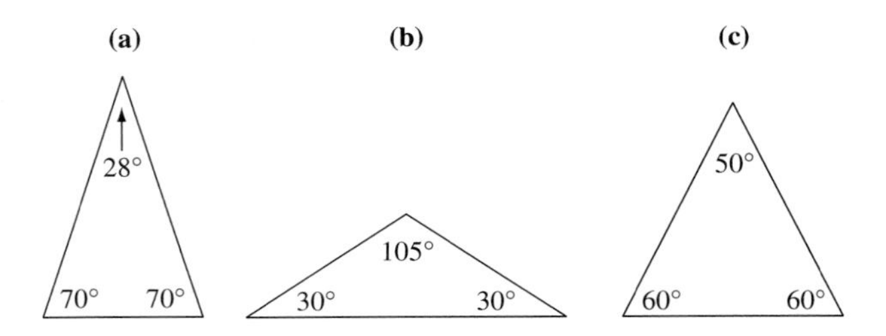

2. Which of the equilateral triangles shown below could be used to tile the hyperbolic plane?

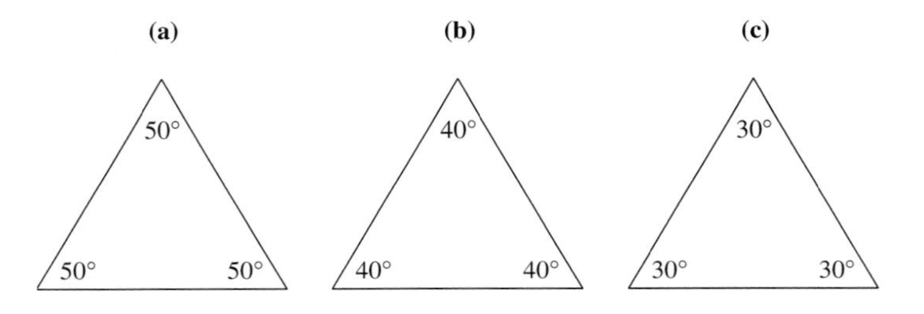

3. In a right triangle, one of the acute angles has measure 35. If the triangle has defect 50, find the measure of the other acute angle.

4. Find the defect of the pentagon having angles of measure 30, 50, 60, 80, and 100.

5. Which of the following are invariants of a circular inversion?
- **(a)** distance
- **(b)** Cross Ratio
- **(c)** circles
- **(d)** a curve that is either a circle or line
- **(e)** Euclidean angle measure
- **(f)** measure of angle between tangents to sides of an h-angle.

6. What are the "lines" of the upper half-plane model for hyperbolic geometry having horizontal boundary ℓ.

7. Under a reflection (inversion) in h-line ℓ, as shown in the diagram here, answer the following:

   (a) To what does the vertical line m transform?

   (b) What is the equation of the image of the h-line m of **(a)** from the information provided in the figure?

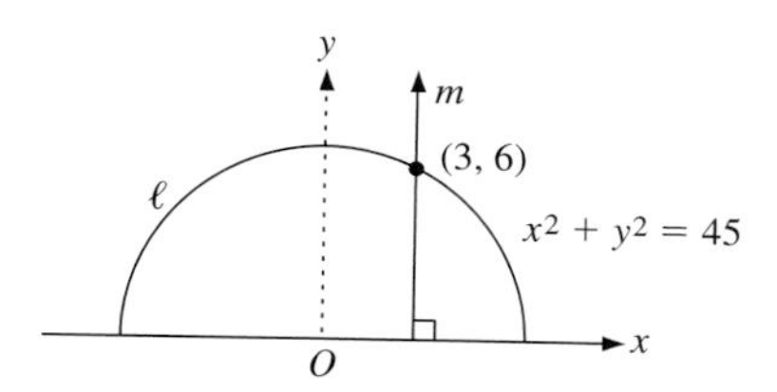

8. Proclus' argument stated that if ℓ passes through A and is parallel to m, and n is any other line through A parallel to m, then the distance PQ becomes infinite as AP becomes infinite, because $PQ > x$ and x becomes infinite from a property of angles and right triangles. This is a contradiction, since PQ cannot become infinite. Therefore, there is only one line through A parallel to ℓ. The error in reasoning committed by Proclus was that he used a property of Euclidean geometry that is not true in hyperbolic geometry. What was it?

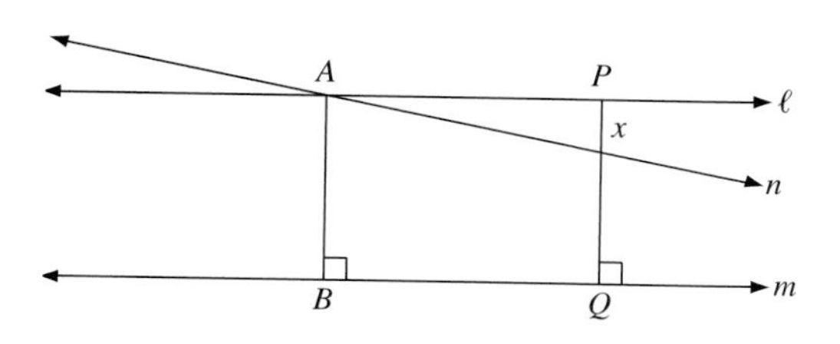

9. A right triangle $\triangle ABC$ is given in hyperbolic geometry with $A = 30$, $C = 90$, and $c = 2$, in standard notation. Find B, accurate to one-tenth degree, then find the angle sum for this triangle.

Formulas Needed:

$$\cosh x = \tfrac{1}{2}(e^{x} + e^{-x})$$
$$\cosh c = \cot A \cot B$$

10. Prove directly, without using defect or its additive properties, that *if* B–D–C and $\triangle ABC$ has angle sum 180, then both triangles $\triangle ABD$ and $\triangle ADC$ have angle sum 180. (***Hint:*** The first step of your proof should be: Suppose that $r + t + u < 180$.)

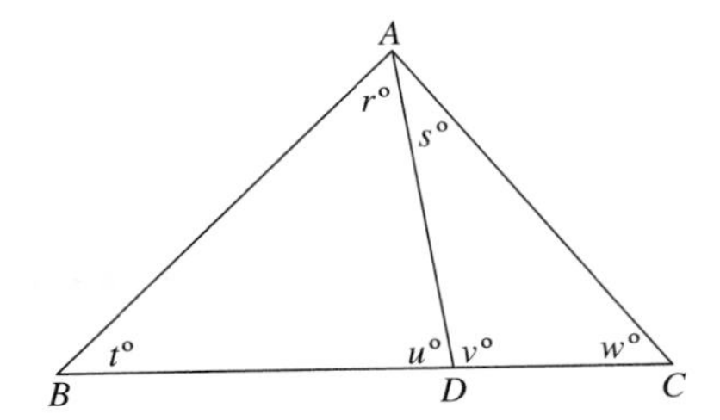

CHAPTER 7

An Introduction to Three-Dimensional Geometry

OVERVIEW

The original axioms for absolute geometry were stated for three-dimensional space, but we have not yet made full use of them. Our development of geometry, beginning with Chapter 3, has taken place in a single, fixed plane. Our purpose here is to take a brief look at the implications that our axiom system has for three-dimensional geometry. In this chapter we will study the basic properties of lines and planes in space, the important shapes of three-dimensional geometry (polyhedra, spheres, cones, and cylinders), and the development of some of the classical formulas for volume. An optional section on spherical geometry appears in which we derive the formulas for spherical trigonometry that closely parallel those of hyperbolic geometry in Section 6.5.

For convenience, we use E^n for Euclidean geometry of dimension n (which can be coordinatized as a geometry whose points are represented by ordered n-tuples of real numbers, or by n-vectors). We denote hyperbolic n-space and elliptic n-space by H^n and S^n, respectively, for $n = 1, 2, 3$; the three geometries all coincide for $n = 1$—the trivial case. (A model for H^3 will be considered in Problem 14, Section 7.5.)

7.1 Orthogonality Concepts for Lines and Planes

Perpendicularity of lines to lines, lines to planes, and planes to planes in 3-space ought to be an altogether straightforward topic. Yet, there are a few surprises in store, and some problems we might not expect. We point out that we have so far defined perpendicularity only for lines that lie in the same plane (in Section 2.8); the other possible concepts for space will require some thought.

NOTE: In geometry and topology we often encounter the concept of **homogeneity** of space: what is true at one point is true at every other point (perhaps only in small neighborhoods of those points). Here, we have an analogous concept: what is true in one plane is true in every other plane. Of course, one new and powerful aspect of our axioms is the validity of the SAS postulate for *triangles lying in different planes* (see Figure 7.1). An interesting problem along these lines is proving that if the Euclidean parallel postulate holds for one plane, it must hold for all other planes. (See Problem 20.)

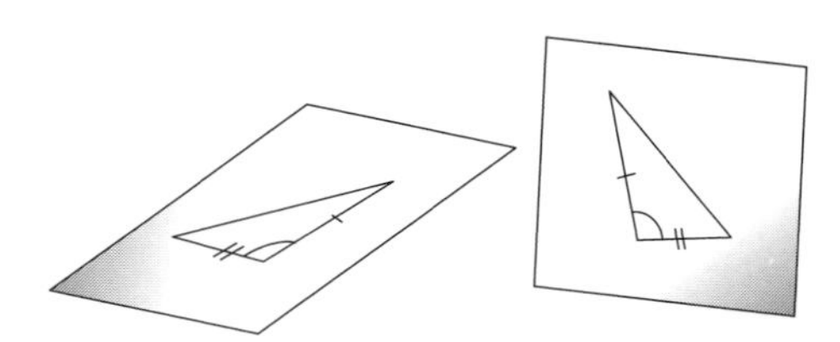

Figure 7.1

LINES PERPENDICULAR TO LINES IN SPACE

We take the same meaning in space for perpendicular lines as we do in plane geometry. That is, two lines are **perpendicular in space** iff they lie in the same plane, and intersect at right angles. Given a point A not on line ℓ there is a unique line through A perpendicular to ℓ (in space) since there is a unique plane containing A and ℓ and a unique perpendicular exists *in this plane*—as indicated (Figure 7.2). However, if A lies *on* line ℓ, then there can be many lines through A perpendicular to ℓ. In Figure 7.3 is illustrated a pair of planes $\boldsymbol{P}_1$ and $\boldsymbol{P}_2$ both containing line ℓ (how do we know they exist?), and in *each* of the two planes there exist lines m_1 and m_2 that are both perpendicular to line ℓ at A.

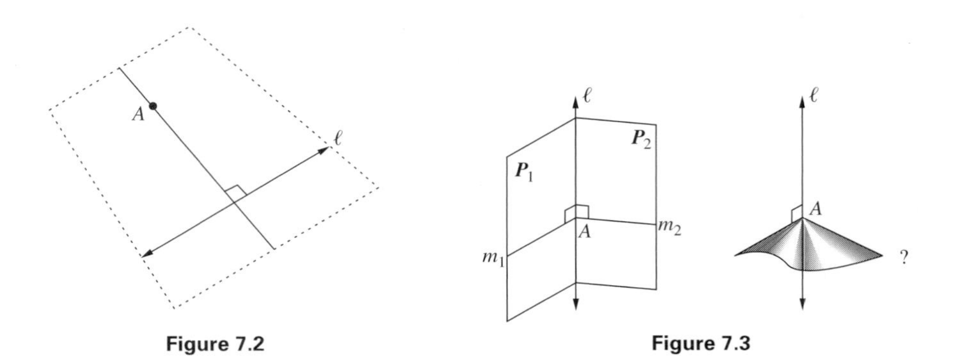

Figure 7.2

Figure 7.3

LINES OR PLANES PERPENDICULAR TO PLANES IN SPACE

The illustration in Figure 7.3 suggests there could exist, theoretically, a nonplanar surface that is generated by the family of lines in space perpendicular to the same line at a point on that line. To refute this notion on the basis of axiomatic geometry (as you are asked to do in Problem 1) is not difficult, but it requires the basic concept of perpendicularity between lines and planes, which we now consider.

DEFINITION: A line ℓ is said to be **perpendicular** to a plane $\boldsymbol{P}$ (and $\boldsymbol{P}$ is **perpendicular** to ℓ) iff ℓ meets the plane at some point A and is perpendicular to all lines in the plane that pass through A. We write, symbolically, $\ell \perp \boldsymbol{P}$ (or $\boldsymbol{P} \perp \ell$). A plane $\boldsymbol{P}$ is **perpendicular** to another plane $\boldsymbol{Q}$ iff $\boldsymbol{P}$ passes through some line that is perpendicular to $\boldsymbol{Q}$, and we write $\boldsymbol{P} \perp \boldsymbol{Q}$ for this condition.

You will be asked to establish in Problem 10 that there always exists a line that is perpendicular to a given plane from a given point not in that plane, while Problem 14 asks for the construction when the point lies in the plane. By merely passing a plane through the perpendicular line (incidence axioms), the *existence* of a plane perpendicular to a given plane through an external, or internal, point is guaranteed. We cannot expect uniqueness here because a given plane can have many planes perpendicular to it at a given point. A construction in geometry will be pursued in the discovery unit at the end of this section, where you can participate in proving the so-called *Fence-Post Property*—the first significant result for E^3.

THEOREM 1: FENCE-POST PROPERTY
If line ℓ is perpendicular to two distinct lines m and n of a plane at some point A, then ℓ is perpendicular to all lines in that plane that pass through A, and the line is perpendicular to the plane at A.

PLANES PERPENDICULAR TO PLANES IN SPACE

We have already defined perpendicularity between two planes: when the first plane contains a line that is perpendicular to the second. Note the asymmetry of this definition—it is by no means obvious that we can turn it around and conclude at this point that if $\boldsymbol{P} \perp \boldsymbol{Q}$, then $\boldsymbol{Q} \perp \boldsymbol{P}$. In fact, this is true, and it is a corollary of Theorem 1.

COROLLARY: If $\boldsymbol{P} \perp \boldsymbol{Q}$ then $\boldsymbol{Q} \perp \boldsymbol{P}$, for any two planes $\boldsymbol{P}$ and $\boldsymbol{Q}$.

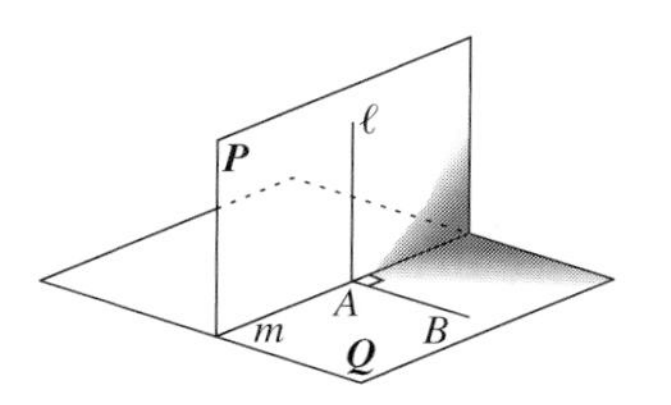

Figure 7.4

PROOF

By definition, $\boldsymbol{P}$ contains a line ℓ that is perpendicular to $\boldsymbol{Q}$ at some point A (Figure 7.4). Let $\boldsymbol{P} \cap \boldsymbol{Q}$ = line m (Axiom I-4). Then at A, and *in plane* $\boldsymbol{Q}$, erect the line $\overleftrightarrow{AB} \perp m$. By Theorem 1, $\ell \perp \overleftrightarrow{AB}$ ($\therefore \overleftrightarrow{AB} \perp \ell$) so $\overleftrightarrow{AB}$ is perpendicular to two lines in $\boldsymbol{P}$, hence to every line in $\boldsymbol{P}$, so that line $\overleftrightarrow{BA} \perp \boldsymbol{P}$ at A. But $\boldsymbol{Q}$ contains line $\overleftrightarrow{BA}$, so by definition, $\boldsymbol{Q} \perp \boldsymbol{P}$.

THEOREM 2: If planes $\boldsymbol{P}$ and $\boldsymbol{Q}$ are perpendicular, and their intersection is line ℓ, then every line in $\boldsymbol{P}$ perpendicualr to ℓ is perpendicular to $\boldsymbol{Q}$ (Figure 7.5).

PROOF

We know that P contains a line $m \perp \boldsymbol{Q}$ at some point $\boldsymbol{T}$. We must show that if A is any other point on ℓ and $\overleftrightarrow{BA} \perp \ell$ in $\boldsymbol{P}$ (that is, $\angle BAT$ is a right angle), then line $\overleftrightarrow{BA} \perp \overleftrightarrow{AC}$ for some other point C in $\boldsymbol{Q}$ not on ℓ, thus $\overleftrightarrow{BA} \perp \boldsymbol{Q}$ by Theorem 1. Locate $C \neq T$ in plane $\boldsymbol{Q}$ so that line $\overleftrightarrow{CT} \perp \ell$ in $\boldsymbol{Q}$. Now $\overleftrightarrow{CT} \perp m$ since $m \perp$ every line in $\boldsymbol{Q}$ passing through T, and $\overleftrightarrow{CT} \perp \ell$. Thus $\overleftrightarrow{CT}$ is perpendicular to every line in $\boldsymbol{P}$ passing through T. Locate B' so that B–A–B' and $BA = AB'$. Then $\triangle BAT \cong \triangle B'AT$ by SAS, making $BT = B'T$. Then, being right triangles, $\triangle BCT \cong \triangle B'CT$ and $BC = B'C$. Thus, $\triangle BCA \cong \triangle B'CA$ by SSS and $\angle BAC \cong \angle B'AC$, or $\overleftrightarrow{BA} \perp \overleftrightarrow{AC}$, as desired.

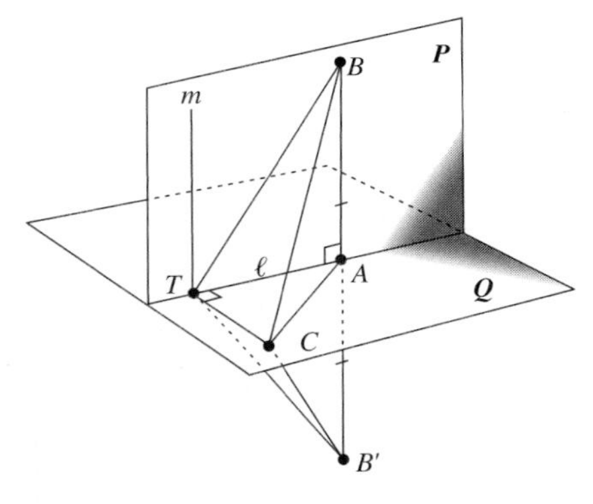

Figure 7.5

We leave the proof of the following result as Problem 18:

COROLLARY: If two planes $\boldsymbol{P}$ and $\boldsymbol{Q}$ are perpendicular to a third plane $\boldsymbol{R}$, their line of intersection ℓ is also perpendicular to $\boldsymbol{R}$.

OUR GEOMETRIC WORLD

If two sheets of plywood are placed at right angles (represented as planes $\boldsymbol{P}$ and $\boldsymbol{Q}$ in Figure 7.6), we can observe their perpendicularity at all points along the common edge. This means that if we slide a carpenter's square along the corner edge, the legs of the square will remain flush to the two plywood sheets. The result of Theorem 2 just proven provides the theoretical basis for this.

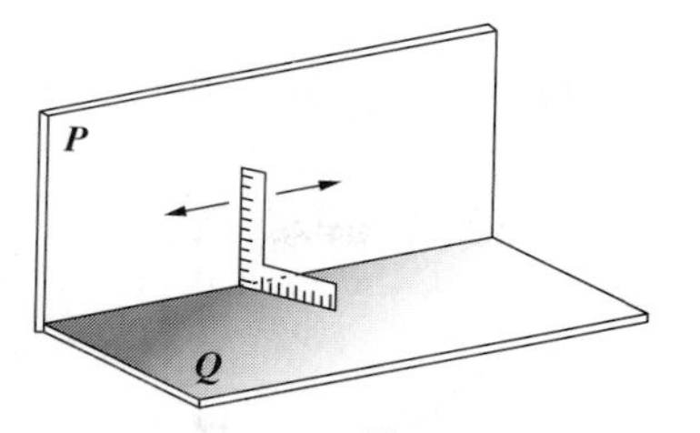

Figure 7.6

DIHEDRAL ANGLE BETWEEN TWO PLANES

The concept of perpendicular planes leads us to a generalization of the concept of angle that applies to space. Let two planes $\boldsymbol{P}$ and $\boldsymbol{P'}$ meet in line ℓ, and on each of these two planes consider a half plane determined by ℓ, denoted $\boldsymbol{H}$ and $\boldsymbol{H'}$. (See Figure 7.7.) The set $\ell \cup \boldsymbol{H} \cup \boldsymbol{H'}$ is then called a **dihedral angle,** denoted $\angle(\boldsymbol{H}, \boldsymbol{H'})$, with the half-planes called **sides,** and line ℓ called its **edge.** This is a kind of three-dimensional angle; any planar cross section of it would be an ordinary angle in the plane.

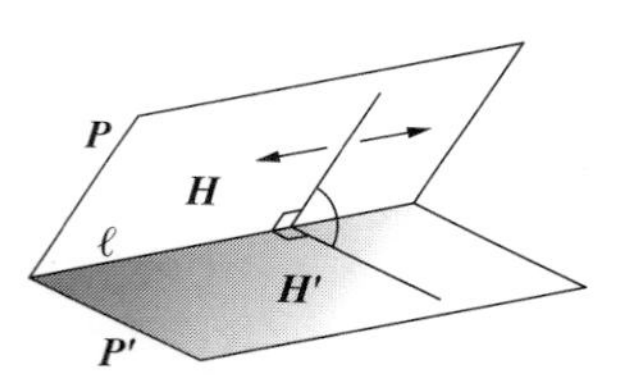

Figure 7.7

It is convenient to have a measure for a dihedral angle, but it may not be clear how to do it at first. A basic property of absolute geometry is that if we start with an angle formed by the rays from any point on the edge of a dihedral angle perpendicular to that edge, and lying in its sides, and then think of "sliding" the angle along the edge of the dihedral angle, the measure of that angle stays constant. Note that this has already been proven in the case of a "right" dihedral angle. The constant value of this angle will then be taken as the **measure** of the dihedral angle, denoted $m\angle(\boldsymbol{H}, \boldsymbol{H'})$.

Proving the property just mentioned makes another interesting problem for you to work on, so we will not spoil it by offering the proof. The usual proof in Euclidean geometry uses properties of parallelograms, but it is just as easy to establish the result for absolute geometry and avoid properties of Euclidean parallelism. This makes it even more interesting. (See Problem 15 for some guidance if you need it.)

EQUIDISTANT LOCUS IN SPACE

Recall that a **locus** of points is merely a *set* of points satisfying a certain property. We continue to use the old and new terminology interchangeably. The last theorem in this section will actually be needed later for the study of spheres.

THEOREM 3: EQUIDISTANT LOCUS THEOREM

The set of all points in E^3 equidistant from two fixed points A and B is a plane that is perpendicular to $\overline{AB}$ at its midpoint (see Figure 7.8).

PROOF

(1) First we prove that if $W \in \boldsymbol{P}$ where $\boldsymbol{P}$ is the plane perpendicular to line segment $\overline{AB}$ at its midpoint M, then point W is equidistant from A and B. In Figure 7.8, consider segment $\overline{WM}$ and the two triangles it forms with $\overline{AB}$; since $\boldsymbol{P} \perp \overline{AB}$ at M, $\overline{AB}$ is perpendicular to all lines in $\boldsymbol{P}$, hence $\overline{AB} \perp \overline{WM}$, and $\triangle AMW \cong \triangle BMW$ by SAS, which implies $WA = WB$, and W is equidistant from A and B.

(2) Let W be equidistant from A and B; we must prove that $W \in \boldsymbol{P}$. In the plane $\boldsymbol{Q}$ determined by line $\overleftrightarrow{AB}$ and point W, W lies on the perpendicular bisector of segment $\overline{AB}$, hence $\overline{AB} \perp \overline{MW}$. But also, line $\overleftrightarrow{AB}$ is perpendicular to the line ℓ of intersection of $\boldsymbol{Q}$ and $\boldsymbol{P}$ since $\overleftrightarrow{AB}$ is perpendicular to every line in $\boldsymbol{P}$. Thus we have two lines, $\overleftrightarrow{MW}$ and ℓ, both perpendicular to AB at M in plane $\boldsymbol{Q}$, hence they coincide, and $W \in \boldsymbol{P}$, ending the proof.

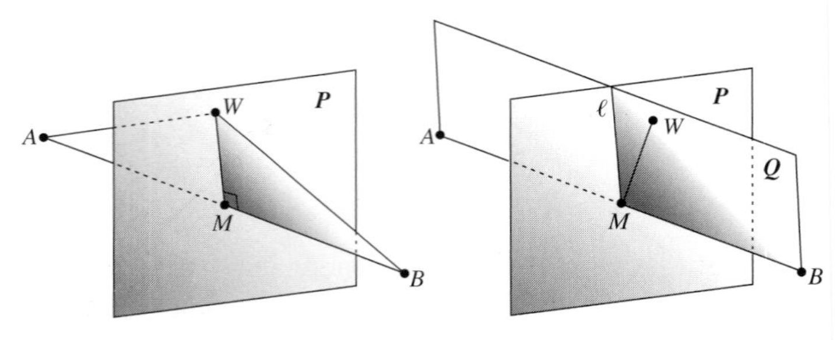

Figure 7.8

Moment for Discovery

Constructing a Perpendicular to a Plane

A steel fence post is being driven into level ground, and we would like it to be perfectly erect, meaning that from all vantage points the post looks straight—the geometric meaning of the term "perpendicular." As all surveyors know, it is only necessary to line the post up in two different directions, using a carpenter's square, say, to do the job (Figure 7.9). The post will then be perpendicular from all other directions.

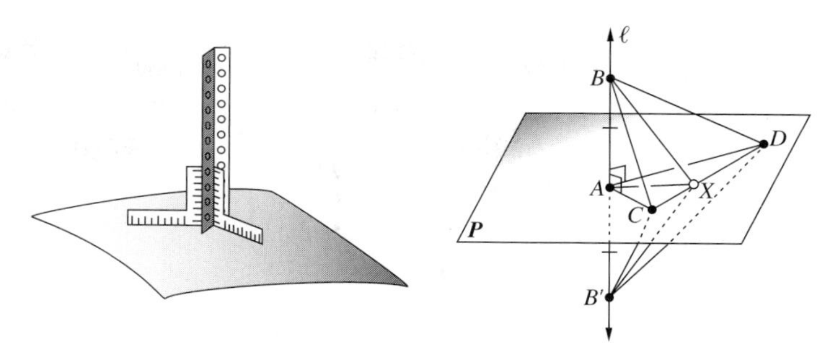

Figure 7.9

Justifying this leads to an interesting problem for three-dimensional geometry. The idea is, we have a line (ℓ) perpendicular to two lines $\overleftrightarrow{AC}$ and $\overleftrightarrow{AD}$ in plane $\boldsymbol{P}$, and we would like to prove line $\ell \perp \overleftrightarrow{AX}$ for every other point X in plane $\boldsymbol{P}$. For convenience, take $X \in \overline{CD}$. Study these steps to see if you can "discover" a proof, filling in all the details as you go:

1. Locate B and B' on ℓ so that B–A–B' and $AB = AB'$. Why is $BC = B'C$? $BD = B'D$?
2. Do you find that $\triangle BCD \cong \triangle B'CD$? Why?
3. Is $\angle BCX \cong \angle B'CX$? What about $\triangle BCX$ and $\triangle B'CX$?
4. Is $BX = B'X$?
5. Can you now prove that line $\ell \perp \overleftrightarrow{AX}$?
6. Is the above proof valid in hyperbolic space as well? (The answer is "yes" if you proved this for absolute geometry and did not use the Euclidean parallel postulate anywhere.)

PROBLEMS (7.1)

GROUP A

1. Prove explicitly that the set of all lines in space perpendicular to a line at some point on that line lies in the same plane.
2. Three mutually orthogonal (perpendicular) planes meet at three lines, and at a single point P. Prove that these three lines of intersection are mutually orthogonal. (Use the corollary of Theorem 2.)
3. A line is perpendicular to an infinite number of lines in a plane. Find an example that shows that this fact does not necessarily imply that the given line is perpendicular to the given plane.
4. Five points A, B, C, D, and E are situated in space such that $m\angle ABC = 120$, $m\angle CBD = 135$, $m\angle ABE = m\angle DBE = m\angle EBC = 90$. Find $m\angle ABD$. (No proof required.)

5. Five points A, B, C, D, and E are situated in space such that $m\angle ABC = m\angle CBE = m\angle DBE = 45$, $m\angle CBD = m\angle ABD = 90$. Find $m\angle ABE$. (***Hint:*** Verify that E lies in the plane of C, B, and D; locate P, Q on rays $\overrightarrow{BA}$, $\overrightarrow{BE}$ with $BP = BQ = \sqrt{2}$ and $\overline{PR} \perp \overline{BC}$. Then $PR = RB = 1$. Find PQ.)

6. The extension of a tripod PQ makes equal angles with the legs at A, B, and C, as shown in the figure. Assuming the legs are of equal length, what must be the relation of line PQ to the plane of A, B, and C? Can you prove it? (See Problem 8 below.)

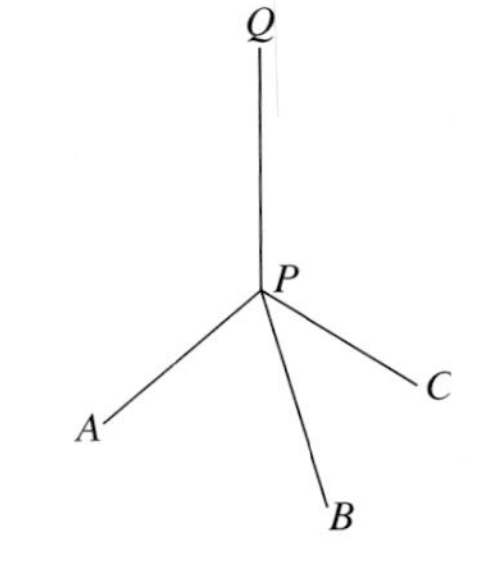

7. Let A, B, C be any three noncollinear points in plane $\boldsymbol{P}$, let O be the circumcenter of $\triangle ABC$ in that plane, and let line $\overleftrightarrow{PO} \perp \boldsymbol{P}$. Prove that

(a) P is equidistant from A, B, and C.

(b) The lines from P through A, B, and C make equal angles with line PO.

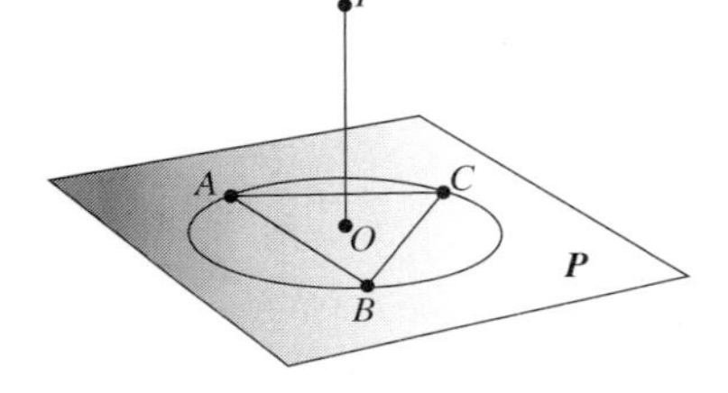

GROUP B

8. Suppose that P is equidistant from three noncollinear points A, B, and C, and does not lie in the plane of A, B, and C. Prove that if O is the circumcenter of $\triangle ABC$, then line $\overleftrightarrow{PO}$ is perpendicular to the plane of A, B, C. (***Hint:*** Use Theorem 3.)

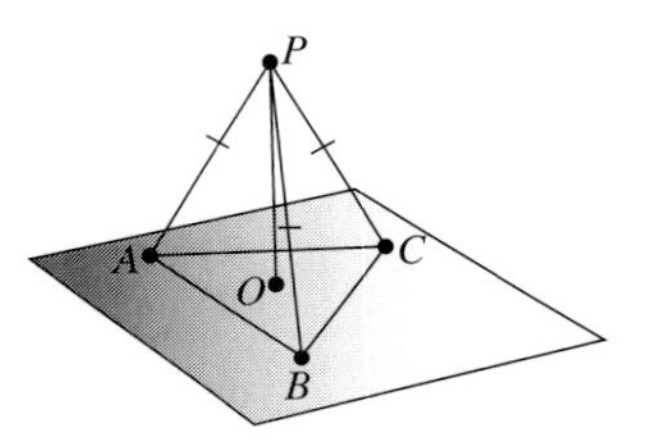

9. Carpenters working in an apartment building that is being renovated have located a ceiling joist (see figure) and have marked a point A on the center of that beam. They want to add a central support beam from the floor 12 ft directly below. If they have a piece of wire 18 ft long, what are some ways they can accurately locate the point B directly below point A if they do not want to use the side walls for a point of reference?

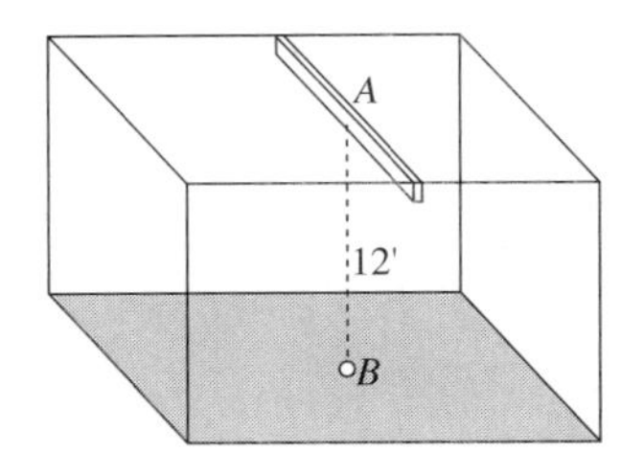

10. Construction of Line Perpendicular to Plane from Given External Point Prove that these steps will yield a line through point A not on plane $\boldsymbol{P}$ that is perpendicular to plane $\boldsymbol{P}$:

(1) Draw any line ℓ in plane $\boldsymbol{P}$.

(2) In the plane $\boldsymbol{Q}$ of A and ℓ, construct the line $\overleftrightarrow{AB} \perp \ell$ (standard Euclidean construction).

(3) In plane $\boldsymbol{P}$, construct line $m \perp \ell$ at B (standard Euclidean construction).

(4) In the plane of A and line m, drop the perpendicular $\overleftrightarrow{AC}$ to line m (standard Euclidean construction). Then line $\overleftrightarrow{AC}$ is the desired perpendicular. (Prove.)

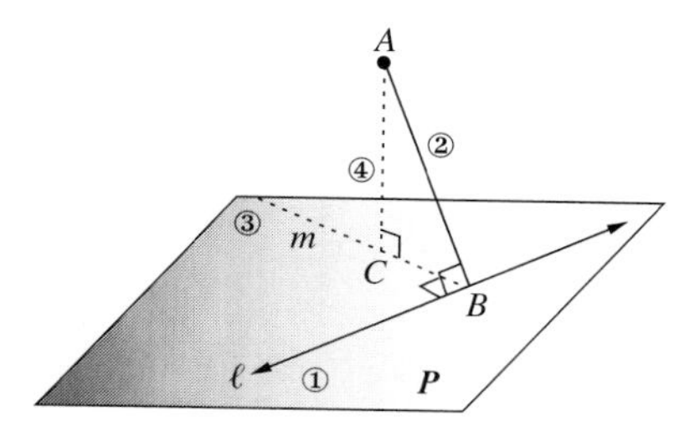

11. If a line and plane have exactly one point in common and the line is not perpendicular to the plane, using the result of Problem 10 and borrowing the result of Problem 12(b), prove there exists a unique plane perpendicular to the given plane containing the given line.

12. (a) Prove that there is only one perpendicular line to any plane $\boldsymbol{P}$ at a point A lying in that plane (assuming one exists; existence of the perpendicular is addressed in Problem 14).

(b) Prove there is only one perpendicular line from an external point A to any plane $\boldsymbol{P}$. (See figure for a hint.)

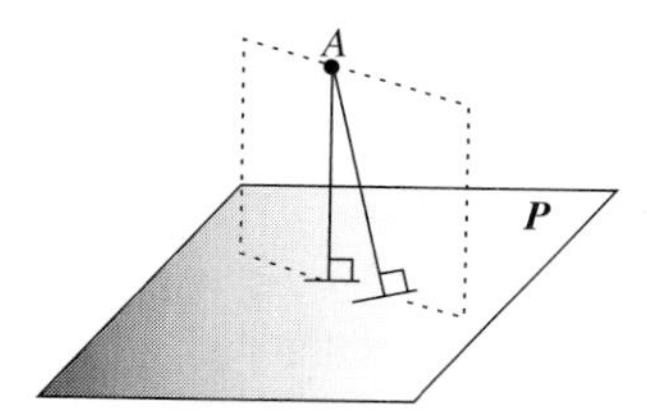

13. Prove that any two lines ℓ and m that are perpendicular to a plane $\boldsymbol{P}$ must lie in the same plane.

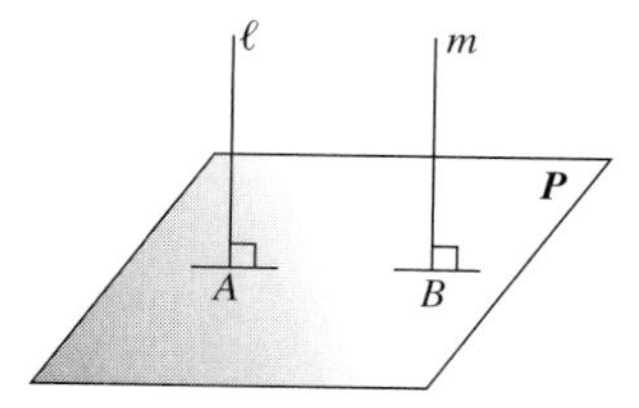

14. Give a Euclidean construction that will yield the line perpendicular to a plane at some point A in that plane, and prove that your construction works.

GROUP C

15. Prove in absolute geometry that perpendicular sections of any dihedral angle are congruent such as $\angle ABC$ and $\angle DEF$ in the figure. (***Hint:*** As shown, assume $AB = DE$ and $BC = EF$. If we can prove $AC = DF$ we will be finished, by SSS. Locate the midpoints L, M, and N of $\overline{AD}$, $\overline{BE}$, and $\overline{CF}$, respectively. Can you prove that $\diamond DLNF \cong \diamond ALNC$?)

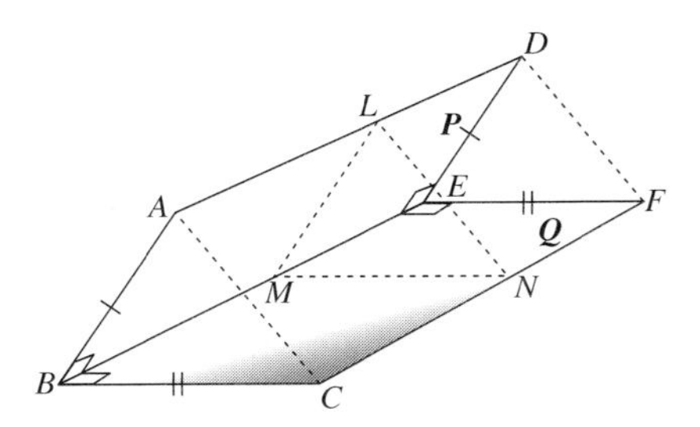

16. In conjunction with Problem 15, prove that any other angle whose sides lie in $\boldsymbol{P}$ and $\boldsymbol{Q}$ and make equal angles with the edge $\boldsymbol{P} \cap \boldsymbol{Q}$ has measure *less than* that of the dihedral angle. (***Hint:*** In the figure, $AB = BC$, $\overleftrightarrow{AB} \perp \ell$, and $\overleftrightarrow{BC} \perp \ell$. Then $AB' > AB$. Use inset in figure, comparing isosceles triangles.)

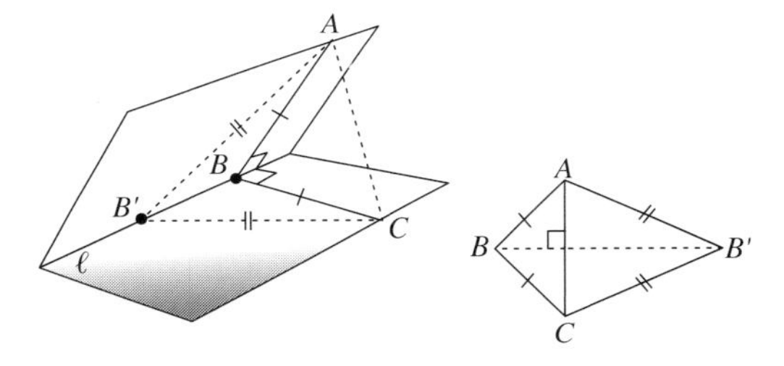

17. Bisector of a Dihedral Angle A half-plane is said to **bisect** a dihedral angle if, taken with the two sides, it forms two dihedral angles having equal measures. Prove that the points of the bisector of a dihedral angle are equidistant from the sides.

18. Give an explicit proof for the corollary to Theorem 2.

19. State and prove the Space-Separation Postulate (establish the desirable properties of the "sides" of any plane, analogous to the properties of half-planes and the sides of a line).

20. Prove that if the Euclidean Parallel Postulate holds in plane $\boldsymbol{P}$ then it holds for any other plane $\boldsymbol{Q}$. (***Hint:*** Use angle sums of triangles.)

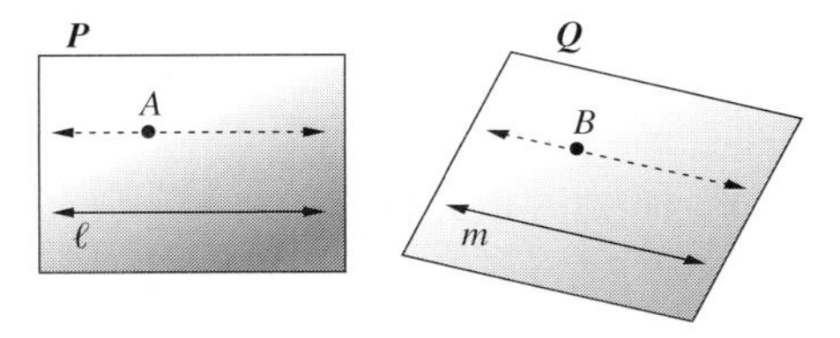

7.2 Parallelism in Space, Prisms, Pyramids, and the Platonic Solids

The *prism* is generated by lines parallel to some fixed line in space varying along a polygonal base. As such, the basis for its properties is the Law of Transitivity for parallel lines in space: *Lines parallel to the same line are parallel to each other.* It is easy to prove transitivity for parallelism in a plane, but somewhat more challenging in space. In the plane, all you have to do is construct a common transversal t to all three lines, as shown in Figure 7.10: If $\ell \parallel m$ and $m \parallel n$ then $m\angle 1 = m\angle 2$ and $m\angle 2 = m\angle 3$, so $m\angle 1 = m\angle 3$ by transitivity of ordinary equality, and therefore, $\ell \parallel n$, as desired. In space, we experience more difficulty. If $\ell \parallel m$ and $m \parallel n$ as indicated in Figure 7.11, we even seem to have trouble showing that lines ℓ and n lie in the same plane!

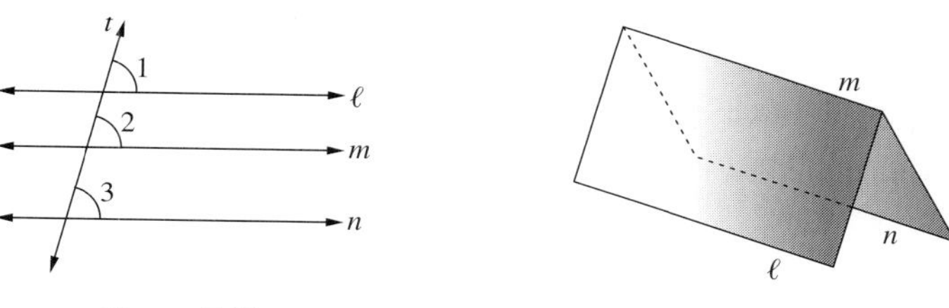

Figure 7.10

Figure 7.11

For convenience, at this point we introduce the notation $\boldsymbol{P}(ABC)$ for the unique plane $\boldsymbol{P}$ passing through three noncollinear points A, B, and C. Similarly, let $\boldsymbol{P}(A, \ell)$ denote the plane that contains point A and line ℓ, $A \notin \ell$.

PARALLELISM FOR LINES AND PLANES IN SPACE

Before we tackle the problem of transitivity, the official definition of parallelism for lines and planes in space will be stated.

DEFINITION: Two lines ℓ and m in E^3 are said to be **parallel** iff they lie in the same plane and do not meet. Two planes are **parallel** iff they do not meet. A line and plane are parallel iff they do not meet. We denote parallelism in all three cases by the usual symbol, $\parallel$.

Since the Transitive Law is valid only for Euclidean geometry, where the Euclidean parallel postulate is assumed, our definition of parallelism applies more directly to Euclidean geometry. However, there is a corresponding definition in hyperbolic space, which we give only for the sake of completeness—we do not intend to do anthing more with hyperbolic parallelism in space at this time.

DEFINITION: In H^3 two rays $\overrightarrow{AB}$ and $\overrightarrow{CD}$ are **asymptotically parallel** iff they lie in the same plane and, in that plane, satisfy the two conditions: (1) $\overrightarrow{AB}$ and $\overrightarrow{CD}$ lie on the same side of line $\overleftrightarrow{AB}$, (2) $\overrightarrow{AB}$ and $\overrightarrow{CD}$ do not meet, and (3) every ray $\overrightarrow{AP}$ in the interior of $\angle BAC$ meets $\overrightarrow{CD}$. A ray is **asymptotically parallel** to a plane iff it is asymptotically parallel to some ray in that plane (Figure 7.12). Two planes are **asymptotically parallel** iff some planar cross section of them consists of two lines containing asymptotically parallel rays (as suggested by the diagram in Figure 7.13).

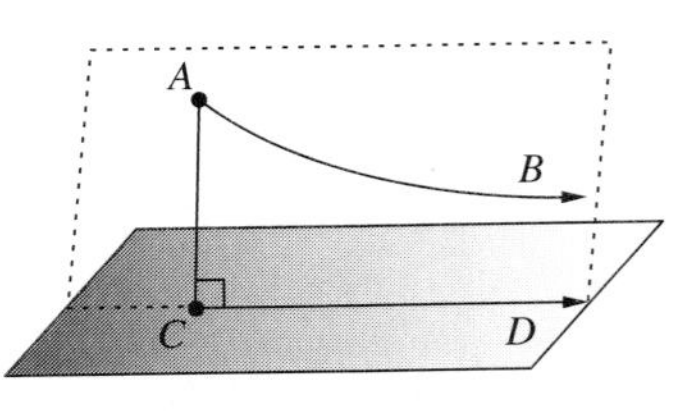

Figure 7.12

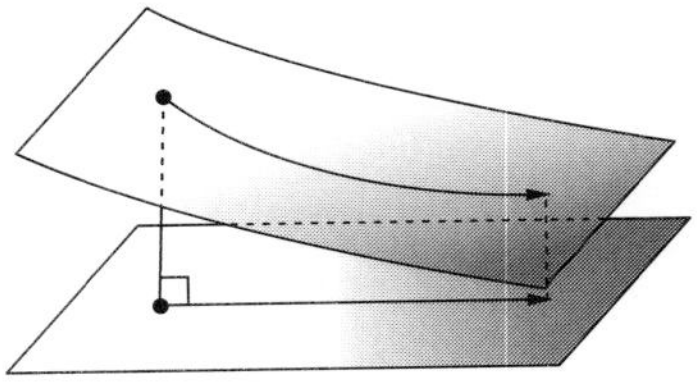

Figure 7.13

LEMMA A: If a line is parallel to a plane then it is parallel to every other line coplanar with it and lying in that plane.

PROOF

If in Figure 7.14 the line m, which is coplanar with ℓ and lies in $\boldsymbol{P}$, were to meet line ℓ at some point A, then ℓ would meet plane $\boldsymbol{P}$ at A.

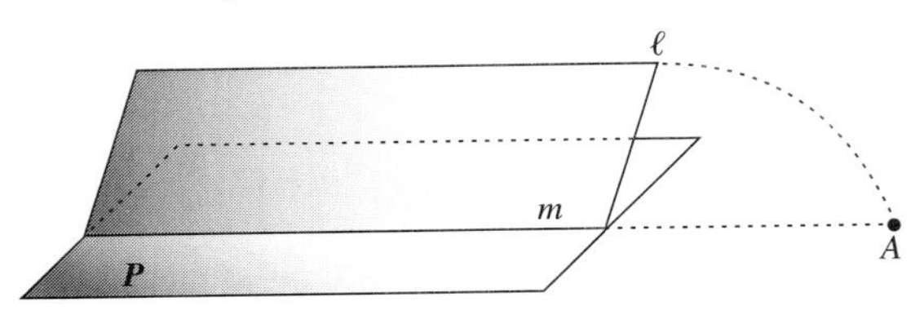

Figure 7.14

The rest of the material in this section is devoted exclusively to E^3 (Euclidean geometry).

THE A-FRAME THEOREM

Here we prove a theorem whose corollary is the transitive law for parallel lines.

THEOREM 1: If two distinct planes each contains exactly one of a pair of distinct parallel lines, then the planes are either parallel or their line of intersection is parallel to each of the given parallel lines.

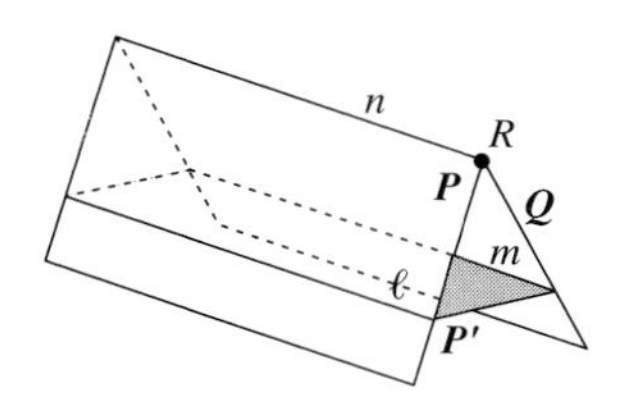

Figure 7.15

PROOF

Suppose $\ell \parallel m$ and planes $\boldsymbol{P}$ and $\boldsymbol{Q}$ contain ℓ and m, respectively. The cases for which there is nothing to prove are $\boldsymbol{P} \parallel \boldsymbol{Q}$, or one of the planes $\boldsymbol{P}$ or $\boldsymbol{Q}$ contains both m and ℓ. So assume otherwise, that $\boldsymbol{P}$ and $\boldsymbol{Q}$ meet at some point R, not in ℓ or m, and hence meet in a line n passing through R (Figure 7.15). We must prove that $n \parallel \ell$ and $n \parallel m$. Suppose $\boldsymbol{P'}$ is the plane of the parallel lines ℓ and m. It follows that $\ell \parallel \boldsymbol{Q}$, for otherwise, ℓ meets $\boldsymbol{Q}$ at a point on both $\boldsymbol{Q}$ and $\boldsymbol{P'}$ (since ℓ lies in $\boldsymbol{P'}$), that is, at a point on $\boldsymbol{P'} \cap \boldsymbol{Q} = m$. $\rightarrow\leftarrow$ By Lemma A, ℓ is parallel to every line in $\boldsymbol{Q}$ that is coplanar with it, or $\ell \parallel n$. In like manner, $m \parallel n$.

COROLLARY: If ℓ, m, and n are each pairwise distinct, and if $\ell \parallel m$ and $m \parallel n$, then $\ell \parallel n$.

PROOF

Suppose that $\boldsymbol{P}$ is the plane of ℓ and m and $\boldsymbol{Q}$ is that of m and n, as in Figure 7.16. Choose A any point on line n and let $\boldsymbol{R}$ be the plane of A and ℓ, with $n' = \boldsymbol{Q} \cap \boldsymbol{R}$. By the previous theorem, since $\boldsymbol{Q}$ and $\boldsymbol{R}$ contain the parallel lines m and ℓ, respectively, their intersection, n', is parallel to both ℓ and m. In particular, $n' \parallel m$ and $n \parallel m$. By the Euclidean parallel postulate, since both n and n' pass through A, $n' = n$; thus since $\ell \parallel n'$, then $\ell \parallel n$, as desired.

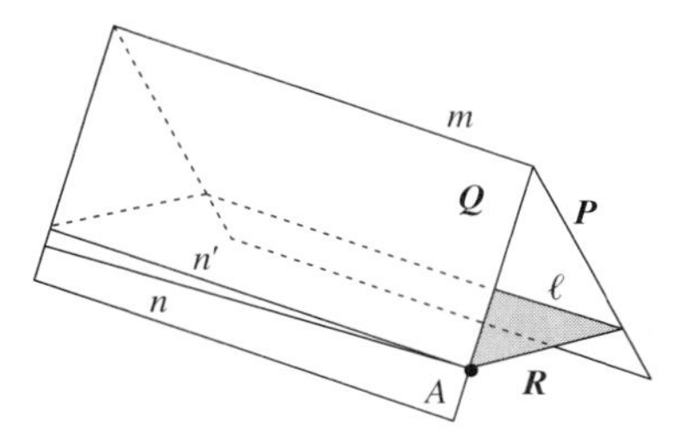

Figure 7.16

BOXES, PARALLELEPIPEDS, PRISMS, AND PYRAMIDS IN E^3

We prefer to call any six-sided figure in E^3 whose faces are quadrilaterals and interiors, as shown in Figure 7.17, a **box** (the technical name: **hexahedron**), which is a generalization of a rectilinear box (with right angles at the corners), just like the convex quadrilateral is a generalization of the parallelogram and rectangle. More generally, any bounded figure that is the intersection of a finite number of **closed half-spaces** (that is, all the points of a plane in E^3 together with the points lying on one side of that plane) is called a **polytope.**

Other basic terms are commonly used when dealing with polytopes: A **convex polyhedron** (or simply, **polyhedron**) is the *boundary* (outer shell) of a polytope. A

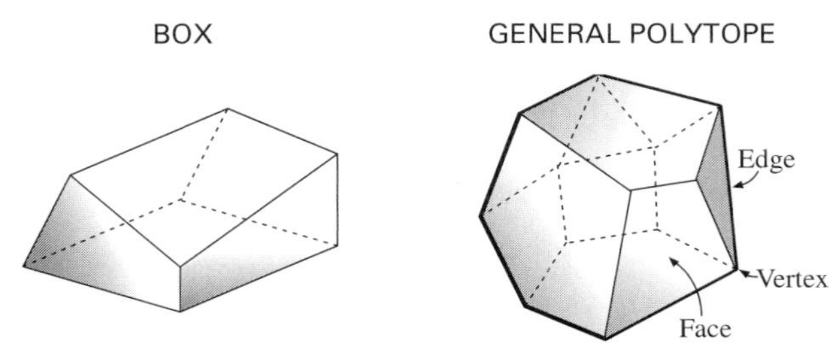

Figure 7.17

face is the set of points of the polytope which lie in a single plane (necessarily a convex polygon and its interior), an **edge** is any side of the polygonal boundary of a face (the intersection of two faces), and the **vertices** are the intersections of any two edges. And, as earlier discussed informally, a **hexahedron** is any polyhedron having six faces, and if each face is a quadrilateral and interior, it is called a **box.**

DEFINITION: A **parallelepiped** is any box whose six sides are each parallelograms (Figure 7.18). If each of the six sides is a rectangle, then the parallelepiped is called a **rectangular box**, and if each side is a square, a **cube**.

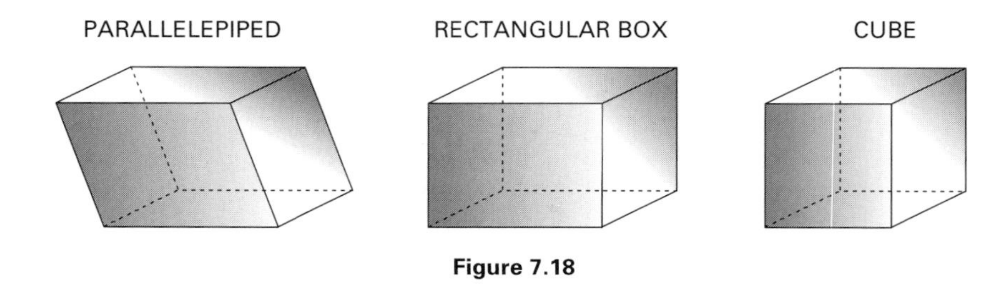

Figure 7.18

Next, we consider the three-dimensional figures known as *prisms* and *pyramids.*

DEFINITION: A **prism** is a polyhedron having two faces lying in parallel planes (one called the **base**, the other the **top**) and the remaining faces (called the **lateral sides**) lying in planes parallel to a fixed line m in E^3. (See Figure 7.19.) If the base of the prism is a triangle, then it is called a **triangular prism**. A **pyramid** is a polyhedron having the property that all faces except one meet at a common point A, called the **apex** (Figure 7.19). These **lateral faces** of the pyramid are necessarily the sides and interiors of certain triangles, and the remaining face (its **base**) is a polygon and interior. If all the faces of a pyramid are triangles and their interiors, it is called a **tetrahedron**.

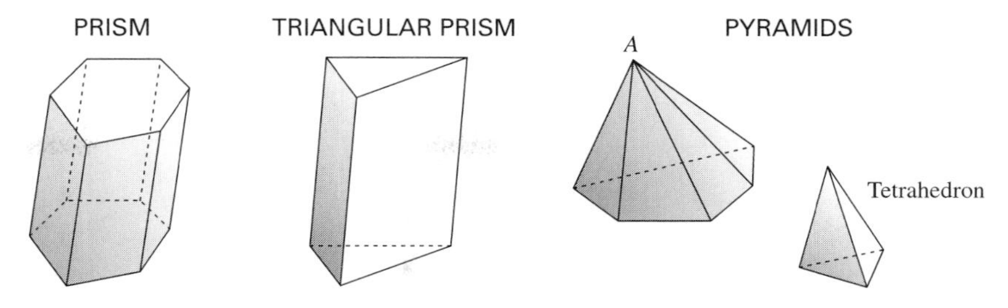

Figure 7.19

LEMMA B. Corresponding edges of the base and top of a prism are parallel.

PROOF

By definition, the base and top of the prism lie in parallel planes, so if corresponding edges are ℓ and m, they belong to the same side of the prism, hence lie in the same plane. Line ℓ cannot intersect the base, hence is parallel to it, so by Lemma A, $\ell \parallel m$.

THEOREM 2: The lateral sides of a prism are parallelograms and interiors.

PROOF

We have already shown that the opposite edges of a face lying in the base and top are parallel; it remains to show that the remaining pairs of opposite edges are parallel. But if ℓ and n are lines containing two edges (Figure 7.20), by hypothesis they are the intersections of two pairs of planes, each parallel to line m, such as $\boldsymbol{P}$ and $\boldsymbol{Q}$ in the figure. It follows that these intersections are also parallel to m, hence, by the Transitive Law, they are parallel to each other. Since opposite sides of each lateral side have been shown to be parallel, the lateral sides are all parallelograms.

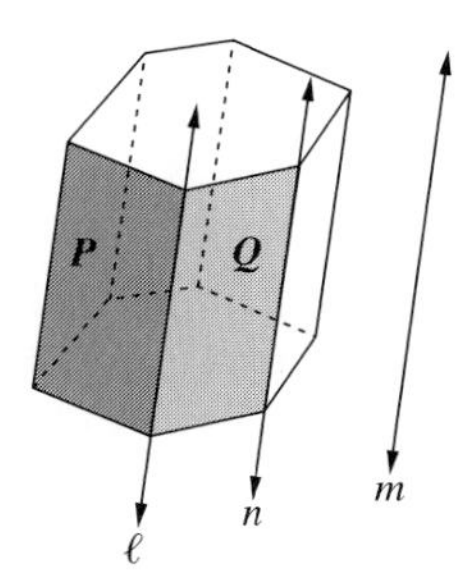

Figure 7.20

COROLLARY: A prism is a parallelepiped iff the base is a parallelogram.

This proof will be left as Problem 17; it remains to show only that the top face is a parallelogram since all other faces are parallelograms by hypothesis and Theorem 2.

DEFINITION: A **prismatic surface** is the set of all points lying in the lines (ℓ) that are parallel to some fixed line m and intersecting a base polygonal path (closed or not) lying in a plane not parallel to m (Figure 7.21). Each such line ℓ passing along the base path is called a **line generator.** A line passing through the geometric center (centroid) of the base (if there is one) and parallel to m is called the **axis** of the prism or prismatic surface. If instead of being parallel to a fixed line in space, the line generators pass through a fixed point C (the **vertex**), the resulting surface is called a **pyramidal** surface. In this case, the **axis** is the line passing through the vertex and the geometric center of the base, if there is one.

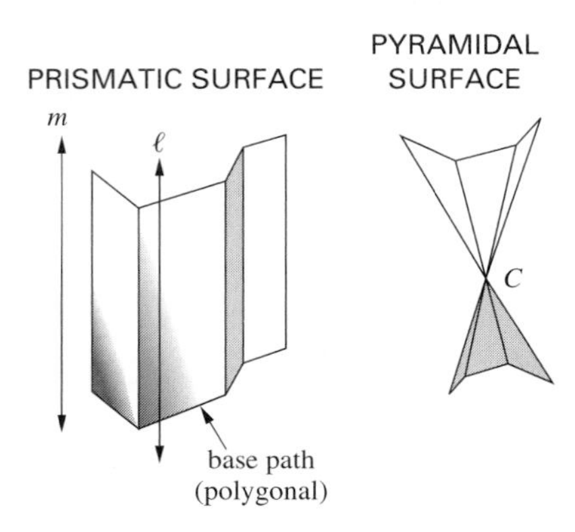

Figure 7.21

Important features of three-dimensional space are reflected by the following interesting phenomena involving a tetrahedron.

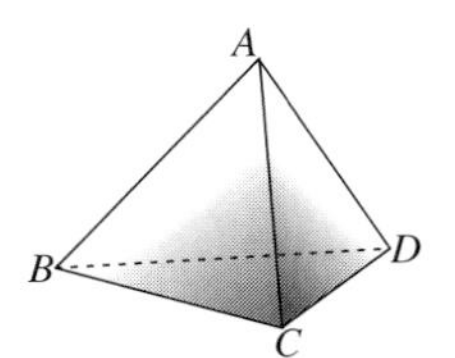

Figure 7.22

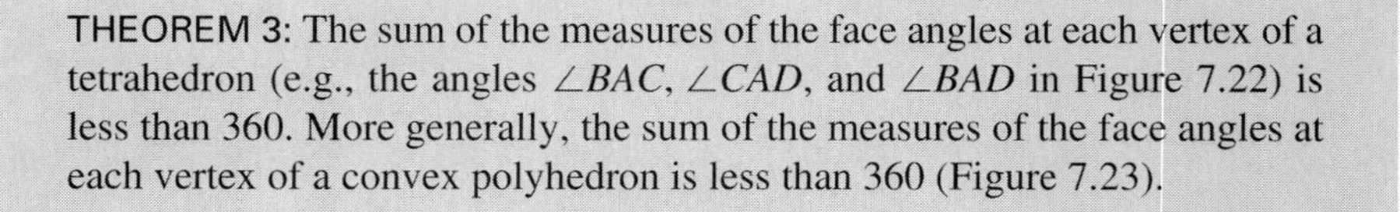

THEOREM 3: The sum of the measures of the face angles at each vertex of a tetrahedron (e.g., the angles $\angle BAC$, $\angle CAD$, and $\angle BAD$ in Figure 7.22) is less than 360. More generally, the sum of the measures of the face angles at each vertex of a convex polyhedron is less than 360 (Figure 7.23).

This theorem will be left as an interesting problem, Problem 18; a hint has been provided in case you want to try it. Theorem 3 can be used to prove a classical theorem dating back to Plato (430–349 B.C.E.), to be mentioned shortly.

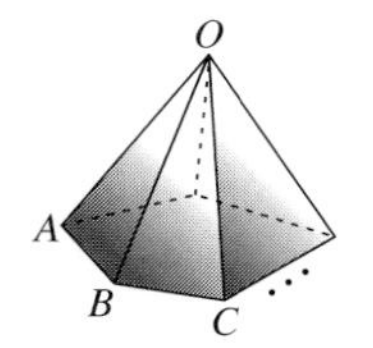

Figure 7.23

THEOREM 4: The measures of the three face angles at a vertex of a tetrahedron satisfy the triangle inequality. That is, the sum of the measures of any two is greater than that of the third. (In Figure 7.22, $m\angle BAC + m\angle CAD > m\angle BAD$.) (See problem 20.)

THE PLATONIC SOLIDS

We end the discussion of geometric solids by mentioning the most famous of all—the **Platonic solids,** so named because Plato used them in his famous works to represent the five elements of the universe: earth, water, air, fire, and the cosmos. These polytopes are characterized by the property that *the faces are regular polygons, each pair congruent,* and there are the same number of faces at each vertex. It is an interesting exercise to prove that these five are the only ones possible, a consequence of Theorem 3; it will be left as a nice problem for you (Problem 20). These solids are displayed in Figure 7.24, along with their corresponding **nets** (planar representations of the surface that can be used to construct physical models).

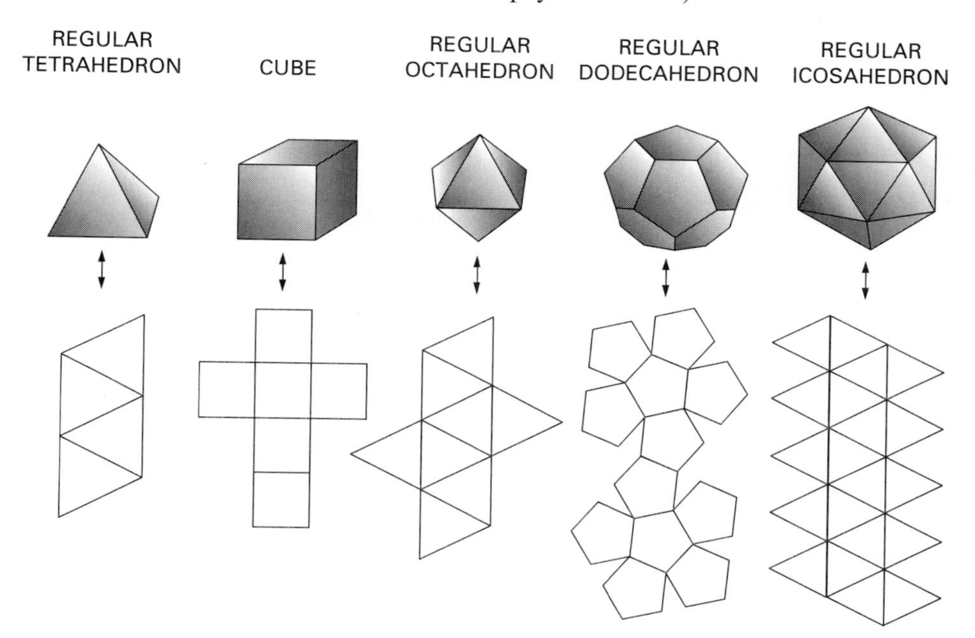

Figure 7.24

EULER'S FORMULA

A famous formula having important applications in both geometry and topology can be your own discovery if you follow the steps in the discovery unit that follows.

Moment for Discovery

Discovering Euler's Formula

Start with a prism, having, say, a pentagon as the base and top, as shown in Figure 7.25.

1. Count the faces, and let that number be F. (Do not forget to count the top and bottom.)
2. Count the edges, and let that number be E.
3. Count the vertices, and let that number be V.
4. Now compute the value of $F - E + V$. This is called the **Euler number** for the prism. Compute this same value, $F - E + V$ for a different prism, one having an octagon for its base and top, as shown in Figure 7.25. What value did you get this time? Do you suspect this number might be the same for all prisms? What number is it?

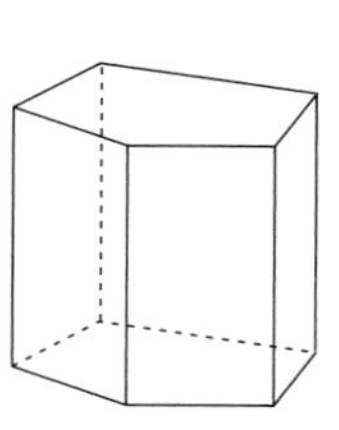
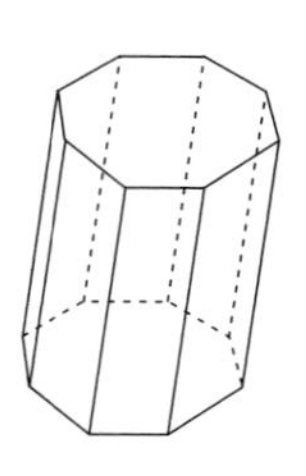

Figure 7.25

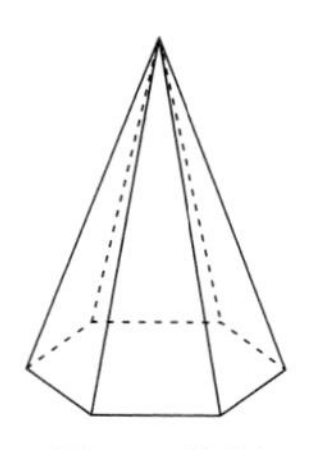

Figure 7.26

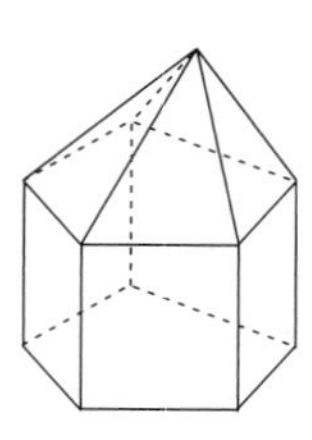

Figure 7.27

5. Perhaps the expression $F - E + V$ has one value for prisms, and a different value for other types of polytopes, thus giving us a characteristic number for each major type of solid. To find out what it is for a pyramid, consider one having a hexagon as base (Figure 7.26). Count the number of faces (F) on the sides and bottom of the pyramid. Calculate E and V for this pyramid.
6. Now, again, find the value of $F - E + V$. Did anything happen?
7. Repeat this experiment for the combination pyramid-prism shown in Figure 7.27.
8. Would you care to make a conjecture concerning all polytopes? Can you prove it? If so, then you have discovered an important theorem of three-dimensional combinatorial geometry. (The proof can be best accomplished through mathematical induction on the number of vertices of a polytope.)

PROBLEMS (7.2)

GROUP A

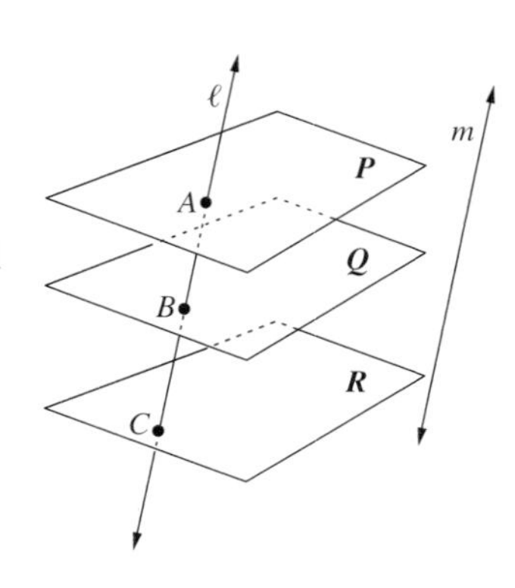

1. If two lines do not intersect and are parallel to the same plane, must they be parallel to each other?
2. Two lines ℓ and m that lie in the same plane are each parallel to a second plane that is not parallel to the first plane. Then $\ell \parallel m$. Prove.
3. Line ℓ pierces three parallel planes $\boldsymbol{P}$, $\boldsymbol{Q}$, and $\boldsymbol{R}$ at points A, B, and C, respectively. If ℓ varies, but remains parallel to a fixed line m that is not parallel to $\boldsymbol{P}$, show that the ratio AB/BC is constant. Generalize.

4. Two parallel lines pierce each of two parallel planes at corresponding pairs of points (P, P') and (Q, Q'). Prove that $\diamond PP'Q'Q$ is a parallelogram.

5. Prove that a plane can be passed through the center of a cube in such a manner that the intersection is a regular hexagon. (See figure for this problem.) You may use the Pythagorean Theorem and trigonometry.

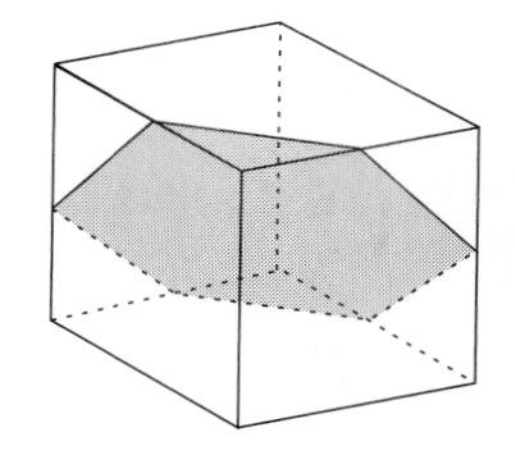

6. What is the length of a diagonal of a cube passing through its center if the sides of the cube are of unit length?

7. Find the altitude $\overline{AE}$ of a regular tetrahedron $ABCD$ having sides of unit length as shown in the figure.

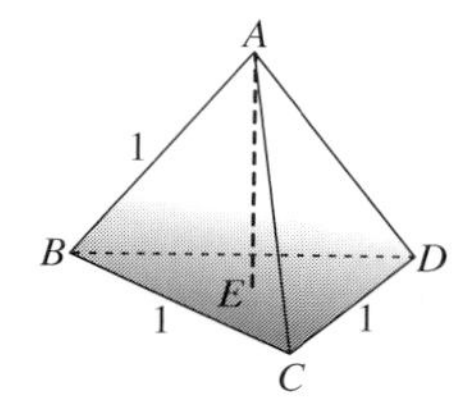

8. Find the altitude $\overline{AF}$ to a square pyramid whose sides are each of unit length.

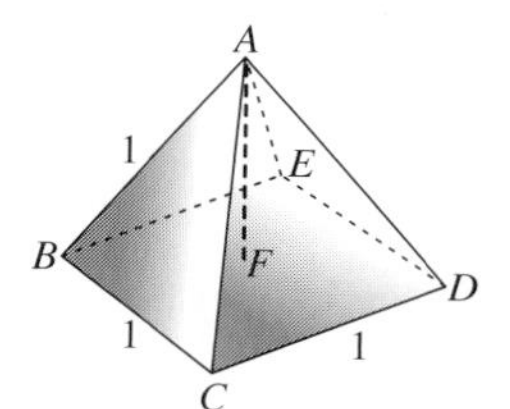

9. What is the measure of the dihedral angle between any two faces of a regular tetrahedron?

10. What is the measure of the dihedral angle between two adjacent faces of

(a) a cube

(b) a regular octahedron?

11. Suppose we place two square pyramids $ABCDE$ and $BCDEF$ with unit sides together to form an octahedron (figure at right), with A and F on opposite sides of the plane of B, C, D. Then all faces are congruent equilateral triangles, hence the octahedron is regular. Prove that A, B, F, and D are coplanar and that $\square ABFD$ is a square congruent to $\square BCDE$.

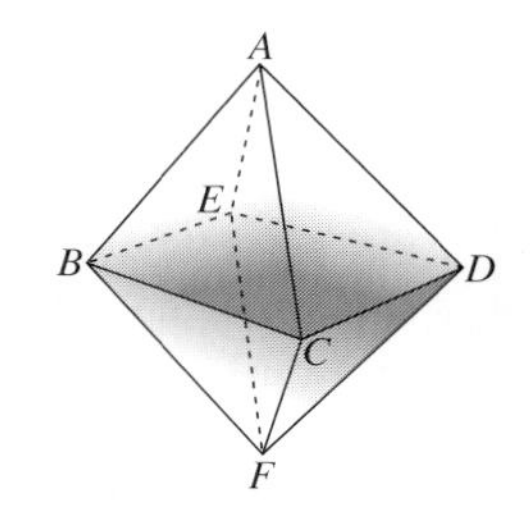

12. Prove or disprove: Two planes perpendicular to the same plane are parallel.

GROUP B

13. A soccer ball is a **semiregular** polyhedron (the faces are regular polygons, but are not all the same). How, and why, does it fit together? Use your knowledge of E^3 to build a convincing argument. (***Hint:*** First step: start with any pentagonal face as base and prove that as one "glues" the five regular hexagons to the edges of this base and turns each of them upward to meet the previous hexagon, the first and last pieces will fit together.)

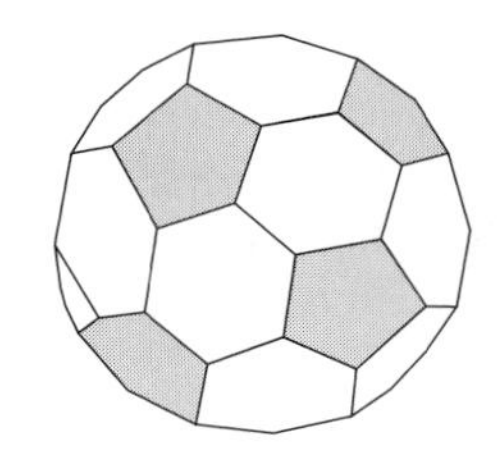

14. Archimedean Solids There are 13 classical solids of a more general type than the five Platonic solids, called **Archimedean solids.** They are the **semiregular** polyhedrons, which are polyhedrons having regular polygons as faces and the same number of each type of polygon meeting at each vertex. Five of these polyhedrons correspond directly to the Platonic solids—they are the **truncated** Platonic solids, formed by slicing off a part of the solid by passing a plane near each vertex. (The construction of the **truncated cube** is shown in figure (1) for this problem; the general procedure is defined in Problem 15.) The "soccer ball" polyhedron introduced in Problem 13 is actually the **truncated icosahedron.**

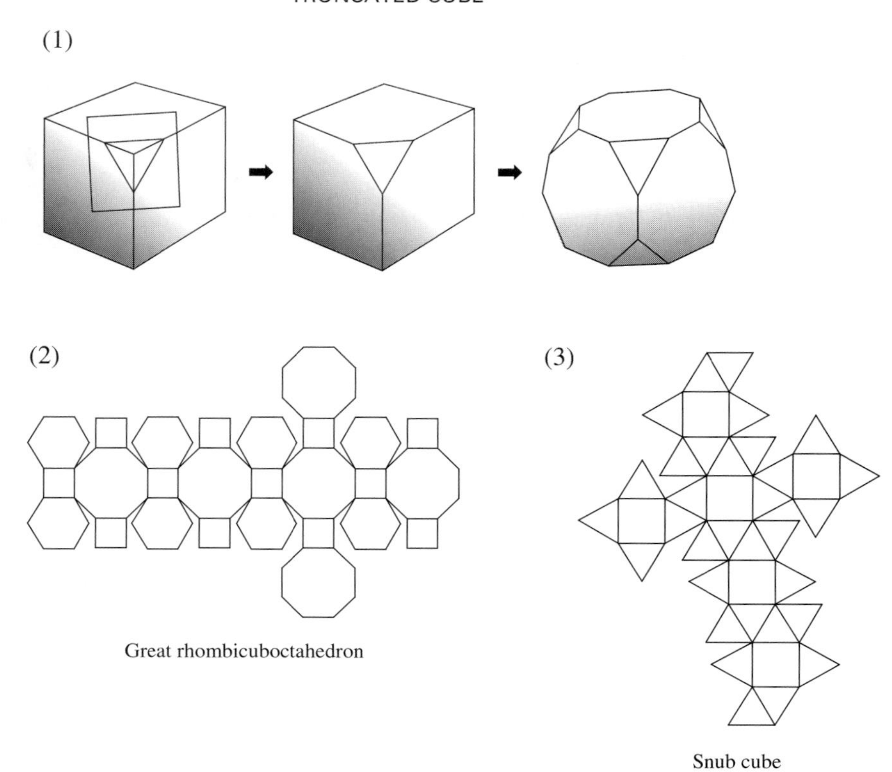

Great rhombicuboctahedron

(a) Show how to construct the truncated icosahedron to yield the "soccer ball" polyhedron. How many faces will it have?

(b) What regular polygons will be the faces of the **truncated octahedron**?

(c) Nets are shown in (2) and (3) for two other nontruncated Archimedean solids. As a special project, enlarge these diagrams on a sheet of poster board, cut them out, and tape them together to form the resulting polyhedrons (technical names given with each diagram).

15. The general truncation of an existing convex polyhedron, called a **truncated polyhedron,** is obtained as follows (see figure, top next page). Let $\boldsymbol{F}$ be a given polyhedron having vertices $A, B, C, D. \ldots$ At vertex A pass a plane $\boldsymbol{P}$ intersecting each edge incident with A and cut off that part of $\boldsymbol{F}$ lying on the same side of $\boldsymbol{P}$ as A, keeping the rest of $\boldsymbol{F}$ and its intersection with $\boldsymbol{P}$, plus the interior of the polygon $\boldsymbol{P} \cap \boldsymbol{F}$, as a new face. Repeat this for the remaining vertices $B, C, D, \ldots$ so that the plane used for each truncation does not interfere with truncations performed previously. Let $\boldsymbol{F}'$ denote the new polyhedron. (If $\boldsymbol{F}$ is regular, the truncating planes can be adjusted so that $\boldsymbol{F}'$ is semi-regular.)

(a) If the vertices of $\boldsymbol{F}$ all have the same **order** n (= number of faces of $\boldsymbol{F}$ that meet the vertex), show that the number of faces, edges, and vertices of $\boldsymbol{F}'$ in terms of those of $\boldsymbol{F}$ is given by

$$F' = F + V, \qquad E' = E + nV, \qquad V' = nV.$$

(b) Using the results of **(a)**, show that

$$F' - E' + V' = F - E + V$$

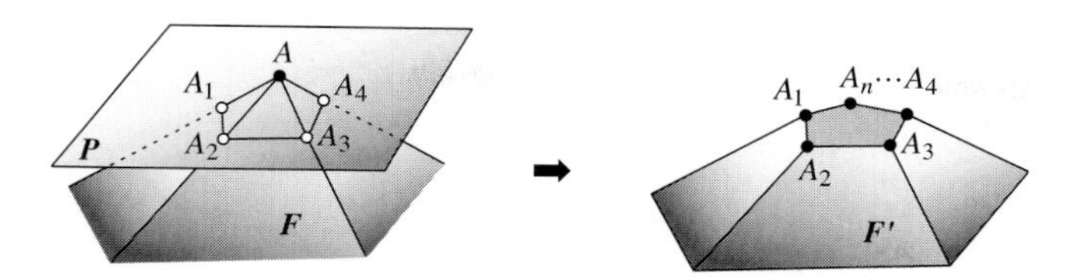

16. Consider the plane $\boldsymbol{B}$ containing the bisector of a dihedral angle, and let the planes containing the sides of the angle be denoted $\boldsymbol{P}$ and $\boldsymbol{Q}$. Set up a correspondence between $\boldsymbol{P}$ and $\boldsymbol{Q}$ by intersecting the perpendiculars to $\boldsymbol{B}$ in the obvious manner (as shown in the figure). Prove that under this special kind of parallel projection, *corresponding triangles are congruent.* (In this case, $\triangle ABC \cong \triangle A'B'C'$.)

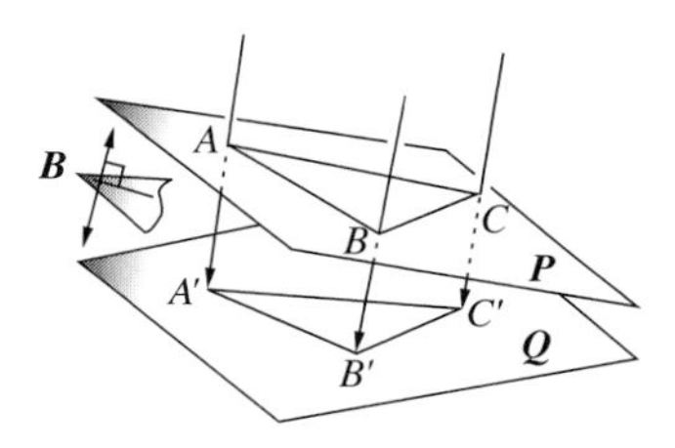

17. Prove the corollary to Theorem 2.

GROUP C

18. Prove that if we take a ray $\overrightarrow{OP}$ in the interior of the "polyhedral angle" $OABCD \cdots YZ$ in Theorem 3 such that the plane perpendicular to $\overrightarrow{OP}$ at Q meets all the given rays at $A, B, C, D, \cdots Y, Z$, then $m\angle AOB < m\angle AQB$, $m\angle BOC < m\angle BQC, \cdots$. (See Problem 16, Section 7.1.) Now sum.

19. Theorem 4 can be proven by a fundamental result from spherical trigonometry: The sum of the lengths of two sides of a spherical triangle is greater than that of the third side. (***Hint:*** Consider A as the center of a sphere, cutting the rays $\overrightarrow{AB}$, $\overrightarrow{AC}$, and $\overrightarrow{AD}$ at B', C', and D', and consider the spherical triangle $\triangle B'C'D'$.)

20. Theorem 3 provides an upper bound (namely 360) for the sum of the angles of any regular polyhedron at each vertex. The following table, properly filled out, will yield a proof that the Platonic solids are the only solids of this type possible (having congruent faces of regular polygons). Complete the table below, and write up the proof in detail. (The first two rows show what is needed; repeat for each polygon type until a negative answer is reached.)

Type of Regular Polygon Used	Number of Faces at Each Vertex	Angle of Each Face Has Measure	Total Angle Measure at Each Vertex	Is This Sum < 360?
Square	3	90	270	Yes
Square	4	90	360	No
Equilateral triangle				

21. More on Euler's Formula Euler's formula is topological in nature. That is, if a polyhedron is "smoothed" out to form a closed, continuous surface, then any convex polyhedron is topologically a sphere, by the process of gradually stretching and shrinking (without tearing) various parts of that surface. The Euler relation $F - E + V = 2$ (found in the above discovery unit) then becomes a characterization of all polytopes topologically equivalent to the sphere. If, however, a polytope is topologically equivalent to the *torus* (doughnut-shaped surface), a different value for $F - E + V$ occurs. Verify this by calculating $F - E + V$ for the polytopes illustrated in the figure; be sure to carefully count all faces, all edges, and all vertices—some of which are understood to exist, but hidden in the diagrams. (Define a **polyhedron** in general to be an edge-to-edge union of (planar) convex polygons and their interiors that is the boundary of some connected solid region in E^3.)

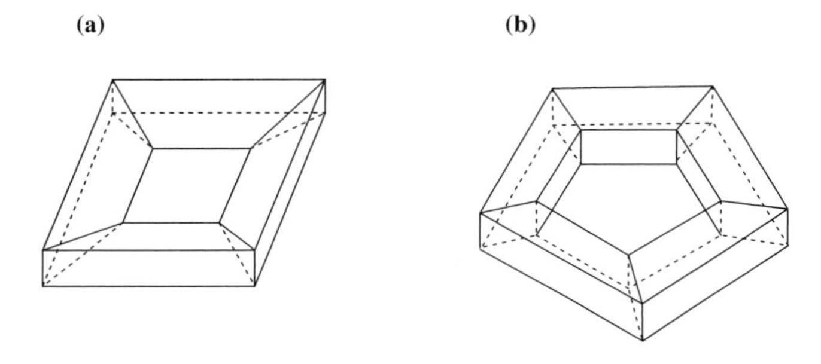

7.3 Cones, Cylinders, and Spheres

The cone and cylinder are generalizations of prismatic and pyramidal surfaces as defined in the previous section. When the polygonal base is replaced by an arbitrary **base curve** we get a **cylindrical surface** if the line generators are parallel or a **conical surface** if the line generators are concurrent at some point (the **vertex**), as illustrated in Figure 7.28. If the base curve is closed (like a circle), the resulting surfaces are called, respectively, a **cylinder** or **cone,** and, if that curve is a circle, a **circular cylinder** or **circular cone.** The **axis** is the line passing through the center of the circle and parallel to the generators of a cylinder or passing through the center and ver-

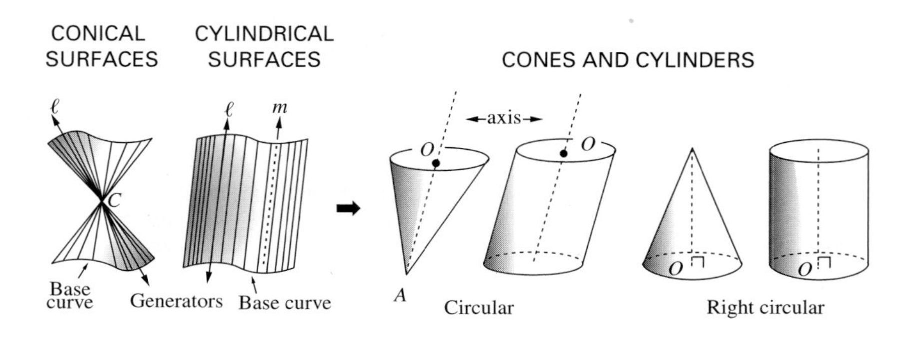

Figure 7.28

tex of a cone. A cone or cylinder is **right circular** when the axis is perpendicular to the plane of the base circle. Note that because the generators of a cone are *lines* and not rays or segments, the cone is an unbounded surface (like the cylinder), having two parts on either side of its vertex that are called **nappes.**

The sphere is the only surface in elementary geometry whose definition involves only the concept of distance in E^3. Consequently, its definition is much simpler than that for the cone or cylinder. Since it is important and is used so frequently, we give a formal definition for it.

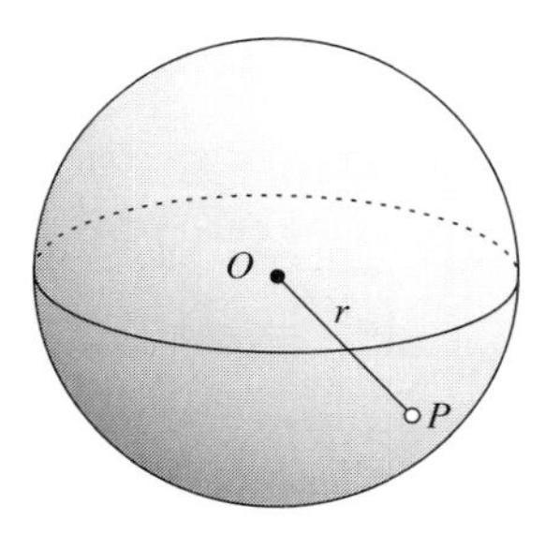

Figure 7.29

DEFINITION: Let O be any point in E^3 and r any positive real. Then, as illustrated in Figure 7.29, a **sphere** with **center** O and **radius** r is the set of all points P in E^3 at a distance r from O. A point Q such that $OQ < r$ is called an **interior point** of the sphere, while a point R for which $OR > r$ is an **exterior point.** A **solid sphere, solid ball,** or simply, **ball** is a sphere together with all its interior points (i.e., for some fixed point O and positive real number r, all points P in E^3 such that $OP \leq r$).

In E^3, a **circle** is defined as the set of all points in a plane lying at a fixed distance from a fixed point in that plane.

THEOREM 1: The intersection of a sphere and plane, if nonempty, is a circle or point. If it is a circle, then its center lies on the intersection of the plane and the line passing through the center of the sphere and perpendicular to the plane.

PROOF

Let $\boldsymbol{S}$ be a sphere having center O and radius r, and $\boldsymbol{P}$ any plane intersecting $\boldsymbol{S}$ (Figure 7.30).

(1) If the plane passes through O, then a point Q lies in the intersection $\boldsymbol{P} \cap \boldsymbol{S}$ iff $Q \in \boldsymbol{P}$ and $QO = r$, that is, iff Q lies on a circle with center O and radius r. Otherwise, let C be the foot of the perpendicular from O to plane $\boldsymbol{P}$, and let $A \in \boldsymbol{P} \cap \boldsymbol{S}$ (by hypothesis), with Q any other variable point of $\boldsymbol{P} \cap \boldsymbol{S}$. Then $OQ = OA = r$. Since $\overline{OC} \perp \boldsymbol{P}$, by HL $\triangle OQC \cong \triangle OAC$ and $CQ = CA =$ constant. Hence, all points $Q \in \boldsymbol{P} \cap \boldsymbol{S}$ belong to the circle $\boldsymbol{K}$ lying in plane $\boldsymbol{P}$, centered at C and having radius $s = CA \geq 0$.

(2) Conversely, assume $Q \in \boldsymbol{K}$, the circle in $\boldsymbol{P}$ centered at C, radius $s = CA$; we must show that $Q \in \boldsymbol{P} \cap \boldsymbol{S}$. Since $CQ = CA$, by SAS $\triangle OQC \cong \triangle OAC$ and $OQ = OA = r$. Hence $Q \in \boldsymbol{S}$, and since $Q \in \boldsymbol{P}$, then $Q \in \boldsymbol{P} \cap \boldsymbol{S}$, finishing the proof that $\boldsymbol{K} = \boldsymbol{P} \cap \boldsymbol{Q}$.

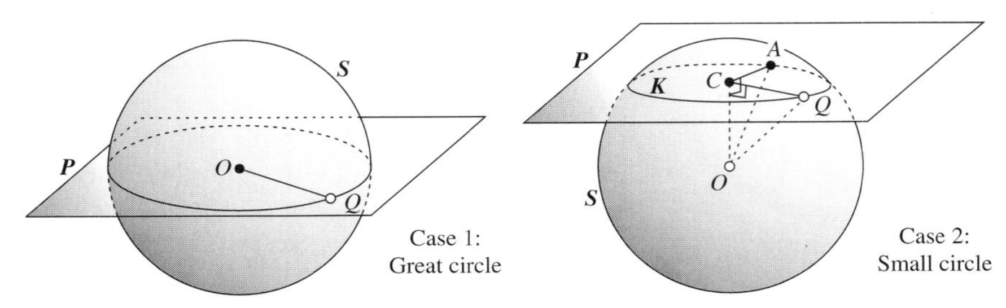

Figure 7.30

DEFINITION: A circle on a sphere whose plane passes through the center of the sphere is called a **great circle;** all others are **small circles.**

We might note that the above theorem is part of the formal basis for taking the unit sphere as a model for spherical geometry, where "lines" are taken as great circles. If A and B are two points that are not the end points of a diameter, then a unique plane is determined by A, B, and O, the center of the sphere, and hence the great circle containing A and B is unique.

It is interesting that both the preceding theorem and the one to follow are theorems of absolute geometry, but the next theorem is much more difficult to prove.

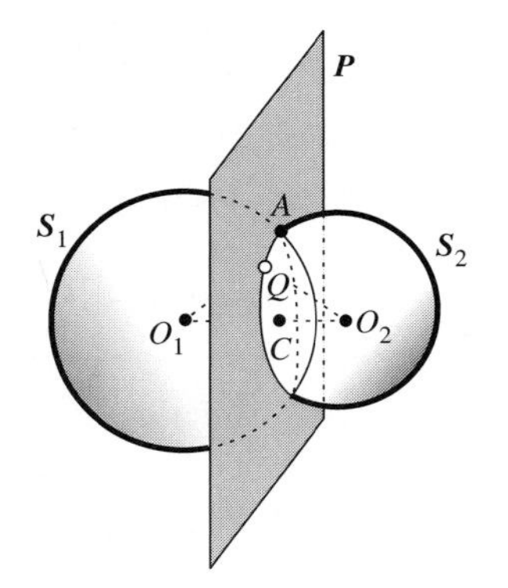

Figure 7.31

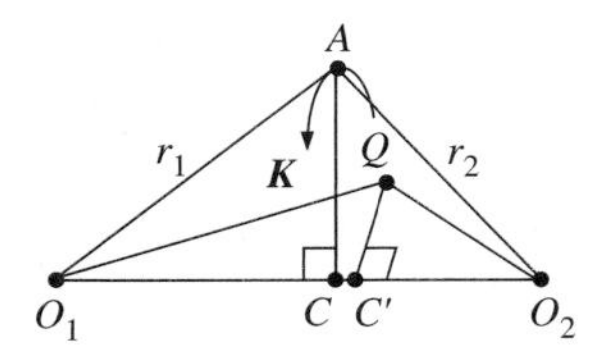

Figure 7.32

THEOREM 2: The nonempty intersection of two distinct spheres S_1 and S_2 is a circle or a point, and if it is the former, the plane of the circle of intersection is perpendicular to the line of centers $\overleftrightarrow{O_1O_2}$ of the two spheres.

PROOF

(1) For sake of argument assume that $\boldsymbol{S_1} \cap \boldsymbol{S_2}$ contains at least two points, and we may then assume that A is a point in $\boldsymbol{S_1} \cap \boldsymbol{S_2}$ not lying on line $\overleftrightarrow{O_1O_2}$ (Figure 7.31). Pass a plane $\boldsymbol{P}$ through A perpendicular to line $\overleftrightarrow{O_1O_2}$, meeting $\overleftrightarrow{O_1O_2}$ at $C \neq A$. By Theorem 1, $\boldsymbol{P} \cap \boldsymbol{S_1}$ and $\boldsymbol{P} \cap \boldsymbol{S_2}$ are circles through A; the two circles must coincide since both have the same center ($= C$) and the same radius ($= CA$). Hence, if that common circle is denoted by $\boldsymbol{K}$, we have $\boldsymbol{K} \subseteq \boldsymbol{S_1} \cap \boldsymbol{S_2}$. It remains to prove that $\boldsymbol{S_1} \cap \boldsymbol{S_2} \subseteq \boldsymbol{K}$.

(2) Suppose that $Q \in \boldsymbol{S_1} \cap \boldsymbol{S_2}$. Let r_1 and r_2 be the radii of $\boldsymbol{S_1}$ and $\boldsymbol{S_2}$. Drop the perpendicular from Q to line O_1O_2 at C' (Figure 7.32). Since $Q \in \boldsymbol{S_1} \cap \boldsymbol{S_2}$, $QO_1 = r_1 = AO_1$ and $QO_2 = r_2 = AO_2$, so that by SSS $\triangle AO_1O_2 \cong \triangle QO_1O_2$, and $\angle AO_1O_2 \cong \angle QO_1O_2$. Hence by HA, $\triangle AO_1C \cong \triangle QO_1C'$ and $O_1C = O_1C'$ or $C' = C$. Therefore, by congruent triangles $QC = AC = s$. To prove that $Q \in \boldsymbol{P}$, note that any line perpendicular to a line that is itself perpendicular to a plane at the point of intersection

must lie in that plane (otherwise, there would be two perpendiculars to a line lying in the same plane). Hence $Q \in \boldsymbol{K}$, as desired, ending the proof that $\boldsymbol{S}_1 \cap \boldsymbol{S}_2 \subseteq \boldsymbol{K}$.

The next theorem is important in the study of the model for hyperbolic space in E^3. It is not provable *in* hyperbolic space (for the reason that in the hyperbolic plane three noncollinear points do not always lie on a circle).

THEOREM 3: Any four noncoplanar points A, B, C, and D determine a unique sphere $\boldsymbol{S}$ passing through them.

PROOF

Let $\boldsymbol{P}$ be the plane of A, B, and C, and locate E, the center of the circumcircle ($\boldsymbol{K}$) of $\triangle ABC$ (Figure 7.33). If ℓ is the perpendicular to $\boldsymbol{P}$ at E, then one can see that every point $F \in \ell$ is equidistant from A, B, and C. Conversely, if F' is any other point in space equidistant from A, B, and C, then by Theorem 3 of Section 7.1, F' lies on the two equidistant loci of A and B, and of B and C, which are planes that intersect along line $\ell \perp \boldsymbol{P}$, hence $F' \in \ell$. Thus, any sphere passing through A, B, and C must have its center on line ℓ; we have only to show that exactly one of those spheres passes through D. Let $\boldsymbol{Q}$ be the plane of line ℓ and point D, and let the line of intesection $\boldsymbol{Q} \cap \boldsymbol{P}$ intersect $\boldsymbol{K}$ at point A'. Then, as shown in Figure 7.33, determine the perpendicular bisector of segment $\overline{A'D}$ in plane $\boldsymbol{Q}$. By previous results in Euclidean geometry, this perpendicular bisector must intersect ℓ at some point O since line $\overleftrightarrow{A'D}$ is not parallel to line $\overleftrightarrow{A'E}$. Then, $OD = OA' = OA$ and it follows that O is equidistant from A, B, C, and D. Hence, we have found the center of the sphere passing through A, B, C, and D, unique by construction.

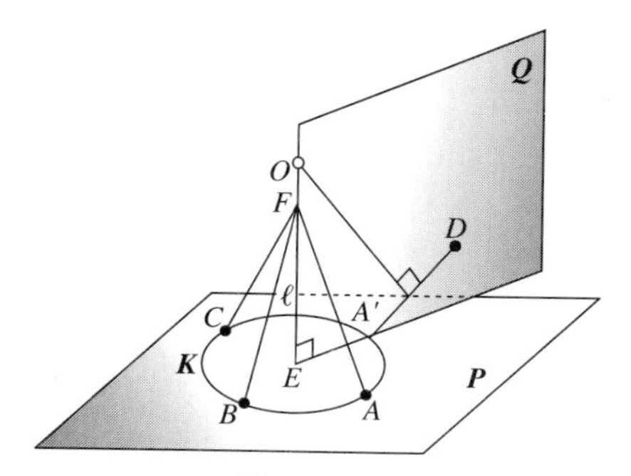

Figure 7.33

TANGENT PLANES AND LINES TO SPHERES

If a plane touches a sphere at just one point, it is called a **tangent** plane, and similarly for a line. Three basic results, relegated to the problem section, are:

- A plane is tangent to a sphere at a point A iff the radius of the sphere to A is perpendicular to the plane at A.
- A line is tangent to a sphere at a point A iff the radius of the sphere to A is perpendicular to the line at A.
- If two lines $\overleftrightarrow{PA}$ and $\overleftrightarrow{PB}$ are tangent to a sphere at A and B from some external point P, then $PA = PB$.

Moment for Discovery

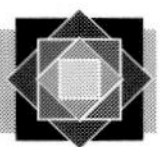

Area Versus Volume Phenomenon

Consider the familiar problem: Among all rectangles inscribed inside the unit circle, find the dimensions of the one having the greatest area. The answer is, naturally, a square having sides $\sqrt{2}$ units in length. We might suppose that the analogous problem for the volume of a cylinder inscribed in a unit sphere has the same answer, namely, when the diameter of the cylinder equals its height (thus, the diagram on the left in Figure 7.34 is the precise cross section of the maximal cylinder). Assuming this to be the case, the cylinder having maximal volume would also have dimensions $\sqrt{2}$ (diameter) by $\sqrt{2}$ (height). Anticipating the volume formulas derived in the next section, the volume of this maximal cylinder would be calculated as follows: $V = \pi r^2 h$, where $r = \frac{1}{2}\sqrt{2}$, $h = \sqrt{2}$.

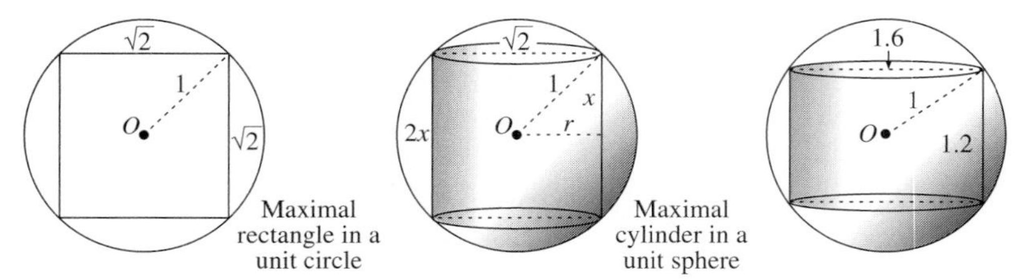

Figure 7.34

1. Use these values to find the volume of this cylinder in terms of π (use three-place accuracy).
2. Consider the inscribed cylinder whose height is slightly less, namely, 1.2 units instead of $\sqrt{2} \approx 1.4$ units. Find the radius r of the cylinder in this case. (See diagram to the far right of Figure 7.34.)
3. Calculate the volume of this inscribed cylinder (with height $= 1.2$) in terms of π. What did you find? Any conclusions? (You should have obtained $V = 0.768\pi$ here.)
4. With $2x$ as the height of an arbitrary inscribed cylinder, find a formula for its volume. [See diagram in the center of Figure 7.34; here, $r \neq \sqrt{2}/2$ but instead, will be some function $f(x)$.]

C 5. Use calculus on the formula in Step 4 to find the correct dimensions of the optimal cylinder (exactly, then approximate by decimal the multiple of π obtained).

6. Draw the cross section of the sphere and maximal inscribed cylinder, and compare it with that of the inscribed square and unit circle. Is the discrepancy noticeable?

PROBLEMS (7.3)

GROUP A

1. Two planes are tangent to a sphere of radius 10. If the dihedral angle of the two planes has measure 60, how far apart are the points of contact, A and B?

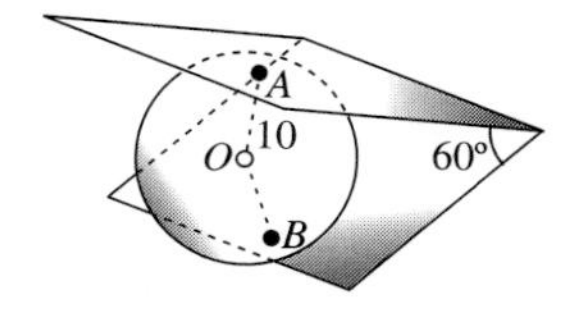

2. A plane is tangent to a sphere iff it is perpendicular to the radius drawn to the point of contact. Prove.

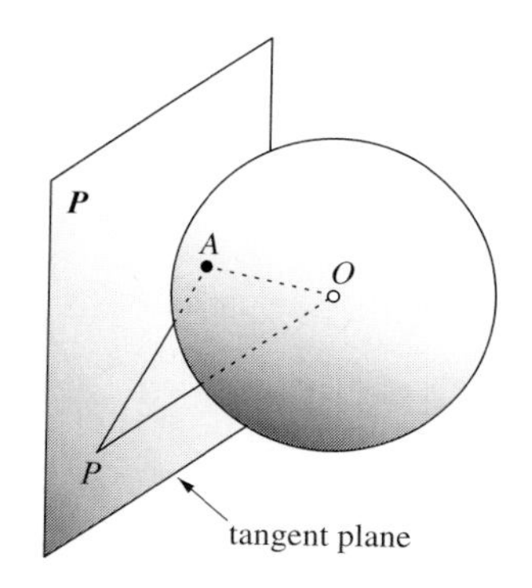

3. In a tetrahedron $ABCD$, regular or not, the six plane bisectors of the dihedral angles determined by each pair of adjacent sides meet at a point equidistant from the sides. Prove. Hence, every tetrahedron has an **insphere**—a sphere that lies inside the tetrahedron and is tangent to all four faces. (See Problem 17, Section 7.1.)

4. Find the radius of the sphere inscribed in
   (a) a cube having unit side,
   (b) a regular tetrahedron having unit side (see Problem 3),
   (c) a regular octahedron having unit side. (***Hint:*** If P is the point of contact of one of the faces $\triangle ABC$, why is $PA = PB = PC$? See Problem 10.)

5. Ten cannon balls of diameter 9 in. are stacked in pyramid fashion, as illustrated. Find the exact height of the stack from ground level. ("Exact" means no decimal approximations are allowed.)

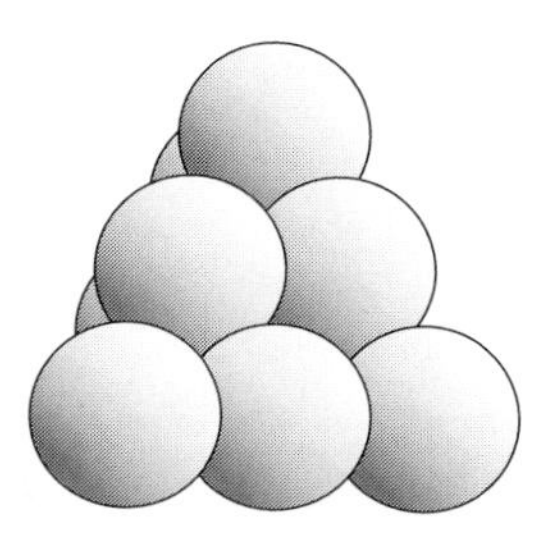

6. Six steel pipes having a diameter of 9 in. are stacked in pyramid fashion, as shown in the figure. What is the exact height of the stack? (How does this compare with the answer to Problem 5?)

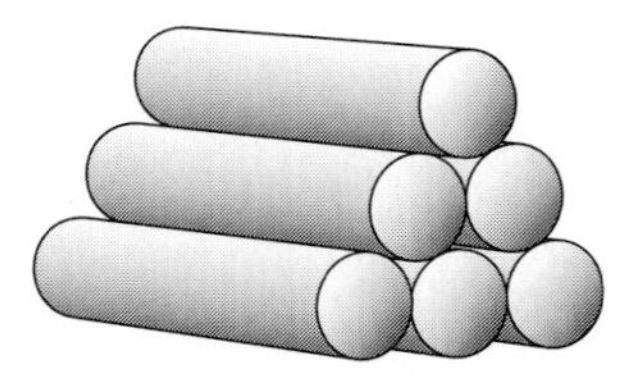

7. A set of four marbles having a diameter of $\frac{3}{4}$ in. is to be packaged inside a plastic spherical container. What is the exact diameter of the smallest container possible?

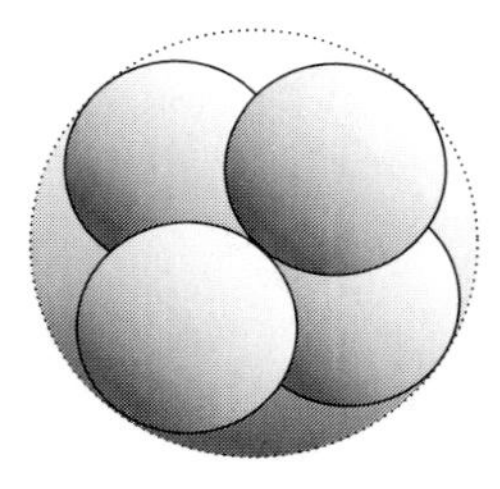

8. A marble having a diameter of $\frac{3}{4}$ in. falls inside a conical paper cup whose diameter and height are both 3 in. How far is the center of the marble from the vertex of the right circular cone representing the cup?

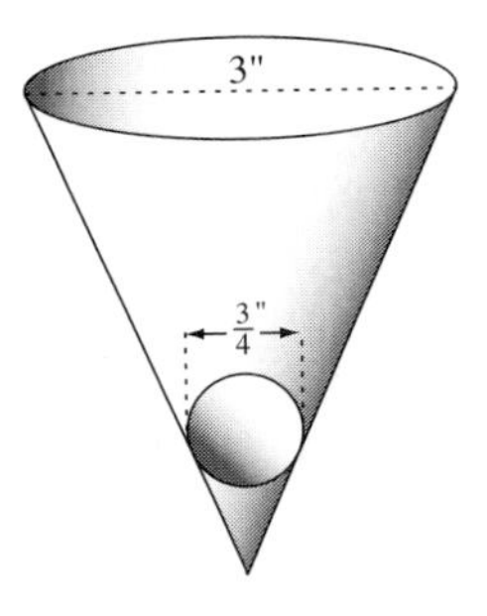

GROUP B

9. Prove that if two planes are tangent to a sphere and the planes are not parallel, the points of contact A and B are equidistant from the line of intersection ℓ of the two planes. Show also that $PA = PB$ for any point P on line ℓ.

10. If lines PA and PB are tangent to a sphere at A and B from some external point P, prove that $PA = PB$.

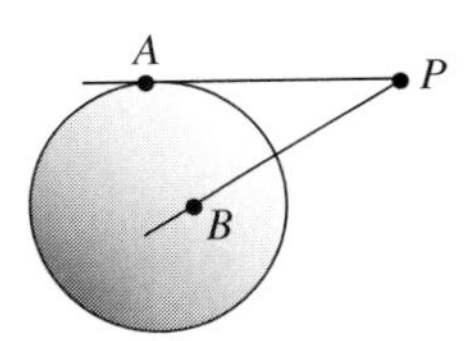

11. If the intersection of a sphere and cone is a circle and no other points, the sphere and cone are said to be **tangent,** and that circle is called the **circle of contact.** Show that a sphere $\boldsymbol{S}$ with center O is tangent to a right circular cone $\boldsymbol{C}$ with vertex A iff (1) point A lies exterior to the sphere, (2) O lies on the axis $\overleftrightarrow{AB}$ of the cone, and (3) each line generator of the cone is tangent to the sphere at a point on the circle of contact. (You may use a property of right circular cones difficult to prove synthetically, that the only circles on such cones are those that are parallel to the base circle with their centers lying on the axis.)

12. In the figure for this problem, two spheres are tangent to the cone and also tangent to a common plane at the points F_1 and F_2. If P is any point of intersection of the plane and cone, prove the characteristic property for an ellipse,

$$PF_1 + PF_2 = \text{constant}.$$

(Show that the value of the constant sum equals $2a = MN$; use result in Problem 9.)

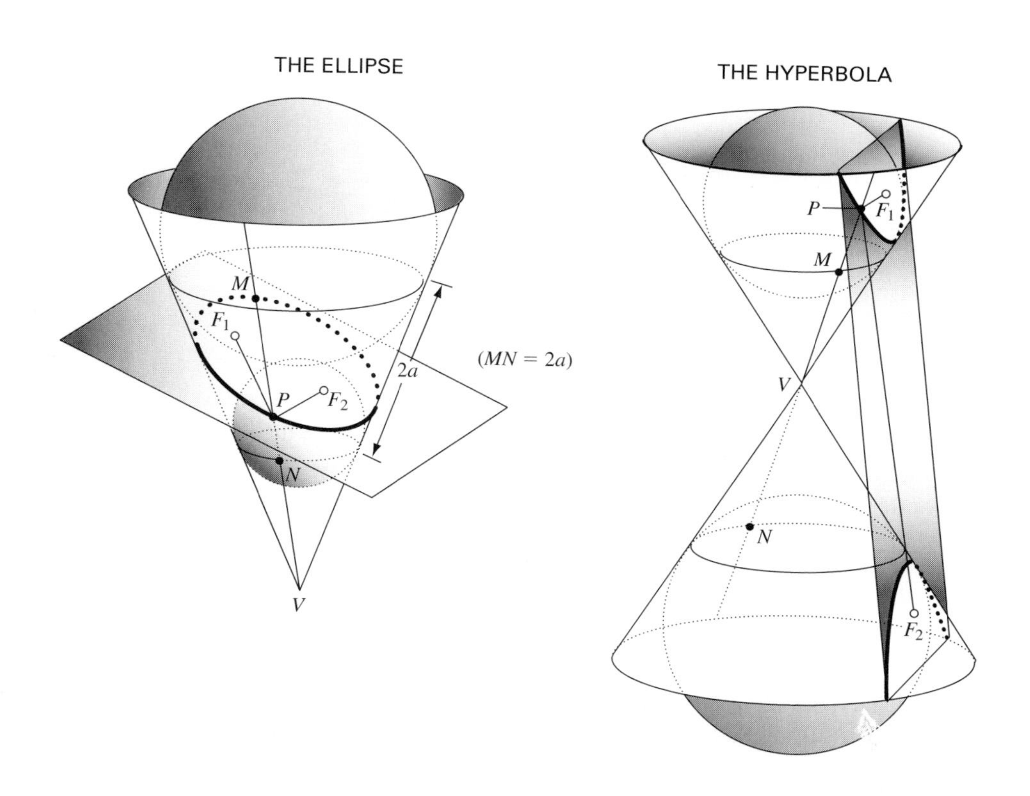

13. As in Problem 12, the two spheres are both tangent to the plane at F_1 and F_2, but the plane is tangent externally to both spheres. This time, prove the characteristic property for a hyperbola,

$$PF_2 - PF_1 = \text{constant},$$

where again the constant is $2a = MN$.

GROUP C

14. Find a way to exhibit the focus-directrix property of a parabola using one sphere tangent to a cone and a plane parallel to one of the cone's line generators.

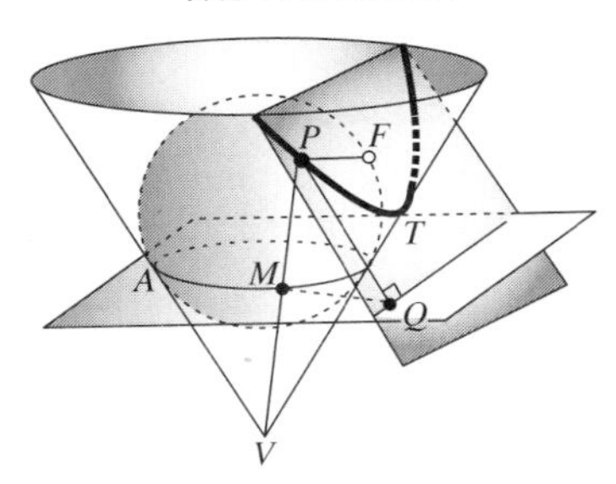

15. Prove that the intersection of a right circular cone and sphere whose center lies on the axis of the cone, if nonempty, is either a circle or the union of two circles. (See Problem 11.)

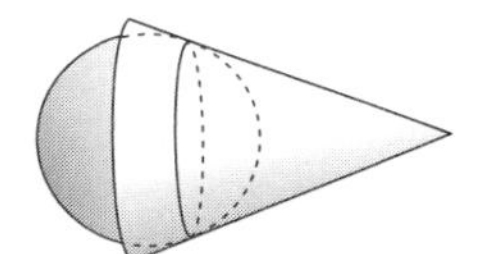

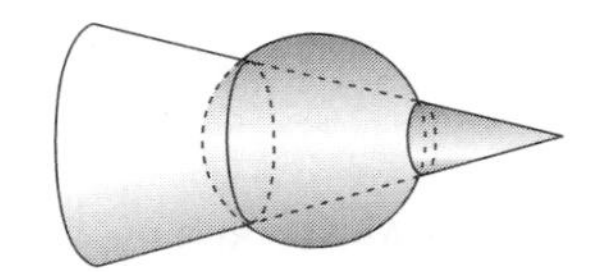

7.4 Volume in E^3

An understanding of the basic concepts in E^3 as introduced in the preceding sections enables us to work with the important topic of volume. We present a modern, axiomatic treatment, where the existence of volume is assumed, obeying desirable properties. Then we will use these properties to derive some of the standard volume formulas. (It might be helpful at this time to review the discussion of area and volume in Section 4.6.)

AXIOMS FOR VOLUME

We assume the same axioms for volume as before, except we shall replace Axiom 4 by the actual formula for the volume of a parallelepiped (height times area of base). As noted earlier (in the problems) we can prove this from the previous list of axioms;[1] assuming this property merely shortens the development.

We consider the following class of three-dimensional objects, whose volumes will be assumed to exist (i.e., **measurable**), to be called **solid regions** (as in Section 4.6):

[1]This is a corollary of Cavalier's Principle and the volume formula $V = Bh$ established in Problem 18, Section 4.6.

(1) Any bounded, convex set in E^3

(2) The complement of a convex set inside a bounded convex set (that is, all the points in a bounded convex set not lying in a convex subset of that convex set)

(3) A finite union or intersection of such sets (1) or (2).

The axioms on volume may now be stated.

1. EXISTENCE POSTULATE	To each region T, there corresponds a real number Vol $T \geq 0$, called its volume.
2. DOMINANCE POSTULATE	If $T_1 \subseteq T_2$ then Vol $T_1 \leq$ Vol T_2.
3. POSTULATE OF ADDITIVITY	If $\text{Vol}(T_1 \cap T_2) = 0$, then $\text{Vol}(T_1 \cup T_2) = \text{Vol } T_1 + \text{Vol } T_2$.
4. UNIT OF MEASURE	The volume of a parallelepiped is the product of its altitude and the area of its base.
5. CAVALIERI'S PRINCIPLE	If all the planes parallel to some fixed plane that meet the regions T_1 and T_2 do so in plane sections having equal areas, whose boundaries are part of the boundaries of the given regions, then Vol T_1 = Vol T_2 (Figure 7.35).

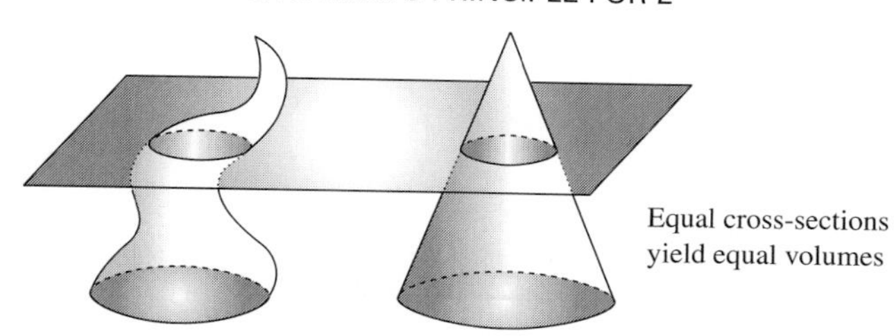

Figure 7.35

Note that there are only five axioms, as compared with six axioms originally given in Section 4.6. The Congruence Postulate was omitted because although it is true, it is simply not very useful in E^3.

VOLUMES OF PRISMS AND CIRCULAR CYLINDERS

We must first take care of a technical detail.

LEMMA: The volume of any planar cross section of a finite union of bounded, convex sets in E^3 is zero.

PROOF

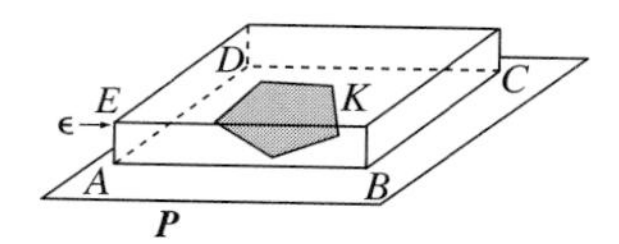

Figure 7.36

Let T be such a finite union, and consider any plane $\boldsymbol{P}$ that cuts T in a set K (Figure 7.36). Since T is bounded, so is K. In fact, K, like T, is also a finite union of bounded, convex sets—all lying in the *same plane*. Thus K is a region and Vol K exists and is nonnegative. We can find a rectangle $\square ABCD$

in $\boldsymbol{P}$ that contains K in its interior, since K is bounded, and Area $\square ABCD = B > 0$. If we construct a segment AE perpendicular to $\boldsymbol{P}$ of arbitrary (small) length ϵ and consider the box $ABCDE$, then

$$K \subseteq \square ABCD \subseteq \text{Box } ABCDE,$$

and by the Dominance Postulate and Axiom 4,

$$0 \leq \text{Vol } K \leq \text{Vol(Box } ABCDE) = B\epsilon.$$

Since ϵ is arbitrary and B and Vol K are constant, it follows that Vol $K = 0$.

We can now obtain the following formula, where B = base area and h = altitude,

(1) VOLUME OF A PRISM $\qquad V = Bh$

We start with a triangular prism whose base has area B and altitude, of length h. In Figure 7.37 is illustrated such a prism $QABC$, with base $\triangle ABC$ having area B, and altitude (length h) perpendicular to the plane $\boldsymbol{P}(ABC)$. In the base plane, construct a parallelogram $ABCD$ having adjacent sides $\overline{AB}$ and $\overline{BC}$, and in the planes $\boldsymbol{P}(QAD)$ and $\boldsymbol{P}(DCS)$ construct parallelograms $AQTD$ and $CSTD$. In this manner, we construct a parallelepiped $QABCD$, as shown in Figure 7.37. This creates a second triangular prism $QACD$ having the same volume as the original prism $QABC$ (by Cavalieri's Principle). That is,

$$V = \text{Vol}(QABC) = \text{Vol}(QACD)$$

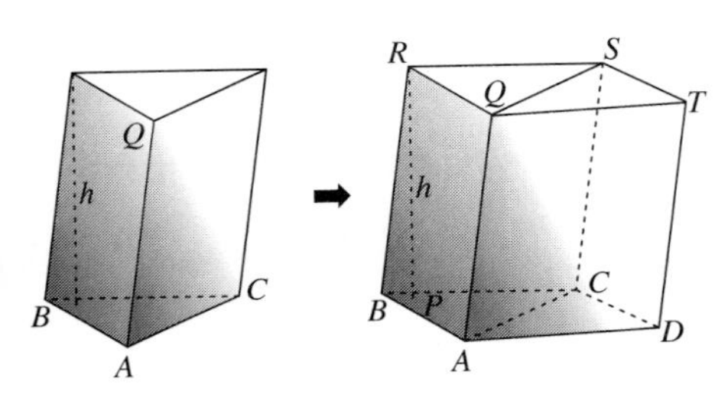

Figure 7.37

Using the theory of area, since $\triangle ABC \cong \triangle ACD$, we obtain Area $\square ABCD = 2$ Area $\triangle ABC = 2B$.

Now by Postulate 4,

$$\text{Vol}(QABCD) = RP \cdot \text{Area } \square ABCD = 2Bh$$

and by the Postulate of Additivity for volume and the preceding lemma [which takes care of the requirement Vol $(QABC \cap QACD) = $ Vol $(\square ACSQ) = 0]$,

$$2Bh = \text{Vol}(QABCD) = \text{Vol}(QABC) + \text{Vol}(QACD) = 2V$$

which gives us the desired result **(1)**.

We extend this formula to any prism in the obvious manner, by subdividing the base polygon (which is convex) into triangles having areas $B_1, B_2, \ldots, B_n$, as shown

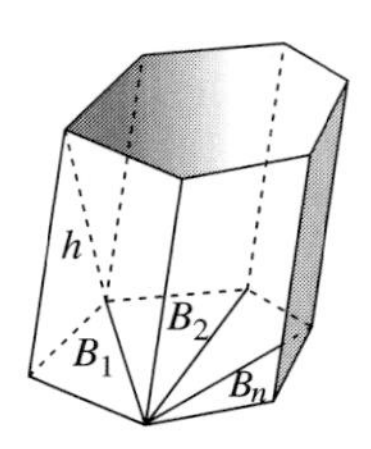

Figure 7.38

in Figure 7.38, which induces a subdivision of the prism into triangular prisms where **(1)** is valid. Thus, again by the Postulate of Additivity and the lemma,

$$V = \text{Vol (Prism)} = B_1h + B_2h + \cdots + B_nh = (B_1 + B_2 + \cdots + B_n)h = Bh$$

where B is the area of the base polygon. This completes the proof of **(1)**.

For the circular cylinder, let $SABD$ be a cylinder with the circle of radius r and center C as base, and altitude $\overline{SE} \perp \boldsymbol{P}$ where $\boldsymbol{P}$ is the plane of the base circle (Figure 7.39). Let a triangle $\triangle PQR$ be determined in plane $\boldsymbol{P}$ having the same area as the circle (i.e., $B = \pi r^2$). (Can you see how to guarantee this?) Let $\overline{PT}$ be the perpendicular to $\boldsymbol{P}$ at P with $SE = TP = h$. It is obvious that a plane section of the cylinder and triangular prism parallel to $\boldsymbol{P}$ are, respectively, congruent to the base of the cylinder and base of the prism, hence have equal areas. By Cavalieri's principle,

$$\text{Volume of prism} = \text{Volume of cylinder} = Bh = \pi r^2 h$$

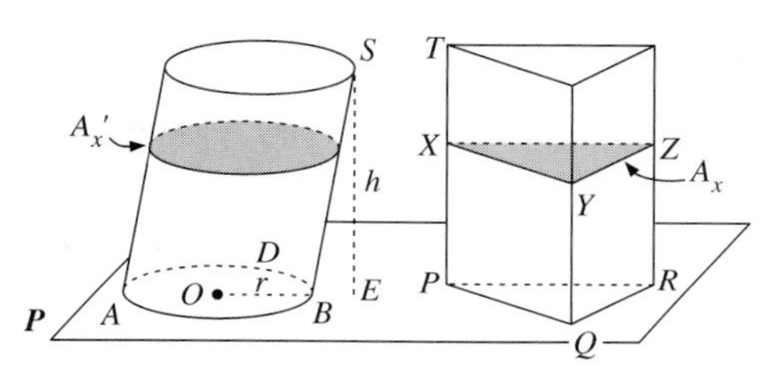

Figure 7.39

The classical result for then follows:

(2) VOLUME OF A CIRCULAR CYLINDER $V = \pi r^2 h$

VOLUMES OF PYRAMIDS AND CIRCULAR CONES

The program for pyramids and cones can be carried out in much the same manner as for prisms and cylinders. As before, we begin with the formula for the volume of a triangular pyramid. It is evident from Cavalieri's principle that any two triangular pyramids having congruent bases and congruent altitudes have the same volume (see Figure 7.40 for a "proof by picture"). Note that by properties of ratios of parallel seg-

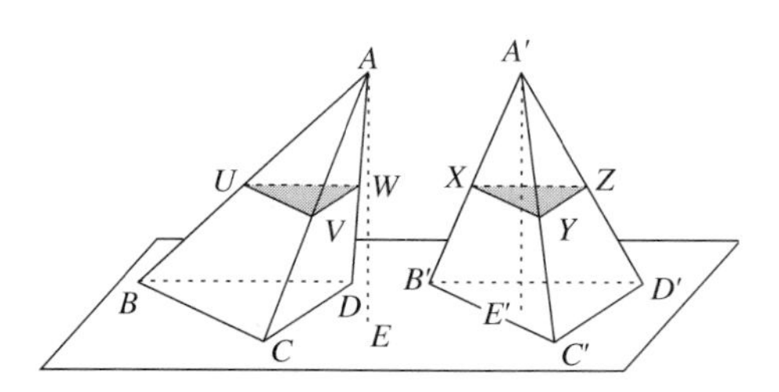

Figure 7.40

ments ($\overline{UV}$ and $\overline{BC}$, etc.), the triangular cross sections have equal areas because $\triangle BCD \cong \triangle B'C'D'$ and $\triangle UVW \cong \triangle XYZ$. This means that we can assume that the given pyramid is a right triangular pyramid, so that two lateral faces lie in planes perpendicular to the base.

From a given right triangular pyramid we can construct a right triangular prism, as shown in Figure 7.41. Two planes can be passed through vertex A that divides the prism into three triangular pyramids as shown in the sequence of diagrams in Figure 7.41. Let the volumes of these three pyramids be designated V_1, V_2, and V_3, where V_1 corresponds to the bottom—and given—pyramid. By choosing the appropriate base and altitude for each pyramid, we can show that $V_1 = V_2 = V_3$: for the top and bottom pyramids, simply use the edges designated h as altitude and the top and base of the prism as bases, which are congruent triangles (area denoted by B). Thus we obtain $V_1 = V_3$. For the top and middle pyramids, we use the edge designated h' as common altitude from vertex A, and the side triangles as bases with area denoted B'. Again, since the triangles are congruent, they have the same area. Thus, $V_2 = V_3$. By additivity of volume, $V_1 + V_2 + V_3 = 3V$ where $V_1 = V$ is the volume of the given right triangular pyramid. Therefore, using the formula already established for a prism, $3V = Bh$ and we have arrived at the following formula for a triangular pyramid having altitude h and base area B:

(3) VOLUME OF A TRIANGULAR PYRAMID $V = \frac{1}{3}Bh$

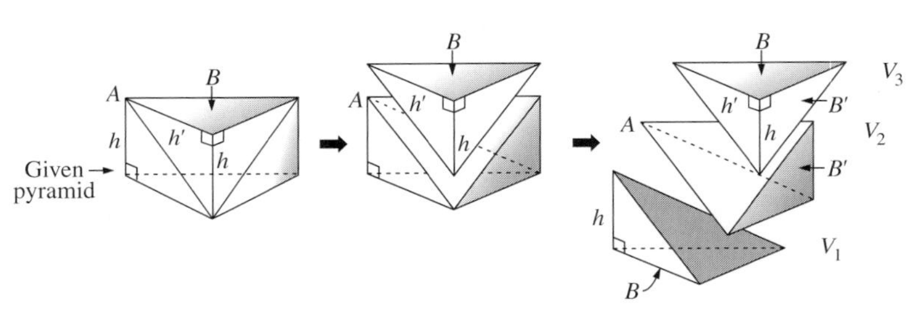

Figure 7.41

The next step is to obtain the formula for the volume V of a circular cone having height h and radius r for the circular base. For convenience, we assume the cone is a right circular cone and that a pyramid with a lateral edge perpendicular to the base and right triangle for base has been constructed so that $\frac{1}{2}B'h' = \pi r^2$, as shown in Figure 7.42. The analysis is similar to that for deducing the volume of a circular

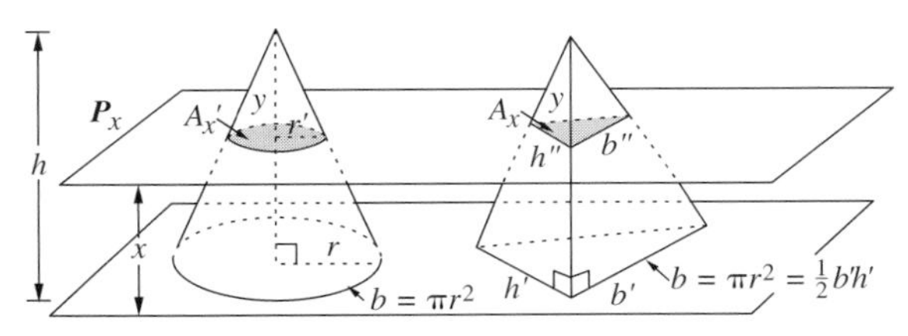

Figure 7.42

cylinder from that of a right prism, so we leave the details for you to fill in (Problem 9). The final result, for a cone having base radius r and altitude h, is:

(4) VOLUME OF A CIRCULAR CONE $V = \frac{1}{3}\pi r^2 h$

THE VOLUME OF A SPHERE, SPHERICAL SEGMENT

We have had a sampling of the usefulness of Cavalieri's principle. It is so powerful that we can deduce the volume of a sphere with relative ease, compared to Archimedes' hard work on the problem. The analysis is illustrated in Figure 7.43. A solid cone with base radius r and height r is removed from a solid circular cylinder with the same radius and height. A plane $\boldsymbol{P}_x$, parallel to the base plane, x units above it, cuts the sphere in a disk of area A_x', and the cylinder minus the cone in an annular ring of area A_x. Using the Pythagorean Theorem for $\triangle OPX$, we have

$$A_x' = \pi s^2 = \pi(r^2 - x^2) = \pi r^2 - \pi x^2$$

$$A_x = \pi r^2 - \pi t^2 \qquad (t/x = r/r = 1 \text{ by similar triangles})$$

$$= \pi r^2 - \pi x^2$$

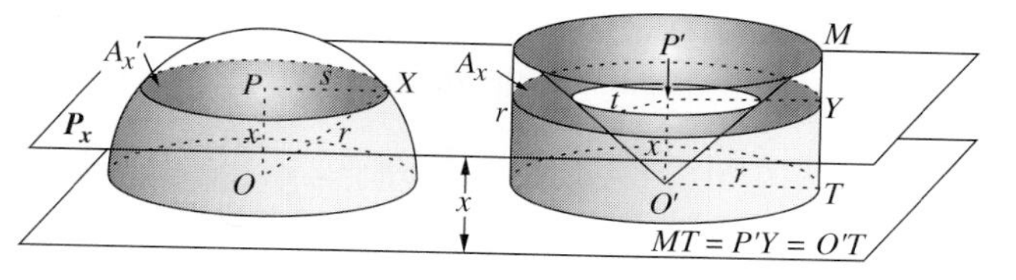

Figure 7.43

Therefore,

$$A_x' = A_x$$

By Cavalieri's principle, with V the volume of the sphere and $\frac{1}{2}V$, the hemisphere,

$$\frac{1}{2}V = \begin{Bmatrix} \text{VOLUME OF CYLINDER} \\ \text{RADIUS } r\text{, HEIGHT } r \end{Bmatrix} - \begin{Bmatrix} \text{VOLUME OF CONE} \\ \text{RADIUS } r\text{, HEIGHT } r \end{Bmatrix}$$

From (**2**) and (**4**), with $h = r$, we obtain $\frac{1}{2}V = \pi r^2 \cdot r - \frac{1}{3}\pi r^2 \cdot r = \frac{2}{3}\pi r^3$, or

(5) VOLUME OF A SPHERE $V = \dfrac{4}{3}\pi r^3$ (radius r)

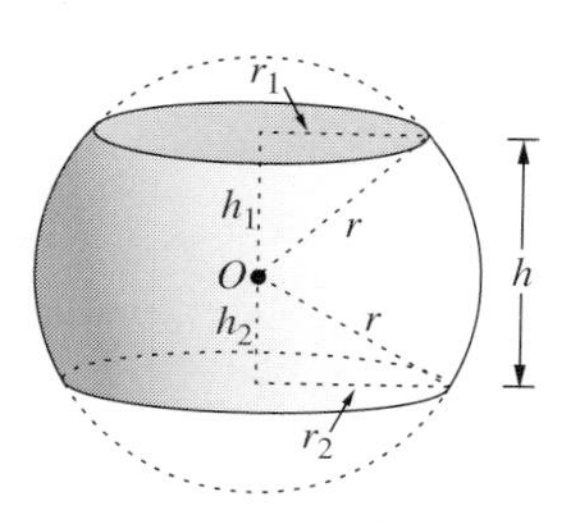

Figure 7.44

The same method can be used to derive the formula for the volume of certain parts of the sphere, such as one important type called a **spherical segment.** This solid is the part of a sphere and interior lying between two parallel planes, whose outer shell is that part of the sphere known as a **zone,** the region between two circles on the sphere in parallel planes. (See Figure 7.44.) The top and bottom of a spherical segment are circular disks—its **bases.** Let the radii of these bases be r_1 and r_2, and let the distance between the parallel places be h, called the **height.** We first obtain a formula for the special case when one of the planes passes through the center of the

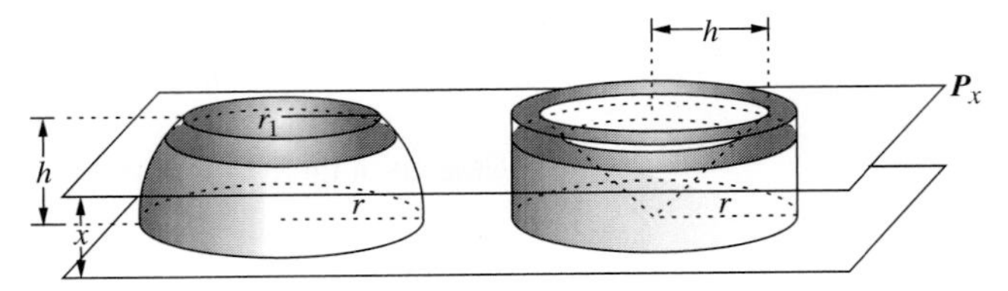

Figure 7.45

sphere (yielding a **hemispherical segment**), as shown in Figure 7.45). We can use Cavalieri's principle as before; the set-up is exactly the same as that of Figure 7.43, when x varies between 0 and h (instead of 0 and r). Since again, $A'_x = A_x$, we have

$$V = \begin{Bmatrix} \text{VOLUME OF CYLINDER} \\ \text{RADIUS } r\text{, HEIGHT } h \end{Bmatrix} - \begin{Bmatrix} \text{VOLUME OF CONE} \\ \text{RADIUS } r\text{, HEIGHT } h \end{Bmatrix}$$

which yields the formula

(6) $$V = \pi r^2 h - \frac{1}{3}\pi h^3$$

To find the formula for the more general spherical segment, having height $h = h_1 + h_2$ and radii r_1 and r_2 of the top and bottom disks, we merely use **(6)** and sum (assuming the bases lie in different hemispheres):

$$\begin{aligned} V_1 + V_2 &= (\pi r^2 h_1 - \tfrac{1}{3}\pi h_1{}^3) + (\pi r^2 h_2 - \tfrac{1}{3}\pi h_2{}^3) \\ &= \pi r^2(h_1 + h_2) - \tfrac{1}{3}\pi(h_1{}^3 + h_2{}^3) \\ &= \pi r^2 h - \tfrac{1}{3}\pi(h_1 + h_2)(h_1{}^2 - h_1h_2 + h_2{}^2) \\ &= \pi r^2 h - \tfrac{1}{3}\pi h(h_1{}^2 - h_1h_2 + h_2{}^2) \\ &= \tfrac{1}{3}\pi h(3r^2 - h_1{}^2 + h_1h_2 - h_2{}^2) \end{aligned}$$

We want to express the final result in terms of the height h and the two radii r_1, r_2. Some algebra tricks and the Pythagorean relations $r^2 = r_1{}^2 + h_1{}^2 = r_2{}^2 + h_2{}^2$ are necessary (as indicated in Figure 7.44). By manipulation and substitution,

$$\begin{aligned} V &= \tfrac{1}{6}\pi h[6r^2 - 2h_1{}^2 + 2h_1h_2 - 2h_2{}^2] \qquad \text{(divided and multiplied by 2)} \\ &= \tfrac{1}{6}\pi h[3(r^2 - h_1{}^2) + 3(r - h_2{}^2) + h_1{}^2 + 2h_1h_2 + h_2{}^2] \\ &= \tfrac{1}{6}\pi h[3r_1{}^2 + 3r_2{}^2 + (h_1 + h_2)^2] \end{aligned}$$

Hence, if the spherical segment has height h and base radii r_1 *and* r_2, we obtain

(7) VOLUME OF A SPHERICAL SEGMENT $$V = \frac{1}{6}\pi h(3r_1{}^2 + 3r_2{}^2 + h^2)$$

This same exact formula results when the spherical segment lies in a single hemisphere, where *subtraction* of two hemispherical segments is necessary, instead of the addition used above. In this case, r_1 and r_2 are still the radii of the top and bottom, but $h = h_1 - h_2$. We leave this as a problem for you (Problem 10).

EXAMPLE 1 A hole having diameter 0.5 cm is drilled from a ball bearing having diameter 1.5 cm. Find the volume of the material removed, and the volume remaining in the bearing. Give an approximate answer accurate to three decimals. (See Figure 7.46.)

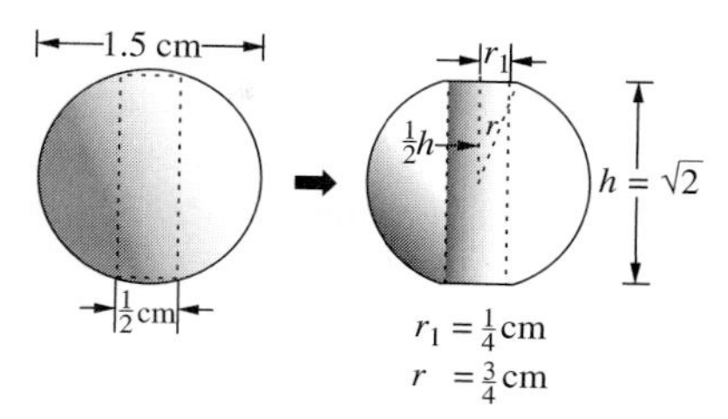

Figure 7.46

SOLUTION

We think of this in terms of removing a right circular cylinder and interior from a spherical segment, whose height would be, by the Pythagorean relation, $2\sqrt{\left(\frac{3}{4}\right)^2 - \left(\frac{1}{4}\right)^2} = \sqrt{2}$. To find the volume of the part of the ball bearing left we subtract the volumes of spherical segment of radii $r_1 = r_2 = \frac{1}{4}$ and height $h = \sqrt{2}$, and a right circular cylinder of radius $\frac{1}{4}$ and height $\sqrt{2}$. Using **(2)** and **(7)**, that value is given by

$$\frac{1}{6}\pi \cdot \sqrt{2}\left[3 \cdot \left(\frac{1}{4}\right)^2 + 3 \cdot \left(\frac{1}{4}\right)^2 + \sqrt{2}^2\right] - \pi \cdot \left(\frac{1}{4}\right)^2\sqrt{2}$$

$$= 19\sqrt{2}\pi/48 - \sqrt{2}\pi/16 = \sqrt{2}\pi/3 \approx 1.481 \text{ cm}^3$$

To find the volume of the material drilled out, we merely subtract the above answer from the volume of the ball bearing itself, which is, from **(5)**, $9\pi/16 \approx 1.767$ cm^3. The volume of the part removed is therefore approximately $1.767 - 1.481 = 0.286$ cm^3. ■

Moment for Discovery

The Golden Bracelet Problem

A jeweler fashions a bracelet made of 14 carat gold whose exterior is a perfect spherical surface. It is constructed by taking as the center of the sphere the point O, as shown in Figure 7.47, and casting the part of the sphere lying outside the

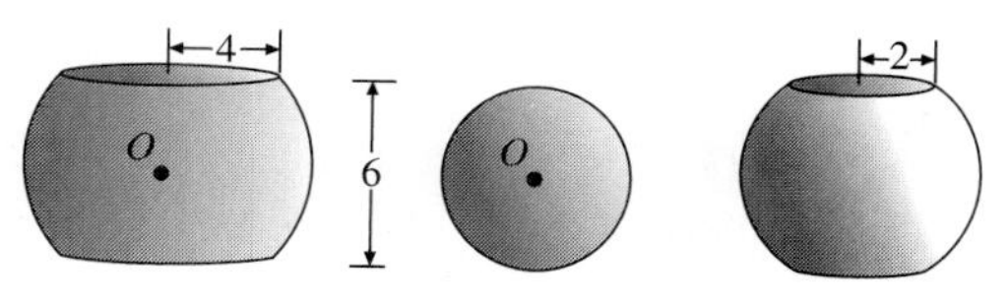

Figure 7.47

right circular cylinder whose axis passes through O. The following calculations devoted to finding how much gold will be used, if correctly computed, may surprise you.

1. Start with an example, as shown in Figure 7.47, where the inside radius $r_1 = 4$ cm, and the width of the bracelet $= h = 6$ cm. Find the volume of the spherical segment formed (you should get $132\pi\text{ cm}^3$), then subtract the volume of the cylinder. Write down your answer. [The formulas **(2)** and **(7)** are used, just as in Example 1.]
2. Find the volume of a sphere whose diameter equals the width of the bracelet (hence $r = 3$ cm). Did anything happen?
3. Now calculate the volume of gold needed for a bracelet having the same width but half the size as the first one, shown in the figure at right. (Here, $r_1 = 2$ cm and $h = 6$ cm.) Did anything happen?
4. Try to prove the phenomenon in general, using formulas **(2)**, **(5)**, and **(7)**, with arbitrary r_1, r, and h.

PROBLEMS (7.4)

GROUP A

1. Find the volume of a regular tetrahedron of unit side.
2. Find the volume of a regular octahedron of unit side.
3. Find the height of the right circular cylinder inscribed in a sphere having unit radius shown in the figure.

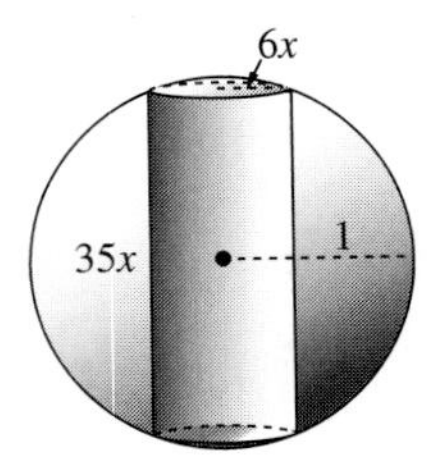

4. The formula for the area of a sphere ($S = 4\pi r^2$—the area of four great circles) can be found by observing that each spherical triangle and approximating Euclidean triangle having as sides the corresponding chords of the sphere can serve as the base of a pyramid whose apex is the center of the sphere. If we sum all these to cover the area of the sphere, we get, in the limit, a relationship between the volume of a sphere and its surface area. Find this relation, and deduce from it the desired formula, using **(5)**.

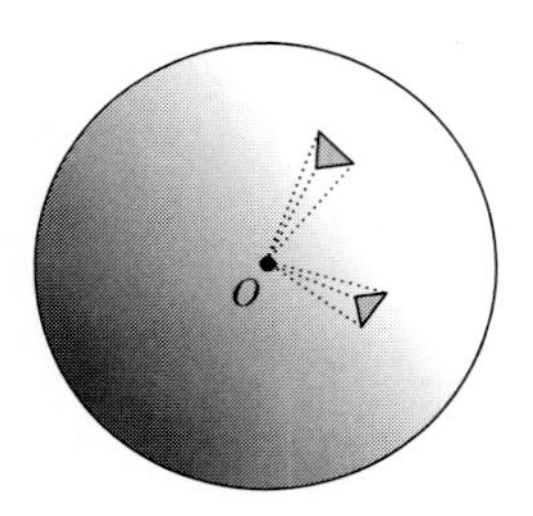

5. Nine tennis balls 3.5 in. in diameter are packed in three layers of three per layer into a cylindrical can having radius $r \approx 3.771$ in. and height $h \approx 9.215$ in. Nine tennis balls can also be packed into a long cylindrical can with the balls in a straight line, having radius $r = 1.25$ in. and height $h = 22.5$ in.

(a) What is the total space wasted using each of the two packing methods, and which wastes the most total space? (Space wasted = volume of can minus volume of balls inside.)

(b) How much material (total area in square inches) is required for the two types of cans used for packaging, and which one requires the most?

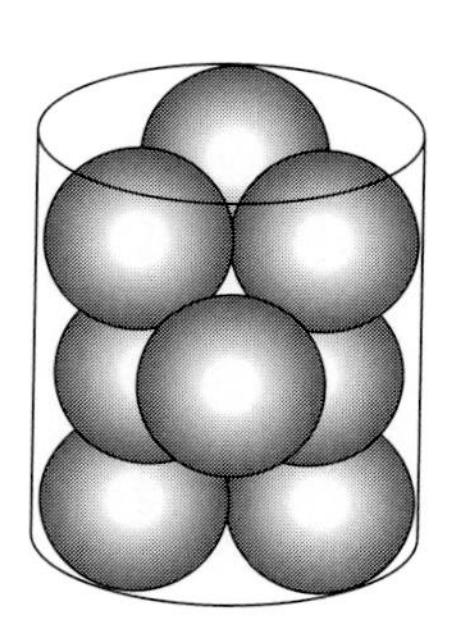

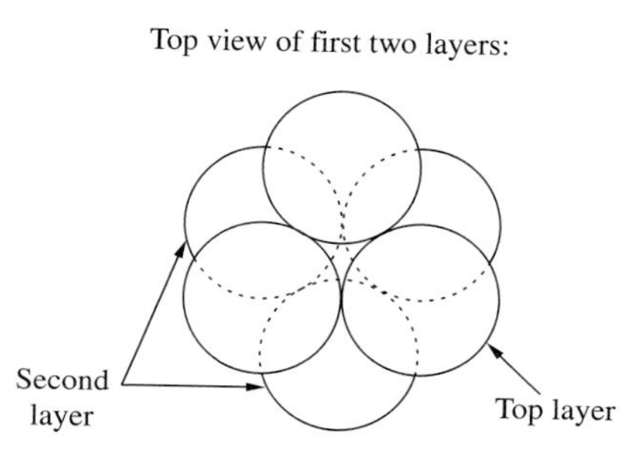

GROUP B

6. Which of the following water tanks will hold the most water when three-fourths full (in terms of the height of the water in the tank)?

(a) A cylindrical tank with radius 20 ft and height 40 ft (height of water in tank = 30 ft); total surface area given by $S = 2\pi rh + 2\pi r^2$.

(b) A spherical tank that has the same total surface area ($S = 4\pi r^2$) as the cylindrical tank in **(a)** counting top and bottom. (***Hint:*** First show that the radius r of the spherical tank must be approximately 24.5 ft, then use **(7)** with $h = \frac{3}{4}(2r)$, $r_1 = 0$, and using the Pythagorean theorem to find r_2.)

7. Deduce Archimedes' relation (engraved on his tomb) between a sphere and its circumscribed cylinder, as shown in the figure, by finding the ratios V/V' (volume) and S/S' (total surface area) in each case.

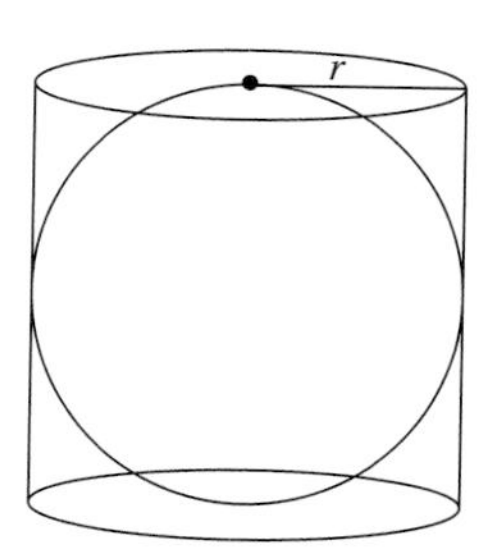

8. Unlike the square, the cube cannot be filled with a finite number of cubes of all different sizes. In the case of a square, a finite number (of squares) is possible. In fact, the optimal number is 21, shown in the figure below. Such a configuration is called a **perfect squared square.** (This is only a recent discovery. It was long thought that a 24-piece tesselation was the optimal one for a square; a long standing theorem states that *at least* 21 squares are required.)

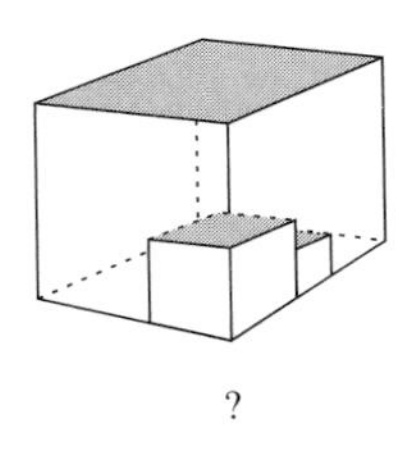

(a) Verify this construction for the square.

(b) Prove that *no finite tesselation is possible* for a cube in E^3 using cubes.

9. Complete the details for the volume of a circular cone, as outlined in the text preceding **(4)**. (***Hint:*** By similar triangles, $r' \neq r = (h - x)x = b''/b'$, etc. Prove that $A_x' = A_x$.)

GROUP C

10. Complete the proof of **(7)** for the case when the bases of the segment lie in the same hemisphere. (***Hint:*** You will need the algebra rule $(h_1 - h_2)(h_1^2 + h_1h_2 + h_2^2) = h_1^3 - h_2^3$ for this case.)

11. A **spherical cap** is the part of a sphere and interior lying on one side of a plane, sometimes called a **segment of one base.** Let the radius of the base be r, and its height, h. Show that the volume is given by the formula $V = \frac{1}{6}\pi h(3r^2 + h^2) = \frac{1}{3}\pi h^2(3R - h)$, where R is the radius of the sphere. (See the figure.) (***Hint:*** From the Pythagorean Theorem can be derived the needed relation $r^2 = 2Rh - h^2$. Verify this as part of your solution.)

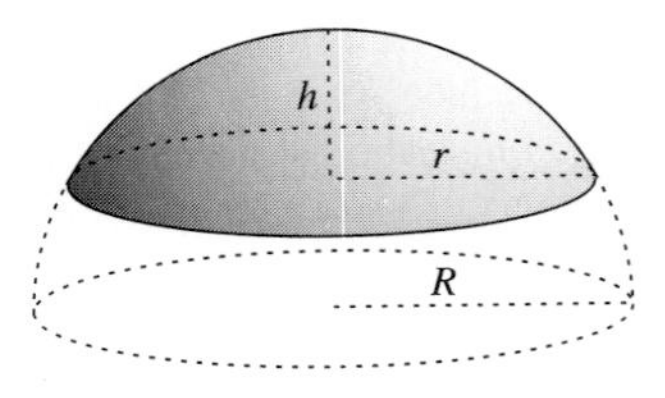

*7.5 Coordinates, Vectors, and Isometries in E^3

As might be expected, the construction of a model for E^3 requires the use of ordered triples of real numbers, (x, y, z), as illustrated in Figure 7.48. Since you are probably already acquainted with the mechanics of a three-dimensional coordinate system, we will just outline the basic procedure, paying close attention to the incidence axioms

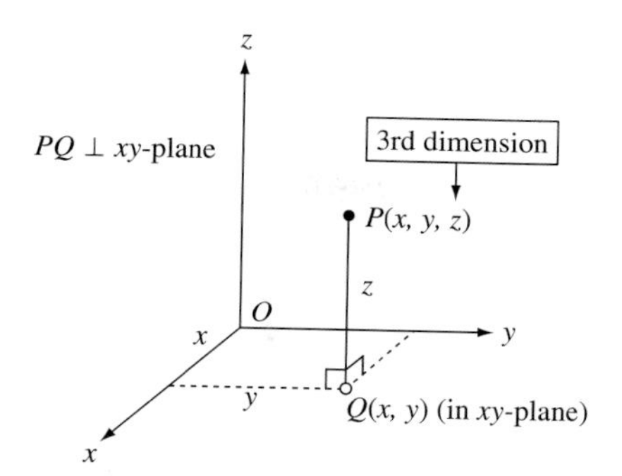

Figure 7.48

involving the interplay between lines and planes, and how all that works out in a three-dimensional coordinate system. We choose to view this as a model, where numbers and equations *represent* geometric objects, instead of rigorously constructing a coordinate system from the geometric objects—as we did for the plane in Section 4.7.

INCIDENCE AXIOMS, DISTANCE AND ANGLE MEASURE

Using coordinates, we construct representations of the undefined terms and basic objects with which the axioms of E^3 are concerned, thereby obtaining a three-dimensional model for axiomatic Euclidean geometry.

POINT (x, y, z) for real numbers x, y, and z

LINE The set of all points (x, y, z) such that

$$x = at + x_0, \qquad y = bt + y_0, \qquad z = ct + z_0$$

for real t (the **parameter**), and a certain triple of constants $[a, b, c]$ not all zero, called the **direction** of the line, and some point (x_0, y_0, z_0) on it (Figure 7.49).

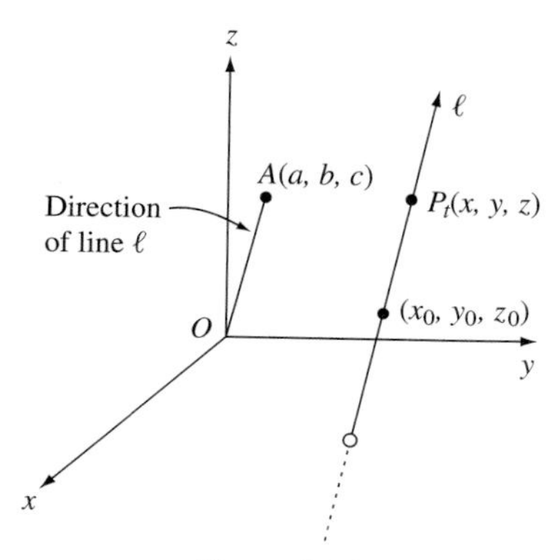

Figure 7.49

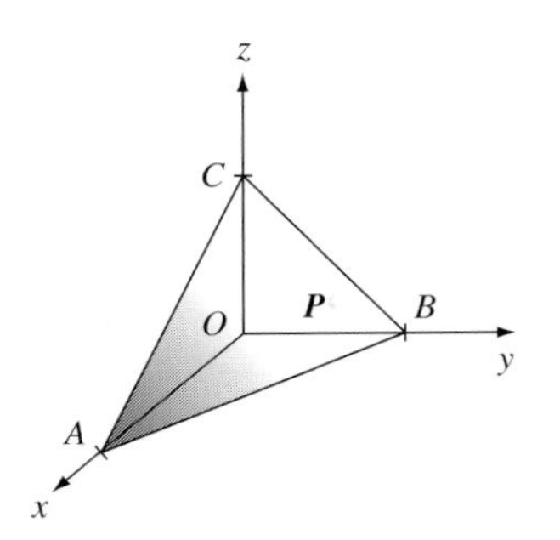

Figure 7.50

PLANE The set of all points (x, y, z) such that

$$ax + by + cz = d$$

for a certain triple of constants $[a, b, c]$ not all zero (called the **normal direction**). If a, b, c and d are all nonzero, the equation can be put into the **intercept form** (Figure 7.50)

$$\frac{x}{A} + \frac{y}{B} + \frac{z}{C} = 1.$$

DISTANCE For any two points $P = (x_1, y_1, z_1)$ and $Q = (x_2, y_2, z_2)$ the distance from P to Q (Figure 7.51) is the number

$$PQ = \sqrt{(x_1 - x_2)^2 + (y_1 - y_2)^2 + (z_1 - z_2)^2}$$

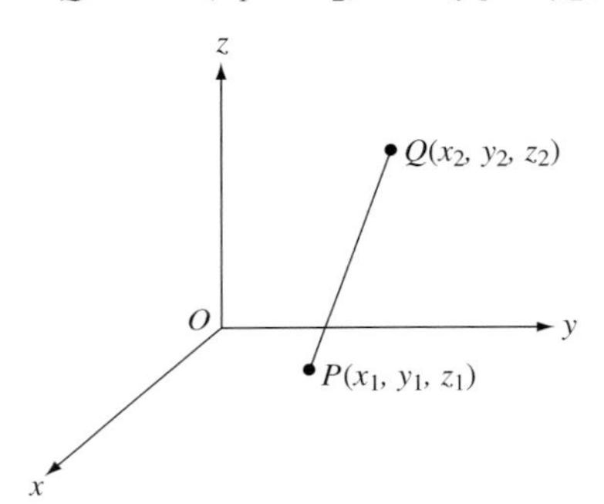

Figure 7.51

ANGLE MEASURE If P, Q, and R have coordinates (x_k, y_k, z_k) for $k = 1, 2, 3$, (Figure 7.52) set

$$(u_1, u_2, u_3) = (x_1 - x_2, y_1 - y_2, z_1 - z_2)$$

and, similarly,

$$(v_1, v_2, v_3) = (x_3 - x_2, y_3 - y_2, z_3 - z_2)$$

Then the angle measure of $\angle PQR$ is defined as the number

$$m\angle PQR = \cos^{-1} \frac{u_1v_1 + u_2v_2 + u_3v_3}{\sqrt{u_1^2 + u_2^2 + v_3^2}\sqrt{v_1^2 + v_2^2 + v_3^2}}$$

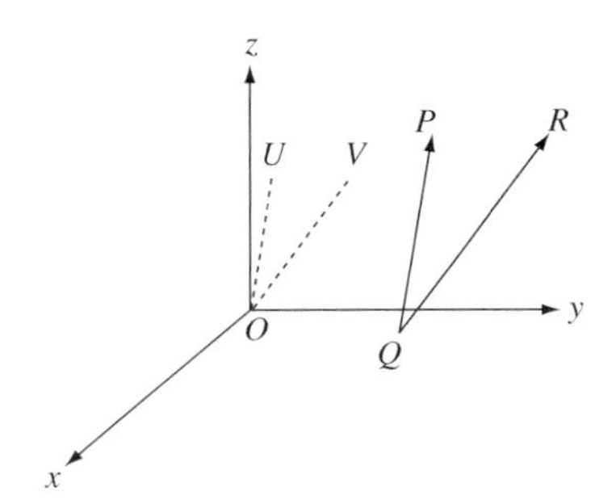

Figure 7.52

WORKING WITH COORDINATES IN GEOMETRY

We now provide two examples to remind you of some of the basic features of three-dimensional coordinate geometry.

EXAMPLE 1 Prove the result directly implied by the axioms that a line ℓ not lying in plane $\boldsymbol{P}$ meets that plane in at most one point, following these steps:

(a) First show this for the example shown in Figure 7.53
Line ℓ: $x = 2t + 2, y = -3t, z = t + 1$ $(x_0 = 2, y_0 = 0, z_0 = 1)$
Plane $\boldsymbol{P}$: $3x - 4y + 2z = 0$
(See Figure 7.53.)

(b) Generalize to the arbitrary case using coordinates.

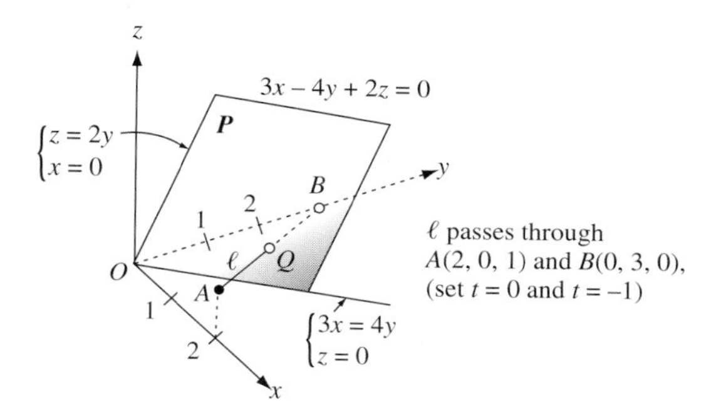

Figure 7.53

SOLUTION

(a) By substitution,

$$3(2t + 2) - 4(-3t) + 2(t + 1) = 0$$
$$6t + 6 + 12t + 2t + 2 = 0$$
$$20t = -8 \rightarrow t = -2/5$$

Thus, the point of intersection is found by setting $t = -2/5$ in the parametric equations for line ℓ; the result is the point $Q(6/5, 6/5, 3/5)$, illustrated in the figure.

(b) In general, we have

$$\ell: x = at + x_0, y = bt + y_0, z = ct + z_0$$
$$\boldsymbol{P}: px + qy + rz = s$$

To determine whether there is a solution, plug in (as before):

$$p(at + x_0) + q(bt + y_0) + r(ct + z_0) = s$$
$$(pa + qb + rc)t = s - px_0 - qy_0 - rz_0$$

There will be a unique solution in t (hence a unique point of intersection) if

$$pa + qb + rc \neq 0$$

and no solution otherwise. (Using vectors, $pa + qb + ac = 0$ implies that ℓ, with direction $[a, b, c]$, is perpendicular to the normal of $\boldsymbol{P}$, with direction $[p, q, r]$, or $\ell \parallel \boldsymbol{P}$.) ■

EXAMPLE 2 Determine in each of the following whether the planes whose equations are given intersect, and if they do, identify the line of intersection by its parametric equations.

(a) $2x - y + z = 1$
$2x - y + z = 2$

(b) $x - y + 2z = 0$
$x + y + z = 2$

SOLUTION

(a) If we subtract the two equations we obtain the contradiction $0 = 1$, so there can be no values for x, y, and z that satisfy both equations simultaneously, hence no point of intersection (the planes are parallel—note the normal direction of the two planes, $[2, -1, 1]$, indicating that they are parallel.)

(b) Write the two equations in the form

$$x - y = -2z$$

$$x + y = 2 - z$$

and solve for x and y in terms of z (which we can essentially take as the parameter t that yields a line of intersection):

Result of Summing:

$$2x = 2 - 3z$$

Result of Subtraction:

$$-2y = -2 - z$$

For convenience, take the parameterization $z = 2t$, which will avoid fractions. Hence we find the line of intersection (in parametric form):

$$\ell\colon x = -3t + 1, y = t + 1, z = 2t \qquad (t \text{ real}).$$

(It can be checked that these coordinates satisfy both equations above for all t, hence the line lies in both planes.) ■

E^3 AS THE VECTOR SPACE V^3

As is customary, we identify vectors with either directions or points in E^3. In practice, we often shift back and forth between the two notions. If the object is to find a certain direction, we use a vector to represent a directed line segment in E^3, while if the object is to find the coordinates of a point, a position vector is used, whose terminal point coincides with the desired point.

We first discuss the actual meaning of the notation V^3—the **vector space of dimension three.** This "space" consists of all possible three-entry columns of real numbers, called **3-vectors,** or just **vectors:**

$$\boldsymbol{v} = \begin{bmatrix} v_1 \\ v_2 \\ v_3 \end{bmatrix}, \quad \text{where } v_1, v_2, v_3 \text{ are real numbers;} \quad \mathbf{0} = \begin{bmatrix} 0 \\ 0 \\ 0 \end{bmatrix} \ \textbf{(zero vector)}$$

(often written in horizontal form as well.)

Define addition and scalar multiplication componentwise in the usual manner (as in two dimensions). A set of k vectors

$$\boldsymbol{v}_1, \boldsymbol{v}_2, \boldsymbol{v}_3, \ldots, \boldsymbol{v}_k$$

is called **linearly independent** iff the only linear combination of those vectors that equals the zero vector is the trivial one:

$$r_1\boldsymbol{v}_1 + r_2\boldsymbol{v}_2 + r_3\boldsymbol{v}_3 + \cdots + r_k\boldsymbol{v}_k = \mathbf{0} \quad \text{iff} \quad r_1 = \cdots = r_k = 0.$$

For any k linearly independent vectors, $\boldsymbol{v}_1, \boldsymbol{v}_2, \boldsymbol{v}_3, \cdots, \boldsymbol{v}_k$, the collection of all linear combinations of those k vectors is called the **span** of $\boldsymbol{v}_1, \boldsymbol{v}_2, \boldsymbol{v}_3, \cdots, \boldsymbol{v}_k$, and is a **subspace of dimension** k. We note that the zero vector $\mathbf{0}$ will belong to any subspace. If $\boldsymbol{M}$ is any set of vectors, the **translate** of $\boldsymbol{M}$ by $\boldsymbol{x}_0$, denoted $\boldsymbol{M} + \boldsymbol{x}_0$, is the set of all vectors $\boldsymbol{v} + \boldsymbol{x}_0$ for all $\boldsymbol{v}$ in $\boldsymbol{M}$. Finally, if

$$\boldsymbol{u} = \begin{bmatrix} u_1 \\ u_2 \\ u_3 \end{bmatrix} \quad \text{and} \quad \boldsymbol{v} = \begin{bmatrix} v_1 \\ v_2 \\ v_3 \end{bmatrix}$$

define

INNER (SCALAR OR DOT) PRODUCT	$\boldsymbol{uv} \equiv \boldsymbol{u} \cdot \boldsymbol{v} = u_1v_1 + u_2v_2 + u_3v_3$
SQUARE NOTATION	$\boldsymbol{u}^2 = \boldsymbol{uv}$ where $\boldsymbol{u} = \boldsymbol{v}$
NORM OF A VECTOR	$\lVert\boldsymbol{u}\rVert = \sqrt{\boldsymbol{u}^2} \equiv \sqrt{\boldsymbol{uu}} \qquad (\boldsymbol{uu} = \boldsymbol{u} \cdot \boldsymbol{u})$

Now we can describe the same geometric concepts as before, only in terms of vectors.

POINT Any 3-vector.

LINE Any one-dimensional subspace or translate thereof.

PLANE Any two-dimensional subspace or translate thereof.

DISTANCE If $p = \boldsymbol{u}$ and $Q = \boldsymbol{v}$ then

$$PQ = \lVert\boldsymbol{u} - \boldsymbol{v}\rVert = \sqrt{(\boldsymbol{u} - \boldsymbol{v}) \cdot (\boldsymbol{u} - \boldsymbol{v})}$$

ANGLE MEASURE If $P = \boldsymbol{p}$, $Q = \boldsymbol{q}$, $R = \boldsymbol{r}$, set $\boldsymbol{u} = \boldsymbol{p} - \boldsymbol{q}$ and $\boldsymbol{v} = \boldsymbol{r} - \boldsymbol{q}$; then $m\angle PQR = \cos^{-1}\dfrac{\boldsymbol{uv}}{\lVert\boldsymbol{u}\rVert\,\lVert\boldsymbol{v}\rVert}$

NOTE: The fact $\cos^{-1}0 = 90°$ gives us the familiar criterion for orthogonality: $\boldsymbol{u} \perp \boldsymbol{v}$ iff $\boldsymbol{uv} = 0$.

Since a line is a translate of some one-dimensional subspace and, as such, is the translate of the linear combinations of a single nonzero vector $\boldsymbol{a}$ (the **direction** of the line), the vector equation for any such line is of the form

$$\boldsymbol{x} = t\boldsymbol{a} + \boldsymbol{x}_0$$

A plane can be characterized by a vector equation of the form

$$\boldsymbol{ax} = d$$

where $\boldsymbol{a}$ is a fixed vector (the **normal** to the plane) and d is a fixed scalar as the following analysis shows: If $\boldsymbol{x}_0$ is some point on the plane, then it must satisfy, by substitution back into the above vector equation, the condition

$$\boldsymbol{ax}_0 = d \to \boldsymbol{ax} = \boldsymbol{ax}_0$$

which can be written as

$$\boldsymbol{a}(\boldsymbol{x} - \boldsymbol{x}_0) = 0$$

Thus, the plane consists of all vectors $\boldsymbol{x}$ such that $\boldsymbol{x} - \boldsymbol{x}_0$ is orthogonal to $\boldsymbol{a}$. It can be shown that such a set of vectors in V^3 is a two-dimensional subspace $\boldsymbol{M}$, hence for each $\boldsymbol{x} - \boldsymbol{x}_0$ in $\boldsymbol{M}$, then $\boldsymbol{x} = \boldsymbol{M} + \boldsymbol{x}_0$, precisely the above description of a plane as the translate of a two-dimensional subspace.

When we say that a plane *passes through* a vector $\boldsymbol{x} = [p, q, r]$ we mean it passes through the *point* $P(p, q, r)$, which is the tip of the position vector $\boldsymbol{OP}$ equivalent to $\boldsymbol{x}$.

EXAMPLE 3

(a) Use vectors to find the equation of the line ℓ passing through $A(4, -2, 2)$ and $B = (2, 6, 8)$, that is, through $\boldsymbol{a} = [4, -2, 2]$ and $\boldsymbol{b} = [2, 6, 8]$.

(b) Find where this line crosses the plane $\boldsymbol{P}$: $3x + 4y - z = 12$, or in vector form, $\boldsymbol{cx} = 12$ where $\boldsymbol{c} = [3, 4, -1]$ and $\boldsymbol{x} = [x, y, z]$.

(c) Find the equation of the unique plane containing ℓ and perpendicular to **P**.

SOLUTION

(a) To find the direction of the line, use a directed line segment

$$\boldsymbol{d} = \boldsymbol{AB} = \boldsymbol{b} - \boldsymbol{a} = \begin{bmatrix} 2 - 4 \\ 6 - (-2) \\ 8 - 2 \end{bmatrix} = \begin{bmatrix} -2 \\ 8 \\ 6 \end{bmatrix}$$

Hence, the vector equation of the line is $\boldsymbol{x} = t\boldsymbol{d} + \boldsymbol{a}$ or, to spell it out with column matrices,

$$\boldsymbol{x} = t\begin{bmatrix} -2 \\ 8 \\ 6 \end{bmatrix} + \begin{bmatrix} 4 \\ -2 \\ 2 \end{bmatrix} \qquad (t \text{ real})$$

(b) Now substitute $\boldsymbol{x}$ into the equation of the plane:

$$\boldsymbol{c}(t\boldsymbol{d} + \boldsymbol{a}) = 12$$

$$t\boldsymbol{cd} + \boldsymbol{ca} = 12 \qquad \text{(dot products)}$$

or, since $\boldsymbol{cd} = [3, 4, -1] \cdot [-2, 8, 6] = 20$ and $\boldsymbol{ca} = [3, 4, -1] \cdot [4, -2, 2] = 2$,

$$20t + 2 = 12 \rightarrow t = \frac{1}{2}$$

$$\therefore \boldsymbol{x} = \frac{1}{2}\boldsymbol{d} + \boldsymbol{a}$$

which yields the point of intersection between ℓ and $\boldsymbol{P}$

$$\boldsymbol{x} = \frac{1}{2}\begin{bmatrix} -2 \\ 8 \\ 6 \end{bmatrix} + \begin{bmatrix} 4 \\ -2 \\ 2 \end{bmatrix} = \begin{bmatrix} 3 \\ 2 \\ 5 \end{bmatrix}$$

that is, the point $C(3, 2, 5)$.

(c) Let the equation of the desired plane $\boldsymbol{Q}$ be, in vector form,

$$\boldsymbol{ex} = k,$$

where $\boldsymbol{e} = [p, q, r]$ and k are to be determined. First, let two points $\boldsymbol{a}$ and $\boldsymbol{b}$ on ℓ corresponding to $t = 0, 1$ lie on $\boldsymbol{Q}$ (forcing $\ell \subseteq \boldsymbol{Q}$). Then, with $\boldsymbol{x} = \boldsymbol{a}$ and $\boldsymbol{x} = \boldsymbol{b}$,

$$\boldsymbol{ea} = \boldsymbol{eb} = k \quad \rightarrow \quad \boldsymbol{e} \cdot \begin{bmatrix} 4 \\ -2 \\ 2 \end{bmatrix} = \boldsymbol{e} \cdot \begin{bmatrix} 2 \\ 6 \\ 8 \end{bmatrix} \quad \rightarrow$$

$$4p - 2q + 2r = 2p + 6q + 8r$$

which simplifies to

$$p - 4q - 3r = 0$$

Next, impose the condition $\boldsymbol{Q} \perp \boldsymbol{P}$. In terms of normals, we have

$$\boldsymbol{ec} = 0 \quad \rightarrow \quad \boldsymbol{e} \cdot \begin{bmatrix} 3 \\ 4 \\ -1 \end{bmatrix} = 0$$

or

$$3p + 4q - r = 0$$

We want to solve the two equations in p, q, r for p and q in terms of r. Summing yields

$$4p - 4r = 0 \quad \rightarrow \quad p = r$$

By substitution, $q = -r/2$ and hence with $r = 2, \boldsymbol{e} = [p, q, r] = [2, -1, 2]$. To find k, we have $\boldsymbol{ea} = k$ or $4p - 2q + 2r = k = 4(2) - 2(-1) + 2(2) = 14$. The desired equation of $\boldsymbol{Q}$ is therefore $\boldsymbol{ex} = 14$, or, in coordinate form, $2x - y + 2z = 14$. ■

OUR GEOMETRIC WORLD

When the astronauts landed on the moon in 1969, they left behind a reflecting "half-box" (Figure 7.54) for the purpose of making accurate measurements using laser beams projected from the Earth. The half-box consists of three mirrors that are mutually orthogonal, an altogether simple design. But it has the interesting property that a laser beam from Earth aimed toward the box will be reflected back to Earth at the exact same spot where the beam originated—regardless of the orientation of the box on the moon and the angle at which the beam strikes it. It is interesting to solve the mathematics of this phenomenon, which can be nicely done using vectors (see Problem 12).

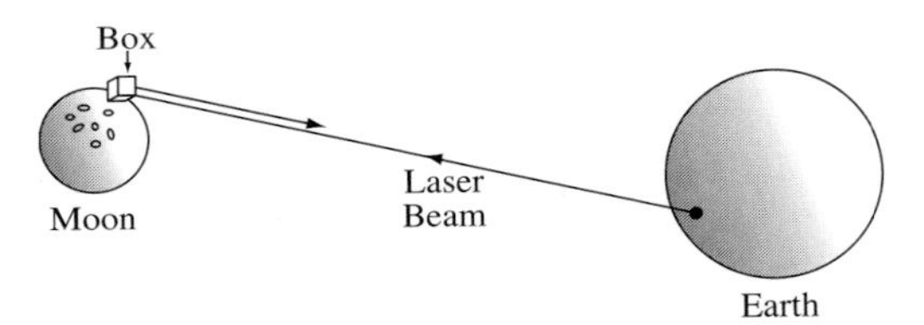

Figure 7.54

EXAMPLE 4 State and prove the Pythagorean theorem in terms of three-dimensional vectors.

SOLUTION

The triangle with vertices $\mathbf{0}$, $\boldsymbol{u}$, and $\boldsymbol{v}$ is a right triangle if $\boldsymbol{uv} = 0$. Therefore, we can state Pythagoras' theorem as follows:

$$\text{If } \boldsymbol{uv} = 0 \text{ then } \|\boldsymbol{u}\|^2 + \|\boldsymbol{v}\|^2 = \|\boldsymbol{u} - \boldsymbol{v}\|^2.$$

PROOF

By definition,

$$\|\boldsymbol{u} - \boldsymbol{v}\|^2 = (\boldsymbol{u} - \boldsymbol{v})^2 = (\boldsymbol{u} - \boldsymbol{v}) \cdot (\boldsymbol{u} - \boldsymbol{v}) = \\ = \boldsymbol{u}^2 - 2\boldsymbol{uv} + \boldsymbol{v}^2 = \|\boldsymbol{u}\|^2 + \|\boldsymbol{v}\|^2$$

(since $\boldsymbol{uv} = 0$). ■

ISOMETRIES IN E^3

Here we briefly explore the coordinate form for a distance-preserving transformation on E^3. For convenience we consider only isometries that fix the origin (0, 0, 0); all the rest can be obtained through translations, exactly as in the situation for plane transformations.

As in E^2, there are special forms of isometries for E^3: Reflection in a plane, space rotations, and direct and opposite isometries, to name a few. Analogies between dimensions two and three are strong enough to give us adequate guidance, and very little explanation should be needed. Of course, it is more convenient to work with matrices and vectors, which we shall do. Our starting point is the matrix form of the most general linear transformation that maps (0, 0, 0) into (0, 0, 0), given first in coordinate form, then in matrix form:

$$\begin{aligned} x' &= a_1x + b_1y + c_1z \\ y' &= a_2x + b_2y + c_2z \\ z' &= a_3x + b_3y + c_3z \end{aligned} \qquad \rightarrow \qquad \begin{bmatrix} x' \\ y' \\ z' \end{bmatrix} = \begin{bmatrix} a_1 & b_1 & c_1 \\ a_2 & b_2 & c_2 \\ a_3 & b_3 & c_3 \end{bmatrix} \begin{bmatrix} x \\ y \\ z \end{bmatrix}$$

The vector form is

$$\boldsymbol{v}' = A\boldsymbol{v},$$

where $\boldsymbol{v}' = [x', y', z']$, $\boldsymbol{v} = [x, y, z]$, and A stands for the 3×3 above matrix. Now we need to guarantee that A be distance preserving. That is, we want

$$\textbf{(1)} \qquad \|\boldsymbol{u}' - \boldsymbol{v}'\| = \|\boldsymbol{u} - \boldsymbol{v}\|, \qquad \text{for all } \boldsymbol{u}, \boldsymbol{v}.$$

Since this also applies to the distance from $\mathbf{0}$ to $\boldsymbol{u}$ and $\boldsymbol{v}$, we must then have $\|\boldsymbol{u}'\| = \|\boldsymbol{u}\|$ and $\|\boldsymbol{v}'\| = \|\boldsymbol{v}\|$. Putting these two conditions together, and squaring out in **(1)** according to the rules for scalar products, we get the necessary—which also turns out to be sufficient—condition for an isometry

$$\boldsymbol{u}'\boldsymbol{v}' = \boldsymbol{u}\boldsymbol{v}, \qquad \text{for all } \boldsymbol{u} \text{ and } \boldsymbol{v}.$$

That is,

$$(A\boldsymbol{u}) \cdot (A\boldsymbol{v}) = \boldsymbol{u}\boldsymbol{v}.$$

In particular, this must be true if $\boldsymbol{u} = [1, 0, 0] = \boldsymbol{v} = [1, 0, 0]$. The above condition then becomes $(A\boldsymbol{u}) \cdot (A\boldsymbol{v}) = 1$. Thus,

$$A\boldsymbol{u} = [a_1, a_2, a_3] = A\boldsymbol{v} \quad \text{and}$$
$$\|A\boldsymbol{v}\| = \|\boldsymbol{v}'\| = \|\boldsymbol{v}\| = 1 \rightarrow a_1^2 + a_2^2 + a_3^2 = 1.$$

That is, the first column of A must be a unit vector. Similarly, if $\boldsymbol{u} = [0, 1, 0]$ and $\boldsymbol{v} = [0, 1, 0]$ then $(A\boldsymbol{u}) \cdot (A\boldsymbol{v}) = \boldsymbol{u}\boldsymbol{v} = 1$ or

$$b_1{}^2 + b_2{}^2 + b_3{}^2 = 1,$$

and the second column of A is a unit vector. It is clear that this pattern continues. If we stagger the "1s" in $\boldsymbol{u}$ and $\boldsymbol{v}$, with $\boldsymbol{u} = [1, 0, 0]$ and $\boldsymbol{v} = [0, 1, 0]$, then we get $(A\boldsymbol{u}) \cdot (A\boldsymbol{v}) = 0$, and thus

$$[a_1, a_2, a_3] \cdot [b_1, b_2, b_3] = 0$$

or

$$a_1b_1 + a_2b_2 + a_3b_3 = 0$$

That is, the dot product of the first and second columns is zero, and hence, as vectors, they are orthogonal. By continuing in this manner, we find:

THEOREM 1 (COORDINATE FORM OF AN ISOMETRY IN E^3)
Suppose that $\boldsymbol{v}' = A\boldsymbol{v}$ is the vector form for a distance-preserving mapping from E^3 to itself that fixes the point **0.** Then A must have the form of an **orthogonal matrix;** that is, the three columns of A are unit vectors in E^3, and they are mutually orthogonal.

EXAMPLE 5 One can readily verify that the following matrix

$$A = \begin{bmatrix} \frac{2}{3} & \frac{1}{3} & \frac{2}{3} \\ \frac{2}{3} & -\frac{2}{3} & -\frac{1}{3} \\ \frac{1}{3} & \frac{2}{3} & -\frac{2}{3} \end{bmatrix}$$

has the desired properties of being orthogonal. Hence, the transformation on E^3 defined by $\boldsymbol{x}' = A\boldsymbol{x} + \boldsymbol{c}$ where $\boldsymbol{c} = [1, -2, -1]$, is an isometry, whose coordinate form is

$$\begin{aligned} x' &= \tfrac{2}{3}x + \tfrac{1}{3}y + \tfrac{2}{3}z + 1 \\ y' &= \tfrac{2}{3}x - \tfrac{2}{3}y - \tfrac{1}{3}z - 2 \\ z' &= \tfrac{1}{3}x + \tfrac{2}{3}y - \tfrac{2}{3}z - 1 \end{aligned} \quad \blacksquare$$

Some of the problems that follow make use of the **unit basis vectors $\boldsymbol{i}, \boldsymbol{j}$, and $\boldsymbol{k}$.** These are the position vectors that lie along the coordinate axes having unit length. They are defined as the vectors

$$\boldsymbol{i} = [1, 0, 0], \quad \boldsymbol{j} = [0, 1, 0], \quad \text{and} \quad \boldsymbol{k} = [0, 0, 1]$$

Because of the algebra of vectors, we have for any three reals x, y, and z:

$$x\boldsymbol{i} + y\boldsymbol{j} + z\boldsymbol{k} = x[1, 0, 0] + y[0, 1, 0] + z[0, 0, 1] = [x, y, z]$$

Thus we observe the significant (and familiar) fact:

Any 3-vector is a unique linear combination of $\boldsymbol{i}, \boldsymbol{j}$, and $\boldsymbol{k}$.

For this reason, the vectors $\boldsymbol{i}, \boldsymbol{j}$, and $\boldsymbol{k}$ are referred to as a **basis** for $\mathbf{V}^3$.

PROBLEMS (7.5)

GROUP A

1. A line has equation $x = t - 1, y = 3t + 1, z = -t$ (t real). Does it intersect both planes $x + y + z = 9$ and $3x - y - 2z = 2$? If so, find the coordinates of the point(s) of intersection.

2. A line has equation $x = t, y = 3t, z = t + 2$, (t real). Find its point of intersection with the line $x = 3s - 7, y = s - 5, z = -6s + 13$ (s real), and show that these two lines are perpendicular.

3. Show that the following parametric forms represent the same two lines:

$$x = 4t - 1, \qquad y = 6t + 3, \qquad z = 2t + 4, \qquad t \text{ real},$$

and

$$x = -2s + 5, \qquad y = -3s + 12, \qquad z = -s + 7, \qquad s \text{ real}.$$

[***Hint:*** Find a replacement for the parameter s in terms of t of the form $s = at + b$, where a and b are constants to be determined. You must then show that this replacement works for all three coordiantes x, y, and z. (For a geometric method, show that the second line passes through $-1, 3, 4$)—a point on the first—then show that the lines are parallel, hence must coincide.)]

4. The line ℓ: $x = t + 1, y = 2t - 1, z = -t + 2$ is oblique to the plane $\boldsymbol{P}$: $3x + y + z = 1$, as illustrated. Find:

(a) the point of interesection of ℓ and $\boldsymbol{P}$

(b) the equation for the unique plane that contains ℓ and is orthogonal to $\boldsymbol{P}$.

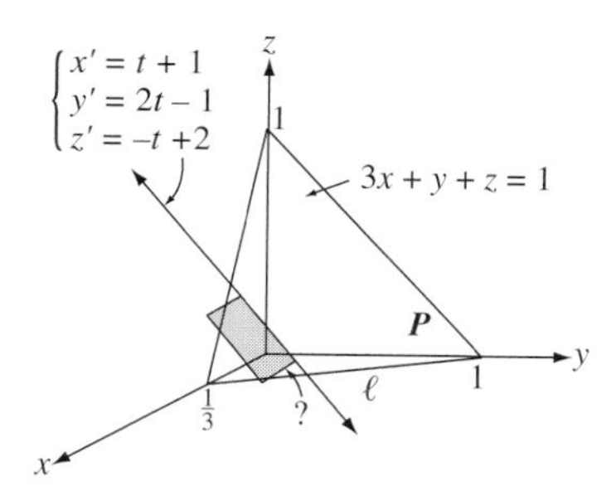

5. A line is given in vector form by $\boldsymbol{x} = t\boldsymbol{a} + \boldsymbol{x}_0$ where $\boldsymbol{a} = \boldsymbol{i} - \boldsymbol{j} + 2\boldsymbol{k}$ and $\boldsymbol{x}_0 = \boldsymbol{j} + 3\boldsymbol{k}$. Show that this line is contained by the plane $\boldsymbol{bx} = -2$ if $\boldsymbol{b} = 3\boldsymbol{i} + \boldsymbol{j} - \boldsymbol{k}$.

6. Using vector formulas only, find the cosine of the angle between

(a) $\boldsymbol{a} = 3\boldsymbol{i} - \boldsymbol{j} + 2\boldsymbol{k}$ and $\boldsymbol{b} = 2\boldsymbol{j} + \boldsymbol{k}$

(b) $\boldsymbol{c} = 3\boldsymbol{i} - \boldsymbol{j} - 2\boldsymbol{k}$ and $\boldsymbol{d} = \boldsymbol{i} + \boldsymbol{j}$.

GROUP B

7. (a) Calculate $\boldsymbol{ab} = \boldsymbol{a} \cdot \boldsymbol{b}$ if $\boldsymbol{a} = \boldsymbol{i} + \boldsymbol{j} - \boldsymbol{k}$ and $\boldsymbol{b} = \boldsymbol{i} - \boldsymbol{j} + \boldsymbol{k}$.

(b) Find the measure of the angle between these two vectors.

8. **(a)** Write the vector equation of the line ℓ passing through $A(3, -1, 2)$ and $B(0, 3, 1)$. (***Hint:*** What does $\boldsymbol{AB}$ or $\boldsymbol{B} - \boldsymbol{A}$ represent?)

(b) Write the vector form of the equation of the plane $\boldsymbol{P}$**:** $x + y - 2z = 4$.

(c) Using vectors, find the unique point of intersection of ℓ and $\boldsymbol{P}$.

9. Find the measure of the dihedral angle formed by the half-planes $z = -\sqrt{2}x + y, z > 0$, and $z = \sqrt{2}x - y, z > 0$ (illustrated in the figure). (***Hint:*** Use normals.)

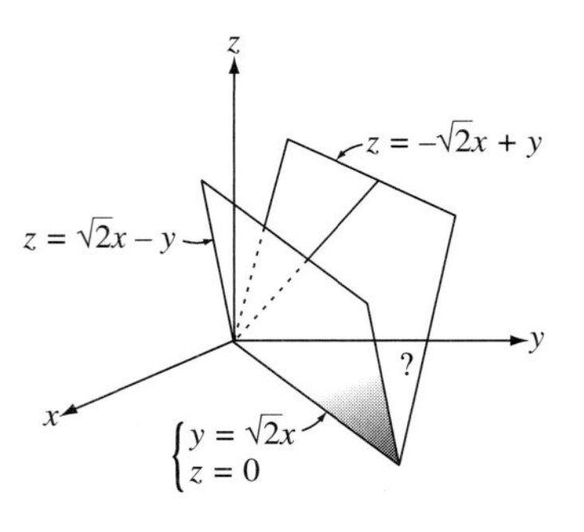

10. **(a)** Show that the following transformation is an insometry by verifying that the matrix A of coefficients is orthogonal:

$$x' = -\frac{7}{11}x + \frac{6}{11}y + \frac{6}{11}z$$

$$y' = \frac{6}{11}x + \frac{9}{11}y - \frac{2}{11}z$$

$$z' = \frac{6}{11}x - \frac{2}{11}y + \frac{9}{11}z$$

(b) By direct calculation, find the distance between $A(1, 1, 0)$ and $B(4, 1, -4)$, the coordinates of the image points A' and B', and the distance between A' and B'.

11. Show that the isometry of Problem 10 is actually a reflection in some plane by

(a) showing that the coordinate equation of the set of all points (x, y, z) that are fixed by the transformation is that of a plane $\boldsymbol{P}$, and

(b) verifying that the midpoint of the segment joining $Q(x, y, z)$ and $Q'(x', y', z')$ lies on $\boldsymbol{P}$ and line QQ' is perpendicular to the plane $\boldsymbol{P}$ you found in part **(a)**.

GROUP C

12. **(a)** Let the vector $\boldsymbol{v} = a\boldsymbol{i} + b\boldsymbol{j} + c\boldsymbol{k}$ represent a ray of light striking the xy-plane. If the xy-plane acts as a mirror, show that the reflected ray is $\boldsymbol{v}_{\mathrm{R}} = a\boldsymbol{i} + b\boldsymbol{j} - c\boldsymbol{k}$.

(b) Show a similar result with respect to the other coordinate planes.

(c) Use the results of **(a)** and **(b)** to prove that a ray of light entering the first octant of the xyz-coordinate system and striking first the xy-plane, then the xz-plane, and finally the yz-plane, will exit parallel to its original path.

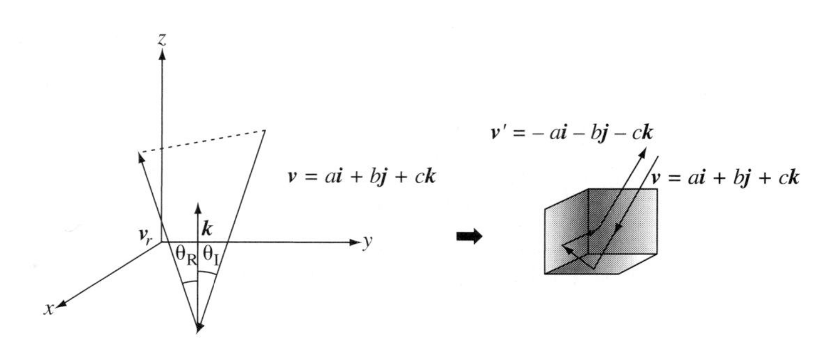

13. It is known that every direct isometry with a fixed point has a line of fixed points. This means that every object turning about a point in space with no outside forces acting must have a "spin axis"—an axis about which is rotates! Illustrate this by finding the spin axis and angle of rotation for the isometry of Example 5.

14. A Model for Three-Dimensional Hyperbolic Geometry Analogous to the upper half-plane model for the hyperbolic plane, coordinate geometry can be used to conveniently define the **upper half-space model** for hyperbolic 3-space. The object of this problem is to explore that model, leaving some details for you to work out. We let $\boldsymbol{H}$ denote the set of all points $P(x, y, z)$ in space such that $z > 0$. The *h-lines* are either semicircles with base diameter on, and lying in a plane perpendicular to, the xy-plane, or vertical rays (parallel to the z-axis) with end points in the xy-plane. The *h-planes* are hemispheres with centers in the xy-plane, or half-planes perpendicular to the xy-plane. Distance and angle-measure are the same as for the upper half-plane model (restricting to a plane for each pair of h-points or pair of vertical h-rays). You should now convince yourself (prove) that (1) *two points determine an h-line*, (2) *three noncollinear points determine an h-plane*, and (3) an h-angle always lies in an ordinary plane perpendicular to the xy-plane. You might find it convenient to work with coordinates: The equation of any sphere with center at $C(h, k, 0)$ is $(x - h)^2 + (y - k)^2 + z^2 + r^2$, which reduces to the form $x^2 + y^2 + z^2 + ax + by = c, z > 0$. (The equation of a vertical half-plane is $ax + by = c$ for constants a, b, c, where a and b are not both zero.) It would take three points lying on such a surface to determine a, b, and c in each case. Explore as many features of hyperbolic 3-space as you can; venture into asymptotic tetrahedrons and a theory of parallelism for h-planes. Compare the synthetic definition given in Section 7.2 for asymptotic parallelism.

THREE-DIMENSIONAL HYPERBOLIC GEOMETRY

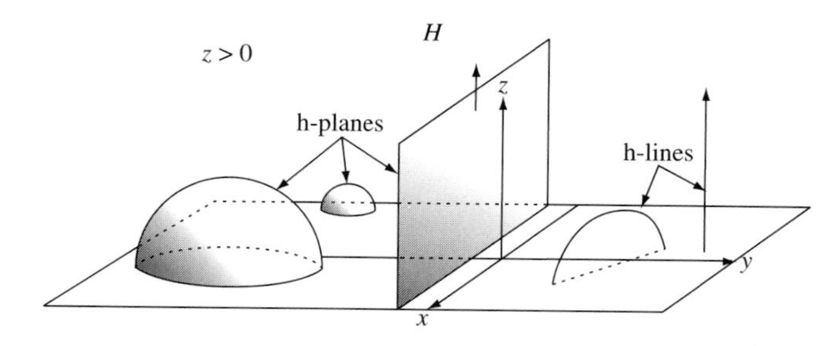

*7.6 Spherical Geometry

Now that we have some experience with three-dimensional geometry, we are in a position to examine the major characteristics of the geometry of the sphere. An axiomatic approach is possible (in fact, this can be found in Appendix D), but here we confine our attention to models, one being the unit sphere $\boldsymbol{S}$—a sphere in E^3 having radius 1. If this sphere were centered at the origin of a three-dimensional coordinate system then its equation would be

$$\boldsymbol{S}: x^2 + y^2 + z^2 = 1$$

LINES AND DISTANCE ON THE SPHERE

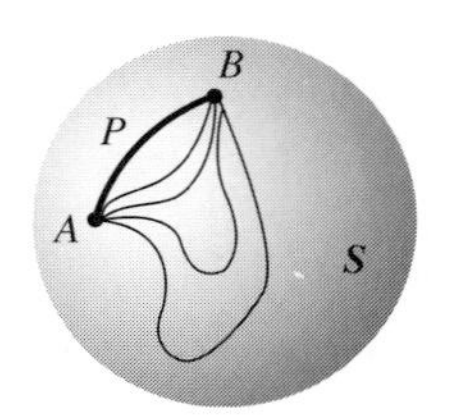

Figure 7.55

First, let us discuss distance and "straight" lines on a spherical surface. Imagine, if you will, a piece of string that passes through two fixed points A and B on $\mathbf{S}$. As the string is gradually pulled tighter and tighter, we know that it will assume the position of a great circle arc $\widehat{APB}$, as illustrated in Figure 7.55. This experiment lends support to the principle that the shortest path between any two points on a sphere is the (minor) great circle arc joining them. Thus, in spherical geometry, the great circles are the analogs of lines in absolute geometry. We define for any two points A and B on $\mathbf{S}$:

$$AB^* = \text{Length of semi-great circle} = \pi \text{ if } A \text{ and } B \text{ are opposite points (poles)}$$

$$AB^* = \text{Length of minor arc } \widehat{AB} \text{ on the great circle passing through } A \text{ and } B\text{, otherwise.}$$

In axiomatic geometry we can let α be the upper bound for distance regardless of the geometry we are studying. Distances on a unit sphere are obviously bounded—the (least) upper bound being given by $AB^* = \pi$ when A and B are polar points. Thus $\alpha = \pi$. (In Euclidean and hyperbolic geometry, $\alpha = \infty$.) If our model for spherical geometry were just our own planet Earth, $\alpha \approx 24{,}900$ (miles). We often use α in place of π to make our discussion apply to any sphere.

In axiomatic geometry it can be proven that if $\alpha < \infty$ then for each given point A in the plane there exists a unique *corresponding* point A' such that $AA' = \alpha$. On a sphere $\mathbf{S}$, this occurs, for example, when A and B are the north and south poles, or so-called **polar** (or **opposite**) points on $\mathbf{S}$. Geometrically, this means that A and B are the end points of a **diameter** of S (defined as the intersection of a line passing through the center of a sphere and the sphere and its interior).

In absolute geometry it was assumed (as an axiom) that two points always determine a unique line. This is not true in $\mathbf{S}$. Let us consider the example of the north and south poles A and B (Figure 7.56). In this case, $\overline{AB}$ is a diameter of $\mathbf{S}$, and any vertical plane passing through O, the center of $\mathbf{S}$, contains this diameter, and the points A and B. This plane intersects $\mathbf{S}$ in a great circle (thus "line"), and there are infinitely many such planes passing through A and B, thus there are an infinite number of

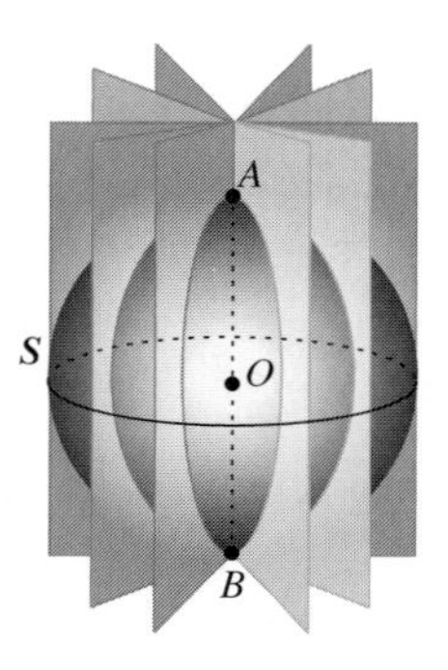

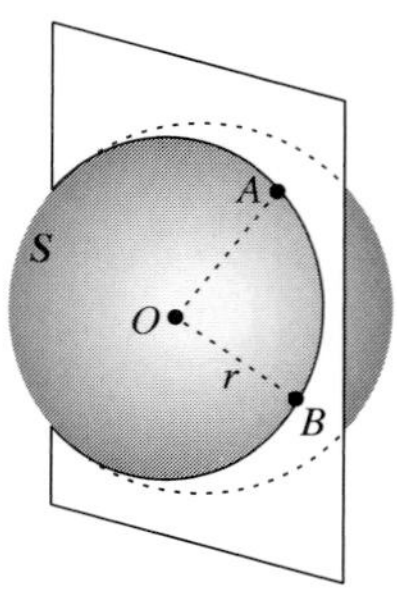

Figure 7.56

"lines" passing through two poles (or opposite points) A and B of $\boldsymbol{S}$. On the other hand, if A and B are not opposite points (when $AB < \alpha$), then O, A, and B are non-collinear and they determine a unique plane. In this case there is only one great circle passing through A and B. We can then be certain that in spherical geometry, *two points determine a line if* $AB^* < \alpha$.

What about betweenness? This is readily determined using distance, as in absolute geometry: A–B–C^* holds iff A, B, and C are distinct points of a great circle on $\boldsymbol{S}$ and $AB^* + BC^* = AC^*$. This evidently occurs whenever B lies on the minor arc $\widehat{AC}$ of the great circle passing through A and C (if $AC^* < \pi$). What if $AC^* = \pi$? Then since there are an infinite number of lines through A and C in this case and they cover all of $\boldsymbol{S}$, it follows that *every point of* $\boldsymbol{S}$ *except A and C lies between A and C*. This shows us that we must be careful about our concept of both distance and betweenness in $\boldsymbol{S}$.

ANGLES AND ANGLE MEASURE

Using the above concept for betweenness we can identify segments, rays, angles, and triangles in $\boldsymbol{S}$: a segment is merely the minor arc of a great circle (like $\widehat{AB}$ in Figure 7.57, where $AB^* < \pi$), a ray is a semi-great circle (like arc $\widehat{CDC'}$ in Figure 7.57, having *two* end points), an angle is the union of two semi-great circles, and a triangle is the union of three (minor) great circle arcs (again as in Figure 7.57).

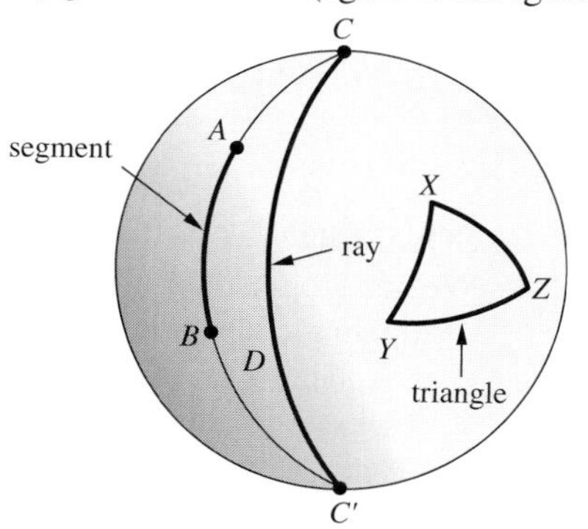

Figure 7.57

Finally, concerning angle measure, we observe that on the surface of the Earth the natural way to measure an angle that has its sides on the Earth's surface is to sight along the sides and use Euclidean angle measure. Thus, $m\angle ABC = m\angle A'BC'$ where rays $\overrightarrow{BA'}$ and $\overrightarrow{BC'}$ are the tangents to arcs $\widehat{BA}$ and $\widehat{BC}$. We can now summarize the components of our model for spherical geometry as follows.

POINT	Any point on the sphere $\boldsymbol{S}$.
LINE	A great circle of $\boldsymbol{S}$.
DISTANCE	The Euclidean length AB^* of minor (or semicircle) arc $\widehat{AB}$ on a great circle passing through A and B.
ANGLE MEASURE	If the tangents in E^3 of sides $\widehat{BA}$ and $\widehat{BC}$ of angle ABC are $\overrightarrow{BA'}$ and $\overrightarrow{BC'}$, then take

$$m\angle ABC^* = m\angle A'BC'$$

NOTE: Following the precedent set in hyperbolic geometry, we use an asterisk following the symbol of a geometric measure, object, or concept to denoting a property of spherical geometry (to distinguish it from the corresponding entities in E^3), and we will start using the terms s-distance, s-segment, . . . for spherical distance, spherical segment, and so on.

EXAMPLE 1 A spherical right triangle on the unit sphere $\boldsymbol{S}$ has vertices $C(0, 0, 1)$ (north pole), $A(1/\sqrt{2}, 0, 1/\sqrt{2})$ (half-way down the longitude line from the north pole to the equator in the xz-plane), and $B(0, 1/\sqrt{2}, 1/\sqrt{2})$ (half-way from the north pole to the equator in the yz-plane), as in Figure 7.58. Find the lengths of the three sides. Is this triangle "isosceles"? Are the base angles congruent?

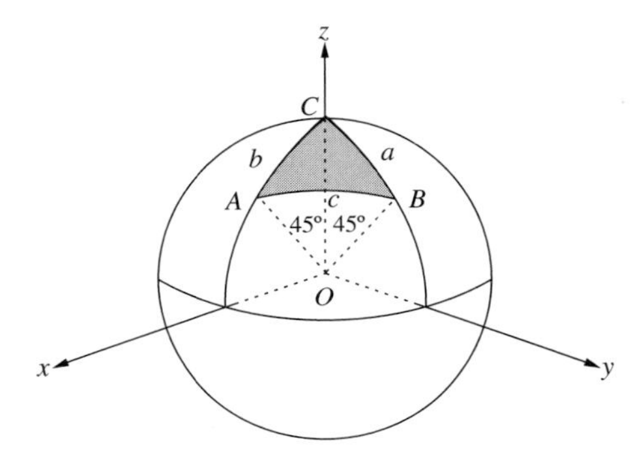

Figure 7.58

SOLUTION

We need to recall how to find the length of arc s of a circle of radius r subtended by a central angle of measure θ; the formula is $s = r\theta$, provided θ is measured in radians. (See Appendix B.) Since $r = 1$, then $s = \theta$ (radian measure). Thus, since in Figure 7.58 we can see that $m\angle BOC = 45$ and $m\angle AOC = 45$, then

$$a = b = \pi/4 \approx 0.785 \qquad \text{(maximum distance} = \alpha \approx 3.14)$$

To find c we use vectors to find the cosine of the angle θ between $\boldsymbol{OA}$ and $\boldsymbol{OB}$, or $m\angle AOB$. Being position vectors, $\boldsymbol{OA} = [1/\sqrt{2}, 0, 1/\sqrt{2}] = (1/\sqrt{2})[1, 0, 1]$ and $\boldsymbol{OB} = [0, 1/\sqrt{2}, 1/\sqrt{2}] = (1/\sqrt{2})[0, 1, 1]$. Then from the formula for the cosine of the angle between two vectors,

$$\cos\theta = \frac{\boldsymbol{OA} \cdot \boldsymbol{OB}}{\|\boldsymbol{OA}\| \, \|\boldsymbol{OB}\|}$$

or, since the factor $1/\sqrt{2}$ cancels out,

$$\cos\theta = \frac{[1, 0, 1] \cdot [0, 1, 1]}{\|[1, 0, 1]\| \, \|[0, 1, 1]\|} = \frac{1 \cdot 0 + 0 \cdot 1 + 1 \cdot 1}{\sqrt{1^2 + 0^2 + 1^2}\,\sqrt{0^2 + 1^2 + 1^2}}$$

$$= \frac{1}{\sqrt{2} \cdot \sqrt{2}} = \frac{1}{2}$$

Thus $\theta = m\angle AOB = 60$ and $c = \pi/3 \approx 1.047$. Yes, the triangle is isosceles, and the base angles are congruent by the symmetry evident in the figure. ■

ISOSCELES TRIANGLE THEOREM

In the preceding example we encountered a property of **S** that agrees with absolute geometry: *The base angles of an isosceles triangle are congruent.* In fact, the proof of this proposition can be patterned after that for absolute geometry using the SAS postulate, which is also true in **S**.

EXAMPLE 2 Find, to the nearest 1/100 unit, the s-distances KS^*, SP^*, and PK^* in **S** if $K = (1, 0, 0)$, $S = (0, 0, 1)$, and P is s-collinear with K and S, with the central angles subtended by arcs SP and PK each of measure 135°. (See Figure 7.59.) Does S lie between K and P?

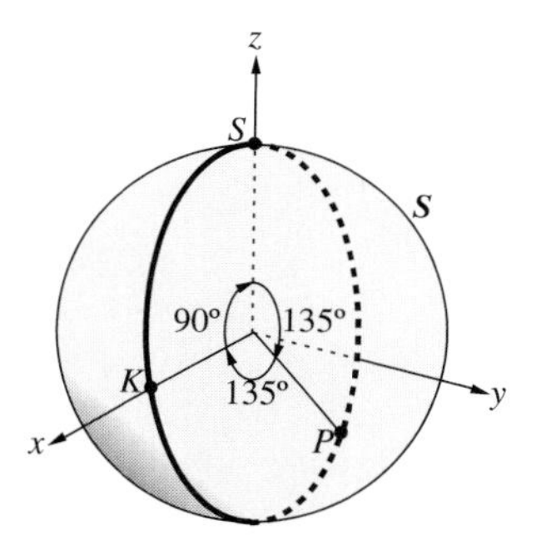

Figure 7.59

SOLUTION

Following the same procedure as in Example 1, we need only change degree measure to radian measure:

$$KS^* = 90(\pi/180) = \pi/2 \approx 1.57 \qquad (\alpha \approx 3.14)$$

$$SP^* = 135(\pi/180) = 3\pi/4 \approx 2.36 = PK^*$$

Since $KS^* + SP^* = 1.57 + 2.36 = 3.93 \neq 2.36 = PK^*$, S does not lie between K and P. In fact, there are no betweenness relations among these three s-collinear points. ■

OUR GEOMETRIC WORLD

Three points on the Earth's surface lying approximately on a great circle are Cape Canaveral (Florida), Cairo (Egypt), and Tokyo (Japan), with distances as follows:

From Cape Canaveral to Cairo: 8000 miles

From Cairo to Tokyo: 7000 miles

If we assume the distance from Cape Canaveral to Tokyo is the sum of the above distances, we would get 15,000 miles—the long way around the circle to Tokyo. The shortest (minor arc) distance is the Earth's circumference minus 15,000, or

$$24{,}900 - 15{,}000 = 9{,}900 \text{ miles}$$

The metric defined in the model $\boldsymbol{S}$ when $\alpha = 24{,}900$ would yield this same answer—the shortest distance between those two points on $\boldsymbol{S}$ is the measure of the minor, not the major, arc joining them.

Figure 7.60

The trigonometric relations for a right triangle in spherical geometry will now be derived. We have chosen a derivation that minimizes algebraic manipulations, but it is valid only for those triangles whose sides are less than $\alpha/2$ in length. However, the formulas obtained turn out to be valid for arbitrary spherical triangles of any size. For convenience, it will be assumed at first that we are working with a sphere $\boldsymbol{S}$ having unit radius.

To further simplify matters, we have placed the given right triangle with one leg on the imaginary equator of $\boldsymbol{S}$, and the other on a longitude line joining the imaginary north and south poles of $\boldsymbol{S}$, as shown in Figure 7.61. We use standard notation throughout, with a, b, and c the spherical lengths of the three sides. Recall that since

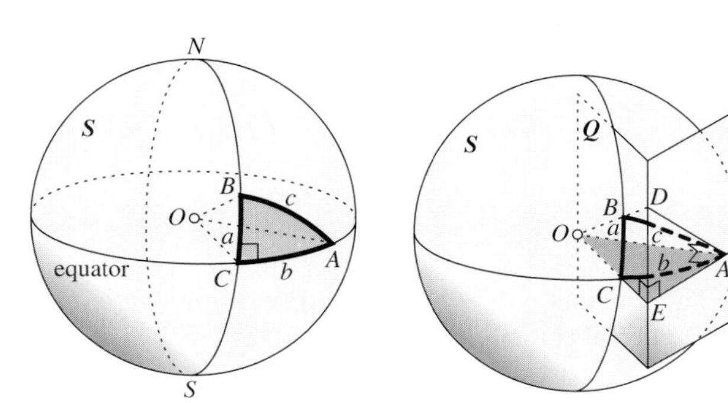

B D O c 1 A O 1 A b C E

Figure 7.61 **Figure 7.62**

S has radius = 1, the central angle measured in radians equals the length of the great circle arc it intercepts. Thus we have:

$$a = m\angle BOC \text{ (in radians)}, \qquad b = m\angle AOC, \qquad \text{and} \qquad c = m\angle AOB.$$

Note that even though we will be making conclusions about spherical geometry, the entire analysis takes place in three-dimensional Euclidean geometry.

SPHERICAL TRIGONOMETRY

Construct the tangent plane $\boldsymbol{P}$ to $\boldsymbol{S}$ at point A, with radius $\overline{OA} \perp \boldsymbol{P}$ (Figure 7.62). Let rays $\overrightarrow{OB}$ and $\overrightarrow{OC}$ meet $\boldsymbol{P}$ at D and E (valid if $AC^*, AB^* < \pi/2$); thus, the plane $\boldsymbol{Q}$ determined by O, B, C meets $\boldsymbol{P}$ in line $\overleftrightarrow{DE}$. The plane OAB determined by $O, A,$ and B contains side $\overline{AB}^*$ and meets $\boldsymbol{P}$ in the line $\overleftrightarrow{AD}$, which must then be tangent to arc $\overset{\frown}{AB}$. Similarly, $\overleftrightarrow{AE}$ is tangent to arc $\overset{\frown}{AC}$. Thus, by definition of spherical angle measure, $A = m\angle BAC^* = m\angle DAE$.

A number of (Euclidean) right triangles are formed that turn out to be very useful. First we have $\overline{OA} \perp \overline{AD}$ and $\overline{OA} \perp \overline{AE}$, forming the two right triangles shown in the inset of Figure 7.62. The equatorial plane OAC contains segment $\overline{OA}$, so it is perpendicular to $\boldsymbol{P}$. But it is also perpendicular to $\boldsymbol{Q}$, which contains side $\overline{BC}^*$ (since $\angle BCA^*$ is a spherical right angle). Thus, both $\boldsymbol{P}$ and $\boldsymbol{Q}$ are perpendicular to plane OAC; by the corollary of Theorem 2, Section 7.1, $\boldsymbol{P} \cap \boldsymbol{Q} \equiv \overline{DE} \perp$ Plane OAC and $\overline{DE} \perp \overline{OE}$ and $\overline{DE} \perp \overline{EA}$. The remainder of the analysis involves only the most elementary trigonometry involving right triangles. We have the following relations:

$$\triangle OAD: \quad \frac{DA}{1} = \tan c \qquad \therefore DA = \tan c$$

$$\frac{1}{OD} = \cos c \qquad \therefore OD = \sec c$$

$$\triangle OAE: \quad \frac{EA}{1} = \tan b \qquad \therefore EA = \tan b$$

$$\frac{1}{OE} = \cos b \qquad \therefore OE = \sec b$$

$$\triangle ODE: \quad \frac{DE}{OD} = \sin\angle DOE = \sin a,\ DE = OD \sin a \qquad \therefore DE = \sec c \sin a$$

$$\frac{OE}{OD} = \cos a,\ OE = OD \cos a \qquad \therefore OE = \sec c \cos a$$

$$\frac{DE}{OE} = \tan a,\ DE = OE \tan a \qquad \therefore DE = \sec b \tan a$$

Hence, in $\triangle ADE$ we have

$$\textbf{(1)} \qquad \sin A = \sin\angle DAE = \frac{DE}{DA} = \frac{\sec c \sin a}{\tan c} = \frac{\sin a}{\cos c \tan c} = \frac{\sin a}{\sin c}$$

and the first relation for spherical triangle ABC emerges. Continuing on,

$$\textbf{(2)} \qquad \cos A = \frac{EA}{DA} = \frac{\tan b}{\tan c}$$

$$\textbf{(3)} \qquad \tan A = \frac{DE}{EA} = \frac{\sec b \tan a}{\tan b} = \frac{\tan a}{\cos b \tan b} = \frac{\tan a}{\sin b}$$

Finally, the two expressions above for OE lead to a relation involving only the sides of the spherical triangle, and could be regarded as the "Pythagorean Theorem" for spherical geometry:

$$OE = \sec b = \sec c \cos a$$

$$(\cos b \cos c)\sec b = (\cos b \cos c) \sec c \cos a$$

$$\textbf{(4)} \qquad \therefore \cos c = \cos a \cos b$$

The formulas we have obtained are collected below for easy reference. (Because $\triangle ABC$ was any right triangle, the relations for sin B, cos B, and $\tan B$ can be obtained by merely changing notation.)

Spherical Trigonometry of the Right Triangle (Unit Sphere)

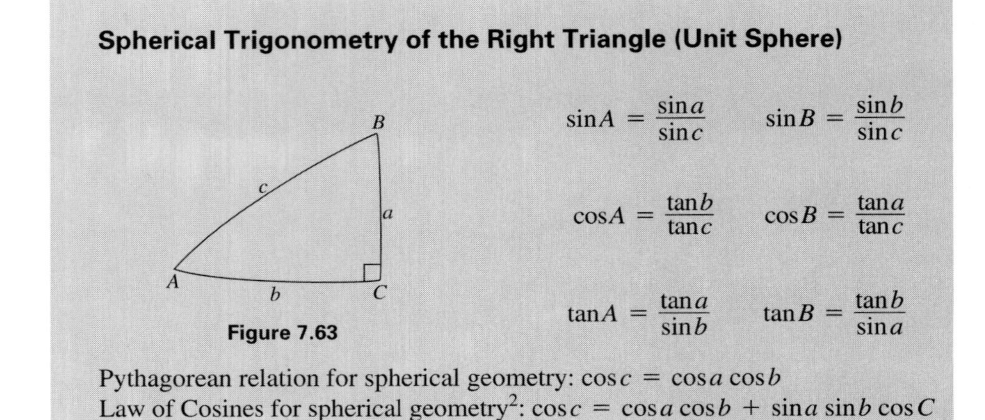

Figure 7.63

$$\sin A = \frac{\sin a}{\sin c} \qquad \sin B = \frac{\sin b}{\sin c}$$

$$\cos A = \frac{\tan b}{\tan c} \qquad \cos B = \frac{\tan a}{\tan c}$$

$$\tan A = \frac{\tan a}{\sin b} \qquad \tan B = \frac{\tan b}{\sin a}$$

Pythagorean relation for spherical geometry: $\cos c = \cos a \cos b$
Law of Cosines for spherical geometry[2]: $\cos c = \cos a \cos b + \sin a \sin b \cos C$

NOTE: It can be shown that these formulas are valid on a sphere of arbitrary radius r, provided a, b, and c are replaced by $a' = a/r$, $b' = b/r$, and $c' = c/r$, respectively. In all these formulas it should be understood that the values for a, b, and c must be in radians.

EXAMPLE 3 Verify the Pythagorean Theorem for spherical geometry in the triangle of Example 1, where the result obtained was

$$a = b = \pi/4 \approx 0.785, \qquad c = \pi/3 \approx 1.047, \qquad C = 90°$$

Then find the other two angles A, B and the angle-sum of $\triangle ABC$, to the nearest one-tenth degree.

---

[2]Established in Problem 17.

SOLUTION

$$\cos c = \cos a \cos b: \qquad \cos \pi/3 = \cos \pi 4 \cos \pi/4 \qquad \text{or}$$
$$\tfrac{1}{2} = (\sqrt{2}/2) \cdot (\sqrt{2}/2)$$

$$\sin A = \frac{\sin a}{\sin c} = \frac{\sin \pi/4}{\sin \pi/3} = \frac{\sqrt{2}/2}{\sqrt{3}/2} = \sqrt{2/3} \quad \rightarrow \quad A = B \approx 54.7^\circ$$

Angle-sum: $A + B + C = 54.7^\circ + 54.7^\circ + 90^\circ = 199.4^\circ$ ■

EXAMPLE 4 Derive the following formula relating the angles of a right triangle to the length of the hypotenuse in spherical geometry, and use it to show that the angle sum of a right triangle is greater than 180. (Compare with Example 3, Section 6.5.)

(5)
$$\cos c = \cot A \cot B$$

SOLUTION

From the formulas for $\tan A$ and $\tan B$, we find

$$\cot A \cot B = \frac{\sin b}{\tan a} \cdot \frac{\sin a}{\tan b} = \frac{\sin b}{\sin a/\cos a} \cdot \frac{\sin a}{\sin b/\cos b} = \cos a \cos b = \cos c$$

We need to prove that $A + B > 90$. From a trigonometric identity,

$$\tan (A + B) = \frac{\tan A + \tan B}{1 - \tan A \tan B}$$

and $\cos c < 1$ for $0 < c < \pi/2$, so we have from **(5)** $\cot A \cot B < 1$, or $\tan A \tan B > 1$. But this makes the denominator of the preceding expression negative when A and B are acute angles. Now, if either A or B is an obtuse or right angle, we would already have $A + B + C > 180$. So assume both $A < 90$, $B < 90$. Then $\tan A$ and $\tan B$ are positive, so the numerator of the preceding expression is positive, while the denominator is negative, hence $\tan (A + B) < 0$ or $A + B > 90$ and therefore $A + B + C > 180$. ■

EXAMPLE 5 Honolulu is 2,091 miles from San Francisco along a great circle that makes an angle of 45° with the line of latitude through that point. The Alaska Peninsula is about an equal distance (2,091 miles) due north of Honolulu.

(a) What is the great circle distance from San Francisco to the Alaska Peninsula? (Use $r = 3{,}960$ for the radius of the earth in miles.)

(b) If Euclidean trigonometry is used for this problem, what discrepancy results? (In Figure 7.64, A = Honolulu, B = San Francisco, C = Alaska Peninsula.)

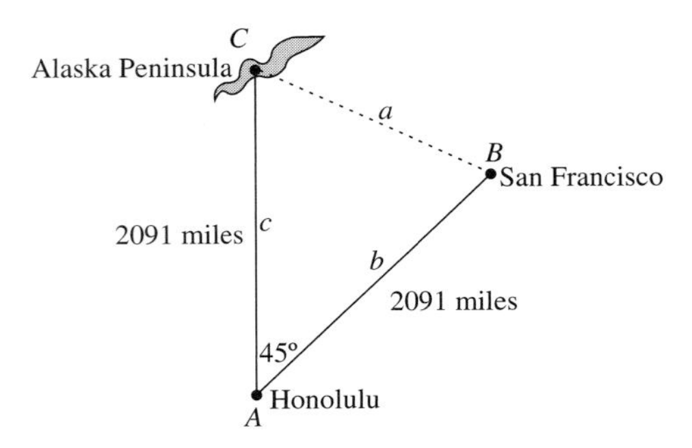

Figure 7.64

SOLUTION

(a) We use the spherical Law of Cosines for a sphere of radius $r = 3960$, with $a' = a/r$, $b' = b/r$, and $c' = c/r$:

$$\begin{aligned}\cos a' &= \cos b' \cos c' + \sin b' \sin c' \cos A \\ &= \cos^2(2091/3960) + \sin^2(2091/3960)\cos 45° \\ &= 0.925650 \quad \rightarrow \quad a' = 0.38804\end{aligned}$$

$$\therefore a' = a/3960 = 0.388804 \rightarrow a = 3960(0.388804) = 1537 \text{ miles.}$$

(b) If $a^2 = b^2 + c^2 - 2bc \cos A$ we have

$$a^2 = 2(2091)^2 - 2(2091)(2091)\cos 45° = 2{,}561{,}222$$

$$\rightarrow a \approx 1600 \text{ miles.}$$

The discrepancy is 53 miles. ■

There are many parallel themes in hyperbolic and spherical geometry. One of them is the angle sum of triangles and the definition of area. For hyperbolic geometry we defined the defect of a triangle as the amount the angle sum is short 180. In spherical geometry, every triangle has angle sum > 180, so we define the **excess** of a triangle as

$$\epsilon(\triangle ABC) = m\angle A + m\angle B + m\angle C - 180$$

Just as in Chapter 6 where we proved that defect was additive, we can also prove here that excess is additive, using practically the same proof. This gives us a reason to take as the definition for area in spherical geometry

$$K = k\epsilon(\triangle ABC) = k(A + B + C - 180)$$

where A, B, and C denote angle measures, and k is some undetermined constant ($k = \pi/180$ is valid for a unit sphere).

Finally, just as in hyperbolic geometry, a virtually identical proof yields:

THEOREM: AAA CONGRUENCE CRITERION FOR SPHERICAL GEOMETRY
If two triangles have the three angles of one congruent, respectively, to the three angles of the other, the triangles are congruent.

COROLLARY: Similar, noncongruent triangles do not exist in spherical geometry.

EXAMPLE 6 In spherical geometry there is only one triangle, up to congruence, having angles of measure 30, 90, and 100. Verify this fact by actually finding the sides of such a triangle (to three-place accuracy) using spherical trigonometry and **(5)**.

SOLUTION
We begin with **(5)**:

$$\cos c = \cot 30° \cot 100° \approx -0.30541 \quad \rightarrow \quad c \approx 1.881$$

With $A = 30$ we have

$$\sqrt{3}/2 = \cos A = \frac{\tan b}{\tan c} = \frac{\tan b}{\tan 1.881} = \frac{\tan b}{-3.11788}$$

$$\text{or} \qquad \tan b = -2.70016$$

$$\therefore b \approx 1.925$$

To find a, use the formula for $\cos c$: $\cos 1.881 = \cos a \cos 1.925$, $a \approx 0.495$ ■

NOTE: $\tan b = -2.70016$ is equivalent to $\tan(b - \pi) = -2.70016$ where $b > 0$.

Moment for Discovery

Area of a Spherical Triangle on a Unit Sphere

The area of a sphere of radius $r = 1$ is 4π square units. The area of a hemisphere is therefore 2π. A **lune** is the region of a sphere bounded by two great semicircles, meeting at two poles. The **angle** of the lune is just the obvious spherical angle formed by the two spherical rays emanating from either pole. A lune L with angle of measure A has area equal to the proportionate amount of L to a hemisphere, which is a "lune" with angle measure 180. That proportional amount is $A/180$. Thus,

$$\frac{\text{Area }(L)}{2\pi} = \frac{A}{180} \qquad \text{or} \qquad \text{Area }(L) = \frac{\pi A}{90} \qquad \text{(square units)}$$

SPHERICAL LUNE

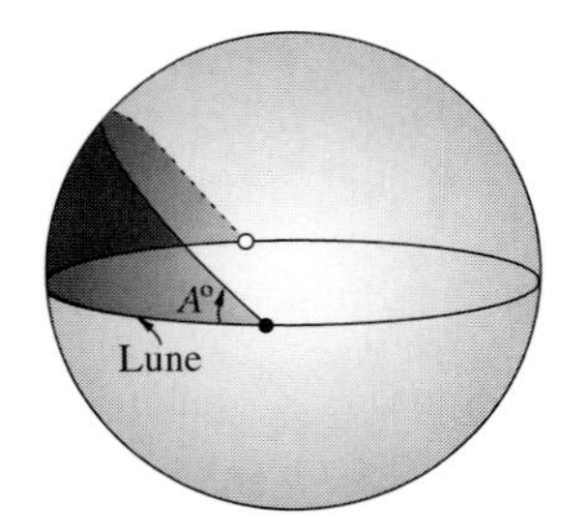

FINDING THE SQUARE UNIT AREA OF A SPHERICAL TRIANGLE

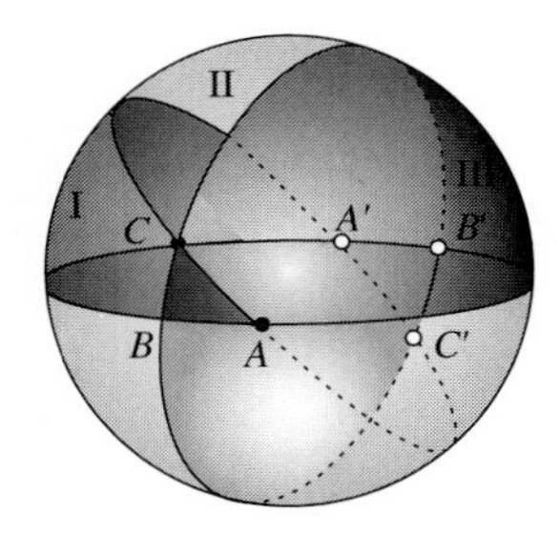

Figure 7.65

Use this formula in the following analysis for the area (as ordinarily understood) of any spherical triangle in terms of square units.

In the figure is indicated the extension of the sides of $\triangle ABC$ to obtain the great circles shown, meeting at the opposite poles A', B', and C'. It is clear that

$$\triangle ABC \cong \triangle A'B'C' \quad \rightarrow \quad K = K'$$

where K and K' are the areas of the triangles. Note that Regions I and III make up the part of the lunes A and B outside $\triangle ABC$, those lunes themselves having areas $\pi A/90$ and $\pi B/90$. Thus,

$$\text{Area (Lune A)} = K + I = \pi A/90$$

where I and K also denote the areas of the figures they represent. Region II is the remainder of the lune of $\angle C' \cong \angle C$ after deleting $\triangle A'B'C'$.

1. Obtain formulas for $K + \text{III}$ and $K' + \text{II} = K + \text{II}$.
2. What type of region is $H \equiv K \cup \text{I} \cup \text{II} \cup \text{III}$? What is its area?
3. Deduce a formula for $K =$ Area $\triangle ABC$. What did you discover? (Use algebra to write expressions for $3K + \text{I} + \text{II} + \text{III}$ and $K + \text{I} + \text{II} + \text{III}$, then solve for K.)

PROBLEMS (7.6)

GROUP A

1. What is the locus of points on the Earth's surface 500 miles due north of the equator? Is it a great circle? If not, what curve is it? Generalize this behavior for any sphere $\mathbf{S}$.
2. Find the excess of
   (a) a spherical triangle $\triangle ABC$ having angles of measure 30, 80, and 150.
   (b) a spherical equilateral triangle $\triangle DEF$ having angles of measure 65 each.

3. Which of the triangles in Problem 2 is the largest? Does "large" mean the same thing in S as it does in Euclidean geometry?

4. Identify the two sides of an s-line ℓ. Use this to sketch the graph and describe the interior of an angle on a sphere S.

5. A spherical triangle $\triangle ABC$ on a unit sphere S has two right angles at B and C, and the length of the included side is $\pi/12 \approx 0.262$. Find the measure of $\angle BAC$ and the excess of the triangle. (***Hint:*** Let $\overline{BC}^*$ lie on the equator and A = north pole.)

6. Use spherical trigonometry to find $A \equiv m\angle BAC$ in Problem 5.

7. A point P moves northward, starting out at the South Pole and moving in such a manner that its direction constantly makes a 45° angle with each line of latitude. Is its locus a great circle or part of a great circle?

8. A circle can be a line in spherical geometry. True or false? If true, where must its center be located?

9. Prove that if B–D–C and ϵ_1 and ϵ_2 are the excesses of $\triangle ABD$ and $\triangle ADC$, then $\epsilon(\triangle ABC) = \epsilon_1 + \epsilon_2$.

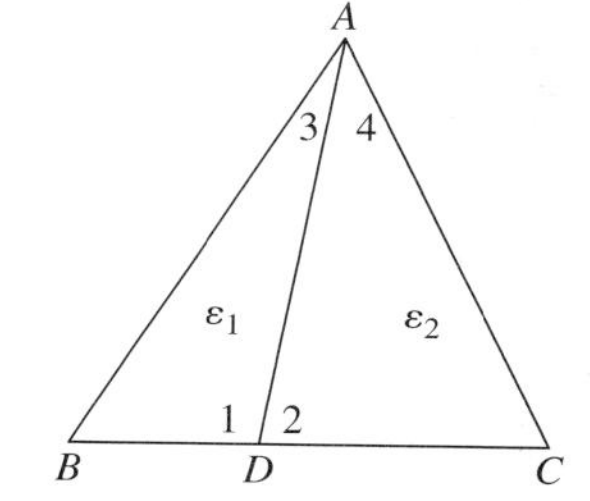

GROUP B

The following problems 10–17 involve spherical trigonometry

10. Given an s-isosceles right triangle $\triangle ABC$ on a unit sphere S with $a = b = \pi/6$. Find c accurate to three places, and the measure of the two nonright angles accurate to one-tenth degree.

11. As in Problem 10, $\triangle ABC$ is an isosceles right triangle, but with acute angles of measure just over 45° (say 45.1°). What size must the sides be, compared with that of the triangle in Problem 9? (***Hint:*** To find the sides see Example 6.)

12. Two great circle arcs on the Earth's surface leading out of London make an angle of 20°. One goes to New York City, a distance of 3,046 miles, the other to Santo Domingo, Dominican Republic, a distance of 4,580 miles. How far is it from New York City to Santo Domingo along a great circle arc? (***Hint:*** See Example 5.)

13. For distances involving several miles or less on the surface of the Earth, why is Euclidean trigonometry a viable tool? Answer by giving your own illustration of some right triangle and the use of $\cos A = b/c$ versus **(2)** above.

14. Find the length of each side of an equilateral triangle having angles each of measure

    (a) 75°

    (b) 150°

    (c) Show that the relationship between the side (of length a) and angle (of measure A) of an equilateral triangle is given by $\sec A = \sec a + 1$.

    (d) What is the maximal length (least upper bound) for a side of an equilateral triangle on the unit sphere?

15. One equilateral triangle on a unit sphere has sides of length 0.3 each, while another has sides of length 2.0 each. Find the measure of each of the angles in each case, accurate to one-tenth degree. Which triangle has the larger excess?

16. There is a unique triangle on the unit sphere whose angles have measure $A = 60°$, $B = 45°$, and $C = 90°$. Find its sides, and estimate your answers to three decimal places.

GROUP C

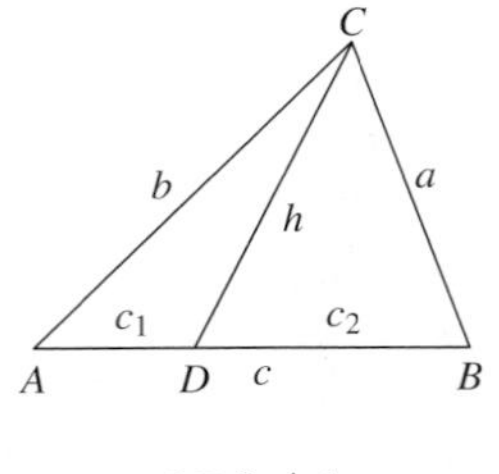

17. Law of Cosines for Spherical Geometry Establish the following formula for a spherical triangle:

$$\cos a = \cos b \cos c + \sin b \sin c \cos A$$

[*Hint:* First write $\cos a = \cos c_2 \cosh h = \cos (c - c_1) \cos h$ and use the addition formula for $\cos(x + y)$, then use $\cos A = \tan c_1 / \tan c$.]

18. Use the Law of Cosines for spherical geometry to prove the SAS Postulate for any spherical model $\boldsymbol{S}$.

19. Prove that each pair of lines in spherical geometry has a common perpendicular.

20. What is the least upper bound of the areas for right triangles on a sphere $\boldsymbol{S}$ having an acute angle of measure 60?

21. Flying "due East" along the equator is well defined and well understood by navigators. Explain why flying "due East" *along a great circle* anywhere else in the world cannot be defined.

22. Using a spherical model, prove that any two lines intersect in spherical geometry.

Chapter 7 Summary

The concept of perpendicularity in space between two planes and between a line and a plane was shown to have natural properties expected to be true, but their proofs frequently led us into some unexpected avenues. We defined the concept of $\boldsymbol{P} \perp \boldsymbol{Q}$ for two planes, but proving $\boldsymbol{Q} \perp \boldsymbol{P}$ turned out to be nontrivial, and required previous significant results (such as the theorem that a line is perpendicular to every line in a plane provided it is perpendicular to just two lines in that plane). The locus property in space that is analogous to the property of the perpendicular bisector of a line segment was established: The locus of points equidistant from two points is the *plane* that bisects the segment joining those two points and perpendicular to it.

Parallelism in space was introduced, where we defined parallelism between a line and a plane, and between two planes. A major result is that parallel lines in space satisfy the Transitive Law for Parallelism: If line ℓ is parallel to line m and m is parallel to n, then line ℓ is parallel to line n. The familiar facts involving rectangular boxes and parallelepipeds were then demonstrated. The more general polyhedron was introduced, and the five Platonic solids were discussed.

Curved surfaces were then introduced: conical surfaces, analogous to pyramidal surfaces formed by planes, and cylindrical surfaces, analogous to prismatic surfaces, led to the familiar concept of cones and cylinders. A sphere, the locus of a point at a constant distance from a fixed point, was introduced, and its interaction with the cone and cylinder discussed. Volume for the solids formed by these surfaces and combinations thereof was shown, with various formulas derived. Included was the volume of a parallelepiped, prism, pyramid, right circular cone, cylinder, and sphere. A formula for a *segment* of a sphere (the part enclosed between two parallel planes) was also derived, and some applications considered.

Three-dimensional coordinates and vectors were introduced, then used to develop spherical geometry from the unit sphere, including the derivation of basic spherical trigonometry, with the formulas for a spherical right triangle turning out to be completely parallel to those of a right triangle in hyperbolic geometry. The *excess* of a spherical triangle (the amount by which the angle sum of a triangle exceeds 180) was discussed briefly, and its additivity leads to a proper concept for area in terms of the angles of a triangle, in agreement with the more familiar concept of area with that involves square units on the surface of the sphere.

Testing Your Knowledge

You are expected to take this test using only the list of axioms, definitions, and theorems in Appendix F as references.

1. Find $m\angle ABE$ if $m\angle ABC = m\angle ABD = m\angle CBD = 90$, $m\angle CBE = 60$, and $m\angle DBE = 30$. You need not prove your answer.

2. A rectangular closet has floor dimensions 5ft by 7.2 ft, and the ceiling is 9.6 ft high. Find the distance *d* from corner to opposite corner along a line not lying in the floor, ceiling, or walls.

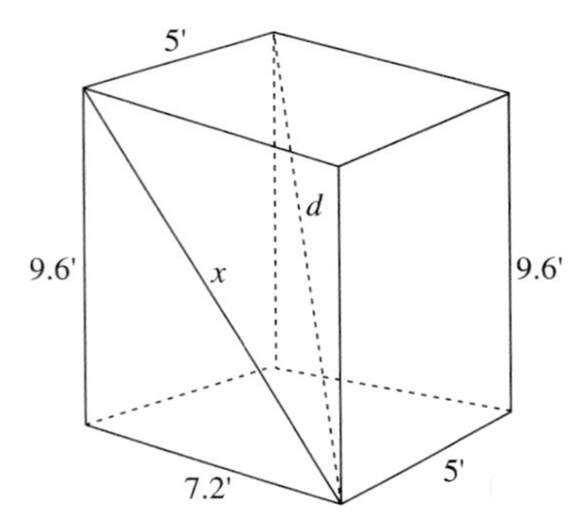

3. Two regular tetrahedrons are joined at a face to form a polyhedron having six congruent equilateral triangles as faces. Tell why the result is *not* a regular polyhedron.

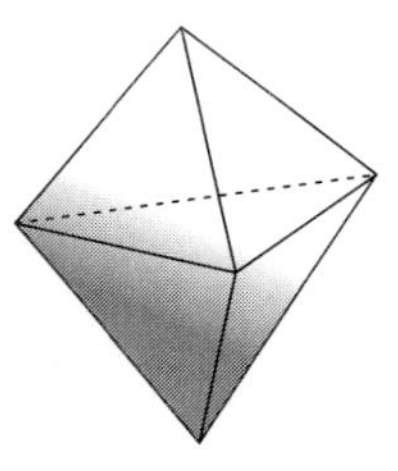

4. How many faces, edges, and vertices does a regular octahedron have?

5. Two perpendicular planes are tangent to a sphere of radius $\sqrt{2}$. How far is the line of intersection of those two planes from the center of the sphere?

6. A conical water tower has a right circular cylinder on its side and a right circular cone at the top. The cone has height equal to its radius. If the radius of the base of the water tower is 10 ft and its total height is 30 ft, find the total volume of the tower. [You will need one or more of the following volume formulas: $V = Bh$, $V = \frac{4}{3}\pi r^3$, $V = \pi r^2 h$, $V = \frac{1}{3}Bh$, $V = \frac{1}{3}\pi r^2 h$, $V = \frac{1}{6}\pi h(3r_1^2 + 3r_2^2 + h^2)$.]

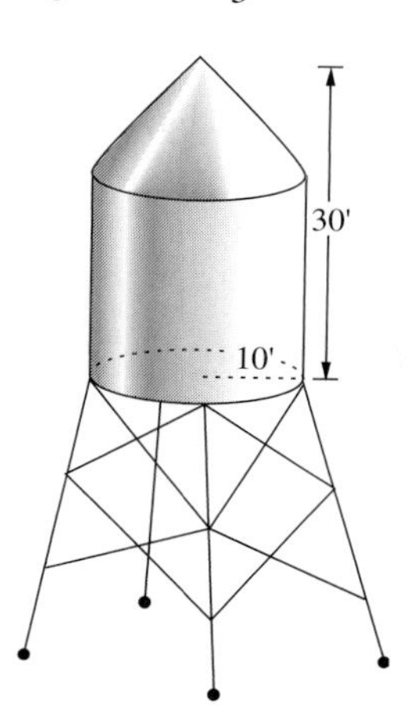

7. Show, on intuitive grounds, that the (lateral) surface area of a right circular cone is one-half the slant height s times the circumference C of the base. Then derive the classical formula for surface area $S = \pi rs$, where r is the radius of the base.

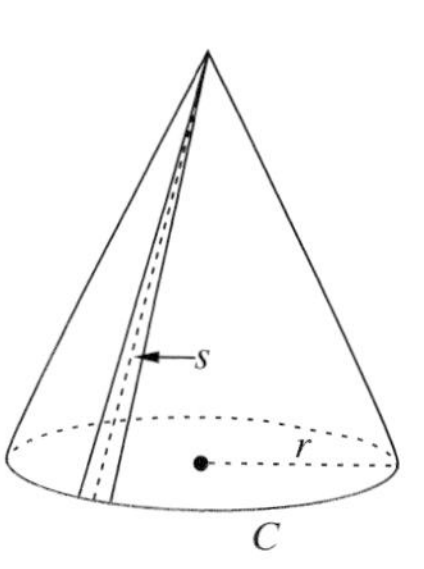

8. The plane $z = 3$ intersects the sphere $x^2 + y^2 + z^2 - 6y - 8z = 25$ in a circle. Find the center and radius of that circle.

9. Find the excess of

(a) a trirectangular spherical triangle (one having three right angles)

(b) a birectangular spherical triangle (two right angles) lying on a unit sphere such that the vertices of those right angles are at a distance $\pi/4$ units apart.

10. Prove that there exists one and only one plane perpendicular to a given line ℓ at some point A on that line. [You must show (1) how to construct such a plane, and (2) prove that there can be only one.]

APPENDIX A

Bibliography

Baravalle, H., *Eighteenth Yearbook*. Washington, D.C.: National Council of Teachers of Mathematics, 1945.

Bonola, R., *Non-Euclidean Geometry*. New York: Dover Publications, 1955.

Burrill, G.F., T.D. Kanold, J.J. Cummins, and L.E. Yunker, *Geometry: Applications and Connections*. New York: Glencoe/McGraw-Hill, 1995.

Burton, D.M., *History of Mathematics: An Introduction,* 2nd ed. New York: McGraw Hill Publishers, 1998.

Coxeter, H.S.M., and S.L. Grietzer, *Geometry Revisited (The New Mathematics Library,* Volume 19). Washington, D.C.: Mathematical Association of America, 1967.

Coxford, A., Z. Usiskin, and D. Hirschorn, *The University of Chicago Study Mathematics Project, Geometry*. Glenview, IL: Scott, Foresman and Company, 1991.

Dudeney, H.E., *Amusements in Mathematics*. New York: Dover Publications, 1958.

Eves, H., *A Survey of Geometry,* Volume One. Boston: Allyn and Bacon, 1963.

Greenberg, M.J., *Euclidean and Non-Euclidean Geometries*. San Francisco: W.H. Freeman and Company, 1993.

Grunbaum, B., and G.C. Shephard, *Tilings and Patterns*. New York: W.H. Freeman and Company, 1987.

Heath, T.L., *The Thirteen Books of Euclid's Elements,* Volume One. New York: Dover Publications, 1956.

Heilbron, J.L., *Geometry Civilized: History, Culture, and Technique*. Oxford: Clarendon Press, 1998.

Hilbert, D., *The Foundations of Geometry*. Chicago, IL: The Open Court Publishing Company, 1902.

Krause, E.F., *Taxicab Geometry: An Adventure in Non-Euclidean Geometry.* New York: Dover Publications, 1987.

Larson, R.E., L. Boswell, and L. Stiff, *Geometry: An Integrated Approach*. Boston: D.C. Heath and Company, 1995.

Lockwood, E.H., "Simson's Line and Its Envelope," *Mathematical Gazette*, **37** (1953), 124–125.

Lockwood, E.H., *A Book of Curves*. New York and London: Cambridge University Press, 1963.

Milnor, J., "Hyperbolic Geometry: The First 150 Years," *The Mathematical Heritage of Henri Poincaré, Proceedings of Symposia in Pure Mathematics of the American Mathematical Society*, **39**, (1996), 25–40.

Moise, E., *Elementary Geometry from an Advanced Standpoint*. Reading, MA: Addison Wesley Longman, 1990.

Trudeau, R., *The Non-Euclidean Revolution*. Boston: Birkhauser, 1995.

Yaglom, I.M., *Geometry Transformations*. New York: Random House, 1962.

APPENDIX B

Review of Topics in Secondary School Geometry

Overview

The first review topic involves one of the simplest of all objects in geometry, the triangle, yet this object is associated with many concepts that are not at all simple, such as congruence, similarity, area, and the Pythagorean Theorem. We shall start with the concept of congruence, then briefly touch on the others, our purpose being merely to remind you of a few things you may need at the beginning of your study. Many topics that are covered in detail in other parts of this book have been omitted here to avoid unnecessary duplication.

B.1 Triangles and Congruence

Recall that a **triangle,** $\triangle ABC$, is determined by any three points A, B, and C that do not lie on a line—its **vertices**—and consists of all the points lying on the segments $\overline{AB}$, $\overline{BC}$, and $\overline{AC}$—its **sides.** The **angles** of $\triangle ABC$ are $\angle A$ ($\angle BAC$), $\angle B$ ($\angle ABC$), and $\angle C$ ($\angle BAC$). Two triangles are said to be **congruent** if the pairs of corresponding sides and angles in the two triangles are congruent (that is, have the *same measure*). We write $\triangle ABC \cong \triangle DEF$, for example, when $\triangle ABC$ and $\triangle DEF$ are congruent under the correspondence $A \leftrightarrow D$, $B \leftrightarrow E$, $C \leftrightarrow F$, and thus, as indicated in Figure B.1 by the usual marks showing congruent segments and angles in geometry,

$$\overline{AB} \cong \overline{DE}, \quad \overline{BC} \cong \overline{EF}, \quad \overline{AC} \cong \overline{DF},$$
$$\angle A \cong \angle D, \quad \angle B \cong \angle E, \quad \text{and} \quad \angle C \cong \angle F$$

This definition of congruence implies a much used concept in general: When any two *figures* (meaning triangles or polygons) are congruent, their corresponding sides and angles are congruent. That is, *corresponding parts of congruent figures are congruent*, to be denoted in this book by CPCF.

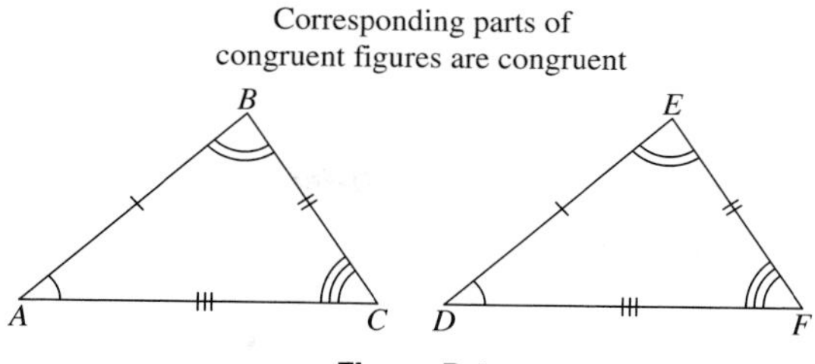

Figure B.1

SAS, ASA, AND SSS POSTULATES

Consider the situation depicted in Figure B.2. In $\triangle ABC$ and $\triangle DEF$ the measures of sides and angles are as follows:

$$AB = 3 = DE, \quad BC = 4 = EF, \quad \text{and} \quad m\angle ABC = 70 = m\angle DEF$$

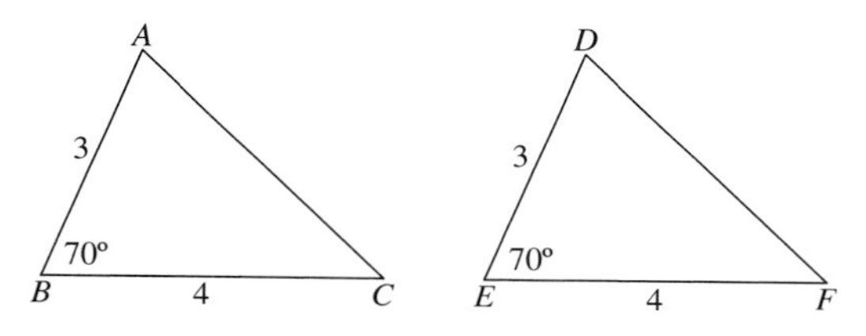

Figure B.2

Then it would follow that the measures of the remaining sides and angles of the two triangles are equal, that is, the *triangles are congruent*. This congruence is due to the **SAS Congruence Criterion** (Side–Angle–Side):

> If two triangles have two sides and the included angle of one congruent, respectively, to the corresponding two sides and included angle of the other, the triangles are congruent.

Similarly, if we had two sets of *angles* congruent, such as $m\angle = 50 = m\angle D$ and $m\angle B = 70 = m\angle E$, as shown in Figure B.3, and the *included sides* congruent,

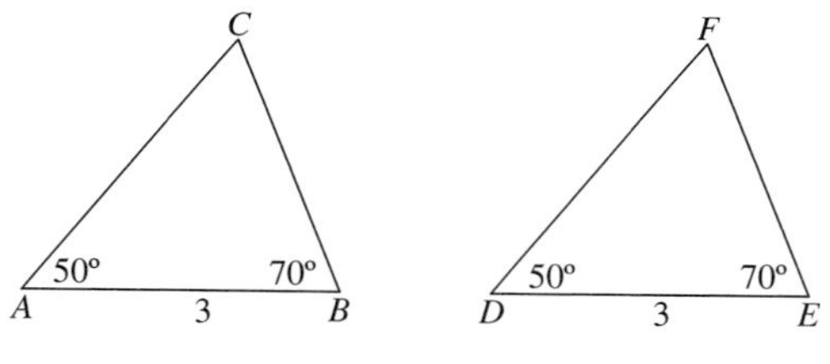

Figure B.3

$AB = 3 = DE$, then again the triangles would be congruent, this time by the **ASA Congruence Criterion** (Angle–Side–Angle). Finally, the last criterion involves the congruence of *all three sides* of two triangles. The **ASA** and **SSS Congruence Criterion** (Side–Side–Side) states:

If two triangles have two angles and the included side of one congruent, respectively, to the corresponding two angles and included side of the other, the triangles are congruent.

If two triangles have their corresponding sides congruent, the triangles are congruent.

ISOSCELES TRIANGLE THEOREM

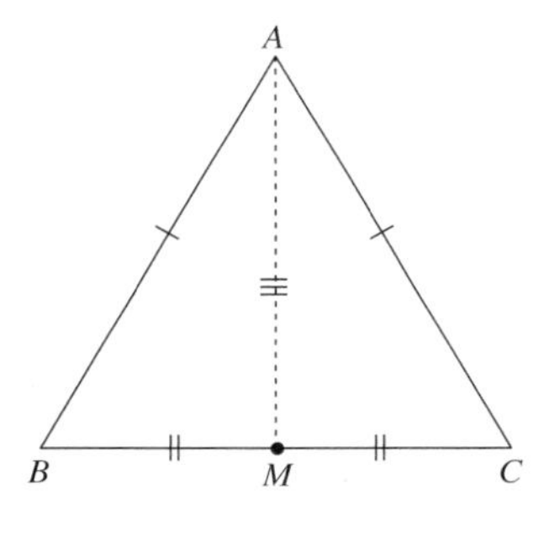

Figure B.4

One of the first applications of the theory of congruent triangles is to establish the basic properties of an **isosceles triangle** (a triangle with two congruent sides). Suppose in $\triangle ABC$ we have $AB = AC$ (remember that congruence of segments is synonymous with equal distances). As shown in Figure B.4, we have constructed the midpoint M of the base $\overline{BC}$ of $\triangle ABC$, and have drawn segment $\overline{AM}$. This forms two triangles, $\triangle ABM$ and $\triangle ACM$. In those triangles we have $\overline{AB} \cong \overline{AC}$ (given), $\overline{BM} \cong \overline{MC}$ (property of midpoint), and $\overline{AM} \cong \overline{AM}$ (Reflexive Law—any object is congruent to itself). Thus, by SSS, $\triangle ABM \cong \triangle ACM$ and $\angle ABM \cong \angle ACM$, and therefore $\angle B \cong \angle C$ in the original triangle. This proves the first half of the **Isosceles Triangle Theorem**:

If two sides of a triangle are congruent, the angles opposite those sides are congruent, and, conversely, if two angles are congruent, the sides opposite are congruent.

AAS CONGRUENCE

Another congruence criterion is readily obtained from the ASA Congruence Criterion by using the fact that in Euclidean geometry the sum of the angles of any triangle equals 180. Thus if two sets of angles in two triangles are congruent, then the measures of the third pair must be the balance of 180 after the first two (equal) angle measures are subtracted, hence are equal and congruent. Thus the ASA Criteria im-

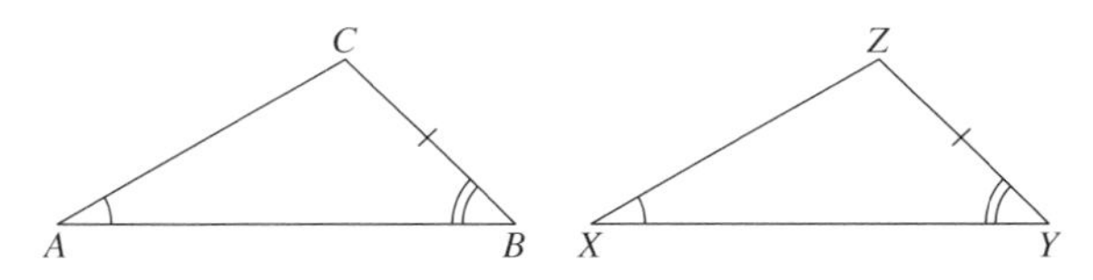

Figure B.5

plies the **AAS Congruence Criterion** (Angle–Angle–Side), illustrated in Figure B.5:

> If two triangles have two angles and a side opposite one of them in one triangle congruent, respectively, to the corresponding two angles and opposite side in the second triangle, the triangles are congruent.

In stating this criterion, you can see how important it is to keep the order of the sides and angles straight. Figure B.6 shows a diagram in which, indeed, two angles and a side opposite one of them in each of the two triangles are congruent, but the triangles are not congruent.

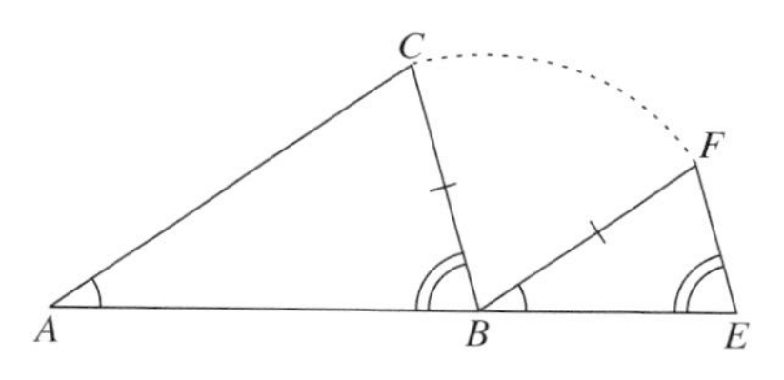

Figure B.6

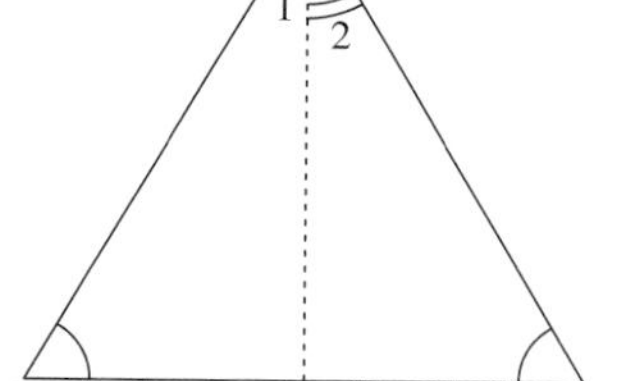

Figure B.7

As an application of the AAS Congruence Criterion, suppose we are given that $\angle B \cong \angle C$ in $\triangle ABC$, as shown in Figure B.7. As indicated, we have constructed the angle bisector $\overline{AD}$ of $\angle BAC$. Then in $\triangle ABD$ and $\triangle ACD$ we would have $\angle 1 \cong \angle 2$, $\angle B \cong \angle C$ (given), and $\overline{AD} \cong \overline{AD}$.

PROBLEM 1 Continue this argument and show that $\overline{AB} \cong \overline{AC}$.

This argument proves the *converse* part of the Isosceles Triangle Theorem. (In such theorems you cannot take the converse for granted; it must be proven separately. For more details on this point, see Section 2.1.) We end this section with two examples of typical geometry problems, the first involving *overlapping* triangles and congruence criteria, the second involving the Angle-Sum Theorem in Euclidean geometry (the sum of the measures of the angles of a triangle is 180).

EXAMPLE 1 In Figure B.8 it is given that $WL = KT$ and $\angle WLT \cong \angle KTL$. Show that $LK = WT$ and $\angle KLT \cong \angle WTL$.

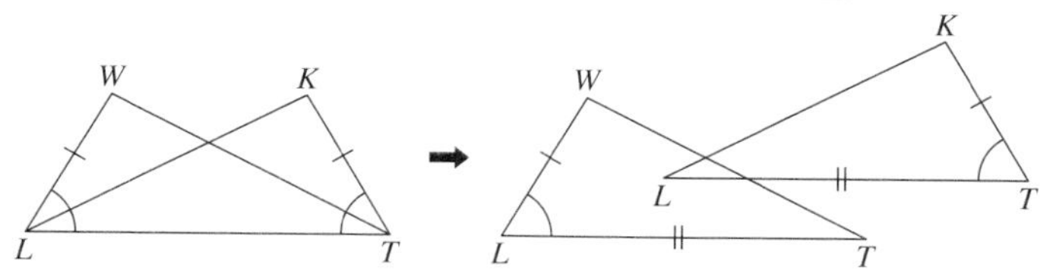

Figure B.8

SOLUTION

It helps if we can visualize the result of pulling the triangles apart, as shown in the diagram to the right in Figure B.8. Because we have $\overline{WL} \cong \overline{KT}$, $\angle L \cong \angle T$, and $\overline{LT} \cong \overline{LT}$, then by SAS $\triangle WLT \cong \triangle KTL$, and by CPCF $\overline{LK} \cong \overline{WT}$ and $\angle KLT \cong \angle WTL$. ■

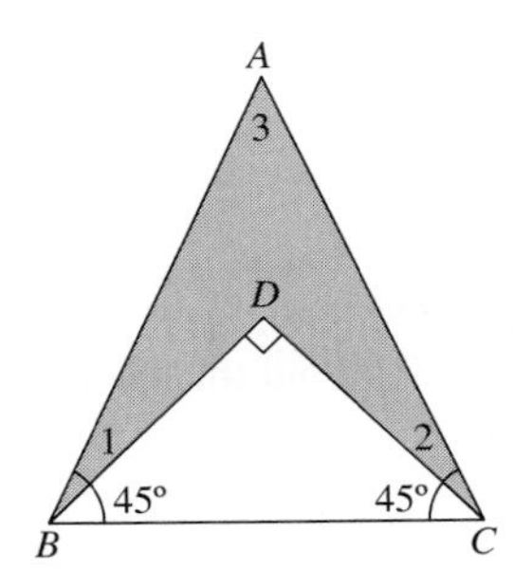

Figure B.9

EXAMPLE 2 A truss is built in the shape shown in Figure B.9, where $AB = AC$, $m\angle BDC = 90$, and $m\angle DBC = m\angle DCB = 45$. Find the sum of the measures of the angles marked 1, 2, and 3. Generalize this result.

SOLUTION

We have:

$$(m\angle 1 + 45) + (m\angle 2 + 45) + m\angle 3 = 180$$
$$m\angle 1 + m\angle 2 + m\angle 3 + 90 = 180$$
$$\therefore m\angle 1 + m\angle 2 + m\angle 3 = 90$$

Generalization: In the above argument, we never used the fact that $AB = AC$, so it apparently works on the triangle of Figure B.10. See if you agree. What changes must be made in the proof? Another generalization is shown in Figure B.11, where we relax the requirement that the right triangle must have acute angles each of which measure 45. In this generalization, a more elegant argument emerges: Extend side $\overline{BD}$ to E on $\overline{AC}$ and use the Exterior Angle Theorem:

$$m\angle 4 = m\angle 1 + m\angle 5 \qquad (m\angle 4 = 90)$$

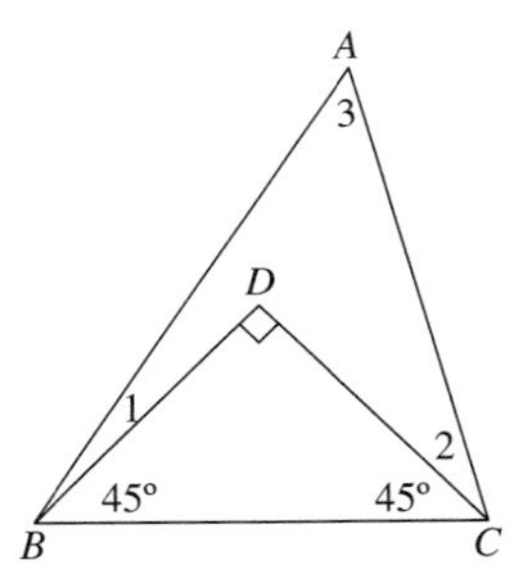

Figure B.10

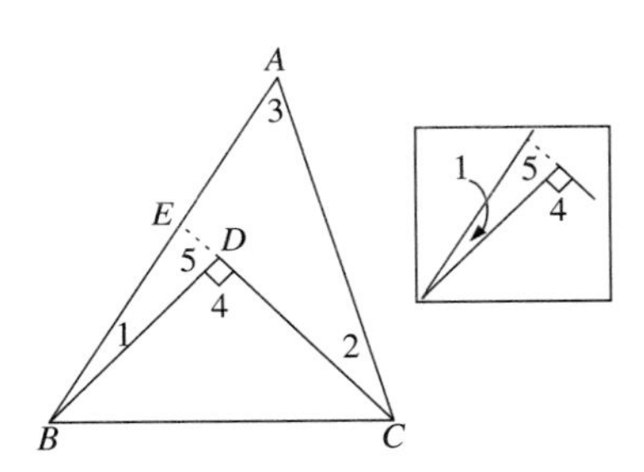

Figure B.11

But, again applying the Exterior Angle Theorem to $\triangle ACE$,

$$m\angle 5 = m\angle 2 + m\angle 3$$

By substitution,

$$90 = m\angle 1 + m\angle 5 = m\angle 1 + (m\angle 2 + m\angle 3)$$

Again the sum of the three angles marked 1, 2, and 3 equals 90. (One further generalization remains; can you find it?) ■

B.2 Congruence in Right Triangles

Right triangles (triangles one of whose angles is a right angle) have their own special congruence criteria, not only because they automatically have one set of angles (the right angles) congruent to start with, but in addition, an SSA criterion is valid that is not valid for general triangles. Recall that the **hypotenuse** of a right triangle is the side opposite the right angle, and the **legs** are the sides adjacent to the right angle.

HA, LA, AND HL CONGRUENCE CRITERIA

If $\triangle ABC$ and $\triangle DEF$ have $m\angle C = m\angle F = 90$, acute angle $B \cong$ acute angle E, and hypotenuse $\overline{AB} \cong$ hypotenuse $\overline{DE}$, as in Figure B.12, then by AAS the triangles would be congruent. This proves the **HA Congruence Criterion**:

> Two right triangles are congruent if they have an acute angle and hypotenuse of one congruent, respectively, to an acute angle and hypotenuse of the other (Figure B.12).

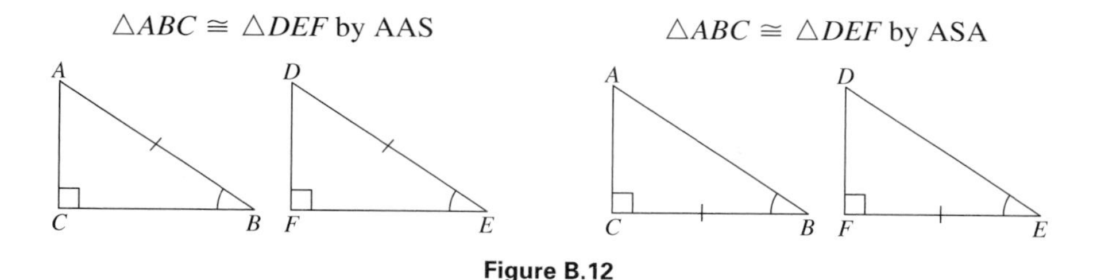

Figure B.12

For other criteria, we could also have a *leg* of one triangle, instead of the hypotenuse, congruent to the corresponding *leg* of the other. Depending on whether those congruent legs were adjacent to or opposite to the congruent acute angles, we could use either the ASA or the AAS Congruence Criterion to establish that the triangles are congruent. This proves the **LA Congruence Criterion**:

> If under some correspondence two right triangles have a leg and acute angle of one congruent, respectively, to the corresponding leg and acute angle of the other, the triangles are congruent.

The final criterion for right triangles is the **HL Congruence Criterion**:

> If two right triangles have a leg and hypotenuse congruent, respectively, to the corresponding leg and hypotenuse of the other, the triangles are congruent.

This criterion requires a special proof; it is not a consequence of any of the previous criteria. As indicated in Figure B.13 segment $\overline{YZ}$ is extended to point W such that $ZW = BC$. The rest of the argument is apparent from the figure; see if you can finish it.

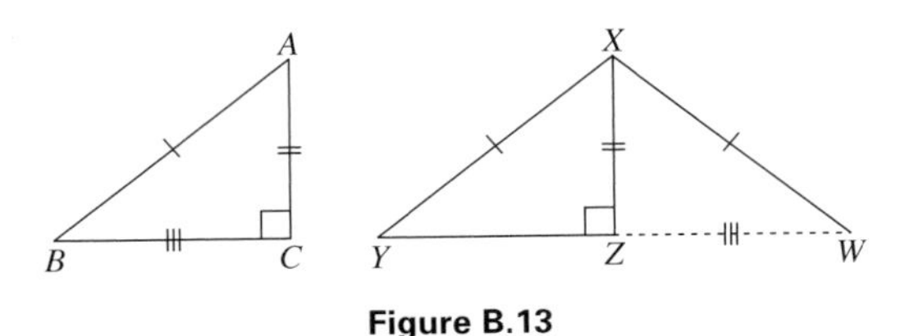

Figure B.13

PROBLEM 2 Finish the argument started in Figure B.13 to show that $\triangle ABC \cong \triangle XYZ$. (***Hint:*** Your argument will ultimately use the Isosceles Triangle Theorem and the HA Congruence Criterion.)

B.3 Circumcircle and Incircle of a Triangle; Locus

The important concepts of perpendicular bisector of a line segment and, dually, bisector of an angle are revealed by the two results that follow. They are stated as *locus* properties. (The **locus** of a point is the "path" or set of points that is determined by that point when it satisfies certain given properties.)

> The locus of a point equidistant from the end points of a line segment is the perpendicular bisector of that line segment.

> The locus of a point equidistant from the sides of an angle is the bisector of that angle.

(Proofs can be found in Section 3.3 and Problem 11, Section 3.6, solved in Appendix E.) These results have straightforward corollaries.

CIRCUMCENTER AND INCENTER OF TRIANGLES

> The perpendicular bisectors of the sides of any triangle are concurrent in a point O that is equidistant from the three vertices, hence is the center of a circle (**circumcircle**) passing through the vertices. (See Figure B.14.)

CIRCUMCIRCLE OF A TRIANGLE

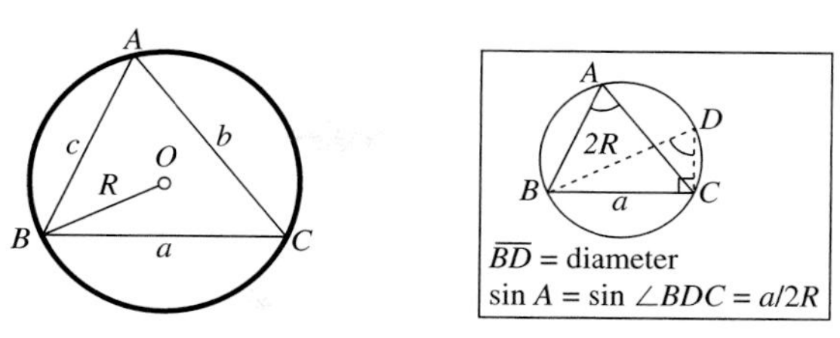

Figure B.14

The bisectors of the angles of any triangle are concurrent in a point I that is equidistant from the three sides of the triangle, hence is the center of a circle (**incircle**) tangent to the sides. (See Figure B.15.)

INCIRCLE OF A TRIANGLE

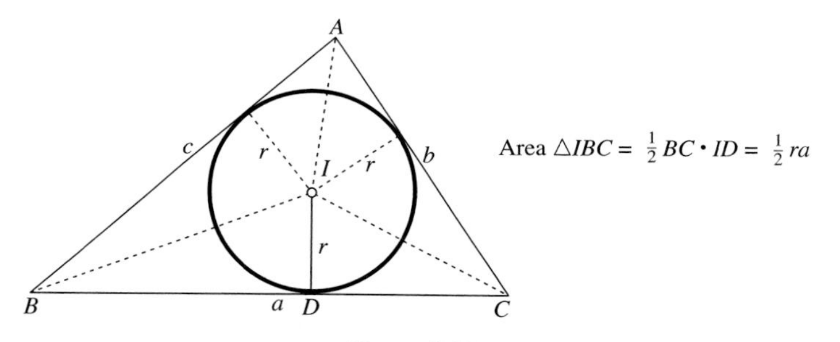

Figure B.15

FORMULAS FOR CIRCUMRADIUS, INRADIUS

It is interesting to derive the classical formulas for the **circumradius** R (radius of the circumcircle) and the **inradius** r (radius of the incircle). The formulas relate these radii to the *area* K of $\triangle ABC$ and the lengths of its three sides a, b, and c in each case:

(1)
$$R = \frac{abc}{4K}$$

(2)
$$r = \frac{2K}{a + b + c}$$

The first equation **(1)** follows from a result established in an optional section of this book, Problem 8, Section 1.4 (see also inset of Figure B.14), which uses the Inscribed Angle Theorem (see below) and trigonometry: $\sin\angle BDC = a/BD = a/2R$, or $R = a/2\sin A$ where $A = m\angle BAC$. All we have to do now is to multiply numerator and denominator by bc and use the area formula $K = \frac{1}{2}bc\sin A$ (or $4K = 2bc\sin A$). The second is shown in Figure B.15, where Area $\triangle IBC = \frac{1}{2}ra$, Area $\triangle IAC = \frac{1}{2}rb$, Area $\triangle IAB = \frac{1}{2}rc$. Sum these three equations: Area

$\triangle ABC = \frac{1}{2}ra + \frac{1}{2}rb + \frac{1}{2}rc = r \cdot \frac{1}{2}(a + b + c) = K. \therefore r = 2K/(a + b + c)$. If s is the **semiperimeter** of $\triangle ABC$ [one-half the perimeter $= (a + b + c)/2$], then **(2)** reduces to simply

$$r = \frac{K}{s} \qquad \textbf{(2')}$$

B.4 Proof Writing

Proving a theorem in geometry (or mathematics in general) amounts to simply writing an explanation of why the theorem is true. There is nothing mysterious about that, but sometimes just coming up with an explanation seems insurmountable at times. Certainly, the art of proving theorems is a learned discipline for which there are a few guidelines, and we shall explore those here, and in Sections 2.1 and 2.2.

First, it is important to have a good recollection of the concepts previously covered. That is why we have collected the results proven in this book in Appendix F, as a quick reference to help you work the problems. Second, you must fully understand the problem you are working on—what is given, and what must be proven. Then you need to think about those results that might have a bearing on the problem. Thus, the first three basic steps in proof writing are

- Understanding previous results,
- Understanding the problem you are working on,
- Putting the two together—observing what is relevant among the previous results.

TWO-COLUMN FORMAT

Another helpful ingredient is understanding how a two-column proof works, for this format can be the way to successful proof writing. It should be pointed out that a paragraph proof is really the goal, and the object of any proof is merely to give an explanation of why something is true, based on previous results. You do not need an outline if you already know *why* the proposition you are trying to prove is true. But if not, then a two-column (T proof) is a possible starting point. Any two-column proof can be mechanically converted to a paragraph proof by just writing down each of the steps in order and placing the reasons after each step.

In a two-column proof, we make certain **statements** or **conclusions** about the concept to be proven, beginning with the hypothesis and ending with what is to be proven. Each of these statements is backed up by a postulate, definition, or *previously proven* theorem. These are the **reasons**, which **justify** each statement. For most students, success in proof writing is a gradual process. So at first the statements of the proofs are provided for you, and all you have to do is to give the reasons. In fact, the first nontrivial proof you are expected to create on your own in the problem sets in this book does not occur until a Group B problem in Section 2.4, or a Group A problem in Section 3.3.

A purely mechanical way to start an outline proof is to list all that is given in a problem as the first (or first several) statements, leaving a blank space in the middle, and placing the desired conclusion at the end. Sometimes it is easier to work backward from the conclusion, writing the statement that just precedes the conclusion. (Often this is not possible, so this practice is probably not worth more than a quick glance.) An area in which this tactic does work frequently is in algebra or in proving trigonometric identities.

EXAMPLE 1 Suppose s is the semiperimeter $[s = \frac{1}{2}(a + b + c)]$ and a, b, and c are the lengths of the sides of any triangle. Prove that $\frac{1}{2}(a + b - c) = s - c$. Thus we are given $\frac{1}{2}(a + b - c)$, and want to prove (or derive from this) $s - c$.

SOLUTION

The first and last steps in outline form are as indicated:

CONCLUSIONS	JUSTIFICATIONS
(1) $s = \frac{1}{2}(a + b + c)$	Given
(2) $\frac{1}{2}(a + b - c)$	Given
—Space to Be Filled in—	
(x) $\therefore s - c$	?

(We do not know the reason for the last step until we have figured out the next-to-last step, and we do not know how many steps our proof will require—the reason for the "x.") In this case, it is easier to work backward, using what we know about s. Observe that:

$$s - c = \frac{1}{2}(a + b + c) - c = \frac{1}{2}a + \frac{1}{2}b + \frac{1}{2}c - c = \frac{1}{2}a + \frac{1}{2}b - \frac{1}{2}c = \frac{1}{2}(a + b - c)$$

Now we know how the outline proof should go:

CONCLUSIONS	JUSTIFICATIONS
(1) $s = \frac{1}{2}(a + b + c)$	Given
(2) $\frac{1}{2}(a + b - c)$	Given
(3) $\frac{1}{2}a + \frac{1}{2}b - \frac{1}{2}c$	Distributive Law
(4) $\frac{1}{2}a + \frac{1}{2}b + (\frac{1}{2}c - c)$	Substitution
(5) $(\frac{1}{2}a + \frac{1}{2}b + \frac{1}{2}c) - c$	Associative Law of Addition
(6) $\frac{1}{2}(a + b + c) - c$	Distributive Law
(7) $\therefore s - c$	Definition of s

(Note that we have made a much bigger thing out of it than the problem warrants, but this shows the process required.) ■

The next example is a more typical geometry problem; it is a challenge for almost anyone, although completely elementary.

EXAMPLE OF "DIFFICULT" PROOF

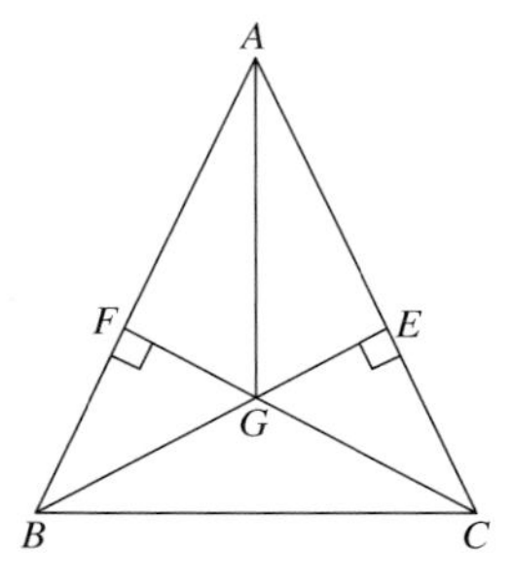

Figure B.16

EXAMPLE 2 Write a proof for the following proposition.
Given: $\triangle ABC$ is isosceles, with $\overline{AB} \cong \overline{AC}$ (Figure B.16). Also, $\overline{BE} \perp \overline{AC}$, $\overline{CF} \perp \overline{AB}$.
Prove: $\overrightarrow{AG}$ bisects $\angle BAC$. (That is, prove that $\angle BAG \cong \angle CAG$.)

SOLUTION

Again, we write down the given items and the last step (desired conclusion), leaving a blank in the middle.

CONCLUSIONS	JUSTIFICATIONS
(1) $\overline{BE} \perp \overline{AC}$	Given
(2) $\overline{CF} \perp \overline{AB}$	Given
(3) $\overline{AB} \cong \overline{AC}$	Given
—Space to Be Filled in—	
(x) $\therefore \angle BAG \cong \angle CAG$.	?

Now comes the hard part. We must fill in the missing steps. Our thought process could take the following form.

(1) About all that is obvious at first is that we need to know something about isosceles triangles. Congruence principles for right triangles might also be involved. (A review of those results might be helpful at this point.)

(2) What kind of strategy might work? Maybe we can show that $\triangle AFG \cong \triangle AEG$.

(3) Look at other pairs of triangles in the figure that might be proven congruent (Figure B.17).

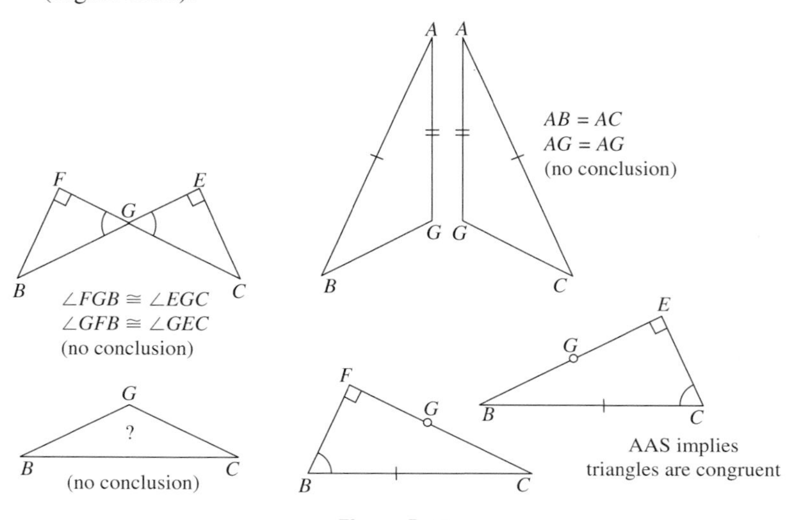

Figure B.17

(4) For each pair of triangles, does the given information actually imply they are congruent? (The response is written beside each diagram in Figure B.17.)

(5) Finally, we find that in the last diagram, we can get somewhere. Using the Isosceles Triangle Theorem, $\angle B \cong \angle C$ and by AAS the triangles would be congruent. But we still need a strategy to pull everything together.

(6) Finding a strategy: Goal 1 would be: Prove $\triangle BCF \cong \triangle CBE$. From the congruent parts we get from that, we might be able to go back to some pairs of triangles we looked at previously and prove something useful. Since we can now prove that $\angle BCF \cong \angle CBE$, $\triangle BGC$—a triangle we previously knew nothing about—can be shown to be isosceles. Does this lead anywhere? If $\overline{BG} \cong \overline{CG}$, then looking at $\triangle ABG$ and $\triangle ACG$ we find that these two triangles will be congruent by SSS. This will finally prove $\angle BAG \cong \angle CAG$—the last step in the outline we started above. Our goals now become:

Goal 1: Prove $\triangle BCF \cong \triangle CBE$.
Goal 2: Prove $\triangle BGC$ is isosceles, with $\overline{BG} \cong \overline{GC}$.
Goal 3: Prove $\triangle ABG \cong \triangle ACG$.

At this point, we now basically know what to write as steps in our proof. Here goes.

	CONCLUSIONS	JUSTIFICATIONS
Goal 1	**(1)** $\overline{CE} \perp \overline{AC}$	Given
	(2) $\overline{CF} \perp \overline{AB}$	Given
	(3) $\overline{AB} \cong \overline{AC}$	Given
	(4) $\overline{BC} \cong \overline{BC}$	Reflexive Property
	(5) $\angle FBC \cong \angle ECB$	Isosceles Triangle Theorem
	(6) $\triangle FBC \cong \triangle ECB$	HA
Goal 2	**(7)** $\angle GBC \cong \angle GCB$	CPCF
	(8) $\overline{BG} \cong \overline{GC}$	Isosceles Triangle Theorem
Goal 3	**(9)** $\overline{AB} \cong \overline{AC}$	Given
	(10) $\overline{AG} \cong \overline{AG}$	Reflexive Property
	(11) $\triangle ABG \cong \triangle ACG$	SSS
	(12) $\therefore \angle BAG \cong \angle CAG$	CPCF ■

B.5 Similar Triangles, Geometric Mean

Suppose instead of having three corresponding *sides* of two triangles congruent we had three pairs of *angles* congruent. This would not guarantee that the triangles are congruent, because two such triangles would be the same *shape* but they might not have the same *size*, as illustrated in Figure B.18. Instead of the sides being congruent, their lengths are **proportional**. That is, the measures of the corresponding sides in the two triangles would be in the *same ratio*:

$$\frac{AB}{XY} = \frac{BC}{YZ} = \frac{AC}{XZ} \quad \text{or} \quad \frac{AB}{BC} = \frac{XY}{YZ} \quad \text{and} \quad \frac{BC}{AC} = \frac{YZ}{XZ}$$

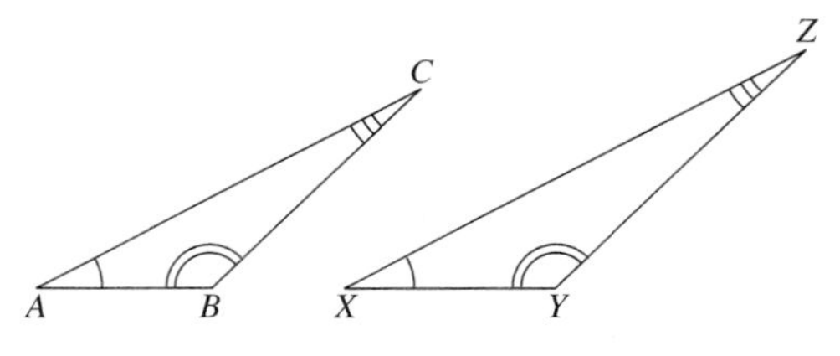

Figure B.18

When this happens in two triangles, they are said to be **similar**. (Recall that in algebra, if $a/b = c/d$ then $ad = bc$ and $a/c = b/d$; that is, in proportions like these, the terms b and c can be interchanged.)

AA, SAS, AND SSS SIMILARITY CRITERIA

The postulate for similarity (which will be proven as a theorem in Section 4.2) is the **AA Similarity Criterion**:

> If two triangles have two pairs (therefore all three pairs) of corresponding angles congruent, the triangles are similar, with corresponding sides in the same ratio.

MIDPOINT-CONNECTOR, SIDE-SPLITTING THEOREMS

Related to this are two quite useful results. The Midpoint-Connector Theorem states that

> The line segment joining the midpoints of two sides of a triangle is parallel to the third side and has length one-half that of the base.

The Side-Splitting Theorem is as follows:

> If a line segment joining points D and E on sides $\overline{AB}$ and $\overline{AC}$ of $\triangle ABC$ is parallel to side $\overline{BC}$, then $AD/DB = AE/EC$.

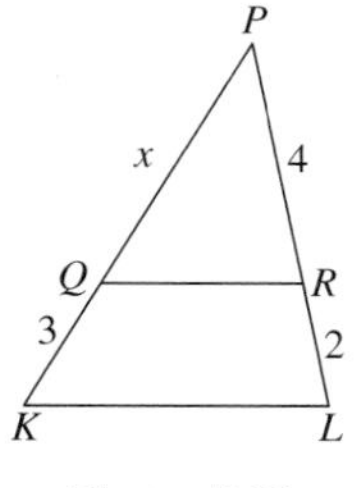

Figure B.19

EXAMPLE 1 In Figure B.19, $\overline{QR} \parallel \overline{KL}$ (therefore $\angle PQR \cong \angle PKL$), with certain measurements indicated. Find x.

SOLUTION
By the Side-Splitting Theorem, $PQ/QK = PR/RL$. That is,

$$\frac{x}{3} = \frac{4}{2} \quad \text{or} \quad 2x = 3 \cdot 4 = 12 \quad \therefore x = 6 \quad \blacksquare$$

GEOMETRIC MEAN

Another important concept is that of the **geometric mean** of two numbers, which is the square root of the product of the two given numbers. That is, a is the geometric mean of b and c iff $a = \sqrt{bc}$. This concept occurs in algebra, geometry, and in nature.

A sequence of numbers $a_1, a_2, a_3, \ldots, a_n$ is a **geometric sequence** iff the ratio of any two consecutive members in the sequence is constant, which is the same as saying that the second of any three consecutive members is the *geometric mean* between the first and third. For example, consider a_5, a_6, and a_7. If $a_5/a_6 = a_6/a_7$ then $a_5a_7 = a_6^2$ or $a_6 = \sqrt{a_5a_7}$.

When two similar triangles have a side in common, the length of the common side is the geometric mean of the lengths of two other sides. For example, in Figure B.20 we have $\triangle ABC \sim \triangle ACD$ (with congruent angles marked) having side $\overline{AC}$ in common. Then $AB/AC = AC/AD$ or $AC^2 = AB \cdot AD$. A special case of this occurs when the angles at B and D are complementary and $m\angle BCD = m\angle B + m\angle D = 90$, or when $\triangle BCD$ is a right triangle (as shown in the diagram that follows). This proves the following theorem.

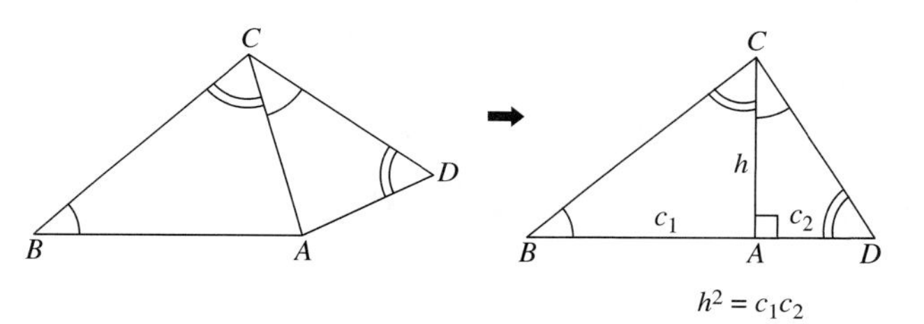

Figure B.20

> The altitude to the hypotenuse of any right triangle is the geometric mean of the segments formed on the hypotenuse.

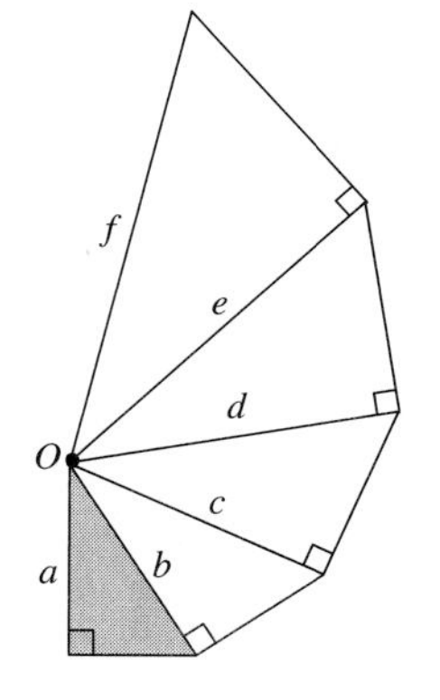

Figure B.21

A geometric sequence has a geometric construction, where the first two members are $a_1 = a$ and $a_2 = b$. Figure B.21 shows a sequence of right triangles, with each subsequent triangle constructed on the hypotenuse of the previous triangle, and all the acute angles at O congruent. If the first of these triangles has sides of length a and b, as shown, then the common sides between right triangles have lengths $a, b, c, d, e, f, \ldots$ which is the given geometric sequence. (Can you show why?)

GEOMETRIC MEAN IN NATURE

EXAMPLE 2 Analogous to Figure B.21, the shell of the chambered nautilus (Figure B.22) outlines a repeated pattern of segment triples such that the second

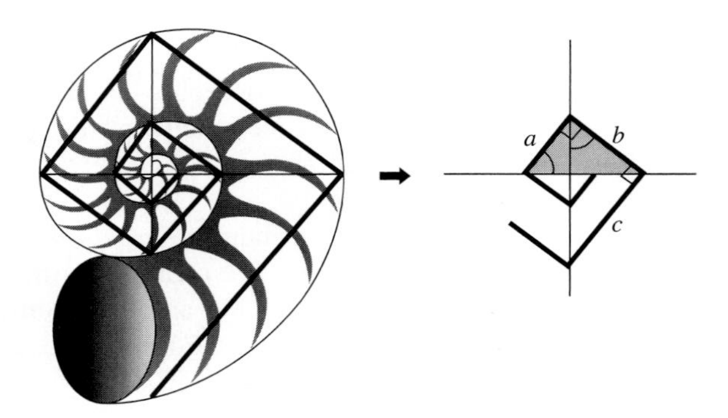

Figure B.22

is the geometric mean of the first and third. The shaded right triangle is similar to the larger right triangle with the vertical hypotenuse (because they have two pairs of congruent angles), so therefore

$$\frac{a}{b} = \frac{b}{c} \quad \text{or} \quad b^2 = ac \quad \blacksquare$$

B.6 Circles

A **circle** is the totality of all points in a plane that lie at a fixed distance r (called the **radius**) from some fixed point O (called the **center**). Any segment $\overline{OP}$ where P is any point on the circle is also called a **radius**, the meaning being clear by context. A **chord** is any segment joining two distinct points of a circle, and a **diameter** is a chord passing through the center O. Further terms associated with circles are exhibited in Figure B.23.

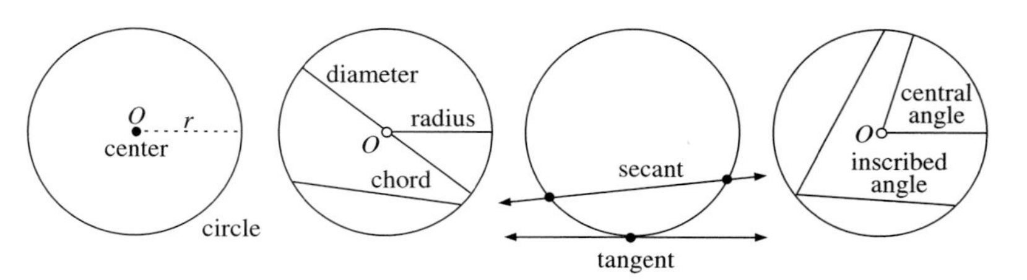

Figure B.23

TANGENT THEOREM

One of the most elementary results about circles is the fact that a tangent is perpendicular to the radius at the point of contact. This, in turn, implies the following useful result.

If $\overline{PA}$ and $\overline{PB}$ are tangents to circle O at A and B, respectively, then $PA = PB$ (Figure B.24).

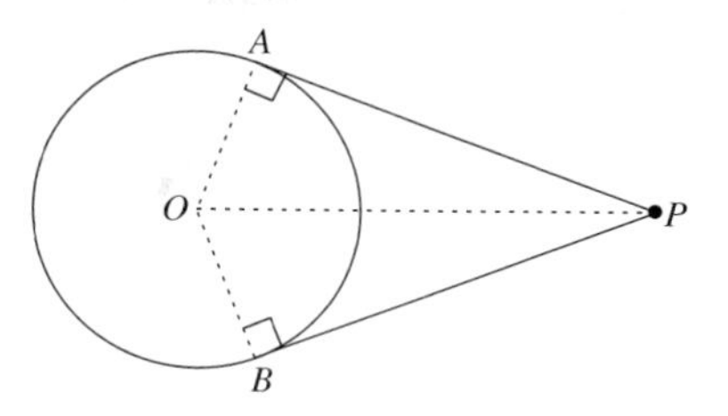

Figure B.24

[PROOF

In right triangles $\triangle PAO$ and $\triangle PBO$, $OA = OB$ and $OP = OP$. By HL, $\triangle PAO \cong \triangle PBO$, and therefore $PA = PB$.]

ARC MEASURE

A **circular arc** is merely some continuous part of a circle, as shown in Figure B.25; the three types of circular arcs are **minor arc**, such as arc $\widehat{AB}$ in Figure B.25, a **semicircle**, such as arc $\widehat{STU}$, and **major arc**, such as arc $\widehat{PQR}$. To measure an arc in a natural way, think of a central angle, with its corresponding inscribed arc, starting to open up (Figure B.26), having measure 45, then 135, then nearly 180 (angle measure cannot exceed 180). As the length of the arc continues to increase, the measure of the corresponding central angle begins to *decrease*, and eventually returns to 0. As this happens, the arc becomes a major arc, however, and, unlike the measure of the corresponding central angle, its measure is growing larger. To compensate for this phenomenon, we must take the measure of any major arc $\widehat{AB}$ to be $360 - m\angle AOB$. (We can let a minor arc be measured by its corresponding central angle, and let a semicircle be defined to have measure 180.)

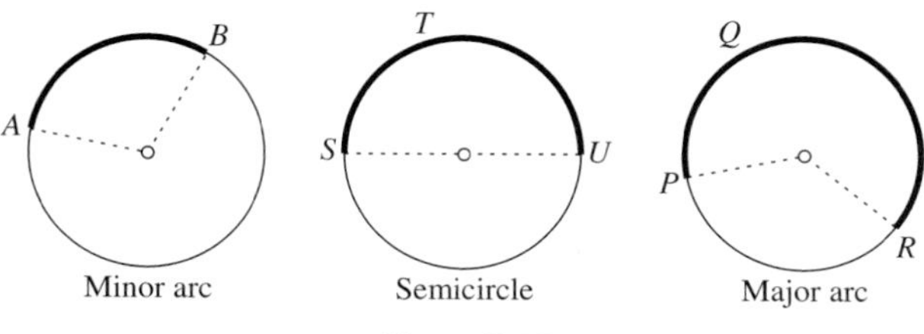

Figure B.25

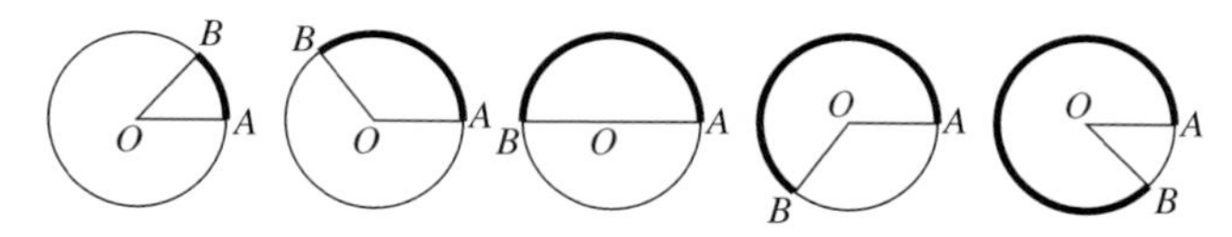

Figure B.26

MEASURE OF INSCRIBED ANGLES

As an ultimate consequence of this definition, the following theorem can be established, as we show how to do in Section 4.5. It is known as the **Inscribed Angle Theorem** (illustrated in Figure B.27):

> Any inscribed angle of a circle has measure equal to one-half that of its inscribed arc.

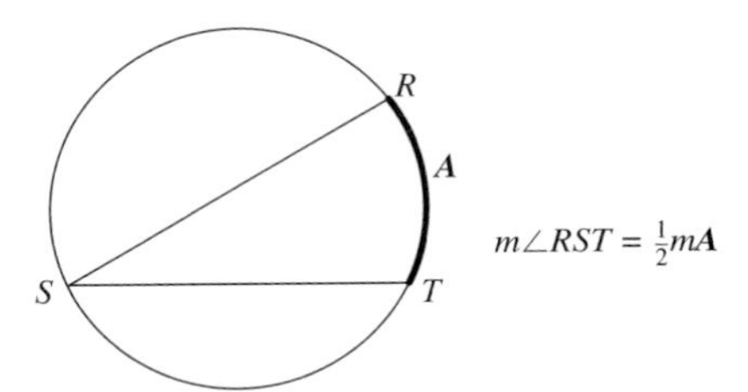

Figure B.27

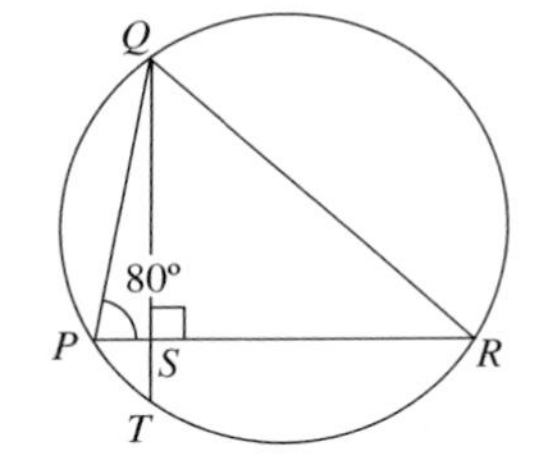

Figure B.28

EXAMPLE 1 $\triangle PQR$ is inscribed in a circle (Figure B.28) and chord $\overline{QT} \perp \overline{PR}$, with $m\angle QPR = 80$. Find $m\widehat{PT}$.

SOLUTION
Because the acute angles of a right triangle are complementary, $m\angle PQT = 10$. $\therefore m\widehat{PT} = 20$. ■

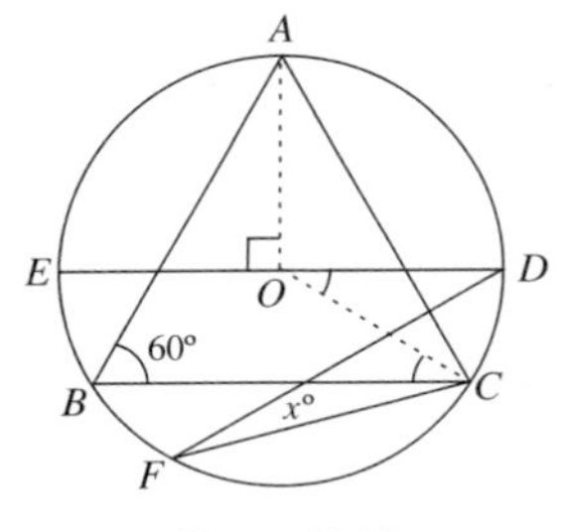

Figure B.29

EXAMPLE 2 Find the measure of $\angle CFD$ (x in Figure B.29) if $\triangle ABC$ is equilateral and $\overline{DE}$ is the diameter of the circle parallel to side $\overline{BC}$.

SOLUTION
Point O is both the circumcenter and incenter of the equilateral triangle, so in $\triangle ABC$ we have $m\angle OCB = m\angle OCA = 30$ and, because $\overline{OC}$ is a transversal for the parallel lines $\overline{ED}$ and $\overline{BC}$, $m\angle DOC = m\angle OCB = 30$. $\therefore m\widehat{CD} = 30$. By the Inscribed Angle Theorem, $m\angle CFD = \frac{1}{2}(30) = 15$. ■

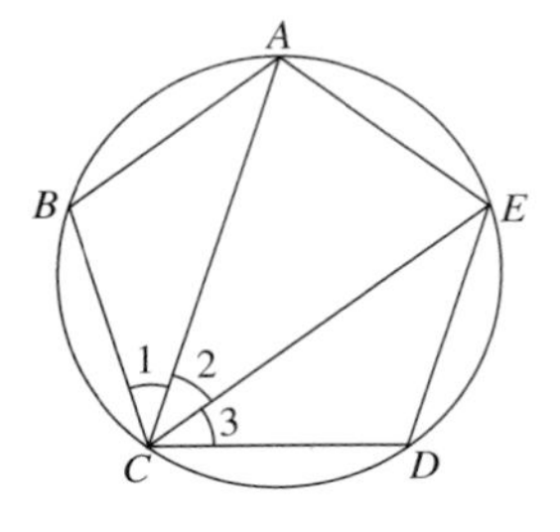

Figure B.30

PROBLEM 3 Regular pentagon $ABCDE$ (sides and angles congruent) is inscribed in a circle. Prove that diagonals $\overline{CA}$ and $\overline{CE}$ are **(a)** congruent, and **(b)** the angle trisectors of $\angle BCD$. (See Figure B.30.) You may use the proposition: *Two arcs of a circle have equal measure if and only if their subtended chords are congruent.*

ANGLE INSCRIBED IN A SEMICIRCLE

The Inscribed Angle Theorem may be specialized to produce an important special case: Let the inscribed angle be $\angle ABC$ where A and C are the end points of a diameter (Figure B.31). Then $m\widehat{ADC} = 180$ (because arcs $\widehat{ABC}$ and $\widehat{ADC}$ are both semicircles), which implies by the Inscribed Angle Theorem $m\angle ABC = \frac{1}{2}(180) = 90$. This proves

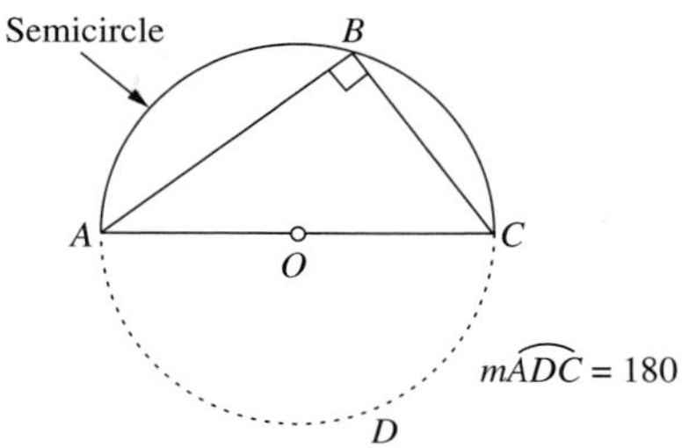

Figure B.31

Any angle inscribed in a semicircle is a right angle.

TWO-CHORD THEOREM

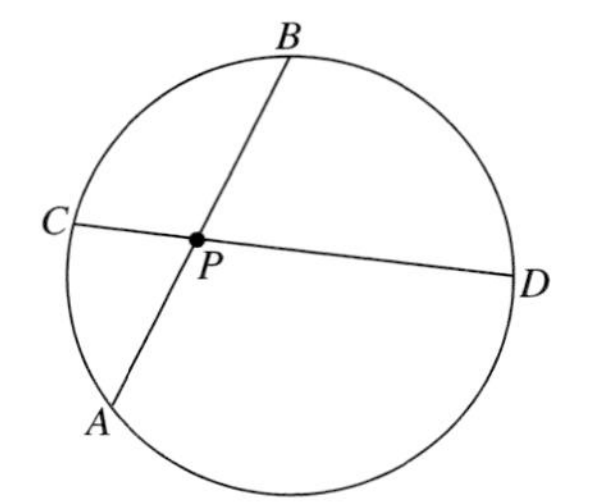

Figure B.32

Another property of circles that is used often in geometry is the **Two-Chord Theorem**:

If $\overline{AB}$ and $\overline{CD}$ are any two chords of a circle intersecting at an interior point P on each segment, then $AP \cdot PB = CP \cdot PD$. (See Figure B.32.)

EXAMPLE 3 In Figure B.33, $DR = 3$, $ER = 4$, $FR = 5$, and $GR = 6$. Find x and y.

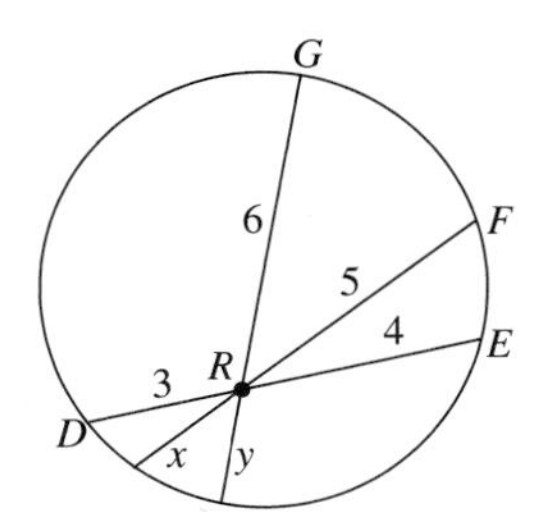

Figure B.33

SOLUTION
By the Two-Chord Theorem,

$$5x = 3 \cdot 4 = 12 \quad \text{or} \quad x = 2.4 \text{ (exact)}$$

$$6y = 12 \quad \text{or} \quad y = 2. \quad \blacksquare$$

METRIC ARC LENGTH

A different kind of arc measure is its *length*, meaning the limit of its approximating inscribed polygonal paths. This would result in a measure compatible with the units of distance (that is, arc length would be measured in inches if distance units were in inches). You may recall that the length of the entire circle of radius r is given by $C = 2\pi r$. A basic result concerning arc length—entirely reasonable without proof—is that the lengths of any two arcs on a circle are in the same ratio as their (degree) measures. That is, in Figure B.34,

$$\frac{s_1}{s_2} = \frac{m\widehat{AB}}{m\widehat{CD}} = \frac{m\angle AOB}{m\angle COD}$$

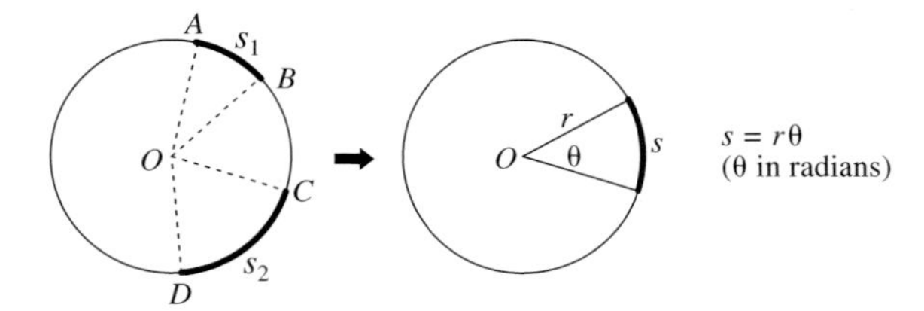

Figure B.34

Thus, if we let $s_1 = s$ for the length of an arbitrary circular arc on a circle of radius r, θ the measure of its central angle and let arc $\widehat{CD}$ be a semicircle, then $s_2 = \pi r$ and, by substitution into the preceding formula,

$$\frac{s}{\pi r} = \frac{\theta}{180}$$

which yields the formula

(1) $$s = \frac{\pi r \theta^\circ}{180}$$

where θ° indicates that θ is measured in degrees.

RADIAN MEASURE

Here we find it desirable to introduce a different measure for angles: the so-called *radian measure*. A **radian** is defined to be a central angle that subtends an arc of length r on a circle of radius r. Because the arc is greater than a chord that spans it, we find that $60^\circ > 1$ radian. As a matter of fact, because circumference $= C = 2\pi r$, there are 2π radians in a complete revolution, or π radians in a semicircle of measure 180°. Thus:

$$\pi \text{ radians} = 180^\circ \qquad 1 \text{ radian} = (180/\pi)^\circ \approx 57.3^\circ$$
$$1^\circ = \pi/180 \text{ radians}$$

Thus, to convert degrees to radians, we multiply by the factor $\pi/180$. Hence $\theta^\circ \cdot (\pi/180)$ or $\pi\theta^\circ/180 = \theta$ is the same as θ° measured in radians. Applying this to **(1)** reduces it to

(1′) $$s = r\theta \qquad (\theta \text{ in radians})$$

B.7 Quadrilaterals, Parallelograms, and Rectangles

A **quadrilateral** is a four-sided figure consisting of the set of points lying on line segments that are joined pairwise to form a closed path, called its **sides**: $\overline{AB}$, $\overline{BC}$, $\overline{CD}$, and $\overline{DA}$. Various types of quadrilaterals can be identified, as illustrated in Figure B.35. An important type of quadrilateral is the **parallelogram**—any quadrilateral

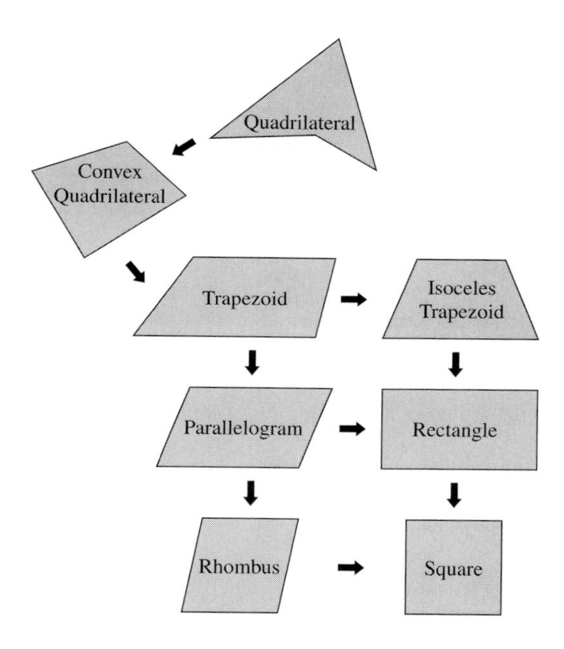

Figure B.35

whose opposite sides lie on parallel lines. As a result of properties of parallel lines, we can obtain the fact that the opposite sides of a parallelogram are congruent, and conversely. This is discussed in Section 4.2 so we need not go into detail at this point. A further fact is that if a quadrilateral has a pair of opposite sides that are congruent and lie on parallel lines, then it is a parallelogram. A parallelogram with a right angle yields the familiar figure called a **rectangle**. The chief properties of a rectangle having sides of length a and b are

- Opposite sides are parallel and congruent
- All four of its angles are right angles
- Diagonals are congruent, having length $\sqrt{a^2 + b^2}$
- Area $= ab$.

PROBLEM 4 Show that if the diagonals of a rectangle are perpendicular, it is a square.

B.8 Trigonometry

The basic definitions for the six trigonometric functions are given below. (See Figure B.36.) They are justified by use of similar triangles, as discussed in Section 4.3.

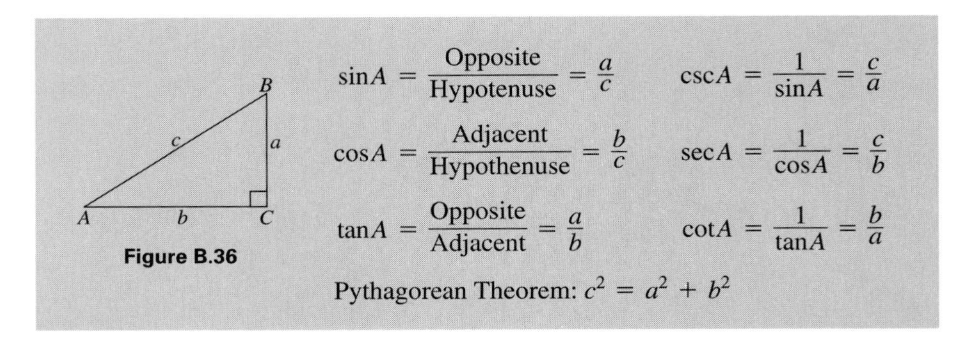

Figure B.36

$$\sin A = \frac{\text{Opposite}}{\text{Hypotenuse}} = \frac{a}{c} \qquad \csc A = \frac{1}{\sin A} = \frac{c}{a}$$

$$\cos A = \frac{\text{Adjacent}}{\text{Hypothenuse}} = \frac{b}{c} \qquad \sec A = \frac{1}{\cos A} = \frac{c}{b}$$

$$\tan A = \frac{\text{Opposite}}{\text{Adjacent}} = \frac{a}{b} \qquad \cot A = \frac{1}{\tan A} = \frac{b}{a}$$

Pythagorean Theorem: $c^2 = a^2 + b^2$

These functions are valid (except where the denominators are zero) for $0 < A < 90$, and may be extended to include all values of A such that $0 \leq A \leq 180$, as in Section 4.3. Certain trigonometric identities are very useful, in particular,

$$\sin^2 A + \cos^2 A = 1$$

$$1 + \tan^2 A = \sec^2 A$$

$$\sin(x + y) = \sin x \cos y + \cos x \sin y$$

$$\cos(x + y) = \cos x \cos y - \sin x \sin y$$

Setting $x = y$ in the last equation produces a so-called **double angle formula** for cosine, and from this emerges the **half-angle formulas** for sine and cosine, developed as follows:

$$\cos 2x = \cos^2 x - \sin^2 x = (1 - \sin^2 x) - \sin^2 x = 1 - 2\sin^2 x$$

Solve for $\sin^2 x$, and replace x by $\frac{1}{2}A$:

$$2\sin^2 \tfrac{1}{2}A = 1 - \cos 2(\tfrac{1}{2}A) \qquad \text{or} \qquad \sin^2 \tfrac{1}{2}A = (1 - \cos A)/2$$

$$\sin \tfrac{1}{2}A = \pm\sqrt{\frac{1 - \cos A}{2}}$$

Similarly,

$$\cos \tfrac{1}{2}A = \pm\sqrt{\frac{1 + \cos A}{2}}$$

$$\tan \tfrac{1}{2}A = \pm\sqrt{\frac{1 - \cos A}{1 + \cos A}}$$

Finally, the **Law of Sines** for $\triangle ABC$ in standard notation is

$$\frac{a}{\sin A} = \frac{b}{\sin B} = \frac{c}{\sin A}$$

and the **Law of Cosines** is

$$a^2 = b^2 + c^2 - 2bc\cos A, \qquad b^2 = a^2 + c^2 - 2ac\cos B,$$
$$c^2 = a^2 + b^2 - 2ab\cos C$$

B.9 Polygons: Area and Perimeter

A **polygon** is the totality of points that lie on segments (its **sides**) that join three or more points taken pairwise (called **vertices**), such that no three points are collinear and no two segments meet except at vertices. An **angle** of a polygon is an angle determined by three consecutive vertices. (See Figure B.37, where a pictorial glossary of a few terms is provided, applied to three types of a seven-sided *heptagon*.) A general polygon with n sides is called an ***n*-gon** (for $n \geq 3$), while for special values of n ($5 \leq n \leq 10$, or $n = 12$) the polygon has special names—respectively, **pentagon**, **hexagon**, **heptagon**, **octagon**, **nonagon**, **decagon**, and **dodecagon**. A **regular** polygon is a convex polygon having congruent sides and congruent angles.

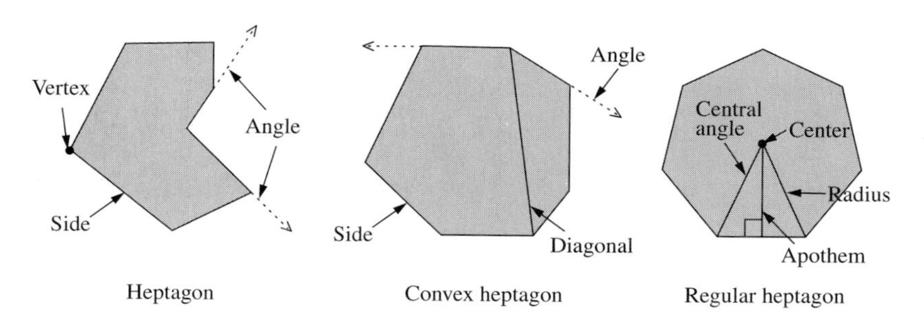

Figure B.37

REGULAR POLYGONS, APOTHEM AND RADIUS

Since a circle can be circumscribed about any regular polygon, the center of the circle and its radius are called the **center** and **radius** of the polygon. An **apothem** of a regular polygon (analogous to the altitude of a triangle) is a line segment drawn from the center of the polygon perpendicular to a side (a term also used for the *length* of that segment). The apothem a of a regular polygon is the altitude of an isosceles triangle with vertex at the center, so the area of this triangle is $\frac{1}{2}as$ where s is the length of any side of the polygon. Hence, the area of a regular polygon with n sides is given by $K = n \cdot (\frac{1}{2}as) = \frac{1}{2}nsa$, and the perimeter is $P = ns$. The following table is a collection of common formulas for regular polygons, including the two we just derived, provided for convenience.

Formulas for Regular n-Sided Polygons
Side $= s$ Apothem $= a$ Radius $= r$

Number of diagonals	$\frac{1}{2}(n^2 - 3n)$
Measure of central angle	$360/n$
Measure of interior angle	$180(n - 2)/n$
Apothem in terms of side	$\frac{1}{2}s\cot(180°/n)$
Side in terms of radius	$2r\sin(180°/n)$
Perimeter (P)	ns
Area	$\frac{1}{2}nsa = \frac{1}{2}Pa$

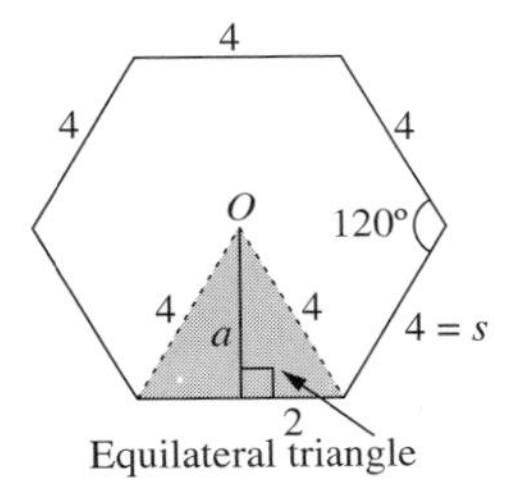

Figure B.38

EXAMPLE 1 In the case of the familiar regular hexagon—the shape often found in nature in the form of honeycombs—suppose each side of the hexagon measures 4 cm, as shown in Figure B.38. We can figure out the length of the apothem in this case, without trigonometry, as follows: $\sqrt{4^2 - 2^2} = \sqrt{12} = 2\sqrt{3}$ cm. We obtain directly the following information (compare with above table):

Number of diagonals $= \frac{1}{2}(36 - 18) = 9$ Apothem $= \frac{1}{2} \cdot 4 \cdot \cot 30° = 2\sqrt{3}$ cm

Measure of central angle $= \frac{360}{6} = 60$ Perimeter $= 6 \cdot s = 6 \cdot 4 = 24$ cm

Measure of each angle $= 180(6 - 2)/6 = 120$ Area $= \frac{1}{2} \cdot 24 \cdot 2\sqrt{3} = 24\sqrt{3}$ cm^2 ■

EXAMPLE 2 A desk ornament has the shape of a regular octagon (Figure B.39). If each side of the ornament measures 1 in., find how high it stands, in both exact and approximate values (accurate to two decimals) and find the area of the face.

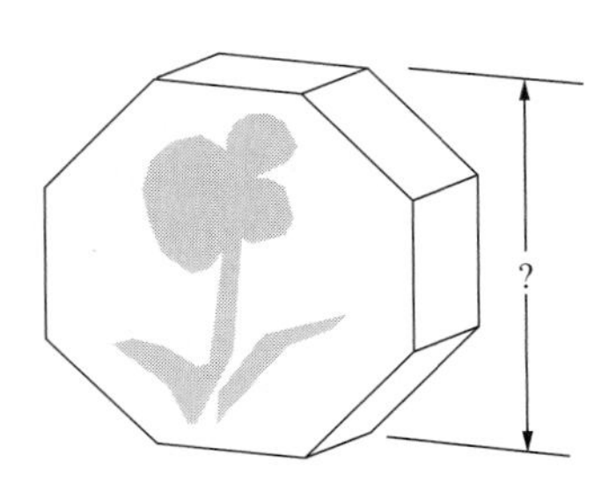

Figure B.39

SOLUTION

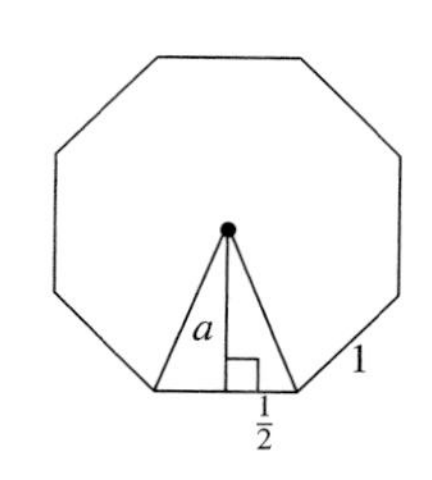

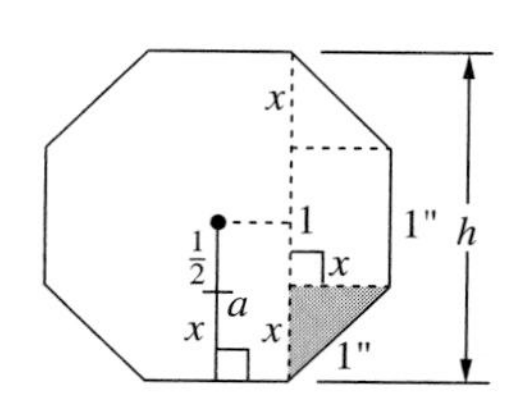

WITH TRIGONOMETRY

$$a = \tfrac{1}{2} \cdot s \cot(180°/8)$$
$$= \tfrac{1}{2} \cot 22.5°$$

$$\therefore \text{Height} = 2a$$
$$= 2(\tfrac{1}{2} \cot 22.5°)$$
$$= \cot 22.5° \approx 2.41 \text{ in.}$$

$$P = 8s = 8 \text{ in.}$$
$$K = \tfrac{1}{2} \cdot 8(\tfrac{1}{2} \cot 22.5°)$$
$$= 2 \cot 22.5° \approx 4.83 \text{ in}^2$$

WITHOUT

(Shaded triangle is isosceles right triangle)

$$x^2 + x^2 = 1^2 \quad \text{or} \quad 2x^2 = 1$$
$$x = \sqrt{2}/2$$
$$\text{Height} = 1 + 2x$$
$$= 1 + 2(\sqrt{2}/2)$$
$$= 1 + \sqrt{2} \approx 2.41 \text{ in.}$$
$$a = \tfrac{1}{2} + x = \tfrac{1}{2}(\sqrt{2} + 1)$$

$$K = \tfrac{1}{2}Pa = \tfrac{1}{2} \cdot 8 \cdot \tfrac{1}{2}(\sqrt{2} + 1) = 2(\sqrt{2} + 1) \quad \blacksquare$$

B.10 Optional Trigonometric Derivation of Heron's Formula

This section shows how to apply a few of the previous equations and identities to derive the famous Heron's Formula for the area of a triangle in terms of the lengths of its sides, a, b, and c:

$$\textbf{(1)} \quad K = \sqrt{s(s-a)(s-b)(s-c)} \qquad \text{where } s = \tfrac{1}{2}(a+b+c)$$

The first part is algebraic, independent of trigonometry or geometry. Since $2s = a + b + c$, then $2s - 2a = a + b + c - 2a = b + c - a$. Hence

$$2s \cdot 2(s-a) = (b+c+a)(b+c-a) = [(b+c)+a][(b+c)-a]$$

Use the algebra factoring law $(D+a)(D-a) = D^2 - a^2$, where $D = b + c$. We obtain

$$4s(s-a) = (b+c)^2 - a^2 = b^2 + 2bc + c^2 - a^2$$

In the same manner,

$$4(s-b)(s-c) = (a-b+c)(a+b-c) = [a-(b-c)][a+(b-c)]$$
$$= a^2 - (b-c)^2 = a^2 - (b^2 - 2bc + c^2)$$

Therefore, we obtain the two equations

$$\textbf{(2)} \qquad 4s(s-a) = 2bc + b^2 + c^2 - a^2$$

$$\textbf{(3)} \qquad 4(s-b)(s-c) = 2bc - b^2 - c^2 + a^2$$

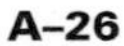

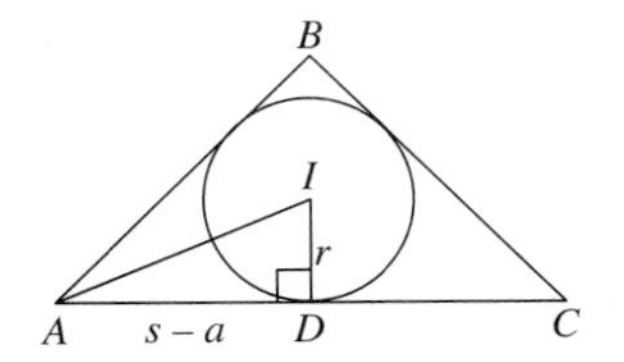

Figure B.40

So far, we have not used any geometry; this was all algebra. But now consider Figure B.40. It is shown in Problem 14, Section 3.8 that $AD = s - a$. Since I is the incenter and lies on all three angle bisectors, then in particular, $\overline{IA}$ bisects $\angle BAC$. Hence in right $\triangle ADI$ we have (using also the half-angle formula for tangent)

$$\tan\tfrac{1}{2}A = \frac{ID}{AD} = \frac{r}{s-a} = \sqrt{\frac{1-\cos A}{1+\cos A}}$$

Square both sides to clear the radical, then multiply numerator/denominator of the expression on the right by $2bc$:

$$\frac{r^2}{(s-a)^2} = \frac{2bc - 2bc\cos A}{2bc + 2bc\cos A}$$

Now we use the Law of Cosines for $\angle A$ in $\triangle ABC$, solving for $2bc\cos A$ so we can use it in the preceding formula:

$$a^2 = b^2 + c^2 - 2bc\cos A \qquad \text{or} \qquad 2bc\cos A = b^2 + c^2 - a^2$$

By substitution,

$$\frac{r^2}{(s-a)^2} = \frac{2bc - (b^2+c^2-a^2)}{2bc + (b^2+c^2-a^2)} = \frac{2bc - b^2 - c^2 + a^2}{2bc + b^2 + c^2 - a^2}$$

But from equations **(2)** and **(3)** we have

$$\frac{r^2}{(s-a)^2} = \frac{4(s-b)(s-c)}{4s(s-a)} \qquad \text{or} \qquad r^2 = \frac{(s-a)^2(s-b)(s-c)}{s(s-a)}$$

Using the formula for r in Section B.3, $r = K/s$, we have

$$\frac{K^2}{s^2} = \frac{(s-a)(s-b)(s-c)}{s} \qquad \text{or} \qquad K^2 = s(s-a)(s-b)(s-c)$$

SOLUTIONS TO PROBLEMS

PROBLEM 1 $\triangle ABD \cong \triangle ACD$ by AAS. $\therefore \overline{AB} \cong \overline{AC}$ by CPCF.

PROBLEM 2 $m\angle XZW = 90 = m\angle C$, $ZW = BC$ (by construction), and $XZ = AC$ (given), so by SAS $\triangle XWZ \cong \triangle ABC$. Then $XW = AB = XY$ and $\triangle XYW$ is isosceles, with $\angle W \cong \angle Y$; $\angle W \cong \angle B$ by CPCF, so $\angle B \cong \angle Y$ and $\triangle ABC \cong \triangle XYZ$ by HA.

PROBLEM 3 **(a)** Since the polygon is regular, $\overline{CB} \cong \overline{CD}$, $\overline{BA} \cong \overline{DE}$, and $\angle B \cong \angle D$. By SAS, $\triangle ABC \cong \triangle EDC$ and $\overline{CA} \cong \overline{CE}$. **(b)** $m\overset{\frown}{BA} = m\overset{\frown}{AE}$ so $m\angle BCA = \frac{1}{2}m\,\overset{\frown}{BA} = \frac{1}{2}m\,\overset{\frown}{AE} = m\angle ACE$. By part (a) $m\angle BCA = m\angle ECD$.

PROBLEM 4 In rectangle $ABCD$ with $\overline{AC} \perp \overline{BD}$ at point E, $AE = EC$ and $BE = BE$, so by SAS $\triangle ABE \cong \triangle CBE$ and $AB = BC$. $\therefore AB = BC = CD = DA$, and $ABCD$ is a square.

APPENDIX C

The Geometer's Sketchpad: *Brief Instructions*

It will be assumed that if you are using *The Geometer's Sketchpad* software, you already have access to a manual on its operations and/or a workbook on examples. It is unnecessary to duplicate that here. We will, however, include some instructions for procedures that are used frequently in this book. In these instructions, as with all the directions given in discovery units and problems, certain key words are used. We list these first, before discussing construction procedures.

SPECIAL TERMS

SELECT

a. To select an object such as a point, segment, or circle, move the arrow so it points to and touches the object to be selected, then left click. The object will be marked by *Sketchpad* as having been selected.

b. To select more than one object at a time, use the procedure of (a), but before you click, hold the Shift key down and select the objects needed.

CHOOSE This term refers to selecting an item from the toolbar commands (at top of screen, and to left of screen). To choose such an item, point and click, then choose a menu item and click again to execute the final command. Occasionally you are given an option to be checked or parameters to be typed in.

CONSTRUCT A term meaning to "draw" an object on your screen without specifications, or with specifications if given. Use the tools in the vertical toolbar at the left side of the screen to locate a point, draw a circle, segment, ray, or line. You must return to the SELECT TOOL at the top of the menu in order to select objects. (An option is to press the Control key while making your selections, again holding down the Shift key, in addition, to make multiple selections.)

SELECT POINT OF INTERSECTION When two objects (segments, lines, circles) cross and you need the point of intersection, point and click at the point of crossing to place the point of intersection into *Sketchpad* memory. (***Note:*** Sometimes this will not work if the point has been previously hidden or if there are three objects crossing at the same point. When this happens, you will have to select the two objects, then choose Point At Intersection from the CONSTRUCT menu.)

LOCATE (SELECT) POINT ON OBJECT To locate (or construct) an arbitrary point on a line, segment, circle, etc., choose POINT TOOL (in the vertical toolbar at left), then place the arrow at object, and left click. A point will appear on the object, and it can be dragged along that object. (Alternate method: Select object, then choose Point On Object from the CONSTRUCT menu.)

HIDE It is often desirable to delete objects of construction not pertinent to the objective of an experiment. Merely select object(s) to be erased and choose Hide (Object) under DISPLAY. (***Note:*** A useful feature of *Sketchpad* is that if you have selected objects of the same type, such as rays, then the Display command will read "Hide Ray(s)," preventing possible errors. Conversely, if you select several points for clearing and this prompt reads "Hide Objects," then you know you have inadvertently picked up an object other than a point that you might not want erased. In case you do inadvertently hide an object you need, you can recover it by choosing EDIT, Undo Hide (Object) before proceeding further.

LABELING In virtually all instructions, our constructions are automatically labeled by *Sketchpad*, and our reference to them will match the labels in *your* sketch if you also use automatic labeling. Before you start construction, choose DISPLAY, then Preferences, then Autoshow Labels (Points), and click OK (labels for other objects will not be used). The first two points you locate in your sketch will now be automatically labeled A, B. If your labels get out of sync with ours and not too much is lost, choose EDIT, Select All, and click. This will erase your construction and will restart the labeling with $A, B, \ldots$.

GEOMETRIC OBJECTS COMMONLY CONSTRUCTED

TRIANGLE Choose SEGMENT TOOL and click with the arrow on your screen for the first vertex, A, of the triangle. Move from A to B holding down the left key on the mouse and by moving the arrow, then release; the second vertex B appears. Move from B to another position, and you will have the third vertex C. Finally, move from C to A to complete the triangle.

ISOSCELES TRIANGLE Construct segment $\overline{AB}$, and, while still selected, choose Point At Midpoint under CONSTRUCT, and while the midpoint is selected, press Shift key and select the segment. With the last two items selected, choose Perpendicular Line under CONSTRUCT. Locate a point D on the perpendicular and draw segments $\overline{DA}$ and $\overline{DB}$. Drag D to adjust (select D and move mouse while holding down the click key).

EQUILATERAL TRIANGLE Construct a segment $\overline{AB}$, with A to the left of B if $\overline{AB}$ is not vertical. Double click on A to make it a center for rotation. Select segment $\overline{AB}$ and point B, then choose Rotate By 60° under TRANSFORM, creating vertex B'. Join B and B' by a segment to complete the triangle. (***Note:*** Objects are always rotated by *Sketchpad* in the *positive mathematical direction* (counterclockwise). To rotate in the clockwise direction, you must prefix the parameter of rotation with a minus sign in the option "Rotate By X degrees" in the Rotate menu under TRANSFORM.)

QUADRILATERAL Use the method of constructing a triangle (above), except that you will use four points A, B, C, and D instead of three.

PARALLELOGRAM

a. The fastest way to construct an arbitrary parallelogram is to first construct a segment, select it and its end points, then choose Copy, then Paste under

EDIT. This will create a segment parallel to the first one; drag it to any desired position and draw in the missing segments. This also works when one side of the parallelogram is already specified.

b. If the parallelogram has two adjacent sides given, $\overline{AB}$ and $\overline{AC}$, select A, then B, and choose Mark Vector "A → B" under TRANSFORM. Select segment $\overline{AC}$ and point C, then choose Translate By Marked Vector under TRANSFORM, creating segment $\overline{BC'}$ parallel and congruent to $\overline{AC}$. Join C and C' by a segment to complete.

SQUARE Construct segment $\overline{AB}$, with A to the left of B, and double click on A. Select segment $\overline{AB}$ and point B, then choose Rotate By 90° under TRANSFORM, creating vertex B'. Double click on B', select segment $\overline{B'A}$ and A, then choose Rotate By 90°, creating vertex A'. Join A' and B to complete.

ANGLE BISECTOR Given $\angle ABC$, select the points A, B, and C, in that order, and choose Angle Bisector under CONSTRUCT. The second point selected is always the vertex of the angle, and the origin of the angle bisector (in this case, B). The wrong order will produce the bisector of the wrong angle.

CIRCLE

a. To create an arbitrary circle without specifications, use CIRCLE TOOL.

b. To construct a circle with a given center A and passing through a given point B, select A, then B, and choose Circle By Center And Point under CONSTRUCT.

c. To construct a circle with a given center A and with a specified geometric radius congruent to a given segment (with neither end point at A), select the given segment and the center A, then choose Circle By Center And Radius under CONSTRUCT.

d. To construct a circle with a given center A and a numerical radius r, translate A horizontally r units to A' (use Translate By Polar Vector: Direction = 0.00°, Magnitude = r inches). Construct the circle with center A and passing through A' as in (b).

ARC

a. If the arc lies on a given circle and its end points are specified (given points on the circle), select the two points *in the counterclockwise direction* on that circle, then the circle, and choose Arc Of Circle under CONSTRUCT.

b. To construct an arc through three points, select the three points in order, and choose Arc Through Three Points under CONSTRUCT. If you get the wrong arc, check the order of the points you specified. They occur on the arc in the same order that they are encountered on the circle, in either the positive, or negative, direction.

OTHER COMMON PROCEDURES

MARK CENTER There are two ways to mark a point for the center of a rotation or dilation. The fastest way is to double click on the point. Also, you can select the point, then choose Mark Center under TRANSFORM.

MARK MIRROR There are two ways to mark a segment, ray, or line as the line of reflection in a transformation. One is to double click on the segment (ray or line). The other is to select the segment or ray, then choose Mark Mirror under TRANSFORM.

AREA, PERIMETER (POLYGONS) To calculate the area of a polygon, select the vertices in consecutive order, and choose Polygon Interior under CONSTRUCT. While the interior is still selected, choose Area (Perimeter) under MEASURE.

AREA, CIRCUMFERENCE (CIRCLES) To calculate the area or circumference of a circle, select the circle, then choose Area (Circumference) under MEASURE.

CREATING SCRIPTS

When you have a procedure that you are going to use repeatedly, sometimes it is advisable to use the Script program in *Sketchpad*. Choose New Sketch and New Script under FILE; *Sketchpad* will place these side by side. Suppose we want to repeat the construction of an equilateral triangle on several different line segments. First, construct one segment $\overline{AB}$. Turn on Record (REC) in the Script Menu and proceed basically as in the above construction of an equilateral triangle: Select point A, then point B (but *not* segment $\overline{AB}$), and rotate B about A 60° to point B'; construct segments $B'A$ and $B'B$. Turn off Record by choosing Stop. For future use, we choose Save under FILE and type out a name for this script, such as Equil-T or Eq-Tri (up to eight characters with no periods or spaces between).

To play the script we created for a given line segment $\overline{PQ}$, select P, then Q, and choose FAST in the Script Menu. An equilateral triangle PQR will quickly appear. You should try this if you do not have much experience with *Sketchpad*. (In this particular example, find out what effect reversing the order of P and Q has on the construction.) Remember, if you closed the script after recording a procedure and you want to use it in a diagram, you must choose Open under FILE and accurately type the name chosen for it. The script will then appear on the screen ready for use.

Any command or series of commands in *Sketchpad* can be recorded in a script, regardless of how involved the procedure is. The construction or procedure is almost instantaneous when it is played back.

Important Note: The success for creating and playing scripts lies in matching up the selected objects with those listed in the Script monitor under Given; these are the ones used to perform the operations in creating the script. They will appear in the exact order in which they were first used in the construction. If care is not taken, unintended objects can appear under Given, requiring extra, unnecessary selections to be made in playing back the script. It is advisable to carefully plan how to use the objects in the construction so as to guarantee that only those you want to select in playing back the script will appear under Given.

APPENDIX D

Unified Axiom System for the Three Classical Geometries

We first present the axioms, then a short discussion will follow indicating how the system works, and, in particular, how axiomatic spherical geometry is developed. However, the complete development is left as an interesting project for the reader; we do not intend to give the full development. Because this system of axioms allows for bounded distance and the study of spherical geometry in addition to the study of hyperbolic and Euclidean geometry, we call the resulting development **neutral geometry**.

UNDEFINED TERMS: *Point, line, plane, space*

AXIOM 1: To each pair of points (A, B) is associated a unique real number, denoted AB, with least upper bound α.

AXIOM 2: For all points A and B, $AB \geq 0$, with equality only when $A = B$.

AXIOM 3: For all points A and B, $AB = BA$.

AXIOM 4: Given any four distinct collinear points A, B, C, and D such that A–B–C, then either D–A–B, A–D–B, B–D–C, *or* B–C–D.

AXIOM 5: Each two points A and B lie on a line, and if $AB < \alpha$, that line is unique.

AXIOM 6: Each three noncollinear points determine a plane.

AXIOM 7: If points A and B lie in a plane and $AB < \alpha$, then the line determined by A and B lies in that plane.

AXIOM 8: If two planes meet, their intersection is a line.

AXIOM 9: Space consists of at least four noncoplanar points, and contains three noncollinear points. Each plane is a set of points of which at least three are noncollinear, and each line is a set of at least two distinct points.

AXIOM 10 (Ruler Postulate): Given line ℓ and two points P and Q on ℓ, the points of ℓ can be placed into one-to-one correspondence with the real numbers x such that $-\alpha < x \leq \alpha$ (called **coordinates**) in such a manner that

(1) points P and Q have coordinates 0 and $k > 0$, respectively

(2) if A and B on the line have coordinates a and b, then

$$AB = |a - b|, \qquad \text{if} \quad |a - b| \leq \alpha$$

$$AB = 2\alpha - |a - b|, \qquad \text{if} \quad |a - b| > \alpha$$

AXIOM 11 (Plane Separation Postulate): Let ℓ be any line lying in any plane $\boldsymbol{P}$. The set of all points in $\boldsymbol{P}$ not on ℓ consists of the union of two subsets H_1 and H_2 of $\boldsymbol{P}$ such that

(1) H_1 and H_2 are convex sets

(2) H_1 and H_2 have no points in common

(3) if A lies in H_1 and B lies in H_2 such that $AB < \alpha$, line ℓ intersects segment $\overline{AB}$.

AXIOM 12: Each angle $\angle ABC$ is associated with a unique real number between 0 and 180 denoted $m\angle ABC$. No angle can have measure 0 or 180.

AXIOM 13 (Angle Addition Postulate): If D lies in the interior of $\angle ABC$, then $m\angle ABD + m\angle DBC = m\angle ABC$.

AXIOM 14 (Protractor Postulate): The set of rays $\overrightarrow{AX}$ lying in a plane and on one side of a given line $\overleftrightarrow{AB}$, including ray $\overrightarrow{AB}$, may be placed into one-to-one correspondence with the real numbers x such that $0 \leq x < 180$ (called **coordinates**) in such a manner that

(1) ray $\overrightarrow{AB}$ has coordinate 0

(2) if rays $\overrightarrow{AC}$ and $\overrightarrow{AD}$ have coordinates c and d, then $m\angle CAD = |c - d|$.

AXIOM 15 (Linear Pair Axiom): A linear pair of angles is a supplementary pair.

AXIOM 16 (SAS Postulate): If two sides and the included angle of one triangle are congruent, respectively, to two sides and the included angle of another, the triangles are congruent.

The major feature of this axiom system is that we have, at the outset, a choice for α. We can assume that either $\alpha = \infty$ (distance is unbounded), or $\alpha < \infty$ (distance is bounded). If we choose $\alpha = \infty$, it is apparent that we will obtain the axiom system for absolute geometry—the geometry we studied in Chapter 3, as the following analysis shows. Many of the above axioms are already, word for word, the same as those for absolute geometry. But some are not. For example, consider how Axiom 5 reads if $\alpha = \infty$: Each two points A and B lie on a line, and *because* $AB < \alpha$, that line is unique. This is the incidence axiom, Axiom I-1. The remaining incidence axioms for neutral geometry (Axioms 6–9) reduce to those of absolute geometry in Chapter 2. The Ruler Postulate, Axiom 10, has that strange condition concerning the case $|a - b| > \alpha$, but if $\alpha = \infty$, this case cannot occur and we are back to the original Ruler Postulate for absolute geometry. Axiom 4 is an extra postulate we do not need because it is a consequence of the Ruler Postulate, so we can just throw it out when $\alpha = \infty$. Hence we conclude that if $\alpha = \infty$, neutral geometry coincides with absolute geometry, and depending on which parallel axiom we choose we can obtain either parabolic (Euclidean) or hyperbolic geometry.

What if our choice were $\alpha < \infty$? Now we are supposed to obtain elliptic (spherical) geometry. We present a sequence of lemmas containing new results, totally different from absolute geometry (as we might expect), that can be observed from the above axioms with very little effort. The goal is to obtain the same theory for triangles having sides of length $< \alpha/2$ as we obtained in absolute geometry. Occasionally these results extend to arbitrary triangles, as in the case of the triangle inequality. Much of the development follows without additional labor; many of Euclid's original arguments employed in absolute geometry can be used here. In what follows, we shall confine ourselves to just a single plane, as we did in our development for absolute geometry.

LEMMA 1: On line ℓ, the points $A[0]$, $B[\alpha]$, $C[\alpha/4]$, $D[-\alpha/4]$, $E[3\alpha/4]$, and $F[-3\alpha/4]$ exist by the Ruler Postulate. If line $m \neq \ell$ passes through A, and H_1 and H_2 are the half-planes of m containing C and D, respectively, then $E \in \overrightarrow{AC}$, $F \in \overrightarrow{AD}$, and $E \in H_1$, $F \in H_2$.

LEMMA 2: Given the same construction as Lemma 1 (with the same points A, B, C, D, E, and F on ℓ), $EF = \alpha/2$ and E–B–F holds. If line m passes through A, then it meets segment $\overline{EF}$ and must also pass through B.

COROLLARY A: For any two points A and B, if $AB = \alpha$ then any line that passes through A must also pass through B.

COROLLARY B: To each point A there corresponds a unique point A' in the plane such that $AA' = \alpha$. (We call A and A' **polar points.**)

LEMMA 3: In $\triangle ABC$, with A' the polar of A, if $AB < \alpha/2$, $AC \leq \alpha/2$, and D is any interior point on $\overline{BC}$, then A–B–A', A–C–A', A–D–A', and $AD < \alpha/2$.

COROLLARY (Exterior-Angle Inequality): In any triangle whose sides are of length less than $\alpha/2$, an exterior angle has measure greater than the measure of either opposite interior angle. Specifically, if $AB < \alpha/2$, $AC \leq \alpha/2$, and A–C–D then $m\angle BCD > m\angle A$ and $m\angle BCD > m\angle B$.

LEMMA 4: If the legs of an isosceles triangle are of length $< \alpha/2$, the base angles are acute.

LEMMA 5: All the inequality theorems of absolute geometry are valid for triangles whose sides are of length less than $\alpha/2$.

LEMMA 6: If $AB \geq \alpha/2$, $BC \geq \alpha/2$, and $AC < \alpha$, then $AB + BC > AC$.

LEMMA 7: If $AB < \alpha/2$, $BC \geq \alpha/2$, and $AC \leq \alpha/2$, then $AB + BC > AC$.

LEMMA 8: If $AB < \alpha/2$, $BC < \alpha/2$, and $AC \leq \alpha/2$, then $AB + BC > AC$.

(***Hint:*** Recall the construction suggested for Problem 25, Section 3.3; for the case $AB + BC < AC$, construct D and E on $\overline{AC}$ so that $AD = AB$ and $CE = CB$, with M the midpoint of $\overline{DE}$, and apply the above corollary, the Exterior Angle Inequality.)

LEMMA 9: If $AB < \alpha/2$, $BC \geq \alpha/2$, and $AC > \alpha/2$, then for C' the polar of C, $AB + AC' > BC'$ and $AB + BC > AC$.

COROLLARY: If $\triangle ABC$ is any triangle (and hence sides are necessarily of length $< \alpha$), then $AB + BC > AC$.

PROBLEM Prove that any two lines ℓ and m in the plane intersect at some point A, and that therefore they intersect at polar points A and B such that $AB = \alpha$.

LEMMA 10: If $\overline{AB} \perp \overline{BC}$ and $AB = \alpha/2$, then $\overline{AC} \perp \overline{BC}$ and $AC = \alpha/2$.

COROLLARY: With the notation and hypothesis of Lemma 10, suppose $A–E–B$ and $A–D–C$ and $EB = DC$. Then $\diamond BCDE$ is a Saccheri Quadrilateral with obtuse summit angles.

The result of the last corollary allows us to establish the Angle Sum Theorem for Spherical Geometry: *The angle sum of any triangle is greater than 180.* The details for this development, as well as other characteristics of spherical geometry that can be established axiomatically, will be left to the reader.

APPENDIX E

Answers to Selected Problems

SECTION 1.1

1. Angle sum equals 360. **3.** Lines $\overleftrightarrow{AB}$ and $\overleftrightarrow{AC}$ are tangent to the circle. **5.** Isosceles triangle (Side-Splitting Theorem); $\rightarrow$ (implies) $AM = MC = MB$. **7. (b)** From **(a)**, $6x + z = 2$, $8x + y + z = 4$, $4x + z = 1$; first and third equations yield $x = \frac{1}{2}$, $z = -1$, and substitution into second equation yields $y = 1 \rightarrow K = \frac{1}{2}B + I - 1$. **9.** Actual proof that reassembled figure is a square: Pivoting about P, Q, and S yields closed figure because P and Q are midpoints and $RS = \frac{1}{2}BC = BC - RS = BR + SC$, or $RS = AR' + AC' = R'S'$. Therefore $XY'Y''X'$ is a rectangle; it will be shown that $X'X = XY'$. Area $(XY'Y''X') =$ Area $\triangle ABC \rightarrow X'X \cdot XY' = (\sqrt{3}/4) \cdot 10^2 = 25\sqrt{3} = QR^2 \rightarrow X'X \cdot XY' = QR^2$. In original figure for problem, $PQ = RS \rightarrow PQRS$ is parallelogram; by congruent triangles, $a = b$ and $RX = QY$. $\therefore XY' = XQ + QY = XQ + RX = QR$ and $X'X \cdot XY' = QR^2 \rightarrow X'X \cdot XY' = XY'^2 \rightarrow X'X = XY'$. **11. (b)** If $m\angle A = 120°$ in $\triangle ABC$, draw lines $\overline{AD}$ and $\overline{AE}$ forming angles of measure 30°, 60°, and 30° at A, then subdivide $\triangle ADE$ by joining vertices to centroid. **13.** $MN = \frac{1}{2}BC = YZ$ and $\overline{MZ} \| \overline{AD} \perp \overline{BC} \| \overline{ZY}$ (Midpoint Connector Theorem) $\rightarrow$ $MNYZ$ is rectangle. Midpoint U of hypotenuse $\overline{NZ}$ of right triangle $\triangle NZY$ is equidistant from N, Y, Z, M (Problem 5) $\rightarrow$ circle centered at $U =$ midpoint of $\overline{XL} = U'$ passes through M, N, Y, Z, L, X. Finally, $U'(U)$ is equidistant from vertices of right triangle $\triangle XDL$, so circle also passes through D, E, and F. **15. (a)** Each angle of polygon $= 128\frac{4}{7}° \rightarrow m\angle BAC = \frac{1}{2}(180° - 128\frac{4}{7}°) = 25\frac{5}{7}°$, $m\angle ABC = m\angle ACB = 77\frac{1}{7}° \rightarrow$ angle measures of $\triangle 3$ are $25\frac{5}{7}°$ (at angle opposite B and C), $51\frac{3}{7}°$ (at B), $102\frac{6}{7}°$ (at C); angle between horizontal diagonal and side bordering pieces 6 and $7 = 102\frac{6}{7}° \rightarrow \triangle(1, 2)$ and $\triangle 3$ will fit in spaces shown (adjacent to dashed lines). $m\angle DAL = 128\frac{4}{7}° - 2(25\frac{5}{7}°) = 77\frac{1}{7}° \rightarrow \triangle 5$ will fit in space shown, and $\overline{LN} \|$ dashed line opposite, forming smaller parallelogram $\rightarrow$ pieces (4, 9), (10, 11), 12 make up rest of larger parallelogram. Remainder of dissection follows from dissection of parallelogram into square.

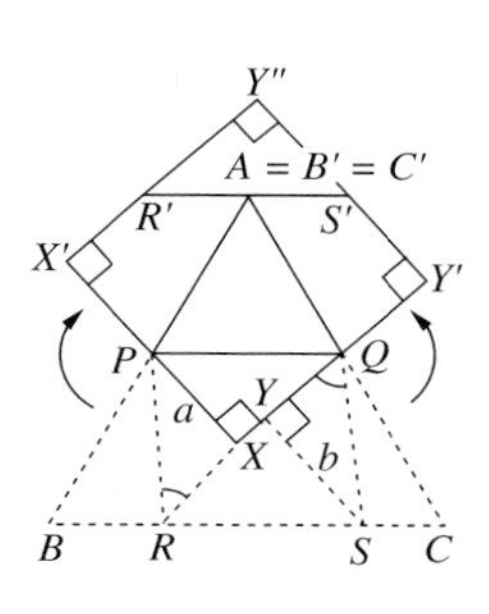

Problem 9

SECTION 1.2

3. $x^2 =$ area of square $= 4 \cdot 6 + 1^2 = 25 \rightarrow x = 5$. **5. (a)** 3.1326286; **(b)** 3.1393502; **(c)** 3.1415576. **9.** Regular inscribed hexagon has side $s_2 = 1$, so regular dodecagon has side $s_3 = \sqrt{2 - \sqrt{4 - 1^2}} = \sqrt{2 - \sqrt{3}}$; s_4 (24 sides) $= \sqrt{2 - \sqrt{4 - s_3^2}} = \sqrt{2 - \sqrt{4 - (2 - \sqrt{3})}} = \sqrt{2 - \sqrt{2 + \sqrt{3}}}$, and s_5 (48 sides) $= \sqrt{2 - \sqrt{4 - s_4^2}} = \sqrt{2 - \sqrt{4 - (2 - \sqrt{2 + \sqrt{3}})}} = \sqrt{2 - \sqrt{2 + \sqrt{2 + \sqrt{3}}}}$. In general, $s_{n+1}(3 \cdot 2^n$ sides$) = \sqrt{2 - \sqrt{2 + \sqrt{2 + \cdots + \sqrt{2 + \sqrt{3}}}}}$ (n radicals); multiply this by $\frac{1}{2}(3 \cdot 2^n)$ for π'_n. **11.** (Refer to construction discussed in Discovery Unit, Section 1.1.) Radius of construction arc $= \sqrt{bh} = \sqrt{8 \cdot 4} = \sqrt{32} \approx 5.7$, and arc meets opposite side at E where $DE = 7$, $EC = 1$. By measuring $\angle AEB$ with a protractor (or other means of analysis), $m\angle AEB = 90° \rightarrow F = E \rightarrow$ construction of Section 1.1 valid (with $F = E$). **13. (a)** $AE = AM + ME = AM + MC = a/2 + \sqrt{a^2 + (a/2)^2} = a/2 + (a/2)\sqrt{5} = \tau a$, so $AE/AD = \tau$. **(b)** Quadratic formula gives $x = \frac{1}{2}[1 \pm \sqrt{1^2 + 4}] = \frac{1}{2}(1 + \sqrt{5}) = \tau$. **(c)** $\tau^2 = \tau + 1 \rightarrow \tau^2/\tau = (\tau + 1)/\tau = 1 + \tau^{-1} \rightarrow \tau^{-1} = \tau - 1$. Next, $\tau^{-2} = (\tau - 1)^2 = \tau^2 - 2\tau + 1 = (\tau + 1) - 2\tau + 1 = 2 - \tau$. **15. (a)** $2/\sqrt{2} = \sqrt{2}$; $1/\sqrt{2 + \sqrt{2}} = 1/\sqrt{2 + \sqrt{2}} \cdot (\sqrt{2 - \sqrt{2}}/\sqrt{2 - \sqrt{2}}) = \sqrt{2 - \sqrt{2}}/\sqrt{(2 + \sqrt{2})(2 - \sqrt{2})} = \sqrt{2 - \sqrt{2}}/\sqrt{2}$; $1/\sqrt{2 + \sqrt{2 + \sqrt{2}}} = 1/\sqrt{2 + \sqrt{2 + \sqrt{2}}} \cdot (\sqrt{2 - \sqrt{2 + \sqrt{2}}}/\sqrt{2 - \sqrt{2 + \sqrt{2}}}) = \sqrt{2 - \sqrt{2 + \sqrt{2}}}/\sqrt{(2 + \sqrt{2 + \sqrt{2}})(2 - \sqrt{2 + \sqrt{2}})} = \sqrt{2 - \sqrt{2 + \sqrt{2}}}/\sqrt{2 - \sqrt{2}}$. **(b)** Product of expressions in

(a) results in $(2/\sqrt{2})\cdot(2/\sqrt{2+\sqrt{2}})\cdot(2/\sqrt{2+\sqrt{2+\sqrt{2}}}) = 2^2\sqrt{2-\sqrt{2+\sqrt{2}}}$. This pattern continues; in general, $2\cdot(2/\sqrt{2})\cdot(2/\sqrt{2+\sqrt{2}})\cdot(2/\sqrt{2+\sqrt{2+\sqrt{2}}})\cdots(2/\sqrt{2+\sqrt{2+\cdots\sqrt{2+\sqrt{2}}}}$ (n radicals) $=$ $2^n\sqrt{2-\sqrt{2+\sqrt{2+\cdots\sqrt{2+\sqrt{2}}}}} = \pi_n$ and converges to π. $\therefore\ \pi/2 = (2/\sqrt{2})\cdot(2/\sqrt{2+\sqrt{2}})\cdot(2/\sqrt{2+\sqrt{2+\sqrt{2}}})\cdots$ and $2/\pi = (\sqrt{2}/2)\cdot(\sqrt{2+\sqrt{2}}/2)\cdot(\sqrt{2+\sqrt{2+\sqrt{2}}}/2)\cdots$. **(c)** $2/\pi =$ $\sqrt{2/4}\,\sqrt{2+\sqrt{2})/4}\,\sqrt{(2+\sqrt{2+\sqrt{2}})/4}\cdots = \sqrt{\tfrac{1}{2}}\sqrt{\tfrac{1}{2}+\tfrac{1}{2}\sqrt{2}/2}\cdot\sqrt{\tfrac{1}{2}+\tfrac{1}{2}\sqrt{2+\sqrt{2}}/2}\cdots$ $\sqrt{\tfrac{1}{2}}\cdot\sqrt{\tfrac{1}{2}+\tfrac{1}{2}\sqrt{\tfrac{1}{2}}}\sqrt{\tfrac{1}{2}+\tfrac{1}{2}\sqrt{\tfrac{1}{2}+\tfrac{1}{2}\sqrt{2}/2}}\cdots = \sqrt{\tfrac{1}{2}}\sqrt{\tfrac{1}{2}+\tfrac{1}{2}\sqrt{\tfrac{1}{2}}}\cdot\sqrt{\tfrac{1}{2}+\tfrac{1}{2}\sqrt{\tfrac{1}{2}+\tfrac{1}{2}\sqrt{\tfrac{1}{2}}}}\cdots$. **17.** If $a = 2.5$, limit $\approx 3.295291 \neq \pi$. Problem 18 addresses the issue in detail. **18.** $\lim_{n\to\infty} 2^n\sqrt{2-\sqrt{2+\sqrt{2+\cdots\sqrt{2+\sqrt{a}}}}} =$ $4\sin^{-1}\tfrac{1}{2}\sqrt{4-a}$ by carefully following analysis leading up to **(4)** and **(5)**, and using relation between chord $s_1(=\sqrt{4-a})$ and its intercepted arc α: $\tfrac{1}{2}s_1 = \sin\tfrac{1}{2}\alpha$. Observe what happens in this limit if $a = 4$; if $a > 4$ the sequence is undefined since $\sqrt{2+\sqrt{2+\cdots+\sqrt{2+\sqrt{a}}}} > 2$. This yields an exact answer to the sequence in Problem 17: $2a\sin^{-1}\tfrac{1}{2}\sqrt{4-a}$. It can be shown that this has the value π *only* when $a = 2, 3$. **19.** By similar triangles, $GB/BE = FB/BD = (BE/BD)/BD \to GB =$ $BE^2/BD^2 = (\tfrac{1}{2})^2/[(\tfrac{7}{8})^2+1] = \tfrac{16}{113} \to CL = 3 + \tfrac{16}{113} - \tfrac{355}{113}$.

SECTION 1.3

1. Perimeter unbounded; area constant because altitude and base of $\triangle ABC$ invariant (perimeter minimum when C is directly above midpoint of $\overline{AB}$). **3.** G invariant = midpoint of arc. (*Proof:* If $m\widehat{CG} = m\widehat{GF}$, $\tfrac{1}{2}m\angle CDG = m\widehat{CG} = m\widehat{GF} = \tfrac{1}{2}m\angle GDF$ so G lies on the bisector of $\angle CDF$ for all points D on arc $\widehat{CF}$). **5.** $x \geq 0$ with $=$ only when D lies on minor arc $\widehat{AC}$. Ptolemy's Theorem: If $ABCD$ is inscribed quadrilateral, $ac + bd = mn$. **7. (a)** Value of $x + y + z$ repeats six times for one revolution of P about circle, with maximum at the three points of tangency, minimum at points half-way between. **(b)** $x^2 + y^2 + z^2 = 5a^2/4$ (constant) where a = side of equilateral triangle. **9.** $\overline{AT}$, $\overline{BU}$, $\overline{CV}$ are perpendicular bisectors of sides of $\triangle PQR$ so meet at circumcenter (point equidistant from vertices). $\therefore$ altitudes of $\triangle ABC$ (same three perpendiculars) are concurrent. **11.** Maximum area when C is above midpoint of $\overline{AD}$ (where altitude is maximum); $AE + ED = ED = AB$ = constant $\to$ perimeter constant. **13.** U is midpoint of $\overline{HO}$ and $HG = 2GO$. **15.** $\triangle KLM$ is similar to original triangle $\triangle ABC$.

SECTION 1.4

5. Let $x = RQ$, $y = AQ$; by Law of Cosines, $x^2 = 2y^2(1-\cos 20°) = 4y^2\sin^2\tfrac{1}{2}(20°)$ or $x = 2y\sin 10°$ and $y = AM\cdot\sec 20° = \tfrac{1}{2}\sec 20° \to x = 2(\tfrac{1}{2}\sec 20°)\sin 10° = \sin 10°\sec 20°$. **7.** The limaçons, with loop. **9.** Midpoint of hypotenuse of right triangles $\triangle APE$, $\triangle APF$ equidistant from vertices $\to$ circle passes through A, E, P, and through A, P, F; $x = AP$ = diameter of circumcircle of $\triangle AEF \to \sin A = EF/x$. But $\sin A = a/2R$ in $\triangle ABC \to EF/x = a/2R$ or $EF = ax/2R$. Similarly with DF and DE. **11.** Equation becomes $c\cdot z + a\cdot x = b\cdot y$; divide both sides by $2R$ and use formulas for sides of pedal triangle to show that $DE + EF = DF$. **13.** Partial derivative with respect to t of $(x/t + y/\sqrt{a^2-t^2} = 1)$ is $-xt^{-2} - \tfrac{1}{2}y(a^2-t^2)^{-3/2}(-2t) = 0$ or $y = x(a^2-t^2)^{3/2}/t^3$; substitute this into equation of line to find x in terms of t: $x/t + x(a^2-t^2)^{3/2}/t^3\sqrt{a^2-t^2} = 1 \to xt^2 + x(a^2-t^2) = t^3 \to t = a^{2/3}x^{1/3}$. Finally, substitute this last result into $y = x(a^2-t^2)^{3/2}/t^3$, and simplify.

SECTION 2.1

1. (a) *If it rains, then John wears his raincoat.* **(d)** *If Mary is unhappy, then it rains.* **3.** Given: Two angles with measure 30. Prove: Angles congruent. Converse: *If two angles are congruent, then they have measure 30.* (False). **5.** Given: Medians of triangle. Prove: Medians concurrent. Converse: *If lines are concurrent, then they are the medians of a triangle.* (True) **7. (a)** Let p = inflation, q = high unemployment; then $p \to q$. **(b)** If there were inflation without high unemployment. **9. (a)** Theorem 1. **(b)** Theorem 2. **(c)** Contrapositive of Axiom 1. **11. (a)** Theorem 1. **(b)** Theorem 2. **(c)** Definition. **13.** (1) p (given); (2) r or s (given); (3) assume r

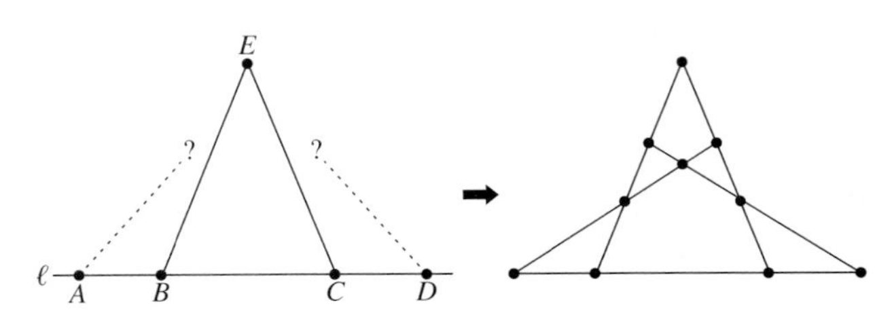

Problem 15

(first logical case); (4) x (Theorem 2); (5) q (Theorem 4); (6) s (second case); (7) y (Theorem 3); (8) $\sim p \rightarrow\leftarrow$ (Theorem 1); (9) $\therefore q$ (Rule of Elimination). **15.** Let ℓ be any line $= \{A, B, C, D\}$. If there were five or more other lines besides ℓ they meet ℓ at A, B, C, and D, forcing two of them to meet at, say A, in violation of Axiom 2. Hence there can exist but four other lines besides ℓ (and there must exist four lines, or Axiom 2 fails). $\therefore$ precisely five lines and only finitely many points (precisely 10) exist. System is categorical. (See figure for model.)

SECTION 2.2

1. A, B, C, D not collinear; area of $\Diamond ACBD$ is one unit square. (See figure.) **7. (b)** Not categorical. **9.** Point O has coordinates $(10, -30)$, so certain sums involving angles used in proof do not follow. **11.** Through "point" *PR* (representing President), draw three "lines" E, P, and N (representing Executive, Program, and Nominating Committees), and locate "points" *VP* (Vice President) and *ST* (Secretary-Treasurer) on E; *VP* must belong to two further "lines," say A and B, as must *ST* (C and D), and A and P must have a member M_4 in common. Continue analysis in this fashion; show that more than seven members will ultimately violate By-Laws. **13. (a)** Think of dabbas as points (represented as dots) and abbas as lines (represented as line segments or curves in plane). Start out with Axioms (1), (2), and (3): let A and B be two (distinct) dabbas, and ℓ_1 the unique abba containing A and B (as shown in figure). By Axiom (4) there exists a dabba C not on ℓ_1 and abbas ℓ_2, ℓ_3 must exist, containing pairs (A, C), (B, C), respectively [Axiom (3)]. Since this construction could be carried out for any two dabbas A and B, we have proven Property (1). Now use Axiom (5) to obtain abba ℓ_4 containing C and "parallel" to ℓ_1, hence there exists $D \neq C$ on ℓ_4; abbas ℓ_5 and ℓ_6 now follow, proving Property (2) and (3). **(b)** *MODEL 1*: Dabbas: $\{A, B, C, D\}$ Abbas: $\ell_1, \ell_2, \ell_3, \ell_4, \ell_5, \ell_6\}$ where $\ell_1 = \{A, B\}$, $\ell_2 = \{A, C\}, \ldots$ (all dabba pairs). *MODEL 2*: See Problem 8, Section 2.3.

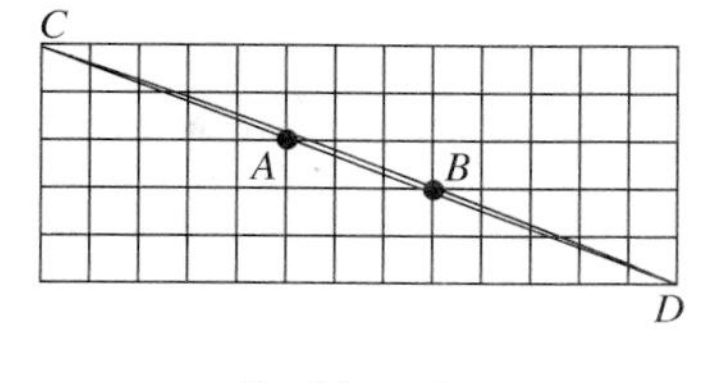

Problem 1

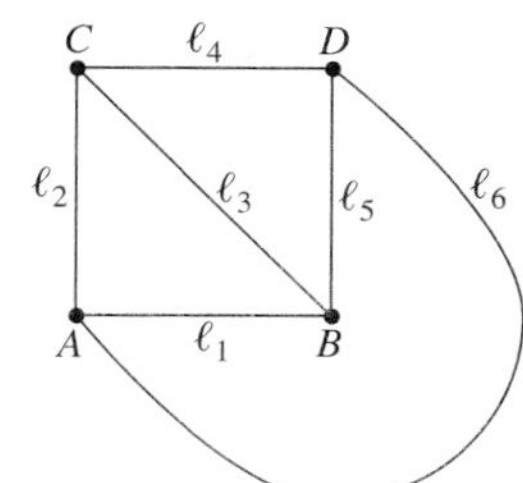

Problem 13

SECTION 2.3

1. See figure. **3.** Axioms I-1, I-2, I-3. **5.** *Lines*: $\{1, 2\}, \{1, n\}, \{2, n\}, n = \text{integer} \geq 3$, $\{3, 4, \ldots, n, \ldots\}$; *Planes*: $\{1, 3, 4, \ldots, n, \ldots\}, \{2, 3, 4, \ldots, n, \ldots\}, \{1, 2, n\}, n \geq 3$. **9.** *Points*: $S = \{A, B, C, D, E\}$; *Lines*: $\{A, B\}, \{A, B, C\}, \{A, D\}, \{A, E\}, \{B, C\}, \{B, D\}, \{B, E\}, \{C, D\}, \{C, E\}, \{D, E\}$; *Planes*: $\{A, B, C, D\}, \{A, B, C, E\}, \{A, D, E\}, \{B, D, E\}, \{C, D, E\}$. **11.** Prerequisite for problem is familiarity with $\mathbf{Z}_p = \{0, 1, 2, 3, \ldots, p-1\}$, where p is a prime and any integer n is *equivalent modulo p* to one of these p numbers. This set is a *finite field*, with so-called *clock arithmetic*: $0 = p, 1 = p + 1, \ldots, 1^{-1} = 1, 2^{-1} = (p-1)/2$ for $p > 2$, and $(p-1)^{-1} = p - 1$ (other inverses have to be worked out individually). In such a system, define *point* as an ordered pair (x, y), and *line* as all points whose coordinates satisfy a linear equation of the form $y = mx + b$ or $x = a$ for constants m, b, a in $\mathbf{Z}_p$. One can use algebra to verify (1) *two points determine a line*, (2) *two lines are parallel iff they are either both vertical lines or have the same slope*, and, from this, (3) Parallel Postulate. Such a system is rich in geometric properties, particularly if $p > 2$ and p is of the form $4m + 3$. **13. (a)** It suffices to prove the Parallel Postulate; consider $\overleftrightarrow{AB}$ and $C \notin \overleftrightarrow{AB}$ in a given plane $\boldsymbol{P}$—work with $D \notin \boldsymbol{P}$ (Axiom I-5), and noncoplanar lines $\overleftrightarrow{AB}$ and $\overleftrightarrow{CD}$. Axiom I-6 yields a unique plane $\boldsymbol{P}'$ containing $\overleftrightarrow{CD}$ parallel to $\overleftrightarrow{AB}$. **(b)** Let ℓ and m be any two lines in $\boldsymbol{P}$, and $\boldsymbol{Q}$ the plane of A and ℓ; by (a), $\boldsymbol{Q}$ contains a line $\ell' \| \ell$ passing through A, and by Axiom I-6, there is a plane $\boldsymbol{P}'$ through ℓ' parallel to m, which will be the unique plane through A parallel to $\boldsymbol{P}$. **(c)** If ℓ has n points $A_1, \ldots, A_n$, and m is any other line containing point $B_1 \notin \ell$, take the parallels to $\overline{A_1B_1}$ passing through $A_2, \ldots, A_n$ to obtain n points on m, then show there can be no others; there are $n + 1$ lines through each point and n lines in each family of parallel llnes, thus $n(n+1)$ lines altogether. **(d)** To count the points in space, let ℓ and m be any two noncoplanar lines; there are n planes containing ℓ, one for each point on m, plus the plane parallel to m, each containing $n^2 - n$ points not counting those on ℓ, so there are $n(n^2 - n) + n^2 = n^3$ points; to count the number of planes, first prove there are $n^2 + n + 1$ planes passing through a fixed point A and n planes in each family of parallel planes $\rightarrow$ there are $n(n^2 + n + 1)$ planes in space. **15. (a)** $\{A, \boldsymbol{K}, U\}, \{\boldsymbol{V}, L, \boldsymbol{H}\}, \{S, \boldsymbol{I}, N\}$. **(b)** $\{\boldsymbol{G}, \boldsymbol{H}, I, P, Q, \boldsymbol{R}, Y, Z, \Sigma\}$ and $\{J, E, \boldsymbol{\Sigma}, Y, \boldsymbol{K}, \boldsymbol{F}, D, Z, L\}$. **(c)** $\{F, \boldsymbol{T}, P, U, Q, D, \boldsymbol{R}, E, S\} \cap \{G, W, \boldsymbol{L}, Q, \boldsymbol{F}, S, \Sigma, M, \boldsymbol{B}\} = \{F, Q, S\}$. **(d)** $\{O, A, Z, V, K, I, E, U, \boldsymbol{P}\}$.

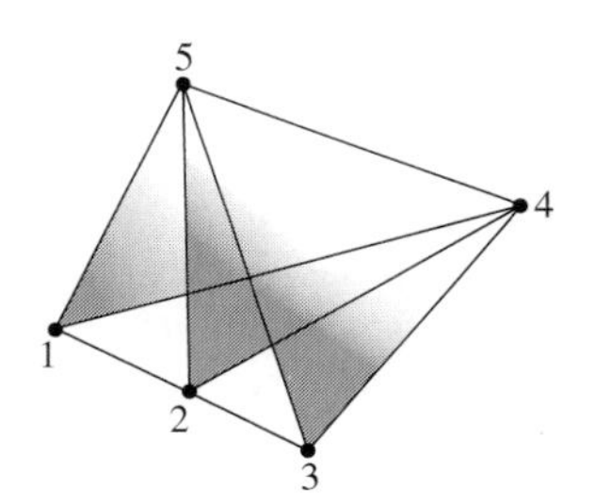

Problem 1

SECTION 2.4

1. C–A–D–B (or B–D–A–C). **3. (a)** B–A–C and B–C–D **(b)** $AC + CD < AD$. **5. (b)** $\overline{BD}$. **(c)** $\overleftrightarrow{BC}$. **(d)** C. **(e)** $\angle ACD$. **7.** $\overrightarrow{CA}, \overrightarrow{CB}, \overrightarrow{AB}, \overrightarrow{CA} = \overrightarrow{CB}$, and $\overrightarrow{AB}$. **9. (a)** $x \geq 3$. **(b)** $-1 \leq x \leq 5$. **(c)** $3 \leq x \leq 5$. **11.** $XY = 15$. **13.** $b = a \pm 8$. **15.** (8) Assume B–A–C (assumption for indirect proof); (9) $BA + AC = BC$ (definition); (10) $BA + (AB + BC) = BC$ [substitution, Step (1)]; (11) $2AB = 0$ (algebra, $AB = BA$); (12) $AB = 0$ and $A = B \rightarrow\leftarrow$ (algebra, $A \neq B$). **17.** From the definition of A–B–C–D, it remains to prove A–C–D $(AC + CD = AD)$: $AC + CD = (AB + BC) + (BD - BC) = AB + BD + BC - BC = AD$. **19.** Suppose $\ell = \overleftrightarrow{AB}$;

by Ruler Postulate, Theorem 3 there exists $C \in \ell$ such that C–A–B. If $C \in \overrightarrow{AB}$ then either $C = A$, $C = B$, A–C–B, or A–B–C $\rightarrow\leftarrow$ (Theorem 1). **21. (a)** D-1, D-2, D-3, but not D-4. **(c)** All real x such that either $x = 2$, $x = 5.5$, $2.5 \le x \le 3$, $3.5 \le x \le 4$, or $4.5 \le x \le 5$. **(d)** If $a < b$ and $a = m$, $b = n$ are integers, then $\overline{ab} = \{m, m + 1, m + 2, \ldots, n\}$. If $a = m$ and $b = n + \theta$, $0 < \theta < 1$, then $\overline{ab} = \{x : x = a, x = b, m + \theta \le x \le m + 1, m + 1 + \theta \le x \le m + 2, \ldots,$ or $n - 1 + \theta \le x \le n\}$. If $a = m + \theta$ and $b = n$, $0 < \theta < 1$, then $\overline{ab} = \{x : x = a, x = b, m + 1 \le x \le m + 1 + \theta, m + 2 \le x \le m + 2 + \theta, \ldots,$ or $n - 1 \le x \le n - 1 + \theta\}$. Finally, if $a = m + \theta$, $b = n + \theta$, $0 < \theta < 1$ and $0 < \phi < 1$, then for $\theta \ge \phi$, $\overline{ab} = \{x : x = a, x = b, m + 1 + \phi \le x \le m + 1 + \theta, m + 2 + \phi \le x \le m + 2 + \theta, \ldots,$ or $n - 1 + \phi \le x \le n - 1 + \theta\}$, and for $\theta < \phi$, $\overline{ab} = \{x : x = a, x = b, m + \phi \le x \le m + 1 + \theta, m + 1 + \phi \le x \le m + 2 + \theta, \ldots,$ or $n - 2 + \phi \le x \le n - 1 + \theta\}$. (Example: Using interval notation, $\overline{(3.7)(6.2)} = \{3.7, 6.2\} \cup [4.2, 4.7] \cup [5.2, 5.7]$.) **(e)** Proof based on least integer notation and its properties: $[x]$ = *least integer n that is* $\ge x$. Basic properties: $[x + y] \le [x] + [y]$ and, for $x < y$, $[x] \le [y]$. Since by definition $d(x, y) = [|x - y|]$, we have $d(x, y) = [|x - y|] = [|x - z + z - y|] \le [|x - z| + |z - y|] \le [|x - z|] + [|z - y|] = d(x, z) + d(z, y)$.

SECTION 2.5

1. 145 **3. (a)** 92. **(b)** 34. **(c)** 43. **(d)** 96.5. **5.** No, using $m\angle PMQ = 120$. Semirigorous proof using betweenness relations evident from figure: Construct ray $\overrightarrow{MN'}$ opposite ray $\overrightarrow{MN} \rightarrow m\angle PMN' = 180 - m\angle PMN = 60$, $m\angle N'MQ = 180 - m\angle NMQ = 60$ (Linear Pair Axiom); using observable fact $\overrightarrow{MP}$–$\overrightarrow{MN'}$–$\overrightarrow{MQ}$, $m\angle PMQ = 60 + 60 = 120$. (Rigorous proof uses results from Section 2.6.) **7. (a)** 126. **(b)** 128. **9.** $\overrightarrow{CF} \perp \overrightarrow{CE}$. **11.** 20. **13.** Let $\angle ABC$ and $\angle ABD$ be linear pair, with $\overrightarrow{BD}$ opposite $\overrightarrow{BC}$, and let $\overrightarrow{BE}$ and $\overrightarrow{BF}$ be the bisectors of those angles (see figure); for rays on A-side of $\overleftrightarrow{BC}$, let $\overrightarrow{BA}$, $\overrightarrow{BF}$ have coordinates a, b (Protractor Postulate). We first show $b > a$: $\overrightarrow{BD}$–$\overrightarrow{BF}$–$\overrightarrow{BA}$ (definition, angle bisector) $\rightarrow m\angle FBD < m\angle ABD$; by Linear Pair Axiom and Protractor Postulate, $b = m\angle FBC = 180 - m\angle FBD > 180 - m\angle ABD = m\angle ABC = a$. $\overrightarrow{BE}$ has coordinate $\frac{1}{2}a$, so $0 < \frac{1}{2}a < a < b$ and $\overrightarrow{BC}$–$\overrightarrow{BE}$–$\overrightarrow{BA}$–$\overrightarrow{BF}$ (dual of Theorem 3, Section 2.4) $\rightarrow$ $m\angle FBE = m\angle FBA + m\angle ABE = \frac{1}{2}(m\angle DBA + m\angle ABC) = \frac{1}{2} \cdot 180 = 90$. **15.** Using Figure 2.37, by definition, pairs $(\angle 1, \angle 3)$ and $(\angle 2, \angle 3)$ are linear pairs, hence are supplementary pairs $\rightarrow m\angle 1 = m\angle 2$ (Theorem 2). **17.** "Locus" of G is arc of circle with diameter $\overline{EF}$ (result of ideas to be discussed in Section 4.5). **19.** By Protractor Postulate, consider ray $\overrightarrow{AC}$ on one side of $\overleftrightarrow{AB}$ having coordinate 90, $\rightarrow m\angle CAB = 90 \rightarrow \overleftrightarrow{AC} \perp \overleftrightarrow{AB}$ (existence established). If second perpendicular m passing through A exists, then two rays on same side of $\overleftrightarrow{AB}$ have same coordinate $\rightarrow\leftarrow$ (this last statement actually depends on Plane Separation Postulate, Section 2.6).

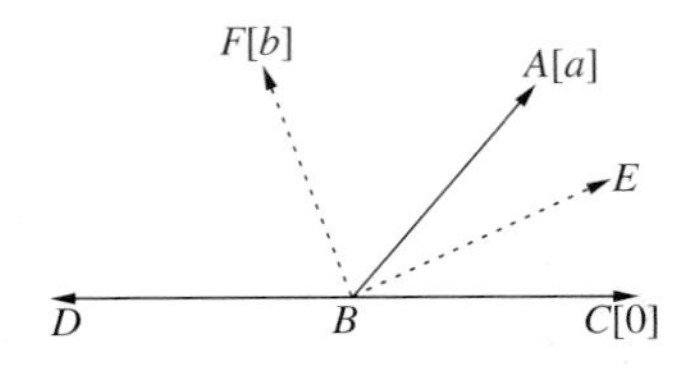

Problem 13

SECTION 2.6

1. (a) $H(P, \overleftrightarrow{TS})$. **(b)** $H(T, \overleftrightarrow{PR})$. **(c)** $H(P, \overleftrightarrow{RT}) \cap H(T, \overleftrightarrow{RP})$. **3. (a)** $H(E, \overleftrightarrow{FG})$. **(b)** $H(F, \overleftrightarrow{DE})$. **5. (a)** 40. **(b)** 160. **(c)** 160. **(d)** None. **7.** Three half-planes: Another half-plane; a parallel strip (points between two parallel lines); interior of angle or interior of triangle; region on one side of a broken line $ABCD$ consisting of $\overrightarrow{BA} \cup \overline{BC} \cup \overrightarrow{CD}$. Four half-planes: Sets mentioned above; interior of (convex) quadrilateral; region lying on one side of a broken line $ABCDE = \overrightarrow{BA} \cup \overline{BC} \cup \overline{CD} \cup \overrightarrow{DE}$. **11.** Ray only: Given $C \in \overrightarrow{AB}$ and $D \in \overrightarrow{AB}$; to prove: $\overline{CD} \subseteq \overrightarrow{AB}$. Let line $\overleftrightarrow{AB}$ have a coordinate system such that $\overrightarrow{AB} = \{x : x \ge 0\}$; if coordinates of C and D are $c < d$, then $c \ge 0$ and $d \ge 0$, $\rightarrow 0 \le c < d \rightarrow$ for any point $P[x] \in \overline{CD}$, $c \le x \le d \rightarrow x \ge 0 \rightarrow P \in \overrightarrow{AB}$ (Theorem 3, Section 2.4). **13.** Given betweenness relations imply that $\overleftrightarrow{DE}$ does not pass through A, B, or $C \rightarrow \overleftrightarrow{DE}$ meets $\overline{AB}$ at F such that A–F–B (Postulate of Pasch). Cases for D, E, F on line $\overleftrightarrow{DE}$ are D–F–E, E–D–F, or D–E–F (desired), so we set out to disprove first two. D–F–$E \rightarrow$ line $\overleftrightarrow{AB} = \overleftrightarrow{BF}$ meets $\overline{CD}$ at A' such that C–A'–D (figure) $\rightarrow \overleftrightarrow{AB}$ meets $\overleftrightarrow{CD}$ in two points A, A' $\rightarrow\leftarrow$ E–D–$F \rightarrow$ line $\overleftrightarrow{AD}$ meets $\overline{BF}$ at C' such that B–C'–$F \rightarrow \overleftrightarrow{AC}$ and $\overleftrightarrow{BF} = \overleftrightarrow{AB}$ have A and C' in common $\rightarrow\leftarrow$ $\therefore$ D–E–F. **17. (a)** Construct ray $\overrightarrow{MQ'}$ opposite ray $\overrightarrow{MQ}$, which then must lie on P-side of $\overleftrightarrow{OR} = H_1$. With coordinate system for rays in H_1 (such that $\overrightarrow{MO}$ has zero coordinate), coordinates of $\overrightarrow{MP}$, $\overrightarrow{MQ'} = 78$, x, where $m\angle Q'MO = 180 - m\angle OMQ = 79$ (Linear Pair Axiom) $\rightarrow |x - 0| = x = 79$. $\therefore m\angle Q'MP = |79 - 78| = 1 \rightarrow m\angle PMQ = 180 - m\angle Q'MP = 179$. **(b)** Here, $x = 79$ as before and $m\angle Q'MP = |79 - 80| = 1 \rightarrow m\angle PMQ = 180 - 1 = 179$. **(c)** $x = 79$; since coordinate of $\overrightarrow{MP} = 79$, $\overrightarrow{MP} = \overrightarrow{MQ'} \rightarrow \overrightarrow{MP}$ opposite $\overrightarrow{MQ}$. **19.** An open half-plane, closed half-plane, or an open half-plane together with an open or closed ray on boundary.

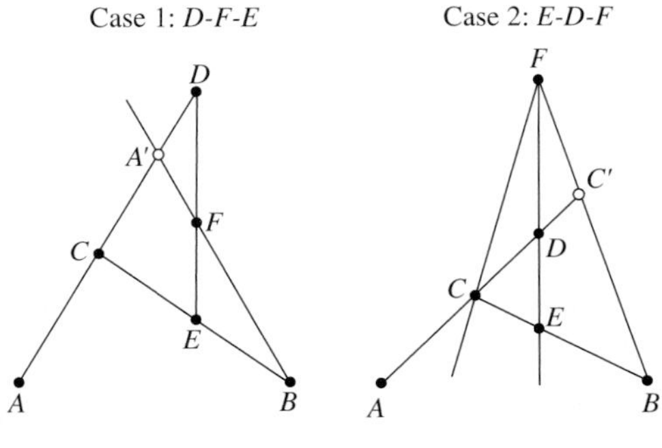

Problem 13

TESTING YOUR KNOWLEDGE (CHAPTER 2)

1. Axiom I-1 (two points determine a line). **2.** Suppose line and plane had second point in common; by Axiom I-3, line would lie in plane, contradicting what was given. $\therefore$ line and plane have only one point in common. **3.** $QT = 3$. **4. (a)** $D[11]$. **(b)** B–C–D, $BC = 8 - 5 = 3$ and $CD = 11 - 8 = 3 = BC$. $\therefore$ C is midpoint of $\overline{BD}$. **5.** $m\angle 1 = 45$. **6.** $\frac{1}{2} \cdot (48 + 115) = 81.5$. **7.** $m\angle SPR = 45$. **8.** $m\angle FAH = 105$. **9.** (See figure.) $H(G, \overleftrightarrow{EF})$, $H(R, \overleftrightarrow{EF})$, or $H(S, \overleftrightarrow{EF})$ and $H(S, \overleftrightarrow{FG})$. **10.** Let $\overrightarrow{AD}$ and $\overrightarrow{BE}$ bisect $\angle BAC$ and $\angle ABC \rightarrow \overrightarrow{AB}$–$\overrightarrow{AD}$–$\overrightarrow{AC} \rightarrow D \in$ Interior $\angle BAC \rightarrow$ ray $\overrightarrow{AD}$ meets side $\overline{BC}$ at F such that B–F–C (Crossbar Theorem). Again, by the Crossbar Theorem, ray $\overrightarrow{BE}$ meets segment $\overline{AF}$ at some point G such that A–G–F. $\therefore$ G $\in$ Interior $\angle ABC$, $\angle BAC$.

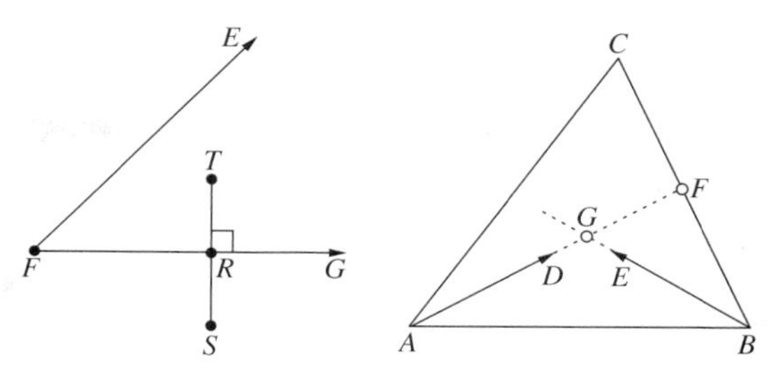

Problem 9 **Problem 10**

SECTION 3.1

1. (a) $\overline{RS} \cong \overline{VW}$, $\overline{RT} \cong \overline{UW}$. **(b)** $ST = RT - RS = UW - VW = UV$. $\therefore$ $\overline{ST} \cong \overline{UV}$. **3.** $\overline{BC} \cong \overline{ZW}$, $\overline{AC} \cong \overline{YZ}$, $\overline{CD} \cong \overline{XZ}$, $\overline{CD} \cong \overline{YW}$, $\overline{BD} \cong \overline{AD}$, $\angle ACD \cong \angle ACB$, $\angle ACD \cong \angle XZY$, $\angle XZY \cong \angle XZW$. (Transitive Law yields several other congruences.) **5.** $\angle XYZ \cong \angle XZY$. **7.** $\triangle ABC \cong \triangle BAC$ (triangles are isosceles with $\overline{AC} \cong \overline{BC}$). **9.** $\triangle GHK$ and $\triangle FDE$, $\triangle CAB$ and $\triangle FDE$. **11. (a)** Identical sets. **(b)** Reflexive Law of Congruence. **(c)** $\overline{AB} \cong \overline{BC} \cong \overline{CA}$ (triangle is equilateral). **13.** Given: $AC + CD + DA = AB + BE + AE \rightarrow (AB + BC) + CD + DA = AB + BE + (AD + DE) \rightarrow BC + CD = BE + DE \rightarrow CD = BE$. **15.** Six: $ABC \leftrightarrow ABC$, ACB, BAC, BCA, CAB, CBA; only one: $ABC \leftrightarrow ABC$. **17.** If $A \neq C$ and $A \neq D$, then C–A–D, and $\overline{CD} = \overline{CA} \cup \overline{AD} \neq \overline{CA}$ (the set $\overline{AD}$ is nonempty). Hence $\overline{CA} \subset \overline{CD}$ (strict inclusion). Similarly, $\overline{AD} \subset \overline{CD}$. $B \in \overline{CD} \rightarrow B \in \overline{CA}$ or $\overline{AD}$. If $B \in \overline{CA}$ then $\overline{AB} \subseteq \overline{CA} \subset \overline{CD} \rightarrow\leftarrow$ If $B \in \overline{AD}$ then $\overline{AB} \subseteq \overline{AD} \subset \overline{CD} \rightarrow\leftarrow$ $\therefore$ $A = C$ or $A = D$. Assume $A = C$; argument similar to preceding shows that $B = D$. If $A = D$ then it follows that $B = C$.

SECTION 3.2

1. $AB^* = BC^* = 6$, $AC^* = 12$, $XY^* = YZ^* = XZ^* = 6$; yes, but not congruent. **3. (a)** $AB^* = 2$, $BC^* = 1 + \sqrt{3} = AC^*$. **(b)** 60. **(c)** Isosceles, equiangular. **5. (a)** Slope $\overleftrightarrow{AC} = (9 - 0) / (-3 - 0) = -3$; Slope $\overleftrightarrow{BC} = (4 - 0) / (12 - 0) = \frac{1}{3} \rightarrow \overleftrightarrow{AC} \perp \overleftrightarrow{BC}$. **(b)** $a = 16$, $b = 12$, $c = 20$. **(c)** $16^2 + 12^2 = 256 + 144 = 400 = 20^2$. $\therefore$ this right triangle satisfies the Pythagorean relation for both Euclidean and Taxicab Metrics. **7.** 18 blocks. **9.** Yes. **11. (a)** $R(1.5, 8.5)$. **(b)** $\sqrt{72.5} \approx 8.515$.

SECTION 3.3

1. (a) $\overline{UV} \cong \overline{UY}$, $\overline{VW} \cong \overline{YX}$, $\overline{UW} \cong \overline{UX}$, $\overline{VX} \cong \overline{WT}$. **(b)** $\overline{RT} \cong \overline{WS}$, $\overline{TM} \cong \overline{MW}$, $\overline{RM} \cong \overline{MS}$. **(c)** $\overline{AC} \cong \overline{CB}$, $\overline{DE} \cong \overline{DF}$. **3.** $x = 5$ (ASA); $y = 6.7$ (ASA or SAS). **5.** $x = 23$ (Isosceles Triangle Theorem); $y = z$ (ASA); $y = z = 90$ (Theorem 3, Section 2.5). **7. (a)** Reflexive Law for Congruence. **(b)** ASA. **(c)** CPCF. **(d)** Definition of congruence. **9.** S and T are both equidistant from Q and R, so $\overleftrightarrow{ST}$ is the perpendicular bisector of $\overline{QR}$. **11.** Construction produces points B and C on sides of angle such that $AB = AC$ and D such that $BD = DC \rightarrow \triangle ABD \cong \triangle ACD$ by SSS $\rightarrow \angle BAD \cong \angle DAC$. **13.** Locus is perpendicular bisector of $\overline{BC}$. (*Proof:* By SAS $\triangle AXC \cong \triangle ABY \rightarrow \angle ACX \cong \angle ABY \rightarrow \angle XCB \cong \angle YBC$ or, by Isosceles Triangle Theorem, $PB = PC$.) **15.** $\overline{PA} \cong \overline{PB} \rightarrow \angle A \cong \angle B$ (Theorem 2); $\overline{AM} \cong \overline{MB}$. $\therefore$ $\triangle PAM \cong \triangle PBM$ (SAS) $\rightarrow \angle AMP \cong \angle BMP$ (CPCF) and $\overline{PM} \perp \overline{AB}$ by Theorem 3, Section 2.5. **19.** If, as shown in figure, D and E trisect segment $\overline{BC}$ and if angles at A were actually of measure 30 each, locate F on $\overline{AC}$ such that $m\angle ADF = 75 = m\angle ADE$. By ASA $\triangle ADF \cong \triangle ADE$ and $DF = DE = DC$. Then $\triangle DCF$ is isosceles. But the angles of $\triangle DCF$ at F and C are of measure 105 and 45 $\rightarrow\leftarrow$ **21. (a)** Construct perpendicular bisector of segment $\overline{AB}$ and choose C any point on it. Then $\triangle CAB$ will be isosceles triangle with base $\overline{AB}$. **(b)** Construct three rays from point O forming three angles of measure 120; on each ray select points A, B, C at same distance from $O \rightarrow \triangle ABC$ is equilateral. **23.** By Crossbar Theorem A–D–$B \rightarrow E$–F–B and C–F–$D \rightarrow C$–E–A; $\angle BFD \cong \angle CFE$ (Vertical Pair Theorem) $\rightarrow \triangle BFD \cong \triangle CFE$ (SAS) $\rightarrow m\angle ADC = 180 - m\angle BDC = 180 - m\angle BEC = m\angle AEB$, and $\angle ABE \cong \angle ACD$ (CPCF), $\overline{BE} \cong \overline{CD} \rightarrow \triangle ACD \cong \triangle ABE$(ASA). $\therefore$ $AD = AE$. **25.** Case (1): If $AB + AC = BC$ then $BC > BA$, so construct D on $\overline{BC}$ so that

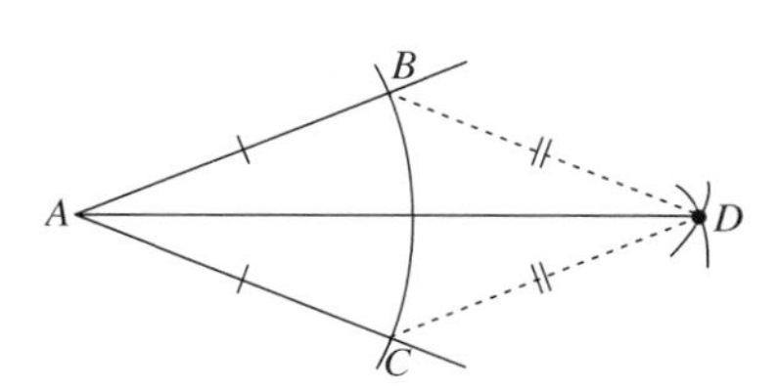

Problem 11

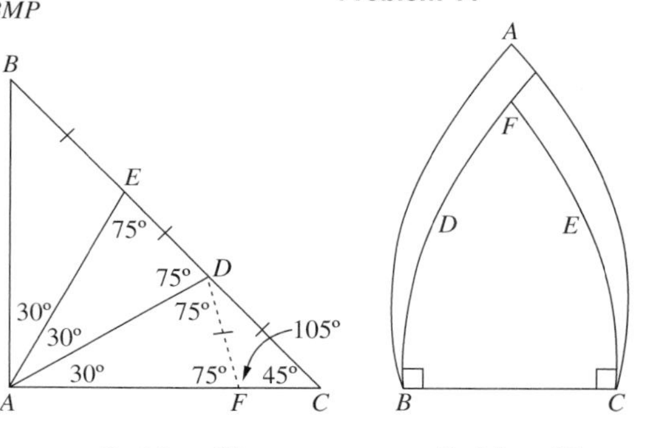

Problem 19 **Problem 25**

$BD = BA$; B–D–$C \rightarrow DC = BC - BD = BC - BA = AC \rightarrow \triangle ABD$, $\triangle ACD$ are isosceles $\rightarrow m\angle BAD = m\angle ADB$, $m\angle CAD = m\angle ADC$; B–D–$C \rightarrow \overrightarrow{AB}$–$\overrightarrow{AD}$–$\overrightarrow{AC} \rightarrow m\angle BAC = m\angle BAD + m\angle DAC = m\angle ADB + m\angle ADC = 180 \rightarrow\leftarrow$ $\therefore AB + AC \neq BC$. Case (2): (See figure.) First show isosceles triangles have acute base angles: If $AB = AC$ and $\angle B$, $\angle C$ are obtuse angles, construct perpendicular rays $\overrightarrow{BD}$ and $\overrightarrow{CE}$ on same side of $\overleftrightarrow{BC}$; they will meet at F by Crossbar Theorem $\rightarrow\leftarrow$ (Problem 24). Now suppose $AB + AC < BC$; construct point D on $\overline{BC}$ with $BD = BA$; $DC = BC - AB > AC \rightarrow E$ with $CE = CD$ and C–A–E exists $\rightarrow \overrightarrow{DE}$–$\overrightarrow{DA}$–$\overrightarrow{DC} \rightarrow m\angle EDC > m\angle ADC = 180 - m\angle ADB > 90 \rightarrow\leftarrow$ $\therefore AB + AC \geq BC$.

SECTION 3.4

1. (2) and (3). **3.** (1) and (3) **5.** $m\angle K < 120$ using exterior angle at L. **7.** $50 < y < 110$. **9.** $90 = m\angle BCD >$ both $m\angle A$, $m\angle B \rightarrow m\angle A < 90$, $m\angle B < 90$. **11.** In figure for Problem 10, suppose $m\angle ACD = 160 = m\angle BCA'$ where A–C–A' (Vertical Pair Theorem); then $m\angle ACB = 20$ by Linear Pair Axiom $\rightarrow$ exterior angles at A and $B > 20$ (Exterior Angle Inequality). **13.** $0 < x < y < z < 90$. **15.** In Figure 3.33, with $\angle A$, $\angle B$ in any $\triangle ABC$, extend $\overline{BA}$ to D such that B–A–$D \rightarrow m\angle 3 > m\angle 2$, so $180 = m\angle 1 + m\angle 3 > m\angle 1 + m\angle 2 \rightarrow m\angle A + m\angle B < 180$. **17.** If K–T–R as in figure for Problem 17, $m\angle WTK > 90 \rightarrow\leftarrow$ (by hypothesis, angles at K and T acute). **19.** $m\angle AMQ > m\angle PAM \rightarrow 2m\angle 1 > 2m\angle 2 \rightarrow m\angle 1 > m\angle 2$. **21.** By result of Problem 20, $x_2 \leq x_1/2$, $x_3 \leq x_2/2$, $x_4 \leq x_3/2, \ldots \rightarrow x_4 \leq (x_2/2)/2 = x_2/4 = (x_1/2)/4 = x_1/8$. In general, $x_n \leq x_1/2^{n-1}$; since $x_1/2^{n-1}$ can be made arbitrarily small by taking n large enough, there exists n such that $m\angle AC_nB = x_n < \epsilon$.

SECTION 3.5

1. $z < y < x$ (Scalene Inequality). **3.** Since other two angles are acute by Exterior Angle Theorem, Scalene Inequality implies side opposite right angle is greater side. **5.** $m\angle B < m\angle DCB < m\angle ACB < m\angle ADC$; $m\angle B < m\angle A = m\angle ACD < m\angle ADC = m\angle BDC$. $\therefore$ $\angle B$ is the smallest, $\angle ADC$ or $\angle CDB$ is largest. **7.** $\angle C$. (*Proof:* Locate D on line containing segment of length 4 such that $CD = 5$; side opposite C in new triangle > 3, so $m\angle C > m\angle B > m\angle A$ by SAS Inequality.) **11. (a)** $y = 31$ by CPCF ($\triangle MCE \cong \triangle MBA$ by SAS); $AE < EC + CA$ by Triangle Inequality, or $2x < y + 35 = 31 + 35 = 66$. **(b)** $AM = x < \frac{1}{2}(66) = 33$. **(c)** $AE + EC > AC \rightarrow 2x + 31 > 35 \rightarrow x > 2$. **13.** By construction indicated in figure for this problem (valid because $m\angle AME = 180 - m\angle AML > 90 > m\angle E$ and $AE > AM$), $m\angle MFE = 180 - m\angle AFM > 90$ so $ME > MF$; since $ME = ML$, $ML > MF \rightarrow m\angle FAM < m\angle MAL$ (SAS Inequality). $\angle DAL \cong \angle MAE$ by CPCF. **15.** Suppose $\triangle ABC$, $\triangle XYZ$ have $AB = XY$, $BC = YZ$, and $AC = XZ$. If $m\angle A > m\angle X$ then by the SAS Inequality, $BC > YZ \rightarrow\leftarrow$ $\therefore m\angle A = m\angle X$ and $\triangle ABC \cong \triangle XYZ$ by SAS. **17.** If $k_1 = k_2 = m\angle BAC + m\angle B + m\angle C$, then $k_1 + k_2 = k_2 + 180 \rightarrow k_1 = 180$.

SECTION 3.6

3. (a) $\triangle BDA \cong \triangle BDC$ by AAS. **(b)** $\triangle BFP \cong \triangle BFQ$ by HA. **(c)** $\triangle BGX \cong \triangle BGY$ by HL. **5.** Given: In right triangles $\triangle ABC$ and $\triangle XYZ$, $AB = XY$ and $BC = YZ$. Construct W on ray $\overrightarrow{XZ}$ so that X–Z–W and $ZW = AC$; $\angle YZW$ is right angle $\rightarrow \angle YZW \cong \angle BCA \rightarrow \triangle YZW \cong \triangle BCA$ (SAS) $\rightarrow YW = AB = YX \rightarrow \triangle XYW$ is isosceles, with $\angle X \cong \angle W \cong \angle A \rightarrow \triangle ABC \cong \triangle XYZ$ (AAS). **7.** No; location of P not unique. **11.** Drop perpendiculars $\overline{ID}$, $\overline{IE}$ to sides (figure) $\rightarrow \triangle ADI \cong \triangle AEI$ by HA ($\angle DAI \cong \angle IAE$ and $AI = AI$) $\rightarrow ID = IE$. Conversely, if $ID = IE$, figure shows how to get a contradiction if $I \notin$ Interior $\angle BAC \rightarrow \overrightarrow{AB}$–$\overrightarrow{AI}$–$\overrightarrow{AC}$; HL $\rightarrow \triangle ADI \cong \triangle AEI$(HL) $\rightarrow$ $\angle DAI \cong \angle IAE \rightarrow I$ lies on bisector of $\angle BAC$. **13.** $m\angle ABC > m\angle ACB \rightarrow m\angle DBC = \frac{1}{2}m\angle ABC \geq \frac{1}{2}m\angle ACB = m\angle ECB \rightarrow$ $m\angle FBC = m\angle FBD + m\angle DBC > m\angle FCB \rightarrow CF > BF$. Construct segment $\overline{CF'} \cong \overline{BF}$ on $\overline{CF}$ and $\angle CF'H \cong \angle BFD \rightarrow \overrightarrow{F'H}$ does not meet $\overline{FG}$ by Exterior Angle Inequality $\rightarrow C$–H–G (Postulate of Pasch) $\rightarrow \triangle FBD \cong \triangle F'CH$ (ASA) $\rightarrow BD = CH < CG < CE \rightarrow\leftarrow$ $\therefore$ $AC \leq AB$. Similarly, $AB \leq AC \rightarrow AB = AC$.

Problem 11

SECTION 3.7

1. Diagonals $\overline{AC}$ and $\overline{BD}$ do not meet. **3. (a)** True. **(b)** False. **5.** $(m\angle P + m\angle 1 + m\angle 3) + (m\angle 4 + m\angle 2 + m\angle R) \le 360$ (Saccheri–Legendre Theorem) $\to m\angle P + 90 + 150 + 50 \le 360$ **7.** *If* $\diamond ABCD$, $\diamond XYZW$ *are convex quadrilaterals with* $\overline{AB} \cong \overline{XY}$, $\angle B \cong \angle Y$, $\overline{BC} \cong \overline{YZ}$, $\angle C \cong \angle Z$, *and* $\angle D \cong \angle W$, *then* $\diamond ABCD \cong \diamond XYZW$. **9. (a)** By SASAS, $\diamond ABCD \cong \diamond EBCF \to m\angle E = m\angle A = 90 = m\angle D = m\angle F$. $\therefore \diamond BEFC, \diamond AEFD$ are rectangles. **(b)** To prove: Angles at B, C are right angles. $\overline{AD} \cong \overline{EF} \to \diamond ABCD \cong \diamond EBCF$ by SASAS $\to \angle ABC, \angle BCD, \angle CBE, \angle BCF$ are right angles (Theorem 3, Section 2.5). **11.** To prove: $m\angle C = 90$ (see figure). $\overline{AC} \cong \overline{BD}$ (corollary of Theorem 2), $\overline{CD} \cong \overline{AB}$ (given) $\to \triangle DCB \cong \triangle ABC$ (SSS) $\to \angle C \cong \angle B =$ right angle. **13.** $\diamond ABCD \cong \diamond EGHF$ (ASASA) $\to AD = EF$. **15.** Given: $\diamond ABCD$, $\diamond XYZW$ with SASSS hypothesis (as marked in figure). $\triangle ABC \cong \triangle XYZ$ (SAS) $\to \angle 1 \cong \angle 3$, $\overline{AC} \cong \overline{XZ}$, $\angle 5 \cong \angle 7 \to \triangle ACD \cong \triangle XZW$ (SSS) $\to \angle 2 \cong \angle 4$, $\angle D \cong \angle W$, $\angle 6 \cong \angle 8$; $\overrightarrow{AB}$–$\overrightarrow{AC}$–$\overrightarrow{AD}$ and $\overrightarrow{XY}$–$\overrightarrow{XZ}$–$\overrightarrow{XW} \to m\angle BAD = m\angle 1 + m\angle 2 = m\angle 3 + m\angle 4 = m\angle YXW$; similarly $m\angle BCD = m\angle YZW$. **17. (a)** $\triangle BMB' \cong \triangle AMQ$, $\triangle CNC' \cong \triangle ANQ$ (HA) $\to m\angle 1 = m\angle 3$, $m\angle 2 = m\angle 4$, $B'M = MQ$, $QN = NC'$. $\therefore m\angle BAC + m\angle ABC + m\angle BCA = (m\angle 1 + m\angle 2) + (m\angle 4 + m\angle BCA) = x + y = 2x$. **(b)** By Saccheri–Legendre Theorem, Angle-sum $\triangle ABC = 2x \le 180 \to x \le 90$. **(c)** $B'C' = B'M + MQ + QN + NC' = 2MQ + 2QN = 2MN$. **19.** Parallelogram with unit sides and angle of measure 60, and unit square. **21.** $m\angle 2 + m\angle 3 + m\angle B \le 180 \to m\angle 2 + m\angle 3 \le 90 = m\angle 1 + m\angle 2 \to m\angle 3 \le m\angle 1 \to AB \le DC$ (SAS Inequality in $\triangle DAC$, $\triangle ABC$). **23.** False: square with unit sides, and rectangle with two opposite sides of unit length.

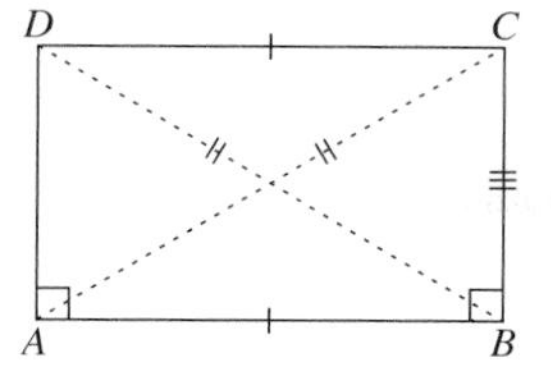

Problem 11

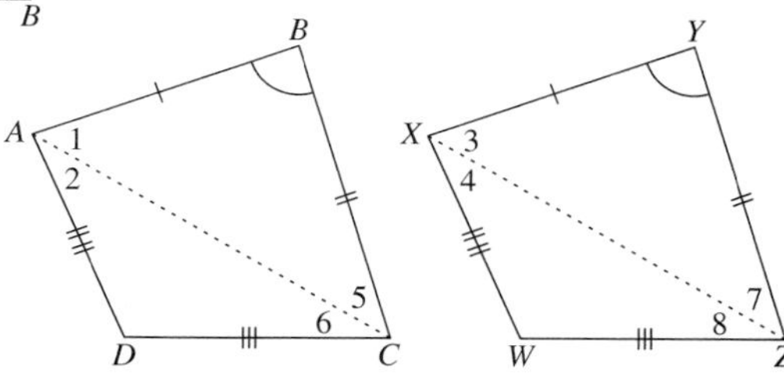

Problem 15

SECTION 3.8

1. Center is equidistant from end points of chord $\to$ it lies on perpendicular bisector (Theorem 2, Section 3.3). **3.** (For minor arcs only) $m\widehat{AB} = m\widehat{CD} \to m\angle AOB = m\angle COD \to \triangle AOB \cong \triangle COD$ (SAS) $\to \overline{AB} \cong \overline{CD}$ (CPCF). **5.** False: in coordinate plane, take circle of radius 5 with equation $x^2 + y^2 = 25$, and consider $\overline{AB}$, $\overline{CD}$ with end points $A(4, 3)$, $B(4, -3)$, $C(2, \sqrt{21})$, $D(2, -\sqrt{21})$; $AB = 6$ but $CD = 2\sqrt{21} \approx 9.17 \ne 2AB$. **11.** (See figure; O = center of circle.) By result of Problem 3, $m\widehat{AB} = m\widehat{CD} \to \angle AOB \cong \angle DOC \to \triangle AOB \cong \triangle DOC \to \angle BAO \cong \angle CDO$; $OA = OD \to \angle OAD \cong \angle ODA \to \angle BAD \cong \angle CDA$. **13.** $RM = RN$, $PS = SQ$, $RS = RS$, $\overrightarrow{RS}$ bisects angles at R and $S \to \diamond RMPS \cong \diamond RNQS$ (SASAS). $\overline{MP} \cong \overline{NQ} \to m\widehat{MP} = m\widehat{NQ}$. **15.** Let $\overleftrightarrow{CD}$ meet $\overline{AB}$ at E; $EA = EC = EB \to AE = EB$. **17.** Line passes through (0, 1), which is rational point, 1 unit from center of circle, having radius 5, so is an interior point. But line meets circle in Euclidean plane at $(\pm 2\sqrt{6}, 1)$ which is not rational. **19.** Let x_0 be any positive real and $\epsilon > 0$. Choose $\delta = \epsilon$. If $|x = x_0| < \delta$ then $PP_0 < \delta$, and $|d(x) - d(x_0)| = |OP - OP_0| = \pm(OP - OP_0)$. By the triangle inequality, $OP - OP_0 \le PP_0$, $OP_0 - OP \le PP_0$ by result of Problem 8, Section 3.5 $\to \pm(OP - OP_0) \le PP_0$ or $|OP - OP_0| \le PP_0 \to |d(x) - d(x_0)| \le PP_0 < \delta = \epsilon$. That is, for $|x - x_0| < \delta$, $|d(x) - d(x_0)| < \epsilon$ and $d(x)$ is continuous. **21.** Let x_0 be any positive real < 360 and $\epsilon > 0$. Construct central angle $\angle P_0O'Q$ such that $P_0Q < \epsilon$, and choose $\delta = m\widehat{P_0Q}$. (See figure.) To show possible, locate D, D' on tangent to circle O' at P_0 such that P_0 is the midpoint of $\overline{DD'}$ and $P_0D = \epsilon$, and draw $\overline{O'D}$ and $\overline{O'D'}$ cutting circle O' at Q, Q' (Secant Theorem). It follows that $P_0Q = P_0Q' < P_0D = \epsilon$. Now if $m\widehat{P_0P} < \delta$, then P falls on either arc $\widehat{Q'P_0}$ or $\widehat{P_0Q}$, and the ray $\overrightarrow{O'P}$ meets $\overline{D'D}$ at some point E with $P_0P < P_0E < \epsilon$ (Scalene Inequality, $\angle P_0PO'$ acute). Then by Triangle Inequality (as in solution to Problem 19), $|d(x) - d(x_0| = |OP - OP_0| = \pm(OP - OP_0) \le PP_0 < \epsilon$ and $d(x)$is continuous. Consider $d(0) = OA = OO' - O'A = d - r' < r$ (given: $d < r + r'$), and $d(180) = OB = OO' + O'B = d + r' > r$ $(r - r' < d)$. By Intermediate Value Theorem, there exists x between 0 and 180 such that $d(x) = r \to OP = r$, so P lies on both circles; elementary construction yields second point Q.

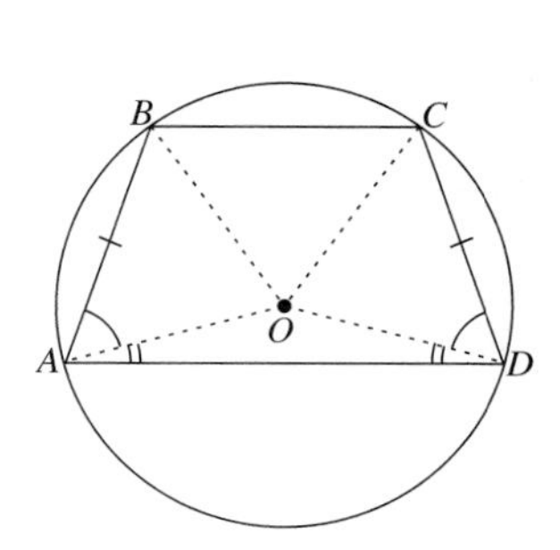

Problem 11

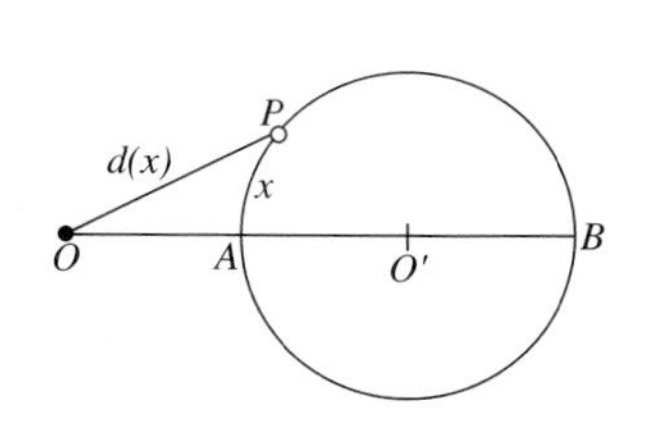

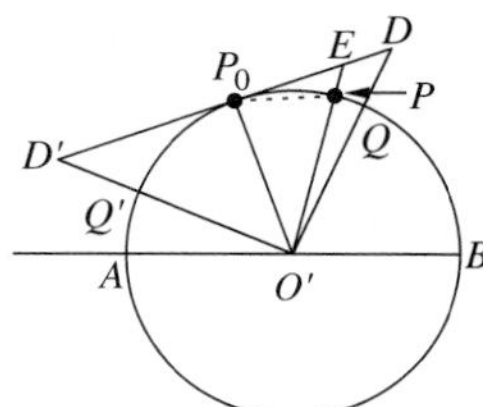

Problem 21

TESTING YOUR KNOWLEDGE (CHAPTER 3)

1. To show independence of SAS Postulate in absolute geometry. **2. (a)** $x = 91$. **(b)** SAS Postulate. **3. (a)** SAS. **(b)** CPCF. **(c)** Vertical Pair Theorem. **(d)** AAS. **(e)** $\overline{AE} \cong \overline{DE}$. **4.** Because $m\angle 1 > m\angle W$ by Exterior Angle Inequality and $m\angle W = m\angle UVW > m\angle 2$ by Angle Addition Postulate. **5. (a)** **6.** (1) $WT = WT$; (2) $\diamond PRTW \cong \diamond QSTW$ (SASAS); (3) $\therefore PR = QS$ (CPCF). **7. (a)** No because they are not congruent. **(b)** Yes. **(c)** $\overline{DE} \cong \overline{RS}$, $\overline{DF} \cong \overline{RU}$, $\angle F \cong \angle U$. **8. (a)** $a = 6$. **(b)** HL Congruence Criterion. **9.** $x = 85$, $y = 150$. **10.** Let tangents at A and B of circle O meet at P; $OA = OB$ and $PO = PO \to \triangle PAO \cong \triangle PBO$ (HL) $\to \angle APO \cong \angle BPO$. $\therefore \overrightarrow{PO}$ bisects $\angle APB$.

SECTION 4.1

1. Property F $\to$ top and middle lines parallel $\to m\angle 3 = m\angle 4$. **3.** $x = 47$, $y = 43$, $z = 47$. **5.** $\angle D$, $\angle B$ complementary (Corollary, Theorem 3) and $\angle AFD$ ($\cong \angle EFC$) complementary to $\angle C \cong \angle B \to \angle D \cong \angle AFD$ (Theorem 2, Section 2.5). **7.** $\overleftrightarrow{DF} \| \overleftrightarrow{AB}$. (*Proof:* $m\angle FDB = \frac{1}{2} m\angle EDB = \frac{1}{2}(m\angle A + m\angle B) = m\angle B$, and Property Z.) **9.** (See figure.) $\ell \| m$ and $\overline{AC}$, $\overline{BD}$ are perpendicular to $m \to m\angle 1 = m\angle 2$ (Property Z); $\overleftrightarrow{AC} \| \overleftrightarrow{BD}$ (Property C) $\to m\angle 3 = m\angle 4$ (Property Z) $\to \triangle ABD \cong \triangle DCA$ (ASA) $\to AC = BD$. **13.** $PR = PS \to m\angle 1 = m\angle 2$ and $PS = PQ \to m\angle 3 = m\angle 4$. Angle sum of $\triangle QRS = m\angle 1 + m\angle 2 + m\angle 3 + m\angle 4 = 2\,m\angle 2 + 2\,m\angle 3 = 2\,m\angle QSR = 180 \to m\angle QSR = 90$. **15.** Draw transversal $t \perp \ell$; because $\ell \| m$, by Corollary D $t \perp m$, $t \perp n \to t \perp$ both n, $\ell \to \ell \| n$ (Corollary C). **17.** (See figure.) $\triangle VWQ$ is equilateral, segment $\overline{QU}$ meets $\overline{PW}$ at R, $\overleftrightarrow{QU}$ is perpendicular bisector of $\overline{VW} \to \overrightarrow{UQ}$ bisects $\angle VUW$, and $RU = RW \to \triangle PRU \cong \triangle QRW$ (AAS) $\to UP = QW = VW$. **19.** $B'C' = 2MC$ established in Problem 17, Section 3.7, angle sum of $\triangle ABC = 180 = 2x$, so $x = 90$ and $m\angle C = m\angle B = 90 \to \diamond BCC'B'$ is rectangle $\to \overleftrightarrow{B'C'} \| \overleftrightarrow{BC}$. By argument of Problem 9, $B'C' = BC \to 2MN = BC \to MN = \frac{1}{2}BC$. **21.** (See figure.) Suppose $\angle ABC$, $\angle XYZ$ have parallel sides ($\overleftrightarrow{AB} \| \overleftrightarrow{XY}$, $\overleftrightarrow{BC} \| \overleftrightarrow{YZ}$). Lines $\overleftrightarrow{BC}$ and $\overleftrightarrow{XY}$ are not parallel (Parallel Postulate and $\overleftrightarrow{YZ} \| \overleftrightarrow{BC}$) $\to$ they meet at some point D. In the case illustrated, by Property F $m\angle 1 = m\angle 2 = m\angle 3 = 180 - m\angle XYZ$. **23.** It must be proven that C, M, and D are collinear. $\triangle AMC \cong \triangle BMD$ by HA $\to \angle AMC \cong \angle BMD$; extend $\overline{CM}$ to D' such that C–M–$D' \to m\angle BMD' = m\angle AMC = m\angle BMD$ by Vertical Pair Theorem. Since a ray $\overrightarrow{MX}$ on D-side of $\overleftrightarrow{AB}$ such that $m\angle BMX = m\angle BMD$ is unique, $\overrightarrow{MD'} = \overrightarrow{MD}$ and $D \in$ line $\overleftrightarrow{CM}$.

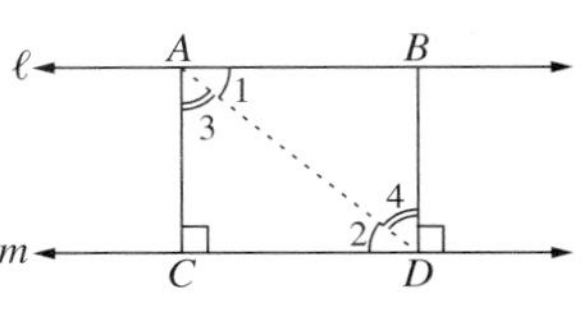

Problem 9

Problem 17

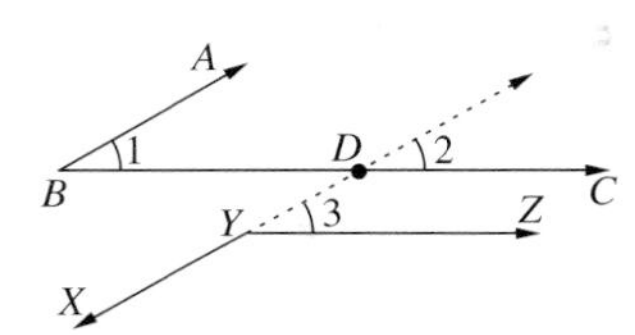

Problem 21

SECTION 4.2

1. $x = 6$, $y = 18$. **3.** $x = 30$, $z = 20$, $y = 56$. (Follow same procedure as Example 2: Draw $\overleftrightarrow{EF} \| \overleftrightarrow{DB}$ forming $\square DEFB \to FB = 42$, $EF = 10$, $FC = y - 42$.) **5. (a)** $KC = AB = 20 \to DK = 8 \to x = 4$, $y = 20$. **(b)** $EF = x + y = 24$. **(c)** Yes. **7.** Inscribe arc with center C on first line, and having radius AB; suppose arc intersects next five lines of paper at D, E, F, G, H. Segment $\overline{CH}$ will also meet those five lines, and by property of parallel projection, all segments formed on $\overline{CH}$ are congruent; because $AB = CD = CH$, each segment on $\overline{CH}$ has desired length. **9.** Opposite and adjacent sides are congruent because it is a parallelogram first, a rhombus second. Answer affirmative because equilateral (convex) quadrilateral is a parallelogram. **11.** Theorem 3: Side Splitting Theorem. **13.** $\overleftrightarrow{ED} \| \overleftrightarrow{AC}$ and $ED = AC$ given, so $\diamond ACDE$ is parallelogram $\to m\angle EDC + m\angle DCA = 180$; $BD = BC$ (CPCF) $\to m\angle BDC = m\angle BCD \to m\angle EDC = m\angle DCA = 90 \to \diamond ACDE$ is rectangle; $m\angle BCD = m\angle ACD - 30 = 60 = m\angle BDC \to \triangle BCD$ is equiangular, hence equilateral $\to AC = BC = CD \to \diamond ACDE$ is square. **15.** $AB = AD \to \angle ABD \cong \angle ADB$; $\overleftrightarrow{AD} \| \overleftrightarrow{BC} \to \angle ADB \cong \angle DBC$ (Z-Property) $\to \angle ABD \cong \angle DBC$ (Law of Transitivity) $\to \overrightarrow{BD}$ bisects $\angle ABC$. **17.** (See figure.) Suppose $\diamond ABCD$ is convex quadrilateral and $\overline{DC} \cong \overline{AB}$, $\overleftrightarrow{DC} \| \overleftrightarrow{AB}$. Draw diagonal $\overline{BD}$; $\angle 1 \cong \angle 2$ (Z-Property), $CD = AB$, $DB = DB \to \triangle CBD \cong \triangle ADB$ (SAS) $\to \angle 3 \cong \angle 4$ (CPCF) $\to \overleftrightarrow{AD} \| \overleftrightarrow{BC}$ (Z-Property). **19.** By construction, $\diamond ABCD$ is equilateral $\to \diamond ABCD$ is a rhombus with $\overleftrightarrow{AD} \| \ell$. **21.** $AE = 17\sqrt{2} \approx 24.041631 \to$ we want a midpoint on $\overline{EB}$ within 25.041631 of A; distances from A to nearest midpoints on either side of point E at each stage are (1) 22.5; (2) 22.5, 33.75; (3) 22.5, 28.125; (4) 22.5, 25.3125; (5) 23.90625, 25.3125; (6) 23.90625, 24.60938. Therefore, sixth bisection produces P on $\overline{EG}$ at a distance 24.60938 from A. **23. (a)** Suppose $\diamond ABCD$ is a trapezoid with $\overleftrightarrow{DC} \| \overleftrightarrow{AB}$, $\overline{AD} \cong \overline{BC}$; construct $\square AECD$ with $E \in \overline{AB}$, extended. $BC = AD = CE \to \triangle CEB$ is isosceles with $\angle CEB \cong \angle CBE$. If $E = B$ then $\diamond ABCD$ is a parallelogram, allowed only if $\diamond ABCD$ is a rectangle $\to \angle A \cong \angle B$; if A–E–B, $\angle DAB \cong \angle CEB \cong \angle CBE$ (F-Property) or $\angle A \cong \angle B$, and if A–B–E then $m\angle DAB = 180 - m\angle CEB = 180 - m\angle CBE = m\angle ABC$ (C-Property) or $\angle A \cong \angle B$. **(b)** Start with the same cases A–E–B, A–B–E and reverse the steps of proof in **(a)**. **(c)** $\triangle DAB \cong \triangle CBA$ by SAS $\to \overline{BD} \cong \overline{AC}$.

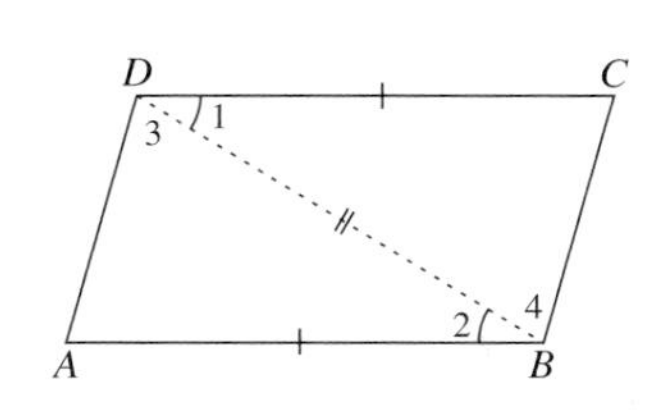

Problem 17

SECTION 4.3

1. $WX = \frac{1}{2}WY$, $WV = \frac{1}{2}WZ$ by definition of midpoint; $\angle XWV \cong \angle YWZ$ (Reflexive Law) $\to \triangle XWV \sim \triangle YWZ$ (SAS Similarity). **3. (a)** Neither $PQ/UV = QR/VW$ nor $PQ/VW = QR/UV \to$ triangles not similar. **(b)** $\triangle PQR \sim \triangle UWV$. **5.** 14 ($\triangle MCN \sim \triangle ACM$). **7. (a)** $\triangle ABC \sim \triangle YXZ$. **(b)** $k = \frac{1}{2}$. **(c)** $m\angle Y = 60$. **9.** $\angle P$ is right angle ($11^2 + 60^2 = 61^2$). **11.** Set $a/b = k = c/d \to a = bk$, $c = dk \to (a + c)/(b + d) = (bk + dk)/(b + d) = k = a/b$. **13.** $x = 9$. **15. (a)** As in Example 2, $AB^2 = BC \cdot BD = BC(BC + CD) = BC(BC + AB) = BC^2 + BC \cdot AB \to AB^2/BC^2 = 1 + AB/BC$ or $x^2 = 1 + x \to x = \tau = AB/BC$. **(b)** $m\angle A = m\angle D \to 2\theta + \phi = 180$, $\theta = 2\phi$ (Euclidean Exterior Angle Theorem) $\to 5\phi = 180$ or $\phi = 36$, $\theta = 72$. **(c)** If M is midpoint of $\overline{BC}$ then $\cos\theta = BM/AB = \frac{1}{2}(BC/AB) = \frac{1}{2}(1/\tau) = 1/(2\tau) \to \sec 72° = 2\tau$; if N is midpoint of $\overline{AD}$ then $\cos\phi = AN/AC = \frac{1}{2}(AD/AB) = \frac{1}{2}(AB/BC) = \tau/2 \to \cos 36° = \tau/2$. **17. (a)** By AA Similarity, $\triangle AEF \sim \triangle ABC \to AF/AC = AE/AB \to AF \cdot AB = AE \cdot AC$. **(b)** $AF/AC = AE/AB$ and $\angle BAE \cong \angle CAF \to \triangle ABE \sim \triangle ACF$ (SAS). **(c)** $AB/AC = AE/AF$ and $\angle FAE \cong \angle CAB \to \triangle AFE \sim \triangle ACB$. **21.** See figure for counterexample. **23.** In $\triangle ABC$, $\triangle XYZ$ suppose $AB = kXY$, $AC = kXZ$, $\angle A \cong \angle X$ with $k > 1$. Construct $\overline{AD} \cong \overline{XY}$ on $\overline{AB}$ and $\overline{AE} \cong \overline{XZ}$ on $\overline{AC}$; $\triangle ADE \cong \triangle XYZ$ by SAS. If $\overleftrightarrow{DE}$ is *not* parallel to $\overleftrightarrow{BC}$, locate E' on $\overline{AC}$ so that $\overleftrightarrow{DE'} \| \overleftrightarrow{BC} \to AD/AB = AE'/AC = AE/AC \to AE' = AE \to E' = E$. $\to\leftarrow$ $\therefore \overleftrightarrow{DE} \| \overleftrightarrow{BC}$ and $\angle ADE \cong \angle B \cong \angle Y \to \triangle ABC \sim \triangle XYZ$ by AA. **25. (a)** $AB = b - a = -(a - b) = -BA$. **(b)** $AB + BC = (b - a) + (c - b) = c - a = AC$. **27.** Stewart's Theorem applied to collinear points A, B, D, with C playing role of P, is $CA^2 \cdot BD + CB^2 \cdot DA + CD^2 \cdot AB + BD \cdot DA \cdot AB = 0$ or, by substitution of $AD = pAB = pc$, $DB = qAB = qc$, with $d = CD$, we have $b^2(-qc) + a^2(-pc) + d^2(c) + (-qc)(-pc)c = 0$, which reduces to $d^2 = pa^2 + qb^2 - pqc^2$.

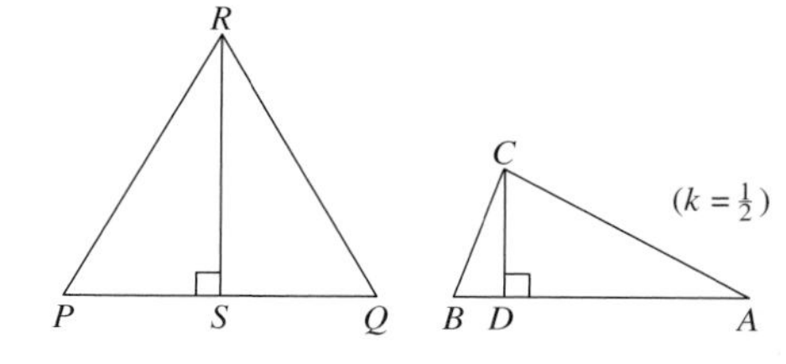

Problem 21

SECTION 4.4

1. (See figure.) **3.** (See figure.) **5.** (See figure.) **7.** $m\angle W = 45$; $m\angle S = m\angle T = 67.5$ **9.** Each point is common vertex of same regular polygons (dodecagon twice, equilateral triangle) having angle measures $150 + 150 + 60 = 360$. **11. (a)** Construction exists iff angle of measure $72/3 = 24$ iff 15-sided regular polygon can be constructed; $n = 15 = 3 \cdot 5$—a product of distinct Fermat primes. **(b)** If angle of measure 8 can be constructed then a 45-sided regular polygon can be constructed; $45 = 3^2 \cdot 5$, which is not of required form. **15.** (See figure.) With $x = MC$ and $\theta = m\angle MCB$, by Law of Sines, $\sin\theta / \sin 45° = \frac{1}{2} \to \sin\theta = \sqrt{2}/4 \to \cos\theta = \sqrt{1 - 2/16} = \sqrt{14}/4 \to \sin\phi = \sin(180° - 45° - \theta) = \sin(45° + \theta) = \sin 45° \cos\theta + \cos 45° \sin\theta = \sqrt{2}/2(\sin\theta + \cos\theta) = \frac{1}{4}(\sqrt{7} + 1)$; $x = x/1 = \sin\phi / \sin\theta = (\sqrt{7} + 1)/\sqrt{2}$, $y = x/\sqrt{2} = \frac{1}{2}(\sqrt{7} + 1)$, $z = \frac{1}{2}(\sqrt{7} - 1) \to MD = y + z = \sqrt{7} \to BD = \sqrt{8} \to m\angle BCD = 90$, $CE = 2y = \sqrt{7} + 1$. $\therefore m\angle ABC \approx 114$, $m\angle MDC \approx 66$, or $m\angle EDC \approx 132$.

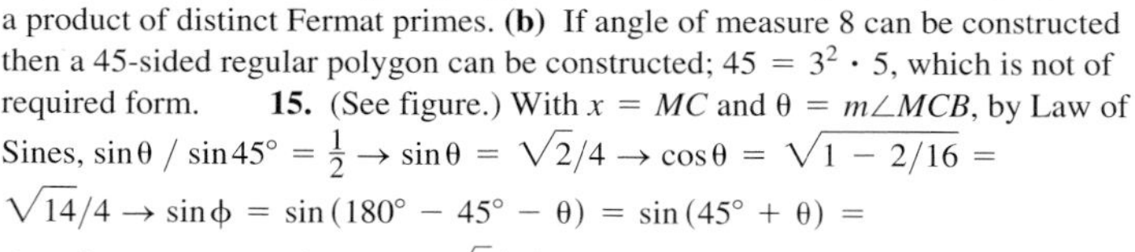

17. $BC = \sqrt{5} \to BF = BE = \frac{1}{2}(\sqrt{5} - 1) = 1/\tau \to \sin\theta = \frac{1}{2}BF/OB = \frac{1}{2}BE = 1/(2\tau) \to \theta = 18 \to m\angle BOF = 36$.

19. If sides of regular pentagon = 1, by Law of Cosines, each diagonal (= side of star) has length $\sqrt{1^2 + 1^2 - 2 \cdot 1 \cdot 1 \cdot \cos 108°} = \sqrt{2 + 2\cos 72°} = 2\sqrt{\frac{1}{2}(1 + \cos 72°)} = 2\cos 36° = \tau$. For angle measure θ of point of star, draw perpendicular from point of star to side of pentagon $\to \sin\frac{1}{2}\theta = \frac{1}{2}/\tau = 1/(2\tau) \to \csc\frac{1}{2}\theta = 2\tau \to \theta = 36$.

21. $n = 24, 30, 32, 34, 40, 48, 51, 60, 64, 68, 80, 85$, and 96.

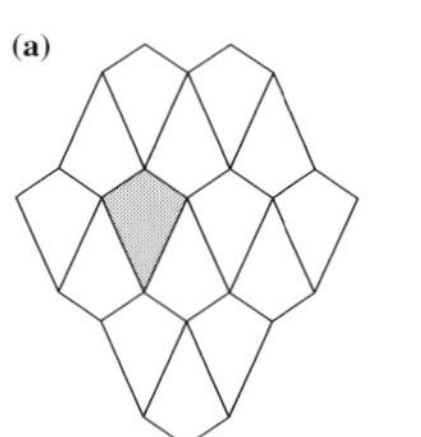

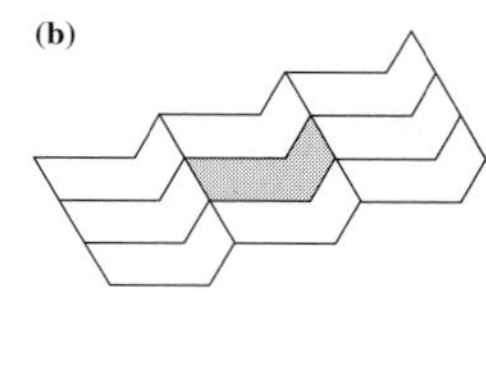

Problem 1

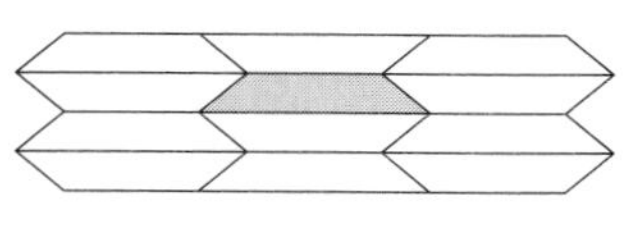

Problem 3

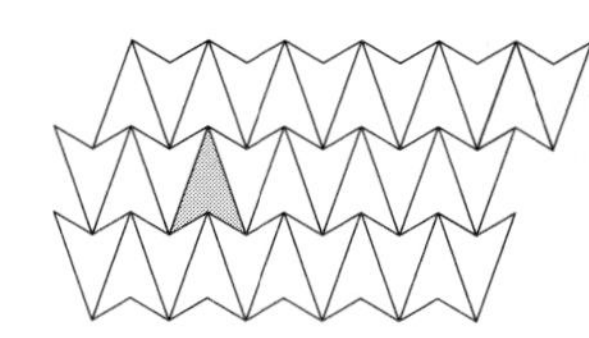

Problem 5

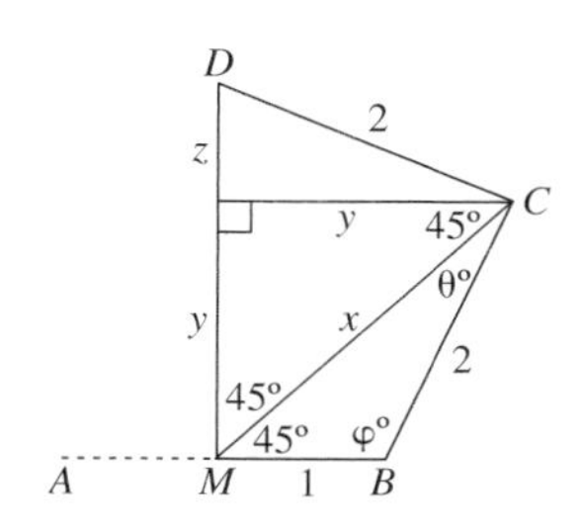

Problem 15

SECTION 4.5

1. (a) $m\angle CAB = 30, m\angle COB = 60, m\angle DAC = 15, m\angle DOC = 30$. **(b)** (1) $m\widehat{BE} = 180 - 20 = 160 \rightarrow m\angle BAE = 80$; (2) $2m\angle BAE + 20 = 180 \rightarrow m\angle BAE = 80$. **3.** 60. **5.** 8. **7. (a)** 30. **(b)** Congruent chords intercept arcs of equal measure. **(c)** $m\angle RPY = \frac{1}{2}m\widehat{RAY} = \frac{1}{2}\cdot 10\cdot 30 = 150$. **9.** $\sqrt{365}$. **11.** Keep measure of $\angle APB < 90$. **13.** 152, 208. **15.** $r = 10\sqrt{2}$ (P lies on $\overline{AB}$). **17.** $12 \sin 20°/\sin 75° \approx 4.249$ (Law of Sines). **19.** Diameter $\overline{AB}$ cuts circle into two semicircles of measure 180; if $m\widehat{AC} = x$ then $m\widehat{BC} = 180 - x \rightarrow m\angle CAT = 90 - m\angle BAC = 90 - \frac{1}{2}(180 - x) = \frac{1}{2}x = \frac{1}{2}m\widehat{AC}$. **21.** $RM = MA = MB = MC \rightarrow$ A, B, C, R are concyclic $\rightarrow R$ bisects arc $\widehat{ARB} \rightarrow m\angle BCR = \frac{1}{2}m\widehat{BR} = \frac{1}{2}m\widehat{RA} = m\angle RCA$. **23. (a)** With $\overline{OP}$ as cevian and $r = OA = OB$ as radius of circle, Power$(P) = OP^2 - r^2 = [(AP/AB)r^2 + (PB/AB)r^2 - (AP/AB)\cdot(PB/AB)\cdot AB^2] - r^2 = PA \cdot PB$. **(b)** $PA \cdot PB =$ Power$(P) = PC \cdot PD$. Theorem 3: Using result of Problem 22(a), $PC^2 =$ Power$(P) = PA \cdot PB$. **25.** $\triangle BOC \cong \triangle COD \rightarrow m\angle AOD = 3m\angle AOC$; $OE = OD \rightarrow m\angle E = m\angle ODE = \theta - \phi$. By Exterior Angle Theorem, $m\angle EAO = m\angle AOD + m\angle ODE = 3(90 - \theta) + \theta - \phi = 270 - 2\theta - \phi$. $\therefore m\angle EOA = 180 - m\angle E - m\angle EAO = \theta + 2\phi - 90$. **27. (a)** $m\angle BXC = m\angle XPB + m\angle PBX = A + \frac{1}{2}(180 - A) = 90 + \frac{1}{2}A =$ constant $\rightarrow X$ varies on circle having $\overline{BC}$ as chord. **(b)** $\triangle PXB$ is equilateral for all P. **29.** Locus is hyperbola (or line if $\mathbf{C}_1$ and $\mathbf{C}_2$ have equal radii).

SECTION 4.6

1. Area of **(a)** = 28.6667; area of **(b)** = 29.4647; area of **(c)** = 29.82333. **3.** 24 sq. units. **5.** Ratio of areas = ratio of AD/AB because altitudes are equal and areas = same constant multiple of AD, AB. **7.** Window designed by carpenter has area = 16 sq. ft. (rectangle) plus $4\pi/2 = 6.28$ sq. ft. (semicircle) = total of 22.28 sq. ft. **(a)** Window with 6 ft. base has area $6\cdot 1 + (\pi\cdot 3^2)/2 = 6 + 9\pi/2 \approx 20.14$ sq. ft. (so far, carpenter is correct). **(b)** If dimensions of rectangular part of window are $5 \times x$, then $10 + 2(x + 2.5) = 20 \rightarrow 2x + 5 = 10 \rightarrow x = 2.5 \rightarrow$ Area $= 5(2.5) + \pi(2.5^2)/2 \approx 22.32$ sq. ft., so carpenter's window does not have maximal area. **9.** Let x = base, y = height of rectangle for any window. Then $2x + 2(y + x/2) = 20 \rightarrow 3x + 2y = 20 \rightarrow y = 10 - 3x/2$. Total area of window $= xy + [\pi(x/2)^2]/2 = x(10 - 3x/2) + \pi x^2/8 \rightarrow A(x) = 10x - (3/2 - \pi/8)x^2$. The derivative is $A'(x) = 10 - (3 - \pi/4)x = 0 \rightarrow x \approx 4.52$ ft.; $y \approx 3.22$ ft. (maximal area ≈ 22.58 sq. ft.). **11.** $L = 2\pi R = 2\pi(r + w/2) \rightarrow wL = 2\pi rw + \pi w^2$; area of washer $= \pi(r + w)^2 - \pi r^2 = 2\pi rw + \pi w^2 = wL$. **13. (a)** 6.3 sq. ft. **(b)** 120 sq. ft. converts to $120/9 = 13.33$ sq. yd. **(c)** Volume of cube with side $s = s^3 \rightarrow$ volume of cube twice as large $= (2s)^3 = 8s^3$ (8 times the volume), which applies to any two similar figures $\rightarrow$ fish twice as long weighs eight times as much. **15.** $A = (3\sqrt{3}/2)s^2$ (hexagon); $A = 2(\sqrt{2} + 1)s^2$ (octagon). **17.** Two x-coordinates for points on left and right boundaries of region lying at distance y units from x-axis are $x = \sqrt{2y - y^2}$; $x = \frac{1}{2}[2y \pm \sqrt{4y^2 - 4(2y^2 - 2y)}] = y + \sqrt{2y - y^2}$. Difference $= y$, so area = that of right triangle = 2 sq. units, as shown in figure. **19.** (1) Base of T_1 = Area $\triangle QTV = \frac{1}{4}B$; Base of T_1' = Area $TVSU = \frac{1}{2}B \rightarrow$ Vol $T_1 = (\frac{1}{4}B)\cdot(\frac{1}{2}h) = \frac{1}{8}Bh$; Vol $T_1' = \frac{1}{2}(\frac{1}{2}B)\cdot(\frac{1}{2}h) = \frac{1}{8}Bh$. $\therefore$ Vol $P = \frac{1}{8}Bh + \frac{1}{8}Bh + 2$ Vol P_1. (2) Apply same analysis to subdivision of pyramid P_1 with base $B_1 = \frac{1}{4}B$ and height $h_1 = \frac{1}{2}h$: Vol $P = \frac{1}{4}Bh + \frac{1}{16}Bh + 4$ Vol P_2. (3) Similarly, Vol $P = \frac{1}{4}Bh + \frac{1}{16}Bh + \frac{1}{64}Bh + 8$ Vol $P_3 \rightarrow$ Vol $P = Bh(\frac{1}{4} + \frac{1}{16} + \frac{1}{64} + \cdots + (\frac{1}{4})^n + \cdots)$. Formula for geometric series: $r + r^2 + r^3 + \cdots + r^n + \cdots = r/(1 - r)$ if $0 < r < 1 \rightarrow$ Vol $P = Bh[\frac{1}{4}/(1 - \frac{1}{4})] = Bh\left(\frac{1}{4}\cdot\frac{4}{3}\right) = \frac{1}{3}Bh$. **21. (a)** Stage 1: $\frac{3}{2}$; stage 2: $\frac{15}{4}$ $(= \frac{3}{2} + \frac{9}{4})$; Stage 3: $\frac{57}{8}$ $(= \frac{3}{2} + \frac{9}{4} + \frac{27}{8})$. **(b)** Stage 1: $\sqrt{3}/16$; Stage 2: $\sqrt{3}/16 + 3\sqrt{3}/64 = 7\sqrt{3}/64$; Stage 3: $\sqrt{3}/16 + 3\sqrt{3}/64 + 9\sqrt{3}/256 = 37\sqrt{3}/256$. **(c)** Perimeter $= \frac{3}{2} + \frac{9}{4} + \frac{27}{4} + \cdots + (\frac{3}{2})^n + \cdots = \infty$; Area $= \sqrt{3}/16(1 + \frac{3}{4} + \frac{9}{16} + \cdots + (\frac{3}{4})^{n-1} + \cdots) = \sqrt{3}/4$; yes. **23.** Stage 1: $1 + 4\cdot(\frac{1}{4}) = 2$; Stage 2: $2 + 12\cdot(\frac{1}{16}) = 2\frac{3}{4} = 2.75$; Stage 3: $2\frac{3}{4} + 36\cdot(\frac{1}{64}) = 3.3125$. **25.** Stage 1: 13; Stage 2: $13^2 + 24 = 193$; Stage 3: $193\cdot 13 + 193\cdot 24 + 24 = 7{,}165$. **27.** At stage n, number of line segments in fractal is 2^{n+1}, each of length $1/2^n \rightarrow$ Perimeter $= 2^{n+1}/2^n = 2$.

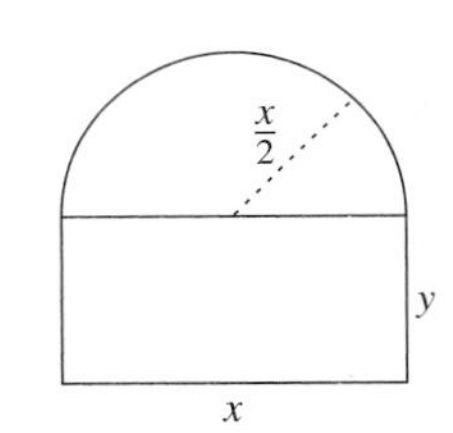

Problem 9

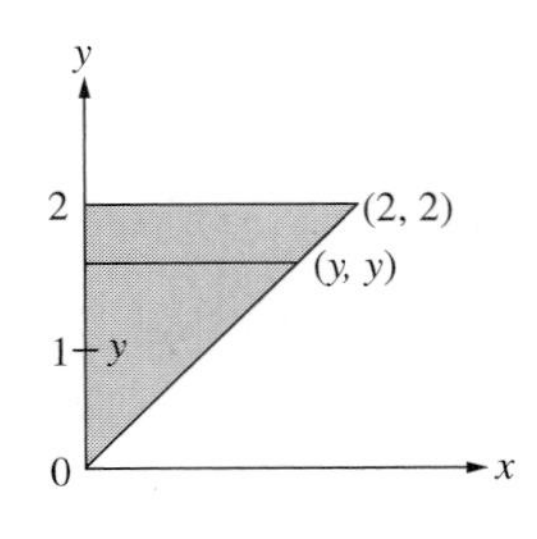

Problem 17

SECTION 4.7

1. (a) $\boldsymbol{u} = [3, 2]$, $\boldsymbol{v} = [6, -9]$, $\boldsymbol{w} = [3, 11]$. **(b)** $\boldsymbol{uv} = 3 \cdot 6 + 2 \cdot (-9) = 0$. **(c)** $\cos B = 1/\sqrt{10}$. **3.** If $AC = BD$ then $\sqrt{(b-2a)^2 + (c-0)^2} = \sqrt{(b-0)^2 + (0-c)^2} \to b^2 - 4ab + 4a^2 + c^2 = b^2 + c^2$ or $4a(-b+a) = 0 \to a = 0\ (a \neq b)$. $\therefore \diamond ABCD$ is a rectangle. **5.** Midpoints of sides are $L[\frac{1}{2}(a+b), 0]$, $M[\frac{1}{2}(b+d), \frac{1}{2}e]$, $N[\frac{1}{2}d, \frac{1}{2}(c+e)]$, $P(\frac{1}{2}a, \frac{1}{2}c) \to$ Slope $\overleftrightarrow{PL} = -c/b =$ Slope $\overleftrightarrow{MN}$, Slope $\overleftrightarrow{PN} = e/(d-a) =$ Slope $\overleftrightarrow{LM} \to \diamond LMNP$ is a parallelogram. **7. *Theorem:*** *For any two orthogonal vectors* $\boldsymbol{u}$ *and* $\boldsymbol{v}$, $\|\boldsymbol{u} + \boldsymbol{v}\|^2 = \|\boldsymbol{u}\|^2 + \|\boldsymbol{v}\|^2$. Proof: $(\boldsymbol{u} + \boldsymbol{v})^2 = (\boldsymbol{u} + \boldsymbol{v}) \cdot (\boldsymbol{u} + \boldsymbol{v}) = \boldsymbol{u} \cdot \boldsymbol{u} + \boldsymbol{u} \cdot \boldsymbol{v} + \boldsymbol{v} \cdot \boldsymbol{u} + \boldsymbol{v} \cdot \boldsymbol{v} = \boldsymbol{u}^2 + \boldsymbol{v}^2 = \|\boldsymbol{u}\|^2 + \|\boldsymbol{v}\|^2$ (because $\boldsymbol{u} \cdot \boldsymbol{v} = 0 = \boldsymbol{v} \cdot \boldsymbol{u}$. **9.** Plane's bearing is due north at $425/(3\frac{1}{8}) = 136$ mph, so $\boldsymbol{v} + \boldsymbol{w} = [r + x, s + y] = [0, 136] \to r + x = 0$ and $s + y = 136$. If $\boldsymbol{v}' = [r', s']$ is the plane's heading for return trip, then $\boldsymbol{v}' + \boldsymbol{w} = [0, -150]$. $\therefore r' + x = 0, s' + y = -150$. Also, magnitudes of $\boldsymbol{v}, \boldsymbol{v}'$ both equal the cruising speed $\to r^2 + s^2 = r'^2 + s'^2 = 145^2$ and $r = r' = -x \to s = \pm s' = -s'$ (evidently, $\boldsymbol{v} \neq \boldsymbol{v}'$) $\to$ second equation in s and y is $-s + y = -150$. Solution is $s = 143$, $y = -7 \to w = [x, y] = [\pm 24, -7] \to$ wind velocity $= 25$ mph and direction is $\approx 16.26°$ south of east, or south of west. **11.** $FE = \sqrt{(x-0)^2 + (y-4)^2} = \sqrt{x^2 + y^2 - 8y + 16} = \sqrt{3y^2 - 12 + y^2 - 8y + 16} = 2\,|y - 1| = 2EG$. If M is midpoint of $\overline{FE}$, then $ME = EG \to \triangle DEM \cong \triangle DEG \to \angle FDM \cong \angle MDE \cong \angle EDG$.

13. Geometric Interpretation: The sum of the squares of the lengths of the diagonals of a parallelogram equals the sum of the squares of the lengths of its sides. **(a) Vector proof:** $\|\boldsymbol{u} + \boldsymbol{v}\|^2 + \|\boldsymbol{u} - \boldsymbol{v}\|^2 = (\boldsymbol{u} + \boldsymbol{v})^2 + (\boldsymbol{u} - \boldsymbol{v})^2 = \boldsymbol{u}^2 + 2\boldsymbol{uv} + \boldsymbol{v}^2 + \boldsymbol{u}^2 - 2\boldsymbol{uv} + \boldsymbol{v}^2 = 2\boldsymbol{u}^2 + 2\boldsymbol{v}^2 = 2\|\boldsymbol{u}\|^2 + 2\|\boldsymbol{v}\|^2$. **(b) Synthetic proof** (using Pythagorean Theorem): In figure, let sides be of length a and b, diagonals c and d, and altitude $h \to c^2 = (a+u)^2 + h^2 = a^2 + 2au + (u^2 + h^2) = a^2 + 2au + b^2$; $d^2 = (a-u)^2 + h^2 = a^2 - 2au + (u^2 + h^2) = a^2 - 2au + b^2 \to c^2 + d^2 = 2a^2 + 2b^2$. **15.** Let G $= (x, y)$. $AG = \frac{2}{3}\boldsymbol{AL} = \frac{2}{3}(\boldsymbol{AB} + \boldsymbol{BL}) = \frac{2}{3}(\boldsymbol{AB} + \frac{1}{2}\boldsymbol{BC}) = \frac{2}{3}\boldsymbol{AB} + \frac{1}{3}\boldsymbol{BC} \to [x - a_1, y - a_2] = \frac{2}{3}[b_1 - a_1, b_2 - a_2] + \frac{1}{3}[c_1 - b_1, c_2 - b_2] \to x - a_1 = \frac{2}{3}(b_1 - a_1) + \frac{1}{3}(c_1 - b_1) \to x = (a_1 + b_1 + c_1)/3$; y coordinate similar. **17.** $A' = (\frac{1}{2}b + \frac{1}{2}\sqrt{3}c, \frac{1}{2}c + \frac{1}{2}\sqrt{3}\,b)$, $B' = (\frac{1}{2}a - \frac{1}{2}\sqrt{3}c, \frac{1}{2}c - \frac{1}{2}\sqrt{3}a)$, $C' = (\frac{1}{2}a + \frac{1}{2}b, -\frac{1}{2}\sqrt{3}b + \frac{1}{2}\sqrt{3}a)$; $AA'^2 = a^2 + b^2 + c^2 - ab + \sqrt{3}bc - \sqrt{3}ac = BB'^2 = CC'^2$. **19.** If $\diamond ABCD$ is as in Problem 5, then $X = (\frac{1}{2}(a+b), \frac{1}{2}(a-b))$, $Y = (\frac{1}{2}(b+d+e), \frac{1}{2}(b-d+e))$, $Z = (\frac{1}{2}(c+d-e), \frac{1}{2}(c+d+e))$, $W = (\frac{1}{2}(a-c), \frac{1}{2}(c-a))$. **(a)** $4XZ^2 = 4YW^2$, Slope $\overleftrightarrow{XZ} = -1/$Slope $\overleftrightarrow{YW}$. **(b)** By coordinate analysis, $M(\frac{1}{2}b, \frac{1}{2}c)$ is common midpoint of $\overline{XZ}, \overline{WY}$ when $d = b - a, e = c$.

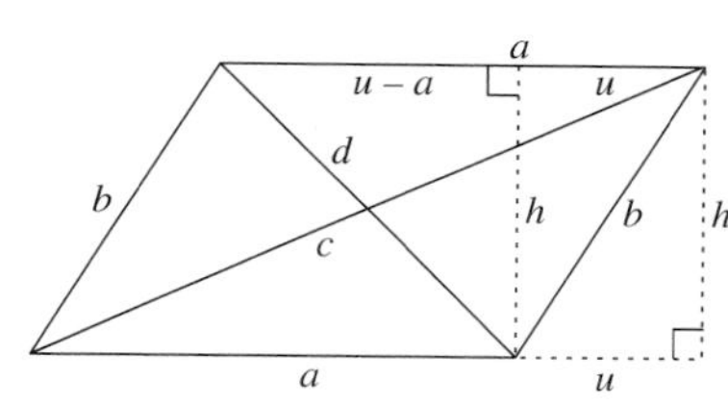

Problem 13

SECTION 4.8

1. 5/3. **3.** Lines meet 18 in. from M. **5.** $= \begin{bmatrix} ABC \\ DEF \end{bmatrix} = (a_1/a_2) \cdot (b_1/b_2) \cdot (c_1/c_2) = (c/b) \cdot (a/c) \cdot (b/a) = cab/bca = 1 \to \overline{AD}, \overline{BE}, \overline{CF}$ concurrent. **7.** By Menelaus' Theorem $\begin{bmatrix} ABD \\ TWX \end{bmatrix} = -1 = \begin{bmatrix} BCD \\ ZTY \end{bmatrix}$

$$\to \left(\frac{AX}{XB} \cdot \frac{BT}{TD} \cdot \frac{DW}{WA}\right) \cdot \left(\frac{BY}{YC} \cdot \frac{CZ}{ZD} \cdot \frac{DT}{TB}\right) = \frac{AX}{XB} \cdot \frac{DW}{WA} \cdot \frac{BY}{YC} \cdot \frac{CZ}{ZD} = (-1)^2 = 1.$$

9. (See figure.) $(BD/DC) \cdot (CE/EA) \cdot (AF/FB) = (y/z) \cdot (z/x) \cdot (x/y) = yzx/zxy = 1$. **11.** $1 = (AE/EC) \cdot (CD/DB) \cdot (BF/FA) = (AE/EC) \cdot 1 \cdot (BF/FA) \to AE/EC = AF/FB$; by converse of Side-Splitting Theorem, $\overleftrightarrow{FE} \| \overleftrightarrow{BC}$. **13.** In $\triangle PAB$, N, B', A' are collinear; in $\triangle PBC$, L, C', B' are collinear, and in $\triangle PCA$, M, A', C' are collinear. By Menelaus' Theorem,

$$\begin{bmatrix} PAB \\ NB'A' \end{bmatrix} = -1 = \begin{bmatrix} PBC \\ LC'B' \end{bmatrix} = \begin{bmatrix} PCA \\ MA'C' \end{bmatrix} \to \text{Product} = -1 = \frac{PA'}{A'A} \cdot \frac{AN}{NB} \cdot \frac{BB'}{B'P} \cdot \frac{PB'}{B'B} \cdot \frac{BL}{LC} \cdot \frac{CC'}{C'P} \cdot \frac{PC'}{C'C} \cdot \frac{CM}{MA} \cdot \frac{AA'}{A'P} = \frac{AN}{NB} \cdot \frac{BL}{LC} \cdot \frac{CM}{MA} = \begin{bmatrix} ABC \\ LMN \end{bmatrix} \to L, M, N \text{ collinear.}$$

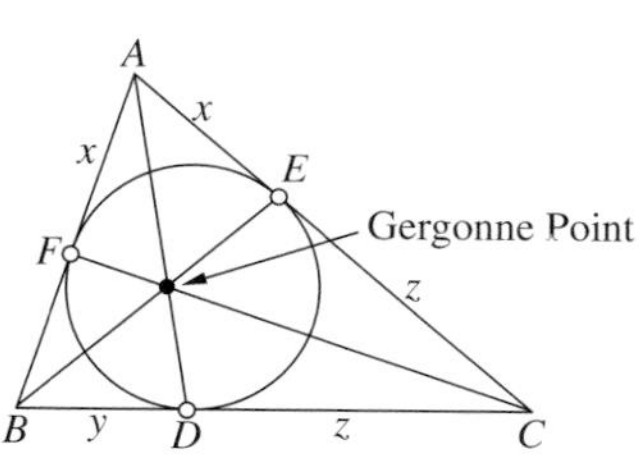

Problem 9

TESTING YOUR KNOWLEDGE (CHAPTER 4)

1. (a) 35. **(b)** 35. **(c)** 63. **2.** $x/4 = 17/5 \to x = 13.6$; $y/3 = 17/5 \to y = 10.2$. **3.** (See figure.) $\angle P \cong \angle W$, $\angle Q \cong \angle T$ (Property Z) $\to$ by AA Criterion $\triangle PQR \sim \triangle WTR$; by definition of similar triangles, $PR/RW = QR/RT$. **4.** $x = 16/25$. **5.** False; p must be a Fermat prime. **6.** (b) and (c). **7. (a)** 140. **(b)** $s = 6 \sin 20°$ or $s = 6 \cos 70°$. **8.** $25\pi/2 + 450 \approx 489.3$ sq. ft. **9.** Let triangle have vertices $A(a, 0)$, $B(b, 0)$, and $C(0, c)$, $a < b$, $c > 0$. Midpoint of $\overline{AB} = M(\frac{1}{2}a + \frac{1}{2}b, 0)$. Use distance formula

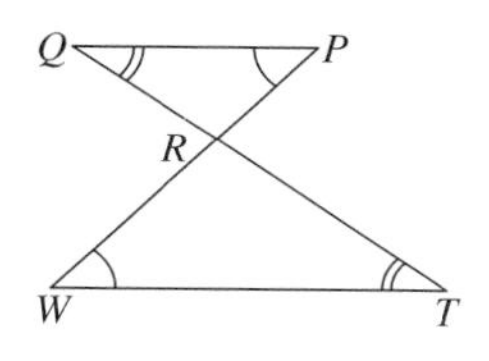

Problem 3

to find MC and set equal to $\frac{1}{2}(b - a) = MA$ to find conditions required on a, b, c. (Condition is $c^2 = -ab$, iff $\angle C$ is right angle.) **10.** (See figure.) By Z-Property, $\angle 1 \cong \angle 2$; by Vertical Pair Theorem, $\angle 3 \cong \angle 4$. Since $\overline{AD} \cong \overline{BC}$, $\triangle AMD \cong \triangle CMB$ (AAS). $\therefore AM = CM$ and $DM = BM$ (CPCF).

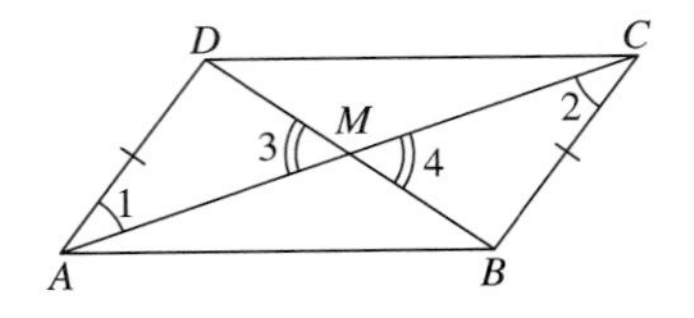

Problem 10

SECTION 5.1

1. (a) $f(1, -1) = (2, 3)$; $f(\frac{1}{2}, \frac{1}{2}) = (1, \frac{3}{2})$; $f(-3, 4) = (-6, -9)$. **(b)** $(2, -3)$ has no preimage; a preimage of $(4, 6)$ is $(2, 1)$; f maps any point of form $(2, y)$ to single point $(4, 6)$. **(c)** No (f is not onto and not one-to-one). **3. (b)** No: f is not onto and not one-to-one. **5.** By substitution, $x' = 1 - y^2$, $y' = \delta y^2 = \delta(1 - x')$ or, when $y \geq 0$, $y' = -x' + 1$ (slope $= -1$) and when $y < 0$, $y' = -(1 - x') = x' - 1$ (slope $= 1$), shown as dashed lines in figure for Problem 5. Yes, f is a transformation (but not linear). *Proof:* (One-to-one) If $(x_1, y_1) \neq (x_2, y_2)$ then $x_1' = x_1^3$, $x_2' = x_2^3 \rightarrow x_1^3 \neq x_2^3$, or if $x_1 = x_2$, $y_1 \neq y_2$, then $y_1' = \delta y^2 \neq \delta' y_2^2 = y'_2$. (Onto) Let (x, y) be a given point; if $(x^{1/3}, \delta\sqrt{|y|}$ is acted upon by f we get $x' = (x^{1/3})^3 = x$ and $y' = \delta(\delta\sqrt{|y|})^2 = \delta|y| = y \rightarrow f(x^{1/3}, \delta y) = (x, y)$. **7.** $y' = -3x'$. No. **9.** $y = 2x + 1$ maps to $(-4x' + 3y') = 2(3x' + 4y') + 1$ or $y' = -2x' - 1/5$ (slope $= -2$); $y = -\frac{1}{2}x + 3$ maps to $y' = \frac{1}{2}x' + 3/5$ (slope $\frac{1}{2}$) $\rightarrow$ image lines perpendicular. **11.** Unique fixed point is $(1.5, 1)$. **13.** Suppose image lines ℓ', m' are perpendicular, but $f^{-1}(\ell') = \ell$, $f^{-1}(m') = m$ are not (as shown in figure). Let n be perpendicular at $P = \ell \cap m$; f preserves perpendicularity, so image of n is ℓ'. It remains to show this is impossible (intuitively clear because f is one to one). Suppose $A \in \ell$, $A \neq P$. Then A' lies on ℓ' and by one-to-one property, $A' \neq P'$. If $B \in n$, $B \neq P$, then f maps B to B' on ℓ' $\rightarrow\leftarrow$ (result of Example 2). **15.** Let $\overline{AC}$ be any segment, and M its midpoint. Construct $\square ABCD$ having $\overline{AC}$ as diagonal; $\overline{BD}$ meets $\overline{AC}$ at M. By result of Problem 14, f maps $\square ABCD$ to $\square A'B'C'D'$, with diagonals $\overline{A'C'}$ and $\overline{B'D'}$ meeting at the midpoint N of $\overline{A'C'}$, $\overline{B'D'}$. But $f(M)$, lying on $\overleftrightarrow{AC}$ and $\overleftrightarrow{BD}$, maps to a point M' on both $\overleftrightarrow{A'C'}$, $\overleftrightarrow{B'D'}$, that is, $M' = N =$ midpoint of $\overline{A'C'}$. **17.** If $A' = A$, $B' = B$, then by result of Problem 15, midpoint M of $\overline{AB}$ is midpoint of $\overline{A'B'} = \overline{AB} \rightarrow M' = M$. All midpoints formed from previous midpoints together with A, B are fixed points also. Arbitrary point C of $\overline{AB}$ is the limit of a sequence of such midpoints, so by continuity, since each of these midpoints is a fixed point, so is C.

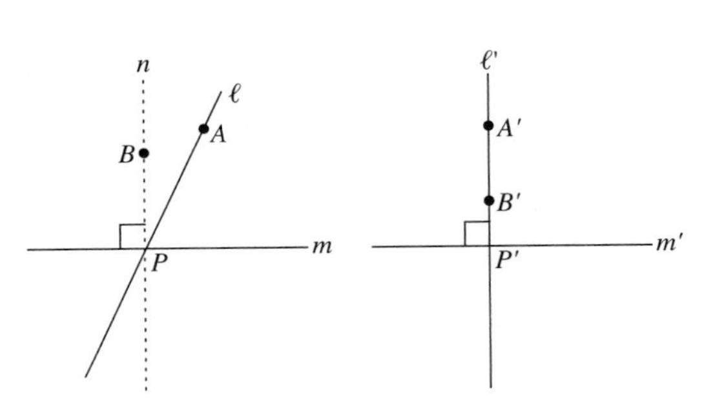

Problem 13

SECTION 5.2

1. M, X. **3.** By definition ℓ is perpendicular bisector of $\overline{CD} \rightarrow Q$ equidistant from $C, D \rightarrow CQ = DQ$. **5.** $y = 11 - 5x$. **7.** gf given by $x'' = 6x + 5y$, $y'' = 10x + 6y$; for fg, rearrange primes, then substitute: $x'' = y' = 6x + 5y$, etc. ($fg = gf$). **9.** For s_ℓ^2, $s_\ell(P) = Q$ where ℓ is perpendicular bisector of $\overline{PQ} \rightarrow s_\ell(Q) = P \rightarrow s_\ell^2(P) = s_\ell(s_\ell(P)) = s_\ell(Q) = P$. **11. (a)** $AP + PB = \sqrt{(1 - t)^2 + (5 - 0)^2} + \sqrt{(t - 8)^2 + (0 - 3)^2}$ or $f(t) = \sqrt{t^2 - 2t + 26} + \sqrt{t^2 - 16t + 73}$ where $1 \leq t \leq 8$; $f'(t) = (t - 1)/\sqrt{t^2 - 2t + 26} + (t - 8)/\sqrt{t^2 - 16t + 73} = 0 \rightarrow (t - 1)^2/(t^2 - 2t + 26) = (t - 8)^2/(t^2 - 16t + 73)$; since $(t - 1)^2 = t^2 - 2t + 1$, first two terms match those of denominator, so by division above becomes $1 - 25/(t^2 - 2t + 26) = 1 - 9/(t^2 - 16t + 73) \rightarrow 9(t^2 - 2t + 26) = 25(t^2 - 16t + 73) \rightarrow 16t^2 - 382t + 1591 = 0$. Roots are $t = 18.5$ (extraneous), $t = 5.375$. **(b)** Reflect $B(8, 3)$ in x-axis to $B'(8, -3)$ and intersect line $\overleftrightarrow{AB'}$ with x-axis: $\overleftrightarrow{AB'}$ has equation $y - 5 = \text{Slope} \cdot (x - 1) = -8(x - 1)/7$; set $y = 0 \rightarrow 8x = 43 \rightarrow x = 5.375$. **13. (a)** $(17, 0)$ or $(8.0, 0)$, approx. (See figure.) **(b)** $(0, 10.8)$ or $(0, 26.8)$ (y-coordinates not exact).

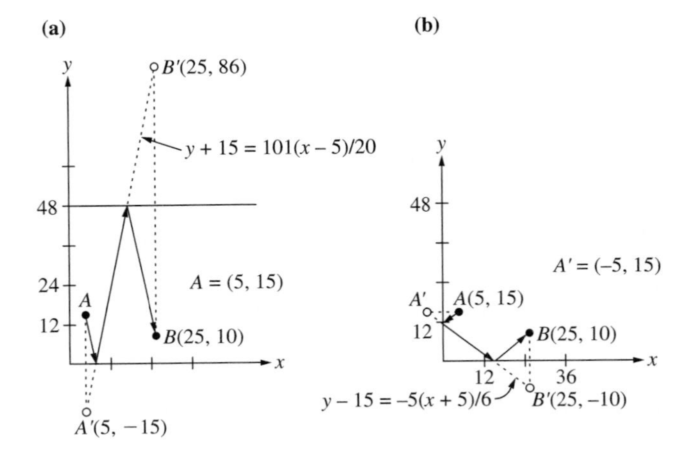

Problem 13

(c) (14.3, 48) (x-coordinate not exact). **15.** At any point on one of sides of orthic triangle and in direction toward either adjacent vertex of that triangle (due to Corollary of Theorem of Problem 20, Section 4.3). **17.** $x = \frac{3}{5}x + \frac{4}{5}y + 2$, $y = \frac{4}{5}x - \frac{3}{5}y - 4 \to -2x + 4y - 10 \to \ell$: $2y = x - 5$; let $P(x,y)$ be given point $\to P' = (3x/5 + 4y/5 + 2, 4x/5 - 3y/5 - 4)$ so midpoint $M = (\frac{1}{2}(x + 3x/5 + 4y/5 + 2), \frac{1}{2}(y + 4x/5 - 3y/5 - 4)) = (4x/5 + 2y/5 + 1, 2x/5 + y/5 - 2)$ whose coordinates satisfy equation of ℓ. Further, using coordinates, Slope $\overleftrightarrow{PP'} = -2$, Slope $\ell = \frac{1}{2}$ so $\overleftrightarrow{PP'} \perp \ell$.

SECTION 5.3

1. (See figure.) **3.** (See figure.) **5. (b)** 3. **(c)** Figure is moved a distance $\sqrt{40} = 2\sqrt{10} \approx 6.32$ units in a direction parallel to line having slope 3. **7.** $x'' = x + 4$, $y'' = y + 8$. **9.** If $P = (x, y)$ and $C(a, b)$ is midpoint of $\overline{PP'}$, $a = \frac{1}{2}(x + x')$, $b = \frac{1}{2}(y + y')$; solving for $x', y' \to x' = -x + 2a$, $y' = -y + 2b$. **11.** (See figure.) **13.** Suppose $x' = x + a$, $y' = y + b$; let $P(p, q)$, $Q(r, s)$ be given $\to P' = (p + a, q + b)$, $Q' = (r + a, s + b) \to P'Q' = \sqrt{[(p + a) - (r + a)]^2 + [(q + b) - (s + b)]^2} = \sqrt{(p - r)^2 + (q - s)^2} = PQ$. **15. (d)** Any point C on perpendicular bisector of $\overline{AB}$ because $CA = CB$. **17.** tr is rotation about P (see figure); $tr = s_m s_\ell{}^2 s_n = s_m s_n$ (center of rotation $= m \cap n = P$). **19.** Nontrivial isometry must be product of one, two, or three reflections (Theorem 2); if it is direct, it must be product of exactly two reflections, $s_\ell s_m$, which is a translation if $m \| \ell$, and a rotation otherwise. **21. (a)** $f = s_\ell$ **(b)** Suppose $t = s_\ell s_m$, where $\ell \| m$. By results of this section we can choose k so that $r = s_k s_\ell \to rt = s_k s_\ell{}^2 s_m = s_k s_m \to rt$ is itself a translation or rotation.

Problem 1

Problem 3

(a)

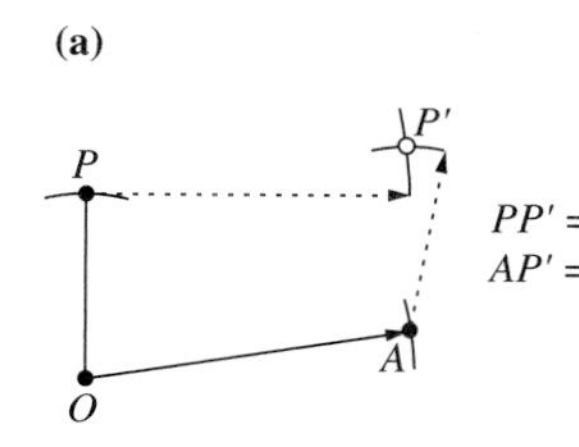

$PP' = OA$
$AP' = OP$

(b)

$PP' = OA$

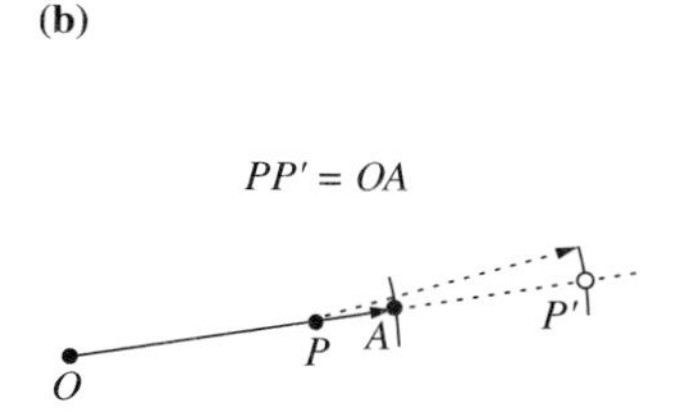

Problem 11

Problem 17

SECTION 5.4

1. (a) 3. **(b)** -2. **3.** $A' = (3, 0)$; $B' = (15, 0)$; $C' = (11, -8)$. $A'B' = 15 - 3 = 12 = 4AB$, $A'C' = \sqrt{128} = 4\sqrt{8} = 4AC$, and $B'C' = \sqrt{80} = 4\sqrt{5} = 4BC \to \triangle ABC \sim \triangle A'B'C'$ (SSS Criterion). **5.** Inverse is $x = -y' - 6$, $y = -x' + 6 \to$ line $y = mx + b$ maps to $-x' + 6 = m(-y' - 6) + b$ or $y' = x'/m - 6 - 6/m + b/m \to m = 1/m$, $b = -6 - 6/m + b/m \to m = \pm 1$ and $b = -6 - 6(\pm 1) + b(\pm 1)$. If $m = 1$, b has no unique solution; hence $m = -1$, $b = -b \to b = 0$, so fixed line is $y = -x$. **7.** $\boldsymbol{CP'} = 5\boldsymbol{CP} \to [x' - 2, y' - 3] = 5[x - 2, y - 3] \to x' = 5x - 8$, $y' = 5y - 12$. **9.** Draw line $\overleftrightarrow{WT}$ and intersect with line $\overleftrightarrow{UV}$. **11.** (See figure: construct line through P' parallel to $\overleftrightarrow{PQ}$ and intersect this with ray $\overrightarrow{AQ}$.) **13.** Given similitude is dilation with fixed point as center (definition of dilation). **15.** Given $\angle ABC$, points A, B, C mapped to A', B', C' such that $A'B' = kAB$, $A'C' = kAD$, $B'C' = kBC \to \triangle A'B'C' \sim \triangle ABC$ by SSS Criterion and $\angle A'B'C' \cong \angle ABC$. **17.** Let f be given similitude $\to d_{1/k}f$ is distance preserving [f takes A, B to A', B' with $A'B' = kAB$, and $d_{1/k}$ takes A', B' to A'', B'' with $A''B'' = (1/k)A'B' = AB$] $\to d_{1/k}f = g =$ product of three or fewer reflections $\to f = d_{1/k}{}^{-1}g = d_k\, g =$ product of dilation and three or fewer reflections. **19.** Let $k = A'B'/AB$, with P, Q given points; $\triangle ABP \sim \triangle A'B'P'$ (AA Criterion) $\to A'P' = kAP$. $\triangle APQ \sim \triangle A'P'Q' \to P'Q'/PQ = A'P'/AP = k \to P'Q' = kPQ$ for all points P, $Q \to$ mapping is similitude. **21.** Let k be proportionality factor between similarity $\triangle ABC \sim \triangle XYZ$; map $\triangle XYZ$ to $\triangle X'Y'Z'$ by dilation $d_{1/k} \to \triangle ABC \cong \triangle X'Y'Z' \to$ there exists isometry d mapping $\triangle ABC$ to $\triangle X'Y'Z' \to$ similitude $d_k d$ maps $\triangle ABC$ to $\triangle XYZ$.

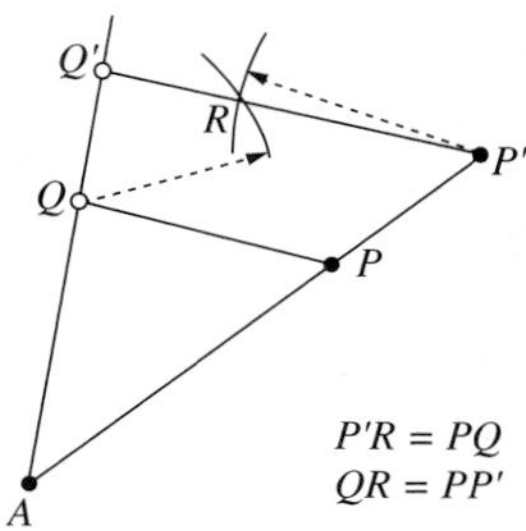

Problem 11

SECTION 5.5

1. (a) $\begin{bmatrix} 2 & 1 \\ -1 & 2 \end{bmatrix} = \begin{bmatrix} a & -\delta b \\ b & \delta a \end{bmatrix}$ a $(= 2, b = -1, \delta = 1)$. **(b)** Dilation d_3 with center $(0, 0)$. **(c)** Translation t_{AB} where $A = (0, 0)$, $B = (-3, 5)$. **3. (a)** $P'(9, 6), Q'(17, 13), R'(-12, -13), S'(-4, -6)$. **(b)** $P'(3, -3), Q'(3, -6), R'(4, 5), S'(4, 2)$. **5.** [Check: $\overline{B'E'} \perp \overline{A'C'}$; $m_1 = \text{Slope } \overleftrightarrow{B'E'} = (4 + 2)/(6 - 10) = -3/2$ and $m_2 = \text{Slope } \overleftrightarrow{A'C'} = 6/9 = 2/3 \to m_1 m_2 = (-3/2)(2/3) = -1$.] **7. (a)** $x' = 0.6x + 0.8y + 0.4$, $y' = 0.8x - 0.6y - 0.8$; $x'' = 0.6x' + 0.8y' + 1.2$, $y'' = 0.8x' - 0.6y' - 2.4$. **(b)** $x'' = x + 0.8$, $y'' = y - 1.6$. **9.** Must show $C(h, k)$ is midpoint of $\overline{PP'}$ for each point $P(x, y)$: Midpoint of (x, y) and $(x', y') = (-x + 2h, -y + 2k)$ has coordinates $\frac{1}{2}(x - x + 2h, y - y + 2k) = (h, k)$. **11. (a)** $k = \sqrt{a^2 + b^2} = \sqrt{12^2 + 5^2} = 13$. **(b)** $\sqrt{4/9 + 5/9} = 1$. Transformation in **(b)**; one in **(a)** direct. **13. (a)** $(x_0 = y_0 = 0.)$ $m' = (c + dm)/(a + bm) = m \to$ if $m = 0$, then $c/a = 0$ or $c = 0$; if $m = 1$, then $(c + d)/(a + b) = 1$ or $d = a + b$; if $m = 2$, then $(c + 2d)/(a + 2b) = 2$ or $2d = 2a + 4b = 2(a + b)$. $\therefore b = 0, d = a$ and matrix is $\begin{bmatrix} a & 0 \\ 0 & a \end{bmatrix}$. **(b)** With $O = (0, 0)$, $P = (x, y)$, $P' = (ax, ay)$, $OP' = \sqrt{a^2x^2 + a^2y^2} = |a|\sqrt{x^2 + y^2} = |a|\, OP$. If $a > 0$ we are finished; if $a < 0$ then $P' = (-bx, -by)$ where $b = -a > 0$ and P' lies on opposite ray of $\overrightarrow{OP}$. **15.** Area of both triangles is 5 sq. units.

17. Suppose matrix of f is $\begin{bmatrix} a & b \\ c & d \end{bmatrix}$ and image of circle $(x - h)^2 + (y - k)^2 = r^2$ is circle. Inverse transformation is f^{-1}: $x = (dx' - by')/e$, $y = (-cx' + ay')/e$ where $e = ad - bc$, which is of the form $x = px' - qy'$, $y = -sx' + ty'$ where $p = d/e$, $q = b/e$, etc; substitution into equation of circle $\to (px' - qy' - h)^2 + (-sx' + ty' - k)^2 = r^2$ or $(p^2 + s^2)x'^2 + (q^2 + t^2)y'^2 - (2pq + 2st)x'y' +$ {linear terms in x', y'} $= r^2 - h^2 - k^2$. This is a circle, so $p^2 + s^2 = q^2 + t^2$ and $2pq + 2st = 0 \to d^2 + c^2 = b^2 + a^2$ and $bd + ac = 0 \to$ by lemma, with roles of b and c reversed, $\begin{bmatrix} a & b \\ c & d \end{bmatrix} = \begin{bmatrix} a & b \\ -\delta b & \delta a \end{bmatrix}$, which is matrix for similitude in either case $\delta = 1, -1$.

SECTION 5.6

1. $fg = [14, 4, 12, 2] = gf$ (matrix exhibited as a vector), and $f^{-1} = [1, -1, -3, 4]$. **3.** Set of dilations given by setting $b = 0$; $\begin{bmatrix} a & 0 \\ 0 & a \end{bmatrix}\begin{bmatrix} b & 0 \\ 0 & b \end{bmatrix} = \begin{bmatrix} ab & 0 \\ 0 & ab \end{bmatrix}$ and $\begin{bmatrix} a & 0 \\ 0 & a \end{bmatrix}^{-1} = \begin{bmatrix} 1/a & 0 \\ 0 & 1/a \end{bmatrix}$ (product, inverse) which are of correct form, so belong to subset. **5.** Cyclic.

7. $\boldsymbol{D}_2 = [e, s_\ell, r_{180}, s_\ell r_{180}]$, where r_{180} has center at M; $\boldsymbol{D}_2$ is the Fours Group. **9.** Coordinate proof: If $A = [a, -\delta b, b, \delta a]$, $B = [c, \epsilon d, d, \epsilon c]$, with $\delta = \pm 1$, $\epsilon = \pm 1$ and $a^2 + b^2 = 1 = c^2 + d^2$, then $AB = [ac - \delta bd, -\delta\epsilon(bc + \delta ad), bc + \delta ad, \delta\epsilon(ac - \delta bd)] \equiv [r, -\delta' s, s, \delta' r]$ where $r = ac - \delta bd$, $s = bc + \delta ad$, and $\delta' = \delta\epsilon$. By algebra, $r^2 + s^2 = a^2c^2 + b^2d^2 + b^2c^2 + a^2d^2 = (a^2 + b^2)\cdot(c^2 + d^2) = 1$; the form of the inverse must also be examined. **13.** (See figure.) Consider $\triangle ABD$, where $m\angle A = \alpha$, $m\angle B = \beta$, and $\therefore m\angle D = \gamma$. By Theorem 1, $r_{2\alpha}[A]r_{2\beta}[B]r_{2\gamma}[C] = r_{-2\gamma}[D]r_{2\gamma}[C]$. If this is not the identity, then $C \neq D$. Let $\ell = \overleftrightarrow{CD}$, and $m, n \parallel$ to lines through C and D making alternate interior angles of measure γ with $\ell \to r_{2\gamma}[C] = s_\ell s_m$ and $r_{2\gamma}[D] = s_\ell s_n \to r_{-2\gamma}[D]r_{2\gamma}[C] = (s_n s_\ell)(s_\ell s_m) = s_n s_\ell{}^2 s_m = s_n s_m =$ translation $(m \| n)$. **15.** By formulas given, direct Lorentz Transformations $\boldsymbol{L}$ are a subset of the generalized Lorentz Group. If $A = \begin{bmatrix} a & b \\ b & a \end{bmatrix}$, $B = \begin{bmatrix} c & d \\ d & c \end{bmatrix}$, and $a^2 - b^2 = 1 = c^2 - d^2$, then $AB = \begin{bmatrix} ac + bd & ad + bc \\ bc + ad & bd + ac \end{bmatrix} \equiv \begin{bmatrix} r & s \\ s & r \end{bmatrix}$ and $A^{-1} = \begin{bmatrix} a & -b \\ -b & a \end{bmatrix}$; by algebra, $r^2 - s^2 = (ac + bd)^2 - (ad + bc)^2 = (a^2 - b^2)(c^2 - d^2) = 1$, so AB is of correct form.

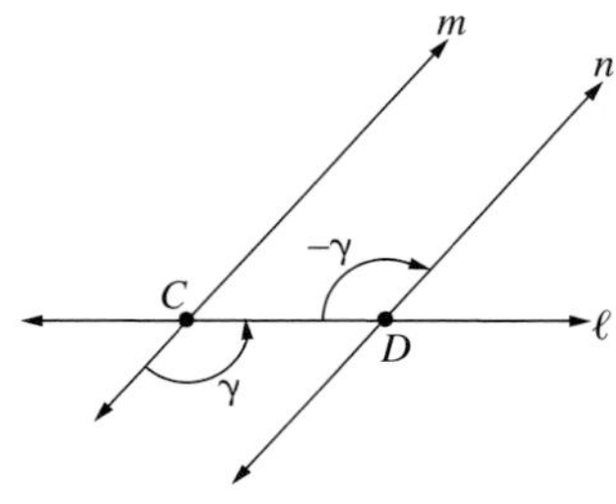

Problem 13

SECTION 5.7

3. Since O lies on perpendicular bisectors of sides of $\triangle ABC$ and sides of $\triangle LMN$ are paralleo to those of $\triangle ABC$, O lies on altitudes of $\triangle LMN \to O =$ orthocenter of $\triangle LMN$; since $d_{-1/2}[G]$ maps A to L, B to M, and C to N, $\triangle ABC$ maps to $\triangle LMN$, H maps to $H' = O$, and lines through G are fixed, so $OG = H'G' = \frac{1}{2}HG$ or $HG = 2GO$. **5.** $P(x, y)$ maps to $P'(x', y')$ where $x' = ax - by + c$, $y' = bx + ay + d$, with $a^2 + b^2 = 1$. Midpoint of $P(x, y)$ and $P'(x', y')$ is $M(x'', y'')$ where $x'' = \frac{1}{2}(x + ax - by + c)$, $y'' = \frac{1}{2}(y + bx + ay + d)$, or $x'' = \frac{1}{2}(a + 1)x - \frac{1}{2}by + \frac{1}{2}c$, $y'' = \frac{1}{2}bx + \frac{1}{2}(a + 1)y + \frac{1}{2}d$, which has the form of a direct similitude $(\delta = 1)$ with dilation factor given by $k^2 = \frac{1}{4}(a + 1)^2 + \frac{1}{4}b^2 = \frac{1}{2}(a + 1)$. If the traced map has same size as original, $f(x, y) = (x'', y'')$ must have $k = 1 \to \frac{1}{2}(a + 1) = 1 \to a = 1, b = 0$ and $x'' = x + \frac{1}{2}c$, $y'' = y + \frac{1}{2}d$ (translation). **7.** Using **(12)**, $P(x, y)$ is first mapped to $P'(x_1, y_1)$ where $x_1 = x\cos(-90°) - y\sin(-90°) = y$, $y_1 = x\sin(-90°) + y\cos(-90°) = -x \rightarrow$ second stake at $C(y, -x)$; $P(x, y)$ mapped to $P''(x_2, y_2)$ where $x_2 - 1 = (x - 1)\cos 90° - (y - 0)\sin 90° = -y$, $y_2 - 0 = (x - 1)\sin 90° + (y - 0)\cos 90° = x - 1 \rightarrow x_2 = 1 - y$, $y_2 = x - 1 \rightarrow$ third stake at $D(1 - y, x - 1)$. Midpoint of $\overline{CD}$ has coordinates given by $[\frac{1}{2}(y + 1 - y), \frac{1}{2}(-x + x - 1)] = (\frac{1}{2}, -\frac{1}{2})$, which forms an isosceles right triangle with $A(0, 0)$ and $B(1, 0)$. **9.** Analogous to Example 1 and works for any odd number of points. **11.** Take copies of $\triangle ABC$ at common point A, filling in with equilateral triangles for the fundamental region of a tessellation. (See figure.) **13.** Starting with $\triangle ABC$ and constructing external isosceles triangles $\triangle QCA$, $\triangle RAB$, $\triangle PCB$ having vertex angles of measure $2p$, $2q$, $2r$, because $p + q + r = 180$ by hypothesis, Corollary C of Theorem 1, Section 5.6 implies product $r_{2p}[P]r_{2q}[Q]r_{2r}[R]$ is identity (since B is fixed point). By Corollary B, $\triangle PQR$ has angle measures p, q, and r, respectively. **15.** Draw any circle tangent to sides of $\angle ABC$ and draw $\overrightarrow{BP}$, where P is given internal point, cutting circle at Q, then let dilation $d_k[B]$ act on circle to map it to one passing through P, where $k = BP/BQ$. **17.** Reflection of line $\overleftrightarrow{BC}$ in point P will intersect ray $\overrightarrow{BA}$ at Q; draw line $\overleftrightarrow{QP}$, cutting ray $\overrightarrow{BC}$ at R. Since $Q = R'$ under this reflection, $PQ = PR$. **19.** Let A be any point on outer circle; rotate innermost circle about A through 60°, and find B, its intersection with middle circle (no solution if B does not exist). Rotate B about point A 60° in reverse direction to find point C; $\triangle ABC$ will be desired triangle.

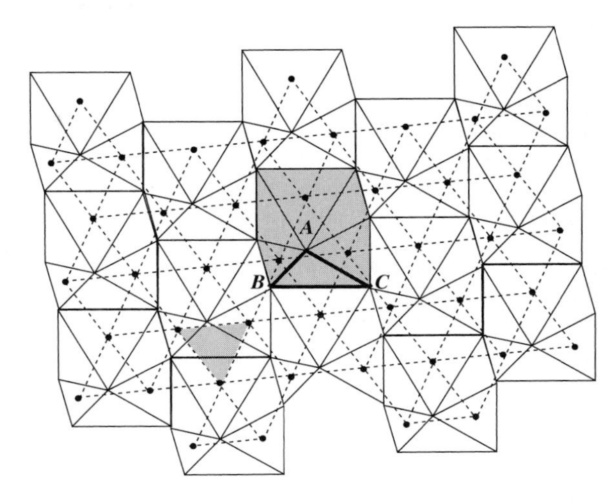

Problem 11

TESTING YOUR KNOWLEDGE (CHAPTER 5)

1. (a) Dilation. **(b)** Neither. **(c)** Similitude, not a dilation. **2.** $\overline{AB} \cong \overline{AC}$ and $\overline{AC} \cong \overline{AD}$, $\angle BAC \cong \angle CAD \rightarrow \triangle BAC \cong \triangle CAD$ (SAS) $\rightarrow \overline{BC} \cong \overline{CD}$. Similarly, $\triangle CAD \cong \triangle DAE$ and $\overline{CD} \cong \overline{DE}$. By Isosceles Triangle Theorem, $m\angle BCD = m\angle BCA + m\angle ACD = m\angle CDA + m\angle ADE = m\angle CDE$. **3.** SSS. **4. (a)** 2. **(b)** 2. **(c)** 1. **(d)** 3. **5.** $a = 1$. [$x = 2x - ay$ and $y = 3x - 2ay - 1$ lead to $x = ay$ and $(a - 1)y = 1$ or $y = 1/(a - 1)$ and $x = a/(a - 1)$ which are fixed points for all a *if* $a \neq 1$.] **6.** (See figure.) Rotation of 90° about origin maps (1, 0) to (0, 1) and translation $x' = x + 1$, $y' = y - 1$ maps (0, 1) to (1, 0), so (1, 0) is a fixed point under the product—the only fixed point. **7. (a)** $x = x' - 4$, $y = y' - 3$. **(b)** $x = 2x' - y' - 1$, $y = 5x' - 3y' - 1$. **8.** $y = 3x$ is one fixed line; all rest are of form $y = 3x + b$. **9. (a)** (3). **(b)** (3). **(c)** (1). **(d)** (2). **(e)** (3). **(f)** (1). **10.** Suppose $\ell \| m$ and ℓ' meets m' at point P. Since a linear transformation is onto, there is a unique point Q such that $Q' = P$. But $Q' \in \ell' \rightarrow Q \in \ell$, and $Q' \in m' \rightarrow Q \in m$, which contradicts $\ell \| m$. $\therefore \ell' \| m'$.

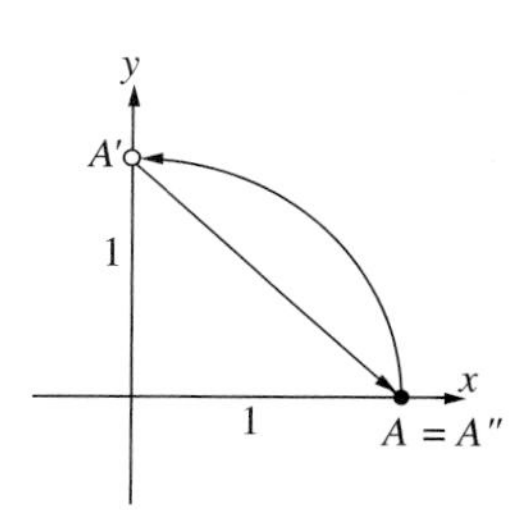

Problem 6

SECTION 6.2

1. See solution for Problem 17, Section 3.5. **3.** Euclid's Fifth Postulate is one way (if $m\angle A + m\angle B < 180$ then lines are not parallel); another is by Property C: If $\overleftrightarrow{AC} \| \overleftrightarrow{BD}$ then $m\angle A + m\angle B = 180 \rightarrow\leftarrow$ **5.** Drop perpendicular $\overleftrightarrow{AD}$ to line $\overleftrightarrow{CQ}$; $\overleftrightarrow{AP} \| \overleftrightarrow{CQ} \rightarrow \overleftrightarrow{AD} \perp \overleftrightarrow{AP}$ (Property C) $\rightarrow \overleftrightarrow{AD} = \overleftrightarrow{AB} = \overleftrightarrow{BD} \rightarrow \overleftrightarrow{BD} \perp \overleftrightarrow{CQ} \rightarrow \overleftrightarrow{BD} = \overleftrightarrow{BC} = \overleftrightarrow{AB}$ (uniqueness of perpendicular from point B). **7.** Suppose $m\angle ABC + m\angle BCD < 180$; if $\overrightarrow{BA'}$ is constructed to make $m\angle A'BC + m\angle BCD = 180$, then $\overleftrightarrow{BA'} \| \overleftrightarrow{CD}$. By Proclus' Postulate, $\overleftrightarrow{BA}$ intersects $\overleftrightarrow{CD}$, as desired. **9.** In Saccheri Quadrilateral $ABCD$ draw $\overline{AC}$; in $\triangle ADC$ and $\triangle CBA$, $\overline{AD} \cong \overline{CB}$, $\overline{AC} \cong \overline{AC} \rightarrow \triangle ADC \cong \triangle CBA$ (LH) $\rightarrow \overline{CD} \cong \overline{AB}$ **11.** Let C–A–C', D–B–D'; if $\angle BAC$ were not a right angle, either $m\angle CAB + m\angle ABD < 180$ or $m\angle C'AB + m\angle ABD' < 180 \rightarrow \overleftrightarrow{AC}$ meets $\overleftrightarrow{BD}$ (Euclid's Fifth Postulate) $\rightarrow\leftarrow$ $\therefore \angle BAC =$ right angle. Similarly, $\angle ACD =$ right angle $\rightarrow \diamond ABCD$ is rectangle $\rightarrow AB = CD$ (Problem 10 or properties of parallelograms). **13.** If $AB \leq MD$, on $\overline{MD}$ construct $\overline{MC'} \cong \overline{AB}$; $\angle C'MC \cong \angle ABM$, $BM = MC \rightarrow \triangle C'MC \cong \triangle ABM$ (SAS) $\rightarrow m\angle C'CM = m\angle AMB > m\angle ACM \rightarrow\leftarrow$ (Exterior Angle Inequality). **15.** Using Hint, $AB = CD \rightarrow AE = AB - EB = CD - GC = DG \rightarrow$ both $\diamond AEGD$, $\diamond GEBC$ are Saccheri Quadrilaterals with $m\angle AEG \leq 90$, $m\angle BEG \leq 90 \rightarrow m\angle AEG = m\angle BEG =$

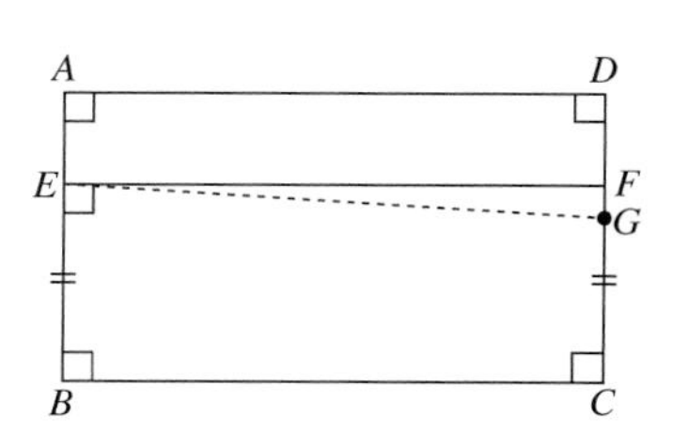

Problem 15

$90 \to \overleftrightarrow{EG} \perp \overleftrightarrow{AB} \to \overleftrightarrow{EG} = \overleftrightarrow{EF} \to G = F$ and $m\angle EFC = m\angle EFD = 90 \to \diamond AEFD, \diamond EBCF$ are rectangles. **17.** Let $\diamond ABCD$ be Lambert Quadrilateral (right angles at A, B, C). Problem 16 $\to \square BEGF$ exists, with B–A–E and B–C–F; let ray $\overrightarrow{CD}$ meet segment $\overline{EG}$ at H, and apply result of Problem 15 to $\diamond BEHC$, then to $\diamond ABCD$. **19.** Double segment $\overline{AP}$ to P_1, let Q_1 be foot of perpendicular from P_1 on line $\overleftrightarrow{AQ}$, and let M be midpoint of $\overline{AQ_1} \to PQ \le PM \le \frac{1}{2}P_1Q_1$, or $P_1Q_1 \ge 2PQ$ (corollary of Theorem 3, Section 3.7). Repeat process, doubling $\overline{AP_1}$ to P_2, with $Q_2 =$ foot of perpendicular from P_2 on line $\overleftrightarrow{AQ} \to P_2Q_2 \ge 2P_1Q_1 \ge 4PQ$; nth step yields $P_nQ_n \ge 2^nPQ \to x$ is unbounded.

SECTION 6.3

1. 15. **3.** Triangles have equal defects, areas if angle measures are equal. Converse not true (individual angles having equal sums in two triangles need not be equal). **5.** $\delta = \delta\,(\triangle PWS) - \delta\,(\triangle WKS) = (180 - 55 - 100 - 10) - (180 - 90 - 70 - 10) = 5$. **7.** $\delta_1 = 180 - 178 = 2$; $\delta_2 = 360 - 354 = 6$; $\delta_3 = 180 - 174 = 6$; $\delta_4 = 360 - 354 = 6 \to \delta_1 + \delta_2 + \delta_3 + \delta_4 = 20$; $\delta(\diamond ABCD) = 360 - 2 \cdot 90 - 2 \cdot 80 = 360 - 340 = 20$. **9.** $\delta(\triangle ABC) = 0 = \delta_1 + \delta_2$; since $\delta_1 \ge 0$ and $\delta_2 \ge 0$, $\delta_1 = \delta_2 = 0$. Yes. **11.** $\delta = 360 - 2 \cdot 90 - 2 \cdot 83 = 14$. (In Problem 10, $\delta = 360 - 353 = 7$.) Reason: line joining midpoints of summit and base of Saccheri Quadrilateral divides it into two congruent Lambert Quadrilaterals. **13. (a)** $(12 - 2)180 - 12\theta = 12 \to \theta = 149$. **(b)** $m\angle AOB = 360/12 = 30$. By additivity of defect, $12\delta = 12$ or $\delta = 1$. **15.** Central angles form five congruent triangles in each pentagon $\to$ any two from either pentagon must have equal angle measures $\to$ triangles congruent (AA Criterion for hyperbolic geometry) $\to$ sides and angles of pentagons congruent. **17. (a)** Answer: No; $\delta > 0 \to 360 - 4m\angle A > 0 \to m\angle A < 90$. **(b)** $k(360 - 4\epsilon) \approx 360k$. **19.** $\delta = (5 - 2)180 - 5m\angle A = 540 - 5m\angle A = 90 \to m\angle A = 90$. Thus, at each vertex, four tiles exactly fit, closing all gaps. Since pentagons are congruent, distance covered by n tiles in a row $= n \cdot$ Diameter, which is unbounded.

SECTION 6.4

1. (See figure.) **3.** Equation of circle has form $(x - 3)^2 + y^2 = 4$, centered at $(3, 0)$, radius 2. $\therefore$ tangent to $x = 5$ at $(5, 0)$ not in $\boldsymbol{H}$. **5.** $AB^* = \ln 3 \approx 1.099$; $CD^*\ \ln\frac{8}{7} \approx 0.134$. **7.** Equation of h-line through A', B' is $x^2 + y^2 = 25k^2$, with end points $M(5k, 0), N(-5k, 0)$; $A'M = B'N, A'N = B'M \to A'B'^* = \ln\dfrac{A'M \cdot B'N}{A'N \cdot B'M} = \ln\dfrac{A'M^2}{A'N^2} = \ln\dfrac{(5k + 3k)^2 + (0 - 4k)^2}{(-5k + 3k)^2 + (0 - 4k)^2} = \ln\dfrac{64k^2 + 16k^2}{4k^2 + 16k^2} = \ln\dfrac{80}{20} = \ln 4$ (constant) $\to \lim A'B'^* = \ln 4 \ne 0$. **9. (a)** Because $H(C, \overleftrightarrow{BA})$ is set of points to right of $\overleftrightarrow{AB}$, $H(A, \overleftrightarrow{BC})$ is set of points in $\boldsymbol{H}$ exterior to circle, and P lies in both sets. **(b)** All h-lines through P that meet $\overrightarrow{BA}^*$ are semicircles through P containing arc $\overline{BC}^*$ in their interiors. **11. (a)** [In Problem 10, $a = b = \ln 2$ and $\overleftrightarrow{AB}^*$ has equation $(x + 24)^2 + y^2 = 1025$.] End points of $\overleftrightarrow{AB}^*$ are given by $(x + 24)^2 = 1025$ or $M(8.016, 0), N(-56.016, 0)$; $c =$

$$AB^* = \left|\ln\frac{AM \cdot BN}{AN \cdot BM}\right| = \ln\sqrt{7.635} \approx 1.016 \to$$

$c^2 \approx 1.032$; $a = b = \ln 2 \approx 0.693 \to a^2 + b^2 \approx 2(0.693)^2 = 0.960 \ne 1.032$. **(b)** $\cosh a = \cosh \ln 2 = 5/4 = \cosh b$; $\cosh c \approx 1.562$ and $\cosh a \cosh b = (5/4)^2 = 1.5625$. **13.** A Saccheri Quadrilateral with acute summit angles at D and E. **17. (a)** Draw semicircle centered at origin, meeting positive x-axis at M and positive y-axis at A, then draw semicircle centered at point on negative x-axis through M, intersecting y-axis at B; $\angle ABM$ is an asymptotic right triangle. **(b)** Three unit semicircles with centers at $(-1, 0)$, $(0, 0)$, and $(1, 0)$ form, at their intersections in $\boldsymbol{H}$ and on x-axis, an isosceles asymptotic triangle. **19. (a)** Ray $\overrightarrow{AP}$ meets ray $\overrightarrow{BD} \to\leftarrow$. **(b)** Since all rays from A interior to $\angle BAC$ meet $\overrightarrow{BD}$, ray $\overrightarrow{AR}$ meets $\overrightarrow{BD}$ at S; $\triangle ABS \cong \triangle ABT \to \angle BAR \cong \angle BAT \cong \triangle BAQ \to$ rays $\overrightarrow{AT}$ and $\overrightarrow{AQ}$ coincide $\to \ell$ meets $\overrightarrow{BD} \to\leftarrow$. **21.** (See figure.) If $m\angle B > m\angle Y$, construct ray $\overrightarrow{BP}$ interior to $\angle AB\Omega$ so that $m\angle ABP = m\angle Y \to$ ray $\overrightarrow{BP}$ meets $\overrightarrow{A\Omega}$ at some point S. On ray $\overrightarrow{X\Gamma}$,

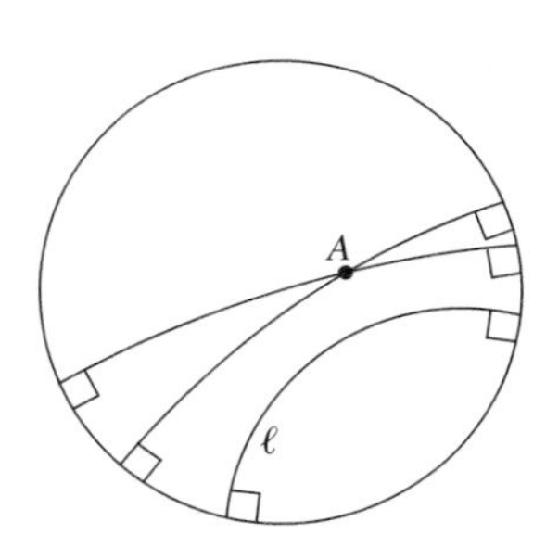

Problem 1

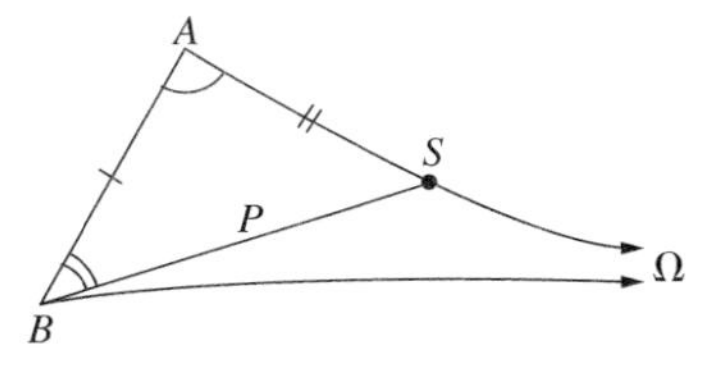

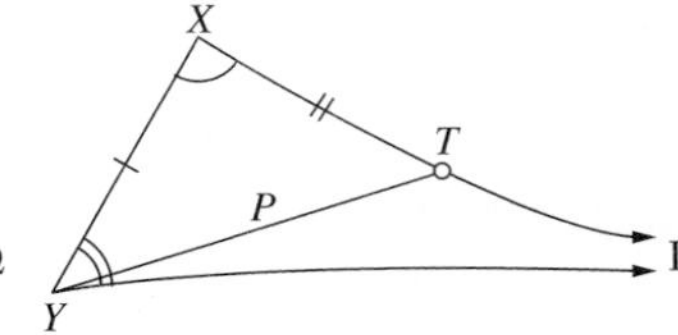

Problem 21

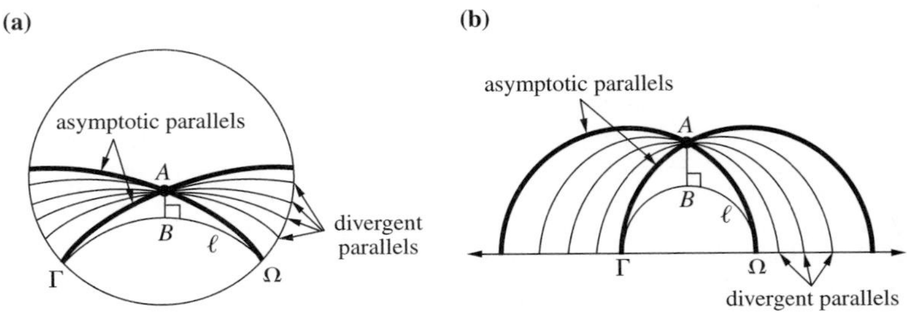

Problem 25

measure off $\overline{XT} \cong \overline{AS} \to \triangle ABS \cong \triangle XYT \to m\angle Y > m\angle XYT = m\angle ABS \to\leftarrow$. **23. (a)** $OD^2 = OP^2 + PD^2 = r^2 + s^2$ becomes, by distance formula and substitution: $(a/2 - c/2)^2 = r^2 + s^2 = (b + a^2/4) + (d + c^2/4)$ reducing to $b + d = -ac/2$. Equations **(10)** and **(11)** then follow. **(b)** Only one. **(c)** Equation of h-line perpendicular to $x^2 + y^2 = 8$ is, by **(11)**, $x^2 + y^2 + cx = -8$, which must pass through $A(3, 2) \to x^2 + y^2 - 7x = -8$. **25.** (See figure.) **27.** Both are possible. **29.** h-line n: $x^2 + y^2 + cx = d$ perpendicular to ℓ: $x^2 + y^2 = r^2$ iff $d = -r^2$; $n \perp m$ iff $d = -b - ac/2$ or $c = 2(r^2 - b)/a$. $\therefore$ n: $x^2 + y^2 + 2(r^2 - b)x/a = -r^2$ is unique h-line perpendicular to both $x^2 + y^2 = r^2$ and $x^2 + y^2 + ax = b$ provided $a \neq 0$. (If $a = 0$, take y-axis.)

SECTION 6.5

1. In **(1)**, let $r^2 = 30 \to x' = 30x/(x^2 + y^2)$, $y' = 30y/(x^2 + y^2)$; then $K' = (5, 5)$, $L' = (2, 4)$. **3.** $A = (0, 1)$ and $B = (1, 0)$; substitution of inverse of **(1)** into equation $x + y = 1$ yields $[x'/(x'^2 + y'^2)] + [y'/(x'^2 + y'^2)] = 1 \to x' + y' = x'^2 + y'^2$. Coordinates $O(0, 0)$, $A(0, 1)$, and $B(1, 0)$ satisfy this equation. **5. (a)** Condition for orthogonality $OD^2 = r^2 + s^2$ becomes $(2 - 6)^2 = 2^2 + (\sqrt{12})^2$ or $(-4)^2 = 16$, which is true $\to \ell \perp m$. **(b)** To save on use of primes, regard given h-lines as images of inversion: ℓ: $x'^2 + y'^2 - 4x' = 0$, m: $x'^2 + y'^2 - 12x' = -24$; substitution of equations of inversion yields $[6x/(x^2 + y^2]^2 + [6y/(x^2 + y^2)]^2 - 4[6x/(x^2 + y^2)] = 0$ $\to x = \frac{3}{2}$ (equation of ℓ'). Similarly, equation of m' is $x^2 + y^2 - 3x = \frac{3}{2}$; ℓ' passes through center of $m' \to \ell' \perp m'$. **9.** $\cosh c = \cosh a \cdot \cosh b = \cosh^2 3 = 101.358 \to c = 5.312$: $\tan A = \tanh 3/\sinh 3 = 0.99505/10.01787 \to A = 5.7° = B$. **11.** $\cosh c = \cosh 3 \cdot \cosh 4 \to c = 6.310$. **13.** For any x, $\cosh(\ln x) = \frac{1}{2}(e^{\ln x} + e^{-\ln x}) = \frac{1}{2}(x + 1/x)$. If $x = \tau$, $\cosh(\ln \tau) = (\tau + \tau^{-1})/2 = \sqrt{5}/2 \to \cosh c = \cosh \tau \cosh \tau = 5/4$; $\cosh(\ln 2) = \frac{1}{2}(2 + \frac{1}{2}) = \frac{5}{4}$, in agreement. $A = B \to \cot^2 A = \cosh c = 5/4 \to \cot A = \sqrt{5}/2 \to A \approx 41.8°$. **15. (a)** $\cos\theta = \tanh a / \tanh y$; $\tanh y = (e^y - e^{-y})/(e^y + e^{-y}) = (1 - e^{-2y})/(1 + e^{-2y})$; $\to$ as $y, x \to \infty$, $\theta \to A$ and $\lim \tanh y = \lim (1 - e^{-2y}) / (1 + e^{-2y}) = (1 - 0)(1 + 0) = 1 \to \cos A = \lim \cos\theta = \tanh a / 1 \to \cos A = \tanh a$. **(b)** $\sin A = \sqrt{1 - \cos^2 A} = \sqrt{1 - \tanh^2 a} = \sqrt{\text{sech}^2 a} = \text{sech}\, a$ (from identity: $\cosh^2 a - \sinh^2 a = 1 \to 1 - \tanh^2 a = \text{sech}^2 a$). $\tan A = \sin A / \cos A = \text{sec}\, a / \tanh a = \text{csch}\, a$ (from definitions for $\text{sech}\, x$, $\tanh x$, and $\text{csch}\, x$). **17.** Defect of pentagon $= (5 - 2)180 - \{\text{Angle-sum}\}$ or $\{\text{Angle-sum}\} = 540 - \delta \to$ each angle has measure $= \frac{1}{5}(540 - \delta) = 360/n$, where δ is defect of pentagon and n is number of tiles about each point (as δ decreases, n increases); values of $360/n$ are 360, 180, 120, 90, 72, 60, ... $\frac{1}{5}(540 - \delta) < \frac{1}{5}(540) = 108$, so we rule out 360, 180, and 120. $\therefore \frac{1}{5}(540 - \delta) = 90$. **19.** $\angle OMC = 90$ (Euclidean angle) $\to \overleftrightarrow{CM}$ tangent to semicircle ℓ at M; let C be center of circle of inversion with radius CM (orthogonal to ℓ). By Property 6 of circular inversion, ℓ maps to itself, M maps to M, and A, B are inverse pairs $\to AM^* = MB^*$. **21.** Because if equidistant locus is a line, there would exist adjacent Saccheri Quadrilaterals having adjacent summit angles (figure), the sum of whose measures $= 180 \to\leftarrow$. **23.** Arc **A** is equidistant locus. Proof: (See figure.) Draw h-line m perpendicular to ℓ centered at G by construction shown. Take this line m as line of reflection (circular inversion); because a circle through G inverts to Euclidean line orthogonal to line of centers (x-axis), and ℓ maps to itself ($\ell \perp m$), $\overleftrightarrow{HF}^*$ maps to vertical line perpendicular to ℓ, that is, y-axis. If $CG = u$, $CD = v$, and r, s are radii of **A** and m, then $r^2 = PD^2 = v^2 + c^2$ where $c = PC$, and $u^2 = GC^2 = GJ^2 + JC^2 = s^2 + c^2 \to r^2 + s^2 = u^2 + v^2 = GD^2$ or $GD^2 = r^2 + s^2 \to \overline{GE} \perp \overline{DE}$. $\therefore$ m is orthogonal to **A** $\to$ arc **A** is self-inverse $\to$ point H maps to point $A \to HF^* = H'F'^* = AB^* =$ const. **25.** $\cosh a = \cosh d \cosh c_2 - \sinh d \sinh c_2 \cos\theta$, $\cosh b = \cosh d \cosh c_1 + \sinh d \sinh c_1 \cos\theta$. To eliminate terms involving $\cos\theta$, multiply both sides of equations by, respectively, $\sinh c_1$ and $\sinh c_2$, then sum. Result: $\sinh c_1 \cosh a + \sinh c_2 \cosh b = \cosh d(\sinh c_1 \cosh c_2 + \cosh c_1 \sinh c_2) = \cosh d \sinh c$, equivalent to desired result on division by $\sinh c$. **27.** Because distances of even several million miles are very small compared to 10^{20}, a right triangle in ***H*** would obey the law $\cosh c = \cosh a \cosh b$ in terms of universe measures, or in terms of local units a', b', c', $\cosh c'/k = \cosh a'/k \cosh b'/k$ (where $k = 10^{20}$). For small x, $\cosh x \approx 1 + \frac{1}{2}x^2$, hence above equation is, but for an incredibly small error, $1 + \frac{1}{2}(c'/k)^2 = [1 + \frac{1}{2}(a'/k)^2] \cdot [1 + \frac{1}{2}(b'/k)^2] = 1 + \frac{1}{2}(a'/k)^2 + \frac{1}{2}(b'/k)^2 + \frac{1}{4}(a'/k)^2 \cdot (b'/k)^2$ or $\frac{1}{2}(c'/k)^2 = \frac{1}{2}(a'/k)^2 + \frac{1}{2}(b'/k)^2 + \frac{1}{4}a'^2b'^2/k^4 \approx \frac{1}{2}(a'/k)^2 + \frac{1}{2}(b'/k)^2 \to \frac{1}{2}c'^2/k^2 \approx \frac{1}{2}a'^2/k^2 + \frac{1}{2}b'^2/k^2$. That is, $c'^2 \approx a'^2 + b'^2$—Euclidean Pythagorean Theorem—precise measurements could not detect error.

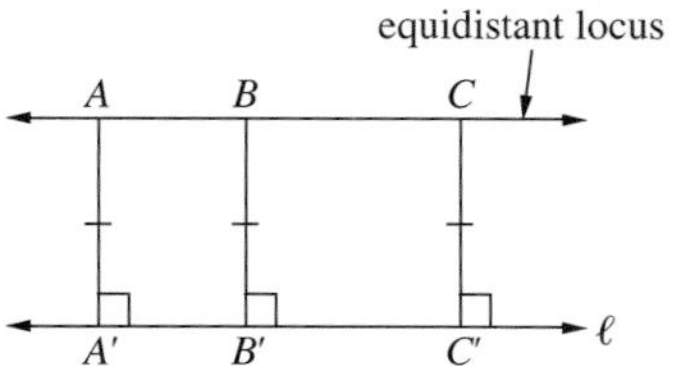

equidistant locus

Problem 21

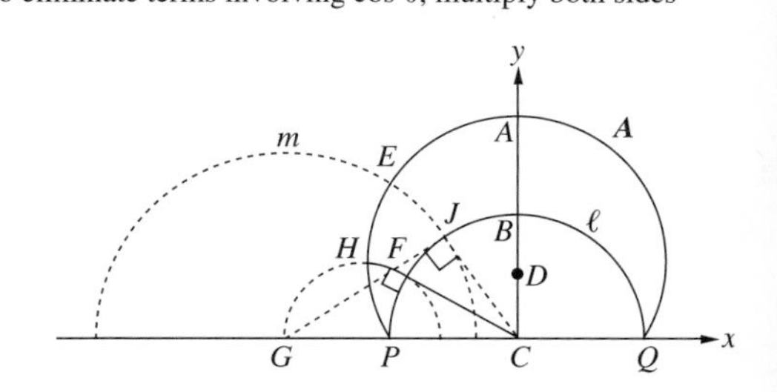

Problem 23

TESTING YOUR KNOWLEDGE (CHAPTER 6)

1. Correct order is (c), (a), and (b) (with defects 10, 12, and 15). **2.** (b) and (c). **3.** $50 = \delta - 180 - 90 - 35 - x \to x = 5$. **4.** $\delta = (5 - 2)180 - 320 = 260$. **5.** (b), (d), and (f). **6.** Semicircles with centers on ℓ (excluding end points) and vertical rays (excluding end points). **7. (a)** Semicircle passing through (3, 6) and (0, 0). **(b)** $x^2 + y^2 + ax = b \to a = -15, b = 0$ (by substitution of coordinates); Answer: $x^2 + y^2 - 15x = 0$. **8.** PQ need not be bounded in hyperbolic geometry. **9.** $\cosh c = \cosh 2 = \cot 30 \cdot \cot B \to \cot B = 2.172104 \to B \approx 24.7°$; angle-sum $= 90° + 30° + 24.7° = 144.7°$. **10.** By Saccheri–Legendre Theorem, $r + t + u \leq 180$ and $s + v + w \leq 180$. If either of these sums were < 180, then total of the two sums < 360 and $m\angle B + (r + s) + (u + v) + m\angle C < 360$ or $m\angle B + m\angle A + 180 + m\angle C < 360 \to$ Angle-sum $\triangle ABC < 180 \to\leftarrow \therefore r + t + u = s + v + w = 180$.

SECTION 7.1

1. If $\boldsymbol{P} \perp \ell$ at A, all lines in $\boldsymbol{P}$ are $\perp$ to ℓ at $A \to$ if $m \perp \ell$ at A, then $m \subseteq \boldsymbol{P}$ or else plane $\boldsymbol{Q}$ of ℓ and m meets $\boldsymbol{P}$ in a second line $\perp$ to ℓ at A (in plane $\boldsymbol{Q}$) $\to\leftarrow$ **3.** Line ℓ can be *in* given plane and other lines can be several parallels in plane $\perp$ to ℓ. **5.** $m\angle ABE = 60$. **7. (a)** O is equidistant from A, B, C and $\overleftrightarrow{PO} \perp \boldsymbol{P}$, so $\triangle POA \cong \triangle POB \cong \triangle POC$ (SAS) $\to PA = PB = PC$. **(b)** From **(a)**, $\angle APO \cong \angle BPO \cong \angle CPO$ (CPCF). **9.** (1) Mark wire at 12 ft. length, place end at A and other end will touch floor at desired point. (2) Extend wire 18 ft. from A to three different points on floor, then mark center of circle through those points. (3) Use wire as bob. **11.** By result of Problem 10, drop line $m \perp$ plane $\boldsymbol{P}$ from $A \in \ell$ and let $\boldsymbol{Q}$ be the plane of ℓ and m. Result of Problem 12(b) $\to \boldsymbol{Q}$ is only plane $\perp \boldsymbol{P}$ containing line ℓ (also uses Theorem 2). **13.** Let $\boldsymbol{Q} = \boldsymbol{P}(A, m)$, which meets $\boldsymbol{P}$ on line $\overleftrightarrow{AB}$; in plane $\boldsymbol{Q}$, let $\ell' \perp \overleftrightarrow{AB} \to \ell' \perp \boldsymbol{P}$ (Theorem 2) $\to \ell' = \ell \subseteq \boldsymbol{Q}$ [result of Problem 12(a)]. **15.** $\Diamond ABED$, $\Diamond CBEF$ are Saccheri Quadrilaterals, so lines $\overleftrightarrow{LM}$ and $\overleftrightarrow{MN}$ are common perpendiculars to $\overleftrightarrow{BE}, \overleftrightarrow{AD}$, and to $\overleftrightarrow{BE}, \overleftrightarrow{CF} \to \overleftrightarrow{BE} \perp \boldsymbol{P}(LMN) \equiv \boldsymbol{R}$ (Theorem 1) $\to \boldsymbol{P} \perp \boldsymbol{R} \to \overleftrightarrow{AD} \perp \boldsymbol{R}$ (Theorem 2) $\to \overleftrightarrow{AD} \perp \overleftrightarrow{LN}, \overleftrightarrow{CF} \perp \overleftrightarrow{LN}, \Diamond ALNC \cong \Diamond DLNF$ (SASAS) $\to \overline{AC} \cong \overline{DF}$ by CPCF. **17.** (See figure.) Let P be any point in the bisecting plane of a dihedral angle $\angle(\boldsymbol{P}, \boldsymbol{Q})$, and drop perpendiculars $\overline{PA}$, $\overline{PB}$ to planes $\boldsymbol{P}$, $\boldsymbol{Q}$; in $\boldsymbol{P}$, draw $\overline{AC} \perp \ell$ and let $\boldsymbol{R} = \boldsymbol{P}\ (APC)$. Show first that $\overline{PB} \subseteq$ plane $\boldsymbol{R}$: Because $\ell \perp \overleftrightarrow{AC}$ (line of intersection of $\boldsymbol{P}$ and $\boldsymbol{R}$), by Theorem 2 $\ell \perp \boldsymbol{R}$; similarly, if $\overleftrightarrow{BC'} \perp \ell$ in $\boldsymbol{Q}$ and $\boldsymbol{R}' = \boldsymbol{P}\ (PBC')$, $\ell \perp \boldsymbol{R}'$. Both $\boldsymbol{R}$ and $\boldsymbol{R}'$ pass through $P \to \boldsymbol{R}' = \boldsymbol{R}$ and $\overline{PB} \subseteq \boldsymbol{R}$; $\ell \perp$ sides of $\angle BCP$ and $\angle ACP \to$ by definition of dihedral bisector, $m\angle ACP = m\angle PCB$ and right triangles $\triangle ACP$, $\triangle PCB$ congruent by $HA \to PA = PB$. (Converse also true.) **19.** Statement: *Given any plane* $\boldsymbol{P}$, *the points of space not lying in* $\boldsymbol{P}$ *belong to two sets* H_1 *and* H_2, *called half-spaces, which satisfy the following properties*: (1) H_1 *and* H_2 *are convex*; (2) H_1 *and* H_2 *have no points in common*; (3) *If* $A \in H_1$ *and* $B \in H_2$ *then* $\boldsymbol{P}$ *intersects segment* $\overline{AB}$ *at some interior point*. Proof Outline: If $\boldsymbol{P}$ is any plane, let D be any point not in $\boldsymbol{P}$, E a point in $\boldsymbol{P}$, and F a point such that D–E–F. Define H_1 as set of all points U in space such that segment $\overline{UF}$ meets $\boldsymbol{P}$, and H_2 set of points V such that segment $\overline{VD}$ meets $\boldsymbol{P}$. If U is point not in plane $\boldsymbol{P}$, it must be shown that $U \in H_1$ or $U \in H_2$ [accomplished by observing plane $\boldsymbol{Q} = \boldsymbol{P}(UDF)$ and using Plane Separation Postulate for $\boldsymbol{P}$ and line $\ell = \boldsymbol{P} \cap \boldsymbol{Q}$]. To show H_1 is convex, let A, B lie in H_1, $C \in \overline{AB}$. By definition of H_1, $\overline{AF}$ and $\overline{BF}$ meet $\boldsymbol{P}$ at A', B'. Apply Plane Separation Postulate in $\boldsymbol{P}\ (ABB')$ to show $\overline{AB}$, therefore C, lies on A-side of line $\overleftrightarrow{A'B'} \to$ line $\overleftrightarrow{A'B'}$ meets segment $\overline{CF}$ at C' such that F–C'–$C \to C \in H_1$, as desired. Convexity of $H_1, H_2 \to H_1$ and H_2 do not meet; use Plane Separation Postulate to prove if $A \in H_1$, $B \in H_2$, then $\boldsymbol{P}$ meets $\overline{AB}$.

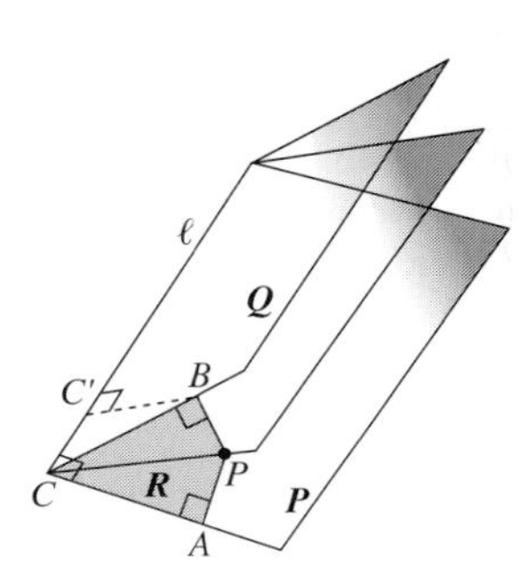

Problem 17

SECTION 7.2

1. No: If ℓ, m intersect at A and lie in plane $\boldsymbol{Q}$ above and parallel to plane $\boldsymbol{P}$, lift m to third plane above $\boldsymbol{Q}$ and parallel to $\boldsymbol{P}$. **3.** If ℓ' meets $\boldsymbol{P}, \boldsymbol{Q}, \boldsymbol{R}$ at A', B', C' and lies in same plane $\boldsymbol{S}$ as ℓ, then $\boldsymbol{S}$ meets $\boldsymbol{P}, \boldsymbol{Q}, \boldsymbol{R}$ at parallel lines $\overleftrightarrow{AA'}, \overleftrightarrow{BB'}, \overleftrightarrow{CC'}$ (or else $\boldsymbol{P}$, $\boldsymbol{Q}$, or $\boldsymbol{R}$ intersect $\to\leftarrow$); by Parallel Projection (Section 4.2) $AB/BC = A'B'/B'C'$. This can then be extended to case when ℓ, ℓ' do not lie in same plane. **5.** (See figure.) Desired hexagon obtained by taking midpoints $A, B, C, \ldots$ of sides of cube. For convenience, let sides of cube have length $2 \to AB = \sqrt{2} = BC$ by Pythagorean Theorem; because $\overline{AK} \perp \overline{KC}$ and $AK = 1 = LC$, $KC = \sqrt{5}$ and $AC = \sqrt{6} \to m\angle ABC = 120$ (Law of Cosines), true for all sides and angles. $\therefore$ hexagon is regular. **7.** $\sqrt{\frac{2}{3}}$. **9.** By Law of Cosines $\cos^{-1}\frac{1}{3} \approx 70.53°$
11. In problem figure, A, B, D, and F are equidistant from E and C, hence lie on perpendicular bisector (plane) of segment $\overline{EC}$ (Theorem 3, Section 7.1). Hence, in $\boldsymbol{P}$, $\Diamond ABFD$ equilateral $\to$ it is a rhombus $\to \triangle BCD \cong \triangle BAD$ (SSS) $\to m\angle BAD = m\angle BCD = 90 \to \Diamond ABFD$ is square. **13.** Start with partial net in a plane consisting of five regular hexagons surrounding one regular pentagon. Gaps between hexagons are congruent, so when hexagons are turned upward to fill in gaps, dihedral angles between hexagons will be congruent, as will those between hexagons and base pentagon. Call this configuration a *base shell*, for convenience. Along trailing edge of shell is found an angle between two adjacent edges of hexagons, and by symmetry through midpoint of edge leading to vertex of base pentagon, this angle is congruent to angle of pentagon $\to$ second base shell congruent to first will exactly fit over part of first shell, and extending it further around. This construction can be repeated until figure must close, forming the desired soccer ball. **15. (a)** One new face added for each original vertex $\to F' = F + V$. At A there are n new edges, and parts of all the original edges are retained; this being true at each of the V original vertices, $E' = E + nV$. For each of the V vertices, n were added and 1 subtracted, hence a net total of $(n - 1)V$ were added to $V \to V' = (n - 1)V + V = nV$. **(b)** $F' - E' + V' = (F + V) - (E + nV) + nV =$

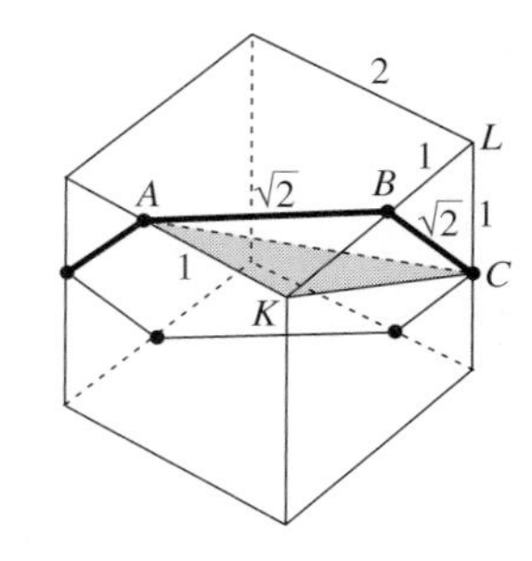

Problem 5

$F - E + V$. **17.** It must be shown that all faces are parallelograms; Theorem 2 takes care of the lateral faces, so this leaves the top $\diamond A'B'C'D'$. If base is $\square ABCD$, two lateral edges parallel $\to$ they are coplanar, and plane containing them intersects base and top at parallel lines (Theorem 2). By Transitive Law, $\overleftrightarrow{A'B'} \| \overleftrightarrow{AB} \| \overleftrightarrow{CD} \| \overleftrightarrow{C'D'}$; similarly, $\overleftrightarrow{AC'} \parallel \overleftrightarrow{B'D'}$ and $\diamond A'B'C'D'$ is parallelogram. **19.** Let sides of spherical $\triangle B'C'D'$ on unit sphere have lengths b, c, and d. Let radius of sphere $= 1 \to m\angle B'A'C' = m\angle BAC = d$. Similarly, $m\angle CAD = b$, $m\angle DAB = c$. From triangle inequality on unit sphere $d + b > c \to m\angle BAC + m\angle CAD > m\angle DAB$. **21. (a)** $(F, E, V) = (12, 24, 12)$. **(b)** $(F, E, V) = (20, 40, 20)$.

SECTION 7.3

1. $10\sqrt{3}$ **3.** By result of Problem 17, Section 7.1, two plane bisectors of a pair of lateral faces meet in a line ℓ, each point of which is equidistant from lateral faces, hence $\ell \subseteq$ third bisecting plane of two lateral faces; plane bisector of dihedral angle formed by base and lateral face will cut line ℓ in some point I equidistance from base and lateral face—true for any lateral face, so I equidistant from all four faces of tetrahedron. **5.** $6\sqrt{6} + 9$. **7.** $3\sqrt{6}/8 + \frac{3}{4}$ in. **9.** Let planes $\boldsymbol{P}$, $\boldsymbol{Q}$ be tangent to sphere O at A, B (figure), and let $\boldsymbol{R} = \boldsymbol{P}(AOB)$. From the result of Problem 2, $\overline{OA} \perp \boldsymbol{P}$ and $\overline{OB} \perp \boldsymbol{Q} \to \boldsymbol{R} \perp \boldsymbol{P}$ and $\boldsymbol{R} \perp \boldsymbol{Q}$. By corollary of Theorem 2, Section 7.1, $\ell \perp \boldsymbol{R}$ at some point $C \to \ell \perp \overline{AC}$, $\ell \perp \overline{BC}$. $OA = OB \to \triangle AOC \cong \triangle BOC$ (HL) $\to AC = BC \to A$ and B are equidistant from ℓ. If $P \in \ell$ then $CP = CP \to \triangle ACP \cong \triangle BCP$ (SAS) $\to AP = BP$. **10.** Let $\boldsymbol{P}$ be plane of $\overleftrightarrow{PA}$, $\overleftrightarrow{PB} \to \boldsymbol{P}$ intersects sphere in circle $\boldsymbol{K}$ (Theorem 1). If $\overleftrightarrow{PA}$ or $\overleftrightarrow{PB}$ not tangent to $\boldsymbol{K}$, they intersect $\boldsymbol{K}$ at a second point, thus sphere, and $\overleftrightarrow{PA}$ or $\overleftrightarrow{PB}$ cannot be tangent lines. Then $PA = PB$ follows from previous result (Corollary of Theorem 2, Section 3.8). **13.** $PF_1 = PM$, $PF_2 = PN \to PF_2 - PF_1 = PN - PM = MN = 2a$.

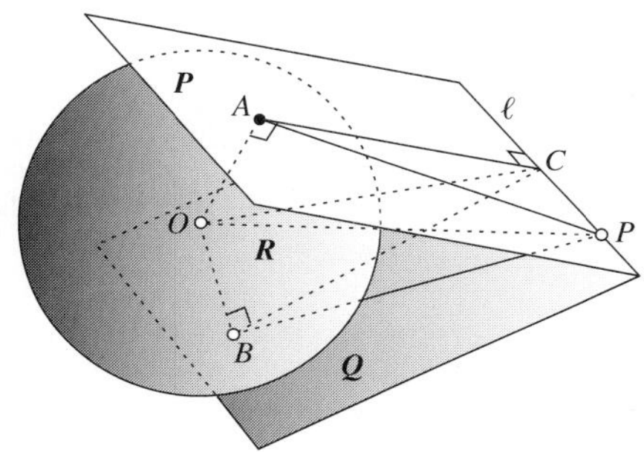

Problem 9

SECTION 7.4

1. $h = \sqrt{\frac{2}{3}} = \sqrt{6}/3$, $B = \sqrt{3}/4 \to V = \frac{1}{3}Bh = \sqrt{2}/12$. **3.** First solve for x: $(6x)^2 + (35x/2)^2 = 1^2 \to x = 2/37 \to h = 70/37$. **5. (a)** $V_1 = \pi(3.771)^2 \cdot (9.215) \approx 411.679$ in.^{3} Volume of 9 tennis balls $= 202.044$ in.^{3} $\to$ wasted space $= 209.635$ in.^{3} (short can). $V_2 = \pi(1.75)^2 \cdot 9 \cdot (3.5) \approx 303.066$ in.^{3} $\to$ wasted space $= 101.022$ in.^{3} (tall can). **(b)** Area (short can) $= 307.689$ in.^{2}; Area (tall can) $= 365.603$ in.^{2} **7.** $V/V' = (4/3\pi r^3)/2\pi r^3 = \frac{2}{3} = S/S'$. **9.** In Figure 7.42, by construction Area (base $\triangle$) $= \pi r^2 = \frac{1}{2}b'h' \to$ by similar right triangles, $y/h = r'/r$, $y/h = h''/h' = b''/b' \to A'_x = \pi r'^2 = \pi r^2(y^2/h^2)$; $A_x = \frac{1}{2}b''h'' = \frac{1}{2}(b'y/h)(h'y/h) = \pi r^2(y^2/h^2) \to A'_x = A_x \to$ volumes $= \to V = \frac{1}{3}Bh = \frac{1}{3}\pi r^2 h$. **11.** Use **(7)** with $r_1 = 0$, $r_2 = r \to V = 1/6\, \pi h(3r_2 + h^2)$; use right triangle formed by segment joining center of sphere to end point of segment labeled $r \to (R-h)^2 + r^2 = R^2 \to r^2 = 2Rh - h^2$.

SECTION 7.5

1. Yes: $(2, 10, -3)$. **3.** Make substitution $s = -2t + 3$. **5.** Let $P = \boldsymbol{x}$ be vector form of point on line $\boldsymbol{x} = t(\boldsymbol{i} - \boldsymbol{j} + 2\boldsymbol{k}) + \boldsymbol{j} + 3\boldsymbol{k} \to x = t$, $y = -t + 1$, $z = 2t + 3$. Substitute into equation of plane: $3x + y - z = -2$. **7. (a)** -1. **(b)** $\cos^{-1}(-\frac{1}{3}) \approx 109.47°$. **9.** $60°$ or $120°$. **11.** Plane of fixed points: $3x = y + z$. **13.** Spin axis: $3\boldsymbol{i} + \boldsymbol{j} + \boldsymbol{k}$; angle of rotation $= \cos^{-1}(-5/6) \approx 146.44°$.

SECTION 7.6

1. Small circle with center at North Pole. Given a "line" ℓ on $\boldsymbol{S}$ (great circle), determine point C half-way around $\boldsymbol{S}$ from any point $A \in \ell$ to opposite side. If P is at constant distance $s < \pi r/2$ units from ℓ, on C-side of ℓ, locus of P will be small circle centered at C having radius $\pi r - s$. **3.** Triangle in **(a)**. Yes; the larger the sides, the larger the angles, and the larger the excess (area). **5.** $m\angle BAC^* = \angle BOC = 180°/12 = 15° \to$ excess $= 15$. **7.** No, because two points of locus are S (South Pole) and E, some point on equator. Great circle determined by these two points is a meridian that makes $90°$ angle with each line of latitude ($\neq 45°$ as required). (Actual curve is spiral-like path starting at South Pole circling upward, but never quite reaching North Pole.) **9.** (See figure.) $\epsilon_1 = m\angle 3 + m\angle B + m\angle 1 - 180$; $\epsilon_2 = m\angle 4 + m\angle 2 + m\angle C - 180$. $\therefore \epsilon_1 + \epsilon_2 = m\angle 3 + m\angle B + m\angle 1 + m\angle 4 + m\angle 2 + m\angle C - 360 = (m\angle 3 + m\angle 4) + m\angle B + m\angle C + (m\angle 1 + m\angle 2) - 360 = m\angle A + m\angle B + m\angle C - 180 = \epsilon(\triangle ABC)$. **11.** $c \approx 0.118$; $a = b \approx 0.083$. **13.** Example: Let $A = 45°$, $C = 90°$, $c = 5.0$ mi. Correct value of b from spherical trigonometry determined as follows: $\tan b' = \tan c' \cos A = \tan(5/3960) \cos 45°$, where $b' = b/3960$, $c' = c/3960$, or $\tan b' = 0.0008\ 9281\ 21 \to b/3960 = 0.0008\ 9281\ 18 \to b = 3.5355\ 3484\ 5$ miles; b as determined from Euclidean

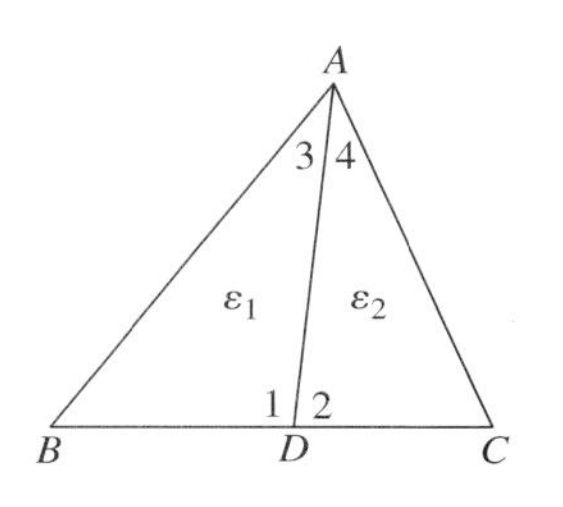

Problem 9

geometry $= c \cos A = 5 \cdot \cos 45° = 3.5355\,3390\,6$ miles. **15.** Use relation of Problem 14: $\sec A = \sec a + 1$. (1) $\sec A = \sec 0.3 + 1 \to A \approx 60.8°$. (2) $\sec A = \sec 2.0 + 1 \to A \approx 135.5°$. **17.** $\cos a = \cos(c_1 - c)\cos h = \cos c_1 \cos c \cos h + \sin c_1 \sin c \cos h = (\cos c_1 \cdot \cos h)\cos c + \sin b \sin c\,[\cos h \sin c_1/\sin b] = \cos b \cos c + \sin b \sin c\,[(\cos b/\cos c_1)\sin c_1/\sin b] = \cos b \cos c + \sin b \sin c \cdot (\tan c_1/\tan b) = \cos b \cos c + \sin b \sin c \cos A$. If D falls on extended sides, the only change in above is replacement of $c_1 - c$ by $c - c_1$, which does not affect proof. **19.** Any two lines (great circles) meet at opposite poles A, A', so locate midpoints M, N such that $AM = MA' = AN = NA' = \pi/2$ on those lines. By SSS $\triangle AMN \cong \triangle A'MN \to \overleftrightarrow{MN} \perp$ both lines. **21.** Because the direction "due East" is at a point not on equator and must be along a latitude.

TESTING YOUR KNOWLEDGE (CHAPTER 7)

1. 90. **2.** $x^2 = 9.6^2 + 7.2^2 = 144$; $d = \sqrt{x^2 + 5^2} = 13$. **3.** Because there are three faces at top and bottom vertex and four at the remaining three. **4.** $F = 8$, $E = 12$, $V = 6$. **5.** 2. **6.** $7000\pi/3 \approx 7{,}330$ cu. ft. **7.** Surface of cone approximated by n very slim isosceles triangles each with altitude approximately s and base x; nx approximates circumference C of base, hence $A \approx n \cdot \frac{1}{2}s \cdot x = \frac{1}{2}sC = \frac{1}{2}s(2\pi r) = \pi rs$. **8.** Center of circle $= (0, 3, 3)$; radius $= 7$. (Projection of circle into xy-plane has equation $x^2 + y^2 - 6x = 40$, $z = 0$). **9. (a)** 90. **(b)** 45. **10.** Let line ℓ be given with point A any point on it. (1) Choose two distinct planes containing ℓ, and in each of those planes construct a line perpendicular to ℓ at A, say lines m, n. The plane $\boldsymbol{P}$ determined by m and n will then be perpendicular to ℓ by Theorem 1, Section 7.1. (2) If second plane $\boldsymbol{Q}$ were perpendicular to ℓ at A, then, since $A \in \boldsymbol{P} \cap \boldsymbol{Q}$, $\boldsymbol{P}$ and $\boldsymbol{Q}$ meet in some line $m \neq \ell$ (because $\ell \perp m$). Choose plane containing ℓ not passing through m; this plane will meet $\boldsymbol{P}$, $\boldsymbol{Q}$ in lines $p, q \to p \perp \ell, q \perp \ell$, which contradicts uniqueness of perpendicular to line in a plane. $\therefore$ the perpendicular plane $\boldsymbol{P}$ at A is unique.

APPENDIX F

Symbols, Definitions, Axioms, Theorems, and Corollaries

SYMBOLS USED IN GEOMETRY

SYMBOL	DESCRIPTION	WHERE DEFINED
$\overleftrightarrow{AB}$	Line determined by A and B	71
$\overline{AB}$	Segment joining A and B	81
$\overrightarrow{AB}$	Ray through B with origin A	81
AB	Distance from A to B	78
$m\overline{AB}$	Measure of segment AB $(=AB)$	85
$A–B–C$	B lies between A and C	79
$A–B–C–D$	Betweenness relation for four points	80
$\overrightarrow{AB}–\overrightarrow{AC}–\overrightarrow{AD}$	Ray $\overrightarrow{AC}$ lies between rays $\overrightarrow{AB}$ and $\overrightarrow{AD}$	91
$\overrightarrow{AB}–\overrightarrow{AC}–\overrightarrow{AD}–\overrightarrow{AE}$	Betweenness relation for four rays	91
$\angle ABC$	Angle with sides $\overrightarrow{BA}$ and $\overrightarrow{BC}$	81
$m\angle ABC$	Measure of $\angle ABC$	90
$\ell \perp m$	Line ℓ is perpendicular to line m	97
$\overline{AB} \perp \overline{CF}$	Segment $\overline{AB}$ is perpendicular to segment $\overline{CD}$	97
$H(A, \overleftrightarrow{BC})$	Half-plane determined by $\overleftrightarrow{BC}$ containing A	108
$\triangle ABC$	Triangle ABC	120
$\diamond ABCD$	Quadrilateral $ABCD$	183
▱$ABCD$	Parallelogram $ABCD$	224
$\square ABCD$	Rectangle, square $ABCD$	186
$\overline{AB} \cong \overline{XY}$	$\overline{AB}$ is congruent to $\overline{XY}$	122
$\angle ABC \cong \angle XYZ$	$\angle ABC$ is congruent to $\angle XYZ$	122
$\triangle ABC \cong \triangle XYZ$	$\triangle ABC$ is congruent to $\triangle XYZ$	123
$\diamond ABCD \cong \diamond XYZW$	$\diamond ABCD$ is congruent to $\diamond XYZW$	184
$\triangle ABC \sim \triangle XYZ$	$\triangle ABC$ is similar to $\triangle XYZ$	237
$\overset{\frown}{AB}, \overset{\frown}{ABC}$	Minor arc $\overset{\frown}{AB}$, major arc $\overset{\frown}{ABC}$	196
$m\overset{\frown}{AB}$	Measure of arc $\overset{\frown}{AB}$	196
$\ell \parallel m$	Line ℓ is parallel to line m	211
$\overline{AB} \parallel \overline{CD}$	Segment $\overline{AB}$ is parallel to segment $\overline{CD}$	211
$\boldsymbol{P}(ABC)$	Plane passing through A, B, C	503
$\boldsymbol{P}(A, \ell)$	Plane containing point A and line ℓ	503

AXIOMS FOR ABSOLUTE GEOMETRY

UNDEFINED TERMS: *Point*, *line*, *plane*, and *space*.

AXIOM I-1: Each two distinct points determine a line.

AXIOM I-2: Three noncollinear points determine a plane.

AXIOM I-3: If two points lie in a plane, then any line containing those two points lies in that plane.

AXIOM I-4: If two distinct planes meet, their intersection is a line.

AXIOM I-5: Space consists of at least four noncoplanar points, and contains three noncollinear points. Each plane is a set of points of which at least three are noncollinear, and each line is a set of at least two distinct points.

AXIOM D-1: Each pair of points A and B is associated with a unique real number, called the **distance** from A to B, denoted by AB.

AXIOM D-2: For all points A and B, $AB \geq 0$, with equality only when $A = B$.

AXIOM D-3: For all points A and B, $AB = BA$.

AXIOM D-4 (Ruler Postulate): The points of each line ℓ may be assigned to the entire set of real numbers x, $-\infty < x < \infty$, called **coordinates**, in such a manner that

(1) each point on ℓ is assigned to a unique coordinate

(2) no two points are assigned to the same coordinate

(3) any two points on ℓ may be assigned the coordinates zero and a positive real number, respectively

(4) if points A and B on ℓ have coordinates a and b, then $AB = |a - b|$.

AXIOM A-1: Each angle $\angle ABC$ is associated with a unique real number between 0 and 180, called its **measure** and denoted $m\angle ABC$. No angle can have measure 0 or 180.

AXIOM A-2 (Angle Addition Postulate): If D lies in the interior of $\angle ABC$, then $m\angle ABD + m\angle DBC = m\angle ABC$. Conversely, if $m\angle ABD + m\angle DBC = m\angle ABC$, then D is an interior point of $\angle ABC$.

AXIOM A-3 (Protractor Postulate): The set of rays $\overrightarrow{AX}$ lying on one side of a given line $\overleftrightarrow{AB}$, including ray $\overrightarrow{AB}$, may be assigned to the entire set of real numbers x, $0 \leq x < 180$, called **coordinates**, in such a manner that

(1) each ray is assigned to a unique coordinate

(2) no two rays are assigned to the same coordinate

(3) the coordinate of $\overrightarrow{AB}$ is 0

(4) if rays $\overrightarrow{AC}$ and $\overrightarrow{AD}$ have coordinates c and d, then $m\angle CAD = |c - d|$.

AXIOM A-4 (Linear Pair Axiom): A linear pair of angles is a supplementary pair.

AXIOM H-1 (Plane Separation Postulate): Let ℓ be any line lying in any plane $\boldsymbol{P}$. The set of all points in $\boldsymbol{P}$ not on ℓ consists of the union of two subsets H_1 and H_2 of $\boldsymbol{P}$ such that

(1) H_1 and H_2 are convex sets

(2) H_1 and H_2 have no points in common

(3) If A lies in H_1 and B lies in H_2, the line ℓ intersects the segment $\overline{AB}$.

AXIOM C-1 (SAS Postulate): If two sides and the included angle of one triangle are congruent, respectively, to two sides and the included angle of another, the triangles are congruent.

DEFINITIONS

SECTION 1.1

median of triangle	segment joining vertex and midpoint of opposite side
altitude of triangle	segment from vertex perpendicular to opposite side (extended)
angle bisector of triangle	segment from vertex to opposite side lying on bisector of angle of triangle
centroid of triangle	point of concurrency of medians
orthocenter of triangle	point of concurrency of altitudes
incenter of triangle	point of concurrency of angle bisectors
circumcenter of triangle	point equidistant from the vertices
circumcircle of triangle	circle passing through the vertices, having the circumcenter as its center
incircle of triangle	circle tangent to the sides, having the incenter as its center
foot of perpendicular	point of intersection of perpendicular with line or segment

SECTION 2.2

independent axioms	axioms that do not logically imply one another
categorical axioms	axiom system that possesses only one model, up to isomorphism

SECTION 2.3

incidence axioms	axioms pertaining to set membership of points to lines or planes, or lines to planes
affine geometry	system of points, lines, planes satisfying incidence Axioms I-1–I-5 and parallel postulate

SECTION 2.4

***B* between *A* and *C* (*A–B–C*)**	if $AB + BC = AC$ and A, B, C are distinct, collinear points
A–B–C–D	when A–B–C, B–C–D, A–B–D, and A–C–D are all valid
segment $\overline{AB}$	set of all points X such that $X = A$, $X = B$, or A–X–B
ray $\overrightarrow{AB}$	set of all points X such that $X = A$, $X = B$, A–X–B, or A–B–X
open segment (ray)	points of segment (ray) excluding end points
interior point of segment (ray)	point of segment (ray) excluding end points (origin)
angle $\angle ABC$	set consisting of rays $\overrightarrow{BA}$ and $\overrightarrow{BC}$ where A, B, and C are noncollinear points
extension of $\overline{AB}$	points on ray $\overrightarrow{AB}$, ray $\overrightarrow{BA}$, or line $\overleftrightarrow{AB}$, whichever applies
midpoint of $\overline{AB}$	point M such that A–M–B and $AM = MB$

SECTION 2.5

$\overrightarrow{AC}$ **between** $\overrightarrow{AB}$, $\overrightarrow{AD}$	if $m\angle BAC + m\angle CAD = m\angle BAD$ for distinct rays $\overrightarrow{AB}$, $\overrightarrow{AC}$, $\overrightarrow{AD}$ (denoted $\overrightarrow{AB}-\overrightarrow{AC}-\overrightarrow{AD}$)
$\overrightarrow{AB}$–$\overrightarrow{AC}$–$\overrightarrow{AD}$–$\overrightarrow{AE}$	when $\overrightarrow{AB}$–$\overrightarrow{AC}$–$\overrightarrow{AD}$, $\overrightarrow{AC}$–$\overrightarrow{AD}$–$\overrightarrow{AE}$, $\overrightarrow{AB}$–$\overrightarrow{AC}$–$\overrightarrow{AE}$, and $\overrightarrow{AB}$–$\overrightarrow{AD}$–$\overrightarrow{AE}$ are all valid
bisector of $\angle ABC$	ray $\overrightarrow{BD}$ between $\overrightarrow{BA}$ and $\overrightarrow{BC}$ such that $m\angle ABD = m\angle DBC$
adjacent angles	two angles having a side in common but no interior points in common
right angle	an angle having measure 90
obtuse angle	an angle having measure greater than 90

perpendicular lines	two lines that contain the sides of a right angle
perpendicular segments (rays)	segments (rays) lying on each of two perpendicular lines
opposite rays	rays $\overrightarrow{BA}$, $\overrightarrow{BC}$ when A–B–C
vertical pair of angles	two angles whose corresponding sides are opposite rays
supplementary pair of angles	two angles whose measures sum to 180
complementary pair of angles	two angles whose measures sum to 90
linear pair of angles	two angles with a common side and opposite rays for the other two sides

SECTION 2.6

convex set	set having property that if points A and B are in that set, $\overline{AB}$ is in the set
half-plane of ℓ	one of the two convex sets H_1, H_2 in Plane Separation Postulate associated with line ℓ
closed half-plane	half-plane together with line that determines it
same side of line	in same half-plane determined by that line
interior point of angle	point in half-planes determined by each side of the angle and containing other side

SECTION 3.1

absolute geometry	development and results provable from 15 axioms previously listed
triangle $\triangle ABC$	set of all points lying on segments $\overline{AB}$, $\overline{BC}$, $\overline{AC}$ (the **sides**)
included side of $\angle A$, $\angle B$	side $\overline{AB}$
included angle of $\overline{AB}$, $\overline{BC}$	angle $\angle B$
opposite side (angle)	in $\triangle ABC$, $\overline{AB}$ and $\angle ACB$, $\overline{BC}$ and $\angle BAC$, or $\overline{AC}$ and $\angle ABC$
congruent segments (angles)	two segments (angles) having equal measures
congruent triangles	two triangles having corresponding sides and angles congruent
SAS Hypothesis	hypothesis in the SAS Postulate for congruence

SECTION 3.2

Taxicab Metric	$PQ = \|x_1 - x_2\| + \|y_1 - y_2\|$ where $P = (x_1, y_1)$, $Q = (y_1, y_2)$ in coordinate plane
taxicab geometry	coordinate geometry with Taxicab Metric replacing Euclidean Metric

SECTION 3.3

isosceles triangle	triangle having two sides congruent
legs of isosceles triangle	the congruent sides
base, base angles of isosceles triangle	included side of angles opposite the legs, angles opposite the legs
vertex, vertex angle of isosceles triangle	vertex, angle opposite base
perpendicular bisector of segment	line perpendicular to segment and bisecting it
kite, dart	four-sided figure $ABCD$ such that $\overline{AB} \cong \overline{AD}$ and $\overline{BC} \cong \overline{CD}$
locus of a point	line, circle, or set of points generated by a point satisfying certain conditions

SECTION 3.4

exterior angle of triangle	angle formed by side and extended side of triangle
angle sum of triangle	sum of measures of the three angles

SECTION 3.6

distance from point to line	length of segment from that point perpendicular to line
equidistant from	lying the same distance from

SECTION 3.7

quadrilateral $\diamond ABCD$	set of all points on segments $\overline{AB}$, $\overline{BC}$, $\overline{CD}$, $\overline{DA}$ **(sides)** such that no three **vertices**, A, B, C, D, are collinear and no two sides meet except at end points
angles of $\diamond ABCD$	$\angle ABC$, $\angle BCD$, $\angle CDA$, and $\angle DAB$
diagonals of $\diamond ABCD$	segments $\overline{AC}$, $\overline{BD}$
adjacent, consecutive vertices of $\diamond ABCD$	the pairs (A, B), (B, C), (C, D), or (A, D)

adjacent sides of $\diamond ABCD$	pairs of sides sharing a common vertex
convex quadrilateral	quadrilateral whose diagonals intersect
congruent quadrilaterals	two quadrilaterals having corresponding sides and angles congruent
rectangle	convex quadrilateral having four right angles
Saccheri Quadrilateral (SQ)	convex quadrilateral $\diamond ABCD$ with two congruent sides perpendicular to third side
base of SQ	segment included by the two right angles
legs of SQ	congruent sides adjacent to base
summit of SQ	side opposite the base
summit angles of SQ	angles opposite base
SQ **associated with** $\triangle ABC$	$\diamond BCC'B'$ such that $\overline{BB'}$ and $\overline{CC'}$ are perpendiculars from B and C to line that passes through midpoints of sides $\overline{AB}$, $\overline{AC}$
Lambert Quadrilateral	convex quadrilateral having three right angles
Hypothesis of Obtuse, Right, or **Acute Angle**	assumption that the summit angles of a Saccheri Quadrilateral are, respectively, obtuse, right, or acute angles

SECTION 3.8

circle (***C***)	set of points in a plane equidistant from a fixed point (**center**)
radius of ***C***	distance from center to any point on ***C*** (numerical), or segment joining center and point on ***C*** (geometric)
chord of ***C***	segment joining two points of circle of ***C***
diameter of ***C***	chord passing through center of ***C***
tangent of ***C***	line having exactly one point (**point of contact**) in common with ***C***
secant of ***C***	line having exactly two points in common with ***C***
arc of ***C***	intersection of ***C*** with closed half-plane $H' \equiv \ell \cup H$
semicircle of *C*	arc $\boldsymbol{C} \cap H'$ when ℓ passes through center of ***C***
minor arc of *C*	arc $\boldsymbol{C} \cap H'$ when H does not contain center
major arc of *C*	arc $\boldsymbol{C} \cap H'$ when H contains center
central angle of ***C***	angle whose sides contain two radii of ***C***
inscribed angle of ***C***	angle whose sides contain two chords of ***C***

SECTION 4.1

parallel lines	lines lying in the same plane that do not intersect
transversal	line that intersects two other lines
alternate interior angles along transversal	a pair of angles two of whose sides lie on the transversal pointing toward each other, with the other two sides lying on the two lines, pointing in opposite directions
corresponding angles along a transversal	a pair of angles two of whose sides lie on the transversal pointing in same direction, with the other two sides lying on the two lines, pointing in same direction

SECTION 4.2

parallelogram	(convex) quadrilateral with opposite sides parallel
rhombus	parallelogram having two adjacent sides congruent
rectangle	parallelogram whose angles are right angles
square	rhombus whose angles are right angles
trapezoid	(convex) quadrilateral with a pair of opposite sides parallel (called **bases**)
legs of trapezoid	sides adjacent to the bases
median of trapezoid	segment joining the midpoints of the legs
isosceles trapezoid	trapezoid with congruent legs and is not an **oblique** parallelogram (a parallelogram having no right angles)

SECTION 4.3

similar polygons	polygons having corresponding angles congruent and corresponding side-lengths in the same ratio
cevian of a triangle	segment joining vertex with point on opposite side (extended)

SECTION 4.4

polygon (*n*-**gon**)	for any positive integer $n \geq 3$, set of all points lying on segments $\overline{P_1P_2}, \overline{P_2P_3}, \overline{P_3P_4}, \ldots, \overline{P_nP_1}$ (**sides**) joining points $P_1, P_2, P_3, \ldots, P_n$ (**vertices**) such that no two sides intersect except at the vertices, and no three consecutive vertices are collinear
angle of polygon	$\angle P_iP_{i+1}P_{i+2}$ for any three consecutive vertices P_i, P_{i+1}, P_{i+2}
convex polygon	polygon for which the interior of any angle contains the rest of the polygon
regular polygon	polygon whose sides and angles are congruent
center, **radius** of regular polygon	center, radius of circle passing through vertices (**circumscribed circle**)
tiling (or **tessellation**) of plane	collection of regions $T_1, T_2, \ldots, T_n, \ldots$, called **tiles** that cover the plane, which have no interior points in common
elementary tiling of order *n*	tessellation each of whose tiles is congruent to one of *n* tiles (called **fundamental regions**)
regular tiling	elementary of order one whose fundamental region is a regular polygon
semiregular tiling	tiling such that (1) it is elementary of order *n*, (2) each of the fundamental regions is a regular polygon, and (3) each vertex (that is, a vertex of a tile) is of the **same type**: the set of tiles and their ordering around each vertex are the same

SECTION 4.7

directed line segment *AB* from *A* to *B*	ordinary segment $\overline{AB}$ with a **direction** (**initial point** *A* and **terminal point** *B*)
vector representation for *AB*	$[c - a, d - b]$, where $A = (a, b)$ and $B = (c, d)$. [Notation: $\mathbf{v}(\boldsymbol{AB})$]

SECTION 4.8

directed distance	distance given by the expression $AB = b - a$, for $A[a]$ and $B[b]$ on a line
linearity number of *D*, *E*, *F* with respect to $\triangle ABC$	defined for *D*, *E*, *F* on $\overleftrightarrow{BC}, \overleftrightarrow{AC}, \overleftrightarrow{AB}$ as the quantity (using directed distance) $\left[\frac{ABC}{DEF}\right] = \frac{AF}{FB} \cdot \frac{BD}{DC} \cdot \frac{CE}{EA}$

SECTION 5.1

transformation	an association (*f*) of a unique point $P' = f(P)$ with a given point *P* in some plane that is one to one
image of *P*	and onto the point *P*′ to which *P* is associated or "mapped"; that is, $f(p)$ under a transformation *f*
linear transformation	one that preserves collinearity
inverse of transformation *f*	mapping f^{-1} such that $f^{-1}(P') = P$ for each point $P' = f(P)$
fixed point of *f*	any point *A* for which $f(A) = A$
identity mapping	transformation for which every point is a fixed point (denoted *e*)

SECTION 5.2

reflection in line ℓ, **line reflection**	transformation $P \leftrightarrow P'$ for which ℓ (called the **line of reflection**) is the perpendicular bisector of segment $\overline{PP'}$ (denoted s_ℓ)
reflection in point *C*, **point reflection**	transformation $P \leftrightarrow P'$ for which *C* is the midpoint of $\overline{PP'}$ (denoted s_c)
isometry (**motion**, **rigid motion**, **Euclidean motion**)	distance-preserving mapping.
direct transformation	one that preserves counterclockwise orientation of triangles
opposite transformation	one that reverses counterclockwise orientation of triangles
product of two mappings *f* and *g*	composition $f \circ g$ defined by $P' = f[g(P)]$ for each point *P*

SECTION 5.3

translation	product of two line reflections s_ℓ and s_m where ℓ and *m* are parallel lines
rotation	product of two line reflections s_ℓ and s_m where ℓ and *m* are intersecting lines
center of rotation	point of intersection of lines ℓ and *m* that define the rotation

SECTION 5.4

glide reflection	product of reflection s_ℓ and translation t_{AB} where $\overleftrightarrow{AB} \parallel \ell$
similitude, **similarity transformation**	linear transformation for which there is a positive constant *k* (**dilation factor**) such that $P'Q' = k\,PQ$ for all pairs of points *P*, *Q*

dilation with center C, dilation factor k transformation d_k that maps any point P to that point P' on line $\overleftrightarrow{CP}$ such that $CP' = k\,CP$ (directed distance assumed)

SECTION 5.6

group, transformation group a set of transformations such that (1) the product fg of any two members f and g belongs to the set (**product closure**) and (2) the inverse f^{-1} of each member f also belongs to the set (**inverse closure**)

order number of distinct elements of a finite group

subgroup subset of transformation group that is itself a transformation group under its operations

general linear group GL(2) set of linear transformations of the plane to itself

SECTION 5.7

motion transformation in the plane that preserves both distance and angle measure

axis of motion line of fixed points, if it exists

nontrivial motion a motion that is not the identity

SECTION 6.3

defect of polygon $P_1P_2P_3 \ldots P_n$ the number $\delta(P_1P_2P_3 \ldots P_n) = 180(n-2) - m\angle P_1 - m\angle P_2 - m\angle P_3 - \cdots - m\angle P_n$

area of polygon $P_1P_2P_3 \ldots P_n$ the number $K = k\delta(P_1P_2P_3 \ldots P_n)$ for some constant k

SECTION 6.5

circular inversion mapping defined in coordinates by $x' = r^2x/(x^2 + y^2)$, $y' = r^2y/(x^2 + y^2)$for some positive constant r

circle of inversion circle whose equation is $x^2 + y^2 = r^2$ where r is the defining constant

SECTION 7.1

perpendicular lines, planes (line to plane): if line meets plane at some point A and is perpendicular to every line in that plane at A; (plane to plane): if first plane contains line that is perpendicular to second

dihedral angle set of all points lying in two closed half-planes having common edge

SECTION 7.2

parallel lines lines lying in same plane that do not meet

parallel planes planes that do not meet

hexahedron six-sided figure, where each side is a quadrilateral and its interior

polytope finite intersection of **closed half-spaces** in E^3 (a plane and all points lying on one side of it)

polyhedron having any number of sides (**faces**) each of which is a polygon and interior (or the *boundary of a polytope*)

edges of polyhedron line segments that are the intersections of any two faces

vertices of polyhedron points of intersection of any two edges

parallelepiped hexahedron having parallelograms and their interiors as faces

rectangular parallelepiped one having rectangles and their interiors as faces

cube rectangular parallelepiped having congruent edges

prism polyhedron having exactly two faces in parallel planes (the **base** and **top**), with remaining faces (the **lateral sides**) lying in planes parallel to a fixed line in E^3

pyramid polyhedron all of whose faces (**lateral sides**), except one (the **base**), meet at a common point (**apex**); lateral faces are necessarily triangles and their interiors

platonic solid a **polytope** whose faces are congruent to one regular polygon, with the same number of faces at each vertex

semiregular polyhedron one having regular polygons (and their interiors) as faces, and having the same number and type of faces at each vertex

SECTION 7.3

circular cylinder set of all points lying on lines (**generators**) that meet a circle (**base circle**) and remain parallel to a fixed line in E^3 not lying in the plane of the circle

circular cone set of all points lying on lines (**generators**) that meet a circle (**base circle**) and pass through a fixed point (**vertex**)

right circular cylinder	circular cylinder whose generators are perpendicular to base plane
right circular cone	circular cone whose **axis** (line joining vertex and center of base circle) is perpendicular to the plane of the circle
sphere	set of all points in E^3 lying at constant distance (**radius**) from a fixed point (**center**)
great circle of sphere	intersection of sphere and plane passing through center
tangent line, **plane** to sphere	line or plane that meets sphere at exactly one point

SECTION 7.4

spherical segment	that part of sphere and its interior lying between two parallel planes

THEOREMS AND COROLLARIES

SECTION 2.3

THEOREM 1: If C and D belong to line $\overleftrightarrow{AB}$ and $C \neq D$, then $\overleftrightarrow{CD} = \overleftrightarrow{AB}$.

THEOREM 2: Two distinct lines meet in at most one point; a line which meets a plane not containing it intersects that plane in exactly one point.

SECTION 2.4

THEOREM 1: If A–B–C then C–B–A, and neither A–C–B nor B–A–C.

THEOREM 2: If A–B–C, B–C–D, and A–B–D hold, then A–B–C–D is true.

THEOREM 3: If $A[a]$, $B[b]$, and $C[c]$ lie on line ℓ, then A–B–C iff $a < b < c$ or $c < b < a$.

THEOREM 4: If C lies on ray $\overrightarrow{AB}$ and $A \neq C$, then $\overrightarrow{AB} = \overrightarrow{AC}$.

THEOREM 5 (Segment Construction Theorem): If $AB < CD$, there exists a unique point E on ray $\overrightarrow{CD}$ such that $AB = CE$ and C–E–D.

COROLLARY: The midpoint of any segment exists, and is unique.

SECTION 2.5

THEOREM 1′ (Referred to but not stated): If the rays $\overrightarrow{AB}$, $\overrightarrow{AC}$, $\overrightarrow{AD}$ have coordinates b, c, and d relative to some half-plane, then $\overrightarrow{AB}$–$\overrightarrow{AC}$–$\overrightarrow{AD}$ iff either $b < c < d$ or $d < c < b$.

THEOREM 1 (Angle Construction Theorem): If $m\angle ABC < m\angle DEF$, there is a unique ray $\overrightarrow{EG}$ such that $m\angle ABC = m\angle GEF$ and $\overrightarrow{ED}$–$\overrightarrow{EG}$–$\overrightarrow{EF}$.

COROLLARY: The bisector of any angle exists and is unique.

THEOREM 2: Angles supplementary (or complementary) to the same angle are congruent.

THEOREM 3: $\overleftrightarrow{BD} \perp \overleftrightarrow{AC}$ at $D \in \overline{AC}$ iff $m\angle ADB = m\angle BDC$.

THEOREM 4: If $A \in \ell$, there exists a unique perpendicular to line ℓ at A.

THEOREM 5 (Vertical Pair Theorem): Vertical angles have equal measures.

SECTION 2.6

THEOREM 1: If H is a half-plane determined by ℓ and $A \in \ell$, $B \in H$, then, except for point A, $\overline{AB} \subseteq H$ and $\overrightarrow{AB} \subseteq H$.

COROLLARY: Let B and F lie on opposite sides of a line ℓ and let A and G be any two distinct points on ℓ. Then segment $\overline{GB}$ and ray $\overrightarrow{AF}$ have no points in common.

THEOREM 2 (Postulate of Pasch): If a line cuts one side of a triangle at an interior point and does not pass through any of the vertices, it also cuts one (and only one) other side.

THEOREM 3:

(1) If $A \in \overrightarrow{BP}$ and $C \in \overrightarrow{BQ}$, then (interior $\overline{AC}$) $\subseteq$ Interior $\angle PBQ$.
(2) If $D \in$ Interior $\angle ABC$, then (interior $\overrightarrow{BD}$) $\subseteq$ Interior $\angle ABC$.

THEOREM 4 (Crossbar Theorem): If $D \in$ Interior $\angle BAC$, then $\overrightarrow{AD}$ meets $\overline{BC}$ at an interior point.

SECTION 3.1

The three laws for any equality relation $\equiv$ are:

REFLEXIVE LAW: For all x, $x \equiv x$;

SYMMETRY LAW: If $x \equiv y$ then $y \equiv x$;

TRANSITIVE LAW: If $x \equiv y$ and $y \equiv z$ then $x \equiv z$.

SECTION 3.3

THEOREM 1 (ASA Theorem): If $\angle A \cong \angle X$, $\overline{AB} \cong \overline{XY}$, and $\angle B \cong \angle Y$, then $\triangle ABC \cong \triangle XYZ$.

THEOREM 2 (Isosceles Triangle Theorem): In $\triangle ABC$, $\overline{AB} \cong \overline{AC}$ iff $\angle B \cong \angle C$.

THEOREM 3 (Perpendicular Bisector Theorem): P lies on the perpendicular bisector of AB iff $PA = PB$.

THEOREM 4 (SSS Theorem): If $\overline{AB} \cong \overline{XY}$, $\overline{BC} \cong \overline{YZ}$, and $\overline{AC} \cong \overline{XZ}$, then $\triangle ABC \cong \triangle XYZ$.

THEOREM 5 (Existence of Perpendicular from an External Point): Given $A \notin \ell$, there exists a unique perpendicular from point A to line ℓ.

SECTION 3.4

THEOREM 1 (Exterior Angle Inequality): The measure of an exterior angle of a triangle is greater than that of either opposite interior angle.

THEOREM 2 (Saccheri–Legendre Theorem): The angle sum of a triangle is less than or equal to 180.

SECTION 3.5

THEOREM 1 (Scalene Inequality): In $\triangle ABC$, $AB > AC$ iff $\angle C > \angle B$.

THEOREM 2 (Triangle Inequality): $AB + BC \geq AC$, with equality only if A–B–C.

COROLLARY (Median Inequality): If M is the midpoint of $\overline{BC}$, $AM < \frac{1}{2}(AB + AC)$.

THEOREM 3 (SAS Inequality): If in $\triangle ABC$ and $\triangle XYZ$, $AB = XY$, $AC = XZ$, but $m\angle A > m\angle X$, then $BC > YZ$. Conversely, if $BC > YZ$, then $m\angle A > m\angle X$.

SECTION 3.6

THEOREM 1 (AAS Congruence Criterion): If $\angle A \cong \angle X$, $\angle B \cong \angle Y$, and $\overline{BC} \cong \overline{YZ}$, then $\triangle ABC \cong \triangle XYZ$.

THEOREM 2 (SSA Theorem): If $\overline{AB} \cong \overline{XY}$, $\overline{BC} \cong \overline{YZ}$, $\angle A \cong \angle X$, and $\triangle ABC \not\cong \triangle XYZ$, then $\angle C$ and $\angle Z$ are supplementary angles.

COROLLARY A: In acute triangles, if $\overline{AB} \cong \overline{XY}$, $\overline{BC} \cong \overline{YZ}$, and $\angle A \cong \angle X$, then $\triangle ABC \cong \triangle XYZ$.

COROLLARY B (HL Theorem): If the hypotenuse and leg of one right triangle are congruent to the hypotenuse and leg of another, the triangles are congruent.

COROLLARY C (HA Theorem): If the hypotenuse and acute angle of one right triangle are congruent to the hypotenuse and acute angle of another, the triangles are congruent.

COROLLARY D (LA Theorem): If a leg and acute angle of one right triangle are congruent to the corresponding leg and acute angle of another, the triangles are congruent.

COROLLARY E (SsA Congruence Criterion): Suppose that in $\triangle ABC$ and $\triangle XYZ$, $\overline{AB} \cong \overline{XY}$, $\overline{BC} \cong \overline{YZ}$, $\angle A \cong \angle X$, and $BC \geq BA$. Then $\triangle ABC \cong \triangle XYZ$.

SECTION 3.7

THEOREM 1 (SASAS Congruence): If convex quadrilaterals $\diamond ABCD$ and $\diamond XYZW$ have $\overline{AB} \cong \overline{XY}$, $\angle B \cong \angle Y$, $\overline{BC} \cong \overline{YZ}$, $\angle C \cong \angle Z$, and $\overline{CD} \cong \overline{ZW}$, then $\diamond ABCD \cong \diamond XYZW$.

THEOREM 2: The summit angles of a Saccheri Quadrilateral are congruent.

COROLLARIES: Let $\diamond ABCD$ be a Saccheri Quadrilateral with base $\overline{AB}$. Then

(1) The diagonals are congruent ($\overline{AC} \cong \overline{BC}$).
(2) If M, N are the midpoints of $\overline{AB}$, $\overline{CD}$, then $\overleftrightarrow{MN}$ is the perpendicular bisector of $\overline{AB}$ and $\overline{CD}$.
(3) If $\angle C$ and $\angle D$ are right angles, $\diamond ABCD$ is a rectangle, and $\overline{CD} \cong \overline{AB}$.

THEOREM 3: The Hypothesis of the Obtuse Angle is not valid in absolute geometry.

COROLLARIES:

(1) If $\diamond ABCD$ is a Saccheri Quadrilateral with base $\overline{AB}$, then $CD \geq AB$.
(2) If M, N are the midpoints of sides $\overline{AB}$, $\overline{AC}$ of $\triangle ABC$, then $MN \leq \frac{1}{2}AB$.

SECTION 3.8

THEOREM 1 (Additivity of Arc Measure): If the union of two arcs is an arc, then the measure of their union equals the sum of the measures of the two arcs.

THEOREM 2 (Tangent Theorem): A line is tangent to a circle iff it is perpendicular to the radius at the point of contact.

THEOREM 3 (Secant Theorem): If a line passes through an interior point of a circle, it is a secant of that circle.

SECTION 4.1

EUCLID'S FIFTH POSTULATE OF PARALLELS: If two lines in the same plane are cut by a transversal so that a pair of interior angles on the same side of the transversal has angle sum less than 180, the lines will meet on that side of the transversal.

AXIOM P-1 (Euclidean Parallel Postulate—Playfair's Version): If ℓ is any line and $P \notin \ell$, there exists a unique line passing through P parallel to ℓ (in the plane of P, ℓ).

THEOREM 1: (Parallelism in Absolute Geometry): If the angles in a pair of alternate interior angles along a transversal of two lines are congruent, the two lines are parallel.

THEOREM 2: If two parallel lines are cut by a transversal, then the angles in each pair of alternate interior angles are congruent.

COROLLARY A (The C Property): Two lines are parallel iff the angles in a pair of interior angles on the same side of the transversal are supplementary.

COROLLARY B (The F Property): Two lines are parallel iff the angles in a pair of corresponding angles along a transversal are congruent.

COROLLARY C (The Z Property): Two lines are parallel iff the angles in a pair of alternate interior angles are congruent.

COROLLARY D: If $t \perp \ell$ and $\ell \parallel m$, then $t \perp m$.

THEOREM 3 (Euclidean Exterior Angle Theorem): The measure of an exterior angle of any triangle equals the sum of the measures of the two opposite interior angles.

COROLLARIES:

(1) The angle sum of any triangle is 180.
(2) The acute angles of any right triangle are complementary.
(3) The sum of the measures of the angles of any convex quadrilateral is 360.
(4) Any Saccheri Quadrilateral or Lambert Quadrilateral is a rectangle.

THEOREM 4 (The Midpoint Connector Theorem): If M and N are the midpoints of $\overline{AB}$ and $\overline{AC}$, respectively, then $\overleftrightarrow{MN} \parallel \overleftrightarrow{BC}$ and $MN = \frac{1}{2}BC$.

COROLLARY: If a line bisects one side of a triangle and is parallel to the second, it also bisects the third side.

SECTION 4.2

THEOREM 1: A diagonal of a parallelogram divides it into two congruent triangles.

COROLLARIES:

(1) The opposite sides of a parallelogram are congruent.
(2) If a convex quadrilateral has opposite sides congruent, it is a parallelogram.
(3) If a convex quadrilateral has a pair of opposite sides both congruent and parallel, it is a parallelogram.
(4) The diagonals of a parallelogram bisect each other.

THEOREM 2 (Midpoint-Connector Theorem for Trapezoids): Let $\diamond ABCD$ be a trapezoid with base $\overline{AB}$, and let M, N be the midpoints of $\overline{AD}$, $\overline{BC}$, respectively. If $\ell \parallel \overleftrightarrow{AB}$ and passes through M, then ℓ passes through N and $MN = \frac{1}{2}(AB + CD)$. Conversely, if ℓ passes through M and N, then $\ell \parallel AB$.

THEOREM 3 (The Side-Splitting Theorem): If $E \in \overline{AB}$, $F \in \overline{AC}$, and $\overline{EF} \parallel \overline{BC}$, then $AE/AB = AF/AC$, or, equivalently, $AE/EB = AF/FC$.

PARALLEL PROJECTION: If A, B, C lie on ℓ, A', B', C' on m, and $\overleftrightarrow{AA'}$, $\overleftrightarrow{BB'}$, $\overleftrightarrow{CC'}$ are parallel lines, then $AB/BC = A'B'/B'C'$.

SECTION 4.3

THEOREM 1 (AA Similarity Criterion): If $\triangle ABC$ and $\triangle XYZ$ have $\angle A \cong \angle X$ and $\angle B \cong \angle Y$, then $\triangle ABC \sim \triangle XYZ$.

THEOREM 2 (SAS Similarity Criterion): If $\triangle ABC$ and $\triangle XYZ$ have $AB/XY = AC/XZ$ and $\angle A \cong \angle X$, then $\triangle ABC \sim \triangle XYZ$.

THEOREM 3 (SSS Similarity Criterion): If $\triangle ABC$ and $\triangle XYZ$ have $AB/XY = BC/YZ = AC/XZ$, then $\triangle ABC \sim \triangle XYZ$.

PYTHAGOREAN THEOREM: If a, b, and c are, respectively, the lengths of the two legs and hypotenuse of a right triangle, then $a^2 + b^2 = c^2$. ***Converse:*** If $a^2 + b^2 = c^2$ then a, b, and c are the lengths of the two legs and hypotenuse of a right triangle.

CEVIAN FORMULA: In $\triangle ABC$, if $\overline{CD}$ is a cevian from vertex C to $D \in \overline{AB}$ and $p = AD/AB = 1 - q$, then $CD^2 = pa^2 + qb^2 - pqc^2$, in standard notation.

SECTION 4.4

THEOREM 2: The angle sum of a convex n-gon is $180(n - 2)$.

COROLLARY A: Each interior angle of a regular polygon has measure $\phi = 180(n - 2)/n$.

COROLLARY B: The sum of the measures of the exterior angles of a convex polygon taken in the same direction is 360.

THEOREM 3 (Theorem of Gauss on Regular Polygons): A regular n-gon may be constructed with the Euclidean tools iff n is either a power of two, or, the product of a power of two and distinct **Fermat primes** of the form $F_m = 2^{2^m} + 1$.

SECTION 4.5

THEOREM 1 (Inscribed Angle Theorem): The measure of an inscribed angle of a circle equals one-half that of its intercepted arc.

THEOREM 2 (Two-Chord Theorem): If chords $\overline{AB}$ and $\overline{CD}$ intersect at an interior point P, then $AP \cdot PB = CP \cdot PD$.

THEOREM 3 (Secant-Tangent Theorem): If a secant $\overleftrightarrow{PA}$ and tangent $\overleftrightarrow{PC}$ meet a circle at the respective points A, B, and C (point of contact), then $PC^2 = PA \cdot PB$.

COROLLARY (Two-Secant Theorem): If two secants $\overleftrightarrow{PA}$ and $\overleftrightarrow{PC}$ of a circle meet the circle at A, B, C, and D, respectively, then $PA \cdot PB = PC \cdot PD$.

SECTION 4.6

THEOREM 1: If R is a rectangle with base of length b units and height of length h units, then Area $R = bh$.

SECTION 4.7

LAW OF COSINES FOR VECTORS: $\boldsymbol{uv} = \|\boldsymbol{u}\|\,\|\boldsymbol{v}\| \cos\theta$, where $\theta = m\angle(\boldsymbol{u}, \boldsymbol{v})$.

THEOREM 1: Any two nonzero vectors $\boldsymbol{u}$ and $\boldsymbol{v}$ are perpendicular iff $\boldsymbol{uv} = 0$.

THEOREM 2 (Triangle Inequality, Vector Form): $\|\boldsymbol{u} + \boldsymbol{v}\| \leq \|\boldsymbol{u}\| + \|\boldsymbol{v}\|$ with equality only when $\boldsymbol{u}$ and $\boldsymbol{v}$ represent two collinear directed segments.

SECTION 4.8

THEOREM 1 (Ceva's Theorem): Cevians $\overline{AD}$, $\overline{BE}$, $\overline{CF}$ of $\triangle ABC$ are concurrent iff $AF/FB \cdot BD/DC \cdot CE/EA = 1$.

THEOREM 2 (Menelaus' Theorem): If points D, E, and F lie on the extended sides of $\triangle ABC$ opposite A, B, and C, respectively, then D, E, and F are collinear iff $AF/FB \cdot BD/DC \cdot CE/EA = -1$.

SECTION 5.1

THEOREM: The inverse of a linear transformation is a linear transformation.

SECTION 5.2

THEOREM 1 (ABCD Property): Reflections are angle-measure preserving, betweenness preserving, collinearity preserving, and distance preserving.

THEOREM 2: The product of an even number of opposite linear transformations is a direct transformation, and the product of an odd number is an opposite transformation.

THEOREM 3 (Translation Mapping): If $\ell \parallel m$, the product of the line reflections s_ℓ and s_m is distance preserving, slope preserving, and maps a line to one that is parallel to it.

SECTION 5.3

THEOREM 1: Given any two congruent triangles $\triangle ABC$ and $\triangle PQR$, there exists a unique isometry that maps the first triangle onto the second.

THEOREM 2 (Fundamental Theorem on Isometries): Every isometry in the plane is the product of at most three reflections, exactly two if the isometry is direct and $\neq e$.

SECTION 5.4

THEOREM 1: A similitude is angle-measure preserving.

COROLLARY (ABC Property): A similitude preserves angle measure, betweenness, and collinearity.

THEOREM 2: A dilation is a similitude, and, accordingly, has the ABC Property.

THEOREM 3: Given any two similar triangles $\triangle ABC$ and $\triangle PQR$, there exists a unique similitude that maps the first triangle onto the second.

COROLLARY (Fundamental Theorem on Similitudes): Any similitude is the product of a dilation and at most three reflections.

SECTION 5.5

COORDINATE FORM OF A SIMILITUDE:

$$\begin{cases} x' = ax - \delta by + x_0, \\ y' = bx + \delta ay + y_0 \end{cases} \qquad \text{(dilation factor } k = \sqrt{a^2 + b^2}\text{)}$$

COORDINATE FORM OF A DILATION [with Center (h, k) and Dilation Factor a]:

$$\begin{cases} x' = ax + x_0, \\ y' = ay + y_0 \end{cases} \qquad (x_0 = h - ah, y_0 = k - ak)$$

COORDINATE FORM OF AN ISOMETRY:

$$\begin{cases} x' = ax - \delta by + x_0, \\ y' = bx + \delta ay + y_0 \end{cases} \qquad (\text{where } a^2 + b^2 = 1)$$

COORDINATE FORM OF A TRANSLATION t_{OA} [where $A = (a,b)$]:

$$\begin{cases} x' = x + a, \\ y' = y + b \end{cases}$$

COORDINATE FORM OF A ROTATION $r_\theta[C]$ [where $C = (h, k)$]:

$$\begin{cases} x' = (x - h)\cos\theta - (y - k)\sin\theta + h, \\ y' = (x - h)\sin\theta + (y - k)\cos\theta + k \end{cases}$$

COORDINATE FORM OF REFLECTION IN ℓ, [where ℓ has Equation $ax + by = c$]:

$$\begin{cases} kx' = (b^2 - a^2)x - 2aby + 2ac, \\ ky' = -2abx - (b^2 - a^2)y + 2bc \end{cases} \qquad (\text{where } k = a^2 + b^2)$$

COORDINATE FORM OF CENTRAL REFLECTION [where $C = (a, b)\ s_c$]:

$$\begin{cases} x' = -x + 2a, \\ y' = -y + 2b \end{cases}$$

SECTION 5.6

THEOREM 1: The product of two rotations $r_{2\alpha}[A]$ and $r_{2\beta}[B]$ is either a translation or rotation. It is a translation iff $\alpha + \beta = 180$, and a rotation otherwise. If $\alpha + \beta = 180$, then $r_{2\alpha}[A]r_{2\beta}[B] = r_{-2\gamma}[C]$ where C is the third vertex of the triangle having base $\overline{AB}$, whose vertices A, B, and C are oriented positively (counterclockwise), and whose angles are α, β, and γ, respectively, each oriented in the positive direction.

COROLLARY A: If $\triangle ABC$ is any triangle in the plane, with vertices oriented in the positive direction, and α, β, and γ the three angles at A, B, and C, respectively, oriented positively, then the product $r_{2\alpha}[A]r_{2\beta}[B]r_{2\gamma}[C]$ is the identity transformation.

SECTION 5.7

AXIOM M (Motion Postulate): There exists a nontrivial motion having any given line as axis.

THEOREM 5 (SAS Congruence Criterion): Suppose that triangles $\triangle ABC$ and $\triangle XYZ$ satisfy the SAS Hypothesis under the correspondence $ABC \leftrightarrow XYZ$, and that the Motion Postulate is assumed. Then $\triangle ABC \cong \triangle XYZ$.

SECTION 6.2

THEOREM 1: If $\triangle ABC$ has angle sum < 180 and D is any point on side $\overline{BC}$, then both $\triangle ABD$ and $\triangle ADC$ have angle sum < 180.

SECTION 6.3

THEOREM 1: The sum of the measures of the angles of any right triangle is less than 180.

COROLLARY: The angle sum of any triangle is less than 180.

THEOREM 2 (AAA Congruence Criterion for Hyperbolic Geometry): If two triangles have the three angles of one congruent, respectively, to the three angles of the other, the triangles are congruent.

SECTION 6.5

PROPERTIES OF CIRCULAR INVERSION: If $\boldsymbol{C}$ is the circle of inversion, then
(1) Points inside $\boldsymbol{C}$ map to points outside $\boldsymbol{C}$, points outside map to points inside, and points on $\boldsymbol{C}$ are self-inverse.
(2) Lines or circles map to lines or circles.
(3) A line through the center O of $\boldsymbol{C}$ is invariant, but the individual points of that line are changed.
(4) A circle through O maps to a line not through O, and the image line is perpendicular to the line that passes through O and the center of the given circle.
(5) Cross-ratio is invariant under circular inversion.
(6) A circular inversion is **conformal**, that is, it preserves curvilinear angle measure (Theorem 2).

THEOREM 3: A reflection exists with respect to any h-line, which preserves h-distance and h-angle measure, and maps h-lines to h-lines, h-segments to h-segments, and h-angles to h-angles.

COROLLARY: The SAS Postulate holds for the upper half-plane model.

SECTION 7.1

THEOREM 1 (Fence-Post Property): If a line is perpendicular to two lines of a plane at their point of intersection A, it is perpendicular to all lines in that plane passing through A.

COROLLARY: For any two planes $\boldsymbol{P}$ and $\boldsymbol{Q}$, if $\boldsymbol{P} \perp \boldsymbol{Q}$ then $\boldsymbol{Q} \perp \boldsymbol{P}$.

THEOREM 2: If planes $\boldsymbol{P}$ and $\boldsymbol{Q}$ are perpendicular, and their intersection is line ℓ, then every line in $\boldsymbol{P}$ perpendicular to ℓ is perpendicular to $\boldsymbol{Q}$.

COROLLARY: If two planes $\boldsymbol{P}$ and $\boldsymbol{Q}$ are perpendicular to a third plane $\boldsymbol{R}$, their intersection ℓ is also perpendicular to $\boldsymbol{R}$.

THEOREM 3 (Equidistant Locus Theorem): The set of all points in E^3 equidistant from two fixed points A and B is a plane that is perpendicular to $\overline{AB}$ at its midpoint.

SECTION 7.2

THEOREM 1 (A-Frame Theorem): If line $\ell \parallel$ line m, $\ell \subseteq$ plane $\boldsymbol{P}$, and $m \subseteq$ plane $\boldsymbol{Q}$, then either $\boldsymbol{P} \parallel \boldsymbol{Q}$ or $\boldsymbol{P}$ and $\boldsymbol{Q}$ intersect in a line n that is parallel to ℓ and m.

COROLLARY: If ℓ, m, and n are each pairwise distinct, and if $\ell \parallel m$ and $m \parallel n$, then $\ell \parallel n$.

THEOREM 2: The lateral faces of a prism are parallelograms.

THEOREM 3: The sum of the measures of the face angles at each vertex of a polyhedron is less than 360.

THEOREM 4: The measures of the three face angles at a vertex of a tetrahedron satisfy the triangle inequality (the sum of any two is greater than the third).

SECTION 7.3

THEOREM 1: The intersection of a sphere and plane $\boldsymbol{P}$, if nonempty, is a circle or point. If a circle, then its center is $\ell \cap \boldsymbol{P}$ where $\ell \perp \boldsymbol{P}$ from the center of the sphere.

THEOREM 2: The nonempty intersection of two distinct spheres is a circle or a point, and if it is the former, the plane of the circle is perpendicular to the line of centers.

THEOREM 3: Any four noncoplanar points lie on a unique sphere.

SECTION 7.4

Formulas for Volume:

PRISM	$V = Bh$	(B = base area, h = altitude)
RIGHT CIRCULAR CYLINDER	$V = \pi r^2 h$	(r = base radius, h = altitude)
RIGHT CIRCULAR CONE	$V = \frac{1}{3}\pi r^2 h$	(r = base radius, h = altitude)
SPHERE	$V = \frac{4}{3}\pi r^3$	(r = radius)
SPHERICAL SEGMENT	$V = \frac{1}{6}\pi h$	$(3r_1^2 + 3r_2^2 + h^2)$

(r_1, r_2 = radii of parallel bases, h = distance between planes of bases)

SECTION 7.5

THEOREM 1 (Coordinate Form of an Isometry): If $\boldsymbol{x}' = A\boldsymbol{x} + \boldsymbol{x}_0$ is the matrix form of a linear transformation f in E^3, then f is an isometry iff A is an **orthogonal** matrix (the three columns of A are unit vectors in E^3 and are pairwise othogonal).

SECTION 7.6

THEOREM (AAA Congruence Criterion for Spherical Geometry): If two triangles have the three angles of one congruent, respectively, to the three angles of the other, the triangles are congruent.

INDEX